Numerical Models
of Oceans *and*
Oceanic Processes

This is Volume 66 in the
INTERNATIONAL GEOPHYSICS SERIES
A series of monographs and textbooks
Edited by RENATA DMOWSKA, JAMES R. HOLTON, and H. THOMAS ROSSBY

A complete list of books in this series appears at the end of this volume.

Numerical Models *of* Oceans *and* Oceanic Processes

LAKSHMI H. KANTHA

University of Colorado
Boulder, Colorado

CAROL ANNE CLAYSON

Purdue University
West Lafayette, Indiana

ACADEMIC PRESS

A Harcourt Science and Technology Company

San Diego San Francisco New York Boston London Sydney Tokyo

Cover photo: A NOAA AVHRR image of the Gulf of Mexico on March 19, 1995. A large anticyclonic warm core eddy shed by the Loop Current and strong wintertime cooling of the water column on the U.S. continental shelf are depicted. It is a challenge to simulate and accurately forecast the processes responsible for these features in numerical ocean models, the subject matter of this book. (Photo courtesy of Dr. Robert Arnone of the Naval Research Laboratory, Stennis Space Center, MS.)

Academic Press
A Harcourt Science and Technology Company
525 B Street, Suite 1900, San Diego, California 92101-4495, USA
http://www.academicpress.com

Academic Press
Harcourt Place, 32 Jamestown Road, London NW1 7BY, UK
http://www.academicpress.com

Library of Congress Catalog Card Number: 99-60406

International Standard Book Number: 0-12-434068-7

PRINTED IN THE UNITED STATES OF AMERICA
00 01 02 03 04 05 MM 9 8 7 6 5 4 3 2 1

To

The Office of Naval Research

on the occasion of its Fiftieth Anniversary

and

The U.S. Navy

Contents

Chapter 1

Introduction to Ocean Dynamics

Chapter 2

Introduction to Numerical Solutions

Chapter 6

Tides and Tidal Modeling

Chapter 7

Coastal Dynamics and Barotropic Models

Chapter 8

Data and Data Processing

Chapter 9

Sigma-Coordinate Regional and Coastal Models

Chapter 10

Multilevel Basin Scale and Global Models

Chapter 11

Layered and Isopycnal Models

Chapter 12

Ice–Ocean Coupled Models

Chapter 13

Ocean–Atmosphere Coupled Models

Chapter 14

Data Assimilation and Nowcasts/ Forecasts

Appendix A

Equations of State

Appendix B

Wavelet Transforms

Appendix C

Empirical Orthogonal Functions and Empirical Normal Modes

Appendix D
Units and Constants

List of Acronyms

AABW	AntArctic Bottom Water
AAM	Atmospheric Angular Momentum
AATSR	Advanced Along Track Scanning Radiometer
ABL	Atmospheric Boundary Layer
ACC	Antarctic Circumpolar Current
ACCE	Atlantic Circulation and Climate Experiment
ADCM	Acoustic Doppler Current Meter
ADCP	Acoustic Doppler Current Profiler
ADEOS	ADvanced Earth Observation Satellite
ADI	Alternate Direction Implicit method
AGCM	Atmosphere General Circulation Model
AIDJEX	Arctic Dynamics Joint EXperiment
ALACE	Autonomous Lagrangian Circulation Explorer (floats)
AMSU	Advanced Microwave Sounding Unit
AOML	Atlantic Oceanographic and Meteorological Laboratory
APE	Available Potential Energy
ATLAS	Automated Temperature Line Acquisition System
ATSR	Along Track Scanning Radiometer
AUV	Autonomous Underwater Vehicle
AVHRR	Advanced Very High Resolution Radiometer
AXBT	Air-deployed eXpendable BathyThermograph
BL	Boundary Layer
BMO	British Meteorological Office
BVP	Boundary Value Problem
CEAREX	Central Eastern ARctic EXperiment
CEOF	Complex Empirical Orthogonal Functions
CFC	ChloroFluoroCarbon
CFD	Computational Fluid Dynamics

CFL	Courant–Friedrichs–Levy
CGCM	Coupled ocean–atmosphere General Circulation Model
CLIVAR	CLImate VARiability and predictability program
CNES	Centre National d'Etudes Spatiales
COADS	Comprehensive Ocean Atmosphere Data Set
COAGCM	Coupled Ocean Atmosphere General Circulation Model
COAMPS	Coastal Ocean Atmosphere Modeling and Prediction System
COARE	Coupled Ocean Atmosphere Response Experiment
COD	Computational Ocean Dynamics
CODAR	Coastal Ocean Doppler Radar
CODE	Coastal Ocean Dynamics Experiment
COIS	Coastal Ocean Imaging Spectrometer
CORE	Continuous Observation of the Rotation of the Earth
CPC	Climate Prediction Center (at NCEP; former Climate Analysis Center)
CPSST	Cross-Product Sea Surface Temperature
CPU	Central Processing Unit
CREAMS	Circulation Research in East Asian Marginal Seas
CrWT	Cross Wavelet Transform
CTD	Conductivity Temperature Depth (profiler)
CUPOM	University of Colorado version of Princeton Ocean Model
CWT	Continuous Wavelet Transform
CZCS	Coastal Zone Color Scanner
DA	Data Assimilation
DFT	Discrete Fourier Transform
DMSP	Defense Meteorological Satellite Program
DOE	Department of Energy
DOM	Dissolved Organic Matter
DWBC	Deep Western Boundary Current
DWT	Discrete Wavelet Transform
EAC	Ensemble Average Closure
ECM	Electromagnetic Current Meter
ECMWF	European Center for Medium-Range Weather Forecasting
EEZ	Exclusive Economic Zone
EEOF	Extended Empirical Orthogonal Functions
EIC	Equatorial Intermediate Current
EKE	Eddy Kinetic Energy
ENM	Empirical Normal Modes
ENSO	El Niño-Southern Oscillation
EOF	Empirical Orthogonal Function
EOS	Earth Observing System
EOS-AM	Earth Observing System AM mission

ERBE	Earth Radiation Budget Experiment
ERICA	Experiment on Rapidly Intensifying Cyclones over the Atlantic
ERM	Exact Repeat Mission
ERS	Earth Resources Satellite
ESA	European Space Agency
EUC	Equatorial UnderCurrent
FCT	Flux-Corrected Transport
FEM	Finite Element Method
FFT	Fast Fourier Transform
FNMOC	Fleet Numerical Meteorology and Oceanography Center
FOCAL	Francais Ocean et Climat dans l'Atlatique equatorial
FT	Fourier Transform
FTCS	Forward in Time, Centered in Space
GAC	Global Area Coverage (4 km resolution AVHRR) data
GALE	Genesis of Atlantic Lows Experiment
GATE	Global Atlantic Tropical Experiment
GCM	General Circulation Model
GCOS	Global Climate Observing System
GEOSAT	GEOdetic SATellite
GEOSECS	Geochemical Ocean Sections Study
GEWEX	Global Energy and Water cycle Experiment
GF	Gigaflops
GFDL	Geophysical Fluid Dynamics Laboratory
GFO GEOSAT	Follow-On
GJ	Gigajoules
GLOSS	GLObal Sea level observing System
GMT	Greenwich Mean Time
GOALS	Global Ocean Atmosphere Land System
GODAE	Global Ocean Data Assimilation Experiment
GOES	Geostationary Orbiting Environmental Satellite
GOOS	Global Ocean Observing System
GPS	Global Positioning System
GTOS	Global Terrestrial Observing System
GTS	Global Telecommunication System
GW	Gigawatts, also gigawords
HIRS	High resolution Infra Red Sounder
HNLC	High Nitrate Low Chlorophyll
HRPT	High Resolution Picture Tube satellite data receivers
ICES	International Council for Exploration of the Sea
IDOE	International Decade of Ocean Exploration
IERS	International Earth Rotation Service

IES	Inverted Echo Sounder
IGBP	International Geosphere Biosphere Programme
IGOSS	Integrated Global Ocean Services System
IMSL	International Math and Science Library
IOC	Intergovernmental Oceanographic Commission
IOP	Intensive Observation Period
IPCC	Intergovernmental Panel on Climate Change
IR	InfraRed
ISCCP	International Satellite Cloud Climatology Project
ITCZ	Inter-Tropical Convergence Zone
IVP	Initial Value Problem
JEBAR	Joint Effect of Baroclinicity and bottom Relief
JGOFS	Joint Global Ocean Flux Study
JIC U.S.	Navy/NOAA Joint Ice Center
JMA	Japan Meteorological Office
JPL	Jet Propulsion Laboratory (at California Institute of Technology)
KC	Kantha-Clayson
KE	Kinetic Energy
LAC	Local Area Coverage (1 km resolution AVHRR) data
LATEX	LouisianA-TEXas
LCE	Loop Current Eddy
LEADEX	LEADs EXperiment
LES	Large Eddy Simulation
LHS	Left-Hand Side
LIDAR	Laser Induced Detection and Ranging
LOD	Length of Day
LOM	Length of Month
LTE	Laplace Tidal Equations
LW	Longwave
MABL	Marine Atmospheric Boundary Layer
MBT	Mechanical BathyThermograph
MCSST	Multi-Channel Sea Surface Temperature
MICOM	Miami Isopycnic COordinate Model
MIZ	Marginal Ice Zone
MIZEX	Marginal Ice Zone EXperiment
MJO	Madden–Julian Oscillations
ML	Mixed Layer
MLD	Mixed Layer Depth
M-O	Monin–Obukhoff
MODE	Mid-Ocean Dynamics Experiment
MODIS	MODerate resolution Imaging Spectroradiometer

MOM	Modular Ocean Model
MOODS	Master Oceanographic Observation Data Set
MW	Megawatts, also Megawords
MY	Mellor–Yamada
NADW	North Atlantic Deep Water
NAO	North Atlantic Oscillation
NASA	National Aeronautics and Space Administration
NAVO	NAVal Oceanographic office
NCAR	National Center for Atmospheric Research
NCDC	National Climate Data Center
NCEP	National Centers for Environmental Prediction (formerly NMC)
NDBC	NOAA Data Buoy Center
NEC	North Equatorial Current
NECC	North Equatorial Counter Current
NEMO	Navy Earth Map Observer
NESDIS	National Environmental Satellite Data and Information Service
NGDC	National Geophysical Data Center
NMC	National Meteorological Center
NMFS	National Marine Fisheries Service
NOAA	National Oceanic and Atmospheric Administration
NODC	National Oceanographic Data Center
NOGAPS	Navy Operational Global Atmosphere Prediction System
NORAPS	Navy Operational Regional Atmosphere Prediction System
NOS	National Ocean Service
NRL	Naval Research Laboratory
NSCAT	NASA SCATterometer
NWP	Numerical Weather Prediction
NWS	National Weather Service
OCTS	Ocean Color and Temperature Sensor
ODE	Ordinary Differential Equation
OGCM	Ocean General Circulation Model
OI	Optimal Interpolation
OLR	Outgoing Longwave Radiation
OML	Oceanic Mixed Layer
OSSE	Observing System Simulation Experiment
OWS	Ocean Weather Station
PAGES	Past Global Changes
PALACE	Profiling Autonomous LAgrangian Circulation Explorer float
PAR	Photosynthetically Available Radiation
PBL	Planetary Boundary Layer

PCA	Principal Component Analysis
PDE	Partial Differential Equation
PDO	Pacific Decadal Oscillation
PE	Potential Energy, also Primitive Equation
PIP	Principal Interaction Pattern
PIRATA	PIlot Research moored Array in the Tropical Atlantic
PMEL	Pacific Marine Environmental Laboratory
POD	Proper Orthogonal Decomposition
PODAAC	Physical Oceanography Distributed Active Archive Center
POES	Polar Orbiting Environmental Satellite
POM	Princeton Ocean Model
POLYMODE	Joint USSR POLYgon and U.S. MODE experiments
POP	Principal Oscillation Pattern; also Parallel Ocean Program
PPM	Piecewise Parabolic Method
PSU	Practical Salinity Units
PVD	Progressive Vector Diagram
QC	Quality Control
QG	Quasi-geostrophic
QH	Quasi-hydrostatic
RAFOS	SOFAR spelled backward
RCM	Rotor-type Current Meter
RCOAM	Regional Coupled Ocean Atmosphere Model
REOF	Rotated Empirical Orthogonal Functions
RHS	Right-Hand Side
RK	Runge–Kutta
RMS	Root Mean Square
ROV	Remotely Operated Vehicle
RV	Research Vessel
SAR	Synthetic Aperture Radar
SAUV	Smart Autonomous Underwater Vehicle
SC	Successive Correction
SCOR	Scientific Committee on Ocean Research
SEASAT	SEA SATellite
SeaWiFS	Sea-viewing WIde Field-of-view Sensor
SEC	South Equatorial Current
SEQUAL	SEasonal response of the EQUatorial AtLantic programme
SHEBA	Surface HEat Budget of the Arctic experiment
SLP	Sea Level Pressure
SMMR	Scanning Multichannel Microwave Radiometer
SOFAR	SOund Fixing and Ranging floats
SOI	Southern Oscillation Index
SOR	Successive Over Relaxation

SSA	Singular Spectrum Analysis
SSH	Sea Surface Height
SSM/I	Special Sensor Microwave Imager
SSS	Sea Surface Salinity
SST	Sea Surface Temperature; Sea Surface Topography in geodesy
STD	Salinity Temperature Depth profiler
SVD	Singular Value Decomposition
SW	Shortwave
TAO	Tropical Atmosphere Ocean array
TC	TOGA/COARE
TF	Teraflops
TFO	TOPEX Follow-On
TOGA	Tropical Ocean Global Atmosphere program
TOPEX	TOPography EXperiment
TOS	TOGA Observing System
TOVS	TIROS Operational Vertical Sounder
T/P	TOPEX/Poseidon
TRMM	Tropical Rainfall Measuring Mission
TVD	Total Variance Diminishing method
UNESCO	United Nations Economic Social and Cultural Organization
UTC	Universal Temps Coordinee (Universal Time Coordinated)
UV	Ultraviolet
VACM	Vector Averaging Current Meter
VMCM	Vector Measuring Current Meter
VOS	Voluntary Observing Ship
TW	Terawatts
WCRP	World Climate Research Program
WD	Wavelet Domain
WFT	Windowed Fourier Transform
WMO	World Meteorological Organization
WOCE	World Ocean Circulation Experiment
WT	Wavelet Transform
WWB	Westerly Wind Burst
WWW	World Wide Web, also World Weather Watch
XBT	eXpendable BathyThermograph
XCP	eXpendable Current Profiler
XCTD	eXpendable Conductivity Temperature Depth

List of Symbols

α	Coefficient of thermal expansion
β	Coefficient of expansion due to salinity
δ, δ_S, δ_M, δ_I,	Western boundary layer thickness, Stommel, Munk, and Inertial
η	Interfacial deflection
ε	Perturbation variable, dissipation rate of TKE
ε_{ijk}	Alternating tensor
δ_{ij}	Kronecker delta function
φ_M	Monin-Obukhoff similarity function
ψ	Stream function
κ	von Karman constant
λ_i	Eigen value
ν, ν_t	Kinematic viscosity, molecular and turbulent.
ω	Frequency
θ	Latitude, fluctuating temperature
ρ	Density
ρ_w, ρ_A, ρ_I	Density of water, air, and ice
σ_{ij}	Ice stress tensor (also turbulent stress tensor)
ζ	Absolute vorticity, sea surface height
τ_{AI}, τ_{IO}, τ_{AO}	Shear stress at air-ice, ice-ocean, and air-ocean interfaces.
ξ_1, ξ_2, ξ_3	Orthogonal curvilinear coordinates
τ_{ij}, τ	Stress tensor
τ^w	Wind stress
τ^b	Bottom stress

$\Delta\rho$	Change in density
Θ	Mean temperature
Φ	Gravitational potential
Φ_t	Tidal potential
Ω	Angular rotation rate of Earth
a	Rossby radius of deformation, radius of Earth, amplitude of the wave
a_1, a_2, a_3	Unit vectors in the three coordinate directions
b	Buoyancy
c_d	Drag coefficient
e_{ij}	Strain rate tensor
f	Coriolis parameter, planetary vorticity
g	Acceleration due to gravity
$\mathbf{g}$	Gravitational body force
h_1, h_2, h_3	Coordinate metrics
k	Wave number
$\mathbf{k}$	Unit vector in the vertical (z) direction
k_T, k_S, k_c	Diffusivity of heat, salt and a passive scalar (kinematic)
n	wave frequency
p	pressure
$\mathbf{q}_T, \mathbf{q}_S$	Kinematic heat flux and salt flux vectors
r	radius of curvature
t	time
u, v, w	Velocity components
$\mathbf{v}$	Velocity vector
x, y, z	Rectangular Cartesian coordinates
x_i, x_2	Rectangular Cartesian coordinates
z	Vertical coordinate
z_i	Inversion height
z_0	Roughness scale
A_M, A_H	Horizontal diffusivity of momentum and scalars (heat, salt, etc.)
A_I	Ice concentration
C	Gravity wave speed
C_e, C_i	External, internal gravity wave speed
C_m	Gravity wave speed of m the baroclinic mode
C_p, C_g	Phase velocity, group velocity
D_M, D_e, D_I	Mixed layer depth, Ekman layer depth, ice thickness
E	Energy anomaly of the eddy
$\mathbf{F}$	Frictional force
F_w	Energy flux due to internal waves

H	Layer depth, bottom depth
K_M, K_H	Vertical diffusivity of momentum and scalars (heat, salt, etc.)
L	Length scale of motions, also meridional extent of the basin
M	Mass anomaly of the eddy
M_x, M_y	Zonal, meridional transports over the water column per unit width
N	Buoyancy frequency
P	Mean pressure
Pr_t	Turbulent Prandtl number
Ri_g, Ri_f	Gradient Richardson number, flux Richardson number
Ri	Bulk Richardson number
Ro, Ro_t	Rossby numbers
S	Mean salinity
S_M	Mixed layer salinity
S_M, S_H	Stability functions in turbulent mixing coefficients
So, Si	Sources and sinks in the conservation equations
T	Mean temperature; also period-of-wave motions
T_i	Inertial period
U, V, W	Velocity components
U_1, U_2, U_3	Velocity components
U_g, V_g	Geostrophic velocity components
U_M	Mixed layer velocity
V_c	Long Rossby wave speed
V_d	Drift velocity of eddies
W	Basin width

Foreword

To appreciate how timely this book is, it is necessary to look back to the early days of ocean modeling. Remarkable progress had been made in numerical weather prediction by Jules Charney and Norman Phillips in the 1940s and 1950s. Joseph Smagorinsky, who had participated in that effort at the Institute for Advanced Study at Princeton, returned to Washington and persuaded the U.S. Weather Bureau, which later became part of the National Atmospheric and Oceanic Administration, to back a visionary effort to build a comprehensive numerical model of global climate, including both the atmosphere and the ocean. The project in its early stages involved less than a dozen scientists and programmers. On the other hand, the supercomputers of the day needed for the work were vast, expensive, and difficult to use. It is not surprising that few other organizations attempted similar research, which required so much in resources and focused on such a long-range goal.

The development of a numerical model of the general circulation of the World Ocean was a priority for Smagorinsky, and the laboratory he organized proved to be an ideal environment for such an undertaking. An enlightened policy on the part of the lab director enabled me to take extensive leaves, during which I returned to the Woods Hole Oceanographic Institution. I found my colleagues there were mildly curious, but they had little active interest in ocean model building because the tools for doing that type of research were only available in a few specialized laboratories. Just a small group interested in geophysical fluid dynamics as it related to oceanography were interested in the rather abstract problems, which the computers of the day were capable of solving.

Today the situation is entirely different. The enormous technical progress in designing and building computers have made modeling widely accessible to laboratories, university departments, and even individuals. Students are adept at using computers before they even begin their university studies. Computer models have become a universally accepted tool for research and education in almost every scientific field. Physical oceanography has evolved from a field that was primarily concerned with the exploration of the World Ocean, to a focus on much more detailed questions involving mechanisms underlying the ocean circulation and its role in climate. A few decades ago most of the observations of the ocean could be summarized in a few printed atlases. Now oceanographers have at their disposal vast archives of data generated by satellites and ships of opportunity. For the first time the satellite altimeter and temperature data allow a "synoptic" view of the ocean, something that atmospheric scientists had taken for granted but was unavailable a few decades ago on the basis of normal observations taken from research vessels. The end of the cold war has unlocked substantial archives of data held by both the United States and the former Soviet Union.

At the same time, many interdisciplinary fields in which numerical models of the ocean circulation can play a useful role are beginning to be explored. The ability of geochemical measurements to detect tiny concentrations of chemicals in the ocean has opened up tremendous possibilities in investigating details of ocean mixing as well as global pathways of tracers injected at the surface. In many of these areas, scientific research and important international policy issues intersect. For example, several modeling groups are attempting to quantify how much of the additional carbon dioxide produced by fossil fuel burning is taken up by the ocean and how this uptake might change with different climate conditions. Modeling groups are trying to refine climatic scenarios for a world with increasing greenhouse gases. This implies understanding how the interaction of the ocean and atmosphere will affect the pattern of rainfall and temperature over the continents. Since the original International Climate Change Panel (IPCC) report in 1990, models of the response of climate to the buildup of greenhouse gases are not considered complete unless they include an active model of ocean circulation.

Advances in coupled models of the atmosphere and ocean have provided a solid justification for observational networks that will include routine subsurface measurements in the ocean, and such a network already exists for monitoring the El Niño area of the equatorial Pacific. In addition to global problems, ocean circulation models have found many important local applications to near-shore areas, where models are needed for both a scientific understanding of coastal circulation as well as the solution to pressing environmental problems involving fisheries management or pollution.

With the large number of important applications in so many areas such as biology, engineering, geochemistry and climate, a wide variety of scientists with diverse backgrounds have become interested in ocean models. Perhaps the one common factor will be familiarity in the use of computers for solving scientific problems. This book by Kantha and Clayson draws from many sources and an extensive background and experience in modeling to provide an understandable teaching tool and guide for research. It summarizes in an understandable way the present approaches to numerically modeling the ocean circulation. The book lays out a path by which a reader could independently access the codes and information from traditional references or Internet sources to begin to use ocean models for his or her own, possibly entirely new, applications.

Kirk Bryan
Princeton, NJ

Preface

Oceans play a pivotal role in our weather and climate. They are also an important source of our food and minerals. Ocean-borne commerce is vital to our increasingly close-knit global economy. Thus, directly or indirectly, oceans are a part of our everyday lives. Yet, we do not understand well the intricate details of their circulation, limits to their biological productivity, their interaction with the atmosphere, or their tolerance to wastes dumped by the ever-increasing human population. Until recently, our capacity to make observations in the oceans was also limited. The tedium and expense of making *in situ* measurements placed a severe limit on our ability to explore oceanic processes. Although sensors orbited on satellites, and carried on freely drifting buoys, autonomous underwater vehicles, and semipermanent strategically located moorings—all triumphant achievements of 20th-century technology—have begun to fill in gaps in our knowledge of how oceans function, they are only a part of the solution.

In the coming century, increasing reliance will therefore be placed on yet another marvel of 20th-century technology: high performance computers. Numerical ocean models that are run on high performance computers, both in a standalone mode and in combination with observed data, will vastly increase our ability to simulate and comprehend oceanic processes, monitor the current state of the oceans, and, to a limited extent, even predict their future state. However, their ability to fulfill modern societal needs will largely depend on the fidelity with which they can simulate oceanic processes. Although the brute power of multiteraflop and even petaflop high performance computers of the coming century will help in this task, they are no substitute for a better understanding and representation of underlying mechanisms. That is why there is a need for a bright, young generation—people who are trained in the use of computers, possess the requisite knowledge base, and are willing to tackle the hard task of deciphering how Nature puts it all together. It is our hope that this book, along with its companion volume, *Small Scale Processes in Geophysical Flows* (Academic Press, 2000), will assist in this task.

Like many other computational sciences, numerical ocean modeling requires a combination of several skills. The first and foremost is an intimate knowledge of ocean dynamics, or at least its salient features. (For modeling the chemical and biological state, it is also essential to have at least some knowledge of the chemical and biological processes.) Unlike descriptive oceanography, dynamical oceanography requires that the student possess adequate skills in mathematics and physics, because the ability to quantify oceanic processes through solution of the underlying conservation equations is a necessity, not an option. The second requirement is the ability to program and interact with computers. Here the younger generations hold an edge. They are more at ease with computers. However, they do require proficiency in at least one programming language (FORTRAN is the current language of choice in scientific computing) and one operating system (UNIX is the current nearly universal system). In addition, familiarity with at least elementary numerical methods is desirable. It is this confluence of requisite skills that is difficult to realize. Seldom does one find a student who possesses all these skills. Our own experience in teaching this subject has been that numerical ocean modeling is best taught along with some ocean dynamics and numerical methods, and this bias has shaped the contents of this volume. Although excellent treatments of ocean dynamics and numerical methods exist, we have endeavored to collect in a single place—for ease of reference and at the risk of being accused of attempting to be all-inclusive— (1) some elementary but salient aspects of the dynamics that one is attempting to simulate on a computer, (2) elementary concepts in numerical solutions of differential equations and numerical analysis, and (3) numerical ocean models. However, topics related to subgrid-scale processes such as oceanic mixing and modeling mixed layers and coupled physical–chemical–biological systems have been covered separately in *Small Scale Processes in Geophysical Flows* (Academic Press, 2000). It is hoped that these two volumes combined will serve as useful references on the topic.

On a subject this vast, it is impractical to provide a thorough and in-depth treatment of each subtopic that could very well require an entire volume of its own. We have, however, tried to provide at least some elementary material on topics with which we felt an ocean modeler should be familiar. Of course, modern references are provided to enable the student to pursue a particular topic in greater depth. The book is written so that a newcomer to the field, provided he or she has the necessary mathematics and physics background and computing skills, can learn the subject with ease and be introduced to some current research topics as well, without having to do an extensive literature survey on his or her own. It is our hope that this book will fascinate and inspire at least a few young people around the world to take up careers in environmental science and engineering and thus contribute to a better understanding of our environment and improve conditions for the population at large, which increasingly will be at the mercy of the vagaries of nature for sustenance in the coming century.

It has been our experience that a subject such as this is best taught using modern multimedia capabilities. Numerical ocean modeling, like other computational sciences, cannot be learned from merely reading a book, no matter how good; it requires a hands-on approach. We therefore hope to make this book electronic. In addition to the regular hard-copy format, we hope to provide the student, eventually, with an electronic supplement that contains the source code, data bases, color graphics, animation packages and sample runs that will enable a more interactive use of the material. As far as is feasible, each chapter will be provided with project-like exercises that are designed to improve the student's skill and understanding. It is one thing to read about, say, coastally trapped waves, but entirely another to actually watch them propagate and witness changes brought on by fluctuatuions in relevant parameters. We hope that this format will make it easier and "more fun" to learn topics that are normally considered quite dry and hard and frankly turn off quite a few brilliant young people.

Finally, in a perhaps overambitious endeavor such as this, mistakes are inevitable, especially in topics on which we are not experts. The mere fact that we could put something like this together on such a vast and intricate field attests to our liberal borrowing (properly attributed, of course) from experts in their individual areas of expertise. We wish to thank them and apologize if we misquoted anyone. We would certainly appreciate hearing about any glaring errors that may have been inadvertently made.

It is our pleasure to acknowledge the contributions of many anonymous reviewers whose comments have greatly improved this book. Particular thanks to Dr. Frank Bryan of the National Center for Atmospheric Research, whose thoughtful and thorough review is greatly appreciated. We thank the many scientists who contributed by sending original figures from their work for inclusion in this text. We also thank the following individuals for helping to prepare many of the final figures for this text: Tristan Johnson, Reed L. Clayson, Jason Hartz, and Rebecca Priddy. Reed L. Clayson also provided valuable editorial assistance.

It was our hope to complete these two books in time for the 50th birthday of the Office of Naval Research, but we severely underestimated the time involved in converting an initial draft to a final peer-reviewed set of chapters. Nevertheless, the principle "better late than never" governs our dedication of these books.

Last but not the least, we thank our very understanding, ever-patient, and tolerant spouses, Kalpana Kantha and Tristan Johnson, for their unflinching support and assistance. LHK thanks Roshan, Vinod, and Kiran for putting up so patiently with an "absentee" father. CAC's infant son, Johann, did his part by consistently getting her up so she could devote productive pre-dawn hours to completing the text.

Lakshmi H. Kantha   Carol Anne Clayson
University of Colorado   Purdue University
Boulder, Colorado   West Lafayette, Indiana
kantha@colorado.edu   clayson@purdue.edu

Prologue

The subject of this treatise is numerical modeling of oceans and oceanic processes, in other words, computational ocean dynamics. Although our emphasis will necessarily be on the oceans, we will also consider the sea ice and the atmosphere in as far as they are essential to the main topic. We will restrict ourselves to physical processes: large scale circulation and density structure. Modeling of chemical and biological processes is discussed briefly in the companion volume *Small Scale Processes in Geophysical Flows* (Academic Press, 2000), which deals primarily with small scale processes responsible for vertical mixing and transport in the oceans.

Ocean modeling and estimation of the oceanic state are becoming increasingly relevant to the understanding and prediction of long-term weather and climate, El Niño, Australasian monsoons, and decadal variability being three very good examples. Modeling the ocean circulation and structure with fidelity is also important to applications such as management of fisheries and pollution control, as well as many naval operations. Our goal is to provide a firm background in principles of ocean modeling and related topics. We will start off with simple models but conclude with comprehensive data-assimilative, state-of-the-art numerical models of the oceans.

It is quite obvious that an ocean modeler must understand what he or she is trying to simulate on a digital computer. We will therefore start off with introductory ocean dynamics in Chapter 1. This is the first leg of the three-legged numerical ocean modeler's chair. Topics treated include broad circulation features such as the basin scale wind-driven gyres, Ekman layers, barotropic and baroclinic flow over topography, fronts, mesoscale variability, and thermohaline circulation. These topics are dealt with in a quasi-analytical fashion simply because a firm foundation in simple yet fundamental concepts in ocean dynamics is indispensable to an ocean modeler. Those already possessing such knowledge can skim through or skip this chapter. Chapter 2 deals with the second leg of the chair: numerical methods. Familiarity with numerical techniques is another prerequisite to ocean modeling. Topics dealt with include methods of solving ordinary and partial differential equations, explicit and implicit methods, steady-state and time-dependent problems, stability of numerical schemes, and spatial and temporal discretization. Examples are given from a simple inertial oscillation problem to a recently developed state-of-the-art nonhydrostatic model from MIT.

We proceed next to the third leg of the chair, the numerical ocean model itself. Chapters 3 to 5 address oceanic processes in the tropics, midlatitudes and high latitudes, respectively, and simple numerical models designed to simulate some of these processes. This division by latitudinal bands is a useful way to look at oceanic processes and their simulation by numerical means. Adjustment processes in the equatorial waveguide and simple reduced gravity models suited to simulating them are described in Chapter 3. Midlatitude processes such as Rossby waves and baroclinic instability, and quasi-geostrophic models suited to simulating midlatitude dynamics are described in Chapter 4. Chapter 5 examines the role of sea ice in high latitude oceans and describes how to model the sea-ice cover. Chapter 6 deals with tides and tidal models. This topic is traditionally of great importance to estuarine and shelf dynamics and models, but it is also becoming increasingly important to modeling the marginal seas and ocean basins. Finally, Chapter 7 explores coastal dynamics and simple barotropic models of the coastal oceans.

Chapters 8 to 11 are devoted to more comprehensive numerical ocean models. We will deal principally with the ocean models that have had considerable history and user experience around the world. This means inevitably those that were developed in North America. Chapter 8 deals with data and data processing—the data needed for the initialization and forcing of these models and the processing of these data and the model output. It is important to realize at the very outset that although all numerical ocean models are based on more or less the same set of governing equations, in one form or other, and therefore might be expected to yield similar results under similar circumstances, in reality, the performance and the fidelity with which oceanic processes are represented can differ widely. This has to do with the fact that when the same continuum equations (that hold exactly in the limit of vanishing grid sizes) are discretized for solutions on a digital computer, the behavior of the resulting discretized forms of the governing equations can differ widely, depending on the horizontal and vertical discretization schemes used, time-stepping methods, and the manner in which certain important terms such as *advection* and *diffusion* are handled. Although horizontal discretization is a very important consideration in numerical models of both the atmosphere and the ocean, it is the vertical discretization that sets most ocean models apart. This is because the bottom topography and its gradient are of overwhelming importance to the ocean circulation and to the horizontal and vertical transport of properties such as temperature and dissolved substances. The choice of the vertical discretization is therefore central to ocean modeling. Unfortunately, at present, no ocean model can cope gracefully with the large topographic gradients found along the continental slope and at many other places in the ocean. High grid resolutions tend to mitigate this problem, and the problem might be solved eventually by bringing raw computer power to bear on it, but the resolutions affordable at present require careful attention to vertical discretization and its consequences.

Each of the three models discussed in Chapters 9 to 11 describe a different approach to vertical discretization while solving the same basic equations governing ocean circulation processes. Chapter 9 describes models formulated in a terrain-following vertical coordinate called the sigma coordinate, patterned after a similar one in the atmosphere (Phillips, 1957). Sigma coordinate models were developed originally for application to shelf and estuarine circulations. Here, the vertical distance z is normalized by the local depth and the number of levels in the vertical is the same at all grid points and independent of the local water column depth. Both the shallow shelf and the deep ocean regions can therefore be resolved without an excessive number of vertical levels, and the kinematic boundary conditions at the bottom can be satisfied correctly. Unfortunately, this is not without penalty, and in regions of large topographic gradients the model has the potential to suffer from truncation errors in the calculation of the horizontal pressure gradient terms in the momentum equations (Haney, 1991; McCalpin, 1994; Mellor *et al.*, 1994). Chapter 10 looks at models formulated in the conventional z-coordinate system, which while free from this problem, suffer from steppy "staircase" discretization of the bottom depth and hence the inexact kinematic boundary condition at the bottom, which can in turn lead to large local vertical velocities (Gerdes, 1993). Here the number of levels varies from grid point to grid point and depends on the local depth. Unless high vertical resolution is provided in the upper layers, shallow shelf regions cannot be modeled properly. Traditionally, z-coordinate models have been applied to ocean basins, pretty much ignoring the shallower shelf regions around their margins.

Both z- and σ-coordinates are Eulerian in nature. The coordinate surfaces are either horizontal or conform to bottom topography. Here, one solves for properties at grid points, whose locations in the vertical are initially specified and held fixed. In other words, the thickness of the vertical "layers" at any grid point is invariant. On the other hand, it is possible to use a semi-Lagrangian approach and let the layer thicknesses vary with time. Here, constant density surfaces (isopycnals) can be used as coordinate surfaces. Chapter 11 considers such layered and isopycnal models. These models are particularly useful when better conservation of properties of water masses in the ocean interior is essential, since diapycnal mixing can be more accurately prescribed and controlled. However, it is inherently difficult to treat the upper mixed layer and other mixing regions in this model (Bleck *et al.,* 1989). The appearance and disappearance of isopycnal surfaces and their intersection with topography are also nontrivial to deal with (Bleck and Boudra, 1986).

Coupled models are described next in Chapters 12 and 13—coupled ice-ocean models suited to application to ice-covered seas in Chapter 12, and coupled ocean-atmosphere models suited to long-term simulations of the state of the coupled ocean-atmosphere system in Chapter 13. Coupled physical-biological models are not treated in this book, but they are discussed briefly in *Small Scale Processes in Geophysical Flows* (Academic Press, 2000). Chapter 14 concludes with a discussion of data assimilation, predictability, and nowcast-forecast systems for the oceans, topics central to the oceanic state estimation and prediction problem of great societal interest. Unlike simulations, which can be carried out without the aid of observational data, predictions require that the model be combined optimally with observational data to provide an estimation of the current state of the ocean, from which a future state can be estimated based on the dynamic constraints provided by a model.

Finally, Appendices A, B, and C present the equations of state, wavelet transforms, and empirical orthogonal functions, respectively. The latter two are becoming increasingly important to analyzing model and observational data. Appendix D is a compilation of information such as physical and dynamical constants useful to ocean modeling.

We have endeavored to present in these books elementary (but important) concepts followed by advanced material on each topic. Given the rapid pace of scientific progress, it is likely that the latter might be subject to change as new knowledge is acquired, but the former, being invariant by definition, should endure. Recent references provided should enable the reader to pursue a particular topic further if need be. The level of treatment here is appropriate to graduate studies. However, some familiarity with fluid flows is desirable. Knowledge of differential equations and computing is essential. The material laid out is useful for teaching a comprehensive, two-semester graduate level course in ocean modeling, although selected topics could be taught in a single semester.

Chapter 1

Introduction to Ocean Dynamics

Oceanography is a relatively young field, barely a century old. Major discoveries such as the reason for the western boundary intensification of currents such as the Gulf Stream and the Kuroshio, and the existence of a deep sound channel in which low-frequency acoustic energy can travel for thousands of kilometers with little attenuation, were not made until the 1940s. Even today, our knowledge of the circulation in the global oceans is rather sketchy and full of holes. However, because of the central role the oceans play in a variety of matters affecting mankind (for example, the climate), this situation is rapidly changing. Satellite remote sensing (including satellite altimetry, infrared, microwave, and ocean color sensors), long-term telemetering arrays, long endurance drifters and gliders, and smart autonomous underwater vehicles are changing radically the way we observe and monitor the global oceans. Numerical ocean modeling is even younger. The very first comprehensive numerical global baroclinic ocean model was formulated by Kirk Bryan in the late sixties (Bryan, 1969). However, the advent of high performance computers has led to a phenomenal growth in the field, especially in the last decade.

Because of the high heat capacity of water (2.5 m of the upper ocean is equivalent to the entire troposphere) and the oceans' large extent (they cover over 70% of the Earth's surface), oceans act as thermal flywheels and moderate our long-term weather. They are also huge reservoirs of CO_2 (containing about 60 times the amount of CO_2 in the atmosphere), and have a long memory (about a millennium), meaning that the residence time is large in the deeper parts of the ocean. Oceans therefore play a crucial role in determining the climatic conditions on our planet on a variety of timescales. Marked changes in the Earth's climate in the past are thought to have been associated with disruptions of the global meridional thermohaline circulation. On shorter timescales of a few years, the variability inherent to the coupled atmosphere–ocean system (primarily in the tropical Pacific) called the El Niño–Southern Oscillation (ENSO) phenomenon causes widespread disruption of the precipitation patterns around the globe. A better understanding of the oceans is also important for other reasons, including defense and commerce needs of nations. Oceans are an important source of protein for many, and a better understanding of their biological characteristics, such as their primary productive capacity, is essential to reversing the decline in fisheries and to a wise management of the oceans' biological resources. The oceans might also be able to supply part of our energy and mineral needs in the coming century.

However, the oceanographers are data-poor in general. Even today, there exist many regions in the southern hemisphere where not a single *in situ* observation of ocean properties has ever been made. Because electromagnetic energy, the backbone of sensors and communications in the atmosphere, does not penetrate deep or propagate far in the ocean, it is also difficult to probe the oceanic interior remotely. The maximum penetration occurs in the blue-green visible range of the spectrum and even then it is limited to a maximum of about 100 m in the clearest waters. Only low frequency acoustic energy is capable of propagating over long distances with very little attenuation. Because of the existence of the sound speed minimum at depth, acoustic energy is trapped in a waveguide and can travel over thousands of kilometers. This makes remote probing by sound, such as acoustic tomography, possible. It is only in the last decade or so that satellite-borne sensors such as infrared radiometers, microwave imagers, and altimeters have begun to fill in the data gaps, especially in the poorly explored southern hemisphere oceans. Since collection of *in situ* data in the oceans is quite expensive, and since satellite-borne sensors provide information on mostly the near-surface layers of the ocean, it is often thought that ocean models are central to understanding the way the oceans function. The hope is that comprehensive ocean models in combination with the sparse *in situ* and the relatively abundant remotely sensed data provide the best means of studying and monitoring the oceans. Herein lies the importance and the promise of ocean models. For estimating the future state of the oceans, i.e., for prediction purposes, numerical ocean models are quite indispensable.

The motions in the ocean are turbulent and span a wide spectrum of spatial and temporal scales. It is important for an ocean modeler to understand what scales are being represented or resolved in the model and what scales are being parameterized, and more importantly what scales are being imperfectly represented. Table 1.5.1 (see also Figure 1.5.1) lists various processes of interest in the oceans (and the atmosphere for completeness) and their corresponding time and length scales, as well as the Rossby numbers associated with them at midlatitude. Before attempting to model oceans and oceanic processes on a digital computer, it is essential to obtain a good understanding of the broad characteristics that one is trying to simulate, as far as is known from observations and analytical means.

The oceanic circulation is a complicated function of the density structure of the water masses composing the ocean basins, the radiative fluxes at its surface, and the forcing (the wind stress and buoyancy fluxes) imposed at the ocean surface by overlying atmosphere (and to some extent the astronomical tide-generating forces due to the Moon and the Sun). The surface forcing contains a variety of temporal (ranging from hourly variations to decadal and beyond) and spatial scales (ranging from kilometer scales associated with a sharp atmospheric front to basin scales). The oceans respond in a very complicated manner to atmospheric forcing and even today we do not understand the details or even the nature of this response over the entire spectrum of forcing, simply because observations spanning the spectrum just do not exist. Understanding oceanic response to this forcing, by necessity, involves appealing to the governing dynamical equations of motion, in conjunction with either simple conceptual or complicated numerical models. This is the domain of dynamical oceanography, and the reader is referred to textbooks such as Gill (1982), Pond and Pickard (1989), Cushman-Roisin (1994), and Mellor (1996a) for a treatment of the subject topic. However, since numerical modeling of the oceans and oceanic processes requires a firm understanding of at least the salient aspects of ocean dynamics, we have attempted to provide a brief summary here in this chapter.

A general idea of the overall structure of the oceans and the broad features of their circulation has been built up over the past century, painstakingly and piece-by-piece, by *in situ* measurements, and a description of these broad features is the domain of descriptive physical oceanography. The reader is referred to textbooks such as Pickard and Emery (1982), Dietrich *et al.* (1980), and Tomczak and Godfrey (1994) for an in-depth treatment of the oceanic structure and circulation. Schmitz (1996a,b) provides a particularly insightful and up-to-date description of the many fascinating features of the large scale circulation in the ocean basins and the interbasin exchanges. The subject matter is however vast and the following attempt to provide a thumbnail sketch does not do adequate justice to the topic.

Careful analyses over the past two decades by, for example, Joe Reid and his colleagues (Reid, 1981; Lynn and Reid, 1968; Mantyla and Reid, 1983) and Schmitz (1996a,b) of the hydrographic data collected over the past century (and archived at NODC), have helped advance our knowledge of the oceanic structure and circulation. Syd Levitus and his co-workers (Levitus, 1982; Levitus and Boyer, 1994; Levitus *et al.*, 1994) have produced an atlas of water mass properties such as temperature, salinity, and oxygen content in the global oceans on monthly climatological timescales. This atlas is available from NODC both in electronic and in atlas forms. The reader is referred to these for recent updates of property distributions in the ocean. Along the same lines, surface forcing has been deduced from ship-based marine surface observations by Hellerman and Rosenstein (1983), and as part of an effort to construct a Comprehensive Ocean Atmosphere Data Set (COADS) by Woodruff *et al.* (1987). These analyses are once again available in digital format and provide a good idea of the surface forcing imposed on the oceans on monthly timescales.

We will deal with these issues later in the book, but for now we merely present maps of the bottom topography (Figure 1.1.1) and the mean surface circulation (Figure 1.1.2) in the three primary basins, the Atlantic, the Pacific and the Indian. They illustrate the broad features of the oceans—such as the deep abyssal plains, the shallow shelves, midocean ridges and island chains, narrow passageways, and straits—and their circulation—such as the intense western boundary currents, the broad gyres spanning the basins, the equatorial current system, and the Antarctic Circumpolar Current connecting the various basins. It is the task of numerical ocean models to reproduce these circulation features and their variability, over the spatial and temporal scales of interest, with as much fidelity as possible.

SALIENT FEATURES

The physical characteristics of the global ocean, in terms of its shape and extent, are determined by tectonic forces that set continents adrift, and create and consume the oceanic crust. Tectonic motions are caused by internal heat flux-driven convection in the Earth's upper mantle, which is believed to extend to depths of over a thousand kilometers. This mantle convection is responsible for the slow drift (of the order of a few to several cm yr^{-1} on average) of the continental (lithospheric) plates that act as though they are floating on top of the upper mantle. Convection also causes creation of oceanic crust at midocean ridges, and its consumption at oceanic trenches due to subduction. In addition, mantle hot spots create island chains such as Hawaii. Consequently, trenches at the edges of subducting tectonic plates such as the Marianas Trench in the western Pacific are the deepest spots in the global ocean. Midocean ridges, island chains, submerged guyots, and seamounts are the shallowest features in the deep parts of the ocean and play an important role in ocean dynamics. The western Pacific in particular is dotted with thousands of seamounts which play an important role in tidal mixing and other aspects of oceanic circulation. Midocean ridges and other topographic features are important to the basin circulation, and it is therefore essential to include them and to resolve them in numerical models. The deep trenches, on the other hand, play a very minor role, and for reasons of efficiency and economy, most ocean circulation models impose a limit on the model depth (a false bottom), roughly equal to the bottom depth in the abyssal oceans, ~5000 m.

Ocean basins are ringed at their margins by shallow continental shelves, less than 200 m deep and ranging in width from a few kilometers to more than a hundred. These shelves are thought to be associated with sedimentary geological processes

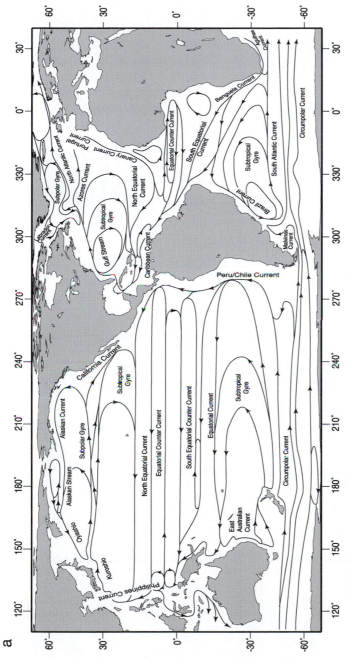

Figure 1.1.2 Broad mean circulation features in the (a) Atlantic and Pacific, and (b) Indian basins. (Based on Tomczak and Godfrey, *Regional Oceanography: An Introduction*, ©1994, Pergamon Press; reprinted by permission of Butterworth Heinemann Publishers, a division of Reed Educational & Professional Publishing Ltd.)

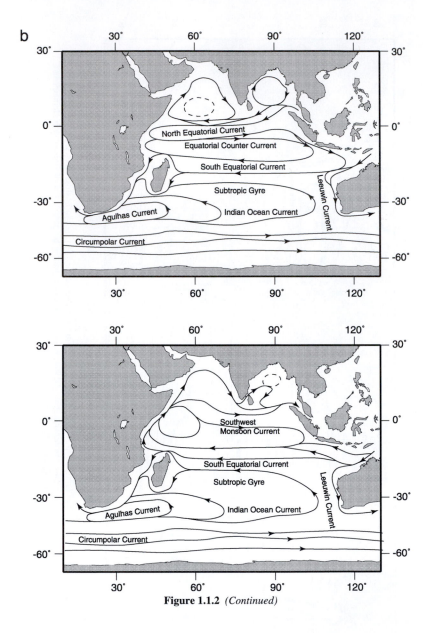

Figure 1.1.2 (*Continued*)

such as those prevalent during the last glacial period, when the sea level was ~125 m lower than at present. Because of the shallow depths involved, circulation on the shelf is forced strongly by winds and astronomical tides, while the density gradients and winds play an important role in the circulation in the deep basin proper. Because of the differing dynamics of the shelf and the basin proper, it is traditional to study them and model them separately, often using different kinds of ocean models. The transition from the shelf to the abyssal ocean is quite abrupt, and consequently these continental slope regions are quite narrow, only a few tens of kilometers wide. Since all ocean models have difficulty at present (because of the resolution affordable on present-day computers) coping with strong topographic changes, continental slope regions are the hardest to model. However, they are becoming increasingly important in the extraction of hydrocarbons, for which the ever-growing human population appears to have an insatiable appetite.

The shape and depth distribution of a basin are important factors in its circulation. From this point of view, the relatively narrow and young Atlantic basin is a strong contrast to the broad Pacific. Tectonic motions continue to broaden the Atlantic and narrow the Pacific. One of the very few geological activities of dynamical importance to the oceans is the submarine earthquakes and slumps that generate destructive tsunamis. Another is thermal vents at midocean spreading centers, whose influence can extend over large distances from the ridge crests. However, only on the timescale of millions of years are the changes in the ocean basin shape and location significant and therefore important to paleo-oceanography. Modeling the circulation in paleo-oceans is both challenging and fascinating. Twice over the past few hundred million years, due to the changing pattern of mantle convection, the continental masses have been alternately together or spread out far apart. This has had profound consequences for the resulting circulation. For example, the agglomeration of continents appears to favor amplification of diurnal tides, whereas the present distribution selectively amplifies the semidiurnal tides.

The three primary basins are connected to one another in the southern hemisphere by the Southern Ocean around the Antarctic subcontinent, while in the northern hemisphere, the Pacific and the Atlantic oceans are connected to the Arctic Ocean by straits. In addition, the Pacific and the Indian Ocean communicate directly through narrow straits in the Indonesian Archipelago. This is the only low latitude connection between two major basins. The flow called the Indonesian Throughflow is of considerable climatic importance, since it is associated with a large heat and salt flux from the Pacific to the Indian Ocean. With the Indian subcontinent forming its northern boundary, the only basin that is limited in its northern extent and hence not directly connected to the Arctic Ocean is the Indian Ocean. The Arctic and the Antarctic polar oceans are quite important from a climatic point of view since they are capped by a sea-ice cover that waxes and wanes with the seasons and modulates the air–sea and radiative fluxes at the air–sea interface. The Southern Ocean helps to redistribute water masses from one basin to another. The Circumpolar Current is the only current that is zonally continuous around the globe.

The western Pacific Ocean is ringed by marginal semienclosed seas, the Bering Sea, the Sea of Okhotsk, the Sea of Japan, the Yellow Sea, the East China Sea, and the South China Sea, whereas the Indian Ocean has the Persian Gulf and the Red Sea, and the Atlantic the Mediterranean Sea, the Caribbean, and the Gulf of Mexico. Semienclosed marginal seas are often dominated by the flow through the narrow straits that connect them to the primary basins. For example, a primary source of variability in the Sea of Japan is the Tsushima Current flowing in through the shallow Korea Strait and out through the Tsugaru and Soya straits. In the Gulf of Mexico, the Loop Current entering through the deep Yucatan Strait and exiting through the Florida Strait is the principal source of variability, since large anticyclonic eddies shed off by this current transport subtropical water masses into the western Gulf. Of the marginal semienclosed seas, the Yellow Sea and the Persian Gulf are relatively shallow (25 and 75 m deep, respectively, on average) and are therefore strongly driven by wind and tidal forcing. On the other hand, seas like the Mediterranean Sea, the Gulf of Mexico, and the Sea of Japan are minibasins. Of the marginal semienclosed seas, only the Bering Sea and the Sea of Okhotsk are significantly affected by sea-ice cover, a substantial fraction of these seas being covered by sea ice during winter.

MEAN CIRCULATION

The horizontal circulation in each primary basin is primarily driven by the surface forcing applied by the overlying atmosphere. This circulation can be divided broadly into equatorial and mid-high latitudinal components. The midlatitudinal circulation in both the Atlantic and the Pacific consists of a basin-wide anticyclonic subtropical gyre with a strong, narrow western boundary current as its western limb, and a broad, sluggish, meandering eastern boundary current as its eastern limb. The wind stress curl drives a slow equatorward transport of water masses in the basin, which is returned poleward by these western boundary currents (the Gulf Stream in the north Atlantic, the Kuroshio in the north Pacific, the Brazil Current in the south Atlantic, and the East Australian Current in the south Pacific), which transport a few tens of Sverdrups (1 Sv = 10^6 m^3 s^{-1}) of subtropical water masses poleward. In the northern hemisphere, the high latitude circulation consists of a cyclonic subpolar gyre that transports colder water masses equatorward. In the southern hemisphere, the Circumpolar Current flowing eastward around the Antarctic continent and spanning all three basins is the predominant high latitude circulation feature. The Southern Ocean constitutes the primary communication path between the three basins, although the Pacific and the Indian basins are connected to each other by narrow passages in the Indonesian archipelago, and the Pacific and the Atlantic are connected to each other indirectly through the Arctic Ocean.

The rate of mechanical work done by the wind on the oceanic general circulation is dominated by the action of the zonal wind stress component on the zonal component of the geostrophic current, principally in the Southern Ocean, but also in the Kuroshio and Gulf Stream/North Atlantic Current regions. The meridonal component of the wind stress is important only in eastern upwelling regions and the broad subtropical gyres act as regions of negative work input. Recent estimates using TOPEX/Poseidon altimetric data (Wunsch 1998) and a global ocean model (Semtner and Chervin 1992) both indicate that the rate of mechanical work done by the atmosphere on the ocean is roughly 900 GW (a global average of 2.8×10^{-3} W m^{-2}), about 25% of that from oceanic tides and very small compared to the 2 PW of net heat exchange at the air-sea interface. Nevertheless, the energy dissipated as a result may be available for cross-isopycnal mixing in the deep oceans and hence could play a very important role in global heat budget. Wunsch (1998) also points out that most of the energy in fluctuating oceanic motions is in the first baroclinic mode (~300 EJ, of which 1.2 EJ is kinetic, the rest potential) and only 2 EJ is in the barotropic motions (divided roughly equally between potential and kinetic energies). This can be compared to the roughly 0.5 EJ present in tidal motions (Kantha 1998). See Lueck and Reid (1984) and Oort et al. (1994) for a fuller discussion of the energetics in the ocean.

The equatorial circulation in each basin consists of a complicated pattern of currents and countercurrents driven primarily by the curl of the wind stress. Superimposed on these are fluctuations on a wide variety of timescales driven by fluctuating wind forcing. Changes in wind forcing generate fast-moving Kelvin and Rossby waves in the equatorial waveguide that in turn produce rapid changes in the equatorial circulation. An eastward flowing Equatorial Undercurrent at a depth of about 100–200 m is a prominent feature of the equatorial circulation. However, detailed differences, having to do with the zonal extent of the waveguide and differences in the prevailing surface forcing, exist between the three basins. Of the three, it is the broad Pacific that plays a dominant role on interannual (year-to-year) changes in global weather, the ENSO being an excellent example [see Webster and Palmer (1997) for a description of the 1997 ENSO]. Southern Oscillation is an oscillation in east-west pressure difference across the equatorial Pacific (between the western tropical Pacific and the southeastern tropical Pacific) and is correlated with El Niño and La Niña phases of ENSO. On the other hand, the North Atlantic Oscillation (NAO) is the north-south pressure difference between the low pressure region near Iceland and the high pressure region near Azores in the subtropics. This difference drives storms across the North Atlantic from west to east during winter. The variability in NAO on interannual to interdecadal scales affects weather patterns and climatic conditions in regions around the North Atlantic, while ENSO appears to have a more global impact.

The circulation in the Indian Ocean differs significantly from that in the other two basins principally because of its limited meridional extent as well as its unique seasonally reversing monsoon winds. These winds give rise to a reversing western boundary current, the Somali Current, which flows northward across the equator during southwest monsoons, but reverses and flows southward during the northeast monsoons. The protruding Indian subcontinent further subdivides the northern part of the basin and radically affects its circulation. Excess evaporation makes the upper layers in the western part, the Arabian Sea, saltier than the rest of the Indian Ocean, while large runoff from rivers draining the Indian subcontinent makes the eastern part, the Bay of Bengal, significantly fresher.

While the tropical Pacific Ocean appears to dominate the weather (and climate) on timescales of 3–5 years, inherent oscillations of the coupled global ocean–atmosphere system are thought to play a prominent role in decadal fluctuations. Analyses of tree ring records in Oregon show an ~17-year peak in the spectra of climatic fluctuations over the past 500 years. Decadal fluctuations also appear prominently in analyses of surface drift in the northeast Pacific over roughly the past century (Ingraham *et al.*, 1998). Latif (1998) discusses interdecadal variability, and its manifestation and dynamics in coupled ocean–atmosphere models.

On even longer timescales, centuries and beyond, it is the north Atlantic that is prominent. During winter, strong storm-induced surface cooling creates dense salty water in its subpolar seas that sinks to the bottom of the basin, flows equatorward, and, after crossing the equator, flows into the Southern Ocean. These waters are returned to the surface by slow upwelling. The resulting meridional thermohaline circulation is important to the long-term climate. This circulation permeates all three basins through the Southern Ocean. Cessation of this "conveyor belt" is thought to trigger radical and rapid changes in midlatitude climate (Broeker, 1997), evidence for which exists in numerous lake and deep ocean sediments, and 110,000-year-long Greenland ice cores containing proxy climatological data from the past. Adkins *et al.* (1998) present evidence for sharp changes in deep sea circulation at the end of the last ice age. Dense water formation occurs also in the Southern Ocean around Antarctica, where Antarctic bottom water (AABW) is formed.

The Indonesian Throughflow is the only low latitude connection between ocean basins at present. Its transport of mass, heat, and salt from the Pacific to the Indian Ocean is important to the circulation and water mass structure of both basins. This heat transport removes a significant fraction of the heat the western Pacific receives and leads to a westward shift of the warm pool and the atmospheric deep convection center. This exerts a significant impact on the tropical and global climate. The throughflow is also an important branch of the global thermohaline circulation. Godfrey (1996) has reviewed the role of the Indonesian Throughflow on the circulation and water mass structure in the Pacific and the Indian oceans. Schneider (1998) describes simulations of the throughflow by the coupled air–sea general circulation model ECHO developed at the Max-Planck Institute at Hamburg and discusses its impact on the global climate.

How the water mass structure is maintained in the global oceans is an important matter (Munk, 1966). The stable stratification that exists in the interior of the oceans at present is due to two processes. One is the cold deep and intermediate water formation in subpolar seas. The bottom water formed in the Greenland and Labrador seas and around the Antarctic continent slowly fills the abyssal oceans with cold, salty water masses. The timescale associated with this process is a few millennia. The slow upwelling that results is counterbalanced by mixing in the ocean interior. The other process is the subduction of water masses (for example, 18°C water in the north Pacific) in the vicinity of the poleward limbs of the sub-tropical gyres that appears at a depth of a few hundred meters in the equatorial regions. This process is called thermocline ventilation (see Pedlosky, 1996) and the timescale associated with it is a few decades. Both these processes are important to the water mass structure and the two complement each other in maintaining the stratification in the ocean interior. Ocean models aiming to simulate variability on multidecadal scales must pay particular attention to thermocline ventilation processes. Climate models investigating long-term changes in Earth's climate must account properly for the thermohaline circulation.

The circulation in the Arctic Ocean is dominated by its perennial ice cover, and consists of an anticyclonic Beaufort Gyre and a Transpolar Drift Stream that transports water masses and sea ice from the vicinity of the Bering Strait into the north Atlantic through the Denmark Strait. This transport of ice and hence freshwater has a strong effect on the air–sea exchange in the northwest Atlantic. Strong wintertime storm forcing of this basin is to some extent moderated by the sea-ice cover.

Ice sheets, occupying 5% of the Earth's land surface, and glaciers are the largest reservoirs of freshwater on Earth, accounting for about 70–80%, with the Greenland sheet accounting for 10% of this, and Antarctica 89%, with the remaining 1% in glaciers and small ice caps (Bindschadler, 1998). Were this all to melt, the global sea level would rise ~70 m. During the previous interglacial epoch, ~125,000 years ago, the ice sheets were slightly smaller and the sea level ~5 m higher. During the last glacial period, ~20,000 years ago, the ice sheet volume was twice the current value and the sea level was ~125 m lower.

Ever since their formation, approximately 3.5 billion years ago, the ocean has played an important role in the geochemical evolution of the Earth. Almost all of the oxygen found in the Earth's atmosphere was created by oxygenic photosynthesis in the ocean by unicellular phytoplankton, with oxygen levels reaching the present day levels roughly 2.2 billion years ago. A massive amount of organic carbon (15 Ptonne) was simultaneously produced and sequestered in sedimentary rocks. Satellites have made it possible to estimate the global annual net primary production (NPP) of carbon, the amount of photosynthetically fixed carbon available to other trophic levels. From ocean color data gathered by the CZCS sensor, Field et al. (1998) estimate that in the present day oceans, NPP is ~49 Gt, roughly one third of which is exported to the deep ocean and hence sequestered from the atmosphere for centuries to millenia (Falkowski et al.1998). The corresponding terrestrial value (deduced from the land vegetation index from AVHRR) is ~56 Gt, comparable to the oceanic value, even though phytoplnkton biomass (~1 Gt) accounts for only 0.2% of the global photosynthetically active primary producer biomass (~500 Gt). This means that the average turnover time scale of biomass is roughly a week in the ocean, compared to 19 years on land (Field et al. 1998). This rapid turnover implies that increased NPP will not result in substantial changes in carbon stored in the phytoplankton biomass, but rather in carbon sequestered through transport of carbon into the oceanic interior. Phytoplankton absorb only 7% of the PAR incident at the ocean surface, while terrestrial plants absorb 31% of the PAR incident on land without permanent ice cover, and the production per unit surface area over land is three times that over the ocean. Maximum NPP is however, similar (1-1.5 kg C m^{-2} yr^{-1}), with upwelling regions being high NPP regions in the ocean and humid tropics on land.

Primary biological production in the ocean is conditional upon the simultaneous availability of solar insolation (the dominant energy source available and essential for photosynthesis), and inorganic nutrients dissolved in the water (to form plant tissue). The latter include inorganic elements carbon, nitrogen, phosphorus and sulfur available in the form of carbon dioxide, nitrates (or molecular nitrogen), phosphates and sulfates dissolved in the water. Primary poductivity is therefore limited to the euphotic zone, where the solar insolation is adequate for photo-synthesis. Solar insolation is the primary limiting factor in high latitudes and responsible for seasonal modulations in productivity. In near-surface waters at low latitudes, the sunlight is usually not a limiting factor, and the growth rate of phytoplankton and hence primary productivity is dictated instead by the scarcest of the inorganic nutrients. Which nutrient is limiting depends on the region under consideration, but in most of the oceans it is nitrogen and secondarily phosphorus. However, certain trace minerals such as dissolved iron are also essential for photosynthesis and the availability of these trace mineral constituents often limits productivity even when the bulk nutrients are plentiful. Vast areas of the global ocean, including Southern Ocean, and the eastern equatorial Pacific are high-nutrient, low-chlorophyll (HNLC) regions, where despite the relative abundance of bulk nutrients, biological productivity is small because of the scarcity of iron (Falkowski et al. 1998, Behrenfeld and Kilber 1999). This opens up the possibility that the productivity in these waters can be increased to the level limited only by the availability of bulk nutrients such as nitrogen, by an artificial augmentation of trace nutrients. This was confirmed by experiments in the equatorial Pacific, where artificial iron enrichment gave rise to spectacular, but temporary phytoplankton blooms (Mullineaux 1999).

MODELING ISSUES

The aspect ratios of the oceans and the atmosphere are small. The average depth of the ocean (troposphere) is ~4 km (10 km) and therefore the vertical scale of motions is O (1 km). For large scale motions, the horizontal scale is O (1000 km). Therefore the ratio of the vertical scale H to the horizontal scale L is O (10^{-3}). From mass continuity, this also means that the ratio of typical vertical velocity W to the horizontal velocity U must also be of the same order (or less because of the inhibition of vertical motions by the ambient stable stratification). Typically W ~ 10^{-4}–10^{-5} m s^{-1} and U ~ 10^{-1} m s^{-1} in the oceans, except in regions of deep convection, where W can reach several centimeters per second. However, while the vertical velocities are small in the oceans, this does not mean they can be neglected. They are very important for processes such as coastal upwelling, thermocline ventilation, ventilation of the deep ocean, and CO_2 cycling in the deep.

Basin scale and global models often ignore or simplify the surrounding shelves and marginal seas, and in most cases, justifiably. Even then, the need to resolve the narrow boundary currents, and the density structure in the vertical, places severe demands on computer resources. Added to this is the difficulty in prescribing accurate surface fluxes, and the necessity to carry out long-term integrations to attain a reasonably equilibrated oceanic state. It is relatively simpler to model semienclosed marginal seas. However, open boundaries at which flow conditions are not known well constitute a problem, although this is not so severe since narrow straits often delineate these boundaries and restrict the interaction of the sea with the basin. On the other hand, marginal seas such as the Labrador Sea, the Greenland Sea, and the Gulf of Maine in the Atlantic, the Arabian Sea and the Bay of Bengal in the Indian Ocean, the Weddell and Ross seas in the Southern Ocean, and continental shelves in general are wide open to the ocean and hence are strongly tied dynamically to the basin. They are best modeled as part of the basin itself. Modeling them separately is a much harder task because of the difficulty in prescribing the variability in lateral boundary conditions on the wide open boundaries on the timescales of interest, although nesting these regional models in a basin-scale model is a viable alternative. The subpolar seas such as the Labrador Sea and the Greenland Sea in the north Atlantic are important sources of deep water. Intermediate water formation occurs in seas like the Sea of Okhotsk. The presence of sea-ice cover during winter in the Bering Sea and the Sea of Okhotsk, as well as the marginal seas in the Arctic and around the Antarctic, means that a coupled sea-ice-ocean model is needed to model the circulation in these seas.

Coupled ocean-atmosphere models are essential for simulations and estimations of the oceanic state over time scales longer than a couple of weeks. Particular attention must be paid to air-sea exchange of momentum, heat and fresh water in these models. Simulations and estimates of the biological (and chemical) state of the oceans requires coupling to an eco-system model. However, in all these cases, the fidelity with which the physical state is represented by the physical model has a profound influence on the coupled models. These issues will be explored in depth in this book.

COMPARISON WITH THE ATMOSPHERE

There are a great many similarities between fluid motions in the oceans and the atmosphere, but there are many differences as well. The latter have to do with the difference in the driving mechanisms and in the scales of motion. It is impossible to list all the similarities and dissimilarities and their causes here. Instead, the reader is referred to excellent texts on geophysical fluid dynamics, the discipline that deals with both the atmosphere and the ocean, such as Gill (1982), Cushman-Roisin (1994), and Pedlosky (1987, 1996). Since quite a bit of what we know about oceanic midlatitude dynamics is derived from work on the atmospheric counterpart, it is also worthwhile to refer to texts on atmospheric dynamics such as Holton (1992).

The principal difference between the oceans and the atmosphere is in the way they are driven. The atmosphere is heated mainly from below, although some bulk heating occurs near the top of and above the tropopause. Solar radiation pretty much passes through the atmospheric envelope (except at certain wavelengths) and heats the ground and the water surfaces, which in turn transfer the heat to the atmosphere. The heat loss, on the other hand, occurs throughout the atmospheric column by radiation to space. The differential radiative heating of continental masses and the oceans drives the large scale atmospheric circulation. This circulation (like the ocean circulation) is predominantly in the horizontal. Narrow regions, where vertical motions penetrate large vertical distances, exist in only localized regions such as the intertropical convergence zones (ITCZ) and cumulonimbus columns. Heat sources and sinks due to condensation of water vapor and evaporation of liquid water have profound effects on atmospheric dynamics, and the resulting variability in cloud cover has a dominant influence on the radiative balance of the atmosphere (and heating of the upper layers of the ocean). A similar mechanism is absent in the oceans. In contrast, the oceans are driven at the surface by atmospheric winds and fluxes, and the solar heating is also confined to a thin layer at the surface. Any heat loss must also occur at the surface.

Another principal difference is in the time and length scales; oceans have a long memory, whereas, except in the stratosphere where chemical constituents can have residence times of many years, there is little direct memory in the atmosphere. Changes in the radiative balance in the atmospheric column (due to clouds and aerosols), and phase conversions involving water vapor and clouds, and the attendant latent heat absorption and release, cause changes in atmospheric circulation, whereas it is the changes in surface forcing that primarily drive the oceanic variability. Slow internal adjustments occur in both through the generation, propagation, and dissipation of planetary Rossby waves, which propagate much more slowly in the oceans. In the oceans, Boussinesq approximation is quite adequate, since the density variations are small. In the atmosphere, the density variations are large, even when adiabatic expansion with height is taken into account, and methods such as the anelastic approximation, which filters out acoustic waves but retains density changes, are often useful for dealing with the bulk of the column.

As far as midlatitude dynamics are concerned, the disparity in length scales between the atmosphere and the oceans is quite important. While the length scale associated with the meridional variation of planetary vorticity is the same in both ($L_\beta = R_E \tan \theta \sim 3500$ km at 30°), the Rossby radius of deformation (a), a measure of the horizontal scale of motions in the fluid column, is about 1000 km in the atmosphere and about 40 km in the midlatitude oceans (for the first baroclinic mode). Also, the intermediate scale $L_I = (a^2 L_\beta)^{1/3}$ is ~1500 km, close to the Rossby radius in the atmosphere, but for the oceans it is ~240 km, much larger than Rossby radius for the oceans. L_I denotes the scales beyond which the standard quasi-geostrophic (QG) equations (see Chapter 4) are not valid (Charney and Flierl, 1981).

Another major difference between the oceans and the atmosphere is the constraint imposed by meridional boundaries in the former and the general lack of obstructions to zonal flow (apart from some midlatitude mountain chains) in the latter. The oceans are divided into various basins, and this leads to western boundary intensification and strong current systems such as the Gulf Stream and Kuroshio. These are unique to the oceans, although intense jet-like flows called jet streams about 1000 km wide (of the order of the Rossby radius) are prevalent along the polar front in the atmosphere.

In summary, it is clear that the circulation in the global oceans is extremely complex, and the task of modeling it with fidelity, over the entire spectrum of its temporal and spatial variability, is both challenging and arduous. Modeling the thermohaline circulation is particularly hard, because of the necessity to avoid spurious modification of deep water masses. Generally speaking, numerical modeling of the oceans and oceanic processes requires meticulous attention to details, from both the dynamical and the numerical points of view. We will discuss the dynamical aspects in the rest of this chapter and the numerical aspects in the next.

1.1 TYPES, ADVANTAGES, AND LIMITATIONS OF OCEAN MODELS

The choice of a particular ocean model or modeling approach depends very much on the intended application, and on the computational and pre- and post-processing capabilities available. A judicious compromise is essential for success.

With this in mind, numerical ocean models can be classified in many different ways [the following is a slightly modified version of that in Kantha and Piacsek (1997)]:

- *Global or regional*. The former necessarily requires high performance computing capabilities, whereas it may be possible to run the latter on powerful modern workstations. Even then, the resolution demanded (grid sizes in the horizontal and vertical) is critical. A doubling of the resolution in a three-dimensional model requires almost an order of magnitude increase in computing (and analysis) resources. It is therefore quite easy to overwhelm even the most powerful high performance computer (and workstation), whether it be a coarse-grained multiple CPU vector processor such as a Cray T-90 or a massively parallel machine such as a Cray T3E, irrespective of whether the model is global or regional. Regional models have to contend with the problem of how to inform the model about the state of the rest of the ocean—in other words, of prescribing suitable conditions along the open lateral boundaries. Often the best solution is to nest the fine resolution regional model in a coarse resolution model of the rest of the basin.
- *Deep basin or shallow coastal*. The prevailing physical processes and the underlying driving mechanisms are essentially different for the two. Circulation in shallow coastal regions is highly variable, driven primarily by synoptic wind and other rapidly changing surface forcing (and near river outflows, by buoyancy differences between the fresh river water and the saline ambient shelf water). Wind mixing at the surface and processes at the bottom are important, and a numerical model that has reliable mixing physics and which resolves the bottom boundary layer is therefore better suited to coastal applications. A model such as the one developed at Princeton University, that employs a bottom-following, sigma-coordinate vertical grid and incorporates an advanced turbulence closure (Blumberg and Mellor, 1987; Kantha and Piacsek, 1993, 1997), may be essential for such applications.

  Deep basins, on the other hand, are comparatively sluggish, and the horizontal density gradients, especially below the wind-mixed upper layers, are a dominant factor in the circulation. The upper mixed layer can often be modeled less rigorously, especially for applications that do not require consideration of air–sea interaction processes. The popular z-level Geophysical Fluid Dynamics Laboratory Modular Ocean Model (MOM2), with or without an upper mixed layer, is a good candidate for modeling the ocean basins (and deep marginal and semienclosed seas) on a variety of timescales ranging from "synoptic" to climatic. The very first global baroclinic ocean model (Bryan, 1969), on which many modern global models are based (Semtner and Chervin, 1992; Semtner, 1995), was a z-level model without an upper mixed layer.

In a z-level model, a number of horizontal levels are defined in the water column and the equations written for the oceanic variables at each level and each point on the model grid in the horizontal, and solved. A hybrid model combining z and σ coordinates is also possible (Gerdes, 1993). These are all Eulerian approaches. Another equally viable approach is a semi-Lagrangian approach (Hurlburt and Thompson, 1980) that divides the ocean vertically into a number of layers and models the variation in properties such as the thickness and density of each layer at each grid point on the horizontal grid. More recently developed isopycnal models (Oberhuber, 1993a,b; Bleck and Smith, 1990) belong to this category. Since mixing in the deep oceans is primarily along isopycnals (isodensity surfaces), isopycnal basin scale models are expected to perform a better job in depicting interior mixing and are better suited to long-term simulations, especially when accurate depiction and preservation of water mass characteristics are essential.

- *Rigid lid or free surface.* Oceanic response to surface forcing can often be divided into two parts: fast barotropic response mediated by external Kelvin and gravity waves on the sea surface, and relatively slower baroclinic adjustment via internal gravity, Kelvin, planetary Rossby, and other waves. On long timescales, it is the internal adjustment that is important to model and it is possible to suppress the external gravity waves by imposing a "rigid lid" on the free surface. This permits larger time stepping of the model, and models used for climatic-type simulations are usually of the rigid lid kind. The very first global ocean model (Bryan, 1969) was a rigid lid model. In such a model, at each time step an elliptic (Poisson) equation for stream function has to be solved. This is difficult to carry out efficiently on vector and parallel processors and for complicated basin shapes (including islands). Also, under synoptic forcing, the convergence of the iterative solver slows down. For these reasons, free surface models, implicit and explicit, are becoming more popular for nonclimatic simulations. A mode-splitting technique must then be employed to circumvent the severe limitation on time stepping that would otherwise be imposed by the presence of fast external modes. To diminish the drawbacks of a rigid lid model, Dietrich *et al.* (1987) and Dukowicz *et al.* (1993) have developed versions in which one works with the pressure on the rigid lid, rather than the barotropic stream function, leaving the domain multiply connected and with better matrix inversion characteristics. With the split-explicit free surface formulation by Killworth *et al.* (1991) and implicit free surface formulation by Dukowicz and Smith (1994), rigid lid models are no longer advantageous from the point of view of either efficiency or dynamics. For shallow water applications, such as storm surge and tide modeling, free surface dynamics must be retained.

- *Hydrostatic, quasi-hydrostatic, or nonhydrostatic.* Most, if not all, large scale circulation models for the oceans (and the atmosphere) are based on the hydrostatic form of the incompressible Navier–Stokes equations. This exploits the fact that the aspect ratio (the ratio of vertical to horizontal scales of motion) of such motions is usually small and that the static stability of the water column in the stably stratified oceans is significant. While they are adequate in modeling gyre scale (≥1000 km) circulation and mesoscale eddies (10–100 km), the hydrostatic approximation breaks down for scales less than the ~10 km typical of small scale processes such as deep convective chimneys, which have horizontal scales of typically 1 km. Hydrostatic approximation also requires neglecting the horizontal component of the Coriolis acceleration for energetic consistency, which means that the angular momentum is only approximately conserved. Approximation where the strict hydrostatic approximation is relaxed to the extent that the horizontal component of the Coriolis acceleration is retained is known as quasi-hydrostatic approximation (White and Bromley, 1995). Fully nonhydrostatic models (Jones and Marshall, 1993; Marshall *et al.*, 1997a,b) can be used at all horizontal scales, but unlike hydrostatic models that require the inversion of only a two-dimensional elliptic equation, they involve the solution of a three-dimensional elliptic equation subject to Neumann boundary conditions, and can therefore be computationally intensive. However, Marshall *et al.* (1997a) show that for large scale circulation applications, the fact that the pressure field is close to hydrostatic can be exploited so that the overhead in solving the nonhydrostatic equations is minimized.
- *Comprehensive or purely dynamical.* Since the density gradients are over-whelmingly important and the density below the upper layers in the global oceans changes very slowly, it is often possible to ignore completely the changes in density with time. The model then becomes purely dynamical and can be used to explore the consequences of changing wind forcing at the surface. Purely dynamical layered models belong to this category and are essentially isopycnal models without the thermodynamic component (Holland and Lin, 1975a,b; Hurlburt and Thompson, 1980). Their principal advantage is that it is often possible to select a limited number of layers in the vertical (as few as two) and still include salient dynamical processes. This enables very high horizontal resolutions necessary for resolving mesoscale features in the oceans to be made. The highest resolution global model at present is the Naval Research Laboratory 1/8° eddy-resolving global model (Metzger *et al.*, 1992) that needs a 16-processor Cray T-90 for multiyear simulations.

Even with modern computing power, it is necessary to sacrifice either vertical or horizontal resolution for many (especially global) simulations. The layered (and isopycnal) models sacrifice vertical resolution, whereas the z-level models, employing a large number of levels in the vertical, are necessarily comparatively coarse-grained in the horizontal. The highest resolution dynamical-thermodynamic z-level model at present is the 1/5° POP model at Los Alamos (Fu and Smith, 1996; Dukowicz and Smith, 1994; see Semtner and Chervin, 1992, for a description of the basic model), and it stretches the capability of a 256-processor CM5 to the limit.

- *With applications to short-term simulations or long-term climate studies*. On climatic timescales, it is extremely important to model correctly the thermohaline circulation driven by the formation of dense deep water masses during strong wintertime cooling in subpolar seas, especially in the Atlantic. Several centuries are needed for a water particle that sank, say, in the north Atlantic, to surface again in the Indian or the Pacific Ocean. Because of this long residence time (or equivalently long memory) of the deep oceans, it is necessary to make multicentury simulations, and irrespective of whether isopycnal or z-level models are employed, the horizontal and vertical resolutions that can be afforded are necessarily coarse (e.g., Boville and Gent, 1998). Accurate ocean simulations on climatic timescales belong to the category of grand challenge problems requiring a multi -teraflop (10^{12} floating point operations per second) computing capability that has been the holy grail of the computer industry in the 1990s.
- *Quasi-geostrophic (QG) or primitive equation based*. In the seventies and early eighties, the limited computing power available then led some to explore simplifications to the governing equations to be solved (Holland,1985). QG models assume that there is a near balance between the Coriolis acceleration and the pressure gradient in the dynamical equations in the rotating coordinate frame of reference in which most ocean models are formulated. This filters out fast gravity waves, and the resulting simplification enables longer time steps to be taken at a given resolution, or equivalently higher vertical and horizontal resolutions for a given computing capability to be achieved. QG models have strong limitations with respect to the accuracy with which some physical processes can be depicted, and are becoming obsolete in the modern high computing power environment. Intermediate models, on the other hand, are in between QG and PE models in complexity and involve retaining higher order terms in the Rossby number expansion of the governing equations (Allen *et al.*, 1990).
- *Barotropic or baroclinic*. In a barotropic model, the density gradients are neglected so that the currents become independent of the depth in the water column. Many phenomena such as tidal sea surface elevation fluctuations and storm surges can be simulated quite adequately by a barotropic model, which is a two-dimensional (in the horizontal) model based on the vertically integrated equations of motion. One of its advantages is that it requires an order of magnitude less computing resources than a comparable baroclinic

model. However, when it is important to model the vertical structure of currents, or the density field, a fully three-dimensional baroclinic model is necessary.

- *Purely physical or physical-chemical-biological.* Often there is a need to model the fate of chemical and biological constituents in the ocean; to do so, it is essential to solve not only the dynamical equations governing the circulation and other physical variables, but also the conservation equations for chemical and biological variables. Modeling the fate of inorganic CO_2 in the oceans, a problem germane to global warming, and modeling the primary productivity in the upper layers of the ocean are two such examples. The former requires solving for at least two more variables, the total CO_2 and alkalinity, whereas the simplest biological model must solve for at least three (but often as high as nine) additional quantities, the nutrient, phytoplankton, and zooplankton concentrations (the so-called NPZ model). The governing equations are transport equations with appropriate source and sink terms, whose parameterization is not always quite straightforward. This implies not only additional complexity but also requires more computing (and data) resources.

- *Process studies-oriented or applications-oriented.* Models that are used to study some salient processes (for example, western boundary currents and gyre circulation) can be considerably simplified and hence made less computationally intensive, since it is often possible to isolate and retain only the relevant physical process, and ignore the rest. Also, such models may not require extensive observational data for model initialization and forcing. They are most often run free in a predictive (or prognostic) mode. On the other hand, applications-oriented models such as those used for ocean prediction purposes require extensive observational data for realistic initialization, forcing, and data assimilation. Assimilation into the model by one means or another of observations of the state of the ocean is indispensable for realistic nowcast/forecast/hindcast applications. Data-assimilative ocean models often employ approaches very similar to those employed by numerical weather prediction (NWP) models in the atmosphere.

- *With and without coupling to sea ice.* Many global ocean models do not at present include sea ice (e.g., Semtner and Chervin, 1992; Semtner, 1995). Some include only an approximate treatment of sea ice. However, comprehensive ice–ocean coupled basin models of the Arctic do exist. For a sea-ice model coupled to a z-level ocean model, see Hibler and Bryan (1987), and to an isopycnal model, see Oberhuber (1993a,b). Sea ice insulates the ocean from the cold atmosphere during winter and mediates the exchange of heat and momentum between the two, and therefore such models involve solving dynamical and thermodynamic equations for the sea-ice cover and its coupling to the underlying ocean. Sea-ice albedo is responsible for the diminished solar insolation penetrating into ice-covered seas during summer. Sea-ice cover therefore plays a very important role in polar and subpolar seas, because of these positive feedback effects it embodies. Climate models concerned with the effect of increasing anthropogenic CO_2 in the atmosphere suggest that global warming could lead to larger increases of temperature in the polar regions and shrinkage of perennial sea-ice cover with potentially catastrophic consequences on coastal communities from the resulting sea level rise due to melting of landfast ice, glaciers, and continental ice sheets.

- *Coupled to the atmosphere or uncoupled.* Finally, for accurate simulation of long timescale processes, it is essential to couple ocean models with atmospheric models. Such coupled models are being increasingly used for such things as forecasting El Niño events. Most often, either the atmosphere or the ocean is highly simplified in such models, although modern high performance computers are enabling comprehensive atmospheric general circulation models (GCMs) to be coupled to comprehensive global ocean models. Applications include simulations of interannual variability (Mechoso *et al.,* 1995) and short-term climate (Boville and Gent, 1998). Truly comprehensive coupled models with applications to long-term climate studies require teraflop computing capability that will be rather routinely available in the coming years.

In this book, we will touch upon many of the above aspects of numerical ocean modeling. However, the field is vast, and the literature voluminous; consequently, all we can do is provide a road map that can help the reader explore a particular aspect. We will attempt to provide the necessary basics of ocean dynamics and numerical modeling to enable a potential ocean modeler to acquire the requisite knowledge base. While Pond and Pickard (1989), Gill (1982), Dietrich *et al.* (1980), and Cushman-Roisin (1994) are excellent starting points for studying ocean dynamics (an advanced treatment can be found in Pedlosky (1979, 1996)), the treatment here is from the point of view of numerical modeling and therefore provides a different perspective on the subject. It is also designed to provide a unified analytical-numerical approach to the subject matter.

1.2 RECENT EXAMPLES

The availability of multigigaflop coarse-grained and massively parallel computers, and abundant satellite remotely sensed data on ocean surface properties, has produced a radical change in numerical ocean modeling. For a long time, the principal limitation in ocean modeling has been the inability to resolve the internal Rossby radius of deformation (see below) crucial to the accurate depiction of salient processes such as mesoscale eddies and western boundary currents. The former are the principal mode of horizontal mixing and transport in the turbulent oceans, and the latter are the underpinnings of the gyre scale circulations in each basin. We are still not at a stage where we can faithfully represent and reproduce these processes in numerical models, but models have reached a stage where they are now "marginally eddy resolving" or "eddy-permitting." Semtner (1995), Stammer *et al.* (1996), and Fu and Smith (1996) describe such models. The resolution attainable is about 1/5° (about 20 km) in comprehensive and 1/8° (about 12 km) in simplified global models, and about 1/16° (about 6 km) in basin scale models. Given the fact that the western boundary currents have widths of less than about 100 km, and mesoscale eddies range in size from about 20 km to hundreds of kilometers, one probably needs at least a 1/20° resolution (about 5 km) to represent the principal ocean dynamics reasonably well. This requires high performance computers with multiteraflop capability. Nevertheless, current ocean models are beginning to be quite realistic.

1.2.1 ALTIMETRY

The advent of satellite precision altimetry, which has enabled sea surface height (SSH) fluctuations to be measured to an accuracy of 2–3 cm rms, has revolutionized oceanic observations (Fu *et al.,* 1994). The NASA/CNES TOPEX/Poseidon (T/P) mission launched in 1992 has been phenomenally successful in measuring SSH fluctuations in the global oceans to an unprecedented degree of accuracy over the past several years. This altimeter samples the ocean along predetermined tracks, with each track revisited every repeat cycle of approximately 9.94 days, and there are now over 6 years of data providing more than 230 realizations. An altimeter measures the distance between itself and the sea surface by measuring the propagation delay in the reflection of its microwave signal by the ocean surface. The dual frequency capability of T/P enables accurate ionospheric corrections to propagation time to be made and a bore-sighted radiometer enables accurate wet tropospheric corrections to be made. Overall, the distance of the sensor from the sea surface can be determined to an accuracy of 2–3 cm. Because of the relatively high orbit (~1350 km from the Earth's surface) and the accuracy with which it can be tracked, the position of the sensor in space (its ephemeres) can also be determined very accurately (2–3 cm rms). These two measurements provide the SSH relative to the Earth's center to an accuracy of about 3–5 cm (Fu *et al.,* 1994).

However, the principal problem is that the major part of this signal is the geoid, the equipotential surface that the sea surface would conform to, were the oceans motionless. This surface cannot be determined accurately at high wavenumbers of interest to mesoscale variability without a dedicated gravimetric mission. However, this stationary geoid signal can be eliminated by averaging the SSH observations over many repeat cycles at points along the track and removing this mean from each of the observations. Unfortunately, this also removes a mean oceanographic SSH signal over the averaging period at each point so that one is left with only the SSH anomaly from that mean. Thus, while SSH variability can be accurately mapped by an altimeter at present, the absolute SSH cannot be. Nevertheless, T/P has provided oceanographers with a convenient means to sample the many important temporal and spatial scales of ocean variability over the global oceans, a feat hitherto impossible to achieve because of the highly scattered and sporadic nature of *in situ* observations. The resulting high quality time series data set has enabled the global ocean variability to be monitored and either compared with or assimilated into high resolution global models. Wunsch and Stammer (1998) discuss the utility of modern satellite altimetry in obtaining a better understanding of the variability of the oceanic circulation.

T/P has also enabled ocean tides to be mapped to an unprecedented level of accuracy in the global oceans [3-cm residual rms error (Andersen *et al.,* 1995, Shum et al. 1997)], making accurate tides available everywhere in the global oceans, not just along the coastline and at a few instrumented locations in the open ocean. Since tides are a major signal in SSH variability, this permits not only accurate subtidal variability to be deduced in the global oceans, but also a better understanding of tides and tidal energetics to be realized. Assimilation of T/P-derived tides into a high resolution global barotropic tidal model (Kantha, 1995b) is one illustration of combining two of the proudest achievements of 20th century technology, high performance computers and precision altimetry, to enable an accurate estimate of the tidal dissipation in the global oceans.

We will present results from both a high resolution 2D global barotropic tidal model and a high resolution 3D global circulation model to illustrate the state of the art in numerical ocean modeling as of 1998.

1.2.2 A 2D Barotropic Model

Kantha (1995b) describes a high resolution (1/5°) nonlinear near-global (the Arctic Ocean excluded) 2D barotropic tidal model that assimilates altimetric tides and measurements from tide gauges (see Chapter 6 on tides) to provide an accurate estimate of tides everywhere in the global oceans. The results of such a model are useful in understanding tidal energetics, a topic of great interest to astronomy and geophysics (Kantha *et al.,* 1995; Munk, 1997; Munk and Wunsch, 1998; Kantha, 1998). The model has 1.4 million grid points and can be integrated for an individual tidal constituent or all the primary ones. Figure 1.2.1 shows the M_2 and K_1 tidal current ellipses derived from the model (see Chapter 6 for more details and results). The ellipses are plotted on a logarithmic scale to enhance the currents in the deep,

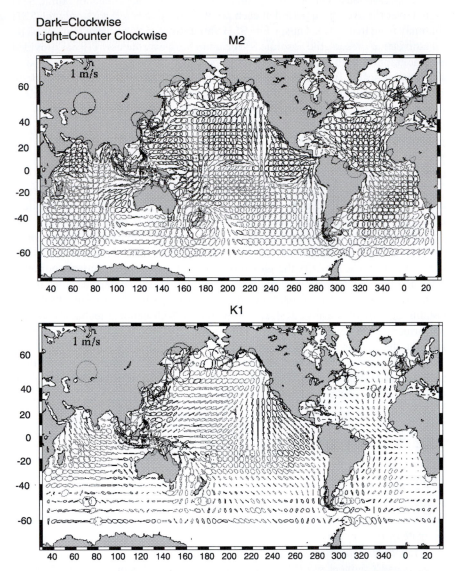

Figure 1.2.1 Tidal current ellipses for M_2 and K_1 tides in the global oceans from a $1/5°$ barotropic tide model. Note the logarithmic scale designed to enhance currents in the deep, which are typically 1 to 2 cm s^{-1}. From Kantha (1995b).

which are typically 1–2 cm s^{-1}, compared to those in shallow water, which are 20–50 cm s^{-1}. Kantha's model results indicate that the lunar tides are dissipating the Earth–Moon gravitational energy at a rate of 3.17 TW and are leading to a slowdown in the Earth's rotation rate by about 2.4 ms per century and an increase in the Moon's orbital radius of about 3.7 m per century. Munk (1997) calls the convergence of the lunar tidal dissipation rate estimates by several diverse methods to about 3.17 TW, one of the triumphs of 20th century science. Kantha and Tierney (1999) have updated these results.

1.2.3 A 3D BAROCLINIC MODEL

Figure 1.2.2, from Fu and Smith (1996), shows the rms SSH variability (with respect to the 1992–1994 mean) derived from a high resolution (also 1/5° on average) global 3D circulation model compared to that derived from T/P altimetry from October 1992 to October 1994. The model is the Parallel Ocean Program (POP) developed at the DOE Los Alamos National Laboratory (Smith *et al.*, 1992; Dukowicz and Smith, 1994), based on the GFDL model (Bryan, 1969). It has 20 vertical levels and over 11 million grid points, and is run on a massively parallel machine (CM5) driven by winds and surface fluxes from ECMWF for the period 1985–1995. This is the most accurate (in comparison with altimetric SSH data) and the highest resolution, fully thermodynamic global ocean model run to date and hence represents the state of the art in numerical ocean modeling. Figure 1.2.2 shows that the model SSH variability is quite realistic compared to that from T/P, while still anemic. The model energy levels are more than a factor of two too low (25 cm^2 vs 64 cm^2 globally). The major western current systems peter out prematurely and do not extend well into the basins as they should. Nevertheless, the model appears to simulate the major ocean current systems quite well (Fu and Smith, 1996), especially features such as the annual cycle. Figure 1.2.3 shows the amplitude and phase of the annual cycle from the model and T/P. The spatial pattern of variability is quite realistic. The phases are reproduced quite well by the model but the amplitudes are smaller. Similar conclusions were reached by Stammer *et al.* (1996) comparing the results from Semtner and Chervin's (1992) global model run at 1.6 times coarser resolution than the POP model. Thus, in general, the current state-of-the-art ocean models are beginning to be quite realistic, a considerable achievement considering the fact that just a decade or so ago, the ocean models could not even reproduce the spatial features of oceanic circulation correctly. While imperfect physics and surface forcing are responsible for some of the residual inaccuracies and the anemic nature of the ocean currents in the model, it

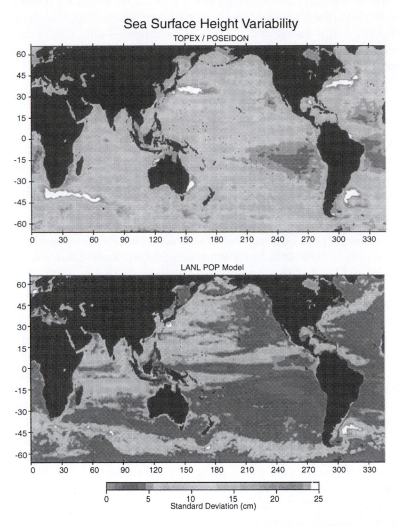

Figure 1.2.2 Standard deviation of the sea surface height from 1992–1994 2-year mean from TOPEX/Poseidon altimetry (top panel) and POP high resolution global model (bottom panel). While the broad features are reproduced by the model, the SSH variability is considerably lower than indicated by observations. From Fu and Smith (1996).

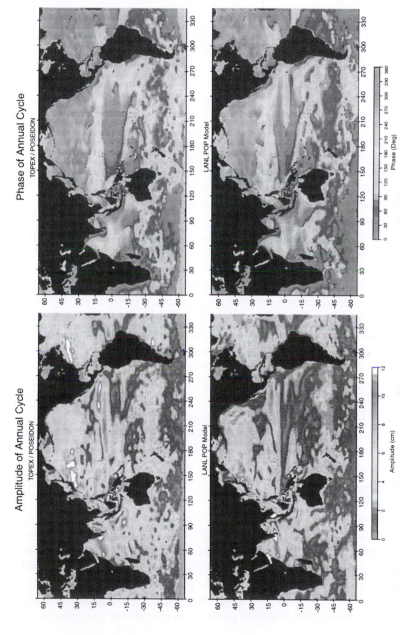

Figure 1.2.3 Amplitude (left) and phase (right) of the annual cycle of SSH variability derived from TOPEX/Poseidon altimetry (top panel) and POP high resolution global model (bottom panel). Note how well the various features are reproduced by the model, despite some major disagreements. From Fu and Smith (1996).

is believed that the resolution is still marginal to resolve the western boundary currents and the mesoscale variability to the fullest extent needed. A quadrupling of the resolution to 1/20° and more accurate physics and surface forcing would be essential to push the model SSH variability close to the observed values. This is one of the many computing grand challenge problems of the coming century.

1.3 GOVERNING EQUATIONS

All numerical ocean models solve one form or other of the same governing equations for oceanic motions, written in the coordinate frame of reference fixed to the rotating Earth. These equations are essentially Navier–Stokes equations (or more appropriately Reynolds-averaged equations for mean quantities, since the flow is invariably turbulent), but with the gravitational buoyancy force (Archimedian force due to density stratification) and Coriolis force (from fictitious accelerations generated due to the noninertial nature of the rotating coordinate frame) prominent in the dynamical balance. In addition, an equation of state relating the density of seawater to its temperature and salinity (and pressure) and conservation equations for temperature and salinity are also solved. In those models that treat turbulence explicitly, equations for turbulence quantities such as the turbulence velocity scale (or equivalently turbulence kinetic energy) and turbulence macroscale are also solved. If chemical or biological components are included, conservation equations for relevant species with appropriate source and sink terms are solved as well.

Global and basin scale models are formulated in spherical coordinates (see Chapter 10), but regional models are usually cast in either uniform or telescoping rectangular Cartesian (latitude–longitude or tilted), or local orthogonal curvilinear, coordinates instead. The coordinate system is right handed, with the origin at the undisturbed mean sea surface, and with the z-axis (named the x_3-axis when tensorial notation is used) in the vertical direction, positive upward (see Figure 1.3.1). The same coordinate system is used for the overlying atmosphere in coupled models. The significance of this is that the definition of static stability involves the same sign for the two media, even though the atmosphere extends over the positive z-axis and the oceans over the negative z-axis. The x-axis (x_1-axis) is usually taken to be in the zonal direction (positive to the east), and the y-axis (x_2-axis) is in the meridional direction (positive to the north). The governing equations of motion are simply statements of conservation of physical quantities associated with fluid particles. For

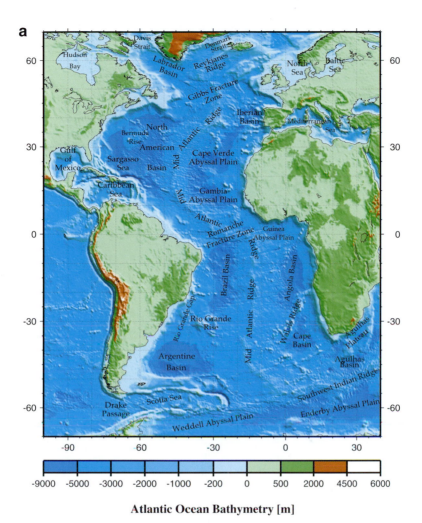

a

Atlantic Ocean Bathymetry [m]

Figure 1.1.1 Bottom topography in the (a) Atlantic, (b) Pacific, and (c) Indian basins derived from the data base of Sandwell *et al.* (1996). Note the various mid-ocean topographic features such as ridges, island chains, and seamounts. These play an important role in the circulation.

b

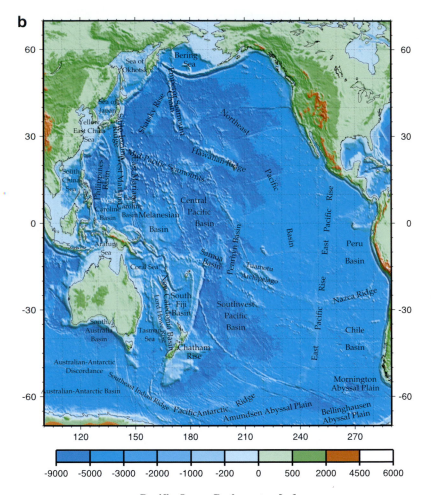

Pacific Ocean Bathymetry [m]

Figure 1.1.1 continued

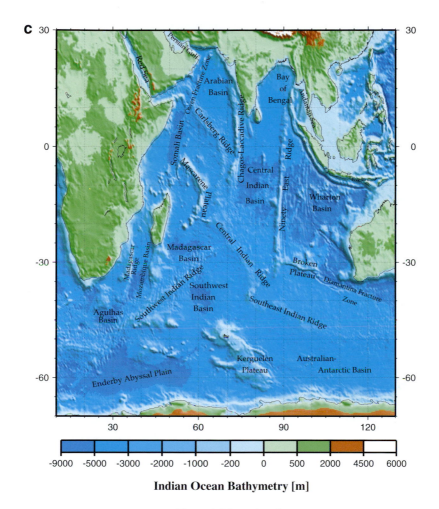

c

Indian Ocean Bathymetry [m]

-9000 -5000 -3000 -2000 -1000 -200 0 500 2000 4500 6000

Figure 1.1.1 continued

Examples of modern massively parallel high performance computers which have the potential for multi-teraflop performance in the coming decades: Cray T3E (top) and SGI Origin 2000 (bottom). Image of the Cray T3E-1200 courtesy of Cray. Cray is a registered trademark of Cray Research, L.L.C., a wholly owned subsidiary of Silicon Graphics, Inc. Image of the Origin 2000 courtesy of SGI. Origin and Origin 2000 are trademarks of Silicon Graphics, Inc.© 1999 Silicon Graphics, Inc. Used by permission. All rights reserved.

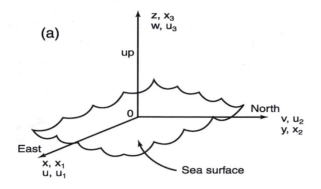

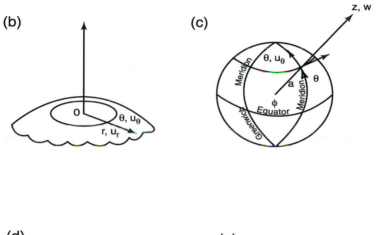

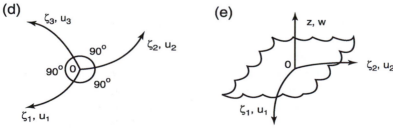

Figure 1.3.1 Coordinate systems used in ocean models: (a) rectangular Cartesian system for regional f and β plane models, (b) cylindrical coordinate system useful for polar oceans, (c) general spherical coordinate system for global models, (d) general orthogonal curvilinear coordinate system, and (e) orthogonal curvilinear coordinate in the horizontal.

the oceans, they are conservation of mass, momentum, heat, and salt. It is the fluid density that is fundamental to its dynamics, no matter which constituents determine it, since it is the density that determines the inertia of the fluid and it is the density variations in a gravity field that determine the forces on the fluid particles. In the oceans, it is the temperature and salinity (or equivalently the amount of dissolved solids) that determine the density, and hence the necessity to carry conservation relations for the two. This also means that one needs an equation of state that relates the fluid density to its constituents (temperature and salinity for the oceans). For the oceans, the equation of state is a complex function of its temperature and salinity, and the ambient pressure, usually given in the form of a sum of a polynomial series (see Appendix A). In the atmosphere, it is the temperature and specific humidity (a measure of the water vapor content) that determine the density, and since water vapor can also be treated as an ideal gas, it is simply the ideal gas law for a mixture of air and water vapor (liquid water and associated phase changes is an added complication). It is often possible to use a simple linear dependence of density on temperature and salinity that suffices for most basin scale physics. But in estuaries and coastal oceans, the ranges of both temperature and salinity are large enough to prevent meaningful linearization around some state. Often, in regions with very little salinity variation, temperature alone is used as a proxy for density. But in regions where temperature or salinity can change independently of one another, it is necessary to carry an equation of state that is a function of both. Since motions take place in a gravitational field, the law of gravitation or an expression for the gravitational potential closes the set of equations.

1.3.1 DIFFERENT FORMS

We will illustrate the governing equations in the Eulerian frame of reference, using vectorial notation for generality. This means that the rate of change of properties is described at a fixed point in space as the fluid is advected past it, unlike the Lagrangian description, where the rate of change following a moving parcel of fluid is described. (The Lagrangian system, while somewhat more natural for fluid motions, is also more complex and is seldom used in ocean models.) The equation of continuity (conservation of mass) can be written as

$$\frac{D\rho}{Dt} + \rho \nabla \cdot \mathbf{v} = 0; \quad \frac{D}{Dt} \equiv \frac{\partial}{\partial t} + \mathbf{v} \cdot \nabla \qquad (1.3.1)$$

where D/Dt denotes the substantial or the total derivative, which is the sum of the rate of change due to temporal changes (the tendency term) and that due to the advection of the fluid past the point (advective term). This form is valid for any compressible fluid and is extensively used in aerodynamics. However, for low speed flows, whose Mach number M (the ratio of the flow velocity U to the speed of sound c_s) is small (M $\ll$ 1), the first term can be neglected and the fluid can be treated as incompressible. Compressibility effects are small up to M ~ 0.3. The sound speed in the ocean is typically 1500 m s^{-1}, and the flow velocities seldom larger than 2–3 m s^{-1}, so that M < 0.002; therefore, to a very good approximation, the flow can be considered to be incompressible. In the atmosphere, M is a bit higher, since the sound speed is ~300 m s^{-1} and flow velocities can reach 60–80 m s^{-1}. But M < 0.25, and therefore compressibility effects can still be ignored. The resulting continuity equation can be written as

$$\nabla \cdot \mathbf{v} = 0 \qquad (1.3.2)$$

We will write this out in various forms for later familiarity. Its forms in simple rectangular Cartesian, cylindrical, and spherical coordinates (see Figure 1.3.1) are

$$\frac{\partial u}{\partial x} + \frac{\partial v}{\partial y} + \frac{\partial w}{\partial z} = 0 \qquad \frac{1}{r}\frac{\partial(u_r r)}{\partial r} + \frac{1}{r}\frac{\partial u_\theta}{\partial \theta} + \frac{\partial w}{\partial z} = 0$$

$$\frac{1}{(R+z)\cos\theta}\frac{\partial u_\phi}{\partial\phi} + \frac{1}{(R+z)\cos\theta}\frac{\partial(u_\theta \cos\theta)}{\partial\theta} + \frac{1}{(R+z)^2}\frac{\partial\left[(R+z)^2 u_r\right]}{\partial z} = 0 \qquad (1.3.3)$$

where $x(x_1)$, $y(x_2)$, and $z(x_3)$ are rectangular coordinates with $u(u_1)$, $v(u_2)$, and $w(u_3)$ being the corresponding velocity components; r, θ, and z are the cylindrical coordinates with , u_r, u_θ, and w being the corresponding velocity components in the radial, tangential, and vertical directions; and ϕ, θ, and z are the modified spherical coordinates (modified because z is used in place of the radius r) with ϕ the longitude, θ the latitude, and u_ϕ, u_θ and u_r the corresponding velocity components in the zonal, meridional, and vertical directions. Note that in applications to the oceans and the atmosphere, the last term can be approximated as $\partial w / \partial z$ since z $\ll$ R, the Earth's radius, and (R + z) approximated as R in the rest of the terms. Modified spherical coordinates are well suited to the oceans and the atmosphere, while unmodified spherical coordinates are best for dealing with the dynamics of Earth's core and the ionosphere. Note also that the equations are written sometimes in terms of colatitude ($\theta' = 90 - \theta$); this form can be obtained simply by replacing $\cos\theta$ by $\sin\theta'$ and $\sin\theta$ by $\cos\theta'$ in all the governing equations of motion. Equation (1.3.2) can be written advantageously in tensorial form, which is compact, especially when applied to the equations for vectorial quantities (e.g., see Kantha, 1989a):

$$\frac{\partial u_i}{\partial x_i} = 0 \qquad i = 1,2 \text{ and } 3 \tag{1.3.4}$$

Einstein's summation convention is invoked whenever repeated indices occur in any term, implying summation over all possible values of the index. When expanded, Eq. (1.3.4) gives the form for the rectangular coordinates in Eq. (1.3.3), except that (x, y, z) and (u, v, w), the familiar forms, are replaced by (x_1, x_2, x_3) and (u_1, u_2, u_3). It is often convenient to separate the horizontal and vertical coordinates in this notation and write instead

$$\frac{\partial u_j}{\partial x_j} + \frac{\partial w}{\partial z} = 0 \qquad j = 1,2 \tag{1.3.5}$$

Because of the thinness of the fluid envelopes surrounding the Earth, the atmosphere, and the oceans, and because of the fact that gravitational forces which act in the vertical direction lead to distinct dynamics in the horizontal and vertical directions, it is often advantageous to make use of these facts by lumping the two horizontal directions together and separating the vertical direction out. The index j assumes only values of 1 and 2.

The most general form for the equations is, of course, the general nonorthogonal curvilinear coordinates, which afford great flexibility in fitting the model grid to complex coastline shapes. While some modelers have ventured to cast the governing equations in this form, this form has not become popular. When such flexibility is desired, a finite-element approach is used instead (e.g., see Le Provost, 1994), since it affords essentially unlimited flexibility in tailoring the model grid to achieve any desired local resolution (see Figure 1.3.2). However, finite-element solutions used to be plagued by $2\Delta x$ noise problems that appear to have been overcome in more recent versions (Lynch and Gray, 1979; Westerink *et al.,* 1992; Lynch *et al.,* 1996). Still, the approach is relatively new compared to the more prevalent finite-difference approach, and its efficacy still largely unproven in geophysical flow calculations, as indeed are many modern approaches such as unstructured grids and adaptive grids widely used in aerodynamical flow calculations. The best compromise between flexibility and the ease of application appears to be afforded by the use of general orthogonal curvilinear coordinates

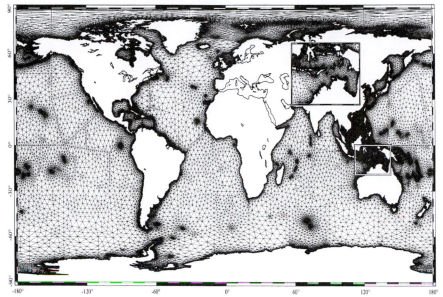

Figure 1.3.2 Finite element grid used by Le Provost *et al.* (1994) in their global barotropic tide model.

(e.g., Kantha and Piacsek, 1993, 1997), in which form the continuity equation can be written as

$$\frac{1}{h_1 h_2 h_3}\left\{\frac{\partial\left(h_2 h_3 u_1\right)}{\partial\xi_1}+\frac{\partial\left(h_3 h_1 u_2\right)}{\partial\xi_2}+\frac{\partial\left(h_1 h_2 u_3\right)}{\partial\xi_3}\right\}=0 \qquad (1.3.6)$$

where ξ_1, ξ_2, ξ_3 are the coordinate directions, and h_1, h_2, and h_3 are the corresponding metrics that define the length of a line element in the coordinate system; $ds = h_1 d\xi_1 \mathbf{a}_1 + h_2 d\xi_2 \mathbf{a}_2 + h_3 d\xi_3 \mathbf{a}_3$, where $\mathbf{a}_1$, $\mathbf{a}_2$, $\mathbf{a}_3$ are unit vectors in the three directions. The metrics h_i are of course functions of the coordinates ξ_i. The forms for the three orthogonal coordinate systems in Figure 1.3.1 (the rectangular Cartesian, cylindrical, and spherical) can be obtained by putting

$$\xi_1 = x_1, \xi_2 = x_2, \xi_3 = x_3 ; h_1 = 1, h_2 = 1, h_3 = 1$$

$$\xi_1 = r, \xi_2 = \theta, \xi_3 = z; h_1 = 1, h_2 = r, h_3 = 1 \qquad (1.3.7)$$

$$\xi_1 = \phi, \xi_2 = \theta, \xi_3 = z; h_1 = (R + z)\cos\theta, h_2 = (R + z), h_3 = 1$$

The general form of Eq. (1.3.6) is seldom necessary. Only the horizontal coordinates need be in orthogonal curvilinear form. This form can be obtained readily from the more general form by putting $\xi_3 = z, h_3 = 1, u_3 = w$, so that the continuity equation becomes

$$\frac{1}{h_1 h_2}\left\{\frac{\partial(h_2 u_1)}{\partial \xi_1}+\frac{\partial(h_1 u_2)}{\partial \xi_2}\right\}+\frac{\partial w}{\partial z}=0 \qquad (1.3.8)$$

1.3.2 TRANSFORMATIONS

We have illustrated the various forms of the governing equations with the simplest possible equation, the incompressible continuity equation. These forms are straightforward, but tedious to derive for more complex equations such as the momentum and scalar conservation relations. One can start with the vectorial form of the governing equations and replace the vectorial operators by their general orthogonal curvilinear forms (e.g., see Appendix 2 of Batchelor, 1979). We are interested in only curvilinear forms in the horizontal directions; therefore, we will present only those forms. Since

$$\frac{\partial a_1}{\partial \xi_2}=\frac{1}{h_1}\frac{\partial h_2}{\partial \xi_1}a_2;\frac{\partial a_2}{\partial \xi_1}=\frac{1}{h_2}\frac{\partial h_1}{\partial \xi_2}a_1$$

$$\frac{\partial a_1}{\partial z}=0;\frac{\partial a_3}{\partial \xi_1}=\frac{\partial h_1}{\partial z}a_1$$

$$\frac{\partial a_2}{\partial z}=0;\frac{\partial a_3}{\partial \xi_2}=\frac{\partial h_2}{\partial z}a_2$$

$$\frac{\partial a_1}{\partial \xi_1}=-\frac{1}{h_2}\frac{\partial h_1}{\partial \xi_2}a_2-\frac{\partial h_1}{\partial z}a_3$$

$$\frac{\partial a_2}{\partial \xi_2}=-\frac{\partial h_2}{\partial z}a_3-\frac{1}{h_1}\frac{\partial h_2}{\partial \xi_1}a_1$$

$$\frac{\partial a_3}{\partial z}=0$$

$(1.3.9)$

the vector operations become

$$\nabla\Theta = \left(\frac{a_1}{h_1}\frac{\partial}{\partial\xi_1} + \frac{a_2}{h_2}\frac{\partial}{\partial\xi_2} + a_3\frac{\partial}{\partial z} \right)\Theta$$

$$\nabla\cdot v = \frac{1}{h_1 h_2}\left(\frac{\partial\left(h_2 u_1\right)}{\partial\xi_1} + \frac{\partial\left(h_1 u_2\right)}{\partial\xi_2} + \frac{\partial\left(h_1 h_2 w\right)}{\partial z} \right)$$

$$\nabla\times v = \frac{a_1}{h_2}\left(\frac{\partial w}{\partial\xi_2} - \frac{\partial(u_2 h_2)}{\partial z} \right) - \frac{a_2}{h_1}\left(\frac{\partial w}{\partial\xi_1} - \frac{\partial(u_1 h_1)}{\partial z} \right)$$

$$+ \frac{a_3}{h_1 h_2}\left(\frac{\partial\left(h_2 u_2\right)}{\partial\xi_1} - \frac{\partial\left(h_1 u_1\right)}{\partial\xi_2} \right)$$

$$\nabla^2\Theta = \frac{1}{h_1 h_2}\left[\frac{\partial}{\partial\xi_1}\left(\frac{h_2}{h_1}\frac{\partial\Theta}{\partial\xi_1} \right) + \frac{\partial}{\partial\xi_2}\left(\frac{h_1}{h_2}\frac{\partial\Theta}{\partial\xi_2} \right) + \frac{\partial}{\partial z}\left(h_1 h_2\frac{\partial w}{\partial z} \right) \right] \qquad (1.3.10)$$

$$u\cdot\nabla v = a_1\left\{ u\cdot\nabla v_1 + \frac{u_2}{h_1 h_2}\left(u_1\frac{\partial h_1}{\partial\xi_2} - u_2\frac{\partial h_2}{\partial\xi_1} \right) + \frac{w}{h_1}\left(u_1\frac{\partial h_1}{\partial z} \right) \right\}$$

$$+ a_2\left\{ u\cdot\nabla v_2 + \frac{w}{h_2}\left(u_2\frac{\partial h_2}{\partial z} \right) + \frac{u_1}{h_1 h_2}\left(u_2\frac{\partial h_2}{\partial\xi_1} - u_1\frac{\partial h_1}{\partial\xi_2} \right) \right\}$$

$$+ a_3\left\{ u\cdot\nabla w + \frac{u_1}{h_1}\left(-u_1\frac{\partial h_1}{\partial z} \right) + \frac{u_2}{h_2}\left(-u_2\frac{\partial h_2}{\partial z} \right) \right\}$$

$$\nabla^2 v = \nabla\left(\nabla\cdot v \right) - \nabla\times\left(\nabla\times v \right)$$

These can be used to derive the equations of motion in any orthogonal coordinate system, starting from their vectorial form. The rate of strain tensor becomes

$$e_{11} = \frac{1}{h_1}\frac{\partial u_1}{\partial\xi_1} + \frac{u_2}{h_1 h_2}\frac{\partial h_1}{\partial\xi_2} + \frac{w}{h_1}\frac{\partial h_1}{\partial z}$$

$$e_{22} = \frac{1}{h_2}\frac{\partial u_2}{\partial\xi_2} + \frac{w}{h_2}\frac{\partial h_2}{\partial z} + \frac{u_1}{h_1 h_2}\frac{\partial h_2}{\partial\xi_1}$$

$$e_{33} = \frac{\partial w}{\partial z}$$

$$e_{12} = \frac{h_2}{2h_1}\frac{\partial}{\partial\xi_1}\left(\frac{u_2}{h_2} \right) + \frac{h_1}{2h_2}\frac{\partial}{\partial\xi_2}\left(\frac{u_1}{h_1} \right) \qquad (1.3.11)$$

$$e_{23} = \frac{1}{2h_2}\frac{\partial}{\partial\xi_2}\left(w \right) + \frac{h_2}{2}\frac{\partial}{\partial z}\left(\frac{u_2}{h_2} \right)$$

$$e_{31} = \frac{h_1}{2}\frac{\partial}{\partial z}\left(\frac{u_1}{h_1} \right) + \frac{1}{2h_1}\frac{\partial}{\partial\xi_1}\left(w \right)$$

1.3.3 GOVERNING EQUATIONS

The most important simplification in ocean modeling is the assumption that the medium is incompressible, which filters out acoustic waves. In studying oceanic (and atmospheric) motions, one is not concerned with the propagation of acoustic waves that arise from compressibility of the medium, unless one is interested in the propagation of acoustic energy itself, as, for example, in acoustic tomography in the global ocean (or sodar probing of the atmosphere). This is particularly important to numerical methods, since the constraints of fast-moving acoustic waves on the time step of integration would otherwise make meaningful long-term integrations impractical. The second most important simplification is ignoring the density changes in the fluid except when gravitational body forces are concerned. This is the Boussinesq approximation equivalent to ignoring mass (inertia) changes due to changes in density. The average potential density of seawater is 1028 kg m^{-3}, and in most of the oceans, this changes by less than 2–3 kg m^{-3}. Only near river mouths and in estuaries does the value approach the freshwater value and even then the error in neglecting the change in density is less than 2–3%. The oceans are weakly (but stably) stratified in the vertical; however, even this weak stratification greatly inhibits vertical motions due to the fact that the motions take place in the presence of a gravitational field, and hence gravitational (buoyancy) forces exert a major influence on the motions. This is the reason for retaining density changes in the gravitational terms while ignoring them in the rest. While it is straightforward to formulate a non-Boussinesq model (Mellor and Ezer 1996), this is unnecessary, since steric effects, namely changes in the sea surface height due to volumetric expansion and contraction of the water column by seasonal heating/cooling, which cannot be modeled by a Boussinesq model can be accounted for separately.

The Earth's rotation rate can be considered constant, even though there is both a small secular change in the length of day on a timescale of centuries due to tidal dissipation ($\sim$2.4 ms cy^{-1}) and short-term changes on timescales of hours to years due to exchanges of angular momentum between the mantle and the atmosphere, the oceans, and the core ($\sim$2 ms peak to peak). The Earth is nearly a spheroid (ellipticity $\sim 1/298$), whose shape arises from the influence of gravitational and centrifugal forces. It is therefore possible to ignore the differences in the polar and equatorial radii and consider the Earth to be spherical, although the equatorial radius is about 0.3% larger than the polar radius. The error in employing spherical instead of spheroidal coordinates is only 0.1% (Hendershott, 1981). Similarly, the altitudinal variations of the gravitational constant (as well as the local variations due to density anomalies in the solid Earth) can be ignored (however, nonspherical shape and nonuniform gravity are important to geodesy), since the depth of the oceans (and the thickness of the atmospheric envelope), an average of 5 km ($\sim$10 km), is small compared to the radius of the Earth, less than 0.08% (0.16%). The effective gravitational acceleration term that appears in the equations on a rotating Earth absorbs the centripetal force term that is a maximum at the equator and zero at the poles. This and the nonspherical shape of the Earth amount to an error of less than 0.5% when a constant but average value of 9.78 m s^{-2} is used for the gravitational acceleration.

The conservation of momentum in an inertial frame of reference with body and frictional forces is

$$\frac{Dv^I}{Dt} = -\frac{1}{\rho}\nabla p + \nabla\Phi^I + \frac{F}{\rho} \qquad (1.3.12)$$

where superscript I is to remind us that the quantities are defined in an inertial frame. The potential term $\nabla\Phi^I$ accounts for the Earth's own gravitational (Archimedian) forces due to density stratification in the fluid and could include astronomical tide-generating forces due to the gravitational fields of the Moon and the Sun. Now, geophysical flows are conveniently dealt with in the frame of reference rotating with the Earth at an angular velocity Ω rad s^{-1}. This is a noninertial frame, and transformation to it (e.g., Pond and Pickard, 1989; Gill, 1982; Cushman-Roisin, 1994) causes two additional terms to appear in the momentum equations, the so-called Coriolis acceleration term $2\Omega \times v$ and the centripetal force term $-\Omega \times (\Omega \times R)$, where R is the radius vector. The latter term can be absorbed into the Earth's gravitational potential (Coriolis force cannot be expressed as the gradient of a potential) and the effective gravitational potential can be written as $g = \nabla\Phi = \nabla\Phi^I - \Omega \times (\Omega \times R)$. It is customary to ignore the variations in the resulting gravitational acceleration g (which amount to 0.5% at the most, due to both this centripetal effect and the Earth's ellipticity) and treat it as a constant. This is true of most real planets, although in laboratory rotating table experiments, the difference between effective and real gravitational acceleration can be substantial (Nezlin and Snezhkin, 1993). The momentum equation can therefore be written as

$$\frac{Dv}{Dt} + 2\Omega \times v = -\frac{1}{\rho}\nabla p + g + \nabla\Phi_t + \frac{F}{\rho} \qquad (1.3.13)$$

Note that an additional term $\nabla\Phi_t$, which is the lunisolar gravitational tidal potential responsible for generating oceanic tides, has been added to Eq. (1.3.12).

The *Coriolis term* is the most important term in the momentum equations for large scale geophysical flows. It is a fictitious force term that appears in a rotating coordinate frame, since unaccelerated straight-line motions in an inertial frame must appear curved in the rotating frame and hence a fictitious force acting perpendicular to both the direction of motion and the axis of rotation is necessary to bring this about in the rotating frame. Because it is a fictitious body force, it does not however do any real work and hence disappears in any energy considerations. The Coriolis terms therefore do not appear in any energy conservation equations. The presence of the Coriolis terms has a profound effect on geophysical flows that sets them apart from flows without any rotation. This, along with the gravitational forces acting due to the stable stratification that prevails most often, and the absence of compressibility effects, distinguishes geophysical flows from aerodynamic flows.

Frictional forces act to dissipate the flow energy. This dissipation rate depends very much on whether the flow is laminar or turbulent, which in turn depends on the magnitude of the flow Reynolds number $Re = UL/\nu$, where ν is the kinematic viscosity, U is the typical flow speed, and L is the scale of the flow. Even for relatively small scale flows in the oceans, $L \sim 10$ km, $U \sim 0.1$ m s^{-1}, $\nu \sim 10^{-6}$ m^2 s^{-1}, and $Re \sim 10^9$. For the atmosphere, the velocities are typically an order of magnitude larger, but so is the kinematic viscosity and hence Re is roughly similar. At these huge Reynolds numbers, the flow is turbulent and consists of random eddying motions superimposed on some statistical mean flow. It is therefore necessary to consider equations for average quantities, rather than instantaneous values. When such averaging is carried out, the nonlinear advection terms introduce additional terms in the momentum equations for the averaged quantities called Reynolds stresses or turbulent stresses (see Chapter 1 of Kantha and Clayson 2000). These are several orders of magnitude larger than molecular stresses and therefore the molecular stress terms can be ignored in the frictional terms. Writing the frictional terms in terms of the turbulent stress tensor τ, the momentum equations become

$$\frac{D\mathbf{v}}{Dt} + 2\Omega \times \mathbf{v} = -\frac{1}{\rho}\nabla p + \mathbf{g} + \nabla\Phi_t + \frac{\nabla\cdot\tau}{\rho} \qquad (1.3.14)$$

In addition, one needs the conservation equations for the temperature T (or more appropriately the potential temperature Θ to take account of adiabatic heating/cooling) and salinity S which, because they are scalar quantities, do not involve any additional terms in the rotating coordinate frame,

$$\frac{D(\Theta,S)}{Dt} = \nabla\cdot(\mathbf{q}_T,\mathbf{q}_S) + So_{T,S} \qquad (1.3.15)$$

where So denotes the source and sink terms, and $\mathbf{q}_T$ and $\mathbf{q}_S$ are the kinematic turbulent heat and salt flux vectors. These are the heat flux vector divided by ρc_p and the salt flux vector divided by ρ. To incorporate pressure effects on density, it is more convenient to deal with potential temperature Θ, the value of the temperature of a fluid parcel brought adiabatically to a reference level z_r, instead of the *in situ* temperature T (in most cases). The equation of state,

$$\rho = \rho(\Theta, S, p) \qquad (1.3.16)$$

completes the set of seven equations for seven quantities, ρ, p, $\mathbf{v}$, Θ, and S, provided the turbulent stress tensor and the turbulent heat and salt flux vectors can be expressed in terms of these quantities. Here indeed lies the principal problem in solving geophysical flows. The turbulent quantities involve higher moments and hence the equations are not closed, if these quantities are retained. It is possible to derive a set of equations for these higher moments, but those equations contain the next higher moments, and so on, ad infinitum (see Chapter 1 of Kantha and Clayson, 2000). This is the so-called closure problem of turbulence that has defied solution for decades. Turbulence closure is still one of the unsolved problems in physics. We will not however deal with the closure problem in this chapter and instead make what is known as the zeroth order approximation (also known as the Boussinesq model in turbulence literature) for these turbulent quantities by assuming that they have the same form as the viscous terms, but replacing the molecular values of diffusivity by constant values known as eddy diffusivities that are several orders of magnitude larger. However, a clear distinction must be made between the horizontal components and the vertical ones, since the effective turbulent coefficients of both the horizontal momentum and heat diffusivities are several orders of magnitude larger than those for the vertical components (the associated gradients are however several orders of magnitude smaller).

We will consider only the rectangular Cartesian latitude–longitude coordinates for simplicity. Using tensorial notation and treating the horizontal coordinates separately from the vertical ones (i and j indices take values of 1 and 2 only), the continuity equation can be written as

$$\frac{\partial U_j}{\partial x_j} + \frac{\partial W}{\partial z} = 0 \qquad (1.3.17)$$

where U_j denotes the horizontal components of mean velocity and W the vertical one, and the momentum equations become

$$\frac{\partial U_i}{\partial t} + \frac{\partial}{\partial x_j}\left(U_j U_i\right) + \frac{\partial}{\partial z}\left(W U_i\right) + (2\Omega \sin \theta)\varepsilon_{i3j} U_j + (2\Omega \cos \theta)W = -\frac{1}{\rho}\frac{\partial P}{\partial x_i}$$

$$+ \frac{\partial}{\partial x_i}\left(\Phi_t\right) + \frac{\partial}{\partial z}\left(K_M \frac{\partial U_i}{\partial z}\right) + \frac{\partial}{\partial x_j}\left(A_M \frac{\partial U_i}{\partial x_j}\right)$$

$$\frac{\partial W}{\partial t} + \frac{\partial}{\partial x_j}\left(U_j W\right) + \frac{\partial}{\partial z}\left(WW\right) - (2\Omega \cos \theta)U_1 = -\frac{1}{\rho}\frac{\partial P}{\partial z} - g \qquad (1.3.18)$$

$$+ \frac{\partial}{\partial z}\left(K_M \frac{\partial W}{\partial z}\right) + \frac{\partial}{\partial x_j}\left(A_M \frac{\partial W}{\partial x_j}\right)$$

where ε_{ijk} is the alternating tensor (equal to 1 if indices occur in cyclic order, -1 if they occur in reverse cyclic order, and 0 if any two of the indices are the same). K_M is the vertical kinematic viscosity (momentum diffusion) coefficient and A_M the horizontal kinematic viscosity coefficient. The vertical diffusivities are considerably smaller than the horizontal ones, typical values being 10^{-5} and 10^{-1} m^2 s^{-1}, respectively, in the deep ocean, and 10^{-2}–10^{-1} and 10^2–10^3 m^2 s^{-1} in the upper mixed layer.

A major uncertainty in numerical models is that associated with accurate pre-scription of these coefficients. These eddy coefficients are not properties of the fluid but are rather properties of the flow itself, and therefore require the flow to be known a priori. This is not possible since the flow solution cannot be obtained unless these coefficients are known. Often, these coefficients are therefore chosen as ad hoc constants in large scale modeling studies. They are not, however, constants. The horizontal coefficients are most often chosen from numerical considerations alone.

1.3.4 APPROXIMATIONS

Even these simplified equations are simply too complex to handle even numerically, and further simplifications are the norm. One such traditional sim-plification exploits the fact that the density variations in the oceans (and the lower atmosphere) are rather small, less than 3% or so, and therefore the density can be considered to be constant, except when body forces due to the motion of a density-stratified fluid in a gravitational field are involved. In other words, changes in the mass or inertia of a fluid parcel due to changes in its density are negligible, while the same changes in density are consequential when a gravitational field is present. This is the so-called *Boussinesq approximation* and involves replacing ρ by a constant reference value ρ_0 (~1035 kg m^{-3}) everywhere except in terms involving the gravitational constant g. Thus under Boussinesq approximation, Eq. (1.3.18) can be written as

$$\frac{\partial U_i}{\partial t} + \frac{\partial}{\partial x_j}\left(U_j U_i\right) + \frac{\partial}{\partial z}\left(W U_i\right) + (2\Omega \sin \theta)\varepsilon_{i3j} U_j + (2\Omega \cos \theta)W = -\frac{1}{\rho_0}\frac{\partial P}{\partial x_i}$$

$$+ \frac{\partial}{\partial x_i}\left(\Phi_t\right) + \frac{\partial}{\partial z}\left(K_M \frac{\partial U_i}{\partial z}\right) + \frac{\partial}{\partial x_j}\left(A_M \frac{\partial U_i}{\partial x_j}\right)$$

$$\frac{\partial W}{\partial t} + \frac{\partial}{\partial x_j}\left(U_j W\right) + \frac{\partial}{\partial z}\left(WW\right) - (2\Omega \cos \theta)U_1 = -\frac{1}{\rho_0}\frac{\partial P}{\partial z} - \frac{\rho}{\rho_0}g$$

$$+ \frac{\partial}{\partial z}\left(K_M \frac{\partial W}{\partial z}\right) + \frac{\partial}{\partial x_j}\left(A_M \frac{\partial W}{\partial x_j}\right)$$

$$(1.3.19)$$

where density ρ is replaced by the reference value ρ_0 except in the term involving gravitational acceleration g in the vertical momentum equation. Note that for a motionless fluid, the terms on the LHS of the vertical momentum equation (W equation) vanish and exact balance between the vertical pressure gradient and the gravitational force, a hydrostatic balance, is the result.

The second simplification that could be made exploits the fact that at certain scales of motion, the aspect ratio of the oceans (and the atmosphere) is small and hence the vertical motions are small and are further inhibited by gravitational forces under stable density stratification. Thus vertical accelerations are small and the vertical frictional forces are also small, and the fluid acts as though it is under static equilibrium as far as vertical motions are concerned. This is the so-called *hydrostatic approximation,* which yields an exact balance between the pressure gradient and the gravitational force in the vertical momentum equation. However, hydrostatic approximation is not always valid, as, for example, in dealing with strong wintertime chimney-like convective flows in subpolar seas such as the Labrador Sea (the atmospheric equivalent is the cumulonimbus activity during a thunderstorm). Fully nonhydrostatic models have to be employed in such situations (Jones and Marshall, 1993).

When the fluid is in motion, the terms on the LHS of the W equation in Eq. (1.3.19) balance the terms on the RHS involving departures from the hydrostatically balanced reference state. For nonhydrostatic effects to be small, the terms on the LHS must be less than either of the deviation terms on the RHS. This requires that the advective timescale U/L, where U is the velocity scale and L is the horizontal length scale of the motions being considered, be short compared to the buoyancy period. Thus a measure of nonhydrostaticity is the ratio $\varepsilon = U/(NL)$, or in terms of the aspect ratio of the fluid $\delta = (H/L)$, where H is the fluid depth, $\varepsilon = \delta F_r$, and $F_r = U/(NH)$ is the Froude number. Clearly, both the aspect ratio and the Froude number must be small for the flow to be in hydrostatic balance. In the main thermocline, $\delta \sim 0.01-0.1$, and Fr < 0.1, so that the flow is very nearly hydrostatic, but for weak stratification and small horizontal scales, hydrostatic approximation breaks down.

Hydrostatic approximation therefore involves neglecting all the terms on the LHS of the W equation in Eq. (1.3.19), including the term involving the horizontal component of the Coriolis acceleration $2\Omega\cos\theta$. This means that a similar term involving $2\Omega\cos\theta$ in the equation for zonal momentum (U equation) in Eq. (1.3.19) must also be neglected. Otherwise, the $2\Omega\cos\theta$ term will not vanish in the equation for the total kinetic energy derived from the U and V equations. This would not be energetically consistent, since the Coriolis accelerations are fictitious, cannot perform any work on the fluid, and hence must vanish identically in any energy equation. The resulting hydrostatic equations are

$$\frac{\partial U_i}{\partial t} + \frac{\partial}{\partial x_j}\left(U_j U_i\right) + \frac{\partial}{\partial z}\left(W U_i\right) + \varepsilon_{i3j} f U_j = -\frac{1}{\rho_0}\frac{\partial P}{\partial x_i} + \frac{\partial}{\partial x_i}\left(\Phi_t\right)$$

$$+ \frac{\partial}{\partial z}\left(K_M \frac{\partial U_i}{\partial z}\right) + \frac{\partial}{\partial x_j}\left(A_M \frac{\partial U_i}{\partial x_j}\right) \qquad\qquad (1.3.20)$$

$$0 = -\frac{1}{\rho_0}\frac{\partial P}{\partial z} - \frac{\rho}{\rho_0}g$$

where $f = 2\Omega\sin\theta$ is the Coriolis term, the component of rotation perpendicular to the ocean surface, θ is the latitude, ρ is the *in situ* density, ρ_0 is the reference density, and g is the gravitational acceleration.

However, neglecting the $(2\Omega\cos\theta)W$ term in the zonal momentum balance becomes untenable as the equator is approached. For example, at the equator, any vertical motion of a fluid parcel involves a change in its distance from the axis of rotation, and strict conservation of its angular momentum requires that this term be retained. Since the vertical velocity W scales like UH/L, this is justified only if the ratio $2\Omega\cos\theta(H/U)$ is small. This is not the case since it can reach values of 0.1 or more near the equator. Clearly, neglecting this term in the zonal momentum balance is more problematic than neglecting acceleration terms in the vertical momentum equation. One possible approach is to retain the term $(2\Omega\cos\theta)U_1$ in the vertical momentum balance (while neglecting the other terms on the LHS) and the $(2\Omega\cos\theta)W$ term in the zonal momentum balance. This is the quasi-hydrostatic (QH) approximation. The QH equations are

$$\frac{\partial U_i}{\partial t} + \frac{\partial}{\partial x_j}\left(U_j U_i\right) + \frac{\partial}{\partial z}\left(W U_i\right) + (2\Omega\sin\theta)\varepsilon_{i3j} U_i + (2\Omega\cos\theta)W = -\frac{1}{\rho_0}\frac{\partial P}{\partial x_i}$$

$$+ \frac{\partial}{\partial x_i}\left(\Phi_t\right) + \frac{\partial}{\partial z}\left(K_M \frac{\partial U_i}{\partial z}\right) + \frac{\partial}{\partial x_j}\left(A_M \frac{\partial U_i}{\partial x_j}\right) \qquad\qquad (1.3.21)$$

$$-(2\Omega\cos\theta)U_1 = -\frac{1}{\rho_0}\frac{\partial P}{\partial z} - \frac{\rho}{\rho_0}g$$

Note that we have further invoked a shallow layer approximation. In spherical coordinates, this implies that since the fluid layer thickness is negligibly small compared to the radius of the Earth R, (R + z) in the governing equations (see Chapter 10) can be replaced by R. However, for the angular momentum to be fully conserved in a QH model, (R + z) should not be approximated by R (Marshall *et al.*, 1997a).

The transport equations for potential temperature Θ and salinity S are

$$\frac{\partial \Theta}{\partial t} + \frac{\partial}{\partial x_j}(U_j \Theta) + \frac{\partial}{\partial z}(W\Theta) = \frac{\partial}{\partial z}\left(K_H \frac{\partial \Theta}{\partial z}\right) + So_\Theta + \frac{\partial}{\partial x_j}\left(A_H \frac{\partial \Theta}{\partial x_j}\right)$$

$$\frac{\partial S}{\partial t} + \frac{\partial}{\partial x_j}(U_j S) + \frac{\partial}{\partial z}(WS) = \frac{\partial}{\partial z}\left(K_H \frac{\partial S}{\partial z}\right) + \frac{\partial}{\partial x_j}\left(A_H \frac{\partial S}{\partial x_j}\right)$$

$$(1.3.22)$$

where K_H is the vertical eddy diffusion coefficient for scalar quantities and A_H the horizontal eddy diffusion coefficient, and So_Θ denotes a volumetric heat source such as that due to penetrative solar heating.

Note that the equation of state (1.3.16) can also be written equivalently in terms of T, S, and p. The equation of state usually employed is the so-called UNESCO equation (Pond and Pickard, 1979; Gill, 1982) and is in the form of polynomial expansions in temperature and salinity (see Appendix A). Note however that mixing considerations in the deep ocean may require that *in situ* temperature (and hence density) be used to evaluate the buoyancy forces in the momentum equations so that local stability of the water column can be computed properly (see Section 10.3).

The above incompressible, Bossinesq, hydrostatic equations are sufficiently accurate for most ocean (and atmosphere) circulation modeling. See Chapter 9 for the general form of these equations in orthogonal curvilinear coordinates. For the spherical coordinates version needed for global ocean modeling, see Chapter 10. Note that in isolated cases such as deep convection (cumulonimbus convection and flow over steep mountains in the atmosphere) and small scale internal wave motions, hydrostatic approximation needs to be relaxed. The horizontal component of Coriolis terms may have to be retained in some cases. For dealing with geophysical flows, it is traditional to consider the above equations applied on a plane tangent to the Earth's surface at the midpoint of that plane, ignore all the curvature effects due to Earth's sphericity, and retain only the latitudinal variation β ($= \partial f/\partial y$) of the Coriolis term f. This so-called *beta-plane approximation* can explain most of the planetary dynamics (a contribution from Carl-Gustaf Rossby). If the latitudinal variation of f is also ignored, one gets the f-plane approximation to the equations, first implemented by Lord Kelvin.

An important point to note is that Eqs. (1.3.19) and (1.3.22) are in their flux-conservative form. This means that the equations for all prognostic variables are in the form

$$\frac{\partial E}{\partial t} + \frac{\partial}{\partial x_j}(F_j) + \frac{\partial}{\partial z}(G) = H \qquad (1.3.23)$$

where E is the prognostic variable, F_j its fluxes in the horizontal direction, G its flux in the vertical direction, and H its source term. These equations are essentially statements of conservation of fluid properties such as mass, momentum, and energy, since in the absence of fluxes at the boundaries of the spatial domain and source terms, global integrals of the quantity would remain unchanged. The divergence form of the spatial gradients illustrates this. Divergence terms can only transport quantities from one point in the spatial domain to another, and on integration over the spatial domain, they vanish. As we shall see in Chapter 2, the use of a staggered grid and the flux-conservative form of the governing equations ensures that the numerical finite-difference equivalents of these equations also preserve these conservation characteristics. Thus, integrals of mass, momentum, scalar concentrations, and energy are conserved globally in the numerical model as well, within, of course, the limitations of a finite precision computer.

By virtue of the hydrostatic approximation, the pressure P anywhere in the fluid column can be determined by

$$\frac{1}{\rho_0} P\left(x_i, z, t\right) = \frac{1}{\rho_0} P_a\left(x_i, t\right) + g\eta\left(x_i, t\right) + \frac{g}{\rho_0} \int_z^0 \rho\left(x_i, z', t\right) dz' \quad (1.3.24)$$

where P_a is the atmospheric pressure and η is the sea surface height.

1.3.5 THERMOBARIC AND CABBELING INSTABILITY

Thermobaric effects (Gill, 1973; Killworth, 1977, 1979; McDougall, 1984, 1987a,b; Garwood *et al.*, 1994; Loyning and Weber, 1997) arise from the increase of the thermal coefficient of expansion of water, especially cold water such as that found in the abyssal regions, with pressure (or equivalently depth). If the temperature stratification in the water column is unstable, a parcel of fluid moving upward will experience increasingly less upward buoyancy forces, while a downward moving one will experience increasingly more downward buoyancy forces, the consequence being that the convective activity in the lower parts is reinforced, while that in the upper parts is weakened. If, in addition, the salinity stratification is stable—since the expansion coefficient for salinity is nearly independent of pressure—it will further weaken the buoyancy forces on the upward moving parcel. If the unstable thermal stratification were to very nearly compensate for the stable salinity stratification, the thermobaric effect, the increase of thermal coefficient of expansion with pressure, could initiate instability, since a parcel moving downward would become less buoyant and continue to move downward. Convection could ensue in the lower parts of the fluid (Gill, 1973; Loyning and Weber, 1997). This process is somewhat similar to the conditional instabilities that occur in moisture-laden atmosphere (see Garwood *et al.*, 1994, for a review).

Thermobaric effects can be illustrated by expanding the equation of state $\rho = \rho$ (T,S,p) in a Taylor series around the mean state (Loyning and Weber, 1997):

$$\rho = \rho_m + \left(\frac{\partial\rho}{\partial T}\right)_m (T - T_m) + \left(\frac{\partial^2\rho}{\partial T^2}\right)_m (T - T_m)^2$$

$$+ \left(\frac{\partial\rho}{\partial S}\right)_m (S - S_m) \qquad (1.3.25)$$

$$+ \left(\frac{\partial\rho}{\partial p}\right)_m (p - p_m) + \frac{1}{2}\left(\frac{\partial^2\rho}{\partial p \partial T}\right)_m (p - p_m)(T - T_m)$$

Here the salinobaric terms, which arise due to the dependence of the salinity expansion coefficient with pressure, are ignored since thermobaric effects overwhelm salinobaric effects. If we now use the definitions for the expansion coefficients,

$$\beta_T = -\frac{1}{\rho_m}\left(\frac{\partial\rho}{\partial T}\right)_m, \beta_S = \frac{1}{\rho_m}\left(\frac{\partial\rho}{\partial S}\right)_m, \beta_p = \frac{1}{\rho_m}\left(\frac{\partial\rho}{\partial p}\right)_m$$

$$\beta_{TB} = \frac{1}{\rho_m}\left(\frac{\partial^2\rho}{\partial p \partial T}\right)_m, \beta_C = \frac{1}{2\rho_m}\left(\frac{\partial^2\rho}{\partial T^2}\right)_m \qquad (1.3.26)$$

we get

$$\rho = \rho_m \begin{bmatrix} 1 - \beta_T (T - T_m) + \beta_S (S - S_m) + \beta_p (p - p_m) \\ + \beta_{TB} (p - p_m)(T - T_m) + \beta_C (T - T_m)^2 \end{bmatrix} \qquad (1.3.27)$$

The fourth term inside the square brackets is the thermobaric term and the last one is the cabbeling term (see McDougall, 1987b). The approximate value for β_T is $7 \times 10^{-6} °C^{-2}$ (Foster, 1972; Killworth, 1977). Near the freezing point of seawater, $-1.7°C$, $\beta_T \sim 2.5 \times 10^{-5} °C^{-1}$ and $\beta_S \sim 8 \times 10^{-4}$ psu^{-1} (Killworth, 1977, 1979; Gill, 1982). Defining a pressure-dependent thermal expansion coefficient, and using the hydrostatic equation to replace pressure $p = p_m - \rho_m g \Delta z$,

$$\hat{\beta}_T = \beta_T - \beta_{TB}(p - p_m) = \beta_T + \beta_{TB}\rho_m g\Delta z = \beta_T + \beta_{T1}\Delta z$$
$$\rho = \rho_m \left[1 - \hat{\beta}_T (T - T_m) + \beta_S (S - S_m) + \beta_p (p - p_m) + \beta_C (T - T_m)^2\right] \qquad (1.3.28)$$

The approximate value for $\beta_{T1} \sim 2.6 \times 10^{-8} °C^{-1} m^{-1}$. In convection problems that involve thermobaric effects (Garwood *et al.*, 1994; Loyning and Weber, 1997), the ratio β_T / β_{T1}, where β_T is the mean value over the water column of depth D, defines an additional length scale (Garwood, 1991) denoting the depth scale of the thermobaric effects. In deep convection problems in cold latitudes, this scale is of the order of 2–3 km. Thermobaric effects appear to be important to deep convection in the Greenland Sea, where the temperature is close to freezing (−1.4°C) and convection can reach depths of 2–3 km. In the Labrador Sea, where similar depths are reached, the typical temperatures are 3°C and hence the thermobaric effects are small. In the Mediterranean Sea, where the temperatures are typically 13–14°C and the depth is 1.5 km, the effects are negligible. In addition to the conventional Rayleigh number Ra that can be defined using the mean value of β_T, one can also define a thermobaric Rayleigh number Ra_{TB}, based on $\beta_{TI}D$ (Loyning and Weber, 1997).

Another instability arises from a nonzero β_C, the quadratic nonlinear dependence of density on temperature called thermostoltic effect. It occurs when a mixture of two water masses of equal density but different T and S characteristics becomes slightly heavier than either. It is caused by the contraction on mixing of waters of different temperatures resulting in a water mass denser than that of either of the parent masses and is thermodynamically irreversible. The resulting instability is called cabbeling instability. See Foster (1972), Killworth (1977), and McDougall (1987b) for a lucid description.

Thermobaric and cabbeling effects must be taken into account in basin scale and global circulation models. In view of the above nonlinearities in the equation of state, in computations of the static stability of the water column in abyssal regions of the ocean, it is often more convenient to use the local form of the equation of state to compute the density changes in the vertical, instead of the potential density based on the potential temperature referred to an arbitrary reference depth. In other words, the change in local *in-situ* density of a fluid parcel when displaced a certain distance in the vertical (usually to the adjacent level in the model) relative to its original value is what determines whether the water column is stably or unstably stratified. This subtle, but rather important point is crucial to depicting the water masses and circulation in the very weakly stratified abyssal depths of the oceans, which are important to aspects such as the thermohaline convection and dense water formation.

1.4 VORTICITY CONSERVATION

An important aspect of geophysical flows is the conservation of potential vorticity, whose derivation is most simply done using vectorial algebra (Muller, 1995). Consider the inviscid momentum equation

$$\frac{d}{dt}v + 2\Omega \times v = -\frac{1}{\rho}\nabla p - \nabla \Phi \qquad (1.4.1)$$

where Φ is the gravitational potential. Taking its curl, one gets an equation for the absolute vorticity,

$$\frac{d}{dt}(\zeta+2\Omega) = (\zeta+2\Omega)\cdot\nabla\mathbf{v} - (\zeta+2\Omega)\nabla\cdot\mathbf{v} - \nabla\left(\frac{1}{\rho}\right)\times\nabla p \qquad (1.4.2)$$

where $\zeta = \nabla\times\mathbf{v}$ is the *relative vorticity* and 2Ω is the *planetary vorticity*. Using the conservation of mass,

$$\frac{d}{dt}\left(\frac{1}{\rho}\right) = \left(\frac{1}{\rho}\right)\nabla\cdot\mathbf{v} \qquad (1.4.3)$$

Eq. (1.4.2) becomes

$$\frac{d}{dt}\left(\frac{\zeta+2\Omega}{\rho}\right) = \frac{\partial}{\partial t}\left(\frac{\zeta}{\rho}\right) + \mathbf{v}.\nabla\left(\frac{\zeta+2\Omega}{\rho}\right)$$

$$= \left(\frac{\zeta+2\Omega}{\rho}\right)\cdot\nabla\mathbf{v} - \frac{1}{\rho}\nabla\left(\frac{1}{\rho}\right)\times\nabla p \qquad (1.4.4)$$

For an incompressible, Boussinesq fluid, this becomes

$$\frac{d}{dt}(\zeta+2\Omega) = \frac{\partial\zeta}{\partial t} + \mathbf{v}.\nabla(\zeta+2\Omega)$$

$$= (\zeta+2\Omega)\cdot\nabla\mathbf{v} - \nabla\left(\frac{1}{\rho}\right)\times\nabla p \qquad (1.4.5)$$

If χ is any property of the fluid that is materially conserved, $\dfrac{d}{dt}\chi = 0$, and then Eq. (1.4.4) gives Ertel's theorem,

$$\frac{d}{dt}\left(\frac{\zeta+2\Omega}{\rho}\cdot\nabla\chi\right) = \frac{1}{\rho}\left[\nabla p\times\nabla\left(\frac{1}{\rho}\right)\right]\cdot\nabla\chi \qquad (1.4.6)$$

where $\zeta_p = \dfrac{\zeta+2\Omega}{\rho}.\nabla\chi$ is known as Ertel's *potential vorticity*. If the term on the RHS is zero, it implies that potential vorticity is a conserved quantity. The spherical coordinate form of potential vorticity is (Muller, 1995)

$$\zeta_p = \zeta_\phi\frac{1}{(R+z)\cos\theta}\frac{\partial\chi}{\partial\phi} + \frac{(\zeta_\theta+2\Omega\cos\theta)}{(R+z)}\frac{\partial\chi}{\partial\theta} + (\zeta_z+2\Omega\sin\theta)\frac{\partial\chi}{\partial z}$$

$$\zeta_\phi = \frac{1}{(R+z)}\frac{\partial w}{\partial\theta} - \frac{\partial v}{\partial z}; \zeta_\theta = \frac{\partial u}{\partial z} - \frac{1}{(R+z)\cos\theta}\frac{\partial w}{\partial\phi} \qquad (1.4.7)$$

$$\zeta_z = \frac{1}{(R+z)\cos\theta}\left(\frac{\partial v}{\partial\phi} - \frac{\partial}{\partial\theta}(u\cos\theta)\right)$$

Since the density is a conserved quantity, under Boussinesq approximation the potential vorticity conservation can be written as

$$\frac{d}{dt}\left(\frac{\zeta + 2\Omega \sin\theta}{\rho_0} \cdot \nabla\rho\right) = 0 \qquad (1.4.8)$$

Note that in the geostrophic limit (Section 1.6), the potential vorticity is ~f (dρ/dz). These aspects are essential to interpreting the numerical model results in terms of water mass advection and transformations (Bryan and Lewis, 1979; Bryan and Sarmiento, 1985; Toggweiler, 1994). Since the deep oceans tend to be adiabatic, conservation of potential vorticity of deep flows is an important requirement for a proper simulation of the meridional circulation in basins like the North Atlantic. For an excellent discussion and recent review of potential vorticity considerations in physical oceanography, see Muller (1995).

1.5 NONDIMENSIONAL NUMBERS AND SCALES OF MOTION

We will next look at some simple limiting cases of these equations for which analytical solutions are possible. But first let us look at some scales and the salient nondimensional quantities. Let U be the typical flow velocity, L be the typical scale of the flow, and T be the typical timescale. Then the ratio of nonlinear advection (inertial) terms to the Coriolis terms in the momentum balance is given by the *Rossby number,*

$$Ro = \frac{U}{fL} \qquad (1.5.1)$$

also known as the Rossby–Kiebel number in Russian literature (Nezlin and Snezhkin, 1993). The significance of this number is that if it is small, the advective terms are small and, for inviscid flows, the balance is primarily between the pressure gradient and the Coriolis forces. The flow is then in near-geostrophic equilibrium. If the balance is exact and the advective terms are negligible, the flow is geostrophic. For midlatitude gyre scale flows in the oceans, $U \sim 10^{-1}$ m s^{-1}, L ~ 10^6 m, f ~ 10^{-4} s^{-1}, and therefore Ro ~ 10^{-3} (very small); the flow is geostrophic. For more energetic processes such as mesoscale eddies, $U \sim 1$ m s^{-1}, L ~ 10^5 m, f ~ 10^{-4} s^{-1}, and therefore Ro ~ 10^{-1} (small but not negligible), and the flow is quasi-geostrophic. In the atmosphere, the velocities are typically an order of magnitude larger, but the scales are also an order of magnitude larger so that the Rossby numbers are similar, though somewhat higher.

The Rossby number is also a measure of the importance of the relative vorticity of the fluid (relative to the rotating coordinate system) vis-à-vis the background planetary vorticity. It can therefore be defined also as the ratio of the relative vorticity to the planetary vorticity:

$$Ro = \frac{\zeta}{f} = \frac{1}{f}\left(k \cdot \nabla x V\right) \qquad (1.5.2)$$

For rotational influence to be dominant, Ro must be small, or equivalently, the time it takes a fluid parcel to traverse a characteristic distance L at velocity U must far exceed the rotation period.

Similarly, a *temporal* Rossby number can be defined based on the timescale of the motions:

$$Ro_t = \frac{T_I}{T} \sim \frac{\omega}{f} \qquad (1.5.3)$$

This is nothing but the ratio of the frequency of motions to the inertial frequency, or equivalently the ratio of the inertial period ($T_I = 2\pi/f$, half a pendulum day at the poles, and one day at 30° latitude) to the period of the motions. The Coriolis parameter f is the natural frequency in a rotating fluid; hence, if the frequency of motion is much higher, the rotation is likely to have little influence on it, whereas if the frequency is small compared to the inertial frequency, the motions are slow enough to be profoundly affected by rotation. Planetary Rossby wave motions, for example, that effect internal adjustments in the oceans (and the atmosphere) have timescales of months to years and are greatly affected by planetary rotation and vorticity, whereas high frequency surface gravity waves with time scales of a few seconds are not in the least bit affected by rotation.

The ratio of the frictional term to the Coriolis term is the *Ekman number*:

$$Ek_t = \frac{K_M}{fL^2} \qquad (1.5.4)$$

This number indicates the relative importance of frictional forces. Away from boundaries, and in the interior of the oceans, frictional terms tend to be negligibly small. They are very important, however, in the turbulent surface and bottom layers. Replacing K_M, the vertical momentum diffusivity, by A_M, the horizontal mixing coefficient, leads to another Ekman number of importance to large scale circulations. Its value is also small, except in regions close to the lateral boundaries—regions of boundary currents.

Geophysical flows are under the action of both rotational and gravitational forces. Even though the prevalent stable ambient stratification is rather small (density changes in the vertical responsible for Archimedian forces on fluid parcels are just a few percent), it has a profound influence on vertical motions and hence on motions in general. To consider stratification effects, it is preferable to consider an idealized stratification condition in which the fluid column is approximated as a layer of thickness H and density ρ_0 overlying another deep layer of density $\rho_0 + \Delta\rho$. The potential energy in such a system is $\Delta\rho gH$. The resulting velocity

scale $C \sim \left(\dfrac{\Delta\rho}{\rho_0} gH\right)^{1/2}$ is the speed of propagation of long wavelength (low wavenumber) disturbances in the system. The length scale a of a flow with this velocity scale that gives a Rossby number of 1 is the scale at which both rotation and stratification are important in the flow:

$$a = \frac{C}{f} \qquad\qquad (1.5.5)$$

This length scale is a very important parameter in geophysical flows and is called the *Rossby radius of deformation*. It is the horizontal extent over which a rotating stratified fluid is affected when disturbed. For example, a Kelvin wave traveling along a coast has a signature extending a few Rossby radii perpendicular to the coast; its influence is small beyond that distance. Kelvin and Rossby waves propagating along the equator have cross-equatorial length scales of a few Rossby radii. The Rossby radius of deformation appears prominently in geostrophic adjustment on an f-plane. It is ~40 km at 30° latitude in the oceans and ~1000 km in the midlatitude atmosphere (roughly the same for Jupiter and Saturn also). The mesoscale variability in the oceans scales with this. It is therefore important to be able to resolve these scales in any numerical ocean model. However, this is still not possible in global models even with modern multigigaflop computers.

In a stably stratified fluid, the characteristic frequency is the frequency at which a fluid parcel displaced vertically from its equilibrium position will oscillate—the buoyancy frequency N. In a continuously stratified fluid with buoyancy frequency N, there exists not one but many discrete modes with propagation speed $C = NH/n$, where n is the vertical mode number. Thus an infinite number of discrete internal Rossby radii can be defined, although only the first few are usually dynamically significant. This is true even if N is not constant in the vertical, in which case, one needs to solve an eigenvalue problem to determine the speed and structure of the modes (see below).

The Rossby radius discussed above is preferably called the internal Rossby radius of deformation since an external Rossby radius a_e based on the external gravity wave speed $C_e = (gH)^{1/2}$ can also be defined. This is called the Rossby–Obukhoff radius in Russian literature. It is usually ~2000 km in the oceans and is about two orders of magnitude larger than the internal radius. At midlatitudes, it is about 3000 km for the atmosphere (~6000 km for Jupiter and Saturn).

In either case, the Rossby radius of deformation denotes the horizontal extent of a disturbance (barotropic/baroclinic) in a rotating fluid. The ratio of this length scale to the length scale of geophysical fluid motions, a/L, is another important nondimensional parameter in geophysical flows.

$$R = \frac{a}{L} = \frac{C}{fL} \qquad\qquad (1.5.6)$$

Note that this is also a Rossby number but defined using the shallow water gravity wave speed C instead of fluid velocity as the velocity scale.

On a rotating planet of radius R, another length scale is also of some importance: the intermediate geostrophic radius $a_{ig} = a\,(R/a)^{1/3}$. This is the scale at which the curvature of the planet is felt strongly, and for a mesoscale feature to last long, its characteristic scale must be less than this scale. Thus the six length scales of great relevance to motion at moderate latitudes are L, a, a_e, a_{ig}, and R. Mesoscale vortices generally have dimension L of the order of Rossby radius a or higher. Long-lived vortex features are much larger in size than the Rossby radius and are more common if R is at least an order of magnitude larger than L; this condition is well satisfied by the giant planets Saturn and Jupiter, leading, for example, to the large Jovian Great Red Spot that has been observed for over 300 years. Features in the Earth's atmosphere seldom last longer than a few weeks. The longest lived eddies in the ocean last several years. These tend to be principally large anticyclones, which behave like Rossby solitons and hence can preserve their structures quite well, whereas cyclones tend to decay rapidly by dispersion (Nezlin and Snezhkin, 1993).

Motions on a rotating planet are inevitably affected by the latitudinal variation of the magnitude of the component of the rotation vector perpendicular to the planet's surface, $f = 2\Omega \sin\theta$, where θ is the latitude. The parameter that describes the variation of planetary vorticity with latitude is $\beta = d\,f/(R\,d\theta)$. A velocity scale, the *Rossby drift speed*, based on the internal Rossby radius a and β can be defined: $V_c = \beta\,a^2$. This is the speed of propagation of nondispersive, midlatitude, linear, long wavelength Rossby waves. Irrespective of whether the feature is cyclonic or anticyclonic, mesoscale features drift westward, in the absence of background currents and other influences such as topographic gradients (see Section 1.15 and Chapter 4). This drift speed V_d is usually $\sim\beta\,r^2$, where r is the characteristic radius.

A characteristic timescale can also be defined: $t_\beta = 2\pi\,(\beta\,a)^{-1}$. Consider a linear Rossby wave packet with a Gaussian distribution of amplitude $[\sim \exp(-r^2/r_p^2)]$ in the radial direction r. The timescale for this packet to disperse such that the maximum velocity in it (determined by geostrophic balance) decreases by a factor e^{-1} is a function of its characteristic radius r_p. This timescale reaches a minimum value roughly equal to t_β when $r_p = a$, the Rossby radius. This is then the dispersion timescale for a linear Rossby wave packet with a characteristic radius equal to the Rossby radius. For the oceans, this timescale is ~80 days; for the atmosphere, it is ~8 days (~15 days for Jupiter). Mesoscale features that have lifetimes an order of magnitude longer than these values can be called long-lived. The lifetime depends to a large extent on the strength of these features, which can be defined as the ratio ω/f, where ω is the absolute value of the angular velocity of their core. Another measure is the ratio of their maximum swirl velocity to their drift speed, V_m/V_d. When this ratio is of the order of 10 or larger, even cyclonic features, which are destined to decay by dispersion spreading, last longer than t_β. Strong anticyclones tend to behave like Rossby solitons, whose dispersion tendency is nearly counterbalanced by nonlinear steepening, and hence can last longer than t_β, even when this ratio is not that large. The lifetime of mesoscale eddies in the ocean is typically several months to more than a year, although some Meddies, subsurface saltwater lenses spawned by the outflow of salty Mediterranean Sea water into the Atlantic, have been tracked for more than 2 years (see Kantha and Clayson, 1999).

A length scale associated with zonal flows with velocity U on a rotating planet, $L_z = (U/2\beta)^{1/2}$, is also an important scale, since this is the meridional scale of features associated with zonal jets. For the Gulf Stream and Kuroshio extensions, L_z is ~160 km, much larger than the Rossby radius a.

Except at low latitudes around the equator, flows in the atmosphere and the oceans are generally baroclinically unstable (see Chapter 4). The result is the generation of meanders and eddies that in turn transport mass, momentum, and heat in the meridional direction. This mechanism is especially important in the atmosphere. The timescale associated with baroclinic instabilities of a flow with characteristic velocity U in a rotating, continuously and uniformly stratified fluid (Eady, 1949; Charney, 1947; Green, 1970; Stone, 1972; Held and Larichev, 1998) is

$$t_{bc} = \frac{N}{f\left|\dfrac{\partial U}{\partial z}\right|} = \frac{Ri^{1/2}}{f} \tag{1.5.7}$$

where $Ri = N^2 / \left|\dfrac{\partial U}{\partial z}\right|^2$ is the gradient Richardson number, which is a measure of the *available potential energy* (APE) stored in the mean stratification.

Available potential energy is an important concept in geophysical flows. It is the potential energy excess available in a given state over and above the theoretical reference state of minimum potential energy reached through a reversible adiabatic process by a rearrangement of the fluid parcels (Oort *et al.*, 1989) [see Huang (1998) for a detailed description of APE in a Boussinesq fluid and its implications]. This is the amount of energy in the system available for possible conversion into kinetic energy (KE) of the flow and hence to drive the motion. All density surfaces are level in the reference state. The flow state however involves a tilt of the density surfaces and this makes the potential energy of the stratified system available for conversion to kinetic energy via baroclinic instabilities. However, this theoretical reference state is hard to ascertain. Instead the Lorenz (1955) approximation, a linearized form of APE, is often used, since no such reference state need be found. This approximation has been applied to the ocean by, for example, Bryan and Lewis (1979), Oort et al. (1989), and Toggweiler and Samuels (1998).

$$ \text{APE} = -\frac{g}{2} \iiint \frac{(\rho - \bar{\rho})^2}{\partial \bar{\rho}_\theta / \partial z} dv \qquad (1.5.8) $$

where ρ is the local in situ density, $\bar{\rho}$ is the mean local in situ density, and $\bar{\rho}_\theta$ is the horizontally averaged potential density at that depth. The rate of change of APE is written assuming both $\bar{\rho}$ and $\bar{\rho}_\theta$ are time invariant, so that

$$ \frac{\partial(\text{APE})}{\partial t} = -g \iiint \frac{(\rho - \bar{\rho})(\partial \rho / \partial t)}{\partial \bar{\rho}_\theta / \partial z} dv \qquad (1.5.9) $$

Above the permanent thermocline, in other words, roughly in the upper kilometer of the water column, the timescale $t_{bc,}$ whose inverse is a measure of the growth rate of baroclinic instabilities, ranges from a few days in regions of western boundary current extensions to about a month [see Chapter 4; see also Stammer (1998)]. It is a few days in the atmosphere.

A length scale associated with baroclinic eddies can be defined using β and t_{bc}: $L_{bc} = (\beta t_{bc})^{-1}$. This length scale is important to the eddy-mediated transfer of scalar properties on a rotating planet. The eddy transfer coefficient $K_{bc} \sim L_{bc}^2 / t_{bc} = \beta^{-2} t_{bc}^{-3}$. Stammer (1998) has used these concepts to estimate the meridional eddy-mediated heat and salt transfer in the global oceans from observed climatological distributions of temperature and salinity and estimates of the eddy transfer coefficient from satellite altimetry.

An important parameter that determines whether the flow is hydrostatic is the ratio of the characteristic advection timescale to the buoyancy period, or equivalently $N_H = U/(NL)$, where L/U is the timescale for advection over a characteristic length scale L. If $N_H \ll 1$, the flow behaves as if it is close to hydrostatic equilibrium. Except in regions of strong wintertime deep convection, the oceans are very nearly hydrostatic.

The ratio of Coriolis to buoyancy frequencies, or equivalently the buoyancy period to inertial period,

$$R_f = \frac{f}{N} = \frac{T_b}{T_I} \qquad (1.5.10)$$

is also an important additional parameter in oceanic motions. For example, the inertial-internal motions are confined by dynamics to the part of the frequency spectrum delineated by frequencies N and f (Chapter 5 of Kantha and Clayson, 1999).

1.5.1 NORMAL MODES

Even when the stratification cannot be idealized (as two-layer or linear), for any arbitrary stratification in the water column, fluid motions can be considered to be a superposition of an infinite sequence of discrete normal modes. These modes can be determined from the solution to the corresponding eigenvalue problem in a flat-bottomed ocean with the prescribed density stratification. Starting from the linearized equations for freely propagating waves, which, for a flat bottom ocean, makes it possible to separate the vertical structure from the horizontal and temporal structure (see Gill, 1982), one gets for the vertical structure

$$\frac{d^2w}{dz^2} + \frac{N^2}{c^2} = 0 \qquad (1.5.11)$$

$$w = 0 \text{ at } z = 0, \ z = -H$$

where w is the vertical velocity and $N(z)$ equal to $\left(-\frac{1}{\rho_0} \frac{d\rho}{dz} \right)^{1/2}$ is the observed buoyancy (Brunt–Vaisala) frequency. The separation constant is $1/c^2$, where c is the internal gravity wave speed. The internal Rossby radii are related to the eigenvalues by $a_i = c/|f|$ except in low latitudes ($|\phi| \leq 5°$), where $a_I = (0.5c/\beta)^{1/2}$. See Emery *et al.* (1984) for the distribution of the Rossby radius in the north Pacific and the Atlantic, and Chelton *et al.* (1998) for a more recent calculation of the gravity wave speed and the radius of deformation in the global oceans. The first mode corresponds to the gravest internal wave motion, the speed of which defines the largest internal Rossby radius. It is often the first one or two modes that are preferentially excited in the oceans (and the atmosphere).

1.5.2 SCALES

The following table (Table 1.5.1) presents the length and time scales associated with physical processes in the oceans and the atmosphere, and Figure (1.5.1) shows a plot of these. As expected, the Rossby numbers decrease with increase in the length and time scales.

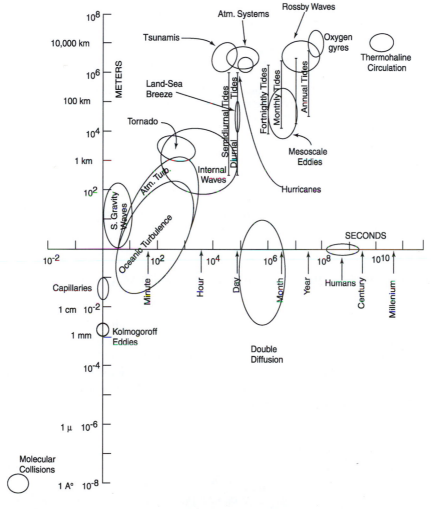

Figure 1.5.1 The range of spatial and temporal scales of motions in the atmosphere and the oceans. The motions span over a 10-decade range in both space and time.

Table 1.5.1 Time and Spatial Scales of Various Oceanic and Atmospheric Phenomena

Process	Length scale	Timescale	Rossby number
Dissipative scales	1–2 mm	~1 s	~10^4
Vertical mixing (ocean)	1–100 m	several minutes	~10^2
Vertical mixing (atm)	1–1000 m	several minutes	~10^2–10^3
Surface waves	1–100	several seconds	~10–10^2
Internal waves	1–10 km	mins–hrs	~10^{-1}–1
Double diffusion	1 cm–10 m	days–weeks	~10
Coastal upwelling	10–20 km	days	~10^{-1}–1
Mesoscale eddies	10–400 km	weeks–months	~10^{-1}
Atm. weather patterns	100–5000 km	days–weeks	~10^{-1}–1
Ocean fronts	5–50 km	weeks	~10^{-1}
Boundary currents	50–100 km	months	~10^{-1}
Basin gyres	2000–15000 km	years	~10^{-2}
Ocean tides	100–1000 km	1/2, 1 day, ...	~10^{-4}–10^{-2}
Tornadoes	few km	<1 hr	~10^3
Hurricanes	500–2000 km	days	~10
Rossby waves	1000 km	months–years	~10^{-3}
Tsunamis	100 km	day	~10^2
Land–sea breeze	10–50 km	1/2 day	~1

Note that the mean free path from kinetic theory gives a length scale of about ~10^{-7} m at sea level for the atmosphere, and the corresponding timescale associated with molecular collisions is ~10^{-4} s.

1.6 GEOSTROPHIC FLOW AND THERMAL WIND

We will consider now some simple limits for which analytical solutions to the governing Boussinesq, hydrostatic equations are possible. We will ignore the tidal potential in the following. Consider a steady, inviscid (Ek = 0) flow at zero Rossby number. Away from the frictional layers at the surface and the bottom, and the lateral boundaries, this is often a good approximation. The nonlinear terms are then zero and the governing equations become

$$\varepsilon_{i3j}fU_j = -\frac{1}{\rho_0}\frac{\partial P}{\partial x_i}$$

$$0 = -\frac{1}{\rho_0}\frac{\partial P}{\partial z} - \frac{\rho}{\rho_0}g$$

(1.6.1)

The flow is called geostrophic. If the density distribution is known, the pressure distribution can be obtained to the level of an unknown constant by integrating the hydrostatic equation; the velocity shear distribution can then be determined from the momentum equation. The utility of the geostrophic approximation is that, knowing the horizontal density distribution, and velocity at any single level in the water column, it is possible to determine the velocity distribution in the entire water column. Assuming the velocity is zero at some depth is one way of deriving velocities in the water column from measured horizontal density gradients, a time-honored method in physical oceanography, although the determination of the *level of no motion* has always been rather ambiguous. Alternatively, if sea level gradients are known and thus the surface velocities, it is then possible to infer the velocities in the water column. The sea surface gradient can in principle be obtained from altimetry, which, in combination with *in situ* measurements, provides a powerful means of deducing the vertical velocity structure and transport in the water column.

Note that the flow is perpendicular to the pressure gradient, with higher pressure to the right (left) in the northern (southern) hemisphere. Thus, the sea level increases by nearly a meter across the Gulf Stream from the coast to the open water. This is also the reason why the flow is cyclonic, counterclockwise (clockwise), around a low pressure system in the atmosphere or a cold core eddy in the ocean and anticyclonic, clockwise (anticlockwise), around a high pressure system in the atmosphere or a warm core eddy in the ocean in the northern (southern) hemisphere. In the oceans, large warm core eddies can have surface elevations almost as high as a meter and give rise to strong anticyclonic swirl velocities of as much as 2 m s^{-1}. The eddies in the Agulhas current south of Africa and the Loop Current eddies in the Gulf of Mexico are two such examples. The well-known example in the atmosphere is that of a hurricane, at whose center the pressure can drop as much as 60 mb, giving rise to strong cyclonic winds often exceeding 60 m s^{-1}.

The reason for needing information on velocity at a single level to deduce the velocity in the water column is simply that the geostrophic balance is a statement about the baroclinic portion of the currents, in other words, the currents generated by horizontal density gradients. The sea level slopes can also be generated by the action of the winds at the surface, and this gives rise to a uniform velocity in the water column, the barotropic component. The sum of the barotropic and baroclinic components gives the total current. The barotropic component is normally not known, without actual current measurements.

The fact that the geostrophic balance provides information on only the vertical variation in currents can be seen by taking the vertical derivative of the horizontal momentum equation in Eq. (1.6.1), interchanging the order of the horizontal and vertical derivatives, and substituting for the vertical gradient of pressure from the hydrostatic equation in Eq. (1.6.1) to get

$$\varepsilon_{i3j}\, f\, \frac{\partial U_j}{\partial z} = \frac{g}{\rho_0}\, \frac{\partial \rho}{\partial x_i} \qquad (1.6.2)$$

Thus, the horizontal gradients of density determine the vertical gradients of horizontal velocities, that is, the vertical shear. This equation is called the *thermal wind* equation, since it was originally applied to the atmosphere, where the density gradients are predominantly due to temperature gradients. In other words, the slope of isotherms or isentropes determines the vertical shear of the winds. Then, knowing the velocity at any point in the column (away from surface and bottom frictional layers), the velocity in the entire column can be derived. It is instructive to write Eq. (1.6.2) in its component form:

$$-f\frac{\partial U_1}{\partial z} = \frac{g}{\rho_0}\frac{\partial \rho}{\partial x_2}; \quad f\frac{\partial U_2}{\partial z} = \frac{g}{\rho_0}\frac{\partial \rho}{\partial x_1} \qquad (1.6.3)$$

It can be seen that the meridional gradient of density determines the vertical shear of the zonal velocity, and the zonal density gradient determines the vertical shear of the meridional velocity. Actually, for strong pressure systems, it is necessary to include the centripetal force terms in the force balance. Then the flow is said to be in cyclogeostrophic balance (see Section 1.15).

If we eliminate pressure from Eq. (1.6.1) by taking the curl of the equation, we get $\partial(fU_j)/\partial x_j = 0$ and since $\partial f/\partial x_1 = 0$, $\beta = \partial f/\partial x_2$, making use of the continuity equation (1.3.17),

$$\beta V = f\frac{\partial W}{\partial z} \qquad (1.6.4)$$

which is simply a statement of vorticity conservation for geostrophic motions (in the absence of external forcing) and is called the *Sverdrup relation*. Any stretching (shrinking) of a fluid element in the vertical direction increases (decreases) its vorticity and must therefore be accompanied by poleward (equatorward) motion of the fluid parcel to feed on the increasing (decreasing) planetary vorticity, since the relative vorticity of geostrophic motion itself is negligible. It is an important constraint to which large scale oceanic circulation must adhere. Vertical integration of Eq. (1.6.4) over the fluid column and imposing the boundary conditions $W = 0$ at the top and the bottom leads to $\beta M_y = 0$, so that the net meridional transport in the water column is zero in the absence of external torques. What happens when the wind stress imposes a torque is discussed in Section 1.9.

It is important to remember that while geostrophic approximation is an excellent diagnostic tool [through Eqs. (1.6.2) to (1.6.4)], much of the dynamics of flows on a rotating plane cannot be so described. Higher order approximations (essentially perturbation expansions in terms of the Rossby number as a small parameter) to the governing equations are essential. Such approximations are called intermediate equations (see Section 4.6). Often, a good approximation to midlatitude dynamics is the so-called quasi-geostrophic approximation that will be dealt with in detail in Chapter 4.

1.7 INERTIAL MOTIONS

Consider a horizontally homogeneous, inviscid flow, away from lateral boundaries. Let the horizontal pressure gradients be zero. From continuity, $\partial w / \partial z = 0$, and since the vertical velocity w is zero at the surface, it must be zero everywhere. The governing equations then reduce, in component form, to

$$\frac{\partial U}{\partial t} - fV = 0; \quad \frac{\partial V}{\partial t} + fU = 0 \tag{1.7.1}$$

The solution is

$$U = V_0 \sin ft; \; V = V_0 \cos ft$$
$$x = x_0 - \frac{V_0}{f} \cos ft; \; y = y_0 + \frac{V_0}{f} \sin ft \tag{1.7.2}$$

The motion described is an inertial oscillation with a period equal to the *inertial period* T_I (half-pendulum day at the poles, 1 day at 30°, and increasing toward the equator). The fluid particles execute a circular motion in the anticyclonic direction, clockwise (counterclockwise) in the northern (southern) hemisphere, with a radius of V_0/f and the center at (x_0, y_0). Vigorous *inertial oscillations* are often a response of the upper ocean to strong and sudden changes in wind forcing. In the atmosphere, inertial oscillations are often evident at night in the atmospheric boundary layer (ABL). Once the solar heating ceases, the turbulence in most of the deep daytime ABL, except close to the ground, decays. The decrease in frictional retardation causes the winds to try to accelerate to their geostrophic values. This leads to characteristic inertial motions over the night, although the time period is normally inadequate for the wind to undergo a complete inertial cycle.

The solutions can be extended to the case of a uniform background current (in a homogeneous fluid) by either adding the pressure gradient terms to Eq. (1.7.1),

$$\frac{\partial U}{\partial t} - fV = -\frac{1}{\rho_0} \frac{\partial P}{\partial x}; \quad \frac{\partial V}{\partial t} + fU = -\frac{1}{\rho_0} \frac{\partial P}{\partial y} \tag{1.7.3}$$

and solving them, or, because of the linearity of the equations, simply adding the resulting geostrophic background currents to the solution (1.7.2):

$$U = U_g + V_0 \sin ft; \; V = V_g + V_0 \cos ft$$
$$x = x_0 + U_g t - \frac{V_0}{f} \cos ft; \; y = y_0 + V_g t + \frac{V_0}{f} \sin ft \tag{1.7.4}$$

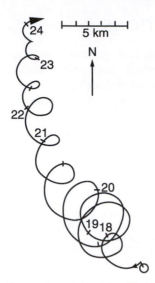

Figure 1.7.1 Progressive vector diagram showing inertial oscillations superposed on a mean current in the Baltic Sea derived from current meter observations of Gustafson and Kullenberg (1933).

These equations describe a particle motion that starts at the initial point (x_0, y_0) and proceeds in the general direction of the background current, but with anticyclonic, clockwise (counterclockwise) looping motions of radius V_0/f around the general direction, in the northern (southern) hemisphere. Drifters deployed in the ocean, and ice floes in the polar seas during summer, very often undergo such motions. Figure 1.7.1 shows an example of such inertial oscillations superposed on a background current calculated from current observations in the Baltic Sea by Gustafson and Kullenberg (1933).

1.8 EKMAN LAYERS

Consider a steady, horizontally homogeneous flow, but one under frictional influence (Ek ≠ 0). Once again, there are no horizontal pressure gradients or horizontal frictional forces because of horizontal homogeneity, and from continuity, there is no vertical motion (w = 0). The governing equations are

$$\varepsilon_{i3j} f U_j = +\frac{\partial}{\partial z}\left(K_M \frac{\partial U_i}{\partial z} \right) \tag{1.8.1}$$

Ekman was the first to study this problem (Ekman, 1905), in response to observations by Sverdrup that the ice floes in the Arctic appeared to drift to the right of the wind, not in the direction of the wind. For tractability, he assumed that K_M is a constant, instead of a function of U_j. This makes the equations linear and analytical solutions feasible. The case he considered was one of a constant wind stress acting over a vertically homogeneous, infinitely deep fluid column. The boundary conditions for this case are

$$K_M \frac{\partial U_j}{\partial z} = \frac{\tau_j^w}{\rho_0} \qquad (z=0)$$

$$U_j \to 0 \qquad (z \to -\infty) \tag{1.8.2}$$

Without loss of generality, we can align the x-axis in the direction of the wind stress and put $\tau_1^w = \tau^w$; $\tau_2^w = 0$. The solution is

$$U(z) = U_s \sin\left(\frac{\pi}{4} + \frac{\pi z}{D_E}\right) \exp\left(\frac{\pi z}{D_E}\right)$$

$$V(z) = -U_s \cos\left(\frac{\pi}{4} + \frac{\pi z}{D_E}\right) \exp\left(\frac{\pi z}{D_E}\right) \tag{1.8.3}$$

$$U_s = \frac{2^{1/2} \pi \tau^w}{f D_E \rho_0}, \quad D_E = \left(\frac{2K_M}{f}\right)^{1/2}$$

A hodograph of the velocity as a function of depth is shown in Figure (1.8.1). Note that the surface velocity U_s is 45° to the right of the wind stress vector (left in the southern hemisphere), and at $z = -D_E$, the flow velocity is actually in a direction opposite to the wind stress. D_E is the *Ekman scale,* the length scale that determines the depth to which the influence of the wind penetrates in a homogeneous fluid. Ekman considered essentially pseudo-laminar solutions, with a value for K_M constant but much larger than the laminar value, to obtain realistic solutions for D_E and U_s. In practice, the flow is turbulent, K_M is a function of the flow itself, and a turbulence closure theory is needed to solve even this simple set of equations. Observationally, the surface velocity has been observed to make a somewhat lesser angle than 45° (10° to 25°), and this angle appears to depend also on the magnitude of the wind stress.

The Ekman form for the Ekman scale and surface velocity is not very useful. Instead one can use scaling arguments to derive these relationships. In a turbulent boundary layer such as the Ekman layer, it is $u_* = \left(\left|\tau^w\right|/\rho_0\right)^{1/2}$, the *friction velocity,* that is the scaling variable. Then, by simple dimensional arguments, D_E is proportional to u_*/f, the value of the proportionality constant depending on how the depth of the layer is defined. The usual value is approximately the same as that of κ

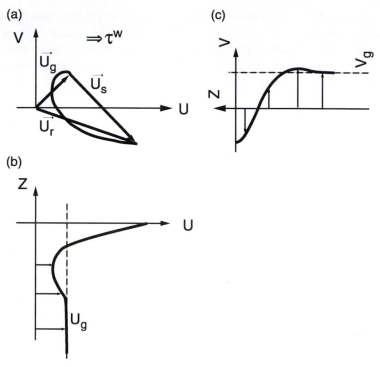

Figure 1.8.1 Ekman current profile for a zonal (eastward) wind stress (in the northern hemisphere): (a) hodograph (plot of zonal current U versus meridional current V), (b) zonal velocity profile (U versus depth z), and (c) meridional velocity profile (V versus depth z). U_g and V_g are geostrophic velocity components. Note the 45° angle the surface velocity makes with the stress vector and the turning of the current to the right of the stress vector with depth.

(= 0.4), the von Karman constant, although values anywhere from 0.2 to 0.6 have been used. The surface velocity must of course scale with u_* so that $U_s \sim u_*$. Careful observations have indeed shown an *Ekman spiral* to exist in the upper layers of the ocean (Price *et al.*, 1987). Schudlich and Price (1998) present detailed observations of Ekman layers during the Long-Term Upper Ocean Study (LOTUS) during both summer and winter. They find that the summer Ekman layer is shallow and its vertical structure and depth subject to strong diurnal modulation. The winter layer is deeper by a factor of two. The mean current does display the characteristic spiral structure but the degree of turning is small compared to the theoretical (laminar) value. They also find that the instantaneous structure of the Ekman layer is rather slab-like, but the long-term mean is a spiral that decays faster in magnitude than it rotates with depth, compared to the theoretical spiral. Careful wind-relative averaging demonstrates that the Ekman transport is in accordance with theoretical results.

A remarkable property of the Ekman flow is that associated with the total transport in the water column, which can be shown by integration in the vertical of Eq. (1.8.1) in its component form and the use of boundary conditions (1.8.2) to be

$$M_x = \int_{-\infty}^{0} U(z)dz = \frac{1}{f}\left(K_M \frac{\partial V}{\partial z}\right)\Big|_{-\infty}^{0} = 0$$

(1.8.4)

$$M_y = \int_{-\infty}^{0} V(z)dz = -\frac{1}{f}\left(K_M \frac{\partial U}{\partial z}\right)\Big|_{-\infty}^{0} = -\frac{\tau^w / \rho_0}{f} = -\frac{u_*^2}{f}$$

This result could have been obtained from the pseudo-laminar solutions (1.8.3), but is far more general, since it does not depend on the value or functional dependence of the mixing coefficient. It is therefore valid for all Ekman layers, whether laminar or turbulent. It is a universal result that has a profound effect on wind-driven ocean circulations. It states that the net *Ekman transport* in the water column is at right angles to the wind stress and to its right (left) in the northern (southern) hemisphere, and depends only on the value of the wind stress and the Coriolis term, and nothing else. Thus a knowledge of the wind stress is all that is needed to derive this transport.

The consequences of Ekman transport are rather profound. This can be seen by integration of the continuity equation (1.3.3) in the vertical:

$$\frac{\partial M_x}{\partial x} + \frac{\partial M_y}{\partial y} + w\Big|_{z=0} - w\Big|_{z=-\infty} = 0 \qquad (1.8.5)$$

Since the vertical velocity at $z = 0$ must vanish, substituting from Eq. (1.8.4) for M_x and M_y,

$$w\Big|_{z=-\infty} = \frac{\partial}{\partial x}\left(\frac{\tau_y^w}{\rho_0 f}\right) - \frac{\partial}{\partial y}\left(\frac{\tau_x^w}{\rho_0 f}\right) = \mathbf{k}\cdot\nabla\times\left(\frac{\tau^w}{\rho_0 f}\right) = \frac{1}{\rho_0 f}\mathbf{k}\cdot\nabla\times\tau^w \qquad (1.8.6)$$

where **k** is the unit vector in the vertical direction and the latitudinal variations of the Coriolis parameter f have been ignored (f-plane approximation). Once again, this result is general, and states that there exists a vertical velocity induced in the deep that is proportional to the curl of the wind stress acting at the surface. In practice, this vertical velocity exists at the base of the turbulent Ekman or mixed layer, and is called *Ekman pumping*. The basis for this can be seen rather easily.

Assume that the wind stress is blowing to the north, but increases from west to east. The resulting Ekman transport is to the east and increases from west to east also. To accommodate this increase in transport, water must upwell from the deep. If there were a decrease in northward wind stress from west to east, downwelling would result.

Wind stress curl (torque)-induced upwelling/downwelling is ubiquitous in the world's oceans and is dramatically evident wherever the wind blowing over the ocean has a jet-like structure (Figure 1.8.2). For example, off the coast of Arabia, during southwest monsoons, the very strong, generally northeastward wind jet causes upwelling to take place to the west of the jet axis and downwelling to its east. This augments the coastal upwelling (see below), brings up the nutrients from below, and increases the primary productivity in the sun-lit upper layers (Brink *et al.,* 1998).

The Ekman transport near the equator also leads to upwelling and downwelling along the equator. The winds are generally easterlies in the equatorial waveguide, and hence there exists an Ekman drift in both the northern and the southern hemispheres that transports surface water away from the equator, inducing up-welling all along the equator (Figure 1.8.2). When, occasionally, winds become westerly, as during strong westerly wind bursts, downwelling takes place. At high latitudes, the presence of sea ice over the ocean also leads to upwelling along the ice edge. The ice surface is rougher than the adjacent water; hence, the drag coefficient is larger over ice than over water. This means that if wind blows parallel to the ice edge, the stress exerted over the ice and hence transmitted through ice to the water below is larger than that over water and a curl of the surface stress exists and leads to upwelling (downwelling) along the ice edge, if the ice edge is to the right (left) of the wind (Figure 1.8.2).

The presence of lateral boundaries also has a profound influence on vertical motions in the ocean. In the northern (southern) hemisphere, if the wind blows along a coast with the coast to its left (right), the Ekman transport induced at right angles and to the right (left) of the wind causes surface waters to be transported away from the coast. Because of the presence of the coast, the velocity perpendicular to the coast must vanish; therefore, the replacement waters must come from underneath. This causes upwelling of colder, deeper waters all along the coast (Figure 1.8.2), which in turn brings nutrients to the euphotic zone and increases the primary productivity in the upper layers. The upwelling regions around the world account for most of the primary productivity, with 90% of the fisheries concentrated in about 10% of the ocean areas that happen to be upwelling regions, mostly along coasts like that off Peru, and during southwest monsoons, off the Arabian and Somali coasts. The high productivity regions show up clearly in ocean color images from a satellite sensor such as the Coastal Zone Color Scanner (CZCS) that NASA orbited on a NOAA satellite in the late seventies. The follow-on to this mission, the SeaWIFS ocean color mission, is also providing excellent color data that should tell

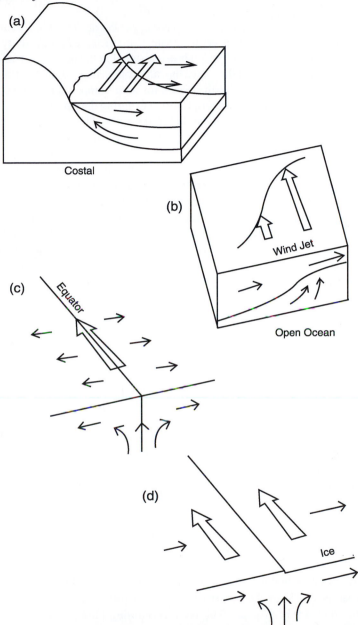

Figure 1.8.2 Upwelling situations in the ocean: (a) coastal upwelling induced by a wind stress parallel to the coast with the coast to its left (in the northern hemisphere), (b) upwelling induced by the curl of the wind stress in the open ocean (for example, off the Arabian coast during summer monsoons), (c) equatorial upwelling due to easterlies, and (d) upwelling along an ice edge (in the northern hemisphere).

66   1 Introduction to Ocean Dynamics

us more about the primary productivity in the upper ocean. This has a bearing not only on fisheries, but also on long-term climate, since increased productivity implies increased carbon uptake, part of which is sequestered in the deep ocean and hence tends to mitigate the effects of the increasing anthropogenic CO_2 in the atmosphere. The cold upwelled waters are also the reason for water temperatures being too cold for swimming off the coast of northern California during the summer, when the prevailing winds blow southward and parallel to the coast. The sea level drops at the coast during upwelling. Also, the sloping of the isotherms causes a subsurface jet in the direction of the wind.

The opposite happens with the wind blowing parallel to the coast but with the coast to its right (left) in the northern (southern) hemisphere. This causes downwelling of water masses, but the effects are not as dramatic as the upwelling case.

Ekman layers also exist at the ocean bottom if there exists a current above the bottom boundary. The solutions for this case can also be obtained readily. For simplicity, align the x-axis in the direction of the flow and consider a velocity U_g in the x direction. With the origin at the bottom, the boundary conditions are

$$U = 0, \ V = 0 \qquad (z = 0)$$
$$U \to U_g, V \to 0 \quad (z \to \infty) \tag{1.8.7}$$

The solution is

$$U(z) = U_g \left[1 - \cos\left(\frac{\pi z}{D_E}\right) \exp\left(-\frac{\pi z}{D_E}\right) \right]$$
$$V(z) = U_g \sin\left(\frac{\pi z}{D_E}\right) \exp\left(-\frac{\pi z}{D_E}\right) \tag{1.8.8}$$

The stress exerted is of considerable interest:

$$\frac{\tau_x^b}{\rho_0} = \frac{\tau_y^b}{\rho_0} = \frac{f}{2} D_E U_g \tag{1.8.9}$$

The corresponding hodograph is shown in Figure (1.8.3a). It can be seen that the bottom stress is 45° to the left (right) of the geostrophic velocity aloft in the northern (southern) hemisphere. This is a bit confusing. However, if one remembers that in the surface Ekman layer, the wind was 45° to the right of the wind stress, the same happens in the *bottom Ekman layer* also. The current veers away to the right of the stress in both cases. This case is of interest in the atmospheric boundary layer, where the stress applied by the wind aloft is at an angle but to its left.

Once again, it is the friction velocity that is a more convenient measure of the depth of the bottom Ekman layer, $D_{Eb} \sim \kappa\, u_{*b}/f$. If the water column is much deeper than the typical Ekman scales, the surface and bottom Ekman layers will be distinct, with a geostrophic current in between. In shallow water on the continental shelf, the two begin to coalesce. When the water is much shallower than the Ekman scale (H $< 0.1\, D_E$), the two Ekman spirals "cancel" each other, and the current and transport in the water column are in the direction of the wind itself. Therefore, the Ekman scale is quite important in considerations of transport in shallow coastal waters also.

The above results are applicable only to neutrally stratified oceans. The Ekman depth denotes the extent of the wind-driven mixing if stratification were not present. However, the presence of stratification can affect the wind-driven mixed layer depth. While in the neutrally stratified case, wind-induced mixing can penetrate to the depth of the neutral Ekman layer, the presence of a strong buoyancy interface such as a seasonal thermocline in the oceans and a strong inversion in the atmosphere tends to greatly inhibit such penetration, and the depth of the buoyancy interface determines the extent of mixing instead.

The solutions indicated by Eq. (1.8.8) are also valid for the atmospheric boundary layer (ABL), and hence it is interesting to obtain the solution for the coupled atmosphere–ocean surface layers. In order to do this, we need to recognize two things: (1) the stress is continuous across the atmosphere–ocean interface (in the absence of transfer of momentum to surface gravity waves), and (2) the velocity must also be continuous across and nonzero, unlike the bottom boundary layer solution above. The modification to the bottom boundary layer solution to take into account an arbitrary geostrophic velocity aloft is a simple coordinate transformation, and similarly, accounting for the bottom velocity is a simple Galilean translation:

$$U^a(z) = \left(U_g^a - U_s^w\right)\left[1 - \cos\left(\frac{\pi\, z}{D_E^a}\right)\exp\left(-\frac{\pi\, z}{D_E^a}\right)\right]$$

$$-\left(V_g^a - V_s^w\right)\sin\left(\frac{\pi\, z}{D_E^a}\right)\exp\left(-\frac{\pi\, z}{D_E^a}\right)$$

$$V^a(z) = \left(V_g^a - V_s^w\right)\left[1 - \cos\left(\frac{\pi\, z}{D_E^a}\right)\exp\left(-\frac{\pi\, z}{D_E^a}\right)\right]$$

$$+\left(U_g^a - U_s^w\right)\sin\left(\frac{\pi\, z}{D_E^a}\right)\exp\left(-\frac{\pi\, z}{D_E^a}\right)$$

(1.8.10)

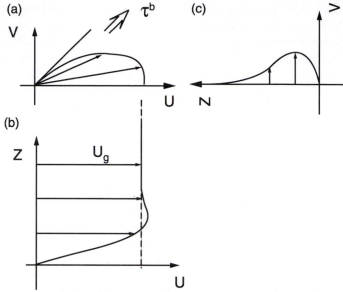

Figure 1.8.3 Ekman current profile for a zonal (eastward) bottom current (in the northern hemisphere): (a) hodograph (plot of zonal current U versus meridional current V), (b) zonal velocity profile (U versus depth z), and (c) meridional velocity profile (V versus depth z). U_g is the geostrophic velocity. Note the 45° angle the bottom stress makes with the geostrophic current vector and the turning of the current to the left of the geostrophic current with depth.

$$\frac{\tau_x^b}{\rho_a} = \frac{f}{2} D_E^a \left[\left(U_g^a - U_s^w \right) - \left(V_g^a - V_s^w \right) \right]$$

$$\frac{\tau_y^b}{\rho_a} = \frac{f}{2} D_E^a \left[\left(U_g^a - U_s^w \right) + \left(V_g^a - V_s^w \right) \right]$$

$$(1.8.11)$$

These stress values can be used to derive the solution for the oceanic layer. Once again it is a straightforward extension of the solution (1.8.3) which can be generalized to include both the wind stress in an arbitrary direction and a geostrophic velocity deep below,

$$U^w(z) = U_g^w + \frac{2^{1/2}\pi}{f D_E^w} \left[\frac{\tau_x^b}{\rho_0} \sin\left(\frac{\pi}{4} + \frac{\pi z}{D_E^w} \right) + \frac{\tau_y^b}{\rho_0} \cos\left(\frac{\pi}{4} + \frac{\pi z}{D_E^w} \right) \right] \exp\left(\frac{\pi z}{D_E^w} \right)$$

$$V^w(z) = V_g^w + \frac{2^{1/2}\pi}{f D_E^w} \left[-\frac{\tau_x^b}{\rho_0} \cos\left(\frac{\pi}{4} + \frac{\pi z}{D_E^w} \right) + \frac{\tau_y^b}{\rho_0} \sin\left(\frac{\pi}{4} + \frac{\pi z}{D_E^w} \right) \right] \exp\left(\frac{\pi z}{D_E^w} \right)$$

$$(1.8.12)$$

with

$$U_s^w = U_g^w + \frac{\pi}{fD_E^w}\left[\frac{\tau_x^b}{\rho_0} + \frac{\tau_y^b}{\rho_0}\right]; V_s^w = V_g^w + \frac{\pi}{fD_E^w}\left[-\frac{\tau_x^b}{\rho_0} + \frac{\tau_y^b}{\rho_0}\right]$$

$$D_E^a = \left(2\frac{K_M^a}{f}\right)^{1/2}; D_E^w = \left(2\frac{K_M^w}{f}\right)^{1/2}$$

(1.8.13)

The Ekman length scales are quite disparate in the two media. If τ is the magnitude of the stress at the air–sea interface, the ratio of friction velocities $u_*^a/u_*^w = (\rho_0/\rho_a)^{1/2} \sim 30$ and this is reflected in the corresponding Ekman scales and the magnitudes of the corresponding velocities: the Ekman scale and velocities are larger by a factor of 30 in the atmosphere. The typical Ekman scale in midlatitudes is about 40 m in the oceans, but 1200 m in the atmosphere. Because of this disparity in velocity scales, it is possible to ignore to a first approximation the terms involving the ocean surface velocities in Eq. (1.8.11), thus decoupling Eq. (1.8.11) from Eq. (1.8.12), although solutions that are somewhat algebraically tedious can be obtained retaining those terms. Figure 1.8.4, from Brown (1990), is a 3D picture of the winds and currents adjacent to the air–sea interface; note that the disparity in magnitudes has been suppressed here deliberately to elucidate the turning of currents (winds) with depth (height).

These solutions are useful for qualitative purposes only since they are pseudo-laminar and depend very much on the mixing coefficient being constant with height or depth. In practice, there is considerable variation in mixing coefficient with height (and depth), and the velocity magnitudes, turning angles, etc., have to be obtained by solving the turbulent set of equations with appropriate closure. Also, geophysical flows are seldom neutrally stratified and the presence of a pycnocline (or an inversion in the atmosphere) often limits the extent of the wind-mixed layers adjacent to the air–sea interface. The turning angle also depends on the nature of stratification. LES studies indicate that the turning angle is small in a convective ABL compared to that in the stably stratified ABL (Moeng and Sullivan, 1994). Nevertheless, the empirical Ekman scale $\kappa u_*/f$ and the Ekman transport $\tau_w/(\rho_0 f)$ are quite accurate and useful as quantitative estimates.

Another interesting aspect pertains to how the currents in the surface Ekman layer spin up. For this, one needs to consider the time-dependent Ekman flow:

$$\frac{\partial U_i}{\partial t} + \varepsilon_{i3j}fU_j = \frac{\partial}{\partial z}\left(K_M \frac{\partial U_i}{\partial z}\right)$$

(1.8.14)

The transient solutions are algebraically tedious (Ekman, 1905). They describe a *Cornu spiral* in the hodograph plane for the surface velocity (Figure 1.8.5).

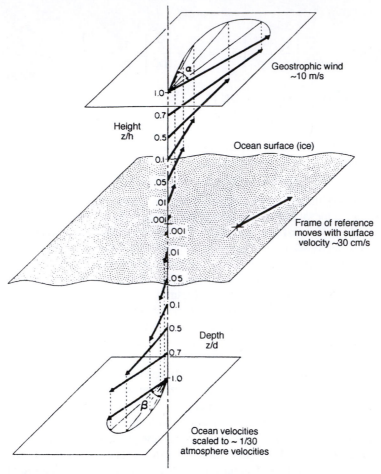

Figure 1.8.4 Ekman profiles in the atmosphere and the ocean. Note that the ocean currents are not drawn to the same scale as atmospheric winds. From Brown (1990).

Oscillations die down in amplitude, and steady state is reached within a few inertial periods. Simple expressions for the vertically integrated transports are, however, possible. By simply vertically integrating the above equation and satisfying the boundary conditions as before,

$$\frac{\partial M_x}{\partial t} - fM_y = \frac{\tau_x^w}{\rho_0}; \frac{\partial M_y}{\partial t} + fM_x = \frac{\tau_y^w}{\rho_0} \qquad (1.8.15)$$

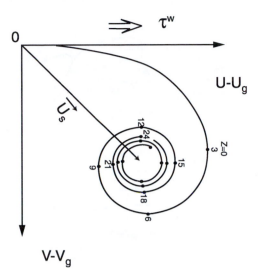

Figure 1.8.5 Hodograph showing how the surface velocity spins up to its steady state value 45° to the right of the wind stress (in the northern hemisphere) when an impulsive wind stress is applied. The numbers denote inertial hours (12 hr is one inertial period).

Aligning the coordinate directions suitably and considering the case of zero τ_x^w, the solution for an impulsively imposed wind stress is

$$M_x = \frac{\tau_y^w}{\rho_0 f}\left[1 - \cos(ft)\right]; \quad M_y = \frac{\tau_y^w}{\rho_0}\sin(ft) \qquad (1.8.16)$$

indicating that the vertically integrated Ekman transports do not decay in time to their steady state values, but simply oscillate about their steady value at the inertial frequency, even while the velocities at any point in the water column tend toward a steady state. A more realistic solution for an impulsively applied wind stress over a stratified ocean can be readily obtained using second-moment turbulence closure and is qualitatively similar (e.g., Mellor, 1996a), even though the pycnocline that strengthens with the deepening of the mixed layer eventually halts further penetration of turbulence into the deep and hence the velocities and transports both tend to have steady oscillatory components.

1.9 SVERDRUP TRANSPORT

Long-term averages of winds at the Earth's surface are primarily zonal (see Figure 1.9.1). As a consequence of various meridional cells in the atmosphere, there is a considerable variation of this zonal wind stress with latitude. Generally, the

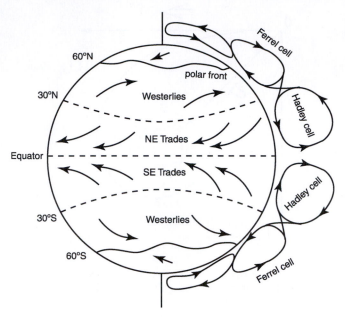

Figure 1.9.1 A sketch of the long-term average winds on the globe showing the trade winds and westerlies, as well as various meridional cells. From Cushman-Roisin (1994).

easterly trade winds prevail in the tropics and westerlies in the midlatitudes. Winds are generally easterly in subpolar regions. An interesting question concerns the kind of oceanic circulation that is driven by these wind patterns. Harold Sverdrup tackled this question in 1947 (Sverdrup, 1947).

Consider a flat-bottomed ocean of depth H driven by wind stress at its surface. For low Rossby numbers typical of large gyre scale motions, the governing equations are linear and in their component form can be written as

$$\frac{\partial U}{\partial x} + \frac{\partial V}{\partial y} + \frac{\partial W}{\partial z} = 0$$

$$-fV = \frac{1}{\rho_0}\left(-\frac{\partial P}{\partial x} + \frac{\partial \tau_x}{\partial z}\right) \tag{1.9.1}$$

$$fU = \frac{1}{\rho_0}\left(-\frac{\partial P}{\partial y} + \frac{\partial \tau_y}{\partial z}\right)$$

The continuity equation can be integrated in the vertical, with boundary conditions W = 0 at z = 0, and z = –H, to yield

$$\frac{\partial M_x}{\partial x} + \frac{\partial M_y}{\partial y} = 0 \qquad (1.9.2)$$

Pressure can be eliminated by cross differentiating the momentum equations and taking the difference:

$$\frac{\partial}{\partial x}(fU) + \frac{\partial}{\partial y}(fV) = \frac{1}{\rho_0}\left(\frac{\partial}{\partial x}\frac{\partial \tau_y}{\partial z} - \frac{\partial}{\partial y}\frac{\partial \tau_x}{\partial z}\right) \qquad (1.9.3)$$

Integration from $z = -H$ to $z = 0$ in the vertical and assuming zero bottom friction gives

$$f\frac{\partial M_x}{\partial x} + \frac{\partial}{\partial y}(fM_y) = \frac{1}{\rho_0}\left(\frac{\partial \tau_y^w}{\partial x} - \frac{\partial \tau_x^w}{\partial y}\right) \qquad (1.9.4)$$

Making use of Eq. (1.9.2), Eq. (1.9.4) can be rewritten as

$$\beta M_y = -\beta\frac{\partial \psi}{\partial x} = \frac{1}{\rho_0}(\mathbf{k}\cdot\nabla\times\tau)$$
$$M_x = \frac{\partial \psi}{\partial y}, M_y = -\frac{\partial \psi}{\partial x} \qquad (1.9.5)$$

where $\beta = \partial f / \partial y$ and (and in 1.9.4) use has been made of the fact that there is no longitudinal variation of f ($\partial f / \partial x = 0$). A stream function has been defined which satisfies the integrated continuity equation in Eq. (1.9.2).

Equation (1.9.5) is the famous *Sverdrup balance* (Sverdrup, 1947) that relates the wind-driven meridional transport in the oceans to the local curl (torque) of the wind stress. Where it is maximum, the meridional transport is also a maximum; where the curl is zero, there is no meridional transport. In the subtropical gyres of the ocean basins (equatorward of about 45°), the prevailing winds and the associated wind stress curl are such that the meridional transport is equatorward, and in the subpolar gyres (poleward of about 45°), the transport is poleward. The latitude of long-term averaged zero wind stress curl tends to delineate the boundary between the two. The Sverdrup balance prevails independent of the density stratification or the vertical structure of the fluid column, or the pattern of wind stress except its local curl. The ability to explain wind-driven circulations in ocean basins is the reason oceanographers are fond of examining the wind stress curl. Sverdrup derived this to explain the North Pacific Counter Current that is opposite to the prevailing winds. The wind stress curl due to mostly zonal easterly trade winds in the equatorial Pacific is responsible for this behavior. Consider only the zonal winds $\tau_x^w(y)$ in a rectangular basin. Then Eq. (1.9.5) can be integrated in the x direction to yield

$$\psi = \psi_0(y) + \frac{1}{\beta}\frac{\partial}{\partial y}\left(\frac{\tau_x^w}{\rho_0}\right) \tag{1.9.6}$$

To satisfy no normal flow conditions along meridional boundaries, stream function ψ must be constant along both western and eastern boundaries. However, there is only one constant at our disposal: ψ_0. It is necessary then to choose either the western or eastern boundary to satisfy the boundary condition. Because of the presence of the intense western boundary current that the Sverdrup solution does not admit, it is better to choose the eastern boundary and this is what Sverdrup did. Imposing $\psi = 0$ at $x = W$, the solution is

$$\psi = \frac{1}{\beta}\frac{\partial}{\partial y}\left(\frac{\tau_x^w}{\rho_0}\right)(x - W) \tag{1.9.7}$$

For a typical wind stress distribution over midlatitude oceans [see Eq. (1.10.4)],

$$\psi = \frac{\pi}{\beta}\frac{W}{L}\frac{\tau_0}{\rho_0}\left(1 - \frac{x}{W}\right)\sin\left(\frac{\pi y}{L}\right) \tag{1.9.8}$$

The scale of the vertically integrated transport is therefore $M_I = \dfrac{\pi}{\beta L}\dfrac{\tau_0}{\rho_0}$ and the actual velocities $U_I = \dfrac{\pi}{\beta L H}\dfrac{\tau_0}{\rho_0}$. For a wind stress of 0.1 N m^{-2} and a midlatitude (30°) value for β of 2×10^{-11} m^{-1} s^{-1}, assuming $W = 1.6\,L$, one gets 24 Sv as the maximum gyre circulation. The transport and velocity scales are about 2.5 m^2 s^{-1} and 5×10^{-3} m s^{-1} for $L = 6000$ km and an effective depth of 500 m. Since the entire transport has to be returned by a western boundary layer, the transport and velocity scales in the western boundary current are larger by a magnitude of about $W/*$, where $*$ is the width of the boundary current: $M_w = \dfrac{\pi}{\beta L}\dfrac{\tau_0}{\rho_0}\dfrac{W}{\delta}$ and $U_w = \dfrac{\pi}{\beta L H}\dfrac{\tau_0}{\rho_0}\dfrac{W}{\delta}$. For a δ value of 100 km and a W of 4000 m, they are 100 m^2 s^{-1} and 0.2 m s^{-1}. The maximum velocities are of course several times this value.

The Sverdrup balance explains the broad features of the equatorial current system and the meridional transports in the midlatitude subtropical and subpolar gyres. Figure 1.9.2, adapted from Hellerman and Rosenstein (1983), shows this

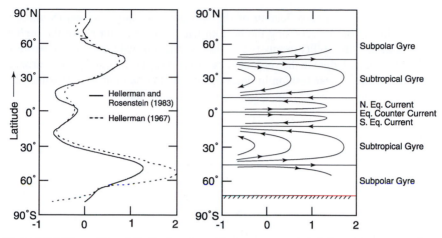

Figure 1.9.2 The meridional distribution of zonal long term mean winds (left) and the corresponding Sverdrup circulation (right). From Hellerman and Rosenstein (1983) and adapted from Pedlosky (1966).

quite well. The various subtropical and subpolar gyres driven by the curl of the zonal wind stress can be seen as well as the equatorial current system. However, the solution is degenerate in the sense it was obtained by ignoring the higher order frictional terms in the governing equations. Consequently, it does not satisfy the crucial lateral boundary condition at the western boundary and hence is unable to explain how the return of these broad equatorward (poleward) meridional flows in the subtropical (subpolar) gyre is accomplished. This is the role of western boundary currents that the Sverdrup solution does not possess.

Pedlosky (1996) discusses in great detail the validity of Sverdrup balance in real ocean basins, with their nonuniform topography and nonzero bottom stress, and of the Sverdrup balance derived for a baroclinic ocean without bottom interaction. The zero bottom stress is a reasonably good approximation in deep oceans, but bottom topographic variations are usually nonnegligible. Nevertheless, the balance appears to hold reasonably well in the real oceans, and certainly for a homogeneous flat-bottomed ocean, and forms the basis of all our theories of wind-driven circulation in the latter.

1.10 WESTERN BOUNDARY INTENSIFICATION (STOMMEL SOLUTION)

By ignoring frictional effects, Sverdrup missed a key part of the wind-driven circulation in the ocean basins, the boundary currents, especially the strong western boundary currents. It was Henry Stommel, in 1948, who provided a simple explanation for the intensification of the western boundary currents (Stommel, 1948). This he did by retaining the bottom friction terms (and hence the higher order terms and the ability to satisfy conditions of no normal flow on both western and eastern boundaries) in the vertical integration of Eq. (1.9.3), and parameterizing them in a simple linear fashion in order to be able to obtain analytical solutions. Using $\tau_x^b = cM_x; \tau_y^b = cM_y$, the equations can then be integrated to give

$$f\frac{\partial M_x}{\partial x}+\frac{\partial}{\partial y}\left(fM_y\right)=\frac{1}{\rho_0}\left(\frac{\partial \tau_y^w}{\partial x}-\frac{\partial \tau_x^w}{\partial y}\right)-c\left(\frac{\partial M_y}{\partial x}-\frac{\partial M_x}{\partial y}\right) \quad (1.10.1)$$

so that Eq. (1.9.5) becomes

$$c\nabla^2\psi+\beta\frac{\partial\psi}{\partial x}=-\frac{1}{\rho_0}\left(k\cdot\nabla\times\tau\right) \quad (1.10.2)$$

This is an elliptic (harmonic) equation and can be readily solved for a closed midlatitude constant depth basin for an assumed wind stress curl distribution, with boundary condition $\psi = 0$ along both the western and the eastern boundary. For a purely zonal wind stress, it can be written as

$$\frac{c}{\beta}\nabla^2\psi+\frac{\partial\psi}{\partial x}=\frac{1}{\beta}\frac{\partial}{\partial y}\left(\frac{\tau_x^w}{\rho_0}\right) \quad (1.10.3)$$

Note that (c/β) multiplying the Laplacian is small, yet cannot be neglected if a western boundary condition of no normal flow has to be satisfied. In the jargon of mathematics, this is a matched asymptotic expansion problem. Note that the RHS is a function only of y and hence the solution is separable in x and y. Stommel assumed a distribution of wind stress typical of midlatitudes over long time scales (see Figure 1.9.2),

$$\tau_x^w=-\tau_0\cos\left(\frac{\pi y}{L}\right);\tau_y^w=0 \quad (1.10.4)$$

which gives zero wind stress curl at y = 0 and L, the southern and northern extents of a closed basin of extent L in the meridional direction (extent W in the zonal direction) and uniform depth H. In the northern half of the basin, we have strong westerlies, flow is from the west, and in the southern half, the flow is easterly. For this situation, he investigated three cases: (1) With f = 0, the solution had east–west symmetry, but the principal balance was between the bottom friction and the wind stress, as can be expected a priori. There was no westward intensification. There

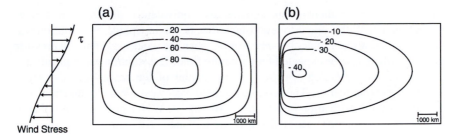

Figure 1.10.1 Streamlines in a flat-bottom rectangular ocean for the Stommel problem (a) on an f-plane and (b) on a β-plane. Note the western intensification in the latter. The idealized wind stress is shown at left.

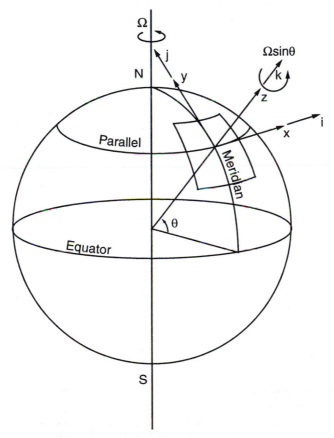

Figure 1.10.2 A sketch of the midlatitude beta plane.

were no pressure gradients at the surface, and equivalently, therefore, no surface deflection. (2) With f equal to a constant, that is, on an f-plane, the solution was similar, but now the primary balance was between the wind stress and the pressure gradient, which is to be expected for a rotating fluid. The solution was equivalent to a high pressure (elevated sea level) in the center of the basin, with clockwise (anticyclonic) circulation. The solution retained east–west symmetry and there was no westward intensification [see Figure 1.10.1a]. (3) With a nonconstant f, that is, on a beta-plane (Figure 1.10.2), the flow was no longer symmetric east to west, and had a strong current along the western boundary (Figure 1.10.1b). The pressure change was confined mostly to the western boundary. Equivalently, the sea level increased eastward at the western boundary to accommodate the northward *western boundary current* (Figure 1.10.1b).

For the wind stress in Eq. (1.10.4), and the canonical values mentioned in Section 1.9, one gets a Sverdrup transport of a maximum of 24 Sv at the central latitude of the basin, decreasing to zero at the northern and southern zonal boundaries. This transport is returned northward in the form of an intense western boundary current.

Solution to Eq. (1.10.3) can be obtained readily by the method of separation of variables by substituting $\psi(x, y) = \dfrac{W}{\beta} \dfrac{\partial}{\partial y}\left(\dfrac{\tau_x^w}{\rho_0}\right)\varphi(x) = \chi(y)\varphi(x)$, where

$$\chi(y) = \frac{W}{\beta} \frac{\partial}{\partial y}\left(\frac{\tau_x^w}{\rho_0}\right) \tag{1.10.5}$$

then $\varphi(x)$ is given by

$$\varphi(x) = \left(\frac{1 - e^{-bW}}{e^{-aW} - e^{-bW}}\right)e^{-ax} - \left(\frac{1 - e^{-aW}}{e^{-aW} - e^{-bW}}\right)e^{-bx} - 1 \tag{1.10.6}$$

where $a = \beta/c$, $b = a\,(1 + \pi^2/(a^2\,W^2))^{1/2}$, and the boundary condition satisfied is $\varphi = 0$ at both $x = 0$ and $x = W$. For the wind stress distribution in Eq. (1.10.4),

$$\chi(y) = \frac{\pi W}{\beta L} \frac{\tau_0}{\rho_0} \sin\left(\frac{\pi y}{L}\right) \tag{1.10.7}$$

and the solution for ψ becomes

$$\psi(x, y) = \frac{\pi W \tau_0}{\beta L \rho_0}\,\varphi(x)\sin\left(\frac{\pi y}{L}\right) \tag{1.10.8}$$

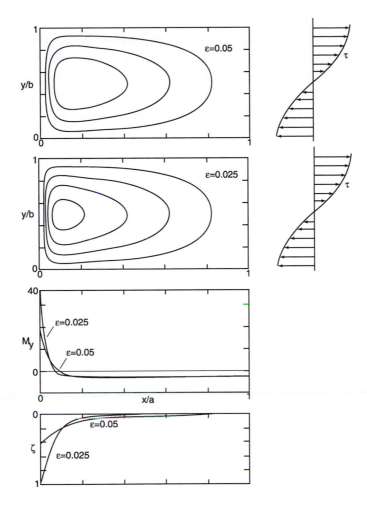

Figure 1.10.3 Stommel solution for two values of the small perturbation parameter: (a) stream function for $\varepsilon = 0.05$, (b) stream function for $\varepsilon = 0.025$, (c) meridional transport across midlatitude plane, and (d) absolute vorticity along midlatitude plane. Adapted from Mellor (1996). Zonal wind stress distribution is shown to the right.

    Stommel's seminal work tied western boundary intensification of currents in midlatitude ocean basins to latitudinal variations in the planetary vorticity. Western boundary currents are unique and central to the basin scale circulation in the oceans. This is a major difference between the circulation in the atmosphere and that in the oceans. The flow in the atmosphere is principally zonal.

The boundary layer thickness or the width of the western boundary current can be derived by balancing the largest term in the Laplacian with the beta term (wind stress contribution is negligible over that width): $c\partial^2\psi/\partial x^2 \sim \beta\partial\psi/\partial x$, which gives $c/\delta^2 \sim \beta/\delta$, so that $\delta \sim c/\beta$. This estimate is useful in obtaining numerical solutions since it is necessary to choose the model grid size so as to resolve this layer to avoid numerical problems. The harmonic equation itself can be solved by any one of the numerous elliptic solvers, including the time-honored successive overrelaxation (SOR) or more modern multigrid techniques (see Chapter 2).

The method of matched asymptotic expansions can be used to obtain an approximate solution to first order in the perturbation variable $\varepsilon = \left(\dfrac{c}{\beta W}\right) \sim \left(\dfrac{\delta}{W}\right)$

following the same procedure as outlined in Section 1.11 below so that

$$\psi(x,y) = \varphi(x)\chi(y); \quad \varphi(x) = \frac{x}{W} - 1 + e^{-ax} \qquad (1.10.9)$$

This solution can also be obtained from Eq. (1.10.6). The corresponding values for the meridional transport M_y and vorticity ζ are

$$M_y = \frac{1}{W}\chi(y)\left\{\frac{e^{-ax}}{\varepsilon} - 1\right\}$$

$$\zeta = -\frac{1}{W^2}\chi(y)\frac{e^{-ax}}{\varepsilon^2} \qquad (1.10.10)$$

Note that the absolute vorticity ζ is negative, but $\partial\zeta/\partial x$ is positive. The solutions are shown in Figure 1.10.3, adapted from Mellor (1996), for the two values of 0.025 and 0.05 for parameter ε, which is the ratio of boundary layer thickness to the width of the basin. Note that in the real ocean, this parameter is about 0.01 to 0.02. At these values, however, advection becomes important and the problem is no longer linear.

1.11 GYRE SCALE CIRCULATION (MUNK SOLUTION)

The Stommel solution is not quite realistic. The inclusion of the bottom friction raises the order of the governing equation and removes the degeneracy in the Sverdrup relation (which is an inviscid solution valid only away from the boundaries; it has no provision for the poleward return of the water masses flowing equatorward), and enables the beta effect to produce a western boundary current.

But the bottom friction is usually negligible in the deep ocean and hence it cannot be very effective. Also, the Stommel solution does not accommodate the broad eastern boundary currents that are part of the midlatitude wind-driven basin scale circulation. Mathematically, this happens because the resulting harmonic equation allows only the condition of no normal flow to the boundary to be satisfied; the order of the equation is not high enough to accommodate the no-slip condition essential to having zero-slip boundary layers adjacent to a boundary. These deficiencies can be removed by appealing to horizontal frictional terms instead, which raise the order of the equations and make it biharmonic and hence capable of accommodating both no-normal-flow and no-slip conditions. This is what Walter Munk did in 1950 (Munk, 1950), introducing the word "gyre" into the oceanographic terminology along the way. The governing equations then become

$$\frac{\partial U}{\partial x} + \frac{\partial V}{\partial y} + \frac{\partial W}{\partial z} = 0$$

$$-fV = \frac{1}{\rho_0}\left(-\frac{\partial P}{\partial x} + \frac{\partial \tau_x}{\partial z}\right) + A_M \nabla^2 U \qquad (1.11.1)$$

$$fU = \frac{1}{\rho_0}\left(-\frac{\partial P}{\partial y} + \frac{\partial \tau_y}{\partial z}\right) + A_M \nabla^2 V$$

which can then be integrated in the vertical as before to yield

$$A_M \nabla^4 \psi - \beta \frac{\partial \psi}{\partial x} = \frac{1}{\rho_0}\left(k \cdot \nabla \times \tau\right) \qquad (1.11.2)$$

This is a biharmonic equation and requires two boundary conditions, $\psi = 0$ and $\partial \psi / \partial n = 0$, corresponding to zero normal flow to the closed boundary as well as no slip at the boundary. The solution can be obtained by matched asymptotic solutions, which incorporate the Sverdrup flow in the interior, a swift, narrow western current and a broad southward eastern boundary current. Munk's solution is more realistic, at least in the context of a flat bottom linear ocean.

The boundary layer thickness or the width of the western boundary current in the Munk solution can be seen by balancing the largest term in the Laplacian with the beta term (wind stress contribution, as before, is largely negligible over that width): $A_M \partial^4 \psi / \partial x^4 \sim \beta \partial \psi / \partial x$,   which   gives $A_M / \delta^4 \sim \beta / \delta$,   so   that $\delta \sim (A_M / \beta)^{1/3}$. A numerical solution can be obtained by converting Eq. (1.11.2) into a set of two coupled harmonic equations that can be solved iteratively (see Chapter 2):

$$A_M \nabla^2 \zeta - \beta \frac{\partial \psi}{\partial x} = \frac{1}{\rho_0} (k \cdot \nabla \times \tau)$$

$$\zeta = \nabla^2 \psi \tag{1.11.3}$$

It is, however, instructive to obtain solutions to the Munk problem analytically, and this can be done using *matched asymptotic expansions*, a routine procedure in boundary layer flows in fluid mechanics. For a simple wind stress distribution (zonal wind stress varying only in the meridional direction), Eq. (1.11.2) becomes

$$\frac{A_M}{\beta} \nabla^4 \psi - \frac{\partial \psi}{\partial x} = -\frac{\chi(y)}{W} \tag{1.11.4}$$

Neglecting the frictional terms once again leads to the Sverdrup solution: $(\partial \psi / \partial x) = \chi(y)$. Integrating this with $\psi = 0$ at $x = W$, the eastern boundary, gives

$$\psi = \left(\frac{x}{W} - 1 \right) \chi(y) \tag{1.11.5}$$

This solution does not of course satisfy any other boundary condition demanded by the flow. One needs to appeal to the full equations (1.11.4) retaining the viscous terms to satisfy the no-slip and no-normal-flow conditions. The problem is akin to the classical boundary layer problem in fluid mechanics and the solution is also similar—the use of matched asymptotic expansions. The formal methodology and mathematics are outlined in, for example, van Dyke (1964), Cole (1968), and Nayfeh (1973), but it is adequate here to outline the method less formally. To do this, we first normalize the equation using the outer length scale W for the spatial variables and $\chi(y)$ for ψ to get

$$\varepsilon \left(\frac{\partial^4 \psi_o}{\partial x_o^4} + \frac{\partial^4 \psi_o}{\partial x_o^2 \partial y^2} + \frac{\partial^4 \psi_o}{\partial y^4} \right) - \frac{\partial \psi_o}{\partial x_o} = -1 \tag{1.11.6}$$

where $x_o = x/W$ is the outer variable and the small parameter $\varepsilon = A_M/(\beta W^3)$ multiplies the viscous terms. Note that the stream function is normalized by $\chi(y)$ as before. As in boundary layer theory, even though the frictional term is small in the bulk of the fluid, it cannot be ignored near the boundary without creating nonphysical degeneracies in the solution. The method of matched asymptotic expansions permits a proper perturbation expansion in the small parameter that still retains the character of the solution. This method was outlined by Ludwig Prandtl himself using an example of a damped spring-mass system with a small mass (Schlichting, 1978). The characteristic nature of such problems is that a small parameter multiplies the highest derivative terms, so that the term cannot be ignored in some parts of the domain even if the parameter is vanishingly small. The gradients in the domain adjust such that the term is comparable to others in that part of the domain even though elsewhere it can be neglected.

The outer solution is found by seeking perturbation series expansions in the small parameter $\varepsilon^{1/3}$: $\psi_o = \psi_o^0 + \varepsilon^{1/3}\psi_o^1 + \varepsilon^{2/3}\psi_o^2 + \cdots$. How do we know what power to use? This parameter is nothing but the ratio of the boundary layer thickness δ to the length scale W in the problem, which, using the estimate we made before, reduces to $\varepsilon^{1/3}$. This can also be deduced formally by looking at the terms in the Eq. (1.11.9) governing the inner solution below. The outer solution must satisfy the outer boundary condition $\psi_o = 0$ at $x_o = 1$. It suffices here to look at the zeroth order term, the limit $\varepsilon \rightarrow 0$. This just gives the nondimensional form of Eq. (1.11.6),

$$\psi_o = (x_o - 1) + o(\varepsilon^{1/3}) \tag{1.11.7}$$

even though this solution can be shown to be actually valid to $O(\varepsilon)$. The inner limit $(x_o \rightarrow 0)$ of this outer solution is -1 to $O(\varepsilon^{1/3})$. Note that it cannot and does not satisfy the inner boundary condition of no normal flow and zero slip at $x_o = 0$. To enable it to do that, we need to insert a boundary layer between the outer solution and the inner boundary condition. Now Eq. (1.11.6) can be rescaled so that the highest derivative terms become of the order one and hence can be retained. This is done by defining an inner variable $x_i = x_o\varepsilon^{-1/3}$, the scaling parameter chosen such that the first term in the parentheses on the LHS of Eq. (1.11.6), which is the highest derivative in the outer variable x_o, becomes of the same order as the term or terms not involving ε. This leads to

$$\left(\frac{\partial^4\psi_i}{\partial x_i^4} + \varepsilon^{2/3}\frac{\partial^4\psi_i}{\partial x_i^2\partial y^2} + \varepsilon^{4/3}\frac{\partial^4\psi_i}{\partial y^4}\right) - \frac{\partial\psi_i}{\partial x_i} = -\varepsilon^{1/3} \tag{1.11.8}$$

The form of this equation of course suggests perturbation expansions of the inner and outer solutions in terms of the small parameter $\varepsilon^{1/3}$:

$$\psi_i = \psi_i^0 + \varepsilon^{1/3}\psi_i^1 + \varepsilon^{2/3}\psi_i^2 + \cdots$$

The inner solution in the limit $\varepsilon \to 0$ can be obtained by taking the limit $\varepsilon \to 0$ of Eq. (1.11.8). We get

$$\frac{\partial^4 \psi_i}{\partial x_i^4} - \frac{\partial \psi_i}{\partial x_i} = 0 \tag{1.11.9}$$

to $O(\varepsilon^{1/3})$. The inner solution must satisfy the inner boundary condition, namely, the conditions of no normal flow and no slip at the western boundary $x_i = 0$: $\partial \psi_i / \partial y = 0 = \partial \psi_i / \partial x_i$. The outer limit ($x_i \to \infty$) of the inner solution must also match the inner limit of the outer solution $\psi_o(0, y) = -1$, since the boundary layer thickness is small in terms of the outer variable. In other words, the asymptotic boundary condition ($x_i \to \infty$) on the inner solution is the asymptotic ($x_o \to 0$) limit of the outer solution. It can be verified that the inner solution satisfying all these conditions is

$$\psi_i = \left[\cos\left(\frac{\sqrt{3}}{2}x_i\right) + \frac{1}{\sqrt{3}}\sin\left(\frac{\sqrt{3}}{2}x_i\right) \right] \exp\left(-x_i/2\right) - 1 + o(\varepsilon^{1/3}) \tag{1.11.10}$$

A solution uniformly valid everywhere can be obtained by simply adding the two solutions, and subtracting their common asymptote,

$$\psi = \psi_i(x_i, y) + \psi_o(x_o, y) - \psi_o(0, y)$$

$$= \left[\cos\left(\frac{\sqrt{3}}{2}x_i\right) + \frac{1}{\sqrt{3}}\sin\left(\frac{\sqrt{3}}{2}x_i\right) \right] \exp\left(-x_i/2\right) + x_o - 1 \tag{1.11.11}$$

to $O(\varepsilon^{1/3})$. In dimensional terms, the solution is

$$\psi = \frac{W}{\beta}\frac{\partial}{\partial y}\left(\frac{\tau_x^w}{\rho_0}\right)\left\{ \left[\cos\left(\frac{\sqrt{3}}{2}\left(\frac{\beta}{A_M}\right)^{1/3}x\right) + \frac{1}{\sqrt{3}}\sin\left(\frac{\sqrt{3}}{2}\left(\frac{\beta}{A_M}\right)^{1/3}x\right) \right] e^{-\left(\frac{\beta}{A_M}\right)^{1/3}\frac{x}{2}} + \frac{x}{W} - 1 \right\} \tag{1.11.12}$$

The corresponding meridional transport M_y and vorticity ζ are

$$M_y = \frac{1}{W} \chi(y) \left\{ \frac{2}{\sqrt{3}} \frac{1}{\varepsilon} e^{-x_i/2} \sin\left(\frac{\sqrt{3}}{2} x_i \right) - 1 \right\}$$

$$\zeta = \frac{1}{W^2} \chi(y) \frac{e^{-x_i/2}}{\varepsilon^2} \left[\frac{1}{\sqrt{3}} \sin\left(\frac{\sqrt{3}}{2} x_i \right) - \cos\left(\frac{\sqrt{3}}{2} x_i \right) \right] \qquad (1.11.13)$$

Note that the absolute vorticity ζ is negative, but $\partial\zeta/\partial x$ is positive.

This is the analytical solution to the Munk problem. Its form is shown in Figure 1.11.1 [adapted from Mellor (1996); see also Pedlosky (1996)] for the two values of 0.025 and 0.05 for parameter ε, which is the ratio of boundary layer thickness to the width of the basin. The solution however does not satisfy the no-slip condition at the eastern boundary and hence does not give an eastern boundary current to this order (not important to the western intensification problem). It is possible, however, to carry out matched asymptotic expansions at the eastern boundary also, but to zeroth order, solutions do not change. At the next order $O(\varepsilon^{1/3})$, however, the no-slip boundary condition can be satisfied, since the error in the mismatch of zeroth order solutions is $O(\varepsilon^{1/3})$. Thus, the Munk solution yielding a narrow western and a broad eastern boundary current can be obtained (see Mellor, 1996, for the full solution).

The method of matched asymptotic expansions can be used for the Stommel problem also. The small parameter here is $\varepsilon = \left(\frac{c}{\beta W} \right) \sim \left(\frac{\delta}{W} \right)$. To obtain an approximate solution to first order in the perturbation variable ε, we follow the same procedure as outlined above. First, the spatial variables in Eq. (1.10.3) are normalized using the outer length scale W and ψ by $\chi(y)$ so that the equation normalized in terms of the outer variable $x_0 = x/W$ is

$$\varepsilon \left(\frac{\partial^2 \psi_o}{\partial x_o^2} + \frac{\partial^2 \psi_o}{\partial y^2} \right) + \frac{\partial \psi_o}{\partial x_o} = 1 \qquad (1.11.14)$$

where $x_o = x/W$ and $\varepsilon = c/(\beta W)$. The outer solution obtained by ignoring ε terms is simply $\psi_o(x_o) = x_o - 1$ to $O(\varepsilon)$, satisfying the boundary condition on the eastern boundary $x_o = 1$, but not that on the western boundary $x_o = 0$. The outer solution at its inner limit, $x_o \to 0$, is -1. The inner solution is obtained by rescaling the spatial variables so that $x_i = x_o \varepsilon^{-1}$, so Eq. (1.10.3) becomes

$$\left(\frac{\partial^2 \psi_i}{\partial x_i^2} + \varepsilon^2 \frac{\partial^2 \psi_i}{\partial y^2} \right) + \frac{\partial \psi_i}{\partial x_i} = \varepsilon \qquad (1.11.15)$$

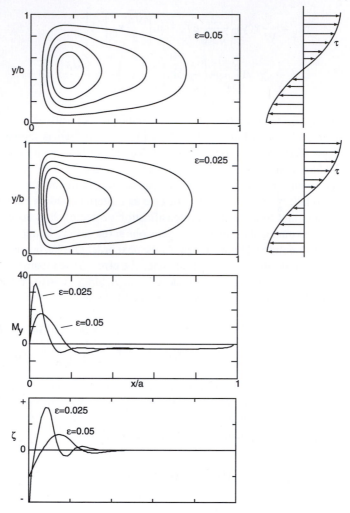

Figure 1.11.1 Munk solution for two values of the small perturbation parameter ε: (a) stream function for ε = 0.05, (b) stream function for ε = 0.025, (c) meridional transport across midlatitude plane, and (d) absolute vorticity along midlatitude plane. Adapted from Mellor (1996). Zonal wind stress distribution is shown to the right.

Note that the scaling retains the largest term, the first term, in the parentheses to balance the second term on the LHS. In the limit of ε → 0,

$$\frac{\partial^2 \psi_i}{\partial x_i^2} - \frac{\partial \psi_i}{\partial x_i} = 0 \qquad (1.11.16)$$

so that the inner solution satisfying the boundary condition at the western boundary
to O(ε) is $\psi_i(x_i) = b\left(e^{-x_i} - 1\right)$. The solution at its outer limit $x_i \to \infty$ is $-b$, and to
match the inner limit of the outer solution b must be 1, so that $\psi(x_i) = e^{-x_i} - 1$.
Matching the inner and outer solution or just adding the outer and inner solutions
and subtracting the common asymptote, -1, one gets the solution that satisfies both
boundary conditions correctly,

$$\psi = x_o - 1 + e^{-x_i} \tag{1.11.17}$$

to O(ε), so that in dimensional terms

$$\psi(x, y) = \varphi(x)\chi(y); \quad \varphi(x) = \frac{x}{W} - 1 + e^{-ax} \tag{1.11.18}$$

    Western intensification can be explained in terms of vorticity conservation. In
order to do this consider Eq. (1.11.3). This equation is simply an equation for
conservation of vorticity ζ and can be integrated across the basin from $x = 0$ to $x = W$ to yield (since $\psi = 0$ at both ends)

$$A_M \left[\frac{\partial \zeta}{\partial x}\bigg|_{x=W} - \frac{\partial \zeta}{\partial x}\bigg|_{x=0} \right] = \int_0^W \frac{(k \cdot \nabla \times \tau)}{\rho_0} \, dx \tag{1.11.19}$$

For southward (northward) Sverdrup transport, or negative (positive) wind stress
curl, the RHS is negative (positive). The return flow in the boundary current is
northward (southward), and zero at the boundary itself but reaching a maximum
away from the boundary so that $\partial \zeta / \partial x$ is positive (negative) at both $x = 0$ and $x = W$.
Thus, the only way total vorticity balance can be maintained is for the value of
$\partial \zeta / \partial x$ at $x = 0$ to be significantly large (and that at $x = W$ to be negligible) enough to
balance the integrated vorticity due to the wind stress curl. This calls for a narrow,
intense western boundary current and a diffuse, weak eastern current, no matter
which direction the Sverdrup transport takes.
    Both the Stommel and Munk solutions possess north–south symmetry because of
the linearity of the solutions. These solutions come in handy in verifying the
numerical solutions for the linear wind-driven circulation problem. However, there
are limitations. The Stommel solution does not satisfy the condition of no slip on
either the western or the eastern boundary. The Munk solution is more realistic in
this respect. Another feature of these solutions is that neglecting advection effects
leads to local vorticity balance at each latitude, which is also not quite realistic. The
linear nature of both the solutions also makes it difficult to obtain narrow western
boundary currents without violating the assumption of linearity.

The relevant Reynolds number is $Re = U_w \delta/A_M = U_w/\beta \delta^2$, typically about 1 for a boundary current of 100 km width and the canonical value of U_w derived in Section 1.9. Based on the maximum velocity, Re can reach several times this value and hence the inertial effects are hardly negligible. This is true of the Stommel solution as well. Thus for plausible values of the friction coefficients, which are essentially unknown tunable parameters, both violate linearity, and if they are chosen not to violate linearity, they give broad anemic western boundary currents. These difficulties prompted a search for solutions in which the nonlinear inertial terms are not neglected. Charney (1955) looked for purely inertial solutions in which the thickness of the western boundary layer is $\delta_I = (U/\beta)^{1/2}$. The solution is similar to the Stommel solution (1.11.16) but with $a = (\beta/U)^{1/2}$. However, this solution is incomplete and valid only for the southern (northern) half of the subtropical (subpolar) gyre (see Pedlosky, 1996). It does not satisfy the no-slip boundary condition at the western boundary, and in order to do that, a viscous sublayer must be embedded within it (Pedlosky, 1996). Note that $Re = (\delta_I/\delta)^2$.

The fully nonlinear vorticity equation contains the vorticity advective term and Eq. (1.11.2) becomes

$$v \cdot \nabla \zeta + \beta \frac{\partial \psi}{\partial x} = -\frac{1}{\rho_0}(k \cdot \nabla \times \tau) - c\,\zeta + A_M \nabla^2 \zeta \qquad (1.11.20)$$

where $v = \frac{1}{H} k \times \nabla \psi$. For the nonlinear case, the north–south symmetry is lost and the maximum velocity shifts northward. Now the nonlinear vorticity advection term can balance the beta term so that $\beta \frac{|\psi|}{\delta} \sim \frac{|\psi|^2}{LH\delta^3}$, where $|\psi|$ denotes the magnitude of the Sverdrup transport $\frac{\tau_0}{\beta \rho_0}$ so that $\delta \sim \beta^{-1} \left(\frac{\tau_0}{\rho_0 LH} \right)^{1/2}$. However, frictional terms are also important and probably of the same magnitude as advection terms.

Solutions to Eq. (1.11.20) require numerical methods. Even then, there are many uncertainties associated with appropriate values for the friction coefficients and the appropriate condition at the western wall, whether slip or no-slip, or some other condition involving subgrid scale motions not explicitly resolved. The discussion of these is beyond the scope of this book (see Pedlosky, 1996, for a thorough expose instead). We will merely present a numerical solution from Ierely (1987) to the nonlinear Munk problem that shows the recirculation in the northwest corner and the north–south asymmetry of the full nonlinear solution (Figure 1.11.2), just to illustrate the profound differences nonlinearity brings in, even for a simple rectangular, flat-bottomed, homogeneous ocean.

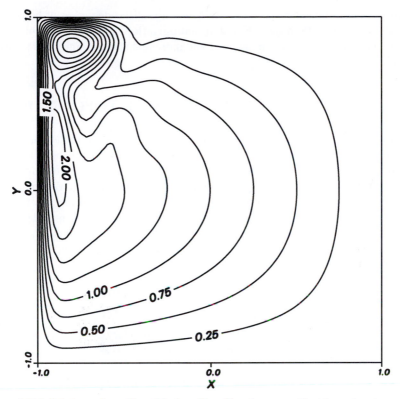

Figure 1.11.2 Solution to the nonlinear Munk problem. Note the corner eddy at the northwest corner of the basin and the pronounced north-south asymmetry. From Ierely (1987).

Thus, while Stommel and Munk solutions are very educational, they are but linear solutions for the simplest case of a homogeneous, flat-bottomed ocean. The real oceans are far more complex. The flow in the boundary currents is quite nonlinear and so nonlinear terms are important. More importantly, the topography is variable, and in concert with the horizontal density gradients, exerts its own torque on the flow. In addition, the wind forcing is seldom as simple as that necessary for analytical tractability. It has a whole spectrum of temporal and spatial scales that cannot be accommodated in analytical models. Finally, thermohaline effects produce density gradients in the oceans that contribute to the transport in the western boundary currents. Analytically, the problem is intractable and numerical solutions are indispensable. Realistic simulation of ocean circulation for realistic basins and realistic surface forcing, including momentum and buoyancy fluxes, is the principal justification for numerical ocean modeling. However, until recently, the lack of adequate computing capability prevented adequate resolution of the narrow western boundary currents and hence a realistic depiction of basin scale circulation. It is only in the 1990s with multigigaflop computing capability that ocean basin models are becoming more and more realistic (Semtner, 1995; Fu and Smith, 1996).

1.12 BAROTROPIC CURRENTS OVER TOPOGRAPHY

Bottom topography plays a very important role in oceanic circulation, which is one reason why the flat-bottomed ocean basins are usually of only academic interest. This influence is especially large for barotropic flows, since the presence of density stratification would otherwise tend to mitigate these effects. We will explore barotropic flows and the effect of topography on them by considering the density to be homogeneous. In this case, there is no horizontal density gradient and hence no vertical shear, and the velocity away from the frictional layers is uniform in the vertical and driven entirely by the sea surface slope. Thus from Eq. (1.3.24), $\overline{\nabla}P = g\overline{\nabla}\eta$, where the gradient operator is two-dimensional and the overbar indicates that only horizontal gradients are considered.

The influence of topography can be seen from retaining the bottom slope in the governing equations. Thus, integrating continuity equation (1.9.1) in the vertical from z = –H to z = η, and making use of the boundary conditions of no flow through the free surface and the impermeable bottom, gives

$$W_{z=\eta} = -v_s \cdot \overline{\nabla}\eta = -U_s(\partial\eta/\partial y) - V_s(\partial\eta/\partial y)$$
$$W_{z=-H} = -v_b \cdot \overline{\nabla}H = -U_b(\partial H/\partial y) - V_b(\partial H/\partial y)$$

$$(1.12.1)$$

We again get Eq. (1.9.2) for the case with nonuniform bottom and free surface deflection, and steady case with $\partial\eta/\partial t = 0$, except that the transports are defined over the entire water column z = –H to z = η. The linear momentum equations in Eq. (1.9.1) for low Rossby number flows can also be integrated similarly to yield

$$-fM_y = -g(H+\eta)\frac{\partial\eta}{\partial x} + \frac{\left(\tau_x^w - \tau_x^b\right)}{\rho_0}$$

$$(1.12.2)$$

$$fM_x = -g(H+\eta)\frac{\partial\eta}{\partial y} + \frac{\left(\tau_y^w - \tau_y^b\right)}{\rho_0}$$

Ignoring the contribution of sea level deflection η to the water column depth, dividing by H, cross differentiating and eliminating η terms, and making use of Eq. (1.12.1), one gets

$$\mathbf{M} \cdot \overline{\nabla}\left(\frac{f}{H}\right) = \mathbf{M}_x \frac{\partial}{\partial x}\left(\frac{f}{H}\right) + \mathbf{M}_y \frac{\partial}{\partial y}\left(\frac{f}{H}\right)$$

$$= \left[\frac{\partial}{\partial x}\left(\frac{\tau_y^w}{\rho_0 H}\right) - \frac{\partial}{\partial y}\left(\frac{\tau_x^w}{\rho_0 H}\right)\right] - \left[\frac{\partial}{\partial x}\left(\frac{\tau_y^b}{\rho_0 H}\right) - \frac{\partial}{\partial y}\left(\frac{\tau_x^b}{\rho_0 H}\right)\right] \quad (1.12.3)$$

where f/H is the potential vorticity. If bottom friction can be neglected (which is fine for the deep oceans), this says that in the absence of external (wind) forcing, a barotropic flow will follow f/H contours in the ocean. The flow and the transport will be parallel to these contours. From considerations of potential vorticity, this is not surprising, since if the relative vorticity is negligible, as should be true for low Rossby number flows, then the potential vorticity of a barotropic flow is simply f/H and this should be conserved following a fluid parcel, in the absence of any external torques. Thus, a fluid parcel must follow f/H contours. Since the change in topography overwhelms by far the change in planetary vorticity, this means that the flow parallels the depth contours.

In the presence of external torques such as those due to the curl of the wind stress divided by H (not just the wind stress), Eq. (1.12.3) says that the flow component perpendicular to the f/H contours depends on this torque. Naturally, it requires a vorticity input to change the potential vorticity and deviate the flow from the f/H contours.

Extension to nonlinear flows is difficult, but some guidance can be obtained by looking at vorticity conservation. In this case, the relative vorticity of the fluid, $\zeta = \mathbf{k} \cdot \nabla \times \mathbf{v} = \dfrac{\partial V}{\partial x} - \dfrac{\partial U}{\partial y}$, is comparable to the planetary vorticity $f = 2\Omega \sin\theta$ and cannot be neglected. Therefore, f needs to be replaced by the total or absolute vorticity $(f + \zeta)$, since in barotropic flows, in the absence of external torques, the quantity that is conserved following a fluid particle is the potential vorticity defined using the total vorticity:

$$\frac{D}{Dt}\left(\frac{f+\zeta}{H}\right) = 0 \quad (1.12.4)$$

In the presence of external torque, this equation generalizes for the steady case:

$$\mathbf{M} \cdot \overline{\nabla}\left(\frac{f+\zeta}{H}\right) = \mathbf{M}_x \frac{\partial}{\partial x}\left(\frac{f+\zeta}{H}\right) + \mathbf{M}_y \frac{\partial}{\partial y}\left(\frac{f+\zeta}{H}\right)$$

$$= \frac{\partial}{\partial x}\left(\frac{\tau_y^w - \tau_y^b}{\rho_0 H}\right) - \frac{\partial}{\partial y}\left(\frac{\tau_x^w - \tau_x^b}{\rho_0 H}\right) \quad (1.12.5)$$

This relationship is the extension of Eq. (1.12.3) to nonzero Rossby numbers.

1.13 BAROCLINIC TRANSPORT OVER TOPOGRAPHY

Thus far, we have dealt with barotropic flows. However, there exist significant horizontal gradients of density in the oceans, especially across fronts such as the western/northern edge of the Gulf Stream. These density gradients act in concert with topographic gradients in the basin. The importance of bottom topography to ocean circulation can be illustrated by application to low Rossby number (geostrophic) baroclinic flow over topography. The governing equations in this case can be written with the help of Eq. (1.3.24) as

$$
-fV = -g\frac{\partial \eta}{\partial x} - \frac{1}{\rho_0}\int_z^0 g\frac{\partial \rho}{\partial x}dz' + \frac{1}{\rho_0}\frac{\partial \tau_x}{\partial z}
$$

$$
fU = -g\frac{\partial \eta}{\partial y} - \frac{1}{\rho_0}\int_z^0 g\frac{\partial \rho}{\partial y}dz' + \frac{1}{\rho_0}\frac{\partial \tau_y}{\partial z}
$$

(1.13.1)

Splitting the density integral into a difference between two integrals, one from −H to 0, and the other from −H to z, Eq. (1.13.1) can also be written as

$$
V = -\frac{1}{f}\left\{-g\left[\frac{\partial \eta}{\partial x} + \frac{1}{\rho_0}\int_{-H}^0 \frac{\partial \rho}{\partial x}dz\right] + \frac{g}{\rho_0}\int_{-H}^z \frac{\partial \rho}{\partial x}dz' + \frac{1}{\rho_0}\frac{\partial \tau_x}{\partial z}\right\}
$$

$$
U = \frac{1}{f}\left\{-g\left[\frac{\partial \eta}{\partial y} + \frac{1}{\rho_0}\int_{-H}^0 \frac{\partial \rho}{\partial y}dz\right] + \frac{g}{\rho_0}\int_{-H}^z \frac{\partial \rho}{\partial y}dz' + \frac{1}{\rho_0}\frac{\partial \tau_y}{\partial z}\right\}
$$

(1.13.2)

Written this way, it is easy to see the various contributions to the velocity in the water column of a fully baroclinic ocean. There are three terms. The first one is the bottom velocity (provided we ignore the bottom Ekman layer or equivalently consider only the velocity well beyond the bottom Ekman layer); it has contributions from both sea level slope and horizontal density gradients. In other words, it includes both the barotropic and a part of the baroclinic contributions. The second term accounts for the departures of velocity in the water column from the bottom velocity due to horizontal density gradients, baroclinicity. The third term accounts for the Ekman component and is prominent near the surface and the bottom. The resulting profile is shown in Figure 1.13.1.

Integrating Eq. (1.13.2) in the vertical from −H to 0 gives

$$
-fM_y = -gH\frac{\partial}{\partial x}\left[\eta + \frac{1}{\rho_0}\int_{-H}^0 g\rho dz\right] + \frac{\partial}{\partial x}\left[\frac{1}{\rho_0}\int_{-H}^0 dz\int_{-H}^z g\rho dz'\right] + \frac{\left(\tau_x^w - \tau_x^b\right)}{\rho_0}
$$

$$
fM_x = -gH\frac{\partial}{\partial y}\left[\eta + \frac{1}{\rho_0}\int_{-H}^0 g\rho dz\right] + \frac{\partial}{\partial y}\left[\frac{1}{\rho_0}\int_{-H}^0 dz\int_{-H}^z g\rho dz'\right] + \frac{\left(\tau_y^w - \tau_y^b\right)}{\rho_0}
$$

(1.13.3)

Cross differentiation to eliminate the terms in the first square brackets (after dividing throughout by H), and making use of the continuity equation (1.9.2) for vertically integrated transports, gives (Kantha *et al.*, 1982; Mellor *et al.*, 1982)

$$\mathbf{M} \cdot \bar{\nabla}\left(\frac{f}{H}\right) = M_x \frac{\partial}{\partial x}\left(\frac{f}{H}\right) + M_y \frac{\partial}{\partial y}\left(\frac{f}{H}\right) = \frac{1}{H^2}\frac{\partial H}{\partial y}\frac{\partial \Phi}{\partial x} - \frac{1}{H^2}\frac{\partial H}{\partial x}\frac{\partial \Phi}{\partial y}$$

$$+ \frac{\partial}{\partial x}\left(\frac{\tau_y^w - \tau_y^b}{\rho_0 H}\right) - \frac{\partial}{\partial y}\left(\frac{\tau_x^w - \tau_x^b}{\rho_0 H}\right) \tag{1.13.4}$$

$$\Phi = \frac{1}{\rho_0}\int_{-H}^{0} dz \int_{-H}^{z} g\rho dz' = -\frac{1}{\rho_0}\int_{-H}^{0} g\rho z dz$$

Equation (1.13.4) for a fully baroclinic flow differs from that for a purely barotropic flow (1.12.3) in that extra terms involving the integral of the first moment of density terms (the first two terms on the RHS) appear in the equations. They were first pointed out by Sarkisyan and co-workers (Sarkisyan and Ivanov, 1971) and by Holland and Hirschman (1972). These are the so-called *Joint Effect of Baroclinicity and Bottom Relief* (JEBAR) or *Joint Effect of Baroclinicity and Bottom Topography* (JEBAT) terms. They represent the torque exerted on the fluid column by the joint action of density and topographic gradients to deviate it from following f/H or planetary vorticity contours (see Mellor *et al.*, 1982; Kantha *et al.* 1982; Huthnance, 1984; Csanady, 1985; Mertz and Wright, 1992; Myers *et al.*, 1996; Cane *et al.*, 1998),

$$\text{JEBAR} = -\left[\frac{1}{H_2}\frac{\partial H}{\partial \psi}\frac{\partial \Phi}{\partial \xi} - \frac{1}{H_2}\frac{\partial H}{\partial \xi}\frac{\partial \Phi}{\partial \psi}\right] = J\left(\Phi, \frac{1}{H}\right) = -\kappa \cdot \nabla \times \left[\frac{\nabla \Phi}{H}\right] \tag{1.13.5}$$

where J is the Jacobian. Ignoring the bottom frictional terms, Eq. (1.13.4) can be written as

$$J\left(\psi, \frac{f}{H}\right) = J\left(\Phi, \frac{1}{H}\right) + \frac{1}{\rho_0}k \cdot \nabla \times \left[\frac{\tau}{H}\right] \tag{1.13.6}$$

The JEBAR term can also be decomposed into a bottom pressure torque and a torque representing depth-averaged pressure (Holland, 1973; Myers *et al.*, 1996).

The JEBAR terms are often more important than the torques due to the wind stress or the bottom friction. They are instrumental in determining the transport in

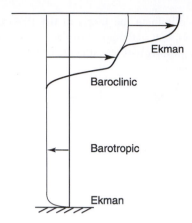

Figure 1.13.1 The velocity profile in the ocean showing the barotropic, baroclinic, and Ekman (top and bottom) components.

oceanic gyres. Using a diagnostic, finite-element, barotropic model of the North Atlantic, Myers *et al.* (1996) found that the transport driven by the bottom pressure torque component of JEBAR dominates their solution except in the subpolar gyre. They found that the inclusion of JEBAR terms was crucial to the proper separation of the Gulf Stream near Cape Hatteras. Even today, the most complex global and basin scale ocean models are plagued with western boundary current separation problems. These currents proceed too far north before separating from the coast. For example, the Gulf Stream penetrates all the way to the Flemish Cap in some models (e.g., Semtner and Chervin, 1992; Semtner, 1995; Marsh *et al.*, 1996), instead of separating at Cape Hatteras at 35° N. This may very well be because the coarse resolution does not yet represent the topography and hence the JEBAR effects well.

Notice that JEBAR terms exist only if topographic gradients exist. They are zero for a flat-bottomed, fully baroclinic ocean and hence do not at all figure prominently in the theory of wind-driven circulation in flat-bottomed baroclinic oceans (which most theoreticians consider for simplicity). For a flat-bottomed ocean, therefore, the horizontal density gradients do NOT alter the total transport in the water column [Eq. (1.12.3)], since although the velocity profiles differ from the barotropic case due to baroclinicity, their integrals will be the same as those for the barotropic case. This is not so when there are topographic gradients. Another way to look at this is to consider JEBAR terms to affect the barotropic potential vorticity balance. Since large scale flows are affected by such a balance, it also means the JEBAR terms also affect the circulation. Greatbatch *et al.* (1991) conclude that the JEBAR term is larger than the wind stress term in the North Atlantic circulation.

The appearance of JEBAR terms also makes the problem intractable analytically, and numerical methods must be employed to solve these equations (see Kantha *et al.,* 1982; Mellor *et al.,* 1982). Realistic numerical models of circulation in the ocean basins invariably account for the basin bottom topography and hence include this term indirectly in the calculations. This is another reason for the need for numerical ocean models. Holland (1973) demonstrated the importance of the JEBAR terms by numerical simulations involving three cases, (1) a constant-depth baroclinic ocean, (2) a variable-depth homogeneous ocean, and (3) a variable-depth baroclinic ocean, and showed that the solutions were fundamentally different in the three cases, with the last case having the largest and most realistic transports. This is due to the action of the JEBAR terms.

Cane *et al.* (1998) argue, however, that the JEBAR term overestimates the influence of topography on oceanic transports. They point out that Godfrey (1989) obtains a realistic global solution with just Sverdrup balance, based on hydrographic data and assuming a level of no motion at 2000 m, totally ignoring topographic variations. They argue that the wind stress is the true external forcing on the ocean, JEBAR terms are not, and the internal structure and hence JEBAR terms adjust to make the Sverdrup balance (1.9.5) hold approximately in the oceans. They suggest that the oceanic transport is pretty much confined to the upper part of the water column and hence is better estimated by the simple Sverdrup balance in a flat bottom ocean than by Eq. (1.13.4). This can be seen by rewriting Eq. (1.13.6) in terms of departures from Sverdrup balance,

$$-\beta\psi_x = \frac{1}{\rho_0}(\mathbf{k}\cdot\nabla\times\tau) + \delta_{Sv}$$

$$\delta_{Sv} = H\left[\text{JEBAR} - (f\nabla\psi + \frac{\tau}{\rho_0})\times\nabla\left(\frac{1}{H}\right)\right] \qquad (1.13.7)$$

$$= \frac{1}{\rho_0}J(p_b, H) = -fw_b$$

where w_b is the vertical component of velocity at the ocean floor due to the divergence of the horizontal components of the geostrophic velocity, and p_b is the pressure at the bottom. Cane *et al.* (1998) postulate that baroclinic effects tend to be such that w_b, and hence δ_{Sv}, is small. This has implications for any diagnostic computations of transport based on observed hydrography, and any numerical calculations that attempt to separate out the barotropic and baroclinic components of the transport. Cane *et al.* (1998) also point out the difficulty of accurately calculating the JEBAR terms from observed hydrography even though the formulation of Mellor *et al.* (1982; see also Kantha *et al.,* 1982) reduces significantly the numerical error in the computation of JEBAR terms.

1.14 COASTAL UPWELLING AND FRONTS

Consider a quiescent coastal ocean with a long, straight coast with a y-axis aligned along the coast and an origin at the coast. Let a wind stress of magnitude τ start blowing parallel to the coast in the positive y direction at $t = 0$. This leads to an Ekman transport away from the coast and hence upwelling along the coast. The basics can be explained by a simple linear reduced gravity model on an f-plane. Such a model describes the motion of an interface between a layer of fluid of depth H overlying an infinitely deep layer. The governing equations are then similar to barotropic shallow water equations except that the gravitational acceleration g is replaced by $g' = g\,(\Delta\rho/\rho_0)$, where $\Delta\rho$ is the density change across the interface. The equations, retaining only the gradients perpendicular to the coastline, are (e.g., Gill, 1982; Cushman-Roisin, 1994)

$$\frac{\partial \eta}{\partial t} + H\frac{\partial u}{\partial x} = 0$$

$$\frac{\partial u}{\partial t} - fv = -g'\frac{\partial \eta}{\partial x} \tag{1.14.1}$$

$$\frac{\partial v}{\partial t} + fu = -\frac{\tau}{\rho_0 H}$$

where η is the interfacial deflection. Note that the wind stress acts as a body force distributed over the depth H of the upper layer. The lower layer is infinitely deep and quiescent. This set can be reduced to a single equation in u,

$$\frac{\partial^2 u}{\partial t^2} + f^2 u - C^2\frac{\partial^2 u}{\partial x^2} = -\frac{f\tau}{\rho_0 H} \tag{1.14.2}$$

where C is the reduced gravity shallow water wave speed $(g'H)^{1/2}$. The solution, subject to the boundary condition $u = 0$ at $x = 0$, is

$$u = \frac{\tau}{f\rho_0 H}\left[1 - \exp(-x/a)\right]$$

$$\eta = \frac{\tau\, t}{\rho_0 C}\exp(-x/a) \tag{1.14.3}$$

$$v = \frac{\tau\, t}{\rho_0 H}\exp(-x/a)$$

where a is the Rossby radius of deformation C/f. The interface deflects upward (and the corresponding sea level deflection is downward, consistent with the water moving offshore). The solution consists of a steady part that gives the Ekman transport and a steadily growing transient that eventually brings the interface to the surface and makes these solutions eventually invalid. Note that the influence of the wind extends to a distance of the order of the Rossby radius away from the coast, reminiscent of the Rossby adjustment problem in midlatitudes (Figure 1.14.1). A coastal jet exists in the upper layer in the same direction as the wind. If the bottom layer were finite, there would also exist an undercurrent in the bottom layer in the opposite direction. Thus for upwelling that occurs along eastern margins of basins due to prevailing equatorward wind stress, an equatorward coastal jet and a poleward undercurrent are generated. The undercurrent is a prominent feature along the California coast, for example.

These solutions are limited since nonlinearity rears its ugly head eventually as the upper layer shrinks. If the wind blows steadily, the interface will surface and begin to move offshore with cold water all along the coast, and the linear solution ceases to be valid long before this happens. The best means of overcoming this limitation is to employ numerical models. In practice, this phenomenon takes the form of jets, squirts, and frontal eddys along this upwelling front, is highly

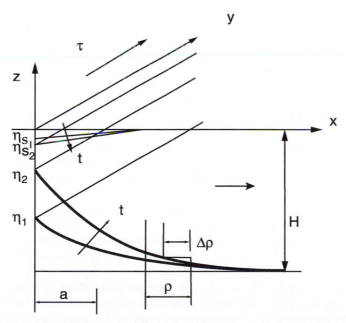

Figure 1.14.1 Coastal upwelling in a reduced gravity model. The evolution of the interface and the sea level with time. Note that the Rossby radius is the cross-shore scale.

unsteady, and is quite complicated to model even numerically. Turbulence closure is necessary to properly account for the wind-driven mixing. The numerical grid must also be small enough to resolve the internal Rossby radius of deformation.

Insight into such coastal *fronts* can be obtained by appealing once again to a simple linear model of a steady, infinitely long front. Let the density differ by $\Delta\rho$ in a surface layer of thickness H(x), with H = 0 at x = 0, the location of the front (with the origin at the front and the y-axis along the front). The velocity component perpendicular to the front u is zero everywhere, and $\partial v / \partial y = 0$. This leads to the vertical velocity being zero by virtue of continuity and the boundary condition at the surface. The governing equation is then simply

$$ fv = g\frac{\partial \eta}{\partial x} + \frac{1}{\rho_0}\int_z^0 g\frac{\partial(\Delta\rho)}{\partial x}dz' \qquad (1.14.4)$$

The solution is

$$ h = H[1 - \exp(-x/a)] $$
$$ \eta = -g'H[1 - \exp(-x/a)] \qquad (1.14.5)$$
$$ v = C\exp(-x/a) $$

where C is the reduced gravity shallow water wave speed $(g'H)^{1/2}$ and a is the Rossby radius C/f. Note once again that the influence of the front extends over a distance perpendicular to the front of the order of the Rossby radius, and the velocity is maximum right at the front and is parallel to the front (Figure 1.14.2). It is a peculiarity of rotating stratified fluids that a steady front such as this can be maintained. Fronts are common in the ocean and the atmosphere: coastal fronts due to upwelling and freshwater runoff, and deep water fronts due to eddies and currents such as the Gulf Stream that transport different water masses into a region creating such fronts. Fronts are also regions of high biological activity and thus important to fisheries.

We looked at a steady infinite front in which the only nonzero velocity component is along the front. We derived solutions assuming the flow is geostrophic. But fronts in both the oceans and the atmosphere, especially the latter, are seldom steady, nor are they infinite in extent. Such fronts can still be treated but within the framework of *semigeostrophic theory*, popular in the atmosphere. This depends simply on the fact that the length scale parallel to the front (y direction) L is normally much larger than the length scale perpendicular to the front (x direction), the Rossby radius a (a « L). The cross-front velocity U is also typically an order of magnitude smaller than the along-front velocity V. In the atmosphere, for example,

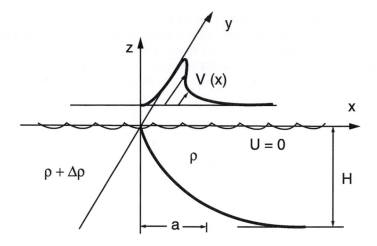

Figure 1.14.2 A sketch of the velocity along a midocean front. Note that the Rossby radius is the cross-front scale.

$L \sim 1000$ km, $a \sim 100$ km, $U \sim 1$ m s^{-1}, and $V \sim 10$ m s^{-1} (Holton, 1992). The governing equations are

$$\frac{Dv}{Dt} - fu = -\frac{\partial p}{\partial x}$$

$$\frac{Du}{Dt} + fv = -\frac{\partial p}{\partial y}$$

$$0 = -\frac{\partial p}{\partial z} - \frac{\rho}{\rho_0} \qquad (1.14.6)$$

$$\frac{\partial u}{\partial x} + \frac{\partial v}{\partial y} + \frac{\partial w}{\partial z} = 0$$

$$\frac{D}{Dt} \equiv \frac{\partial}{\partial t} + u\frac{\partial}{\partial x} + v\frac{\partial}{\partial y} + w\frac{\partial}{\partial z}$$

Using an advection time scale of U/L, the ratio of the advective terms to the Coriolis terms in the equation for cross-front momentum (x direction) is U^2/L divided by fV equal to Ro (U/V) (where Ro = U/fL < 1), much smaller than 1. Thus, these advective nonlinear terms can be ignored and the along-front velocity is to avery good approximation (1% in the above case) in geostrophic balance. However, in the along-front (y direction) momentum equation, the ratio is VU/L divided by fU equal to Ro (V/U) of the order of unity, and the inertial terms cannot be neglected; geostrophy is not even approximately valid. Separating the velocity into geostrophic and ageostrophic components, to a good approximation $v = v_g$ and $u = u_g + u_a$, with u_g and u_a of the same order. The resulting equations for inviscid flow in the vicinity of a front are called semigeostrophic equations (Holton, 1992):

$$\frac{Dv_g}{Dt} - fu_a = 0$$

$$fv_g = -\frac{\partial p}{\partial y}; fu_g = -\frac{\partial p}{\partial x}$$

$$0 = -\frac{\partial p}{\partial z} - \frac{\rho}{\rho_0}$$

$$\frac{\partial u_a}{\partial x} + \frac{\partial w}{\partial z} = 0$$

$$\frac{D}{Dt} \cong \left[\frac{\partial}{\partial t} + \left(u_g \frac{\partial}{\partial x} + v_g \frac{\partial}{\partial y}\right)\right] + \left(u_a \frac{\partial}{\partial x} + w \frac{\partial}{\partial z}\right)$$

(1.14.7)

The difference from the quasi-geostrophic equations is that the advection terms are based on the full velocity, the sum of the geostrophic as well as the ageostrophic components (second parentheses), not just the former.

Nonlinearity is an inseparable aspect of coastal upwelling and frontal processes; hence these processes are best handled by numerical models. It is essential to ensure that the model resolution in the cross-shore and cross-front direction is adequate to resolve the Rossby radius of deformation in order to do a good job of simulating these numerically. While this is straightforward to do in a limited domain model of a front or upwelling region, it is prohibitively expensive to do so in basin scale and global models. Consequently, in such models these processes may not be well resolved, and consequently may be poorly represented.

1.15 MESOSCALE EDDIES AND VARIABILITY

In the early years of oceanography, contributions by Sverdrup, Ekman, Stommel, and Munk helped greatly advance our understanding of the broad aspects of oceanic circulation. Observational campaigns were also undertaken in an attempt to map the water mass structure and the general circulation in various ocean basins. However, it was not until the late sixties that repeated observations at specific sites revealed the startling fact that the majority of the kinetic energy in the oceans resides not in the steady gyres, but in eddies with scales of the order of the Rossby radius that are ubiquitous. MODE (MODE Group, 1978) and POLYMODE observations that were subsequently undertaken in the 1970s in the Atlantic greatly improved our understanding and appreciation of oceanic mesoscale variability (see Robinson, 1983). Instead of being sluggish, the oceans (certainly the regions above the permanent thermocline) have constantly changing circulation with vigorous embedded mesoscale eddy activity. These mesoscale eddies and the tremendous variability they generate greatly complicate the task of measuring and mapping the various properties in the ocean. They certainly make the task of modeling and prediction of the oceans very difficult if not impossible. It is therefore important for an ocean modeler to understand mesoscale variability, especially since the Holy Grail of ocean modelers has always been the fine horizontal resolution needed to resolve and reproduce these features in their basin scale and global model simulations. The computing power available at present is making this dream a reality and the ability to interpret model-generated variability requires that we understand mesoscale eddies and the associated variability.

Mesoscale variability generated by rings shed off energetic ocean currents such as the Gulf Stream in the Atlantic, the Kuroshio in the Pacific, the Aghulas Current off the southern tip of Africa, the Brazil retroflection, and the Loop Current in the Gulf of Mexico is well-known. The streams are hydrodynamically unstable; the resulting meanders develop into closed loops pinching off rings (Olson, 1991). Depending on which side of the stream the loops occur, the rings can be cyclonic or anticyclonic. For example, anticyclonic rings are spawned on the northern side of the Gulf Stream, and cyclonic rings on the southern side. The Loop Current sheds only anticyclonic rings. The lifetime of these rings is determined not only by their characteristics, such as size and strength, but also by the direction they are carried toward by the background currents. The anticyclonic warm core rings shed off the Gulf Stream travel southwestward along the mid-Atlantic Bight slope and are rapidly reabsorbed by the Gulf Stream. Consequently, their lifetimes are limited to a few months, even though they could conceivably last longer if this geographical constraint were absent. The cold core rings, on the other hand, drift southwestward in the Sargasso Sea and are also absorbed by the Gulf Stream eventually, but last much longer (up to 1.5 years), in spite of the fact they are cyclones. Since weak cyclones are destined to decay rapidly by dispersion (the timescale is roughly that for a linear Rossby wave packet to disperse; see Section 1.5), the long life of the Gulf Stream cyclonic rings may be due to the fact that they are strong to begin with, and by virtue of conservation of total vorticity, their transport to lower latitudes by background circulation, where the planetary vorticity is less, tends to intensify them.

The large anticyclonic Loop Current eddies have a lifetime of only about a year on the average, since their westward drift brings them in contact with the Mexican shelf, where interaction with topography tends to attenuate them rapidly. Mesoscale eddies in the broad oceanic gyres are both cyclonic and anticyclonic, and their size

(50–300 km) tends to be somewhat larger than the internal Rossby radius (40–50 km). Their strength, as gauged by the ratio ω/f, where ω is their rotation speed, is not very strong and they generally drift westward (counter to the direction of planetary rotation) at speeds close to their drift speed V_d.

Another example of long-lived mesoscale features in the ocean is the sub-surface lens. Meddies spawned by the salty subsurface Mediterranean outflow into the Atlantic at a depth of around 1000 m are mostly anticyclonic, and with a large value for the ratio of the swirl velocity to the drift velocity, strongly nonlinear, and therefore can last for many years.

An example of large mesoscale features generated by the action of winds is the anticyclones in the eastern Pacific off the Gulfs of Tehuantepec and Papagayo (Willett, 1996). They are several times the Rossby radius, and their rotation velocity is several times their drift velocity. They drift westward for a few months until they are absorbed by the equatorial current system. In the atmosphere, intense vortices such as hurricanes and tornadoes are also quite common. To deal with these motions, it is necessary to employ cylindrical coordinates with the origin at the center of these vortex-like features. Then the principal balance in the radial direction is a slightly modified form of geostrophic balance, *cyclogeostrophic balance*, and the flow is called the gradient flow (or gradient wind in the atmosphere). In these flows, the frictional and tangential acceleration forces are negligible, so there exists a primary balance between the horizontal pressure gradient, the Coriolis force, and the centrifugal force:

$$-fv - \frac{v^2}{r} = -\frac{1}{\rho_0}\frac{\partial p}{\partial r}$$

(1.15.1)

$$-\frac{\rho}{\rho_o} = \frac{1}{\rho_0}\frac{\partial p}{\partial z}$$

Therefore

$$\frac{\partial v}{\partial z} = -g\frac{\frac{\partial}{\partial r}\left(\frac{\rho}{\rho_0}\right)}{\left(f + \frac{2v}{r}\right)}$$

(1.15.2)

The importance of the centrifugal term depends on the magnitude of the Rossby number $Ro = V/fR$, where R is the radius of curvature. This is nothing but the ratio of the centrifugal force to the Coriolis force. For low Rossby numbers, of course, the second term representing centrifugal forces is small. The relevant length scales here are the radius of the eddy and the Rossby radius of deformation a. Another relevant parameter is the pressure drop or rise at the center relative to the ambient.

For the same pressure change then, the balance that prevails depends very much on the ratio of the radius to the Rossby radius, R/a. If R/a is small, the vortex is tightly wound, the swirl velocities are large, the Rossby number is large, and the balance is simply between the centrifugal force and the pressure gradient. The Coriolis term is negligible (see Figure 1.15.1). This is typical of strong vortices, such as tornadoes in the atmosphere. This balance requires necessarily a pressure drop at the vortex center to have the pressure gradient oppose the centrifugal forces. Tornadoes therefore have low pressure centers no matter in which direction they rotate, cyclonic or anticyclonic. The balance that prevails is called cyclostrophic balance.

Such intense vortices do not occur in the oceans. Most oceanic eddies are under cyclogeostrophic balance where all three terms are roughly of similar magnitude. This comes about because the ratio R/a is of the order unity. This also means that the Rossby number, while small, is not negligible. Around a low pressure system (cold core eddies), the Coriolis and the centrifugal forces are both outward and the pressure gradient balances both, while around a high pressure system, the pressure gradient and centrifugal force are outward and are balanced by the Coriolis force (Cushman-Roisin, 1994). When R/a is large compared to unity, the centrifugal forces can be neglected and geostrophic balance prevails.

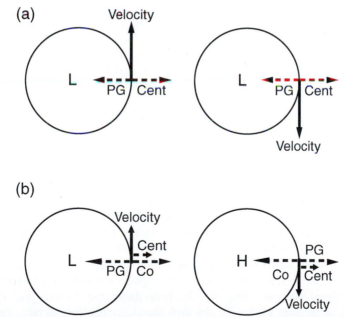

Figure 1.15.1 (a) Cyclostrophic balance in strong cyclonic (left) and anticyclonic (right) eddies (in the northern hemisphere). The principal balance is between the pressure gradient and the centrifugal force (the Coriolis force is negligible). (b) Cyclogeostrophic balance in cyclonic (left) and anticyclonic (right) eddies (in the northern hemisphere).

The simplest model for oceanic mesoscale vortices is a quasi-geostrophic, inviscid, shallow water model. Substituting $H = H_o + h$, where H_0 is the undisturbed layer depth and $h(x,y,t)$ is its perturbation, and $f = f_0 + \beta y$, the midlatitude beta-plane approximation, Eq. (1.12.4), the potential vorticity conservation equation for a single-layer reduced gravity fluid (see also Chapter 3), can be rewritten as

$$\left(1 + \frac{h}{H_0}\right)\left(\frac{D\zeta}{Dt} + \beta v\right) - \left(\frac{\zeta + f_0 + \beta y}{H_0}\right)\frac{Dh}{Dt} = 0 \qquad (1.15.3)$$

Under quasi-geostrophic approximation,

$$U \sim -\frac{g'}{f_0}\frac{\partial h}{\partial y}; V \sim \frac{g'}{f_0}\frac{\partial h}{\partial x}; \frac{Dh}{Dt} \sim \frac{\partial h}{\partial t} \qquad (1.15.4)$$

the equation for the total derivative of vorticity $\zeta \sim \frac{g_e}{f_0}\nabla^2 h$ becomes

$$\frac{D\zeta}{Dt} = \left(\frac{\partial}{\partial t} + U\frac{\partial}{\partial x} + V\frac{\partial}{\partial y}\right)\zeta = \frac{g'}{f_0}\frac{\partial}{\partial t}\nabla^2 h + \frac{g'^2 H_0}{f_0^2}J\left(h, \nabla^2 h\right)$$
$$\qquad (1.15.5)$$
$$J\left(h, \nabla^2 h\right) = \frac{\partial}{\partial x}(h)\frac{\partial}{\partial y}\left(\nabla^2 h\right) - \frac{\partial}{\partial y}(h)\frac{\partial}{\partial x}\left(\nabla^2 h\right)$$

Substituting Eq. (1.15.4) in Eq. (1.15.5) and using Rossby radius a to normalize x and y, H_0 to normalize h, $1/f_0$ to normalize t, and $C = (g'H_0)^{1/2}$ to normalize velocities, one gets the nondimensional equation (see Nezlin and Snezhkin, 1993)

$$\frac{\partial}{\partial \hat{t}}(\hat{\nabla}^2\hat{h} - \hat{h}) = \hat{V}_c\frac{\partial\hat{h}}{\partial\hat{x}} + \hat{V}_c\hat{h}\frac{\partial\hat{h}}{\partial\hat{x}} + \hat{J}(\hat{h}, \hat{\nabla}^2\hat{h}) = 0 \qquad (1.15.6)$$

where carets denote nondimensional quantities and J is the Jacobian. The first two terms describe linear Rossby wave dynamics (see Chapter 4). In the linear limit, cyclones and anticyclones behave similarly (changing h to −h does not alter the equation). The last two terms are however nonlinear. The first of these is similar to the nonlinearity in shallow water gravity wave equations that gives rise to solitons and is called the scalar nonlinearity. It also introduces cyclone–anticyclone asymmetry, since changing h to −h changes the equation. Only if h > 0 (for an anticyclone) can this term balance linear dispersion giving rise to monopolar Rossby solitons that can propagate with little dispersion. In that case, the drift velocity is larger than V_c, the Rossby drift speed, so that there is no radiation of linear Rossby waves and hence no decay. The drift speed increases with maximum amplitude h_0 of the soliton: $V_d = V_c (1 + \alpha h_0)$, with $\alpha \sim 0.2$ (Nezlin and Snezhkin 1993), akin to surface gravity wave solitons whose propagation speed also increases in a similar fashion with amplitude (see Chapters 5 and 6 of Kantha and Clayson, 2000). The equation (without the last term) is very much similar to the classical shallow surface gravity wave Korteveg–deVries (KdV) equation that permits, once again because of the asymmetry of the nonlinear term, only elevation (and not depression) solitons.

Omitting the scalar nonlinearity term leads to the classical Charney–Obukhoff equation for nonlinear quasi-geostrophic vortices, with only the second nonlinear term, called the vector nonlinearity, present in the equation. These equations allow for a dipolar soliton, a cyclone–anticyclone pair called a modon. For a nearly axisymmetric vortex, the ratio of the vector to scalar nonlinearity is proportional to square of the ratio of the Rossby radius to the vortex radius, so that scalar nonlinearity dominates for large vortices, but vector nonlinearity dominates for small vortices. For a fascinating description of Rossby solitons and their implications for mesoscale features in the oceans and planetary atmospheres, see Nezlin and Snezhkin (1993).

An important property of an isolated eddy, whether cyclonic or anticyclonic, is its drift velocity with respect to the ambient waters. This drift velocity is important simply because it is the means by which an enormous quantity of water masses associated with the eddy are transported across the oceans and mixed into different water masses. Therefore, eddies effect transport and mixing, and it is of interest to know the rate at which this is done; for this purpose, the drift velocity of the eddy must be calculated. Isolated eddies in a quiescent ocean move westward due to the difference in rotation rate in their northern and southern regions (with respect to the eddy center). This is due to the latitudinal variation of planetary vorticity, the beta effect, and hence it is called the beta drift.

Bjerknes was the first to explain why pressure disturbances always tend to propagate westward. Consider a high pressure center on a beta-plane; the associated velocities are anticyclonic, clockwise. Assuming a totally geostrophic flow in the eddy (see Hendershott, 1981), it is easy to see that the flow between a given pair of isobars is larger (in the northern hemisphere) in the southern half of the eddy, compared to the northern half, since the geostrophic velocity is inversely proportional to the Coriolis parameter. This causes convergence at the western edge of the eddy and a divergence at the eastern edge. Therefore the pressure drops at the eastern edge and increases at the western edge, and the high pressure pattern propagates west. A similar argument applies for a low pressure center, except that the rotation is reversed. Beta drift can be quite large close to the equator and cause fast westward migration of oceanic eddies. At the latitude of the Gulf of Tehuantepec, the beta drift can amount to about 11 cm s^{-1} (Willett, 1996). Laccadive eddies in the southern Arabian Sea at a latitude of ~10° travel west at a speed of about 18 cm s^{-1} (Lopez and Kantha, 1998; Bruce *et al.,* 1998).

On the oceanic side, several attempts have been made to quantify the beta drift using dynamical equations for an isolated eddy. Pioneering work by Nof (1981) on isolated eddies has been followed up by Killworth (1983) and more recently Benilov (1996). Nof (1983) has examined the drift characteristics of isolated nonlinear cyclonic and anticyclonic eddies on a beta-plane, using power series expansions to the integrated form of the exact governing equations of a two-layer reduced gravity fluid in a frame of reference moving with the eddy. The eddy is assumed to move westward at a steady speed with no change in shape or structure, under the assumption $\varepsilon = \beta a/f < Ro = V_m/fr_e$, the Rossby number, with r_e characterizing the eddy size. The equations of motion for the interior of the eddy are integrated over its volume and combined with those of the fluid exterior to the eddy, and a perturbation expansion in ε is used to provide the translation speed. We will not go into the details of the derivation, for which the reader is referred to Nof (1983), who indicates that the westward beta drift is given by

$$\frac{V_d}{V_c} = \frac{\int_0^{r_e} 2\pi\, r\, dr \int_{r_e}^r V_s(r')\, [H+h(r')]\, dr'}{f\, a^2 \left[\int_0^{r_e} h(r')\, 2\pi\, r'\, dr' \right]} + o\left(\varepsilon^2\right) \qquad (1.15.7)$$

where $h(r)$ is the deflection of the interface from the undisturbed depth of H, $V_s(r)$ is the swirl distribution, and r_e is the eddy size. For cyclonic eddies, the numerator is positive, the denominator is negative, and the drift is therefore negative and westward; for anticyclonic eddies, it is exactly the other way and the drift is still westward. For cyclonic eddies, the drift velocity increases with the eddy size, but decreases with its amplitude, while for anticyclonic eddies, it decreases with size but increases with its strength. Anticyclonic eddies travel faster than the Rossby wave speed and the cyclonic ones travel slower. The eddy is carrying its entire mass within $r = r_e$ westward; the fluid outside is simply displaced northward and southward and left behind (Nof, 1983).

Assuming either $h(r)$ or $V_s(r)$ provides the other from the cyclostrophic balance equation, which for a reduced gravity model is

$$fV_s + \frac{V_s^2}{r} = \frac{g\Delta\rho}{\rho_0}\frac{dh}{dr} \qquad (1.15.8)$$

For a linear swirl distribution to the edge of the eddy r_e, the drift speed for an anticyclonic eddy is given by

$$\frac{V_d}{V_c} = \frac{-2\left[1 + \dfrac{h_c}{3H}\right]}{1 + \left[1 - 8\left(\dfrac{a}{r_e}\right)^2 \dfrac{h_c}{H}\right]} + o\left(\varepsilon^2\right) \qquad (1.15.9)$$

where h_c is the deflection of the pycnocline at the eddy center (maximum) from the undisturbed state and H is the undisturbed depth of the upper layer. For a cyclonic eddy, the same expression holds, except the sign of the term involving the amplitude is switched. Note that the normalized eddy amplitude is bounded by unity for a cyclonic eddy, but by 0.125 $(r_e/a)^2$ for an anticyclonic one and can be larger than unity. Nof (1983) also calculates the drift velocities for cyclonic and anticyclonic Gulf Stream (GS) rings by assuming a parabolic velocity distribution, and finds the results differ by less than 10% from the case of linear velocity distribution, indicating a weak dependence of the beta drift velocity with the structural details of the eddy. Figure (1.15.2), from Nof (1983), shows the beta drift for a cyclonic eddy with $h_c/H = 1$, and an anticyclonic one with an identical structure but $H = 0$. Also shown are Rossby wave speeds for a half-wavelength in both the meridional and zonal directions of the eddy diameter.

For a typical cold core GS ring in the Sargasso Sea, $\beta = 2.10^{-11}$ m^{-1} s^{-1}, g' = 0.017 m s^{-2}, f = 8.10^{-5} s^{-1}, r_e = 100 km, H = 600 m, V_m = 1.6 m s^{-1} at 60 km radius, Ro = 0.31, and the drift speed is 1.8 cm s^{-1}. Such a ring transports 10^{13} m^3 of water masses westward at the rate of about 1.6 km d^{-1} (Nof, 1983). For a typical warm core GS ring in the mid-Atlantic Bight, $\beta = 2.10^{-11}$ m^{-1} s^{-1}, g' = 0.015 m s^{-2}, f = 8.10^{-5} s^{-1}, r_e = 80 km, H = 500 m, h_c = 50 m, V_m = 1.6 m s^{-1} at 80 km radius, Ro = 0.23, and the drift speed is 1.6 cm s^{-1}. These values are several times less than observed values, which are typically ~5 cm s^{-1}. This could be due to several factors, including the absence of the background currents in Nof's model. To a first-order approximation, the background current can be simply added to the self-induced drift velocity to obtain the total drift velocity.

Nof (1981, 1983) ignored the beta-induced correction to the velocity field, and his model predicts steady translation speeds even for those eddies whose velocity is less than the Rossby wave speed and hence might radiate energy and decay. Benilov (1996) corrected this problem and arrived at a formula for the eddy drift speed using reduced gravity approximation,

$$\frac{V_d}{V_c} = \frac{\displaystyle\int_0^\infty \left\{r\left[H + h(r)\right]V_s(r)\right\}2\pi\, r\, dr}{f\, a^2 \displaystyle\int_0^\infty \left\{r\dfrac{dh}{dr}\right\}2\pi\, r\, dr} \qquad (1.15.10)$$

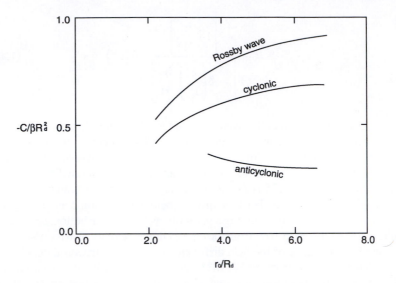

Figure 1.15.2 Beta drift of cyclonic and anticyclonic eddies. From Nof (1983).

which reduces to an elegant form

$$\frac{V_d}{V_c} = 1 + \frac{E}{MC^2} \qquad (1.15.11)$$

where M and E are mass and energy anomalies associated with the eddy:

$$M = \int_0^\infty \{\rho h(r)\} 2\pi r dr$$

$$E = \frac{1}{2} \int_0^\infty \{g\Delta\rho h^2(r) + \rho[H + h(r)]V_s^2(r)\} 2\pi r dr \qquad (1.15.12)$$

For an anticyclonic eddy, $M > 0$, and hence the drift speed is larger than the Rossby wave speed. For a cyclonic eddy, $M < 0$, and the drift velocity can be westward (at speeds less than the Rossby wave speed) or eastward. Benilov (1996) indicates that those with westward drift are not permissible and in reality might radiate energy and decay rapidly.

A particular case of Eq. (1.15.1), where the pressure gradient vanishes, the balance is between the Coriolis force and the centrifugal force: $V = f R$ and the Rossby number is unity. This is an inertial jet and can only occur if the rotation of the jet is anticyclonic. A good example of such an inertial jet in the atmosphere is in the Gulf of Tehuantepec. The Sierra Madre forms a mountain chain approximately

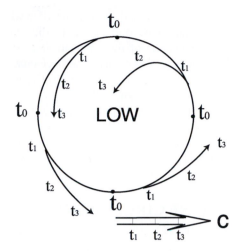

Figure 1.15.3 Particle trajectories in an eddy on a mean background current.

1500 m high off the west coast of Mexico, except for three gaps in the chain, one about 40 km wide near this gulf. During winter, high pressure systems moving into the Gulf of Mexico on the eastern side of the Sierra Madre give rise to a strong pressure drop across this gap, through which wind pours across at a speed often exceeding 20–30 m s^{-1} into the Gulf of Tehuantepec. In the gap itself, the Coriolis force is balanced by the pressure gradient across the gap, but once released from the constraints of the gap, the cross-axis pressure gradient vanishes, and the wind jet curves to the right at a radius just enough for the centrifugal force to balance the Coriolis force. The resulting anticyclonic flow is thought to be responsible for spinning up an anticyclonic eddy in that gulf, which then drifts westward and is eventually absorbed by the equatorial current system in the eastern Pacific (Willett, 1996).

Equation (1.15.1) holds only in the absence of friction and advection by a mean current or wind. Both can be significant in the atmosphere. The presence of friction at the ground causes the tangential velocities to be brought to zero, decreasing the centrifugal force on the fluid parcels as the ground is approached. But the pressure gradient remains unchanged. The consequence is that the parcels move inward in the boundary layer, irrespective of the direction of rotation. This radial convergence gives rise to an axial (vertical) velocity away from the ground.

Advection by mean flow has a slightly different effect. In half the eddy, where advection augments the swirl velocity, fluid parcels move outward and away from the center, whereas on the half-portion, where the advective velocity opposes the swirl velocity, fluid parcels tend to move inward toward the eddy center (Figure 1.15.3). For a westward propagating anticyclonic eddy, the southern half has fluid parcels moving away from the center and the northern half toward it. Since the swirl velocity of eddies decreases rapidly beyond a certain radius, the advective effects become important in the outer parts of the eddy; in an eddy with strong swirl velocities, a buoy well within the radius of maximum velocity tends to stay within the eddy for long periods of time, whereas that far beyond has a tendency to be entrained away from the eddy by the ambient flow.

Graef (1998) points out that the beta-plane shallow water, reduced gravity equations used in many studies mentioned above omit terms related to the Earth's radius, in effect having to do with the curvature of the Earth's surface. He suggests a more accurate set based on Muller (1995). However, this does not appear to change the results.

1.16 THERMOHALINE CIRCULATION AND BOX (RESERVOIR) MODELS

As seen earlier, wind-driven circulation is largely responsible for the strong western boundary currents and vigorous gyre scale circulations in ocean basins. Naturally, the time scales associated with the variability of this kind of circulation are the scales associated with the wind itself, principally seasonal and interannual. In wind-driven ocean circulation theories, the density effects are largely ignored for the most part or the density field in the bulk of the ocean is considered to be invariant. Consequently, the variability in oceanic circulation due to slow and small but nonnegligible changes in the oceanic density field is ignored. Over long timescales of interest to climatic studies, of the order of millennia and beyond, it is the changes in the density structure of the basins that are of primary importance. This change is brought about by the meridional advection of water masses and the associated vertical advection and diffusion of heat and salt in the water column, these processes being driven by the heat and salt fluxes at the surface. At present, this *meridional thermohaline circulation* is driven primarily by wintertime cooling and deep water formation in the subpolar seas of the North Atlantic and the Greenland and Labrador seas. These cold, dense water masses sink to the bottom and flow equatorward, cross the equator, and are advected into the Indian and Pacific oceans, where they slowly surface again and flow poleward again. Figure 1.16.1 (from Schmitz, 1995) shows a schematic of this circulation as originally depicted by Broecker (1987, 1991). This circulation, often called the thermohaline conveyor belt (Broecker, 1991), has been updated recently by Schmitz (1995) based on the latest observations and model results. The timescale associated with it, the timescale for a particle to complete one complete circuit, is typically a millennium.

While only 15 to 30 Sv of water mass transports is involved (comparable to that involved in the gyre scale horizontal wind-driven circulations, but less than the inertial circulations of the western boundary currents that occur on scales less than the basin scale), this meridional thermohaline circulation is of critical im-

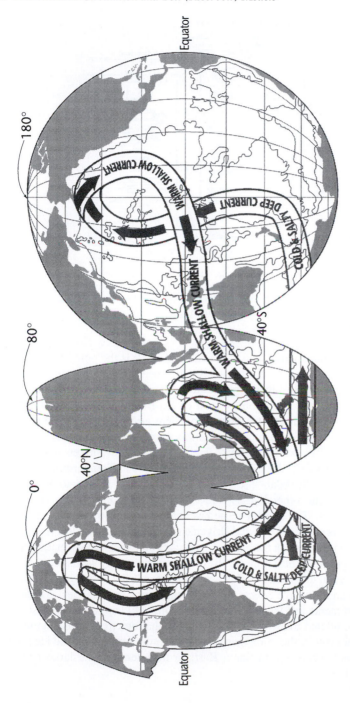

Figure 1.16.1 Meridional thermohaline circulation (the conveyor belt) in the global oceans, the two-layer version suggested by Broecker. From Schmitz (1995).

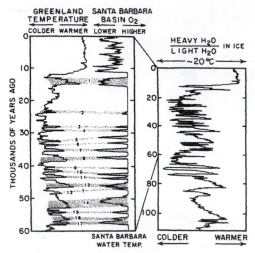

Figure 1.16.2 Oxygen isotope records in Greenland ice cores and Santa Barbara marine deposits showing large and abrupt changes in the recent geological past. (Reprinted with permission from Broeker, W. S., The great ocean conveyor, in *Global Warming: Physics and Facts.* Copyright 1992 American Association for the Advancement of Science.)

portance to climatic changes and the oceanic biota. It is the poleward transport of the excess heat gained in the tropics by the oceans and the atmosphere that moderates the climate in mid to high latitudes. Satellite measurements indicate that under present climatic conditions, a total of 5.5 PW of heat must be transported poleward across 35° latitude by the atmosphere and the oceans (Trenberth and Solomon, 1994). The North Atlantic accounts for nearly two-thirds of this, roughly 1.2 PW across 25° N, while the North Pacific accounts for the remainder of about 0.8 PW (Bryden and Hall, 1980; Bryden *et al.*, 1991; Trenberth and Solomon, 1994). Of the North Atlantic transport, roughly half is due to thermohaline convection, and the other half is due to wind-driven circulation and western boundary currents. The North Pacific has no deep water formation and hence the transport there is primarily by wind-driven circulation (Engelhardt, 1996). Thus the thermohaline circulation in the North Atlantic accounts for approximately one-third of the global poleward heat transport by ocean currents. Any weakening, cessation, or rapid change in the pattern of this circulation can bring about drastic changes in climate and can have large effects on the oceanic biota, such as species extinctions. Deep sea sediments and ice cores contain records of changes in the past climate, thought to be related to these changes in thermohaline circulation. Figure 1.16.2, from Broeker (1997), shows oxygen isotope records from Greenland ice sheet and marine deposit records from Santa Barbara that demonstrate large and abrupt changes in climate.

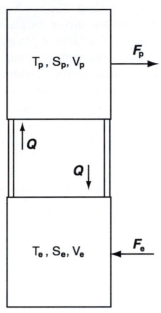

Figure 1.16.3 Two-box model of the thermohaline circulation, the two boxes representing the low latitude (bottom) and high latitude (top).

The meridional gradient in solar isolation is largely responsible for the meridional circulation. The net heat loss in the polar oceans cools the surface waters and makes them dense enough to sink to the bottom and flow equatorward and, after upwelling slowly in the equatorial oceans, gain heat near the surface and flow poleward again. This circulation involves great asymmetry in that the sinking regions are rather narrow and limited, while the upwelling regions are broad and extensive. Also, the temperature gradients are principally confined to the permanent thermocline, with the deep waters being nearly isothermal. This simple picture is however greatly complicated by salinity effects. At present, there exists excess precipitation at high latitudes and excess evaporation at low latitudes. This haline forcing resulting from local imbalances in precipitation and evaporation acts in the direction opposite to thermal forcing, since the freshwater flux in the polar oceans tends to counteract the cooling effect on the buoyancy of surface parcels and hence can weaken the thermohaline circulation or even reverse it. Such reversals can be rather sudden and have been called thermohaline catastrophes. Excellent recent reviews of thermohaline processes can be found in Weaver and Hughes (1992), Marotzke (1994, 1996), Whitehead (1995), and Park (1996).

The best way to simulate and understand thermohaline circulation is through comprehensive three-dimensional coupled models of the global atmosphere and oceans integrated for tens of centuries (Boville and Gent, 1998). However, inadequacies in computing power and imperfections in model physics do not yet permit this approach to be fully exploited. The next best method is through three-dimensional ocean circulation models driven by plausible wind and buoyancy forcing at the surface. This approach is also at present rather limited, since the computing resources needed to run even a relatively coarse resolution global ocean model over tens of centuries (needed for equilibration) of simulation and experimentation are still quite prohibitive. Besides, most global ocean models still have relatively imperfect physics related to the dense and deep water formation. At present, therefore, highly simplified models such as zonally averaged two-dimensional models in the meridional plane and simple box models are being used to investigate the plausible variability of thermohaline circulation. The simplest such conceptual model is the box model consisting of well-mixed reservoirs in communication with one another, pioneered by none other than Henry Stommel in the 1960s (Stommel, 1961). While such a box model of thermohaline circulation is so highly simplified as to bear little resemblance to the actual thermohaline circulation in the global oceans, it is nevertheless a very useful tool for understanding the basic physical mechanisms involved in the variability, especially the thermohaline catastrophes brought on by sudden reversals in the direction of the meridional circulation. The principles can be illustrated by a two-box thermohaline circulation model (Figure 1.16.3) consisting of two well-mixed reservoirs representing the equatorial and polar regions of the ocean connected to each other.

Let the temperature, salinity, and volume be T_P, S_p, and V_p in the polar reservoir and T_e, S_e, and V_e in the equatorial reservoir (Stommel, 1961; Thual and McWilliams, 1992; Park, 1996). Let F_p and F_e be the salt fluxes at the surface and let T_p^a, T_e^a and S_p^a, S_e^a be the reference (or air) temperatures and salinities. Let Q be the magnitude of the volume transport between the two reservoirs, and τ_T, τ_S be the damping (or restoring) timescales that tend to restore both the polar and the equatorial reservoir to some reference conditions. The governing equations for salinity and heat conservation in the two reservoirs are given by

$$V_p \frac{\partial S_p}{\partial t} = V_p F_p + |Q|(S_e - S_p)$$

$$V_e \frac{\partial S_e}{\partial t} = V_e F_e - |Q|(S_e - S_p)$$

$$V_p \frac{\partial T_p}{\partial t} = \frac{V_p}{\tau_T}(T_p^a - T_p) + |Q|(T_e - T_p)$$ \hspace{1cm} (1.16.1)

$$V_e \frac{\partial T_e}{\partial t} = \frac{V_e}{\tau_T}(T_e^a - T_e) - |Q|(T_e - T_p)$$

The first term in each equation represents the change due to local air–sea exchanges, and the second that due to the horizontal advection brought about by the meridional transport between the two reservoirs. Parameterization of both these terms is controversial, and the results depend to some extent on how that is done. Traditionally, following Stommel (1961), the volume transport has been parameterized by assuming a frictional balance, in other words, assuming that the frictional dissipation (vv_{zz}) balances the meridional pressure gradient (p_y / ρ_0) (v being the meridional velocity and z the vertical coordinate), so that the volume transport is proportional to the buoyancy difference between the reservoirs,

$$Q = Q_f = -c_f \frac{\left(\rho_e - \rho_p\right)}{\rho_0} = c_f \left[\alpha(T_e - T_p) - \beta(S_e - S_p)\right] \qquad (1.16.2)$$

where a linear equation of state of the form $\rho = \rho_0(1 - \alpha T + \beta S)$ has been used to close the equations; α, β are the coefficients of thermal and saline expansion. However, there are indications that it may be more appropriate to use geostrophic balance conditions to derive this transport (Bryan, 1991; Whitehead, 1995; Park, 1996). Consider a basin with meridional boundaries driven by a meridional temperature gradient. In this situation, a circulation is set up in the basin such that the zonal temperature gradient $\partial T / \partial x$ is of the same order as the meridional gradient $\partial T / \partial y$, and the thermal wind balance provides a scale for the resulting meridional velocity V_g (e.g., see Park, 1996),

$$V_g \sim \frac{g\alpha\Delta T\delta_T}{fL} \qquad (1.16.3)$$

where L is the meridional scale, and δ_T is the thermal boundary layer thickness. Since the meridional transport induces a vertical velocity into the thermal boundary layer of magnitude,

$$W_g \sim \frac{V_g \delta_T}{L} \qquad (1.16.4)$$

which must be such that the vertical advection of heat is balanced by the diffusion across the thermocline, $wT_z \sim kT_{zz}$,

$$W_g \sim \frac{k}{\delta_T} \qquad (1.16.5)$$

Equations (1.16.3) to (1.16.5) give

$$V_g \sim \left[\frac{\kappa(g\alpha\Delta T)^2}{f^2 L}\right]^{1/3} ; \quad \delta \sim \left[\frac{\kappa f L^2}{g\alpha\Delta T}\right]^{1/3}$$

$$Q_g \sim V_g\delta \sim \left[\frac{\kappa^2 Lg\alpha\Delta T}{f}\right]^{1/3} \sim (\alpha\Delta T)^{1/3} \qquad (1.16.6)$$

Equation (1.16.4) is also consistent with classical beta-plane thermocline theories that assume a linear geostrophic vorticity balance on the beta-plane $\beta v = f w_z$, so that $\beta V \sim f W / \delta$, which is equivalent to Eq. (1.16.4).

Extending to the case with both meridional temperature and salinity gradients and using the linear equation of state,

$$Q = Q_f = -c_g \left[\frac{(\rho_e - \rho_p)}{\rho_0} \right]^{1/3} = c_g \left[\alpha(T_e - T_p) - \beta(S_e - S_p) \right]^{1/3} \quad (1.16.7)$$

The coefficients c_f and c_g are chosen usually to yield a volume transport of about 10 Sv for present conditions of 20°C temperature difference and 2 psu salinity difference between the two reservoirs. When the thermal mode is active, with sinking in the polar reservoir and upwelling in the equatorial one, Q is positive; when the haline mode dominates, with sinking in the equatorial region and upwelling in the polar, Q is negative. Equations (1.16.2) and (1.16.7) can be written as

$$Q_{f,g} = c_{f,g} \left[\alpha(T_e - T_p) - \beta(S_e - S_p) \right]^{n_{f,g}} ; \ n_f = 1, \ n_g = 1/3 \quad (1.16.8)$$

Two types of boundary conditions are used to derive the salt flux. Traditionally, damping to reference values has been used, in analogy with the heat flux:

$$F_{p,e} = \frac{1}{\tau_S} \left(S_{p,e}^a - S_{p,e} \right) \quad (1.16.9)$$

There are reasons to suspect that this may not be appropriate for salt fluxes, since the evaporation and precipitation processes depend largely on thermal conditions. Instead, the freshwater and hence the salt flux should depend on the meridional gradient of temperature and can be parameterized as (Marotzke, 1996; Park, 1996)

$$F_p = \frac{\gamma S_0}{2\varepsilon_w d} \left(T_e - T_p \right)^m = -F_e \quad (1.16.10)$$

where $m \sim 3.5$ (Park, 1996), γ is the atmospheric transport efficiency, ε_w is the ratio of ocean area to catchment area of the basin, d is the box depth, and S_0 is the average salinity. For $m = 0$, this condition is the same as prescribing a fixed value of the excess precipitation.

Since it is the differences in temperature and salinity between the two reservoirs that occur prominently, it is possible to deal with quantities $\Delta T = \left(T_e - T_p \right)$ and $\Delta S = \left(S_e - S_p \right)$. Then the volume transport can be written as

$$\frac{Q_{f,g}}{c_{f,g}(\alpha\Delta T)} = \left(1 - \frac{\beta\Delta S}{\alpha\Delta T}\right)^{n_{f,g}} \tag{1.16.11}$$

where $n = 1$ for frictional control and $1/3$ for geostrophic control.

The nature of solutions can be illustrated by looking at a simpler subset of the governing equations when the damping of temperature is infinitely fast so that $\tau_T = 0$ and the temperatures of the two reservoirs attain the reference temperatures and remain there. The temperature equations drop out (Haidvogel and Bryan, 1993; Marotzke, 1994) and ΔT becomes fixed. Then an equation for R_b, the ratio of buoyancies due to salinity and temperature differences, can be derived:

$$\frac{\partial}{\partial \hat{t}}(R_b) = \varepsilon - r(1 - R_b)^n R_b$$

$$\tilde{Q} = r(1 - R_b)^n \tag{1.16.12}$$

$$R_b = \frac{\beta\Delta S}{\alpha\Delta T}; \quad \varepsilon = \frac{\beta F V_p}{c_{f,g}(\alpha\Delta T)^2}; \quad \hat{t} = \left(\frac{V_p}{2c_{f,g}\alpha\Delta T}\right)t; \quad r = \left(\frac{V_p + V_e}{2V_p}\right)$$

where $\tilde{Q}$ is the normalized volume transport. This equation admits multiple equilibria for particular values of ε, the normalized salt flux. For the frictional control case ($n = 1$) with equal reservoir volumes ($r = 1$), three steady state solutions exist (Haidvogel and Bryan, 1993),

$$R_b = \frac{1}{2}\left\{(1 + \sqrt{1 - 4\varepsilon}), (1 - \sqrt{1 - 4\varepsilon}), (1 + \sqrt{1 + 4\varepsilon})\right\}$$

$$\tilde{Q} = 1 - R_b \tag{1.16.13}$$

and multiple equilibria are possible for $\varepsilon < 0.25$. Figure 1.16.4 shows the volume transport $\tilde{Q}$ plotted as a function of the salt flux ε. Of the three possible solutions above, the first one is unattainable because it is unstable. The second solution corresponds to the thermal mode ($\tilde{Q} > 0$), where the thermal forcing dominates and overcomes that due to salinity forcing, but this region is confined to $0.5 \leq \tilde{Q} \leq 1.0$. The flow is strong with sinking in the polar reservoir and upwelling in the equatorial one, corresponding to the current North Atlantic conditions. The third solution ($\tilde{Q} < 0$) corresponds to the haline mode, where the salinity contribution overwhelms that due to the thermal gradient. The flow is weak but reversed, and sinking occurs in the equatorial reservoir and upwelling in the polar. As ε increases monotonically from zero beyond 0.25, the solution must switch suddenly from the thermal mode to the haline mode at 0.25. This is known as thermohaline catastrophe, and there is evidence in deep sea sediments that such sudden transitions have occurred in the geologic past.

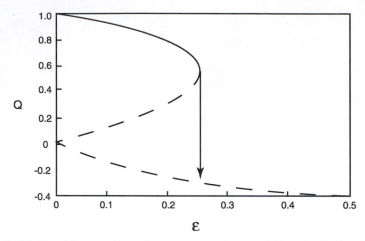

Figure 1.16.4 Plot of the normalized volume transport against the salt Xux, showing the two solution states in the simplest two-box model. The dashed portion is unstable so that transitions between the full curve to the dash-dot curve indicating a catastrophic change are possible.

The physical basis of a thermohaline catastrophe is quite simple. As the freshwater flux (equivalently ε) increases, the meridional buoyancy gradient driving the transport between the reservoirs decreases, the circulation becomes sluggish, and the polar oceans become even fresher so that the salinity gradient increases even more and the circulation becomes even more sluggish. A stage is reached where the circulation is so weak that it cannot remove the salinity anomalies and the salinity gradient overcomes the thermal gradient and the circulation reverses.

Welander (1986) extended the two-reservoir model to three reservoirs taking into account the southern polar region and showed that there now can exist four stable equilibrium states, with two symmetric solutions, one with sinking in the polar reservoirs and upwelling in the equatorial one, and the other with upwelling in the polar reservoirs and sinking in the equatorial one. There are two asymmetric solutions as well corresponding to sinking in one of the polar regions and upwelling elsewhere.

The general case of Eq. (1.16.1) has been dealt with by Park (1996) for both the frictional and geostrophic transport, and the restoring and interactive salinity flux cases. Equation (1.16.1) can be reduced to two equations in nondimensional form,

$$\frac{\partial}{\partial \hat{t}}(\Delta \hat{T}) = \Delta \hat{T}^{a} - \Delta \hat{T}(1 + \hat{Q})$$

$$\frac{\partial}{\partial \hat{t}}(\Delta \hat{S}) = \hat{F} - \hat{Q}\Delta \hat{S}$$

(1.16.14)

where

$$\left(\Delta\hat{T},\Delta\hat{T}^a\right)=\left[\left(\frac{1}{V_e}+\frac{1}{V_p}\right)\alpha c_f\tau_T\right]\left(\Delta T,\Delta T^a\right); \ \hat{t}=t/\tau_T$$

$$\Delta\hat{S}=\left[\left(\frac{1}{V_e}+\frac{1}{V_p}\right)\beta c_f\tau_T\right]\Delta S; \ \hat{Q}=\left[\left(\frac{1}{V_e}+\frac{1}{V_p}\right)\tau_T\right]Q \qquad (1.16.15)$$

$$\hat{F}=\left[\left(\frac{1}{V_e}+\frac{1}{V_p}\right)\beta c_f\tau_T^2\right]\left(F_e-F_p\right)$$

with

$$\hat{Q}_f=\Delta\hat{T}-\Delta\hat{S}$$

$$\hat{Q}_g=\lambda\left(\Delta\hat{T}-\Delta\hat{S}\right)^{1/3}; \ \lambda=c_g c_f^{-1/3}\tau_T^{2/3}\left(\frac{1}{V_e}+\frac{1}{V_p}\right)^{2/3} \qquad (1.16.16)$$

for volume transport conditions corresponding to frictional and geostrophic conditions, respectively, and

$$\hat{F}=\xi\left(\Delta\hat{S}^a-\Delta\hat{S}\right); \ \xi=\frac{\tau_T}{\tau_S}$$

$$\hat{F}=\Delta\hat{T}^m\left[\frac{\gamma\hat{S}_0\tau_T}{\varepsilon_w d}\right]\left[\left(\frac{1}{V_p}+\frac{1}{V_e}\right)c_f\alpha\tau_T\right]^{-m} \qquad (1.16.17)$$

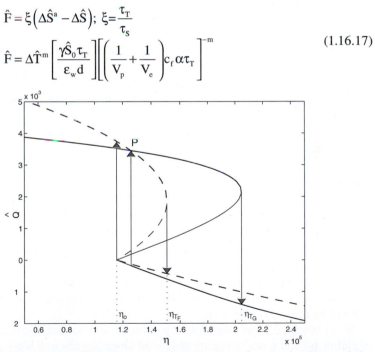

Figure 1.16.5 Plot of the normalized volume transport against the salt flux, showing the two solution states in the more comprehensive two-box model. The solid curve is for the geostrophic and the dashed curve for the frictional model. From Park (1996).

for freshwater flux conditions at the surface for restoring and interactive boundary conditions, respectively. The value of λ in Eq. (1.16.16) is determined by equating the values of $\hat{Q}$ for the two cases for present day conditions: $\lambda = 0.00228$.

Multiple equilibria occur for $\xi \ll 1$. Thual and McWilliams (1992) show that for $\xi = 0.002$, the two-box frictional model yields bifurcation results similar to those of a zonally averaged frictional model. Figure 1.16.5 shows the normalized transport $\hat{Q}$ against the imposed normalized freshwater flux $\Delta \hat{S}^a$ for an imposed meridional temperature gradient $\Delta \hat{T}^a$ of 0.00577 (from Park, 1996) for both frictional and geostrophic transport conditions but with restoring salinity boundary conditions. These conditions correspond to dimensional units of $\Delta T = 20°C$, $Q = 10$ Sv, and $\tau_T = 100$ days. The lower branches of the curve correspond to the haline mode, and the upper ones to the thermal mode. The middle ones are unstable and hence unattainable. In both frictional and geostrophic cases, as the salinity forcing $\Delta \hat{S}^a$ increases, the salinity gradient increases, while the temperature gradient does not change much because of the strong negative feedback. The increase in salinity gradient weakens the density gradient and hence the thermohaline circulation $\hat{Q}$, until the catastrophic transition at $\Delta \hat{S}_2^a$ from the thermal mode to the haline mode. Conversely, as the salinity gradient decreases from the states representing the haline mode, the buoyancy forcing also weakens and circulation becomes more sluggish, until the circulation stops when the salinity gradient cancels the temperature gradient exactly and the buoyancy gradient becomes zero. Any further decrease in the salinity gradient causes a catastrophic transition to the thermal mode at $\Delta \hat{S}_1^a$. Point P denotes the present conditions in the North Atlantic. Thus, while the frictional model indicates that a small change in haline forcing (freshwater flux) from present conditions can bring about a catastrophic change to the haline mode, the geostrophic model indicates that that is quite unlikely. The results for interactive haline forcing conditions are qualitatively similar to those for the restoring forcing conditions.

Figure 1.16.6 (from Park, 1996) shows the temporal evolution of the non-dimensional volume transport and heat flux as a function of nondimensional time for an increase in $\Delta \hat{S}^a$ of 6 to 8% across $\Delta \hat{S}_2^a$ at t = 0, all other conditions being the same as above. The catastrophic transition to haline mode occurs in a short span of a few hundred years for the geostrophic case, whereas the frictional case takes considerably longer (time is normalized by the restoring timescale of 100 days). The heat flux virtually ceases as the circulation stops during the transition, before being restored by reversed circulation.

While box models are educative, it is difficult to simulate the crucial processes of narrow bottom water formation and broad upwelling via well-mixed reservoirs. For this reason, the 2D meridional plane models and fully 3D global ocean models are being increasingly used to study thermohaline circulation. In addition, the global

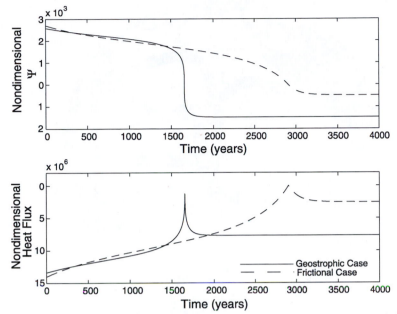

Figure 1.16.6 Temporal evolution of the volume transport (top) and heat flux (bottom) during a catastrophic transition. From Park (1996).

thermohaline circulation pattern is quite complex and involves interbasin exchanges that cannot be studied by simple two- and three-box models. Schmitz (1995) has provided a thorough survey of global thermohaline circulation and a summary map (Figure 1.16.7) that exhibits the complex nature of this circulation. Fully 3D models [see Sarmiento (1992) and Toggweiler (1994) for a description of these models and a discussion of the model results], despite limitations imposed by the need for long integration times, are best suited for simulation of the details of this complex circulation.

1.17 NUMERICAL MODELS

We have so far considered analytical and semi-analytical solutions to simplified subsets of the full equations that nevertheless clarify many salient aspects of the oceanic processes. Numerical ocean models, on the other hand, can and do employ the full set of Equations (1.3.16) and (1.3.18) to (1.3.20). However, these equations need to be supplemented by initial and boundary conditions on the prognostic variables. These depend on the nature of the solution sought.

The oceans are driven by momentum, heat, and salt fluxes at the air–sea interface. The boundary conditions at the sea surface ($z = \eta$) are therefore

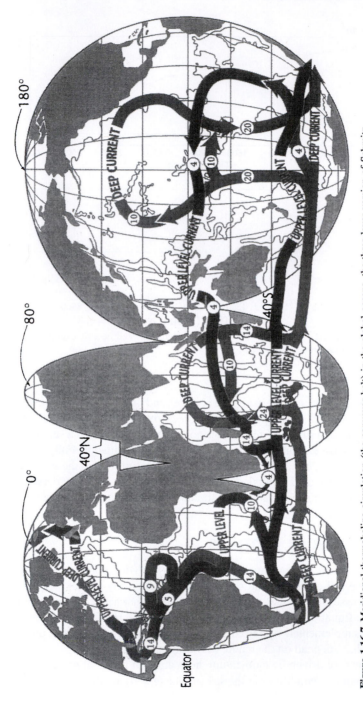

Figure 1.16.7 Meridional thermohaline circulation (the conveyor belt) in the global oceans, the three-layer version of Schmitz (1995).

$$K_M \left(\frac{\partial U_i}{\partial z} \right) = (\tau_{oi})$$

$$K_H \left(\frac{\partial}{\partial z}(\Theta, S) \right) = (q_H, q_S)$$

$$W = \frac{\partial \eta}{\partial t} + U_j \frac{\partial \eta}{\partial x_j} \qquad (1.17.1)$$

where τ_{0i} is the kinematic shear stress acting at the free surface due to the action of winds and waves (taken as nearly equal to the kinematic wind stress) and $q_{H,S}$ are the heat and salt fluxes (kinematic). The term q_H is determined by the net heat balance at the air–sea interface due to short-wave and long-wave solar heating, backradiation by the ocean surface, and the turbulent sensible and latent heat exchanges, and q_S by the difference between evaporation and precipitation. Accurate parameterization of these air–sea fluxes has been the subject of intense research for several decades (for example, the 1992 multinational Tropical Ocean Global Atmosphere/Coupled Ocean Atmosphere Response Experiment, or TOGA/COARE). For more details, see Chapter 4 of Kantha and Clayson (2000).

The conditions at the ocean bottom ($z = -H$) are of no mass transfer through the ocean bottom:

$$W = -U_j \frac{\partial H}{\partial x_j}$$

and

$$K_M \left(\frac{\partial U_i}{\partial z} \right) = (\tau_{bi})$$

$$K_H \left(\frac{\partial}{\partial z}(\Theta, S) \right) = (0, 0)$$

$$(1.17.2)$$

The last of the above conditions implies no heat or salt transfer through the ocean bottom (geothermal fluxes at the ocean bottom are negligible, except perhaps in the immediate vicinity of midocean spreading centers). The bottom stresses are usually parameterized using quadratic drag law with $c_d \sim 0.0025$,

$$\tau_{bi} = c_d |U_{bi}| U_{bi} \qquad (1.17.3)$$

or by assuming that the lowest model grid point falls within the logarithmic law region and using the well-known logarithmic relationship between the mean velocity and the friction velocity to derive the drag coefficient (e.g., Blumberg and Mellor, 1987; Kantha and Piacsek, 1993). Some ocean models (e.g., Wunsch *et al.*, 1997; Haidvogel *et al.*, 1997) use a linear drag term with the drag proportional to the velocity ($\tau_{bi} = \alpha U_{bi}$, where $\forall$ has the dimensions of velocity). However, the proportionality coefficient is ill-determined and becomes essentially a tuning parameter.

When additional quantities such as turbulence velocity and length scales are modeled explicitly, corresponding conservation equations need to be solved along with appropriate boundary conditions at the ocean surface and bottom (Blumberg and Mellor, 1987; Kantha and Piacsek, 1993; Kantha and Clayson, 1994). The same holds for modeling chemical and biological quantities.

If the lateral boundary is a closed boundary, then it is straightforward to apply the lateral boundary condition; the component of velocity perpendicular to the boundary is zero and there is no lateral mass, heat, or salt flux through the boundary. If it is open, on the other hand, as is usual in regional models of coastal and marginal seas, it is necessary to prescribe open boundary conditions. This is a difficult problem since complete information on various flow properties must be specified and this depends to a large extent on how well the flow at the boundary is known. The best strategy is to nest the model in a coarser resolution model of the basin or a global model. In many cases, this is not feasible and hence it is not possible to inform the model about what the rest of the ocean is doing. The best strategy under these conditions is some form of Sommerfeld radiation boundary condition on dynamical quantities, which ensures that disturbances approaching the boundary from the inside are radiated out and not bottled up (Blumberg and Kantha, 1985; Roed and Cooper, 1986; Kantha *et al.*, 1990). This is usually of the form

$$\frac{\partial \varsigma}{\partial t} + C \frac{\partial \varsigma}{\partial x_n} = 0 \qquad (1.17.4)$$

where ς is a variable such as the sea surface height, n denotes direction normal to the boundary, and C is the phase speed of the approaching disturbance. Proper prescription of C is important to the success of the radiative boundary condition and has been the subject of much research (Orlanski, 1976; Blumberg and Kantha, 1985).

If there is inflow at the lateral boundary, temperature and salinity of the incoming flow must be prescribed. If there is outflow, on the other hand, these quantities are simply advected out:

$$\frac{\partial (\Theta, S)}{\partial t} + U_n \frac{\partial (\Theta, S)}{\partial x_n} = 0 \qquad (1.17.5)$$

We will deal with these aspects in greater detail in later chapters.

Note that the horizontal mixing terms in Equations (1.3.16) and (1.3.18) to (1.3.20) corresponding to unresolved subgrid scale processes have been parameterized simply as Laplacian diffusion terms, as is often the practice. A more rigorous form for these terms can be found in Blumberg and Mellor (1987). The values for these coefficients are most often chosen as constants in a rather ad hoc manner based on purely numerical considerations. While vertical mixing coefficients K_M and K_H can be rigorously modeled by turbulence closure theories (e.g., Kantha and Clayson, 1994; Mellor and Yamada, 1982), there does not exist a similar approach for these terms. One approach, widely used in atmospheric modeling, is that due to Smagorinsky (1963) and is similar to the classical mixing length theory of turbulence. Here the mixing coefficient is assumed to be proportional to the mean strain rate, so that

$$A_M = C\left(\Delta x_1\right)\left(\Delta x_2\right)\left[\left(\frac{\partial U_i}{\partial x_j}+\frac{\partial U_j}{\partial x_i}\right)\left(\frac{\partial U_i}{\partial x_j}+\frac{\partial U_j}{\partial x_i}\right)\right]^{\frac{1}{2}}; \quad A_H \sim A_M \quad (1.17.6)$$

where C is the Smagorinsky coefficient, with a value of around 0.04, and Δx_1 and Δx_2 are the grid sizes. This approach assumes that the subgrid scales fall within the Kolmogoroff inertial subrange, an assumption not always satisfied. A practical consequence of using this model is that strong horizontal shear is accompanied by strong horizontal mixing, which tends to smear out thermal fronts. A more general approach is to assume that the mixing coefficients are a sum of a constant background value and the Smagorinsky value given by Eq. (1.17.6) and to choose the values assigned to each appropriately (e.g., Kantha, 1995b).

Another approach is to assign a constant cell Reynolds number $R_N = \frac{|U_j||\Delta x_j|}{A_M}$ and determine the value of the mixing coefficient thusly. Some modelers have used a biharmonic form ($\nabla^4 \equiv \frac{\partial^2}{\partial x_k \partial x_k}\frac{\partial^2}{\partial x_k \partial x_k}$) to model these terms (O'Brien, 1985). In this form, the terms serve principally to control the so-called $2\Delta x$ noise in the numerical solutions, since this form of friction is scale-sensitive.

Suffice it to say that modeling horizontal diffusion terms is still rather ad hoc.

Haney (1971) suggested that the prescription of the total heat flux at the surface take the form $Q = Q_0 + (\partial Q/\partial T)(T_a - T_s)$, where Q_0 contains contributions from the downward SW solar heat flux at the ocean surface and the LW backradiation and latent heat flux that would result (sensible heat flux would be zero), were the air–sea temperature difference $(T_a - T_s)$ to be zero. The second term contains contributions to the backradiation, and latent and sensible heat fluxes per degree excess of air–sea temperature difference. The values for Q_0 and $\partial Q/\partial T$ can be computed from past heat flux observations and are strong functions of latitude, with typical values for Q_0 being ~300 ly day^{-1}, and for $\partial Q/\partial T$, ~90 ly day^{-1} °C^{-1}, at 15° latitude. The values depend on a variety of factors including the prevailing cloud cover and wind stress intensity. However, this form permits long-term climatological-type simulations to be carried out with only the air temperature needing to be prescribed and without the burden of having to carry out the detailed heat flux calculations at the air–sea interface. The Haney (1971) parameterization can also be written as $Q = \gamma(T_a^* - T_s)$, where T_a^* is the apparent equilibrium temperature of the atmosphere (Chu *et al.,* 1998) and not the actual air temperature, and can be computed from climatological heat fluxes.

A form similar to the Haney (1971) parameterization is often used even under synoptic forcing. Since the parameters that influence the heat (and salt) flux, such as the cloud cover and wind stress intensity, are seldom known precisely, forcing an ocean model with fluxes derived from NWP analyses or even from observations can lead to a slow drift in the upper layer temperatures and salinities away from observed values. Even small imbalances in the heat and salt fluxes can have large cumulative effects and result in simulations that drift far from reality. Consequently, to control such drifts, T_A in the Haney (1971) parameterization is replaced by T_R, a reference temperature, which is usually a climatological average value of SST (monthly or seasonal value) for climatological simulations, and several-day composite MCSST value for synoptic simulations: $Q = Q_0 + (\partial Q/\partial T)(T_R - T_S)$. Such a prescription has the effect of damping the model SST toward the reference value and controlling its drift, yet permitting shorter timescale fluctuations to be simulated, the degree of damping being dependent on the chosen value of $\partial Q/\partial T$. An alternative is to simply damp the SST toward the reference value with the damping time constant chosen appropriately.

To summarize, in this chapter, we have dealt with the very broad and very basic outlines of oceanic circulation and processes. After discussing numerical techniques in Chapter 2, we will revisit these topics, but with more specificity and discuss how to obtain the relevant numerical solutions.

Chapter 2

Introduction to Numerical Solutions

Numerical ocean modeling is simply a special case of computational fluid dynamics (CFD) which, in its most general form, involves solutions of a coupled set of nonlinear partial differential equations (PDEs) governing the time-dependent behavior of properties of a fluid flowing in three-dimensional space and acted upon by various forces. Therefore, the governing equations are invariably one form or another of Navier–Stokes equations, supplemented by conservation equations for relevant scalar properties such as fluid temperature, salinity, turbulent energy, phytoplankton, and nutrient concentrations with appropriate source and sink terms. More appropriately, since most practical situations involve high Reynolds number turbulent flows, ensemble-averaged Reynolds-type equations derived from Navier–Stokes equations are almost always used instead. Here lies one of the principal conceptual difficulties in CFD: how to parameterize the unknown turbulent stress and diffusion terms that result, in other words, the turbulence closure problem. Enormous efforts have been expended in trying to find acceptable solutions to this problem [see Chapter 1 of Kantha and Clayson (2000) for a discussion of the problem]. Ignoring this for the moment, the task of solving the resulting PDEs in the most efficient manner possible requires careful attention to the nature of the flow, the available computer resources, and the simplifications that can be made without adversely affecting the solutions sought. While the enormous increase in the raw computing power and concomitant decrease in computing costs over the past few decades have made realistic CFD solutions to many practical flow problems feasible, algorithmic improvements have also contributed significantly. The principal task in CFD is to find the optimum solution technique to the fluid flow on hand, and this requires an intimate knowledge of the nature of the flow under consideration.

The principal differences between CFD in different fields of science, such as aerodynamics, geophysical fluid dynamics, and astrofluid dynamics, are related to the differences in the underlying equations. The fluid can be considered to be incompressible in most geophysical fluid flows. Mach number M, the ratio of typical flow velocity, U, to the speed of sound, c_s, is small (less than 0.002 in the oceans and 0.2 in the atmosphere). However, this simplification is more than counterbalanced by the fact that the fluid is stratified (density is not constant) and hence under the action of Archimedean (buoyancy) forces in a gravitational field. Because of the planetary rotation, the equations are usually written in a rotating (noninertial) frame of reference, giving rise to fictitious body forces that nevertheless have a dominant effect on the flow (Chapter 1). With the exception of very high viscosity flows in the Earth's mantle, most geophysical flows (the atmosphere and the oceans) occur at very high Reynolds numbers and are invariably turbulent.

In contrast, while buoyancy forces and ambient rotation can most often be ignored, compressibility is often the most important aspect of aerodynamic flows, and aerodynamic CFD often involves simulating supersonic/hypersonic (M > 1) flows with as much fidelity as possible in replicating the shock structure. However, the task of reproducing sharp changes across a shock without unphysical oscillations or too much smearing of the gradients is somewhat similar to the task of simulating sharp thermal and salinity fronts in ocean models. The principal advantage aerodynamical CFD enjoys is the feasibility of obtaining accurate verification data through either wind tunnel or in-flight measurements under controlled conditions. Consequently, highly advanced CFD techniques such as 3D unstructured grids, adaptive grids (where the computational grid is fitted to the flow and changed as needed during the computation), and shock-capturing schemes have been routinely used. In contrast, techniques used in the oceans and the atmosphere, especially the former, tend to be rather conservative, simply because one often does not have the luxury of verification data under controlled conditions to assess the efficacy of the schemes. The flows also tend to be more complex and less well understood. CFD in astrofluid dynamics involves even a more complicated set of equations that include not only the effects of rotation, stratification, compressibility, and high Reynolds numbers, but also body forces due to the flow of ionized fluid in a magnetic field (magneto-hydrodynamical forces), self-gravitation of the massive bodies of fluid, forces exerted by electromagnetic radiation, and massive sources of heat generated by fusion. Validation data under controlled conditions are nonexistent.

Ocean modeling therefore bears much resemblance to, and often borrows advanced techniques from, conventional CFD, and it is therefore useful for an ocean modeler to have some familiarity with CFD. Many excellent CFD texts exist, but the two-volume set by Fletcher (Fletcher, 1988a,b) is exemplary. Fletcher (1988a) is an excellent introduction to elementary CFD concepts and techniques, whereas Fletcher (1988b) deals with advanced methods. The two, combined, form an excellent modern reference source for CFD. In contrast, the two-volume set by Hoffman and Chiang (1993) deals with similar subjects, but from a more cookbook-type approach and is therefore useful to beginners. Anderson's (1995) combined treatment of aerodynamics and CFD is useful for one not familiar with aerodynamics (see also Wendt, 1992). Advanced treatments of specialized topics in aerodynamical CFD can be found in Holt (1988) and Peyret and Taylor (1983). Canuto *et al.* (1988) is an excellent reference on spectral methods in CFD. Press *et al.* (1992) is a valuable reference on numerical methods in general. Haltiner and Williams (1980) contains a good description of numerical techniques applied to atmospheric flows. O'Brien (1985) is a collection of papers from a NATO workshop on ocean modeling, and provides a good idea of the state of the subject in the early 1980s.

In analogy with CFD, ocean and atmospheric modeling can be termed "computational ocean dynamics" (COD) and "computational atmosphere dynamics" (CAD), or both can be grouped under "computational geophysical fluid dynamics" (CGFD). However, there is no consensus at present for such terminologies.

2.1 INTRODUCTION

The advent of digital computers in the late fifties ushered in a new era in solving practical fluid flow problems. The advances in computing since then have been phenomenal. Since 1960, the era of the rather primitive IBM704, there has been a roughly six orders of magnitude increase in the speed of the central processing unit (CPU), so that in modern machines such as the CRAY T90, the clock speed of 2 to 3 ns is rapidly approaching the physical limit of about 1 ns for conventional silicon wafer technology (new technologies such as Josephson junctions hold the promise of 10-ps switching speeds). This, combined with parallel processing of CPU instructions, is pushing the peak (theoretical) performance of individual CPUs above the gigaflop (GF, 10^9 floating point operations per second) limit. For example, an individual Cray T90 processor now reaches 1.8 GF. Consequently, a multiple CPU machine such as the 16-processor Cray T90 is now capable of sustained multigigaflop performance, and a CFD problem that would have taken a century to complete on an IBM704 takes less than an hour on a T90. Computing costs have dropped in a similar fashion, so what would now cost a few hundred dollars would have been prohibitively expensive, millions of dollars, three decades ago. The cost of computer memory has also dropped dramatically so that instead of being able to store only a few thousand words in core memory, it is now possible to store close to 1 Gw (~10^9 words), thus permitting large practical flow problems to be solved efficiently. The increased computer speed and memory, and decreased computing costs, combined with algorithmic improvements in solution methods, have made it possible to do truly large scale computations on high performance computers. The individual workstations have also kept pace, and many computing-intensive calculations that would have required high performance computers just a decade ago can be carried out now on modern workstations such as the DEC Alpha and Silicon Graphics desktop machines, or even a modern PC. See Figure 2.1.1 for the evolution of computing since 1955.

2.1.1 ARCHITECTURE

Scalar architecture of the sixties meant that a CPU would carry out one instruction for each clock cycle. Since CPU tasks essentially break down into simple operations such as fetching contents of a memory location into a CPU storage register, performing arithmetic operations on words stored in registers, and restoring the contents of a register into memory, arithmetic operations on each element of an array would take several clock cycles. However, vector (or pipelined) CPUs, such as the Cray-1S of the seventies, carried out several instructions concurrently for each clock cycle, thus operating on several elements of an array at a time, each at a different stage of the arithmetic operation at any given time. Thus, for the same clock speed, the speed of computation of a vector processor would be several times that of a scalar processor. As long as care was taken not to include conditional statements and the like that would force the CPU to process array elements one at a time, and the computations consisted primarily of array operations, faster computations would be possible. However, the physical limit on CPU speed is still the bottleneck, and the performance level at present is still in the GF range for even the fastest vector CPU. Multiple CPUs working on the same problem is the only way to overcome this limitation, and modern vector-processing machines such as the CRAY T90 have therefore several CPUs (16 in this case) and multitasking capability to push the sustained performance to the multi-GF range.

Massively parallel computers with hundreds or even thousands of extremely fast vector CPUs are necessary to breach the teraflop (TF, 10^{12} flops) barrier. These machines are expected to be the high performance computers of choice in the coming century, and operate in either the SIMD (single instruction multiple data) or MIMD (multiple instruction multiple data) mode. In SIMD machines,

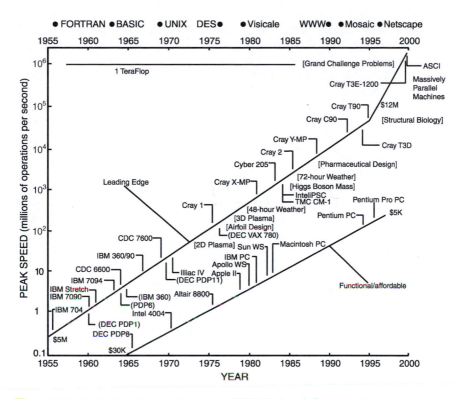

Figure 2.1.1 Evolution of computing since 1955. Both mainframes and workstations are shown. Adapted from Brenner (1996).

the same operation is carried out synchronously on multiple elements of an array, so that during operations on a scalar, most of the CPUs are idle. In MIMD machines, each CPU operates on a different part of the CFD problem, so that dividing the tasks between CPUs (load balancing) and carrying out the sequential operations characteristic of computational methods that may require one CPU to wait until the results are available from another (interprocessor communication) are nontrivial programming tasks (see Haidvogel *et al.*, 1997 for discussion of adaptation of their spectral element model to the parallel processing environment). For machines with distributed (not shared) memory, this also means that a large communication bandwidth between processors is necessary so that the data and computational results can be rapidly transmitted from one CPU to another as needed. Consequently, the increased speed from parallel processing by N CPUs is far below a factor of N, and depends crucially on the nature of the CFD problem, the computer architecture, and the efficiency of the parallel code. Nevertheless, for very large "Grand Challenge"-type computational problems that include realistic ocean and climate models, massively parallel computers are the only solution feasible. CPU speeds are at present slowly nearing the 1 ns (1 GHz) mark, with Cray T3E-1200 having a clock speed of 600 MHz and a single processor peak performance of 1.2 GF. For a 1000 CPU machine, this translates to over 1 TF peak performance. The current top performer is an SGI ASCI (Accelerated Strategic Computation Initiative) machine consisting of 5040 (500 MHz) processors with a peak performance rating of 2.52 TF. This machine has the potential to achieve a sustained performance level of over 1 TF for some selected very large problems (Dongarra 1999; see Van der Steen 1999 for an overview of high performance computers).

Federal programs such as the Accelerated Strategic Computing Initiative (ASCI) have the goal of building a 3- to 4-TF machine by the turn of the century, and reaching a 10-TF capability by the year 2003 or so. It remains to be seen what sustained performance level can be attained by then on practical CFD and ocean modeling problems. Meanwhile, fast CPUs such as the Pentium III that are the workhorses of the PC industry are being put together to produce massively parallel computers with multi-GF capability unheard of even with high performance computers just a decade ago. Modern workstations such as the DEC Alpha now provide a desktop computing capability equivalent to a fraction of a Cray T90 CPU, enabling not only sophisticated pre- and postprocessing for ocean models, but also the solving of CFD problems that would have required a high performance computer just a few years ago. Progress is rapid, and it is hard to anticipate the advances that will be made even on a time scale of a few years.

2.1.2 COMPUTATIONAL ERRORS

Numerical computations involve two types of errors: *round-off errors* and *truncation errors*. The former result from the inability of a computer to represent a floating point number to infinite precision. Thus floating point numbers are inevitably rounded off and represented only approximately in arithmetic operations. Thus, while integer arithmetic is exact on any computer, the accuracy of the floating point operations depends on the word length used and how the number is represented in the machine—what fractions of the word are assigned to the mantissa and the exponent. In 32-bit (word length is 4 bytes) computers of the past, numbers could be represented only to roughly 7 significant decimal digits, but most modern CFD work is carried out on 64-bit machines such as the CRAY T90, CRAY T3E, and DEC Alphas, which permit a number to be represented to about 15 significant decimal digits. Machine accuracy, which is indicative of the precision with which an arithmetic floating point operation can be carried out on a computer, is defined as the lowest number that can be added to 1.0 and still yield an answer other than one. For a 32-bit machine, this is around 10^{-8}, and for a 64-bit machine, around 10^{-16}. However, it is possible to perform double precision (and any arbitrary precision) arithmetic on a machine of any given basic precision, using software instructions, the penalty being a significant slowdown in computations. In any case, round-off errors are unavoidable in any numerical operation on a digital computer, and in sequential operations on an operand they tend to accumulate. One has very little control on the magnitude of the cumulative round-off error, except by the use of expensive higher precision arithmetic that slows down the computations. Many CFD calculations demand 64-bit precision, for example, multiyear runs of ocean models involving keeping track of the salinity of water parcels.

The larger word length of modern CFD computers also means that a larger dynamic range of numbers can be supported. For example, while an ANSI/IEEE 754 standard 32-bit floating point number can range from $\pm 1.2 \times 10^{-38}$ to $\pm 3.4 \times 10^{+38}$ (any number beyond these limits produces the underflow and overflow terminations), 64-bit numbers can be between $\pm 2.2 \times 10^{-308}$ and $\pm 1.8 \times 10^{+308}$. On a Cray vector processor, the numbers can go from roughly 10^{-2467} to 10^{2465}.

Truncation errors arise from the need for discretization in computing on a digital computer. A variable continuous in space can only be represented on a digital computer at preselected discrete points in space. The ordinary and partial derivatives of a variable at such discrete points have to be represented in terms of the values of the variable at that grid point and its neighbors. This is done using Taylor series expansions. The value of a function y(x) at the neighboring grid point j+1 (x = x_0+h, where h is the grid spacing) can be expressed as an infinite series that contains terms involving the function y(x) and its derivatives at the grid point j (x = x_0):

$$y_{j+1} = y_j + \left.\frac{dy}{dx}\right|_j h + \left.\frac{d^2y}{dx^2}\right|_j \frac{h^2}{2!} + \cdots$$

$$(2.1.1)$$

Expansions such as these can be used to derive expressions for any order derivative in terms of the grid point values of the function itself, by truncating the expansions suitably. For example, the first derivative at grid point j (x = x_0) can be written as

$$\left.\frac{dy}{dx}\right|_j = \frac{y_{j+1} - y_j}{h} + \varepsilon$$

$$\varepsilon = \frac{h}{2!} \left.\frac{d^2y}{dx^2}\right|_j + \cdots = o(h)$$

$$(2.1.2)$$

The derivative is now represented in terms of finite differences between the values of the variable at two adjacent grid points. ε is the truncation error, and in this case, it is of order h. This finite difference approximation is therefore called a first order accurate scheme. An nth order accurate scheme would have a truncation error of $o(h^n)$. A finite-difference approximation of any accuracy to any order derivative can be obtained using Taylor series expansions at neighboring grid points. Higher derivatives and higher order accurate schemes naturally require more grid points. For example, a second-order accurate scheme for the first derivative requires values at three successive points and can be written as

$$\left.\frac{dy}{dx}\right|_j = \frac{y_{j+1} - y_{j-1}}{2h} + o(h^2)$$

(2.1.3)

Because of symmetry, y_j does not appear explicitly. Similarly, a second-order accurate scheme can be obtained for the second derivative:

$$\left.\frac{d^2y}{dx^2}\right|_j = \frac{y_{j+1} - 2y_j + y_{j-1}}{h^2} + o(h^2)$$

(2.1.4)

Once again, because of symmetry, a second-order accurate scheme for the second derivative results from the use of only three points, whereas an asymmetric scheme of the same accuracy would have required values at four grid points. Symmetric finite differences such as in Eqs. (2.1.3) and (2.1.4) are called centered differences. Equation (2.1.2) is asymmetric and is called a forward differencing (or a downwind scheme, assuming that the flow is in the direction of increasing x). In contrast, the following first-order accurate scheme is a backward difference (or an upwind scheme):

$$\left.\frac{dy}{dx}\right|_j = \frac{y_j - y_{j-1}}{h} + o(h)$$

(2.1.5)

Higher order accurate schemes can be derived by using more adjacent grid points. For example, a centered fourth-order accurate finite-difference scheme for the first derivative is

$$\left.\frac{dy}{dx}\right|_j = \frac{-y_{j+2} + 8y_{j+1} - 8y_{j-1} + y_{j-2}}{12h} + o(h^4)$$

(2.1.6)

Similar expressions can be derived for partial derivatives of a function of two or more variables using corresponding Taylor series expansions.

Truncation error is present even if the computer were to have infinite precision, that is, zero round-off error. It is also cumulative. A consequence of truncation errors is that CFD calculations that involve PDEs and ODEs are only approximate. In theory, the truncation error is under the control of the modeler, since reducing the grid size (equivalently, increasing the model resolution) or going to a higher order scheme reduces the truncation error. By letting h → 0, the truncation error of any order scheme can be eliminated, in theory. In practice, since doubling the resolution in a numerical model requires an order of magnitude increase in computing time and manyfold increase in memory requirements, reducing truncation errors by increasing the resolution is not always practical. Even if it were possible to make h arbitrarily small to minimize truncation errors, on a finite precision computer, reducing h causes increased round-off errors, since the same calculation will now involve a correspondingly larger number of arithmetic operations. Thus, since the total computational error in any CFD calculation is the sum of round-off and truncation errors, in practice, as h is decreased, the total error decreases initially because of more rapidly decreasing truncation error, up to a point. Any further increase causes the total error to increase, since the increase in round-off error begins to overwhelm any further decrease in the truncation error. In most cases, the choice of the model resolution is dictated by available computer resources, and often it is impractical to perform a convergence test by carrying out calculations at even two different model resolutions.

Using a higher order accurate scheme is often a means of reducing truncation error without increasing model resolution, but it is rare to see CFD calculations go beyond a fourth-order accurate spatial differencing scheme. Most CFD calculations use second-order accurate spatial differences, since specifying boundary conditions then is far simpler than in a fourth-order accurate scheme. Most ocean models are also formulated using first- or second-order spatial differences. The preferred temporal difference scheme is also mostly first-order or second-order accurate. If a lower truncation error is desirable, it is simpler to increase the grid resolution and tolerate the additional computational burden rather than reformulate an existing complex code using higher order differences. However, increasing attention is being given to formulating new models using higher order spatial (Bender, 1996; Haidvogel and Beckmann, 1997) and temporal (Durran, 1991; Holland *et al.,* 1998) differencing schemes.

The choice between the use of a high-order scheme at a particular grid resolution and the use of a low-order scheme at a high grid resolution to reduce truncation errors and increase the accuracy of the calculations involves principally a trade-off between increased complexity and increased computational requirements. Obviously, dynamical considerations dictate the minimum grid resolutions desirable. For example, the need to properly resolve the Rossby radius of deformation determines the grid resolution required in eddy-resolving ocean models. A higher order scheme cannot fill in the missing dynamics and compensate for the deficiencies due to inadequate grid resolution.

However, assuming that the grid chosen enables the dynamical processes to be represented, a higher order scheme can generally represent those processes better. Along the same lines, a reduction in truncation errors by an increase in grid resolution implies greater memory and CPU time requirements, compared to only increased CPU time requirements with a higher order accurate scheme. The vector processor architecture being what it is, CPU time requirements are often not a strong function of the accuracy of the scheme. All this would argue for as high an order scheme as possible to be employed, constrained only by the complexity of the lateral boundary condition prescription. But how high is optimal?

A detailed analysis of the order and resolution considerations in an ocean model can be found in Sanderson (1998). Briefly, the computational cost of a differencing scheme of order n (and hence truncation error $\varepsilon = a\Delta^n$, where Δ is the grid size and a is a proportionality constant dependent on the characteristics of the problem and the scheme, and a chosen model domain) can be written in most cases as $c = b \, n^m \, \Delta^{-d}$, where d is the problem dimension (usually 4 for a time-dependent three-dimensional oceanic process) and b is a proportionality constant dependent on the algorithm. The coefficient m is close to 1 for any optimal implementation. Ignoring the implications of increased round-off errors in a higher resolution model for simplicity, for a desired truncation error ε, the relative computational costs of two differencing schemes of order n_1 and n_2 is

$$\frac{c_2}{c_1} = \left(\frac{n_2}{n_1}\right)^m \left(\frac{\Delta_1}{\Delta_2}\right)^d = \left(\frac{n_2}{n_1}\right)^m \left(\frac{\varepsilon}{a}\right)^{d(n_1^{-1} - n_2^{-1})}$$

(2.1.7)

Clearly, if we ignore the increased complexity of boundary conditions (and in some cases burden of recoding), in terms of the computational burden, a higher order scheme is preferable. Since the product of the truncation error and the computational cost is proportional to $n^m \Delta^{n-d}$, it can be seen that truncation error can be reduced far more quickly by increasing m rather than decreasing Δ, as long as n < d. Sanderson (1998) finds that in general, as long as n < d, grid refinement is not a good strategy to bring down computational errors. Therefore, schemes higher than second order are beneficial to ocean modeling. As a corollary, implementing a scheme of order much higher than d is not very cost effective, the practical limit being perhaps d+1. This means that it is seldom advantageous to go beyond fourth-order differences for oceanic problems. Even then, it might be simpler to difference only the most important terms in the governing equations with a higher order scheme, such as the pressure gradient (McCalpin, 1994) and Coriolis terms. This is the primary reason for the increasing use of third-order time differencing (Durran, 1991) and fourth-order spatial differencing schemes in ocean models (Haidvogel and Beckmann, 1997).

Ocean modelers should be aware of the above limitations to the speed and accuracy of numerical computations.

2.2 ORDINARY DIFFERENTIAL EQUATIONS

While numerical ocean modeling invariably involves solving partial differential equations of one form or another, many of the modeling problems involve solving ordinary differential equations (ODEs). Box models of thermohaline circulation discussed in Section 1.15 and biological ecosystem models without advection and diffusion effects are typical examples. These problems are governed by a set of ordinary differential equations, usually nonlinear and usually coupled. No matter what the order of the individual equations involved, the set can always be reduced to a system of N first-order ordinary differential equations for N dependent variables.

Even during the solution of PDEs, when the spatial discretization is implemented, whether it is through finite differences, finite elements, or semispectral methods (see discussion below), the resulting equations are ODEs in time. Then any of the schemes to be discussed below can be brought to bear on the problem, consistent only with the accuracy desired and memory and CPU requirements. Since in most ocean modeling problems, the evolution of the physical system from a known initial state is of interest, these problems are called initial value problems. Initial value problems involve solving for the future state of the system, namely, the values for the N variables at some finite time t, given the initial state, the values for these variables at t = 0. The solution involves dividing the time interval into a number of small steps and advancing the solution over each time step. The accuracy, efficiency, and stability of the chosen scheme depend naturally on the step size chosen.

The time-honored method for the solution of an initial value problem involving a set of coupled nonlinear first-order ordinary differential equations is the fourth-order Runge–Kutta (RK) method. However, there exist more efficient methods such as the highly efficient adaptive step-size embedded fifth-order RK and the Bulirsch–Stoer (BS) methods. Both these methods make calculations that indicate the magnitude of the truncation error for the adopted step size. This information is then used to decrease the step size at the current step to obtain a predetermined accuracy and optimize the step size for the next step. These are presented and described in Press *et al.* (1992). While we will describe the principles behind these advanced methods briefly, we will present only the conventional RK method and the Adama–Bashforth predictor–corrector schemes, since they suffice for most nonrepetitive applications.

2.2.1 RUNGE–KUTTA METHOD

Consider first a single first-order ordinary differential equation for a single dependent variable y(t) of the form

$$\frac{dy}{dt} = f(y, t)$$

$$(2.2.1)$$

Depending on the functional form of f(y,t), the equation can be linear or nonlinear. The value y (0) is known and the value of y(T) is desired. The interval t = 0 to t = T is divided into N subintervals (n = 1, N+1), and then the problem is the following: if y^n is known, how to determine y^{n+1} efficiently and accurately, the step size being $\Delta t = t_{n+1} - t_n$. The classical explicit (or forward) Euler method involves evaluating the value of the derivative in Eq. (2.2.1) at the initial point t_n only. This first-order accurate method is neither very accurate nor very stable. The fourth-order RK method involves evaluating the derivative at not only the initial point but also twice more at the midpoint and once at the end point t_{n+1} of the interval, and taking a weighted mean of the four values for the derivative to advance the solution from t_n to t_{n+1}:

$$k_1 = f(t_n, y^n)$$

$$k_2 = f(t_n + 0.5\,\Delta t, y^n + 0.5\,\Delta t\,k_1)$$

$$k_3 = f(t_n + 0.5\,\Delta t, y^n + 0.5\,\Delta t\,k_2)$$

$$k_4 = f(t_n + \Delta t, y^n + \Delta t\,k_3)$$

$$y^{n+1} = y^n + \frac{\Delta t}{6}\left(k_1 + 2k_2 + 2k_3 + k_4\right) + o(\Delta t^5)$$

$$(2.2.2)$$

Note that each new estimate of the derivative is based on the previous best estimate. If the function f involves only the independent variable t, the method is equivalent to the classical Simpson's formula for integration of a function f(t). The truncation error in each step is $o(\Delta t^5)$, but the cumulative truncation error over the entire interval is $o(\Delta t^4)$. The method is quite reliable and robust, though not the most efficient. Extension to a system of N equations,

$$\frac{dy_m}{dt} = f_m(y_1, y_2 \cdots, y_m, t); \quad m = 1, ..., N$$

$$(2.2.3)$$

is straightforward [see Section 2.10 for application to the thermohaline catastrophe problem involving a set of two coupled first-order nonlinear ordinary differential equations in Eq. (1.15.14)].

If instead of using an arbitrary predetermined fixed step size, the step size can be controlled adaptively during the solution to obtain desired accuracies, the efficiency of the RK solution can be greatly improved. Adaptive step size control requires an estimate of the truncation error to be obtained during the solution at each time step. This can be achieved by obtaining two solutions at each step, one with full step and the other with two half steps. Since the dependence of the truncation error on the step size is known, the difference between the two solutions, one with resolution h and the other with resolution h/2, is a convenient estimate of the truncation error and enables one to determine what step size should have been taken to obtain a solution with desired predetermined accuracy. If the predetermined accuracy is not met at the current step, the solution can be repeated with the optimum step size determined. If it is exceeded, the step size can be increased to the optimum in the next step. This simple principle of local extrapolation enables one to take small steps where needed but larger ones where the desired accuracy does not warrant small steps, and greatly improves the computational efficiency.

Another method to estimate the truncation error is to use embedded RK formulas, where the same functions can be used to evaluate the solution to two different accuracies. Fehlberg's method (Press *et al.,* 1992) uses the difference between the solutions from a fifth-order RK formula and an embedded fourth-order one to estimate the truncation error, and then use it to adapt the step size to obtain the desired accuracy. Press *et al.* (1992) present this highly efficient RK method and the associated code which should be used if the efficiency of the RK solution is a critical issue.

RK methods are robust and best when the nature of solutions is not known a priori to be well behaved. For well-behaved functions without any singularities in the integration interval, a different class of even more efficient integration methods are available. These are called Bulirsch–Stoer methods (Press *et al.,* 1992), and they involve extrapolation to the limit of zero step size at each time step. This is accomplished by obtaining several solutions at each step but with different step sizes. By treating these solutions as analytical functions of the step size, either a polynomial or a rational function extrapolation can be used to determine what the solution would be if the step size were infinitesimally small. The extrapolation yields both the solution and the associated error estimate, which can then be used to refine the solution by increasing the number of solutions obtained at each step. Obviously, the number of solutions to be obtained at each step, the step sizes to be used, and the extrapolation method all determine the efficiency of the method. The step sizes to be used are determined by a predetermined sequence, but the number of solutions to obtain at each step is determined by a process of iteration, starting with two, where the extrapolated solution is checked for desired error characteristics. If the solution obtained is not satisfactory, another solution is obtained with the next (reduced) step size,

Richardson's extrapolation is applied, and the error reexamined. The process is continued until the desired error estimate is realized, or terminated and the whole procedure repeated with a reduced overall step size. As in the embedded RK method above, adaptive step size control is used to control the overall step size. Press *et al.* (1992) describe this method also, in great detail, and present the associated code, which should be used if one suspects a priori that the solutions might not have singularities in the domain and high efficiency is critical.

A method that has been used successfully in time discretization of geophysical problems is the third-order Adams–Bashforth scheme (Durran, 1991; Haidvogel *et al.,* 1997):

$$\hat{y}^{n+1} = y^n + \frac{\Delta t}{12}\left[23\left(\frac{dy}{dt}\right)^n - 16\left(\frac{dy}{dt}\right)^{n-1} + 5\left(\frac{dy}{dt}\right)^{n-2}\right] + o(\Delta t^4)$$

$$(2.2.4)$$

This method overcomes some of the limitations of second-order RK and other schemes that are routinely used for time differencing of PDEs in ocean and atmosphere models. It can also be used as part of a predictor–corrector scheme, where the predictor step gives a first guess value of y^{n+1} based on Eq. (2.2.4), where the caret indicates the first guess value, which is then used to obtain the updated value for y^{n+1} with a corrector step,

$$y^{n+1} = y^n + \frac{\Delta t}{12}\left[5\left(\frac{d\hat{y}}{dt}\right)^{n+1} + 8\left(\frac{dy}{dt}\right)^n - \left(\frac{dy}{dt}\right)^{n-1}\right] + o(\Delta t^4)$$

$$(2.2.5)$$

thus avoiding the necessity to solve an implicit equation. Repetition of this procedure by iteration to a desired accuracy at each time step further improves the solution. For problems that involve complex expressions for the time derivatives, which is certainly the case for time-dependent multidimensional geophysical problems, these predictor–corrector methods often outperform Bulirsch–Stoer-type schemes (Press *et al.,* 1992). However, they are also very costly and therefore seldom used.

The methods discussed above are "explicit" schemes. When there are two or more dependent variables, there arises the possibility that there is more than one timescale involved in the problem. In that case, the time step must be small enough to resolve the smallest of them, even though accuracy considerations may permit a much larger one, or the scheme can become unstable. The equations are then called stiff, and require "implicit" schemes that will allow larger steps to be taken (consistent only with the accuracy desired) while maintaining stability. Implicit versions of RK and BS schemes are available (see Press *et al.,* 1992). Here we will illustrate the problem using a simple linear first-order equation: $dy/dt = f(y) = -\alpha y$ ($\alpha > 0$). Use of the explicit Euler scheme gives

$$y^{n+1} = y^n + \Delta t (dy / dt)^n = (1 - \alpha \Delta t) y^n \tag{2.2.6}$$

which is unstable for $\Delta t > 2/\alpha$. Since $|y^n|$ increases monotonically with time step n, the cure is simply an implicit scheme, where the derivative is evaluated at time step n+1 instead of n (the backward Euler scheme),

$$y^{n+1} = y^n + \Delta t (dy / dt)^{n+1} \rightarrow y^n / (1 + \alpha \Delta t) \tag{2.2.7}$$

for which y^n decreases monotonically with n, as it should according to the original equation, no matter what the step size is. Implicit schemes are, however, difficult to implement for a general set of equations, but using the first term of the Taylor series expansion for the derivative at time step n+1, the implementation becomes simpler. For example, the above equation can be solved by this semi-implicit method (although not needed in this case),

$$y^{n+1} = y^n + \Delta t \left[f(y)\big|_{y^n} + \left(\frac{\partial f}{\partial y} \right)_{y^n} (y^{n+1} - y^n) \right] \tag{2.2.8}$$

which gives the same result in this case as the fully implicit method.

There may be occasions where the dependent variables are not all known at the initial point, but only a few are, with the remainder known at an end (or intermediate) point. These problems are known as two-point boundary value problems. For example, we may know the phytoplankton concentration at the initial point but not the zooplankton concentration, which may be known only at the final point in time. Such problems are readily solved using the initial value problem approach, by guessing the values of the unknown variables at the initial point and obtaining the solutions at the end point. By a process of iteratively adjusting the guessed values until the solutions for the unknown variables match their known values at the end point, the solution can be completed. This is, however, beyond the scope of this book and the reader is referred to Press *et al.* (1992) for description of, and the software for, efficient solutions to two-point boundary value problems.

2.3 PARTIAL DIFFERENTIAL EQUATIONS

In CFD and numerical ocean modeling, as in many other subjects involving computation of fluid flow fields, we deal primarily with partial differential equations of one form or another. These equations are generally nonlinear [meaning there occur terms involving products of dependent variable(s) and/or their derivatives], and defy analytical solutions; hence the need for numerical methods of solution. For fluid flow problems governed by some form of Navier–Stokes equations, the nonlinear advection terms are the principal source of difficulty. However, these terms involve first-order derivatives and are not the highest order terms in the equations. It is usually the diffusion terms involving second derivatives that are of highest order, and hence it is these terms that determine the nature of the equations and therefore the method suitable for their solution. This can be illustrated by a simple linear second-order PDE for a single dependent variable $\varphi(x,t)$ in two independent variables t and x, of which t could be either space or time:

$$A(x,t)\frac{\partial^2 \varphi}{\partial t^2} + B(x,t)\frac{\partial^2 \varphi}{\partial t \partial x} + C(x,t)\frac{\partial^2 \varphi}{\partial x^2}$$

$$+D(x,t)\frac{\partial \varphi}{\partial t} + E(x,t)\frac{\partial \varphi}{\partial x} + F(x,t)\varphi + G(x,t) = 0$$

$$(2.3.1)$$

The nature of this equation depends only on the terms involving the highest order derivatives, the first three. The rest of the terms do not matter (even if the lower order terms were nonlinear). Mathematically, there exist two cardinal directions in x–t space defined by solutions to the characteristic equation:

$$A\left(\frac{dx}{dt}\right)^2 - B\left(\frac{dx}{dt}\right) + C = 0$$

$$(2.3.2)$$

These directions, called characteristic directions, are given by

$$\left(\frac{dx}{dt}\right) = \frac{B \pm \sqrt{B^2 - 4AC}}{2A}$$

$$(2.3.3)$$

If both characteristics are real and distinct ($B^2 > 4AC$), the equation is hyperbolic; if they are real but coincident ($B^2 = 4AC$), the equation is parabolic; and if they are complex-valued ($B^2 < 4AC$), it is elliptic.

Physically, characteristics define the lines along which information propagates in the physical (x–t) domain. Wave propagation problems are governed by hyperbolic equations which have real characteristics, supersonic flows being a typical example. Steady-state problems governed often by elliptic equations have imaginary characteristics, Stommel and Munk problems being classical examples. The real or complex-valued nature of the characteristics is important to the nature of the problem and therefore the method to be devised to solve it.

For practical flow problems involving more than one dependent variable (say, n) and a set of coupled PDEs, the governing equations can be converted to a set of first-order PDEs,

$$\frac{\partial \Phi}{\partial t} + \mathbf{A}(x, ..., t)\frac{\partial \Phi}{\partial x} + \mathbf{B}(x, ..., t)\frac{\partial \Phi}{\partial y} + ... + P(\Phi, x, ...) = 0$$

(2.3.4)

where Φ is a vector containing the dependent variables, matrices $\mathbf{A}$ and $\mathbf{B}$ are functions of the dependent variables x, y, ..., and time t, and P is another vector. The nature of this problem depends on the eigenvalues of the matrices $\mathbf{A}$ and $\mathbf{B}$. If the eigenvalues of $\mathbf{A}$ are real and distinct, the problem is hyperbolic in x and t—if complex valued then it is elliptic in x and t. Similarly, $\mathbf{B}$ governs the nature of solutions in y–t space.

The subset of Eq. (2.3.4) involving only the independent variables x and y defines a steady-state problem in two dimensions, whose characteristic directions are determined by the solution of the equation (Hoffman and Chiang, 1993)

$$\left| \mathbf{A}(x, y)m_x + \mathbf{B}(x, y)m_y \right| = 0$$

(2.3.5)

which is a nth-order algebraic equation with n solutions for the normals to the characteristic surfaces, that is, the characteristic directions m_y / m_x. If all the characteristics are real, the system is hyperbolic. If they are all complex valued, the system is elliptic.

2.3.1 CONSISTENCY, CONVERGENCE, AND STABILITY

Well-posedness of the problem (as defined by Hadamard) is also quite important. In mathematical problems relevant to CFD, this means that not only the solution must exist and be unique, but it must also depend continuously on auxiliary data. In other words, the governing equations alone do not determine the solution uniquely. It is the equations along with the auxiliary conditions (initial and boundary conditions) that render the solution unique. Numerical problems in general require that consistency, convergence, and stability conditions also be satisfied. This means that the discrete approximation to the governing differential equation must be consistent in the sense that one must be able to recover the original differential equation by letting the discretization intervals go arbitrarily to zero. This ensures that the scheme approximates the differential equation under study, not some other equation. Consistency can be checked by reversing the discretization process by substituting Taylor series expansions for the adjacent grid points and requiring the time step Δt and the grid sizes Δx, Δy, ..., go to zero in an arbitrary fashion. The result should be the original PDE. Usually one uses Taylor series expansions to derive the finite difference approximations to the governing differential equations. Therefore, if the discretization interval is made to go to zero, the truncation error vanishes and one recovers the original differential equation. Therefore, consistency is usually not a problem. Only rarely does one encounter a scheme where that may be a problem, a rare example being the Dufort–Frankel scheme for parabolic equations.

The computational scheme must also be convergent in the sense that as the discretization interval is reduced, the numerical solution should converge to the true solution. Normally, convergence is hard to prove theoretically for complex systems, although straightforward in principle to check empirically by repeating the calculations at different grid sizes (limited only by the increased cumulative round-off errors). Because of the resource requirements, this is seldom done. Therefore, it is hard to assure convergence. However, a theorem by Lax comes to the rescue. The Lax equivalence theorem states that if a scheme is consistent and stable, it is also convergent. For a consistent scheme, it is therefore (necessary and) sufficient to check only the stability of the scheme. The reasoning is as follows. If a scheme is consistent, one can recover the governing PDE by reversing the discretization process that uses Taylor series expansions. Now, if the scheme is also stable, successive refinements of the grid should enable convergence of the finite-difference solution to the exact solution, at least in theory, since truncation errors decrease monotonically to zero. In practice, round-off errors intervene. Nevertheless, one relies on this theorem and seldom explicitly checks convergence by empirical means. Most often, the model resolution is chosen based simply on available computer resources and stability requirements.

2.3.1.1 Stability

The computational algorithm must also be stable and not susceptible to numerical problems arising from accumulation and amplification of round-off errors. Any spontaneous perturbations to the solution must not have a tendency to be amplified at any point during the procedure. Otherwise, inevitable computational errors get amplified and swamp the real solution, making the results meaningless. This often imposes a limit on the time step allowed and decreases the efficiency of the scheme. It is absolutely essential to assure the stability of a discretization scheme. The most commonly used method is the von Neumann scheme. Its applicability is, however, restricted to linear systems and to only the interior points in the domain. Local discretization must be invoked for nonlinear problems.

The method due to von Neumann depends on the fact that since the equations are linear, the error in the solution, namely the difference between the numerical solution and the true solution, is also governed by the same equation as the numerical solution itself. Now if we think of the error as being composed of components with different wavelengths $n\Delta x$, and if we investigate under what conditions all of those components stay bounded in time, we can arrive at a condition for stability of the scheme. L. F. Richardson, who attempted the very first numerical weather prediction during the First World War (Richardson, 1922), used an unstable scheme (and an ill-suited form of governing equations), so that his predictions, though pioneering in every way, were not useful. It was Courant, Friedrichs, and Lewy, who in 1928 derived the time step constraints for the stability of an explicit scheme, and Carl-Gustaf Rossby's work in the 1930s that made it possible to use for atmospheric prediction the vorticity conservation equation (instead of the primitive equation then plagued by inertia-gravity wave problems) that filters out troublesome gravity waves. However, successful numerical weather prediction did not become feasible until the advent of electronic computers in the late 1940s (Richardson had estimated it would take 64,000 human "computers" just to keep up with the weather, let alone forecast it, on a global basis!), when Charney, Fjortoft, and von Neumann used the vorticity conservation equation and the first electronic computer at Princeton (ENIAC) to make the very first successful forecast (albeit for only the geopotential at one level).

In 1D problems governing the temporal evolution of the variable u(x), von Neumann's technique involves substitution of $u_j^n = \xi^n \exp(jk\Delta x), k=1,2,...,$ where ξ $(=\xi^{n+1}\xi^{-n})$ is a complex number that denotes the amplitude of the kth spatial mode, and is called the amplification factor. If ξ stays bounded, that is, $|\xi| \leq 1$ for all possible values of the wavenumber k, computational errors either decay or stay unamplified and the solution is stable. This criterion is quite straightforward to check, although algebraically messy. Note that ξ is in general complex. For a system of equations with more than one dependent variable, as is usual in CFD, von Neumann's method leads to an amplification matrix ξ, instead of a scalar amplification factor. Then it is necessary to seek the eigenvalues λ_m of the matrix, where m is the number of variables. The stability condition becomes $|\lambda_m| \leq 1$, for all m, for all possible values of wavenumber k.

The von Neumann method can be illustrated using the 1D linear advection equation [see Eq. (2.5.3) below]. The exact solution to this equation is $u = f(x - ct)$, where $u = f(x)$ indicates the initial condition. The initial form propagates unchanged as time goes on, in the positive x direction at the speed of advection c. Using a time step of Δt and a grid size of Δx, it is easy to see that the true solution traverses a grid if $c\Delta t = \Delta x$, or if the Courant number, $C = c\Delta t/\Delta x$, is equal to unity. Intuitively, it is clear that if one is seeking numerical solutions, it would be impossible to propagate the solution from grid point to grid point any faster than the true solution itself is propagating. In other words, C cannot exceed

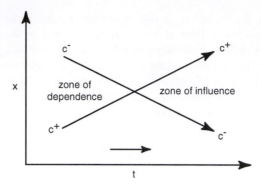

Figure 2.3.1 The zones of dependence and influence in hyperbolic system x-t space.

unity without the solutions becoming unphysical. More generally, the computational domain of dependence for the point under consideration must contain the physical domain of dependence (Figure 2.3.1). In practice, this manifests itself by the solutions becoming unstable to random round-off errors, leading to a failure of the numerical scheme.

Mathematically, this can be shown by considering the solution to be a superposition of Fourier modes of different wavenumbers and ensuring none of the modes amplify with time. Because of the linear nature of the equation, the error also satisfies the same equation, and therefore, this also means that errors in the numerical solution stay bounded. This can be illustrated with the advection equation (2.5.3). Consider a single mode of wavenumber k:

$$u(x,t) = \text{Re}[U(t)\exp(ikx)] \tag{2.3.6}$$

Substitution into Eq. (2.5.3) gives the oscillation equation,

$$\frac{\partial U}{\partial t} + (ikc)U = 0 \tag{2.3.7}$$

the solution of which is $U(t) = \text{Re}[U(0)\exp(-ikct)]$. Combining Eqs. (2.3.6) and (2.3.7), it is easy to see that the solution is the initial form propagating unchanged. Now consider the finite-difference equivalent of Eq. (2.3.6):

$$u_j^n = \text{Re}[U^{(n)}\exp(ijk\Delta x)] \tag{2.3.8}$$

Using centered (leapfrog) time differencing in Eq. (2.3.7),

$$U^{(n+1)} = U^{(n-1)} - iC\sin(k\Delta x)U^{(n)} \qquad (2.3.9)$$

If we define an amplification factor ξ by $U^{(n+1)} \equiv \xi U^{(n)}$, the stability of the numerical solution requires that $|\xi|^n = |U^{(n)}|/|U^{(0)}|$ be bounded, or equivalently $\ln|\xi| \leq o(\Delta t/t)$ or $|\xi| \leq 1 + o(\Delta t/t)$ for small $\Delta t/t$. This is the von Neumann necessary condition for stability. However, in the face of random round-off and other computational errors, the sufficient condition is $|\xi| \leq 1$. Since ξ is generally complex, this essentially means that in the complex (Reξ – Imξ) plane, for stability of the scheme, all possible values of ξ must lie within a circle of unit radius centered at the origin. Equation (2.3.9) gives the amplification factor for the leapfrog-centered spatial difference scheme:

$$\xi = -iC\sin(k\Delta x) \pm \left[1 - \left(\frac{C}{2}\right)^2 \sin^2(k\Delta x)\right]^{1/2}$$
$$(2.3.10)$$

The sufficient condition for stability $|\xi| \leq 1$ is satisfied if $|C\sin(k\Delta x)| \leq 1$ for all admissible values of $k\Delta x$. Since $k\Delta x$ can assume values from 0 to 2π, the maximum possible value of $\sin(k\Delta x)$ is 1, and hence $C \leq 1$ is required for the stability of the scheme. The scheme is conditionally stable.

While the von Neumann method is simple conceptually, in practice, it is valid only in the interior of the domain and does not take into account the influence of the boundary conditions. The matrix method does, and it also depends on expressing the solution at time step n+1 in terms of that at time step n:

$$\mathbf{U}^{(n+1)} = \mathbf{A}\mathbf{U}^{(n)} \qquad (2.3.11)$$

The solution at all grid points in the spatial domain is accounted for. $\mathbf{U}$ is now a vector of length N, where N is the number of grid points, and $\mathbf{A}$ is a N×N matrix. The form for $\mathbf{A}$ can be readily obtained from the difference equations along with the boundary conditions. It is clear that $\mathbf{U}^{(n)} = \mathbf{A}^n\mathbf{U}^{(0)}$, and therefore $\mathbf{A}$ is the amplification matrix. The stability of the scheme depends on the properties of this matrix. For the solution to be bounded, it is sufficient that no eigenvalue of this matrix exceed unit magnitude, i.e., $|\lambda_n| \leq 1$, n = 1, ..., N, where the N eigenvalues are given by roots of the characteristic equation: $|\mathbf{A} - \lambda\mathbf{I}| = 0$. In principle, they can exceed unity by a small value of O($\Delta t/t$), just as in Neumann's analysis above. Also, if the amplifying eigenmode is not present in the initial condition, the solution cannot go unstable. But in practice, round-off errors during the solution make it essential to assure that none of the eigenvalues exceed unit magnitude. The procedure can be illustrated using the diffusion equation [see Eq. (2.5.21) below]. Using the two-level FTCS (forward in time and centered in space) scheme for simplicity,

$$\left(\frac{u_j^{n+1} - u_j^n}{\Delta t}\right) = v\left(\frac{u_{j+1}^n - 2u_j^n + u_{j-1}^n}{\Delta x^2}\right) + o(\Delta x^2, \Delta t)$$

(2.3.12)

the $(N-1) \times (N-1)$ amplification matrix $\mathbf{A}$ becomes

$$A = \begin{bmatrix} 1-2D & D & \cdots & 0 \\ D & 1-2D & & \vdots \\ \vdots & & \cdots & D \\ 0 & \cdots & D & 1-2D \end{bmatrix}$$

(2.3.13)

where $D = v\Delta t/\Delta x^2$ and boundary conditions $u_0^n = u_N^n = 0$ have been used. This matrix is tridiagonal with elements a (= D), b (= 1–2D), and c (=D), and its eigenvalues are given by (Haltiner and Williams, 1980)

$$\lambda_n = b + 2(ac)^{1/2}\cos(n\pi/N) = 1 - 4D\sin^2(n\pi/2N) \quad (2.3.14)$$

All the eigenvalues are bounded if $D \le 1/2$, and therefore the scheme is conditionally stable.

Both the von Neumann and the matrix methods are best suited to linear equations. The third method that is suitable for nonlinear equations also is the energy method, where the sum of the squares of the dependent variable (equivalent to energy) $\sum_j (u_j^n)^2$ is examined for boundedness. If the energy is bounded, then the solution at each individual point u_j^n is also bounded and stability assured. This can be illustrated for the FTUS (forward in time and upstream differencing) of the advection equation (2.5.3), for which this method is simpler, compared to other schemes (Mesinger and Arakawa, 1976; Haltiner and Williams, 1980),

$$\left(\frac{u_j^{n+1} - u_j^n}{\Delta t}\right) + c\left(\frac{u_j^n - u_{j-1}^n}{\Delta x}\right) = o(\Delta x, \Delta t)$$

(2.3.15)

which gives

$$u_j^{n+1} = (1-C)u_j^n + Cu_{j-1}^n$$
$$\sum_j (u_j^{n+1})^2 = (1-C)^2 \sum_j (u_j^n)^2 + C^2 \sum_j (u_{j-1}^n)^2$$
$$+ 2(1-C)C\sum_j (u_j^n)(u_{j-1}^n)$$

(2.3.16)

Using cyclic boundary conditions leads to $\sum_j (u_j^n)^2 = \sum_j (u_{j-1}^n)^2$. This and the use of Schwarz's inequality $\sum ab < (\sum a^2)^{1/2}(\sum b^2)^{1/2}$ in Eq. (2.3.16), which requires that $(1-C)C > 0$, leads to

$$\sum_j (u_j^{n+1})^2 \leq [(1-C)^2 + 2(1-C)C + C^2]\sum_j (u_j^n)^2 = \sum_j (u_j^n)^2$$

(2.3.17)

Thus, as long as the substitution of the inequality holds, the scheme is stable. This of course requires $C < 1$, the CFL condition (described below).

2.3.2 ELLIPTIC, HYPERBOLIC, AND PARABOLIC SYSTEMS

A well-posed problem for an elliptic equation involves proper boundary conditions at the periphery of the physical domain in which solutions are sought. These boundary conditions can be either Dirichlet conditions, where the value of the dependent variable(s) is prescribed on the boundary, or Neumann conditions, where its derivative (usually normal to the boundary) is prescribed. In the latter case, the solution can be determined to an unknown constant of integration. The boundary conditions can also be mixed (or Robin) conditions involving a linear combination of the dependent variable and its derivative. The number of boundary conditions needed depends on the order of the equation. Harmonic equations involving only second-order derivatives need only one condition on the dependent variable. For example, in steady-state fluid flow problems governed by the elliptic Laplace's equation,

$$\nabla^2 \varphi \equiv \left(\frac{\partial^2}{\partial x^2} + \frac{\partial^2}{\partial y^2} \right) \varphi(x, y) = 0$$

(2.3.18)

or the Poisson equation in which the RHS is replaced by $f(x, y)$, it is sufficient to prescribe φ *or* its derivative along the boundary. On the other hand, it is necessary to prescribe two conditions for a biharmonic equation

$$\nabla^4 \varphi \equiv \left(\frac{\partial^4}{\partial x^4} + 2\frac{\partial^2}{\partial x^2}\frac{\partial^2}{\partial y^2} + \frac{\partial^4}{\partial y^4} \right) \varphi(x, y) = f(x, y)$$

(2.3.19)

Both φ *and* its derivative need to be prescribed along the boundary.

Hyperbolic equations involve some sort of propagation with little or no dissipation. A disturbance at a given point in the physical domain travels at a finite speed through the domain along the two characteristics emanating "downstream" from that point, and therefore its effect can only be felt inside the wedge-shaped domain bounded by the two characteristics. Similarly, a point in the physical domain can only be influenced by the wedge-shaped region "upstream" of it, bounded by the two characteristics (Figure 2.3.1; this becomes clearer if one thinks of a supersonic flow in the positive x direction). Because of this directionality, solutions can be obtained from known upstream conditions and step-by-step integration of the governing equations by marching downstream at suitable increments governed by a stability criterion involving the speed of propagation of disturbances (c) in the medium. Hyperbolic problems involving time and space are initial value problems, and solutions require integration, step-by-step, forward in time from known initial conditions, seeking solutions that satisfy the boundary conditions demanded in the physical domain. The time step allowed for solution is limited by the speed of propagation of infinitesimal disturbances (sound speed in supersonic flows). Physically, this means for a chosen grid size, one cannot propagate numerical solutions forward in time faster than a disturbance itself would propagate across the grid: the time step Δt is bounded by $\Delta x/c$, where Δx is the grid size in physical space. Equivalently, the Courant number, $C = c\Delta t / \Delta x$, must be less than unity. More precisely, the computational domain of dependence must contain the physical domain of dependence of the point being time stepped. This is the famous Courant–Friedrichs–Lewy (CFL) condition. It constrains the efficiency of solutions to time marching problems involving PDEs, in explicit time stepping schemes that determine solutions at the next time step $(t + \Delta t)$ one physical point (in x) at a time.

It is possible to circumvent this limitation by using implicit time stepping methods, which however require obtaining solutions simultaneously at all physical points (for all x) at the next time step. Thus, while the time step is unconstrained by stability considerations and limited only by the accuracy of the transient solutions one demands, the solution usually requires "inverting" a large but sparsely populated matrix. One seldom "inverts" very large matrices by brute-force inversion methods such as conventional Gaussian elimination; instead, one uses iterative techniques that start from an initial guess and converge to the correct solution. Which method one uses depends very much on the desire for simplicity and demand for efficiency. Often in problems involving second-order differences of second derivative terms in the equations (problems involving mixing in the vertical, for example), the resulting matrix takes the form of a tridiagonal matrix or something equally simple, for which fast algorithms are available. Otherwise, it is a question of whether a fast matrix solver can be employed and whether the problem permits fast convergence of solutions in the iterative scheme used. It is also a matter of the computer architecture available. Massively parallel computer architecture is inherently less well suited to solving problems involving global dependence, because of the intercommunication required of the different processors. Explicit methods involving only local dependences exploit this architecture more efficiently.

Parabolic problems are typical of problems in which the diffusion of a property in the spatial domain with increase in time is involved. Diffusion of heat and viscous diffusion of momentum and vorticity are classical examples. The tendency of the solutions is to smooth out eventually any initial discontinuities because of the diffusive nature of the physical process (whereas in hyperbolic problems, the lower order nonlinear advection terms can actually form discontinuities even if the initial state did not have any). The steady-state limit of time-dependent problems involving more than one spatial direction and governed by a parabolic PDE reduces to an elliptic equation.

Thus, both parabolic and hyperbolic problems are "time-dependent" problems, whereas the elliptic problem is a "steady-state" problem. Numerically, the solution techniques for parabolic problems are not much different from the ones for hyperbolic problems. Both are time-dependent initial value problems (IVPs), and this is all that matters. However, the methods employed for elliptic problems are much different, since these are steady-state boundary value problems (BVPs), requiring solutions in the entire physical domain subject to some conditions at the boundaries of the domain, and it is usual to seek an iterative solution starting from an initial guess and converging to the true solution. In an elliptic problem, a disturbance at any point in the physical domain is felt instantaneously at all other points in the domain; therefore a solution must be sought simultaneously at all points in the domain, unlike time-dependent parabolic and hyperbolic problems, where it is possible to seek solutions one physical point at a time. Numerically, this means converting the elliptic PDE to a set of algebraic equations and therefore "inverting" a large matrix. In practice, iterative techniques are used to obtain the solution and here once again the efficiency in terms of the speed of convergence of solutions is crucial.

When seeking numerical solutions, the distinction between IVPs and BVPs often becomes blurred, since both can be solved using iterative matrix solvers. In fact it is possible to convert a BVP to an IVP by adding terms involving the time derivative of the dependent variable and integrating to a steady asymptotic state starting from a guessed initial state. This is called the pseudo-transient method. In fact, iterative relaxation methods for a BVP can be thought of as equivalent to iteration in time, in other words an IVP.

2.4 ELLIPTIC EQUATIONS AND STEADY-STATE PROBLEMS

Consider the elliptic equation (2.3.18). Using a centered difference scheme for the derivatives, which results in a second-order accurate scheme,

$$\frac{\partial \varphi}{\partial x} = \frac{\varphi_{i+1,j} - \varphi_{i-1,j}}{2\,\Delta x} + o(\Delta x^2); \quad \frac{\partial^2 \varphi}{\partial x^2} = \frac{\varphi_{i+1,j} - 2\varphi_{i,j} + \varphi_{i-1,j}}{2\,\Delta x^2} + o(\Delta x^2)$$

(2.4.1)

Eq. (2.3.18) reduces to the set of algebraic equations

$$a_{i,j}\varphi_{i+1,j} + b_{i,j}\varphi_{i-1,j} + c_{i,j}\varphi_{i,j+1} + d_{i,j}\varphi_{i,j-1} + e_{i,j}\varphi_{i,j} = f_{i,j} \quad i = 1, I; \; j = 1, J \quad (2.4.2)$$

By substituting $l = j + (i-1)J$; $l = 1, (I \cdot J)$, this set of equations can be converted to a single matrix equation

$$\mathbf{A} \cdot \mathbf{X} = \mathbf{B} \qquad (2.4.3)$$

where $\mathbf{A}$ is the $(N \times N)$ coefficient matrix; $\mathbf{X}$ is a vector of length N, containing the values of the unknown variable φ at all $N = I \cdot J$ grid points starting from $i = 1$, $j = 1$, to $i = I$, $j = J$; and $\mathbf{B}$ is a vector of the same length containing the forcing terms on the RHS of Eq. (2.4.2). The matrix $\mathbf{A}$ is large but sparse, principally tridiagonal but with fringes.

Many CFD problems reduce ultimately to the solution of a matrix equation such as Eq. (2.4.3). An efficient method of "inversion" is therefore of considerable importance, especially since the matrix size is usually very large (for a problem involving a 100×100 grid, the matrix size is $10^4 \times 10^4$). The matrix is however usually sparse, as in the solution of the Laplace equation, and simplifications are therefore possible.

The two principal methods of solution of the set of algebraic equations (2.4.3) are direct methods and iterative methods. Direct methods such as Gauss and Gauss–Jordan elimination are computationally intensive, and for large, dense matrices subject to large round-off errors. They are seldom used in large 3D CFD problems. In those cases where matrices are sparse and banded, special versions of Gauss elimination are extensively used. But in the most general case, the method of choice is usually the iterative method, where the solution is started from an initial guess and relaxed to the true solution using efficient acceleration techniques.

2.4.1 DIRECT SOLVERS

One possible method for solving Eq. (2.4.3) directly is the use of conventional Gauss or Gauss–Jordan elimination procedures that eliminate all the sub- and superdiagonal elements of the matrix by systematic normalization and subtraction. Gaussian forward elimination of subdiagonal elements followed by back substitution is also possible. Press *et al.* (1992) discuss these. The elimination procedure requires at each step division by an element, called the pivot. For the method to be successful and accurate, it is essential to select the largest possible element as the pivot. This often requires rearranging the rows and columns at each step. While exchanging rows to bring the largest element to the desired position does not change the solution, exchanging columns scrambles it. The former is called partial pivoting, and is simple to implement and most often is sufficient. The latter is called full pivoting and requires additional bookkeeping, since the scrambled solution has to be reconstituted properly. Choosing as a pivot that element which would have been the largest if the original equations had been normalized to make their largest coefficient unity is often used and is called implicit pivoting. These methods, however, require a priori knowledge of the RHS vector **B**. See Press *et al.* (1992) for a Gauss–Jordan elimination routine with full pivoting.

2.4.1.1 LU Decomposition

A preferred direct method is LU decomposition, because of the simple nature of solution of the resulting triangular set: substitution. Irrespective of the structure of the matrix **A**, whether sparse (meaning many of the elements are zero), dense (few or no elements are zero), or banded (only elements on and in the vicinity of the diagonal are nonzero), this requires that the matrix **A** be factored into lower triangular **L** and upper triangular **U** matrices: **A** = **L.U**. It follows then that **A.X** = (**L.U**) **X** = **L** (**U.X**) = B, so that one can solve **L.Y** = B first by simple forward substitution, made possible because of the structure of **L**, to find the vector **Y**, and then use **Y** to solve **U.X** = Y and obtain X by simple back substitution made possible by the structure of **U**. In other words, one solves **U.X** = **L**$^{-1}$B. The task is then to perform LU decomposition efficiently. Press *et al.* (1992) list one such algorithm. Let

$$
\begin{bmatrix} a_{11} & a_{12} & a_{13} \\ a_{21} & a_{22} & a_{23} \\ a_{31} & a_{32} & a_{33} \end{bmatrix} = \begin{bmatrix} \alpha_{11} & 0 & 0 \\ \alpha_{21} & \alpha_{22} & 0 \\ \alpha_{31} & \alpha_{32} & \alpha_{33} \end{bmatrix} \begin{bmatrix} \beta_{11} & \beta_{12} & \beta_{13} \\ 0 & \beta_{22} & \beta_{23} \\ 0 & 0 & \beta_{33} \end{bmatrix} \quad (2.4.4)
$$

To find coefficients α_{ij} and β_{ij}, one writes out $N+N^2$ equations:

$$
\begin{aligned}
\alpha_{ii} &= 1 \quad (i=1, \ldots, N) \\
\alpha_{i1}\beta_{1j} + \ldots &= a_{ij} \quad (i=1, \ldots, N; \; j=1, \ldots, N)
\end{aligned} \quad (2.4.5)
$$

These equations are solved readily by rearranging in a particular order, using what is known as Crout's algorithm (Press *et al.*, 1992):
For each j = 1, 2, ..., N:

For each i = 1, 2, ..., j: $\beta_{1j} = a_{1j}$ and $\beta_{ij} = a_{ij} - \sum\limits_{k=1}^{i-1} \alpha_{ik}\beta_{kj}$; i = 2,..., j

For each i = j+1, j+2, ..., N: $\alpha_{ij} = \dfrac{1}{\beta_{jj}}\left[a_{ij} - \sum\limits_{k=1}^{j-1} \alpha_{ik}\beta_{ki} \right]$

Go to next j.

Pivoting, namely, rearranging the rows and columns of a matrix to select the largest possible matrix element as the element to perform normalization of the other elements of the matrix, is essential to Crout's algorithm. However, only partial pivoting, where only rows are interchanged, is adequate. Since interchanging rows of a matrix during its inversion does in no way affect the solution, this is much simpler to implement than full pivoting, which scrambles the solution [see Press *et al.* (1992) for a good discussion of this topic]. Once the elements of **L** and **U** are known, first, vector Y is obtained by forward substitution,

$$y_1 = \frac{b_1}{\alpha_{11}}\,;\; y_i = \frac{1}{\alpha_{ii}}\left[b_i - \sum\limits_{j=1}^{i-1}\alpha_{ij}y_j \right]\; (i = 2, ..., N)$$

(2.4.6)

and then, the unknown vector X is obtained from back substitution:

$$x_N = \frac{y_N}{\beta_{NN}}\,;\; x_i = \frac{1}{\beta_{ii}}\left[y_i - \sum\limits_{j=i+1}^{N}\beta_{ij}x_j \right]\; (i = N - 1, N - 2, ..., 1)$$

(2.4.7)

Both Press *et al.* (1992) and Fletcher (1988a) list FORTRAN codes to perform LU decomposition (LUDCMP and FACT, respectively), followed by forward and backward substitutions (LUBKSB and SOLVE, respectively). Press *et al.* use the more efficient Crout's algorithm for LU decomposition. These implementations are meant for only dense matrices. They are wasteful in terms of both storage requirements and computational time for sparse matrices since zero elements will be stored and operated on. There are implementations specifically designed for sparse matrices to minimize the number of arithmetic operations and storage. Press *et al.* (1992) provide a particularly simple version (TRIDAG) optimized for tridiagonal matrices that combines the factorization and substitution (no pivoting is employed since it is almost always unnecessary).

One advantage of LU decomposition is that it does not require a priori knowledge of the RHS vector B. Therefore, once the matrix is factored, it can be used for efficient repeated solutions for different B vectors, using the substitution code. This also means that for a large set of equations, serious loss of precision caused by accumulation of round-off errors can be corrected and machine precision restored by iterated improvement (Press *et al.,* 1992). Let X_c be the corrected solution, X_w be the erroneous solution, and δX be the error. Since $\mathbf{A}.X_c = B$, multiplying $\mathbf{A}$ and X_w will not yield B, but B+ δB. Therefore, one can solve $\mathbf{A}. \ \delta X = \delta B = (\mathbf{A}.X_w - B)$ by mere substitution to obtain the correction δX needed without the need to factorize matrix $\mathbf{A}$ again. Press *et al.* (1992) provide an algorithm that does this.

2.4.1.2 Thomas Algorithm

The use of three-point finite differences (symmetric second-order accurate schemes) for systems with a single unknown variable, or a finite-element method with linear interpolation, leads to matrices with tridiagonal structures. Use of split-step methods (see Section 2.5) to solve multidimensional problems also generates tridiagonal systems. Line relaxation iterative methods also require solution of tridiagonal matrices (see Section 2.4.2). Hence, efficient algorithms for tridiagonal matrices are essential to CFD. A particularly efficient (but because of its recursive nature, difficult to vectorize and parallelize) algorithm is the Thomas algorithm. It consists of two steps. The first is known as forward sweep, which manipulates the matrix into an upper triangular form with unit elements along the diagonal (the vector B is also modified in the process to B′). This is followed immediately by a backward sweep, which involves simple back substitution to obtain the unknown vector X. It is essentially an optimized Gaussian elimination method that uses recursive relations for solution.

Let $a_i x_{i-1} + b_i x_i + c_i x_{i+1} = d_i$ be the tridiagonal matrix equation. Expanded,

$$\mathbf{A} \cdot X = \begin{bmatrix} b_1 & c_1 & 0 & 0 & . & 0 & 0 & 0 \\ a_2 & b_2 & c_2 & 0 & . & 0 & 0 & 0 \\ 0 & a_3 & b_3 & c_3 & . & 0 & 0 & 0 \\ . & . & . & . & . & . & . & . \\ . & . & . & . & . & a_{N-1} & b_{N-1} & c_{N-1} \\ . & . & . & . & . & 0 & a_N & b_N \end{bmatrix} \begin{bmatrix} x_1 \\ x_2 \\ x_3 \\ . \\ x_{n-1} \\ x_N \end{bmatrix} = \begin{bmatrix} d_1 \\ d_2 \\ d_3 \\ . \\ d_{N-1} \\ d_N \end{bmatrix} = D \quad (2.4.8)$$

The matrix $\mathbf{A}$ is particularly sparse and only the diagonal, subdiagonal, and superdiagonal are populated. There is no need to store the N×N matrix in its entirety; only three vectors a_i, b_i, and c_i of length N need to be stored. And the solution takes O(N) operations instead of O $(N^3/3)$ for the general LU decomposition.

The Thomas algorithm uses the regression relationship

$$x_i = \alpha_i - \frac{c_i}{\beta_i} x_{i+1}$$

$$(2.4.9)$$

where the coefficients α_i, β_i are to be determined during the forward sweep. Changing i to i–1 in Eq. (2.4.9) and substituting into the ith equation, $a_i x_{i-1} + b_i x_i + c_i x_{i+1} = d_i$, and rearranging gives

$$x_i = \left(\frac{d_i - a_i \alpha_{i-1}}{b_i - (a_i c_{i-1} / \beta_{i-1})} \right) - \left(\frac{c_i}{b_i - (a_i c_{i-1} / \beta_{i-1})} \right) x_{i+1}$$

(2.4.10)

so that

$$\alpha_i = \frac{d_i - a_i \alpha_{i-1}}{\beta_i} ; \quad \beta_i = b_i - \frac{a_i c_{i-1}}{\beta_{i-1}} \quad (i = 2, ..., N)$$

(2.4.11)

These are recursive relations that require the values of α_1 and β_1, which can be determined from the first equation $b_1 x_1 + c_1 x_2 = d_1$:

$$\alpha_1 = d_1/\beta_1 \text{ and } \beta_1 = b_1$$

(2.4.12)

This completes the forward sweep. All the coefficients in the recursive relationship (2.4.9) are known. The backward sweep starting from element N provides the solution sought. Substitution of Eq. (2.4.9) into the last equation, $a_N x_{N-1} + b_N x_N = d_N$, gives

$$x_N = \left(\frac{d_N - a_N x_{N-1}}{b_N} \right) = \left(\frac{d_N - a_N (\alpha_{N-1} - c_{N-1} x_N / \beta_{N-1})}{b_N} \right)$$

(2.4.13)

so that

$$x_N = \left(\frac{d_N - a_N \alpha_{N-1}}{\beta_N} \right) = \alpha_N$$

(2.4.14)

Knowing x_N from Eq. (2.4.14), x_i for i = N–1, N–2, ..., 1, can be determined from Eq. (2.4.9), since the coefficients are known from Eqs. (2.4.11) and (2.4.12).

As in all Gaussian elimination techniques, serious round-off errors could result from ill conditioning of the matrix. To avoid this, it is necessary that $|b_i| > |a_i| + |c_i|$. The Thomas algorithm is particularly simple to code; Fletcher (1988a) lists one.

Use of higher order accurate schemes for systems involving a single unknown variable generate matrix equations with bandwidth larger than tridiagonal ones. However, the Thomas algorithm can be generalized to deal with these matrices, which are still sparse. Fletcher (1988a) discusses and lists a generalized Thomas algorithm for pentadiagonal matrices. If there is more than one unknown variable and a system of equations instead of just one, the solution procedure often leads to a block tridiagonal matrix in which instead of scalars, M×M submatrices form the diagonal, subdiagonal, and superdiagonal elements of a tridiagonal matrix. The Thomas algorithm can be generalized to this situation as well (Fletcher, 1988a) and requires $O(NM^3)$ operations.

For elliptic equations solved on simple uniform rectangular grids, the matrix elements are constants, and in this case direct solvers making use of a procedure called cyclic reduction followed by Fourier series representation are quite efficient, since modern fast Fourier transform (FFT) techniques can be brought to bear. These are, however, beyond our scope here and the reader is referred to Fletcher (1988a) and Press *et al.* (1992) for a discussion of these methods.

2.4.2 ITERATIVE SOLVERS AND RELAXATION METHODS

For very large sparse matrices characteristic of CFD and ocean modeling problems, the most common method is the iterative or relaxation method. It consists of splitting the matrix $\mathbf{A}$ in Eq. (2.4.3) into two parts, one of which is easily invertible ($\mathbf{E}$) and the other is the remainder ($\mathbf{F}$): $\mathbf{A} = \mathbf{E} + \mathbf{F}$. The equation becomes

$$\mathbf{E} \cdot \mathbf{X} = -\mathbf{F} \cdot \mathbf{X} + \mathbf{B} \qquad (2.4.15)$$

Relaxation methods involve successive iterations of the form

$$\mathbf{X}^{n+1} = -\mathbf{E}^{-1}\mathbf{F} \cdot \mathbf{X}^n + \mathbf{E}^{-1}\mathbf{B} = \mathbf{M} \cdot \mathbf{X}^n + \mathbf{N} \qquad (2.4.16)$$

or equivalently

$$\mathbf{X}^{n+1} = \mathbf{X}^n + \mathbf{E}^{-1}\mathbf{R}^n; \quad \mathbf{R}^n = \mathbf{A}\mathbf{X}^n - \mathbf{B} \qquad (2.4.17)$$

starting from an initial guess $\mathbf{X}^1$. $\mathbf{R}^n$ is the vector of residuals at the iteration step n. The norm of this tends to zero as the solution converges. $\mathbf{M} = \mathbf{E}^{-1}\mathbf{F}$ is called the iteration matrix. The speed of convergence depends very much on the property of this matrix, specifically its spectral radius σ, which is the magnitude of the largest eigenvalue λ_j of the matrix. The eigenvalues are the roots of the algebraic equation obtained by zeroing the determinant,

$$|\mathbf{M} - \lambda\mathbf{I}| = 0 \qquad (2.4.18)$$

where $\mathbf{I}$ is the identity matrix. Then $\sigma = \max_j |\lambda_j|$; $j = 1, I - J$. For real, symmetric matrices, all eigenvalues are real.

The magnitude of the spectral radius of $\mathbf{M}$ determines the convergence properties. If $\sigma < 1$, convergence is assured. This can be seen by writing the error $\mathbf{e}^n = \mathbf{X}_t - \mathbf{X}^n$. The error is then governed by the equation $\mathbf{e}^{n+1} = \mathbf{M} \cdot \mathbf{e}^n = \mathbf{M}^n \cdot \mathbf{e}^0$ so that the error in the initial guess $\mathbf{e}^0$ approaches zero and the solution converges if $\mathbf{M}^n$ approaches zero. This is assured if the spectral radius is less than unity. Often this translates to the requirement that matrix $\mathbf{A}$ be diagonally dominant. The smaller the value of the spectral radius, the faster the convergence.

2.4.2.1 Jacobi and Gauss–Seidel Methods

For illustrative purposes, consider the Laplace equation, and for convenience, assume the same grid sizes in the x and y directions. Then Eq. (2.4.2) becomes

$$\varphi_{i+1,j} + \varphi_{i-1,j} + \varphi_{i,j+1} + \varphi_{i,j-1} - 4\varphi_{i,j} = 0 \qquad (2.4.19)$$

The classical point Jacobi (or Richardson) method consists of the iteration scheme

$$\varphi_{i,j}^{n+1} = \varphi_{i,j}^n + 0.25(\varphi_{i+1,j}^n + \varphi_{i-1,j}^n + \varphi_{i,j+1}^n + \varphi_{i,j-1}^n - 4\varphi_{i,j}^n)$$

$$(2.4.20)$$

If the corresponding matrix $\mathbf{A}$ can be decomposed into a strictly lower triangular matrix $\mathbf{L}$ (with only elements below the diagonal nonzero), a diagonal matrix $\mathbf{D}$ (with only elements along the diagonal nonzero) and a strictly upper triangular matrix $\mathbf{U}$ (with only elements above the diagonal nonzero), then the Jacobi method is equivalent to

$$\mathbf{X}^{n+1} = -\mathbf{D}^{-1}(\mathbf{L}+\mathbf{U}) \cdot \mathbf{X}^n + \mathbf{D}^{-1}\mathbf{B} \qquad (2.4.21)$$

or equivalently

$$x_j^{n+1} = \frac{1}{A_{jj}}\left(B_j - \sum_{\substack{k \\ k \neq j}}^{N} A_{jk}x_k^n \right)$$

$$(2.4.22)$$

For Dirichlet boundary conditions, the spectral radius of the iteration matrix $\mathbf{M} = \mathbf{D}^{-1}(\mathbf{L}+\mathbf{U})$ is, for the general case of nonequal grid sizes,

$$\sigma_J = \frac{\cos(\pi/I) + (\Delta x/\Delta y)^2 \cos(\pi/J)}{1 + (\Delta x/\Delta y)^2}$$

$$(2.4.23)$$

where Δx and Δy are grid sizes. This value is usually close to unity even though smaller than unity; hence, convergence in the Jacobi method is impractically slow. For I, J »1, assuming $\Delta x = \Delta y$, $\sigma_J \sim 1- (\pi^2/2N)$, where $N = I\,J$, the total number of grid points. The convergence rate, defined as the number of iterations needed to reduce the error by an order of magnitude, is $-(\log \sigma_J)^{-1}$ and is therefore ~0.5N. Since each iteration takes 5N operations, it takes $2.5N^2$ operations to reduce the error by an order of magnitude.

The weighted Jacobi method, a simple variant of Eq. (2.4.20), is given by

$$\mathbf{X}^{n+1} = [(1-\omega)\mathbf{I} - \omega\mathbf{D}^{-1}(\mathbf{L} + \mathbf{U})] \cdot \mathbf{X}^n + \omega\mathbf{D}^{-1}\mathbf{B}$$

$$(2.4.24)$$

where ω can be chosen to accelerate the convergence somewhat.

The point Gauss–Seidel (or Liebmann) method involves a slight variation from Jacobi method,

$$\varphi_{i,j}^{n+1} = \varphi_{i,j}^n + 0.25(\varphi_{i+1,j}^n + \varphi_{i-1,j}^{n+1} + \varphi_{i,j+1}^n + \varphi_{i,j-1}^{n+1} - 4\varphi_{i,j}^n)$$

$$(2.4.25)$$

equivalent to

$$\mathbf{X}^{n+1} = -(\mathbf{L}+\mathbf{D})^{-1}\mathbf{U} \cdot \mathbf{X}^n + (\mathbf{L}+\mathbf{D})^{-1}\mathbf{B} \qquad (2.4.26)$$

or

$$x_j^{n+1} = \frac{1}{A_{jj}}\left(B_j - \sum_{k=1}^{j-1} A_{jk}x_k^{n+1} - \sum_{k=j+1}^{N} A_{jk}x_k^n\right)$$

$$(2.4.27)$$

The iteration matrix $\mathbf{M} = (\mathbf{L+D})^{-1}\mathbf{U}$ has a spectral radius equal to the square of the Jacobi spectral radius. Therefore, convergence is twice as fast but still impractically slow. Convergence is guaranteed for both the Jacobi and the Gauss–Seidel methods.

Relaxation methods can be point relaxation methods as above, where grid points are updated one at a time at each iteration cycle. Using point Gauss–Seidel relaxation as an example, in Eq. (2.4.25) the only unknown point is (i,j) on the LHS. This is therefore a point iterative method. Thinking of the grid points as if arranged on a checkerboard, it is easy to see that the solution at the red (even) points depends only on the black (odd) points and vice versa. Thus, it is possible to update the red and black set of points alternately, by updating even or odd points in each halfsweep of every relaxation cycle. This red–black point relaxation (or odd–even ordering) is the one most used.

Relaxation can also be line relaxation, where the points are relaxed an entire row or column at a time. Now if Eq. (2.4.25) is rewritten slightly as

$$-0.25\varphi_{i-1,j}^{n+1} + \varphi_{i,j}^{n+1} - 0.25\varphi_{i+1,j}^{n+1} = \varphi_{i,j}^n + 0.25(\varphi_{i,j+1}^n + \varphi_{i,j-1}^{n+1} - 4\varphi_{i,j}^n)$$

a solution can be obtained at all i points for each j, i.e., row by row, by solving the tridiagonal matrix that results. Similarly, the solution can also be obtained column by column. This is the line Gauss–Seidel relaxation and is twice as fast as the point Gauss–Seidel relaxation; the convergence rate is considerably improved by bringing in information at each step of the solution from points other than the immediate neighbors.

2.4.2.2 Successive Overrelaxation (SOR)

If one recognizes the fact that these relaxation methods essentially add a correction to the previous solution, it is easy to understand how the venerable successive overrelaxation (SOR) method works. SOR has been the workhorse of elliptic solvers until replaced recently by modern multigrid and conjugate gradient methods. The advantage of SOR shows up when using the point Gauss–Seidel method, where instead of the correction term at each iteration step, one overcorrects,

$$\varphi_{i,j}^{n+1} = \varphi_{i,j}^{n} + 0.25\alpha(\varphi_{i+1,j}^{n} + \varphi_{i-1,j}^{n+1} + \varphi_{i,j+1}^{n} + \varphi_{i,j-1}^{n+1} - 4\varphi_{i,j}^{n})$$

(2.4.28)

equivalent to

$$\mathbf{X}^{n+1} = \mathbf{X}^{n} - \alpha(\mathbf{L}+\mathbf{D})^{-1}\mathbf{R}^{n};\ \mathbf{R}^{n} = \mathbf{A}\mathbf{X}^{n} - \mathbf{B}$$

(2.4.29)

or

$$x_j^{n+1} = \frac{\alpha}{A_{jj}}\left(B_j - \sum_{k=1}^{j-1} A_{jk}x_k^{n+1} - \sum_{k=j+1}^{N} A_{jk}x_k^{n}\right) + (1-\alpha)x_j^{n}$$

(2.4.30)

This is the weighted average of the Gauss–Seidel value and the previous value. α is the relaxation parameter, and for convergence, $0<\alpha<2$. Overrelaxation, $\alpha>1$, is necessary for faster convergence than the usual Gauss–Seidel method. The optimal choice for the relaxation parameter is $\alpha = 2\left(1+\sqrt{1-\sigma_j^2}\right)^{-1}$, for which the spectral radius is $\sigma = \sigma_j^2\left(1+\sqrt{1-\sigma_j^2}\right)^{-2}$ (Press *et al.,* 1992). Note that this is only asymptotically valid and is attained only after $O(J)$ iterations. In initial stages, the error actually grows. For the Dirichlet example used in the Jacobi and Gauss–Seidel methods, the number of iterations needed to reduce the error by an order of magnitude is $\sim 0.33\ N^{1/2}$ for $\Delta x = \Delta y$ and $I = J = N^{1/2}$. Obviously, the choice of this parameter is critical, since the rate of convergence is best in a narrow window around this optimum and deteriorates rapidly away from this window. However, for a general problem, it is difficult to determine the optimum value a priori and some experimentation is essential. Point and line, and Jacobi and Gauss–Seidel relaxations can all be overrelaxed, but the optimum value of the relaxation parameter is 1 for the point Jacobi method. In other words, the regular Jacobi method is as optimal as it gets.

It is possible to choose a different value of α at each iteration step, since the above optimal value is not optimal in the beginning. When α is kept constant, the SOR method is called stationary overrelaxation. When it is changed from iteration to iteration, it is called nonstationary overrelaxation Using a particular sequence for α during iteration accelerates the convergence significantly. One such method is called the Chebyshev acceleration. Press *et al.* (1992) describe this method which changes the α in each halfsweep in a way that assures that the norm of the error decreases monotonically. The sequence of α values start from 1, ending up with the optimum value: $\alpha_0 = 1$, $\alpha_1 = (1 - 0.5\sigma_J^2)^{-1}$, ..., $\alpha_{I+1} = (1 - 0.25\sigma_J^2 \alpha_I)^{-1}$, ..., $\alpha_\infty = \alpha_{opt}$. A slight modification of SOR called symmetric SOR, where matrix $\mathbf{E}$ in Eq. (2.4.17) is $(\alpha\mathbf{L}+\mathbf{D})\mathbf{D}^{-1}(\mathbf{U}+\mathbf{D})/(2\alpha-\alpha^2)$, allows the SOR method to be accelerated by Chebyshev acceleration (Fletcher, 1988a). SOR can also be either point SOR or line SOR, the latter being $\sqrt{2}$ faster than the former at the same value of the optimal relaxation parameter (Hirsch, 1988).

An important issue is the ease with which a relaxation scheme can be vectorized and parallelized on modern high performance computers. For example, a regular line Jacobi relaxation is easier to vectorize than the line Gauss–Sidel SOR, and therefore, even though less efficient, it may perform better. Just like the red–black scheme in point relaxation methods, Jacobi line relaxation can be done on alternating lines during successive sweeps. This is called zebra line relaxation and can be done by rows or columns. Zebra line relaxation is well suited to modern computers.

2.4.3 PRECONDITIONED CONJUGATE GRADIENT METHOD

Some means of accelerating convergence is desirable when using SOR. As seen before, the rate of convergence depends on the eigenvalues λ of the iteration matrix. Each eigenvalue denotes the factor by which that particular eigenmode of the error is reduced in one iteration. Naturally, the one with the largest modulus determines the convergence rate. The ratio $|\lambda_{max}|/|\lambda_{min}|$, the condition number of the matrix, is indicative of the bandwidth of the eigenvalue spectrum. If it is much larger than unity, the eigenvalues close to λ_{max} dominate and slow the convergence rate. One way to improve convergence is to precondition the matrix by multiplying it with a preconditioning matrix, so that the preconditioned matrix has an eigenvalue spectrum with a smaller condition number, and smallest possible value for λ_{max}.

Another technique to accelerate convergence is the conjugate gradient method. It involves minimizing the function

$$f(\mathbf{x}) = \frac{1}{2}\mathbf{x}^T \cdot \mathbf{A} \cdot \mathbf{x} - \mathbf{x}^T \cdot \mathbf{B}$$

$$(2.4.32)$$

which is equivalent to equating its gradient to zero:

$$\nabla f = \mathbf{A} \cdot \mathbf{x} - \mathbf{B} = 0 \tag{2.4.33}$$

which is the same as solving Eq. (2.4.3). Thus, the method is reduced to minimizing a function in n-dimensional space. This method is called the conjugate gradient method and is applicable to only symmetric, positive-definite matrices ($\mathbf{z}^T\mathbf{A}\mathbf{z} > 0$ for any $\mathbf{z}$), as is the usual case for elliptic equations. The method consists of the steps (Fletcher, 1988a)

1. $\mathbf{x}^{n+1} = \mathbf{x}^n + \lambda^n \mathbf{P}^n; \quad \mathbf{R}^{n+1} = \mathbf{R}^n - \lambda^n \mathbf{U}^n$

2. $\sigma^{n+1} = (\mathbf{R}^{n+1})^T \mathbf{R}^{n+1}; \quad \alpha^{n+1} = \sigma^{n+1} / \sigma^n$

3. $\mathbf{P}^{n+1} = \mathbf{R}^{n+1} + \alpha^{n+1} \mathbf{P}^n; \quad \mathbf{U}^{n+1} = \mathbf{A} \cdot \mathbf{P}^{n+1}$

4. $\lambda^{n+1} = \sigma^{n+1} / \left[(\mathbf{P}^{n+1})^T \mathbf{U}^{n+1} \right]$

$$(2.4.34)$$

where $\mathbf{P}^n$ constitutes a search direction (which is initially equal to the initial residual $\mathbf{R}^1$). The choice of σ^{n+1} in Eq. (2.4.34) ensures that these search directions are mutually orthogonal. The choice of λ^n in Eq. (2.4.34) ensures that the function (2.4.32) is minimized in the $\mathbf{P}^n$ direction. At each step, the method tends to eliminate contribution of an eigenvector in the series expansion of the residual R based on eigenfunctions of A. In the absence of round-off errors, this method produces a robust algorithm that takes no more than N iterations. If some eigenvalues of A are bunched together, it takes even less. However, round-off errors tend to reduce the effectiveness of this technique.

If the equations are rewritten so that matrix $\mathbf{A}$ is transformed to a matrix close to an identity matrix, this can be avoided. This is called preconditioning, and preconditioned conjugate gradient methods are a very efficient means of solving finite-difference equations resulting from PDEs (Fletcher, 1988a; Press *et al.,* 1992). Preconditioning reduces the spread of the eigenvalues so that the convergence is faster. The method is equivalent to solving instead of Eq. (2.4.3),

$$\mathbf{A}' \cdot \mathbf{x}' = \mathbf{B}' \tag{2.4.35}$$

where

$$\mathbf{A'} = \mathbf{WE}^{-1}\mathbf{AW}^{-1}; \ \mathbf{x'} = \mathbf{W} \cdot \mathbf{x}; \ \mathbf{B'} = \mathbf{WE}^{-1}\mathbf{B} \qquad (2.4.36)$$

The matrix $\mathbf{W}$ is chosen so that $\mathbf{W}^T\mathbf{W} = \mathbf{E}$. Different choices of $\mathbf{E}$ and hence $\mathbf{W}$ give rise to different schemes. The preconditioned conjugate gradient scheme then boils down to

1. $\mathbf{x}^{n+1} = \mathbf{x}^n + \lambda^n \mathbf{P}^n$

2. $\mathbf{S}^{n+1} = \mathbf{S}^n - \lambda^n \mathbf{U}^n; \hat{\mathbf{R}}^{n+1} = \mathbf{WS}^{n+1} = \mathbf{WE}^{-1}\mathbf{R}^{n+1}$

3. $\sigma^{n+1} = (\hat{\mathbf{R}}^{n+1})^T \hat{\mathbf{R}}^{n+1}; \ \alpha^{n+1} = \sigma^{n+1} / \sigma^n$

4. $\mathbf{P}^{n+1} = \mathbf{S}^{n+1} + \alpha^{n+1}\mathbf{P}^n; \ \hat{\mathbf{P}}^{n+1} = \mathbf{WP}^{n+1}$

5. $\mathbf{U}^{n+1} = \mathbf{E}^{-1}\mathbf{A} \cdot \mathbf{P}^{n+1}; \ \hat{\mathbf{U}}^{n+1} = \mathbf{WU}^{n+1}$

6. $\lambda^{n+1} = \sigma^{n+1} \big/ \left[(\hat{\mathbf{P}}^{n+1})^T \hat{\mathbf{U}}^{n+1} \right]$

$$(2.4.37)$$

where $\mathbf{S}$ and $\hat{\mathbf{R}}$ represent pseudo-residuals. As before, convergence is indicated by the value of σ^{n+1}.

2.4.4 MULTIGRID METHODS

It is possible to accelerate the solution beyond what is possible by SOR, using modern multigrid methods. Relaxation methods can be thought of as essentially methods that iteratively damp the error field resulting from the incorrect initial guess. The rate of convergence depends therefore on how exactly the error is damped by the relaxation technique. It appears that relaxation is quite efficient in attenuating the high wavenumber components of the error field. A few iterations usually suffice to attenuate wavenumber components close to the grid resolution wavenumber $(2\Delta x)^{-1}$. However, the low wavenumbers get attenuated very slowly, and after a few iterations, further decrease of total error is consequently slow. Multigrid methods accelerate the convergence of the relaxation methods by a clever exploitation of the fact that high frequency components of error are quickly eliminated during relaxation, and that a low frequency error at one grid resolution appears to be a high frequency error to a coarser grid. Instead of solving the problem on a single grid, they solve it successively on multiple grids with grid resolutions that are multiples of each other.

The multigrid technique can be illustrated using a simple 1D boundary value problem (Briggs, 1987):

$$\left(\frac{d^2}{ds^2} - \sigma\right)x(s) = b(s); \quad 0 \le s \le 1$$

$$x(0) = x(1) = 0$$

$$(2.4.38)$$

Using the centered second-order accurate scheme, the corresponding finite-difference equation becomes

$$x_{j-1} - (2 + \sigma h^2)x_j + x_{j+1} = h^2 b_j \qquad (1 \le j \le n)$$

$$x_1 = x_n = 0$$

$$(2.4.39)$$

Equation (2.4.39) is of course a $(n-2 \times n-2)$ tridiagonal system for solutions at interior grid points $j = 2, n-1$, and can be solved efficiently using, for example, the Thomas algorithm. More generally, the system of linear equations resulting from a finite-difference approximation to a multidimensional elliptic problem is of the form given by Eq. (2.4.3). If y_j denotes an approximate solution, and $e_j = x_j - y_j$, the error, the error is governed by the residual equation $\mathbf{A} \cdot \mathbf{E} = \mathbf{R} = \mathbf{B} - \mathbf{A} \cdot \mathbf{Y}$. Without loss of generality, for illustrative purposes, one can put $\mathbf{B} = 0$. Then the exact solution $\mathbf{X} = 0$, and the error is simply $\mathbf{E} = -\mathbf{Y}$.

Consider an initial guess for y (or equivalently error) consisting of different wavenumbers m =1 to n:

$$y_j = \sum_{m=1}^{n} \sin\left(2\pi \frac{m}{n} j\right)$$

$$(2.4.40)$$

The application of, say, the weighted Jacobi method for 100 iterations shows that the higher the wavenumber, the more rapid the attenuation. The wavenumber component n is attenuated quite rapidly, while wavenumber 1 decays only very slowly. Also while the first few iterations bring down the total error considerably, principally due to suppression of high wavenumber components, further decrease of error is slow. High frequency components are quickly eliminated, but low frequency components remain. Thus, after the first few iterations, convergence slows down dramatically. The slower elimination of low frequency components of error is what slows convergence in all iterative schemes. Relaxation methods are efficient smoothers. But while they are effective in suppressing noisy components in the initial error, they leave the smoother components essentially unchanged. The rate of convergence is not much affected by weighting factors, and is actually worsened by any decrease in grid spacing h. Thus, attempts to obtain more accurate solutions by refining the grid only worsen the situation.

The guiding principle behind multigrid methods is the recognition that during relaxation, oscillatory components (m » n/2) are damped strongly, while smooth components (m « n/2) are not (Brandt, 1977; Briggs, 1987; Hackbusch, 1985; McCormick, 1987). Assume that the oscillatory components of error have been eliminated on a particular grid by a few relaxation sweeps. Since smooth components remain relatively unchanged, there is little advantage to iterating further. But since what appears to be a smooth component to a fine grid appears to be oscillatory to a coarser grid, relaxation on a coarser grid instead would eliminate these, some at least, more rapidly. Thus, it is advantageous to move to a coarser grid, when relaxation on the current grid stalls, and carry out relaxation sweeps there.

The process can start from the grid on which the solution is desired. After a few iterations, when the convergence slows, the residual is transferred to a coarser grid. Since the low frequency components of the error on the fine grid appear as though they are of higher frequency to the coarser grid, a few iterations on the coarser grid are more efficient at attenuating some of these error components than many more continued iterations on the fine grid itself. This improved residual estimate (from relaxation of the residual equation) is then transferred back to the fine grid to obtain a better error estimate and hence a more accurate solution. This process of successive solutions on the fine and coarse grids can be continued to convergence. This approach is called *coarse grid correction*. This process can also be continued until the coarsest possible grid (2×2) is reached or can be stopped at some intermediate point.

Another strategy is to start from a coarse grid, make relaxation sweeps there, and use the solution as an initial guess on a finer grid and continue going to finer grids until the desired resolution is reached, repeating the entire cycle until convergence. This is the *nested iteration*. Both strategies require ways of correctly and efficiently transferring information between grids, coarse to fine and fine to coarse, the former called prolongation and the latter called restriction. A simple linear interpolation often works best for prolongation, and simple injection where the coarse point just assumes the value of the corresponding fine grid point works well (although a weighted average of surrounding fine grid points also works well) for restriction. A simple two-grid coarse grid correction scheme follows (repeated to convergence):

Relax on $A^h X^h = B^h$ on fine grid Ω^h p times with initial guess Y^h

    Compute $R^{2h} = I_h^{2h} R^h$ the residual on coarse grid Ω^{2h}

        Relax on $A^{2h} E^{2h} = R^{2h}$ on coarse grid Ω^{2h} p times with initial guess $E^{2h} = 0$

    Correct fine grid solution $Y^h \leftarrow Y^h + I_{2h}^h E^{2h}$

Relax on $A^h X^h = B^h$ on fine grid Ω^h p times with new initial guess Y^h

Here the superscript h denotes the original grid, 2h, the coarse grid with mesh size twice that of the original. This two-grid method is quite efficient and the asymptotic convergence rate is independent of the grid size. The amount of work scales with the total number of grid points $N = n^d$, where n is the number of grid points in one of the d problem dimensions. This is to be compared to $N^{3/2}$ for the optimal SOR, and $N^{5/4}$ for the conjugate gradient technique.

2.4.4.1 V-Cycle

To exploit the advantages of smoothing properties of coarser grids even more, more than two grids may be employed:

Relax on $A^hX^h=B^h$ on fine grid Ω^h p times with initial guess Y^h
  Compute $R^{2h} = I_h^{2h} R^h$ the residual on coarse grid Ω^{2h}
   Relax on $A^{2h}E^{2h}=R^{2h}$ on coarse grid Ω^{2h} p times with initial guess $E^{2h}=0$
     Compute $R^{4h} = I_{2h}^{4h} R^{2h}$ the residual on coarse grid Ω^{4h}
      Relax on $A^{4h}E^{4h}=R^{4h}$ on coarse grid Ω^{4h} p times, initial guess $E^{4h}=0$
       Compute $R^{8h} = I_{4h}^{8h} R^{4h}$ the residual on coarse grid Ω^{8h}

              ....

          Relax on $A^{Lh}E^{Lh}=R^{Lh}$ on coarsest grid Ω^{Lh}

              ....

         Correct fine grid solution $Y^{4h} \leftarrow Y^{4h} + I^{4h}_{8h} E^{8h}$
       Relax on $A^{4h}X^{4h}=B^{4h}$ on fine grid Ω^{4h} q times with initial guess Y^{4h}
     Correct fine grid solution $Y^{2h} \leftarrow Y^{2h} + I^{2h}_{4h} E^{4h}$
   Relax on $A^{2h}X^{2h}=B^{2h}$ on fine grid Ω^{2h} q times with new initial guess Y^{2h}
  Correct fine grid solution $Y^h \leftarrow Y^h + I^h_{2h} E^{2h}$
Relax on $A^hX^h=B^h$ on fine grid Ω^h q times with new initial guess Y^h

Here the superscript 4h denotes the coarse grid with mesh size four times that of the original, and so on. This method is called the V-cycle multigrid method and is highly efficient. There are many other variants of the multigrid technique possible, such as the W-cycle. The efficiency $O(n^d)$ is comparable to that of the best FFT-based direct solvers $O(n^d \log n)$. However, it is important to note that SOR cannot be used in multigrid methods since the high frequency attenuating capability crucial to multigrid technique is much inferior to that of, say, the regular relaxation methods such as Gauss-Seidel. The black–red point relaxation is often optimal for use in a multigrid method. For more details and software, the reader is referred to Fletcher (1988a) and Press *et al.* (1992). Note that both FFT-based methods and multigrid methods are more complicated in their applicability to complex-shaped domains such as the ocean basins.

   Note that while relaxation can be applied to the original equation or the residual equation in linear systems, it has to be applied to the original equation if the system is nonlinear. For more details the reader is referred to Brandt (1977), Hackbusch (1985), Briggs (1987), Fletcher (1988a), and Press *et al.* (1992).

2.4.5 PSEUDO-TRANSIENT METHOD

The iterative technique for steady-state BVPs is generally equivalent to time marching a corresponding time-dependent problem to its asymptotic state. To see this add the time derivative to the Laplace equation (2.3.18) to get a parabolic IVP:

$$\frac{\partial \varphi}{\partial t} = k\nabla^2 \varphi$$

(2.4.41)

This equation can be solved to its steady state from an arbitrary initial state, subject to the same boundary conditions as the original BVP, so that the asymptotic steady state corresponds to the solution of the original BVP. Using an explicit FTCS (forward in time, centered in space) scheme (and assuming $\Delta x = \Delta y$), one gets

$$\varphi_{i,j}^{n+1} = \varphi_{i,j}^{n} + \frac{k\Delta t}{\Delta x^2}(\varphi_{i+1,j}^{n} + \varphi_{i-1,j}^{n} + \varphi_{i,j+1}^{n} + \varphi_{i,j-1}^{n} - 4\varphi_{i,j}^{n})$$

(2.4.42)

where n denotes the current time, and Δt is the time step. For stability, $\frac{k\Delta t}{\Delta x^2} \leq 0.25$ (see Section 2.5). Using a value of 0.25 gives exactly the same equation as the Jacobi iterative scheme does [Eq. (2.4.20)], except that n now refers to the time level and not the iteration.

This technique of artificially adding a time-dependent term to convert the BVP to an IVP and seeking the asymptotic steady-state solution is called a pseudo-transient method. It is an effective means of overcoming problems in solving a complex steady-state problem, where the equations may change their nature from hyperbolic to elliptic in parts of the domain. The technique was used successfully in the early sixties to solve the then nearly intractable steady-state problem of high Mach number flow around a blunt reentry body, where the flow decelerates to subsonic conditions immediately around the stagnation point, before accelerating back to supersonic conditions. Consequently, the domain contains embedded elliptic regions, and the position of the shock front is unknown a priori. The complications of patching solutions in different domains, across boundaries that must be determined as part of the solution, can be circumvented by converting the steady-state problem to an IVP and marching forward from an estimated initial state to the asymptotic steady state. A prime advantage of solving the equivalent unsteady problem to asymptotic state is that the equation is parabolic and powerful splitting techniques for multidimensional time-dependent problems (see below) such as alternate direction implicit (ADI) techniques can be pressed into service.

2.5 TIME-DEPENDENT PROBLEMS

Time-dependent problems can be hyperbolic propagation/advection type or parabolic diffusion type, or can have elements of both. Many numerical aspects related to time-dependent problems can be illustrated using a simple one-dimensional linear advection–diffusion equation

$$\frac{\partial u}{\partial t} + c\frac{\partial u}{\partial x} = v\frac{\partial^2 u}{\partial x^2}$$

(2.5.1)

and its nonlinear counterpart

$$\frac{\partial u}{\partial t} + u\frac{\partial u}{\partial x} = v\frac{\partial^2 u}{\partial x^2}$$

(2.5.2)

the Burger's equation. Equation (2.5.2) is a degenerate form of the 1D Navier–Stokes equation (pressure forces important to fluid flows are ignored) but embodies aspects related to formation of discontinuities (shocks), or more appropriately high gradient regions. It is important to represent these regions accurately in CFD problems, and this, combined with the fact that analytical solutions can be found to Eq. (2.5.2), has made it a favorite topic in CFD.

The importance of advection–diffusion equation (2.5.1) stems from the fact that it is the one-dimensional counterpart of a conservation equation for a scalar property of a fluid without of course any sources or sinks. It can describe, for example, the ozone concentration in the atmosphere or the concentration of a pollutant in the ocean. It is therefore important that the solution technique preserve important flow properties. First of all, unphysical negative concentrations cannot be allowed to develop. Secondly, the integral over the model domain of the property must be conserved, so that there is no artificial numerical source or sink of the material. Finally, sharp gradients that exist must not be overly smoothed by the numerical method. These conditions require considerable ingenuity in the solution of Eqs. (2.5.1) and (2.5.2), especially when the diffusion term is absent ($v = 0$). In this case, the system is hyperbolic.

2.5.1 ADVECTION EQUATION AND HYPERBOLIC SYSTEMS

Consider Eq. (2.5.1) without the diffusion terms, the hyperbolic problem:

$$\frac{\partial u}{\partial t} + c\frac{\partial u}{\partial x} = 0$$

(2.5.3)

The exact solution is $u = G(x - ct)$, where $u\,(t = 0) = G(x)$ is the initial condition. An initial disturbance propagates in the positive x direction without any change in shape, or equivalently is advected at a constant velocity c in the positive x direction. Solution along the characteristic (line drawn in x–t space with slope dx/dt = c) drawn from any point (x, 0) remains constant so that the solution at any point x_1 at time t_1 is equal to the solution at $x = x_1 - ct_1$ at t = 0. The problem then is how to accurately solve this equation numerically.

If we use the forward in time and centered in space (FTCS) differencing scheme,

$$\frac{u_j^{n+1} - u_j^n}{\Delta t} + c\frac{u_{j+1}^n - u_{j-1}^n}{2\Delta x} = 0 + o(\Delta t, \Delta x^2)$$

$$u_j^{n+1} = u_j^n - (1/2)C(u_{j+1}^n - u_{j-1}^n); \quad C = c\Delta t / \Delta x$$

$$(2.5.4)$$

This scheme is explicit, and first-order accurate in time and second-order accurate in space. C is the Courant number. Explicit schemes allow one to advance the solution to the next time level, one spatial grid point at a time, and are quite simple to implement. Using the von Neumann stability analysis and substituting $u_j^n = \xi^n \exp i(jk\Delta x)$ in Eq. (2.5.4), we get for the amplification factor $\xi = 1 - iC \sin(k\Delta x)$ so that $|\xi| = [1 + C^2 \sin^2(k\Delta x)]^{1/2} \geq 1$ for any value of k, no matter what the Courant number is. This scheme is unconditionally unstable.

Consider instead a first-order scheme in space, the upwind scheme (also called donor cell scheme), which uses backward differences (assuming c > 0):

$$\frac{u_j^{n+1} - u_j^n}{\Delta t} + c\frac{u_j^n - u_{j-1}^n}{\Delta x} = 0 + o(\Delta t, \Delta x)$$

$$u_j^{n+1} = (1 - C)u_j^n + Cu_{j-1}^n$$

$$(2.5.5)$$

The amplification factor is $|\xi| = [1 + 4C(C-1)\sin^2(k\Delta x / 2)]^{1/2} \leq 1$. The upwind scheme is therefore conditionally stable. Using Taylor series expansions in (2.5.5) and eliminating $\partial^2 u / \partial t^2$ using Eq. (2.5.3), it can be seen that Eq. (2.5.5) is equivalent to

$$\frac{\partial u}{\partial t} + c\frac{\partial u}{\partial x} = -\frac{\Delta t}{2}\frac{\partial^2 u}{\partial t^2} - \frac{c\Delta x}{2}\frac{\partial^2 u}{\partial x^2} = \frac{c\Delta x}{2}(1 - C)\frac{\partial^2 u}{\partial x^2}$$

$$(2.5.6)$$

In other words, the upwind scheme solves the advection–diffusion equation instead of the pure advection equation. The artificial, purely numerical, diffusion (or viscosity) introduced by the scheme is equal to $\nu = 0.5(1-C)C\Delta x^2/\Delta t$. This numerical damping is zero only if $C = 1$ (then the solution is also exact), and is responsible for suppressing numerical instabilities and making numerical solution possible. Many other schemes follow the same strategy, the magnitude of the artificial diffusion introduced being different for different schemes. For example, the Lax scheme

$$\frac{u_j^{n+1} - (u_{j+1}^n + u_{j-1}^n)/2}{\Delta t} + c\frac{u_{j+1}^n - u_{j-1}^n}{2\Delta x} = 0 + o(\Delta t, \Delta x^2)$$

(2.5.7)

has amplification factor $|\xi| = [\cos^2(k\Delta x) + C^2\sin^2(k\Delta x)]^{1/2} \le 1$ and is conditionally stable, but is equivalent to solving the advection–diffusion equation with artificial diffusion of $\nu = \Delta x^2/(2\Delta t)$. The leapfrog scheme (centered and second-order accurate in both time and space),

$$\frac{u_j^{n+1} - u_j^{n-1}}{2\Delta t} + c\frac{u_{j+1}^n - u_{j-1}^n}{2\Delta x} = 0 + o(\Delta t^2, \Delta x^2)$$

(2.5.8)

is neutrally stable. This can be seen by substituting $u_j^n = \xi^n \exp i(jk\Delta x)$, which results in a quadratic equation for the amplification factor, whose solution is $\xi_{1,2} = -iC\sin(k\Delta x) \pm \sqrt{1 - C^2\sin^2(k\Delta x)}$. For $C \le 1$, the scheme is neutrally stable. For $C > 1$, it is unstable. Because three time levels are used instead of two, the solution consists of two modes, one physical and the other a spurious computational mode, and the computational mode must be suppressed by the addition of explicit temporal smoothing or a viscosity term. Stability analysis can also be done by substituting $v_j = u_j^{n-1}$, so that Eq. (2.5.8) reduces to a two-equation set:

$$u_j^{n+1} = v_j^n + C(u_{j+1}^n - u_{j-1}^n)$$

$$v_j^{n+1} = u_j^n$$

(2.5.9)

Making the usual substitution, one gets

$$\begin{bmatrix} \xi_1 \\ \xi_2 \end{bmatrix}^{n+1} = \begin{bmatrix} -2iC\sin(k\Delta x) & 1 \\ 1 & 0 \end{bmatrix} \begin{bmatrix} \xi_1 \\ \xi_2 \end{bmatrix}^n$$

(2.5.10)

The stability condition is that the spectral radius (the magnitude of the eigenvalue largest in magnitude) of the 2×2 amplification matrix ξ $[=\xi^{n+1}(\xi^n)^{-1}]$ on the RHS of Eq. (2.5.10),

$$\lambda_{1,2} = -iC\sin(k\Delta x) \pm \sqrt{1 - C^2 \sin^2(k\Delta x)} \, ,$$

be ≤ 1, which yields the same result as above. For a system of m equations (as, for example, in a linear 1D shallow water wave propagation equation, which has two coupled equations), a similar procedure can be applied that leads to an m × m amplification matrix and m eigenvalues and the von Neumann condition for stability is that the eigenvalue with the largest magnitude (the spectral radius of the amplification matrix) must be bounded by unity.

An advantage of the leapfrog scheme is that no artificial diffusion is introduced by the scheme. However, this comes at the penalty of having to suppress the spurious computational mode. One technique to accomplish this is the use of a temporal filter called the Asselin–Robert filter that will be discussed in the next section. Another method that can be used is a two-step predictor–corrector-type leapfrog-trapezoidal scheme:

$$u_j^* = u_j^{n-1} - C(u_{j+1}^n - u_{j-1}^n)$$

$$u_j^{n+1} = u_j^n - \frac{C}{2}\left[\frac{1}{2}(u_{j+1}^n + u_{j+1}^*) - \frac{1}{2}(u_{j-1}^n + u_{j-1}^*)\right]$$

$$(2.5.11)$$

This however doubles the number of computations per step. The scheme is still second-order accurate but the trapezoidal step involved in taking the average of the first guess value at time level n+1, and the value at time level n, strongly damps out the computational mode in the leapfrog scheme. This scheme is also conditionally stable, i.e., stable for $C \leq 1$.

Other multistep methods are also popular in solving hyperbolic problems. The two-step Lax–Wendroff scheme uses an intermediate time level to obtain an interim solution:

$$\frac{u_{j+1/2}^{n+1/2} - (u_{j+1}^n + u_j^n)/2}{\Delta t/2} + c\frac{u_{j+1}^n - u_j^n}{\Delta x} = 0$$

$$\frac{u_j^{n+1} - u_j^n}{\Delta t} + c\frac{u_{j+1/2}^{n+1/2} - u_{j-1/2}^{n+1/2}}{\Delta x} = 0$$

$$(2.5.12)$$

The scheme is second-order accurate (in both space and time), and is conditionally stable, i.e., stable for $C \leq 1$. It introduces an artificial numerical viscosity equal to $\nu = 0.5C^2\Delta x^2/\Delta t$.

Note that in all the above schemes, the spatial differences were evaluated at the previous time level or levels, so that the unknown values at time level n+1 could be evaluated one spatial grid point at a time. In other words, the resulting equations contain only one unknown grid point j at time level n+1. The time stepping scheme is explicit and can be readily extended to multidimensional problems. It is also possible to evaluate the spatial differences at time level n+1, but this results in equations that contain the unknown grid point j as well as its neighboring points at time level n+1. This means that the solution at time level n+1 must be obtained simultaneously at all spatial grid points and cannot be obtained grid point by grid point. Time stepping scheme becomes implicit. The unknown value at each grid point is coupled to that at the neighboring point, a matrix equation replaces the simple algebraic relation of an explicit method, and the matrix needs to be inverted to advance the solution to the next time level. Implicit time differencing schemes are therefore computationally more intensive, but have the advantage of being unconditionally stable, meaning that arbitrarily large time steps can be taken consistent only with the accuracy demanded. Extension to more than one spatial dimension requires, however, a technique such as split time stepping, as we will see in the next section in connection with the diffusion equation.

2.5.1.1 FCT Schemes

None of the above schemes is completely satisfactory. The first-order upwind scheme smears out the solution over the neighboring grid points; gradients are excessively smoothed out. However, while the scheme is too diffusive, no unphysical oscillations develop. In second-order schemes such as the two-step Lax–Wendroff scheme, local ripples appear, especially near regions of sharp gradients. These local extrema are unacceptable since they lead to negative concentrations locally. The second-order schemes however do not overly smear sharp gradients. Thus, first-order schemes suffer from too much numerical diffusion while second-order schemes suffer from too much numerical dispersion. The former smears out gradients, while the latter produces oscillations. The reason is that the leading order truncation term in a first-order scheme involves $\partial^2 u / \partial x^2$, an even-order spatial derivative, which is diffusive, whereas in a second-order scheme it is $\partial^3 u / \partial x^3$, an odd-order derivative, which is dispersive.

Spurious oscillations and negative values cannot be tolerated when solving advection equations governing the transport and diffusion of tracers such as the greenhouse gases in the atmosphere and dissolved chemical substances in the ocean. Too much smearing of an initial concentration gradient is also undesirable. The problem is particularly serious in long-term climate-type simulations.

A time-honored method to reduce oscillations and eliminate negative values in a second-order scheme is to introduce additional, explicit damping terms, either second order or fourth order. Thus, for the two-step Lax–Wendroff scheme, which is equivalent to

$$u_j^{n+1} = u_j^n - (C/2)(u_{j+1}^n - u_{j-1}^n) + (C^2/2)(u_{j+1}^n - 2u_j^n + u_{j-1}^n)$$

$$(2.5.13)$$

one adds a dissipation term, either $v_2 \Delta x^2 (\partial^2 u / \partial x^2)$ or $-v_4 \Delta x^4 (\partial^4 u / \partial x^4)$, in other words, $v_2(u_{j+1}^n - 2u_j^n + u_{j-1}^n)$ or $-v_4(u_{j+2}^n - 4u_{j+1}^n + 6u_j^n - 4u_{j-1}^n + u_{j-2}^n)$, to the RHS of Eq. (2.5.12). The latter, a biharmonic-type damping term, is especially effective in damping high frequency oscillations. These terms also stabilize the solutions so that larger time steps can be taken. Some smearing of gradients is however unavoidable. Traditionally, second-order leapfrog in time and central differencing in space have been extensively employed in ocean models (for example, the POM in Chapter 9, and the GFDL model in Chapter 10), in combination with diffusivities large enough to suppress spurious oscillations. However, this may lead to excessive smearing of scalar properties such as temperature and salinity, which in turn can lead to incorrect dynamical solutions, especially in long simulations (Gerdes *et al.*, 1991; Hecht *et al.*, 1995).

Flux-corrective transport (FCT) schemes developed by Boris and Book (1973; see also Zalesak, 1979; Fletcher, 1988a) combine the best of the diffusive low order and dispersive high order schemes to arrive at an optimum solution with minimum diffusion and dispersion. They can be thought of as two-step schemes in which in the first step, a smooth ripple-free interim solution is obtained, and in the second step, any excessive smoothing in the first step is removed by adding an antidiffusive term, a term with negative diffusivity. For the Lax–Wendroff scheme, this means adding a diffusive term in the first step and an antidiffusive term in the second:

$$u_j^* = u_j^n - (C/2)(u_{j+1}^n - u_{j-1}^n) + (v_2 + C^2/2)(u_{j+1}^n - 2u_j^n + u_{j-1}^n)$$

$$u_j^{n+1} = u_j^* - v_2^*(u_{j+1}^* - 2u_j^* + u_{j-1}^*)$$

$$(2.5.14)$$

Appropriate selection of the positive and negative diffusivities gives an optimum solution. However, it is important to limit (clip) the antidiffusion introduced so that it does not lead to oscillations. This is the most crucial step of an FCT scheme. For the above Lax–Wendroff scheme, Boris and Book (1973, 1975) recommend

$$v_2 = (1 + 2C^2)/6, \quad v_2^* = (1 - C^2)/6$$

$$(2.5.15)$$

Consider the flux-conservative form of the equation, where the fluxes at neighboring half points are $F_{j-1/2} = c\,(u_j - u_{j-1})$ and $F_{j+1/2} = c\,(u_{j+1} - u_j)$. Basically, FCT schemes can be considered to consist of the following steps (Zalesak, 1979):

1. Determining the antidiffusive fluxes at these half-points, which are equal to the differences between some low order scheme diffusive but ripple-free fluxes and some high order scheme dispersive fluxes: $A_{j\pm1/2} = F^H_{j\pm1/2} - F^L_{j\pm1/2}$.

2. Limiting or clipping these antidiffusive fluxes $A^C_{j\pm1/2} = C_{j\pm1/2} A_{j\pm1/2}$ ($0 \leq C_{j\pm1/2} \leq 1$).

3. Adding these clipped antidiffusive fluxes to the low order scheme solution,

$$u^*_j = u^n_j - \frac{\Delta t}{\Delta x}(F^L_{j+1/2} - F^L_{j-1/2})$$

$$u^{n+1}_j = u^*_j - \frac{\Delta t}{\Delta x}(A^C_{j+1/2} - A^C_{j-1/2})$$

$$(2.5.16)$$

to get the final ripple-free flux-limited or flux-corrected solution.

The flux limiting step is crucial. The original Boris and Book (1973) prescription for limiting antidiffusive fluxes is

$$A^c_{j+1/2} = \text{sgn}(A_{j+1/2}) \max \left\{ 0, \min \left[\begin{array}{l} |A_{j+1/2}|, \\[2mm] \text{sgn}(A_{j+1/2})\dfrac{\Delta x}{\Delta t}(u^*_{j+2} - u^*_{j+1}), \\[2mm] \text{sgn}(A_{j+1/2})\dfrac{\Delta x}{\Delta t}(u^*_j - u^*_{j-1}) \end{array} \right] \right\}$$

$$A^c_{j-1/2} = \text{sgn}(A_{j-1/2}) \max \left\{ 0, \min \left[\begin{array}{l} |A_{j-1/2}|, \\[2mm] \text{sgn}(A_{j-1/2})\dfrac{\Delta x}{\Delta t}(u^*_{j+1} - u^*_j), \\[2mm] \text{sgn}(A_{j-1/2})\dfrac{\Delta x}{\Delta t}(u^*_{j-1} - u^*_{j-2}) \end{array} \right] \right\}$$

$$\text{sgn}\,\varphi = \varphi / |\varphi|; \text{ equal to } +1 \text{ if } \varphi \geq 0 \text{ and } -1 \text{ if } \varphi < 0 \qquad (2.5.17)$$

Extension to more than one dimension, while straightforward, requires that split time stepping (see next section) be employed. Otherwise, the existence of fluxes in more than one direction requires that this fact be recognized in clipping the antidiffusive fluxes in any particular direction. Zalesak (1979) presents a modified FCT scheme that does so. In two dimensions, this implies that all four antidiffusive fluxes $A^C_{i+1/2, j}$, $A^C_{i-1/2, j}$, $A^C_{i, j+1/2}$, and $A^C_{i, j-1/2}$ into a grid cell act in concert to ensure that the solution in that cell is ripple-free,

$$u_{i,j}^{*} = u_{i,j}^{n} - \frac{\Delta t}{\Delta x \Delta y}(F_{i+1/2,j}^{L} - F_{i-1/2,j}^{L} + G_{i,j+1/2}^{L} - G_{i,j-1/2}^{L})$$

$$u_{i,j}^{n+1} = u_{i,j}^{*} - \frac{\Delta t}{\Delta x \Delta y}(A_{i+1/2,j}^{C} - A_{i-1/2,j}^{C} + A_{i,j+1/2}^{C} - A_{i,j-1/2}^{C})$$

$$(2.5.18)$$

where F and G indicate fluxes across the cell faces. This means that the following six quantities be computed and used in limiting fluxes:

$$P_{i,j}^{+} = \max(0, A_{i-1/2,j}) - \min(0, A_{i+1/2,j}) + \max(0, A_{i,j-1/2}) - \min(0, A_{i,j+1/2})$$

$$Q_{i,j}^{+} = \frac{\Delta x \Delta y}{\Delta t}(u_{i,j}^{max} - u_{i,j}^{*})$$

$$R_{i,j}^{+} = \{\min(1, Q_{i,j}^{+}/P_{i,j}^{+}) \text{ if } P_{i,j}^{+} > 0; \ 0 \text{ if } P_{i,j}^{+} = 0$$

$$P_{i,j}^{-} = \max(0, A_{i+1/2,j}) - \min(0, A_{i-1/2,j}) + \max(0, A_{i,j+1/2}) - \min(0, A_{i,j-1/2})$$

$$Q_{i,j}^{-} = \frac{\Delta x \Delta y}{\Delta t}(u_{i,j}^{*} - u_{i,j}^{min})$$

$$R_{i,j}^{-} = \{\min(1, Q_{i,j}^{-}/P_{i,j}^{-}) \text{ if } P_{i,j}^{-} > 0; \ 0 \text{ if } P_{i,j}^{-} = 0$$

$$(2.5.19)$$

Note that $P_{i,j}^{+}$ and $P_{i,j}^{-}$ represent the sum of all antidiffusive fluxes into and away from grid cell (i,j). The coefficients in $A_{i+1/2,j}^{C} = C_{i+1/2,j} A_{i+1/2,j}$ and $A_{i,j+1/2}^{C} = C_{i,j+1/2} A_{i,j+1/2}$ (bounded by values 0 and 1) are chosen according to

$$C_{i+1/2,j} = \begin{cases} \min(R_{i+1,j}^{+}, R_{i,j}^{-}) \text{ if } A_{i+1/2,j} \geq 0 \\ \min(R_{i,j}^{+}, R_{i+1,j}^{-}) \text{ if } A_{i+1/2,j} < 0 \end{cases}$$

$$C_{i,j+1/2} = \begin{cases} \min(R_{i,j+1}^{+}, R_{i,j}^{-}) \text{ if } A_{i,j+1/2} \geq 0 \\ \min(R_{i,j}^{+}, R_{i,j+1}^{-}) \text{ if } A_{i,j+1/2} < 0 \end{cases}$$

$$(2.5.20)$$

with similar expressions for $C_{i-1/2,j}$ and $C_{i,j-1/2}$. The maximum and minimum values of the variable above and below which the grid point value is not allowed to fall are chosen according to

$$u_{i,j}^{a} = \max(u_{i,j}^{n}, u_{i,j}^{*})$$

$$u_{i,j}^{max} = \max(u_{i-1,j}^{a}, u_{i+1,j}^{a}, u_{i,j}^{a}, u_{i,j-1}^{a}, u_{i,j+1}^{a})$$

$$u_{i,j}^{b} = \min(u_{i,j}^{n}, u_{i,j}^{*})$$

$$u_{i,j}^{min} = \min(u_{i-1,j}^{b}, u_{i+1,j}^{b}, u_{i,j}^{b}, u_{i,j-1}^{b}, u_{i,j+1}^{b})$$

$$(2.5.21)$$

Zalesak (1979) recommends that this algorithm be used instead of the original Boris and Book (1973, 1975) scheme for multidimensional problems. The scheme is finding increasing use in modeling tracers in the ocean, where it is important to avoid spurious overshoots and undershoots. For example, Gerdes (1993) has implemented such a scheme in the s-coordinate version of the GFDL ocean model (see Chapter 10).

2.5.1.2 TVD Schemes

FCT schemes are two-step schemes that are easy to implement for an existing scheme but difficult to implement as implicit formulations. A total variance diminishing (TVD) scheme (Harten, 1983; Fletcher, 1988a) is a scheme that builds on a first-order scheme by adding on antidiffusive fluxes that are limited to assure TVD. First-order schemes are monotone schemes, in that an initially monotonic solution will stay monotonic with time and will not develop any local extrema due to spurious oscillations that destroy monotonicity. However, they are also highly diffusive and, without impractically fine resolutions, smear sharp gradients. One remedy is to use a TVD scheme that not only is free from spurious oscillations, but also achieves second-order accuracy in at least regions where the solution varies smoothly. The total variance is defined as $TV(u) = \int |\partial u / \partial x| dx$, and in a numerical solution, it is $TV(u^n) = \sum |u_{j+1}^n - u_j^n|$, so that if $TV(u^{n+1}) \leq TV(u^n)$, the scheme is TVD. In a TVD scheme, the antidiffusive flux is added, but care is taken to see that TVD condition is satisfied. In this sense, the TVD scheme is similar to the FCT scheme.

The basis of TVD schemes can be seen by examining the Lax–Wendroff scheme, which can be interpreted as adding an antidiffusive flux to the first-order upwind scheme. The upwind scheme (2.5.5) is TVD but far too diffusive. The Lax–Wendroff scheme effectively adds an antidiffusive flux proportional to C^2 to the upwind scheme which has a diffusion proportional to C, such that

$$u_j^{n+1} = u_j^n - C(u_j^n - u_{j-1}^n) - (F_{j+1/2}^n - F_{j-1/2}^n);$$

$$F_{j-1/2}^n = \left(\frac{C}{2} - \frac{C^2}{2}\right)(u_j^n - u_{j-1}^n), F_{j+1/2}^n = \left(\frac{C}{2} - \frac{C^2}{2}\right)(u_{j+1}^n - u_j^n)$$

$$(2.5.22)$$

However, it adds too much antidiffusion and is therefore oscillatory. If the antidiffusive fluxes added could be chosen to be as large as feasible without violating TVD, one would have a viable TVD scheme. This is done by suitably multiplying the Lax–Wendroff corrective fluxes by a function $\phi(r_j)$ that depends on the ratio of contiguous gradients $r_j = (u_j - u_{j-1})/(u_{j+1} - u_j)$,

$$u_j^{n+1} = u_j^n - C(u_j^n - u_{j-1}^n) - (F_{j+1/2}^c - F_{j-1/2}^c);$$

$$F_{j-1/2}^c = \phi(r_{j-1})\left(\frac{C}{2} - \frac{C^2}{2}\right)(u_j^n - u_{j-1}^n), F_{j+1/2}^c = \phi(r_j)\left(\frac{C}{2} - \frac{C^2}{2}\right)(u_{j+1}^n - u_j^n)$$

(2.5.23)

with function $\phi(r_j)$ equal to min $(2, r_j)$ for $r_j > 1$; min $(2r_j, 1)$ for $1 \geq r_j > 0$; and zero for $r_j \leq 0$ (Fletcher, 1988a).

Leonard (1979) has developed schemes such as Quadratic Upstream Interpolation for Convective Kinematics (QUICK) that are third-order accurate. Holland *et al.* (1998) use a variant of this third-order upwind scheme, which is identical to that of Farrow and Stevens (1995), in the ocean part of the NCAR Climate system model. This scheme can be illustrated using a simple 1D version of the advection equation for a tracer ϕ:

$$\frac{\partial \phi}{\partial t} + \frac{\partial}{\partial x}(u\phi) = 0$$

(2.5.24)

For a uniform grid with grid spacing Δx, the conventional centered space difference scheme would be

$$\frac{\partial \phi_i}{\partial t} + \frac{u_{i+1/2}}{2\Delta x}(\phi_{i+1} + \phi_i) - \frac{u_{i-1/2}}{2\Delta x}(\phi_i + \phi_{i-1}) = 0$$

(2.5.25)

The third-order upwind scheme would give instead (Holland *et al.*, 1998)

$$\frac{\partial \phi_i}{\partial t} + \frac{u_{i+1/2}}{16\Delta x}(-\phi_{i+2} + 9\phi_{i+1} + 9\phi_i - \phi_{i-1}) - \frac{u_{i-1/2}}{16\Delta x}(-\phi_{i+1} + 9\phi_i + 9\phi_{i-1} - \phi_{i-2})$$

$$-\frac{|u_{i+1/2}|}{16\Delta x}(-\phi_{i+2} + 3\phi_{i+1} - 3\phi_i + \phi_{i-1}) + \frac{|u_{i-1/2}|}{16\Delta x}(-\phi_{i+1} + 3\phi_i - 3\phi_{i-1} + \phi_{i-2}) = 0$$

(2.5.26)

This is equivalent to solving

$$\frac{\partial \phi}{\partial t} + \frac{\partial}{\partial x}(u\hat{\phi}) + \frac{\partial}{\partial x}\left[\left(\frac{|u|\Delta x^3}{16}\right)\frac{\partial^3 \phi}{\partial x^3}\right] = 0$$

(2.5.27)

where $\hat{\phi}$ is the value of ϕ at the cell walls, obtained from quadratic interpolation. This scheme can be contrasted to the first-order upwind scheme which is equivalent to solving

$$\frac{\partial \varphi_i}{\partial t} + \frac{u_{i+1/2}}{2\Delta x}(\varphi_{i+1} + \varphi_i) - \frac{u_{i-1/2}}{2\Delta x}(\varphi_i + \varphi_{i-1})$$

$$-\frac{|u_{i+1/2}|}{2\Delta x}(\varphi_{i+1} - \varphi_i) + \frac{|u_{i-1/2}|}{16\Delta x}(\varphi_i - \varphi_{i-1}) = 0$$

(2.5.28)

or equivalently

$$\frac{\partial \varphi}{\partial t} + \frac{\partial}{\partial x}(u\hat\varphi) - \frac{\partial}{\partial x}\left[\left(\frac{|u|\Delta x}{2}\right)\frac{\partial \varphi}{\partial x}\right] = 0$$

(2.5.29)

where $\hat\varphi$ is the value of φ at the cell walls, obtained from linear interpolation. For nonuniform grids (see Section 2.6), the expression (2.5.26) is more complicated, but can be written as (Holland *et al.*, 1998)

$$\frac{\partial \varphi_i}{\partial t} + \frac{u_{i+1/2}}{\Delta x_i}\left(\frac{\Delta x_{i+1}\varphi_i + \Delta x_i\varphi_{i+1}}{\Delta x_{i+1} + \Delta x_i} - D_{i+1/2}\right)$$

$$-\frac{u_{i-1/2}}{\Delta x_i}\left(\frac{\Delta x_i\varphi_{i-1} + \Delta x_{i-1}\varphi_i}{\Delta x_{i-1} + \Delta x_i} - D_{i-1/2}\right) = 0$$

(2.5.30)

where

$$D_{i+1/2} = \frac{\Delta x_i \Delta x_{i+1}}{\Delta x_{i-1} + 2\Delta x_i + \Delta x_{i+1}}\left(\frac{\varphi_{i+1} - \varphi_i}{\Delta x_{i+1} + \Delta x_i} - \frac{\varphi_i - \varphi_{i-1}}{\Delta x_i + \Delta x_{i-1}}\right) \quad \text{if } u_{i+1/2} > 0$$

$$= \frac{\Delta x_i \Delta x_{i+1}}{\Delta x_i + 2\Delta x_{i+1} + \Delta x_{i+2}}\left(\frac{\varphi_{i+2} - \varphi_{i+1}}{x_{i+2} + \Delta x_{i+1}} - \frac{\varphi_{i+1} - \varphi_i}{\Delta x_{i+1} + \Delta x_i}\right) \quad \text{if } u_{i+1/2} < 0$$

$$D_{i-1/2} = \frac{\Delta x_{i-1}\Delta x_i}{\Delta x_{i-2} + 2\Delta x_{i-1} + \Delta x_i}\left(\frac{\varphi_i - \varphi_{i-1}}{\Delta x_i + \Delta x_{i-1}} - \frac{\varphi_{i-1} - \varphi_{i-2}}{\Delta x_{i-1} + \Delta x_{i-2}}\right) \quad \text{if } u_{i-1/2} > 0$$

$$= \frac{\Delta x_{i-1}\Delta x_i}{\Delta x_{i-1} + 2\Delta x_i + \Delta x_{i+1}}\left(\frac{\varphi_{i+1} - \varphi_i}{\Delta x_{i+1} + \Delta x_i} - \frac{\varphi_i - \varphi_{i-1}}{\Delta x_i + \Delta x_{i-1}}\right) \quad \text{if } u_{i-1/2} < 0$$

(2.5.31)

If a leapfrog scheme is used for time discretization, for stability, the diffusion terms must be lagged, i.e., evaluated at time step n−1. A predictor–corrector scheme (Farrow and Stevens, 1995) such as the Adams–Bashforth scheme is an alternative, but more expensive. Holland *et al.* (1998) compare the three schemes above and assert that the third-order upwind scheme removes local extrema in tracer concentrations (but not completely) while keeping the diffusion reasonably small. It is also less costly computationally than FCT schemes outlined earlier, although strict monotonicity is not preserved.

2.5.1.3 Nonlinear Advection

The nonlinear advection equation can give rise to the formation of a discontinuity from an initially smooth disturbance. Without an explicitly introduced or numerically introduced viscosity term, the scheme becomes unstable. The two popular methods for the solution of the nonlinear advection equation are the explicit two-step Lax–Wendroff and the explicit predictor–corrector MacCormack schemes. For the linear advection problem, the MacCormack scheme popular in aerodynamic CFD is equivalent to the Lax–Wendroff scheme. To preserve integral conservation properties in the finite-difference equivalent of the differential equation, nonlinear equations such as this are always written in flux-conservative form before a solution is attempted:

$$\frac{\partial E}{\partial t} + \frac{\partial F}{\partial x} = 0; \; E = U, \; F = u^2 / 2$$

(2.5.32)

Otherwise, the round-off and truncation errors can quickly lead to violation of conservation laws. The two-step Lax–Wendroff scheme becomes

$$E_{j+1/2}^{n+1/2} = \left(\frac{E_j^n + E_{j+1}^n}{2} \right) - \frac{\Delta t / 2}{\Delta x} (F_{j+1}^n - F_j^n)$$

$$E_j^{n+1} = E_j^n - \frac{\Delta t}{\Delta x} (F_{j+1/2}^{n+1/2} - F_{j-1/2}^{n+1/2})$$

(2.5.33)

The use of half grid points in both time and space assures a second-order accurate scheme in both time and space. The scheme is conditionally stable (C $\leq$ 1), with C determined by the maximum possible value of u.

In the MacCormack scheme, an interim solution is predicted at the n+1 time level at all spatial grid points, based on forward (backward) spatial differences at time level n. This solution is then corrected by using the backward (forward) spatial difference at time level n+1, based on the predicted solutions:

$$\overline{E}_j^{n+1} = E_j^n - \frac{\Delta t}{\Delta x} (F_{j+1}^n - F_j^n)$$

$$E_j^{n+1} = E_j^n - \frac{\Delta t}{\Delta x} \left(\frac{F_{j+1}^n - F_j^n}{2} + \frac{\overline{F}_j^n - \overline{F}_{j-1}^n}{2} \right) = \left(\frac{E_j^n + \overline{E}_{j+1}^n}{2} \right) - \frac{\Delta t}{\Delta x} \left(\frac{\overline{F}_j^n - \overline{F}_{j-1}^n}{2} \right)$$

(2.5.34)

The use of alternating forward and backward differencing assures second-order accuracy. The stability condition is the same as the two-step Lax–Wendroff scheme.

Both these schemes can be easily extended to multidimensions. However, both suffer from underdamping and develop oscillations, which can be seen from propagating a sharp discontinuity. The normal remedy is the addition of a linear damping term as before, or the use of a FCT scheme. The use of a FCT scheme is actually quite straightforward since it can be considered a correction applied to the solutions given by the MacCormack or Lax–Wendroff schemes, and consists of the following steps (Fletcher, 1988a):

1. Computing diffusive fluxes: $\nu_{j+1/2} = (1/6) + (1+3)(u_{j+1/2}^n \Delta t / \Delta x)^2$

$$F_{j+1/2}^d = \nu_{j+1/2}(E_{j+1}^n - E_j^n)$$

2. Computing antidiffusive fluxes: $\nu'_{j+1/2} = (1/6) - (1/6)(u_{j+1/2}^{n+1} \Delta t / \Delta x)^2$

$$F_{j+1/2}^{ad} = \nu'_{j+1/2}(E_{j+1}^{n+1} - E_j^{n+1})$$

3. Computing the diffused solution: $\quad \hat{E}_j^{n+1} = E_j^{n+1} + (F_{j+1/2}^d - F_{j-1/2}^d)$

4. Computing the differences: $\delta\hat{E} = \hat{E}_{j+1}^{n+1} - \hat{E}_j^{n+1}$

5. Limiting antidiffusive fluxes: $S = \text{sgn}(F_{j+1/2}^{ad})$

$$F_{j+1/2}^c = S \max\{0, \min[S\delta\hat{E}_{j-1/2}, \left|F_{j+1/2}^{ad}\right|, S\delta\hat{E}_{j+3/2}]\}$$

Computing the desired solution:

$$E_j^{n+1} = \hat{E}_j^{n+1} - (F_{j+1/2}^c - F_{j-1/2}^c)$$

(2.5.35)

Fletcher (1988a) lists software to carry out these steps and provides illustration of its application to modeling a shock.

A TVD scheme for the nonlinear equation can be expressed as

$$E_j^{n+1} = E_j^n - \frac{\Delta t}{\Delta x}(F_{j+1/2}^c - F_{j-1/2}^c)$$

$$F_{j+1/2}^c = 0.5(F_j^n + F_{j+1}^n) - 0.5s(F_{j+1}^n - F_j^n) + 0.5\phi(r_j)[s - C_{j+1/2}](F_{j+1}^n - F_j^n)$$

$$s = \text{sgn}(C_{j+1/2}); r_j = (E_{j+1-s} - E_{j-s})/(E_{j+1} - E_j)$$

(2.5.36)

where the scheme accounts for the flow direction and computes the ratio of contiguous gradients from upstream.

The nonlinear advection–diffusion equation has also been investigated by Chang (1995) and Wang (1995) using an entirely new approach called the conservation element/solution element technique, which is capable of resolving a shock-like discontinuity in a single mesh point without any oscillations. It is equivalent in performance to TVD schemes (Harten, 1983), but is much simpler and more efficient and is designed to obey local and global conservation laws (Wang, 1995).

In the oceans and the atmosphere also, one encounters regions with strong gradients, such as oceanic thermal and salinity fronts, and atmospheric fronts, and it is important that these be not smeared out excessively. FCT and TVD schemes can play an important role in simulating these, providing accurate solutions of scalar constituents being transported by the fluid flow. See Rood (1987) for a discussion of advection algorithms used in atmospheric models.

2.5.2 DIFFUSION EQUATION AND PARABOLIC SYSTEMS

Consider Eqs. (2.5.1) and (2.5.2) without the advection terms:

$$\frac{\partial u}{\partial t} = v \frac{\partial^2 u}{\partial x^2}$$

(2.5.37)

This equation is linear and parabolic. It describes diffusion with time of a substance or property in space starting from a prescribed initial distribution: $u(x,0) = F(x)$. The principal outcome is smearing of gradients with time. It can be readily solved using an FTCS scheme:

$$\frac{u_j^{n+1} - u_j^n}{\Delta t} = v \left[\frac{u_{j+1}^n - 2u_j^n + u_{j-1}^n}{\Delta x^2} \right] + o(\Delta t, \Delta x^2)$$

$$u_j^{n+1} = u_j^n + D(u_{j+1}^n - 2u_j^n + u_{j-1}^n); \quad D = v\Delta t / \Delta x^2 \qquad (2.5.38)$$

This scheme is explicit and first-order accurate in time and second-order accurate in space. D is the diffusion number. Using the von Neumann stability analysis and substituting $u_j^n = \xi^n \exp i(jk\Delta x)$, we get for the amplification factor $\xi = 1 - 4D\sin^2(k\Delta x/2)$, so that $|\xi| \leq 1$ for all wavenumbers if $D \leq 1/2$. This scheme is therefore conditionally stable, and easy to implement. The stability constraint imposes a limit on the time step that can be taken, which can be avoided by the use of an implicit scheme, where the spatial differences are taken at time level n+1, instead of n:

$$\frac{u_j^{n+1} - u_j^n}{\Delta t} = \nu \left[\frac{u_{j+1}^{n+1} - 2u_j^{n+1} + u_{j-1}^{n+1}}{\Delta x^2} \right] + o(\Delta t, \Delta x^2)$$

(2.5.39)

This leads to a tridiagonal system:

$$-Du_{j-1}^{n+1} + (1 + 2D)u_j^{n+1} - Du_{j+1}^{n+1} = u_j^n$$

(2.5.40)

The solution at time level n+1 must be obtained simultaneously at all grid points in the domain by inversion of the matrix. In this case, an efficient tridiagonal solver can be used. However, this increased complexity of the solution is accompanied by increased efficiency. Neumann stability analysis gives for the amplification parameter $\xi = [1 + 4D \sin^2(k\Delta x / 2)]^{-1}$, so that $|\xi| \leq 1$ for any value of D. This scheme is therefore unconditionally stable, and an arbitrarily large time step can be taken. In practice, if one is interested in the details of the transient solution to the IVP, the truncation error now determines what time step to take, not the stability constraint. If one is not interested in how the solution evolves in time, but seeks the asymptotic steady state in the quickest way possible, time steps can be larger. The scheme is first-order accurate in time, and second-order accurate in space.

In contrast, an explicit leapfrog scheme, which is second-order accurate in time, gives

$$\frac{u_j^{n+1} - u_j^{n-1}}{2\Delta t} = \nu \left[\frac{u_{j+1}^n - 2u_j^n + u_{j-1}^n}{\Delta x^2} \right] + o(\Delta t^2, \Delta x^2)$$

(2.5.41)

The equation for the amplification parameter is now quadratic and yields two roots:

$$\xi = \frac{1}{2} \left[p \pm (p^2 + 4)^{1/2} \right]; \quad p = 8D \sin^2(k\Delta x / 2) \quad .$$

The positive sign denotes the physical mode and the negative sign the computational mode. Since p is positive for all positive nonzero values of D, the computational mode has $|\xi| \geq 1$ for any value of D. This scheme is therefore unconditionally unstable. This is the scheme that L. F. Richardson, the pioneer in weather prediction, attempted to use in the 1920s, without success. This deficiency could have been remedied very simply, had he known about the stability aspect of finite-difference formulations, by simply evaluating the spatial derivative at time step n−1. If the diffusion terms are thus lagged, one gets for the amplification factor $\xi = \pm(1-p)^{1/2}$; the scheme is stable.

There are two important methods that are commonly used for solution of parabolic equations. Both are second-order accurate in time and space, and both are unconditionally stable. One is the Crank–Nicolson scheme, where the time derivative is evaluated at the n+1/2 point in time, and the spatial derivative is taken as the average of the points surrounding this in (x, t) space so as to make the scheme second-order accurate in both time and space:

$$\frac{u_j^{n+1} - u_j^n}{\Delta t} = \frac{v}{2}\left[\frac{u_{j+1}^n - 2u_j^n + u_{j-1}^n}{\Delta x^2} + \frac{u_{j+1}^{n+1} - 2u_j^{n+1} + u_{j-1}^{n+1}}{\Delta x^2}\right] + o(\Delta t^2, \Delta x^2)$$

(2.5.42)

The RHS is the average of the RHS of the FTCS and implicit schemes. The scheme is implicit and leads to a tridiagonal system:

$$Du_{j-1}^{n+1} - 2(1+D)u_j^{n+1} + Du_{j+1}^{n+1} = -[Du_{j-1}^n + 2(1-D)u_j^n + Du_{j+1}^n]$$

(2.5.43)

Stability analysis yields for the amplification factor $\xi = [1 - 2D \sin^2(k\Delta x/2)]$ $[1 + 2D \sin^2(k\Delta x/2)]^{-1}$, so that $|\xi| \leq 1$ for any value of D. This scheme is therefore unconditionally stable and highly desirable for solving diffusion problems.

The second scheme, the Dufort–Frankel scheme, is unconditionally stable, even though explicit. It is a three-time-level scheme, second-order accurate in time and space, where the u_j point in the spatial derivative is replaced by the average of its neighbors,

$$\frac{u_j^{n+1} - u_j^{n-1}}{2\Delta t} = v\left[\frac{u_{j+1}^n - (u_{j+1}^n + u_{j-1}^n) + u_{j-1}^n}{\Delta x^2}\right] + o(\Delta t^2, \Delta x^2)$$

(2.5.44)

so that

$$u_j^{n+1} = \frac{2D}{1+2D}(u_{j+1}^n + u_{j-1}^n) + \frac{1-2D}{1+2D}u_j^{n-1}$$

(2.5.45)

However, care must be taken since the scheme is not always consistent.

2.5.2.1 Multidimensional Diffusion Problems

These can be solved using either implicit or explicit methods. The explicit method for the two-dimensional diffusion problem

$$\frac{\partial u}{\partial t} = \nu \left[\frac{\partial^2 u}{\partial x^2} + \frac{\partial^2 u}{\partial y^2} \right]$$

(2.5.46)

is

$$\frac{u_{i,j}^{n+1} - u_{i,j}^{n}}{\Delta t} = \nu \left[\frac{u_{i+1,j}^{n} - 2u_{i,j}^{n} + u_{i-1,j}^{n}}{\Delta x^2} + \frac{u_{i,j+1}^{n} - 2u_{i,j}^{n} + u_{i,j-1}^{n}}{\Delta y^2} \right] + o(\Delta t, \Delta x^2, \Delta y^2)$$

$$u_{i,j}^{n+1} = u_{i,j}^{n} + D(u_{i+1,j}^{n} - 2u_{i,j}^{n} + u_{i-1,j}^{n}) + D'(u_{i,j+1}^{n} - 2u_{i,j}^{n} + u_{i,j-1}^{n})$$

(2.5.47)

where $D' = D (\Delta x/\Delta y)^2$. Using von Neumann's method, the stability condition can be shown to be $D \le \frac{1}{2} \left(\frac{\Delta y^2}{\Delta x^2 + \Delta y^2} \right)$. Thus, for $\Delta x = \Delta y$, the limit on time step is half that of the 1D problem with the same grid size Δx. The effective grid size is half that of the 1D problem. Extension to 3D causes the time step to be reduced to one-third the 1D value.

An unconditionally stable, fully implicit scheme is obtained as before in the 1D problem by using the n+1 time level for spatial derivatives, but one gets a pentadiagonal system

$$Du_{i-1,j}^{n+1} + D'u_{i,j-1}^{n+1} - [1 + 2(D + D')]u_{i,j}^{n+1} + D'u_{i,j+1}^{n+1} + Du_{i+1,j}^{n+1} = u_{i,j}^{n}$$

(2.5.48)

that is harder to invert.

2.5.2.2 Split-Step Methods

However, it is far simpler to use fractional time step or split-step methods that lead to solution of tridiagonal systems instead. The former splits the multidimensional problem into a series of one-dimensional problems. Use of the Crank–Nicolson scheme with this approach gives, for the first step, in which y derivatives are ignored,

$$\frac{u_{i,j}^{n+1/2} - u_{i,j}^{n}}{\Delta t / 2} = \frac{v}{2} \left[\frac{u_{i+1,j}^{n+1/2} - 2u_{i,j}^{n+1/2} + u_{i-1,j}^{n+1/2}}{\Delta x^2} + \frac{u_{i+1,j}^{n} - 2u_{i,j}^{n} + u_{i-1,j}^{n}}{\Delta x^2} \right]$$

$$(2.5.49)$$

In the next step, x derivatives are ignored:

$$\frac{u_{i,j}^{n+1} - u_{i,j}^{n+1/2}}{\Delta t / 2} = \frac{v}{2} \left[\frac{u_{i,j+1}^{n+1/2} - 2u_{i,j}^{n+1/2} + u_{i,j-1}^{n+1/2}}{\Delta y^2} + \frac{u_{i,j+1}^{n+1} - 2u_{i,j}^{n+1} + u_{i,j-1}^{n+1}}{\Delta y^2} \right]$$

$$(2.5.50)$$

This scheme is unconditionally stable and second-order accurate in time and space and requires a tridiagonal matrix to be inverted at each fractional step. For 3D problems, one uses three fractional steps, each with a time step of $\Delta t/3$.

However, a more popular technique is the alternate direction implicit (ADI) method, where instead of ignoring the derivative in the other direction, the derivative is retained but in explicit form so that the resulting matrix is still tridiagonal. In the first step, the x direction is made implicit, while the y direction is kept explicit. In the subsequent step, the x direction is made explicit, but the y direction is implicit,

$$\frac{u_{i,j}^{n+1/2} - u_{i,j}^{n}}{\Delta t / 2} = \frac{v}{2} \left[\frac{u_{i+1,j}^{n+1/2} - 2u_{i,j}^{n+1/2} + u_{i-1,j}^{n+1/2}}{\Delta x^2} + \frac{u_{i,j+1}^{n} - 2u_{i,j}^{n} + u_{i,j-1}^{n}}{\Delta y^2} \right]$$

$$(2.5.51)$$

$$\frac{u_{i,j}^{n+1} - u_{i,j}^{n+1/2}}{\Delta t / 2} = \frac{v}{2} \left[\frac{u_{i+1,j}^{n+1/2} - 2u_{i,j}^{n+1/2} + u_{i-1,j}^{n+1/2}}{\Delta x^2} + \frac{u_{i,j+1}^{n+1} - 2u_{i,j}^{n+1} + u_{i,j-1}^{n+1}}{\Delta y^2} \right]$$

$$(2.5.52)$$

leading to two tridiagonal systems to be solved at each time step:

$$Du_{i-1,j}^{n+1/2} - 2(1+D)u_{i,j}^{n+1/2} + Du_{i+1,j}^{n+1/2} = -[D'u_{i,j-1}^{n} + 2(1-D')u_{i,j}^{n} + D'u_{i+1,j}^{n}]$$

$$D'u_{i,j-1}^{n+1} - 2(1+D')u_{i,j}^{n+1} + D'u_{i,j+1}^{n+1} = -[Du_{i-1,j}^{n+1/2} + 2(1-D)u_{i,j}^{n+1/2} + Du_{i+1,j}^{n+1/2}]$$

$$(2.5.53)$$

Stability analysis gives for the amplification factor

$$\xi = \frac{[1 - 2D\sin^2(k_x \Delta x/2)]}{[1 + 2D\sin^2(k_x \Delta x/2)]} \cdot \frac{[1 - 2D'\sin^2(k_y \Delta y/2)]}{[1 + 2D'\sin^2(k_y \Delta y/2)]} \leq 1$$

(2.5.54)

so the method is unconditionally stable. Extension to three dimensions is straightforward, with two directions kept explicit at each fractional step of $\Delta t/3$.

2.6 FINITE-DIFFERENCE (GRID POINT) METHODS

Now that we have a basic understanding of CFD techniques, let us look at how discretization in the horizontal directions is carried out in ocean models. There are principally three ways to solve the governing equations in the horizontal directions: (1) finite-difference, (2) spectral or spectral transform, and (3) finite-element methods. The zonal and meridional continuity of the atmospheric envelope makes it inherently well adapted to spectral and spectral transform methods, which are therefore the preferred methods in the atmosphere (Holton, 1992; Haltiner and Williams, 1980). The availability of fast transforms makes this technique feasible, since transformation from spectral to physical space and back is necessary for data insertion and calculation of nonlinear terms. On the other hand, because of the meridional boundaries, and their complicated shape, the ocean basins are inherently ill suited to spectral techniques. Finite-difference methods are more prevalent here. Finite-element methods (FEMs) originated in the coastal engineering community are becoming increasingly popular because of their nearly unlimited flexibility in adapting the grid locally to any desired resolution. However, the finite-difference technique is the current commonly used method for basin scale and global ocean modeling and this is the method we concentrate upon in this book.

Various numerical aspects of using finite-difference methods to solve for oceanic (atmospheric) flows can be explored by considering two simple cases: the classical one-dimensional advection diffusion equation and the shallow water equation. These were fully explored 20 years ago by Akio Arakawa and his co-workers (Mesinger and Arakawa, 1976), and the following is based mostly on their work as well as that of others such as Haltiner and Williams (1980) and Holton (1992).

Consider first the advection–diffusion equation (2.5.1). Using Taylor series expansions at adjacent grid points j−1 and j+1,

$$u_{j+1} = u_j + \left(\frac{\partial u}{\partial x}\right)_j \Delta x + \left(\frac{\partial^2 u}{\partial x^2}\right)_j \frac{\Delta x^2}{2!} + \left(\frac{\partial^3 u}{\partial x^3}\right)_j \frac{\Delta x^3}{3!} + o(\Delta x^4)$$

$$u_{j-1} = u_j - \left(\frac{\partial u}{\partial x}\right)_j \Delta x + \left(\frac{\partial^2 u}{\partial x^2}\right)_j \frac{\Delta x^2}{2!} - \left(\frac{\partial^3 u}{\partial x^3}\right)_j \frac{\Delta x^3}{3!} + o(\Delta x^4)$$

(2.6.1)

and by suitable combinations of u_{j-1} from u_{j+1}, we get

$$\left(\frac{\partial u}{\partial x}\right)_j = \frac{u_{j+1} - u_j}{\Delta x} + o(\Delta x)$$

$$\left(\frac{\partial u}{\partial x}\right)_j = \frac{u_j - u_{j-1}}{\Delta x} + o(\Delta x)$$

$$\left(\frac{\partial u}{\partial x}\right)_j = \frac{u_{j+1} - u_{j-1}}{2\Delta x} + o(\Delta x^2)$$

$$\left(\frac{\partial^2 u}{\partial x^2}\right)_j = \frac{u_{j+1} - 2u_j + u_{j-1}}{\Delta x^2} + o(\Delta x^2)$$

$$(2.6.2)$$

The first two finite-difference approximations are first-order accurate, while the last two are second-order accurate. The first one involves forward differences, the second one backward differences, and the third one centered differences. Converting the spatial derivatives in Eq. (2.6.1) to finite differences (second-order accurate) leads to the following differential-difference equation involving the time derivative:

$$\frac{du_j}{dt} = -c\left(\frac{u_{j+1} - u_{j-1}}{2\Delta x}\right) + v\left(\frac{u_{j+1} - 2u_j + u_{j-1}}{\Delta x^2}\right) + o\left(\Delta x^2\right) \qquad (2.6.3)$$

In order to advance the solutions in time, it is necessary to evaluate the RHS. If we denote time step n+1 as the future, n as the present, and n−1 as the past, then it is clear that if the RHS is evaluated at time step n or n−1, the solution at time step n+1 at grid point j can be readily obtained, based on its value at previous time steps n and n−1. The scheme is explicit. On the other hand, if any of the terms on the RHS are evaluated at time step n+1, then the solution requires simultaneous solutions at more than one grid point and the scheme is implicit or semi-implicit, depending on whether or not all terms on the RHS are evaluated at time step n+1. There are principally two ways to finite difference the time derivative:

$$\left(\frac{du}{dt}\right)_j^n = \frac{u_j^{n+1} - u_j^n}{\Delta t} + o(\Delta t)$$

$$\left(\frac{du}{dt}\right)_j^n = \frac{u_j^{n+1} - u_j^{n-1}}{2\Delta t} + o(\Delta t^2)$$

$$(2.6.4)$$

The first one is a two-time-level scheme, whereas the second one is a three-time-level scheme (in this case a leapfrog scheme, because the time derivative does not involve the present time step n). Multilevel schemes are more accurate, although care must be taken in using them. Since most numerical solutions involve integration forward in time from known initial conditions over many time steps, it is essential to have a scheme that is as accurate as possible in time differencing. The leapfrog scheme is second-order accurate and hence is heavily used. Three-level schemes generally involve grid point values at n−1, n, and n+1 time steps, since the RHS of Eq. (2.6.3) can involve any or a combination of the values at these time steps. How exactly the time differencing is done and how the terms on the RHS are evaluated determine the method and difficulty of solution as well as its accuracy, stability, and other numerical characteristics. Consider the FTCS scheme:

$$u_j^{n+1} = u_j^n - (C/2)(u_{j+1}^n - u_{j-1}^n) + D(u_{j+1}^n - 2u_j^n + u_{j-1}^n)$$

$$C = c\Delta t / \Delta x; \quad D = v\Delta t/\Delta x^2$$

$$(2.6.5)$$

Using von Neumann stability analysis, we get for the amplification factor

$$\xi = 1 - iC \sin(k\Delta x) - 2D[1 - \cos(k\Delta x)]$$

$$(2.6.6)$$

Geometrically, $|\xi|$ is the length of the line from origin to an ellipse with semimajor and semiminor axes of 2D and C, and centered at (1−2D, 0) in x–y space. This ellipse must lie within the unit circle for stability. Figure 2.6.1 shows the amplification factor plotted in polar and Cartesian coordinates for various values of the two parameters. This condition leads to the equation ($\theta = k\Delta x$)

$$\left(D + \frac{C^2}{4D}\right) + \left(-D + \frac{C^2}{4D}\right)\cos\theta \le 1; \quad 0 \le \theta \le 2\pi$$

$$(2.6.7)$$

The necessary and sufficient condition for the amplification factor $|\xi| \le 1$ is (e.g., Hirsch, 1988)

$$C^2 \le 2D \le 1 \qquad (2.6.8)$$

so that

$$\frac{c^2\Delta t}{v} \le 2 \text{ and } \frac{v\Delta t}{\Delta x^2} \le \frac{1}{2} \qquad (2.6.9)$$

such that the time step must be less than the smaller of those given by these two conditions. The scheme is unconditionally unstable when $v = 0$. Viscosity stabilizes an otherwise unstable central difference scheme for advection. Equation (2.6.9) implies $(c\Delta t / \Delta x) \leq 1$, the classic Courant–Friedrichs–Lewy condition for stability. However, this is a necessary but not sufficient condition. The cell Reynolds number, $R_c = C/D = c\Delta x/v$, must satisfy $2C \leq R_c \leq (2/C)$, so that stable calculations can be performed when values of R_c are higher than 2. Incorrect derivation of the von Neumann condition has led to the often incorrect assertion in literature that $R_c \leq 2$ for stability.

Usually the CFL condition $\Delta t \leq \Delta x / c$ is the more restrictive condition that needs to be satisfied in this case. However, the condition $\Delta t \leq \Delta x^2 /(2v)$ can be more restrictive if the viscosity v is large, as in some physical problems. Using an implicit scheme for the diffusion terms instead,

$$u_j^{n+1} = u_j^n - (C/2)(u_{j+1}^n - u_{j-1}^n) + D(u_{j+1}^{n+1} - 2u_j^{n+1} + u_{j-1}^{n+1}) \quad (2.6.10)$$

the amplification factor becomes

$$\xi = \{1 - iC \sin(k\Delta x)\} / \{1 + 2D[1 - \cos(k\Delta x)]\} \quad (2.6.11)$$

Since the denominator is always ≥ 1, and the numerator is always ≤ 1, it is easy to see that the stability condition is unconditionally satisfied irrespective of the value of C or D. In other words, making the diffusion terms implicit permits the severe time step limitation of an explicit scheme to be sidestepped.

With leapfrog time differencing and the diffusion term evaluated at the previous time step,

$$\left(\frac{u_j^{n+1} - u_j^{n-1}}{2\Delta t}\right) = -c\left(\frac{u_{j+1}^n - u_{j-1}^n}{2\Delta x}\right) + v\left(\frac{u_{j+1}^{n-1} - 2u_j^{n-1} + u_{j-1}^{n-1}}{\Delta x^2}\right) + o(\Delta x^2, \Delta t^2)$$

$$u_j^{n+1} = u_j^{n-1} - C(u_{j+1}^n - u_{j-1}^n) + 2D(u_{j+1}^{n-1} - 2u_j^{n-1} + u_{j-1}^{n-1}) \quad (2.6.12)$$

The amplification factor is

$$\xi^2 = 1 - i2C\xi \sin(k\Delta x) - 4D[1 - \cos(k\Delta x)] \quad (2.6.13)$$

Recall that the leapfrog scheme with centered spatial differences is conditionally stable for the advection equation, but unstable for the diffusion equation. This is why in the mixed case of both advection and diffusion in Eq. (2.6.6) the advection and diffusion terms are finite differenced the way they are,

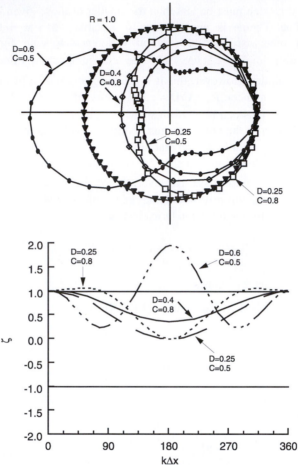

Figure 2.6.1 Ampification factor for the advection-diffusion equation for the leapfrog scheme (in polar and Cartesian coordinates).

with diffusion terms lagged one time step (to n−1) to assure stability, while advection terms are evaluated at the present time step n. This explicit (second-order accurate in both time and space) formulation is often preferable to the ones mentioned previously.

Spatial differencing has the tendency to introduce artificial *numerical* (or computational) *dispersion* into wave propagation that was not present in the original equations. This can be seen by looking at the characteristics of the differential-difference equation of the advection equation. Finite differencing only the spatial derivatives,

$$\frac{du_j}{dt} + c\frac{u_{j+1} - u_{j-1}}{2\Delta x} = 0$$

(2.6.14)

and expanding in terms of Fourier modes, $u_j(t) = \text{Re}[U(t)\exp i(jk\Delta x)]$, one gets

the oscillation equation, $\dfrac{du}{dt} + ik\hat{c}U = 0$, with $\hat{c} = c\dfrac{\sin(k\Delta x)}{k\Delta x}$ as the phase speed

of the numerical modes. Their group velocity is $\hat{c}_g = c\cos(k\Delta x)$. While the original PDE had a nondispersive wave with constant phase and group velocities equal to c, the waves in the numerical solution are dispersive, with their phase speed dependent on the wavenumber and decreasing from the theoretical value of c at very long wavelengths to zero at $k\Delta x = \pi$, for the wavenumber corresponding to the shortest resolvable wavelength $2\Delta x$, which is incidentally stationary, that is, has zero phase velocity! The group velocity also decreases monotonically, becomes zero for a $4\Delta x$ wave, and reverses sign and continues to decrease to −c for a $2\Delta x$ wave. Thus an initial disturbance disperses instead of propagating intact. This artificial numerical dispersion is a serious problem and affects the fidelity of numerical solutions. Higher order schemes mitigate some of these problems, but no method is satisfactory at wavelengths close to $2\Delta x$. If such wavenumbers are important, the only solution is to increase the grid resolution.

Extension of these aspects to two dimensions is straightforward. For example, the CFL condition for the leapfrog scheme for the advection equation can be shown to be $C \leq 2^{-1/2}$ for grids with equal spacing in the two directions.

We next consider the shallow water equation. So far we have dealt with PDEs with only one dependent variable. For application to geophysical flows, extension to more than one variable is essential. This brings in considerations such as staggered grids that are best illustrated with the simplest possible problem. Consider a linear 1D shallow water equation for a nonrotating fluid of constant depth H:

$$\frac{\partial \eta}{\partial t} + H\frac{\partial u}{\partial x} = 0$$

$$\frac{\partial u}{\partial t} + g\frac{\partial \eta}{\partial x} = 0$$

(2.6.15)

This has solutions of the form $u, \eta \sim \exp i(kx - \sigma t)$, which satisfy the dispersion relation $c = \sigma/k = (gH)^{1/2}$. The phase speed is independent of the wavenumber; the shallow water waves are nondispersive. An initial disturbance retains its form as it propagates with time in the x direction. The group velocity $c_g = \partial \sigma / \partial k = c$.

Now consider numerical solutions. Using centered spatial differencing, the differential-difference equations can be written as

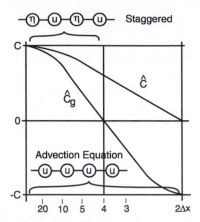

Figure 2.6.2 Illustration of numerical dispersion. Plot of the phase and group velocities as a function of the wavenumber. Note that the $2\Delta x$ wave is stationary and its group velocity is actually negative.

$$\frac{\partial \eta_j}{\partial t} = -H\frac{u_{j+1} - u_{j-1}}{2\Delta x} ; \quad \frac{\partial u_j}{\partial t} = -g\frac{\eta_{j+1} - \eta_{j-1}}{2\Delta x}$$

$$(2.6.16)$$

which have solutions of the form $u_j, \eta_j \sim \exp i(kj\Delta x - \sigma t)$, which satisfy the dispersion relation $\hat{c} = \sigma/k = c\dfrac{\sin(k\Delta x)}{k\Delta x}$. Once again, the phase speed in the numerical solution is not independent of the wavenumber. The group velocity c_g = $\partial \sigma / \partial k = c\ \cos\ (k\Delta x)$. The numerical waves have become dispersive. Finite differencing in space has caused computational dispersion in this case as well. As the wavenumber increases, the phase speed decreases monotonically from c, the true value at $k\Delta x = 0$, to zero at $k\Delta x = \pi$. The shortest resolvable two-grid interval wave of wavelength $2\Delta x$ is stationary. The corresponding group velocity decreases from c at $k\Delta x = 0$ to zero at $k\Delta x = \pi/2$, and to $-c$ at $k\Delta x = \pi$, the shortest resolvable wavelength (see Figure 2.6.2).

While the results are similar to those for the advection equation, there is a very important difference. The shallow water equation has two dependent variables, and it is possible to calculate the solutions on a staggered grid, with variables carried at alternate points, but discretization interval Δx remains unchanged for each variable, as shown in Figure 2.6.2 (in a nonstaggered grid, both variables are carried at the same grid points). This halves the computational time without altering the truncation error! A further advantage is that wavenumbers $k\Delta x > \pi/2$ are eliminated so that the large phase speed errors and negative group velocities are no longer a problem. This is one reason for the extensive use of staggered grids in solving fluid flow problems using finite-difference methods.

2.6.1 STAGGERED GRIDS

Extension to two-dimensional flows is straightforward but best done including rotation. The governing equations, in both primitive variables u and v, and in terms of divergence $\delta = \left(\dfrac{\partial u}{\partial x} + \dfrac{\partial v}{\partial y} \right)$ and vorticity $\zeta = \left(\dfrac{\partial v}{\partial x} - \dfrac{\partial u}{\partial y} \right)$, are

$$\frac{\partial \eta}{\partial t} + H\left(\frac{\partial u}{\partial x} + \frac{\partial v}{\partial y} \right) = 0$$

$$\frac{\partial u}{\partial t} - fv + g\frac{\partial \eta}{\partial x} = 0$$

$$\frac{\partial v}{\partial t} + fu + g\frac{\partial \eta}{\partial y} = 0$$

$$\frac{\partial \eta}{\partial t} + H\delta = 0$$

$$\frac{\partial \delta}{\partial t} - f\zeta + g\left(\frac{\partial^2 H}{\partial x^2} + \frac{\partial^2 H}{\partial y^2} \right) = 0$$

$$\frac{\partial \zeta}{\partial t} + f\delta = 0$$

$$(2.6.17)$$

For simplicity, consider the grid size to be the same in both directions (d = Δx = Δy). As discussed in Chapter 1, an important length scale is the Rossby radius of deformation a = c/f, and the ratio d/a of the grid size d to this length scale is an important parameter.

Now there are five possible arrangements of the three dependent variables η, u, and v: grids A to E, of which only A is nonstaggered (Figure 2.6.3). These are called *Arakawa grids*. The shortest distance between grid points with the same variable is d. All these grids have the same number of dependent variables per unit area; hence, the computation time requirement is similar. But the nature of the solutions is quite different. Note that grid E can be obtained by a 45° rotation from grid B and has the same density of points if the grid spacing is chosen as $2^{1/2}$d. Note that except for grid E, for which it is $(2d)^{1/2}$, the shortest resolvable wave is 2d.

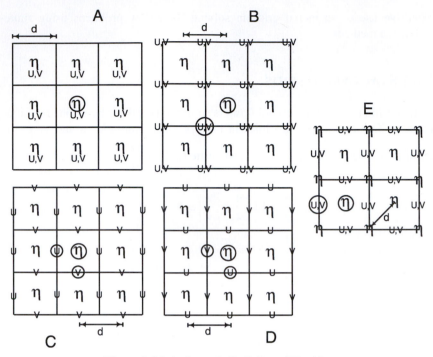

Figure 2.6.3 Arakawa A, B, C, D, and E grids.

How well a numerical scheme does in representing the exact differential equations (apart from the truncation error and accuracy questions) depends on how well each term in the governing equations involving spatial gradients can be calculated at the chosen resolution. Randall (1994) provides a particularly lucid discussion of this aspect. Ideally, in the calculation of tendency terms at a given grid point, anything that involves averaging values at the surrounding grid points should be avoided. The nonstaggered A grid permits ready evaluation of Coriolis terms in the u and v equations since u and v are collocated, but the pressure gradient (the spatial gradient of η) calculation involves averaging, and so does the divergence calculation in the continuity equation. The averaging process alters the pressure gradient in each direction (for example, $\partial\eta/\partial x$ requires averaging in the x direction). The B grid also permits ready evaluation of Coriolis terms, since u and v are collocated, but involves averaging in both pressure gradient and divergence calculations, with a very important difference. The calculation for the gradient in one direction involves averaging in the other direction (for example, $\partial\eta/\partial x$ requires averaging in the y direction) and therefore, changes in that direction are preserved. This is very important for geostrophic adjustment. The C grid is eminently suited to calculating the pressure gradients and divergence, without any averaging, but the Coriolis terms are compromised, because of the averaging required, since u and v are not collocated. This means that only the waves that are less affected by Coriolis forces are well represented. If the grid resolution is such that the smallest waves resolved are relatively insensitive to Coriolis forces, then the C grid can be expected to perform well. The D grid is ideally suited to enforcing the geostrophic balance, but requires averaging in the calculation of the pressure gradient, divergence, and Coriolis terms. The E grid permits calculation of all the Coriolis terms, the pressure gradient terms, and the divergence without any averaging. However, its 1D form is exactly the same as that of an A grid (but with the grid spacing reduced by a factor of $2^{1/2}$), with all three variables η, u, and v collocated. Thus it cannot do a good job on propagations in the principal coordinate directions.

To illustrate these aspects, consider wave-like solutions with wavenumber k_x in the x direction and k_y in the y direction: $\exp(ik_x x + ik_y y)$. The exact dispersion relation and the dispersion relation in various Arakawa grids are (Randall, 1994)

$$\left(\frac{\sigma}{f}\right)^2 = 1 + \left(\frac{a}{d}\right)^2 \left[(k_x d)^2 + (k_y d)^2\right] \qquad \text{- Exact}$$

$$= 1 + \left(\frac{a}{d}\right)^2 \left[\sin^2(k_x d) + \sin^2(k_y d)\right] \qquad \text{- Grid A}$$

$$= 1 + 2\left(\frac{a}{d}\right)^2 \left[1 - \cos(k_x d)\cos(k_y d)\right] \qquad \text{- Grid B}$$

$$= 4\left(\frac{a}{d}\right)^2 \left[\sin^2\left(\frac{k_x d}{2}\right) + \sin^2\left(\frac{k_y d}{2}\right)\right]$$

$$+ \frac{1}{4}[1 + \cos(k_x d) + \cos(k_y d) + \cos(k_x d)\cos(k_y d)] \quad \text{- Grid C}$$

$$= \left(\frac{a}{d}\right)^2 \left[\cos^2\left(\frac{k_x d}{2}\right)\sin^2(k_y d) + \cos^2\left(\frac{k_y d}{2}\right)\sin^2(k_x d)\right]$$

$$+ \frac{1}{4}[1 + \cos(k_x d) + \cos(k_y d) + \cos(k_x d)\cos(k_y d)] \quad \text{- Grid D}$$

$$= 1 + 2\left(\frac{a}{d}\right)^2 \left[\sin^2\left(\frac{k_x d}{\sqrt{2}}\right) + \sin^2\left(\frac{k_y d}{\sqrt{2}}\right)\right] \qquad \text{- Grid E}$$

$$= 1 + 4\left(\frac{a}{d}\right)^2 \left[\sin^2\left(\frac{k_x d}{2}\right) + \sin^2\left(\frac{k_y d}{2}\right)\right] \qquad \text{- Grid Z}$$

$$(2.6.18)$$

We have also added Randall's Z grid, similar to Arakawa's A grid, but with the primitive variables u and v replaced by the vorticity and divergence and collocated with η. The exact dispersion relation shows that both the phase speed $\sigma/|\mathbf{k}|$ and the group speed $|\partial\sigma/\partial\mathbf{k}|$ are nonzero at all wavenumbers, where $\mathbf{k}$ is the wavenumber vector. The phase speed increases monotonically with the wavenumber. These dispersion characteristics are important to geostrophic adjustment processes, and a numerical scheme should preserve these as far as possible. It is important to note that since the group velocity of the waves is never zero, local accumulation of energy is not possible during the adjustment process.

For simplicity, consider the case when the gradients in the y direction are negligible ($k_y = 0$, $k_x = k$). This case can be used to study geostrophic adjustment, a proper simulation of which requires that the inertia-gravity wave propagation must be correctly simulated. The exact dispersion relation is $(\sigma/f)^2 = 1 + k^2 a^2$, but finite differencing on various grids gives (see also Mesinger and Arakawa, 1976)

$$\left(\frac{\sigma}{f}\right)^2 = 1 + \left(\frac{a}{d}\right)^2 \sin^2(kd) \qquad \text{- Grid A}$$

$$= 1 + 4\left(\frac{a}{d}\right)^2 \sin^2\left(\frac{kd}{2}\right) \qquad \text{- Grid B}$$

$$= \cos^2\left(\frac{kd}{2}\right) + 4\left(\frac{a}{d}\right)^2 \sin^2\left(\frac{kd}{2}\right) \qquad \text{- Grid C}$$

$$= \cos^2\left(\frac{kd}{2}\right) + \left(\frac{a}{d}\right)^2 \sin^2\left(\frac{kd}{2}\right) \qquad \text{- Grid D}$$

$$= 1 + 2\left(\frac{a}{d}\right)^2 \sin^2\left(\frac{kd}{\sqrt{2}}\right) \qquad \text{- Grid E}$$

$$= 1 + 4\left(\frac{a}{d}\right)^2 \sin^2\left(\frac{kd}{2}\right) \qquad \text{- Grid Z}$$

$$(2.6.19)$$

For the numerical solutions, the wave speed is now a function of both the ratio a/d and kd. Figure 2.6.4 shows a plot of the dispersion relation for a/d = 2. Grid A has negative group velocities for waves with a wavelength below 4d. The 2d wave, the shortest resolved, has a zero group velocity and behaves like an inertial oscillation. The consequence is that the solutions are extremely noisy and require filtering (Kalnay-Rivas *et al.*, 1977). Grid B, on the other hand, has zero group velocity reached for 2d waves only. Grid C has a similar behavior for a/d = 2. But for a/d < 1/2, the frequency decreases monotonically and departs early from the true solution, whereas for a/d > 1/2, it increases monotonically. For a/d = 1/2,

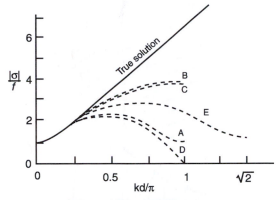

Figure 2.6.4 Dispersion relation for the 1-D shallow water wave for various Arakawa grids for a/d = 2.

the group velocity is zero for all k. Thus, if the C grid has a resolution fine enough to resolve the Rossby radius it does well; if not, the performance suffers. Grid D has negative group velocities below a wavelength of 4d like grid A, but the 2d wave is stationary. Grid E behaves similar to B, except that negative group velocities are reached for wavelength $2^{3/2}d$ and inertial oscillations result for wavelength $2^{1/2}d$. The results for B and E are equivalent, since one can be obtained from the other by a counterclockwise coordinate rotation by 45°. Clearly, all grids have potential problems with geostrophic adjustment since zero group velocities occur, but Arakawa grids A and D (and to some extent E) are particularly inadequate in terms of computational dispersion and therefore seldom used, although for nested models, especially movable ones, unstaggered A grids are more convenient (Kurihara *et al.,* 1998; see also Ginis *et al.,* 1998).

Most ocean models use either an Arakawa B or an Arakawa C staggered grid for horizontal discretization. For the fully two-dimensional case of Eq. (2.6.17), Figure 2.6.5 shows the true solution in wavenumber space and solutions for grids B and C for a/d = 2. The dispersion characteristics are better for grid C than for grid B, with no wave having a negative group velocity. However, if the grid does not resolve the Rossby radius (a/d < 1), grid C loses this advantage and grid B is preferable. This is one reason the very first comprehensive global ocean model was formulated by Kirk Bryan on a B grid. The available computer resources were not adequate then to permit the Rossby radius (typically 40 km in midlatitudes) to be resolved. Even today, many global models, such as MOM2, are still cast on Arakawa B grids (Semtner, 1995). However, with modern multigigaflop computers, finer grid resolutions are becoming possible and global ocean models (POP, for example) are being increasingly cast in Arakawa C grids. Atmospheric models, before the advent of spectral methods, used the

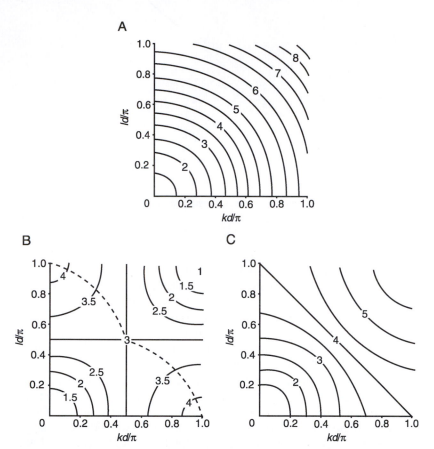

Figure 2.6.5 Dispersion relation for the 2-D shallow water wave for Arakawa B(B) and C(C) grids compared to the true solution (A) for a/d = 2.

Arakawa C grid, since a grid size less than the Rossby radius (~1000 km) was affordable.

Randall (1994) advocates the use of Z grids. No averaging whatsoever is required in solving for η, δ, and ζ on an unstaggered grid. Figure 2.6.6, from Randall (1994), shows the dispersion relation at two values of a/d, 0.1 and 2. The Z grid behaves very nearly like the exact dispersion at both values of a/d. Since it is unstaggered, collapsing to one dimension does not make any difference. The C grid is particularly bad at an a/d of 0.1, and the B grid has moderate problems at both values. According to Randall (1994), these properties had been anticipated earlier (e.g., Neta and Williams, 1989).

The CFL condition for a barotropic (or reduced gravity) calculation for an ocean of effective depth H_e and effective gravity g_e depends on the advective

a/d=2   a/d=0.1

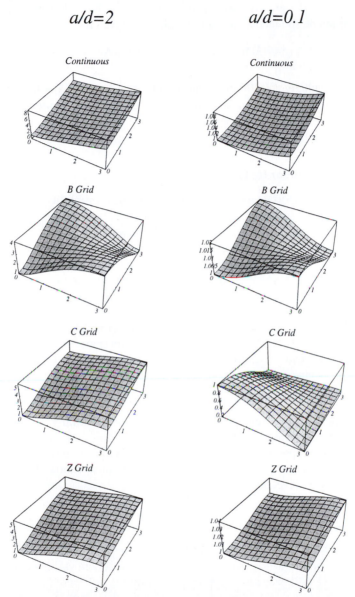

Figure 2.6.6 Dispersion relation for continuous shallow water equations and finite difference schemes on Arakawa B, C, and Randall Z grids, the left column for a/d = 2, and the right for a/d = 0.1. The horizontal coordinates are normalized wavenumbers kd and id, and the vertical coordinate is frequency normalized by f. From Randall (1994).

velocity U and the gravity wave speed $c = (g_e H_e)^{1/2}$. For arbitrary values of Δx and Δy, it can be shown that the CFL condition is given by

$$(c + |U|)\Delta t < \Delta x_e; \quad \Delta x_e = [\Delta x^{-2} + \Delta y^{-2}]^{-1/2}$$

$$(2.6.20)$$

For the atmosphere, $c \sim 100$ m s^{-1} and $U \sim 60$ m s^{-1} or less. For the oceans, $c \sim$ 2–3 m s^{-1} and $U \sim 2$ m s^{-1} or less for internal gravity wave modes. For external gravity wave modes, in the deep ocean, $c \sim 220$ m s^{-1}, and for the atmosphere, the Lamb wave has a roughly similar propagation speed, 300 m s^{-1}.

2.6.1.1 Nonlinear Instability

Finally, it is important to be aware of the nonlinear instability that arises from the nonlinear interaction (through advection terms) of modes with high wavenumbers close to the shortest resolvable wavelength of $2\Delta x$. These are too short to be represented on the grid and their energy is falsely aliased into lower wavenumbers. Repeated aliasing over the course of integration leads to a rapid buildup of energy in the $2\Delta x$ to $4\Delta x$ wavelengths and eventual instability. One strategy to prevent this from happening is to periodically remove these wavelengths by selective filtering and hence prevent a spurious energy buildup. Another is to add an artificial diffusion ($A\nabla^2 u$) term. Mesinger and Arakawa (1976) suggest that it may be better to control the spurious energy inflow to these short modes instead of suppressing their amplitude. This can be done by finite-difference schemes that conserve certain integral properties of the original differential equations. The important quantities to conserve are the mean kinetic energy and the mean square vorticity or enstrophy.

When dealing with the primitive form of the governing equations, it is important to make sure that the particular form used in finite differencing leads automatically to preservation of the conservation properties of the original equation. This can be illustrated by looking at the conservation of energy in the simple nonlinear advection equation,

$$\frac{\partial u}{\partial t} + u\frac{\partial u}{\partial x} = 0$$

$$(2.6.21)$$

and the corresponding equation for kinetic energy:

$$\frac{\partial}{\partial t}\left(\frac{u^2}{2}\right) = -u^2\frac{\partial u}{\partial x}$$

$$(2.6.22)$$

The integral of the KE over the domain depends only on the boundary conditions:

$$\frac{d}{dt}\int_a^b (u^2/2)dx = -(u_b^3 - u_a^3)/3$$

(2.6.23)

If $u_a = u_b = 0$, there is no flux of energy into the domain, and the mean KE is conserved. Now consider the centered space difference form of the differential-difference equation:

$$\frac{\partial u_j}{\partial t} = -u_j \frac{u_{j+1} - u_{j-1}}{2\Delta x}$$

(2.6.24)

The expression for total KE is obtained by multiplying by u_j and summing over all grid points except the end points:

$$\frac{d}{dt}\sum_{j=2}^{N-1} \frac{u_j^2}{2}\Delta x = -\frac{1}{2}\sum_{j=2}^{N-1}\left(u_j^2 u_{j+1} - u_j^2 u_{j-1}\right)$$

(2.6.25)

The term in parentheses on the RHS is not of the form $(A_{j+1} - A_j)$, and hence the RHS does not necessarily sum to zero; therefore, this form is not energy conserving. However the form

$$\frac{\partial u_j}{\partial t} = -\frac{\left(u_{j+1} + u_j + u_{j-1}\right)\left(u_{j+1} - u_{j-1}\right)}{6\Delta x}$$

(2.6.26)

is, since the RHS of

$$\frac{d}{dt}\sum_{j=2}^{N-1} \frac{u_j^2}{2}\Delta x = -\frac{1}{2}\sum_{j=2}^{N-1}\left[u_j u_{j+1}\left(u_{j+1} + u_j\right) - u_{j-1}u_j\left(u_j + u_{j-1}\right)\right]$$

(2.6.27)

is in proper form and hence sums to zero (Haltiner and Williams, 1980).

While some sort of averaging is necessary to assure conservation of the energy in a nonstaggered grid, the use of staggered grids and equations in flux-conservative form is one way of assuring such conservation (see equations in Chapters 6 and 9 for examples). This is especially important for nonlinear terms in the equations since aliasing tends to build up spurious energy at high wavenumbers as described above. Flux form for advective terms then assures that their finite-difference equivalents are automatically in the form $(A_{j+1} - A_j)$, so that mass, momentum, and kinetic energy conservation properties of the original equations are not compromised when the equations are discretized. Most primitive equation-based finite-difference schemes using staggered grids conserve the mean kinetic energy, not necessarily the enstrophy, and hence there is a need for an artificial diffusion term when nonlinear terms are present.

Mesinger and Arakawa (1976) suggest that if the finite-difference scheme conserved both mean kinetic energy and enstrophy, this would prevent the spurious inflow of energy to these nonlinear modes and hence the buildup. It would also obviate the need for artificial damping. This can be readily seen from the vorticity advection equation, which can be written without the diffusion terms as

$$\frac{\partial}{\partial t}\zeta + \mathbf{v}.\nabla\zeta = 0$$

or since $\zeta = \nabla^2\psi$ and $\mathbf{v} = \mathbf{k}\times\nabla\psi$

$$\frac{\partial}{\partial t}\nabla^2\psi + J(\nabla^2\psi, \psi) = 0$$

$$(2.6.28)$$

where J is the Jacobian, which can be evaluated numerically in one of the three following ways:

$$J^0(p, q) = \frac{\partial p}{\partial x}\frac{\partial q}{\partial y} - \frac{\partial p}{\partial y}\frac{\partial q}{\partial x}$$

$$J^1(p, q) = \frac{\partial}{\partial y}\left(q\frac{\partial p}{\partial x}\right) - \frac{\partial}{\partial x}\left(q\frac{\partial p}{\partial y}\right)$$

$$J^2(p, q) = \frac{\partial}{\partial x}\left(p\frac{\partial q}{\partial y}\right) - \frac{\partial}{\partial y}\left(p\frac{\partial q}{\partial x}\right)$$

$$(2.6.29)$$

Mathematically the three forms are equivalent. Numerically, with a centered difference scheme, they can all be finite differenced to second-order accuracy. But only when the Jacobian is evaluated as the average of the three Jacobians above does the scheme conserve the average kinetic energy, the average vorticity, and the average square vorticity or enstrophy (Arakawa, 1966; Mesinger and Arakawa, 1976). This Jacobian is called the Arakawa Jacobian and use of this to finite difference nonlinear advection terms ensures that no buildup of spurious energy at large wavenumbers, and hence no nonlinear instability, occurs (Arakawa, 1970; Arakawa and Lamb, 1977, 1981). Arakawa and Hsu (1990) have generalized this to the barotropic shallow water equations, and derived a family of finite-difference schemes that ensures potential vorticity dissipation with little or no energy dissipation.

2.6.2 TIME DIFFERENCING AND FILTERING

Once a chosen spatial differencing scheme (whether finite-difference, finite-element, or spectral) is implemented in an ocean or atmosphere model, the PDE governing the evolution in time of a prognostic variable reduces to an ODE of the form

$$d\varphi_{i,j} / dt = F \qquad (2.6.30)$$

where F denotes the terms on the right-hand side that are known from the previous time steps as long as the method is explicit. The most popular time differencing scheme is the three-time-level leapfrog scheme, since it provides a second-order accuracy in time with only one function evaluation each time step. It is neutral stabilitywise in that it neither amplifies nor damps linear oscillatory motion. Its most serious shortcoming is the time-splitting instability that develops in nonlinear problems because of the presence of the undamped purely computational mode in addition to the physical mode. When applied to nonlinear equations, there is a tendency for the computational mode to amplify slowly, generating a $2\Delta t$ noise. One way around this problem is to take an occasional step with a two-time-level scheme, which discards the oldest time level and hence eliminates the computational mode. Another is to use a time filter of the form

$$\varphi_f^n = \varphi^n + \frac{\alpha}{2}(\varphi^{n+1} - 2\varphi^n + \varphi_f^{n-1})$$

$$(2.6.31)$$

where $\varphi^{n+1} = \varphi_f^{n-1} + 2\Delta F(\varphi^n)$ at each time step. Subscript f denotes the filtered variable. This is called the Asselin, or more appropriately the Asselin–Robert, filter (Asselin, 1972; see also Robert, 1966). This filter damps high frequencies selectively and is efficient in suppressing the computational mode. This method is very commonly used in both atmospheric spectral and ocean finite-difference models. Various values are used for α, anywhere from 0.1 to 0.3, the usual value being around 0.1. However, the use of the Asselin–Robert filter tends to degrade the accuracy of the scheme, making it less than $O(\Delta t^2)$ accurate, the exact degree of degradation depending on the chosen value of α.

As we saw earlier, a two-level scheme avoids the problem of the computational mode, but is only first-order accurate. The second-order Runge–Kutta methods such as Huen's method require two function evaluations each time step, and also lead to slowly growing instabilities. The fourth-order Runge–Kutta is highly accurate and often used, but it requires four function evaluations per time step and imposes far higher storage requirements compared to two-time-level schemes. Multistep predictor–corrector schemes, such as the Adams–Bashforth schemes, are one alternative. Unfortunately, the second-order Adams–Bashforth scheme is not only subject to the same instability as the second-order Runge–Kutta schemes, but is subject to large phase speed errors (Mesinger and Arakawa, 1976). This prompted Durran (1991) to examine a third-order Adams–Bashforth scheme that appears to have all the advantages of leapfrogging, such as requiring only one function evaluation each time step and high accuracy, but is not subject to time splitting. However, it requires 50% more storage space (for the same grid size), which is not necessarily a handicap on modern computers, which routinely make available a large core memory. If the grid spacing can be increased by a suitable factor, the same storage space requirement can be realized at a modest decrease in spatial resolution (~12% for 3D problems).

An Adams–Bashforth scheme of Nth order involves the use of

$$\varphi^{n+1} = \varphi^n + \Delta t \sum_{m=0}^{N-1} a_m F(\varphi^{n-m})$$

(2.6.32)

where the coefficients a_m are determined by substituting Taylor series expansions into Eq. (2.6.32) and requiring terms less than $O(\Delta t^N)$ to cancel (Gear, 1971; Durran, 1991). The first-order scheme, which is the Euler scheme, requires $a_0 = 1$; the second-order scheme requires $a_0 = 3/2$, $a_1 = -1/2$. The third-order scheme gives

$$\varphi^{n+1} = \varphi^n + \Delta t \left[\frac{23}{12} F(\varphi^n) - \frac{16}{12} F(\varphi^{n-1}) + \frac{5}{12} F(\varphi^{n-2}) \right] + o(\Delta t^4)$$

(2.6.33)

To assess the characteristics of this scheme vis-à-vis the Asselin-filtered leapfrog scheme, Durran (1991) followed the traditional approach that uses the oscillation equation (Mesinger and Arakawa, 1976):

$$d\varphi / dt = i\omega\varphi$$

(2.6.34)

In the limit $(\omega\Delta t) \ll 1$, for the Asselin–Robert-filtered leapfrog scheme, the amplification factor $|A| = |\varphi^{n+1}/\varphi^n|$ is $[0.5\alpha(2-\alpha)^{-1}](\omega\Delta t)^2 + O(\omega\Delta t)^4$ for the physical mode, but it is $(1-\alpha) + O(\omega\Delta t)^2$, relatively independent of the time step, for the computational mode. The filter introduces a second-order error in the amplitude of the physical mode whose impact depends on the magnitude of α. The phase speed of the physical mode,

$$1 + [0.1667(1+\alpha)(1-0.5\alpha)^{-1}(\omega\Delta t)^2],$$

is only second-order accurate and increases with α. On the other hand, for the third-order Adams–Bashforth scheme, the amplification factor is $1 - 0.375(\omega\Delta t)^4$, and the phase speed is $1 - 0.375(\omega\Delta t)^4$. Clearly, the latter scheme is more accurate. Stability requires that $(\omega\Delta t)$ be less than 0.724 for the Adams–Bashforth scheme, but for the leapfrog one, it is decreased below 1 by filtering, and is 0.816 for $\alpha = 0.4$. Durran's (1991) Table 1 provides a nice summary of the characteristics of various other time differencing schemes including these two. In general, second-order time differencing schemes have fourth-order amplitude errors and second-order phase speed errors. A third-order scheme reduces the phase speed error further to fourth order.

Kurihara (1965; see also Kurihara *et al.*, 1998) uses a predictor–corrector method based on the leapfrog scheme but without the Asselin–Robert time filtering,

$$\varphi^{n+1} = \varphi^n + \Delta t[(1-\gamma)F(\varphi^*) + \gamma F(\varphi^n)]$$

$$\text{where} \quad \varphi^* = \varphi^{n-1} + 2\Delta t F(\varphi^n)$$

$$(2.6.35)$$

where γ is a weight chosen to damp unwanted noise. This scheme damps the computational mode without degrading the accuracy. For $\gamma = 0.5$, the amplification factor is $1-0.25(\omega\Delta t)^4$ and the phase speed is $1-0.083(\omega\Delta t)^2$.

Also, the leapfrog scheme is unstable when applied to diffusion–dissipation-type problems (unless the diffusion terms are lagged and evaluated at time step n−1 instead of n), whereas the third-order Adams–Bashforth scheme can be. When applied to the friction equation (Mesinger and Arakawa, 1976),

$$d\varphi / dt = -c\varphi \qquad (2.6.36)$$

the third-order Adams–Bashforth is stable for $(c\Delta t)$ less than 0.545, whereas the lagged leapfrog involving evaluation of the RHS at time step n-1 is stable for $(c\Delta t)$ less than 0.5.

When solving PDEs such as the advection equation, the time step is constrained by the Courant number criterion, and the spatial discretization is related to the time discretization by the Courant number $C = c\Delta t/\Delta x$. Therefore, $\omega\Delta t = C(k\Delta x)$, and the temporal filtering that the Asselin–Robert filter introduces is equivalent to a second-order spatial filter.

In multiply nested models, the A grid is often the grid of choice, and this can lead to large errors in phase speeds of short waves of wavelength less than $4\Delta x$. In addition, if a second-order scheme is used and the Asselin–Robert filter is not, unwanted high frequency noise can develop in nonlinear problems. To suppress high frequency "noise" and preserve low frequency signals, Kurihara *et al.* (1998) and Ginis *et al.* (1998) use a two-step scheme for time integration for any prognostic variable,

$$\varphi^* = \varphi^n + \Delta t[LF^n + HF^n + DF^n]$$

$$\varphi^{n+1} = \varphi^* + \Delta t\{[(1-\gamma_1)LF^n + \gamma_1 LF^*] + [(1-\gamma_2)HF^n + \gamma_2 HF^*] + DF^n\}$$

$$(2.6.37)$$

where LF denotes low frequency terms due to advection, HF denotes high frequency terms due to the pressure gradient and Coriolis acceleration, and DF denotes the fluxes due to vertical transport and diffusion, and horizontal diffusion terms. By suitable choice of γ_1 (~0.5) and γ_2 (~2.5), low frequency modes important to the flow evolution are preserved and the high frequency noise is suppressed.

Filtering is a time-honored technique to suppress high frequency components in either space or time. In modeling nonlinear systems, it is essential to have either implicit diffusion generated from the numerical scheme (such as in upwind differencing schemes) or explicit diffusion (as when using a leapfrog scheme) to suppress the accumulation at the grid scale of energy cascaded from larger scales down the wavenumber spectrum, which manifests itself as $2\Delta x$ noise and in severe cases can lead to model blowup. Occasional use of spatial filters during time integration acts as artificial diffusion and suppresses these high wavenumber modes. However, its use tends to also suppress lower wavenumbers. This can be illustrated with a 1D example. A single application of a simple three-point Laplacian smoother

$$\varphi_j^f = \varphi_j + \gamma(\varphi_{j+1} - 2\varphi_j + \varphi_{j-1})$$

$$(2.6.38)$$

suppresses waves of wavelength $2\Delta x$ by 100%, but also those of wavelength $4\Delta x$ by 50% and $6\Delta x$ by 25%, when $\gamma = 0.5$. This undesirable effect of the smoother can be controlled by making use of the fact that Eq. (2.6.38) acts as a desmoother if γ is negative and less than -0.5. Therefore, if a desmoothing filter is applied to a field smoothed by a smoother, the longer wavelengths can be more or less restored. For $\gamma = -0.56$, the wavelengths $4\Delta x$ and $6\Delta x$ decrease by only 22 and 4% from their unsmoothed values by the sequential application of a smoother-desmoother. This strategy is employed by Kurihara *et al.* (1998) and Ginis *et al.* (1998) in their multiply nested models that employ A grid schemes and hence are inherently noisy.

2.6.3 COMPUTATIONAL GRIDS

For the most part, global ocean models employ a longitude (λ)–latitude (θ) grid in spherical coordinates. The complicated shapes of the ocean basins and intervening land masses do not lend themselves readily to a boundary-adaptive grid, unless the constraint on the spatial grid is relaxed, and unstructured grids such as those used in finite-element models (see Section 2.8) are used. Within the constraints of structured grids, a nonuniform longitude–latitude grid can be used, often with a variable grid spacing in one or both directions. In such nonuniform grids, $\Delta\lambda = f(\lambda$ only) and $\Delta\theta = g(\theta$ only), where f and g are slowly varying functions [equivalently in the x–y domain, $\Delta x = f(x)$ and $\Delta y = g(y)$]. For example, the need to resolve wave motions in the equatorial waveguide is addressed by a finer resolution in the meridional direction in the waveguide (say $1/6°$), changing gradually to a coarser resolution (say $1/2°$) in mid to high latitudes. Such grids are called telescoping grids. The grid size in the zonal direction is usually kept constant, since there is no advantage to be gained in general by varying the grid size, because of the complicated zonal extent of the basins and the associated boundary currents.

In regional models, the use of an unstructured, orthogonal curvilinear or a telescoping grid may often be unavoidable, since local processes may need to be modeled as accurately as possible and very high resolution uniform grids would be prohibitive. Unstructured grids widely used in solving conventional CFD problems (such as flow around an aircraft that requires fitting a 3D grid to its complicated shape) offer the most flexibility. Finite-element ocean models are therefore becoming increasingly popular, but proper care is essential in laying out such a grid; attention to rate of change of the mesh size and the resulting impact on the accuracy and efficiency of solutions is necessary. Numerical grid generation, whether for an unstructured grid finite-element model or a structured grid orthogonal curvilinear one, is itself a complex undertaking (e.g., see Thompson *et al.,* 1985; Hoffman and Chiang, 1993; Le Provost *et al.,* 1994; Lynch *et al.,* 1996).

The use of a telescoping grid requires some caution. It is essential that the grid spacing be changed gradually. Otherwise, spurious reflection of waves could occur at the grid boundaries, leading to a severe distortion of transient adjustment processes central to a time-dependent integration. The rule of thumb is that the ratio of grid sizes of adjacent grids must be bounded, kept within a value of say 1.1. The more gradual the change, the better, within practical constraints. While finite-element models afford unlimited flexibility in adapting the local grid resolution to local requirements, it is important to keep in mind that they are not immune to the need for a gradual change in grid spacing. The use of orthogonal curvilinear grids in some models (Haidvogel *et al.,* 1991b; Kantha and Piacsek, 1993; Blumberg and Mellor, 1987) offers similar flexibility, though much more limited than that in a finite-element model due to the structured nature of the grid, and is also subject to similar constraints. In addition, the ratio of grid sizes in the two orthogonal directions in both telescoping and curvilinear grids must also kept bounded, and within a reasonable value. The rule of thumb is that a ratio of about 3–5 suffices. When using nested grids to better resolve local processes, it is essential also to cap the ratio of the coarse to finegrid sizes to within 3–5, to prevent spurious reflections at the boundaries of the nests. There has never been much rigorous research into these aspects of finite-difference grids, so a more rigorous guidance is not possible.

Another point to remember in finite differencing on nonuniform grids is that to preserve the accuracy of the scheme chosen, it is essential to properly take into account the change in grid size. This can be easily done using Taylor series expansions.

$$u_{j+1} = u_j + \left(\frac{\partial u}{\partial x}\right)_j \left(\frac{\Delta x_j + \Delta x_{j+1}}{2}\right) + \frac{1}{2!}\left(\frac{\partial^2 u}{\partial x^2}\right)_j \left(\frac{\Delta x_j + \Delta x_{j+1}}{2}\right)^2$$

$$+ \frac{1}{3!}\left(\frac{\partial^3 u}{\partial x^3}\right)_j \left(\frac{\Delta x_j + \Delta x_{j+1}}{2}\right)^3 + \cdots$$

$$u_{j-1} = u_j - \left(\frac{\partial u}{\partial x}\right)_j \left(\frac{\Delta x_{j-1} + \Delta x_j}{2}\right) + \frac{1}{2!}\left(\frac{\partial^2 u}{\partial x^2}\right)_j \left(\frac{\Delta x_{j-1} + \Delta x_j}{2}\right)^2$$

$$+ \frac{1}{3!}\left(\frac{\partial^3 u}{\partial x^3}\right)_j \left(\frac{\Delta x_{j-1} + \Delta x_j}{2}\right)^3 + \cdots$$

$$(2.6.39)$$

where index j denotes a cell of size Δx_j. By suitable combinations of u_{j-1} from u_{j+1}, we get

$$\left(\frac{\partial u}{\partial x}\right)_j = \frac{2(u_{j+1} - u_j)}{\Delta x_{j+1} + \Delta x_j} + o(\Delta x_{av})$$

$$\left(\frac{\partial u}{\partial x}\right)_j = \frac{2(u_j - u_{j-1})}{\Delta x_j + \Delta x_{j-1}} + o(\Delta x_{av})$$

$$\left(\frac{\partial u}{\partial x}\right)_j = \left(\frac{u_{j+1} - u_j}{\Delta x_{j+1} + \Delta x_j} + \frac{u_j - u_{j-1}}{\Delta x_j + \Delta x_{j-1}}\right) + o(\Delta x_{av}^2)$$

$$\left(\frac{\partial^2 u}{\partial x^2}\right)_j = \frac{8}{\Delta x_{j+1} + 2\Delta x_j + \Delta x_{j-1}}\left(\frac{u_{j+1} - u_j}{\Delta x_{j+1} + \Delta x_j} - \frac{u_j - u_{j-1}}{\Delta x_j + \Delta x_{j-1}}\right) + o(\Delta x_{av}^2)$$

$$(2.6.40)$$

The use of a uniform or nonuniform longitude–latitude grid in global models is problematic near the pole in the northern hemisphere (the convenient presence of the Antarctic Continent capping the South Pole provides welcome relief in the southern hemisphere). The Arctic Ocean caps the North Pole, and it is often essential to include it in global ocean modeling, because of its interaction with the North Atlantic. The convergence of meridians at the Poles has always been a source of numerical difficulties never satisfactorily resolved. In addition to leading to excessive resolution and thus waste of computing resources, in explicit schemes, time stepping constraints become severe. There are essentially six solutions possible: (1) Terminate the global model at a latitude of about 65° N, and exclude the Arctic. Many global models do this, because of the exclusion of sea-ice cover and physics in those models. This, however, leads to open lateral boundaries with their attendant problems. (2) Thin out the grid as the pole is approached, by omitting every other one; this is seldom satisfactory. (3) Apply a zonal spatial filter to damp out instabilities (Arakawa and Lamb, 1977); but this introduces undesirable spurious nonlocal forcing that can affect solutions near the pole and elsewhere. Filters are also very expensive at high resolutions (Wehner and Covey, 1995). This solution is also seldom satisfactory. (4) Split the model domain into polar and nonpolar parts, with the polar part modeled on a rotated spherical grid, with model North Pole moved away from the real North Pole (often to be relocated on the equator), with the two solutions sutured appropriately. Eby and Halloway (1994), for example, have proposed such a scheme. However, this is really equivalent to two separate calculations. While it carries the advantage that existing global ocean codes can be used with minimum changes, matching the two grids and solutions leads to inevitable discontinuities and other complexities. Semtner (1987), Hibler and Bryan (1987), and Hakkinen *et al.* (1997) used rotated spherical grids to model the Arctic Ocean, with the computational pole relocated such that the computational equator passes through the geographic pole. (5) Use a more general orthogonal curvilinear coordinate system that locates grid singularities conveniently on continents that are masked out in any ocean calculation. For example, Smith *et al.* (1997) have used a dipole grid with poles located in Canada and Asia. Murray (1996) has proposed the use of the so-called Pan Am grid that permits the grid spacing in the zonal direction to be increased toward the North Pole! While quite attractive from a numerical point of view, such a grid is yet to be employed by a global model, because of the extensive recoding needed for the commonly used global ocean models (MOM2, POP, and MICOM—see Chapters 9–12). (6) Use a spectral model (see Section 2.7). While spherical harmonics are natural for computations on a sphere and eliminate the "pole problem," they are not problem-free. Experience with atmospheric GCMs has shown that advection calculation for scalars tends to be poor, even leading to negative values for the scalars (Randall *et al.*, 1998). Abandonment of spectral methods in favor of a semi-Lagrangian advection scheme leads to numerical efficiency (Bates *et al.*, 1993), but does not conserve advected quantities exactly.

An innovative technique proposed in the 1960s (Sadourny *et al.*, 1968) is the use of the finite-difference approach on a spherical geodesic grid (Thuburn, 1997), a technique with potential benefits to both ocean and atmospheric GCMs.

Another modern technique widely used in CFD is adaptive grids. For example, in CFD calculations of a high Mach number flow around a blunt body, proper resolution of shock fronts and similar features requires that the grid be adapted in shape to the shock front, with high resolution in the direction perpendicular to the shock front. Since the location of the shock is not known a priori, an iterative method is needed, where the grid is first laid out to conform to an approximate estimate of the shock front location, but is then continuously refined and adapted to the calculated front as the calculation proceeds, until a satisfactory solution emerges through convergence. Such adaptive grids are also useful in ocean modeling to resolve time-dependent sharp flow features such as the Gulf stream front and eddies, especially since the locations of these features may change with time.

When structured grids are used, nesting a finer resolution grid around a region of interest within a coarser and larger grid to better resolve local processes is an attractive possibility (e.g., Kurihara *et al.,* 1998; Ginis *et al.,* 1998). However, this requires matching the finer solution with the coarser one, either through one-way passive or two-way interactive schemes (see Section 2.10). Such nesting is widely used in atmospheric models (see Zheng *et al.,* 1986) for mesoscale forecasts of selected regions; for example, the operational NOAA/NCEP hurricane model developed at GFDL uses two-way multiple movable Arakawa A grid nests, with the inner 1/6° (5° × 5° domain) and 1/3° (11° × 11° domain) resolution grids moving within a stationary 1° outer grid (Kurihara *et al.,* 1998). The technique is also being increasingly used in ocean modeling (Spall and Holland, 1991; Oey and Chen, 1992; Fox and Maskell, 1995; Ginis *et al.,* 1998). Most nested models are finite-difference models; for a nested spectral model see Cocke (1998), who describes its application to hurricane predictions.

In one-way nesting, the coarser grid model results are saved along the open boundaries of the finer grid model domain, and the fine grid model is then integrated with these boundary conditions. The inherent assumption is that there is no feedback from processes in the smaller domain on those in the larger domain. In cases where the dynamical interactions between the two is important, two-way nesting is used, with information from one mesh to the other transferred across a narrow region of overlap between the two meshes (Oey and Chen, 1992; Ginis *et al.,* 1998). At the end of each coarse mesh integration time step, the open boundary conditions on the fine mesh are specified by spatial interpolation, and the fine mesh model is integrated for a similar time step (usually the coarse mesh time step is a multiple of the fine mesh one) and the interior points of the coarse mesh are updated by suitable averaging of the fine mesh properties. If the interior grid points of the coarse mesh are a subset of the fine mesh grid points, the latter is particularly simple. However, the principal difficulty is the instabilities and noise resulting from mismatch at the interface between the two grids, due to false reflection or aliasing of disturbances propagating from the fine mesh to the coarse one (Perkins *et al.,* 1997) (see Section 2.10). Another problem is that conservation of properties is not enforced at the interface between the two grids (Fox and Maskell, 1995) and so integrations are possible only for short intervals, where this does not matter. Kurihara *et al.* (1979, 1989,

1998) have developed a nesting technique that avoids these problems and successfully applied it to hurricane forecasts in the atmosphere. Ginis *et al.* (1998) describe its adaptation to layered ocean models.

2.7 SPECTRAL (SPECTRAL TRANSFORM) METHOD

Even though the spectral method is not widely used in ocean modeling, its heavy use in atmospheric modeling, and often in modeling the atmosphere in coupled ocean–atmosphere models, makes it important to be knowledgeable about the method. For a more complete treatment, see Washington and Parkinson (1986), Haltiner and Williams (1980), and Holton (1992).

Spectral methods represent the spatial variations of the dependent variables in terms of a truncated series of orthogonal basis functions, such as trigonometric functions in rectangular space and spherical harmonics (Legendre functions) in spherical space, and then solve for the evolution of the coefficients of these basis functions in time. Finite differences, on the other hand, are local approximations. Thus, in a spectral method, the solution at any spatial point depends on all other points in the model domain, whereas in the latter, it depends only on the neighboring points. Global basis functions are also a more efficient means of representing spatial variations. These make the spectral method physically and computationally more attractive, wherever feasible. The zonal and meridional continuity of the atmosphere (meaning that no zonal or meridional boundaries exist) makes it ideally suited to the spectral method. This is the principal reason that there exist few atmospheric models that use finite differences. However, global dependence makes it harder to adapt these methods to a massively parallel computing environment. Finite-difference (or grid point) methods are better suited. It is interesting to see if this spurs further development and use of finite-difference models for the atmosphere in the future. On the other hand, the oceans are zonally (and in some places meridionally) discontinuous and the basin shapes complicated, and hence finite-difference models are the ones most widely used, closely followed by finite-element models in recent years.

For a box-shaped ocean basin with closed boundaries (or open boundaries subject to periodic boundary conditions), a double Fourier series (sine and cosine trigonometric functions) is a natural basis function. If the governing equations were linear, there would be no question as to the superiority of the spectral method, especially at low resolutions where the finite-difference methods are affected greatly by numerical dispersion problems discussed above. However, because of the ubiquitous presence of nonlinear advective terms, spectral methods require the evaluation of a huge number of terms involving the interaction between the various spectral modes. Since the number of such terms is proportional to the square of the number of modes, and the number of modes increases with the increase in spatial resolution, this evaluation becomes inefficient and impractical for high resolution models. This is the principal reason spectral methods were not popular until the 1970s when Steve Orszag and others discovered a way to avoid this bottleneck. The solution to the bottleneck consists of transforming the model results from wavenumber spectral space to physical grid point space at each time step, evaluating the nonlinear advective terms in the physical space, and transforming these troublesome terms back into spectral space before time stepping the solutions. This procedure was made feasible by the discovery of fast, efficient Fourier transform techniques by Cooley and Tuckey in the mid-1960s, since otherwise the transform and inverse transform from spectral to physical space and back would form a major time-consuming step. Since observational data are available only at grid points in physical space, and since all nowcast/forecast models involve data assimilation in one form or another, such transformations have to be efficient for spectral transform methods to be successful in practical applications.

For the thin atmospheric envelope that surrounds the Earth, spherical harmonics are the natural basis functions. Spherical harmonics are an exponentially convergent approximation to variations on a sphere (finite differences on latitude–longitude grids are only algebraically convergent) and this is quite attractive to solving problems on a sphere. While spectral transform methods are the current norm for global atmospheric models both of the NWP (ECMWF T-106 model, for example) and climate type (see Boer *et al.,* 1992; Gates, 1992; Mechoso *et al.,* 1995), and grid point methods for regional models (NCAR MM-5, for example), the advent of massively parallel computers and the need for very high resolutions and continuous data assimilation have refocused interest on grid point methods. However, grid point methods using a regular latitude–longitude grid must make special provisions for handling the near-pole regions, where because of the grid convergence due to convergence of meridians, time step limitations can become extremely severe otherwise (Haltiner and Williams, 1980). Spectral methods are also naturally conducive to conserve important integral quantities such as the kinetic energy and the mean square vorticity, whereas special care must be exercised to assure this in a finite-difference model. Since spherical harmonics are eigenfunctions of the Laplace operator on a sphere, the inversion of linear elliptic operators in an implicit time stepping scheme is straightforward (Haidvogel *et al.,* 1997). However, no FFT-type fast, efficient technique exists for computing discrete spherical harmonic transforms, which, computed at present by direct, less efficient methods, dominate the calculations at high resolutions (Haidvogel *et al.,* 1997). Since the number of operations is proportional to the number of grid points N in a grid point model, but n log n (where n is the number of modes) in a spectral model (Holton, 1992), it is natural that increasingly fine resolutions affordable on modern computers tend to make grid point methods more and more attractive. Spectral techniques are also difficult to implement on computers with distributed memory because of their global nature. Also, enhancements in model resolution in localized regions such as around mountain ranges are nearly impossible in the spectral approach, whereas techniques such as nesting and telescoping make local mesh refinement possible in grid point methods. In any case, complex, irregular geometry and the presence of narrow western boundary currents leading to highly laterally inhomogeneous flows render spherical-harmonics-based spectral methods unattractive to the oceans.

Current operational atmospheric spectral models have all the variables available in both spectral and physical space at every time step, and calculations are done in both spectral and physical space. Computations of physical processes involving such things as radiative transfer and convection are performed in physical space. Data assimilation is also done in physical space.

Any scalar variable Φ can be expanded in spherical harmonics based on latitude θ and longitude ϕ,

$$\Phi = \sum_m \sum_n \Phi_n^m P_n^m(\mu) \exp(im\phi); \quad \mu = \sin\theta \qquad (2.7.1)$$

$$P_n^m(\mu) = (1-\mu^2)^{m/2} \frac{d^m}{d\mu^m} P_n(\mu); \quad P_n(\mu) = \frac{1}{2^n n!} \frac{d^n \left[(\mu^2-1)^n\right]}{d\mu^n} \qquad (2.7.2)$$

where P_n^m is the associated Legendre polynomial of the first kind of degree n and order m, normalized to unity, and $|m| \le n; m = 0, \pm 1, \pm 2, \pm 3 ..., n-1, 2, 3.....$ The first three of these are

$$P_1^1(\mu) = (1-\mu^2)^{1/2}; \quad P_2^1(\mu) = 3\mu(1-\mu^2)^{1/2}; \quad P_3^1(\mu) = 3(1-\mu^2)$$

$$(2.7.3)$$

Index m denotes the zonal wavenumber, the number of waves around a latitude circle, and $n - |m|$ denotes the meridional wavenumber, or the number of nodes in the latitudinal direction, excepting the poles. P_n is the Legendre polynomial:

$$P_0(\mu) = 1; \quad P_1(\mu) = \mu; \quad P_2(\mu) = \frac{1}{2}(3\mu^2 - 1) ... \qquad (2.7.4)$$

Figure 2.7.1 shows the spatial pattern of some of the spherical harmonics. Since the spherical harmonics are truncated at some number, this number defines the resolution of the model. There are two possible truncations: (1) triangular, where M = N equal to some integer (for example, the ECMWF T106 model, where 106 is the maximum number of modes retained—the resolution is 1.2° in both directions), and (2) rhomboidal, where $N = |m| + M$. The two truncations

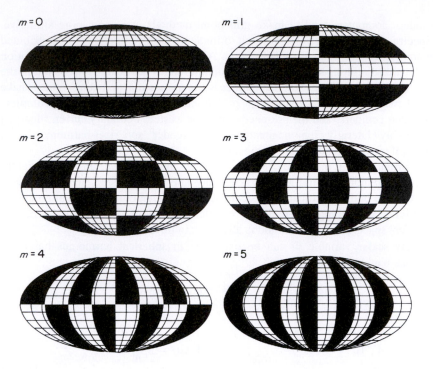

Figure 2.7.1 Patterns of antisymmetric and symmetric spherical harmonic functions used as basis functions in the spectral solution method for n = 5. From Baer (1972).

are shown in Figure 2.7.2. Triangular truncation provides nearly the same horizontal resolution in the zonal and meridional directions and is preferred for high resolution models. Rhomboidal truncation provides the same meridional resolution for every zonal wavenumber and is advantageous for low resolution models.

Washington and Parkinson (1986) illustrate the spectral method using the simple shallow water wave model for an ocean of uniform depth H, covering the globe [see also Holton (1992), who uses the barotropic vorticity equation],

$$\frac{\partial^2 \eta}{\partial t^2} = gH \frac{1}{R^2} \left\{ \frac{\partial}{\partial \mu} \left[(1-\mu^2) \frac{\partial}{\partial \mu} \right] + \frac{1}{(1-\mu^2)} \frac{\partial^2}{\partial \phi^2} \right\} \eta$$

(2.7.5)

which yields for the coefficients η_n^m

$$\frac{d^2 \eta_n^m}{dt^2} = -\sigma^2 \eta_n^m \,; \; \sigma^2 = \frac{n(n+1)gH}{R^2}$$

(2.7.6)

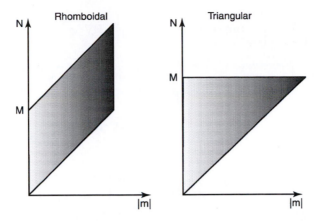

Figure 2.7.2 Rhomboidal and triangular truncation showing which modes are retained. From Simmons and Bengtsson (1984), reprinted with the permission of Cambridge University Press.

where R is the Earth's radius. This equation can be readily solved and the solution built up by summation of the various harmonics.

Washington and Parkinson (1986) also illustrate the spectral transform method using shallow water equations. For dynamical calculations, instead of the velocity components, the prognostic variables are the vorticity and divergence, which are scalar quantities that can be readily integrated in spectral space. Since von Helmholtz's discovery in the 19th century that any vector field can be decomposed into a solenoidal, nonrotational component and a rotational component, $\mathbf{v} = \mathbf{k} \times \overline{\nabla}\psi + \nabla\varphi$, solving for the vorticity, $\zeta = \mathbf{k}.(\overline{\nabla} \times \mathbf{v}) = \nabla^2\psi$, and horizontal divergence, $D = \overline{\nabla} \cdot \mathbf{v} = \nabla^2\varphi$, is equivalent to solving for the horizontal velocity field $\mathbf{v}$ ($\overline{\nabla}$ is the horizontal gradient operator, ψ is the stream function, and φ is the velocity potential). Thus, instead of the primitive equations

$$\frac{\partial \eta}{\partial t} = \frac{H}{R\cos\theta}\left\{\frac{\partial u}{\partial \phi} + \frac{\partial}{\partial \theta}(v\cos\theta)\right\}$$

$$\frac{\partial u}{\partial t} - fv = \frac{-g}{R\cos\theta}\frac{\partial \eta}{\partial \phi}$$

$$\frac{\partial v}{\partial t} + fu = \frac{-g}{R}\frac{\partial \eta}{\partial \theta}$$

$$(2.7.7)$$

that grid point models use, we have

$$\frac{\partial \eta}{\partial t} = \frac{-1}{R^2(1-\mu^2)} \left\{ \frac{\partial}{\partial \phi}(u\eta) + (1-\mu^2)\frac{\partial}{\partial \mu}(v\eta) \right\} - H\nabla^2\varphi$$

$$\frac{\partial}{\partial t}(\nabla^2\psi) + f\left(\nabla^2\varphi + \frac{v}{R\mu}\right) = \frac{-1}{R(1-\mu^2)} \left\{ \frac{\partial}{\partial \phi}(u\nabla^2\psi) + (1-\mu^2)\frac{\partial}{\partial \mu}(v\nabla^2\psi) \right\}$$

$$\frac{\partial}{\partial t}(\nabla^2\varphi) - f\left(\nabla^2\psi - \frac{u}{R\mu}\right) = \frac{1}{R(1-\mu^2)} \left\{ \frac{\partial}{\partial \phi}(v\nabla^2\psi) + (1-\mu^2)\frac{\partial}{\partial \mu}(u\nabla^2\psi) \right\}$$

$$- \nabla^2\left(\frac{u^2 + v^2}{2(1-\mu^2)} + g\eta\right)$$

$$u = -\frac{(1-\mu^2)}{R}\frac{\partial \psi}{\partial \mu} + \frac{1}{R}\frac{\partial \varphi}{\partial \phi} \, ; \quad v = \frac{(1-\mu^2)}{R}\frac{\partial \varphi}{\partial \mu} + \frac{1}{R}\frac{\partial \psi}{\partial \phi}$$

$$(2.7.8)$$

The variables η, ψ, φ, u, and v are expanded in spherical harmonics (the property that the Laplacian of a harmonic is proportional to the harmonic itself is useful). When products of two variables are required, transformation to physical space of each variable is made and the product is transformed back to spectral space. A similar approach holds for rectilinear space, where the basis functions are much simpler trigonometric functions.

Finally, spectral methods can also be used for discretization in the vertical. Instead of finite differences, basis functions could be used in the vertical also. Haidvogel *et al.* (1991b; see also Hedstrom, 1995) have developed a semispectral model which uses a spectral collocation method with Chebyshev functions as basis functions in the vertical. The horizontal discretization is, however, staggered finite difference. Moisan *et al.* (1995) describe its recent application to coastal upwelling off California.

2.8 FINITE-ELEMENT METHODS

Finite-element methods (FEMs) originated in solid mechanics in the 1960s. Their phenomenal success there led to their extension to areas such as fluid flows in the 1970s. However, detailed analysis (e.g., Pinder and Gray, 1977) showed that while the method had great potential, it was also plagued by severe computational problems and high computational costs. It was not until the 1980s that the method began to enjoy a renaissance, when it was discovered (e.g., Lynch and Gray, 1979) that the transformation of the continuity equation to a higher order wave equation not only eliminated the computational problem, but also made both implicit and explicit time stepping computationally efficient. These advances have made FEMs quite competitive in comparison to the finite-difference approach (see Lynch, 1983, for a review) and its application to both 2D barotropic (Foreman *et al.,* 1993; Le Provost *et al.,* 1994) and, more recently, 3D baroclinic (e.g., Lynch *et al.,* 1996) problems feasible.

The principal advantage of FEMs is that they are inherently better adapted to complex geometries and flows with large spatial inhomogeneities. Finite-element techniques belong to the class of Galerkin methods, where the variable is expanded in terms of basis functions of certain spatial structures. Spectral methods can also be grouped under Galerkin methods; however, while spectral methods use orthogonal global basis functions, finite-element methods use basis functions that are polynomials and nonzero only in a limited domain. Consider a 1D problem (Haltiner and Williams, 1980) governed by an operator operating on the dependent variable $u(x)$:

$$\frac{\partial}{\partial t} u(x,t) + L[u(x,t)] = f(x,t)$$

$$(2.8.1)$$

The usual procedure is to represent the dependent variable in the form of a truncated series:

$$u(x,t) = \sum_{j=1}^{N} u_j(t)\varphi_j(x)$$

$$(2.8.2)$$

The problem then reduces to a set of N coupled ordinary differential equations that can be readily solved:

$$\sum_{j=1}^{N} \frac{du_j}{dt} \int_a^b \varphi_i(x)\varphi_j(x)dx + \int_a^b \varphi_i(x)L\left[\sum_{j=1}^{N} u_j(x)\varphi_j(x)\right]dx$$

$$= \int_a^b \varphi_i(x)f(x,t)dx \qquad i = 1, 2, \dots N$$

$$(2.8.3)$$

The method can be illustrated for the advection equation, where $L \equiv c\, \partial / \partial x$:

$$\frac{\partial u}{\partial t} + c\frac{\partial u}{\partial x} = 0$$

$$(2.8.4)$$

Choose a tent-shaped basis function, nonzero up to only the adjacent grid points:

$$\varphi_j(x) = \begin{cases} 0 & x < (j-1)\Delta x \\ [x - (j-1)\Delta x] / \Delta x & (j-1)\Delta x < x < j\Delta x \\ [(j+1)\Delta x - x] / \Delta x & j\Delta x < x < (j+1)\Delta x \\ 0 & x > (j+1)\Delta x \end{cases}$$

(2.8.5)

Substitution in Eq. (2.8.4) gives, after some algebraic manipulation,

$$\frac{1}{6}\left(\frac{du_{I+1}}{dt} + 4\frac{du_I}{dt} + \frac{du_{I-1}}{dt}\right) + c\left(\frac{u_{I+1} - u_{I-1}}{2\Delta x}\right) = 0 \quad \text{for} \quad j = I$$

(2.8.6)

Thus, in contrast to finite-difference methods, the time derivative is a weighted average over several points, which improves the accuracy quite a bit. Using leapfrog time differencing,

$$[(u_{I+1}^{n+1} - u_{I+1}^{n-1}) + 4(u_I^{n+1} - u_I^{n-1}) + (u_{I-1}^{n+1} - u_{I-1}^{n-1})] + 6\frac{c\Delta t}{\Delta x}(u_{I+1}^n - u_{I-1}^n) = 0$$

(2.8.7)

Neumann stability analysis indicates that the Courant number C has to be less than $3^{-1/2}$. However, while this restriction on time step is more stringent than a second-order leapfrog finite-difference scheme (C < 1), it is comparable to that of the fourth-order leapfrog scheme C < 0.73. The performance is better than either two. The numerical phase speed is given by

$$\hat{c} = \frac{1}{k\Delta t}\sin^{-1}\left[\frac{c\Delta t}{\Delta x}\frac{3\sin(k\Delta x)}{2 + \cos(k\Delta x)}\right] \xrightarrow[\frac{c\Delta t}{\Delta x} < 1]{} \frac{c}{k\Delta x}\frac{3\sin(k\Delta x)}{2 + \cos(k\Delta x)}$$

(2.8.8)

The numerical group velocity is

$$\hat{c}_g \xrightarrow[\frac{c\Delta t}{\Delta x} < 1]{} \frac{[1 + 2\cos(k\Delta x)]}{[2 + \cos(k\Delta x)]^2}$$

(2.8.9)

Equations (2.8.8) and (2.8.9) illustrate the basic advantages and disadvantages of finite-element methods. On the one hand, the numerical dispersion is far less severe than that in the second-order leapfrog finite differencing method, and even better than the fourth-order scheme. However, the $2\Delta x$ wave is still stationary, so that the high wavenumbers are adversely impacted. In fact, the group velocity for this wave is $-3c$, negative and three times faster than that given by regular finite-difference models. This means that small scale noise propagates rapidly in FEMs and leads to inaccurate solutions with severe artificial near-2Δx modes.

For 2D problems, the tent function is replaced by a three-noded triangular element, with the values of zero on two of the vertices and increasing linearly to one at the third, the nodal point. We will not go into details here but refer the reader to Pinder and Gray (1977).

Early FEM models applied to study circulation in water bodies were plagued with these severe spurious modes that caused unphysical node-to-node oscillations and required an inordinate amount of artificial dissipation to suppress them (Gray, 1982). This problem prevented widespread application of FEMs in the seventies and early eighties. A reformulation using the wave-continuity equation approach (Lynch and Gray, 1979) appears to have overcome the above difficulty, and has led to a resurgence of FEMs (e.g., Westerink and Gray, 1991; Westerink *et al.*, 1992, 1994). The technique consists of substituting for the continuity equation a second-order wave equation that apparently has the desired effect of smoothing out these spurious high frequency components (Westerink *et al.*, 1992). The phenomenal success of finite-difference schemes has been predominantly due to the use of staggered spatial grids with the ability to propagate 2Δx modes. FEM models using the wave-continuity formulation appear now to do the same and yield solutions that look very similar to the staggered grid finite-difference models (Westerink and Gray, 1991; Westerink *et al.*, 1994).

To illustrate the wave-continuity approach, consider the simple linear 1D shallow water equations (2.6.15). These equations in their unaltered primitive form cannot be used in a FEM model, for reasons discussed above. Instead, the time derivative of the continuity equation is combined with the spatial derivative of the momentum equation and a constant factor α times the continuity equation to yield the wave-continuity equation, which is then solved with the momentum equations (in nonconservative form), so that Eq. (2.6.15) is replaced by

$$\frac{\partial^2 \eta}{\partial t^2} + \alpha \frac{\partial \eta}{\partial t} + H \frac{\partial}{\partial x}\left(\alpha u - g \frac{\partial \eta}{\partial x}\right) = 0$$

$$\frac{\partial u}{\partial t} + g \frac{\partial \eta}{\partial x} = 0$$

$$(2.8.10)$$

α is a tunable numerical constant on the order of 10^{-4} (Lynch *et al.*, 1996). These equations can then be discretized spatially using finite elements. The resulting equations can be solved in the time domain using standard time stepping techniques or in the spectral domain. A solution is first obtained for the elevation, followed by computation of currents using the momentum equations.

The attractiveness of FEMs in affording nearly unlimited flexibility in providing high resolution "local" resolutions, wherever needed, can be seen from the grid used by Le Provost *et al.* (1994) to compute global barotropic tides (Figure 1.3.2). They used a spectral FEM based on the wave-continuity equation by assuming the presence of a dominant frequency. Le Provost and Vincent (1986) conducted sensitivity studies of the precision of the FEM for oceanic tides and concluded that higher order polynomials and a mesh size that is at least 1/15th the wavelength being resolved are essential for accuracy. In addition, a much higher level of grid refinement may be needed to predict currents more accurately than sea surface elevations in a FEM tidal model (Westerink *et al.*, 1992).

The FEM has been applied extensively to 2D barotropic problems such as tides and storm surges. Only in recent years has it been applied to 3D baroclinic problems (e.g., Ip and Lynch, 1994; Lynch *et al.*, 1996). A principal problem appears to be mass conservation. While globally this is assured, the wave-continuity equation approach may not conserve mass locally. Additionally, it is not clear how FEMs handle scalars, such as dissolved gases and salinity in 3D baroclinic problems, where properties such as conservation and positive definiteness are of great importance.

2.8.1 SPECTRAL ELEMENT APPROACH

Iskandarani *et al.* (1995) and Haidvogel *et al.* (1997) describe an approach that attempts to combine the advantages of global spectral and local finite-element techniques—exponential convergence of spectral techniques and the resultant accuracy; lack of the pole problem in spectral methods; flexibility of FEMs to represent complex geometry; and ready implementation of local mesh refinement and scalability on massively parallel computers of FEMs. The approach, called the spectral element method (SEM), has its origins in engineering fluid flows (Fischer, 1989), but has been successfully applied to 2D barotropic problems (Ma, 1993; Iskandarani *et al.*, 1995; Wunsch *et al.*, 1997). Haidvogel *et al.* (1997) describe its application to reduced gravity simulation of wind-driven currents in the Pacific based on a telescoping global SEM grid shown in Figure 2.8.1. The computations in a SEM are performed on an unstructured set of conforming, quadrilateral, isoparametric elements encompassing the model domain, each of which is mapped into a unit square in the computational domain (ξ, ζ). Within each of these elements, the dependent variables are approximated by expansions in polynomials or test functions. For example, the velocity vector $\mathbf{u}$ and the elevation η are expanded in terms of Legendre Cardinal functions,

$$ h_i(\xi) = \frac{-(1 - \xi^2) L'_{N-1}(\xi)}{N(N-1)L_{N-1}(\xi_i)(\xi - \xi_i)}, \quad i = 1, \dots N $$

$$ (2.8.11) $$

where N is the number of nodes in each direction per element, L_{N-1} is the Legendre polynomial of degree $(N-1)$, ξ_I are its N roots, and the prime denotes its derivative. The expressions for **u** and η are

$$\mathbf{u}(\xi, \zeta, t) = \sum_{j=1}^{N} \sum_{i=1}^{N} \mathbf{u}_{i,j}(t) h_i(\xi) h_j(\zeta)$$

$$\eta(\xi, \zeta, t) = \sum_{j=1}^{N-2} \sum_{i=1}^{N-2} \eta_{i,j}(t) \hat{h}_i(\xi) \hat{h}_j(\zeta) \qquad (2.8.12)$$

where hats denote corresponding interpolation functions for η (equivalently pressure). The reason for the N−2 number of nodes for η and hence the limit N−2 in the summation for η is because the degree of interpolation function for pressure must be 2 less than that for velocity on a staggered mesh, to avoid spurious modes. Multiplying the governing equations $\partial \mathbf{u}/\partial t = \mathbf{X}$ by $h_i(\xi) h_j(\zeta)$ and $\partial \eta/\partial t = y$ by $\hat{h}_i(\xi) \hat{h}_j(\zeta)$, and substituting for **u** and η from Eq. (2.8.12) and integrating over the spectral element, one gets a set of ordinary differential equations for $\mathbf{u}_{i,j}$ and $\eta_{i,j}$:

$$M_{i,j;k,l} \frac{d}{dt} \mathbf{u}_{i,j} = \int_A \mathbf{X}_{i,j} h_i(\xi) h_j(\zeta) dA; \quad M_{i,j;k,l} = \int_A h_i(\xi) h_j(\zeta) h_k(\xi) h_l(\zeta) dA$$

$$\hat{M}_{i,j;k,l} \frac{d}{dt} \eta_{i,j} = \int_A y_{i,j} \hat{h}_i(\xi) \hat{h}_j(\zeta) dA; \quad \hat{M}_{i,j;k,l} = \int_A \hat{h}_i(\xi) \hat{h}_j(\zeta) \hat{h}_k(\xi) \hat{h}_l(\zeta) dA$$

$$(2.8.13)$$

where M and $\hat{M}$ are diagonal matrices, easy to invert. Even explicit integration of Eq. (2.8.13) requires inversion of these matrices. Any time stepping scheme is possible, but Haidvogel *et al.* (1997) use a third-order Adams–Bashforth explicit scheme. Writing each equation in Eq. (2.8.13) in the generic form M (dx/dt) = r, this scheme takes the form (Gear, 1971)

$$\frac{x^{n+1} - x^n}{\Delta t} = M^{-1} \left(\frac{23}{12} r^n - \frac{16}{12} r^{n-1} + \frac{5}{12} r^{n-2} \right)$$

$$(2.8.14)$$

Note that the third-order scheme requires values known at two previous time levels, unlike the leapfrog scheme that requires only one. These computations are performed for each element, and the contribution of different elements to the system of equations is assembled; only the vector r of the right-hand sides in Eq. (2.8.13) needs to be treated thusly. The limiting time step is governed by the CFL criterion: $\Delta t \sim L/(CN^2 M)$, where L is the length scale, C is the wave speed, and M is the number of elements.

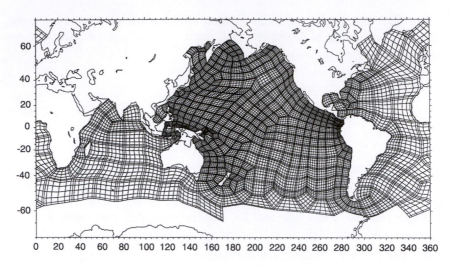

Figure 2.8.1 The model grid used by Haidvogel *et al.* (1997) in their spectral element model of the Pacific Ocean. Each of the 448 spectral elements has a 7×7 mesh.

The governing equations are solved element by element. One or a group of contiguous elements can therefore be assigned to a different processor. Because the time stepping is explicit and hence local, interprocessor communication is needed only when updating collocation points on the edges of the element that lie on the boundary of the processor's domain and hence are shared with other processors. The elements communicate only with adjacent elements at their common boundaries (this permits efficient adaptation to parallel computers), with neighboring elements sharing common points, at which the terms in the equations are obtained by weighted averaging. Thus, global spherical harmonics are replaced by local polynomial transforms that assure fast convergence, at the elemental level. The quadrilateral elements permit easy local mesh refinement as can be seen from Figure 2.8.1. Thus the flexibility of FEM is retained. Since the number of elements M and the size of spectral transforms N (and hence the order N^3 computations) can be individually controlled, considerable flexibility exists. However, the optimum combination of M and N for efficiency, and accuracy, is an important unresolved question.

The reader is referred to Haidvogel *et al.* (1997) and other references cited above for more details on this rather novel technique. While attractive from a numerical point of view, until some experience is built up with the model over time, especially in 3D baroclinic flows, it will not be clear what the major advantages and disadvantages of this method are when it comes to accurate representation of physical processes in the oceans.

2.9 PARAMETERIZATION OF SUBGRID SCALE PROCESSES

Even if our computer models were perfect, digital representation and finite computing resources would still place a limit on how well we can reproduce physical processes. The finite grid size compels us to suitably parameterize those processes that cannot be explicitly modeled—subgrid scale processes, primarily turbulent mixing and diffusion processes in the water column that affect water mass properties. It is not surprising that incorrect parameterization of subgrid scale processes can have an adverse effect on the computed oceanic state (e.g., Holland, 1989). This is typical of all CFD problems. The hope has always been that if one could model explicitly the energy-containing large eddies in the flow, then the inevitable errors in parameterizing the smaller scale processes would not seriously degrade the modeled state. This is the philosophy behind large eddy simulations (LES's) in turbulent flow computations. Ocean models that can resolve at least a part of the spectrum of mesoscale eddies that carry most of the kinetic energy in the oceans are called eddy-resolving models. Given that these eddies scale with the internal Rossby radius of deformation, typically a few tens of kilometers, it is easy to see why fully eddy-resolving capability has not yet been realized, at least in global ocean models.

Ocean models tackle a much more complex problem than LES models do. LES is usually applied to a fully turbulent flow and has met with a measure of success in simulating vertical mixing in flows such as that in a convective ABL [see, for example, Chapter 1 of Kantha and Clayson (2000)]. However, the grid resolutions needed (on the order of meters in the vertical and horizontal directions) make it impossible to use LES to compute vertical mixing in ocean models. Vertical mixing is therefore invariably parameterized, or simulated using less resource-intensive second-moment turbulence closure (Kantha and Clayson, 2000, Chapter 2). However, vertical mixing in the stably stratified interior of the oceans away from the fully turbulent regions near the surface and the bottom is highly intermittent and very poorly understood. Even the sources of this mixing are not clearly known. It is thought that small scale internal waves fed by the cascade of energy down the spectrum from tides and winds play a major role in mixing the interior (Munk and Wunsch, 1998). Mixing at lateral boundaries of ocean basins and double-diffusive processes may also play a role. We do not know how to model intermittent turbulence correctly, and even if we could, it would be impractical to model the small scales involved explicitly. Therefore, unlike vertical mixing in the upper and benthic boundary layers, which can be at least approximately simulated by simple turbulence models, vertical mixing (more appropriately diapycnal mixing) in the interior is inevitably parameterized in ocean models.

Constant [the mixing coefficient in the abyssal oceans appears to be roughly constant at 10^{-5} m^2 s^{-1} (Kunze and Sanford, 1996)] or ambient stability-dependent values that are a function of the local buoyancy frequency N are employed for the vertical (or equivalently diapycnal) eddy diffusion coefficients K_M and K_H. In the upper layers, away from the upper mixed layer, Richardson-dependent parameterizations are used (e.g., Pacanowsky and Philander, 1981). Ocean models are sensitive to the choice of the values and functional forms for these. Gargett *et al.* (1989) found that the poleward meridional heat flux in the z-level GFDL model was considerably modified by the use of $K_H \sim N^{-1}$ instead of a constant value. Other quantities involving deep circulation are also sensitive to the diffusivities of heat and salt used (Gargett and Halloway, 1992; Zhang *et al.*, 1998). Understanding and interpreting such sensitivity through model simulations is made difficult by the interdependence of horizontal and vertical mixing parameterizations (or equivalently isopycnal and diapycnal mixing in those models that use isopycnal mixing schemes and in isopycnal models—see Chapters 10 and 11). Some tests have been made in non-eddy-resolving (in the horizontal) basin scale models (Gargett *et al.*, 1989). The long-term integrations (hundreds of years) needed to achieve an equilibrium state in the deep oceans make sensitivity studies with even marginally eddy-resolving models of ocean basins impractical.

Horizontal (or equivalently isopycnal) mixing parameterization is even more problematic. The models are clearly sensitive to it (e.g., Holland, 1985, 1989). On the one hand, horizontal mixing occurs over a wide spectrum of scales, and even if the larger mesoscale eddies can be modeled, there is still a need to parameterize the clearly physical (dye patches released in the ocean do spread horizontally) horizontal mixing processes that take place at smaller scales. On the other hand, horizontal mixing has been regarded as an unwanted but purely numerical artifact of finite horizontal resolutions needed to prevent numerical instabilities in a nonlinear model, with the implicit implication that it can be reduced to inconsequential values as the grid resolution becomes finer and finer. However, how significant the neglected mixing would be is not known, and no studies have been carried out to resolve this issue. Meanwhile, modelers use either the harmonic or the biharmonic form for horizontal mixing, the latter being quite effective in damping the $2\Delta x$ noise. The coefficients are empirically selected to be the smallest possible values that provide "smooth" solutions, without a numerical instability. Mean deformation rate-dependent values

(Smagorinsky, 1963) or a combination of different values are also often employed. See Muller and Henderson (1989) and Muller and Halloway (1989) for a thorough discussion of parameterization of small scale processes (see also Kantha and Clayson, 1999) and their effect on numerical ocean models.

2.10 LATERAL OPEN BOUNDARY CONDITIONS

The need to prescribe boundary conditions on open lateral boundaries is the Achilles heel of regional models, whether they are of the ocean, the atmosphere, or coupled. Warner *et al.* (1997; see also Miyakoda and Rosati, 1977; Phillips, 1979; Davies, 1983; Gustafsson, 1990) discuss in detail the limitations to simulation and forecast skill of limited area atmospheric models (LAMs) due to the necessity to prescribe lateral boundary conditions (LBCs) from a coarse mesh model, usually global. Many of these apply to regional ocean and coupled ocean–atmosphere models as well. The major differences have to do with the slower advection speeds and the longer timescales associated with the oceans. Also, there does not yet exist an accurate operational global ocean model in which to embed a regional ocean model.

LAMs are finding increasing use at NWS for regional forecasts, by small nations for air quality and agricultural studies, and by the military for operational use. A major advantage of a LAM is the ability to model region-critical processes more accurately and at a much higher resolution than is possible in a coarse mesh global model. However, a major and unavoidable limitation comes through the LBCs. Errors in LBCs propagating/advecting into the modeled domain have a major impact on the evolution of the atmosphere inside the domain. The errors are of two kinds: (1) The information prescribed on the lateral boundaries usually comes from a coarse mesh model with a poorer horizontal (and possibly vertical) resolution and less precise parameterization of physical processes. (2) Numerical techniques used to interface the coarse and fine meshes can introduce errors. Transient gravity-inertia modes generated at the lateral boundaries due to mismatches in physics and other aspects of the two meshed models can also lead to spurious noise inside the modeled domain. Even if the excitation of gravity modes can be avoided, contamination of interior solutions by advection of errors on a longer time scale cannot be. There have been cases where the improvements inherent to a LAM were completely swamped by LBC errors propagating into the domain (Warner *et al.,* 1997). It is therefore extremely important to chose a proper domain in modeling a limited region and limiting the period of simulation appropriately. Generally, the larger the domain surrounding the domain of interest the better, since it takes longer for the errors from the lateral boundaries to propagate to the region of interest, and therefore the forecast skill can be retained longer. Thus the simulation skill is critically dependent on the choice of the LAM domain. The domain chosen must not only be large enough to place the region of interest as far away as possible from inflow lateral boundaries, but it should also include regions crucial to simulating local processes more accurately. The domain size is limited by available computational resources, and therefore in practice the domain choice is reflected in the time over which LAM forecasts are useful.

While hindcasts benefit from the possibility of using analyses to provide LBCs, the forecast skill of the coarse mesh model is unavoidably the most severe limiting factor in the forecast skill of a LAM. In general, the time for advection of errors in LBCs to the region of interest limits the time period over which a LAM retains useful skill. Initialization also plays a crucial role. Generally, if the interior forcing is overwhelmingly stronger than the external forcing introduced through the lateral boundaries, LAMs could prove useful over a much longer period. It is also possible to move the fine mesh with the feature of interest in order to make the best use of the finer and better parameterization possible. NOAA/NCEP operational hurricane models use triple nesting, with a 160-km coarse stationary outer mesh, but 80- and 40-km inner meshes that move with the hurricane (Kurihara *et al.,* 1979; Ginis *et al.,* 1997). The NCAR MM5 model also has facilities for nesting a movable fine grid.

It is often, but not always, possible to make simultaneous simulations on both the coarse mesh and an embedded fine mesh, with each providing LBCs to the other during the simulation (Harrison and Elsberry, 1972; Phillips and Shukla, 1973; Zheng *et al.,* 1986). This interactive or two-way nesting (Phillips and Shukla, 1973; Zheng *et al.,* 1986) is often preferable to one-way or parasitic nesting (Shapiro and O'Brien, 1970; Anthes, 1974), where LBCs are derived from a prior coarse mesh simulation or analysis. Perkey and Maddox (1985) demonstrate the improvement possible from the feedback for a convective precipitation system. Choosing the coarse and fine mesh grids to overlap and have common grid points is a useful strategy, since the feedback to the coarse mesh is simplified. However, unless the coarse mesh is global, it is still necessary to use one-way nesting to provide LBCs to the coarse mesh model from the global model.

Nested nowcast/forecast models are relatively rare for the oceans. This is partly due to the fact that unlike the atmosphere, where global forecast models are routinely run by several NWP centers around the world, no global or even basin scale ocean model is yet operational, although as of this writing, several efforts are underway. Thus regional models are applied to seas where the influence of external forcing is limited or small. Excellent examples are marginal semienclosed seas with straits restricting communication with the outside: the Mediterranean Sea, Sea of Japan, Red Sea, etc. For regions with extensive lateral boundaries, simple climatological average conditions are often used. Naturally, this limits the model skill. Two-way nesting is also a fairly recent phenomenon for ocean models (Oey and Chen, 1992; Fox and Maskell, 1995).

Irrespective of whether the nesting is one-way or two-way, a numerical scheme should avoid exciting gravity-inertia wave-induced noise at the lateral boundaries. In one-way nesting, often the LBCs are relaxed to coarse mesh conditions. Radiation or wave-absorbing sponge conditions are often necessary to prevent reflection of outward propagating disturbances back into the domain (Engquist and Majda, 1977; Miller and Thorpe, 1981; Israeli and Orzag, 1981; Raymond and Kuo, 1984; Hedley and Yau, 1988). Normal strategy is to prescribe or relax LBCs to coarse mesh conditions at inflow boundaries and use advection/radiation BCs on the outflow boundaries.

For two-way nesting, it is necessary to interpolate the coarse grid solution to the fine grid and filter the fine mesh solution that feeds back to the coarse mesh. Simple linear interpolation and averaging surrounding grid points, respectively, are often adequate, although propagation of noise from one mesh to another due to inevitable mismatches is a common problem (Ookochi, 1972; Kurihara *et al.,* 1979; Skamarock *et al.,* 1989; Peggion, 1994; Perkins *et al.,* 1997). Because the forecast skill depends so much on the dominant dynamical process prevalent in the domain, the choice of the domain, the LBC implementation scheme, the type of nesting, the complexity of large scale flow, and various other factors, it is advisable to perform sensitivity studies over a wide range of conditions before relying on the skill of a LAM.

Because of the importance of LBCs in both atmospheric and oceanic regional models, we will describe them briefly. Remember, a constituent property ϕ is simply advected around by a fluid and hence it is simpler to prescribe LBCs for them. If the flow is into the domain, then the value of the property being advected in must be prescribed: $\phi = \phi_d(t)$. If the flow is out of the domain, the property is simply advected out: $\partial \phi / \partial t + u_n \partial \phi / \partial n = 0$, where u_n is velocity normal to the boundary. Often, the value of the property is damped to a reference (say, climatological) value: $\partial \phi / \partial t = [\phi_r(t) - \phi]/T_d$, where T_d is the damping timescale; the smaller the value of T_d, the more strongly the property is held close to the reference value. Often, for lack of a better alternative, the temperature and salinity in ocean models are damped to climatological averages at lateral boundaries. Damping is also useful for prescribing the values at the ocean–atmosphere interface. For example, when precipitation is unknown, the surface salinities may be damped to their climatological values.

It is the dynamical variables such as the SSH and the velocities that are harder to prescribe, since they are affected by both advection and wave motions. Consider a scalar dynamical prognostic variable in a regional model, for example, η, the SSH. There are basically five different ways of prescribing its value on the lateral open boundaries:

1. Clamped LBC. Here the value of the variable is prescribed on the boundaries. This is the classic Dirichlet condition. The prescribed values can come from observations or another model, and can be time dependent.

$$\eta = \eta_d(t) \tag{2.10.1}$$

$$R = \alpha R^n + (1-\alpha)R^{n+1} \tag{2.10.8}$$

The major disadvantage of clamping is that it ignores internal variability and waves impinging on the boundary are reflected back into the domain, causing unwanted noise.

2. Zero gradient. This is the classic Neumann condition.

$$\partial\eta / \partial n = 0 \tag{2.10.2}$$

This can, however, lead to unrealistic interior solutions (Chapman, 1985; Blumberg and Kantha, 1985).

3. Newtonian damping. The variable can also be damped to a reference value. For example, instead of prescribing the tidal values of SSH on a lateral boundary, the SSH can be damped to tidal values.

$$\partial\eta / \partial t = [\eta_r(t) - \eta] / T_d \tag{2.10.3}$$

This affords a little more flexibility to the boundary in adjusting to internal variability. Newtonian damping is also used often as boundary conditions at the top of the atmosphere to prevent gravity waves from being reflected back into the domain.

4. Sponge boundaries. Here the fluid at the boundary is made highly viscous to strongly damp out any fluctuations. The lateral viscosity (A_M, A_H) is artificially increased to accomplish this. The viscosity is gradually increased, either linearly or exponentially, as the boundary is approached, so as not to cause discontinuous changes near the boundary. Sponge conditions, while effective in damping out any waves impinging on the boundary from inside, do not permit external influences to be propagated in either.

5. Sommerfeld radiation conditions. By far the most commonly used is the Sommerfeld radiation condition (Sommerfeld, 1979). It is similar to the advection condition mentioned above, except that the wave velocity is used instead of the advective velocity. At, say, the eastern boundary,

$$\partial\eta/\partial t + c\partial\eta/\partial x = 0 \qquad (2.10.4)$$

This allows a monochromatic wave $\eta(x - ct)$ propagating east toward that boundary to exit undisturbed through the boundary [the sign in Eq. (2.10.4) should be negative for the western boundary]. While elegant in principle, in practice its use is beset with problems. One major problem is that the phase velocity of the disturbance may not be known a priori. It is important to prescribe the correct value of c to prevent generation of noise at the boundary that might corrupt the interior solutions. Orlanski (1976) proposed that the value be computed immediately adjacent to the boundary from known interior solutions and this value then be used in Eq. (2.10.4). Thus if IM denotes the boundary,

$$\eta_{IM}^{n+1} = \eta_{IM}^{n} - R(\eta_{IM}^{n} - \eta_{IM-1}^{n})$$

$$R = c\Delta t / \Delta x = -(\eta_{IM-1}^{n+1} - \eta_{IM-1}^{n})/(\eta_{IM-1}^{n} - \eta_{IM-2}^{n})$$

$$(2.10.5)$$

for a first-order time difference scheme. If leapfrogging is used, to prevent numerical instability η^{n} on the RHS must be replaced with the average of η^{n+1} and η^{n-1} so that (Orlanski, 1976)

$$\eta_{IM}^{n+1} = [(1-R)\eta_{IM}^{n} + 2R\eta_{IM-1}^{n}]/(1+R)$$

$$R = c\Delta t / \Delta x = -(\eta_{IM-1}^{n+1} - \eta_{IM-1}^{n-1})/(\eta_{IM-1}^{n+1} + \eta_{IM-1}^{n-1} - 2\eta_{IM-2}^{n})$$

$$(2.10.6)$$

R must be less than unity for stability of the scheme. Equations (2.10.5) and (2.10.6) are applicable strictly to 1D situations, although in this form it is used in 2D models as well. Extension to 2D is straightforward:

$$\eta_{IM,j}^{n+1} = [(1-R_x)\eta_{IM,j}^{n} + 2R_x\eta_{IM-1,j}^{n} + 2R_y(\eta_{IM,j}^{n} - \eta_{IM,j-1}^{n})](1+R_x)$$

$$R_x = c\Delta t / \Delta x = -(\eta_{IM-1,j}^{n+1} - \eta_{IM-1,j}^{n-1})/(\eta_{IM-1,j}^{n+1} + \eta_{IM-1,j}^{n-1} - 2\eta_{IM-2,j}^{n}) \qquad (2.10.7)$$

$$R_y = c\Delta t / \Delta y = -(\eta_{IM-1,j}^{n+1} - \eta_{IM-1,j}^{n-1})/(\eta_{IM-1,j}^{n+1} + \eta_{IM-1,j}^{n-1} - 2\eta_{IM-1,j-1}^{n})$$

In practice, the phase speed computed can be noisy and spiky, especially during the passage of crests and troughs, and this error in turn propagates into the interior. One remedy that Kantha et al. (1990) propose is to recursively filter the value of R, by storing the previous values and taking the weighted average of the past and present:

$$R = \alpha R^{n} + (1-\alpha)R^{n+1} \qquad (2.10.8)$$

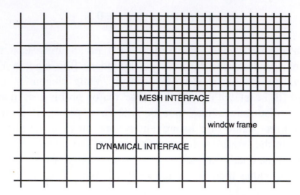

Figure 2.10.1 Postion of the dynamical interface. The mesh interface is surrounded by a window frame of two coarse mesh width. The outer rim of the frame is the dynamical interface.

This appears to lead to much smoother solutions. Even then, care must be exercised in the use of Eq. (2.10.4), since in general, the impinging waves may not be monochromatic or may be propagating at an angle to the boundary, and the wave speed may not be a function of only the local gradients. These problems can often be reduced by taking an average along the boundary for the calculated phase speed using values at the neighboring grid points. Still the difficulty of calculating the phase speed accurately and the adverse impact of incorrect calculation has prompted some modelers to use a fixed value for R (a value of unity is often used).

When it is necessary to radiate out an outgoing wave, but still be able to prescribe an incoming wave (as, for example, in tidal calculations), Eq. (2.10.4) can be modified (Kantha, 1985),

$$\partial \eta / \partial t + c_o \partial \eta / \partial x = [(c_o + c_i)/ c_i]\partial \eta_i / \partial t$$

$$(2.10.9)$$

where subscripts i and o refer to incoming and outgoing waves. This situation arises in two-way nesting. Perkins *et al.* (1997) discuss the need to radiate out the difference $\eta_I - \eta_O$ between the inner and the outer mesh solutions at the boundaries of a nested fine mesh model using a scheme similar to Eq. (2.10.9) in order to reduce the noise at the boundary of a nested grid model and make long-term integrations possible.

For a split-mode model (Chapter 9), it is necessary to radiate out both the external and the internal modes. The external mode gravity wave speed is known so that R = $(gH)^{1/2}\Delta t/\Delta x$ works satisfactorily, even though averaging along the boundary is recommended. The interior mode is however problematic, since the internal baroclinic mode is a function of ambient stratification and not always known a priori. Also since its speed is often comparable to advective speeds, it is necessary to use the sum $(c_i + u)$ instead of c_i. The first-mode baroclinic wave speed is roughly 2.5 m s^{-1} in the global oceans.

The use of overlap between the fine and coarse meshes is a strategy that avoids some of the problems such as nonconservation of properties in a two-way nested models. This strategy has been successfully employed for modeling hurricanes (Kurihara *et al.*, 1998). An overlap called the window frame, equal to two outer grids (Figure 2.10.1), separates the dynamical interface from the physical mesh interface, so that the mesh interface is less susceptible to any numerical noise that may result from dynamical coupling. The coarse domain integration is made first taking into account values in the window frame, thus preserving values and fluxes at the dynamical interface. The results are used during the integration of the fine mesh model, with values from the coarse mesh temporally interpolated to the fine mesh time steps, thus conserving transports of properties during the dynamical coupling between the two models. The open boundary prescription is carried out in two steps. In the first step, the boundary grid values are obtained by interpolation from the interior, and in the second step, these values are damped toward reference values, which are obtained during model integration and not prescribed. Thus for any prognostic variable,

$$\varphi_b^{n+1} = (1 - \gamma_1)\varphi_b^n + \gamma_1\varphi_b^r$$
$$\varphi_b^r = \gamma_2\left(\varphi_{b-1} + \varphi_b^o - \varphi_{b-1}^o\right) + (1 - \gamma_2)\varphi_b^o$$

$$(2.10.10)$$

where subscript b denotes the boundary grid point, and b−1 the inner grid point next to the boundary; superscript o denotes values from the larger scale model, and r the reference value. The damping factor γ_1 and the weighting factor γ_2 are chosen empirically. For details of this strategy and its performance, see Kurihara *et al.* (1989).

2.11 COMPUTATIONAL ISSUES

Ocean models make a large demand on computer resources: CPU time, core memory, and disk storage. Because ocean eddies are much smaller than weather systems, the resolution needed is therefore much finer. Fine resolution also forces one to take smaller time steps in explicit models due to CFL constraints. For explicit free surface models, the time step is limited by the speed of the fast moving surface gravity waves, and one has to take a large number of small time steps to integrate over a simulation or forecast period (mode splitting helps alleviate this problem). For implicit ocean models, which filter out these gravity modes, the CPU time requirement is governed by the rate of convergence of the iterative method used to solve the resulting Poisson equation.

Explicit model codes are usually readily vectorizable and parallelizable, and generally need few additional arrays to store the auxiliary variables that may be needed to speed up the computations. In contrast, the vectorization/ parallelization of the Poisson/Helmholtz solvers associated with implicit codes is usually a nontrivial problem, and for some schemes that have been used up to now on serial machines, not at all feasible. The extra work resulting from the iterative or matrix inversion solution can often increase the total CPU time so that it is comparable to, or even exceeds, that for explicit codes, especially on vector/parallel computers. In addition, there are almost always extra arrays needed during this stage of the computations. The two-color version of the successive overrelaxation and the conjugate gradient methods are two techniques that are well suited to vectorization and parallelization in implicit codes.

In the early days of high performance computers, the extent of core memory available was usually so small that all the arrays needed for computations in ocean models, especially global ones, would not fit within the core, and elaborate methods were employed to make efficient use of high speed disks to transfer arrays into and out of the core as needed. GFDL models (MOM2, for example) still retain such an architecture. With high speed memory becoming much cheaper, modern high performance computers have core memories measured in gigawords, and many ocean models can now reside in memory, although the need for out-of-core models has not totally vanished, especially for very high resolutions and global coverage. In-core models such as the Los Alamos POP are, however, better suited to efficient massive parallelization than the out-of-core ones such as MOM2, because of the considerable disk input/output (I/O) involved.

Disk/tape storage requirements for storing ocean model results are also often in tens to hundreds of gigabytes, and depend on the length of the simulation, and how often and how many variables are required to be stored for later analyses. Disk storage and postprocessing requirements often constrain the temporal resolution and the details of the analyses carried out on the results of an ocean model.

Data-assimilative ocean models require even more resources than the free-running ones, the additional memory and CPU time requirements depending very much on the method of assimilation. It is not unusual for assimilation to more than double the CPU time requirements, even for simple OI-type schemes. Methods such as Kalman filters and adjoint methods are even more demanding. Generally, data assimilation on massively parallel computers requires considerable investment of time and effort for efficient implementation.

Here are some typical CPU time and memory requirements for some large ocean models and for diverse computers, to cover a spectrum of configurations and to familiarize the reader with resource requirements of computational ocean modeling. The 1/8° 6-layer NRL global model (2051 × 1145 × 6 grid) requires 1.8 Gw of memory and 2 CPU hr per month of simulation on a 256-node CM5-

E. The 1/16° Pacific model (1977 × 1313 × 6 grid) requires 385 Mw and 17 single-processor CPU hr per month on a Cray C-90. The 1/5° global explicit barotropic tidal model discussed in Chapter 1 (1801 × s; 729 grid) employs a time step of 13 s, assimilates 4000 data points every time step, and requires 65 Mw of memory and 12 single-processor CPU hr for a 10-day simulation on a 16-processor Cray C-90, assimilating 4000 data points every time step. A 15-level northern hemisphere Arctic ice–ocean model (360 × 360 × 15 grid) requires 40 Mw of memory and 25 CPU hr per month on a Cray C-90. A 30-level sigma-coordinate model of the eastern Pacific (163 × 229 × 30 grid) requires 42 Mw of memory and 4 CPU hr per month on Cray C-90. The small Gulf of Mexico sigma-coordinate nowcast/forecast model (85 × 86 × 22 grid) discussed in Chapter 14 requires 30 Mw of memory and 6 CPU hr for a month-long simulation in the nowcast mode and 4 CPU hr in the forecast mode, the additional time requirements for the simple OI-based data assimilation being in this case about 50%. An idea of the storage requirements can be obtained from the fact that even this model required 2 gigabytes to store the model output at 5-day intervals for a 10-year-long simulation without any data assimilation, and postprocessing of this output required numerous hours on a powerful workstation.

2.12 EXAMPLES

We will next look at some examples illustrating the art of numerical solutions starting from the simplest to the complex, where some of the concepts discussed earlier are put to use.

2.12.1 INERTIAL OSCILLATIONS

Consider inertial oscillations discussed in Section 1.7, for which analytical solutions are available for comparing with numerical solutions. These are governed by two very simple, linear, coupled ODEs, with dependent variables U and V, and independent variable t. We have to decide on the horizontal grid configuration, grids A, B, or C (Figure 2.6.3). In both A and B grids, U and V are collocated and the location of the grid point can be indexed i, j. We could use a first-order scheme for time differencing, but this is seldom done, since the accuracy is so poor. If we use the simplest possible explicit second-order accurate leapfrog time differencing scheme instead, as is done in many ocean and atmosphere models that solve a more complete set of equations, denoting the time step by n, we get

$$\frac{U_{i,j}^{n+1} - U_{i,j}^{n-1}}{2\Delta t} - f_{i,j} V_{i,j}^n = 0$$

$$\frac{V_{i,j}^{n+1} - V_{i,j}^{n-1}}{2\Delta t} + f_{i,j} U_{i,j}^n = 0$$

$$(2.12.1)$$

Clearly, to resolve the process well, one must choose $\Delta t \ll 2\pi/f$. Equation (2.12.1) is energetically consistent, since if we multiply the first equation by $U_{i,j}^n$ and the second by $V_{i,j}^n$ and add the two, the Coriolis terms would vanish as they should. This is why the B grid is so useful when rotation is important. If we chose the C grid instead, we have the option of using either

$$\frac{U_{i,j}^{n+1} - U_{i,j}^{n-1}}{2\Delta t} - 0.25\,(f_{i,j} V_{i,j}^n + f_{i-1,j} V_{i-1,j}^n + f_{i-1,j+1} V_{i-1,j+1}^n + f_{i,j+1} V_{i,j+1}^n) = 0$$

$$\frac{V_{i,j}^{n+1} - V_{i,j}^{n-1}}{2\Delta t} + 0.25\,(f_{i,j} U_{i,j}^n + f_{i+1,j} U_{i+1,j}^n + f_{i+1,j-1} U_{i+1,j-1}^n + f_{i,j-1} U_{i,j-1}^n) = 0$$

$$(2.12.2)$$

or

$$\frac{U_{i,j}^{n+1} - U_{i,j}^{n-1}}{2\Delta t} - 0.0625\,(f_{i,j} + f_{i-1,j} + f_{i-1,j+1} + f_{i,j+1})$$

$$(V_{i,j}^n + V_{i-1,j}^n + V_{i-1,j+1}^n + V_{i,j+1}^n) = 0$$

$$\frac{V_{i,j}^{n+1} - V_{i,j}^{n-1}}{2\Delta t} + 0.0625\,(f_{i,j} + f_{i+1,j} + f_{i+1,j-1} + f_{i,j-1}) \qquad (2.12.3)$$

$$(U_{i,j}^n + U_{i+1,j}^n + U_{i+1,j-1}^n + U_{i,j-1}^n) = 0$$

Equation (2.12.2) is energetically consistent, while Eq. (2.12.3) is not, except in the limit of constant f. Remember also that the leapfrog scheme requires damping the computational mode. One approach is to use temporal smoothing in the form of an Asselin-Robert filter:

$$\tilde{U}_{i,j}^n = U_{i,j}^n + \alpha(U_{i,j}^{n+1} + \tilde{U}_{i,j}^{n-1} - 2U_{i,j}^n)$$

$$\tilde{V}_{i,j}^n = V_{i,j}^n + \alpha(V_{i,j}^{n+1} + \tilde{V}_{i,j}^{n-1} - 2V_{i,j}^n)$$

$$(2.12.4)$$

Proceeding to the next time step requires

$$\tilde{U}_{i,j}^{n-1} = \tilde{U}_{i,j}^n,\ \tilde{U}_{i,j}^n = U_{i,j}^{n+1};\ \tilde{V}_{i,j}^{n-1} = \tilde{V}_{i,j}^n,\ \tilde{V}_{i,j}^n = V_{i,j}^{n+1} \qquad (2.12.5)$$

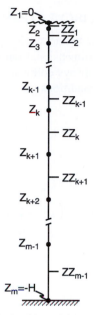

Figure 2.12.1 Staggered vertical grid used for computing the density gradient.

For increased accuracies, one could use the fourth-order RK (but we are still faced with the spatial averaging problem of a C grid). Omitting the i, j indexes for convenience, we have

$$k_1 = -fV^n$$

$$m_1 = fU^n$$

$$k_2 = -f(V^n + 0.5\,\Delta t\, m_1)$$

$$m_2 = f(U^n + 0.5\,\Delta t\, k_1)$$

$$k_3 = -f(V^n + 0.5\,\Delta t\, m_2)$$

$$m_3 = f(U^n + 0.5\,\Delta t\, k_2)$$

$$k_4 = -f(V^n + \Delta t\, m_3)$$

$$m_4 = f(U^n + \Delta t\, k_3)$$

$$U^{n+1} = U^n + \frac{\Delta t}{6}(k_1 + 2k_2 + 2k_3 + k_4) + o(\Delta t^5)$$

$$V^{n+1} = V^n + \frac{\Delta t}{6}(m_1 + 2m_2 + 2m_3 + m_4) + o(\Delta t^5)$$

$$(2.12.6)$$

2.12.2 THERMOHALINE CIRCULATION

Consider the two-box thermohaline model presented in Section 1.16. The dependent variables are the nondimensional temperature and salinity ΔT and ΔS (omitting the tilde), which are always collocated. The algorithm for the case of frictional control and restoring conditions (see Section 1.16) is

$$k_1 = \Delta T^a - (\Delta T)_i (1.0 + Q_i); \quad Q_i = (\Delta T)_i - (\Delta S)_i$$

$$m_1 = F_i - Q_i (\Delta S)_i; \quad F_i = \xi[\Delta S^a - (\Delta S)_i]$$

$$k_2 = \Delta T^a - [(\Delta T)_i + k_1 h/2][1.0 + \hat{Q}_i]; \hat{Q}_i = [(\Delta T)_i + k_1 h/2] - [(\Delta S)_i + m_1 h/2]$$

$$m_2 = \hat{F}_i - \hat{Q}_i [(\Delta S)_i + m_1 h/2]; \quad \hat{F}_i = \xi[\Delta S^a - \{(\Delta S)_i + m_1 h/2\}]$$

$$k_3 = \Delta T^a - [(\Delta T)_i + k_2 h/2][1.0 + \hat{Q}_i]; \hat{Q}_i = [(\Delta T)_i + k_2 h/2] - [(\Delta S)_i + m_2 h/2]$$

$$m_3 = \hat{F}_i - \hat{Q}_i [(\Delta S)_i + m_2 h/2]; \quad \hat{F}_i = \xi[\Delta S^a - \{(\Delta S)_i + m_2 h/2\}]$$

$$k_4 = \Delta T^a - [(\Delta T)_i + k_3 h][1.0 + \hat{Q}_i]; \hat{Q}_i = [(\Delta T)_i + k_3 h] - [(\Delta S)_i + m_3 h]$$

$$m_4 = \hat{F}_i - \hat{Q}_i [(\Delta S)_i + m_3 h]; \quad \hat{F}_i = \xi[\Delta S^a - \{(\Delta S)_i + m_3 h\}]$$

$$(\Delta T)_{i+1} = (\Delta T)_i + \frac{h}{6}(k_1 + 2k_2 + 2k_3 + k_4) + o(h^5)$$

$$(\Delta S)_{i+1} = (\Delta S)_i + \frac{h}{6}(m_1 + 2m_2 + 2m_3 + m_4) + o(h^5)$$

$$(2.12.7)$$

where we have used h for Δt. Transitions from either thermal to haline mode or vice versa can be investigated using Eq.(2.12.7).

2.12.3 NORMAL MODES

Now consider the eigenvalue problem (1.5.11) governing the normal modes in the ocean. Although this formulation is for an infinite flat-bottomed ocean, as long as the scale of variation of the ocean depth and density stratification is much larger than the radius of deformation, these equations can be used for real oceans. Naturally, one has to use observed densities, obtained from temperature and salinity profiles from CTD casts and an equation of state. Because of the sparsity of such data, all available data are binned into discrete boxes [say, 1° × 1°, as did Chelton *et al.* (1998)] and averaged. The normal practice is to decimate the casts into observations at standard predefined depths (Levitus,

1982; Levitus and Boyer, 1994; Levitus *et al.*, 1994). Thus one is left with density values at standard, nonuniformly spaced vertical depths z_k (k = 1, M) (see Figure 2.12.1). Denoting the midpoints between standard depths z_k and z_{k+1} by zz_k (k = 1, M−1),

$$N^2\big|_{zz_k} = -\frac{g}{\rho_0}\left(\frac{\rho(z_k) - \rho(z_{k+1})}{z_k - z_{k+1}}\right) \tag{2.12.8}$$

where $\rho(z_k)$ is the *in situ* density at z_k adjusted adiabatically to zz_k. This gives a better estimate of the stability of the water column than the potential density referenced to some arbitrary depth (usually the sea surface), because of the thermobaric effects. The second derivative d^2w/dz^2 can be computed using

$$\frac{d^2w}{dz^2}\bigg|_{zz_k} = \frac{1}{\Delta z_k}\left[\left(\frac{w_{k-1} - w_k}{\Delta zz_{k-1}}\right) - \left(\frac{w_k - w_{k+1}}{\Delta zz_k}\right)\right] \tag{2.12.9}$$

$$\Delta z_k = z_k - z_{k+1}; \quad \Delta zz_k = zz_k - zz_{k+1}$$

so that Eq. (1.5.8) can be written as

$$\left(\frac{1}{\Delta z_k \Delta zz_{k-1}}\right)\frac{1}{N_k^2} w_{k-1} - \left(\frac{1}{\Delta z_k \Delta zz_{k-1}} + \frac{1}{\Delta z_k \Delta zz_k}\right)\frac{1}{N_k^2} w_k + $$
$$\left(\frac{1}{\Delta z_k \Delta zz_k}\right)\frac{1}{N_k^2} w_{k+1} = -\frac{1}{c^2} w_k \tag{2.12.10}$$

In matrix notation we have $\mathbf{A}\cdot\mathbf{w} = \lambda\mathbf{w}$, a matrix eigenvalue problem that yields M−1 values for eigenvalues λ_m and eigenfunctions w_m, corresponding to M-1 baroclinic modes (barotropic mode is excluded). Matrix $\mathbf{A}$ reduces to a tridiagonal form once values for w_0 and w_M are substituted using the boundary conditions at z = 0 and z = −H. This problem can be readily solved using one of the standard routines and use can be made of the fact that $\mathbf{A}$ is sparse and tridiagonal (see Press *et al.*, 1992).

2.12.4 GYRE SCALE CIRCULATION

Now consider the Stommel problem, which is governed by the harmonic equation (1.10.2). Using second-order centered differences,

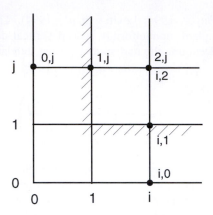

Figure 2.12.2 Boundary conditions for the Munk problem.

$$c\left(\frac{\psi_{i+1,j} - 2\psi_{i,j} + \psi_{i-1,j}}{\Delta x^2} + \frac{\psi_{i,j+1} - 2\psi_{i,j} + \psi_{i,j-1}}{\Delta y^2}\right) +$$

$$\beta\left(\frac{\psi_{i+1,j} - \psi_{i-1,j}}{2\Delta x}\right) = f_{i,j}$$

(2.12.11)

so that the coefficients in Eq. (2.1.5) become

$$a_{i,j} = c + \frac{\beta\Delta x}{2}, b_{i,j} = c - \frac{\beta\Delta x}{2}, c_{i,j} = c\frac{\Delta x^2}{\Delta y^2} = d_{i,j}$$

$$e_{i,j} = -2c\left(1 + \frac{\Delta x^2}{\Delta y^2}\right), f_{i,j} = \left[\frac{\tau_0}{\rho_0}\frac{\pi}{L}\sin\left(\frac{\pi y}{L}\right)\Delta x^2\right]_{i,j}$$

(2.12.12)

SOR in Press *et al.* (1992) can be used to solve this equation subject to the boundary condition $\psi = 0$ along the boundary. It is important to select grid size Δx to resolve the western boundary layer, which has a thickness on the order c/β.

For $\beta = 2 \times 10^{-11}$ m^{-1} s^{-1}, c = 10^{-6} s^{-1}, L = W = 4000 km, the boundary layer scale is 100 km and it is appropriate to choose a grid size of, say, 25 km to resolve the western boundary current. Kinematic wind stress magnitude is $\tau_0 / \rho_0 = 10^{-4}$ m^2 s^{-2}, a reasonable value. These values lead to a southward Sverdrup transport of 16 Sv, which is returned northward in the western boundary current.

For the Munk problem, we need to solve Eq. (1.11.2) subject to the conditions $\psi = 0$ and $\partial \psi / \partial n = 0$ along the boundary, satisfying both no-normal-flow and no-slip conditions. This is however done by converting this biharmonic equation in stream function ψ into two coupled harmonic equations, one for ψ and another for vorticity ζ:

$$A_M \left(\frac{\zeta_{i+1,j} - 2\zeta_{i,j} + \zeta_{i-1,j}}{\Delta x^2} + \frac{\zeta_{i,j+1} - 2\zeta_{i,j} + \zeta_{i,j-1}}{\Delta y^2} \right) -$$

$$\beta \left(\frac{\psi_{i+1,j} - \psi_{i-1,j}}{2\Delta x} \right) = -f_{i,j} \tag{2.12.13}$$

$$\left(\frac{\psi_{i+1,j} - 2\psi_{i,j} + \psi_{i-1,j}}{\Delta x^2} + \frac{\psi_{i,j+1} - 2\psi_{i,j} + \psi_{i,j-1}}{\Delta y^2} \right) = \zeta$$

The second boundary condition in terms of the gradient of ψ must, however, be converted to a condition on vorticity ζ. This can be done as follows. Consider the western boundary (Figure 2.12.2). The numerical grid extends from $i = 1$ to I. Imagine a fictitious grid $i = 0$ outside the western boundary. From the boundary conditions, we know that $\psi_{1,j} = 0$ and $\left. \frac{\partial \psi}{\partial x} \right|_{i,j} = 0$ so that $\psi_{2,j} = \psi_{0,j}$. At $i = 1$,

$\frac{\partial^2 \psi}{\partial y^2} = 0$ so that $\zeta = \frac{\partial^2 \psi}{\partial x^2} = \frac{\psi_{0,j} - 2\psi_{1,j} + \psi_{2,j}}{\Delta x^2}$. This gives $\zeta_{1,j} = \frac{2\psi_{2,j}}{\Delta x^2}$. Values

for vorticity on other boundaries can be derived similarly. The procedure is to successively solve the stream function and vorticity equations subject to the above boundary conditions each iteration step using SOR or some other similar method, until the solution converges to desired accuracy. The accelerated SOR routine presented in Press *et al.* (1992) can be used to solve the biharmonic Munk problem. The value of $A_M \sim 10^4$ m^2 s^{-1} once again gives a 100-km boundary layer. However, the nature of the boundary layer is different in that the velocity profile is in the form of a jet, with a maximum away from the western boundary, whereas for the Stommel case, the maximum velocity is right at the boundary. This is of course due to no-slip condition being satisfied in the former but not the latter.

The Stommel and Munk problems can be combined in a single equation,

$$A_M \nabla^4 \psi - c \nabla^2 \psi - \beta \frac{\partial \psi}{\partial x} = \frac{1}{\rho_0} (\mathbf{k} \cdot \nabla \times \tau) \tag{2.12.14}$$

which can be solved the same way as the Munk problem. One gets the Munk solution when $c = 0$ and the Stommel solution when $A_M = 0$. By imposing a wind stress of the form $-\tau_0 \cos (n\pi y/L)$, one can obtain multiple gyres. For n=2, double gyre circulation crudely representative of the subtropical and subpolar gyres in the basins is obtained.

2.12.5 ADVECTION PROBLEMS

Problems associated with solving the advection equation for a tracer can be illustrated by considering the advection equation (2.5.3). The various finite-difference schemes, in order, are the leapfrog with second- and fourth-order spatial differences, and the third-order Adams–Bashforth with second- and fourth-order spatial differences.

$$\frac{u_j^{n+1}-u_j^{n-1}}{2\Delta t}+c\left(\frac{u_{j+1}^n-u_{j-1}^n}{2\Delta x}\right)=0$$

$$\frac{u_j^{n+1}-u_j^{n-1}}{2\Delta t}+c\left[\frac{4}{3}\left(\frac{u_{j+1}^n-u_{j-1}^n}{2\Delta x}\right)-\frac{1}{3}\left(\frac{u_{j+2}^n-u_{j-2}^n}{4\Delta x}\right)\right]=0$$

$$\frac{u_j^{n+1}-u_j^{n}}{\Delta t}+c\left[\frac{23}{12}\left(\frac{u_{j+1}^n-u_{j-1}^n}{2\Delta x}\right)-\frac{16}{12}\left(\frac{u_{j+1}^{n-1}-u_{j-1}^{n-1}}{2\Delta x}\right)+\frac{5}{12}\left(\frac{u_{j+1}^{n-2}-u_{j-1}^{n-2}}{2\Delta x}\right)\right]=0$$

$$\frac{u_j^{n+1}-u_j^{n}}{\Delta t}+c\left\{\begin{array}{l}\frac{4}{3}\left[\frac{23}{12}\left(\frac{u_{j+1}^n-u_{j-1}^n}{2\Delta x}\right)-\frac{16}{12}\left(\frac{u_{j+1}^{n-1}-u_{j-1}^{n-1}}{2\Delta x}\right)+\frac{5}{12}\left(\frac{u_{j+1}^{n-2}-u_{j-1}^{n-2}}{2\Delta x}\right)\right]\\[10pt]-\frac{1}{3}\left[\frac{23}{12}\left(\frac{u_{j+2}^n-u_{j-2}^n}{4\Delta x}\right)-\frac{16}{12}\left(\frac{u_{j+2}^{n-1}-u_{j-2}^{n-1}}{4\Delta x}\right)+\frac{5}{12}\left(\frac{u_{j+2}^{n-2}-u_{j-2}^{n-2}}{4\Delta x}\right)\right]\end{array}\right\}=0$$

(2.12.15)

Note that the Asselin–Robert filter involves the use of Eq. (2.6.31). Durran (1991) presents solutions to the fourth-order spatial-difference equations in the periodic domain $0 \leq x \leq 1$, and with $\Delta x = 1/32$, $c = 1/4$. The initial condition is

$$u(x) = [64(x-0.5)^2-1]^2 \text{ for } 0.375 \leq x \leq 0.625; 0 \text{ otherwise}$$

This is equivalent to advecting a Gaussian-shaped initial distribution repeatedly through the domain, the unit of time being the time needed to advect it through the domain once, $\Delta x/u$. Figure 2.12.3 shows the leapfrog and Adams–Bashforth solutions compared to the exact solution for two values of the Courant

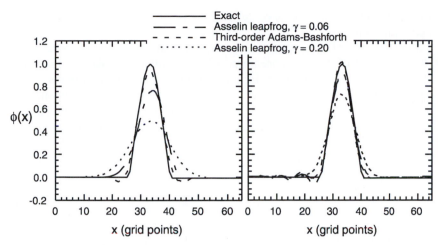

Figure 2.12.3 Comparison of an exact solution to the advection equation with Adams-Bashforth and Asselin-Robert-filtered leapfrog time differencing in a fourth-order finite-difference scheme at a nondimensional time 3, for Courant number of (left) 0.5 and (right) 0.2.

number C, 0.5 and 0.2. The third-order Adams–Bashforth scheme outperforms the Asselin–Robert-filtered leapfrog scheme in every case. The Asselin–Robert filter damps out the signal considerably, the damping being larger for larger values of C. The same results but with double the spatial resolution show significant improvement from increased spatial resolution.

2.12.6 MIT NONHYDROSTATIC GLOBAL OCEAN MODEL

The most recent effort at constructing a better ocean model is that of the group at MIT led by John Marshall (Marshall *et al.,* 1997a,b). This model has been specifically designed for efficient implementation on massively parallel computers and uses the full incompressible, Boussinesq, nonhydrostatic Navier–Stokes equations on a sphere, instead of the conventional hydrostatic version. As is the norm for global (deep-ocean) models, a rigid lid is imposed to filter out fast gravity waves, but this necessitates the solution of a three-dimensional elliptic equation for pressure, which is normally so intensive computationally as to be prohibitive for anything but regional deep convection studies. But clever exploitation of the salient properties of the solution and use of efficient numerical techniques have enabled the model to be used globally with little additional overhead in computing the flow in the vast majority of the oceanic regions where the conditions are very close to hydrostatic. It is instructive to examine the numerical aspects of this model, as it represents the state-of-the-art in ocean modeling.

Marshall *et al.* (1997a) start with the set of governing equations given by Eqs. (10.2.1) and (10.2.2) for an incompressible, stratified, Boussinesq fluid on a rotating sphere. However, the tendency, advective, and diffusion terms are not neglected in the equation for the vertical velocity. The meridional components of the Coriolis acceleration are also retained, and for accurate conservation of angular momentum on a spherical Earth, thin shell approximation is not invoked (R+z is not replaced by R). Retaining the full form for the equation for vertical velocity makes the pressure an additional prognostic variable governed by a 3D Poisson equation, with homogeneous Neumann boundary conditions,

$$\nabla^2 p = \xi_p; \quad \nabla p \cdot n = 0 \qquad (2.12.16)$$

where n is the unit normal and ∇^2 is a three-dimensional Laplacian operator in spherical coordinates. The RHS, ξ_p, is complicated and will not be written out (instead, see Marshall *et al.*, 1997a).

The time differencing used is the quasi-second-order Adams–Bashforth scheme, which avoids the computational mode problem of the leapfrog scheme. Thus each prognostic variable φ (u, v, w, T, and S) is evaluated according to

$$\frac{\varphi^{n+1} - \varphi^n}{2\Delta t} = \left[\left(\frac{3}{2} + \chi \right) \xi_\varphi^n - \left(\frac{1}{2} + \chi \right) \xi_\varphi^{n-1} \right] \qquad (2.12.17)$$

where ξ_φ is the RHS in each equation (except for any terms involving pressure). $\chi \sim 0.1$.

Spatial differencing uses a control volume approach on a C grid (see Chapter 9 for a sigma-coordinate model with a similar implementation). The C grid has difficulties in dealing with the Coriolis term at resolutions coarser relative to the Rossby radius, since spatial averaging can lead to excitation of a checkerboard mode in the horizontal divergence field. The A and B grids avoid this, but solving the pressure equation in a B grid involves a nine-point stencil (in the horizontal direction) and is more complex (Dukowicz *et al.*, 1993), whereas the solution on a C grid involves only a five-point stencil that consists of only the point and its four immediate neighbors. The differencing scheme conserves the variance of T and S, and energy, but not enstrophy.

Solving Eq. (2.12.16) for the three-dimensional total pressure field in the domain is computationally intensive and the practicality of the model for large applications depends on the efficiency with which it can be solved. Convergence can be slow even with accelerated techniques for solving elliptic equations. Instead Marshall *et al.* (1997a,b) split the pressure into three components: the pressure on the rigid lid p_s, the hydrostatic component of the pressure p_h, and the nonhydrostatic component p_{nh}. The hydrostatic component can be readily calculated from the usual diagnostic equation for the vertical pressure gradient term that has only the departures of buoyancy from the state of rest on the RHS. Of the remaining two, p_s is governed by a 2D Poisson equation, while p_{nh} is governed by the more difficult 3D Poisson equation. For the hydrostatic and quasi-hydrostatic models, it is sufficient to solve for p_s. Then the pressure anywhere in the water column is the sum of p_s and p_h, and prognostic equations for the horizontal velocity components involving the horizontal gradients of pressure can be solved. The vertical velocity is then "diagnosed" from the continuity equation. In the nonhydrostatic model, it is necessary to solve for both p_s and p_{nh}. Without going into too many details (but see Marshall *et al.*, 1997a,b), the problem of finding the pressure in the water column reduces to solving the following matrix equations:

$$\mathbf{A}_{2D} \cdot \mathbf{p}_s = \mathbf{B}_{2D} \tag{2.12.18}$$

$$\mathbf{A}_{3D} \cdot \mathbf{p}_{nh} = \mathbf{B}_{3D} \tag{2.12.19}$$

Both $\mathbf{A}_{2D}$ (size $N_x N_y \times N_x N_y$) and $\mathbf{A}_{3D}$ (size $N_x N_y N_z \times N_x N_y N_z$) are symmetric and positive-definite matrices well suited for the use of the conjugate gradient solver. N_x, N_y, and N_z denote the number of grid points in the zonal, meridional, and vertical directions, with N_x and N_y typically in the hundreds and N_z in the tens. Matrix $\mathbf{A}_{3D}$ is huge (typically $10^5 \times 10^5$ to $10^6 \times 10^6$), and "inverting" Eq. (2.12.19) efficiently with minimum storage and CPU requirements is a great challenge, especially on massively parallel computers, where overhead represented by communication between different processors must be minimized.

The matrix $\mathbf{A}_{2D}$ is sparse with five diagonal elements, with the leading diagonal being dominant. The matrix $\mathbf{A}_{3D}$ is also sparse but has seven diagonals composed of $N_x \times N_y$ submatrices of dimension $N_z \times N_z$, one for each vertical column of the ocean. Of these, the diagonal blocks $\mathbf{D}$ arise from the $\partial^2/\partial z^2$ terms and the off-diagonal blocks $\mathbf{d}$ arise from the second-order horizontal derivative terms, ∇_h^2.

$$\mathbf{A}_{3D} = \begin{bmatrix} \mathbf{D}_1 & \mathbf{d} & . & & \mathbf{d} & & & \\ \mathbf{d} & \mathbf{D}_2 & \mathbf{d} & . & & \mathbf{d} & & \\ & \mathbf{d} & \mathbf{D}_3 & \mathbf{d} & . & & . & \\ & & & & & & & \mathbf{d} \\ \mathbf{d} & & . & \mathbf{d} & \mathbf{D}_i & \mathbf{d} & . & \\ & . & & & & & & \\ & & . & & & & & \\ & & . & \mathbf{d} & . & . & \mathbf{d} & \mathbf{D}_{N_x N_y} \end{bmatrix} \tag{2.12.20}$$

Since the aspect ratio of the ocean is small, the elements $\mathbf{d}$ are smaller than elements $\mathbf{D}$ by a factor of $\sim(\Delta z/\Delta x)^2$, and therefore $\mathbf{A}_{3D}$ is dominated by its diagonal elements $\mathbf{D}$. Using the preconditioned conjugate gradient method then involves using as preconditioner a diagonal matrix consisting of the inverse of these elements $\mathbf{D}^{-1}$ as diagonal elements. In the hydrostatic limit, elements $\mathbf{d}$ are zero, and the preconditioning matrix $\mathbf{K}_{3D}$ will be an exact inverse of $\mathbf{A}_{3D}$. When nonhydrostatic effects are present but small, as is usual in the majority of the ocean, convergence is rapid. For inverting $\mathbf{A}_{2D}$, a similar strategy could be employed, but Marshall *et al.* (1997b) use even a better approximation for the preconditioning matrix $\mathbf{K}_{2D}$.

The preconditioned conjugate gradient solver exploits the sparseness and diagonal dominance of matrix $\mathbf{A}$. The efficiency of the preconditioned conjugate gradient method depends very much on the choice of the preconditioning matrix. In many CFD problems, matrix $\mathbf{A}$ is sparse and diagonally dominant. Then choosing the approximate inverse of $\mathbf{A}$, a block-diagonal matrix consisting of the inverse of the diagonal matrix elements that make up the dominant block-diagonal terms of $\mathbf{A}$, as the preconditioner is a logical choice. Choose a matrix $\mathbf{K}$ that is an approximate inverse of $\mathbf{A}$ so that $\mathbf{C} = \mathbf{I} - \mathbf{KA}$ is small, where $\mathbf{I}$ is the identity matrix. Then premultiplying the original matrix equation $\mathbf{A}\ \mathbf{p} = \mathbf{B}$ by $\mathbf{K}$ (omitting subscripts for clarity), one gets $(\mathbf{I}-\mathbf{C})\ \mathbf{p} = \mathbf{KB}$. Then the iterative scheme in solving this equation is $\mathbf{p}^{n+1} = \mathbf{p}^n + \mathbf{b}^n$, where $\mathbf{b}^n = \mathbf{K}\ \mathbf{r}^n$ is called the search direction and $\mathbf{r}^n = \mathbf{B} - \mathbf{A}\ \mathbf{p}^n$ is the residual vector. The values of $\mathbf{r}$ and $\mathbf{b}$ for the next iteration are given by the Richardson iteration procedure: $\mathbf{r}^{n+1} = \mathbf{r}^n - \mathbf{A}\ \mathbf{b}^n$, $\mathbf{b}^{n+1} = \mathbf{K}\ \mathbf{r}^{n+1}$. The conjugate gradient method consists of a particular optimum choice of the search direction, $\mathbf{b}^{n+1} = \mathbf{K}\ \mathbf{r}^{n+1} + \beta\ \mathbf{b}^n$, so that the search directions during consecutive iterations are the conjugate of one another (or mutually orthogonal): $\mathbf{A}\ \mathbf{b}^{n+1} \cdot \mathbf{b}^n = 0$. A parameter α is also be chosen to minimize the magnitude of the residual vector $\mathbf{r}^n$. The algorithm consists then of the scheme

$$\mathbf{r}^0 = \mathbf{B} - \mathbf{Ap}^0; \mathbf{b}^0 = \mathbf{Kp}^0$$

$$\alpha = [\mathbf{r}^n \cdot \mathbf{Kr}^n][\mathbf{b}^n \cdot \mathbf{Ab}^n]^{-1}$$

$$\mathbf{p}^{n+1} = \mathbf{p}^n + \alpha \mathbf{b}^n; \mathbf{r}^{n+1} = \mathbf{r}^n - \alpha \mathbf{Ab}^n$$

$$\beta = [\mathbf{r}^{n+1} \cdot \mathbf{Kr}^{n+1}][\mathbf{r}^n \cdot \mathbf{Kr}^n]^{-1}$$

$$\mathbf{b}^{n+1} = \mathbf{Kr}^{n+1} + \beta \mathbf{b}^n$$

Iterate if $\left|\mathbf{r}^{n+1} \cdot \mathbf{r}^{n+1}\right| > \varepsilon$

$$(2.12.21)$$

where ε is the desired accuracy. The choice of $\mathbf{K}_{3D}$ as a block-diagonal matrix consisting of inverses of elements $\mathbf{D}$, and therefore finding the product $\mathbf{K} \cdot \mathbf{r}$ in Eq. (2.12.21) involves finding the inverse $\mathbf{D}^{-1}$ or equivalently solving for $\mathbf{x}$ in $\mathbf{D} \mathbf{x} = \mathbf{r}$. Here Marshall *et al.* (1997b) exploit the fact that $\mathbf{D}$ is tridiagonal and can be easily LU decomposed so that the solution for $\mathbf{x}$ involves $O(N_z)$ operations instead of $O(N_z^2)$. It is this choice of preconditioner $\mathbf{K}_{3D}$ which is a good approximation to the inverse of $\mathbf{A}_{3D}$ and which, because of the implementation that decomposes the domain into vertical columns to ensure that all cells in a column reside on the same processor, does not require interprocessor communication, that makes the nonhydrostatic model practical for global ocean applications. The convergence of the scheme for p_{nh} is quite rapid if the departure from the hydrostatic condition is small, as is the norm for most of the oceans. Marshall *et al.* (1997a) present three applications of the MIT nonhydrostatic ocean model:

1. Simulation of a laboratory experiment on convection in a rotating fluid ($127 \times 127 \times 19$) with cell sizes of $0.5 \times 0.5 \times 1.5$ cm and a time step of 0.05 s—here the convergence of the 3D Poisson solver takes 60 iterations and the 2D solver 450 iterations, but the total speedup over a 3D Poisson solver that would not decompose the pressure field into its components but solves for the total pressure instead is a factor of 5.

2. Simulation of mixed layer instability ($200 \times 119 \times 19$) with cell sizes of $250 \times 250 \times 100$ m and a time step of 120 s—here the 3D solver takes 38 iterations, and the 2D solver 250, but the speedup is 6.

3. Global ocean at $1°$ resolution ($360 \times 180 \times 20$) with cell sizes of $100 \times 100 \times 0.1$ km and a time step of 20 min—here the 3D solver takes 1 iteration to converge, the 2 D takes 450, and the overall speedup is a remarkable 20. It is this speedup that makes the model practical for global ocean simulations.

For more details, the reader is referred to the original papers cited above and the references therein. It is clear that this model (among others) represents the current state-of-the-art in numerical modeling of oceans and oceanic processes. Many of the concepts discussed in earlier sections are put to practical use in this model.

This concludes our discussion of numerical methods. We will now turn our attention to specific oceanic regions and processes.

Chapter 3

Equatorial Dynamics and Reduced Gravity Models

The oceans are driven primarily at the surface. This surface forcing (winds and buoyancy fluxes) contains a variety of temporal and spatial scales. An important question that ocean dynamicists try to answer is: How do the oceans respond to forcing at various temporal and spatial scales? If there is a change somewhere in the ocean basin, how does the rest of the basin recognize that and adjust to it? These questions are in the realm of dynamical oceanography. The field was pioneered by people like Bjerknes, Ekman, Sverdrup, Rossby, Stommel and Munk barely half a century ago. As we saw in Chapter 1, these pioneers employed various simplifications to the complex nonlinear governing equations that nevertheless explained certain aspects of ocean dynamics. For example, Ekman elucidated the response of the upper ocean to wind stress and showed how rotation affects the currents. Stommel showed how the westward intensification of ocean currents is a consequence of the rotation on a spherical planet, namely, the change of planetary vorticity with latitude. But it was Carl Gustaf Rossby who laid down the foundations of planetary dynamics by simplifying the rotating sphere to a simple tangent plane with changing planetary vorticity (β-plane) that can be used to explain most of the low-frequency changes that occur in the oceans and the atmosphere.

In the next three chapters, we will explore the transient response of the oceans to external forcing and discuss simple numerical models designed to capture salient aspects of the dynamical processes involved. In doing so, we will

247

subdivide he oceans by latitude: equatorial, mid-latitudes and high-latitudes. Given the change in latitude of planetary vorticity, and the meridional gradients in thermal forcing on Earth, such a division is quite logical.

3.1 OCEANIC DYNAMICAL RESPONSE TO FORCING

Imagine an ocean in equilibrium. Now imagine that it is forced hard in some localized region that perturbs its equilibrium. Further assume that the disturbance contains a spectrum of frequencies and scales. The question is, how does the ocean respond?

It does so by generation and propagation away from the region of disturbance of various waves. A pebble tossed into a pond is a loose analogy, but in the oceans, these radiated waves are affected crucially by the fact that we are on a rotating planet. Rotation puts stringent constraints on these waves. It is easy to see that if the time scale (the period) of a wave is comparable to or much larger than the rotational time scale, the inertial period ($\sim 2\pi/f$), then planetary rotation must affect that wave. High-frequency gravity waves we are used to seeing on the surface of a pond or a lake (periods less than a minute or so) are unaffected by rotation, but gravity waves of very low frequency (such as Poincare waves) are. Such rotational gravitational waves travel fast across the basin and communicate changes (certain portions of it) rather quickly.

The existence of boundaries affects how disturbances propagate in the ocean basins. Kelvin waves are a prime example. These waves are dependent on the presence of rotation for their existence, yet they travel at a speed unaffected by rotation, their phase speed being the same as that of the shallow water wave [$C = (gH)^{1/2}$]. Rotation, however, constrains them to the vicinity of oceanic boundaries. Their amplitude is maximum at the boundary, and they are confined to offshore distances of the order of the mid-latitude Rossby radius of deformation ($a_M = C/f$). They propagate with the coast to the right (left) in the northern (southern) hemisphere. The salient timescale is the inertial period T_M proportional to $1/f$.

The equatorial waveguide also entertains Kelvin waves, which exist there because of an "imaginary boundary" at the equator. Equatorial Kelvin waves are also confined, to distances on the order of the equatorial Rossby radius of deformation [$a = (C/\beta)^{1/2}$] in the cross-equatorial direction. Their existence depends on the fact that planetary vorticity varies with latitude ($\beta = \partial f/\partial y \neq 0$). Furthermore, they are restricted to propagating eastward in the waveguide. The salient timescale is the equatorial inertial period $T \sim (\beta C)^{-1/2}$. The phase speed c, which determines these parameters, depends on the ambient density stratification. For example, for a uniformly stratified ocean with buoyancy frequency N_0 and depth D_0, for the first baroclinic mode, the mid-latitude radius of deformation $a_M \sim N_0D_0/f$, the equatorial radius of deformation $a \sim (N_0D_0/\beta)^{1/2}$, and the equatorial inertial period $T \sim (\beta N_0 D_0)^{-1/2}$.

Barotropic, mid-latitude Kelvin waves travel at $\sim$200 to 220 m s^{-1} around the basin and take a day or so to travel around the north Pacific. The phase speed of baroclinic Kelvin waves is however two orders of magnitude smaller (less than 3 m s^{-1}) and, hence, they take correspondingly longer times. Nevertheless, compared to planetary Rossby waves, they travel fast and are the first harbingers of change. Kelvin waves are especially important in the equatorial waveguide. It is the propagation of Kelvin and Rossby waves that brings about changes in the equatorial waveguide. Kelvin waves propagate relatively fast and therefore communicate changes quite rapidly from one end of the basin to the other. In a timescale of the order of 5 to 10 weeks, they can communicate changes across ocean basins; for example, the first mode baroclinic equatorial Kelvin wave, traveling at $\sim$2.4 m s^{-1} can cross the huge Pacific from west to east in a mere 2.5 months.

An important point about Kelvin waves is that they are dispersionless. Their phase speed C_p is independent of their frequency ω ($C_p = \omega/k$, where k is the wave number). If a disturbance created in the equatorial ocean were to consist of a spectrum of spatial scales (and hence ω), all of the Kelvin waves would propagate to the east at the same speed and therefore the entire disturbance would arrive more or less unchanged. If, on the other hand, the phase speed were a function of frequency, different parts would arrive at different times, lower frequencies earlier. In other words, the original disturbance would get "dispersed." Dispersion is a crucial aspect of wave propagation because in a dispersive wave, while the phases propagate at the phase speed C_p, the energy in the wave group (or equivalently the disturbance) travels at the group velocity ($C_g = \partial\omega/\partial k$), and it is the energy propagation that is important to the dynamics. For nondispersive waves $C_p = C_g$, and Kelvin waves are nondispersive.

We finally come to planetary or Rossby waves. These depend on planetary vorticity f being nonconstant. It is the change of planetary vorticity with latitude that gives rise to Rossby waves (Kelvin waves can exist on an f plane but Rossby waves can exist only on a β plane). They generally have low frequencies and propagate very slowly across the ocean basins. Their phase propagation is always westward. They take months to many years to cross the ocean basins and are extremely important to the adjustment of the thin oceanic (and atmospheric) envelope on the globe to low-frequency forcing, such as seasonal and interannual. Rossby waves are in general dispersive, although long Rossby waves are nondispersive. In fact, a peculiarity of planetary Rossby waves is that for certain frequencies, the phases could be propagating westward, while the group velocity and hence the energy propagation is eastward!

What do these waves really do when they arrive at a point and communicate changes that occurred elsewhere? They effect changes in the ambient velocity field. For example, a downwelling equatorial Kelvin wave propagating eastward in the equatorial waveguide depresses the thermocline and induces an eastward fluid velocity in its wake. This induced velocity advects warm western Pacific water masses toward the eastern Pacific. It is very important to realize that a different water mass needs to be advected into a region (discounting local changes) to change the local water masses. This can only be done by advection, a slow process in the oceans, since advective speeds are typically smaller (less than 1 m s^{-1}) than the wave speeds, outside western boundary currents. Wave motions cause changes in prevailing currents, which in turn bring about changes in water masses.

Most planetary adjustment processes can be explained by accounting for the Kelvin and Rossby waves generated by changes in forcing on the corresponding tangent β plane. Equatorial dynamics is particularly simpler because the existence of the waveguide defines the direction of propagation of disturbances to be principally zonal. Nevertheless, the governing equations are nonlinear and involve effects of both rotation and stratification. Simplifications are essential to understanding physical processes, although for accurate depiction, numerical solutions of the full equations are necessary.

Every dynamical system has its own preferences as far as its response to disturbances to its equilibrium state. This response can be divided into two groups: forced and free. Consider a simple forced pendulum. When set in motion, in the absence of external forcing, it oscillates at its natural frequency. However, if the pendulum is forced steadily at some frequency, it responds with swinging at that frequency, irrespective of what its natural frequency is. The amplitude depends on the amplitude of forcing. If, on the other hand, the forcing frequency is close to its natural frequency, it responds vigorously. The resulting amplitude is limited only because of the dissipation in the system. The amplitude will be such that the energy input by the forcing balances dissipation. If the disturbance contains a spectrum of frequencies, the response is highest at its natural frequency but small elsewhere. In other words, systems with their own natural frequencies are rather selective in their response.

Wave propagations are not involved in the dynamics of a pendulum, but they are essential to the response of fluid bodies. If the surface of a pond is disturbed, gravity waves radiate out from the region of disturbance and slosh around the pond until the disturbance is eventually dissipated. The frequencies and wave numbers generated by the disturbance are critically dependent on the natural frequencies of the system. The derivation of these response characteristics is an eigenvalue problem. Dynamics dictate that a definite relationship exists between frequencies and wave numbers ($k=2\pi/\lambda$, $\lambda \sim$ wavelength), the so-called dispersion relationship that needs to be satisfied by the system. For gravity waves on a pond, ignoring surface tension, the dispersion relationship is $\omega^2 = gk \tanh(kH)$, where H is the fluid depth, k the wave number, and ω is the frequency. Any wave that the transient disturbance creates must satisfy this condition. In general, the waves that radiate out will consist of a spectrum in the frequency-wave number regime constrained by this equation. Note that in general, the group velocity $C_g = d\omega/dk$ is not equal to the phase velocity $C_p = \omega/k$. These waves are dispersive. This is the reason why the original disturbance is broken up into a group of waves, each arriving at a distant location at a different time.

For deep water waves,

$$\omega^2 = gk$$

$$C_p = \omega/k = \sqrt{g/k} \qquad\qquad (3.1.1)$$

$$C_g = \frac{1}{2}\sqrt{g/k} = \frac{C_p}{2}$$

The wave energy travels in the same direction as the wave crests, but at half the speed. Lower wave numbers (longer waves, lower frequencies) travel faster and therefore arrive earlier than higher wave numbers. If one looks at a group of such waves carefully, one can see individual crests propagating to the front of the group and disappearing, while new crests appear to be generated spontaneously at the back of the group. This is because $C_p > C_g$ (the opposite happens for capillary waves for which $C_p < C_g$). However, if the water is shallow,

$$\omega^2 = gH\,k^2$$

$$C_p = \omega/k = (gH)^{1/2} \qquad\qquad (3.1.2)$$

$$C_g = (gH)^{1/2} = C_p$$

Shallow water waves are nondispersive. All the waves in the group will travel at the same speed irrespective of the frequency, and the entire pulse will arrive at the same time.

The dispersion relation can be derived by postulating small disturbances that leads to linearized versions of the governing dynamical equations. For surface gravity waves, the linearized equations are

$$\frac{\partial u}{\partial x} + \frac{\partial w}{\partial z} = 0$$

(3.1.3)

$$\rho \frac{\partial u}{\partial t} = -\frac{\partial p}{\partial x}, \rho \frac{\partial w}{\partial t} = -\frac{\partial p}{\partial z}$$

where u and w are the perturbation velocities in the direction of wave motion and in the vertical direction, respectively; p is the perturbation pressure. These equations reduce to $\nabla^2 p = \left(\dfrac{\partial^2}{\partial x^2} + \dfrac{\partial^2}{\partial z^2} \right) p = 0$, a simple Laplacian equation. The resulting flow must satisfy appropriate boundary conditions at the surface (linearized, valid only for small η, where η is the surface perturbation) and at the bottom:

$$w = \frac{\partial \eta}{\partial t} \text{ and } p = \rho g \eta \qquad \text{at } z = 0$$

$$w = 0 \text{ at } z = -H$$

(3.1.4)

Now we look for wavelike disturbances: $\eta = \text{Re}\left[A\, e^{i(kx - \omega t)} \right]$, where A is the amplitude. This represents a wave with its crest traveling in the positive x direction at speed $C_p = \omega/k$; wave length $\lambda = 2\omega/k$ and the wave period $T = 2\pi/\omega$. Equations (3.1.3) and (3.1.4) determine the wave motions permissible by the dynamics. This is an eigenvalue problem, because the solution has to satisfy the governing equations as well as the boundary conditions. Only particular (ω, k) combinations (eigenvalues) are permissible:

$$\omega^2 = gk \tanh(kH)$$

(3.1.5)

This is the dispersion relation. The linearized (small perturbation) governing equations of motion with appropriate auxiliary conditions tell us what kind of wave motions are permitted by the system. This principle holds for any dynamical system, no matter how complex, including propagation in the equatorial waveguide. We will not, however, attempt to deal with equatorial dynamics in great detail, since excellent treatments exist. Philander (1990) in particular, does an excellent job of describing equatorial dynamics and its consequences to ENSO. He also cites an extensive bibliography on the topic that is reasonably current (the late 1980s). Here follow some basics that an ocean modeler should know.

3.2 GOVERNING EQUATIONS

In the equatorial ocean, both f and β are nonzero, and there is density stratification in the vertical as well. The full nonlinear equations under hydrostatic and Boussinesq approximations are

$$\frac{Du}{Dt} - fv = -\frac{1}{\rho_0}\frac{\partial P}{\partial x} + \frac{\partial}{\partial z}\left(K_M \frac{\partial u}{\partial z} \right)$$

(3.2.1)

$$\frac{Dv}{Dt} + fu = -\frac{1}{\rho_0}\frac{\partial P}{\partial y} + \frac{\partial}{\partial z}\left(K_M \frac{\partial v}{\partial z} \right)$$

where

$$\frac{D}{Dt} = \frac{\partial}{\partial t} + u\frac{\partial}{\partial x} + v\frac{\partial}{\partial y} + w\frac{\partial}{\partial z}$$

are the horizontal momentum equations, and the hydrostatic balance gives

$$0 = -\frac{1}{\rho_0}\frac{\partial P}{\partial z} + \frac{\rho}{\rho_0} g$$

(3.2.2)

Continuity equation is

$$\frac{\partial u}{\partial x} + \frac{\partial v}{\partial y} + \frac{\partial w}{\partial z} = 0$$

(3.2.3)

In addition, we need an equation for changes in density ρ. This is provided by the equation of state that relates density to changes in temperature and salinity. In general, the equation of state is nonlinear and requires separate conservation relations for temperature and salinity. However, linearization around a base state allows the two conservation equations to be combined into a single conservation relation for density:

$$\frac{D\rho}{Dt} = \frac{\partial}{\partial z}\left(K_H \frac{\partial \rho}{\partial z} \right)$$

(3.2.4)

Now consider small perturbations from a state of rest, where the hydrostatic balance prevails:

$$\rho(x,y,z,t) = \bar{\rho}(x,y,z) + \rho'(x,y,z,t)$$
$$u,v,w(x,y,z,t) = u', v', w'(x,y,z,t)$$
$$p(x, y, z, t) = \bar{p}(x, y, z) + p'(x,y,z,t)$$

(3.2.5)

The density perturbation equation becomes (dropping all primes)

$$\frac{\partial \rho}{\partial t} - N^2 \frac{w}{g} = \frac{\partial}{\partial z}\left(K_H \frac{\partial \rho}{\partial z}\right)$$

(3.2.6)

where N is the Brunt-Vaisala (buoyancy) frequency $= \left(-\dfrac{g}{\rho_0}\dfrac{d\bar{\rho}}{dz}\right)^{1/2}$. This is the

natural frequency of oscillation of a fluid parcel displaced in the vertical, under the action of restoring gravitational forces. The rest of the linearized (small disturbance) equations are

$$\frac{\partial u}{\partial x} + \frac{\partial v}{\partial y} + \frac{\partial w}{\partial z} = 0$$

$$\frac{\partial u}{\partial t} - fv = -\frac{1}{\rho_0}\frac{\partial p}{\partial x} + ...$$

$$\frac{\partial v}{\partial t} + fu = -\frac{1}{\rho_0}\frac{\partial p}{\partial y} + ...$$

(3.2.7)

$$0 = -\frac{1}{\rho_0}\frac{\partial p}{\partial z} + \frac{\rho}{\rho_0}g$$

Note that (3.2.6) and (3.2.7) involve perturbation quantities, but primes have been dropped from u', v', w', p', ρ' for convenience. If we assume that $K_M = K_H = A/N^2$, for purely analytical convenience, it is possible (McCreary, 1981a,b; Philander, 1990) to seek solutions of (3.2.7) of the form

$$u = \Sigma_m u_m(x,y,t)R_m(z)$$
$$v = \Sigma_m v_m(x,y,t)R_m(z)$$
$$w = \Sigma_m w_m(x,y,t)S_m(z)$$
$$p = \rho_0 g\Sigma_m \eta_m(x,y,t)R_m(z)$$
$$\rho = \Sigma_m \rho_m(x,y,t)S_m(z)$$

(3.2.8)

Substituting $R_m = -gh_m \partial S_m / \partial z$, where h_m is the constant of separation, and dropping the subscript m from all the variables for convenience, the equations for each mode m become

$$\frac{\partial \eta}{\partial t} + h\left(\frac{\partial u}{\partial x} + \frac{\partial v}{\partial y}\right) = -\frac{A\eta}{gh}$$

$$\frac{\partial u}{\partial t} - fv = -g\frac{\partial \eta}{\partial x} - \frac{Au}{gh}$$

$$\frac{\partial v}{\partial t} + fu = -g\frac{\partial \eta}{\partial y} - \frac{Av}{gh}$$ (3.2.9)

$$\frac{\partial^2 S}{\partial z^2} + \frac{N^2}{gh}S = 0$$

The last of these equations is the vertical structure equation. The significance of h is that it is related to the vertical scale of motion and determines the gravity wave speed $C = (gh)^{1/2}$ that is permitted by the system. To determine the values of h (and hence C) the system permits, we need to solve the eigenvalue problem that consists of the previous equations and the boundary conditions at the surface and the bottom. For a flat bottom ocean,

$$w = S = 0 \text{ at } z = -H_b$$
(3.2.10)

$$w - \frac{1}{\rho_0 g}\frac{\partial p}{\partial t} = S - h\frac{\partial S}{\partial z} = 0 \text{ at } z = 0$$

For baroclinic modes, the surface boundary condition can be simplified to that of a rigid lid: $w = S = 0$. Thus, dynamically, their SSH signature is unimportant (except as a diagnostic for detection by tide gauges and altimeters). The vertical structure equation and the boundary conditions constitute an eigenvalue problem that determines for a given $N(z)$; that is, for a given density structure of the ocean, the eigenvalues for h,

$$h_m \ (m=0,1,2,...)$$ (3.2.11)

where m is the vertical mode number. This is an infinite but discrete set, although normally only the first few modes are significant. The values of h_m and $C_m = (gh_m)^{1/2}$ decrease as m increases. Thus, a set of wave motions with phase speeds decreasing with mode number m is permitted by the system. Mode $m = 0$ is the barotropic mode, which is independent of the ambient stratification, and its equivalent depth h is equal to the depth of the fluid column H_b. The speed of propagation is that of a shallow water gravity wave $C_0 = (gH_b)^{1/2}$. The vertical structure of the barotropic wave (say, S_0) is uniform with depth and has no zero crossings. Modes $m \geq 1$ are baroclinic modes and they depend on the ambient stratification. They will be different for different $N(z)$ or $\bar{\rho}(z)$ distributions. These modes will have m zero crossings in the vertical. For example, S_1 (and W_1) will have one zero crossing. Figure 3.2.1 from Philander (1990) shows the typical $\bar{\rho}(z)$ distribution representative of the tropical ocean and the associated vertical modal structure. Note that N^2 reaches a maximum in the thermocline, which exists typically at depths of 100 to 200 m in the equatorial oceans.

If we ignore friction and put A=0 in Eq. (3.2.9). each baroclinic mode m satisfies the equations

$$\frac{\partial \eta}{\partial t} + h_m \left(\frac{\partial u}{\partial x} + \frac{\partial v}{\partial y} \right) = 0$$

$$\frac{\partial u}{\partial t} - fv + g\frac{\partial \eta}{\partial x} = 0 \qquad\qquad (3.2.12)$$

$$\frac{\partial v}{\partial t} + fu + g\frac{\partial \eta}{\partial y} = 0$$

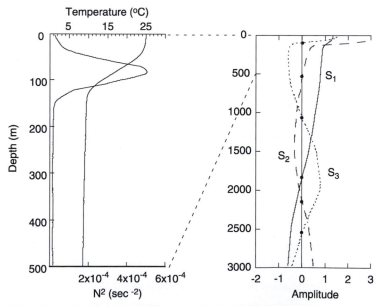

Figure 3.2.1   Typical temperature and buoyancy frequency distributions in the tropical oceans (left), and the corresponding vertical structure of the first three baroclinic modes (right). From Philander (1990).

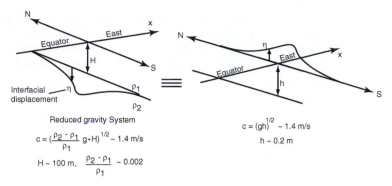

Figure 3.2.2 Sketches illustrating the equivalence between a two-layer ocean with a quiescent bottom layer (left) and an equivalent barotropic ocean with a depth two orders of magnitude shallower (right).

These are precisely the shallow water equations for free barotropic motions on an ocean of depth h_m! η is the free surface deflection. Each baroclinic mode in the equatorial waveguide can therefore be considered equivalent to a barotropic mode on an ocean of depth h_m (Figure 3.2.2); h_m is therefore known as the equivalent depth. Note that at the equator, $f=\beta y$ where $\beta \sim 2\Omega/R = 2.29 \times 10^{-11}$ $m^{-1}s^{-1}$; R is the Earth's radius, Ω its angular rotation rate. On the equatorial β plane, the important dynamical quantities are (1) $C_m = (gh_m)^{1/2}$, the speed of propagation of the shallow water gravity wave; (2) $a_m = (C_m/\beta)^{1/2}$, the equatorial Rossby radius of deformation, and (3) $T_m = (\beta C_m)^{-1/2}$, the equatorial inertial period. A similar analysis holds for mid-latitudes, but there the thermocline is much deeper (~800 to 1000 m).

The various modes, their equivalent depths, inertial periods, and radii of deformation are shown in Table 3.3.1, adapted from Philander (1990). Note that the equivalent depths of baroclinic modes are very small compared to the ocean depth. Note also that the mid-latitude Rossby radius of deformation, $a_M = C/f$ (10 to 50 km for the first mode) is an order of magnitude smaller, while the mid-latitude inertial period, $T_M = 2\pi/f$ (0.5 to 3 days) is comparable.

The significance of the preceding modal expansions is that the adjustment of the vertically stratified equatorial ocean to changing forcing can be considered to be via excitation of Kelvin and their corresponding Rossby wave modes. These are permissible wave motions in the equatorial wave mode. To a first approximation, each of these modes can be treated independently, and the resulting dynamical consequences can be explored simply by looking at the corresponding shallow water equations. For example, Chapter 3 of Philander (1990) deals with the second baroclinic mode and its consequences to equatorial adjustment. Within the linear approximation, the total response of the waveguide can be considered to be a superposition of the most significant modes. What modes will be preferentially excited (since they are an infinite set) depends on the structure of the surface forcing. Wind stress can often be treated as a body force distributed in the upper layer and if it is further expressed as vertical modes, then the projection of the forcing function onto a mode is what excites that particular mode. Thus, responses to changes in remote forcing in the waveguide are a series of appropriate Kelvin wave modes (an infinite number) and their associated Rossby modes. Thus, for example, if winds shift west of a particular longitude, to the east of it Kelvin waves induce changes, lowest mode arriving first, followed successively by higher modes. In the equatorial Pacific, it appears that the first baroclinic mode is preferentially excited and therefore plays a very important role in oceanic adjustment.

If we substitute $h_m = (g'/g) H_m$, and $\eta = (g'/g) \eta_I$, (3.2.12) can be written as:

$$\frac{\partial \eta_I}{\partial t} + H_m \left(\frac{\partial u}{\partial x} + \frac{\partial v}{\partial y} \right) = 0$$

$$\frac{\partial u}{\partial t} - fv + g' \frac{\partial \eta_I}{\partial x} = 0 \qquad (3.2.13)$$

$$\frac{\partial v}{\partial t} + fu + g' \frac{\partial \eta_I}{\partial y} = 0$$

These are the governing equations for motion in a two-layer stably stratified fluid, with an infinite bottom layer, the so-called one and a half layer reduced gravity model (see Chapter 1, Section 4.1). The depth of the upper layer is H_m, g' is the reduced gravity, and η_I is the interfacial deflection; $C_m^2 = g' H_m = g\, h_m$. Thus each baroclinic mode m can also be considered to be governed by a one and a half layer reduced gravity model equations with an upper layer depth H_m and a value for the reduced gravity g' equal to $g\, h_m/H_m = g\, \Delta\rho/\rho_0$, where $\Delta\rho$ is the change in density across the interface between the two layers.

In the presence of wind forcing, the wind stress τ (or, more appropriately, the portion relevant to that particular mode) can be treated as a body force distributed over the upper layer H_m and the governing equations (once again dropping the subscript m for convenience) become

$$g' \frac{\partial \eta}{\partial t} + C^2 \left(\frac{\partial u}{\partial x} + \frac{\partial v}{\partial y} \right) = 0$$

$$\frac{\partial u}{\partial t} - fv + g' \frac{\partial \eta}{\partial x} = \frac{\tau_x}{H} \qquad (3.2.14)$$

$$\frac{\partial v}{\partial t} + fu + g' \frac{\partial \eta}{\partial y} = \frac{\tau_y}{H}$$

This set of equations can be reduced to a single equation for the meridional velocity component v:

$$\left(v_{xx} + v_{yy} \right)_t + \beta v_x - \frac{1}{C^2} v_{ttt} - \frac{f^2}{C^2} v_t = F$$

$$F = \frac{f^2}{C^2} \left(\frac{\tau_x}{H} \right)_t - \frac{1}{C^2} \left(\frac{\tau_y}{H} \right)_{tt} + \frac{1}{H} \frac{\partial}{\partial x} \left(\frac{\partial \tau_x}{\partial y} - \frac{\partial \tau_y}{\partial x} \right)$$

(3.2.15)

where the subscripts denote derivatives. The response of the equatorial waveguide to wind forcing can be considered to be a superposition of various baroclinic modes excited by the wind, with each mode satisfying Eq. (3.2.15). Steady state solution to this equation is given by (dropping $\partial / \partial t$ terms)

$$\beta v_x = \frac{1}{H} (\nabla \times \tau)_x$$

$$\int v dz = vH = \frac{1}{\beta} |\nabla \times \tau|$$

(3.2.16)

These are the Sverdrup balance equations that relate the meridional transport to the curl of the wind stress (see Chapter 1). They can be invoked to explain the banded structure of the equatorial currents.

3.3 EQUATORIAL WAVES

Adjustments in the equatorial waveguide to changes in forcing are through the generation, propagation, reflection, and interaction of gravity, Kelvin and Rossby wave motions. A downwelling (upwelling) Kelvin wave is generated and radiated to the east, while a slower-moving upwelling (downwelling) Rossby wave propagates to the west. Once these waves reach the meridional continental boundaries at the end of the equatorial waveguide, their energy is reflected and/or leaked into other forms of wave motion such as coastal Kelvin waves. Fortunately, they can be studied with a simple one and a half layer reduced gravity model. Let us go back to (3.2.15), drop the forcing terms

$$\left(v_{xx} + v_{yy} \right)_t + \beta v_x - \frac{1}{C^2} v_{ttt} - \frac{\beta^2 y^2}{C^2} v_t = 0$$

(3.3.1)

and seek wave solutions of the form: $v(x, y, t) = V(y) \exp i(kx - \omega t)$:

$$V_{yy} + \left(\frac{\omega^2}{C^2} - k^2 - \frac{\beta k}{\omega} - \frac{\beta^2 y^2}{C^2} \right) V = 0 \qquad (3.3.2)$$

Only the real part need be considered. Now seek solutions that satisfy $V \to 0$ as y $\to \pm\infty$. They are

$$V = D_m \left(\frac{y}{a} \right) \exp i(kx - \omega t)$$

$$D_m \left(\frac{y}{a} \right) = 2^{-m/2} H_m \left(\frac{y}{\sqrt{2}a} \right) \exp \left(-\frac{y^2}{4a^2} \right)$$

$\qquad (3.3.3)$

D_m is a parabolic cylinder function and H_m Hermite polynomial of order m. Substitution of this form into the governing equation provides the dispersion relation

$$\left(\frac{\omega}{C} \right)^2 - k^2 - \frac{\beta k}{\omega} = (2m + 1) \frac{\beta}{C} \qquad (3.3.4)$$

that contains all permissible forms of wave motions including Kelvin waves, Yanai waves, Rossby waves, and inertia-gravity (Poincare) waves in the waveguide.

However, it is more instructive (Potemra *et al.*, 1991) to start from the governing equations for freely propagating baroclinic equatorial waves, Eq. (3.2.13) and deal with quantities normalized by the gravity wave speed C, the equatorial radius of deformation $a = (C/\beta)^{1/2}$ and the equatorial inertial period $T = (\beta C)^{-1/2}$:

$$t = T(t')$$

$$x, y = a(x', y')$$

$$\eta = h(\eta')$$

$$u, v, C_p, C_g = C(u', v', C_p', C_g')$$

$$\omega = T^{-1}(\omega')$$

$$k = a^{-1}(k')$$

$\qquad (3.3.5)$

Primes denote normalized quantities. We will deal henceforth with normalized quantities only and drop primes for convenience.

$$\frac{\partial \eta}{\partial t} + \left(\frac{\partial u}{\partial x} + \frac{\partial v}{\partial y} \right) = 0$$

$$\frac{\partial u}{\partial t} - yv + \frac{\partial \eta}{\partial x} = 0 \qquad (3.3.6)$$

$$\frac{\partial v}{\partial t} + yu + \frac{\partial \eta}{\partial y} = 0$$

This equation, though simple looking, entertains a rich spectrum of wave motions that are crucial to equatorial dynamics (Potemra et al., 1991). The dispersion relation is (Eq. 3.3.4 gives the dimensional version)

$$\omega^2 - k^2 - \frac{k}{\omega} = 2m + 1; \qquad m = -1, 0, 1, 2... \qquad (3.3.7)$$

Value $m = -1$ gives Kelvin waves, $m = 0$ Yanai waves, $m = 1, 2$ ... give Rossby and gravity waves. The meridional structure of these waves is expressed in terms of Hermite functions $Y_m(y)$:

$$Y_m(y) = \frac{H_m(y) \exp\left(-y^2 / 2\right)}{\left(2^m m! \pi^{1/2}\right)^{1/2}}; \quad m = 0, 1, 2, 3... \qquad (3.3.8)$$

where $H_m(y)$ are Hermite polynomials:

$$H_0(y) = 1$$

$$H_1(y) = 2y$$

$$H_2(y) = 4y^2 - 2$$

$$H_3(y) = 8y^3 - 12y \qquad (3.3.9)$$

$$H_4(y) = 16y^4 - 48y^2 + 12$$

$$H_5(y) = 32y^5 - 160y^3 + 120y \quad ...$$

Plot of (3.3.7) for various m is the single most important piece of information on equatorially trapped waves, whether in the oceans or the atmosphere. Figure

Table 3.3.1

Values of the Equivalent Depth, the Gravity Wave Speed, the Equatorial Rossby Radius, the Equatorial Inertial Period, and the Spectral Gap for Various Modes in the Pacific Ocean, the Indian Ocean, and the Tropical Atmosphere.

m	h_m (m)	C_m (m/s)	$(a_m$ km)	T_m (days)	Spectral gap(days)	Type
			Pacific			
0	4000	200	2955	0.17	0.6–3.6	Barotropic
1	0.6	2.4	325	1.56	5.8–32.6	1st baroclinic
2	0.2	1.4	247	2.04	7.5–42.2	2nd baroclinic
3	0.08	0.88	197	2.62	9.7–54.8	3rd baroclinic
4	0.04	0.63	165	3.14	11.6–65.6	4th baroclinic
			Indian			
2	0.28	1.67	270	1.87	6.9–39.1	2nd baroclinic
			Atmosphere			
1	370	60	1620	0.31	1.1–6.5	1st baroclinic

3.3.1 shows the dispersion relation for the ocean, and Figure 3.3.2 compares them for the ocean and the atmosphere.

We find four types of waves: (1) inertia-gravity waves, (2) Kelvin waves, (3) Yanai (mixed Rossby-gravity) waves, and (4) Rossby waves. At high frequencies $(\omega > 1)$, inertia-gravity waves can exist. At low frequencies $(\omega < 1)$, Rossby (planetary) waves result. Between these, there exists a gap in the frequency spectrum $(\omega T \sim 0.3$ to $\omega T \sim 1.7$ or equivalently between periods $T_p \sim 3.7$ T to $T_p \sim 20.9$ T). Yanai waves and Kelvin waves fill this gap. Table 3.3.1 also shows the spectral gap for various modes.

3.3.1 KELVIN WAVES

For m = −1, $\omega = k$ satisfies the dispersion relationship:

$$C_p = C_g = 1$$
$$\eta = \pi^{1/4} Y_0(y)\cos\ (kx - \omega t) = \exp\left(-y^2/2\right)\cos\ (kx - \omega t) \qquad (3.3.10)$$
$$u = \eta;\ v = 0$$

These are Kelvin waves; they are nondispersive. Both phase and group velocities are positive; the propagation is eastward only. They do not have to be periodic. They can be in the form of a pulse. They are generated by abrupt changes in wind forcing—sudden relaxation or strengthening of zonal winds. They are *symmetric* about the equator, Gaussian in shape with a meridional scale

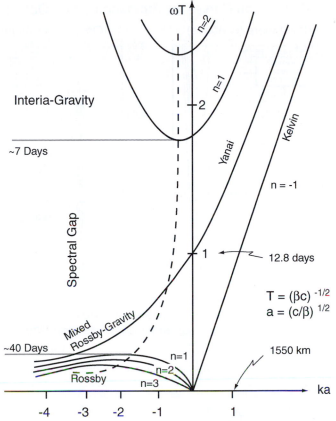

Figure 3.3.1 Dispersion relationship for equatorial wave motions showing the inertia-gravity, Yanai, Kelvin, and Rossby modes. This relationship holds for all modes with appropriate values for the Rossby radius a and the equatorial inertial period T. Wavelength corresponding to ka = 1 is 2040 km for the first and 1550 km for the second baroclinic mode in the Pacific. Period corresponding to $\omega T = 1$ is 9.8 days for the first and 12.8 for the second mode. Adapted from Cane and Sarachik (1976).

proportional to the Rossby radius. The height extremum occurs at the equator. The cross-equatorial structure of the pycnocline deflection due to Kelvin waves is shown in Figure 3.3.3. There are no meridional currents associated with these waves, only zonal currents that also peak at the equator (Figure 3.3.4).

These solutions can also be obtained from Eq. (3.3.6) by putting v = 0 a priori and seeking wavelike solutions propagating in the zonal direction. Because they are unaffected by rotation, the energy of equatorial Kelvin waves is equally

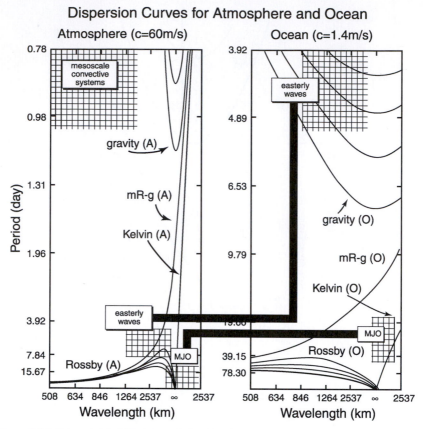

Figure 3.3.2 Comparison of dispersion relationships for the atmospheric and the oceanic equatorial wave guides. Note the strong mismatch in motions in the two waveguides. Note that the oceanic mode is the second mode. For the first mode, whose speed is 2.4 s^{-1}, the right-hand figure must be rescaled appropriately. MJO stands for Madden–Julian oscillations. (Figure provided by P. Webster.)

partitioned between kinetic and potential forms. The total energy density per unit

length $\quad E = 2\int_{-\infty}^{\infty} \frac{1}{2}\rho_0 g\overline{\eta^2}\,dy = 4\int_{0}^{\infty} \frac{1}{2}\rho_0 g\frac{1}{2}\left[\eta_o \exp\left(\frac{-y^2}{2a^2}\right)\right]^2 dy$, and the eastward

energy flux is

$$E = \frac{\sqrt{\pi}}{2}\rho_0 g\eta_0^2 a;\ F_e = EC_g = \frac{\sqrt{\pi}}{2}\rho_0 g\eta_0^2 aC \qquad (3.3.11)$$

where η_0 is the wave amplitude right at the equator, where it is the maximum.

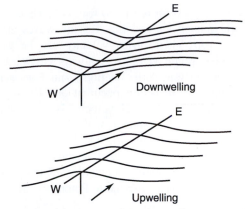

Figure 3.3.3 The cross-equatorial structure of the thermocline corresponding to upwelling and downwelling equatorial Kelvin waves.

3.3.2 YANAI WAVES

Here $m = 0$. There are two solutions; of these the first one, $\omega = -k$, gives solutions that are unbounded in the y direction and hence not physical. The other solution corresponds to the planetary Rossby-gravity mode: an inertia-gravity wave at high frequencies ($\omega > 1$), but a planetary wave at low frequencies ($\omega < 1$).

$$k = \omega - \frac{1}{\omega} \quad \left(\omega = \frac{1}{2}\left[\left(k^2 + 4\right)^{1/2} - k\right] \right)$$

$$C_p = \frac{\omega^2}{\omega^2 - 1} ; C_g = \frac{\omega^2}{\omega^2 + 1} \qquad (3.3.12)$$

$$\eta = -\frac{\omega}{2^{1/2}} \, Y_1(y) \sin(kx - \omega t)$$

$$u = \eta; \; v = Y_0(y) \cos(kx - \omega t)$$

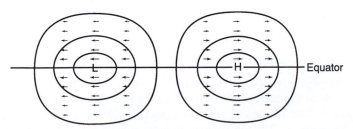

Figure 3.3.4 The structure of the currents associated with an equatorial Kelvin wave. Note that the currents are zonal.

C_g is always positive. $C_p > 0$ for $\omega > 1$; < 0 for $\omega < 1$. These waves behave like Kelvin waves at high frequencies and like Rossby waves at low frequencies. They fill the spectral gap. Their phase velocity can be eastward (+ve) or westward (−ve), but their group velocity is always eastward (+ve). They are *antisymmetric* about the equator, series of highs and lows across the equator. The maximum and minimum in height occur at one Rossby radius from the equator. At the equator, the meridional velocity is maximum, but the zonal velocity is zero. Flow is across the equator, with alternating direction every half wavelength, from northward to southward and back. Typically, Yanai waves have periods of 26 days in the Indian Ocean, velocity of 0.1 to 0.3 m/s (Kindle and Thompson, 1989). They are readily detected in IR imagery as well because of ∼ 1° C SST anomalies associated with them (Potemra et al., 1991).

3.3.3 ROSSBY WAVES

Here $m \geq 1$, but $\omega < 1$, that is, the frequencies are much smaller than the equatorial inertial frequency. Then the first term in the dispersion relation can be neglected:

$$\omega = \frac{-k}{k^2 + (2m+1)}; \qquad m = 1, 2 \ldots$$

$$C_p = \frac{-1}{k^2 + (2m+1)}; C_g = \frac{k^2 - (2m+1)}{\left(k^2 + (2m+1)\right)^2}$$

$$\eta_m = \frac{1}{2}\left[\frac{(2m)^{1/2} Y_{m-1}(y)}{\omega + k} - \frac{(2(m+1))^{1/2} Y_{m+1}(y)}{\omega - k} \right] \sin(kx - \omega t) \qquad (3.3.13)$$

$$u_m = -\frac{1}{2}\left[\frac{(2m)^{1/2} Y_{m-1}(y)}{\omega + k} + \frac{(2(m+1))^{1/2} Y_{m+1}(y)}{\omega - k} \right] \sin(kx - \omega t)$$

$$v_m = Y_m(y) \cos(kx - \omega t)$$

There exist an infinite number of discrete modes $m = 1, 2 \ldots$. For a given mode, $C_g < 0$ (westward energy propagation) for $|k| < (2m+1)^{1/2}$; $C_g > 0$ (eastward energy propagation) for $|k| > (2m+1)^{1/2}$. C_p is always negative, phases always propagate westward. *Odd* modes ($m = 1, 3, 5 \ldots$) have *symmetric* structure about the equator. *Even* ($m = 2, 4 \ldots$) modes are *antisymmetric* about the equator. The maximum and minimum in height are away from the equator, not at the equator. The cross-equatorial structure of symmetric and asymmetric Rossby waves is shown in Figure 3.3.5 (from Philander, 1990). The Kelvin wave

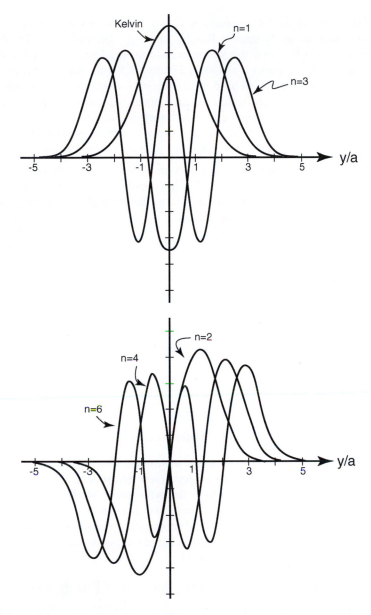

Figure 3.3.5 Meridional structure of symmetric and antisymmetric Rossby modes. Note that the odd modes are symmetric while even modes are antisymmetric with respect to the equator. The Kelvin wave is also shown. Note that the first and second modes have their maximum signature about a Rossby radius away from the equator. The cross-equatorial scale is the equatorial Rossby radius of deformation. From Philander (1990).

structure is also shown for comparison. Figure 3.3.6 shows the current structure associated with an antisymmetric Rossby wave. An alternating pattern of highs and lows is evident. Note the strong cross-equatorial currents, unlike Kelvin waves which Rossby waves is shown in Figure 3.3.5 from Philander (1990). The Kelvin wave structure is also have no meridional currents associates with them. The alternating southward/northward meridional velocities are also character-istic of Yanai waves, which are also antisymmetric about the equator.

The phases (or crests) of Rossby waves always propagate westward. However, the group velocity $C_g = \partial\omega / \partial k$ is negative for low values of ka but positive for high values. This means long Rossby waves propagate energy westward, short ones eastward. The division depends on m and ka. In this sense they are similar to mid-latitude Rossby waves that have dispersion relation (for zonal propa-gation) in dimensional terms:

$$\frac{\omega}{\beta a_M} = \frac{-ka_M}{1+k^2 a_M^2} \qquad (3.3.14)$$

where a_M = C/f is the mid-latitude radius of deformation. The only difference is that they have, in addition, a nonzero meridional propagation velocity. Long mid-latitude Rossby waves are nondispersive and have a phase speed and group velocity given by $C_p = C_g = -\beta\, a_M^2$.

For very long Rossby waves, the second term in the denominator can be neglected and in dimensional terms

$$\frac{\omega}{\beta a} = \frac{-ka}{(2m+1)} \qquad (3.3.15)$$

and

$$C_p = C_g = -\beta a^2 / (2m+1) \qquad (3.3.16)$$

In other words, very long Rossby waves are not dispersive. These long Rossby waves play a crucial role in equatorial adjustment. These, along with Kelvin waves, are the most germane to this adjustment. Mode m = 1 propagates energy westward at 1/3 Kelvin wave speed, m = 2 at 1/5, m = 3 at 1/7, and so on. Thus they are slower than the eastward propagating Kelvin waves but they are the only way the western regions realize changes are occurring to the east, since Kelvin waves cannot propagate westward.

The zonal energy flux due to Rossby waves is $g'\int_{-\infty}^{\infty} \eta u\ dy$ in dimensional

quantities. Normalizing by $C^3 a$, the nondimensional value of zonal energy flux is

$$F_e = \frac{1}{4}\left[\frac{m+1}{(\omega-k)^2} - \frac{m}{(\omega+k)^2}\right] \qquad (3.3.17)$$

Note that $F_e = EC_g$, where E is the energy density of waves:

$$E = \frac{1}{2}\int_{-\infty}^{\infty}\left[\left(u^2 + v^2\right) + \eta^2\right] dy \qquad (3.3.18)$$

Note the appearance of C_g in the expression for the energy flux; energy propagates at the group velocity. The governing equations can be reduced to

$$\frac{\partial E}{\partial t} + C_g\frac{\partial E}{\partial x} = 0 \qquad (3.3.19)$$

which is nothing but conservation of wave energy density E.

3.3.4 INERTIA-GRAVITY (POINCARE) WAVES

Here $m \geq 1$, but $\omega > 1$, that is the frequencies are much higher than the equatorial inertial frequency. Then the third term in the dispersion relation can be neglected. These are inertia-gravity waves, gravity waves whose frequency is so low that they are affected by planetary rotation effects. Their structure is the same as that of Rossby waves (see Figure 3.3.4).

$$\omega = k^2 + (2m+1); \qquad m=1, 2, \ldots$$

$$C_p = \pm\left(1 + \frac{2m+1}{k^2}\right)^{1/2}; C_g = \frac{1}{C_p} = \pm\left(1 + \frac{2m+1}{k^2}\right)^{-1/2}$$

$$\eta_m = \frac{1}{2}\left[\frac{(2m)^{1/2}\,Y_{m-1}(y)}{\omega+k} - \frac{\left(2(m+1)\right)^{1/2}\,Y_{m+1}(y)}{\omega-k}\right]\sin(kx-\omega t) \qquad (3.3.20)$$

$$u_m = -\frac{1}{2}\left[\frac{(2m)^{1/2}\,Y_{m-1}(y)}{\omega+k} + \frac{\left(2(m+1)\right)^{1/2}\,Y_{m+1}(y)}{\omega-k}\right]\sin(kx-\omega t)$$

$$v_m = Y_m(y)\cos(kx-\omega t)$$

Both C_p and C_g can be +ve (eastward) or −ve (westward). But phase and energy propagation are both in the same direction.

Thus, Rossby and inertia-gravity waves can propagate energy eastward or westward. Rossby waves do so at long periods greater than, say, 40 days, gravity waves at short periods less than, say, 7 days. As mentioned before, there exists a big spectral gap between Poincare and Rossby waves in the $\omega - k$ space, which is filled to some extent by the Yanai and Kelvin waves, which, however, can only propagate energy eastward; there are no free wave motions in the gap that can propagate energy westward.

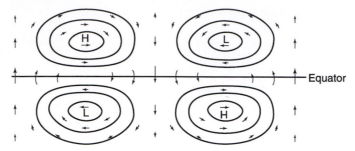

Figure 3.3.6 Current structure associated with an even mode (antisymmetric) Rossby wave. Note the alternating north-south pattern of meridional currents.

When a westward propagating symmetric odd mode Rossby wave is reflected from a western meridional boundary, the associated mass flux is usually returned by an eastward Kelvin wave of the same frequency (frequency is conserved during reflection). However, it returns only roughly 50% of the energy incident; the rest are returned by short eastward propagating Rossby waves. For an antisymmetric even mode Rossby wave incident, no mass flux is involved, so no Kelvin waves are reflected, only short Rossby waves. Similarly, when a Kelvin wave impinges on an eastward meridional boundary, the net mass flux is brought to zero two ways: poleward traveling Kelvin waves and westward propagating Rossby (symmetric) waves. If Yanai or Kelvin waves in the spectral gap are incident on an eastern boundary, there is *no* wave to return the energy and mass back to the west. Energy piles up and goes into poleward coastally trapped Kelvin waves!

In altimetric records, a high or low symmetric anomaly bump at the equator traveling east corresponds to a Kelvin pulse. Antisymmetric highs and lows traveling east and west correspond to Yanai waves, if in the spectral gap. Symmetric or antisymmetric Rossby waves traveling westward at slower speeds than Kelvin pulses traveling eastward may also be seen. Inertia-gravity waves travel at high speeds and because the altimeter repeat cycle period (~10 days for TOPEX) is typically longer than their period (<7 days); they are not properly sampled by the altimeter.

A good way to detect Kelvin wave propagation along the equator in a time series of horizontal distribution maps such as altimetric SSH anomaly maps from each repeat cycle or model-derived interfacial deflection fields is to plot an $x - t$ (Hovmuller) diagram along the equator. However, Rossby waves are best seen in $x - t$ diagrams plotted at a distance roughly one to two Rossby radii away from the equator. Antisymmetric Rossby (and Yanai) wave signatures in these fields can be enhanced by looking at $\eta_+ - \eta_-$, (instead of η_+ or η_- separately), where subscripts plus and minus denote the values north and south of the equator, respectively. Similarly, symmetric waves are enhanced by the sum $\eta_+ + \eta_-$.

3.4 EQUATORIAL CURRENTS

Now let us look at equatorial currents driven by zonal winds. Away from the boundaries (eastern and western), the response can be analyzed using shallow water Eq. (3.2.15) by zeroing $\partial / \partial x$ and retaining wind forcing:

$$v_{tt} + f^2 v - C^2 v_{yy} = -f \frac{\tau_x}{H}; \quad f = \beta y \qquad (3.4.1)$$

Here once again, the most important parameters are the equatorial inertial period T and the equatorial Rossby radius of deformation a. For times longer than the equatorial inertial period, the balance is

$$fv = -\frac{\tau_x}{H} \qquad (3.4.2)$$

sufficiently far away from the equator. This sets up an Ekman drift at distances larger than the Rossby radius (y > a) and the mass conservation therefore requires upwelling (or downwelling) near the equator. Near the equator, for t >T, the first term can be neglected and the resulting solution is

$$v = -\frac{\tau_x}{H} T^{-1/2} Q$$

$$u = \frac{\tau_x}{H}\left(1 - \frac{y}{a}\right) Q t \qquad (3.4.3)$$

This is the equatorial (Yoshida) jet, confined to about a radius of deformation on either side of the equator, with an associated meridional Ekman drift (Figure 3.4.1). However, the jet keeps accelerating with time. This is not what happens in reality. It is an artifact of the infinite zonal extent. In reality, the meridional boundaries intervene and impose u = 0, through waves generated at those boundaries (a Kelvin wave at the western and Rossby wave at the eastern boundary), which propagate inward and stop the acceleration of the jet, resulting in steady state conditions.

Now we have most of the components of equatorial dynamics that can be used to explain the response of the equatorial ocean to changes in forcing. If, for example, easterly winds suddenly start blowing over the entire basin, this sets up an accelerating equatorial jet, poleward Ekman drift and upwelling all along the equator. However, this does not satisfy the zero zonal velocity condition at meridonal boundaries. These are imposed by the Kelvin wave from western and Rossby waves from eastern boundaries that establish zonal pressure gradients, arrest the acceleration of the jet, and lead to steady state conditions. These wave motions consist of successively increasing mode numbers. Barotropic mode arrives first, the first baroclinic Kelvin mode next, and so on. Only when all the

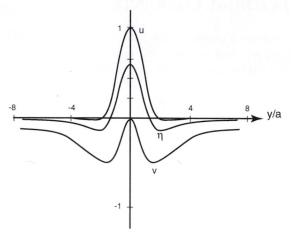

Figure 3.4.1   The meridional structure of the Yoshida jet. The thermocline displacement and the zonal and meridional velocities all have a cross-equatorial scale of a Rossby radius.

modes have arrived is a steady state established. This steady state is one of no motion, where the wind stress and the zonal pressure gradient exactly balance. However, in the real ocean, not all the modes make it, since higher modes tend to be more severely dissipated and therefore do not contribute greatly to the adjustment process. Thus, the jet is not totally canceled by the zonal pressure gradient.

It appears that most of the equatorial adjustment is accomplished by one mode, the first baroclinic mode (Kelvin and Rossby). Note that Philander (1990) asserts it is the second mode that is dominant; altimetric and other observations suggest that it is the first mode that is preferentially excited in the tropical Pacific. Thus, as the westward jet is set in motion and is being accelerated by the easterlies, the first baroclinic modes arrive and cancel most, but not all, of the acceleration of the jet. The vertical structures of the surface jet and the baroclinic modes may not coincide exactly. According to Philander (1990), the discrepancies set up the Equatorial Undercurrent (Figure 3.4.2 from Philander, 1990). This eastward current, maintained by the eastward pressure gradients due to easterlies, is a very important feature of the equatorial ocean (Figure 3.4.3). It responds vigorously to changes in zonal winds blowing at the surface and may, in fact, be regarded as the "pulse" of the equatorial dynamics. In the Pacific, it is about 300 km wide, straddles the equator, and is often continuous over distances of over 10,000 km. It transports 30–40 sv of eastward. It is deep in the western Pacific (100 to 150 m) and shoals somewhat

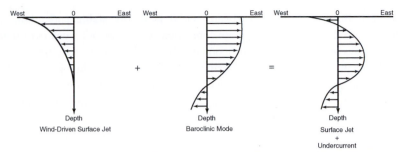

Figure 3.4.2 Superposition of the wind-driven jet and the baroclinic mode giving rise to the EUC. From Philander (1990).

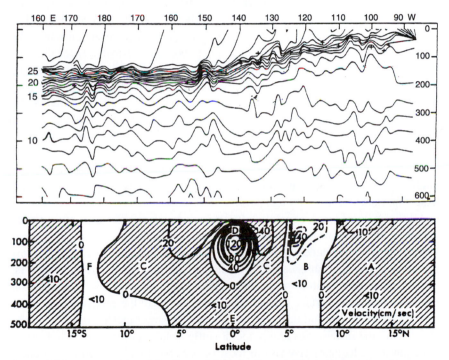

Figure 3.4.3 The structure of the Equatorial Under Current. The isotherms along the equator showing the thermocline (top); the EUC is concentrated around the thermocline. The zonal component of the velocity across a meridional section (bottom): A. North Equatorial Current, B. North Equatorial CounterCurrent, C. South Equatorial Current, D. Equatorial UnderCurrent, E. Intermediate Equatorial Current, and F. South Equatorial CounterCurrent. (From Moore, D. W., and S. G. H. Philander, Modeling of the tropical ocean circulation, in *The Sea,* Copyright Wiley 1977. Reprinted by permission of John Wiley & Sons Inc.)

in the east. Because it is an eastward jet, it is highly stable and retains its structure, even though it may meander somewhat (with a 2- to 3-week period and ~2000 km wavelength) about the equator. But overall it is the most robust feature of equatorial dynamics. During an ENSO event, this jet can often disappear because of the weakening of easterlies and/or onset of strong westerlies. In the Indian Ocean, because the winds reverse periodically, a steady eastward undercurrent is not always present.

The principal dynamical balance in the equatorial oceans is quite complex. Only a fully nonlinear model can delineate different regions with different balances (Philander, 1990) and is important in explaining some anomalous features of the undercurrent. We have looked at highly simplified linear physics of low amplitude wave motions with a horizontally homogeneous density structure and in the absence of background currents. Many of these neglected effects are important. For example, since higher mode waves tend to travel at phase speeds comparable to or smaller than prevailing mean currents, the influence of currents on wave propagation is important. Normally, mean currents lead to a Doppler shift of kU in frequency, but there are additional complications (see Philander, 1990). Also, we have assumed equatorial trapping and vertical standing modes. Neither of these is necessarily satisfied all the time. Some disturbances can travel vertically and some adjustment can take place with meridional propagation of information. Then there are possible nonlinear effects—a Kelvin pulse that has a different phase speed than a linear one. Also, winds don't always blow zonally. Response to meridional (cross-equatorial) winds is also quite important (Philander, 1990). The only way to consider all these important effects is through numerical models.

All things considered, equatorial dynamics is much simpler than mid-latitude dynamics. We have nearly deterministic processes in the waveguide with rapid (compared to mid-latitudes) adjustment to changes in external forcing. The most important is the near-absence of baroclinic instabilities. The current system is, for the most part, stable, with some rather dynamically less consequential instabilities of some surface currents (Legeckis waves in the eastern Pacific). On the other hand, in mid-latitudes, a zonal jet can be baroclinically unstable. In the atmosphere, the instabilities of a predominantly zonal jet on a rotating sphere (β plane physics adequate) give rise to most of our "weather" in the mid-latitudes. Such instabilities, by their very nature, are rather difficult to predict. They were not even understood well until Eady and Charney showed how they are germane to the mid-latitude atmosphere some 40 years ago. Mid-latitude dynamics and simple models to elucidate them will be the subject of the next chapter. But first, we will describe a simple numerical model that captures the essence of equatorial processes.

3.5 REDUCED GRAVITY MODEL OF EQUATORIAL PROCESSES

Comprehensive numerical circulation models that can simulate processes in the equatorial oceans, including those discussed earlier in this chapter, will be described in Chapters 9 to 11. Here, we will confine ourselves to describing the simplest model that can do so, the reduced gravity model. A reduced gravity model is very useful for understanding adjustment processes in the equatorial waveguide. The governing equations for the one and a half layer reduced gravity model are the same as those for a barotropic tidal model (see Chapters 6 and 7), except for the reduced value of the gravitational acceleration. In their most general form, they can be written in orthogonal curvilinear coordinates and in flux-conservative form as

$$\frac{\partial \eta}{\partial t} + \frac{1}{h_1 h_2}\left[\frac{\partial}{\partial \xi_1}\left(h_2 u_1 D\right) + \frac{\partial}{\partial \xi_2}\left(h_1 u_2 D\right)\right] = 0 \qquad (3.5.1)$$

$$\frac{\partial}{\partial t}\left(u_1 D\right) + \frac{1}{h_1 h_2}\left[\frac{\partial}{\partial \xi_1}\left(h_2 u_1^2 D\right) + \frac{\partial}{\partial \xi_2}\left(h_1 u_1 u_2 D\right)\right]$$

$$-f u_2 D + \frac{u_1 u_2 D}{h_1 h_2}\frac{\partial h_1}{\partial \xi_2} - \frac{u_2^2 D}{h_1 h_2}\frac{\partial h_2}{\partial \xi_1} = -g' D \frac{1}{h_1}\frac{\partial \eta}{\partial \xi_1} + \tau_1 + DF_1 \qquad (3.5.2)$$

$$\frac{\partial}{\partial t}\left(u_2 D\right) + \frac{1}{h_1 h_2}\left[\frac{\partial}{\partial \xi_1}\left(h_2 u_1 u_2 D\right) + \frac{\partial}{\partial \xi_2}\left(h_1 u_2^2 D\right)\right]$$

$$+ f u_1 D + \frac{u_1 u_2 D}{h_1 h_2}\frac{\partial h_2}{\partial \xi_1} - \frac{u_1^2 D}{h_1 h_2}\frac{\partial h_1}{\partial \xi_2} = -g' D \frac{1}{h_2}\frac{\partial \eta}{\partial \xi_2} + \tau_2 + DF_2 \qquad (3.5.3)$$

where the last term in each momentum equation contains the horizontal viscosity terms. Assuming a simple harmonic form for these terms (for a more appropriate form, see Section 5.6),

$$F_{1,2} = \frac{1}{h_1 h_2}\left[\frac{\partial}{\partial \xi_1}\left(A \frac{h_2}{h_1}\frac{\partial u_{1,2}}{\partial \xi_1}\right) + \frac{\partial}{\partial \xi_2}\left(A \frac{h_1}{h_2}\frac{\partial u_{1,2}}{\partial \xi_2}\right)\right] \qquad (3.5.4)$$

where A is the coefficient of horizontal viscosity. Term η is the interfacial deflection; u_1 and u_2 are the two components of horizontal velocity in ξ_1 and ξ_2 coordinate directions, respectively; and h_1 and h_2 are the metrics of the orthogonal curvilinear coordinate system ξ_1, ξ_2, respectively. D is the total depth of the upper layer $= H + \eta$, where H is its undisturbed thickness. Also, τ_1 and τ_2 are the wind stress components. Note that the equations are fully nonlinear. These equations can be readily solved in the equatorial waveguide to study the generation, propagation, reflection, interaction, and dissipation of equatorial wave motions. Note that this same set of equations can be used to study phenomena such as the annual Rossby wave propagation in mid-latitudes. By proper choice of x_1, x_2 and h_1, h_2, any orthogonal coordinate system can be represented. For a spherical coordinate system, $\xi_1 = \phi$, $\xi_2 = \theta$, $h_1 = R \cos \theta$, and $h_2 = R$, where R is the radius of the Earth, ϕ is the longitude, and θ is the latitude. For rectangular Cartesian coordinates, $\xi_1 = x$, $\xi_2 = y$, $h_1 = h_2 = 1$. Even though full spherical coordinates are desirable for accurate simulation of processes in the equatorial waveguide, we shall use the equatorial beta plane for simplicity:

$$\frac{\partial \eta}{\partial t} + \left[\frac{\partial}{\partial x}(uD) + \frac{\partial}{\partial y}(vD) \right] = 0 \tag{3.5.5}$$

$$\frac{\partial}{\partial t}(uD) + \left[\frac{\partial}{\partial x}(u^2 D) + \frac{\partial}{\partial y}(uvD) \right] - f\,vD = -g'D\frac{\partial \eta}{\partial x} + \tau^x$$
$$+ D\left[\frac{\partial}{\partial x}\left(A\frac{\partial u}{\partial x} \right) + \frac{\partial}{\partial y}\left(A\frac{\partial u}{\partial y} \right) \right] \tag{3.5.6}$$

$$\frac{\partial}{\partial t}(vD) + \left[\frac{\partial}{\partial x}(uvD) + \frac{\partial}{\partial y}(v^2 D) \right] + fuD = -g'D\frac{\partial \eta}{\partial y} + \tau^y$$
$$+ D\left[\frac{\partial}{\partial x}\left(A\frac{\partial v}{\partial x} \right) + \frac{\partial}{\partial y}\left(A\frac{\partial v}{\partial y} \right) \right] \tag{3.5.7}$$

where u and v are the velocity components in the x and y coordinate directions. Quantity f is the Coriolis parameter and is a function of latitude.

Let us use the notation $i+1/2$ to denote the location half a grid to the east, and $i-1/2$ to denote that half a grid to the west of the location of the variable being finite-differenced; similarly $j+1/2$ and $j-1/2$ denote locations half a grid to the north and to the south respectively. Then the continuity equation, Eq. (3.5.5), becomes :

$$\frac{d\eta_{i,j}}{dt} = -\left[\frac{(uD)_{i+1/2,j} - (uD)_{i-1/2,j}}{\Delta x_{i,j}} \right] - \left[\frac{(vD)_{i,j+1/2} - (vD)_{i,j-1/2}}{\Delta y_{i,j}} \right] \tag{3.5.8}$$

By defining the spatial gradient operators of any quantity φ such that

$$\delta_x \varphi = \left(\varphi_{i+1/2,j} - \varphi_{i-1/2,j}\right) / \left(x_{i+1/2,j} - x_{i-1/2,j}\right)$$

$$= 2\left[\varphi(x + \Delta x_e / 2, y) - \varphi(x - \Delta x_w / 2, y)\right] / \left[\Delta x_e + \Delta x_w\right]$$

$$\delta_y \varphi = \left(\varphi_{i,j+1/2} - \varphi_{i,j-1/2}\right) / \left(y_{i,j+1/2} - y_{i,j-1/2}\right)$$

$$= 2\left[\varphi(x, y + \Delta y_n / 2) - \varphi(x, y - \Delta y_s / 2)\right] / \left[\Delta y_n + \Delta y_s\right]$$

(3.5.9)

the finite-difference equations can be written in a more compact form. For example, the continuity equation (3.5.8) can be written as (omitting the index i,j)

$$\frac{d\eta}{dt} = -\delta_x \left(uD\right) - \delta_y \left(vD\right) \tag{3.5.10}$$

Similarly, the momentum equations become

$$\frac{d}{dt}\left(uD\right) = -\left[\delta_x \left(u^2 D\right) + \delta_y \left(uvD\right)\right] + fvD - g'D\delta_x \left(\eta\right) + \tau^x$$

(3.5.11)

$$+D\left\{\delta_x \left[A\delta_x \left(u\right)\right] + \delta_y \left[A\delta_y \left(u\right)\right]\right\}$$

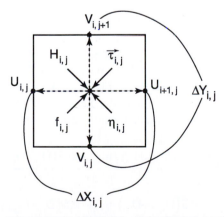

Figure 3.5.1 Staggered C grid computational cell indexed i,j showing the location of different variables.

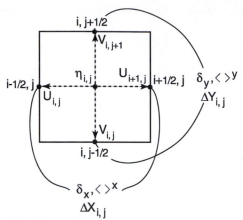

Figure 3.5.2 Staggered C grid computational cell indexed i,j showing the location of elevation η.

$$\frac{d}{dt}(vD) = -\left[\delta_x\left(uvD\right) + \delta_y\left(v^2D\right)\right] - fuD - g'D\delta_y\left(\eta\right) + \tau^y$$

$$(3.5.12)$$

$$+D\left\{\delta_x\left[A\delta_x\left(v\right)\right] + \delta_y\left[A\delta_y\left(v\right)\right]\right\}$$

The first decision to make is the choice of the grid system. We will use the C grid, since it is unlikely that affordable grid sizes would be larger than the Rossby radius. The naming convention followed is shown in Figure 3.5.1; $\eta_{i,j}$ is at the center of the grid indexed (i,j) of size Δx and Δy in the x and y directions, respectively; the u velocity indexed (i,j) is displaced to the west by $\Delta x/2$, and the v velocity indexed (i,j) is displaced to the south by $\Delta y/2$. Note that the locations $x+\Delta x_e/2$, $x-\Delta x_w/2$ $y+\Delta y_n/2$, and $y-\Delta y_s/2$ should be interpreted as half a grid point to the east, west, north, and south, respectively, with respect to the location of the quantity being finite-differenced. Figures 3.5.2 to 3.5.4 show these locations for η, u and v, respectively. Thus the Equations (3.5.10) to (3.5.12) become (note that the bottom depth $D_{i,j}$, and wind stress $\tau_{i,j}$ are collocated with $\eta_{i,j}$)

$$\frac{d\eta_{i,j}}{dt} = -\left[\frac{0.5\left(D_{i+1,j}+D_{i,j}\right)u_{i+1,j} - 0.5\left(D_{i,j}+D_{i-1,j}\right)u_{i,j}}{\Delta x_{i,j}}\right]$$

$$-\left[\frac{0.5\left(D_{i,j+1}+D_{i,j}\right)v_{i,j+1} - 0.5\left(D_{i,j}+D_{i,j-1}\right)v_{i,j}}{\Delta y_{i,j}}\right]$$

$$(3.5.13)$$

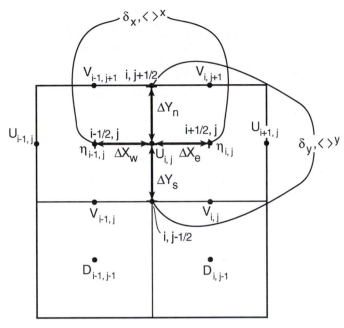

Figure 3.5.3 Staggered C grid computational cell indexed i,j showing the location of the u component of velocity.

$$\frac{d}{dt}\left[0.5u_{i,j}\left(D_{i-1,j}+D_{i,j}\right)\right] = -0.5\left[\frac{\left(u_{i+1,j}+u_{i,j}\right)^2 D_{i,j} - \left(u_{i,j}+u_{i-1,j}\right)^2 D_{i-1,j}}{\left(\Delta x_{i,j}+\Delta x_{i-1,j}\right)}\right]$$

$$-\frac{0.125}{\left(\Delta y_{i-1,j}+\Delta y_{i,j}\right)}\left[\begin{array}{c}\left(D_{i,j+1}+D_{i,j}+D_{i-1,j+1}+D_{i-1,j}\right)\left(v_{i,j+1}+v_{i-1,j+1}\right)\left(u_{i,j+1}+u_{i,j}\right)\\ -\left(D_{i,j}+D_{i,j-1}+D_{i-1,j}+D_{i-1,j-1}\right)\left(v_{i,j}+v_{i-1,j}\right)\left(u_{i,j}+u_{i,j-1}\right)\end{array}\right]$$

$$+0.0625\left(f_{i,j}+f_{i-1,j}\right)\left(v_{i,j}+v_{i,j+1}+v_{i-1,j}+v_{i-1,j+1}\right)\left(D_{i,j}+D_{i-1,j}\right)$$

$$-0.5g'\left(D_{i,j}+D_{i-1,j}\right)\left(\eta_{i,j}-\eta_{i-1,j}\right)+0.5\left(\tau_{i,j}^x+\tau_{i-1,j}^x\right)$$

$$+0.5\left(D_{i,j}+D_{i-1,j}\right)\left\{\frac{A_{i,j}\left[u_{i+1,j}-u_{i,j}\right]/\Delta x_{i,j}-A_{i-1,j}\left[u_{i,j}-u_{i-1,j}\right]/\Delta x_{i-1,j}}{0.5\left(\Delta x_{i,j}+\Delta x_{i-1,j}\right)}\right\}$$

$$+\frac{\left(D_{i,j}+D_{i-1,j}\right)}{\left(\Delta y_{i,j}+\Delta y_{i-1,j}\right)}\left\{\begin{array}{c}\left(\dfrac{A_{i,j}+A_{i,j+1}+A_{i-1,j+1}+A_{i-1,j}}{\Delta y_{i,j}+\Delta y_{i,j+1}+\Delta y_{i-1,j+1}+\Delta y_{i-1,j}}\right)\left[u_{i,j+1}-u_{i,j}\right]\\ -\left(\dfrac{A_{i,j-1}+A_{i,j}+A_{i-1,j}+A_{i-1-1,j}}{\Delta y_{i,j-1}+\Delta y_{i,j}+\Delta y_{i-1,j}+\Delta y_{i-1,j-1}}\right)\left[u_{i,j}-u_{i,j-1}\right]\end{array}\right\}$$

$$(3.5.14)$$

$$\frac{d}{dt}\left[0.5v_{i,j}\left(D_{i,j}+D_{i,j\text{-}1}\right)\right]=-0.5\left[\frac{\left(v_{i,j+1}+v_{i,j}\right)^2 D_{i,j}-\left(v_{i,j}+v_{i,j\text{-}1}\right)^2 D_{i,j\text{-}1}}{\left(\Delta y_{i,j}+\Delta y_{i,j\text{-}1}\right)}\right]$$

$$-\frac{0.125}{\left(\Delta x_{i,j}+\Delta x_{i,j\text{-}1}\right)}\left[\begin{array}{l}\left(D_{i+1,j}+D_{i,j}+D_{i+1,j\text{-}1}+D_{i,j\text{-}1}\right)\left(u_{i+1,j}+u_{i+1,j\text{-}1}\right)\left(v_{i,j}+v_{i+1,j}\right)\\-\left(D_{i,j}+D_{i\text{-}1,j}+D_{i,j\text{-}1}+D_{i\text{-}1,j\text{-}1}\right)\left(u_{i,j}+u_{i,j\text{-}1}\right)\left(v_{i\text{-}1,j}+v_{i,j}\right)\end{array}\right]$$

$$-0.0625\left(f_{i,j}+f_{i,j\text{-}1}\right)\left(u_{i,j}+u_{i+1,j}+u_{i,j\text{-}1}+u_{i+1,j\text{-}1}\right)\left(D_{i,j}+D_{i,j\text{-}1}\right)$$

$$-0.5g'\left(D_{i,j}+D_{i,j\text{-}1}\right)\left(\eta_{i,j}-\eta_{i,j\text{-}1}\right)+0.5\left(\tau^y_{i,j}+\tau^y_{i,j\text{-}1}\right)$$

$$+\frac{\left(D_{i,j}+D_{i,j\text{-}1}\right)}{\left(\Delta x_{i,j}+\Delta x_{i,j\text{-}1}\right)}\left\{\begin{array}{l}\left(\dfrac{A_{i,j}+A_{i+1,j}+A_{i+1,j\text{-}1}+A_{i,j\text{-}1}}{\Delta x_{i,j}+\Delta x_{i+1,j}+\Delta x_{i+1,j\text{-}1}+\Delta x_{i,j\text{-}1}}\right)\left[v_{i+1,j}-v_{i,j}\right]\\[12pt]-\left(\dfrac{A_{i\text{-}1,j}+A_{i,j}+A_{i,j\text{-}1}+A_{i\text{-}1,j\text{-}1}}{\Delta x_{i\text{-}1,j}+\Delta x_{i,j}+\Delta x_{i,j\text{-}1}+\Delta x_{i\text{-}1,j\text{-}1}}\right)\left[v_{i,j}-v_{i\text{-}1,j}\right]\end{array}\right\}$$

$$+0.5\left(D_{i,j}+D_{i,j\text{-}1}\right)\left\{\frac{A_{i,j}\left[v_{i,j+1}-v_{i,j}\right]/\Delta y_{i,j}-A_{i,j\text{-}1}\left[v_{i,j}-v_{i,j\text{-}1}\right]/\Delta y_{i,j\text{-}1}}{0.5\left(\Delta y_{i,j}+\Delta y_{i,j\text{-}1}\right)}\right\}$$

$$(3.5.15)$$

It is clear that a lot of careful bookkeeping is involved in formulating the finite difference equivalents. The averaging of four surrounding velocities involved in computing the Coriolis terms in (3.5.14) and (3.5.15) is what makes the C grid inaccurate in simulating geostrophic adjustment processes when the grid size exceeds the Rossby radius of deformation. In contrast, the pressure gradient terms do not involve any averaging so that superinertial processes are well depicted.

Equations (3.5.13) to (3.5.15) are the C-grid forms of Eqs. (3.5.10) to (3.5.12), respectively. Note that we have used second-order central differences. Similar expressions can be written down for fourth-order spatial differences, but they are far more complex and seldom used. The order of spatial differencing to be used involves tradeoffs between increased accuracy (for the same grid size), on the one hand, and increased complexity of finite difference equations and the boundary conditions, and increased computational burden, on the other.

So far, we have ignored temporal differencing. Equations (3.5.10) to (3.5.12), or equivalently Eqs. (3.5.13) to (3.5.15) for the C grid, are a set of first-order, coupled, nonlinear, ordinary differential equations that can be solved by the methods outlined in Chapter 2. A choice has to be made now as to the time differencing scheme to be used. This depends on the accuracy desired. In time-

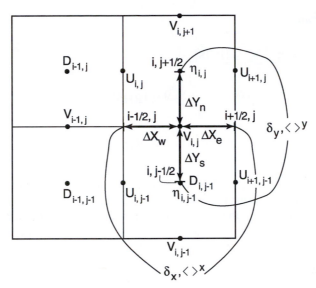

Figure 3.5.4   Staggered C grid computational cell indexed i,j showing the location of the v component of velocity.

dependent problems, one would like to take as large a time step as feasible. This usually rules out two time level schemes, since they are only first-order accurate. Three time level schemes yield $O(\Delta t)^2$ accuracy and are widely used, although as we saw in Section 2.6, with the second-order leapfrog scheme, time-splitting manifests itself in nonlinear problems. We also saw in Section 2.6, the possibility of using third-order Adams-Bashforth schemes. Ultimately, it all depends on various tradeoffs involving storage requirements, the CPU time available, and the complexity.

A choice has also to be made between implicit and explicit schemes. Fully implicit schemes are far too complex for this type of problem. They are also not very practical since they require inversions of complicated matrices. However semi-implicit schemes, where only some of the terms on the RHS are evaluated at time level n+1, are possible. Explicit time schemes, where the terms on the RHS are evaluated at either time level n or n-1, are the simplest. If the terms on the RHS can be evaluated at time steps n-1 and/or n, the previous time levels, solutions at time level n+1 can be readily obtained. The explicit leapfrog scheme is widely used in combination with Asselin-Roberts filter to suppress the computational mode and prevent time-splitting, so that Equations (3.5.10) to

(3.5.12) become

$$\eta_{i,j}^{n+1} = \eta_{i,j}^{n-1} + 2\Delta t \left\{ RHS \right\}_{\eta}$$

$$u_{i,j}^{n+1} = \left[u_{i,j}^{n-1} \left(D_{i,j}^{n-1} + D_{i-1,j}^{n-1} \right) + 4\Delta t \left\{ RHS \right\}_{u} \right] / \left(D_{i,j}^{n+1} + D_{i-1,j}^{n+1} \right) \qquad (3.5.16)$$

$$v_{i,j}^{n+1} = \left[v_{i,j}^{n-1} \left(D_{i,j}^{n-1} + D_{i,j-1}^{n-1} \right) + 4\Delta t \left\{ RHS \right\}_{v} \right] / \left(D_{i,j}^{n+1} + D_{i,j-1}^{n+1} \right)$$

where $\{ RHS \}_{\eta,u,v}$ denote the RHS of Eqs. (3.5.10) to (3.5.12), respectively. Note that for the leapfrog scheme to be stable, the dissipation (horizontal viscosity and any frictional) terms in the momentum equations must be lagged—that is, evaluated at time level n−1. The rest of the terms on the RHS in Eq. (3.5.16) are evaluated at time level n. Note that D = H + η, and therefore, η equation is solved first, so that D^{n+1} is known, before solving for the velocity components u and v. Asselin-Roberts filter is applied after integration at each time step:

$$\left(\eta_{i,j}^{n} \right)_{f} = \eta_{i,j}^{n} + \gamma \left[\left(\eta_{i,j}^{n-1} \right)_{f} - 2\eta_{i,j}^{n} + \eta_{i,j}^{n+1} \right]$$

$$\left(u_{i,j}^{n} \right)_{f} = u_{i,j}^{n} + \gamma \left[\left(u_{i,j}^{n-1} \right)_{f} - 2u_{i,j}^{n} + u_{i,j}^{n+1} \right] \qquad (3.5.17)$$

$$\left(v_{i,j}^{n} \right)_{f} = v_{i,j}^{n} + \gamma \left[\left(v_{i,j}^{n-1} \right)_{f} - 2v_{i,j}^{n} + v_{i,j}^{n+1} \right]$$

The value of γ most often used is 0.1. However, as we saw in Section 2.6, the use of Asselin-Roberts filter degrades the accuracy of the time-differencing scheme. The use of third-order Adams-Bashforth scheme (Durran, 1991) can provide higher accuracy but at the expense of additional storage needed for prognostic variables at the time level n−2.

The CFL condition dictates that the time step Δt be less than the minimum value of $1/2[(\Delta x^{-2} + \Delta y^{-2})^{-1}(g'H)^{-1/2}]$ in the domain. This constraint can be relaxed by making the time-differencing scheme semi-implicit by evaluating the pressure gradient terms involving the horizontal gradients of η in Eqs. (3.5.11) and (3.5.12) at time level n+1. Since during the sequential solution, η^{n+1} is known before u^{n+1} and v^{n+1} are evaluated, such a scheme does not involve any matrix inversions.

Although B grid is not necessary for this problem, it is instructive to examine the finite-differencing involved. Figure (3.5.5) shows the location of variables and indexing on the B grid. Figures (3.5.6) and (3.5.7) show the locations of the half-grid points for η, and u and v, respectively. The finite-difference forms for the B grid are

$$\frac{d\eta_{i,j}}{dt} = -\left[\frac{0.25\left(D_{i+1,j}+D_{i,j}\right)\left(u_{i+1,j}+u_{i+1,j+1}\right)-0.25\left(D_{i,j}+D_{i-1,j}\right)\left(u_{i,j}+u_{i,j+1}\right)}{\Delta x_{i,j}}\right]$$

$$-\left[\frac{0.25\left(D_{i,j+1}+D_{i,j}\right)\left(v_{i+1,j+1}+v_{i,j+1}\right)-0.25\left(D_{i,j}+D_{i,j-1}\right)\left(v_{i+1,j}+v_{i,j}\right)}{\Delta y_{i,j}}\right]$$

$$(3.5.18)$$

$$\frac{d}{dt}\left[0.25u_{i,j}\left(D_{i-1,j}+D_{i,j}+D_{i,j-1}+D_{i-1,j-1}\right)\right]=$$

$$-0.5\left[\frac{\left(u_{i+1,j}+u_{i,j}\right)^2\left(D_{i,j}+D_{i,j-1}\right)-\left(u_{i,j}+u_{i-1,j}\right)^2\left(D_{i-1,j}+D_{i-1,j-1}\right)}{\left(\Delta x_{i,j}+\Delta x_{i-1,j}+\Delta x_{i,j-1}+\Delta x_{i-1,j-1}\right)}\right]$$

$$-\frac{0.5}{\left(\Delta y_{i-1,j}+\Delta y_{i,j}+\Delta y_{i-1,j-1}+\Delta y_{i,j-1}\right)}\left[\begin{array}{l}\left(D_{i,j}+D_{i-1,j}\right)\left(v_{i,j+1}+v_{i,j}\right)\left(u_{i,j+1}+u_{i,j}\right)\\-\left(D_{i,j-1}+D_{i-1,j-1}\right)\left(v_{i,j}+v_{i,j-1}\right)\left(u_{i,j}+u_{i,j-1}\right)\end{array}\right]$$

$$+0.0625\left(f_{i,j}+f_{i-1,j}+f_{i,j-1}+f_{i-1,j-1}\right)\left(D_{i,j}+D_{i-1,j}+D_{i,j-1}+D_{i-1,j-1}\right)v_{i,j}$$

$$-0.5g'\left(D_{i,j}+D_{i-1,j}+D_{i,j-1}+D_{i-1,j-1}\right)\left[\frac{\left(\eta_{i,j}+\eta_{i,j-1}-\eta_{i-1,j}-\eta_{i-1,j-1}\right)}{\left(\Delta x_{i,j}+\Delta x_{i-1,j}+\Delta x_{i,j-1}+\Delta x_{i-1,j-1}\right)}\right]$$

$$+0.25\left(\tau_{i,j}^x+\tau_{i-1,j}^x+\tau_{i,j-1}^x+\tau_{i-1,j-1}^x\right)$$

$$+\frac{\left(D_{i,j}+D_{i-1,j}+D_{i,j-1}+D_{i-1,j-1}\right)}{\left(\Delta x_{i,j}+\Delta x_{i,j-1}+\Delta x_{i-1,j}+\Delta x_{i-1,j-1}\right)}\left\{\begin{array}{l}\left(A_{i,j}+A_{i,j-1}\right)\left(u_{i+1,j}-u_{i,j}\right)/\left(\Delta x_{i,j}+\Delta x_{i,j-1}\right)\\-\left(A_{i-1,j}+A_{i-1,j-1}\right)\left(u_{i,j}-u_{i-1,j}\right)/\left(\Delta x_{i-1,j}+\Delta x_{i-1,j-1}\right)\end{array}\right\}$$

$$+\frac{\left(D_{i,j}+D_{i-1,j}+D_{i,j-1}+D_{i-1,j-1}\right)}{\left(\Delta y_{i,j}+\Delta y_{i,j-1}+\Delta y_{i-1,j}+\Delta y_{i-1,j-1}\right)}\left\{\begin{array}{l}\left(A_{i,j}+A_{i-1,j}\right)\left(u_{i,j+1}-u_{i,j}\right)/\left(\Delta y_{i,j}+\Delta y_{i-1,j}\right)\\-\left(A_{i,j-1}+A_{i-1,j-1}\right)\left(u_{i,j}-u_{i,j-1}\right)/\left(\Delta y_{i,j-1}+\Delta y_{i-1,j-1}\right)\end{array}\right\}$$

$$(3.5.19)$$

$$\frac{d}{dt}\left[0.25v_{i,j}\left(D_{i-1,j}+D_{i,j}+D_{i,j-1}+D_{i-1,j-1}\right)\right]=$$

$$-\frac{0.5}{\left(\Delta x_{i,j-1}+\Delta x_{i,j}+\Delta x_{i-1,j-1}+\Delta x_{i-1,j}\right)}\left[\begin{array}{l}\left(D_{i,j}+D_{i,j-1}\right)\left(v_{i+1,j}+v_{i,j}\right)\left(u_{i+1,j}+u_{i,j}\right)\\-\left(D_{i-1,j}+D_{i-1,j-1}\right)\left(v_{i,j}+v_{i-1,j}\right)\left(u_{i,j}+u_{i-1,j}\right)\end{array}\right]$$

$$-0.5\left[\frac{\left(v_{i,j+1}+v_{i,j}\right)^{2}\left(D_{i,j}+D_{i-1,j}\right)-\left(v_{i,j}+v_{i,j-1}\right)^{2}\left(D_{i,j-1}+D_{i-1,j-1}\right)}{\left(\Delta y_{i,j}+\Delta y_{i-1,j}+\Delta y_{i,j-1}+\Delta y_{i-1,j-1}\right)}\right]$$

$$-0.0625\left(f_{i,j}+f_{i-1,j}+f_{i,j-1}+f_{i-1,j-1}\right)\left(D_{i,j}+D_{i-1,j}+D_{i,j-1}+D_{i-1,j-1}\right)u_{i,j}$$

$$-0.5g'\left(D_{i,j}+D_{i-1,j}+D_{i,j-1}+D_{i-1,j-1}\right)\left[\frac{\left(\eta_{i,j}+\eta_{i-1,j}-\eta_{i,j-1}-\eta_{i-1,j-1}\right)}{\left(\Delta y_{i,j}+\Delta y_{i-1,j}+\Delta y_{i,j-1}+\Delta y_{i-1,j-1}\right)}\right]$$

$$+0.25\left(\tau_{i,j}^{y}+\tau_{i-1,j}^{y}+\tau_{i,j-1}^{y}+\tau_{i-1,j-1}^{y}\right)$$

$$+\frac{\left(D_{i,j}+D_{i-1,j}+D_{i,j-1}+D_{i-1,j-1}\right)}{\left(\Delta x_{i,j}+\Delta x_{i,j-1}+\Delta x_{i-1,j}+\Delta x_{i-1,j-1}\right)}\left\{\begin{array}{l}\left(A_{i,j}+A_{i,j-1}\right)\left(v_{i+1,j}-v_{i,j}\right)/\left(\Delta x_{i,j}+\Delta x_{i,j-1}\right)\\-\left(A_{i-1,j}+A_{i-1,j-1}\right)\left(v_{i,j}-v_{i-1,j}\right)/\left(\Delta x_{i-1,j}+\Delta x_{i-1,j-1}\right)\end{array}\right\}$$

$$+\frac{\left(D_{i,j}+D_{i-1,j}+D_{i,j-1}+D_{i-1,j-1}\right)}{\left(\Delta y_{i,j}+\Delta y_{i,j-1}+\Delta y_{i-1,j}+\Delta y_{i-1,j-1}\right)}\left\{\begin{array}{l}\left(A_{i,j}+A_{i-1,j}\right)\left(v_{i,j+1}-v_{i,j}\right)/\left(\Delta y_{i,j}+\Delta y_{i-1,j}\right)\\-\left(A_{i,j-1}+A_{i-1,j-1}\right)\left(v_{i,j}-v_{i,j-1}\right)/\left(\Delta y_{i,j-1}+\Delta y_{i-1,j-1}\right)\end{array}\right\}$$

$$(3.5.20)$$

Note that the Coriolis terms do not involve any averaging of the velocities because of the colocation of u and v, but the pressure gradient terms involve averaging η. This is the reason geostrophic adjustment processes are well depicted by the B grid when the grid size is larger than the Rossby radius. On the other hand, because of the averaging involved in computing the pressure gradients, superinertial processes are not well simulated.

It is now clear that the exact form for the terms in Eqs. (3.5.10) to (3.5.12) depends on the particular grid chosen. If we define spatial averaging operators

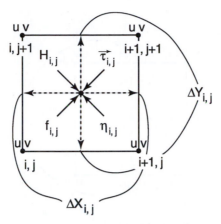

Figure 3.5.5 Staggered B grid computational cell indexed i,j showing the location of the v component of velocity.

such that

$$\langle\varphi\rangle^x = \left(\varphi_{i+1/2,j} + \varphi_{i-1/2,j}\right)/2$$
$$= 0.5\left[\varphi(x + \Delta x_e/2, y) + \varphi(x - \Delta x_w/2, y)\right]$$

$$\langle\varphi\rangle^y = \left(\varphi_{i,j+1/2} + \varphi_{i,j-1/2}\right)/2$$
$$= 0.5\left[\varphi(x, y + \Delta y_n/2) + \varphi(x, y - \Delta y_s/2)\right] \qquad (3.5.21)$$

$$\langle\varphi\rangle^{xy} = \left(\varphi_{i+1/2,j+1/2} + \varphi_{i-1/2,j+1/2} + \varphi_{i+1/2,j-1/2} + \varphi_{i-1/2,j-1/2}\right)$$
$$= 0.25\left[\begin{array}{c}\varphi(x + \Delta x_e/2, y + \Delta y_n/2) + \varphi(x - \Delta x_w/2, y + \Delta y_n/2) \\ +\varphi(x + \Delta x_e/2, y - \Delta y_s/2) + \varphi(x - \Delta x_w/2, y - \Delta y_s/2)\end{array}\right]$$

Eqs. (3.5.11) and (3.5.12) can be written as

$$\frac{d}{dt}\left(u\langle D\rangle^x\right) = -\left[\delta_x\left(u^2 D\right) + \delta_y\left(uvD\right)\right] + \langle f\rangle^x \langle v\rangle^{xy}\langle D\rangle^x - g'\langle D\rangle^x \delta_x(\eta) + \langle\tau^x\rangle^x$$
$$+D\left\{\delta_x\left[A\delta_x(u)\right] + \delta_y\left[A\delta_y(u)\right]\right\}$$

$$(3.5.22)$$

$$\frac{d}{dt}\left(v\langle D\rangle^y\right) = -\left[\delta_x\left(uvD\right) + \delta_y\left(v^2 D\right)\right] - \langle f\rangle^y \langle u\rangle^{xy}\langle D\rangle^y - g'\langle D\rangle^y \delta_y(\eta) + \langle\tau^y\rangle^y$$
$$+D\left\{\delta_x\left[A\delta_x(v)\right] + \delta_y\left[A\delta_y(v)\right]\right\}$$

$$(3.5.23)$$

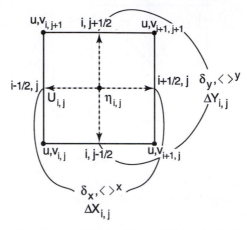

Figure 3.5.6 Staggered B grid computational cell indexed i,j showing the location of elevation η.

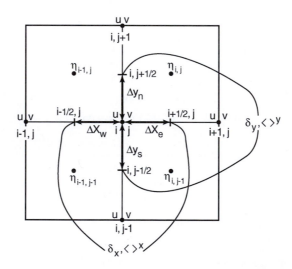

Figure 3.5.7   Staggered B grid computational cell indexed i,j showing the location of the u and v components of velocity.

for the C grid and as

$$\frac{d}{dt}\left(u\langle D\rangle^{xy}\right) = -\left[\delta_x\left(u^2D\right)+\delta_y\left(uvD\right)\right]+\langle f\rangle^{xy} v\langle D\rangle^{xy} - g'\langle D\rangle^{xy}\delta_x(\eta)+\langle \tau^x\rangle^{xy}$$
$$+\langle D\rangle^{xy}\left\{\delta_x\left[A\delta_x(u)\right]+\delta_y\left[A\delta_y(u)\right]\right\}$$

(3..5.24)

$$\frac{d}{dt}\left(v\langle D\rangle^{xy}\right) = -\left[\delta_x\left(uvD\right)+\delta_y\left(v^2D\right)\right]-\langle f\rangle^{xy} u\langle D\rangle^{xy} - g'\langle D\rangle^{xy}\delta_y(\eta)+\langle \tau^y\rangle^{xy}$$
$$+\langle D\rangle^{xy}\left\{\delta_x\left[A\delta_x(v)\right]+\delta_y\left[A\delta_y(v)\right]\right\}$$

(3.5.25)

for the B grid. Note that for the sake of convenience, the expressions in square and curly brackets have not been written down. Their form can, however, be gleaned from Eqs. (3.5.14) and (3.5.15) for the C grid and Eqs. (3.5.19) and (3.5.20) for the B grid.

For a nonuniform Cartesian coordinate grid, Δx can only be a function of y, so that the index Δx_j should suffice to describe the grid size in the x direction. Similarly Δy_i is adequate to describe the grid size in the y direction. Nevertheless, we have used $\Delta x_{i,j}$ and $\Delta y_{i,j}$ so that the finite-difference formulations could be easily translated to a fully orthogonal curvilinear grid for which Δx and Δy are both functions of both x and y. The finite-differencing is similar for the general orthogonal coordinates, except that Eqs. (3.5.1) to (3.5.4) must be used as starting points.

It is also advantageous to cast the finite difference equations in control-volume form, by recognizing the fact that $h_1\partial\xi_1 = \partial x$ and $h_2\partial\xi_2 = \partial y$ and multiplying all the governing equations by ∂x and ∂y before carrying out finite differencing. Recognizing that in finite-difference form, ∂x and ∂y become Δx and Δy, and ξ_1 is independent of ξ_2 and vice versa, and defining finite-difference operators (not gradient operators that were defined in Eq. 3.5.9),

$$\delta_x\varphi = (\partial\varphi/\partial\xi_1)\,\Delta\xi_1 = \varphi(\xi_1+\Delta\xi_1/2) - \varphi(\xi_1-\Delta\xi_1/2)$$
$$\delta_y\varphi = (\partial\varphi/\partial\xi_2)\,\Delta\xi_2 = \varphi(\xi_2+\Delta\xi_2/2) - \varphi(\xi_2-\Delta\xi_2/2)$$

(3.5.26)

Equations (3.5.1) to (3.5.3) become (after substituting $u_1 = u$, and $u_2 = v$)

$$(\Delta x\Delta y)\frac{\partial\eta}{\partial t} = -\left[\delta_x\left(\Delta yuD\right)+\delta_y\left(\Delta xvD\right)\right]$$

(3.5.27)

$$(\Delta x\Delta y)\frac{\partial}{\partial t}(uD) = -\left[\delta_x\left(\Delta y\,u^2D\right)+\delta_y\left(\Delta x\,uvD\right)\right]$$
$$+f\,vD\left(\Delta x\Delta y\right)-\left\{uvD\delta_y\left(\Delta x\right)-v^2D\delta_x\left(\Delta y\right)\right\}$$
$$-g'D\left(\Delta y\right)\delta_x\left(\eta\right)+\left(\Delta x\Delta y\right)\tau^x +D\left[\delta_x\left(A\frac{\Delta y}{\Delta x}\delta_x u\right)+\delta_y\left(A\frac{\Delta x}{\Delta y}\delta_y u\right)\right]$$

(3.5.28)

$$
\begin{aligned}
(\Delta x \Delta y)\frac{\partial}{\partial t}(vD) &= -\left[\delta_x\left(\Delta y\ uvD\right)+\delta_y\left(\Delta x\ v^2 D\right)\right]\\
&\quad - fuD\left(\Delta x \Delta y\right)-\left\{uvD\delta_x\left(\Delta y\right)-u^2 D\delta_y\left(\Delta x\right)\right\}\\
&\quad -g'D\left(\Delta x\right)\delta_y\left(\eta\right)+\left(\Delta x \Delta y\right)\tau^y + D\left[\delta_x\left(A\frac{\Delta y}{\Delta x}\delta_x v\right)+\delta_y\left(A\frac{\Delta x}{\Delta y}\delta_y v\right)\right]
\end{aligned}
$$

$$(3.5.29)$$

Equations (3.5.27) to (3.5.29) are similar in form to Eqs. (3.5.10) to (3.5.12) in rectangular Cartesian coordinates, and the exact form for the terms in these equations also depends on the grid chosen. Note the appearance in the momentum equations of extra terms indicated by curly brackets. These terms are identically zero in the rectangular Cartesian coordinate system. They can be absorbed into the Coriolis terms by defining a pseudo-Coriolis parameter f_p:

$$f_p = f - \left\{u\delta_y\left(\Delta x\right)-v\delta_x\left(\Delta y\right)\right\}/\left(\Delta x \Delta y\right)\qquad(3.5.30)$$

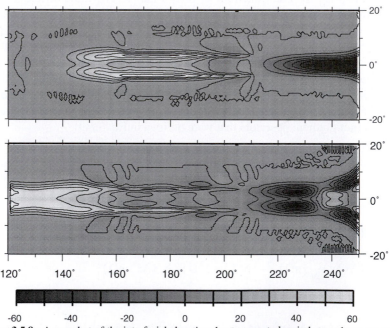

Figure 3.5.8 A snapshot of the interfacial elevation due to a westerly wind stress burst symmetric about the equator. The stress was turned on for a few days and then turned off impulsively (top, 30 days after wind stress turned off; bottom, 130 days after wind stress turned off). Note the Kelvin and first mode Rossby waves in the waveguide.

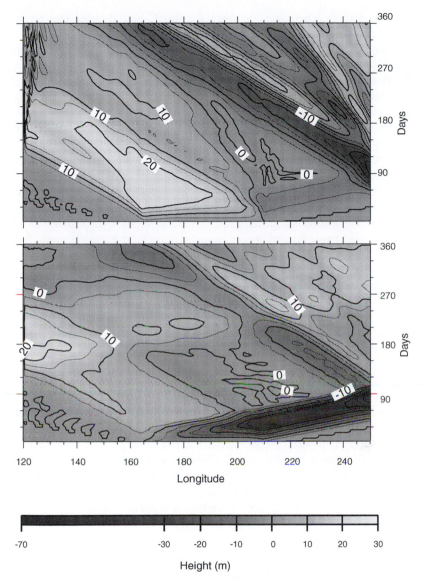

Figure 3.5.9 An x-t (Hovmuller) plot along 3°N (top) and the equator (bottom) showing the various propagations in the waveguide and their reflections at the meridional boundaries.

With this substitution, the general orthogonal curvilinear form of the equations become very much similar to the Cartesian form, except that Eqs. (3.5.27) to (3.5.29) are in control volume form (the operators define spatial differences, not spatial gradients). Their specific form for the C and B grids can be easily deduced and will not be given here.

Figure (3.5.8) shows an application of the reduced gravity model to simulating equatorial processes. The model uses the C grid at 1/3 degree resolution. The domain extends from 30 °S to 30 °N and encompasses the entire equatorial Pacific. The value of H (100 m here) and the value of the reduced gravity can be chosen consistent with either the first (C ~ 2.4 m s^{-1}) or second (C ~ 1.4 m s^{-1}) mode. Realistic coastlines are used, including the complex island chain in the tropical western Pacific. Radiation boundary conditions are employed at the north and south zonal boundaries to radiate the poleward propagating coastal Kelvin waves out of the domain.

Wave motions are generated by switching on an easterly wind stress in the central portion of the domain for 30 days and then switching it off. By controlling the meridional structure of this wind stress, it is possible to excite symmetric odd mode Rossby waves or antisymmetric even mode ones. A wind stress symmetric about the equator generates a strong Kelvin wave propagating to the east and a strong first mode Rossby wave traveling more slowly to the west. Figure (3.5.8) shows a snapshot of the interfacial deflections after the wind has been turned off. The Kelvin and Rossby modes can be seen. At the eastern boundary, reflection of the Kelvin wave energy in the form of a series of westward propagating Rossby wave modes and leakage into poleward-propagating coastally trapped Kelvin waves can be clearly seen in the simulations. Reflection of the Rossby mode at the western boundary involves eastward propagating Kelvin and Rossby modes. The contrasting behavior of high-frequency gravity modes and low-frequency Rossby modes excited by the wind pulse is noteworthy. x – t plots along carefully chosen zonal transects show the wave propagations quite clearly. A transect along the equator (Figure 3.5.9) displays the Kelvin wave propagations rather well, whereas a transect at 3 °N shows the first mode Rossby waves. The propagation speeds are in agreement with theoretical values (for example 2.4 m s^{-1} for the first mode Kelvin wave and 0.8 m s^{-1} for the first mode Rossby wave). Animation of the modeled interfacial deflection fields can show clearly the various aspects of generation, propagation, and reflection of waves in the waveguide.

Chapter 4

Midlatitude Dynamics and Quasi-Geostrophic Models

Incompressible, hydrostatic, Boussinesq equations for a stratified fluid in a gravitational field on a rotating sphere are adequate to describe most geophysical flows. Hough in 1897 was the first to investigate motions on a spherical Earth by expansions in spherical harmonics. However, these solutions are rather complex and not until Carl-Gustaf Rossby proposed in 1937 a simple tangent β plane approximation did it become possible to obtain a deeper understanding of midlatitude dynamics. Rossby, in an attempt to explain the movement of midlatitude atmospheric weather patterns, approximated the equations in rectangular Cartesian coordinates on a local tangent plane by ignoring all metric terms in the equations and retaining only the linearized meridional variation of the Coriolis term. This β plane approximation was a stroke of genius that enabled simple analytical solutions to be obtained and a better understanding of midlatitude dynamics to be reached. This is the simplest possible extension of the f-plane introduced by Lord Kelvin in 1879, which still retains the latitudinal variation of planetary vorticity crucial to slow internal adjustments in midlatitude oceans (and the atmosphere). In most of the following, we ignore dissipation and consider inviscid flows on a midlatitude beta plane.

It is often useful to think of the oceanic circulation as consisting of two principal components. One component is the wind-driven and wind-influenced relatively vigorous motions that are predominantly in the horizontal. These consist of midocean basin-scale gyres, the swift and narrow western boundary currents that become predominantly zonal when they leave the coast, undergo instabilities and generate eddy motions. The other, the thermally driven deep convection in subpolar seas and the resulting meridional circulation, which recirculates water vertically in the water column, is slow and sluggish, with timescales of a few hundred to thousand years, but nevertheless important to long-term climate. Of the two, it is the gyrelike circulations, with associated thermocline ventilation processes that are central to midlatitude dynamics.

4.1 LINEAR MOTIONS

Consider dissipationless, unforced, incompressible, hydrostatic, Boussinesq motions in rectangular Cartesian coordinates on a midlatitude beta plane. If we consider these motions as perturbations to a basic motionless background, the equations for the perturbation quantities can be written as

$$\frac{\partial u}{\partial x} + \frac{\partial v}{\partial y} + \frac{\partial w}{\partial z} = 0$$

$$\frac{du}{dt} - fv = -\frac{\partial p}{\partial x}$$

$$\frac{dv}{dt} + fu = -\frac{\partial p}{\partial y} \qquad (4.1.1)$$

$$0 = -\frac{\partial p}{\partial z} - g\frac{\rho}{\rho_0}$$

Here y is the meridional and x the zonal direction; p is the kinematic pressure; u, v, and w are velocity components in the zonal, meridional and vertical directions; ρ is the density and ρ_0 is the reference density. $f = f_0 + \beta y$, where $\beta = d f / dy$. Typical values for f and β are 10^{-4} s^{-1} and 2×10^{-11} m^{-1} s^{-1}. Note that the quantities in Eq. (4.1.1) are perturbations to a basic motionless background, but a simple background flow can be accommodated. For example, the time derivative can be approximated by $d / dt = \partial / \partial t + U \partial / \partial x$ for a uniform zonal flow.

These equations contain a rich variety of physical processes relevant to midlatitudes and will form the basis of discussions that follow. First, one needs to decide on how one handles the density stratification in the vertical. There are two approaches. The first is to approximate the continuous density stratification by a series of discrete layers, each of uniform density. The second is to assume a simple, analytically tractable stratification, for example, a uniform stratification with depth. Both these methods allow analytical solutions to be obtained. For an arbitrarily general stratification, one must employ numerical methods.

Consider a layered model for stratification first. Let the fluid column be divided into N layers, with thickness of kth layer being $h_k = H_k + \eta_{k-1} - \eta_k$, where H_k is the mean undisturbed thickness of layer k and η_k is the deflection of the interface at the bottom of the kth layer (k increases from 1 at the top to N at the bottom). The hydrostatic equation can then be integrated in the vertical to yield

$$\overline{\nabla} p_k = g\rho_k \overline{\nabla}\eta_k + g\sum_{i=1}^{k-1}\rho_i \overline{\nabla} h_i \qquad (4.1.2)$$

where $\overline{\nabla}$ refers to the horizontal gradients only. Integrating the continuity equation in the vertical over each layer and using the kinematic conditions at the top and bottom of each layer,

$$w_k = -\frac{d}{dt}\eta_k \qquad (4.1.3)$$

where w_k is the vertical velocity at the bottom of the kth layer, we get

$$\frac{d}{dt}(h_k) = -\frac{d}{dt}(\eta_k - \eta_{k-1}) = \frac{\partial(u_k h_k)}{\partial x} + \frac{\partial(v_k h_k)}{\partial y} \qquad (4.1.4)$$

Because of the absence of dissipation and external forcing, these equations are equivalent to conservation of potential vorticity in each layer:

$$\frac{d}{dt}\left(\frac{f+\zeta_k}{h_k}\right) = 0 \; ; \; \zeta_k = \frac{\partial v_k}{\partial x} - \frac{\partial u_k}{\partial y} \qquad (4.1.5)$$

A special set of these equations for a two-layer fluid is of particular interest, since this is often the limit that can be handled with ease by analytical models. The governing equations for this case are

$$\frac{dh_1}{dt} = \frac{d}{dt}(\eta_0 - \eta_1) = -\left[\frac{\partial}{\partial x}(h_1 u_1) + \frac{\partial}{\partial y}(h_1 v_1)\right]$$

$$\frac{du_1}{dt} - fv_1 = -g\frac{\partial\eta_0}{\partial x}$$

$$\frac{dv_1}{dt} + fu_1 = -g\frac{\partial\eta_0}{\partial y}$$

$$\frac{dh_2}{dt} = \frac{d\eta_1}{dt} = -\left[\frac{\partial}{\partial x}(h_2 u_2) + \frac{\partial}{\partial y}(h_2 v_2)\right]$$

$$\frac{du_2}{dt} - f v_2 = -g\left[\left(1 - \frac{\Delta\rho}{\rho_0}\right)\frac{\partial\eta_0}{\partial x} + \frac{\Delta\rho}{\rho_0}\frac{\partial\eta_1}{\partial x}\right]$$

$$\frac{dv_2}{dt} + f u_2 = -g\left[\left(1 - \frac{\Delta\rho}{\rho_0}\right)\frac{\partial\eta_0}{\partial y} + \frac{\Delta\rho}{\rho_0}\frac{\partial\eta_1}{\partial y}\right]$$ (4.1.6)

$$\Delta\rho = \rho_2 - \rho_1; \quad h_1 = H_1 + \eta_0 - \eta_1; \quad h_2 = H_2 + \eta_1 - \eta_B$$

where η_0 is the free surface deflection and η_B is the displacement due to nonuniform bottom depth. Thus, the two-layer equations correspond to retaining the barotropic and the first baroclinic mode. It is also the simplest subset that permits the influence of the bottom topography on baroclinic fluid motions.

A particular subset of these equations, in which the thickness of the bottom layer goes to infinity and the bottom layer becomes quiescent ($u_2 = v_2 = 0$), is called the reduced gravity equation. In general, reduced gravity terminology is applied to any number of layers as long as the bottom layer is infinitely deep and quiescent. Since in layered models the bottom topography is submerged in the bottom layer, this also means that such an approximation removes all the effects of bottom topography and eliminates the barotropic mode. The 1-1/2 layer reduced gravity subset of Eq. (4.1.6) can be obtained by letting $h_2 \rightarrow \infty, u_2 \rightarrow 0, v_2 \rightarrow 0$, which gives

$$\eta_0 = \left(\frac{\Delta\rho}{\rho_0}\right)\eta_1\left(1 - \frac{\Delta\rho}{\rho_0}\right)^{-1}$$ (4.1.7)

which states that the free surface deflection is proportional to the interfacial deflection, but one or two orders of magnitude smaller. Remember, this corresponds to the baroclinic mode only, since the barotropic mode is absent. Using Eq. (4.1.7) to eliminate η_0 in the governing equations for the upper layer, we get

$$\frac{dh_1}{dt} + \frac{\partial}{\partial x}(h_1 u_1) + \frac{\partial}{\partial y}(h_1 v_1) = 0$$

$$\frac{du_1}{dt} - f v_1 = -g'\frac{\partial\eta_1}{\partial x}$$ (4.1.8)

$$\frac{dv_1}{dt} + f u_1 = -g'\frac{\partial\eta_1}{\partial y}$$

where $g' = (\Delta\rho/\rho_0)g$ is the so-called reduced gravity. These equations correspond to the first baroclinic mode in the fluid. Note that these equations are the same as those for the barotropic mode, except that the reduced gravity g' replaces gravitational acceleration g. η_1 is the interfacial deflection. Since η_0 is negligible compared to η_1, and both deflections are small compared to the mean layer thickness, the continuity equation in Eq. (4.1.8) can be written in the form

$$\frac{d\eta_1}{dt} + H_1\left(\frac{\partial u_1}{\partial x} + \frac{\partial v_1}{\partial y}\right) = 0 \qquad (4.1.9)$$

Using the approximate form of Eq. (4.1.7) in which the denominator is put equal to unity, so that $\eta_0 = (\Delta\rho/\rho_0)\eta_1$, these equations can also be recast in the form

$$\frac{d\eta_0}{dt} + H_e\left(\frac{\partial u_1}{\partial x} + \frac{\partial v_1}{\partial y}\right) = 0$$

$$\frac{du_1}{dt} - fv_1 = -g\frac{\partial\eta_0}{\partial x} \qquad (4.1.10)$$

$$\frac{dv_1}{dt} + fu_1 = -g\frac{\partial\eta_0}{\partial y}$$

which are the equations for a barotropic mode on an ocean with its depth equal to the equivalent depth $H_e = (\Delta\rho/\rho_0)H_1$. The relevant parameters for such a problem are

Gravity Wave Speed $C = (gH_e)^{1/2}$
Rossby radius of deformation $a = C/f$
Inertial period $T = 2\pi/f$

Figures 4.1.1 and 4.1.2 from Chelton *et al.* (1998) show the first mode gravity wave speed and the Rossby radius of deformation in the global oceans. The shallow water gravity wave speed above is also the speed at which a mid-high latitude Kelvin wave travels with the coast to the right (in the northern hemisphere), and the Rossby radius of deformation is the distance over which its influence is felt away from the coast. The first baroclinic Kelvin wave speed is approximately 2.5 m s^{-1}, so that $D_e \sim 64$ cm, and the motions are equivalent to barotropic motions on an ocean of 64 cm depth! The Rossby radius of deformation is about 40 km at 30° latitude (see Chelton *et al.*, 1998; Emery *et al.*, 1984).

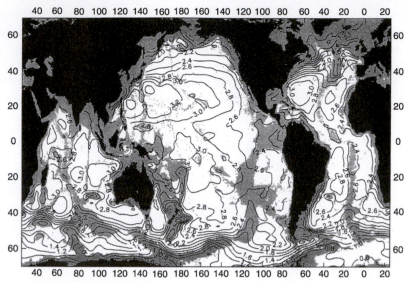

Figure 4.1.1 The distribution of the speed of the first baroclinic gravity wave in the global oceans (From Chelton et al. 1998).

The attractiveness of the reduced gravity equations is that by suitably varying the equivalent depth, any baroclinic mode can be dealt with. Many of the slow adjustment processes involving Rossby (planetary) wave generation and propagation can be studied using this simplified reduced gravity set of equations. The linear form of the equations derived earlier is well suited to obtaining analytical solutions. In addition, the full nonlinear form is the same as the comprehensive set of equations for barotropic motions in the ocean (but with a reduced equivalent depth), and numerical methods can be used to obtain solutions involving any baroclinic mode. We will consider analytical solutions for reduced gravity modes before discussing the full two-layer solutions. For simplicity, we ignore the subscripts, but remember that the following solutions apply to any mode including the barotropic mode, provided the equivalent depth is chosen appropriately.

4.1.1 INERTIA-GRAVITY (SVERDRUP/POINCARE) WAVES

If we seek wavelike solutions to Eq. (4.1.10) of the general form $\exp i(k_x x + k_y y - \omega t)$, it is easy to see that the equations demand that the following dispersion relation be satisfied:

$$\omega^2 = f^2 + C^2 k^2 = f^2(1 + k^2 a^2); \quad k^2 = k_x^2 + k_y^2 \tag{4.1.11}$$

This can be seen by substituting $(\eta, u, v) = (\eta', u', v') \exp i(k_x x + k_y y - \omega t)$ in Eq. (4.1.10) and demanding nontrivial solutions for primed quantities, which leads to the requirement that the determinant of the coefficients of these primed quantities vanish. This in turn leads to the above dispersion relationship. The wave motions satisfying Eq. (4.1.11) are inertia-gravity waves, waves whose frequency is so low that they are affected by planetary rotation. Rotation has made the normally nondispersive shallow water gravity waves dispersive:

$$C_p = C\left[1 + (ka)^{-2}\right]^{1/2} ; C_g = C\left[1 + (ka)^{-2}\right]^{-1/2} \tag{4.1.12}$$

Both the phase and energy propagation are in the same direction, although at different velocities. These waves are also called Sverdrup waves. They are often referred to as Poincare waves (although they were actually first discussed by Lord Kelvin), even though Poincare waves refer to the superposition of incident

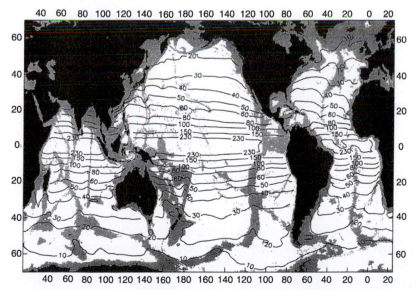

Figure 4.1.2 The distribution of the Rossby radius of deformation of the first baroclinic mode in the global oceans (From Chelton et al. 1998).

and reflected Sverdrup waves at a straight coast. They have a lower bound imposed on their frequency. Their frequency (period) must be greater (smaller) than the inertial frequency (period), and therefore they are also called superinertial waves. Long Poincare waves (ka ~ 0) are essentially inertial oscillations, while waves with wavelength shorter than the Rossby radius travel at the speed of a shallow water wave C, in other words, they do not feel the effect of rotation. Fluid particles under the action of these waves trace an elliptic path in the clockwise (counterclockwise) direction in the northern (southern) hemisphere, with the ratio of major to minor axis of ω/f. Like regular gravity waves, they are specularly reflected at a boundary (angle of incidence equal to angle of reflection), but unlike gravity waves, with a phase shift. A nonzero β is not essential for the existence of these waves; they can exist on a f-plane.

4.1.2 KELVIN WAVES

Kelvin waves are the simplest possible solution to Eq. (4.1.10). However, they require a lateral boundary, since the prime requirement and the reason for simplification is the requirement that the velocity normal to the boundary be zero. Lord Kelvin was the first to point this out in 1879. Aligning the x-axis along the boundary and y-axis perpendicular to it, we can put v = 0, and by cross-differentiating the continuity and u-momentum equations, we get

$$\frac{\partial^2}{\partial t^2}(\eta, u) = C^2 \frac{\partial^2}{\partial x^2}(\eta, u)$$

$$fu = -g \frac{\partial \eta}{\partial y}$$

(4.1.13)

The first equation of this set is the classic wave equation for linear, non-dispersive waves and admits solutions of the form $(\eta, u) \sim F(x \pm Ct)$. It is easy to verify that the solution is

$$\eta = -D \exp(-y/a) F(x - Ct)$$

$$u = C \exp(-y/a) F(x - Ct); \quad v = 0$$

(4.1.14)

Note that only the waves traveling in the positive x direction are admissible. Those that travel in the negative x direction lead to solutions that increase exponentially with increase in y, that is, away from the boundary, and hence are not physically permissible. This just means that for positive (negative) f, meaning in the northern (southern) hemisphere, only Kelvin waves propagating with the coast to the right (left) can exist. Their influence is felt to a distance of the order of the Rossby radius away from the boundary. Therefore, Kelvin waves are trapped by the boundary and, needless to say, a lateral boundary such as a coast is necessary for their existence. Note also that Kelvin waves are nondispersive:

$$C_p = C_g = C \qquad (4.1.15)$$

Both the phase and group speeds are equal to the shallow water wave speed. A remarkable feature of Kelvin waves is that their speed is the same as that of a shallow water wave; in other words, even though they depend on rotation for their existence, their phase speed is independent of the exact value of rotation! The only way rotation is felt is in the extent of their influence across the basin. The stronger the rotation, the more tightly they are trapped by the basin boundary. In the limit of zero rotation, the Rossby radius becomes infinite and the Kelvin wave reduces to an ordinary shallow water gravity wave traveling parallel to the coast. Kelvin waves do not have to be periodic. They can be in the form of a pulse. They are generated by abrupt changes in wind forcing—sudden relaxation or strengthening of winds. If a disturbance such as a wind burst generates a Kelvin wave pulse, it tends to travel essentially intact around the basin.

Because they are unaffected by rotation, the energy of Kelvin waves is also equally partitioned between kinetic and potential forms. The total energy density

per unit length $\quad E = 2\int_0^\infty \frac{1}{2}\rho_0 g \overline{\eta^2} dx = 2\int_0^\infty \frac{1}{2}\rho_0 g \frac{1}{2}\left[\eta_o \exp(-x/a)\right]^2 dx$, and the

energy flux along the coast are

$$E = \frac{1}{4}\rho_0 g \eta_0^2 a; \quad F_e = E C_g = \frac{1}{4}\rho_0 g \eta_0^2 a C = \frac{1}{4f}\rho_0 g^2 \eta_0^2 D_e \qquad (4.1.16)$$

where η_0 is the wave amplitude right at the coast, where it is the maximum. Since a decreases with latitude, to maintain the same energy flux, the amplitude of a Kelvin wave must increase as it propagates poleward.

Since the average ocean depth is about 4 km, the phase speed of a barotropic Kelvin wave is ~200 m s^{-1}; such a wave can travel ~9000 km in half a day. In shallow seas such as the North Sea, with an average depth of 50 m, the propagation is slower; there, a Kelvin wave created by a wind burst off England can reach Norway in about a day. However, the most important waves as far as oceanic adjustment to external forcing is concerned are the baroclinic Kelvin waves. The first baroclinic mode has a typical speed of 2.5 m s^{-1} and a Rossby radius of deformation of a few tens of kilometers (depending on the latitude; the typical midlatitude value is 40 km) and takes weeks to travel substantial distances in ocean basins. Nevertheless, they are much faster than midlatitude Rossby waves, which take months to years to cross an ocean basin. Note also that Kelvin waves can exist on an f-plane, whereas latitudinal variation of f is essential for the existence of Rossby waves. Their energy, however, gets more (less) concentrated as they propagate poleward (equatorward), since for the same gravity wave speed, the Rossby radius decreases with increase in latitude.

Propagation of tides often takes the form of barotropic Kelvin waves propagating along the right coast. The interference pattern of Kelvin waves propagating in opposite directions in a channel gives rise to amphidromic points at which the sea level displacement is zero and around which the wave crests and troughs rotate in the cyclonic, counterclockwise (clockwise) direction in the northern (southern) hemisphere.

4.1.3 PLANETARY ROSSBY WAVES

By far the most important to long term midlatitude adjustment processes are Rossby or planetary waves that are generated on a beta plane $f = f_0 + \beta y$, typical midlatitude values for f_0 and β being 8×10^{-5} s^{-1} and 2×10^{-11} m^{-1} s^{-1}. Unlike Kelvin and Poincare waves, which travel so fast that they are relatively unaffected by slow basin scale oceanic circulation, Rossby waves travel so slowly that the background current might exert an important influence; therefore, it is instructive to retain a background geostrophic zonal current $U = -(g/f_0)(\partial H_e / \partial y)$ in the following derivation (Kessler, 1990). This leads to a set of equations in a reference frame moving with velocity U, which are a small modification to the governing equations shown in Eq. (4.1.10):

$$\frac{\partial \eta}{\partial t} + H_e \left(\frac{\partial u}{\partial x} + \frac{\partial v}{\partial y} \right) + v \frac{\partial H_e}{\partial y} = 0$$

$$\frac{\partial u}{\partial t} - fv = -g \frac{\partial \eta}{\partial x} \qquad (4.1.17)$$

$$\frac{\partial v}{\partial t} + fu = -g \frac{\partial \eta}{\partial y}$$

These equations can be reduced to a single equation for η:

$$\left(1 - a^2 \nabla^2\right) \frac{\partial \eta}{\partial t} - \left(\beta a^2 + U\right) \frac{\partial \eta}{\partial x} = 0; \quad \nabla^2 \equiv \frac{\partial^2}{\partial x^2} + \frac{\partial^2}{\partial y^2} \qquad (4.1.18)$$

Assuming wave motions of the form $\exp i(k_x x + k_y y - \omega t)$, we get a dispersion relation in the moving frame of reference:

$$\omega_m = -\frac{\left(\beta a^2 + U\right)k_x}{\left(1+k^2a^2\right)} \tag{4.1.19}$$

where the effective beta is now $\beta + U/a^2$. In the stationary frame of reference, we get

$$\omega = \omega_m + Uk = -\frac{\left(\beta a^2 - Uk^2 a^2\right)k_x}{\left(1+k^2a^2\right)}$$

$$C_{px} = -\frac{\left(\beta a^2 + U\right)}{\left(1+k^2a^2\right)} + U$$

$$C_{py} = -\frac{\left(\beta a^2 + U\right)k_x k_y}{k_y^2\left(1+k^2a^2\right)} \tag{4.1.20}$$

$$C_{gx} = \frac{\left(\beta a^2 + U\right)\left[\left(k_x^2 - k_y^2\right)a^2 - 1\right]}{\left(1+k^2a^2\right)^2} + U$$

$$C_{gy} = \frac{2\left(\beta a^2 + U\right)\left(k_x a\right)\left(k_y a\right)}{\left(1+k^2a^2\right)^2}$$

Note that unlike equatorial Rossby waves (that propagate only in the westerly direction in the equatorial waveguide), there is no condition on the wave number or frequency as long as the dispersion relationship is satisfied. The Rossby modes can therefore have a continuous spectrum, unlike the equatorial waveguide, which allows for the existence of only discrete modes, albeit an infinite number of them for each baroclinic mode. Also, the midlatitude Rossby waves can have velocity components in the meridional direction. The speed of the equatorial Rossby waves is much larger as well. In fact, their speed is proportional to but less than the gravity wave speed. For example, the first baroclinic equatorial Rossby wave travels at one third the gravity wave speed C.

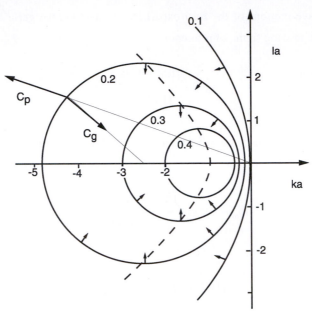

Figure 4.1.3 Dispersion relationship for mid-latitude planetary Rossby waves. Frequency contours are in units of βa. The wavenumber components are normalized by $1/a$. Each circle corresponds to a single frequency, frequency increasing as the radius decreases. The phase velocity is directed away from the origin, whereas the group velocity is directed towards the center of each circle. No Rossby waves exist for frequencies higher than $0.5\beta a$ (adapted from Gill 1982).

Let us first look at the Rossby modes with $U = 0$. In this case

$$\omega = \frac{-\beta a^2 k_x}{1 + k^2 a^2}$$

$$C_{px} = \frac{-\beta}{\left(1 + k^2 a^2\right)}; \quad C_{py} = \frac{-\beta k_x k_y}{k_y^2 \left(1 + k^2 a^2\right)} \qquad (4.1.21)$$

$$C_{gx} = \frac{\beta a^2 \left[\left(k_x^2 - k_y^2\right) a^2 - 1\right]}{\left(1 + k^2 a^2\right)^2}; \quad C_{gy} = \frac{2\beta a^2 \left(k_x a\right)\left(k_y a\right)}{\left(1 + k^2 a^2\right)^2}$$

The dispersion relation is shown in Figure (4.1.3) from Gill (1982). These waves are transverse waves, meaning that the particle velocity in the fluid due to these waves is perpendicular to their phase velocity. Note that the phase velocity in the zonal direction is always negative—that is, westward—irrespective of the value of k_x. However, the group velocity, the propagation velocity of the energy

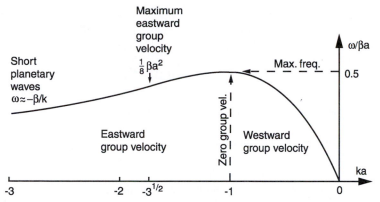

Figure 4.1.4 Rossby wave dispersion relationship; frequency normalized by βa. Note the eastward group velocity for short waves (from Gill, 1982).

associated with these waves, can be positive or negative—that is, eastward or westward. For waves propagating in the zonal direction ($k_y a = 0$), for $k_x a > 1$, the group velocity is eastward; for $k_x a < 1$, the group velocity is westward. Thus, long Rossby waves propagate their energy westward, the short ones eastward (Figure 4.1.4) The property of short waves, namely the possibility of phase and group velocities being in opposite directions, is quite fascinating. Long Rossby waves are nondispersive.

The westward phase propagation of Rossby wave motions can be understood by appealing to vorticity conservation (Gill, 1982; Holton, 1992). Consider a flat bottom ocean of homogeneous fluid on a beta plane. Since H is constant, barotropic Rossby waves conserve absolute vorticity and owe their existence to the meridional gradient of planetary vorticity (baroclinic waves, on the other hand, conserve potential vorticity and exist because of the potential vorticity gradient). Just as gravitational forces act to restore a fluid parcel displaced vertically in a stably stratified fluid leading to vertical oscillations about its rest position and internal wave motions, the meridional gradient of absolute vorticity provides the restoring force for particles displaced in the meridional direction, leading to westward propagating Rossby wave motions. Consider a string of fluid parcels originally at rest in the zonal direction along a latitude circle. If a sinusoidal displacement in the meridional direction is imposed upon this string, because of conservation of total vorticity ($f+\zeta$), a northward (southward) displacement would produce a negative, anticyclonic (positive, cyclonic) vorticity in the fluid. The meridional velocities induced (Figure 4.1.5) advect the particles southward west of the southern tip of the sinusoidal displacement and northward west of the northern tip. This displaces the wave westward. The fluid particles oscillate back and forth, perpendicular to the direction of phase

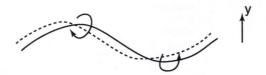

Figure 4.1.5 Westward propagation of disturbances on a beta plane.

propagation (that is, in the meridional direction), a property typical of wave motions in the interior of an incompressible fluid.

For a typical midlatitude synoptic-scale disturbance in the atmosphere, the zonal wavelengths are typically ~6000 km, and the corresponding westward nondispersive barotropic Rossby wave speed is 8 m s^{-1}. However, these disturbances invariably travel eastward, since they are advected by westerly zonal winds of much larger magnitude. Such free modes are not very important compared to the topographically trapped forced modes excited by mountain chains such as the Himalayas and Rockies as the atmosphere traverses the mountain range (Holton, 1992).

The ratio of the kinetic energy density to the potential energy density of a midlatitude Rossby wave is given by

$$R = \frac{KE}{PE} = \frac{\rho g^2 k^2 \eta_0^2 H_e / 4f^2}{\rho g \eta_0^2 / 4} = (ka)^2 \qquad (4.1.22)$$

This means that short Rossby waves have most of their energy in their motions, while the long waves that are more important generally to midlatitude adjustment processes have most of their energy in the form of potential energy. The total energy density multiplied by the group velocity is the energy flux. Also, Rossby waves are not reflected specularly at a straight coast, but it is the group velocity that gets reflected specularly.

The interesting limit of long Rossby waves can be obtained from Eq. (4.1.18) simply by neglecting the Laplacian term, which is equivalent to ignoring the tendency terms in the momentum equations, but not in the continuity equation:

$$\frac{\partial \eta}{\partial t} - \left(\beta a^2\right)\frac{\partial \eta}{\partial x} = 0 \qquad (4.1.23)$$

This is nothing but an equation for a linear nondispersive wave propagating westward such that

$$C_p = C_g = -\beta a^2 \qquad (4.1.24)$$

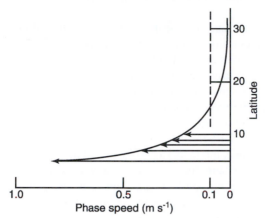

Figure 4.1.6 Long mid-latitude first baroclinic Rossby wave speed as a function of the latitude. Also shown are the first five equatorial Rossby wave modes and their turning latitudes (from Gill,1982).

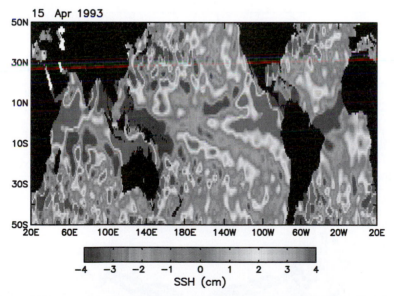

Figure 4.1.7 SSH from altimetric observation showing the characteristic speedup of the westward propagating Rossby wave with decreasing latitude (figure provided by D. Chelton).

These long Rossby waves are perfectly geostrophic but divergent. Their phase speed is one of the most important parameters in midlatitude ocean response problems.

Semiannual, annual, and multiyear Rossby waves traveling across the ocean basins are ubiquitous in altimetric measurements of sea surface height. Barotropic Rossby waves are relatively fast, able to traverse ocean basins in the matter of a week or so. But baroclinic Rossby waves are relatively slow and take years to traverse the basins, although in the equatorial waveguide, they travel fast enough to traverse the basin in a matter of months. Since the Rossby radius a increases dramatically as the equator is approached, their speed increases as well. For example, the long Rossby wave speed (both phase and group) is a mere 1.5 cm s^{-1} at 30° latitude, but increases to 22 cm s^{-1} at 10°. Figure 4.1.6 from Gill (1982) shows this dramatic speedup. Also shown are the first five equatorial Rossby modes and their turning latitudes. This dramatic increase in propagation speed with decreasing latitude is evident in the SSH signature of a baroclinic Rossby wave propagating westward in a basin in both altimetric observations (Figure 4.1.7) as well as numerical model results.

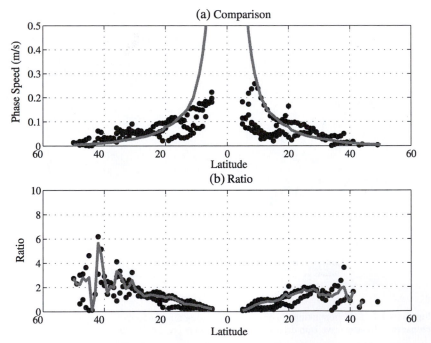

Figure 4.1.8   Linear long Rossby wave speed compared to the speed estimated from altimetric observations (from Fox 1997, see also Chelton and Schlax 1996).

It is important to note that shorter Rossby waves are indeed affected by the background current. Short waves $\left(k_x^2 - k_y^2\right)a^2 > 1$ with eastward group velocity have their Doppler shift enhanced, while the westward propagating waves $\left(k_x^2 - k_y^2\right)a^2 < 1$ have their Doppler shift reduced, with no Doppler shift in the limit of very long, westward propagating nondispersive waves.

Curiously enough, the effective beta appears to be much higher, and altimetry indicates (Chelton and Schlax, 1996) that these waves travel much faster than indicated by the relationship discussed earlier for linear waves in a stationary medium (Figure 4.1.8). As can be seen from the limit of ka = 0 in (4.1.20), this speed is unaffected by any background motion either. In the long wave limit, the Doppler shift represented by the final U term in the expression for the westward phase and group velocities is exactly balanced by the change due to the pycnocline slope! This has been called the non-Doppler effect (see, e.g., Kessler, 1990). However, since the basic shallow water wave speed depends on the local equivalent depth, the meridional changes in the pycnocline depth due to the background zonal current do affect the Rossby radius and hence the Rossby wave speed at any latitude, compared to the case with uniform equivalent depth or, equivalently, the resting ocean.

Killworth and Edwards (1997) have attempted to explain the higher effective beta of Rossby waves in midlatitude oceans. Charney and Flierl (1981) also show that solitary waves travel at a speed higher than the linear wave speed. The Korteweg-De Vries equation describes a class of finite amplitude solutions, including cnoidal and solitary waves (e.g., see the review by Miura, 1976; see also Kantha and Clayson, 1999). The latter result from an exact balance between dispersion and nonlinear steepening so that the wave maintains its shape as it propagates. The propagation speed depends on the wave amplitude. Charney and Flierl (1981) provide solutions for solitary Rossby waves both when $a < L < L_I$ and when $L > L_I$. In the latter case, the solution is in terms of Airy functions Ai:

$$\eta = \text{Ai}(Y)F(x - ct)$$

$$\frac{L^2}{a^2}\frac{\partial F}{\partial t} = \frac{\partial^3 F}{\partial x^3} + \frac{U}{fL}\frac{L^4}{a^4}\left[\int_{Y_0}^{\infty}\text{Ai}^3 \Big/ \int_{Y_0}^{\infty}\text{Ai}^2\right]\frac{\partial}{\partial x}F^2$$

$$Y = Y_0 + \left(2\frac{\beta L}{f}\frac{L}{a}\right)^{1/3}\left(y + \frac{\pi}{2}\right); Y_0 = -2.3381 \tag{4.1.25}$$

$$C_p = -\beta a^2\left[1 + \pi\frac{\beta L}{f}\frac{L}{a} + \left(2\frac{\beta L}{f}\frac{L}{a}\right)^{1/3}Y_0\right]$$

While analytical solutions such as Eq. (4.1.25) are useful, it is desirable to seek solutions to the fully nonlinear reduced gravity equations by numerical means. In the real world, nonlinear motions prevail in the oceans and the atmosphere, and this is one reason for the utility of numerical models.

4.1.4 TOPOGRAPHIC ROSSBY WAVES

Rossby waves owe their existence to changes in planetary vorticity with latitude. A similar effect can also be produced by changes in the depth of the water column in a rotating fluid. These are called topographic Rossby waves, and they can be studied by considering the more general case of a two-layer QG flow (Veronis, 1981). Note that QG approximation filters out the inertia-gravity waves from the equations (see Section 4.5), and solutions can thus be obtained only for the Rossby modes. Consider the linearized, potential vorticity conservation equation for each layer:

$$\frac{\partial}{\partial t}\zeta_k + \beta v_k - \frac{f}{H_k}\left(\frac{\partial}{\partial t} + u_k\frac{\partial}{\partial x} + v_k\frac{\partial}{\partial y}\right)h_k = 0$$

$$\zeta_k = \frac{\partial v_k}{\partial x} - \frac{\partial u_k}{\partial y} \qquad\qquad k=1,2$$

(4.1.26)

For linear waves,

$$\left(u_1\frac{\partial}{\partial x} + v_1\frac{\partial}{\partial y}\right)h_1 = 0$$

$$\left(u_2\frac{\partial}{\partial x} + v_2\frac{\partial}{\partial y}\right)(h_2 + \eta_2) = 0$$

(4.1.27)

Also
$$\frac{\partial}{\partial t}h_1 = \frac{\partial}{\partial t}(\eta_0 - \eta_1); \quad \frac{\partial}{\partial t}h_2 = \frac{\partial}{\partial t}\eta_1$$

(4.1.28)

Using Eq. (4.1.27) and Eq. (4.1.28) in Eq. (4.1.26) and making use of $\eta_B \ll H_2$,

$$\frac{\partial}{\partial t}(\eta_0 - \eta_1) - a_e^2\left(\frac{\partial^2}{\partial x^2} + \frac{\partial^2}{\partial y^2}\right)\frac{\partial}{\partial t}\eta_0 - \beta a_e^2\frac{\partial}{\partial x}\eta_0 = 0$$

$$\frac{\partial}{\partial t}\eta_1 - a^2\left(\frac{\partial^2}{\partial x^2} + \frac{\partial^2}{\partial y^2}\right)\frac{\partial}{\partial t}\hat{\eta} - \beta a^2\frac{\partial}{\partial x}\hat{\eta}$$

$$+\frac{g}{f}\left(\frac{\partial \eta_B}{\partial y} - \frac{\partial \hat{\eta}}{\partial x}\right) = 0$$

$$\hat{\eta} = \left(1 - \frac{\Delta\rho}{\rho_0}\right)\eta_0 + \frac{\Delta\rho}{\rho_0}\eta_1$$

(4.1.29)

where a_e is the barotropic and a the baroclinic Rossby radii of deformation. Substituting $(\eta_0, \eta_1) \sim \exp i(k_x x + k_y y - \omega t)$ in Eq. (4.1.29) leads to the dispersion relation

$$\left[1 + k^2 a_e^2 \left(1 + k^2 a^2 \frac{H_2}{H}\right)\right]\omega^2$$

$$+ \left\{\beta a_e^2 \left(1 + 2k^2 a^2 \frac{H_2}{H}\right) + \beta_t a_e^2 \left(1 + k^2 a^2\right)\frac{H_2}{H}\right\} k_x \omega$$

$$+ k_x^2 a_e^2 a^2 \frac{H_2}{H} \beta(\beta + \beta_t) = 0;$$

(4.1.30)

$$H = H_1 + H_2; \quad a_e^2 = \frac{gH}{f^2}; \quad a^2 = \frac{g'H_1}{f^2}$$

$$\beta_t = \frac{f}{H_2 k_x}\left(k_x \frac{\partial \eta_B}{\partial y} - k_y \frac{\partial \eta_B}{\partial x}\right)$$

Note the appearance of topographic beta in the dispersion relation. If the depth of the fluid column changes only in the meridional direction, it just modifies β. If the depth varies in the zonal direction also, it defines an effective north. For long waves ($ka \to 0$), one solution to the dispersion relation becomes

$$\omega_e = -\frac{\beta a_e^2 k_x \left(1 + \frac{\beta_t}{\beta}\frac{H_2}{H}\right)}{\left(1 + k^2 a_e^2\right)}$$

(4.1.31)

This is just the barotropic Rossby wave influenced by topography. If β_t is large, the second term in the numerator dominates and we have topographic Rossby waves. Generally, if depth increases poleward, β_t is positive and the topographic beta effect reinforces the regular beta. If it is negative, it overrides and can even cause eastward propagation. The second solution is the baroclinic mode and is given by

$$\omega = -\beta a^2 k_x \frac{\left(\dfrac{H_2}{H} + \dfrac{\beta_t}{\beta}\right)}{\left(1 + \dfrac{\beta_t}{\beta}\right)} \tag{4.1.32}$$

In the limit of both strong ($\beta_t > \beta$) and weak ($\beta_t < \beta$) topographic gradients, this relation yields the conventional long wavelength nondispersive Rossby waves, the only difference being the factor H_2/H. For weak gradients, the equivalent ocean depth is $(\Delta\rho/\rho_0)H_1(H_2/H)$, and for strong gradients it is $(\Delta\rho/\rho_0)H_1$.

For short waves ($ka \gg 1$), the dispersion relation becomes

$$k^4\omega^2 + (2\beta + \beta_t)k^2 k_x \omega + k_x^2 \beta(\beta + \beta_t) = 0 \tag{4.1.33}$$

yielding a nondivergent baroclinic Rossby wave confined to the upper layer, $\omega = -\beta k_x / k^2$, and a barotropic mode, $\omega_e = -(\beta k_x / k^2)(1 + \beta_t / \beta)$. For strong topographic gradients, the barotropic mode is confined to the bottom layer and hence the mode is bottom trapped.

4.2 CONTINUOUS STRATIFICATION

The other extreme to layered approximation is that of continuous stratification, and this complements the layered analysis quite well. Once again the starting point for these investigations is the QG pseudo-potential vorticity conservation equation for the water column with uniform stratification (pseudo because conservation is only on projection onto a horizontal plane), which can be written as (Charney and Flierl, 1981; Veronis, 1981)

$$\left(\frac{\partial^2}{\partial x^2} + \frac{\partial^2}{\partial y^2} + \frac{1}{S^2}\frac{\partial^2}{\partial z^2}\right)\frac{\partial p}{\partial t} + \beta\frac{\partial p}{\partial x} = 0 \tag{4.2.1}$$

where p is the perturbation pressure and $S = N/f_0$ is the stratification parameter, N being the buoyancy frequency. If the upper boundary conditions are made that of a rigid lid, $w \sim \dfrac{\partial p}{\partial z} = 0$, then the barotropic Rossby radius becomes infinite. Assuming a bottom with a uniform meridional slope α_y, the bottom boundary condition becomes $w = \alpha_y v$ or

$$\frac{1}{S^2}\frac{\partial^2 p}{\partial z \partial t} + H\beta_t \frac{\partial p}{\partial x} = 0 \qquad (4.2.2)$$

where H is the depth and $\beta_t = f_0 \alpha_y / H$ is the topographic beta. Seeking wavelike solutions for p leads to the eigenvalue problem

$$\frac{1}{S^2}\frac{\partial^2 \hat{p}}{\partial z^2} - \left(k^2 + k_x \frac{\beta}{\omega}\right)\hat{p} = 0$$

$$\frac{\partial \hat{p}}{\partial z} = 0 \text{ at } z=0; \quad \frac{\partial \hat{p}}{\partial z} = \frac{H\beta_t S^2 k_x}{\omega} p \text{ at } z = -H \qquad (4.2.3)$$

which provides for an infinite but discrete set of wave modes.

Solutions of the form $\hat{p} \sim \cos(mz/H)$ (not valid for pure topographic waves for which $\beta = 0$) then yield the dispersion relation for each mode m:

$$\omega = -\frac{\beta a^2 k_x}{m^2 + k^2 a^2}; \quad m \tan m = \frac{\beta_t a^2 k_x}{\omega}$$

$$\frac{\beta}{\beta_t} m \tan m = -\left(m^2 + k^2 a^2\right) \qquad (4.2.4)$$

where the Rossby radius is now defined as

$$a = SH = \frac{NH}{f_0} = \frac{(g'H)^{1/2}}{f_0}; g' = g\frac{\rho_{bot} - \rho_{top}}{\rho_0} \qquad (4.2.5)$$

These waves always travel to the west no matter what the value of bottom slope is, and whether topographic beta aids or opposes regular beta. When the two reinforce each other, the response becomes more barotropic; in the other case, it becomes more baroclinic (Veronis, 1981).

Solutions of the form $\hat{p} \sim \cosh(nz/H)$ (not valid for pure Rossby waves for which $\beta_t = 0$) yield the dispersion relation for each mode n:

$$\omega = -\frac{\beta a^2 k_x}{k^2 a^2 - n^2}; \quad n \tanh n = -\frac{\beta_t a^2 k_x}{\omega}$$

$$\frac{\beta}{\beta_t} n \tanh n = k^2 a^2 - n^2 \qquad (4.2.6)$$

Pure topographic Rossby waves therefore have the dispersion relation

$$\omega = -\frac{\beta_t a^2 k_x}{ka \tanh(ka)} \qquad (4.2.7)$$

For $ka \ll 1$ (weak stratification or long waves), $\omega = -\beta_t k_x / k^2$. These are topographic barotropic Rossby waves. For $ka \gg 1$ (strong stratification or short waves), $\omega = -\alpha N k_x / k$. These motions are bottom trapped or bottom intensified; the current amplitudes increase toward the bottom. There is no phase difference at different depths in the water column; the motion is columnar, although of different intensities at different depths. The frequency of these topographic Rossby waves is proportional to the meridional bottom slope and must be lower than $N\alpha$. If the bottom slopes only in the meridional direction, then these waves travel parallel to isobaths, with the shallow water to the right. The motion is nondivergent and the velocity components are either in or out of phase.

Topographic Rossby waves are ubiquitous along the continental slopes of ocean basins (Hamilton, 1990). However, it is necessary to consider a more general stratification $N^2(z)$ to interpret these observations. Charney and Flierl (1981) provide solutions in the case of uniform slope in the zonal and meridional directions for the barotropic and first two baroclinic modes. The results are qualitatively similar to the uniform stratification case. When topographic gradients are weak, they act as effective beta, due to vortex stretching and shrinking; the effect is weaker for the baroclinic modes than barotropic mode as expected. For the stratification they used, the topographic effect for the first baroclinic mode is 40% of the barotropic mode. When topographic beta opposes the regular beta, an eastward traveling, bottom-trapped wave exists with a trapping scale and phase speed of

$$H_t = \frac{H}{\beta k_x}\left(\beta_{ty} k_x - \beta_{tx} k_y\right); C_p = \frac{\beta H_t^2 N_{z=-H}^2}{f^2} \qquad (4.2.8)$$

For strong topographic gradients, the modes are surface trapped and have large vertical shear. When topographic beta opposes the regular beta, an eastward traveling, surface-trapped wave exists with a trapping scale and phase speed in the limit of $L < a$ of

$$H_t = \frac{f}{kN_{z=-H}}; C_p = \frac{k_y \beta_{tx} - k_x \beta_{ty}}{H_t k_x k^2} \qquad (4.2.9)$$

These results are more accurate compared to either the two-layer or constant stratification cases discussed earlier. For example, the trapping scale is much larger for constant stratification since bottom stratification is weaker than the average value and consequently the phase speeds much larger (Charney and Flierl, 1981).

4.3 GEOSTROPHIC ADJUSTMENT AND INSTABILITIES

Geophysical flows are dominated by rotation effects. This has profound influence on how the medium adjusts when perturbed. In the absence of rotation, when a stably stratified fluid at rest is disturbed, internal waves are radiated out but the resulting motions are eventually dissipated by friction to return the medium to a state of rest. In the presence of rotation, the final state is not one of rest but that of geostrophic equilibrium. The adjustment occurs in a timescale of the order of rotation timescale (the inertial period) and the final state contains more potential energy than that of a state of rest, yet rotation does not allow this minimum energy state to be attained. This final state may itself be unstable, and conversion of this available potential energy (APE) into flow kinetic energy might occur under certain conditions. The question of geostrophic adjustment and the instability of geostrophic/quasigeostrophic flows is the subject of this section.

4.3.1 GEOSTROPHIC ADJUSTMENT

Rossby (1938) was one of the first to consider the adjustment problem in a rotating fluid. This adjustment can be illustrated by considering geostrophic adjustment on a f-plane. Imagine a fluid initially at rest in the domain ($-\infty < x < \infty$, $-\infty < y < \infty$) undergoing a small perturbation in surface elevation imposed at time t=0 of the form $\eta = -\eta_0 \, \text{sgn}(x)$. The problem posed is homogeneous in the y-direction and so the governing shallow water equations become

$$\frac{\partial \eta}{\partial t} + \frac{\partial}{\partial x}(uh) = 0$$

$$\frac{\partial u}{\partial t} + u\frac{\partial u}{\partial x} - fv = -g\frac{\partial \eta}{\partial x}$$

$$\frac{\partial v}{\partial t} + u\frac{\partial v}{\partial x} + fu = 0$$

(4.3.1)

The problem can also be expressed in terms of conservation of potential vorticity

$$\zeta_p = \frac{f + \partial v / \partial x}{h} = \frac{f}{h_0} \qquad (4.3.2)$$

where h_0 is the initial layer depth. The final steady state is not one of rest but one of geostrophic balance with the elevation antisymmetric w.r.t $x = 0$ and velocity at right angles to the elevation gradient:

$$\eta = \eta_0 \left[-1 + \exp(-x/a) \right] \quad x > 0$$

$$= \eta_0 \left[1 - \exp(x/a) \right] \qquad x < 0 \qquad (4.3.3)$$

$$u = 0; \ v = -\frac{g\eta_0}{fa} \exp(-|x|/a)$$

This final steady state is shown in Figure 4.3.1 and can also be derived by considering the potential vorticity conservation equation, Eq. (4.3.2), alone. Rotation confines the deformation of the sea surface height to the vicinity of $x = 0$ to a distance on the order of the Rossby radius of deformation. The peculiar property of rotation that imposes a nonhomogeneous final steady state is responsible for the existence of "discontinuities" such as fronts in the oceans (and the atmosphere). Such fronts cannot exist in the absence of rotation, which can be seen by putting a, the Rossby radius, to infinity in Eq. (4.3.3). Fronts with motions predominantly along and confined to the distance on the order of the radius of deformation are ubiquitous in both the oceans and the atmosphere.

The change in potential energy and the kinetic energy per unit length from the initial perturbed state of elevated sea surface, but no motion, to the final geostrophic steady state are 3/2 and 1/2 times ($\rho_0 g a \eta_0^2$). Only one third of the potential energy is released and converted to KE (unlike the case without rotation where all the PE is converted to KE). The rest is radiated away by Poincare waves, which of course get dissipated during propagation. Interestingly enough, the final steady state is also one of minimum total energy or equivalently the maximum possible energy loss even for the most general case of multilayer arbitrary potential vorticity distribution as long as we are dealing with uniformity in one direction (in this case y-direction) (see Cushman-Roisin, 1994).

The transient solutions involve more algebra and will not be presented (see Gill, 1982, instead). The fluid reaches this state by radiation of Poincare waves. The first waves to arrive at any point in the domain are the short gravity waves traveling at the speed of the shallow water wave unaffected by rotation. In the wake of the front they constitute, arrive Poincare waves that eventually produce a geostrophic steady state. The adjustment timescale within a distance of the

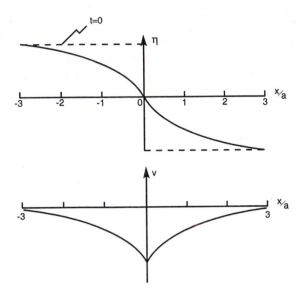

Figure 4.3.1 Geostrophic adjustment on an f - plane: the asymptotic state is one of geostrophic balance with elevation and geostrophic current distributions as shown. The initial SSH perturbation is also shown.

Rossby radius of deformation of the origin is itself on the order of the rotation timescale, or the inertial period. However, the adjustment does not take place monotonically, but overshooting occurs with resulting damped oscillatory return to steady state, with oscillations occurring at the inertial frequency. These oscillations are produced by long waves whose frequency is inertial but whose group velocity is not exactly zero, so that the energy does disperse away leading to a steady state eventually (Gill, 1982).

We considered a highly idealized linearized barotropic problem to illustrate the basic principles of adjustment. Adjustment in a finite basin involves both Poincare and Kelvin waves, barotropic and baroclinic. Adjustment on a beta plane is even more complex and the approach to the final state necessarily involves the generation and radiation of Rossby waves and considerations of the ocean basin as a whole, but the principles are similar and we will not deal with these problems here. They are best dealt with using numerical methods.

Note that if we replace g by the appropriate reduced gravity g', we get the corresponding equations for a baroclinic mode. The baroclinic adjustment processes are therefore similar except that the process is slower, and the Rossby radius and hence the extent of deformation are much smaller.

4.3.2 INSTABILITIES

Another aspect of importance in midlatitude dynamics is the instability of the western boundary currents such as the Gulf Stream, once they depart the coast and begin their predominantly zonal phase. Freed from the constraints of topography of the basin margins, these zonal currents are free to meander, undergo instabilities, and shed off eddies. These instabilities bear some resemblance to more extensively studied instabilities of zonal jets in the atmosphere, where the generation and movement of weather patterns in midlatitudes depend very much on such instabilities. Considerable effort has been devoted by atmospheric scientists, starting with Charney (1947) and Eady (1949), to a study of baroclinic instabilities of an atmospheric zonal jet on a midlatitude beta plane. These are also relevant to ocean currents, and we will summarize the findings briefly for completeness. However, the literature is vast and the topic quite difficult to summarize in a short review.

There are two principal types of instabilities that are important to large-scale geophysical flows: barotropic and baroclinic. These are associated with intrinsic instability of jetlike motions and are therefore dynamical instabilities, as opposed to static instabilities that result from density differences brought on in a gravitational field by heating/cooling. Consequently, they require appealing to dynamical equations to sort out aspects such as the condition for the onset and the growth rate of instabilities. The basic flow that undergoes instability has both horizontal and vertical shear. Instabilities essentially extract energy from this mean flow. Barotropic instability is associated with the horizontal shear of the mean flow, and the instabilities feed on the KE of the mean flow. Baroclinic instability, on the other hand, is associated with the vertical shear of the mean flow, and the instabilities grow at the expense of the potential energy of the mean flow instead. They often also effect conversion of available potential energy (APE) into KE of the mean flow as well. This is simply due to the fact that the existence of vertical shear requires the existence of horizontal density gradients by virtue of thermal wind balance requirements and hence sloping of the isodensity surfaces relative to the horizontal. The basic flow then has APE that can be converted into KE, since the basic state may not be one of lowest energy state. In other words, the flow might be unstable.

Investigating barotropic and baroclinic instabilities requires considerations of growth of perturbations to the basic flow, the realm of a venerable field of fluid mechanics with history dating back to the times of Lord Kelvin and Lord Rayleigh (see, e.g., Swinney and Gollub, 1981). The principal complication compared to the classic stability analysis in fluid mechanics is the presence of gravitational potential and rotation. Linear analysis is the simplest since then the perturbations to a simple mean flow state may be considered small and the governing equations for the perturbed quantities made linear for tractability.

Such analysis provides the condition for the onset of instabilities and the initial growth rates, as well as the modes that grow fastest. Linear analysis cannot, however, indicate the final state of the flow that has undergone instabilities. For this, one has to appeal to either the full nonlinear equations and resort to numerical methods or carry out higher order perturbation expansions. Either way, finite amplitude stability analysis is harder and outside the scope of this review. We will concentrate instead on linear analysis of instabilities of a zonal jet on a midlatitude beta plane.

4.3.2.1 Barotropic Instability

Consider the governing inviscid equations for a homogeneous fluid contained between two horizontal flat surfaces. Since w must vanish at both surfaces, it vanishes identically everywhere and drops out of the equations. Because of the absence of density gradients, the horizontal velocities are uniform in the vertical:

$$\frac{\partial u}{\partial x} + \frac{\partial v}{\partial y} = 0$$

$$\frac{du}{dt} - fv = -\frac{\partial p}{\partial x}$$

$$\frac{dv}{dt} + fu = -\frac{\partial p}{\partial y} \tag{4.3.4}$$

$$\frac{d}{dt} \equiv \frac{\partial}{\partial t} + u\frac{\partial}{\partial x} + v\frac{\partial}{\partial y}$$

Now consider the basic state of simple zonal flow $u = U(y)$, $v = 0$. The corresponding pressure is given by geostrophic balance $fU(y) = -\partial P/\partial y$. Now consider perturbations to this mean flow: u', v', and p'. Substituting $u = U + u'$, $v = v'$, $p = P + p'$ in Eq. (4.3.4) and ignoring the small nonlinear terms involving products of perturbation quantities, one gets a linear set of equations for perturbation quantities:

$$\frac{\partial u'}{\partial x} + \frac{\partial v'}{\partial y} = 0$$

$$\frac{\partial u'}{\partial t} + U\frac{\partial u'}{\partial x} + v'\frac{\partial U}{\partial y} - fv' = -\frac{\partial p'}{\partial x} \tag{4.3.5}$$

$$\frac{\partial v'}{\partial t} + U\frac{\partial v'}{\partial x} + fu' = -\frac{\partial p}{\partial y}$$

Defining a perturbation stream function ψ, a single equation can be derived:

$$\left(\frac{\partial}{\partial t} + U\frac{\partial}{\partial x}\right)\nabla^2\psi + \left(\beta - \frac{d^2 U}{dy^2}\right)\frac{\partial\psi}{\partial x} = 0 \tag{4.3.6}$$

Now look for wavelike solutions that can decay or amplify in the zonal direction: $\psi(x, y, t) = \phi(y)\exp ik(x - Ct)$, where C is complex valued. Substituting this in Eq. (4.3.6) we get an equation for ϕ:

$$\frac{d^2\phi}{dy^2} - k^2\phi + \left(\frac{\beta - \left(d^2 U / dy^2\right)}{U - C}\right)\phi = 0 \tag{4.3.7}$$

This equation is a subset of the classic Taylor-Goldstein equation in fluid mechanics and needs to be solved subject to boundary conditions at the lateral boundaries, which for simplicity we assume are closed: $\phi_{y=0} = \phi_{y=L} = 0$. We now investigate conditions under which the phase speed can be complex and its imaginary part positive, corresponding to growing wave modes. Note the appearance of U-C in the denominator. The location y_c at which $U(y_c) = C$ is known as the critical level. Its significance is that around the critical level, a transfer of energy from the mean flow to the wave flow can occur due to the interaction of the wave flow with the mean flow vorticity gradients. This can cause the wave flow to grow at the expense of the mean flow. The flow is then said to be unstable. Equation (4.3.7) along with the two boundary conditions at the lateral zonal boundaries of the domain (the zonal mean flow is unbounded in the zonal direction) constitutes an eigenvalue problem, meaning that nontrivial solutions exist for only specific values of C. Analytical solutions are possible only for some simple U(y) distributions, but following Rayleigh's lead, useful conditions can be derived by looking at integrals over the domain. A necessary condition for instability can be shown to be that $\left(\beta - d^2 U / dy^2\right)$ must vanish somewhere in the domain. An even stronger condition is that in addition to this quantity vanishing at least once in the domain, the quantity $\left(U - U_0\right)\left(\beta - d^2 U / dy^2\right)$, where U_0 is the value of U where this happens, must be positive in a finite portion of the domain (Cushman-Roisin, 1994). Bounds on the growth rate can be shown to be (Pedlosky, 1987)

$$U_{min} - \frac{\beta L^2}{2(\pi^2 + k^2 L^2)} < C_r < U_{max};$$

$$\left(C_r - \frac{U_{min} + U_{max}}{2}\right)^2 + C_i^2 \leq \left(\frac{U_{max} - U_{min}}{2}\right)^2 + \frac{\beta L^2 (U_{max} - U_{min})}{2(\pi^2 + k^2 L^2)} \tag{4.3.8}$$

where k is the zonal wave number and L is the meridional extent. The first inequality states that the wave must travel at a speed less than the maximum flow speed and a critical level must exist (on an f-plane). The second inequality reduces to the famous semicircle theorem by Louis Howard (Howard, 1961) for the f-plane, namely that in the C_r - C_i plane, the vector C must lie within the semicircle in the northeast quadrant with center at $[0.5(U_{min}+U_{max}),0]$ and a radius of $0.5(U_{max}-U_{min})$.

Linear theory does not indicate any limits to the growth of unstable perturbations. In reality, these disturbances do not grow beyond a bound. For horizontally sheared flows, the final state is a mean flow with a weaker shear and vortices embedded in the shear region.

4.3.2.2 Baroclinic Instability

Geophysical flows are never purely barotropic; hence, baroclinicity plays a dominant role in instabilities. These problems were first considered by Eady (1949) for an f-plane and Charney (1947) for a beta plane. Since then, because of the importance of the problem to midlatitude weather patterns, much work has been done in extending the results (Green,1970; Held and Larichev, 1996; Pedlosky 1987, 1996; Stone, 1972). Two-layer flows have also been considered (see Holton, 1992; Pedlosky, 1987). The Eady problem is the simplest to describe. Consider the governing inviscid QG equations for a uniformly stratified fluid contained between two horizontal rigid flat surfaces on an f-plane. QG dynamics are adequate to describe these instabilities. The basic flow whose instability is under question is uniform in the horizontal but vertically sheared, opposite the case considered for barotropic instability of horizontally sheared but vertically uniform flow. The flow velocity is zero at the bottom, but a Galilean transformation to any uniform velocity makes no difference to the results. Conservation of pseudo-potential vorticity ζ_p (Charney and Flierl, 1981) gives (note that beta is zero)

$$\left(\frac{\partial}{\partial t} + v \cdot \nabla\right)\zeta_p = 0; \quad \zeta_p = \nabla^2 \psi + \frac{\partial}{\partial z}\left(\frac{N^2}{f^2}\frac{\partial \psi}{\partial z}\right) + \beta y \tag{4.3.9}$$

where ψ is the stream function. The basic mean flow has $U = U_{MZ}/H$, $V = 0$, $\Psi = -\dfrac{U}{H}yz$; $Q = 0$. Substituting for each variable, a mean flow value and a perturbation, and linearizing the resulting equation for perturbation quantities (by ignoring the nonlinear terms in perturbation quantities), one gets an equation for the perturbation stream function:

$$\left(\frac{\partial}{\partial t} + \frac{Uz}{H}\frac{\partial}{\partial x}\right)\left(\nabla^2\psi + \frac{N^2}{f^2}\frac{\partial^2\psi}{\partial z^2}\right) = 0 \qquad (4.3.10)$$

Seeking wavelike solutions, $\psi = A(z)\exp i(k_x x + k_y y - \omega t)$, we get

$$-i\left(\omega - \frac{Uz}{H}k_x\right)\left[\frac{f^2}{N^2}\frac{d^2A}{dz^2} - k^2A\right] = 0 \qquad (4.3.11)$$

Let $A(z) = B\cosh(nz) + C\sinh(nz)$, where $n = Nk/f$, and using no flow boundary conditions ($w = 0$) at $z = 0$ and H, (4.3.11) becomes

$$\left(\omega - \frac{Uz}{H}k_x\right)\frac{dA}{dz} + \frac{U}{H}k_x A = 0 \qquad (4.3.12)$$

Equation (4.3.12) constitutes an eigenvalue problem that has the solution

$$\left(\frac{C}{U} - \frac{1}{2}\right)^2 = \frac{1}{4} + \frac{1}{(nH)^2} - \frac{\coth(nH)}{nH} \qquad (4.3.13)$$

Complex-valued solutions exist for C for $nH < 2.3994$, one of the two complex-valued roots having a positive imaginary part and hence growing exponentially. Since the Rossby radius of deformation $a = NH/f$, this condition corresponds to $ka < 2.399$, or the wavelength exceeding 2.619 times the deformation radius. All modes satisfying this condition are unstable and grow, but the fastest growing mode corresponds to $k_y = 0$ and $k = k_x = 1.6062/a$, or the wavelength of 3.912 a (Cushman-Roisin, 1994; Gill 1982). The value of a is 1000 km in the atmosphere so that this corresponds to a wavelength of ~ 4000 km. The maximum growth rate is given by 0.3098 (f/N) dU/dz.

A timescale associated with baroclinic instability can be defined as

$$t_{bc} = \frac{N}{f\left|\dfrac{\partial U}{\partial z}\right|} = \frac{Ri^{1/2}}{f} \qquad (4.3.14)$$

where $Ri = N^2 / \left|\dfrac{\partial U}{\partial z}\right|^2$ is the gradient Richardson number, which is a measure of the available potential energy (APE) stored in the mean stratification. APE is the difference in potential energy between the state of the fluid and a reference state, which is the state of minimum potential energy that can be reached through a reversible adiabatic process (Huang, 1998). All density surfaces are level in the reference state. The flow state, however, involves a tilt of the density surfaces and this makes the potential energy of the stratified system available for conversion to kinetic energy via baroclinic instabilities. This timescale is ~2 days in the atmosphere. In the ocean, above the permanent thermocline, t_{bc} ranges from a few days in regions of western boundary current extensions to about a month (Stammer, 1998). A length scale can also be defined based on β and t_{bc}: $L_{bc} = (\beta\, t_{bc})^{-1}$. This length scale is important to the eddy-mediated transfer of scalar properties such as heat on a rotating planet.

The Eady flow has zero potential vorticity for the basic flow. Application to beta plane by Charney (1947) involved a meridional gradient of potential vorticity of the basic flow equal to beta. Surprisingly, the results are remarkably similar qualitatively although there are quantitative differences. The constant in the growth rate is now reduced by 8% to 0.286 but is independent of beta. Therefore the e-folding scale of the instability is still roughly 2 days. However, Charney's case involves no artificial lid at the tropopause and is unbounded at the top. The scales for the fastest-growing mode are set by the value of beta and the basic flow conditions (Gill, 1982):

$$k_x = 0.79\left(\frac{\beta N}{f}\right) \bigg/ \left(\frac{dU}{dz}\right); \quad H_R = \frac{f}{k_x N} = 1.26\frac{f}{\beta}\left(\frac{f}{N^2}\frac{dU}{dz}\right) \qquad (4.3.15)$$

where H_R is the vertical scale of the flow, the so-called Rossby height. The term in the last parentheses denotes the slope of the isentropes. H_R has a value of 10 km, the height of the tropopause; hence, these scales are very similar to those found by Eady. Observed cyclonic and anticyclonic disturbances in the atmosphere have zonal and meridional wavelengths close to that given above (~ 4000 km) and e-folding rates of 2 to 3 days.

The structure of the most unstable mode is given in Figures 4.3.2 and 4.3.3 for both the Eady and Charney problems (from Gill, 1982). These disturbances

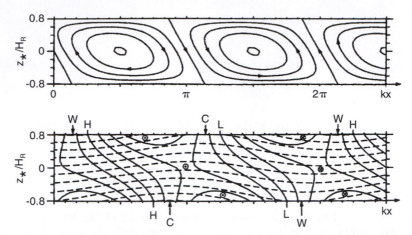

Figure 4.3.2   The Eady problem (f - plane baroclinic instability): the distribution of properties of the most unstable wave in the x-z plane; x is the zonal and z the verical direction. There is no y-dependence; (top) shows the ageostrophic part of the streamfunction, and (bottom) shows contours of normal velocity (solid) and isentropes (dotted). H and L show highs and lows of the geopotential, and W and C show the warmest and coldest points on each boundary. At all levels, flow poleward is warmer than flow southward leading to a net poleward heat flux (from Gill 1982).

transport heat poleward, with warm air flowing poleward at the steering level where the phase speed of the wave is equal to the wind speed. The large cyclones and anticyclones generated by baroclinic instability of the zonal flow are the primary mechanism for poleward heat flux in the atmosphere, whereas it is the western boundary currents that carry the burden in the ocean.

More general necessary conditions for instability for arbitrary potential vorticity distribution are (see Cushman-Roisin, 1994) either 1. The meridional gradient of potential vorticity of the basic flow must change sign somewhere in the domain, or 2. The sign of the meridional gradient of potential vorticity of the basic flow must be opposite to that of the velocity gradient at the top, or 3. This sign must be same as that of the velocity gradient at the bottom. If none of these conditions is satisfied, flow is stable. Any meridional slope of the underlying surface can be either stabilizing or destabilizing, depending on whether it reinforces planetary beta or opposes it.

Application of these instability theories to oceanic flows is both beneficial and problematic. First of all, these theories indicate that the preferred scale of instability is tied to the Rossby radius, which is 40 km in midlatitide oceans. The scale of the unstable wave is ~160 km, consistent with observations. Because of low shearing rates, the growth rates are also small (Gill, 1982). This is also consistent with the meandering and eddy-shedding scales of zonal streams such

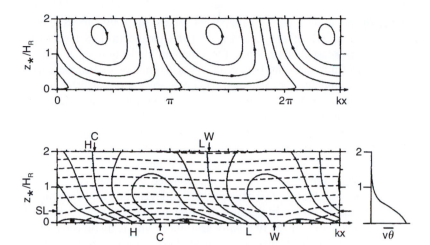

Figure 4.3.3 The Charney problem (β - plane baroclinic instability): the distribution of properties of the most unstable wave in the x-z plane; x is the zonal and z the vertical direction. There is no y-dependence; (top) shows the ageostrophic part of the streamfunction, and (bottom) shows contours of normal velocity (solid) and isentropes (dotted). W and C show the warmest and coldest points on each boundary. At all levels, flow poleward is warmer than flow southward leading to a net poleward heat flux. Poleward heat flux is shown to the right. Note that the flow extends vertically beyond what is shown (from Gill 1982).

as the Gulf Stream and Kuroshio extensions. However, the eddies created transport little heat toward the poles.

The instabilities of weatern boundary current extensions, such as the Gulf Stream and Kuroshio extensions are mixed in that the resulting perturbations have their energy increase at the expense of both the horizontal and vertical shear of the jet, to varying degrees depending on the ratio of the internal Rossby radius to the size of the jet itself (jet halfwidth). Since the unstable wavelengths scale with these length scales, it is obvious that when this ratio is close to one, mixed barotropic and baroclinic instability can prevail. Since the midlatitude radius of deformation is typically 40 km and the jet halfwidth is also typically 40 to 60 km, it is clear that the instabilities tend to be of the mixed type. In the atmosphere, however, instabilities tend to be principally baroclinic and are associated often with strong fronts, generated for example when cold continental air masses collide with warm oceanic air masses. The result is often explosive cyclogenesis, of the type often observed off the east coast of continents during winter.

As can be expected, in the ocean, meanders develop, propagate, and grow, eventually pinching off rings that then propagate away or are reabsorbed quickly or eventually by the stream. The length scale associated with meanders is $L_M \sim (U/2\beta)^{1/2}$, as can be expected from simple dimensional reasoning. The importance of this scale is that, for longer meanders, the flow curvature is small at the crest and trough of the meander, and the meanders then tend to travel westward like a Rossby wave, since the effect of deflection on a beta plane induces a westward translation. For meanders with smaller scales, the curvature effects dominate and the meanders propagate eastward (see Cushman-Roisin, 1994). This scale is typically 150 to 200 km in the ocean, and the meanders are usually smaller than this and therefore tend to propagate eastward.

The principal reason that ocean modelers should be aware of the instabilities of oceanic flows is quite simple. Producing realistic flow instabilities and eddy motions in a numerical model does depend very much on the choice of model parameters. Large horizontal viscosities necessitated by coarse grid resolutions tend to damp out instabilities. For example, the anticyclonic Loop Current eddies shed off by the Loop Current in the Gulf of Mexico appear to be due primarily to the barotropic instability of the first baroclinic mode and realistic eddy shedding depends very much on the proper choice of the horizontal viscosity parameter in the model. On the other hand, in eddy-resolving ocean models, it is possible to choose model parameters that give explosive and hence unrealistic growth of baroclinic instabilities. Some layered models of the Gulf Stream simulate far too large a growth rate for Gulf Stream meanders. It is therefore essential to understand the implications of barotropic and baroclinic instabilities for successful modeling of mesoscale variability in the oceans. A more thorough discussion of these aspects can be found in Gill (1982), Pedlosky (1987), Holton (1992), and Cushman-Roisin (1994).

4.4 SPINUP

Another aspect of importance in basin scale and global ocean modeling is the question of spinup. It is important to assure that the modeled ocean has reached a state of equilibrium under the applied forcing before analyzing the model results. Simple as this may seem, it is often difficult to assure that this condition is met, especially for some basin-scale and global OGCMs. These models are so resource-intensive that it is not always possible to achieve a true statistical equilibrium, especially for climatic applications that may require the model to be run for hundreds of years.

It is in principle possible to initialize an ocean model with homogeneous water masses and run it under, say, climatological surface forcing until the oceanic state attains some resemblance to the true state involving horizontal and vertical gradients in density. But the timescale for the meridional thermohaline circulation is literally hundreds of years and therefore it would require a prohibitive amount of computing resources for such a spinup, even if it could be assured that the model physics is adequate to accurately represent the processes involved. Besides, unless one is interested in oceanic states other than the present, as for example in paleo-ocean studies, it is unnecessary to do so. Instead, the model is usually initialized with the present observed state of the oceanic water mass structure and integrated forward until the circulation is consistent with the prescribed water mass structure. This is the primary reason for the popularity of climatological databases such as Syd Levitus's seasonal hydrography of the world's oceans (Levitus, 1982; Levitus and Boyer, 1994; Levitus *et al.,* 1994). Levitus hydrography is extensively used by ocean modelers to initialize their model oceans. It is the implicit assumption, in most cases, that since the deep oceans change very slowly and only on timescales of millenium or so, the oceanic state derived by averaging over the past several decades of observations is adequate to represent the current state. This would be true if the observations were reasonably complete in terms of spatial coverage. However, the sparsity of the deep oceanic data and the resulting inadequacy of spatial coverage make the inferred oceanic state imperfect at best.

The question of importance to ocean modelers is simply this: How long does the model ocean initialized with observed hydrography, started from rest, and driven by repeating (or constant) surface forcing take to attain an equilibrium state, where the circulation is in dynamical balance with the prescribed deep density field? Similarly, once the ocean is "spun up," how long does it take for the model ocean to adjust to changes in forcing conditions? Answers to these questions can of course be derived empirically. Often, the practice is to integrate the model and monitor quantities such as the total KE and PE to see when they attain an equilibrium state. However, some idea of the spinup timescales can be obtained by appealing to simple dynamical considerations involving well-known slow adjustment processes. It is clear that the deep ocean will require realistic intermediate and deep water formations and hundreds of years to adjust to changes in surface forcing (adjustment of the upper ocean above the main thermocline itself is quicker, of the order of 50 years or so). This is the domain of climate modelers and, of necessity, they employ highly simplified ocean models that can be integrated for such long periods. Obtaining realistic deep water formation in simplified OGCMs is a big challenge since these processes occurring in subpoar seas require realistic simulation of deep convection processes, which in turn require high horizontal resolutions and more comprehensive physics that make the model prohibitively expensive.

For investigations of oceanic variability on shorter timescales such as interannual changes such as due to ENSO phenomena, it is adequate to assure that the upper ocean (say above the main thermocline) is in reasonable equilibrium. In the equatorial waveguide, this means assuring that the Kelvin and Rossby waves produced by the surface forcing have had a chance to propagate across the model basin and bring about changes in the currents that in turn effect changes in water masses. The first baroclinic Rossby wave mode travels at roughly 0.8 m s^{-1} on the average in the Pacific, the widest ocean basin (roughly 18,000 km wide) and hence takes 9 months to cross it from one end to the other. The corresponding Kelvin wave takes 3 months to do so. However, spinup involves higher baroclinic modes that travel slower (but are damped somewhat faster), and therefore the spinup time needed for the model ocean to reach an equilibrium state is a few years. This time is much smaller for narrower basins such as the Atlantic and the Indian. The adjustment timescale to changes in surface forcing is similar.

The timescale for spinup is much longer in the midlatitude oceans. The most salient quantity here is the speed of the long nondispersive Rossby mode (βa^2). At 30° latitude, this is about 2.5 cm s^{-1} and the Pacific is about 12,000 km wide there, and hence it takes 11 years for the Rossby wave to cross the Pacific! Jacobs *et al.* (1994) have detected the Rossby waves generated by the 1983 ENSO event impacting the western Pacific region in 1993 TOPEX altimetric SSH signal. For many high resolution basin models, such a spinup time is rather prohibitive. This is one principal reason why simple layered models have enjoyed such popularity in basin and global high resolution eddy-resolving ocean models. As far as midlatitude dynamics is concerned, it is the North Indian Ocean that is the quickest to spinup. The subcontinent of India divides it into two subbasins, the Arabian Sea and the Bay of Bengal, the former being the widest about 2500 km. Using a latitude of 20°, the approximate northern extent of the Indian Ocean, the Rossby mode speed is about 7 cm s^{-1} and hence it takes slightly more than a year for the Rossby mode to propagate across the Arabian Sea. The equatorial Indian Ocean itself is 6000 km wide and hence it takes the first baroclinic mode about 3 months to cross it from end to end. Consequently, high resolution modeling of the Indian Ocean using models with comprehensive physics is much simpler for the North Indian Ocean (and the Atlantic) than for the Pacific.

The scenario involved in spinup from rest is as follows. Assume an impulsive application of surface wind forcing to an ocean basin at rest, initialized with observed density fields. This generates Kelvin and Rossby modes. The motion at any longitude increases with time until the westward propagating long baroclinic Rossby waves generated at the eastern boundary arrive, inducing a steady Sverdrup flow in their wake. By the time these Rossby waves reach the western boundary, a steady state will have been reached everywhere except at the western boundary itself, where the adjustment depends on the propagation and dissipation of slow moving reflected Rossby waves. The spinup timescale is essentially the time for Rossby waves to traverse the basin from west to east. Since there are an infinite number of such modes, the spinup time is larger than that indicated by that for the first mode alone, but for all practical purposes, since higher modes also tend to be more heavily damped, it is the lowest modes, the first and probably the second, that need to be considered. The surest way is to monitor the basin energetics for attainment of an equilibrium state.

We have ignored barotropic modes in the preceding discussion. It is the barotropic mode that effects changes in Ekman pumping due to changes in wind forcing. Therefore for mere adjustment of Sverdrup balance in the upper layers, the timescale is that of the passage of the barotropic Rossby mode across the basin, which is of the order of days.

One way to attain faster spinup is through data assimilation. If the true state of the ocean corresponding to the applied forcing is known, it is possible to force the model ocean toward the observed state by nudging it toward the observed state during spinup. While this is straightforward in principle, it is harder to implement, simply because the observed state may not be complete enough. The most important quantity in this respect is the SSH. If the SSH is known and the model density field can be adjusted during spinup such that the model SSH is nudged toward the observed value, faster spinup is possible. Fortunately, the advent of satellite altimetry has made it possible to derive information on the SSH necessary to do this. This works especially well for barotropic models, since there is no need then to adjust the density field. An example is given by Kantha (1996), who used observed tidal SSH fields to spin up a high resolution model of barotropic tides in the global oceans.

Bryan (1984; see also Pacanowski, 1995) used a robust diagnostic method, to effect a faster spinup in the deep. The method (see Chapter 10) depends on artificially speeding up the spinup of deeper waters; this can, however, lead to severe distortions in dynamics, whose effect on the final spunup state is not clear.

4.5 QUASI-GEOSTROPHIC MODELS

As we saw earlier, incompressible, hydrostatic, Boussinesq equations for a gravitationally stratified fluid on a rotating sphere are adequate to describe most geophysical flows. In fact, this is the set of equations that are solved in a numerical ocean general circulation model (OGCM). Comprehensive all-inclusive physics is essential for universal applicability of such a model and for real-life applications such as nowcasting and forecasting the four-dimensional oceanic state. Unfortunately, such models are also resource intensive, especially when run on a global scale, and tax even the most modern high performance computers, so that their use is necessarily rather selective. Also because of the complexity, it is also difficult to decipher the results from such a model. For the purposes of understanding the dynamical processes involved, it is often preferable to appeal to a simpler subset of the governing equations that contain only the physics that are necessary to study such processes, but are sufficient in the sense they exclude those that are not central to the processes being studied. Clearly, the simplification must be done with care and the limitations of the model must be kept in mind in interpreting the results.

Quasi-geostrophic (QG) approximation on a midlatitude beta plane is one such approximation that has proved quite useful for understanding the dynamical processes involved in the large-scale ocean circulation away from the equator and outside the continental shelves of the principal ocean basins. QG models are far more efficient than PE (primitive equation based) models (Holland, 1985), since they filter out the troublesome fast gravity waves, and this permits integrations to be carried out for hundreds of years necessary often for statistical and dynamical equilibrium to be attained. Such luxury is unaffordable for PE-based OGCMs (Semtner, 1995). QG models enjoyed popularity in the 1970s and 1980s, principally because even with the limited supercomputing capabilities available then, they enabled high resolution, eddy-resolving simulations of basin-scale ocean circulation to be carried out. This eddy-resolving capability not only enables baroclinic instabilities central to gyre-scale circulation to be more accurately depicted, but it permits more realism to be attained overall, since, because of the small grid sizes possible, the subgrid scale diffusivities needed from numerical considerations do not have to be so large as to make the oceans diffusion dominated. Instead of parameterizing the effects of eddies, QG models enabled them to be modeled explicitly. In this sense, QG models have the same common objective as LES models in turbulence. However, with the advent of multigigaflop high performance computers of the 1990s that for the very first time have enabled marginally eddy-resolving global simulations to be made with PE models (Fu and Smith, 1996; Semtner, 1995), and because of the inapplicability of QG models to equatorial regions, which are a crucial component of the ocean-atmosphere system, QG models have become increasingly irrelevant in recent years. Nevertheless, because of the insight they provide into midlatitude dynamics, QG dynamics and QG models deserve continued study and use. We will describe here the highly versatile QG model developed by Bill Holland (and Julliana Chow) of NCAR (Chow and Holland, 1986). Comprehensive full-physics PE models are described in Chapters 9 to 11.

QG approximation enables gravity waves (permitted by PE equations and mostly unimportant to slow long timescale adjustment in geophysical flows) to be filtered out and hence an order of magnitude improvement in integration time step (that is no longer constrained by gravity wave speed-governed CFL criterion) to be attained. It has a long history dating back to the early 1950s. Since the method allows often for simple analytical models to be constructed, it has often been the principal tool of many geophysicists (see Pedlosky, 1979; Veronis, 1981) and some ocean modelers (Holland, 1985). Pedlosky's book is the principal authoritative treatment of this approach in geophysical fluid dynamics, but the essentials can be illustrated by application to shallow water equations (Holland, 1985).

4.5.1 GOVERNING EQUATIONS

Consider dissipationless, unforced, shallow water equations in rectangular Cartesian coordinates for barotropic flow on a midlatitude beta plane:

$$\frac{\partial \eta}{\partial t} + \frac{\partial (uD)}{\partial x} + \frac{\partial (vD)}{\partial y} = 0$$

$$\frac{\partial u}{\partial t} + u\frac{\partial u}{\partial x} + v\frac{\partial u}{\partial y} - fv + g\frac{\partial \eta}{\partial x} = 0 \qquad (4.5.1)$$

$$\frac{\partial v}{\partial t} + u\frac{\partial v}{\partial x} + v\frac{\partial v}{\partial y} + fu + g\frac{\partial \eta}{\partial y} = 0$$

Here y is the meridional and x the zonal directions; $f = f_0 + \beta y$, where $\beta = d f / dy$, and D is the layer depth. For a barotropic mode, D is also the depth of the water column and η is the free surface deflection. (These equations are also applicable to baroclinic modes, provided g′, the reduced gravity, replaces g. Then η is the interfacial deflection.) The layer thickness can be written as D = H + η - H_B, where H is the mean thickness of the layer, H_B is the perturbation to it due to the bottom depth variations. It is convenient to assume an f-plane (β =0) for the following derivation. It is also convenient to deal with normalized quantities:

$$t = \left(\frac{L}{U}\right)(t')$$

$$x, y = L\left(x', y'\right) \qquad (4.5.2)$$

$$\eta = N\left(\eta'\right)$$

$$u, v = U\left(u', v'\right)$$

We will deal henceforth with normalized quantities only, but we will drop primes for convenience.

$$\varepsilon F\left(\frac{\partial \eta}{\partial t} + u\frac{\partial \eta}{\partial x} + v\frac{\partial \eta}{\partial y}\right) - \left[u\frac{\partial}{\partial x}\left(\frac{H_B}{H}\right) + v\frac{\partial}{\partial y}\left(\frac{H_B}{H}\right)\right]$$

$$+\left(1 + \varepsilon F\eta - \frac{H_B}{H}\right)\left(\frac{\partial u}{\partial x} + \frac{\partial v}{\partial y}\right) = 0$$

$$\varepsilon\left(\frac{\partial u}{\partial t} + u\frac{\partial u}{\partial x} + v\frac{\partial u}{\partial y}\right) - v + \frac{\partial \eta}{\partial x} = 0$$

$$\varepsilon\left(\frac{\partial v}{\partial t} + u\frac{\partial v}{\partial x} + v\frac{\partial v}{\partial y}\right) + u + \frac{\partial \eta}{\partial y} = 0$$

(4.5.3)

where $\varepsilon = U/(fL)$ is the Rossby number, and $F = f^2L^2/(gD) = (L/a)^2$ is the Froude number. Note that $N = fUL/g$ to render the pressure gradient term of the same order as the Coriolis term. The term F is assumed to be of order one. The term $\delta = \left(\frac{\partial u}{\partial x} + \frac{\partial v}{\partial y}\right)$ is the divergence and $\zeta = \left(\frac{\partial v}{\partial x} - \frac{\partial u}{\partial y}\right)$ is the vorticity. Eq. (4.5.3) is equivalent to an equation for the conservation of potential vorticity ζ_p:

$$\frac{\partial \zeta_p}{\partial t} + u\frac{\partial \zeta_p}{\partial x} + v\frac{\partial \zeta_p}{\partial y} = 0; \quad \zeta_p = (1 + \varepsilon\zeta)\left(1 + \varepsilon F\eta - \frac{H_B}{H}\right)^{-1}$$

(4.5.4)

Approximation $\varepsilon = 0$ gives the geostrophic approximation. Approximation $\varepsilon \ll 1$ enables perturbation expansions to be made in terms of ε. If only the zeroth geostrophic and first-order terms are retained, one gets the QG model. It is also possible to retain higher order terms and obtain what are known as intermediate models, which are in-between QG and PE models in terms of accuracy. Expanding all the variables in terms of ε,

$$\eta = \eta_0 + \varepsilon\eta_1 + \cdots$$
$$u, v = (u, v)_0 + \varepsilon(u, v)_1 + \cdots$$

(4.5.5)

and assuming $H_B/H = \varepsilon\eta_B \sim O(\varepsilon)$, the zeroth- order fields are geostrophic and divergenceless:

$$\delta_0 = \frac{\partial u_0}{\partial x} + \frac{\partial v_0}{\partial y} = 0$$

$$v_0 = \frac{\partial \eta_0}{\partial x} \tag{4.5.6}$$

$$u_0 = -\frac{\partial \eta_0}{\partial y}$$

The first-order equations are

$$F\left(\frac{\partial \eta_0}{\partial t} + u_0 \frac{\partial \eta_0}{\partial x} + v_0 \frac{\partial \eta_0}{\partial y} \right) - \left(u_0 \frac{\partial \eta_B}{\partial x} + v_0 \frac{\partial \eta_B}{\partial y} \right)$$

$$+ \left(\frac{\partial u_1}{\partial x} + \frac{\partial v_1}{\partial y} \right) = 0$$

$$\left(\frac{\partial u_0}{\partial t} + u_0 \frac{\partial u_0}{\partial x} + v_0 \frac{\partial u_0}{\partial y} \right) - v_1 + \frac{\partial \eta_1}{\partial x} = 0 \tag{4.5.7}$$

$$\left(\frac{\partial v_0}{\partial t} + u_0 \frac{\partial v_0}{\partial x} + v_0 \frac{\partial v_0}{\partial y} \right) + u_1 + \frac{\partial \eta_1}{\partial y} = 0$$

Formulated this way, it is easy to see the limitations imposed by QG approximation. Three principal conditions need to be satisfied:

$$\varepsilon = \frac{U}{fL} \ll 1; \quad \frac{H_B}{H} \ll 1; \quad \frac{\eta}{H} \ll 1 \tag{4.5.8}$$

The first condition (Rossby number must be small) prevents the QG model from being accurate for strongly inertial jets; slow basin scale circulation is their best forte. It also makes it invalid for equatorial dynamics. The second condition (the bottom slope must be small) prevents the QG model from being applied to regions of strong topographic changes, such as continental slopes. Thus QG models are best for nearly flat-bottomed ocean basins. The third condition (the interfacial slope must be small) is not much of a problem for barotropic modes but is clearly a major limitation for baroclinic modes, where the displacement of the isopycnal (density) surfaces can be quite large compared to the undisturbed layer thickness. Also QG models are best applied to adiabatic problems with no heat diffusion since thermal forcing is difficult to accommodate. Of course, gravity wave physics, which characterizes fast adjustment processes that may often be important to the dynamics, is lost. However, their efficiency and simplified dynamics more than counterbalance these limitations.

The zeroth-order fields can be solved by forming the vorticity conservation equation,

$$\frac{D\zeta_0}{Dt} = \frac{\partial\zeta_0}{\partial t} + u_0\frac{\partial\zeta_0}{\partial x} + v_0\frac{\partial\zeta_0}{\partial y} = -\left(\frac{\partial u_1}{\partial x} + \frac{\partial v_1}{\partial y}\right) = -\delta_1$$

$$\zeta_0 = \left(\frac{\partial v_0}{\partial x} - \frac{\partial u_0}{\partial y}\right) = \nabla^2\eta_0$$

(4.5.9)

leading to nothing but a statement of QG potential vorticity conservation of a fluid parcel:

$$\frac{D}{Dt}\left(\nabla^2\eta_0 - F\eta_0 + \eta_B\right) = 0$$

$$\text{or } \frac{\partial}{\partial t}\left(\nabla^2\eta_0 - F\eta_0\right) = -J\left(\eta_0, \nabla^2\eta_0 + \eta_B\right)$$

(4.5.10)

where the derivative D/Dt is the substantial derivative or the derivative following a fluid parcel and $J(a,b) = \left(\dfrac{\partial a}{\partial x}\dfrac{\partial b}{\partial y} - \dfrac{\partial b}{\partial x}\dfrac{\partial a}{\partial y}\right)$ is the Jacobian.

Equation (4.5.10) also holds the means to extend QG formulation to fully three-dimensional rotating stratified geophysical flows on a spherical Earth:

$$\frac{D}{Dt}\left[\nabla^2\psi + \frac{\partial}{\partial z}\left(\frac{1}{S(z)}\frac{\partial\psi}{\partial z}\right) + f\right] = 0$$

(4.5.11)

where ψ is the QG stream function and $S(z)$ is the static stability parameter depending on background gravitational stratification, equal to $N(z)/f$, where N is the buoyancy frequency. The effect of forcing and friction can be represented by a nonzero term on the RHS. This equation forms the basis of eddy-resolving QG models formulated by Holland (1978, 1985), which will be dealt with in this section.

Numerical implementation of the solution to Eq. (4.5.11) involves dividing the water column into N layers (see Figure 4.5.1 from Holland, 1985) and

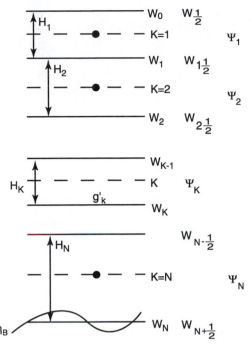

Figure 4.5.1 Discretization in the vertical in Holland's QG model detailing the properties of various layers. (Redrawn from O'Brien, J. J., *Advanced Physical Oceanographic Numerical Modeling*, figure 2, p. 211, © 1985 Reidel with kind permission from Kluwer Academic Publishers.)

deriving potential vorticity conservation relation for each layer:

$$\frac{D}{Dt}\left[\nabla^2\psi_k + \frac{f_0}{H_k}(\eta_k - \eta_{k-1}) + f\right] = R_k \qquad (4.5.12)$$

where the term R_k represents the wind forcing and frictional terms; ψ_k is the QG streamfunction for each layer k (k increases from 1 at the top to N at the bottom) and η_k is the interfacial deflection at its bottom; $\eta_0 = 0$ corresponding to a rigid lid and $\eta_N = \eta_B$ corresponding to the perturbation of the bottom layer thickness by topography; D_k is the undisturbed layer thickness. For rectangular Cartesian coordinates (beta plane), Eq. (4.5.11) can also be cast in the form

$$\frac{\partial}{\partial t}\left(\nabla^2\psi_k\right) = J\left[\left(f + \nabla^2\psi_k\right), \psi_k\right] + \frac{f_0}{H_k}(w_k - w_{k-1})$$

$$+\delta_{k1}\frac{\nabla\times\tau}{H_1}-\delta_{kN}c\nabla^2\psi_N+\left(A_H\nabla^4\psi_k-A_{BH}\nabla^6\psi_k\right)$$

$$w_k=\frac{f_0}{g_k'}\left\{J\left[\left(\psi_k-\psi_{k+1}\right),\left(\frac{H_{k+1}}{H_k+H_{k+1}}\psi_k+\frac{H_k}{H_k+H_{k+1}}\psi_{k+1}\right)\right]-\frac{\partial}{\partial t}\left(\psi_k-\psi_{k+1}\right)\right\}$$

$$k=1,N$$

$$(4.5.13)$$

where $J(E,F)=\dfrac{\partial E}{\partial y}\dfrac{\partial F}{\partial x}-\dfrac{\partial E}{\partial x}\dfrac{\partial F}{\partial y}$ is the Jacobian, and x is the zonal and y the

meridional coordinate, and δ_{ij} is the Kronecker delta; w_k is the vertical velocity

at the bottom of layer k. Quantity τ is the wind stress and the term involving τ appears only in the first layer; it leads to Ekman pumping at the base of the first layer and is therefore equivalent to a body force acting on the upper layer; c is a dimensional bottom friction coefficient with units of inverse time and provides for the drag on the bottom layer. Coefficients A_H and A_{BH} are coefficients of horizontal diffusion, corresponding to harmonic and biharmonic forms commonly used in numerical modeling; these provide for parameterization of subgrid scale dissipation and are chosen by numerical considerations. The biharmonic form is efficient at suppressing the $2\Delta x$ noise. $g_k'=g\left(\rho_{k+1}-\rho_k\right)/\rho_0$

is the reduced gravity corresponding to the interface at the bottom of layer k; ρ_k is the density of layer k and ρ_0 is the reference density. The model can be run on a f-plane with $f=f_0$ or on a beta plane with $f=f_0+\beta y$. The zonal and meridional components of QG velocity are given by

$$u_k=-\frac{\partial\psi_k}{\partial y};v_k=\frac{\partial\psi_k}{\partial x}\qquad(4.5.14)$$

Equation (4.5.13) needs to be solved with boundary conditions at the top ($w_0=0$) and the bottom [$w_N=J(\psi_N,\eta_B)$] of the water column. Note that like all layered model formulations, the bottom topography is buried in the bottom-most layer and the upper layers feel the topography only indirectly through interaction with the bottom layer. The lateral boundary conditions involve no slip at the boundary so that the tangential component of velocity $u^t_k\sim\partial\psi_k/\partial n$, where n is the unit normal to the boundary, must vanish.

We will not go into the aspects of discretization in the horizontal. The equations are elliptic and a variety of Poisson solvers can be employed to solve them (see Chapter 2). For a box ocean model, FFT methods are the most efficient.

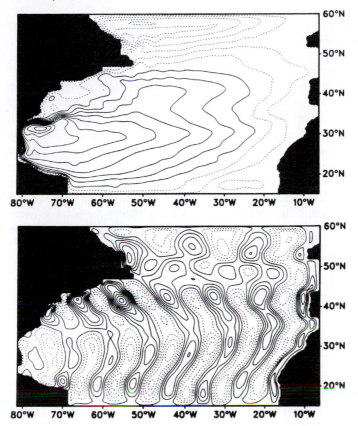

Figure 4.5.2 Results from QG model of the north Atlantic (from Holland 1985): the long term mean circulation showing the Gulf Stream and the subtropical gyre (top) and the thermocline depth anomaly showing the annual period first baroclinic mode Rossby waves propagating across the basin from east to west. (Reprinted from O'Brien, J. J., *Advanced Physical Oceanographic Numerical Modeling*, figures 16 and 17, p. 223, © 1985 Reidel with kind permission from Kluwer Academic Publishers.)

4.5.2 APPLICATIONS

Note that QG formulation is ideally suited to the study of slow oceanic adjustment brought on by Rossby wave generation and propagation in a basin. For application of QG model to a variety of oceanic flows, including baroclinic instability of zonal jets, such as the Gulf Stream, see Holland (1978, 1985) and the reference list cited therein. For a comparison between QG and PE models, see Semtner and Holland (1978). Here, we will present just one example from Holland (1985). Figure 4.5.2 shows a simulation of the North Atlantic, using a

$1/4°$ model of the basin. The top panel shows the time-averaged stream function, whereas the bottom shows the annual Rossby waves marching westward to effect the slow changes due to the seasonal variability in the wind forcing. See Holland (1985) for other applications of QG models.

QG models break down in regions of strong topographic changes. On the other hand, intermediate models (Allen *et al.,* 1990; Gent and McWilliams, 1982; McWilliams and Gent, 1980) have the robustness and dynamical simplicity of QG models, yet retain validity for flows over $O(1)$ topographic changes, such as the continental slope. They filter out gravity waves and permit solutions to be obtained more efficiently than PE models. Intermediate models are still small Rossby number approximations to the PE models, but unlike QG equations, most of them retain the full continuity equation. An insightful survey of various intermediate barotropic models, including the balance equations can be found in Allen *et al.* (1990, see also Barth *et al.,* 1990), who have made a systematic study of their capability to simulate linear ageostrophic coastally trapped waves, such as Kelvin and continental shelf waves. They conclude that intermediate models are more accurate than QG models in such situations. Nevertheless, enormous increases in computing power available have made both QG and intermediate models of principally academic interest.

Chapter 5

High Latitude Dynamics and Sea-Ice Models

High latitude oceans and seas are an integral part of the global climate system and are an especially sensitive indicator of global climatic changes, since their perennial and seasonal ice covers are rather sensitive to changes in the climate system. It is the presence of the sea-ice cover that is the most distinguishing feature of the high latitude oceans. The ice cover mediates exchanges between the atmosphere and the ocean. It insulates the ocean from the cold atmosphere during winter, preventing excessive heat loss from the oceans. Thus, it affects both the dynamics and thermodynamics of high latitude oceans. During summer, the higher albedo of the ice and its snow cover reduce the absorption of solar radiation by the ocean. These thermodynamic effects in turn have an effect on the state of the ocean and the ice cover itself. Most of the dynamical aspects of polar oceans can be explained by f-plane dynamics, since beta effects are quite small. The principal complication arises from the presence and interaction of ice cover and the close coupling of the ice dynamics to the ocean dynamics.

5.1 SALIENT FEATURES OF ICE COVER

The exchange of momentum between the atmosphere and the ocean is through the ice cover, and since ice can entertain internal stresses and behaves like a complex viscous-plastic (possibly elastic) medium on large scales of interest to ocean dynamics, ice plays a central role in the dynamics of the coupled media. It

337

is impossible to explore every aspect of sea-ice-covered ocean here; the literature is vast. Instead we will explore some salient aspects and refer the reader instead to existing reviews. *The Geophysics of Sea Ice,* edited by Norbert Untersteiner (Untersteiner, 1986), is a collection of articles on various aspects of sea ice and is an excellent place to start exploring sea ice and its effects on the dynamics and thermodynamics of ice-covered waters. *Sea Ice Processes and Models* edited by Robert Pritchard (Pritchard, 1980) is an earlier post-AIDJEX (Arctic Ice Dynamics Joint Experiment) summary of our knowledge. More recent summaries can be found in the two-volume set edited by Walker Smith (1990) and the Nansen centennial volume *The Polar Oceans and Their Role in Shaping the Global Environment*, edited by Ola Johannessen, Robin Muench, and Jim Overland. In addition, a collection of articles on studies on marginal ice zones (MIZ) and CEAREX can be found in special sections of the *Journal of Geophysical Research* (e.g., Muench *et al.,* 1991); the *Journal of Physical Oceanography* also contains results from current and past studies on ice-covered oceans.

Sea ice cover is an important aspect of subpolar seas and shelves around the Arctic in the northern hemisphere (Figure 5.1.1). In the southern hemisphere, the sea ice cover around the continent of Antarctica waxes and wanes with the seasons (Figure 5.1.2). The dynamical and thermodynamic exchanges between the atmosphere and the ocean are particularly interesting near the ice edge, in the so-called marginal ice zone (MIZ). For example, when the wind is essentially parallel to the ice edge, the roughness contrast between the ice and the open water leads to differing stresses being applied on the two sides of the ice edge resulting in intense upwelling/downwelling at the ice edge. On the other hand, if the wind is off-ice, during winter, the intense heat loss from the ocean to the cold atmosphere generates a convective internal boundary layer that affects the state of both the atmospheric and the oceanic boundary layers. The ice edge is also associated with a sharp salinity front due to a layer of brackish water immediately underneath created by melting of ice near the ice edge. This front is often unstable and generates meanders and eddies that are seen in SAR and visible imagery (ice floes act as flow tracers).

The perennial sea-ice cover in the interior of the Arctic is composed of large ice floes with a spectrum of sizes, typical size being a few kilometers. These floes are driven by the winds and ocean currents, and their motion is influenced by the bumping and grinding collisions between the floes as they move in response. These floe-floe interactions on a wide spectrum of floe scales is the most important, fascinating as well as frustrating part of the dynamics of sea-ice-covered oceans. This is simply because it is not always clear how to account for these collisions and their effect on ice motions (and hence the ocean circulation) over the entire spectrum of floe sizes. Traditionally, these interactions have been characterized by incorporating internal ice stresses into the dynamical equations

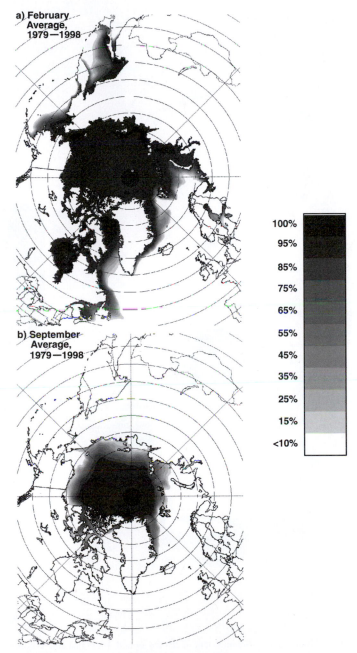

Figure 5.1.1 The minimum and maximum extent of Arctic sea ice and its concentration (generated using the Bootstrap Algorithm and provided by J.C. Comiso).

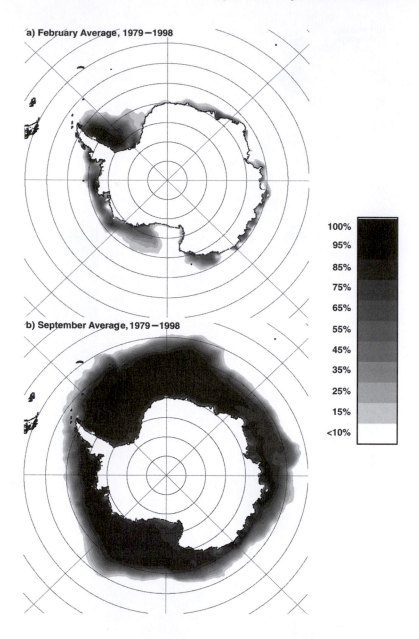

Figure 5.1.2 The minimum and maximum extent of sea ice around Antarctica and its concentration (generated using the Bootstrap Algorithm and provided by J.C. Comiso).

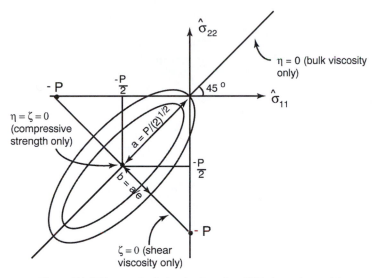

Figure 5.1.3 Viscous-plastic ice rheology from Hibler's sea-ice model.

for ice motion, with ice treated as a continuum and with a ice rheology postulated a priori. This lumps all such uncertainties into terms involving ice rheology. It appears, however, that at least on large spatial scales characterizing the basin scale circulation in the Arctic, Bill Hibler's viscous-plastic rheology (Hibler, 1979) gives meaningful results for ice motion and thickness distributions in the Arctic. The principal premise of this rheology is simple: sea-ice cover can not sustain tensile forces as well as it can the compressive forces. The rheology makes note of this. Sea ice is treated as a fluid continuum, but with a complex constitutive relationship. While at low strain rates, ice behaves as a Newtonian viscous fluid, when strain rates exceed a certain value, it begins to flow like a plastic medium, that is, the stress becomes independent of the strain rate. The resulting rheology is shown in Figure 5.1.3. It describes an ellipse in the negative principal stress quadrant.

On smaller spatial scales and in the vicinity of the marginal ice zones, where the floe sizes are small, the appropriate rheology to use is not clear (but see Flato and Hibler, 1992). Fortunately, the importance of ice internal stresses decreases quite rapidly with the decrease in ice concentration, and ice stresses can often be (but not always) ignored in the dynamical equations for the ice in the MIZ.

The underlying ocean responds in turn to the ice motion, and the circulation that is set up is determined by the effective shear stress applied by ice floes at the ice-ocean interface. Since ice can sustain internal stresses, the stress applied to the ocean and hence the circulation depends very much on the rheological properties of sea ice. When the ice motion piles up ice against a coast, its

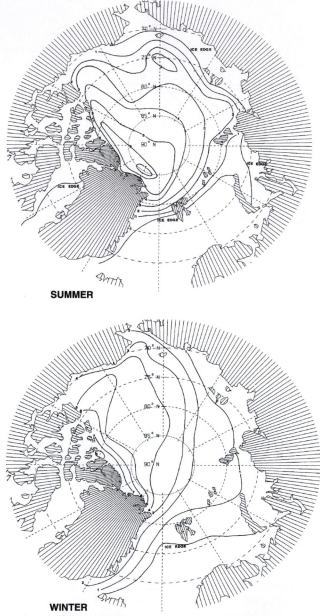

Figure 5.1.4 Mean ice draft in summer (top) and winter (bottom) derived from submarine observations. Reprinted from *Cold Regions Science and Technology*, 13, Bourke, R H. and R. P. Garrett, Sea ice thickness distribution in the Arctic ocean, 259–280, 1987, with permission from Elsevier Science.

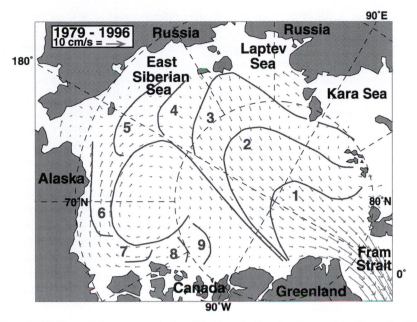

Figure 5.1.5 Observed long-term mean ice drift. Note the Beaufort Gyre and the Transpolar Drift Stream. The numbers indicate the average number of years for the ice in that location to exit the Arctic through Fram Strait. From Pfirman *et al.* (1997).

thickness builds up and the wind stress is balanced principally by internal stresses in ice, and therefore no stress is applied to the underlying ocean. When ice is under divergence, large leads open up and the internal ice stresses can often be neglected, and under these conditions, the wind stress is transferred to the water below. The average thickness of the sea ice cover is ~3 m in the Arctic with thinner ice on the Siberian shelves, but the ice can be thicker than 6 m toward the Canadian archipelago (Figure 5.1.4). In general, the ice near the periphery of the Arctic (and that close to the Antarctic continental masses) is less mobile than that in the central Arctic. The large scale ice motion in the Arctic consists of an anticyclonic gyre centered in the Beaufort Sea and a transpolar drift that transports ice from the vicinity of the Bering Straits and Siberian shelves to the Fram Straits and out into the North Atlantic (Figure 5.1.5). The large scale ocean circulation reflects this (Barry *et al.*, 1993). Analysis of sea ice trajectories based on historical drift data from the International Arctic Buoy Program (Pfirman *et al.*, 1997) shows that Kara Sea is a major contributor of ice to the Barents Sea and the southern limb of the Transpolar Drift Stream. Aagard (1989) and Barry *et al.* (1993) are excellent, albeit descriptive (nondynamical) reviews of the Arctic circulation and physical oceanography, the latter being the most recent.

Other important processes prevalent in both the Arctic and its surrounding seas, and around Antarctica, are important to climate. Greenland and Labrador Seas are regions of deep water formation in winter, and this deep water is an integral part of the thermohaline circulation in the global oceans. These waters sink to the bottom of the Atlantic and cross the equator and eventually surface in the tropical Pacific, the whole journey taking about a millennium (Schmitz, 1995; see also Chapter 1). This is what accounts for the long "memory" of the global oceans. At times in the distant past, warming episodes have melted the glaciers on the North American Continent with the resulting fresh water layer in the North Atlantic leading to a cessation of deep convection. The consequences have been catastrophic to marine life, and the climate change has been marked (Broeker, 1997). The seas around Antarctica are regions of formation of the Antarctic Bottom Water (AABW). Siberian shelves and the Sea of Okhotsk are regions of formation of intermediate water. For a better understanding of the longer thermohaline meridional circulation timescales in the global climate, it is necessary to understand and model processes related to dense water formation in these ice-covered seas.

Tidal motions are quite small in the Arctic itself (Kowalik and Proshutinsky, 1993 and 1995; Lyard, 1997); the sea-ice cover further inhibits tidal motions (tides are, however, large in the Antarctic marginal seas, the Weddell and Ross Seas; see Kantha, 1995). Therefore, it is the winds that are the principal driving mechanism for the ocean circulation in the Arctic. Tides are, however, significant in its surrounding seas such as the Sea of Okhotsk and the Bering Sea (Kantha and Piacsek, 1993). As far as large scale dynamics is concerned, because of the near-zero values of beta, the polar latitudes are not as interesting as mid and low latitudes that can sustain planetary waves and the rich spectrum of processes associated with them. However, processes associated with f-plane dynamics, such as Kelvin waves and continental shelf waves, become prominent in the Arctic and around Antarctic. These processes are strongly affected by the capping sea-ice cover and the absence of meridional boundaries (unlike lower latitudes). Unfortunately, because of the difficult logistics involved in making any measurements in the harsh Arctic and Antarctic environments, especially during the bitterly cold and hazardous dark winter months, when the influence of the capping ice cover is the greatest on these processes, much remains unknown.

The few observations that are available have been made from drifting ice camps in the Beaufort Sea and the central Arctic. These observations show the sea ice drift in the central Arctic to be composed of a long-term average of a few cm s^{-1} overall, with inertial oscillations and synoptic scale fluctuations due to wind events superimposed on it. The Arctic buoy program (Untersteiner and

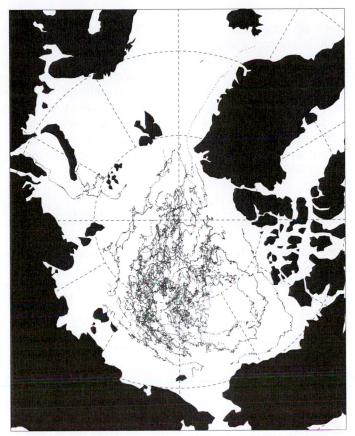

Figure 5.1.6 Trajectories of ice stations in the Arctic showing the Beaufort Gyre and the Transpolar Drift Stream. From Barry *et al.* (1993).

Thorndyke, 1982) has deployed drifting buoys periodically in the Arctic, and their trajectories indicate a fascinating spectrum of ice motions in the Arctic (see Figure 5.1.6 from Barry *et al.,* 1993).

Of course, the most important parameter in large scale dynamics is the internal Rossby radius of deformation a. In the polar seas, because of the large value of f and small value of the gravity wave speed, the Rossby radius is rather small, on the order of a few to 10 kilometers, compared to 40 km typical of midlatitudes. This has implications to modeling the Arctic. For those processes such as Kelvin waves, the spatial resolution needed to resolve them is quite high.

Currently, atmospheric GCMs are not very skillful when it comes to simulation and prediction of the polar atmosphere. One reason is the very crude treatment of sea ice. While climate models (Boville and Gent, 1998) attempt to include some representation of ice dynamics, NWP models do not; instead the sea-ice extent is prescribed and held fixed. Since the atmosphere-ice-ocean interactions that impact the evolution of the polar atmosphere are a function of both the dynamics and thermodynamics of ice cover, it is essential to not only include them but also account for the two-way interactions between the atmosphere and the underlying sea-ice-covered ocean. Given the importance of the reservoir of cold polar air to midlatitude weather, and the importance of sea-ice cover to long-term climate, attempts are underway to incorporate more realistic sea ice and atmosphere-ice-ocean interactions in atmospheric GCMs (Randall *et al.*, 1998).

5.2 MOMENTUM EQUATIONS FOR SEA ICE

The unique feature of high latitude dynamics is the presence of sea-ice cover and hence it is necessary to solve for the state of sea ice as well. The governing momentum equations for the sea-ice cover, treated as a two-dimensional (but heavily fractured) continuum (Kantha and Mellor, 1989b; Mellor and Kantha, 1989), are

$$\frac{\partial}{\partial t}(D_I U_{Ii}) + \frac{\partial}{\partial x_j}(D_I U_{Ij} U_{Ii}) - \varepsilon_{3jk} f_3 D_I U_{Ik} = -g D_I \frac{\partial \eta_o}{\partial x_i}$$

$$+ \frac{A_I}{\rho_I}(\tau_{AIi} - \tau_{IOi}) + F_i \qquad (5.2.1)$$

$$D_I = A_I h_I; \qquad F_i = \frac{1}{\rho_I} \frac{\partial \sigma_{ij}}{\partial x_j}$$

where A_I is the ice concentration, the area fraction of ice, D_I is the average ice thickness ($D_I = A_I h_I$, where h_I is the ice floe thickness), and U_{Ii} are the velocity components of ice and τ_i is the shear stress. Subscripts AI and IO denote the air-ice and ice-ocean interfaces; σ_{ij} is the internal ice stress, whose divergence F_i is the additional force term in the momentum balance. η_0 is the sea level and the term in the momentum equation due to its slope can often be approximated using geostrophy:

$$-g D_I \frac{\partial \eta_o}{\partial x_i} = -\varepsilon_{3jk} f_3 D_I U_{gk} \qquad (5.2.2)$$

where U_{gi} is the geostrophic velocity in the water column below.

The term τ_{AI} is the stress applied by the atmosphere and can be related to W_i, the geostrophic wind aloft often by a simple bulk formula:

$$(\tau_{AI1}, \tau_{AII}) = \rho_a c_d (W_k W_k)^{1/2} \left[(W_1 \cos\alpha - W_2 \sin\alpha), (W_1 \sin\alpha + W_2 \cos\alpha) \right]$$

$$(5.2.3)$$

where c_d is the drag coefficient dependent on the roughness of the sea ice, the pack ice being much smoother ($c_d \sim 10^{-3}$) than rafted ice at the edge of the ice pack ($c_d \sim 2 - 3 \times 10^{-3}$), and α the turning angle, the angle between the geostrophic velocity vector and the shear stress. During AIDJEX, c_d was found to be 0.97×10^{-3} and $\alpha \sim 25°$ (Coon, 1980). Note that these drag coefficients are much smaller than the usual 10 m values over sea ice (Overland, 1985).

The term τ_{IO} is the stress exerted on the water by ice and can also be parameterized similarly:

$$
(\tau_{IO1}, \tau_{IO2}) = \rho_w c_{dw} \left[(U_k - U_{gk})(U_k - U_{gk}) \right]^{1/2}
$$

$$
\left\{
\begin{array}{l}
\left[-(U_1 - U_{g1})\cos\beta + (U_2 - U_{g2})\sin\beta \right], \\[6pt]
\left[-(U_1 - U_{g1})\sin\beta - (U_2 - U_{g2})\cos\beta \right]
\end{array}
\right\}
\qquad (5.2.4)
$$

McPhee (1980) used $c_{dw} = 0.0055$ and $\beta = 23°$ to successfully simulate the drift of the AIDJEX ice camps.

A more accurate computation of the stresses at the air-ice and ice-ocean interfaces requires modeling of the turbulent atmospheric boundary layer above (for example, Kantha and Mellor, 1989a) and the oceanic mixed layer below the sea ice (Kantha and Mellor, 1989b).

Of the various terms in the momentum balance Eq. (5.2.1), which can also be written compactly as

$$\frac{D}{Dt}(D_I U_{Ii}) = F_{AI} - F_{IO} + F_c + F_s + F_i \qquad (5.2.5)$$

where D/Dt denotes the total derivative, the most important are the atmospheric stress, F_{AI}, the ocean drag, $-F_{IO}$, and the Coriolis force, F_c, when ice concentrations are small (<90%), and the ice internal stresses, F_i, when concentrations exceed 90%. The tendency terms are usually three orders of magnitude smaller than the wind stress, and nonlinear advection terms can often be ignored (Oberhuber, 1993a). However, the oceanic surface tilt term, F_s, even though small, is important to long-term simulations.

Treating the sea ice cover as a continuum is only valid at spatial scales larger than the individual floes; nevertheless, it makes it much easier to model sea ice cover and its interactions with the underlying ocean.

5.3 CONSTITUTIVE LAW FOR SEA ICE (ICE RHEOLOGY)

The most important term in the sea-ice dynamical equations is the internal stress term σ_{ij}. To make any progress in modeling sea ice, its rheology must be postulated. This is hard to do since the sea-ice cover in polar and subpolar seas consists of a collection of ice floes of a variety of sizes and thicknesses constantly bumping and grinding against the neighboring floes as they move in a general direction. Moreover, the characteristics of the floes themselves change as the marginal ice zone is approached; the floes tend to become smaller (on the order of tens to hundreds of meters), whereas in the central Arctic their size is measured in kilometers.

The difficulty of postulating a constitutive relationship that is valid over the entire range of observed sea-ice states ranging from the rafted pancake ice in the MIZ to the pack ice in the central Arctic consisting of individual floes of tens of kilometers in size, and extending over spatial scales of a few tens of meters at the ice edge to tens of kilometers in the interior of the ice pack, is quite self-evident. Our current understanding of sea ice-behavior and ice internal stresses has improved considerably as a result of the Sea Ice Mechanics Initiative sponsored by the Office of Naval Research (Weeks and Timco, 1998). The program sought to develop physically based constitutive and fracture models, through observations of floe-floe interactions and stresses in the pack ice. The results were published in a special section of the *Journal of Geophysical Research* (Volume 103, number C10, 21,737–21,925, 1998). Overland *et al.* (1998) found that sea ice in Beaufort Sea in the Arctic behaves like a granular plastic material, with ice moving in large aggregates of 20 to 200 km separated by narrow shear zones, consistent with granular behavior. Shear failure of ice is rather common under compression.

The ice in the central Arctic is easily susceptible to ridging by collisions between floes; in other words, energy is expended through the ridge formation process. One thing we know for sure about sea-ice cover is that it is quite weak in tension and hence is easily ripped apart by tensile forces to form leads, long linear openings. Cracks open up in the floes in the interior and leads form. We also know that ice resists compressive forces, the compressive strength being a strong function of the ice thickness. While thin ice yields easily under compression leading to ridge and rubble formation and thickening of ice, thick ice can resist compression quite well. We also know from the behavior of ice near coastlines that it behaves like a plastic medium under an applied shear stress and begins to flow once certain deformation values are reached. All this means that even though individual ice floes are solid elastic media, on spatial scales much larger than the size of the individual floes, the rheological behavior is quite complex. Simple Newtonian (linear) viscous rheology with only shear viscosity, as used in early ice dynamics studies (Campbell, 1965), does not suffice, since there is no resistance to convergence. A generalized linear viscous law with both shear and bulk viscosities (and hence compressive strength) may work well far from shore, but close to the shore, the effective viscosities are nonlinear. To overcome this deficiency, plastic rheology, where the state of ice is independent of the magnitudes of the stress and strain rates, a highly nonlinear behavior has been proposed. But a pure plastic rheology is hard to implement, and it is also necessary to allow for cases when the ice is not flowing plastically to accommodate ice behavior both near and far from a shore.

Whether or not internal ice stresses are an important component of the momentum balance depends very much on the state of the sea ice itself. Intuitively, during summer, one would expect even the pack ice in the central Arctic to be unable to support significant internal stresses in its highly fragmented state with wide open leads between ice floes. On the other hand, during winter, when leads are few and far in between (and most often covered by thinner ice), internal ice stresses should be important. Convincing evidence for this behavior can be found in McPhee (1980), who showed that it was possible to simulate the drift speed of the manned ice camps during AIDJEX quite accurately during summer, even when internal ice stresses were completely ignored. On the other hand, during winter, large errors could be seen in the drift of the same camps, when the internal ice stress terms were ignored in the momentum balance (Figure 5.3.1). Inertial oscillations are also more evident in ice motion during summer than during winter (McPhee, 1980). At the edge of the ice pack, with ice broken into small floes by wave action and with large openings between floes, and ice often in the form of pancake ice, once again one should expect internal stresses to play a small part in the ice dynamics. Overall, observations suggest that the internal ice stresses, in addition to being a strong function of the ice thickness, should also be a strong function of the ice concentration or compactness (the area fraction of an elemental area covered by ice).

In a solid medium, a constitutive law relates the stress in the medium to the strain, Hooke's linear law for an elastic medium being an excellent example. On the other hand, in a fluid medium, the stress is a function of the strain rate, resulting in the classical Navier-Stokes equations for a Newtonian viscous fluid. When the stress exceeds a certain value, called the yield stress, some elastic solids begin to flow plastically, meaning that the stress becomes independent of the strain (and the strain rate). This is for simple ideal plastic behavior (in case the material hardens as it flows, the stress once again becomes a function of the

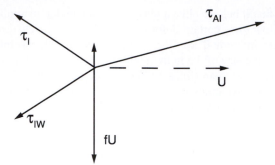

Figure 5.3.1 Momentum balance on an ice floe.

strain, but with a different proportionality constant). Based on extensive observations during AIDJEX, Coon *et al.* (1974) and Coon (1980; see also Colony and Rothrock, 1980, and Pritchard, 1980) proposed an elastic-plastic constitutive law for sea ice in the shape of a squished teardrop in the negative quadrant of the principal stress space (a shape somewhat similar to the elliptical yield curve for Hibler's rheology). The stress is constrained to lie within the space defined by the teardrop, whose size is a strong function of the ice thickness. The sea ice is supposed to behave as an elastic medium if the stress lies within the yield curve, while flowing as an ideal plastic once the stress state reaches the yield curve. Pritchard (1980) showed that with proper estimation of the yield strength, this law simulates the observed ice velocity field in the near-shore region of the Beaufort Sea during the winter period of AIDJEX.

Coon *et al.* (1998) discuss an anisotropic elastic-plastic constitutive law for sea ice. In sea-ice models so far, sea ice has been considered to be an isotropic medium, requiring only two stress and two deformation invariants, with the isotropic behavior modeled by a yield function with a preselected, fixed shape in the two-dimensional parameter space of the stress invariants, Hibler's viscous-plastic rheology being a typical example, with sea ice regarded as plastic material for stresses above the failure and viscous below. During summer, there is little pack ice stress and the dynamics are little affected, and the constitutive law is largely irrelevant. However, wintertime pack ice with few or no active leads is inherently anisotropic, and Coon *et al.* (1998) describe a rheology characterized by three stress components and three deformation rates, two of these normal and one shear. However, such very general anisotropic rheology is yet to be used in sea ice models, although Hibler and Schulson (1997) have proposed and Geiger *et al.* (1998) have used a simpler anisotropic rheology of truncated ellipse, which is Hibler's (1979) ellipse confined only to the negative quadrant of the principal stress space. This eliminates tensile stress completely so that the sea ice cannot support any tensile stress whatsoever.

5.3.1 VISCOUS-PLASTIC ICE RHEOLOGY

For numerical modeling purposes, however, it is preferable to have the simplest possible constitutive law that encapsulates the sea-ice behavior yet is numerically tractable. Bill Hibler made a seminal contribution to sea-ice modeling when he proposed a viscous-plastic rheology for sea ice in the central Arctic (Hibler, 1979, 1980a, b, c). He assumed that the sea ice can be treated as a two-dimensional continuum, an assumption essential for simplicity in modeling sea ice, and that it is isotropic. The latter assumption leads to a much simpler characterization of the stress in terms of the strain rate invariants. Now, for an isotropic continuum, the principal axes of its stress tensor must coincide with that of its strain rate tensor and the constitutive relationship relating the two must be invariant to any rotation of the coordinate system or interchange of the axes (Schlichting, 1978, page 55). This requires that the principal stress be of the form $\hat{\sigma}_{ij} = a\hat{\epsilon}_{ij} + b\hat{\epsilon}_{kk} + c$, or equivalently, $\sigma_{ij} = a\epsilon_{ij} + b\delta_{ij}\epsilon_{kk} + c$.

Hibler (1979) postulates

$$\sigma_{ij} = 2\eta\epsilon_{ij} + \left[(\zeta - \eta)\epsilon_{kk} - (P/2) \right]\delta_{ij}$$

$$\epsilon_{ij} = \frac{1}{2}\left[\frac{\partial U_{li}}{\partial x_j} + \frac{\partial U_{lj}}{\partial x_i} \right] \qquad (5.3.1)$$

$$\zeta \equiv \zeta(\epsilon_{ij}, P); \quad \eta \equiv \eta(\epsilon_{ij}, P)$$

where σ_{ij} is the two-dimensional stress tensor and ϵ_{ij} is the strain rate (deformation) tensor. P is the pressure (or ice strength) term, and is a strong function of the ice thickness and compactness. ζ is the (nonlinear) bulk viscosity, and η is the (nonlinear) shear viscosity; both are functions of the pressure and deformation tensor. The pressure and viscosity terms are given by (with modifications to ensure energy conservation, Hibler and Ip, 1995; see also Geiger *et al.*, 1998)

$$P_m = P^* D_I \exp[-C(1 - A_I)]; \quad C \sim 20, \ P^* \sim 2.75\text{x}10^4 \, \text{Nm}^{-2}$$

$$\zeta = \frac{P_m}{2\Delta}, \eta = \frac{\zeta}{e^2}; \ e \sim 2 \qquad (5.3.2)$$

$$\Delta = \max\left\{ \left[\epsilon_{kk}^2 + \frac{1}{e^2}\left(2\epsilon_{ij}\epsilon_{ij} - \epsilon_{kk}^2 \right) \right]^{1/2}, \epsilon_0 \right\}; \ \epsilon_0 \sim 2\text{x}10^{-9}\text{s}^{-1}$$

where e is the ratio of the major axis to the minor axis of the elliptic yield curve, or equivalently the ratio of compressive strength to the shear strength of ice. The bound on the deformation rate and hence the stresses is to approximate the rigid states inside the yield curve. The maximum ice strength P_m (equal to ice pressure for high strain rates) is a strongly nonlinear function of the ice concentration because of the large value of C. There is therefore a rapid decrease in ice internal stresses when A_I falls below 85%. Note that it is also a linear function of ice thickness D_I. Overland and Pease (1988) found that a quadratic dependence on thickness gave better results for coastal processes, but the quadratic form is seldom used in modeling pack ice in the interior.

The above formulation allows for plastic flow of sea ice at deformation rates larger than $\sim 10^{-4}$ day^{-1} to be modeled within the framework of a fluid flow, which allows linear viscous behavior at smaller values. The compressive strength is a strong function of the ice compactness and thickness. Both shear and bulk viscosities are allowed for and made functions of the pressure and deformation rates. Consequently, the sea ice strongly resists compression and shear, but allows dilation with little stress. Hibler's rheology is the most commonly used rheology for ice modeling purposes and corresponds to an isotropic viscous-plastic continuum with an elliptic yield curve in the negative quadrant of the principal stress space (Figure 5.1.3). While it is possible to device alternative constitutive relationships (Coon, 1980, Coon et al. 1998, for example), this rheology appears to work well and yields realistic results for the Arctic sea-ice cover, including its thickening towards the Canadian Archipelago and the transpolar ice drift (see Hibler, 1984, for a detailed discussion of the rheological behavior of sea ice and modeling ice rheology in numerical models).

It is also important to remember the inherent limitations of the particular rheology used when interpreting the results of a numerical simulation. Viscous-plastic law appears to have stood the test of time, at least for coarse grid simulations of sea ice in the Arctic. Holland et al. (1993) have made an extensive study of sea-ice behavior in a model using standard rheology. Other rheologies have been proposed recently, using an analogy to a cavitating flow, for example (Flato and Hibler, 1992). This cavitating fluid rheology is equivalent to ignoring the shear stresses and hence its shape in the principal axis domain collapses from Hibler's (1979) ellipse to a line along the major axis of the ellipse.

Since any proposed rheology is an empirical postulate with little rigorous theoretical basis, the ultimate test is how well it simulates the known behavior of sea ice cover in ice-covered polar and subpolar seas. Geiger et al. (1998) carried out a detailed study of ice drift and deformation in the Weddell Sea and compared the results of the ice model to observations. They found that the viscous-plastic rheology gave the best results (cavitating fluid rheology was found inaccurate), confirming that some shear resistance is essential to realistic rheology. Ice deformation proved to be far more difficult to predict than ice drift, exhibiting more sensitivity to internal ice parameters. None of the ice models performed satisfactorily with respect to ice deformation.

The fact that Eq. (5.3.1) and Eq. (5.3.2) define an ellipse in the principal stress space can be seen as follows. The principal strain rates are given by

$$\hat{\varepsilon}_{11} = \frac{1}{2}\left[\Theta + \left(\Theta^2 - 4\Phi\right)^{1/2}\right]; \hat{\varepsilon}_{22} = \frac{1}{2}\left[\Theta - \left(\Theta^2 - 4\Phi\right)^{1/2}\right]; \hat{\varepsilon}_{12} = \hat{\varepsilon}_{21} = 0$$

$$(5.3.3)$$

where $\Theta = \varepsilon_{11} + \varepsilon_{22}$ and $\Phi = \varepsilon_{11}\varepsilon_{22} - \varepsilon_{12}^2$ are the two invariants associated with the strain rate tensor or equivalently $\varepsilon_{I} = \hat{\varepsilon}_{11} + \hat{\varepsilon}_{22}$ and $\varepsilon_{II} = \hat{\varepsilon}_{11} - \hat{\varepsilon}_{22}$. It is easy to see from Eq. (5.3.1) that the principal stresses are given by (the principal values of any tensor π_{ij} are given by the roots λ_1 and λ_2 of the characteristic equation of π_{ij}: $\left|\pi_{ij} - \lambda\delta_{ij}\right| = 0$)

$$\hat{\sigma}_{11} = \zeta\varepsilon_{I} + \eta\varepsilon_{II} - (P/2); \hat{\sigma}_{22} = \zeta\varepsilon_{I} - \eta\varepsilon_{II} - (P/2); \hat{\sigma}_{12} = \hat{\sigma}_{21} = 0$$

$$(5.3.4)$$

Since

$$\Delta = \max\left\{\left[\varepsilon_{I}^2 + \frac{\varepsilon_{II}^2}{e^2}\right]^{1/2}, \varepsilon_0\right\}; \zeta = \frac{P}{2\Delta} \text{ and } \eta = \frac{P}{2\Delta e^2}$$

$$(5.3.5)$$

Hibler's constitutive law for sea ice can be written as

$$\left(\hat{\sigma}_{11} + \hat{\sigma}_{22} + P\right)^2 + e^2\left(\hat{\sigma}_{22} - \hat{\sigma}_{11}\right)^2 = \min\left\{P^2\left[1, \left(\frac{\varepsilon_{I}^2 + \varepsilon_{II}^2/e^2}{\varepsilon_0^2}\right)\right]\right\}$$

$$(5.3.6)$$

Since the equation for an ellipse in the principal stress space (see Figure 5.3.2) is

$$\left(x + a_0\right)^2 + \frac{a^2}{b^2}\left(y - b_0\right)^2 = a^2;$$

$$x = \left(\hat{\sigma}_{11}\cos\theta + \hat{\sigma}_{22}\sin\theta\right); y = \left(\hat{\sigma}_{22}\cos\theta - \hat{\sigma}_{11}\sin\theta\right)$$

$$(5.3.7)$$

it is easy to see that for $b_0 = 0$, and $\theta = 45°$, we get

$$\left(\hat{\sigma}_{11} + \hat{\sigma}_{22} + 2^{1/2}a_0\right)^2 + \frac{a^2}{b^2}\left(\hat{\sigma}_{22} - \hat{\sigma}_{11}\right)^2 = 2a^2$$

$$(5.3.8)$$

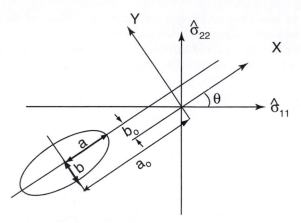

Figure 5.3.2 The elliptic yield curve in principal stress space.

Thus, we have a yield curve in the form of an ellipse oriented at $45°$ to the principal stress axes and centered at $(-P/2, -P/2)$, with eccentricity e and the semimajor axis equal to either $P/\sqrt{2}$ or $(P/\sqrt{2})\left[\left(\varepsilon_I^2 + \varepsilon_{II}^2/e^2\right)/\varepsilon_0^2\right]^{1/2}$. The stresses in the modeled sea ice fall on either one of the two elliptic yield curves.

The above constitutive relationship is quite general and several interesting limits are worth exploring. If $\eta = 0$ (no shear viscosity, only bulk viscosity), it is easy to see from Eq. (5.3.8) that the stress-strain law is simply a straight line inclined at an angle of $45°$ to the principal axes (see Figure 5.3.2):

$$\hat{\sigma}_{11} = \zeta\varepsilon_I - (P/2); \quad \hat{\sigma}_{22} = \zeta\varepsilon_I - (P/2) \rightarrow \hat{\sigma}_{11} = \hat{\sigma}_{22} \qquad (5.3.9)$$

If $\zeta = 0$ (no bulk viscosity, only shear viscosity), the law is a straight line in the negative quadrant (see Figure 5.3.2):

$$\hat{\sigma}_{11} = \eta\varepsilon_{II} - (P/2); \quad \hat{\sigma}_{22} = -\eta\varepsilon_{II} - (P/2) \rightarrow \hat{\sigma}_{11} + \hat{\sigma}_{22} = -P \qquad (5.3.10)$$

If both viscosities are zero, it is just a point at the center of the ellipse. Roed and O'Brien (1983) used a constitutive law that ignored both viscosities (many earlier sea-ice models, such as that by Parkinson and Washington, 1979, ignored ice internal stresses altogether assuming free drift instead). Semtner (1987) used the zero shear viscosity limit ($\eta = 0$) of Hibler's formulation to relax the time step constraints imposed by large shear viscosities. This is similar to the cavitating fluid model of Flato and Hibler (1992). However, ignoring shear viscosities for numerical expediency involves compromises to the physics of sea-ice rheology (Geiger *et al.* 1998). Overall, the full formulation by Hibler is to be preferred for general applicability.

It is also interesting to note that the constitutive relationship for a general Newtonian viscous fluid is a subset of Eq. (5.3.1), with no η term inside the square brackets and $\zeta = -\eta$ (by Stoke's hypothesis) and with P and η being constants. The relationship is the same as in Eq. (5.3.10) but with constant coefficients. Not surprisingly, the law traces the same straight line as the shear viscosity-only limit of Hibler's rheology (except that the coefficients are constants and not functions of macro properties of the medium).

Hakkinen (1987) proposed a constitutive law of the form

$$\sigma_{ij} = \phi_1 \varepsilon_{ij} + \left[-\frac{\phi_1}{2} \varepsilon_{kk} - \phi_0 \right] \delta_{ij}$$

$$\phi_0 = \rho_1 D_1 \mu_0 \exp\left[-C(1 - A_1) \right] \quad (\Theta \leq 0)$$

$$\phi_1 = \rho_1 D_1 \mu_1 \exp\left[-C(1 - A_1) \right] \exp\left[-\gamma \Theta |\Psi| \right]$$

$$\Theta \equiv \text{tr}(\varepsilon_{ij}) = (\varepsilon_{11} + \varepsilon_{22}); \quad \Psi \equiv \frac{1}{2} \left[(\text{tr } \varepsilon_{ij})^2 - \text{tr}(\varepsilon_{ij} \varepsilon_{ij}) \right] = (\varepsilon_{11} \varepsilon_{22} - \varepsilon_{12}^2)$$

$$\mu_0 = 1 \, \text{Nkg}^{-1}, \mu_1 = 10^4 \, \text{m}^2 \text{s}^{-1}, C = 15, \gamma = 3 \times 10^8 \text{s}^3$$

$$(5.3.11)$$

In Hibler's notation, this is equivalent to $\zeta = 0$ (e = 0), with

$$\eta = \frac{P}{2\Delta}, \quad \Delta = \frac{2\mu_0}{\mu_1} \exp\left[\gamma \Theta |\Psi| \right]; \quad C = 15, P^* = 2\rho_1 \mu_0 = 1820 \text{Nm}^{-2}$$

$$(5.3.12)$$

It is therefore equivalent to shear viscosity-only rheology and for tension ($\Theta > 0$), pressure terms are put to zero (but not the shear viscosity), whereas Hibler's formulation provides for a small but nonzero tensile strength for ice.

The viscous-plastic rheology has been the workhorse of ice dynamics modeling since it was first formulated (Hibler, 1979), and a better understanding of its behavior and sensitivity to various parameters has emerged since then (e.g., Steele *et al.,* 1997). Hibler (1979) originally selected a P^* value of 5000 N m^{-2}, which has since then been increased by a factor of 5.5 so that the ice velocities from the dynamical model are in better tune with values observed by buoys (Hibler and Walsh, 1992). In a careful analysis of force balance in the ice model coupled to an ocean model of the Arctic, Steele *et al.* (1997) show the exquisite sensitivity of the resulting ice velocity and thickness distributions to the choice of P^*. While the value in Eq. (5.3.2) gave the most realistic depiction of both the ice transport in and out of the Arctic and the ice thicknesses along the north American side, decreasing it by a factor of 5 led to unrealistically large ice velocities, and unrealistically large ice thicknesses along north America. The internal ice stresses became negligible and the ice was in essentially free drift. On the other hand, increasing it by a factor of 5 led to virtual cessation of ice motion, with ice thickness distributions tending to approach the pure thermodynamic equilibrium limit, which they do if $P^* \to \infty$. The balance is then principally like land-fast ice, between the wind stress and the internal stress gradient. The dependence on increasing P^* is, however, highly nonlinear. Note that while most sea-ice models assume the pressure gradient to be proportional to the average ice thickness in Eq. (5.3.1), formulations such as Rothrock (1975) and Hibler (1980) consider a more complex function based on a thickness distribution (see also Flato and Hibler, 1995).

Steele *et al.* (1997) also confirm that the tendency terms in the force balance Eq. (5.2.1) are negligible. All three terms—the pressure gradient, the shear viscosity, and bulk viscosity—contribute significantly to the internal stress gradient, which is important to the force balance on ice in the Arctic, except during summer, where the ice is in free drift in most of the Arctic. The balance is roughly between the air stress, the water drag, and the internal stress gradient on timescales of a few weeks and beyond (Figure 5.3.1), while on daily timescales, the balance is between the air stress and water drag or air stress and internal stress gradient, depending on the location and time. The transition between the two appears to occur around the synoptic timescales (5 to 7 days).

Hibler and Schulson (1997) have recently proposed truncating the original Hibler (1989) ellipse. Their modification of a truncated elliptical yield curve is equivalent to

$$\eta = \min\left(\frac{\zeta}{e^2}, \eta_0\right); \; \eta_0 = (P/2 - \zeta \varepsilon_{kk})/\varepsilon_s$$

$$\varepsilon_s = \left[\left(\varepsilon_{11} - \varepsilon_{22}\right)^2 + 4\varepsilon_{12}^2\right]^{-1/2}$$

(5.3.13)

where ε_s is the maximum shear rate. The bound on the shear stress ensures that there is no tensile stress. In addition, to ensure that there is no stress when the strain rate is zero, Geiger *et al.* (1998) put $P = 2\Delta\zeta_0$, where $\zeta_0 = \min [P_m/2\Delta, \zeta_{max}]$. When plastic flow occurs, P is constant equal to P_m, but for $\zeta = \zeta_{max}$, the stress state lies on a geometrically similar but smaller yield curve (Geiger *et al.*, 1998).

It is not clear if viscous-plastic rheology is appropriate for the MIZ. The characteristics of the ice cover are basically different here than in the central Arctic, where this rheology has been successfully employed. Gutfraind and

Savage (1997) have shown that the results of discrete floe simulations of ice motion in a marginal ice zone are in general agreement with the ice motions derived from a viscous-plastic model, but with the elliptical yield curve replaced by Mohr-Coulomb rheology, where the yield curve is a circle.

5.3.2 ELASTIC-VISCOUS-PLASTIC ICE RHEOLOGY

While Hibler's viscous-plastic law is the well-accepted standard in numerical modeling of sea ice dynamics, it suffers from numerical disadvantages. This can be seen (Hunke and Dukowicz, 1997) by rewriting Eq. (5.3.1):

$$\frac{1}{2\eta}\sigma_{ij} + \frac{1}{4\zeta}\left[\left(1 - \frac{\zeta}{\eta}\right)\sigma_{kk} + P\right]\delta_{ij} = \varepsilon_{ij} \qquad (5.3.14)$$

Now, the reason for Hibler's imposition of a lower bound Eq. (5.3.2) on the strain rate becomes clear. Without that, the effective viscosities become infinite in the limit of zero strain rate and Eq. (5.3.14), the ideal viscous-plastic law, becomes singular. In this limit, when the ice pack moves as a rigid solid, Hibler's bounds allow it to creep very slowly as a viscous fluid (Hibler, 1986). However, the maximum possible value of shear viscosity η is set at $\sim 0.6 \times 10^8$, and this imposes a stringent time step restriction for stability in an explicit numerical scheme, when the ice is nearly rigid (Chapter 2). Typical time step is ~ 1 s for a 100 km grid (Ip *et al.*, 1991). Implicit schemes, such as SOR (Hibler, 1979) and line relaxation (Holland *et al.*, 1993; Oberhuber, 1993a), are therefore used to circumvent this limitation and take steps as long as a day, but they suffer from slow convergence properties for large matrices typical of high resolution models. Implicit methods are also ill suited to parallel computers. Another shortcoming is that for accurate transient behavior, even in implicit methods, the time step must resolve adequately the shortest timescales in the applied forcing. In this case, the time step must be of the order of minutes for, say, daily forcing. Hibler (1979) and Zhang and Hibler (1997) use a predictor-corrector scheme to step forward in time the ice momentum equations, while solving iteratively the resulting equations at each time step by SOR. The two components of the momentum equations are decoupled by evaluating the cross derivatives at time step n (instead of n+1), and this permits row-by-row solution of the U equations and column-by-column solutions of V equations. This, however, results in inaccuracies in ice drift, and Hunke and Dukowicz (1997) present an alternative scheme involving a less efficient backward Euler scheme for time marching but an efficient preconditioned conjugate gradient method for iteration that does not split the strain rate tensor.

In an effort to overcome the severe time step limitations imposed by the full viscous-plastic law, Flato and Hibler (1990, 1992) proposed a zero shear viscosity rheology called a cavitating fluid model. This is equivalent to letting e → ∞. This is solely for numerical expediency and has since been shown to be of doubtful validity. Steele *et al.* (1997) show that shear viscous forces in ice are a significant part of the internal stress gradient and hence the force balance in the Arctic, as well as the energy balance. Observations also suggest a value of 2 to 3 for e (Stern *et al.*, 1995).

In an effort to overcome the limitations and still retain the essence of viscous-plastic behavior, Hunke and Dukowicz (1997) propose an elastic-viscous-plastic law that permits efficient explicit time stepping (and hence suitable to multiprocessor architectures) and accurate transient behavior. This is accomplished simply by adding an elastic term to Eq. (5.3.12) to obtain a prognostic equation for the stress tensor:

$$\frac{1}{E}\frac{\partial \sigma_{ij}}{\partial t} + \frac{1}{2\eta}\sigma_{ij} + \frac{1}{4\zeta}\left[\left(1 - \frac{\zeta}{\eta}\right)\sigma_{kk} + P\right]\delta_{ij} = \varepsilon_{ij} \qquad (5.3.15)$$

Unlike the earlier AIDJEX-derived elastic-plastic law constructs, this is done for purely numerical expediency with parametric values chosen solely for numerical efficiency and not for accurate elastic behavior. In the limit of large η and ζ, the law reduces to an elastic law,

$$\frac{1}{E}\frac{\partial \sigma_{ij}}{\partial t} = \varepsilon_{ij} \qquad (5.3.16)$$

where E is the Young's modulus, but the viscous-plastic behavior Eq. (5.3.14) is retained in the steady state limit. Thus, the numerical behavior is controlled by Eq. (5.3.16) when the ice is nearly rigid and this permits large time steps to be taken in an explicit scheme. Unlike for pure viscous-plastic constitutive law, any readjustment to time-dependent forcing involves fast elastic waves and is rapid, hence the transient response of ice tends to be more accurate even with larger time steps. Hunke and Dukowicz (1997) solve Eq. (5.3.15), holding η and ζ fixed, obtaining the stress tensor σ_{ij}, which is then used to solve the momentum equation, Eq. (5.2.1). The viscosities are updated every N time steps.

Following Hunke and Dukowicz (1997), the time-stepping aspects can be illustrated by the analysis of a simplified 1-D version of Eqs. (5.2.1) and

(5.3.15), neglecting σ_{22} and σ_{12}, and absorbing the $P/4\zeta$ term into $\sigma_{11} = \sigma$:

$$\rho_I A_I D_I \frac{\partial U_I}{\partial t} = \frac{\partial \sigma}{\partial x} + (\tau_{AI} - \tau_{IO})$$

$$\frac{1}{E} \frac{\partial \sigma}{\partial t} + \frac{\sigma}{\zeta} = \frac{\partial U_I}{\partial x}$$

(5.3.17)

In the limit $E \rightarrow \infty$, one gets the viscous-plastic model

$$\rho_I A_I D_I \frac{\partial U_I}{\partial t} = \zeta \frac{\partial^2 U_I}{\partial x^2} + (\tau_{AI} - \tau_{IO})$$

(5.3.18)

with its timescale $T_v = (\rho_I A_I D_I / \zeta) \Delta x^2$. In the limit $\zeta \rightarrow \infty$, one gets the elastic model

$$\frac{\partial^2 U_I}{\partial t^2} = C_e^2 \frac{\partial^2 U_I}{\partial x^2} + \frac{1}{\rho_I A_I D_I} \frac{\partial}{\partial t} (\tau_{AI} - \tau_{IO})$$

(5.3.19)

where $C_e = (E / \rho_I A_I D_I)^{1/2}$ is the elastic wave speed, and the elastic timescale is T_e $= \Delta x / C_e = (\rho_I A_I D_I / E)^{1/2} \Delta x$. By choosing the value of E appropriately, T_e can be made several orders of magnitude larger than T_v. For the most general 2-D elastic-viscous-plastic case, Neumann stability analysis shows that an explicit scheme is stable as long as $\Delta t \leq 2^{1/2} T_e$, irrespective of the value of T_v, contrary to the viscous-plastic law, which would require a highly stringent $\Delta t \leq T_v/2$. To obtain accurate transient response to forcing, Hunke and Ducowicz (1997) show that the time split $N = T_e/T_v$ is optimum. They choose for the general 2-D problem

$$0 < E < \frac{2\rho_I A_I D_I}{(\Delta t_e)^2} \min(\Delta x^2, \Delta y^2); \quad (\Delta t_e)^2 = \sqrt{2} T_v \Delta t_v$$

(5.3.20)

Thus given a suitable choice of T_v and Δt_v, Δt_e is known and the corresponding value of E ($E \rightarrow \infty$ as $\Delta t_e \rightarrow 0$). Note that $E \rightarrow 0$ as $A_I \rightarrow 0$. The procedure is to select the time steps Δt_v and Δt_e, and the corresponding value of E, and subcycle over $N = \Delta t_v/\Delta t_e$ steps, the prognostic momentum and ice internal stress equations holding viscosities constant, before updating them. Hunke and Ducowicz (1997) claim that to produce an accurate transient response in a 2-D test problem, the implicit viscous-plastic code requires time steps of the order of 60 s and hence CPU times of the order of hours, whereas the elastic-viscous-plastic explicit code with typical values of $\Delta t_v = 21600$ s and $\Delta t_e = 216$ s takes less than a minute to produce similar accuracies. More significantly, the CPU time is still less than half that for the viscous-plastic code with a traditional time step of 1 day. It remains to be seen if passage of time reinforces its use, but it is noteworthy that the addition of a time-dependent term to change the nature of the relationship from a diagnostic to prognostic equation is somewhat similar in philosophy to converting a harder elliptic steady-state problem to a time-steppable parabolic non-steady-state problem in conventional CFD (see Chapter 2).

In Section 5.6, we present yet another approach, the application of a generalized Thomas algorithm by Kantha and Mellor (1989b), to solving the ice dynamical equations.

5.4 CONTINUITY EQUATIONS FOR SEA ICE

It is important to note that since the rheological properties of sea ice are strong functions of the ice compactness and thickness, a sea ice model must also allow for accurate computation of these quantities. Even the simplest model for sea ice must allow for the possibility of open water (or equivalently thin ice). This is usually done by considering each elemental area to be composed of ice floes of thickness h_I and open water, the area concentration or compactness of ice being A_I and therefore the average ice thickness is $D_I = h_I A_I$. Equations are then needed for A_I and D_I:

$$\frac{\partial}{\partial t}(A_I) + \frac{\partial}{\partial x_j}(A_I U_{Ij}) = \frac{\rho_0}{\rho_I} E(1 - A_I) w_{AO} / (D_I / A_I) \quad (0 \le A_I \le 1)$$

$$\frac{\partial}{\partial t}(D_I) + \frac{\partial}{\partial x_j}(D_I U_{Ij}) = \frac{\rho_0}{\rho_I} \left[A_I (w_{AI} + w_{IO}) + (1 - A_I) w_{AO} \right] (0 \le A_I \le 1)$$

$$(5.4.1)$$

The second equation is simply a statement of conservation of mass. The terms on the RHS are source and sink terms due to freezing and melting brought on by thermodynamic exchanges with the atmosphere and the ocean. w denotes the ice growth rate, with subscripts AI, AO, and IO denoting atmosphere-ice, atmosphere-ocean, and ice-ocean interfaces. The first equation is empirical, patterned after the mass conservation equation, and was first proposed by Nikiferov (1957). Nevertheless it is remarkable in that it captures the ice behavior quite well. When ice is diverging, A_I decreases according to this equation and open water is created. Under convergence, A_I increases and when it reaches a value of unity, any further convergence does not increase A_I and hence the ice thickness increases, simulating ridging processes.

In modeling the dynamics of sea ice cover, the above model corresponding to a single thick ice category co-existing with open water (or equivalently very thin ice) is the simplest and most tractable. However, recent studies have highlighted the importance of regarding the sea ice to consist of several ice categories and thicknesses (or even a continuous thickness distribution) to more accurately simulate sea ice thermodynamics and hence air-ice-ocean interactions. In principle, Eqs. (5.2.1) and (5.4.1) can be extended to multiple ice categories and thicknesses by keeping track of each of these separately. However, this necessarily involves conversion of one category and thickness to another, implying interaction terms in each set of equations that are hard to model without excessive empiricism. There also arise questions on how to handle floe-floe interactions and rheological properties of a multiple-thickness sea ice cover.

5.5 RESPONSE OF SEA ICE TO STORM PASSAGE

Ice growth and decay both due to thermodynamics at the ocean and atmospheric interfaces, as well as that due to dynamical processes of dilation and ridging, can be dealt with by solving Eq. (5.4.1) together with the Eq. (5.2.1). However, the computation of the source and sink terms in Eq. (5.4.1) requires solution of the coupled ice-ocean system, which is the subject of Chapter 12. Here we ignore thermodynamical aspects and deal only with the dynamical processes and their influence on the ice properties.

The Arctic Buoy program showed that the sea ice in the Arctic behaves differently to the passages of low and high pressure systems. Basically, the behavior depends on whether the winds are cyclonic or anticyclonic. For a low pressure system with cyclonic winds, the ice moves away from the center and the resulting divergence thins the ice and decreases the ice concentration. Consequently, the ice is pretty much in free drift and internal ice stresses are negligible around the storm center. The ice drift is also at an angle to the isobars (Figure 5.5.1, top). On the other hand, for a high pressure system, with anticyclonic winds, the ice is driven toward the center, compacted and the thickness increased. Ice stresses become an important part of the momentum balance and near-solid body rotation around the storm center is the result (Figure 5.5.1, bottom).

A good idea of the response of sea ice to atmospheric forcing and its differing behavior during convergence and divergence can be obtained by examining the response of a box ocean covered by sea ice of a certain thickness and concentration to passage of low and high pressure systems through the domain. This involves the solution of Eqs. (5.2.1) and (5.4.1), the latter without any source or sink terms on the RHS. The numerical details of solution are explained

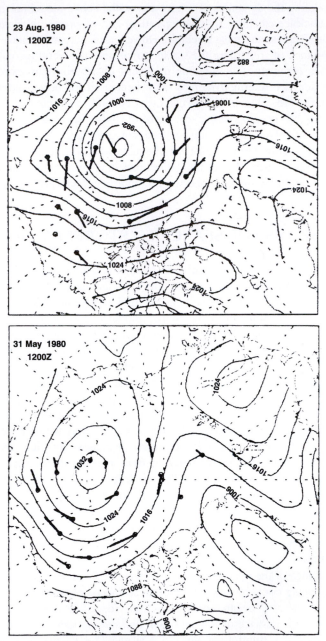

Figure 5.5.1 Arctic buoy trajectories during the passage of a low pressure (cyclonic) system (top) and a high pressure (anticyclonic) system (bottom). From Thorndike and Colony (1980).

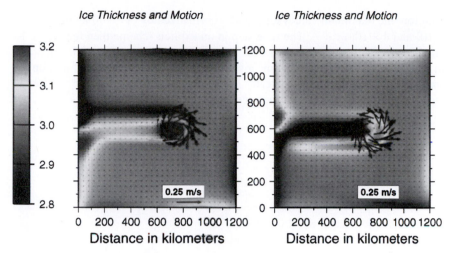

Figure 5.5.2 Ice thickness distribution during the passage of a low-pressure (cyclonic) system (right) and a high pressure (anticyclonic) system (left).

below. Figure 5.5.2 shows the ice thickness distribution for the two cases during the passage of the storm. For cyclonic storms, the divergence of ice leads to thinning of ice underneath the storm and free drift prevails, while for anticyclonic ones, ice convergence leads to thicker ice in the center and a solid-body like motion of sea ice around the center.

5.6 NUMERICS

Solving for ice thickness and compactness (concentration) requires solving the dynamical equations with ice rheology involving internal stress terms. Internal stresses can build up to large values and lead to severe time step limitations on explicit numerical ice models. Making the diffusion terms implicit permits the severe time step limitation of an explicit scheme to be sidestepped (see Chapter 2). This strategy is essential for solving ice dynamical equations, which involve extremely large viscosities for the interior ice pack. However, this advantage has a penalty of its own: the solution scheme is more complex, since it now involves the inversion of a matrix. For 1-D problems, a tridiagonal matrix is involved and efficient schemes for inversion are available. For 2-D problems, the solution requires an iterative solver, which will be outlined below.

For sea ice, it is possible to estimate the time step limitation imposed by large viscosities in ice. For a 3 m thick ice with a compactness of unity, $D_I = 3$ m, $P = P^* D_I = 3 \times 10^4 \, \text{N m}^{-1}$. The minimum value of deformation rate is $2 \times 10^{-7} \, \text{s}^{-1}$ and therefore the maximum values for η and ζ normalized by $\rho_I \sim 900$ kg m^{-3} are $\sim 2 \times 10^7$ and 8×10^7 m^2 s^{-2}. The time step in an explicit scheme then for, say, a 10 km grid size is about 2 s. Now consider the speed of propagation of disturbances in ice. This can be obtained by linearizing the governing equations of motion for small disturbances around a steady state. For simplicity consider motion in one dimension only and the ice to be of constant thickness. Let $U_I = 0 + U_I', A_I = A_I^0 + A_I'$, $P = P^0 + P'$. Then the perturbation quantities obey

$$\frac{\partial}{\partial t}(A_I') + A_I^0 \frac{\partial}{\partial x}(U_I') = 0$$

$$\frac{\partial}{\partial t}(U_I') + \frac{1}{\rho_I D_I^0} \frac{\partial}{\partial x}(P') = 0 \qquad (5.6.1)$$

$$P' = P^* D_I^0 C \exp\left[-C(1 - A_I^0)\right] A_I'$$

which yields a wave equation,

$$\frac{\partial^2}{\partial t^2}(U_I') - c^2 \frac{\partial^2}{\partial x^2}(U_I') = 0$$

$$(5.6.2)$$

$$c^2 = \frac{P^* C A_I^0}{\rho_I} \exp\left[-C(1 - A_I^0)\right]$$

The maximum value of c occurs for ice compactness of unity. Substituting for various ice parameters, $c \sim 15$ m s^{-1} and therefore the CFL criterion would give a limit of 8 minutes for a 10 km grid 2-D ice model. The viscosity constraint is therefore far more severe.

5.6.1 GOVERNING EQUATIONS IN ORTHOGONAL CURVILINEAR COORDINATES

We will present the governing equations for sea ice in orthogonal curvilinear coordinates, which is the form used in the ice-ocean coupled model of Mellor and Kantha (1989), Kantha and Mellor (1989b), Hakkinen, Mellor, and Kantha (1992), and Hakkinen and Mellor (1990, 1992). We will then discuss the solution technique. A review of ice-ocean coupled models can be found in Hakkinen (1990) and Mellor and Hakkinen (1995) (see also Chapter 12). The governing equations are

$$\frac{\partial}{\partial t}(A_I) + \frac{1}{h_1 h_2}\left[\frac{\partial}{\partial \xi_1}(A_I U_I h_2) + \frac{\partial}{\partial \xi_2}(A_I V_I h_1)\right] =$$

$$\frac{\rho_w}{\rho_I} E(1 - A_I) w_{AO} / (D_I / A_I)$$

$$+ \frac{1}{h_1 h_2}\left[\frac{\partial}{\partial \xi_1}\left(A_H \frac{h_2}{h_1}\frac{\partial A_I}{\partial \xi_1}\right) + \frac{\partial}{\partial \xi_2}\left(A_H \frac{h_1}{h_2}\frac{\partial A_I}{\partial \xi_2}\right)\right] \qquad (0 \le A_I \le 1)$$

$$(5.6.3)$$

$$\frac{\partial}{\partial t}(D_I) + \frac{1}{h_1 h_2}\left[\frac{\partial}{\partial \xi_1}(D_I U_I h_2) + \frac{\partial}{\partial \xi_2}(D_I V_I h_1)\right] =$$

$$\frac{\rho_w}{\rho_I}\left[A_I(w_{AI} + w_{IO}) + (1 - A_I)w_{AO}\right]$$

$$+ \frac{1}{h_1 h_2}\left[\frac{\partial}{\partial \xi_1}\left(A_H \frac{h_2}{h_1}\frac{\partial D_I}{\partial \xi_1}\right) + \frac{\partial}{\partial \xi_2}\left(A_H \frac{h_1}{h_2}\frac{\partial D_I}{\partial \xi_2}\right)\right]$$

$$\frac{\partial}{\partial t}(D_I U_I) + \frac{1}{h_1^2 h_2}\left[\frac{\partial}{\partial \xi_1}(D_I U_I U_I h_1 h_2) + \frac{\partial}{\partial \xi_2}(D_I U_I V_I h_1^2)\right]$$

$$- \frac{1}{h_1^2}\frac{\partial h_1}{\partial \xi_1}(D_I U_I U_I) - \frac{1}{h_1 h_2}\frac{\partial h_2}{\partial \xi_1}(D_I V_I V_I) - f D_I V_I = -\frac{g D_I}{h_1}\frac{\partial \eta_0}{\partial \xi_1}$$

$$+ \frac{A_I}{\rho_I}(\tau_{AIx} - \tau_{IOx}) + \frac{F_x}{\rho_I}$$

$$\frac{\partial}{\partial t}(D_I V_I) + \frac{1}{h_1 h_2^2}\left[\frac{\partial}{\partial \xi_1}(D_I U_I V_I h_2^2) + \frac{\partial}{\partial \xi_2}(D_I V_I V_I h_1 h_2)\right]$$

$$- \frac{1}{h_1 h_2}\frac{\partial h_1}{\partial \xi_2}(D_I U_I U_I) - \frac{1}{h_2^2}\frac{\partial h_2}{\partial \xi_2}(D_I V_I V_I) + f D_I U_I = -\frac{g D_I}{h_2}\frac{\partial \eta_0}{\partial \xi_2}$$

$$+ \frac{A_I}{\rho_I}(\tau_{AIy} - \tau_{IOy}) + \frac{F_y}{\rho_I}$$

where

$$F_x = \frac{1}{h_1 h_2}\left[\frac{\partial}{\partial\xi_1}(\sigma_{11}h_2) + \frac{\partial}{\partial\xi_2}(\sigma_{12}h_1) + \sigma_{12}\frac{\partial h_1}{\partial\xi_2} - \sigma_{22}\frac{\partial h_2}{\partial\xi_1}\right]$$

$$F_y = \frac{1}{h_1 h_2}\left[\frac{\partial}{\partial\xi_2}(\sigma_{22}h_1) + \frac{\partial}{\partial\xi_1}(\sigma_{12}h_2) + \sigma_{21}\frac{\partial h_2}{\partial\xi_1} - \sigma_{11}\frac{\partial h_1}{\partial\xi_2}\right]$$

$$\sigma_{ij} = 2\eta\varepsilon_{ij} + \left[(\zeta - \eta)\varepsilon_{kk} - \frac{P}{2}\right]\delta_{ij} \qquad (5.6.4)$$

$$\varepsilon_{11} = \frac{1}{h_1}\frac{\partial U_I}{\partial\xi_1} + \frac{V_I}{h_1 h_2}\frac{\partial h_1}{\partial\xi_2}; \varepsilon_{22} = \frac{1}{h_2}\frac{\partial V_I}{\partial\xi_2} + \frac{U_I}{h_1 h_2}$$

$$\frac{\partial h_2}{\partial\xi_1}$$

$$\varepsilon_{12} = \frac{h_2}{2h_1}\frac{\partial}{\partial\xi_1}\left(\frac{V_I}{h_2}\right) + \frac{h_1}{2h_2}\frac{\partial}{\partial\xi_2}\left(\frac{U_I}{h_1}\right)$$

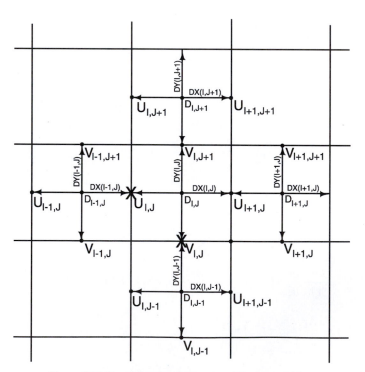

Figure 5.6.1 C grid showing the location of the ice variables.

These equations can be cast in finite difference form (see Figure 5.6.1 for the grid):

$$\Delta x\; \Delta y\frac{\partial}{\partial t}(A_I) + \Delta_x\left(A_I U_I \Delta y\right) + \Delta_y\left(A_I V_I \Delta x\right) =$$

$$\Delta x\; \Delta y\left[\frac{\rho_w}{\rho_I}E(1-A_I)w_{AO}/(D_I/A_I)\right]$$

$$+ \;\Delta_x\left(A_H\Delta y\frac{\Delta_x(A_I)}{\Delta x}\right) + \Delta_y\left(A_H\Delta x\frac{\Delta_y(A_I)}{\Delta y}\right) \qquad (0 \le A_I \le 1)$$

$$\Delta x\; \Delta y\frac{\partial}{\partial t}(D_I) + \Delta_x\left(D_I U_I \Delta y\right) + \Delta_y\left(D_I V_I \Delta x\right) =$$

$$\Delta x\; \Delta y\frac{\rho_w}{\rho_I}\left[A_I\left(w_{AI}+w_{IO}\right)+(1-A_I)w_{AO}\right]$$

$$+ \;\Delta_x\left(A_H\Delta y\frac{\Delta_x(D_I)}{\Delta x}\right) + \Delta_y\left(A_H\Delta x\frac{\Delta_y(D_I)}{\Delta y}\right)$$

$$(5.6.5)$$

$$\Delta x\; \Delta y\frac{\partial}{\partial t}(D_I V_I) + \Delta_x\left(D_I U_I V_I\Delta y\right) + \Delta_y\left(D_I V_I V_I\Delta x\right) + \overline{f}D_I U_I\Delta x\; \Delta y =$$

$$-gD_I\Delta x\Delta_y\left(\eta_0\right) + \Delta x\; \Delta y\;\frac{A_I}{\rho_I}\left(\tau_{AIy}-\tau_{IOy}\right)$$

$$+\frac{1}{\rho_I}\left[\Delta_y\left(\sigma_{22}\Delta x\right)+\sigma_{12}\Delta_x\left(\Delta y\right)+\Delta_x\left(\sigma_{12}\Delta y\right)-\sigma_{11}\Delta_y\left(\Delta x\right)\right]$$

$$\overline{f}=f-\left[\frac{U_I\Delta_y\left(\Delta x\right)-V_I\Delta_x\left(\Delta y\right)}{\Delta x\; \Delta y}\right]$$

$$\sigma_{11}=2\eta\left[\frac{\Delta_x(U_I)}{\Delta x}+\frac{V_I}{\Delta x}\frac{\Delta_y(\Delta x)}{\Delta y}\right]-\frac{P}{2}$$

$$+(\zeta-\eta)\left[\frac{\Delta_x(U_I)}{\Delta x}+\frac{V_I}{\Delta x}\frac{\Delta_y(\Delta x)}{\Delta y}+\frac{\Delta_y(V_I)}{\Delta y}+\frac{U_I}{\Delta y}\frac{\Delta_x(\Delta y)}{\Delta x}\right]$$

$$\sigma_{12}=\eta\left[\frac{\Delta y}{\Delta x}\Delta_x\left(\frac{V_I}{\Delta y}\right)+\frac{\Delta x}{\Delta y}\Delta_y\left(\frac{U_I}{\Delta x}\right)\right]$$

$$(5.6.6)$$

where

$$\Delta x = h_1 d\xi_1; \quad \Delta y = h_2 d\xi_2$$

$$\Delta_x (\cdots) = \frac{\partial}{\partial \xi_1}(\cdots)d\xi_1; \quad \Delta_y (\cdots) = \frac{\partial}{\partial \xi_2}(\cdots)d\xi_2 \qquad (5.6.7)$$

Writing

$$\tau_{IOx} = \rho_w C_{HU}(U_I - U_o); \quad \tau_{IOy} = \rho_w C_{HU}(V_I - V_o) \qquad (5.6.8)$$

and using the staggered C grid, and the leapfrog scheme for time differencing and centered differencing for spatial derivatives, taking care to evaluate the ice internal stress terms at time step n+1 for assuring an implicit scheme, the U momentum equation can be written in finite difference form as

$$A_{xy} U_{i,j}^{n+1} + A_{xp} U_{i+1,j}^{n+1} + A_{xm} U_{i-1,j}^{n+1} + A_{yp} U_{i,j+1}^{n+1} + A_{ym} U_{i,j-1}^{n+1} =$$

$$R_2 U_{i,j}^{n-1} - \Pi_{i,j}^n$$

$$A_{xy} = R_1 + R_3 + R_4 - R_5 + R_6 + R_7 - R_8 + R_9 + R_{11} + R_{12} + R_{14}$$

$$A_{xp} = -R_4 - R_5 + R_{15}; \quad A_{xm} = -R_6 + R_7 - R_{15}$$

$$A_{yp} = -R_8 - R_{10}; \quad A_{ym} = R_9 - R_{13}$$

$$(5.6.9)$$

where

$$R_1 = \left(\frac{D_{i,j}^{n+1} + D_{i-1,j}^{n+1}}{2}\right) \frac{\Delta x \, \Delta y}{2\Delta t}; \quad R_2 = \left(\frac{D_{i,j}^{n-1} + D_{i-1,j}^{n-1}}{2}\right) \frac{\Delta x \, \Delta y}{2\Delta t}$$

$$R_3 = \frac{\rho_w}{\rho_I}\left(\frac{A_{i,j}^{n+1} C_{HUi,j}^n + A_{i-1,j}^{n+1} C_{HUi-1,j}^n}{2}\right) \Delta x \, \Delta y$$

$$R_4 = (\zeta + \eta)_{i,j}^n \frac{\Delta Y_{i,j}}{\Delta X_{i,j}}; \quad R_5 = \frac{1}{4}(\zeta - \eta)_{i,j}^n \frac{\Delta Y_{i+1,j} - \Delta Y_{i-1,j}}{\Delta X_{i,j}}$$

$$R_6 = (\zeta + \eta)_{i-1,j}^n \frac{\Delta Y_{i-1,j}}{\Delta X_{i-1,j}}; \quad R_7 = \frac{1}{4}(\zeta - \eta)_{i-1,j}^n \frac{\Delta Y_{i,j} - \Delta Y_{i-2,j}}{\Delta X_{i-1,j}}$$

$$R_8 = \frac{1}{4}\left(\eta_{i,j}^n \frac{\Delta X_{i,j}}{\Delta Y_{i,j}} + \eta_{i-1,j}^n \frac{\Delta X_{i-1,j}}{\Delta Y_{i-1,j}}\right)\left(\frac{\Delta X_{i,j+1} + \Delta X_{i-1,j+1} - \Delta X_{i,j-1} - \Delta X_{i-1,j-1}}{\Delta X_{i,j} + \Delta X_{i-1,j} + \Delta X_{i,j+1} + \Delta X_{i-1,j+1}}\right)$$

$$R_9 = \frac{1}{4}\left(\eta_{i,j}^n \frac{\Delta X_{i,j}}{\Delta Y_{i,j}} + \eta_{i-1,j}^n \frac{\Delta X_{i-1,j}}{\Delta Y_{i-1,j}}\right)\left(\frac{\Delta X_{i,j+1} + \Delta X_{i-1,j+1} - \Delta X_{i,j-1} - \Delta X_{i-1,j-1}}{\Delta X_{i,j} + \Delta X_{i-1,j} + \Delta X_{i,j-1} + \Delta X_{i-1,j-1}}\right)$$

$$R_{10} = \frac{1}{2}\left(\frac{\eta_{i,j}^{n}\dfrac{\Delta X_{i,j}^{2}}{\Delta Y_{i,j}} + \eta_{i-1,j}^{n}\dfrac{\Delta X_{i-1,j}^{2}}{\Delta Y_{i-1,j}} + \eta_{i,j+1}^{n}\dfrac{\Delta X_{i,j+1}^{2}}{\Delta Y_{i,j+1}} + \eta_{i-1,j+1}^{n}\dfrac{\Delta X_{i-1,j+1}^{2}}{\Delta Y_{i-1,j+1}}}{\Delta X_{i,j+1} + \Delta X_{i-1,j+1}} \right)$$

$$R_{11} = \frac{1}{2}\left(\frac{\eta_{i,j}^{n}\dfrac{\Delta X_{i,j}^{2}}{\Delta Y_{i,j}} + \eta_{i-1,j}^{n}\dfrac{\Delta X_{i-1,j}^{2}}{\Delta Y_{i-1,j}} + \eta_{i,j+1}^{n}\dfrac{\Delta X_{i,j+1}^{2}}{\Delta Y_{i,j+1}} + \eta_{i-1,j+1}^{n}\dfrac{\Delta X_{i-1,j+1}^{2}}{\Delta Y_{i-1,j+1}}}{\Delta X_{i,j} + \Delta X_{i-1,j}} \right)$$

$$R_{12} = \frac{1}{2}\left(\frac{\eta_{i,j}^{n}\dfrac{\Delta X_{i,j}^{2}}{\Delta Y_{i,j}} + \eta_{i-1,j}^{n}\dfrac{\Delta X_{i-1,j}^{2}}{\Delta Y_{i-1,j}} + \eta_{i-1,j-1}^{n}\dfrac{\Delta X_{i-1,j-1}^{2}}{\Delta Y_{i-1,j-1}} + \eta_{i,j-1}^{n}\dfrac{\Delta X_{i,j-1}^{2}}{\Delta Y_{i,j-1}}}{\Delta X_{i,j} + \Delta X_{i-1,j}} \right)$$

$$R_{13} = \frac{1}{2}\left(\frac{\eta_{i,j}^{n}\dfrac{\Delta X_{i,j}^{2}}{\Delta Y_{i,j}} + \eta_{i-1,j}^{n}\dfrac{\Delta X_{i-1,j}^{2}}{\Delta Y_{i-1,j}} + \eta_{i-1,j-1}^{n}\dfrac{\Delta X_{i-1,j-1}^{2}}{\Delta Y_{i-1,j-1}} + \eta_{i,j-1}^{n}\dfrac{\Delta X_{i,j-1}^{2}}{\Delta Y_{i,j-1}}}{\Delta X_{i,j-1} + \Delta X_{i-1,j-1}} \right)$$

$$R_{14} = 2\left[(\zeta+\eta)_{i,j}^{n} + (\zeta+\eta)_{i-1,j}^{n} \right]\frac{\left(\Delta Y_{i,j} - \Delta Y_{i-1,j}\right)\left(\Delta Y_{i,j} - \Delta Y_{i-1,j}\right)}{\left(\Delta Y_{i,j} + \Delta Y_{i-1,j}\right)\left(\Delta X_{i,j} + \Delta X_{i-1,j}\right)}$$

$$R_{15} = 2\left[(\zeta-\eta)_{i,j}^{n} + (\zeta-\eta)_{i-1,j}^{n} \right]\frac{\left(\Delta Y_{i,j} - \Delta Y_{i-1,j}\right)}{\left(\Delta X_{i,j} + \Delta X_{i-1,j}\right)}$$

$$(5.6.10)$$

and

$$\Pi_{i,j}^{n} = \Delta_{x}\left(D_{I}^{n}U_{I}^{n}U_{I}^{n}\Delta y\right) + \Delta_{y}\left(D_{I}^{n}U_{I}^{n}V_{I}^{n}\Delta x\right) - \bar{f}D_{I}^{n}V_{I}^{n}\Delta x\Delta y + gD_{I}^{n}\Delta y\Delta_{x}\left(\eta_{0}^{n}\right)$$

$$-\frac{\rho_{w}}{\rho_{I}}A_{I}^{n}C_{HU}^{n}U_{o}^{n}\Delta x\Delta y - \frac{1}{\rho_{I}}A_{I}^{n}\tau_{AIx}^{n}\Delta x\Delta y - \eta\frac{\Delta y}{\Delta x}\Delta_{x}\left(\frac{V_{I}^{n}}{\Delta y}\right)\Delta_{y}\left(\Delta x\right)$$

$$-\Delta_{x}\left[(\zeta+\eta)^{n}\frac{V_{I}^{n}}{\Delta x}\Delta_{y}\left(\Delta x\right) + (\zeta-\eta)^{n}\Delta_{y}\left(V_{I}^{n}\right)\right] - \Delta_{y}\left[\eta^{n}\Delta y\Delta_{x}\left(\frac{V_{I}^{n}}{\Delta y}\right)\right]$$

$$+\left[(\zeta+\eta)^{n}\frac{1}{\Delta y}\Delta_{y}\left(V_{I}^{n}\right) + (\zeta-\eta)^{n}\frac{V_{I}^{n}}{\Delta x\Delta y}\Delta_{y}\left(\Delta x\right)\right]\Delta_{x}\left(\Delta y\right) + \Delta y\Delta_{x}\left(\frac{P}{2}\right)$$

$$\Delta x\,\Delta y = \frac{1}{4}\left(\Delta X_{i,j} + \Delta X_{i-1,j}\right)\left(\Delta Y_{i,j} + \Delta Y_{i-1,j}\right)$$

$$(5.6.11)$$

We will not write down the finite differences above explicitly since the expressions are algebraically tedious. The form of these can be readily seen from the code. Note, however,

$$\Delta_x(\cdots) = \frac{2\left[(\cdots)_{i,j} - (\cdots)_{i-1,j}\right]}{\Delta X_{i,j} + \Delta X_{i-1,j}}; \quad \Delta_y(\cdots) = \frac{2\left[(\cdots)'_{i,j} - (\cdots)'_{i-1,j}\right]}{\Delta Y_{i,j} + \Delta Y_{i-1,j}}$$

(5.6.12)

where the primes are to remind us that the values must be evaluated at points marked by crosses in Figure 5.6.1, not at the grid centers.

The V momentum equation can be written in finite difference form as

$$A_{xy}V_{i,j}^{n+1} + A_{xp}V_{i+1,j}^{n+1} + A_{xm}V_{i-1,j}^{n+1} + A_{yp}V_{i,j+1}^{n+1} + A_{ym}V_{i,j-1}^{n+1} =$$
$$R_2 V_{i,j}^{n-1} - \Pi_{i,j}^n$$
$$A_{xy} = R_1 + R_3 + R_4 - R_5 + R_6 + R_7 - R_8 + R_9 + R_{11} + R_{12} + R_{14}$$
$$A_{xp} = -R_8 - R_{10}; \quad A_{xm} = R_9 - R_{13}$$
$$A_{yp} = -R_4 - R_5 + R_{15}; \quad A_{ym} = -R_6 + R_7 - R_{15}$$

(5.6.13)

where

$$R_1 = \left(\frac{D_{i,j}^{n+1} + D_{i,j-1}^{n+1}}{2}\right)\frac{\Delta x \, \Delta y}{2\Delta t}; \quad R_2 = \left(\frac{D_{i,j}^{n-1} + D_{i,j-1}^{n-1}}{2}\right)\frac{\Delta x \, \Delta y}{2\Delta t}$$

$$R_3 = \frac{\rho_w}{\rho_I}\left(\frac{A_{i,j}^{n+1}C_{HUi,j}^n + A_{i,j-1}^{n+1}C_{HUi,j-1}^n}{2}\right)\Delta x \, \Delta y$$

$$R_4 = (\zeta + \eta)_{i,j}^n \frac{\Delta X_{i,j}}{\Delta Y_{i,j}}; \quad R_5 = \frac{1}{4}(\zeta - \eta)_{i,j}^n \frac{\Delta X_{i,j+1} - \Delta X_{i,j-1}}{\Delta Y_{i,j}}$$

$$R_6 = (\zeta + \eta)_{i,j-1}^n \frac{\Delta X_{i,j-1}}{\Delta Y_{i,j-1}}; \quad R_7 = \frac{1}{4}(\zeta - \eta)_{i,j-1}^n \frac{\Delta X_{i,j} - \Delta X_{i,j-2}}{\Delta Y_{i,j-1}}$$

$$R_8 = \frac{1}{4}\left(\eta_{i,j}^n \frac{\Delta Y_{i,j}}{\Delta X_{i,j}} + \eta_{i,j-1}^n \frac{\Delta Y_{i,j-1}}{\Delta X_{i,j-1}}\right)\left(\frac{\Delta Y_{i+1,j} + \Delta Y_{i+1,j-1} - \Delta Y_{i-1,j} - \Delta Y_{i-1,j-1}}{\Delta Y_{i,j} + \Delta Y_{i,j-1} + \Delta Y_{i+1,j} + \Delta Y_{i+1,j-1}}\right)$$

$$R_8 = \frac{1}{4}\left(\eta_{i,j}^n \frac{\Delta Y_{i,j}}{\Delta X_{i,j}} + \eta_{i,j-1}^n \frac{\Delta Y_{i,j-1}}{\Delta X_{i,j-1}}\right)\left(\frac{\Delta Y_{i+1,j} + \Delta Y_{i+1,j-1} - \Delta Y_{i-1,j} - \Delta Y_{i-1,j-1}}{\Delta Y_{i-1,j} + \Delta Y_{i,j} + \Delta Y_{i,j-1} + \Delta Y_{i-1,j-1}}\right)$$

$$R_{10} = \frac{1}{2}\left(\frac{\eta_{i,j}^n \frac{\Delta Y_{i,j}^2}{\Delta X_{i,j}} + \eta_{i,j-1}^n \frac{\Delta Y_{i,j-1}^2}{\Delta X_{i,j-1}} + \eta_{i+1,j}^n \frac{\Delta Y_{i+1,j}^2}{\Delta X_{i+1,j}} + \eta_{i+1,j-1}^n \frac{\Delta Y_{i+1,j-1}^2}{\Delta X_{i+1,j-1}}}{\Delta Y_{i+1,j} + \Delta Y_{i+1,j-1}}\right)$$

$$R_{11} = \frac{1}{2}\left(\frac{\eta_{i,j}^n \frac{\Delta Y_{i,j}^2}{\Delta X_{i,j}} + \eta_{i,j-1}^n \frac{\Delta Y_{i,j-1}^2}{\Delta X_{i,j-1}} + \eta_{i+1,j}^n \frac{\Delta Y_{i+1,j}^2}{\Delta X_{i+1,j}} + \eta_{i+1,j-1}^n \frac{\Delta Y_{i+1,j-1}^2}{\Delta X_{i+1,j-1}}}{\Delta Y_{i,j} + \Delta Y_{i,j-1}}\right)$$

$$R_{12} = \frac{1}{2}\left(\frac{\eta_{i,j}^n \frac{\Delta Y_{i,j}^2}{\Delta X_{i,j}} + \eta_{i,j-1}^n \frac{\Delta Y_{i,j-1}^2}{\Delta X_{i,j-1}} + \eta_{i-1,j-1}^n \frac{\Delta Y_{i-1,j-1}^2}{\Delta X_{i-1,j-1}} + \eta_{i-1,j}^n \frac{\Delta Y_{i-1,j}^2}{\Delta X_{i-1,j}}}{\Delta Y_{i,j} + \Delta Y_{i,j-1}}\right)$$

$$R_{13} = \frac{1}{2}\left(\frac{\eta_{i,j}^n \frac{\Delta Y_{i,j}^2}{\Delta X_{i,j}} + \eta_{i,j-1}^n \frac{\Delta Y_{i,j-1}^2}{\Delta X_{i,j-1}} + \eta_{i-1,j-1}^n \frac{\Delta Y_{i-1,j-1}^2}{\Delta X_{i-1,j-1}} + \eta_{i-1,j}^n \frac{\Delta Y_{i-1,j}^2}{\Delta X_{i-1,j}}}{\Delta Y_{i-1,j} + \Delta Y_{i-1,j-1}}\right)$$

$$R_{14} = 2\left[(\zeta+\eta)_{i,j}^n + (\zeta+\eta)_{i,j-1}^n\right]\frac{(\Delta X_{i,j} - \Delta X_{i,j-1})(\Delta X_{i,j} - \Delta X_{i,j-1})}{(\Delta X_{i,j} + \Delta X_{i,j-1})(\Delta Y_{i,j} + \Delta Y_{i,j-1})}$$

$$R_{15} = 2\left[(\zeta-\eta)_{i,j}^n + (\zeta-\eta)_{i,j-1}^n\right]\frac{(\Delta X_{i,j} - \Delta X_{i,j-1})}{(\Delta Y_{i,j} + \Delta Y_{i,j-1})}$$

$$(5.6.14)$$

and

$$\Pi_{i,j}^n = \Delta_x\left(D_I^n U_I^n V_I^n \Delta y\right) + \Delta_y\left(D_I^n V_I^n V_I^n \Delta x\right) + \overline{f}D_I^n U_I^n \Delta x \Delta y + gD_I^n \Delta x \Delta_y\left(\eta_0^n\right)$$

$$-\frac{\rho_w}{\rho_I} A_I^n C_{HU}^n V_o^n \Delta x \Delta y - \frac{1}{\rho_I} A_I^n \tau_{AIy}^n \Delta x \Delta y - \eta\frac{\Delta x}{\Delta y}\Delta_y\left(\frac{U_I^n}{\Delta x}\right)\Delta_x\left(\Delta y\right)$$

$$-\Delta_y\left[(\zeta+\eta)^n \frac{U_I^n}{\Delta y}\Delta_x\left(\Delta y\right) + (\zeta-\eta)^n \Delta_x\left(U_I^n\right)\right] - \Delta_x\left[\eta^n \Delta x \Delta_y\left(\frac{U_I^n}{\Delta x}\right)\right] \qquad (5.6.15)$$

$$+\left[(\zeta+\eta)^n \frac{1}{\Delta x}\Delta_x\left(U_I^n\right) + (\zeta-\eta)^n \frac{U_I^n}{\Delta x \Delta y}\Delta_x\left(\Delta y\right)\right]\Delta_y\left(\Delta x\right) + \Delta x \Delta_y\left(\frac{P}{2}\right)$$

$$\Delta x\ \Delta y = \frac{1}{4}\left(\Delta X_{i,j} + \Delta X_{i,j-1}\right)\left(\Delta Y_{i,j} + \Delta Y_{i,j-1}\right)$$

Note that

$$\Delta_x\left(\cdots\right) = \frac{2\left[\left(\cdots\right)'_{i,j} - \left(\cdots\right)'_{i-1,j}\right]}{\Delta X_{i,j} + \Delta X_{i,j-1}}; \quad \Delta_y\left(\cdots\right) = \frac{2\left[\left(\cdots\right)_{i,j} - \left(\cdots\right)_{i,j-1}\right]}{\Delta Y_{i,j} + \Delta Y_{i,j-1}}$$

(5.6.16)

where the primes are to remind us once again that the values must be evaluated at points marked by crosses in Figure 5.6.1, not at the grid center.

5.6.2 SOLUTION TECHNIQUE

These equations can be solved using a generalization of the Gaussian elimination for a tridiagonal matrix (Thomas algorithm). Consider a finite difference equation of for a 1-D problem of the form

$$A_{xy}F_i + A_{xp}F_{i+1} + A_{xm}F_{i-1} = A_i$$

(5.6.17)

Substitute

$$F_i = P_iF_{i-1} + Q_i$$

(5.6.18)

Change the indices in Eq. (5.6.18) to get F_{i-1} and eliminate this term in Eq. (5.6.17) to derive the following recursive relations for P_i and Q_i:

$$P_i = \frac{-A_{xp}}{A_{xy} + A_{xm}P_{i-1}}$$

$$Q_i = \frac{A_i - A_{xm}Q_{i-1}}{A_{xy} + A_{xm}P_{i-1}}$$

(5.6.19)

Knowing the boundary condition at i = 1 (whether Dirichlet or Neumann), P_1 and Q_1 can be evaluated and using Eq. (5.6.19); P_i and Q_i can be evaluated for all i, sweeping forward from i =1 to i = IM. From the boundary condition at i = IM, F_{IM} is known and Eq. (5.6.18) can then be used to solve for all i, starting from i = IM on the sweep back. Thus the solution is obtained in two sweeps, one forward and one back (no iteration is needed).

For a 2-D problem, a similar procedure can be used, except that several sweeps are needed in an iterative scheme (developed by Dr. James Herring of

Dynalysis of Princeton, Princeton, New Jersey) to obtain the solution. The convergence is, however, fast. Consider the finite difference equation for the 2-D problem:

$$A_{xy}F_{i,j} + A_{xp}F_{i+1,j} + A_{xm}F_{i-1,j} + A_{yp}F_{i,j+1} + A_{ym}F_{i,j-1} = A_{i,j} \tag{5.6.20}$$

Substitute

$$F_{i,j} = P_{i,j}F_{i+1,j} + Q_{i,j}F_{i,j+1} + R_{i,j} \tag{5.6.21}$$

Change the indices in Eq. (5.6.21) to get $F_{i-1,j}$ and $F_{i,j-1}$ and eliminate these terms in Eq. (5.6.20) to derive the following recursive relations for $P_{i,j}$, $Q_{i,j}$ and $R_{i,j}$:

$$P_{i,j} = \frac{-A_{xp}}{A_{xy} + A_{xm}P_{i-1,j} + A_{ym}Q_{i,j-1}}$$

$$Q_{i,j} = \frac{-A_{yp}}{A_{xy} + A_{xm}P_{i-1,j} + A_{ym}Q_{i,j-1}} \tag{5.6.22}$$

$$R_{i,j} = \frac{A_{i,j} - A_{xm}\left(Q_{i-1,j}F_{i-1,j+1} + R_{i-1,j}\right) - A_{ym}\left(P_{i,j-1}F_{i+1,j-1} + R_{i,j-1}\right)}{A_{xy} + A_{xm}P_{i-1,j} + A_{ym}Q_{i,j-1}}$$

The procedure is to start the solution at the southwest corner $i = 1$, $j = 1$ from known values of P, Q, and R (which can be found for either Dirichlet or Neumann boundary conditions on F) along the southern and western boundaries. Use of Eq. (5.6.22) then provides values of P, Q, and R for all i and j in the forward sweep to the northeast corner $i = IM$, $j = JM$. Knowing the value of F along the northern and eastern boundaries (Neumann or Dirichlet), one sweeps back to the southwest corner using Eq. (5.6.21) to obtain F for all i and j. However, it takes a few iterations for the values on the boundaries to be felt inside the domain and therefore several sweeps are necessary for the solution. Convergence is fast, however. The procedure also sidesteps the severe limitation due to large ice viscosities on the time step in an explicit scheme.

Chapter 6

Tides and Tidal Modeling

The oceanic tide refers to the rhythmic rise and fall of sea level with time, made evident at a coast by the periodic advancing and receding of the waters from the shore. At locations where the tidal range (the difference between the high and low waters) is large— as, for example, in the Bay of Fundy—it exposes and submerges features along the coast. In estuaries with large tides, such as the Severn River in Great Britain, the Tsientang River in China, and the Amazon estuary, the tide travels up the estuary as a spectacular bore. The tides arise due to the gravitational effects of the Sun and the Moon; even in prehistoric times, people living along seashores must have been aware of the tidal phenomenon and its intimate association with the various phases of the moon. There is evidence that the people of the Indus Valley civilization in India, nearly 4500 years ago, knew about tides and made use of this knowledge in the construction of their docks (Pugh, 1987). The Macedonian army of Alexander the Great, used to the small tides in the Mediterranean Sea, must have marveled at the large tides at the mouth of the Indus river. Tidal variations in sea level have been recorded at some coastal stations, harbors, and ports for centuries, for use by ship-borne commerce. But only recently has it been possible to observe and model tides accurately everywhere in the global oceans by a powerful combination of highly accurate satellite altimeters that measure the sea level fluctuations to within 3 to 5 cm accuracy (Fu *et al.*, 1994) and comprehensive numerical tidal models running on powerful multigigaflop supercomputers (for example, Kantha, 1995b; Le Provost *et al.*, 1998, Tierney *et al.*, 1999).

Tides slowly dissipate the gravitational energy resident in the Earth-Moon system to the tune of about 3.17 TW (slightly larger than 2.92 TW, the entire electric power generation capacity in the world in 1995; other interesting comparisons: energy loss in an earthquake is ~0.3 TW; in a typical solar storm, ~1.5 TW; heat flux from outer core to the lower mantle, ~5.7 TW; geothermal

375

heat flux, ~30 TW; solar insolation on Earth, ~1.7×10^5 TW). The resulting torque between the Earth and the Moon decreases the angular momentum of the Earth, and the Earth spins down ever so slightly leading to a monotonic increase in the length of the day (LOD). This was first pointed out by none other than Immanuel Kant in 1754 and later by Lord Kelvin. To conserve the angular momentum of the Earth/Moon system, this loss of angular momentum due to the despinning of the Earth is made up by an increase in the orbital angular momentum of the Moon. By virtue of Kepler's law, this is brought about by an increase in the orbital radius of the Moon but a decrease of the lunar orbital velocity, leading to an increase in the length of the month (LOM). These secular trends in LOD (and LOM) and the resulting Earth-Moon evolution are important to astronomy and geophysics. Recent analysis of laminated tidal sedimentary records from Utah, Indiana, Alabama, and Australia (Sonett *et al.* 1996), produced by the semimonthly spring and neap tides during the Proterozoic era 900 million years ago, indicates that the year consisted then of 481 days, each about 18 hr long. This is equivalent to an average LOD increase of 2.4 ms cy^{-1}. These observations are consistent with modern measurements (Dickey *et al.*, 1994; Hide and Dickey, 1991; Munk, 1997; Wahr, 1988) of the increase in the orbital radius of the Moon of about 3.7 cm year^{-1} by lunar laser ranging (LLR), and decrease in the Moon's orbital velocity to the tune of about 25 arc-seconds per century squared that corresponds to an increase in LOM of 34.6 ms cy^{-1} (these values being equivalent to a tidal dissipation value of 3.05 TW and LOD increase of 2.1 ms cy^{-1}).

   The evolution of the Earth-Moon system since its formation is a topic of great interest to astronomers (Kagan, 1997; Kagan and Sundermann, 1996). This evolution depends very much on the history of tidal dissipation. Gerstenkorn (1955) hypothesized that about 2.5 billion years ago, the Moon might have approached the Earth closely enough to have had catastrophic effects on both planets. This distance is close to the Roche limit (about 2.89 times the Earth's radius for a liquid and 2.80 for a solid satellite), which is the distance at which tidal forces generated by the primary planet overwhelm the self-gravitational forces of the satellite and cause it to be pulled apart. At such small separation distances, the oceanic tides on Earth would have reached amplitudes of over a kilometer, with enough energy dissipation to boil the oceans off in a few decades and cause runaway greenhouse warming of the planet to more than 1500 K by the enormous water vapor in the atmosphere (Munk, 1968). The dissipation would also have melted part of the mantles of both planets. Yet the sedimentary records on Earth, and lunar rocks that are more than 4 billion years old (Lee *et al.*, 1997), indicate that no such event happened. The Earth and the Moon were probably close only at the birth of the solar system, at which time oceans did not exist. Oceans appear to have formed about a billion years later by accretion of water from impacting cometary objects. Once formed, the oceans are thought to have assumed the principal role of dissipating lunar tidal energy.

If the current theories are right, the Moon formed at a distance of about the Roache limit, from accretion of the ejecta in the disk surrounding the Earth that was created by a giant off-center impact by a Mars-sized object, roughly 4.5 billion years ago (Ida *et al.*, 1997; Lissauer, 1997,). Tidal dissipation has been causing it to move away ever since. Since currently the dissipation in the oceanic tides is about 20 times that in the solid Earth tides, this discrepancy can be explained only by more accurate computations of tidal dissipation in the past, including that in the paleo-oceans (Hansen, 1982; Kagan and Sundermann, 1996; Webb, 1982), accounting properly for the effect of plate tectonics and the shape of the continents on the normal mode and resonance characteristics of the oceanic response to tidal forcing. Careful investigations of the response characteristics of the paleo-oceans (e.g., Kagan, 1997; Kagan and Sundermann, 1996; Webb, 1982) indicates that the shape and location of the oceans have an important effect on which tidal constituents are amplified and which are not. Changes in the cellular convective patterns in the Earth's mantle in its geological past have led to alternate consolidation of continents into a supercontinent (such as Pangea) and their breakup and drift. The former tends to amplify the diurnal constituents, whereas the latter tends to amplify the semidiurnal ones (Kagan and Sundermann, 1996). Tidal dissipation in the principal oceanic semidurnal M_2 tide itself might have ranged from a minimum of 0.35 TW to a maximum of 4.5 TW over the past 600 million years. The length of a sidereal day might itself have dipped to as low as 12 hr in the very distant past. The reader is referred to Kagan and Sundermann (1996) for a fascinating account of the paleo-ocean tides and their effect on Earth-Moon evolution. The problem of reconciling the tidal dissipation over the past 4.5 billion years with observations remains unresolved.

While the tidal dissipation occurring in the global oceans is now well known from geophysical measurements such as lunar laser ranging (Dickey *et al.*, 1994) and very long baseline interferometry, and tracking of satellites (Ray *et al.*, 1994b), accurate identification of the regions of intense tidal dissipation in the oceans and its quantification are still quite elusive (Kantha, 1995; Munk, 1997; Kantha, 1998). A majority of the dissipation occurs on shallow shelves like the Patagonian and European shelves, and it is still difficult to measure or compute tides and tidal dissipation accurately enough on various shallow shelves around the world. An accurate determination of the oceanic tides and the associated load tides in the solid Earth is also germane to understanding the inner workings of the Earth's core (Wahr, 1988).

One reason for the modern interest in tides lies in being able to compute them accurately enough so that they can be subtracted out of the altimetric signals, both in the open oceans and in the marginal and coastal seas. Modern satellite-borne altimeters such as the NASA/Centre National d'Etudes Spatiales(CNES) TOPEX/ POSEIDON (T/P) altimeter have achieved such a high level of precision (2 to 3 cm rms), their orbits can be so precisely determined (2 to 3 cm rms), and the various sources of error in determining the microwave signal propagation time so accurately accounted for (Fu *et al.*, 1994) that the uncertainty in determining the tidal sea surface height (SSH) has become one of the few major remaining sources of error in altimetry (Molines *et al.*, 1994; Schlax and Chelton, 1994), the others being the departures of the ocean from a purely inverse barometric response to atmospheric pressure forcing (Kantha *et al.*, 1994; Ponte, 1994,) and the electromagnetic bias produced by the asymmetry of large surface waves in high sea states (Eifouhaily *et al.*, 1999).

Tidal phenomenon was one of the very first geophysical problems that became amenable to theoretical analyses, once Sir Isaac Newton laid down the law of gravitation in 1687. Since then, famous names like Bernoulli, Laplace, Kelvin, Munk, and Cartwright have been associated with tidal theories. After Newton explained the semidiurnal nature of tides in *Principia*, it took another half a century for Daniel Bernoulli to attempt to quantify tides through the equilibrium theory of tides, and nearly a century before Marquis de Laplace laid down a rigorous foundation for modern tidal research in the form of separation of tides into different species and formulation of the hydrodynamical equations that govern tidal motions. Nearly another century elapsed before Lord Kelvin initiated the harmonic analysis of tides and made it possible for tides to be predicted well in advance from observations in the past. Among modern contributors to tidal science, Walter Munk, David Cartwright, and their collaborators have been particularly prominent.

Tidal fluctuations of sea level along the world's coasts are quite well known and well measured. However, only recently have we been able to measure tides in the open ocean (from satellites) and to model them accurately enough in the open ocean for many oceanographic applications. Tides are not only fascinating as a regularly occurring, highly repeatable phenomena in a chaotic geophysical world, but they are also of considerable practical importance. For a basic treatment of tidal phenomena, the reader is referred to Defant (1961), Dietrich *et al.* (1980), Pugh (1987), and Pond and Pickard (1989). For an excellent survey of oceanic tides, Cartwright (1977, 1993), Hendershott (1981), and Reid (1990) are quite valuable. For a good idea of the status of the subject, the various technical articles in Parker (1991) serve as excellent starting points. Ray (1993) and Shum et al. (1997) review the accuracy of current global tidal models specifically for application to modern altimetry. A recent reference on tides is the special section on the geophysical evaluation of the T/P mission in *Journal of Geophysical Research*, 99 (C12) 24,369-25,062, which contains several articles on global ocean tides (see, e.g., Le Provost *et al.*, 1994; Ma *et al.*, 1994;

Schrama and Ray, 1994; also see Kantha, 1995). Melchior (1981) is still an excellent reference source for solid Earth tides. Wahr (1988) describes the influence of oceanic tides on Earth's rotation (see also Dickey, 1995). The classic treatise "The Rotation of the Earth" by Munk and MacDonald (1960; see also Lambeck, 1980) constitutes a landmark in this field. It brought together for the first time concepts from diverse fields such as oceanography, paleo-oceanography, solid Earth geophysics, tectonics, astronomy, and planetology to bear upon a comprehensive discussion of the topic. A meeting on tides held in late 1996 at the Royal Society of London, in honor of the 70th birthday of tidalist David Cartwright, brought together many tidalists from around the world. The proceedings of that meeting (published as a special issue on *Tidal Science in Progress in Oceanography,* Vol. 40, 1997, edited by R.D. Ray and P. L. Woodworth) provide a fascinating glimpse into the status of tides and tidal research as of 1997. Cartwright himself has authored a treatise on the history of tidal research (Cartwright, 1998). For a modern perspective on tides, see Kantha (1998). The brief review of tides here in this chapter is designed to be a mathematical complement to the excellent but somewhat descriptive treatment of tidal phenomena by Pugh (1987).

The principal reason for the interest of geophysicists in oceanic tides is because a faithful depiction of oceanic tides enables more accurate inferences to be made about the Earth itself. Geodetists are interested in the influence of tides on Earth's gravity field and its rotation. Astronomers are interested principally because of the effects of tides on the ephemerides of the Moon and the consequent implications. Oceanographers study tides partly because they are ubiquitous in their measurements. Though often treated as a mere contaminant and a nuisance, a careful study of oceanic tides enables one to better understand the response of the global oceans to external forcing.

6.1 DESCRIPTION OF TIDES

Oceanic tides are mostly the result of the gravitational attraction between the Earth and the Moon, the astronomical body closest to the Earth, and that between the Earth and the Sun, the most massive heavenly body in our planetary system. As we shall see, the gravitational forces that raise tides are proportional to the ratio of the mass of the heavenly body to that of the Earth and to the cube of the ratio of the Earth's radius to the distance between the two bodies. Therefore tides due to other planetary bodies are negligible (the contribution of Venus, the planet nearest Earth, is 5.4×10^{-5} of that of the Moon). We will first provide a descriptive explanation for the observed tidal characteristics and follow it up with a proper mathematical development.

To understand tides, it is necessary to understand the orbital characteristics of the satellite Moon around the Earth and that of the Earth around the Sun (Figure

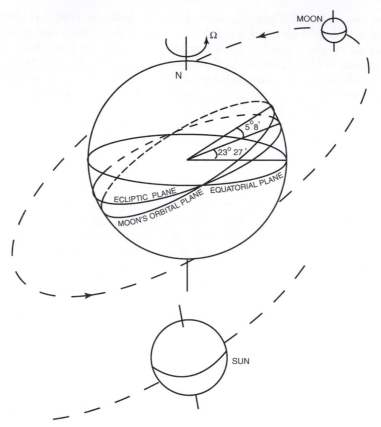

Figure 6.1.1 Sketch showing the orbital characteristics of the Earth-Moon-Sun system.

6.1.1). While the Earth rotates around its axis once every 23 h 56 m 4 s (sidereal day, 86 164 s, is equal to 0.997 269 solar day), the Earth and the Moon rotate around a common center once every 27 d 7 h 43 m 14.9 s (27.3217 d, 1 sidereal month), the direction of rotation being the same as the Earth's rotation (anticlockwise looking down on the north pole). This center is located 4671 km from the Earth's center of mass and 379 730 km from Moon's center of mass, the distance between the two centers being 384 405 km on the average. Due to eccentricity, the perigee of Moon's orbit is 10% closer than the apogee. In addition, the plane of the Moon's orbit around the Earth is inclined at an angle between 18.5 and 28.5° to the equatorial plane of the Earth, with this declination modulated over a period of 18.613 years. The Earth itself orbits around the Sun every 365 d 5 h 49 m 12 s (365.2422 d, 1 tropical year), with the ecliptic plane, the plane of Earth's orbit around the Sun inclined at 23° 27' to the equatorial plane. Since the spin axis of Earth does not change much during the year, the apparent declination (the latitude relative to the equatorial plane) of the Sun varies between -23.5 to $23.5°$ during the year. This is responsible for the seasons on Earth. Also our Gregorian calendar is tied to vernal equinox, the time when the Sun crosses the equator (zero declination), rather than the perigee (perihelion) of Earth's orbit around the Sun. The lunar orbit is inclined at $5°\,08'$ to the ecliptic plane, and this angle (or equivalently the lunar nodal point) varies with a period of 18.613 years. Its perigee itself varies with a period of 8.861 years (perihelion has a period of 20 942 years). All these orbital variations have an effect on the resulting tides.

Sir Isaac Newton was the first to explain the predominantly semidiurnal nature of tides on Earth whose rotation rate is diurnal. To understand this, it is convenient to consider the Earth to be nonrotating and concentrate initially on just the motion of the Earth and the Moon about each other around a common center. Now ignore the inclination of the Moon's orbit and consider the mutual rotation to be in the plane of the Earth's equator. Let R_E be the Earth's radius, R being the distance between the centers of the Earth and the Moon. If we neglect for the moment the eccentricity of the Moon's orbit, every point on the Earth then describes a circle with radius equal to the distance between the center of rotation and the Earth's center, and this requires that there be a centripetal acceleration of the same magnitude at all points on the Earth, directed toward and parallel to the Earth-Moon axis. This is, of course, provided by the gravitational attraction between the two bodies. The centrifugal force balances the gravitational force exactly at the center of the Earth, but because of the differing distances between each point on Earth and the Moon's center of gravity, the gravitational force over the Earth is nonuniform (Figure 6.1.2). The uniform centripetal acceleration is equal to the average gravitational force (per unit mass) $F = GM_M/R^2$ ($G \sim 6.67 \times 10^{-11}$ N m kg^{-2}, MM $\sim 7.38 \times 10^{22}$ kg, R $\sim$ 384 405 km). However, the sublunar point A, the point on Earth's surface facing the moon directly, is closer to the moon by a distance R_E, and hence the gravitational force there is $GM_M/(R-R_E)^2$ and antilunar point B, the point on the other side of the Earth away from the Moon, is $GM_M/(R+R_E)^2$. At points C and D, at $90°$, it is $GM_M/(R^2+R_E^2)$. Thus the residual force at each point can be calculated making use of the fact that $a = R_E/R$ ($R_E \sim 6371$ km, $R_M \sim 1738$ km) is a small number (1/60.34) and hence these expressions can be expanded in terms of a. It is traditional to retain only the first term, which is $O(a^2)$, although higher order terms are necessary for better accuracy.

The excess force at both A and B directed away from the Earth's center in both cases is $[GM_M/(R-R_E)^2 - GM_M/R^2] \sim F[2a +O(a^2)]$. The forces at C and D have components tangential and perpendicular to the Earth's surface of $GM_M/(R^2+R_E^2)$ multiplied by $R/(R^2+R_E^2)^{1/2}$ and $R/(R^2+R_E^2)^{1/2}$, respectively, and hence the residual force components are 1.5 Fa^3 and 1.5 Fa^2 in the tangential and perpendicular directions respectively. The residual or excess tide generating

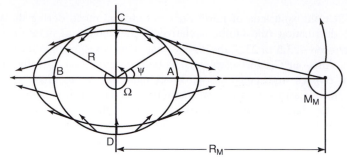

Figure 6.1.2 Residual gravitational forces acting on an ocean of uniform depth covering the Earth.

gravitational force at each point on Earth's surface can be similarly resolved into components parallel and perpendicular to the Earth's surface. The perpendicular component just modulates the gravitational acceleration, and, since F/g = $(MM/M_E)a^2$, this modulation is $O(a^4$ to $a^3)$, 10^{-5} to 10^{-7} of the magnitude of the gravitational acceleration g and can be easily neglected. It is the tangential component, even though even smaller, that is important in producing tides, since it is of the same order as other horizontal forces on a fluid mass. It is called the tractive force and moves water around the globe (Figure 6.1.3). If the water masses on a globe covered everywhere by water were to move instantly

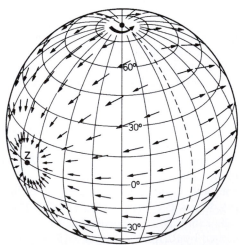

Figure 6.1.3 Tidal tractive forces due to Moon's gravitation on a globe covered by water. (From Dietrich. G. K., *et al.*, Tidal phenomenon, in *General Oceanography*. Copyright © 1980 Gebruder Borntraeger Verlag. Reprinted by permission of John Wiley & Sons, Inc.)

in response to this force, there would be a bulge facing the Moon at A, another at B, and flattening at points C and D. The tractive force is zero at A and B, but reaches a maximum between the sublunar (A) and antilunar (B) points, and the poles (C and D).

So far, we have neglected other aspects, such as the eccentricity of the lunar orbit and the Earth's rotation. If the rotation of the Earth were to be reinstated now, it is easy to see that since the bulges of the water at A and B are tied to the Earth-Moon axis, as long as the water responds instantly (the so-called equilibrium tide approximation), the Earth would rotate under these bulges so that an observer on the surface of the Earth would experience two bulges of water and the accompanying lows passing through each day. Thus on a diurnally rotating planet, there would be semidiurnal tides and two highs and two lows every day. If the Moon were stationary with respect to the Earth, the period associated with such a tide would be exactly half the sidereal day, approximately 12 hr. Since the Moon is also rotating in the same direction as a point on Earth's surface, the effective period of the bulges is half a lunar day, 12.421 hr. This is the period of the basic lunar tide M_2, subscript 2 referring to the semidiurnal period. M_2 is the dominant tidal constituent. Thus, at any point on the Earth's surface, the gravitational forces pile up water into a bulge that slowly moves around with the Moon, while the Earth rotates under this slowly moving bulge. Since the period of this tide is greater than 12 hr by 25.24 min, each day, the tides arrive 50.47 min later than the previous day. Newton and Bernoulli were the first to explain the semidurnal nature of tides; this was a shining example of the success of the law of gravitation in explaining a real, complex phenomenon on Earth. However, this picture is highly simplified and the equilibrium tides provide a very inaccurate estimate of oceanic tides (but do well for solid Earth tides), as we shall see shortly.

Now consider the nuances of the Moon's orbit. Since the Moon's orbital plane is inclined with respect to the Earth's equatorial plane, the north-south symmetry with respect to the equator cannot be expected. To an observer on the Earth's surface, the high tides will be larger if the Moon travels directly overhead than otherwise. Therefore, each day, the two high waters would differ from each other, except when the Moon is directly over the equator during the lunar month. The difference between the two high waters at any point is the diurnal inequality, which is zero when the Moon is passing directly over the equator. This inequality depends on the latitude of the observational point, and the fraction of the lunar month, and can be represented as a superposition of the semidiurnal tide M_2 with a diurnal component that has the same angular rotation rate as M_2. The diurnal component is composed of two subcomponents, K_1 and O_1 (the former having a contribution from Sun as well), and is a maximum when the Moon is at its northern- or southern- most declination every half lunar month and zero when it is over the equator. Since the declination of the Moon has a period of one tropical (sidereal) month (27.3217 days), the overall effect is to

TABLE 6.1.1

Important Tidal Constituents and Their Characteristics (α Is the Elasticity Factor and β Is the Load Factor)

Species	Darwinian symbol	Doodson number	Period (hr or day)	Speed (deg/hr)	Frequency (cy/day)	Eqbm tide Amp (m)	Alpha	Beta	Name
	M_0	055.555	infinity	—	0	0.013465			Constant lunar tide
	S_0	055.555	infinity	—	0	0.006247			Constant solar tide
	Mf	075.555	(13.6608)	1.0980331	0.07320220	0.042017			Declination tide to M_0-lunar fortnightly
	Mm	065.455	(27.5546)	0.5443747	0.03629165	0.022191			Elliptical tide of first order to M_0-lunar monthly
0 (LP)	Ssa	057.555	(182.6211)	0.0821373	0.00547582	0.019542	0.693	0.953	Declination tide to S_0-solar semiannual
	Msm	073.555	(14.7653)	1.0158958	0.06772638	0.004239			Variation tide to M_0
	Sa	056.554	(365.2425)	0.0410667	0.00273778	0.003104			Elliptical tide of first-order to S_0-solar annual
	Msf	063.655	(31.8119)	0.4715211	0.03143470	0.003678			Evection tide to M_0
	Mt	085.455	(9.1329)	1.6424070	0.10949385	0.008049			Ter-mensual tide
	K_1	165.550	23.9345	15.0410686	1.00273791	0.142408	0.7364		Diurnal principal declination tide
	O_1	145.555	25.8193	13.9430356	0.92953571	0.101266	0.6950		Diurnal principal lunar tide
	P_1	163.555	24.0659	14.9589314	0.99726209	0.047129	0.7059		Diurnal principal solar tide
	Q_1	135.655	26.8684	13.3986609	0.89324406	0.019387	0.6946		Elliptical tide of first-order to O_1
	M_1	155.655	24.8332	14.4966939	0.96644626	0.007965	0.6962		Elliptical tide of first-order to K_1
	J_1	175.455	23.0985	15.5854433	1.03902956	0.007965	0.6911		Elliptical tide of first-order to K_1
	OO_1	185.555	22.3061	16.1391017	1.07594011	0.004361	0.6925		Diurnal declination tide of second-order
	ρ_1	137.455	26.7231	13.4715145	0.89810097	0.003685	0.6948		Evection tide to O_1
	σ_1	127.555	27.8484	12.9271398	0.86180932	0.003098	0.6930		Variation tide to O_1
1 (D)	π_1	162.556	24.1321	14.9178647	0.99452431	0.002754	0.7027	0.940	Large elliptical solar
	$2Q_1$	125.755	28.0062	12.8542862	0.85695241	0.002564	0.6930		Smaller elliptical lunar
	ϕ_1	167.555	23.8045	15.1232060	1.00821373	0.002028	0.6657		Second-order solar
	θ_1	173.655	23.2070	15.5125898	1.03417265	0.001526	0.6784		Evectional
	χ_1	157.455	24.7091	14.5695476	0.97130317	0.001522	0.6994		Smaller evectional
	τ_1	147.555	25.6681	14.0251730	0.93501153	0.001325	0.6956		
	Ψ_1	166.554	23.8693	15.0821354	1.00547569	0.001132	0.5285		Smaller elliptical solar
	S_1	164.556	24.0000	15.0000020	1.00000013	0.001116	0.7126		Radiational

	Symbol	Doodson							Description
	M_2	255.555	12.4206	28.9841042	1.93227362	0.244102			Semidiurnal principal lunar tide
	S_2	273.555	12.0000	30.0000000	2.00000000	0.113572			Semidiurnal principal solar tide
	N_2	245.655	12.6583	28.4397295	1.89598197	0.046735			Large elliptical tide of first-order to M_2
	K_2	275.555	11.9672	30.0821373	2.00547582	0.030875			Semidiurnal declination tide to M_2
	ν_2	247.455	12.6260	28.5125831	1.90083888	0.008877			Large evection tide to M_2
	μ_2	237.555	12.8718	27.9682084	1.86454723	0.007463			Large variation tide to M_2
2 (SD)	L_2	265.455	12.1916	29.5284789	1.96856526	0.006899			Small elliptical tide of first-order to M_2
	T_2	272.556	12.0164	29.9589333	1.99726222	0.006636	0.693	0.953	Large elliptical tide of first-order to S_2
	$2N_2$	235.755	12.9054	27.8953548	1.85969032	0.006184			Elliptical tide of second-order to M_2
	ε_2	227.655	13.1273	27.4238337	1.82825558	0.001804			
	λ_2	263.655	12.2218	29.4556253	1.96370835	0.001800			Smaller evectional
	η_2	285.455	11.7545	30.6265121	2.04176747	0.001727			
	R_2	274.554	11.9836	30.0410667	2.00273778	0.000950			Smaller elliptic solar
3 (TD)	M_3	355.555	8.2804	43.4761563	2.89841042	0.003198			Ter-diurnal principal lunar tide
	$2SM_2$		11.6070	31.0158958	2.0677264				Compound tide from M_2, S_2
	M_4		6.2103	57.9682084	3.8645472				Overtide from M_2
	MS_4		6.1033	58.9841042	3.9322736				Compound tide from M_2, S_2
(SW)	S_4		6.0000	60.0000000	4.0000000				Overtide from S_2
	M_6		4.1402	86.9523127	5.7968208		0.693		Overtide from M_2
	$2MS_6$		4.0924	87.9682084	5.845472			1.000	Compound tide from M_2, S_2

Based on Dietrich *et al.* (1980) and Desai (1996).

produce a lunar tidal potential with semidiurnal, diurnal, fortnightly, and monthly components. In other words, the orbital tilt or declinational variation is responsible for producing both the diurnal and long-period components, both of which would be zero, were there no tilt (although the interaction of the Sun and the Moon would still produce a fortnightly modulation). Long period lunar (and solar) tides are also called declinational tides.

The Moon's orbital plane is inclined to the ecliptic plane by $\sim 5^\circ$ 09' and precesses with a 18.61-year period (the Metonic cycle) so that the Moon's maximum declination varies between 18° 18' to 28° 36'. The result is a slow but strong 18.61-year modulation of lunar (and only lunar) tides, which is taken into account by nodal factors computed each year and used to correct the amplitude and phase of the lunar tides. For example, the amplitude modulation due to the Metonic cycle is 1 ± 0.115 for K_1, and 1 ± 0.037 for M_2.

Similarly, the eccentricity of the Moon's orbit (0.0549) is such that the distance between the centers of the Earth and the Moon at perigee is 0.9 times that at apogee. Since, as we saw earlier, the tide producing tractive forces are $O(a^3)$, the ratio of the tractive forces at perigee to those at apogee is $(0.9)^{-3} \sim 1.3$. This monthly (lunar) variation is represented by N_2, with a period slightly different than that of M_2, the two reinforcing each other at perigee but opposing each other at apogee. An additional complication is that the position of the lunar perigee itself rotates slowly with a period (8.847 years) different from that of the Metonic cycle.

So far we have ignored the tides due to the Sun's gravitational forces. While the mass of the Sun (1.991×10^{30} kg) is 2.7×10^7 times that of the Moon (7.38×10^{22} kg; mass of Earth $- 5.977 \times 10^{24}$ kg), because it is 389 times farther (149.5×10^6 km), its tidal force is only $2.7 \times 10^7 / (389)^3 \sim 0.459$ times that of the moon. Nevertheless, both lunar and solar tides are important in the world's oceans. The Earth and the Sun rotate about a common center once a year and the gravitational forces between the two create two bulges along the Earth-Sun axis. The same arguments therefore lead to a semidiurnal tide S_2 of exactly 12 hr period due to the Sun, and the diurnal inequality contribution to K_1 from the Sun due to the inclination of the ecliptic plane to the Earth's equatorial plane. Because of the annual period of the declination of the Sun, the overall effect is to produce a solar tidal potential with a semidiurnal, diurnal, semiannual, and annual components. The eccentricity of the Earth's orbit around the Sun (0.0168) also affects the tides, although the effect is much smaller than that due to the eccentricity of the Moon's orbit around the Earth (0.0549). The principal lunar and solar tidal frequencies are shown in Table 6.1.1.

When lunar and solar tractive forces reinforce each other, as, for example, when the Sun and the Moon are on the same or opposite sides of the Earth (full Moon and the new Moon), the tides are the largest and are called high spring tides. The high tides are the highest and the low tides are the lowest, and the tidal range, the difference between the high and the low, is the largest. When they counteract each other, as when they are in quadrature, $90°$ to each other, the tides are low and are called low neap tides. In the former, M_2 and S_2 reinforce, and in the latter oppose each other. Since the tidal tractive force due to the Sun is roughly half that of the Moon, the ratio of spring to neap tides is nearly three (Cartwright, 1997). Newton was also the first one to explain this spring-neap tidal cycle. In addition, because of the solar declination, when the Sun crosses the equatorial plane at the time of the spring and autumn equinoxes, the semidiurnal solar tidal forcing is at its maximum; therefore, the corresponding fortnightly spring tide is the highest.

6.2 FORMULATION: TIDAL POTENTIAL

While the above description is simplified enough to explain the basic underlying mechanisms, quantitative deductions require a mathematical approach to deriving the lunisolar gravitational potential. The following is based on Lambeck (1988), Cartwright (1993), and Desai (1996). For gravitational purposes, the heavenly bodies can be considered to be point masses. It is convenient to consider a geocentric coordinate system when considering Earth's tides. In this coordinate system, the gravitational potential at any point P with coordinate $\mathbf{r}$ due to a mass at another point with coordinate $\mathbf{r}_M$ (Figure 6.2.1) is

$$\Phi = \frac{GM_m}{|\mathbf{r} - \mathbf{r}_M|} \tag{6.2.1}$$

If ψ is the angle between the two vectors,

$$|\mathbf{r} - \mathbf{r}_M| = R\left[1 + \hat{a}^2 - 2\hat{a}\cos\psi\right]^{1/2}$$
$$|\mathbf{r}| = r, |\mathbf{r}_M| = R; \ \hat{a} = r/R \tag{6.2.2}$$

Expanding in a Taylor series in terms of $\hat{a}$ (< 1),

$$\frac{1}{|\mathbf{r} - \mathbf{r}_M|} = \frac{1}{R}\sum_{n=0}^{\infty} \hat{a}^n P_n(\cos\psi) \tag{6.2.3}$$

where P_n are conventional Legendre functions (polynomials) of degree n given by

$$nP_n(x) = (2n-1)xP_{n-1}(x) - (n-1)P_{n-2}(x) \qquad n=2,3 \ldots$$
$$P_0(x) = 1, P_1(x) = x, P_2(x) = (3x^2 - 1)/2, P_3(x) = (5x^3 - 3x)/2 \tag{6.2.4}$$

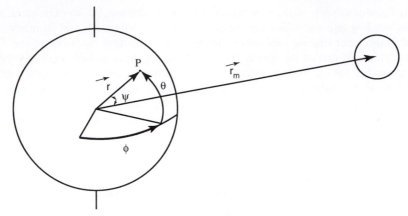

Figure 6.2.1 Sketch defining the angles and distances involved in deriving the tidal potential. (Provided by S. Desai.)

so that the tidal potential at the Earth's surface becomes a sum,

$$\Phi = \frac{GM_M}{R} \sum_{n=0}^{\infty} \left(\frac{R_E}{R} \right)^n P_n(\cos\psi)$$

$$= g \sum_{n=0}^{\infty} K_n P_n(\cos\psi)$$

$(6.2.5)$

where

$$g = \frac{GM_E}{R_E^2}, \quad K_n = R_E \frac{M_M}{M_E} \left(\frac{R_E}{R} \right)^{n+1}$$

$(6.2.6)$

so that

$$K_n g = \frac{GM_M}{R} \left(\frac{R_E}{R} \right)^n.$$

Here R_E is the equatorial radius of the Earth, g is the equatorial gravitational acceleration, and M_E is the mass of the Earth. This expression applies to solar tidal potential as well with appropriate modifications. The mean values for K_2 and K_3 (based on the mean value of R) are 0.358 373 m and 0.005 946 m for the Moon, and 0.164 570 m and 0.000 700 m for the Sun, respectively. K_3 is small but non-negligible for the Moon, but negligibly small for the Sun. K_0 and K_1 are irrelevant to tides. The term n = 0 is constant; hence, the gradient of the potential, the force is zero. The term n = 1 is constant for all points on the Earth and its gradient with respect to ($R_E\cos\psi$), GM_M/R^2, is responsible for the orbital motions of the Moon around the Earth (and the Earth around the Sun). It does not contribute to rotational perturbations or deformations of the Earth (Desai, 1996). The remaining terms (n = 2, 3 ...) add up to zero when summed over a spherically symmetric Earth (the oblateness can be neglected in computing the lunisolar gravitational potential, because of the large distances R_M, R_s involved), but are the ones responsible for Earth's deformations and the lunisolar torques acting on Earth. For example, the most dominant tide-producing term is that for n = 2:

$$\Phi = (GM_M / R)(R_E / R)^2 P_2(\cos\psi) = gK_2(3\cos^2\psi - 1)/2 \qquad (6.2.7)$$

The gradient of these terms is the tidal (tractive) force:

$$F = \nabla\Phi = \frac{GM_M}{R} \nabla \left[\sum_{n=0}^{\infty} \left(\frac{R_E}{R} \right)^n P_n(\cos\psi) \right] \qquad (6.2.8)$$

Making use of spherical geometry and expressing angle ψ in terms of the Greenwich longitude ϕ and latitude θ of the point on Earth under consideration and the corresponding values ϕ_M and θ_M for the Moon (or Sun) center of mass:

$$\cos\psi = \sin\theta\sin\theta_M + \cos\theta\cos\theta_M\cos(\phi - \phi_M) \qquad (6.2.9)$$

θ_M is the lunar (solar) declination, and $(\phi - \phi_M)$ is the hour angle of the point. Decomposing Legendre functions in ψ into those in ϕ and θ,

$$\Phi(r,\phi,\theta) = \frac{GM_M}{R} \sum_{n=2}^{\infty} \sum_{m=0}^{n} \left(\frac{R_E}{R} \right)^n (2 - \delta_{0m}) \frac{(n-m)!}{(n+m)!} P_{nm}(\sin\theta) \qquad (6.2.10)$$
$$P_{nm}(\sin\theta_M)\cos m(\phi - \phi_M)$$

where P_{nm} are associated Legendre polynomials of degree n and order m given by the recursion relation (with $P_{00}(x) = 1$)

$$P_{n,n}(x) = (2n-1)\left(1 - x^2\right)^{1/2} P_{n-1,n-1}(x)$$
$$P_{n+1,n}(x) = (2n+1)xP_{n,n}(x) \qquad (6.2.11)$$
$$P_{n,m}(x) = \left[(2n-1)xP_{n-1,m}(x) - (n+m-1)P_{n-2,m}(x)\right]/(n-m)$$

For example, since $P_{20}(x) = (3x^2-1)/2$, $P_{21}(x) = 3x(1-x^2)^{1/2}$, and $P_{22}(x) = 3(1-x^2)$, the expression for the second degree potential (n=2) is

$$\Phi_2(r,\phi,\theta) = K_2 g \begin{bmatrix} \dfrac{1}{4}\left(3\sin^2\theta - 1\right)\left(3\sin^2\theta_M - 1\right) + \dfrac{3}{4}\sin 2\theta \sin 2\theta_M \\[2mm] \cos\left(\phi - \phi_M\right) + \dfrac{3}{4}\cos^2\theta \cos^2\theta_M \cos 2\left(\phi - \phi_M\right) \end{bmatrix} \qquad (6.2.12)$$

Since it is the term $(\phi - \phi_M$, the hour angle) involving ϕ_M that determines the associated frequency, and ϕ_M has a frequency of a cycle per lunar (and solar) day, of the three terms in the square brackets, the second corresponds to diurnal tides, and the third to semidiurnal tides. Because of the low frequency of a few cycles per lunar month (solar year) associated with the multiplier, the first term corresponds to long period tides (Cartwright, 1993). These are the tidal species first pointed out by Laplace. If the Earth were rigid and comprised of a uniform-depth ocean that responded instantaneously to the above tidal potential, the resulting tide, called the equilibrium tide, would have an amplitude equal to Φ_2/g. However, the Earth is far from rigid, and lunisolar tidal forces generate tides in the solid Earth as well, which reduce the equilibrium tidal amplitude to about 0.693 of this value.

It is easy to see from Eq. (6.2.12) that the long period tides are produced due to changes in lunar (solar) declination. They are independent of the longitude of the point and hence are zonally uniform. They have a maximum amplitude at the poles and are zero at a latitude of 35° 16'. The diurnal species has a maximum amplitude when the lunar (solar) declination is maximum. It is modulated at twice the frequency of variation of this declination. The spatial structure involves both meridional and zonal variations, with maximum amplitude reached at 45° latitude and zero at the equator and the poles. The semidiurnal species is also modulated in a similar fashion, but the amplitude reaches a maximum when the declination is zero. The spatial structure involves both meridional and zonal variations, with maximum amplitude at the equator and zero at the poles. Since the equilibrium tides represented by Eq. (6.2.12) have an amplitude of Φ_2/g, it is easy to see that the semidiurnal equilibrium tide has a maximum amplitude of 0.75 $K_2 = 0.2688$ m, much smaller than the observed value.

The expression for the third-order potential has a similar form, but has an extra term involving $\cos 3(\phi - \phi_M)$. Thus one gets similar, but smaller, contributions plus another species, a terdiurnal tide with a frequency of three cycles per day.

Because the denominator (n+m)! becomes very large for large degree and order, it is more convenient to write Eq. (6.2.11) in terms of normalized associated Legendre functions. The normalizing factor is chosen such that

(Desai ,1996; note that this is different from that in Lambeck, 1988):

$$\bar{P}_{nm} = a_{nm}P_{nm} = (-1)^m \left[\frac{(2n+1)}{4\pi} \frac{(n-m)!}{(n+m)!} \right]^{1/2} P_{nm} \qquad (6.2.13)$$

so that

$$\Phi(r,\phi,\theta) = \frac{GM_M}{R} \sum_{n=2}^{\infty} \sum_{m=0}^{n} \left(\frac{R_E}{R} \right)^n \frac{a_{nm}^2}{2n+1} P_{nm}(\sin\theta) P_{nm}(\sin\theta_M) \cos m(\phi-\phi_M)$$

$$= \sum_{n=2}^{\infty} \sum_{m=0}^{n} \bar{P}_{nm}(\sin\theta) \left[b_{nm}(t) \cos m\phi + c_{nm}(t) \sin m\phi \right]$$

$$b_{nm}(t) = \frac{4\pi GM_M}{R} \frac{(2-\delta_{0m})}{2n+1} \bar{P}_{nm}(\sin\theta_M) \cos(m\phi_M) \left(\frac{R_E}{R} \right)^n$$

$$c_{nm}(t) = \frac{4\pi GM_M}{R} \frac{(2-\delta_{0m})}{2n+1} \bar{P}_{nm}(\sin\theta_M) \sin(m\phi_M) \left(\frac{R_E}{R} \right)^n$$

$$(6.2.14)$$

This gives the tidal potential at any point on the surface of the Earth at time t. Note that the coefficients b_{nm} and c_{nm} are functions of time and dependent only on the ephemeris of the Moon as seen in geocentric coordinates, but they are independent of the coordinates of the observation point.

Since the once-per-month and once-per-year orbital motions of the Moon and the Sun have low frequencies compared to the frequency of rotation of the Earth, these coefficients have principal frequencies centered around m cycles per day, with the orbital motions modulating these principal frequencies. The second-degree coefficients (n = 2) have largest amplitudes concentrated in three frequency bands: m = 0, 1, and 2, the long period, diurnal, and semidiurnal bands. The higher degree coefficients also contribute additional bands centered around m = n cycles per day, but of a much smaller value. Long period band m = 0 is independent of the longitude of the heavenly body and has principal frequencies at fortnightly (Mf) and semiannual (S_{sa}) frequencies. The diurnal and semidiurnal bands have a bandwidth of about 0.3 cy d^{-1} centered at 1 and 2 cy d^{-1}, while the long period tides have frequencies from 0 to 0.15 cy d^{-1} but are, however, an order of magnitude smaller than the high frequency tides (Desai, 1996).

The infinite series can of course be truncated since $a_M \sim (60.3)^{-1}$ and $a_S \sim (23481)^{-1}$. Truncating at n = 2 or 3 suffices for lunisolar tidal potential. For n = 2, $(M_S/R_S^3) = 0.4605 (M_M/R_M^3)$, but for n = 3, $(M_S/R_S^4) = 0.0012 (M_M/R_M^4)$, and therefore the n = 3 term is usually neglected in the solar tidal potential (Cartwright and Edden, 1973; Cartwright and Taylor, 1971), but Wenzel and

Zurn (1990) use the expansions of Tamura (1987) to show that for modern high precision gravimeters it is necessary to retain the $n = 3$ solar and $n = 4$ lunar terms.

The coefficients $b_{nm}(t)$ and $c_{nm}(t)$ are predictable centuries in advance, since the motions of heavenly bodies in the solar system can be predicted very accurately. The ephemerides of the Sun and the Moon in geocentric coordinates (the distance from Earth's center, and the longitude and latitude with respect to the ecliptic) can be defined by five astronomical arguments (angles) and expressed as an infinite series of these angles. The tidal potential can be expanded into a harmonic series composed of functions whose arguments are linear combinations of these five astronomical arguments and the sixth involving the hour angle of the Moon. Darwin (1886) and Doodson (1921) were the first ones to expand the tidal potential into a harmonic series. More recently, Cartwright and Taylor (1971) and Cartwright and Edden (1973) used spectral techniques to expand the lunar tidal potential to $n = 3$ and solar to $n = 2$, followed by Tamura (1987) who went to the next higher degree for both, which adds two additional significant digits and many more spectral components to Cartwright and Taylor's expansions (Desai, 1996).

The combined luni-solar tidal potential is

$$\Phi(r,\phi,\theta) = \frac{4\pi GM_M}{R_E} \sum_{n=2}^{\infty} \sum_{m=0}^{n} \left(\frac{R_E}{L_M}\right)^{n+1} \frac{2-\delta_{0m}}{2n+1} \overline{P}_{nm}(\sin\theta)$$

(6.2.15)

$$\left[\begin{array}{l} \left(\dfrac{L_M}{R}\right)^{n+1} \overline{P}_{nm}(\sin\theta_M)\cos m(\phi-\phi_M) + \\[3mm] \dfrac{M_S}{M_M}\left(\dfrac{L_M}{L_S}\right)^{n+1}\left(\dfrac{L_S}{R_S}\right)^{n+1} \overline{P}_{nm}(\sin\theta_S)\cos m(\phi-\phi_S) \end{array}\right]$$

where L_M and L_S are the major axes of the Moon's orbit around Earth and Earth's orbit around the sun, and R_S is the distance to the Sun. This can be written as

$$\Phi(r,\phi,\theta) = g\sum_{n=2}^{\infty}\sum_{m=0}^{n}\sum_{j} H_{nmj}\,\overline{P}_{nm}(\sin\theta)\cos(m\phi+\omega_{nmj}t+\alpha_{nmj})$$

$$\omega_{nmj}t+\alpha_{nmj} = d_1\tau + (d_2-5)s_0 + (d_3-5)h_0 + (d_4-5)p_0 + (d_5-5)n_s$$

(6.2.16)

$$+(d_6-5)p_s - \chi_{nm}\frac{\pi}{2}$$

$$\chi_{nm} = 1 \text{ (for odd } n+m\text{); } 0 \text{ (for even } n+m\text{)}$$

where d_i are integers that constitute the Doodson argument number $d_1d_2d_3d_4d_5d_6$; d_1 defines the tidal species (0, 1, and 2 cy d^{-1}) and equals the order m of the spherical component from which it is derived; d_1d_2 define the group number (each group separated by about 1 cy mo^{-1}); and $d_1d_2d_3$ define the constituent number (each constituent separated by 1 cy year^{-1}). The origin of each tidal component can be deduced from the Doodson number. For example, M_2 with 2τ originates from the Moon, S_2 with argument $2\tau_s = 2$ ($\tau + s_0 - h_0$) originates with the Sun. The time dependence of the coefficients H_{nmj} is very small and is usually ignored (Cartwright and Taylor, 1971; see also Lambeck, 1988) so that all the temporal variability is concentrated in the cosine term.

The Tamura (1987) development has 209 constituents in the long period band, 343 in the diurnal and 277 in the semidiurnal bands, derived from degree 2 tidal potential, compared to 104, 162, and 119 of Cartwright and Edden (1973). Table 6.1.1 shows a compilation of the arguments, phases, the origin, and the magnitude of the term H_{nmj} in the tidal potential of each component. Note that the values for H_{nmj} are the current best estimates and are from Desai (1996). See Desai (1996) and the original sources, Cartwright and Taylor (1971) and Cartwright and Edden (1973) for more details of the development leading to Table 6.2.1. Tamura's (1987) development leads to values that differ from the above by less than ~0.1%. Some of the H_{nmj} terms are negative and are either tabulated thus (Desai, 1996) or they are quoted as positive as in Table 6.2.1 but with the values of the phases that need to be added to those in Eq. (6.2.16) to make the quoted values negative also listed. Figure 6.2.2 shows the spectral distribution of degree 2 diurnal and semidiurnal tides from Cartwright and Edden (1973).

The amplitude of the equilibrium tide is Φ/g, and for the most important term in the tidal potential, namely that of degree two, it can be written as

$$\zeta_2 = \frac{\Phi_2}{g} = \sum_j H_{20j}\,\overline{P}_{20}(\sin\theta)\cos(\omega_{20j}t + \alpha_{20j}) + \sum_j H_{21j}\,\overline{P}_{21}(\sin\theta)\cos(\phi + \omega_{21j}t + \alpha_{21j})$$

$$+ \sum_j H_{22j}\,\overline{P}_{22}(\sin\theta)\cos(2\phi + \omega_{22j}t + \alpha_{22j}) \tag{6.2.17}$$

By virtue of Eq. (6.2.13),

$$\zeta_2 = \frac{\Phi_2}{g} = \sum_j A_{20j}\left(\frac{3}{2}\cos^2\theta - 1\right)\cos(\omega_{20j}t + \alpha_{20j}) + \sum_j A_{21j}\,\sin 2\theta\,\cos(\phi + \omega_{21j}t + \alpha_{21j})$$

$$+ \sum_j A_{22j}\,\cos^2\theta\cos(2\phi + \omega_{22j}t + \alpha_{22j}) \tag{6.2.18}$$

$$A_{20j} = \left(\frac{5}{4\pi}\right)^{1/2} H_{20j};\ A_{21j} = \frac{3}{2}\left(\frac{5}{24\pi}\right)^{1/2} H_{21j};\ A_{22j} = 3\left(\frac{5}{96\pi}\right)^{1/2} H_{22j}$$

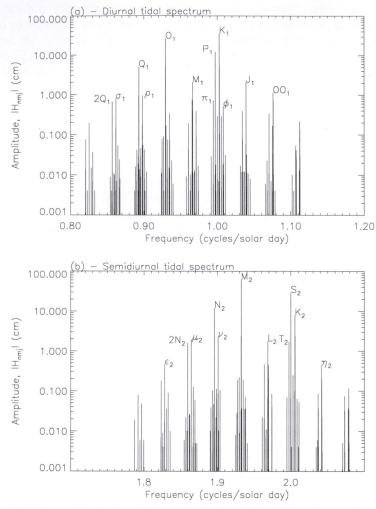

Figure 6.2.2 Spectral distribution of diurnal (top) and semidiurnal (bottom) tides. From Desai (1996).

Table 6.1.1 shows the various constituents, their Doodson numbers, equilibrium amplitudes A_{nmj}, and other characteristics. The ter-diurnal tide ($n=3$, $m=3$) and nonlinear shallow water tides are also shown.

Not all the tidal constituents are named; only the principal ones. The most important semidiurnal tide in the global oceans is M_2, the lunar tide, and the most important diurnal one is K_1, which is a declinational tide with contributions from both the Moon and the Sun. In most cases, it is adequate to consider the largest four semidiurnal (M_2, S_2, N_2, K_2), four diurnal (K_1, O_1, P_1, Q_1) and three long-period tides (Mf, Mm, S_{sa}). Note that the semidiurnal constituents have a subscript of 2, and diurnal ones 1 (ter-diurnal one, 3, and long period tides, none). The solar semidiurnal tide S_2 has contributions from meteorological forcing due to day-night changes in the solar heating of the atmosphere and the oceans. So do the semiannual S_{sa} and annual S_a, which are dominated by seasonal changes in solar heating and consequent changes in sea level. Lunar long period tides, the fortnightly Mf and monthly Mm, are dominated by gravitational forcing. M_2 would be the only lunar tide if the Moon's orbit were circular and in the equatorial plane. N_2 and Q_1 arise due to the ellipticity of Moon's orbit. K_2 arises from both lunar and solar declinations, while O_1 arises from lunar and P_1 from solar declinations.

Computing the tides on Earth requires the complicated motion of the Earth-Moon-Sun system to be described suitably. The two most common reference systems are the equatorial and ecliptic. In the equatorial system, the basic reference circle is the Earth's equator projected upon the celestial sphere and the declinations of astronomical bodies are measured with respect to the plane through the equator. In the ecliptic system, the celestial extension of the ecliptic plane, the plane of the apparent motion of the Sun, is used as the reference. In both systems, the angular distances around these planes are measured eastward relative to a point referred to as the vernal equinox (the two points of intersection of the ecliptic plane and the equatorial plane define the vernal and autumnal equinoxes, which are more or less fixed with respect to the stellar background, even though they precess slowly against the stellar background with a period of about 26000 years). This is called the right ascension of the body, which along with its declination determines the position of the body in space against the stellar background.

The orbit and position of a heavenly body in either system can be defined by six parameters (see Pugh, 1987, for an excellent description of these for the Earth-Moon-Sun system). Following the usual convention (e.g., Doodson, 1921), these six astronomical arguments (angles) are as follows:

τ —the mean lunar time = H +180° (= H+12 h)

s_0— the mean longitude of the Moon

h_0— the mean longitude of the Sun

p_0— the mean longitude of lunar perigee

n_S— the negative mean longitude of the ascending lunar node

p_s— the mean longitude of the solar perigee (perihelion)   (6.2.19)

TABLE 6.2.1

Argument and Phase, Doodson Number for Important Tidal Constituents;
the Amplitude Denotes the Magnitude in the Harmonic Expansion;
Rank Is According to Amplitude within Each Species

Species (S)	Doodson No.	i	j	k	l	m	Name	Tidal potential/g (m)	Rank	Origin	Argument*	Phase
							Longperiod					
	055.555	0	0	0	0	0	M_0	0.21487		M		
	055.555	0	0	0	0	0	S_0	0.09969		S		
	056.554	0	1	0	0	−1	Sa	0.00492	7	S	$+0s + 1h + 0p - 1p_s$	
	057.555	0	2	0	0	0	Ssa	0.03098	3	S	$+0s + 2h + 0p$	
0	063.655	1	−2	1	0	0	Msm	0.00672	5	M	$+1s - 2h + 1p$	
	065.455	1	0	−1	0	0	Mm	0.03518	2	M	$+1s + 0h - 1p$	
	073.555	2	−2	0	0	0	Msf	0.00583	6	M	$+2s - 2h + 0p$	
	075.555	2	0	0	0	0	Mf	0.06661	1	M	$+2s + 0h + 0p$	
	085.455	3	0	−1	0	0	Mt	0.01276	4	M	$+3s + 0h - 1p$	
							Diurnal					
	125.755	−3	0	2	0	0	$2Q_1$	0.00664	11		$-4s + 1h + 2p$	$-\pi/2$
	127.555	−3	2	0	0	0	σ_1	0.00802	9	M	$-4s + 3h + 0p$	$-\pi/2$
	135.645	−2	0	1	−1	0		0.00947			$-3s + 1h + 1p$	$-\pi/2$
	135.655	−2	0	1	0	0	Q_1	0.05019	4	M	$-3s + 1h + 1p$	$-\pi/2$
	137.445	−2	2	−1	−1	0		0.00180			$-3s + 3h - 1p$	$-\pi/2$
	137.455	−2	2	−1	0	0	ρ_1	0.00954	8	M	$-3s + 3h - 1p$	$-\pi/2$
	145.535	−1	0	0	−2	0		0.00152			$-2s + 1h + 0p$	$\pi/2$
	145.545	−1	0	0	−1	0		0.04946			$-2s + 1h + 0p$	$-\pi/2$
1	145.555	−1	0	0	0	0	O_1	0.26216	2	M	$-2s + 1h + 0p$	$-\pi/2$
	145.755	−1	0	2	0	0		0.00171			$-2s + 1h + 2p$	$\pi/2$
	147.555	−1	2	0	0	0	τ_1	0.00343	16		$-2s + 3h + 0p$	$\pi/2$
	155.455	0	0	−1	0	0		0.00741			$-1s + 1h - 1p$	$\pi/2$
	155.655	0	0	1	0	0	M_1	0.02062	5	M	$-1s + 1h + 1p$	$\pi/2$
	155.665	0	0	1	1	0		0.00414			$-1s + 1h + 1p$	$\pi/2$
	157.455	0	2	−1	0	0	χ_1	0.00394	14		$-1s + 3h - 1p$	$\pi/2$
	162.556	1	−3	0	0	1	π_1	0.00713	10		$+0s - 2h + 0p + 1p_s$	$-\pi/2$
	163.545	1	−2	0	−1	0		0.00137			$+0s - 1h + 0p$	$\pi/2$
	163.555	1	−2	0	0	0	P_1	0.12201	3	S	$+0s - 1h + 0p$	$-\pi/2$
	164.556	1	−1	0	0	1	S_1	0.00289			$+0s - 0h + 0p + 1p_s$	$\pi/2$
	165.545	1	0	0	−1	0		0.00730			$+0s + 1h + 0p$	$-\pi/2$
	165.555	1	0	0	0	0	K_1	0.36867	1	M, S	$+0s + 1h + 0p$	$\pi/2$
	165.565	1	0	0	1	0		0.05002			$+0s + 1h + 0p$	$\pi/2$
	166.554	1	1	0	0	−1	Ψ_1	0.00293	17		$+0s + 2h + 0p - 1p_s$	$\pi/2$
	167.555	1	2	0	0	0	ϕ_1	0.00525	12		$+0s + 3h + 0p$	$\pi/2$
	173.655	2	−2	1	0	0	θ_1	0.00395	14		$+1s - 1h + 1p$	$\pi/2$
	175.455	2	0	−1	0	0	J_1	0.02062	6	M	$+1s + 1h - 1p$	$\pi/2$
	175.465	2	0	−1	1	0		0.00409	13		$+1s + 1h - 1p$	$\pi/2$
	183.555	3	−2	0	0	0		0.00342	17		$+2s - 1h + 0p$	$\pi/2$
	185.355	3	0	−2	0	0		0.00169			$+2s + 1h - 2p$	$\pi/2$
	185.555	3	0	0	0	0	OO_1	0.01129	7	M	$+2s + 1h + 0p$	$\pi/2$
	185.565	3	0	0	1	0		0.00723			$+2s + 1h + 0p$	$\pi/2$

(Continues)

TABLE 6.2.1 (*Continued*)

Species (S)	Doodson No.	i	j	k	l	m	Name	Tidal potential/g (m)	Rank	Origin	Argument*	Phase
							Semidiurnal					
	225.855	−3	0	3	0	0		0.00180			$-5s + 2h + 3p$	
	227.655	−3	2	1	0	0	ε_2	0.00467	10		$-5s + 4h + 1p$	
	235.755	−2	0	2	0	0	$2N_2$	0.01601	9	M	$-4s + 2h + 2p$	
	237.555	−2	2	0	0	0	μ_2	0.01932	6	M	$-4s + 4h + 0p$	
	238.554	−2	3	0	0	−1		0.00130			$-4s + 5h + 0p$	
	244.656	−1	−1	1	0	1		0.00102			$-3s + 1h + 1p$	π
	245.645	−1	0	1	−1	0		0.00451			$-3s + 2h + 1p$	π
	245.655	−1	0	1	0	0	N_2	0.12099	3	M	$-3s + 2h + 1p$	
	246.654	1	1	1	0	−1		0.00113			$-3s + 3h + 1p$	
	247.455	−1	2	−1	0	0	ν_2	0.02298	5	M	$-3s + 4h - 1p$	
	248.454	1	3	−1	0	−1		0.00106			$-3s + 5h - 1p$	
	253.755	0	−2	2	0	0		0.00190			$-2s + 0h + 2p$	π
	254.556	0	−1	0	0	1		0.00218			$-2s + 1h + 0p$	π
	255.545	0	0	0	−1	0		0.02358			$-2s + 2h + 0p$	π
2	255.555	0	0	0	0	0	M_2	0.63194	1	M	$-2s + 2h + 0p$	
	256.554	0	1	0	0	−1		0.00193			$-2s + 3h - 0p$	
	263.655	1	−2	1	0	0	λ_2	0.00466	11		$-1s + 0h + 1p$	π
	265.455	1	0	−1	0	0	L_2	0.01786	7	M	$-1s + 2h - 1p$	π
	265.655	1	0	1	0	0		0.00447			$-1s + 2h + 1p$	
	265.665	1	0	1	1	0		0.00197			$-1s + 2h + 1p$	
	272.556	2	−3	0	0	1	T_2	0.01718	8	S	$+0s - 1h + 0p + 1p_s$	
	273.555	2	−2	0	0	0	S_2	0.29402	2	S		
	274.554	2	−1	0	0	−1	R_2	0.00246			$+0s + 1h + 0p - 1p_s$	π
	274.555	2	−1	0	0	0		0.00305	13		$+0s + 1h + 0p$	
	275.545	2	0	0	−1	0		0.00102			$+0s + 2h + 0p$	π
	275.555	2	0	0	0	0	K_2	0.07993	4	M, S	$+0s + 2h + 0p$	
	275.565	2	0	0	1	0		0.02382			$+0s + 2h + 0p$	
	275.575	2	0	0	2	0		0.00259			$+0s + 2h + 0p$	
	283.655	3	−2	1	0	0		0.00086			$+1s + 0h + 1p$	
	285.455	3	0	−1	0	0	η_2	0.00447	12		$+1s + 2h - 1p$	
	285.465	3	0	−1	1	0		0.00195			$+1s + 2h - 1p$	
							Ter-diurnal					
3	355.555	0	0	0	0	0	M_3	0.00828		M	$-3s + 3h + 0p$	
							Shallow water					
							$2SM_2$				$+2s - 2h + 0p$	
							M_4				$-4s + 4h + 0p$	
							MS_4				$-2s + 2h + 0p$	
							S_4					
							M_6				$-6s + 6h + 0p$	
							$2SM_6$				$-4s + 4h + 0p$	

* Argument = $(i - S)s + (j + S)h + k^*p + m^*p_s$
Based on Dietrich *et al.* (1980) and Desai (1996).

where H is Moon's Greenwich hour angle. Each of these six arguments can be computed accurately for centuries in advance from cubic expressions of time reckoned from some epoch, such as the beginning of year 1900 (Melchior, 1981; Pugh, 1987; see also Eq. 6.8.9). The rates of change of these six quantities define the six fundamental frequencies (periods) of astronomical motions related to the orbital motions of the Earth-Moon-Sun system (Pugh, 1987):

$$
\begin{aligned}
\omega_1 & \quad - \dot{t} \quad - \quad 1.0351 \text{ dy} \ - \ 14.4921°\text{hr}^{-1} \\
\omega_2 & \quad - \dot{s}_0 \ - \ 27.3217 \text{ dy} \ - \ 0.5490°\text{hr}^{-1} \\
\omega_3 & \quad - \dot{h}_0 \ - 365.2422 \text{ dy} \ - \ 0.0411°\text{hr}^{-1} \\
\omega_4 & \quad - \dot{p}_0 \ - \quad \ 8.8500 \text{ yr} \ - \ 0.0046°\text{hr}^{-1} \\
\omega_5 & \quad - \dot{n}_s \ - \ 18.6100 \text{ yr} \ - \ 0.0022°\text{hr}^{-1} \\
\omega_6 & \quad - \dot{p}_s \ - \quad 20\,942 \text{ yr} \ - \ 0.000002 \text{ hr}^{-1}
\end{aligned}
\qquad (6.2.20)
$$

The frequency ω_1 corresponds to the mean lunar day, ω_2 to the sidereal month, and ω_3 to the tropical year; ω_4 represents the motion of Moon's perigee, ω_5 represents the regression of lunar node, and ω_6 represents the motion of Sun's prihelion. To this can be added the frequency ω_0, corresponding to the mean solar day (1 day, $15°$ hr^{-1}). By convention, these frequencies are in radians per unit time, while applications use degrees per hour (Pugh, 1987). The Doodson numbers in Eq. (6.2.16) (see also Tables 6.1.1 and 6.2.1) that label each tidal constituent are determined by the angular velocity of that particular constituent, which is a linear combination of these fundamental frequencies:

$$
\omega_c = d_1\omega_1 + (d_2 - 5)\omega_2 + (d_3 - 5)\omega_3 + (d_4 - 5)\omega_4 + (d_5 - 5)\omega_5 + (d_6 - 5)\omega_6 \quad (6.2.21)
$$

Note that this labeling of each constituent is in terms of ω_1 corresponding to a lunar day and is preferred by most tidalists (Cartwright and Edden, 1973; Doodson, 1921). It is also possible to label the same constituents with respect to the solar day (Schureman, 1941) using the relationship $\omega_1 = \omega_0 - \omega_2 + \omega_3$.

The frequency corresponding to a sidereal day (0.9973 day) is $\omega_s = \omega_0 + \omega_3 = \omega_1 + \omega_2$. The synodic period (29.5307 day) governing the lunar phases (new Moon to full Moon to new Moon) determines the 14.7654 day spring-neap cycle of tides that corresponds to the frequency difference $\omega_2 - \omega_3$ (also equal to $\omega_0 - \omega_1$). The spring-neap tides are themselves modulated by tidal force variations at frequency $\omega_2 - \omega_4$ or equivalently the anomalistic period (27.5546 day), corresponding to the closest approach of the Moon to Earth (perigee to perigee).

6.3 BODY, LOAD, ATMOSPHERIC, AND RADIATIONAL TIDES

6.3.1 BODY (SOLID EARTH) TIDES

It is not only the ocean that responds to tidal forcing. The atmosphere and the solid Earth do as well. The gravitational tides in the atmosphere are very small. Therefore, the tides in the atmosphere are driven principally by solar diurnal heating of the atmospheric column. Even these are rather small (equivalent to about 1 cm of water or less), and, except for some applications, normally neglected. The solid Earth responds to tidal forcing as an elastic body and not like a rigid body, and so these body tides also have a rich spectrum of components. The solid Earth tides are quite important for a variety of applications related to gravity, such as the analysis of modern gravity meters, precision tracking of satellites and the Global Positioning System, as well as the calculation of oceanic tides. As far as their effect on oceanic tides is concerned, their principal influence is to reduce the amplitude of the equilibrium tides the ocean feels, since the Earth deforms in the same direction as the oceans under the tidal forcing by the Moon and the Sun. Because the natural frequencies are an order of magnitude larger than tidal frequencies, the response of the solid Earth to tidal forcing is nearly instantaneous so that the spatial distribution of the Earth's body tides is the same as that of the tidal forcing. The tidal bulges are displaced about 0.21° from the sublunar and antilunar points, but this can be neglected unless one is interested in the resulting torques on the Earth and the Moon and the consequent deceleration of the Earth's rotation and increase in the Moon's orbital momentum. In other words, equilibrium tides result. The consequence is a reduction in the effective amplitude of the tidal potential felt by the oceanic equilibrium tides, with respect to the Earth's surface. This elasticity factor must be accounted for in numerical models of oceanic tides. The gravitational potential is simply multiplied by this elasticity factor. Another aspect of this elasticity of solid Earth is that the Chandler wobble of the Earth's rotation axis has a period of 14.4 months, compared to a value of 10 months, Euler value that would result were the Earth to be rigid, the ratio of the latter to the former being roughly the same as the elasticity factor (see Section 6.11).

To compute tides in the solid Earth, it is traditional to consider an elastic, oceanless Earth (the effect of the ocean loading is to produce the so-called load tides, which will be mentioned later) and solve the linearized governing equations for the deformation, assuming a constitutive law (that relates the stress to the strain). The deformation vector $u = (u_r, u_\phi, u_\theta)$, denoting the displacement of a point on Earth can be written as (e.g., Desai, 1996)

$$(u_r, u_\phi, u_\theta) = \left[\frac{H(r)}{g} \Phi(r, \phi, \theta), \frac{L(r)}{g \sin \theta} \frac{\partial}{\partial \phi} \Phi(r, \phi, \theta), \frac{L(r)}{g} \frac{\partial}{\partial \theta} \Phi(r, \phi, \theta) \right] \quad (6.3.1)$$

These displacements themselves increase the tidal potential so that the total potential at the deformed surface is $[1 + K(r)] \; \Phi(r,\phi,\theta)$. The proportionality constants $H(r)$, $K(r)$, and $L(r)$ on a spherical surface with a radius equal to the mean radius of the Earth are written in terms of Love numbers h and k introduced by Love in 1909, and number l introduced by Shida in 1912 (Desai, 1996; Melchior, 1981), which describe the elastic response of the Earth to tidal forcing:

$$h = H(R), k = K(R), l = L(R). \hspace{2cm} (6.3.2)$$

Each spherical component of the tidal potential of degree n and order m has a corresponding set of Love numbers. However, the dependence on m is so small that it is usually ignored, and Love numbers are considered only functions of degree n. In application to oceanic tide calculation, it is the radial deformations of the solid Earth that are important; this appears as the elasticity factor in the governing tidal equations:

$$\alpha_n = (1 + k_n - h_n) \hspace{2cm} (6.3.3)$$

In other words, the tide potential Φ_n produces a solid Earth deformation $\zeta_s = h_n \Phi_n/g$. ζ_s is known as the body tide. It increases the potential by an amount $k_n\Phi_n$ at the Earth's surface. But since a particle on the surface is displaced from R_E to $R_E+ \zeta_s$, there is a local decrease in the potential by $h_n\Phi_n$, so that the effective tide potential is $(1 + k_n - h_n) \Phi_n = \alpha_n \Phi_n$. The k_n term denotes the self-attraction effect. The numerical values depend very much on the exact model used for the Earth. Wahr's values (Wahr, 1981) are widely used in most of the work related to tides (see Desai, 1996). For the second-degree tidal potential, $k_2 \sim 0.302$, $h_2 \sim 0.609$, $l_2 \sim 0.085$ so that $\alpha_2 \sim 0.693$ ($h_3 \sim 0.290$, $k_3 \sim 0.093$, $l_3 \sim 0.015$ for the third-degree potential). However, since there is a sharp resonance of Earth's core near one cycle per day, the Love numbers are frequency dependent for diurnal tides (see Table 6.3.1). In particular, the values of k_2 and h_2 for K_1 are reduced significantly: $k_2 \sim 0.256$, $h_2 \sim 0.520$, so that $\alpha_2 \sim 0.736$.

While the amplitude of body tides is given by $h_2\Phi_2/g$, typically a few tens of centimeters, the observed change in gravitational acceleration is $-2 (1 - 1.5 k_2 + h_2) \Phi_2/R_E$, on the order of 100 µgal (1 gal $= 10^{-2}$ m s^{-2}), about 10^{-7} times the unperturbed value. It is difficult to observe the solid Earth tidal displacements from ground-based instruments, since the instrument itself is displaced, but sensitive gravimeters can detect the associated changes in the gravitational acceleration, and tiltmeters, the associated tilt of the Earth's surface with respect to the gravitational direction (Wahr, 1981; see also Melchior, 1981).

There are several sets of Love numbers: the first set described above corresponds to the response of the Earth to applied potential without any loading of the surface. The second set due to the response to the applied load are the

TABLE 6.3.1
Love Numbers and the Elasticity Factors for Principal
Tidal Constituents

Constituent	k	h
Q_1	0.298	0.603
O_1	0.298	0.603
P_1	0.287	0.581
K_1	0.256	0.520
N_2	0.302	0.609
M_2	0.302	0.609
S_2	0.302	0.609
K_2	0.302	0.609

From Foreman *et al.* (1993).

load Love numbers, k_n', h_n', and l_n' (see Section 6.3.2), and the third due to the response to horizontal frictional forces such as due to ocean currents are shear Love numbers, k_n'', h_n'', and l_n''. Of these nine, only six are independent: k_n, h_n, l_n, h_n', l_n', and l_n'' (see Lambeck, 1988); for example, $k_n' = k_n - h_n$.

6.3.2 LOAD TIDES

The oceanic tide also affects the solid Earth below. The loading due to oceanic tide deforms it elastically. These ocean tide-induced solid Earth tides are called load tides. The loading effects extend to continents as well and are important to gravimetry. As a crude approximation, however, if ζ is the oceanic tide, the load tide can be taken as $0.07 \, \zeta$ in magnitude and exactly in antiphase with the oceanic tide, and the gravitational potential due to it is $0.05 \, g \, \zeta$ (Cartwright, 1993). However, a more accurate method is to calculate equivalent load Love numbers, using a Green's function approach and an Earth model. Needless to say, the ocean tides must be known accurately to compute the load tides, which, in turn, affect the ocean tides through modification of the effective tidal potential. It is traditional to decompose the ocean tides into spherical harmonics of various degrees and compute the Love numbers for each. If ζ_n (θ,ϕ,t) is the ocean tide, then the load tide is $\gamma_n \, h_n' \, \zeta_n$ and the associated tidal potential $g \, (1 + k_n') \, \gamma_n \, \zeta_n$, where k_n' and h_n' are the load Love numbers, and $\gamma_n = 3 \, (\rho / \rho_E)/(2n+1) = 0.5628/(2n+1)$; ρ and ρ_E are mean densities of the ocean and the Earth ($k_2' \sim -0.307$, $h_2' \sim -1.101$, $l_2' \sim 0.023$; $k_3' \sim -0.197$, $h_3' \sim -1.068$, l_3' ~ 0.070). Farell (1972) and Dahlen (1976) are two early calculations of load Love numbers. Pagiatakis (1990) is the most recent (Cartwright, 1993; see also Lambeck, 1988).

For calculation of oceanic tides, the appropriate method is to use an iterative scheme in which the calculated global ocean tides are used to compute the load tides and the induced tidal potential, which is then used to recalculate the ocean tides. However, the calculation of load tides can be rather tedious, since a large number of terms to a very high degree need to be included for accuracy and the calculation made by global integration. It is therefore traditional to use instead the crude approximation for the effective tidal potential due to load tides mentioned above, even though it is not valid over shallow ocean areas. Another alternative is to use load tides calculated from earlier tidal models (for example, Ray, Sanchez, and Cartwright, 1988, who use GEOSAT altimeter-derived ocean tides to calculate load tides).

Load tides define the deformation of the solid Earth, including regions over land, due to the fluctuating load of the oceanic tides above and the opposing effect of the gravitational attraction of the solid Earth by the mass of the oceans (Kantha *et al.,* 1995). They are a small fraction of the ocean tides and are roughly $180°$ out of phase with them. The phase shift is simply due to the fact that the load tide is of opposite sign to the ocean tide. The load tides can be computed using spherical harmonic decomposition of oceanic tides and the load Love numbers (Farell, 1972; Francis and Mazzega, 1990; Munk and MacDonald, 1960). Spherical harmonics to degree 180 and the load Love numbers of Farrell (1972) have been used by Kantha *et al.* (1995) to compute load tides over the global oceans. Figure 6.3.1 shows the load tides for M_2 and K_1 from Kantha *et al.* (1995). Comparison with those calculated from the Desai and Wahr (1995) tidal model and Cartwright *et al.* (1991) load tide results derived from Geosat altimetry shows similar values for both the amplitude and phase in the open ocean. Only in the margins of the ocean basins, especially in regions of high tides such as the Patagonian shelf and the Irish Sea, are substantial differences present.

Accurate computation of load tides is important to many applications such as gravimetry, the global positioning system, and geophysics. While load tide results derived from the venerable Schwiderski (1980) tidal model are still quite adequate for many applications, submillimeter accuracies are highly desirable for geophysical applications such as inferring the workings of the Earth's core. As in most applications, tides are just a nuisance and need to be taken out as accurately as possible so that the much smaller residual signals can be interpreted properly, and geophysical applications are no exception. Figure 6.3.2 shows the gravity anomalies resulting from M_2 load tides computed more accurately (using Green's function approach) from the global ocean model of Kantha (1995b). The anomalies are plotted at various stations where gravimeter data are available. The differences between the various models are shown in the form of anomaly vectors, with in-phase and quadrature components.

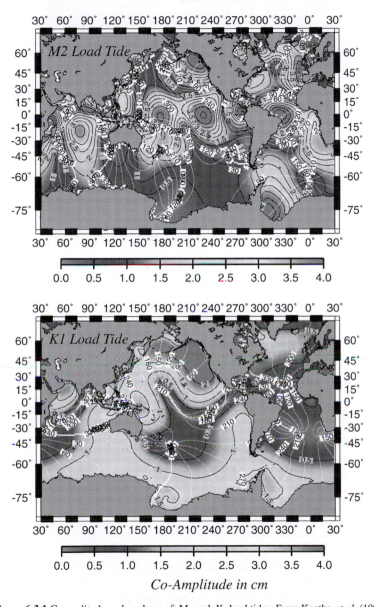

Figure 6.3.1 Coamplitude and cophase of M_2 and K_1 load tides. From Kantha *et al.* (1995).

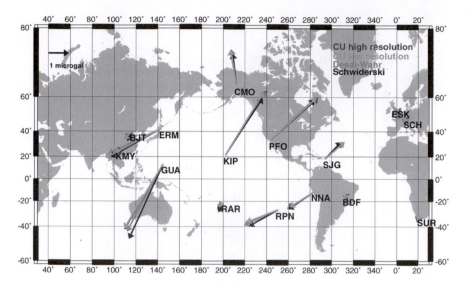

Figure 6.3.2 Gravity anomalies at selected sites resulting from M_2 load tides calculated from different tidal models.

Body tides and load tides have to be taken into account when deducing oceanic tides from altimetric measurements. An altimeter measures the geocentric tide ζ_g, which includes contributions from body tides ζ_b and load tides ζ_l : $\zeta_g = \zeta + \zeta_b + \zeta_l$. It is therefore necessary to calculate and subtract the body and load tides as accurately as possible to deduce oceanic tides from altimetric measurements, even if it is assumed that the altimetric tides are completely free from various analysis, instrument, and measurement-related errors.

6.3.3 ATMOSPHERIC TIDES

Tides also occur in the atmosphere, but they are due to both gravitational and thermal forcing, with the latter being the dominant. They are most evident in surface pressure measurements but can also be seen in surface temperature and pressure fluctuations and winds aloft. The pressure wave is most pronounced at the semidiurnal frequency (exactly 12 hr) leading the corresponding temperature wave by about 10 hr, while the temperature wave is predominantly diurnal (exactly 24 hr), consistent with the diurnal solar heating. Tidal winds in the middle and upper atmosphere are well documented and modeled (e.g., Forbes and Hagan, 1988), but only recently has it become possible to observe them in the troposphere using UHF and VHF radar wind profilers (Whiteman and Bian,

1996). Atmospheric tides bear some resemblance to internal tides in the ocean, in that they are also internal waves with tidal frequencies, but are very ineffectively excited by astronomical tidal forcing. Instead, the solar thermal forcing is the most efficient. Atmospheric tides excited by gravitational forcing are virtually uniform in the vertical, while those excited by thermal forcing exhibit vertical structure. In midlatitudes, the weak surface pressure signals due to tides are swamped by the large meteorological disturbances. The tidal surface pressure signals are relatively larger in the tropics (~1.2 hPa (mb), about 1 cm amplitude in equivalent water level fluctuations) and more easily discernible, but decrease toward the poles (~0.2 hPa at $60°$ latitude).

The pressure fluctuations produced by atmospheric tides are felt by the ocean and since the ocean tends to adjust to changes in atmospheric pressure like an inverse barometer, with a rise (fall) in sea level of ~0.98 cm for every mb drop (rise) in atmospheric pressure, oceanic tides of equivalent amplitude are also generated. The oceanic tide most affected is S_2. Observations clearly show that the amplitude and phase of S_2 are anomalous compared to its neighboring, gravitationally forced M_2 and N_2, suggesting the importance of radiational tidal forcing at this frequency. The ratio of radiational part of S_2 to the gravitational part is anywhere from 0.076 to 0.16, with lower values in the tropics and an increase toward high latitudes (Pugh, 1987). In addition, in some coastal regions with strong land/sea diurnal breezes generated by the land/sea heating contrast, a diurnal S_1 tide of a few centimeters in amplitude might also be present.

Excellent reviews of atmospheric tides can be found in Chapman and Lindzen (1970), Lindzen (1979, 1990), and Whiteman and Bian (1996). While their influence on middle and upper atmosphere is of great interest to geophysicists, their importance to troposphere dynamics derives largely from the observed correlation between the divergences due to tidal winds and semidurnal precipitation maxima in the tropics and subtropics (Whiteman and Bian, 1996). For ocean modelers, they are of interest principally because of the contrast they provide to oceanic tides. We include a brief review here for completeness.

Atmospheric tides as seen at the surface of the Earth are primarily S_2, solar semidiurnal, even though solar internal heating of the atmosphere is predominantly diurnal. While it was known since Laplace's time that solar thermal forcing was responsible, the precise mechanism eluded even people like Lord Kelvin. Only in the 1960s did it become clear that the absorption of solar radiation by water vapor in the troposphere (about one third) and ozone in the upper stratosphere (about two thirds) accounted for most of the semidiurnal pressure fluctuations observed at the surface, far more than that due to surface heating and gravitational forcing (Lindzen, 1990). Second, since vertically propagating diurnal modes cannot exist poleward of about $30°$ latitude, over most of the globe, the strong diurnal tides excited by the diurnal thermal forcing at higher levels of the atmosphere are trapped there and are not felt at the surface. Equatorward of $30°$ latitude, vertically propagating waves suffer from destructive interference (Whiteman and Bian, 1996). No similar restrictions exist for semidiurnal tides over most of the globe; therefore, the surface tidal signals are predominantly semidiurnal. The surface pressure perturbation, consisting of two maxima and two minima around the globe, follows the apparent motion of the Sun from east to west, with the maxima occurring about 2.3 hr before local solar noon and midnight. The associated pressure gradients in the atmospheric column produce strong semidiurnal wind fluctuations in the middle and upper atmosphere, and associated geomagnetic effects in the ionosphere. Internal gravity waves created in the upper stratosphere by solar heating and amplifying as they propagate upward can create tidal winds of as much as 20 m s^{-1} amplitude in the mesosphere and the lower thermosphere. These aspects are important to propagation of electromagnetic signals from orbiting satellites. Tidal winds are, however, much smaller in the troposphere, less than 1 m s^{-1} as deduced from VHF radar profilers (see Whiteman and Bian, 1996).

Surface pressure observations indicate that the dominant atmospheric tide is the semidiurnal S_2 (period of 12 hr) followed by the diurnal S_1 (period of 24 hr). Lunar tide L_2, of gravitational origin equivalent to oceanic M_2, is 20 times smaller than S_2, with a maximum amplitude of 0.09 hPa at the equator. S_2 and S_1 can be represented (in hPa) by (Lindzen, 1990; see also Haurwitz and Cowley, 1973)

$$S_2 \sim P_s \cos^3 \theta \sin\left(2\pi\frac{t}{12}+\phi_1^*\right)+0.0425\left(3\sin^2\theta-1\right)\sin\left(2\pi\frac{t_U}{12}+\phi_2^*\right)$$

$$\sim P_s \cos^3\theta\cos\left[2\pi\frac{t_U}{12}-\phi_1\right]+0.0425\left(3\sin^2\theta-1\right)\cos\left(2\pi\frac{t_U}{12}-\phi_2\right)$$

$$S_1 \sim 0.593\cos^3\theta\sin\left(2\pi\frac{t}{24}+\phi_3^*\right)$$

$$\sim 0.593\cos^3\theta\cos\left(2\pi\frac{t}{24}-\phi_3\right)$$

$$P_s = 1.160 \text{ hPa}, \ t = t_U + 24\lambda/360$$

$$\phi_1^* = 2.758 \ (158°), \ \phi_2^* = 2.059 \ (118°), \phi_3^* = 0.209 \ (12°)$$

$$\phi_1 = \frac{\pi}{2}-\phi_1^*-\frac{\pi}{180}2\lambda, \ \phi_2 = \frac{\pi}{2}-\phi_2^*, \ \phi_3 = \frac{\pi}{2}-\phi_3^*-\frac{\pi}{180}\lambda$$

$$(6.3.4)$$

where t is the local mean solar time (LMST) and t_U the universal (Greenwich) time in hours and λ is the east longitude. The terms containing starred quantities follow the atmospheric convention. The first term in S_2 is by far the most dominant term. Therefore, at any instant, two complete pressure waves surround the Earth, with maximum amplitudes occurring at the equator. The pressure wave moves from east to west following the apparent path of the Sun, with pressure maxima occurring at 0944 and 2144 LMST at all longitudes.

Atmospheric tides can be described quite well by linearized hydrostatic equations for the atmosphere written in a transformed vertical coordinate $z = -\ln(p/p_s)$, where p is the pressure. This leads to Laplace's tidal equations with forcing provided by the absorption by ozone in the middle atmosphere and water vapor in the troposphere (see Lindzen, 1990; Platzman, 1991). The solution consists of many modes with different equivalent depths, the principal semidiurnal tide S_2 having an equivalent depth of 7.85 km, with two-thirds contribution from O_3 (see Lindzen, 1990, for details). However, a simple diagnostic model for tidal wind fluctuations can be obtained by appealing to the linear Laplace tidal equations (see below), with only the time-dependent, Coriolis, and pressure gradient terms retained in the momentum balance (see Section 6.4) with pressure gradients prescribed from Eq. (6.3.4).

The corresponding wind fluctuations (above the boundary layer) can be derived by (Chapman and Lindzen, 1970; see also Whiteman and Bian, 1996)

$$u \sim V_s\left(1+1.5\sin^2\theta\right)\sin\left(2\pi\frac{t}{12}+\phi_1^*+\pi\right) \sim V_s\left(1+1.5\sin^2\theta\right)\cos\left(2\pi\frac{t}{12}-\phi_u\right)$$

$$v \sim -2.5V_s\sin\theta\sin\left(2\pi\frac{t}{12}+\phi_1^*+\frac{\pi}{2}\right) \sim -2.5V_s\sin\theta\cos\left(2\pi\frac{t}{12}-\phi_v\right)$$

$$V_s = \frac{P_s}{\rho a \Omega} \sim 0.2\ \text{ms}^{-1},\ \phi_u = -\frac{\pi}{2}-\phi_1^*-\frac{\pi}{90}\lambda,\ \phi_v = -\phi_1^*-\frac{\pi}{180}\lambda$$

$$(6.3.5)$$

The u and v components are in quadrature, 180° and 90° out of phase with pressure, respectively. Poleward of 20° latitude, the two components are of roughly equal amplitude so that the tidal wind vector rotates clockwise twice a day on a near-circular path. Around the equator, the winds are nearly zonal. The zonal velocity maximum occurs at 0344 and 1544 LMST, leading the pressure maximum by 6 hr, and the meridional velocity maximum occurs at 0044 and 1244 LMST in the northern (9 hr lead time), and 0644 and 1844 LMST in the southern (3 hr lead time) hemisphere. More important, the tidal winds converge along a particular meridian 44 min after local sunrise and sunset, and diverge 44 min after local noon and midnight. These convergences and divergences may play a role in local precipitation patterns. At higher levels both semidiurnal and diurnal tides are evident in the velocity fluctuations. See Whiteman and Bian (1996) for a description of atmospheric semidiurnal tidal winds and their measurements by radar profilers.

Figure 6.3.3 from Lindzen (1990) shows the coamplitude and cophase of the S_2 tide at Earth's surface (in hPa). Figure 6.3.4 from Whiteman and Bian (1996) shows the pressure and wind perturbations at 0000 UTC due to S_2. Note that Eq. (6.3.4) describes the pressure pattern over the globe quite well, except in the vicinity of major mountain ranges and the two poles.

According to Platzman (1991), M_2 tide in the atmosphere is maintained almost entirely by the M_2 ocean tide, with a flux of about 10 GW. This loss is, however, quite inconsequential to the oceanic M_2 tide, whose dissipation in the ocean itself is about 2.4 TW.

6.3.4 RADIATIONAL TIDES

So far we have considered only the gravitational ocean tides. Observations show that the solar semidiurnal, diurnal, and annual tides have amplitudes and phases different from those expected from their neighbor constituents. This has to do with the diurnal and annual solar heating of the atmosphere and the ocean. The component most affected is S_2, whose anomalous portion is about 17% of the gravitational tide, and the driving force is the atmospheric S_2 tide, which causes pressure fluctuations at the ocean's surface. The upper layers of the ocean are also heated and cooled diurnally by the Sun and the resulting thermal expansion and contraction cause fluctuations in sea level on the order of a centimeter—these are called radiational tides and have the same frequency as S_2. The component S_1 also has radiational contribution and in fact its gravitational part is quite negligible (Cartwright, 1993). The annual tide S_a is also affected greatly by seasonal heating and cooling processes in both the ocean and the atmosphere. It is possible to account for these radiational tides by defining a radiational pseudo-potential (Cartwright, 1977; Munk and Cartwright, 1966; Pugh, 1987). The potential is zero at night and varies as the cosine of the zenith angle during the day. Like the gravitational tidal potential (see Section 6.2), it can be expanded in Legendre functions in terms of a_S. Neglecting terms of the order of a_S^2,

$$\Phi_R = S\left(R_S / \overline{R}_S\right)\left[(1/4) + (1/2)P_1(\cos\psi) + (5/16)P_2(\cos\psi) + ...\right] \qquad (6.3.6)$$

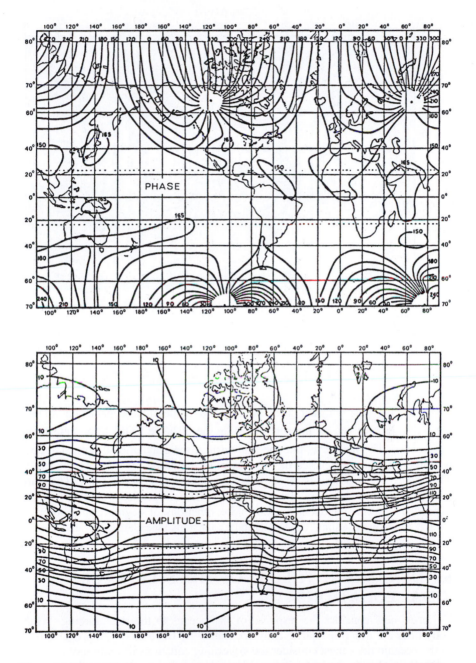

Figure 6.3.3 Coamplitude and cophase of atmospheric semidiurnal S_2 tides. From Lindzen (1990).

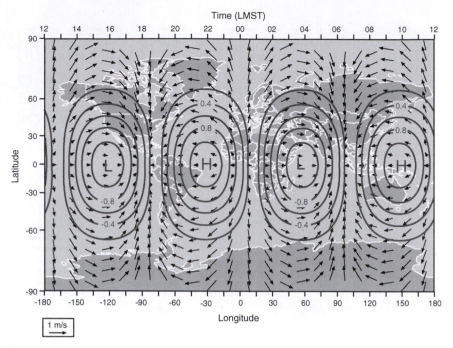

Figure 6.3.4 Pressure and wind perturbations due to atmosperic S_2 tide. From Whiteman and Bian (1996).

where S is the solar constant (1380 W m^{-2}) and the potential depends on the heat per unit area received by the surface. R_S is the distance to the Sun, and $\bar{R}_S$ its mean value. The first term, like the first term of the gravitational potential, does not produce tides; it denotes the net day/night radiation received by the Earth. The second term generates large daily and annual radiational tides. The third term is similar to the gravitational term. These Legendre functions can be expanded in terms of associated Legendre functions of degree n and order m, similar to the gravitational tidal potential expansions (see Cartwright and Tayler, 1971).

6.4 DYNAMICAL THEORY OF TIDES: LAPLACE TIDAL EQUATIONS

Let us now consider the response of the oceans to the tidal forcing. While body tides are described sufficiently accurately by Love theory, a proper model of the oceanic tides must consider the governing equations of motion of the fluid under the action of the body forces due to the applied tidal potential and allow for the shape of the basins, their varying depths, the dissipation due to bottom friction, and, in some cases, the nonconstant density of the water masses. Here lies the difficulty of the subject, since complex nonlinear fluid dynamics are involved. The dynamical theory of tides was initiated by Laplace in 1776, and tidal computations based on dynamical theory were made recently by Schwiderski (1980), Le Provost *et al.* (1994), Kantha (1995b) and Tierney *et al.* (1999), among others.

Laplace considered a rotating, homogeneous ocean on a spherical surface, but linearized the nonlinear governing equations for tractability. This is, of course, equivalent to assuming a zero Rossby number for the flow. He also considered the water depth to be shallow compared to the scale of the horizontal tidal motions (small aspect ratio oceans) and derived a set of shallow water equations on a rotating Earth including the body force terms due to the applied tidal potential. These were modified to include the frictional forces by Pekeris and Accad (1969) and the effects of ocean self-attraction and load deformation of the bottom by Hendershott (1981). With these modifications, Laplace's tidal equations (LTE) become (see Hendershott, 1977, 1981)

$$\frac{\partial \eta}{\partial t} + \frac{1}{R \cos \theta} \left[\frac{\partial}{\partial \phi} (u_\phi H) + \frac{\partial}{\partial \theta} (u_\theta H \cos \theta) \right] = 0$$

$$\frac{\partial u_\phi}{\partial t} - 2\Omega u_\theta \sin \theta = -\frac{g}{R \cos \theta} \frac{\partial}{\partial \phi} \left(\eta_g - \frac{\Gamma}{g} \right) + F_\phi \qquad (6.4.1)$$

$$\frac{\partial u_\theta}{\partial t} + 2\Omega u_\phi \sin \theta = -\frac{g}{R} \frac{\partial}{\partial \theta} \left(\eta_g - \frac{\Gamma}{g} \right) + F_\theta$$

where F_ϕ and F_θ are frictional terms involving the components of bottom stress divided by the fluid density, τ_b / ρ_0; η is the oceanic tide referred to the Earth's surface and η_g is the geocentric ocean tide, which includes the ocean bottom deflection η_e due to solid Earth tide η_b and load tide η_l. Γ is the total tidal potential that includes Φ, the direct astronomical contribution from the Sun and the Moon, and the contributions from the solid Earth body tide, self-attraction of the ocean tide and the load tide

$$\Gamma = \Phi + k_2 \Phi + \phi_l \text{ with } \phi_l = \sum_n (1 + k'_n) g \gamma_n \eta_n$$

$$\eta_e = h_2 \frac{\Phi}{g} + \eta_l \text{ with } \eta_l = \sum_n h'_n \gamma_n \eta_n \qquad (6.4.2)$$

where k's and h's are Love numbers. Note that only the spherical harmonic of degree 2 is retained in the tidal potential terms, whereas the load tide terms are contributions summed up over all degrees n, with η_n denoting the contribution of the ocean tide of degree n. Therefore,

$$
\eta_g - \Gamma / g = \left[\eta - \sum_n (1 + k'_n - h'_n) \gamma_n \eta_n \right] - (1 + k_2 - h_2) \frac{\Phi}{g}
$$
$$
= \beta \eta - \alpha_2 \zeta
$$
(6.4.3)

The coefficient α_2 accounts for the augmentation of the tide generating potential by yielding of the solid Earth (k_2 term) and its decrease due to the geocentric solid Earth tide (h_2 term) and its value is 0.693 (see Section 6.3.1). The coefficient β accounts for the loading effect due to ocean tides. It requires knowledge of the load tides, produced by ocean tides, which in turn requires that the distribution of ocean tides over the global oceans be known. Hence, prescription of β requires a priori knowledge of the ocean tides, which cannot be computed accurately without knowing β. One approach is an iterative technique, where the ocean tides are calculated with an initial guess for β, and the calculations are updated with a refined value estimated from load tides computed from the ocean tides from the previous iteration, using Green's function approach:

$$
\beta = 1 - \frac{\eta_1}{\eta} = 1 - \frac{1}{\eta} \iint_{\text{ocean}} \eta (\phi', \theta') G (\phi', \theta' | \phi, \theta) \cos\theta' d\theta' d\phi'
$$
(6.4.4)

The Green's function formulation for calculation of the load tides is due to Farrel (1972). In routine tidal modeling, the value of β is, however, seldom calculated using Farrell's Green functions. Instead, an approximation is used by assuming that the ocean load tide due to deformation of the solid earth and weight of the tidal column is in exact antiphase with the ocean tide and a fixed fraction of it. The value is chosen as 0.940 and 0.953 for diurnal and semidiurnal constituents (Foreman *et al.*, 1993). This approach simplifies tidal modeling considerably, and the errors introduced by this approximation are not overly large for most applications. However, β is in general not a constant and Ray (1998) suggests that serious errors are incurred by assuming that it is, especially for non-data-assimilative tidal models. Fortunately, load tides in the global ocean are now known a priori to a reasonable degree of accuracy (Ray and Sanchez, 1989) and these can be prescribed and $\beta\eta$ replaced by $\eta - \eta_1$. A procedure where one starts from this (or a constant β approximation), calculates the load tides from the resulting ocean tides, and iterates to convergence (Tierney *et al.*, 1999) is preferable for applications such as precision altimetry and inferring the effect of tides on gravimetric measurements.

Although $\alpha_2 = 0.693$ is a good approximation for all tidal constituents and was the value used by Schwiderski (1980) in his tidal model, the Love numbers are actually frequency dependent (Table 6.3.1). Constituents P_1 and K_1 are especially affected by the narrow "free core nutation" resonance at approximately 1.0056 cy d^{-1} (Wahr, 1981). This effect arises from the motion of the Earth's liquid core relative to the outer crust. Accurate tidal modeling must account for this frequency dependence of α.

The tides obtained by solving Eq. (6.4.1) depend very much on the shape and depth distribution of the ocean basin, as well as the frictional forces, and are considerably different than the equilibrium tides. For example, because of the inertia of water masses, the spring and neap tides at any point lag, by a day or two, the corresponding phases of the Moon, and this delay is called the age of tides (Murty and El-Sabh, 1986). Similarly, the high tides follow after the Moon reaches its perigee. At places, where the frequency of oscillation of the water body is close to a tidal frequency (the normal modes for the ocean have periods ranging from 8 to 80 hr; Platzman, 1981), resonance conditions cause large tides, whereas at places called amphidromic points, tides vanish. The response of the oceans to tidal forcing consists of forced long wave motions, composed of Kelvin and Poincare gravity waves, and planetary and topographic Rossby waves. Of these, Kelvin and Poincare waves are relevant to short period tides, whereas long period tides involve the low frequency Rossby waves. Kelvin waves are coastally trapped within a distance on the order of the Rossby radius of deformation $a = C/f$ and travel at the shallow water gravity wave speed of $C = (gH)^{1/2}$, with the coast to the right (left) in the northern (southern) hemisphere. Poincare waves are gravity waves that are affected by rotation and can travel across basins. Rossby waves are generated due to latitudinal variation of planetary vorticity, and long waves travel at a speed of βa^2 (Chapter 1). Because of the complex fluid dynamical processes involved, the solution is best obtained using a computer, even though the LTE are linear.

If the natural period of oscillation of a bay is close to the tidal period, then the tides are greatly amplified relative to those at the mouth. This happens in the Bay of Fundy, for example. The conditions for resonance in a long narrow channel of length L and depth H is that the channel be a quarter-wave oscillator: $L \sim (1/4) (gH)^{1/2}T$, where T is the tidal period. For $H \sim 100$ m, this critical length is ~ 350 km for the M_2 tide. The Bay of Fundy, with an average depth of 100 m and length of 300 km, is fairly close to resonance with M_2. More accurate calculations, taking into account the Earth's rotation and the variable topography, show that the natural period in the Bay is 13.3 hr, close to the semidiurnal band of tidal frequencies so that a large tidal response can be expected, as indeed observed.

The energy flux across a vertical section is in general $u(H + \eta)(\rho g\eta + \rho u^2/2)$ per unit width of the tide. For small amplitude waves ($\eta/H \ll 1$) in a channel, however, $u = C\eta/H$, so that the energy flux becomes $\rho gHu\eta$. This has interesting consequences. For a tide whose elevation and current differ in phase by $\varphi_\eta - \varphi_u$, the energy flux averaged over a tidal cycle is $0.5\rho g H u_0 \eta_0 \cos(\varphi_\eta - \varphi_u)$, where subscript 0 denotes amplitudes. For a progressive wave, the phase difference is zero and hence the energy flux is $0.5\rho g H u_0 \eta_0 \sim (0.5\rho g \eta_0^2)C$, where the term in parentheses is the energy density averaged over the tidal cycle. For a standing wave, in the absence of bottom friction, the phase difference is $\pi/2$, and hence the net energy flux over the tidal cycle is zero. For a wave propagating across a discontinuous bottom, as, for example, from the deep ocean onto a shelf, energy flux considerations demand that the ratio of amplitudes of the transmitted wave to the incident wave be $2[1+(C_s/C_d)]^{-1}$ so that the amplitude for the reflected wave is $[1-(C_s/C_d)][1+(C_s/C_d)]^{-1}$. For typical shelf depth of 200 m and deep water depth of 4000 m, these ratios are 1.635 and 0.635, so that the tide on the shelf is considerably amplified.

Tides at a point can be either predominantly semidiurnal, predominantly diurnal, or mixed. Their nature is determined by the form ratio $F = (K_1 + O_1)/(M_2 + S_2)$, the ratio of the sum of the amplitudes of the two most important diurnal tides to the sum of the amplitudes of the two most important semidiurnal tides (it is possible to define the ratio including all eight of the diurnal and semidiurnal tides, but the difference is small). If $F < 0.25$, the tides are predominantly semidiurnal (see Figure 6.4.1), if $0.25 < F < 1.5$, the tides are mixed, but mainly semidiurnal, if $1.5 < F < 3.0$, the tides are mixed, but mainly diurnal, and if $F > 3.0$, the tides are predominantly diurnal (Dietrich, 1980; Pond and Pickard, 1989). Figure 6.4.2 shows this ratio for the global oceans. For the primarily semidiurnal tides, the mean range at spring tide is $2(M_2 + S_2)$, and for primarily diurnal tides, the mean spring range is $2(K_1 + O_1)$.

Equation (6.4.1) ignores horizontal motions of the solid Earth (equivalent to assuming Shida numbers l_n are zero). If these are retained instead (John Wahr, personal communication), because of the bottom boundary condition of zero mass flow through the boundary, the vertical integration of the continuity equation introduces additional terms on the RHS of the first equation in Eq. (6.4.1):

$$\frac{\partial \eta}{\partial t} + \frac{1}{R\cos\theta}\left[\frac{\partial}{\partial \phi}(U_\phi H) + \frac{\partial}{\partial \theta}(U_\theta H \cos\theta)\right] = \frac{1}{R\cos\theta}U_\phi^b\frac{\partial H}{\partial \phi} + \frac{1}{R}U_\theta^b\frac{\partial H}{\partial \theta} \qquad (6.4.5)$$

where U_ϕ^b, U_θ^b are the components of the horizontal velocity of the moving bottom, which can be written as

$$U_\phi^b = \frac{\partial}{\partial t}\left[\frac{l_n}{\cos\theta}\frac{\partial}{\partial \phi}\left(\frac{\Phi_n}{g}\right)\right]; \quad U_\theta^b = \frac{\partial}{\partial t}\left[l_n\frac{\partial}{\partial \theta}\left(\frac{\Phi_n}{g}\right)\right] \qquad (6.4.6)$$

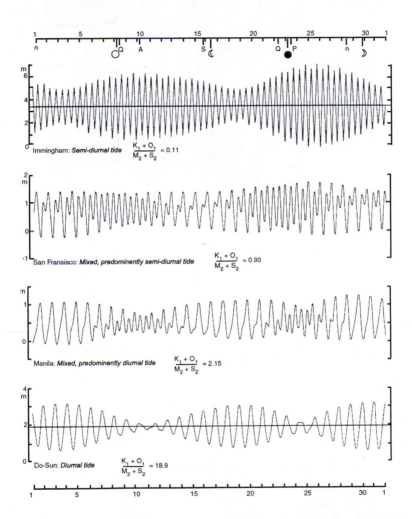

Figure 6.4.1 Tidal records of sea surface height showing predominantly semidiurnal (top), predominantly diurnal (bottom), and mixed (middle) tides. (From Dietrich, G. K., *et al.*, Tidal phenomenon, in *General Oceanography*. Copyright © 1980 Gebruder Borntraeger Verlag. Reprinted by permission of John Wiley & Sons, Inc.)

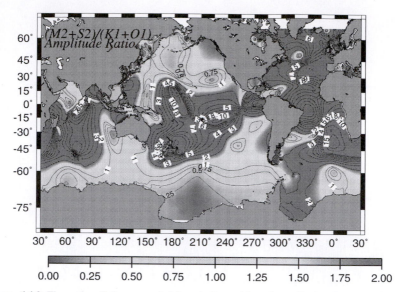

Figure 6.4.2 The ratio of the sum of M_2 and S_2 semidiurnal amplitudes to that of K_1 and O_1 diurnal tides in the global oceans. The tide is predominantly semidiurnal except in regions like the Sea of Okhotsk. From Kantha (1995).

If we retain only second-degree terms and redefine the sea surface elevation as

$$\hat{\eta} = \eta - \eta_H; \quad \eta_H = l_2\left(\frac{1}{R\cos^2\theta}\frac{\partial\zeta}{\partial\phi}\frac{\partial H}{\partial\phi} + \frac{1}{R}\frac{\partial\zeta}{\partial\theta}\frac{\partial H}{\partial\theta}\right) \quad (6.4.7)$$

we get

$$\frac{\partial\hat{\eta}}{\partial t} + \frac{1}{R\cos\theta}\left[\frac{\partial}{\partial\phi}(U_\phi H) + \frac{\partial}{\partial\theta}(U_\theta H\cos\theta)\right] = 0$$

$$\frac{\partial U_\phi}{\partial t} - 2\Omega U_\theta\sin\theta = -\frac{g}{R\cos\theta}\frac{\partial}{\partial\phi}(\beta\hat{\eta} - \alpha\zeta) + F_\phi - \frac{g}{R\cos\theta}\frac{\partial}{\partial\phi}(\beta\eta_H)$$

$$\frac{\partial U_\theta}{\partial t} + 2\Omega U_\phi\sin\theta = -\frac{g}{R}\frac{\partial}{\partial\theta}(\beta\hat{\eta} - \alpha\zeta) + F_\theta - \frac{g}{R}\frac{\partial}{\partial\theta}(\beta\eta_H)$$

$$(6.4.8)$$

Equation (6.4.8) is very much similar to Eq. (6.4.1), except for the appearance of extra forcing terms on the RHS of the momentum equations that depend on the bottom slopes (only degree 2 terms are important). For a flat-bottomed ocean, these terms disappear. The maximum effect occurs over strong topographic changes such as continental slopes and seamounts. Nevertheless, the ratio of these terms to the tidal potential term is such that the correction is likely to be a few percent at best.

6.5 EQUILIBRIUM THEORY OF TIDES

Newton (and Bernoulli) put forth the equilibrium theory of tides. They considered the Earth to be covered with water of uniform depth and density. This ocean is also supposed to respond instantaneously to the tidal forcing; hence, the two bulges on the Earth-Moon axis move with the Moon and are always directly under the Moon, with no lag or lead, while the Earth rotates underneath. The response is similar to that of the elastic Earth. Note that the equilibrium tides can be derived simply by putting all terms in Eq. (6.4.1), except those containing the tidal potential and the tidal elevation, to zero. This is a static solution, equivalent to assuming that the induced velocities are infinitesimally small. Thus there is no kinetic energy in equilibrium tides, only potential energy. Retaining only the $n = 2$ term in the tidal potential, which is the most dominant,

$$\beta\eta = (1 + k_2 - h_2)\frac{\Phi}{g} = \alpha_2\zeta \quad \rightarrow \quad \eta \sim 0.693\frac{\Phi}{\beta g} \tag{6.5.1}$$

These are called self-consistent equilibrium tides (Agnew and Farell, 1978, Desai 1996) since load-induced deformations and self-gravitation of ocean tides are also accounted for. If the approximation $\alpha \sim 1$ and $\beta \sim 1$ is used, classical equilibrium tides for a rigid Earth result:

$$\zeta = \frac{\Phi}{g} = \frac{K_2}{2}\left(3\cos^2\psi - 1\right); \quad K_2\big|_{M,S} = R_E\left(\frac{M}{M_E}\right)_{M,S}\left(\frac{R_E}{R}\right)^3_{M,S} \tag{6.5.2}$$

The decomposition of this of course leads to

$$\zeta = C_0(t)\left(3\sin^2\theta - 1\right) + C_1(t)\sin 2\theta + C_2(t)\cos^2\theta$$

$$C_0(t) = \frac{1}{4}K_2\left(3\sin^2\theta_M - 1\right)$$

$$C_1(t) = \frac{3}{4}K_2\sin 2\theta_M\cos\left(\phi - \phi_M\right) \tag{6.5.3}$$

$$C_2(t) = \frac{3}{4}K_2\cos^2\theta_M\cos 2\left(\phi - \phi_M\right)$$

This classical equilibrium theory explains the semidiurnal nature of oceanic tides and the daily inequality, but the predicted amplitude (from Eq. 6.4.2) is a maximum of 0.2688 m for the semidiurnal, 0.189 m for the diurnal, and 0.057 m for the long-period lunar tides, and 0.46 times these values for the solar ones. Because of the solid Earth tides and their effects, these amplitudes are multiplied by the elasticity factor α (0.693). These values are far smaller than the observed tides, which can reach several meters in range at some places, and the phases of the tides are in considerable disagreement with the phases predicted by equilibrium theory. There can be usually a lead or lag of several hours in high water with respect to the passage overhead of the celestial bodies. This is simply due to the fact that the inertia of the water masses is ignored in the theory and the influence of continental masses dividing the global oceans into distinct basins is also ignored. Friction is also ignored. The water is free to adjust and is assumed to adjust instantaneously to the applied force so that a static equilibrium prevails. This condition is seldom satisfied for short period tides, and even for long period tides, such as the fortnightly and monthly tides, the equilibrium assumption is only approximately valid (solid Earth tides are nearly in equilibrium as we saw before). Even longer period semiannual and annual tides, the 14-month pole tide, and the 18.6-year nodal tide are probably well approximated by equilibrium theory.

6.6 TIDAL ANALYSIS: ORTHOTIDES

Tides are well measured and well known along much of the coastlines of the world. It is in the deep basins that they were not accurately known until the advent of satellite altimetry. From a record of sea surface elevation of sufficient length at a point, it is possible to deduce the tidal components at that point. Generally, the longer the time series, the better the accuracy of the extracted components. The two methods that are used for tidal analysis are the harmonic analysis and spectral methods. The harmonic method introduced by Lord Kelvin in 1870 assumes that the tides can be regarded as superposition of different harmonics whose frequencies are known from astronomy but whose amplitudes and phases are to be determined from the record itself by a least-squares fitting process. The information can then be used for tidal predictions. However, it is important that the tidal signal be adequately sampled; the sampling period must be a small fraction of the period of the signal. Half-hourly or hourly sampling is adequate for most short-period tides, although higher frequency nonlinear tides such as M_4 may benefit from more frequent sampling. Coastal tide gauges and bottom pressure gauges have their sampling rates set to obtain hourly or half-hourly values. On the other hand, altimetric measurement of tides involves sampling at the altimetric repeat cycle period, which is on the order of days (~10 days for TOPEX). In that case, the sampling period is several times the period of the high frequency tide itself, and a least square fitting of the tidal signal implies extracting the signal at the corresponding aliasing frequency. For example, in extracting M_2 signal from TOPEX records, one is actually fitting a sine wave of period equal to $T_{TP}[\text{Min} (X, 1-X)]$, where $X = [(T_{TP}/T_{M2}) - \text{INT}(T_{TP}/T_{M2})]^{-1}$, approximately 61 days, to the record.

The length of record needed to extract different components depends primarily on the closeness in frequency of the components that are to be extracted. Rayleigh separation criterion dictates that it is necessary to have a minimum record length of $2\pi (f_1-f_2)^{-1} = T_1T_2/(T_1-T_2)$ to separate two signals with periods (radian frequencies) $T_1 (f_1)$ and $T_2 (f_2)$. Naturally, the more closely spaced the two frequencies are—that is, the closer the two periods T_1 and T_2 are to each other—the longer the record length needed. Normally, 369 days of hourly data are needed to extract 20 to 30 constituents with adequate separation of closely spaced constituents, although it is possible to use as short a record as 29 days to extract fewer constituents. However, noise contamination also plays an important role. In the presence of noise, the longer the record, the better the extraction of any particular signal. This is especially true of long period tides such as the lunar monthly Mm, since their amplitudes are small (1 to 2 cm) and their signals are buried in large noise contributed by atmospheric fluctuations at similar frequencies. Consequently, record lengths of many years are needed to extract meaningful long period tidal signals (Miller *et al.,* 1993; Wunsch, 1967).

While this time-honored technique works very well, it does not take advantage of the fact that the tidal admittance function, which is the ratio of the tidal response to tidal forcing, is generally smooth. Tidal constituents that are closely spaced to one another and hence have essentially the same admittance function are treated independently of one another. While there is nothing wrong with this approach, it is possible to take advantage of the inherent smoothness of the tidal response function to make accurate tidal predictions with a smaller number of constituents (Groves and Reynolds, 1975; Munk and Cartwright, 1966). Generally speaking, at any point, both the ratio of the observed tidal amplitude to the equilibrium amplitude and the tidal phase with respect to the equilibrium tide vary smoothly with tidal frequency. Therefore, determining the functional form of the complex admittance function for the ocean, which is the ratio of the Fourier transform of the output to that of the input and hence a function of frequency, enables tides to be determined. Such spectral methods treat the tidal record like any other time series and using Fourier transforms, transform the signal into frequency domain. The frequency (in Hz) analyzed thus ranges from the Nyquist frequency $(2 \Delta t)^{-1}$ to T^{-1}, where T is the length of the record and Δt is the sampling interval.

The response methods also facilitate separation of gravitational and radiational tides. Shorter records can be used to determine the tidal constants accurately. Munk and Cartwright (1966) introduced a better response analysis method, improved later by Groves and Reynolds (1975). This method uses a convolution formalism and treats the ocean tide $\eta(t)$ as a weighted sum of the past, present, and future values of the forcing, the equilibrium tidal potential. This is equivalent to using a Fourier series of a certain period to predict the tidal admittance function; this period should be greater than twice the bandwidth of individual bands of each species (Desai, 1996). For diurnal and semidiurnal species, the bandwidth is 0.3 cy d^{-1}; therefore, the period is 1.67 days. Munk and Cartwright (1966) chose 2 days for their analysis and this has since then been adhered to. Starting from Eq. (6.2.9) for the tidal potential, the equation is rewritten as (Cartwright and Ray, 1990)

$$\Phi(\phi,\theta,t) = g\sum_{n=2}^{\infty}\sum_{m=0}^{n}\left[b_{nm}(t)F_{nm}(\phi,\theta) + c_{nm}(t)G_{nm}(\phi,\theta)\right] \qquad (6.6.1)$$

where a_{nm} and b_{nm} are real and imaginary parts of a complex coefficient $C_{nm}(t)$, computed from the ephemerides of the Sun and the Moon. The tide at any point can be written as a weighted sum:

$$\eta(t) = \int_{-\infty}^{\infty} w(t')C(t-t')dt' \equiv \sum_{k=-N}^{N} w(k\tau)C(t+k\tau) \qquad (6.6.2)$$

The complex admittance is then

$$Z(\omega) = \int_{-\infty}^{\infty} w(t')\exp(-i\omega t')dt' \equiv \sum_{k=-N}^{N} w(k\tau)\exp(-i\omega k\tau) \qquad (6.6.3)$$

from which the amplitude and phase of the tide can be calculated. The higher the value of N, the better-resolved the variations of the admittance function in frequency space, with $w(0)$ only giving a flat response independent of the frequency. Since the weighting function $w(k\tau)$, the system response at $t = 0$ to a unit impulse at $t = k\tau$, decreases rapidly with time lag, a value of τ of 2 days and N of 2 is found to be generally adequate (Munk and Cartwright ,1966). For the weak third harmonic (terdiurnal tide), $w(0)$ is sufficient. Formally,

$$\eta(\phi,\theta,t) = \sum_{n=2}^{\infty}\sum_{m=0}^{n}\eta_{nm}(\phi,\theta,t)$$

$$= \sum_{n=2}^{\infty}\sum_{m=0}^{n}\sum_{k=-K}^{K}\left[b_{nm}(t-k\Delta t)U_{nm}(k,\phi,\theta) + c_{nm}(t-k\Delta t)V_{nm}(k,\phi,\theta)\right]$$

$$(6.6.4)$$

where K is a small integer (usually 1 here) and $\Delta t = 2$ days. If we define

$$b_{nm}(t) = \cos(\sigma_{nm}t); \quad c_{nm}(t) = \sin(\sigma_{nm}t)$$

$$\eta_{nm}(t) = R_{nm}\cos(\sigma_{nm}t - \varphi_{nm}) \tag{6.6.5}$$

$$X_{nm}(\sigma_{nm}) = R_{nm}\cos(\varphi_{nm}); \quad Y_{nm}(\sigma_{nm}) = -R_{nm}\sin(\varphi_{nm})$$

then the tidal admittance can be written as

$$Z_{nm}(\sigma_{nm}) = X_{nm}(\sigma_{nm}) + iY_{nm}(\sigma_{nm})$$

$$= \sum_{k=-K}^{K}\left[U_{nm}(k,\phi,\theta) + iV_{nm}(k,\phi,\theta)\right]\exp(-i\sigma_{nm}k\Delta t) \tag{6.6.6}$$

Groves and Reynolds (1975) pointed out the nonorthogonality of terms in Eq. Eq. (6.6.1) and suggested instead expansions in terms of orthotides of the form

$$\eta(\phi,\theta,t) = \sum_{m=1}^{2}\sum_{i=1}^{3}\left[U_i^m(\phi,\theta)P_i^m(t) + V_i^m(\phi,\theta)Q_i^m(t)\right] \tag{6.6.7}$$

Note that only $m = 1$ and 2 are retained (diurnal and semidiurnal species, not long period species $m = 0$) and only degree $n = 2$ is considered ($n = 3$ and higher are small and can be omitted). The coefficients P_i^m and Q_i^m can be written as

$$P_1^m(t) = p_{00}^m b_{2m}(t)$$

$$P_2^m(t) = p_{10}^m b_{2m}(t) + p_{11}^m b_{2m}^+(t)$$

$$P_3^m(t) = p_{20}^m b_{2m}(t) + p_{21}^m b_{2m}^+(t) - q_{21}^m c_{2m}^-(t)$$

$$Q_1^m(t) = p_{00}^m c_{2m}(t) \tag{6.6.8}$$

$$Q_2^m(t) = p_{10}^m c_{2m}(t) + p_{11}^m c_{2m}^+(t)$$

$$Q_3^m(t) = p_{20}^m c_{2m}(t) + p_{21}^m c_{2m}^+(t) - q_{21}^m b_{2m}^-(t)$$

The coefficients p_{ii} and q_{ii} have been computed by Groves and Reynolds (1975) from harmonic expansion of $b_{nm}(t)$ to assure orthogonality:

$$b_{2m}(t) = \cos(\sigma_{2m}t)$$

$$b_{2m}^+(t) = b_{2m}(t+\Delta t) + b_{2m}(t-\Delta t) = \cos(\sigma_{2m}t)\, C_1(\sigma_{2m})$$

$$b_{2m}^-(t) = b_{2m}(t+\Delta t) - b_{2m}(t-\Delta t) = -\sin(\sigma_{2m}t)\, S_1(\sigma_{2m})$$

$$c_{2m}(t) = \sin(\sigma_{2m}t)$$

$$c_{2m}^+(t) = c_{2m}(t+\Delta t) + c_{2m}(t-\Delta t) = \sin(\sigma_{2m}t)\, C_1(\sigma_{2m})$$ $\qquad$ (6.6.9)

$$c_{2m}^-(t) = c_{2m}(t+\Delta t) - c_{2m}(t-\Delta t) = \cos(\sigma_{2m}t)\, S_1(\sigma_{2m})$$

where

$$C_1(\sigma_{2m}) = 2\cos(\sigma_{2m}\Delta t);\ S_1(\sigma_{2m}) = 2\sin(\sigma_{2m}\Delta t)$$

The tidal admittances can be written as (omitting for clarity the subscript 2 corresponding to the degree)

$$X_i^m(\sigma_i^m) = p_{00}^m U_1^m + \left[p_{10}^m + p_{11}^m C_1(\sigma_i^m)\right] U_2^m + \left[p_{20}^m + p_{21}^m C_1(\sigma_i^m) + q_{21}^m S_1(\sigma_i^m)\right] U_3^m$$

$$Y_i^m(\sigma_i^m) = p_{00}^m V_1^m + \left[p_{10}^m + p_{11}^m C_1(\sigma_i^m)\right] V_2^m + \left[p_{20}^m + p_{21}^m C_1(\sigma_i^m) + q_{21}^m S_1(\sigma_i^m)\right] V_3^m$$

where

$$\eta(t) =) \sum_{m=1}^{2} \sum_{i=1}^{3} A_i^m \cos(\sigma_i^m t - \varphi_i^m)$$

(6.6.10)

$$A_i^m \cos\varphi_i^m = (-1)^m \frac{\Phi_i^m}{g} X_i^m(\sigma_i^m);\ A_i^m \sin\varphi_i^m = (-1)^m \frac{\Phi_i^m}{g} Y_i^m(\sigma_i^m)$$

These 12-term forms of the above expressions are valid for only three tidal constituents for each species (m = 1, 2), a total of six. While they can be easily generalized to a larger number, they suffice for most applications.

From data on diurnal and semidiurnal tides near Argentina, Alcock and Cartwright (1978) discovered that the first six orthoweights are fairly independent of the next four (which were significantly smaller than the first six) and accounted for most of the "wiggliness" of the diurnal and semidiurnal admittance functions. It is a common practice, therefore, to use the above form in tidal analysis. However, orthoweights up to 12 as calculated by Desai (1996) are presented in Table 6.6.1. Note that these are slightly different than the ones

TABLE 6.6.1
Orthotide Constants up to 12

i	Order	Coefficient	Lag	Semidiurnal (cm^{-1})	Diurnal (cm^{-1}) Without FCN	Diurnal (cm^{-1}) With FCN
1	0	p_{00}	0	0.019805	0.029575	0.028486
2	2	p_{10}	0	0.089759	0.139752	0.138607
		p_{11}	1	-0.063291	-0.079882	-0.078478
3	4	p_{20}	0	0.344734	0.595555	0.594826
		p_{21}	1	-0.163116	-0.300197	-0.300076
		q_{21}	1	0.091581	0.150533	0.150138
4	6	p_{30}	0	0.207953	0.570171	0.568212
		p_{31}	1	-0.548335	-0.577160	-0.575941
		p_{32}	2	0.446491	0.299021	0.298688
		q_{31}	1	-0.420687	-0.174150	-0.174357
5	8	p_{40}	0	3.923634	2.602574	2.594211
		p_{41}	1	-1.823687	-1.795240	-1.788498
		p_{42}	2	-0.142089	0.502043	0.499594
		q_{41}	1	2.105056	0.743674	0.742103
		q_{42}	2	-0.812342	-0.522503	-0.520660
6	10	p_{50}	0	5.795867	1.446212	1.447651
		p_{51}	1	-2.207406	-1.454041	-1.454278
		p_{52}	2	-1.675517	1.204397	1.202675
		p_{53}	3	0.987998	-0.482360	-0.481132
		q_{51}	1	4.520375	-0.522994	-0.520581
		q_{52}	2	-2.580833	0.400683	0.398598
7	12	p_{60}	0	9.716074	7.339055	7.316722
		p_{61}	1	-6.656145	-5.585597	-5.566378
		p_{62}	2	1.493624	2.135412	2.123411
		p_{63}	3	0.306761	-0.227880	-0.223932
		q_{61}	1	4.534568	2.671913	2.667511
		q_{62}	2	-4.017822	-2.684204	-2.678739
		q_{63}	3	1.068871	1.013172	1.008757

From Desai (1996).

originally calculated by Groves and Reynolds (1975), and the diurnal values have also been computed taking into account the free core resonance. Groves and Reynolds present values up to order 24, but as we said before, there is seldom a need to go beyond order 4. Note that $C_k(\sigma_i^m) = \cos(k\sigma_i^m \Delta t)$ and $S_k(\sigma_i^m) = \sin(\sigma_i^m k\Delta t)$. Table 6.6.2 shows the orthoweights for the long period tides.

The utility of these relationships is that once the primary constituents for each high frequency tidal species (say M_2, S_2, and K_2 for $m = 2$, and K_1, O_1, and P_1 for $m = 1$) are known from a model or observations, the 12 orthotides U_i^m, V_i^m

424 6 Tides and Tidal Modeling

TABLE 6.6.2
Long-Period Orthotide Constants

Tidal band	Doodson number of included components	U_{10}^{20} (cm^{-1})
M_m	062.646 to 068.454	0.392441
M_f	071.755 to 077.575	0.195191
M_t	080.656 to 095.375	0.982441

From Desai (1996).

(i = 1, 3; m = 1, 2) can be calculated from the above expressions and used to compute the possible values for all other constituents of the two species. Thus, the tidal sea level can be calculated by a summation over all significant constituents. For example, the coamplitude/cophase distributions of the six primary tides obtained from the global model of Kantha (1995b) have been used to extend the results to a total of 30 semidiurnal and 30 diurnal constituents. After including the 12 long period tides using equilibrium tide approximation, the orthotide approach has been used to calculate the tidal SSH at any point on the globe at any given time for subtraction from altimetric signals (see http://www.cast.msstate.edu/Altimetry). An added advantage of the orthotide approach is that there is no need to consider the nodal factors; they are automatically taken into account.

The orthotide approach depends very much on the tidal admittance functions being slowly varying and smooth functions in frequency space; this is what makes it possible to calculate the tidal response to forcing at the neighboring frequencies, once the admittance of a few salient constituents within the frequency band are known (see Figure 6.2.1). Given the fact that the oceans have a set of free modes of oscillation and resonances are to be expected, it is surprising that the admittance is indeed smooth. The reason, however, is not clear. It is possible that frictional damping of the free modes is sufficient to ensure smoothing out of the peaks in the admittance function. This holds at least in deep oceans, where the tides are remarkably linear. Care must be taken in applying this approach to coastal and marginal seas, where resonances might make the assumptions approximate at best. Nevertheless, the orthotide approach works remarkably well and is often preferred for estimating the linear tides in the global oceans (nonlinear shallow water tides are more difficult to handle).

6.7 TIDAL CURRENTS

Tidal elevations are usually specified in terms of tidal coamplitude A and cophase G relative to either Greenwich (Greenwich phase lag) or local time:

$$\eta = \text{Re}\left\{Ae^{i\sigma t}e^{-iG}\right\} = A\cos(\sigma t - G)$$
$$= (A\cos G)\cos(\sigma t) + (A\sin G)\sin(\sigma t)$$

(6.7.1)

where G is the Greenwich phase lag and A is the amplitude. Note that the oceanographic convention is to use a phase lag, hence the negative sign associated with G. However, geophysicists often use +G in describing sinusoidal geophysical signals referred to Greenwich, and this often causes confusion in deducing the geophysical implications of oceanic signals such as tides on geophysical parameters (akin to the practice of geophysicists and geodetists using θ for colatitude instead of latitude). The values A cos G and A sin G are known as in-phase and quadrature components of the tide. An equivalent description is in terms of the real part X (in-phase component) and imaginary part Y (out-of-phase component) of the complex tidal admittance, which is defined as

$$X = (\Phi/g)^{-1} A \cos G$$
$$Y = - (\Phi/g)^{-1} A \sin G$$
(6.7.2)

It is, however, common for many tidal tables to refer the tides to a local time. The local phase lag g is related to G by $g = G - 15^{\circ} t_Z$, where t_Z is the local time zone, expressed in hours west of Greenwich, and all the values are in degrees.

Tidal currents can also be specified by the amplitude and Greenwich phase lag of the two components of the tidal current vector, each of which can be determined independently from harmonic (or response) analysis. However, they are best represented in the form of a tidal ellipse, instead of the individual components. Four parameters completely describe the tidal ellipse and hence the tidal current for a single constituent: the amplitudes U_{MA} and U_{MI} of the major and minor axes, the orientation ϑ of the major axis (counterclockwise from east), and the time of maximum current φ, the phase. If the eastward and northward tidal current components are $U_E \cos(\sigma t - \varphi_E)$ and $U_N \cos(\sigma t - \varphi_N)$, then the tidal ellipse parameters are given by

$$U_{MA} = U_{CC} + U_C; U_{MI} = U_{CC} - U_C$$

$$\vartheta = 0.5 \tan^{-1} \left[\frac{2(A_1 A_2 + B_1 B_2)}{A_1^2 + B_1^2 - (A_2^2 + B_2^2)} \right]$$
(6.7.3)

$$\varphi = 0.5 \tan^{-1} \left[\frac{2(A_1 B_1 + A_2 B_2)}{A_1^2 + A_2^2 - (B_1^2 + B_2^2)} \right]$$

where

$$A_1 = U_E \cos\varphi_E; A_2 = U_N \cos\varphi_N$$
$$B_1 = U_E \sin\varphi_E; B_2 = U_N \sin\varphi_N$$

$$U_{CC} = 0.5\left[(A_1 + B_2)^2 + (A_2 - B_1)^2\right]^{1/2} \qquad (6.7.4)$$

$$U_C = 0.5\left[(A_1 - B_2)^2 + (A_2 + B_1)^2\right]^{1/2}$$

A_1, A_2 are the in-phase components of the two tidal components, and B_1, B_2 are the quadrature parts of the two. The tip of the current vector sweeps an ellipse with half-major and half-minor axes given by U_{MA} and U_{MI}, once every tidal cycle, the angle of the vector with north at any time t being

$$\theta = \frac{\pi}{2} - \tan^{-1}\left[\frac{U_N \cos(\sigma t - \varphi_N)}{U_E \cos(\sigma t - \varphi_E)}\right] \qquad (6.7.5)$$

If U_{MI} is positive, the current vector rotates counterclockwise; if negative it rotates clockwise. The maximum excursion of a fluid particle in a tidal cycle is UT/π, 14.2U km for M_2. If $U_{CC} = U_C$, the ellipse degenerates to a line and the tidal current simply reverses periodically, as, for example, in a narrow channel.

It is also possible to consider the resultant current vector as a superposition (Figure 6.7.1) of two current vectors of lengths U_{CC} and U_C, each rotating at an angular speed σ, but U_{CC} rotating counterclockwise, while U_C rotates clockwise. At time t = 0, the U_{CC} and U_C vectors make angles φ_{CC} and φ_C counterclockwise from the East:

$$\varphi_{CC} = \tan^{-1}\left[\frac{(A_2 - B_1)}{(A_1 + B_2)}\right]; \varphi_C = \tan^{-1}\left[\frac{(A_2 + B_1)}{(A_1 - B_2)}\right] \qquad (6.7.6)$$

When U_{CC} and U_C vectors are aligned, the current is maximum and along the semimajor axis of the current ellipse of length ($U_{CC}+ U_C$); when opposed, it is a minimum and along the semiminor axis |$U_{CC} - U_C$|. If $U_{CC} > U_C$ (or $0 < \varphi_N - \varphi_E < \pi$) the total current vector rotates counterclockwise, and if $U_C > U_{CC}$ (or $\pi < \varphi_N - \varphi_E < 2\pi$) it rotates clockwise. If $U_C = U_{CC}$, (or $\varphi_N - \varphi_E = 0, \pi$), the current is rectilinear. The direction and phase of the semimajor axis are 0.5 ($\varphi_N + \varphi_E$), positive counterclockwise from east.

Figure 6.7.2 shows examples of a predominantly semidiurnal tidal current from Pond and Pickard (1989).

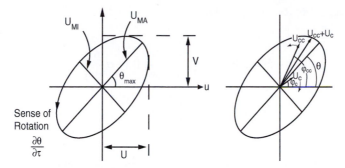

Figure 6.7.1 Tidal current ellipse (left) and decomposition into clockwise and counterclockwise rotating current vectors (right).

High-precision satellite altimetry enables one to determine only the tidal elevations. To derive the corresponding tidal currents, it is necessary to appeal to the governing equations of motion. In principle, this can be done by a barotropic tidal model assimilating the observed altimetric tidal elevations (for example, Kantha, 1995b). However, an approximate estimate can be made by appealing to the momentum equations of the LTE (Eq. 6.5.1) and ignoring the frictional

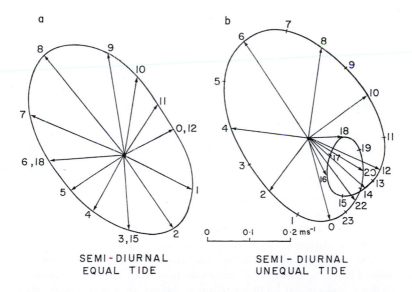

Figure 6.7.2 Tidal current ellipse for a semi-diurnal (left) and unequal (right) tide. (From Pond, S. and Pickard, G. L., *Introductory Dynamical Ocenography*, © 1989 Pergamon Press; reprinted by permission of Butterworth Heinemann Publishers, a division of Reed Educational & Professtional Publishing Ltd.)

terms (Desai, 1996; Ray *et al.*, 1994b):

$$\frac{\partial u_\phi}{\partial t} - 2\Omega u_\theta \sin\theta = -\frac{g}{R\cos\theta}\frac{\partial}{\partial\phi}\left(\beta\eta - \alpha\frac{\Phi}{g}\right)$$

$$\frac{\partial u_\theta}{\partial t} + 2\Omega u_\phi \sin\theta = -\frac{g}{R}\frac{\partial}{\partial\theta}\left(\beta\eta - \alpha\frac{\Phi}{g}\right)$$

(6.7.7)

With η known from altimetry, the RHS is known, including the influence of ocean loading and self-attraction term β. Considering a single constituent (frequency ω) and the velocity components as complex quantities, whose real and imaginary parts provide currents in phase and in quadrature to the tidal potential,

$$i\omega u_\phi - 2\Omega u_\theta \sin\theta = -\frac{g}{R\cos\theta}\frac{\partial}{\partial\phi}\left(\beta\eta - \alpha\frac{\Phi}{g}\right)$$

$$i\omega u_\theta + 2\Omega u_\phi \sin\theta = -\frac{g}{R}\frac{\partial}{\partial\theta}\left(\beta\eta - \alpha\frac{\Phi}{g}\right)$$

(6.7.8)

The absence of nonlinear and frictional terms makes the determination of the tidal current components straightforward. However, these equations are valid only in the deep ocean, where currents are small and friction two to three orders of magnitude less than the other terms in the LTE. The method breaks down, therefore, in shallow water. Also, there are singularities at critical latitudes where the tidal frequency equals the inertial frequency. These critical latitudes occur near the equator for long period tides, $\sim 30°$ latitude for diurnal tides, and near the poles ($\sim 75°$) for semidiurnal tides. This is a numerical artifact since the RHS does not necessarily vanish at these critical latitudes also, and hence interpolation to bridge across these latitudes is essential (Chao *et al.*, 1996a; Ray *et al.*, 1994b). Since the oceanic tides change not only the moment of inertia tensor of the Earth due to elevation changes, but also its angular momentum due to tidal current changes, the LOD calculations require knowledge of tidal currents also. While the above method enables approximate estimation in deep water from known tidal sea level variations (say from altimetry), extension to shallow water and critical latitudes requires the use of a high resolution data-assimilative tidal model (Kantha, 1995b; Le Provost *et al.*, 1994).

Figure 1.2.1 shows tidal current ellipses for M_2 and K_1 components in the global oceans from Kantha (1995b).

6.8 GLOBAL TIDAL MODELS

Tidal modeling has a rich history. For a recent overview, see Shum *et al.,* 1997 (see also Parker, 1991; Ray, 1993; and the special T/P sections in the *Journal of Geophysical Research,* December 1994 and December 1995). The following comes from Kantha (1995b).

It is widely accepted that, as long as sufficient resolution is provided and some degree of tuning (such as of bottom friction) is permitted, it is possible to obtain tidal elevations from barotropic tidal models. It is for this reason that finite element barotropic models have long been popular with coastal engineers. Nevertheless, as is clear from recent work using very high resolution, finite element models in the Gulf of Mexico (Westerink *et al.,* 1992), the shelf off British Columbia (Foreman *et al.,* 1993), and the Mediterranean Sea, it is still difficult to obtain tides accurately in regional tidal models, especially the phases, without fine-tuning of inputs such as boundary forcing and/or bottom friction. Accurate bottom topography is indispensable. As far as global tidal models are concerned, lack of understanding of tidal energy sinks in the deep basins along midocean ridges by conversion to baroclinic tides and inadequate resolution of the narrow shelves around the ocean basins, which leads to incorrect representation of energy loss there by bottom friction, appear to be primarily responsible for the fact that hydrodynamic tide models running free from constraints of observed data have difficulty producing tides accurate enough for use in applications such as altimetry. Inaccurate load tide distribution also contributes to errors in the ocean tides derived from such models. This is the primary reason why assimilation of observed tides either by altimetry or from tide gauges is indispensable at present to obtain results consistency with the accuracy demanded by modern precision altimetry.

There have been many attempts to derive tides in the global oceans (e.g., Accad and Pekeris, 1978; Parke and Hendershott, 1980; Schwiderski, 1980; Zahel, 1991). The very first attempts at modeling global tides ignored the shallow water regions and provided instead for the flux of tidal energy around the periphery of the ocean basins (Accad and Pekeris, 1978). Parke and Hendershott (1980) were among the very first to obtain accurate tidal maps for M_2 and K_1 in the global oceans. However, a milestone in tidal modeling was reached when Schwiderski (1980) was able to compute all eight important short period and the two long period lunar tides, using a $1°$ resolution global model assimilating data from about 1700 tide gauges. Since then, tidal modeling has become a cottage industry, with numerous models under development, spurred principally by the TOPEX (T/P) mission (and its follow-on Jason-1), one of whose goals was to obtain an accurate knowledge of global ocean tides.

The modern global tidal models can be divided into three groups. The first group, starting with the pioneering work of Cartwright and Ray (1990) on Geosat altimetric data (see also Cartwright, 1991; Cartwright and Ray, 1991), involves analysis of altimetric data to extract various tidal signals or, more appropriately, extraction of energy at the aliasing frequency of the various tidal constituents, since the temporal sampling interval is much larger than the tidal period for diurnal and semidiurnal tides. While for a precision altimeter such as TOPEX, the orbit determination errors are not a major factor in extracting tides, some subtidal energy at the aliasing frequency is invariably folded into the extracted tidal signal. This error tends to decrease with an increase in the record length. Cartwright and Ray's results were a significant improvement over the only available global tides at the time—namely, those derived by Schwiderski (1980, 1983)—but suffered from inaccuracies of Geosat data, the unsuitability of the orbital period to extract tides accurately, and the short record length. The T/P mission was designed to minimize aliasing errors (especially for M_2), and its very high accuracy enables a much better extraction of tidal SSH, which is confirmed by the current global models based on altimetric data analyses. The three tidal models from NASA Goddard Space Flight Center (GSFC) (Ray et al., 1994a; Sanchez and Pavlis, 1996; Schrama and Ray, 1994), the University of Texas (Ma et al., 1994), and the University of Colorado (Desai and Wahr, 1994, 1995) belong to this category and are comparable in performance to one another. They offer significant improvements over Cartwright and Ray (1990) (see also Cartwright et al., 1991) tides from Geosat. The Sanchez and Pavlis (1996) model (GSFC) uses Proudman function expansions (see Sanchez et al., 1992) to deduce the eight primary constituents, while the Ray et al. (1994a) model (RSC) uses orthotide formulation. The Schrama and Ray (1994) model (hereinafter referred to as SR) is based on corrections to Schwiderski's model. The Ma et al. (1994) model (CSR) is based on the work of Eanes (1994). The Desai and Wahr (1994) model (DW) used 2 years of TOPEX data (69 cycles) to derive all eight primary constituents, although the current version (Desai and Wahr, 1995) has used 79 cycles.

The second group consists of pure hydrodynamic models run without any data assimilation. The Grenoble finite element model (Le Provost et al., 1994) (see also Le Provost et al., 1991) belongs to this category (although its latest version, FES95.2 [Grenoble] utilizes results from an empirical altimetric tide model). While such models are independent of and hence free from altimetric biases and errors, their performance depends, to a large extent, on the model resolution, the degree to which energy sinks are properly represented, and the degree of fine-tuning of various model parameters. Finite element models have an advantage in terms of resolution but nevertheless suffer from inaccuracies of the underlying bathymetric data base. While good performance can therefore be expected in the open ocean, accurate predictions are not assured in shallow coastal regions, as recent works cited above demonstrate. The Le Provost et al. (1994) model divides the global ocean into subregions for numerical reasons and hence has to contend with open boundaries and the necessity to infer tides on these open boundaries. Also, their method involves a quasi-linear frictional formulation and assumes that the M_2 component is the dominant one. This assumption is not always satisfied; there are regions (for example, the Mediterranean Sea and the Sea of Okhotsk) where the diurnal tide K_1 is the most dominant and not the semidiurnal M_2 tide. Still, the model manages to exhibit a high degree of fidelity as indicated by the rms errors when compared to pelagic gage data. Also, it is one of the only two global models with a high enough resolution in shallow regions to provide meaningful results in coastal and marginal seas. The purely hydrodynamic model results have been updated recently with the FES98 version, which dispenses with division of the domain into separate parts and instead derives the solution over the entire global domain including the Arctic. These results confirm the difficulty of obtaining accurate tides without data constraints.

The third group involves the combined use of a dynamical model and observed tidal data, pioneered by Schwiderski (1980, 1983). Given the inevitable inaccuracies in dynamical modeling, data assimilation alone holds the promise of overcoming them without an inordinate amount of tuning of the model parameters. Schwiderski assimilated about 1700 coastal and island tide gage records into a coarse resolution (1°) dynamical model to produce one of the very first reliable estimates of global tides. He did not have access to altimetry-derived tides, however, and it appears that his data constraints were not adequate to provide better than 10-cm accuracy, even in the open ocean. The coarse resolution also implied inaccurate tides in shallow water and marginal seas and inaccurate representation of shallow water tidal energy sinks. The inverse model of Ebert *et al.* (1994) (OSU) also combines a hydrodynamic model and T/P altimetric crossover analyses to derive accurate tides in the global oceans. However, the methodology involves linearization of the governing equations and the frictional term, and the resolution is not fine enough for shallow coastal regions. It performs very well in most of the global oceans. More recently, the Grenoble finite element model FES98 has been rerun with assimilation of data from tide gauges alone.

Given the inaccuracies in bathymetric and other data bases essential for accurate tidal models, assimilation of observed tidal heights from coastal tide gages and in the open ocean, the same methodology as used by Schwiderski (1980) is one of the most effective means of deducing the tidal elevations. This is the methodology employed by Kantha (1995b). This model retains full nonlinear physics and bottom friction characterization. Tidal component data from 525 quality-controlled coastal tide gages and altimetric tides derived from TOPEX data by Desai and Wahr (1994) and subsampled to 3° resolution (3406 points) are assimilated into the model. A simple approach to assimilation is taken so that most of the computing resources can be devoted to calculating the dynamical consequences and not the details of data assimilation. Tierney *et al.* (1999) have used tides derived more accurately from longer T/P records and better quality-controlled tide gauges to derive global tides from a $1/4°$ model using the same methodology.

Shum *et al.* (1997) provide a more detailed discussion of these TOPEX-based global tidal models, and a CD-ROM containing the results from these models is available from NASA Jet Propulsion Laboratory. Here the Kantha (1995b) model is described in detail.

The governing equations for the barotropic tidal model are the full three-dimensional (3-D) baroclinic equations for continuity and momentum, integrated vertically over the water column. The governing equations for continuity and momentum are written here in orthogonal curvilinear coordinates:

$$\frac{\partial \eta}{\partial t} + \frac{1}{h_1 h_2}\left[\frac{\partial}{\partial x_1}\left(h_2 \bar{u}_1 D\right) + \frac{\partial}{\partial x_2}\left(h_1 \bar{u}_2 D\right)\right] = 0 \qquad (6.8.1)$$

$$\frac{\partial}{\partial t}\left(\bar{u}_1 D\right) + \frac{1}{h_1 h_2}\left[\frac{\partial}{\partial x_1}\left(h_2 \bar{u}_1^2 D\right) + \frac{\partial}{\partial x_2}\left(h_1 \bar{u}_1 \bar{u}_2 D\right)\right]$$

$$-f\bar{u}_2 D + \frac{\bar{u}_1 \bar{u}_2 D}{h_1 h_2}\frac{\partial h_1}{\partial x_2} - \frac{\bar{u}_2^2 D}{h_1 h_2}\frac{\partial h_2}{\partial x_1}$$

$$= -gD\frac{\partial}{\partial x_1}(\beta\eta - \alpha\zeta) + \left(\frac{\tau_{01} - \tau_{b1}}{\rho_0}\right) - \frac{gD^2}{\rho_0}$$

$$\left[\frac{\partial}{\partial x_1}\int_{-1}^{0}d\sigma\int_{\sigma}^{0}\rho d\sigma' - \left(\frac{\sigma}{D}\frac{\partial D}{\partial x_1} + \frac{1}{D}\frac{\partial \eta}{\partial x_1}\right)\int_{-1}^{0}\rho d\sigma'\right]$$

$$+ \frac{gD}{\rho_0}\frac{\partial D}{\partial x_1}\int_{-1}^{0}d\sigma\int_{\sigma}^{0}\sigma'\frac{\partial \rho}{\partial \sigma'}d\sigma' + D\bar{F}_1 \qquad (6.8.2)$$

$$\frac{\partial}{\partial t}\left(\bar{u}_2 D\right) + \frac{1}{h_1 h_2}\left[\frac{\partial}{\partial x_1}\left(h_2 \bar{u}_1 \bar{u}_2 D\right) + \frac{\partial}{\partial x_2}\left(h_1 \bar{u}_2^2 D\right)\right]$$

$$+ f\bar{u}_1 D + \frac{\bar{u}_1 \bar{u}_2 D}{h_1 h_2}\frac{\partial h_2}{\partial x_1} - \frac{\bar{u}_1^2 D}{h_1 h_2}\frac{\partial h_1}{\partial x_2}$$

$$= -gD\frac{\partial}{\partial x_2}(\beta\eta - \alpha\zeta) + \left(\frac{\tau_{02} - \tau_{b2}}{\rho_0}\right) - \frac{gD^2}{\rho_0}$$

$$\left[\frac{\partial}{\partial x_2}\int_{-1}^{0}d\sigma\int_{\sigma}^{0}\rho d\sigma' - \left(\frac{\sigma}{D}\frac{\partial D}{\partial x_2} + \frac{1}{D}\frac{\partial \eta}{\partial x_2}\right)\int_{-1}^{0}\rho d\sigma'\right]$$

$$+ \frac{gD}{\rho_0}\frac{\partial D}{\partial x_2}\int_{-1}^{0}d\sigma\int_{\sigma}^{0}\sigma'\frac{\partial \rho}{\partial \sigma'}d\sigma' + D\bar{F}_2 \qquad (6.8.3)$$

where the last term in each momentum equation contains the horizontal diffusion terms as well as the so-called dispersion terms. Term η is the free surface deflection, $\bar{u}_1$ and $\bar{u}_2$ are the two components of horizontal velocity in x_1 and x_2 directions, respectively, and h_1 and h_2 are the metrics of the orthogonal coordinate system x_1, x_2, respectively. For a spherical coordinate system employed in this model, $h_1 = R\cos\theta$, $h_2 = R$, where R is the radius of the Earth and $x_1 = 1$, $x_2 = \theta$, the latitude. The quantity f is the Coriolis curvilinear parameter $(2\Omega\sin\theta)$, and D is the total depth of the water column $= (H + h)$, where H is the bottom depth. Also, τ_0 is the wind stress, and τ_b is the bottom friction. The terms in large brackets arise from gradients of the density field, which can be ignored since we need consider only a homogeneous fluid for the barotropic tidal model. These terms will be important for the baroclinic version. Note also that we ignore the wind stress terms and use a quadratic drag law for the bottom friction terms, where the bottom friction coefficient c_d can be prescribed a priori:

$$[\tau_{b1}, \tau_{b2}] = c_d\,\rho_0 \left[\bar{u}_1^2 + \bar{u}_2^2\right]^{1/2} [\bar{u}_1, \bar{u}_2] \qquad (6.8.4)$$

The drag coefficient is chosen as 0.0025. In barotropic models some degree of tuning this coefficient is traditional. However, Kantha (1995b) chose to hold this value fixed in assimilative model runs, since data assimilation tends to minimize the need for such tuning.

The first terms on the right-hand side of Eqs. (6.8.2) and (6.8.3) are the tidal potential terms. Note that the equations are fully nonlinear. Not only are the advective terms retained in the momentum equations, but also no assumption is made as to the magnitude of free surface deflection vis-a-vis the local depth of the water column. The equations are therefore appropriate for simulating compound tides as well. Also, because of the presence of nonlinear terms, parameterization of subgrid-scale dissipation is of some importance. The model uses a Laplacian formulation for horizontal mixing, but the viscosity coefficient consists of two terms: the first one is a constant chosen a priori to represent a "background" value (200 m²s⁻¹); the second, based on Smagorinsky's (1963) nonlinear formulation, makes the viscosity dependent on the mean strain rate. The Smagorinsky constant K can also be selected a priori (0.02; the usual value is 0.04). Most of the dissipation is, however, provided by bottom friction in shallow water in a vertically integrated model such as this:

$$\bar{F}_1 = A_H \nabla^2 \bar{u}_1 \; ; \quad F_2 = A_H \nabla^2 \bar{u}_2$$

where

$$A_H = A_H^B + A_H^S$$

$$A_H^S = K\left(h_1 \partial \xi_1\right)\left(h_2 \partial \xi_2\right)$$

$$\left[\left(\frac{\partial \overline{u}_1}{h_1 \partial \xi_1}\right)^2 + \left(\frac{\partial \overline{u}_2}{h_2 \partial \xi_2}\right)^2 + 0.5\left(\frac{\partial \overline{u}_1}{h_2 \partial \xi_2} + \frac{\partial \overline{u}_2}{h_1 \partial \xi_1}\right)^2\right]^{1/2}$$ (6.8.5)

where K is the Smagorinsky constant.

The most noteworthy aspect of these equations is the tidal potential term on the right-hand side that represents astronomical forcing by the celestial bodies, the Moon and the Sun. As we saw in Section 6.3, it is traditional to expand the tidal potential terms in frequency space and consider it as a sum of several terms representing different tidal constituents. The frequencies of these constituents are determined by the trajectories of the Moon and the Sun with respect to the Earth. The amplitudes correspond to that given by equilibrium tidal theory. The most important ones are the so-called primary constituents, namely, the four semidiurnal constituents M_2, S_2, N_2, and K_2, the four diurnal constituents, K_1, O_1, P_1, and Q_1, and the three long-term constituents Mf, Mm, and Ssa, although it is possible to identify about 26 constituents that have equilibrium tidal amplitudes exceeding a centimeter or so (see Tables 6.2.1 and 6.2.2).

The equilibrium tidal potential ζ can be written as (see Section 6.2)

$$\zeta(\phi,\theta,t) = \sum_{m,j} B_{mj} f_{mj}(t_0) L_m(\theta) \cos\left[2\pi(t-t_0)/T_{mj} + m\phi + u_{mj}(t_0)\right]$$ (6.8.6)

where ϕ is the longitude (measured eastward from Greenwich in radians), θ is the latitude (measured north of equator in radians), and t is the universal time. Here t_0 is the reference time (say 0000 UT on January 1, 1975). Subscript m = 0, 1, and 2 refers to tidal species (0 is long-term declinational, 1 is diurnal, and 2 is semidiurnal). B_{mj} is the amplitude of the jth constituent of the mth species, and T_{mj} is the corresponding period:

$$L_0(\theta) = \frac{3}{2}\cos^2\theta - 1$$
$$L_1(\theta) = \sin 2\theta$$ (6.8.7)
$$L_2(\theta) = \cos^2\theta$$

The terms f_{mj} and u_{mj} are the so-called nodal factors (Doodson, 1927). These arise from the modulation of the amplitude and phases of astronomical forcing by the 18.6-year variations in lunar orbit and need to be considered for any accurate calculation of tides at any given time. Note that these affect only the lunar constituents and not the solar ones. When the Moon's declination reaches a maximum of 28° 36' (e.g., November 1987), f reaches a minimum for M_2 of 0.963; when it reaches a minimum of 18° 18', 9.3 years later, f becomes 1.037. Table 6.8.1 (from Pugh, 1987) gives the approximate nodal factors for the major lunar constituents. Note that the nodal factors for nonlinear shallow water tides

TABLE 6.8.1
Nodal Modulation Terms for Important Lunar Tidal Constituents

	f	u
M_m	$1.000 - 0.130\cos(N)$	$0.0°$
M_f	$1.043 + 0.414\cos(N)$	$-23.7°\sin(N)$
Q_1, O_1	$1.009 + 0.187\cos(N)$	$10.8°\sin(N)$
K_1	$1.006 + 0.115\cos(N)$	$-8.9°\sin(N)$
$2N_2, \mu_2, \nu_2, N_2, M_2$	$1.000 - 0.037\cos(N)$	$-2.1°\sin(N)$
K_2	$1.024 + 0.286\cos(N)$	$-17.7°\sin(N)$

N = 0 March 1969, November 1987, June 2006, and so on, at which time the diurnal terms have maximum amplitudes, but M_2 is at a minimum. M_2 has maximum equilibrium amplitudes in July 1978, March 1997, October 2015, and so on.
From Pugh (1987).

are multiples or products of the linear tides; for example, for M_4, the amplitude is the square of that for M_2 and the phase is twice that of M_2.

The tidal model can be run with any number of constituents with appropriate Doodson (1927) nodal factors for that particular year to compute the total tide from all relevant constituents. However, the methodology followed here does not require the use of nodal factors, since we simulate the primary constituents one component at a time and use the resulting tidal response characteristics to compute the total tides using the orthotide approach of Groves and Reynolds (1975) and Cartwright and Ray (1990). Following Schwiderski (1980) in writing the equilibrium tide,

$$\zeta(\phi,0,t) = \sum_{m,j} B_{mj} L_m(\theta) \cos\left(\sigma_{mj} t + m\phi + \chi_{mj}\right) \qquad (6.8.8)$$

where B_{mj} and L_m are same as before, $\sigma_{mj} = 2\pi/T_{mj}$, and χ_{mj} is the astro-nomical argument (see Dietrich *et al.*, 1980), which is a function of the lunar and solar ephemerides h_0, s_0, p_0, n_s, and p_s. Here h_0, s_0, p_0, n_s, and p_s are relative to Greenwich midnight of day D (usually in degrees) and can be expanded in powers of universal time T, reckoned in units of a Julian century (36525 mean solar days) from a certain epoch, such as midnight at Greenwich meridian on 0/1 January 1900 (Pugh, 1987). On the basis of the work of astronomers over the past, it is possible to determine these in degrees with extraordinary precision (Melchior, 1981; Pugh, 1987; Schwiderski, 1980) by

$$
\begin{aligned}
h_0 &= 279.69668 + 36000.768925485\,T + 3.03.10^{-4}\,T^2 \\
s_0 &= 270.434358 + 481267.88314137\,T - 0.001133\,T^2 + 1.9.\,10^{-6}\,T^3 \\
p_0 &= 334.329653 + 4069.0340329575\,T - 0.10325\,T^2 - 1.2.\,10^{-5}\,T^3 \qquad (6.8.9) \\
n_s &= 259.16000 - 1934.14\,T + 0.0021\,T^2 \\
p_s &= 281.22083 + 1.71902\,T + 0.00045\,T^2 + 3.0.10^{-6}\,T^3
\end{aligned}
$$

where

T	$(27392.500528 + 1.0000000356 \ D)/36525;$
D	$d + 365 \ (y - 1975) + Int \ [(y - 1973)/4];$
d	Day number of the year (d = 1 for January 1);
y	Year ($\geq$ 1975);
Int	Integral part of argument (to account for the leap year);
t	Universal standard time (UT) in seconds.

(6.8.10)

The epoch used here is the beginning of 1900 as in Table 4.2 of Pugh (1987). Thus, it is possible to compute χ_{mj} at any location and time. The corresponding ocean tide will be

$$\eta = \sum_{m,j} \eta_{mj} \cos\left(\sigma_{mj} \, t + \chi_{mj} - \delta_{mj}\right)$$

(6.8.11)

where δ_{mj} is the Greenwich phase in radians. It is also called the cophase, and η_{mj} is the coamplitude of the tide.

The term α, the factor multiplying the tidal potential ζ, is the effective Earth elasticity factor taken as 0.69 by Schwiderski (1980) and Hendershott (1981). Wahr (1981) has shown that for diurnal constituents the Love numbers are frequency dependent and are affected by the free core nutation of the Earth's axis, which has a nearly diurnal frequency. Therefore, α is slightly different for different tidal constituents (e.g., Foreman *et al.*, 1993). This frequency dependence (see Table 6.3.1) is especially important for the diurnal component K_1, whose equilibrium amplitude is increased by nearly 10%.

The factor β accounts rather crudely for the effect of load tides in the dynamical equations. Following conventional practice (Foreman *et al.*, 1993; Pekeris and Accad, 1969), the load tide due to the deformation of the ocean bottom under the varying load of the ocean tides can be considered to be in antiphase with the oceanic tide and a small fraction of it. Foreman *et al.* (1993) recommend values of 0.940 and 0.953 for β for diurnal and semidiurnal constituents. A more accurate method would be to use the load tides computed using the ocean tide model results in an iterative scheme to improve the ocean tide and load tide results globally. This may have some bearing on the accuracy of ocean tides in deep water, but in shallow water, the nonlinear components are likely to have a greater impact on tides.

The inadequacy of altimetric tide measurements alone over coastal oceans around the margins of the primary basins is quite self-evident. The smaller spatial scales of tidal variability in these regions introduce serious errors. It is here that a dynamical model with proper physics can fill the gap. A high-resolution, nonlinear, barotropic global tidal model combining observations with dynamics can essentially help "extrapolate" altimetric tides on to the shelf. Coastal tide gages would also be helpful in "interpolating" the tides between the open ocean and the coast. The strategy is very similar to that of Schwiderski (1980, 1983), except that we now have the benefit of accurate tides over the deep oceans derived from altimetry and modern computing resources to attain high resolution. The results of such a model are useful not only in many coastal applications such as pollution tracking, commerce, and defense needs, but also are important to shallow water altimetry and many geophysical applications.

Accurate tides are essential for deducing the subtidal circulation in the global oceans from altimetry, both in the deep ocean and in shallow coastal and semienclosed seas. It is often easy enough to mistake residual tidal energy at the corresponding aliased frequency as a true subtidal oceanic signal, as some earlier Geosat analyses did (see Schlax and Chelton, 1994, for a discussion of tidal aliasing errors). Given the fact that roughly 75% of the sea surface height variability is due to oceanic tides, residual tides have the potential to seriously contaminate the subtidal signals. One indication of the accuracy of tides is the degree of reduction of variance of crossover differences resulting from subtraction of tidal signals. The crossover difference rms value for T/P altimetry is roughly 40 cm over the deeper parts of the global oceans without removal of any ocean tides (Andersen *et al.*, 1995). Removing tides using Schwiderski's (1980, 1983) tidal model results in the reduction of the rms value to about 11 to 12 cm, on the average, while the use of Cartwright and Ray's (1990) model improves this value by around 0.5 cm. These values are used as benchmarks to determine the accuracy of more recent global tidal models (Shum *et al.*, 1997). Current T/P tidal models achieve a further reduction of about 1 cm to give a crossover variance of about 10 cm. Molines *et al.* (1994) discuss this aspect in great detail and explain why the tidal correction is a major one in altimetry.

Figure 6.8.1 shows an example of the application of barotropic equations to the problem of deducing the tidal SSH and currents in the global oceans. The reader is referred to Kantha (1995b) for details, but briefly, the equations are cast in spherical coordinates and the tidal potential terms are expressed as a sum of a series containing various tidal components (the atmospheric forcing terms are zeroed out for this application). The resulting equations are solved on a $1/5°$ latitude-longitude C grid covering the global oceans (excluding the Arctic) for each tidal component. The bottom depths over the model grid are derived from a digital data base (ETOP05 from NOAA) containing world topography at $1/12°$ resolution. However, for the results to be accurate enough for certain applications such as altimetry, inevitable errors that result from inaccurate knowledge of bottom depths and friction coefficients have to be offset by data assimilation.

The tidal SSH data derived from the currently operational NASA/CNES TOPEX/Poseidon precision altimeter (Desai and Wahr, 1995) have been assimilated into the model as well as those from coastal tide gages around

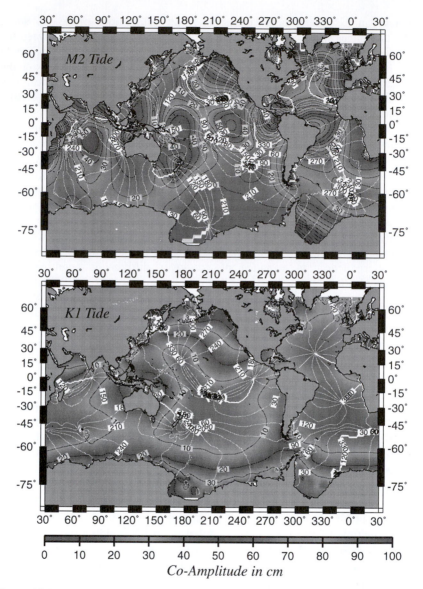

Figure 6.8.1 Coamplitude and cophase of M_2 and K_1 tides in the global ocean. From Kantha (1995).

the world's coastlines. A simple data assimilation scheme has been used where at each time step, the model-predicted SSH is replaced by a weighted sum of the model SSH and the observed SSH, with weights determined a priori. The result

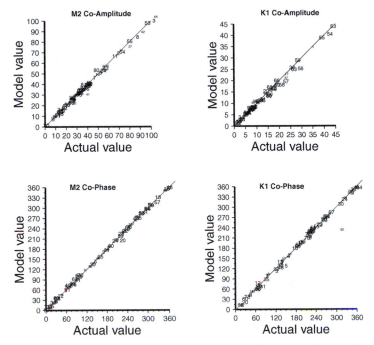

Figure 6.8.2 Comparison of the coamplitudes and cophases of modeled M_2 and K_1 tides with observations at pelagic tide gages at which high quality tidal observations are available.

is tidal SSH that is accurate to within a few centimeters over the global oceans, including shallow coastal, and semienclosed seas. This information is useful for many applications, such as an accurate determination of the subtidal SSH variability from altimetric data, gravimetry, and determination of tidal dissipation. The model grid is 1800×729, the domain extending to the Bering Strait in the north and the Antarctic continent in the south, and takes nearly 15 CPU hr for a 10-day simulation on a CRAY C-90. The finite difference formulations are very similar to those presented for the reduced gravity in Chapter 3 and the barotropic model in Chapter 7 (see also Chapter 9), and will not therefore be repeated here.

Figure 6.8.2 shows the accuracy attained by this data-assimilative tidal model in the form of scatter plots that compare modeled and observed tides from an independent set of accurate tide and bottom pressure gages over the global oceans. More details on Kantha's model results can be found on the Web at http://www.cast.msstate.edu/Tides2D Internet site.

6.9 REGIONAL TIDAL MODELS

There is often a need for accurate high resolution tides and tidal currents in a small region of the global oceans, usually a coastal ocean or a marginal sea, often semienclosed. While the tidal elevations may very well be known accurately along many coastlines from tide gage measurements, the dearth of open ocean bottom pressure gages makes it harder to infer them elsewhere in the region. The logical solution is a regional tidal model. Also a model is often the only means of deducing tidal currents and dissipation of tidal energy by bottom-friction in the region. Implementation of such a model is straightforward, using the set of equations (6.8.1) to (6.8.3). For such regional models, with length scales less than, say, 1000 km, it is possible to put β equal to unity and ignore the loading effects. It is preferable to retain full nonlinearity, if the region is shallow. The most important input to such barotropic tidal models is the bottom topography. It is absolutely essential to have accurate bottom depths at the chosen grid resolution. It is then possible to obtain accurate tides by some fine-tuning the bottom friction coefficient. The grid resolution must be chosen commensurate with the spatial scales as indicated by topography. The tidal wavelength $\lambda = (gH)^{1/2}T$, where H is the bottom depth and T is the tidal period. For H ~ 10 m, λ ~ 430 km for semidurnal and 640 km for diurnal tides. Therefore 10-20 km resolution is usually adequate to resolve even nonlinear components such as M_4.

Tides in semienclosed seas (and bays) without a narrow (and/or shallow) opening to the outside (for example, the Persian Gulf) are usually driven by the tides outside the region. The tides are then the so-called co-oscillating tides. When co-oscillating tides dominate, direct astronomical forcing of tides can be neglected. In regions with highly restricted access to the outside (for example, the Red Sea), direct forcing is quite important. For example, in the eastern Mediterranean Sea, the tides are principally forced by the local tidal potential. While the tidal potential terms are included in the equations of motion, Eqs. (6.8.2) and (6.8.3), the effect of co-oscillation has to be prescribed through the conditions on the open boundaries.

The most straightforward approach is to prescribe the observed tides or tides from a global model along the open boundaries. This is called one-way nesting. With the advent of altimetry and accurate global tidal models, it is now possible to prescribe such boundary conditions quite accurately anywhere in the global oceans. Equations (6.8.1) to (6.8.3) can then be solved subject to

$$\eta(t)\big|_B = \eta_B(t) \tag{6.9.1}$$

Often, such a "clamped" boundary generates noisy solutions, especially if the dissipation is low; in that case it is possible to provide some flexibility by damping the boundary to the observed tide:

$$\left.\frac{\partial\eta}{\partial t}\right|_B = \frac{\eta_B(t)-\eta}{T_d} \tag{6.9.2}$$

By choosing the damping timescale T_d properly, it is possible to get smoother solutions. Yet another approach is to allow any internal disturbances to radiate out so that a final equilibrium state can be readily achieved. The radiation condition is usually of the form

$$\frac{\partial\eta}{\partial t} + C\frac{\partial\eta}{\partial n} = 0 \tag{6.9.3}$$

where C is the phase speed of the disturbance $(gH)^{1/2}$, and n is the unit normal positive in the outward direction. It is possible to combine this with the continuity equation $\partial\eta/\partial t + \partial M/\partial n = 0$, where M is the transport normal to the boundary, to obtain

$$M + C\eta = \text{constant} \tag{6.9.4}$$

as an alternative form of radiation boundary condition. However, it is simply not enough to radiate disturbances out at an open boundary; the incoming tide has to be prescribed as well. Kantha (1985) suggests

$$\frac{\partial\eta}{\partial t} + C\frac{\partial\eta}{\partial n} = \frac{2\partial\eta_i}{\partial t} \tag{6.9.5}$$

with η_i corresponding to incoming tides. Once again, since η_i is often not known a priori, some experimentation is necessary to obtain observed tides on the open boundary.

For examples of tides and tidal models in marginal and semienclosed seas, see http://www.cast.msstate.edu/Tides2D. Traditional treatises on the subject (Defant, 1961; Dietrich *et al.*, 1980; Pugh, 1987) contain detailed descriptions of the barotropic tides in important marginal seas around the world, such as the Yellow Sea, the Mediterranean Sea, and the Red Sea (see also Parker, 1991). Kowalik and Polyakov (1998) is an excellent example of the application of a barotropic model to the tidal science of a marginal sea, in this case, the Sea of Okhotsk.

6.10 GEOPHYSICAL IMPLICATIONS

The geophysical impact of oceanic tides is large (Dickey *et al.*, 1994; Hide and Dickey, 1991; Munk and MacDonald, 1960; Ray *et al.*, 1994; Wahr, 1988). Tidal sea surface elevation fluctuations, both short and long term, modulate the moment of inertia tensor of the earth, and the tidal current fluctuations cause

Length-of-Day Variations

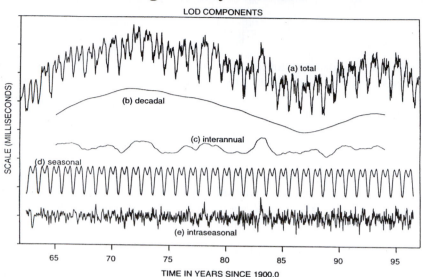

Figure 6.10.1 LOD fluctuations over a wide spectrum of temporal scales. (Reprinted from Dickey, J. O., Earth rotation variations from hours to centuries, in *Highlights of Astronomy,* figure 2a, p. 19, © 1995 with kind permission from Kluwer Academic Publishers.)

exchange of angular momentum between the ocean and the solid Earth, leading to length of day (LOD) fluctuations that are clearly discernible in geophysical measurements of the rotation rate of the Earth (Figure 6.10.1 from Dickey, 1995). While the major part of the variability in the length of day is due to the exchange of momentum between zonal atmospheric motions and the solid Earth, Ray *et al.* (1994) showed that a substantial part of the signal is due to oceanic tides. They were able to compute the influence of semidiurnal and diurnal ocean tides on the LOD, using Cartwright and Ray (1990) tidal solutions for tidal sea levels and currents inferred from these heights using Laplace tidal equations. The computations show remarkably good agreement with observational data (Figure 6.10.2). Long-period tides are major contributors to the LOD fluctuations (Kantha *et al.,* 1998).

Another major influence of oceanic tides is on the braking of the Earth-Moon system due to tidal dissipation of its gravitational energy (Munk, 1997; Munk and MacDonald, 1960). Here the atmosphere has a negligible influence. Lunar laser ranging (LLR) shows a 3.7 cm year^{-1} increase in the semimajor axis of the Moon's orbit (Dickey *et al.,* 1994; Hide and Dickey, 1991) corresponding to a dissipation of 3.05 TW and a LOD increase of 2.1 ms cy^{-1}. Satellite tracking data also clearly show the perturbations in the orbits of artificial satellites due to tidal modulation of the geopotential. All these have been used to deduce the global rate of energy dissipation by oceanic tides (Schrama and Ray, 1994).

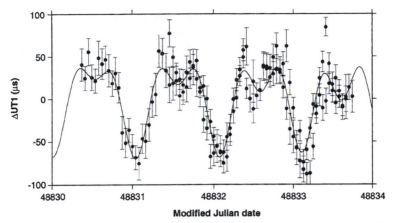

Figure 6.10.2 Comparison of observed variations in UT1 with those predicted using Laplace tidal equations and tidal SSH derived from GEOSAT. (Reprinted with permission from R. D. Ray, D. J. Steinberg, B. F. Chao, and D. E. Cartwright, Diurnal and semidiurnal variations in the Earth's rotation rate induced by oceanic tides, *Science* **264**, 830–832, 1994b. Copyright 1994 American Association for the Advancement of Science.)

According to values quoted by Schrama and Ray, these observations suggest a range of values for M_2 tidal dissipation of about 2.41 to 2.54 TW, with corresponding values for S_2, K_1, and O_1 being 0.35 to 0.37, 0.33 to 0.34, and 0.19 to 0.22 TW. The total dissipation rate due to tides, including that dissipated in solid Earth, amounts to about 3.75 TW (Kantha *et al.,* 1995), of which 3.17 TW is due to the Moon (M_2, N_2, O_1, and Q_1, some from K_1) and 0.58 TW due to the Sun (S_2, K_2, K_1, and P_1). According to geological sedimentary records, the average value for LOD increase has been ~ 2.4 ms cy^{-1} for the past 900 million years, with 481 days in a year and, since the length of the year (LOY) does not change much, with LOD then being ~18 hr (Sonett, 1996). This also implies that the Moon was then ~38 000 km closer to the Earth. If the current rates prevail, the Moon will be farther by roughly the same value in another billion years and the LOD will be ~30 hr, an answer, albeit highly delayed, to modern working day pressures.

6.10.1 TIDAL DISSIPATION AND LOD

The very first estimate of 1.4 TW by Jeffreys (1920) for tidal dissipation for the most important component M_2 was based on G. I. Taylor's estimate of bottom friction-induced dissipation in the Irish Sea (Taylor, 1919). Munk and MacDonald (1960) estimated the power input from the Moon correctly at 2.5 TW, but this value was revised to as high as 4 to 5 TW in the 1970s, before settling back to 2.5 TW in recent years. According to the well-known tidalist, Walter Munk, the convergence of tidal dissipation estimates to 2.5 TW± 0.05

TW is one of the triumphs of 20th century science (Munk, 1997). Of this, only 0.1 TW is due to dissipation in solid Earth, confirmed recently by careful analysis of T/P records by Ray *et al.* (1996), with oceanic tides constituting the lion's share.

Tidal despinning of the Earth and the cumulative effect of the LOD increase over time lead to significant displacement of the time and longitude at which an eclipse is observed (Munk, 1997; Munk and MacDonald, 1960). This was first noted by the astronomer Halley in 1695. Eclipses have had profound religious significance to many civilizations and hence are often meticulously recorded. Since Babylonian times 30 centuries ago, the accumulated time difference (assuming an average rate of 1 ms cy^{-1}) is nearly 3 hr; this leads to changes in both the time and longitude at which eclipses are observed. From careful analysis of ancient records of lunar and solar eclipses, equinoxes and lunar occultations, it is possible to deduce the lunar secular deceleration rate and hence the tidal dissipation rate (Kagan and Sundermann, 1996). While uncertainties in interpreting the ancient records have plagued the reliability of these estimates in the past (Munk and MacDonald, 1960), most recent analyses (for example, Ciyuan and Yau, 1990) have yielded a dissipation value of 3.17 TW, remarkably close to estimates by other methods and the value from Kantha's numerical model.

Telescopic observations of occultations of stars in the lunar path, the longitude of solar eclipses, and Mercury's passage across the solar disk can also be used to deduce the lunar orbital deceleration and hence the tidal dissipation rate. Following the pioneering work by Spencer Jones (Jones, 1939), whose estimates based on lunar occultations lead to a value for tidal dissipation of 2.76 TW, more recent and reliable estimates by Morrison (1978) from 250-year data on Mercury's passage yield a value of 3.17 TW, again remarkably close to other estimates. Finally, tidal perturbations of the ephemerides of artificial satellites can also be used to estimate the tidal dissipation rates and recent estimates applying this method to more than 30 satellites (Marsh *et al.*, 1990) also yield a value of 3.17 TW. The fourth method, the lunar laser ranging, yields a value of 3.07 TW (Bursa, 1990). See Kagan and Sundermann (1996) for a fascinating survey and history of tidal dissipation estimates; their table, reproduced here as Table 6.10.1, shows a complete list of estimates by various methods.

But for the tidal friction, the mean ocean (solid Earth) tidal bulges corresponding to the second spherical harmonic of the ocean (solid Earth) tides would be exactly at the sublunar and antilunar points and there would be no retarding pull of the Moon on the bulges (and hence no despinning of the Earth) and no corresponding pull from the bulges on the Moon (and hence no change in the Moon's orbital momentum). Since the Earth's angular rotation velocity is greater than that of the Moon's orbital angular velocity, the tidal friction displaces the bulges so that they lead the Moon by an angle that is proportional to the tidal dissipation (Figure 6.10.3). Since the sum of the angular momentum

TABLE 6.10.1

Astronomical Estimates of Lunar Acceleration and Corresponding Lunar Tidal Dissipation

Type of observation	Author(s)	$-\dot{n}''/\text{cy}^2$	$-\dot{E} \times 10^{12}$ W
Ancient observations	Fotheringham (1920)	30.8	3.76
of lunar and solar	De Sitter (1927)	37.7 ± 4.1	4.60 ± 0.50
eclipses, equinoxes, and	Newton (1970)	41.6 ± 4.3	5.08 ± 0.52
lunar occultations		42.3 ± 6.1	5.16 ± 0.74
	Stephenson (1972)	34.0 ± 2.0	4.15 ± 0.24
	Muller and Stephenson (1975)	37.5 ± 5.0	4.58 ± 0.61
	Muller (1975)	34.5 ± 3.0	4.21 ± 0.37
		30.4 ± 3.0	3.71 ± 0.37
	Muller (1976)	30.0 ± 3.0	3.66 ± 0.37
	Stephenson (1978)	30.0 ± 3.0	3.66 ± 0.37
	Newhall *et al.* (1983)	26.21 ± 2.0	3.20 ± 0.24
	Ciyuan and Yau (1990)	26.0	3.17
Telescopic observations	Spencer Jones (1939)	22.4 ± 0.88	2.76 ± 0.11
of lunar occultations,	Van Flandren (1970)	52.0 ± 16.0	3.34 ± 1.95
solar longitude, and	Osterwinter and Cohen (1972)	38.0 ± 8.0	4.64 ± 0.98
Mercury's passage	Morrison (1973)	42.0 ± 6.0	5.12 ± 0.73
across the solar disk	Morrison and Ward (1975)	26.0 ± 2.0	3.17 ± 0.24
	Van Flandren (1978, cited by Morrison, 1978)	36.0 ± 4.0	4.39 ± 0.49
	Morrison (1978)	26.0 ± 2.0	3.17 ± 0.24
	Van Flandren (1981)	21.4 ± 2.6	2.61 ± 0.32
Variations in	Cazenave (1982)	21.9 ± 1.6	2.67 ± 0.20
satellite orbit	Marsh *et al.* (1988)	25.4 ± 0.6	3.10 ± 0.07
elements	Christodoulidis *et al.* (1988)	25.3 ± 0.6	3.08 ± 0.07
	Cheng *et al.* (1990)	24.8 ± 0.8	3.03 + 0.10
	Marsh *et al.* (1990)	26.0 ± 0.5	3.17 ± 0.06
Laser sounding of	Calame and Mulholland (1978)	24.6 ± 1.6	3.00 ± 0.20
the Moon	Williams *et al.* (1978)	23.8 ± 4.0	2.90 ± 0.49
	Ferrari *et al.* (1980)	23.8 ± 3.1	2.90 ± 0.38
	Dickey *et al.* (1983)	25.3 ± 1.2	3.09 ± 0.15
	Dickey (1987, cited by Burša, 1987)	24.9 ± 1.0	3.04 ± 0.12
	Newhall *et al.* (1990)	24.9 ± 1.0	3.04 ± 0.12
	Burša (1990)	25.17 ± 0.8	3.07 ± 0.10

From Kagan and Sundermann (1996).

due to Earth's rotation about its axis $C_E \Omega$ ($C_E \sim 8.118 \times 10^{37}$ kg m^2) and that due to the orbital motions of the Earth and the Moon about their common center $M_M M_E (M_M + M_E)^{-1} R^2 n$ must remain constant (ignoring the very small angular momentum of the Moon's rotation about its axis), the associated tidal torque L leads to a decrease of the Earth's rotational angular momentum and a

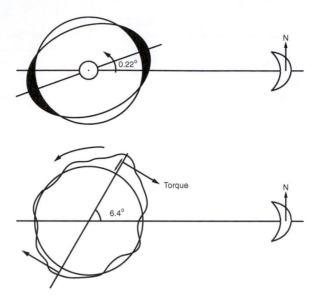

Figure 6.10.3 Sketch showing the mean solid Earth (top) and ocean (bottom) tidal bulges displaced by 0.22° and 6.4°, respectively, form the sublunar point. These are responsible for gravitational torques that decrease the Earth's spin momentum and increase the lunar orbital momentum. From Munk (1997) and Cartwright (1993).

corresponding increase in the Moon's orbital angular momentum so that

$$L = -C\frac{d}{dt}(\Omega) = \frac{M_M M_E}{(M_M + M_E)}\frac{d}{dt}(R^2 n) = -\frac{M_M M_E R^2}{3(M_M + M_E)}\frac{dn}{dt} \quad (6.10.1)$$

after using Kepler's third law, $n^2 R^3 = G (M_M + M_E)$, where G is the universal gravitational constant, to substitute for dR/dt:

$$\frac{dR}{dt} = -\frac{2R}{3n}\frac{dn}{dt} \quad (6.10.2)$$

Thus, the Earth's spin momentum is transferred to lunar orbital momentum (Munk, 1997) by the torque exerted by the tidal bulges. Both Ω and n decrease with time so that both LOD and LOM (length of the lunar month) increase. The total energy of the Earth-Moon system, which is the sum of the kinetic energy of the orbital motion of the Earth and the Moon around their common center, $0.5 \, M_M M_E (M_M + M_E)^{-1} R^2 n^2$, the kinetic energy of the Earth's rotation, $0.5 \, C_E \Omega^2$, and the potential energy of the Moon, $- G\, M_M\, M_E\, R^{-1}$, is

$$E = \frac{1}{2}\left[C_E \Omega^2 - G\frac{M_M M_E}{2R} \right] \tag{6.10.3}$$

This energy decreases with time due to tidal dissipation with the dissipation rate D_S $(= -dE/dt)$ being equal to $L\,(\Omega - n)$; therefore, the change in the lunar angular orbital velocity n and the change in Earth's angular rotation velocity Ω are related to the tidal dissipation by (Kagan and Sundermann, 1996; Munk, 1997; Munk and MacDonald, 1960)

$$D_s = -\frac{R^2 M_M M_E}{3(M_M + M_E)}(\Omega - n)\frac{dn}{dt} = -C_E(\Omega - n)\frac{d\Omega}{dt}$$

$$\Omega = \frac{2\pi}{T_d},\, n = \frac{2\pi}{T_m},\, \frac{d\Omega}{dt} = -\frac{2\pi}{T_d^2}\frac{dT_d}{dt},\, \frac{dn}{dt} = -\frac{2\pi}{T_m^2}\frac{dT_m}{dt} \tag{6.10.4}$$

where $M_{M,E}$ refers to the masses of the Moon and the Earth, R is the distance between the two, Ω is the angular velocity of the Earth, T_d is a sidereal day (86,164 s), and T_M is the lunar month (27.53 days). D_S is the dissipation rate.

For the present day Earth-Moon configuration, dissipation in terawatts is $D_S = -0.1224\,(dn/dt) = 1.53\, dT_d/dt$, where dn/dt is in arc-seconds per century squared and dT_d/dt is the change in LOD in ms cy^{-1}. For a total lunar dissipation value of 3.17 TW (of which about 0.15 TW is due to solid Earth tides), dn/dt has a value of -25.9" cy^{-2}, very close to the consensus value of -25" cy^{-2} from astronomical observations. Many modern estimates based on telescope data and Babylonian eclipse records (see a tabulation by Kagan and Sundermann, 1996; see also Munk, 1997), as well as global model results (Kantha *et al.*, 1995), agree with this value. The rate of change of LOD is 2.07 ms cy^{-1}. By Kepler's third law, the Moon recedes from Earth according to Eq. (6.10.2). For the current configuration, $dR/dt \sim -0.149\,(dn/dt) \sim 1.215\, D_S$ in units of m cy^{-1}, where dn/dt is in arc-seconds per century squared, and D_S is in TW, so that for dn/dt ~ 25.9" cy^{-2}, $dR/dt \sim 3.86$ m cy^{-1}.

The displacement of the solid Earth tidal bulges from their equilibrium position is about $0.22°$ and the displacement of the ocean bulges (the ocean having many large local bulges and indents) is $6.4°$ in the mean (Munk, 1997), the ratio of the two being about 29. Tides due to the Sun have a similar effect on the Sun, but the angular acceleration is eight orders of magnitude smaller than that of the Moon (Munk and MacDonald, 1960) and negligible for all practical purposes. Solar tides can cause changes in the Earth's orbit, but do not affect its rotational characteristics much. Only lunar tides are important to the despinning of the Earth.

6.10.2 TIDAL ENERGETICS

It is instructive to examine the equation for conservation of barotropic tidal energy. Consider for simplicity a rectangular Cartesian coordinate system and linear governing equations, without horizontal diffusion terms:

$$\frac{\partial \eta}{\partial t} + \frac{\partial}{\partial x_i}(u_i D) = 0$$

$$\frac{\partial}{\partial t}(u_j D) - \varepsilon_{j31}(u_1 D) = -gD\frac{\partial}{\partial x_j}(\beta\eta - \alpha\zeta) - \frac{\tau_j}{\rho_0} \tag{6.10.5}$$

(i, j = 1, 2). Multiplying the momentum equation by $\rho_0 u_j$ and the continuity equation by $\rho_0 g\eta$ and summing up, the energy conservation equation becomes

$$\frac{\partial T}{\partial t} + \frac{\partial}{\partial x_i}(F_i) = W - D_b$$

$$T = K + P; \quad K = \frac{\rho_0}{2}D\, u_j u_j; \quad P = \frac{\rho_0}{2}\beta g\eta^2 \tag{6.10.6}$$

$$F_i = \rho_0 gDu_i(\beta\eta - \alpha\zeta), W = \rho_0 g\alpha\zeta\frac{\partial\eta}{\partial t}, D_b = u_j\tau_{bj}$$

where F is the energy flux, W is the rate of work done by astronomical tidal forces, D_b is bottom dissipation rate, K is the kinetic energy, P is the potential energy, and T is the total energy. Note that if the nonlinear and the horizontal friction terms were retained, one would also have terms indicating dissipation by horizontal friction and flux of kinetic energy. For deep oceans, these terms can be neglected. The most important terms are the second term on the LHS, the flux divergence term, and the first term on the RHS, the work done by astronomical forces on the ocean. The second term on the RHS, the bottom dissipation, is nearly negligible in the deep, but dominant in shallow water. Taking averages over a complete tidal cycle, the tendency terms drop out and we are left with

$$\left\langle \frac{\partial F_i}{\partial x_i} \right\rangle = \langle W \rangle - \langle D_b \rangle \tag{6.10.7}$$

This states that the flux divergence in an elemental area is simply the difference between the rates of work done and the bottom dissipation, all quantities averaged over a tidal cycle. When there is excess work done in the area, the excess energy is exported. Where bottom dissipation rate exceeds the rate of work done (as in shallow marginal seas) or the work done is negative, there has to be a convergence of tidal power from elsewhere. Averaged over the whole basin, of course, the divergence term drops out and the equation becomes a simple statement of balance between the total work done by the astronomical forces and the bottom dissipation of tidal energy. However, the patterns of distribution of the work done and bottom dissipation are quite interesting. The work done is principally in certain deep ocean regions, but most dissipation occurs in shallow marginal seas. The tidal currents transport the power deposited in the deep oceanic regions to the shallow marginal seas. While there are negative-work-done regions, where the ocean transfers energy to the moon (and the sun), overall net work is done on the oceans to the tune of about 3.75 TW, of which ~0.25 TW is the work done on the ocean bottom.

While there is now a definite consensus on the work done by tidal forces, the mechanisms and regions of dissipation are still uncertain. The ratio of the areas of the ocean regions shallower than 200 m (0.21×10^{14} m^2) and those deeper than 200 m (3.12×10^{14} m^2) is about 0.07. However, the tidal currents in deep water are almost an order of magnitude smaller, and since the bottom dissipation is proportional to the cube of the tidal velocities, conventional thinking is that even though more than 95% of the work done by tidal forces is in the deep oceans, they account for less than 2 to 3% of the tidal dissipation (see the Table 2 of Kagan and Sundermann, 1996, for example), and most of the dissipation occurs in shallow seas. However, this ignores the conversion of barotropic tidal energy into internal tides over seamounts and midocean ridges, which is now thought to constitute a significant fraction of the tidal energy balance (see Section 6.12). It also ignores highly dissipative turbulent mixing resulting from tidal currents around island chains and around submerged flanks of seamounts. These two mechanisms together might account for about 1 TW of tidal dissipation in the deep oceans.

The rate of work done by the tidal gravitational forces can be computed if the tidal SSH is known (Hendershott, 1977, 1981). The expression above is only the work done on the ocean. The total rate of work done, however, must include the power transferred to induced bottom deflections (Cartwright and Ray, 1990; Hendershott ,1972) and dissipated. The rate of work done by tidal potential and hydrostatic pressure forces on the ocean with an elastic bottom averaged over a tidal period can therefore be written as (Kagan and Sundermann, 1996)

$$\langle \dot{W}_o \rangle = \rho_0 \iint\limits_{ocean} \left[\left\langle \left(\Gamma - \frac{p_b}{\rho_0} \right) \frac{\partial \eta}{\partial t} \right\rangle + g \left\langle \left(1 + \frac{\rho_a}{\rho_0} \right) \eta \frac{\partial \eta_e}{\partial t} \right\rangle \right] dA \quad (6.10.8)$$

where angle brackets denote average over a tidal cycle; η is the ocean surface deflection relative to the bottom whose tidal deflection itself is η_e; ρ_0 and ρ_a are densities of water and air; p_b is the barometric tide ($p_s - g\, \rho_a\, \eta_e$), the pressure at the surface due to atmospheric tide. Since p_b does not exceed 1 mb (10^2 Pa), $p_b/g\rho_0$ does not exceed 10^{-2} m; hence, this term can be neglected. Similarly, the term involving ρ_a/ρ_0, which is $\sim 10^{-3}$, can also be neglected (Platzman, 1991) to yield expressions given by Hendershott (1979, 1981) and Cartwright and Ray (1990):

$$\langle \dot{W}_o \rangle = \rho_0 \iint\limits_{\text{ocean}} \left[\left\langle \Gamma \frac{\partial \eta}{\partial t} \right\rangle + g \left\langle \eta \frac{\partial \eta_e}{\partial t} \right\rangle \right] dA \qquad (6.10.9)$$

From Eq. (6.4.2),

$$\Gamma = (1 + k_2)\Phi + \phi_1; \quad \eta_e = h_2 \Phi / g + \eta_1 \qquad (6.10.10)$$

where the second term on the RHS is much smaller than the first, and rate of work done can be written as

$$\langle \dot{W}_o \rangle = \rho_0 \iint\limits_{\text{ocean}} \left[\alpha \left\langle \Phi \frac{\partial}{\partial t} (\eta - \eta_1) \right\rangle + \left\langle \frac{\partial \eta}{\partial t} (\phi_1 - g\eta_1) \right\rangle \right] dA \qquad (6.10.11)$$

where $\alpha = (1 + k_2 - h_2)$. Using orthogonality of spherical functions, to a slightly lower accuracy, because $\eta_1 \sim -\varepsilon\eta$ and $\phi_1 \sim \varepsilon' g\eta$ ($\varepsilon \sim 0.075, \varepsilon' \sim 0.05$),

$$\langle \dot{W}_o \rangle = \rho_0 \alpha (1 + \varepsilon) \iint\limits_{\text{ocean}} \left[\left\langle \Phi \frac{\partial \eta}{\partial t} \right\rangle \right] dA \qquad (6.10.12)$$

Note that this includes the fraction ε of the work done and dissipated in the mantle due to the load tides induced by the ocean tides. Kagan and Sundermann (1996) further subdivide this into the fraction denoting work done by the tidal potential on the ocean, $(1 + k_2)/\alpha$, and the fraction denoting the tidal energy flux to the lithosphere at the ocean-lithosphere interface due to work done by hydrostatic pressure at the ocean bottom on the lithosphere, $-h_2/\alpha$ (also their expression ignores factor ε). The rate of work done on atmospheric gravitational tides can be similarly written as (Kagan and Sundermann, 1996; Platzman, 1991)

$$\langle \dot{W}_a \rangle = \rho_0 \iint\limits_{\text{atm}} \left[\left\langle \Gamma \frac{\partial}{\partial t} \left(\frac{p_b}{g\rho_0} - \mathfrak{M} \right) \right\rangle + g \left\langle \left(\frac{p_b}{g\rho_0} - \mathfrak{M} \right) \frac{\partial}{\partial t} (\eta + \eta_e) \right\rangle \right] dA$$

$$(6.10.13)$$

where $\gamma = 1$ over the ocean and 0 over land. The first term is due to work done by tidal potential and the second term is that due to hydrostatic pressure. Recent estimates by Kagan and Sundermann (1996; see also Platzman, 1991) indicate that the total rate of work done by the Moon on atmospheric gravitational tides and dissipated in the atmosphere is about 12 GW. This is of negligible consequence to Earth-Moon evolution since the total rate of work done on the ocean is 300 times more, 3.17 TW.

The work done on solid Earth tides (Kagan and Sundermann, 1996) is

$$\langle \dot{W}_E \rangle = \rho_E \iint \left[\left\langle \Phi \frac{\partial \eta_e}{\partial t} \right\rangle \right] dA$$

$$= \frac{2\pi}{9} N_{22} \rho_E R_E^2 (1 + k_2) h_2 \left(\frac{Do^2}{g} \right) (\Omega - n) \sin \delta \tag{6.10.14}$$

where $Do = 0.75 GM_E R_E^2 R^{-3} \sim 2.621 m^2\ s^2$ is the Doodson constant, $N_{nm} = 2(n+m)!\ [(2n+1)(n-m)!]^{-1}$ is the normalization factor for the spherical harmonic of degree n and order m, and δ is the phase-lag angle. The last term, $\sin \delta$, can be approximated by Q^{-1}, where Q, the quality factor, is

$$Q = \langle T \rangle \sigma / \langle D_S \rangle = 2\pi \langle T \rangle / \langle D_S \rangle t_p \tag{6.10.15}$$

where $t_p(\sigma)$ is the tidal period (frequency), T is the total energy, and $D_S(\sigma)$ is the dissipation rate (equal to the rate of work done). Estimates of Q for the solid Earth vary widely but fall within the range 100 to 1000, so that the tidal dissipation rate in the solid Earth cannot exceed about 200 GW (Kagan and Sundermann, 1996) and is therefore an order of magnitude less than that in oceanic tides. This is why ocean tides dominate in any discussion of tidal dissipation of the energy in the Earth-Moon system.

Quality factor Q provides an idea of how rapidly the energy in the system is dissipated. It is indicative of whether the system is heavily or lightly damped. If its value is less than 2π, it means that the system is heavily damped and the energy resident in the system will run down in less than a cycle. If its value is much larger than 2π, the system is lightly damped and its response to external forcing might involve resonant modes. Hendershott (1981) quotes a value of around 17 for M_2 oceanic tide and about 25 for worldwide semidiurnal tides. Based on theoretical considerations of a multimode resonant system, Kagan and Sundermann (1996) estimate a value of roughly 20 for semidiurnal and 10 for diurnal modes. Kantha's model (Kantha, 1998) indicates a global value for Q of roughly 23; thus, if the tidal forcing were to stop, the energy in M_2 tides would be dissipated in about 3.7 tidal cycles, or roughly 46 hr. For diurnal tides, Q is roughly 12, so the decay time is 1.8 cycles (44 hr). Clearly, the ocean responds vigorously to semidiurnal tidal forcing with excitation of resonant modes and amplification of the astronomical tidal forcing. In fact, the total energy in M_2 (392 PJ) is ~13 times the value it would be (30.5 PJ) if the oceanic response were tepid and close to equilibrium. On the other hand, long-period tides, the lunar fortnightly Mf and the lunar monthly Mm have Q values close to 2π and hence are heavily damped. The total energy resident in them is close to the equilibrium value. The ocean responds as a heavily damped system to barotropic forcing at these low frequencies (Kantha, 1998; Kantha *et al.*, 1998; Wunsch *et al.*, 1997).

Since the shallow oceans comprise a small percentage of the area of the global oceans, their contribution to the work done is small enough to be negligible. It is therefore possible to derive the power input, even from global tidal models that are inaccurate in coastal oceans and semienclosed seas, provided accurate tidal sea level is available in the open ocean. Assuming the work done is balanced entirely by the dissipation of oceanic tides, the corresponding values for tidal dissipation can be inferred. Kantha's global model (Kantha, 1995b) gives values of 2.57, 0.41, 0.12, 0.03, 0.38, 0.19, 0.04, and 0.01 TW for M_2, S_2, N_2, K_2, K_1, O_1, P_1, and Q_1 respectively, including the work done on the bottom (divide by 1.075 to get just the oceanic dissipation). Thus, nearly 3.75 TW of power is being put into the ocean by the lunisolar tidal forcing and dissipated! The expression in Eq. (6.10.12) has been used to compute these values from the model results. These values are in excellent agreement with those deduced by Schrama and Ray (1994) using tides derived from 1 year of T/P data and the values inferred from satellite tracking data. The M_2 dissipation is also in agreement with the values derived from LLR measurements (Dickey *et al.*, 1994).

While the shallow coastal parts of the oceans play a small role in the tidal power input into the global oceans, they play an overwhelmingly large part in dissipating the tidal energy deposited, since the tidal dissipation due to bottom friction is large. It is for computing the locations and values of tidal dissipation in the global oceans that global models that are accurate in shallow water are essential. Tidal energy has to be transported from regions where it is deposited into the global oceans, into shallow regions where it is dissipated. It is therefore useful to look at the distribution of power fluxes as well. It is here that accurate tidal currents are important, and it has been difficult to deduce currents accurately without dynamical models of tides, accurate enough to provide a good idea of tidal currents in shallow water. In addition to the total value of tidal dissipation, the geographic distribution of the major energy sinks has been of interest to tidalists since Taylor (1919). Jeffreys (1920) thought incorrectly that Bering Sea accounted for a majority of tidal dissipation. It has now been possible to map the major tidal energy sinks using a global tidal model valid in shallow waters and assimilating altimetric tides (Kantha *et al.*, 1995), and, in

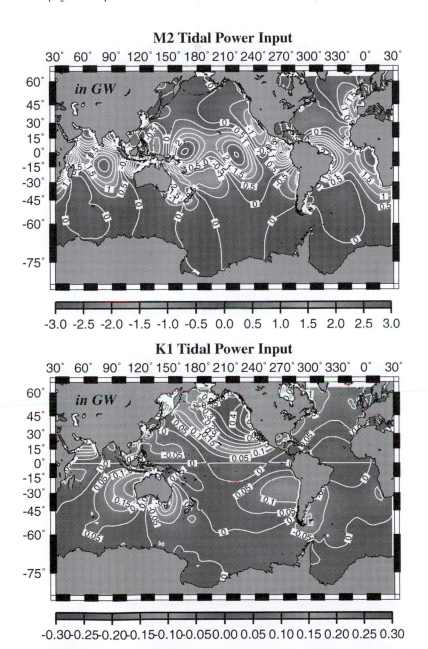

Figure 6.10.4 Power input distribution in each 2° square box for M_2 and K_1 tides. Note that the power input is predominantly in low latitudes.

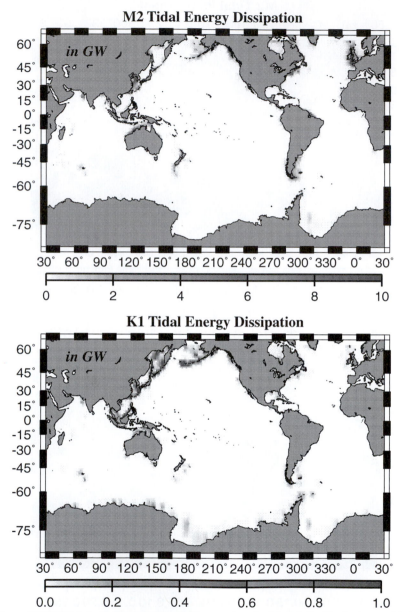

Figure 6.10.5 Tidal dissipation rate in each 2° square box for M_2 and K_1 tides by bottom friction. Note that the dissipation is predominantly in shallow marginal seas.

actuality, its contribution as an energy sink is small (~70 GW) compared to the European shelf (~300 GW), Patagonian shelf (~160 GW), Yellow Sea (~150 GW?) for the semidiurnal, and Sea of Okhotsk (~100 GW) for the diurnal tides. Nearly 95% of the 2.1 TW of frictional dissipation of M_2 tidal energy occurs in 5% of the global oceans, while 91% of the global oceans account for less than 2%. This is natural since in the deep oceans, with an average depth of 4 km, the

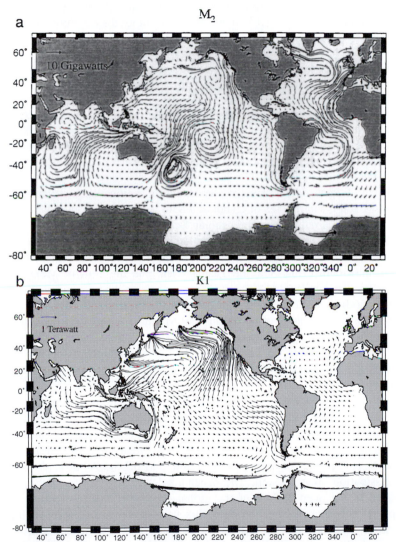

Figure 6.10.6 Tidal power fluxes in each 2° square box for M_2 and K_1 tides. Note the large fluxes from high power input regions into high dissipation regions.

current velocities are of the order of 2 cm s^{-1} and the bottom friction of the order of 2×10^{-5} W m^{-2}, whereas in shallow seas, the currents often reach 1 m s^{-1} and the bottom frictional dissipation values of 2.5 W m^{-2}. Kantha's barotropic model (1995), however, indicates only 2.1 TW (out of the 2.4 TW) of M$_2$ dissipation due to bottom friction in shallow coastal and marginal seas.

In addition to the total values of power input (and hence tidal dissipation) in global oceans, it is interesting to look also at the spatial pattern of the power input distribution. Figure 6.10.4 shows the power input distributions in 2 degree square boxes (and dissipation due to bottom friction) for the two primary tides, M$_2$ and K$_1$. The semidiurnal tidal energy is deposited principally in the equatorial belt between 30°S and 30°N. The patterns are roughly similar for M$_2$ and S$_2$. The two positive lobes in the South Pacific are equally strong for M$_2$, while the region of positive work is stronger in the western Pacific for S$_2$. Significant power is being input into both these tides in the northeast Atlantic off Europe. Similarly, negative work is being done on semidiurnal tides in the Arabian Sea, the Bay of Bengal, and south of the Indonesian Archipelago in the Indian Ocean and off the Mexican coast, between Australia and New Zealand, and off Philippines in the Pacific.

The K$_1$ diurnal tidal energy is deposited in regions such as around the northeast Pacific and the Sea of Okhotsk. There is also significant positive work being done on diurnal tides in the Arabian Sea, the Bering Sea, southeast Indian Ocean off the west coast of Australia, and the regions in the southeast Pacific off the west coast of South America around 30°S. There is negative work being done in the South China Sea, the Tasman Sea, and the Gulf of Mexico. The patterns are very similar for O$_1$.

These findings are consistent with those of Cartwright and Ray (1991), who also note the concentration of power input of M$_2$ and S$_2$ south of the equator (see also Zahel, 1980). They also point out that the work done outside the 66°S to 66°N band is negligible, so that the current model should provide reasonable estimates for these quantities in the global oceans, even though the Arctic is excluded.

Figure 6.10.5 shows the corresponding distributions of tidal energy dissipation by bottom friction in the global oceans for M$_2$ and K$_1$. As expected, the principal regions of dissipation are the shallow coastal oceans such as the Patagonian shelf, the Yellow Sea, and the Irish/Scottish shelf. However, regions around New Zealand, off northwest Australia, and off the Amazon show up as high dissipation areas as well. The oceanic tidal fluxes transmit the energies deposited in the deep regions of the global oceans to shallow regions where dissipation is taking place. The power flux vectors shown in Figure 6.10.6 for M$_2$ and K$_1$ illustrate this point quite well. Large flux convergences into high regions of dissipation in Figure 6.10.5 are noteworthy. For example, the convergence of K$_1$ tidal energy into the Bering Sea and the Sea of Okhotsk is consistent with the large dissipation taking place there. Similarly, there is a large convergence of M_2 energy onto the European shelf, for example. Convergence onto the Antarctic continent of both M_2 and K_1 power is noteworthy.

The mechanical energy resident in tides is itself of considerable interest. With tidal solutions from a data-assimilative tidal model (for example, Kantha, 1995b), it is possible to compute these quantities reliably (Table 6.10.2). Kantha's model (Kantha, 1998) indicates that about 390 PJ of M_2 tidal energy is resident in the global oceans (excluding the Arctic), with an overall ratio of potential to kinetic energy of about 0.62. The energy resident in S_2, K_1, and O_1 is 65, 56, and 37 PJ, respectively.

The 390 PJ of energy in M_2 barotropic tides translates to roughly 1153 J m^{-2} in terms of energy density on the average (distributed over an oceanic area of 3.4×10^{14} m^2, excluding the Arctic), although in regions of large tides, the value can exceed ten times this value. The total energy resident in all semidiurnal tides is 480 PJ (dissipation rate of 3.13 TW) and all tides is close to 580 PJ, with an energy density of 1717 J m^{-2}. The PE/KE ratio is around 0.63 for semidiurnal tides and 0.46 for diurnal tides. Unlike, for example, high frequency surface gravity waves, equipartition of energy does not prevail.

Table 6.10.2 summarizes tidal energetics as derived from the global models. Internal and long-period tides are also included.

6.11 CHANGES IN EARTH'S ROTATION

Both the rate of rotation and the position of the rotation axis of the Earth undergo changes with time. The Earth's rotation axis is inclined at an angle of 23.5° to the celestial pole and rotates about it once every 26,000 years. This precession is caused by the torque exerted by the Sun and the Moon on the equatorial bulge of the nonradially symmetric Earth. Superimposed on this slow precession are short-period oscillations of the Earth's rotation axis relative to inertial space (celestial reference coordinate system) called its nutational motions, the largest of which has an amplitude of 9" and a period of 18.7 years. The motion of the rotation axis when referred to an Earth-fixed reference frame (terrestrial coordinate system) is called the polar motion or wobble. Polar motion and nutation are different descriptions of the same process. However, geodetic measurements, which observe motion of the Earth's surface relative to celestial reference sources, cannot distinguish between the two. Traditionally, nutations are taken to be motions with diurnal periods, while wobble is taken as mostly due to excitation with timescales much longer than a day. However, retrograde (clockwise motions as viewed from above the North Pole opposite to Earth's rotation direction; prograde motions are counterclockwise) diurnal polar motion

TABLE 6.10.2
Tidal Energetics

Barotropic tides

Comp	Power* (GW)	KE (PJ = 10^15 J)	TE	PE/KE	Q	Period (hr/dy)	Power (Relative to M₂)	TE (Relative to M₂)	Eq PE (PJ)	TE Den+ (J m⁻²)
M_2	2572.6	242.0	392.0	0.62	23.03	12.42	1.0	1.0	30.49	1153
S_2	410.3	38.0	65.0	0.71	24.81	12.00	0.159	0.165	6.60	191
N_2	117.8	11.6	18.8	0.62	23.78	12.66	0.046	0.048	1.12	55
K_2	30.1	3.1	5.4	0.74	28.12	11.97	0.012	0.014	0.49	16
K_1	378.7	36.0	55.5	0.54	11.57	23.93	0.147	0.220	9.71	163
O_1	193.5	27.0	37.4	0.39	14.12	25.82	0.075	0.146	4.91	110
P_1	40.6	5.6	7.7	0.38	14.69	24.07	0.016	0.024	1.06	23
Q_1	8.2	1.1	1.5	0.43	12.82	26.87	0.003	0.008	0.18	4
Mf	0.369	0.240	0.381	0.59	5.90	(13.66)	$O(10^{-4})$	$O(10^{-3})$	0.245	1.13
Mm	0.022	0.013	0.049	2.91	6.20	(27.55)	$O(10^{-6})$	$O(10^{-4})$	0.068	0.14
All 4 SD	3130.8	294.7	481.2	0.633	23.22	(Based on M_2 freq)	1.217	1.227	38.70	1415
All 4 D	621.0	69.7	102.1	0.46	12.89	(Based on K_1 freq)	0.241	0.397	15.86	300
Lunar	3170.4	308.7	491.7	0.593	$M_2, N_2, O_1, Q_1, Mf, Mm + 0.68(K_2, K_1)$		1.232	1.372	43.95	1446
Solar	581.7	56.1	92.2	0.643	$S_2, P_1 + 0.32(K_2, K_1)$		0.226	0.264	10.92	271
All	3752.1	364.8	583.9	0.600			1.458	1.637	1.861	1717

Baroclinic tides:

Comp	Diss. (GW)	TE (PJ)	TE Den[x] $J\,m^{-2}$	Q	Alt. TE (PJ)	Alt. TE Den[x] $J\,m^{-2}$
M_2	360	50	167	19.5	50	167
S_2	57	8	27	20.4	?	?
K_1	42	15	50	26.0	?	?
All 4 S	440	61	203	19.5	?	?
All 4 D	81	29	97	26.1	?	?
All	521	90	300	?	?	?

° Excluding the Arctic Ocean.

* Includes solid Earth and load tides; divide by 1.075 to get values for ocean tides only.

+ Energy expected if tides were in equilibrium.

+ Based on global ocean area of 3.4×10^{14} m² (excluding the Arctic).

x Based on ocean area of 3×10^{14} m² (equatorward of 60° latitude).

excited by ocean tides is indistinguishable from nutations defined in the traditional sense. In addition, there exists a nearly diurnal (23.88-hr period) retrograde polar motion due to excitation of resonant motions of the Earth's core relative to the mantle, called free core nutation. The most conspicuous long-period wobble of great importance to geophysics is the 14.4 month Chandler wobble. In addition, the redistribution of masses due to tidal motions also gives rise to polar motion or wobble at tidal periods. The Earth's nutational (and equivalently polar) motions are of great interest to geophysicists, since they contain important information about its deviations from rigidity (Chao and Ray, 1997; Desai, 1996; Gross et al., 1997).

Changes in the rotation rate are reflected in the LOD and universal time fluctuations (with respect to the atomic clock time), while the misalignment of rotation axis with axis of symmetry causes wobbling motions. In the absence of external torques, the angular momentum of the Earth system, comprising the atmosphere, oceans, the mantle, and crust and the core, is a conserved three-dimensional quantity (Figure 6.11.1). If we ignore the lunar torque that reduces the Earth's spin momentum over long timescales, this means that changes in the angular momentum of the atmosphere and the oceans (and the Earth's liquid core) must be reflected in corresponding changes in the rotation vector of the mantle. Changes in the axial component of the vector are reflected in the

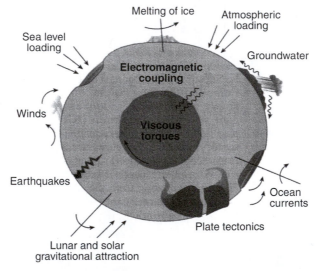

Figure 6.11.1 Sketch showing the forces (indicated schematically by arrows) acting on the mantle that can change its rotation. On seasonal and subseasonal scales, it is the angular momentum exchange between the mantle and the fluid envelopes, the oceans and the atmosphere, and the hydrological redistributions of water and ice that affect both the Earth's rotation rate and its polar motions. Core-mantle coupling and plate tectonics operate on much longer timescales. Adapted from Lambeck (1980).

rotation rate and those in the two equatorial components in the motion of the pole relative to the crust, the wobble. The Earth's variable rotation can be measured by several different techniques: (1) very long baseline interferometry (VLBI), which uses a global network of radio telescopes that simultaneously track extra-galactic radio sources; (2) satellite laser ranging (SLR), which uses laser pulses from a network of stations reflected from a satellite with a highly stable orbit such as laser geodynamic satellite (LAGEOS); (3) lunar laser ranging (LLR), which uses arrays of reflectors on the Moon; and (4) global positioning system (GPS), which is similar to VLBI but uses the GPS constellation of satellites. These observations are collected and distributed by the International Earth Rotation Service (IERS), while the Climate Analysis Center of the NCEP provides the atmospheric contribution as determined from the products from major meteorological centers around the world (Dehant *et al.,* 1997; Salstein *et al.,* 1993). Several excellent reviews of the Earth's rotation exist (Chao, 1994; Eubanks, 1993; Herring, 1991; Hide and Dickey, 1991; Lambeck, 1980; Munk and MacDonald, 1960; and Wahr, 1988).

VLBI uses radio signals from distant radio sources such as quasars received simultaneously by multiple radio telescopes separated by long baselines. The difference in arrival times of radio wave fronts can be used to determine the three-dimensional orientation of the Earth and hence its three-dimensional rotation vector as a function of time. Changes in rotation about the polar axis expressed as changes in universal time UT1 (which is the polar motion-corrected universal time, UT, as determined by overhead passage of stars and whose derivative gives LOD), and the changes in the orientation of the rotation axis can be measured with sufficient precision to resolve many geophysical processes such as ocean and solid Earth tides, atmospheric and oceanic circulation changes, and ENSO processes. The Continuous Observation of the Rotation of the Earth (CORE) program aims to measure UT1 to a precision of 3 to 5 μs hourly (1 to 2 μs daily), and polar motion to 80 to 130 micro arc-second hourly (30 to 50 μs daily) by 2001 (Clark *et al.,* 1998). This level of precision will enable more accurate inferences to be made about geodynamical processes and geophysical parameters such as the anelasticity of the Earth's mantle. Intensive VLBI campaigns such as Cont94, Cont95, and Cont96 have made accurate hourly measurements of UT1 and polar motion that are well suited to inferring the effects of oceanic tides and other high frequency processes on Earth's rotation. Figure 6.11.2 from Clark *et al.* (1998), who describe the intensive VLBI campaigns and the CORE program, shows the magnitude and temporal scales of the geophysical processes that affect the Earth's rotation shown in Figure 6.11.1.

Consider now the effect of tides on the Earth's rotation rate or, equivalently, the length of day, LOD (Agnew and Farrell, 1978; Chao and Ray, 1997; Desai, 1996; Dickman and Nam, 1995). The effect of tides is twofold. First, the fluctuations in the sea level due to tides bring about changes in the moment of inertia tensor of the Earth. The second effect is the motion effect due to tidal currents, of roughly similar magnitude. Considering the core, mantle, oceans, and the atmosphere together, conservation of angular momentum of the total system implies that changes in tidal currents must effect corresponding changes in the rotation rate of the mantle. Changes in inertia tensor also cause changes in the Earth's rotation. These changes, referred to as UT1 or LOD changes, can now be measured to a precision of a few microseconds (Dickey, 1994, and Dickman and Nam, 1995, for example). The major signal in LOD or UT1 is that due to the ever-changing winds in the atmosphere (Dehant *et al.*, 1997) on a variety of timescales, from intraseasonal 40-to-60 day oscillations to ENSO-related wind anomalies and quasi-biennial stratospheric changes, with as much as a 2 ms change in LOD between the boreal summer and winter seasons (see Salstein *et al.*, 1993, for plots showing excellent correlation on certain timescales between geodetic measurements of LOD changes and the atmospheric contribution to them). However, long-term changes due to seasonal adjustment of oceanic masses, ENSO events, and long-period ocean tides, and short-term changes due to semidiurnal and diurnal tides are also discernible. Accurate elimination of tidal effects enables inferences to be made about Earth's anelasticity at tidal frequencies (for example, Desai, 1996). There are also contributions to angular momentum changes due to changes in continental storage of water, ice, and snow and transfer of momentum between the faster-rotating inner core and the mantle through viscous, electromagnetic, and other coupling mechanisms. See Wahr (1986, 1988), Eubanks (1993), Desai (1996), Dickman and Nam (1995), and Chao and Ray (1997) for details.

The equilibrium state of the Earth has approximate rotational symmetry and hence to a zeroth order, only two moments of inertia, equatorial (A) and polar (C), are adequate to describe the state of steady rotation with an angular velocity of Ω rad s^{-1} (1 cycle per sidereal day) and an angular momentum of $C\Omega$. The inertia tensor is diagonal (in reality, the off-diagonal terms are small but nonzero and the two equatorial moments of inertia also differ by a small amount, but these are about 1% of the value of C-A and hence readily ignored). If tidal deformations cause changes in the inertia tensor, the new values of the inertia tensor and the rotation vector are

$$\omega(t) = \Omega\left[1 + m_1(t), m_2(t), m_3(t)\right]$$

$$c(t) = \begin{bmatrix} A + c_{11}(t) & c_{12}(t) & c_{13}(t) \\ c_{21}(t) & A + c_{22}(t) & c_{23}(t) \\ c_{31}(t) & c_{32}(t) & C + c_{33}(t) \end{bmatrix} \tag{6.11.1}$$

The time-dependent values are small perturbations. Quantities m_1 and m_2 define the angular offsets in x_1 and x_2 directions, respectively, of the rotation vector

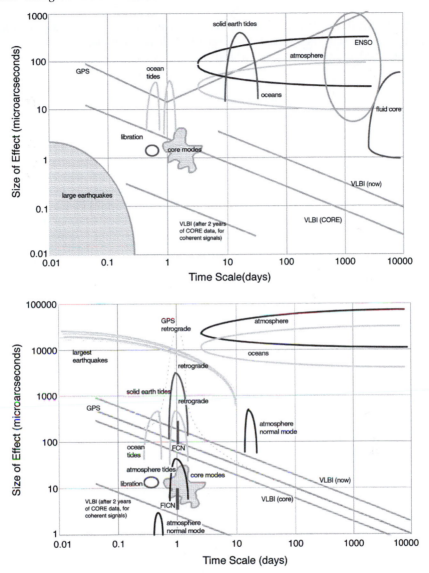

Figure 6.11.2 Sketches showing the magnitudes and timescales of various processes contributing to the changes in Earth's rotation, both its rotation rate (top) and polar motion (bottom). The accuracy levels of current VLBI observations, as well as those expected in the future due to the CORE program, are also indicated. From Clark *et al.* (1998).

from its equilibrium position; m_3 denotes changes in the rotation rate seen as changes in the universal time UT1 or in LOD. It is traditional to combine the first two by defining complex quantities:

$$m(t) = [m_1(t) + im_2(t)]$$

$$c(t) = [c_{13}(t) + ic_{23}(t)]$$

(6.11.2)

If we define a reference rotation pole, then nutations are motions of this reference pole with respect to the inertial space, and polar motions are the motions of the nonaxial rotation axis of the Earth with respect to the reference pole. Thus nutation parameters give the location of the reference pole in inertial frame, and the polar motion parameters give the location of the reference pole in the terrestrial frame (Eubanks, 1993). Now, if we adopt the polar motion gauge in which the reference rotation pole is fixed in space, m(t) and m_3 are governed by the linearized equations of conservation of angular momentum that can be written as (Desai, 1996; Eubanks, 1993; Wahr, 1986, 1988)

$$\left(m + \frac{i}{\Omega_C} \frac{d}{dt} m \right) = \left(\chi - \frac{i}{\Omega} \frac{d}{dt} \chi \right); \quad \chi = \chi_1 + i\chi_2$$

$$\frac{d}{dt} m_3 = -\frac{d}{dt} \chi_3$$

(6.11.3)

where χ denotes nondimentional excitation functions to be defined shortly. Ω_C is the complex Chandler wobble frequency given by

$$\Omega_C = \sigma_C \left(1 + \frac{i}{2Q} \right)$$

$$\sigma_C = \Omega \frac{C - A}{A} \frac{k - k_2}{k} \left(1 - \frac{k_2}{k} \frac{C - A}{A} \right) = \sigma_E \frac{k - k_2}{k} \left(1 - \frac{k_2}{k} \frac{\sigma_E}{\Omega} \right)$$

(6.11.4)

$$\text{where} \quad \sigma_E = \Omega(C - A)/A; \quad k = \frac{3(C - A)G}{\Omega^2 R_E^5}$$

R_E is the Earth's radius and σ_E is the Euler frequency corresponding to the normal mode of a rigid Earth (period of 306 days); σ_C is the frequency of the true normal mode given by the frequency of the Chandler wobble (period of 434.45 sidereal days) that accounts for the yielding of the mantle due to load-induced and rotational deformations; k is a secular Love number (~0.93608); and k_2 (effective value of 0.34784) accounts for rotational tidal deformations, corrected for the Chandler wobble pole tide. Q is the quality factor (~50 to 200) that accounts for the anelasticity of the Earth at Chandler frequency. If the excitation were to remain constant, Chandler wobble would decay with a characteristic timescale of $2Q/\sigma_C$, about 10 to 20 years. Anelasticity effects (nonzero Q) can therefore be neglected for high frequency fluctuations in the Earth's rotation induced by processes such as short-period tides.

By convention, the mean position of the rotation axis over a certain interval (1900 to 1905) is taken as the x_3 axis. x_1 axis is taken perpendicular to this along the Greenwich meridian, and x_2 axis is along 90°E. If the polar motion is defined as $p = X_p + i(-Y_p)$, where X_p and Y_p are positions on the Earth's surface of the rotation axis with respect to the reference pole, then (Eubanks ,1993),

$$m = \left(1 - \frac{i}{\Omega}\frac{d}{dt}\right)p \qquad (6.11.5)$$

and therefore,

$$p + \frac{i}{\Omega_C}\frac{dp}{dt} = \chi \qquad (6.11.6)$$

Thus, knowing the complex-valued excitation function (or chi-function) χ enables polar motion to be determined. For polar motions with timescales much shorter than the Chandler wobble decay timescale, Ω_C can be replaced by σ_C. By convention, positive χ_2 is along 90°E, while x_{2p} is taken as negative along along 90°E, explaining the minus sign in the expression for p (negative sign creates a right-handed coordinate system). Eq. (6.11.6) holds at frequencies far from the free core resonance frequency (nearly diurnal retrograde frequency). The observed polar motion at nearly diurnal and nearly semidiurnal prograde and nearly semidiurnal retrograde frequencies appear to be dominated by short-period ocean tides, while long-period tides (especially Mf) appear to contribute significantly to polar motion at fortnightly and monthly periods (Gross *et al.,* 1997).

The parameters m_3 and χ_3 are related to LOD change and UT1 through

$$m_3 = -\frac{\Delta(\text{LOD})}{\text{LOD}} = \frac{d}{dt}(\text{UT1} - \text{TAI})$$

(6.11.7)

$$\Delta\text{UT1} = \text{UT1} - \text{TAI} = \int_0^t \chi_3(\tau)\, d\tau$$

where UT1 is the universal time and TAI is the reference atomic clock time. LOD is taken as 1 solar day. Observations are, however, reported as the angle p_3 through which the Earth has rotated during (UT1-TAI) (Eubanks, 1993)

$$p_3(t) = -\Omega \int_0^t \frac{\Delta(\text{LOD})}{\text{LOD}}\, dt = \Omega(\text{UT1} - \text{TAI})$$

(6.11.8)

The rotational variations are a function of changes in the moment of inertia due to mass redistributions in the atmosphere and the oceans, the so-called mass effect. They are also affected by momentum exchanges between the Earth's mantle with these fluid media (neglecting exchanges with the core that are thought to be responsible primarily for decadal fluctuations and beyond). The latter, called the motion effect, can be computed from the tangential stress exerted on the crust by the winds and ocean currents and the differential pressure across mountains and midocean ridges. However, it is more reliable instead to compute changes in the angular momentum of the atmospheric and the oceanic motions. Thus the dimensionless excitation functions (written originally for the atmosphere by Barnes *et al.,* 1983) for barotropic motions in a homogeneous ocean can be written as (Eubanks, 1993)

$$\left[\chi_1^{ms}, \chi_2^{ms}\right] = -\frac{1.098\rho_a R_E^4}{(C-A)} \int_0^{2\pi} \int_{-\pi/2}^{\pi/2} \eta \, \sin\phi \, \cos^2\phi[\cos\lambda, \sin\lambda] d\phi \, d\lambda$$

$$\left[\chi_1^{mo}, \chi_2^{mo}\right] = -\frac{1.5913\rho_a R_E^3}{(C-A)\Omega} \int_0^{2\pi} \int_{-\pi/2}^{\pi/2} (H+\eta) \, \cos\phi$$

$$\times \begin{bmatrix} (U_\lambda \sin\phi\cos\lambda - U_\phi \sin\lambda), \\ (U_\lambda \sin\phi\sin\lambda + U_\phi \cos\lambda) \end{bmatrix} d\phi \, d\lambda$$

$$\chi_3^{ms} = \frac{0.753\rho_a R_E^4}{C_m} \int\limits_0^{2\pi} \int\limits_{-\pi/2}^{\pi/2} \eta \; \cos^3\phi d\phi \; d\lambda$$

$$\chi_3^{mo} = \frac{0.998\rho_a R_E^3}{\Omega C_m} \int\limits_0^{2\pi} \int\limits_{-\pi/2}^{\pi/2} (H+\eta) \; U_\lambda \; \cos^2\phi d\phi \; d\lambda$$

$$(6.11.9)$$

where ρ_a is the average density of the oceans (1035 kg m^{-3}), η is the sea surface deviation from rest position, and H is the ocean depth. Note that C_m is used in calculating χ_3, since the core is considered to be decoupled at these high frequencies.

For the atmosphere, a similar set applies:

$$\left[\chi_1^{ms}, \chi_2^{ms}\right] = -\frac{1.098 R_E^4}{(C-A)g} \int\limits_0^{2\pi} \int\limits_{-\pi/2}^{\pi/2} p_s \; \sin\phi \; \cos^2\phi [\cos\lambda, \sin\lambda] d\phi \; d\lambda$$

$$\left[\chi_1^{mo}, \chi_2^{mo}\right] = -\frac{1.5913\rho_a R_E^3}{(C-A)\Omega g} \int\limits_0^{2\pi} \int\limits_{-\pi/2}^{\pi/2} \int\limits_{p_s}^{p_t} \cos\phi \left[\begin{array}{c} (U_\lambda \sin\phi\cos\lambda - U_\phi \sin\lambda), \\ (U_\lambda \sin\phi\sin\lambda + U_\phi \cos\lambda) \end{array} \right] dp \; d\phi \; d\lambda$$

$$\chi_3^{ms} = \frac{0.753 R_E^4}{C_m g} \int\limits_0^{2\pi} \int\limits_{-\pi/2}^{\pi/2} p_s \; \cos^3\phi \; d\phi \; d\lambda$$

$$\chi_3^{mo} = \frac{0.998 R_E^3}{\Omega C_m g} \int\limits_0^{2\pi} \int\limits_{-\pi/2}^{\pi/2} \int\limits_{p_s}^{p_t} U_\lambda \; \cos^2\phi \; dp \; d\phi \; d\lambda$$

$$(6.11.10)$$

where p_s is the surface pressure and p_t is the pressure at the top of the atmosphere (while the density is small in the stratosphere, the large velocities there make it essential to include the stratosphere in angular momentum computations). If one assumes that the oceans behave in an inverse barometric fashion to pressure forcing, then changes in the weight of the atmospheric column due to pressure changes are instantaneously compensated by changes in the height of the ocean column so that the pressure on the ocean bottom remains unchanged. Thus the mass excitation functions need be integrated only over land, except that compensation must be made for shifts of atmospheric mass from over land masses to over the oceans and vice versa, by allowing for changes that occur in average surface pressure over the ocean, which does not effect any changes in sea level and therefore no changes in the inertia tensor occurs. This is the so-called IB assumption. The other extreme is to assume that the ocean surface acts as a rigid lid and integrate over both land and oceans. This is unrealistic. In reality, the situation is in between, but closer to the IB approximation, since the oceans respond like an inverted barometer to pressure forcing except at periods shorter than about a week.

For tidal motions, the excitation functions $\chi_{1,2,3}$ are periodic and of the form $\overline{\chi}_{1,2,3} \exp\{i(\sigma t - \varphi_{1,2,3})\}$ and therefore the polar motions (in radians) and UT1 fluctuations (in seconds) can be written in the form

$$(X_p, Y_p) = (\overline{X}_p, \overline{Y}_p) \exp\left\{i\left[\sigma t + (\varphi_x, \varphi_y)\right]\right\} = \frac{\sigma_C}{\sigma - \sigma_C} (-\overline{\chi}_1, \overline{\chi}_2) \exp\left\{i\left[\sigma t - (\varphi_1, \varphi_2)\right]\right\}$$

$$\Delta UT1 = \frac{\overline{\chi}_3}{\sigma} \exp\left\{i\left(\sigma t - \varphi_3 - \frac{\pi}{2}\right)\right\}$$

$$(6.11.11)$$

Note that by convention, the phase of polar motions is defined as a Greenwich phase lead, hence the positive sign associated with φ_x and φ_y.

The effect of mantle anelasticity (finite Q) is negligible at these frequencies and Q^{-1} has been put to zero in deriving Eq. (6.11.11). However, the effect of nearly diurnal retrograde free core nutation (FCN) is not negligible for tidally induced retrograde polar motions. Equations (6.11.6) and (6.11.11) do not account for FCN effects on polar motion. Correction factors that account for FCN (Desai, 1996; Gross, 1993; Sasao and Wahr, 1981) are

$$1 + 0.095\left(\frac{\sigma_C - \sigma}{\sigma_{FCN} - \sigma}\right) \text{ and } 1 + 0.00055\left(\frac{\sigma_C - \sigma}{\sigma_{FCN} - \sigma}\right) \quad (6.11.12)$$

for mass and motion, respectively, where $\sigma_{FCN} = [1 + (1/433.2)]\Omega_o$, corresponding to a period of 23.88 hr; Ω_o corresponds to the sidereal day.

Since a periodic vector quantity with its sinusoidal components of different amplitudes and phases can be looked upon as superposition of two counterrotating vectors (the sum of which describes an ellipse), the polar motion is usually decomposed into prograde and retrograde motions:

$$p = PM_{pro} e^{i\sigma t} e^{i\varphi_{pro}} + PM_{ret} e^{-i\sigma t} e^{i\varphi_{ret}}$$

$$PM_{pro} = 0.5\left[\left(X_p^{in} - Y_p^{qd}\right)^2 + \left(X_p^{qd} + Y_p^{in}\right)^2\right]^{1/2}; \varphi_{pro} = \tan^{-1}\left(-\frac{X_p^{qd} + Y_p^{in}}{X_p^{in} - Y_p^{qd}}\right) \quad (6.11.13)$$

$$PM_{ret} = 0.5\left[\left(X_p^{in} + Y_p^{qd}\right)^2 + \left(X_p^{qd} - Y_p^{in}\right)^2\right]^{1/2}; \varphi_{ret} = \tan^{-1}\left(\frac{X_p^{qd} - Y_p^{in}}{X_p^{in} + Y_p^{qd}}\right)$$

where

$$X_p^{in} = \overline{X}_p \cos\varphi_x = PM_{pro} \cos\varphi_{pro} + PM_{ret} \cos\varphi_{ret}$$

$$X_p^{qd} = \overline{X}_p \sin\varphi_y = -PM_{pro} \sin\varphi_{pro} + PM_{ret} \sin\varphi_{ret}$$

$$Y_p^{in} = \overline{Y}_p \cos\varphi_y = -PM_{pro} \sin\varphi_{pro} - PM_{ret} \sin\varphi_{ret}$$

$$Y_p^{qd} = \overline{Y}_p \sin\varphi_y = -PM_{pro} \cos\varphi_{pro} + PM_{ret} \cos\varphi_{ret}$$

The major and minor axes of the ellipse described by the tidal polar motion are $(PM_{pro}+PM_{ret})$ and $(PM_{pro}-PM_{ret})$.

Fluctuations in Earth's rotation due to oceanic tides can be deduced from both VLBI observations and global tide models. Desai (1996) and Chao and Ray (1997) tabulate the observed (Dickman, 1993; Gipson, 1996; Herring and Dong, 1994; Sovers, *et al.*, 1993) and model-derived amplitudes and phases of the prograde and retrograde polar motions for the primary semidiurnal and diurnal tides, and Desai (1996) does the same for long-period lunar tides (Gross *et al.*, 1997, tabulate the chi-functions for long-period lunar tides). Table 6.11.1 shows VLBI values from Gipson (1996; see also Gross *et al.*, 1997; Chao and Ray, 1997) and model-derived values from the tidal models of Kantha (1995b) and Kantha *et al.* (1998). Since Kantha's model does not include the Arctic, values deduced from Schwiderski's model for the Arctic have been included. The M_2 value includes the nearly 2 µs libration (Chao and Ray, 1997; Ray *et al.*, 1997). The agreement between the model and observations is reasonable, especially in view of the fact that the error in hourly VLBI observations of $\Delta UT1$ exceeds 5 µs and that of polar motions exceeds 100 µs. Since Earth's rotation is a stringent global check on the accuracy of a tidal model, this suggests that data-assimilative global tidal models have become quite skillful.

For M_2 and S_2, the amplitudes are 18.6 and 8 µs; for K_1 and O_1, the values are roughly 19 and 22 µs, with most contribution coming from currents for both diurnal and semidiurnal tides. The long-period tides Mf and Mm have amplitudes of ~115 µs each, with a small but important contribution from currents. The closeness of the long-period tides to equilibrium tide values is indicated by the fact that the corresponding equilibrium values are 108 and 115 µs for Mf and Mm (equilibrium phase is $270°$). Figure 6.11.3 shows Earth's rotation observations made during the intensive VLBI campaign Cont94, filtered to remove low frequency motions, compared to the contributions of the eight principal short-period tides from Kantha's tidal model. The agreement is remarkably good, suggesting that at these frequencies it is the tidal motions that dominate Earth's rotation variations.

For further reading on the fascinating subject of Earth's rotation variations and its implications, see the reviews by Wahr (1986, 1988) and Eubanks (1993).

TABLE 6.11.1
Earth Rotation Values from VLBI Observations

Const.	ΔUT1		Prograde polar motion		Retrograde polar motion	
	Amp(μs)	Phs(deg)	Amp(μas)	Phs(deg)	Amp(μas)	Phs(deg)
M_2	18.6 (18.9)	236 (234)	62 (99)	110 (133)	257 (302)	272 (259)
S_2	8.0 (8.1)	264 (265)	14 (37)	89 (103)	128 (148)	303 (291)
N_2	3.7 (5.1)	239 (246)	16 (15)	108 (145)	46 (43)	269 (260)
K_2	2.9 (2.2)	283 (276)	15 (9)	104 (93)	18 (40)	346 (295)
K_1	18.6 (19.1)	29 (34)	166 (164)	63 (64)	(11053)	(128)
O_1	22.2 (22.1)	37 (47)	148 (137)	74 (79)	(135)	(276)
P_1	5.8 (6.2)	25 (32)	51 (53)	60 (68)	(881)	(134)
Q_1	5.6 (5.7)	31 (40)	33 (28)	81 (86)	(44)	(307)
Mf	111 (115)	281 (295)	41 (75)	320 (280)	53 (46)	307 (347)
Mm	126 (115)	275 (280)	52 (49)	286 (181)	79 (52)	36 (99)
Ssa*	(813)	(270)	(131)	(184)	(131)	(4)
Sa*	(306)	(270)	(179)	(184)	(179)	(4)

* Ssa and Sa values are for equilibrium gravitational tides.
From Gipson (1996) and Gross *et al.* (1997), and from tidal models (in parentheses)
of Kantha (1995b) and Kantha *et al.* (1998).

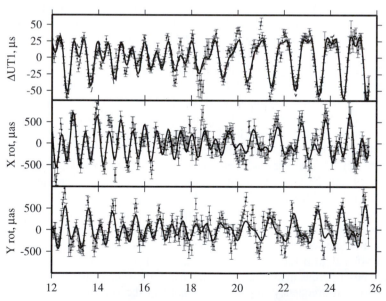

Figure 6.11.3 Comparison between hourly VLBI measurements of ΔUT1 (in ms), and polar motion indicated by X and Y (in mas) during the intensive campaign Cont94 with those deduced from the global tidal model results of Kantha (1995). The low frequency variability and the atmospheric contribution have been removed from VLBI observations. The retrograde diurnal contribution has also been removed from polar motions. The thick line indicates the model and the thin line the observations. The x-axis denotes days in January since the start of 1994.

6.12 BAROCLINIC (INTERNAL) TIDES

Internal tides, which involve vertical oscillations of isotherms in the oceans, are ubiquitous. In fact, they have long been a nuisance to oceanographers, since they "contaminate" their measurements. However, they are all but invisible to the naked eye, since even though the amplitude of isotherm excursions can reach several tens of meters, the corresponding sea surface height fluctuations reach only a few centimeters. This is the primary reason internal tides did not receive any attention until 200 years ago (Wunsch, 1975), while barotropic tides must have been noticed since prehistoric times (Cartwright, 1998). Until recently, their detection required *in situ* instrumentation such as thermistors or current meters. However, with the advent of precision altimetry, they appear to be identifiable in altimetric records (Kantha and Tierney, 1997; Ray and Mitchum, 1996). Kantha and Tierney (1997) estimate that global baroclinic tidal energy may be as much as ~90 PJ; 15% of the barotropic energy and the dissipation of tidal energy through the generation and dissipation of baroclinic tides is also around 15% of the power input into barotropic tides, ~600 GW. Internal tides by definition involve vertical shear and hence are important to mixing in the oceans. Since baroclinic tides are an important source of deep ocean internal waves (surface winds and waves are the others) and mixing, they assume a very important role in deep mixing and ocean dynamics. In fact, indications are that internal tides may be a dominant source of deep ocean mixing and a significant component in thermocline maintenance (Munk and Wunsch, 1998). The following comes from Kantha and Tierney (1997).

The astronomical tide generating forces are body forces that act uniformly throughout the water column. It is therefore difficult to see how they would generate internal tides directly. However, barotropic tidal currents flowing over topographic changes in the ocean have the potential to vertically displace the isotherms and generate internal tides. The consequent transfer of even a small fraction of the energy resident in barotropic tides to baroclinic tides is important to the oceans. It is now thought that midocean ridges and seamounts may constitute prime generation sites (Morozov, 1995). Were the oceans flat-bottomed, no internal tides could occur. Wunsch (1975) in his seminal review points out that internal tides are closer in many ways to atmospheric (internal) tides than the barotropic tides in the oceans; their dynamics are very much similar to inertial-internal gravity waves in the atmosphere and the oceans.

Internal tides arising from the density structure in the water column also play an important role in determining the structure of the tidal currents. It is estimated that as much as 15% of the power input into global barotropic tides is transferred to internal baroclinic tides along midocean ridges, seamounts, and the edges of continental shelves, which is then either dissipated in shallow water or scattered into other modes and dissipated eventually in the deep ocean. While the energy in internal tides is usually a fraction of that in barotropic tides, the currents are of similar magnitude. See Wunsch (1975), Hendershott (1981), Pugh (1987), Wunsch *et al.*, (1997), and Kantha and Tierney (1997) for a description of internal tides.

Internal waves can be generated in a stratified fluid when fluid parcels are forced to undergo vertical deflections under gravity by their passage over topographic changes. Flow over mountains provides a spectacular example of such internal waves, and has long been studied for its importance in the vertical transfer of momentum in the atmosphere. A similar mechanism is also operative in the oceans. Oscillating tidal currents due to barotropic tides can interact with ocean bottom topography such as ocean ridges, narrow straits, and shelf edges to generate internal tides and internal solitons. The resultant transfer of energy from long barotropic tidal scales to much shorter internal motions, which constitutes a cascade in wavenumber space of energy to higher wavenumbers, is important to the mixing and energetics of the interior of the oceans. Since the frequency is known, it is possible to estimate the wavelength of the resulting internal motions. Since internal tides are dominated by lowest vertical modes (Wunsch, 1975), assuming a first mode baroclinic wave speed of 2.5 m s^{-1}, this indicates a wavelength at generation of an internal tide of approximately 110 km for the semidiurnal and 210 to 230 km for the diurnal tides. Higher modes are even shorter.

The above values are consistent with the correlation scales of 100 to 200 km observed for midocean internal tides; while these are shorter than the corresponding values for barotropic tides, they are still large compared to the ocean depth and internal tides can therefore be regarded as long waves. A second aspect of the disparity in length scales between the external and internal tides is that the internal tides are not necessarily phase-locked to the barotropic tides, although they owe their existence to them. Also, waves of such small scale can be expected to be affected significantly by background currents, and in general baroclinic tides are distorted, frequency-shifted, and more rapidly dissipated as they propagate away from their source. Thus, one can expect significant internal tide energies only in the immediate vicinity of topographic features, although there is evidence that the first mode internal tide can propagate many hundreds (Hendry, 1977) to one or two thousands of kilometers (Schott, 1977) before its energy is either dissipated or transferred by nonlinear interactions to other modes. Unlike barotropic tides, which have line spectra, internal tides therefore tend to have a narrow but finite bandwidth in spectral space. This is not only because of the efficient transfer in spectral space made possible by resonant nonlinear triad interactions, but also because of the influence of density and current fluctuations in the background medium.

While surface tides account for a large fraction (90%) of the differential travel time, internal tides are also readily observed in reciprocal acoustic tomography since they affect substantially the travel time of acoustic energy in the ocean. Dushaw *et al* . (1995) describe and quantify internal tides observed in

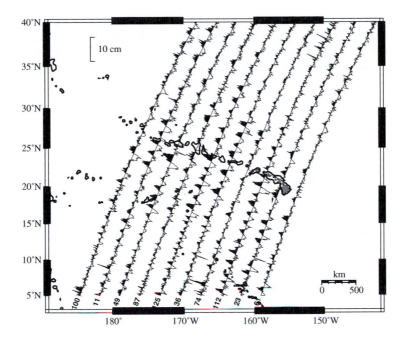

Figure 6.12.1 High pass filtered along-track M2 signals across Hawaiian ridge showing SSH manifestations of internal tides. From Ray and Mitchum (1996).

their 1987 reciprocal tomography experiment in a relatively quiet region of the North Pacific. They were able to detect large scale, phase-locked M_2 internal tides generated by the Hawaiian Ridge, 2000 km north of the ridge. Chiswell (1994) reported *in situ* observations of internal tides with amplitudes as high as 12 m at the Hawaiian Ocean Time Series site. Ray and Mitchum (1996) report that these phase-locked baroclinic tides are also seen in precision altimetry from TOPEX, even though their surface manifestation is a modulation of the SSH signal by a few centimeters at a wavelength of roughly 120 km (Figure 6.12.1). Along-track altimetric tidal solutions are replete with short wavelength features near sharp topographic changes that are clearly a surface manifestation of internal tides in the vicinity of sharp topographic changes. Figure 6.12.2 shows the global distribution of M_2 along-track tidal signals in ascending tracks. These were obtained by careful band-pass filtering of collinear differences between along-track signals two tidal cycles apart (to enhance the detection of M_2) to retain only the 70 to 200 km wavelengths and subjecting the resulting time series to both harmonic analyses to extract the M_2 component. This narrow-band filtering in both frequency and wavenumber band highlights the first mode M_2 baroclinic tides, which tend to have wavelengths of around 110 km. While some mesoscale and noise contamination is inevitable (see Kantha and Tierney, 1997), the large signals apparent in regions like the Hawaiian island chain in the Pacific and the Mascaren Ridge near Madagascar in the Indian Ocean are quite realistic. To illustrate possible noisy regions, along-track signals extracted in a similar fashion are also shown for N_2 (Figure 6.12.2). These should be principally noise and mesoscale signals, since the N_2 tidal signatures should be all but invisible to the altimeter. Comparison of M_2 and N_2 distributions in Figure 6.12.2 provides a good idea of the extent of corruption of M_2 baroclinic tide signals.

While the altimetric ground tracks do not always intersect the tidal crests at right angles, the resulting distributions are a reasonable indication of the baroclinic tides in the global oceans. The square of the SSH signal is proportional to the energy density. The global average from altimetry is 0.16 cm^2 for M_2 corresponding to an average internal tide amplitude of about 4 m, even though the amplitudes can reach as high as 80 m in some regions (Morozov, 1995). Observations cited by Morozov indicate that the amplitudes can exceed 100 to 120 m around the Mascaren Ridge near Madagascar in the Indian Ocean during spring tides. Since

$$E_d = \rho \frac{\Delta\rho}{\rho} \overline{ga^2} = \rho g \overline{a_s^2} \left(\frac{\Delta\rho}{\rho} \right)^{-1} \quad J\,m^{-2} \qquad (6.12.1)$$

where a is the amplitude of the internal wave, a_s is the amplitude of the sea surface perturbation due to it and $\Delta\rho/\rho$ is the density jump at the buoyancy interface, this corresponds to 50 PJ of M_2 baroclinic tidal energy in the global oceans, taking $\Delta\rho/\rho$ as 10^{-3}. The overbar denotes the average over a tidal cycle.

Accurate modeling of internal tides in the global oceans requires a high resolution fully three-dimensional baroclinic tidal model, perhaps assimilating altimetric tides. No such model exists at present. However, simple models of internal tide generation have been developed by Rattray et al. (1996), and Baines (1982). Baines (1982) examined internal tide generation at the continental shelf break and identified several important source regions around the world, including the mouth of the Amazon, the Irish shelf, and the Northeast American shelf. He also showed that the contribution of all these regions to tidal dissipation was very small, ~16 GW, with a peak rate of less than 400 to 500 W m^{-1}. This squelched any interest in baroclinic tides as a major factor in the dissipation of global tidal energy, until recent observations, both in situ (Dushaw et al., 1995) and altimetric (Ray et al., 1996), showed otherwise. Morozov (1995) has reexamined the internal tide generation problem, based on Baines' theory, and has shown that midocean ridges are strong internal tide generation regions and may account for as much as 20,000 W m^{-1} of tidal dissipation. His scaling, however, leads to an excessive estimate: 1.1 TW of dissipation in semidiurnal baroclinic tides alone.

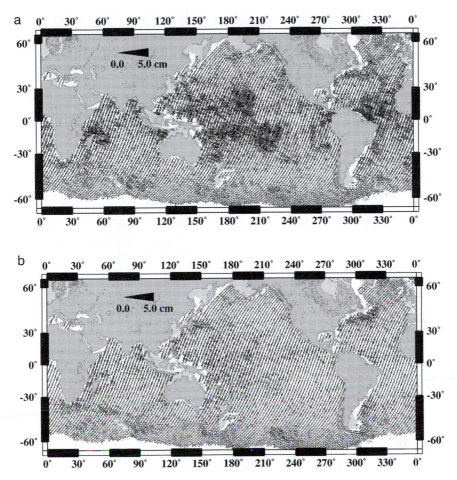

Figure 6.12.2 Global distribution of (a) M_2 and (b) N_2 internal tides along ascending tracks. N_2 is shown to give an idea of the noise level since N_2 amplitudes should be small enough to be indiscernible in altimetric records.

Kantha and Tierney (1997) consider a linear ridge with barotropic tidal current sloshing over it at frequency σ. Let the stratification be assumed to consist of a homogeneous layer of depth h bounded by a buoyancy interface with a density jump of $\Delta\rho$ across it. Let N be the buoyancy frequency in the fluid below. N corresponds to the stratification below the main thermocline with a timescale of typically 1 hr. Let σ be the tidal frequency, corresponding to a period of 12.42 hr for M_2. The ratio of the vertical (Cg_v) and horizontal (Cg_h) group velocities of internal waves generated by tides and propagating into the deep ocean is proportional to the ratio of these two timescales (0.12). These vertically propagating internal waves are the primary sink for the energy in the baroclinic tides, and the energy flux E_f in these waves, which is the energy density E_d in the tides multiplied by the vertical component of the group velocity C_{gv}, is the dissipation rate. It can be easily shown (Baines, 1982, for example) that the momentum equation for the baroclinic tides consists of a body force term proportional to the local gradient of the bottom depth ∇H. No baroclinic tides will be generated if ∇H is zero. It is the component of barotropic velocity vector $\mathbf{U}$ perpendicular to the ridge that is responsible for displacing the isotherms in the vertical and generating the baroclinic tides; therefore, the baroclinic velocities must scale with this component of barotropic velocity. Density stratification is also necessary. This suggests that an appropriate modification to Baines' theory would be to let the resulting baroclinic velocity u be proportional to $(\mathbf{U} \cdot \nabla H/H)\lambda$, where λ is the wavelength, which can be taken as $2\pi C/\sigma$, where the horizontal phase velocity C is $[(\Delta\rho/\rho)gh]^{1/2}$ for the first mode interfacial waves running along the buoyancy interface. The energy density E_d and the vertical energy flux E_f can be written as

$$E_d = \alpha\rho u^2 H = \alpha\rho\left[\left(\mathbf{U} \cdot \frac{\nabla H}{H}\right)\lambda\right]^2 H = \alpha\rho\left[\left(\mathbf{U} \cdot \frac{\nabla H}{H}\right)\frac{2\pi C}{\sigma}\right]^2 H \quad \text{J m}^{-2}$$

$$E_f = E_d C_{gv} = E_d C\frac{C_{gv}}{C_{gh}} = E_d C\left(\frac{\sigma^2 - f^2}{N^2 - \sigma^2}\right)^{1/2} \quad \text{W m}^{-1} \qquad 6.12.2)$$

$$C = \left(\frac{\Delta\rho}{\rho}gh\right)^{1/2}$$

where α is an undetermined proportionality constant. Since the tidal current varies in direction with time, its average over the tidal cycle in the direction perpendicular to the ridge must be taken. Kantha and Tierney (1997) used the above relationships and the barotropic currents from the 1/5° barotropic tidal model of Kantha (1995b) to compute the energy density and the dissipation rate of baroclinic tides in the global oceans in waters deeper than 1000 m (since the theory is not strictly valid for sloping continental shelves). A value of 2.5 m s^{-1} (accurate to ± 20%; see Chelton *et al.,* 1997) was taken as representative of the global average of the long-wave speed C. The value of α was chosen as 0.55 to yield a total energy in the first mode M_2 baroclinic tides in waters deeper than 1000 m of 50 PJ as indicated by altimetric measurements. The corresponding dissipation rate is 360 GW. Table 6.10.2 tabulates the values for different components. The corresponding values are 8 PJ and 57 GW for S_2. A simple scaling to the entire semidiurnal band yields a value 61 PJ and 440 GW.

Applying the same methodology to K_1 tides yields 29 PJ and 81 GW, and 29 PJ and 81 GW for the diurnal band. Thus, the total energy resident in the global first mode baroclinic tides is likely to be around 90 PJ, corresponding to an energy density of 300 J m^{-2}. The total dissipation rate is 521 GW. Kantha and Tierney (1997) suggest that this is an underestimate for a variety of reasons, and the likely value is about 600 GW, about 17% of the estimated 3.5 TW rate of work done by the Sun and the Moon on ocean tides. This value is, however, subject to an uncertainty of perhaps a factor of 2, which will take some time to resolve through *in situ* and altimetric observations. Figure 6.12.3 shows the possible locations of large internal tides in the global oceans. It also shows the amplitude of the M_2 baroclinic tides. Large amplitudes are evident near midocean topographic features such as ridges and island chains.

The fraction of the energy input into ocean by baroclinic tides has long been a subject of considerable interest to oceanography, and so is the amount of energy resident in internal tides. The average energy in M_2 barotropic tides is approximately 1100 J m^{-2} (about 12 TJ in a one degree by one degree box), although this value can reach as high as 10,000 J m^{-2}. The energy resident in baroclinic tides is thought to be 10 to 50% of this (Hendershott, 1981; Wunsch, 1975). Dushaw's measurements indicated a point value of as high as 26%, while the range-averaged value was about 8%. Taking a value of 20% as reasonable as an average, this means that, on the average, 220 J m^{-2} of energy is resident in M_2 internal tides. Dushaw *et al.* (1995) observed a net energy of 315 J m^{-2} in first-mode internal tides with about 250 and 200 J m^{-2} in the semidiurnal and M_2 bands, with the ratio of kinetic to potential energy of 0.26 to 0.35, close to the theoretical value. On the other hand, the dissipation rate due to internal tides is estimated to amount to roughly 10 to 12% of the tidal power input (Hendershott, 1981; Wunsch, 1975), about 0.25 to 0.3 TW for M_2 and 0.4 TW overall. Considering the energy in the deep-sea internal waves, given as 3800 J m^{-2} by the Garett-Munk spectrum, which appears to hold quite well throughout the midlatitude oceans, the energy in the barotropic tides (1717 J m^{-2}) and baroclinic tides (~300 J m^{-2}) is comparatively much less (Munk, 1997). According to Munk (1997; see also Munk and Wunsch, 1998), an energy flux (dissipation rate) of ~ 200 GW is needed to maintain the vertical momentum diffusivity rates of ~10^{-5} m^{2}s^{-1} observed by modern microstructure measurements in the deep ocean (Gregg and Sanford, 1988). Internal tides dissipate far more and might very well be a principal source of internal waves and mixing in most of the deep ocean. Munk (1966) and Munk and Wunsch (1998) indicate that to maintain the structure of the permanent thermocline in face of the 25 Sv deep water formation in the subpolar regions requires an energy flux of 1.9 TW. Clearly, internal tides cannot provide all but a significant fraction of this. Wunsch (1999) quotes estimates that indicate that winds might put in 900 GW into general circulation of the oceans, which is of the same order

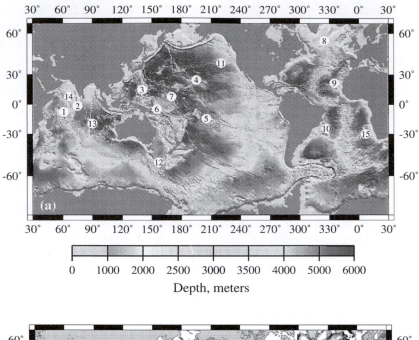

Depth, meters

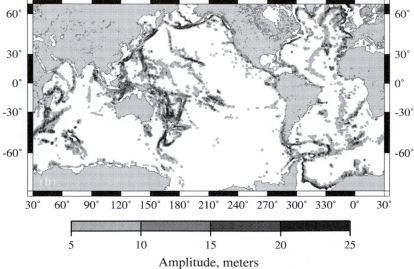

Amplitude, meters

Figure 6.12.3 (a) The location of large internal tides in the global oceans and (b) the amplitude of M2 baroclinic tides derived using the currents from the barotropic model; only amplitudes larger than 5 m are shown. (Reprinted from Kantha, L. H., and C. C. Tierney, Global baroclinic tides, *Progress in Oceanography* **40**, 163–178. Copyright 1998 with permission from Elsevier Science.)

as that dissipated through internal tides. If this is true, the energetics of internal tides are as important as that of the ocean currents!

Another aspect of internal (and external) tides is that they cannot propagate freely beyond a latitude corresponding to $\sigma = f$. They are evanescent beyond this critical latitude, which is roughly $30°$ for diurnal and $75°$ for the semidiurnal tides. But the Rossby radius of deformation, which determines the meridional scale of decay, is several thousand kilometers for surface tides and hence there is no effective trapping (Wunsch, 1975). On the other hand, this scale is a few tens of kilometers for internal tides; therefore, any internal tidal energy that exists poleward of these latitudes is trapped over the generating topography. Another salient aspect of internal tides appears to be their intermittency (Wunsch, 1975); unlike surface tides, they appear to appear out of and disappear into the background noise. Also, *in situ* measurements of semidiurnal internal tides are more common than those of diurnal ones.

Strong tidal flow over sills can give rise to internal solitons. During the flood tide, the isopycnals are elevated slightly, but during the ebb tide they get severely depressed. This depression is often released, resulting in a train of internal wave solitons. The Sulu and Andaman Seas and the Straits of Gibraltar are prime regions of such IW solitons, although they have been observed elsewhere. IW solitons are evident in 10 Hz altimetric records, and Pinkel *et al.* (1996) made an intensive study of 50 m amplitude solitons and encountered unexpectedly in the tropical Western Pacific during the TOGA/COARE observations. A train of solitons, consisting of 3 to 7 solitons, with isopycnals depressed by tens of meters (with corresponding SSH bumps of several centimeters) and trough-to-trough distances of 3 to 7 km are observed in the Sulu Sea (Apel *et al.*, 1985). They are also ordered, with the largest soliton of the train propagating the fastest. The dissipative timescale for the solitons varies from a few tens of hours in deeper waters to a few hours in shallow water. Solitons generated by semidiurnal tides have their energy dissipated within the semidiurnal period. Despite their spectacular nature and large power densities $(30\,000\text{W m}^{-1})$, Munk (1997) suggests that their contribution to tidal dissipation, an upper bound of about 30 GW, is relatively insignificant to the tidal energy balance of the global oceans. The same is true of the spectacular bores, such as the 8 m high Pororoca in the Amazon river estuary whose roar can be heard at distances of 20 km and yet which dissipates less than 1 GW (Munk, 1997). Spectacular tides in bays and inlets such as Bay of Fundy, while very impressive, suffer the same fate, with the Bay of Fundy contributing about 20 GW and all such coastal indentations amounting to less than 100 GW (Munk, 1997). It is the very covert baroclinic tides in the deep ocean that appear to be a major dissipation mechanism for global tides, next only to bottom dissipation in shallow seas, accounting for ~ 600 GW of tidal dissipation out of the 3750 GW supplied by the Sun and the Moon! Table 6.10.2 summarizes internal tidal energetics as derived from global models.

6.13 LONG-PERIOD TIDES

Long-period ocean tides, principally the lunar fortnightly (Mf), the lunar monthly (Mm), the solar semiannual (Ssa), the solar annual (Sa), the 14-month pole tide, and the 18.6-year nodal tide, have much smaller amplitudes than the short-period semidiurnal and diurnal ocean tides. The largest, Mf, has amplitudes of less than 2 to 3 cm in most cases. The long-period ocean tides are therefore nearly insignificant to overall tidal energetics, but nevertheless they are quite important to geophysics and oceanography. Because of the low frequencies involved, it is often assumed that these tides are close to equilibrium and, in fact, for some applications the equilibrium approximation suffices. However, the departures from equilibrium of these tides are important from the point of view of the response of the global oceans to low frequency forcing. Specifically, an important question in oceanography has always been whether external forcing excites vigorous resonant basin modes or a generally weak damped response in the global oceans. An evaluation of the quality factor Q that indicates how fast the energy resident at a particular frequency decays is important to answering this question.

The long-period ocean tides need to be known precisely for GPS and precision altimetry and for many geophysical applications such as evaluating the contribution of the long-period ocean tides to tidal variations in the Earth's rotation rate. Accurate evaluations of the ocean tide contributions to the length of day (LOD) variations will allow improved constraints on the anelasticity of the Earth's mantle at these long periods (Desai, 1996; Wahr and Bergen, 1986). The following comes from Kantha *et al.* (1998).

The oceanic response to long-period forcing has been controversial since Laplace's time (Laplace, 1776), and people like Kelvin, Darwin, Hough, Rayleigh, Poincare, Proudman, and Wunsch have been involved with this ongoing debate (see Wunsch, 1967, and Miller *et al.*, 1993, for a history of research on long-period tides and thorough reviews of the problem). Specifically, the question of whether they can be approximated by their equilibrium solutions has been important since then. Because of their low amplitudes that make their measurements highly susceptible to contamination by noise, and because of the long record lengths needed to separate them effectively from the large background noise, until recently it has been rather difficult to obtain reliable observational data to decide this issue. An additional complication is the fact that the Mf frequency is the difference in the frequencies of the semidurnal tides M_2 and K_2, and the Mm frequency is the difference between M_2 and N_2. Since M_2 is the strongest tide in the ocean, there is always a question of the nonlinear contribution of semidiurnal frequencies to Mf and Mm, which must doubtless be large in some places compared to the direct astronomical forcing. A time-honored method to resolve this issue has been to examine MSf frequency, which is the difference between M_2 and S_2; the absence of significant energy at MSf is an indication that nonlinear contributions from semidiurnal tides are unimportant (Wunsch, 1967). Wunsch *et al.* (1997) discuss the dynamics of the long-period tides, including the 14-month pole tide that is about 5 mm in amplitude and is driven by the wobble of the Earth's axis and not the gravitational forces of heavenly bodies.

Wunsch (1967) made a careful analysis of the then available observations and concluded that there exist significant departures from equilibrium values, of as much as $30°$ in phase and 60% in amplitude, for both Mf and Mm. This is in accord with the conclusions of Proudman (1959) that while the pole tide is in equilibrium, and the semiannual Ssa and annual Sa tides will probably be in equilibrium, the fortnightly and monthly tides are probably not. Wunsch also used quasi-geostrophic (QG) approximations to Laplace's tidal equations in an idealized, rectangular, flat-bottomed Pacific basin to show that the resonant Rossby wave modes of the order of 1000 km in wavelength govern the nonequilibrium portion of the dynamics of these tides. Because of the low frequencies involved (and unlike short-period tides), the oceanic response to long-period tidal astronomical forcing is often assumed to be similar to its response to long-term wind forcing, that is, governed by QG dynamics. Rossby waves, western intensification, and other features that are central to long-term, wind-driven circulation in the global ocean might therefore be important to long-period tides also (Hendershott, 1981). In fact, Hendershott comments that, were the long-period ocean tides of significant amplitudes to have been of interest to tidalists, discoveries such as the beta plane and western intensification of currents might very well have been made much earlier.

Wunsch's analytical solutions are limited by the artificial nature of the basin configuration, and it is not clear if the short Rossby modes can account for the basin scale response. Schwiderski's (1982) data-assimilative numerical model for these tides showed a basin scale response. Meridional distributions of the zonally averaged tidal admittances for Mf and Mm from GEOSAT observations by Ray and Cartwright (1991) appear smooth and lack small scale features. A recent analysis of tide gauges in the the Pacific Ocean by Miller *et al.* (1993) shows that while the Mf tide has phase lags of 10 to $40°$ and the Mm tide is slightly closer to equilibrium than the Mf tide, their admittances appear to lack short scale, Rossby wavelike features. Based on sensitivity studies using a global finite element barotropic tidal model, they suggest that the departures from equilibrium can be explained by remotely forced gravity waves propagating from the Arctic through the Atlantic into the Pacific. More recently, using a spectral element numerical model that included the Arctic, Wunsch *et al.* (1997) conclude that while the Arctic does not appear to play a predominant role in the overall global long-period tidal dynamics, both Rossby wave and gravity wave responses exist in their hydrodynamic model solutions. Kantha *et al.* (1998) have used a $1°$ global hydrodynamic model to compute the lunar long-period tides and their influence on LOD (UT1) fluctuations.

One possible solution to long-period tides that is often used as an approximation is the self-consistent equilibrium approximation (Agnew and Farrell, 1978). Note that these tides have no meridional gradients, although in practice, departures from equilibrium of these tides are substantial and introduce such gradients.

Schwiderski (1980, 1982) was one of the first to model the long-period lunar Mf and Mm tides in the global oceans. Among more recent attempts to model these tides, those by Miller *et al.* (1993), Wunsch *et al.* (1997), and Kantha *et al.* (1998) are noteworthy. Miller *et al.* (1993) used a global finite element barotropic tidal model and concluded that the departures from equilibrium can be explained by remotely forced gravity waves from the Arctic propagating through the Atlantic into the Pacific. Wunsch *et al.* (1997) also used a finite element global barotropic model but with considerable idealization, such as the use of a rigid bottom, linearized bottom friction and artificial frequencies (14 and 28 days for Mf and Mm, respectively). Their principal conclusion is that the ocean behaves like a heavily damped system to tidal forcing at these low frequencies, with a Q value of around 4.6 to 4.9. Their model shows the oceanic response to consist of some gravity wave response superimposed on a principally Rossby wave response. Neither of these two models involves data assimilation.

Kantha *et al.* (1998) assimilated Mf and Mm tides estimated empirically from T/P altimetric records (see Desai, 1996; Desai and Wahr, 1995) into a one-degree barotropic tidal model covering the global oceans except the Arctic. Figure 6.13.1 shows the global distribution of the Mf and Mm tides so obtained. The results show more structure, especially around topography, than the smooth results of Schwiderski (1982). They also confirm the significant departures from equilibrium of these tides. However, Mm tide is closer to equilibrium than the Mf.

Extraction of long-period ocean tides from tidal records has always been problematic because the weak signals tend to be swamped by the meteorological noise in the records, and long record lengths and careful analyses are essential for meaningful results. The lack of accurate observations of long-period tides helped fuel the early debate on whether the Mf and Mm tides are close to equilibrium, until careful analyses of available observations first by Wunsch (1967) and more recently by Miller *et al.* (1993) have laid this issue to rest. Both GEOSAT (Ray and Cartwright, 1991) and T/P (Desai, 1996) altimetry confirm their conclusions. However, because of the smallness of the signals (1 to 2 cm amplitude) compared to the accuracy of the sensor (2 to 3 cm rms, Fu *et al,.* 1994), altimetric measurements are also likely to be plagued by the same problem. This is why the data-assimilative model of Kantha *et al.* (1998) provides a more accurate description, from which the following characteristics emerge.

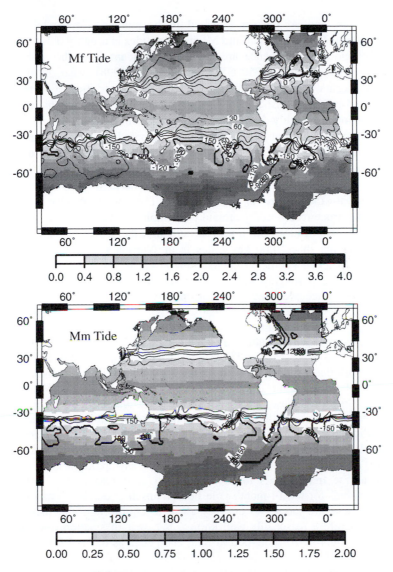

Figure 6.13.1 Coamplitude and cophase plots for Mf and Mm. From Kantha *et al.* (1998).

The energy resident in Mf and Mm, their dissipation rates, and their Q values and PE/KE ratio are of some importance to deciphering the response of the global oceans to low frequency forcing. Steady wind-driven circulations in the oceans are thought to bear close resemblance to long-period tides, especially Mm, since the physics in the two cases, the western intensification, midlatitude

Rossby waves, and so on are expected to be similar (Hendershott, 1981; Wunsch, 1967). The work input into Mf and Mm tides is roughly 0.344 GW and 0.022 GW globally and hence are rather insignificant from the global tidal dissipation point of view (3750 GW total; Kantha, 1998). Also, most of this dissipation occurs near lateral boundaries, with bottom friction accounting for less than 0.4%, in contrast to the Wunsch *et al.* (1997) model, in which the bottom friction is roughly 50%. The total energy in Mf and Mm, 0.381 and 0.049 PJ, respectively, is also quite negligible compared to the 580 PJ resident in short-period tides (Kantha, 1998). The PE/KE ratios are 0.59 and 2.9, respectively. The ratio of PE/KE of Mf to that of Mm is roughly 4.9; this is consistent with what is expected for long Rossby waves whose PE/KE ratio is proportional to the square of their period. The model results also show considerable structure especially around topographic features, so it is likely that Rossby wave propagations are involved, although the scales of these may not correspond to the short 1000 km scales from Wunsch's highly idealized flat-bottomed, rectangular Pacific basin model (Wunsch, 1967).

The value of Q is roughly 5.8 to 5.9 for Mf indicating that the energy in this tide is renewed once every 0.9 cycles or once every 12.6 to 12.8 days or so, in contrast with Q value of 23 for M_2, whose energy is renewed roughly every 3.7 cycles or about 46 hr, and 11.6 for K_1, whose decay scale is 1.8 cycles or about 44 hr. The corresponding values are 5.8 to 6.2 for Mm and hence the decay timescale is 25.4 to 27.2 days. These low Q values for both Mf and Mm are consistent with that (~5.5) inferred from the response of the basin circulation to 5-day synoptic forcing (as summarized in a review paper by Wunsch and Stammer, 1997) and those derived by Wunsch *et al.* (1997) from a spectral element global model with a rigid bottom. The barotropic tidal model-derived Q values indicate a nearly monotonic decrease in the value of Q as one goes from semidurnal (23) to diurnal (13) to Mf (5.9) and Mm (6.2) long-period tides. However, it is not clear if the values are likely to asymptote or continue their decrease as the frequency is further lowered. Unfortunately, it is difficult to deduce Q values for the 14-month pole tide (Wunsch *et al.,* 1997) and the 18.6 year-tide. These are known to be very close to equilibrium and their energy content is therefore reasonably well known, but their dissipation rates are not. Table 6.10.2 summarizes long-period tidal energetics as derived from Kantha's global model (Kantha, 1998; Kantha *et al.,* 1998).

The long-period tides Mf and Mm are only a few centimeters in amplitude, but their importance is in their geophysical impact. They show up prominently in the observations of LOD (UT1) fluctuations. The amplitude of these LOD fluctuations is an order of magnitude larger than those due to short-period tides, and it plays a prominent role in inferring geophysical properties such as the anelasticity of the Earth (Desai, 1996). Both the Earth's moment of inertia tensor and its zonal angular momentum are affected by these tides; hence, accurate estimates of both the SSH and the currents induced by these tides are needed to infer their effects on Earth's rotation. The resulting perturbation in UT1 has an amplitude of 114.9 ms and a phase of 294.9° for Mf and 115.0 ms and a phase of 279.7° for Mm (the corresponding T/P values are 110.4 ms and a phase of 295.8° for Mf and 114.7 ms and a phase of 276.3° for Mm).

More accurate determination of lunar long-period tides enables a better determination of the contributions of the anelasticity of the Earth's mantle to UT1 at these frequencies. The small anelasticity signal buried in the large tidal signals requires considerable precision in the computation of long-period ocean tides and its effects on UT1 fluctuations, including both the mass and motion contributions. Kantha *et al.* (1998) results as well as the empirical T/P model estimates of the contribution of anelasticity of the Earth's mantle to UT1 fluctuations at these frequencies are in approximate agreement with the current geophysical models. See Desai (1996) for detailed discussion of this issue.

Solar semiannual and annual tides, the pole tide and the nodal tide have such low frequencies that it is unlikely that assuming them to be in equilibrium is far from accurate. The first two are strongly driven by radiational forcing due to solar heating of the oceans and the atmosphere during the course of the year. The amplitudes of Ssa and Sa are typically 2 to 4 cm in the global oceans, even though Sa can reach values exceeding 10 cm in some locations. The gravitational contribution to these tides is well described by the equilibrium approximation

$$\zeta_{S_{sa}} = 0.019542\ \alpha\left(\frac{3}{2}\cos^2\theta - 1\right)\cos(2h_0)$$

$$\zeta_{S_a} = 0.003104\ \alpha\left(\frac{3}{2}\cos^2\theta - 1\right)\cos(h_0 - p_s)$$

(6.13.1)

These amplitudes are typically less than a centimeter, and it is clear that the gravitational contribution is overwhelmed by the radiational one. The small nodal tide has an approximate amplitude of

$$\zeta_N = -0.018\alpha\left(\frac{3}{2}\cos^2\theta - 1\right)\cos n_0$$

(6.13.2)

and a period of 18.61 years.

Finally, tides are also induced by variations in the centrifugal force and these are called rotational tides. The most important rotational tide from geophysical point of view is the Chandler pole tide with a typical amplitude less than 0.5 cm, driven by the 434.45 sidereal day Chandler wobble. An annual pole tide is driven by seasonal variations in polar motion, and length-of-day tide by fluctuations in the rotation rate. Because of its low frequency, the pole tide is also very nearly in equilibrium, and hence its value is given by

(Lambeck, 1980, 1988)

$$\zeta_{CP} = -0.699 \frac{\Omega^2 R_E^2}{g} \alpha \left(m_1 \cos\phi + m_2 \sin\phi \right) \sin\theta \cos\theta \qquad (6.13.3)$$

However, it is very difficult to discern the pole tide and the nodal tide in tidal records because of the extremely low signal to noise ratio (Wunsch et al., 1997). Were it ever to become possible to extract these from tidal records accurately enough, several interesting deductions such as the anelasticity of the Earth and Q of the oceans at these low frequencies could be made.

6.14 SHALLOW WATER TIDES AND RESIDUAL CURRENTS

For deep ocean basins, the bottom frictional forces are usually small and relatively unimportant. Consequently, the dynamical theory of Laplace yields a solution for each tidal forcing component (partial tide) that is independent of another. The resulting astronomical tides are linear, and tidal height at any point can be obtained by simple superposition of the different linear astronomical tides. But in shallow water such as continental shelves, the bottom friction is important and in some cases, as in estuaries, the amplitude of the tide is comparable to the water depth. The presence of nonlinear advection terms and the quadratic bottom friction give rise to higher harmonics of astronomical tides. The frequencies of these shallow-water tides is a linear combination of those of the linear tides or multiples of the fundamental frequency. For example, if the primary tidal frequency is ω_1, nonlinearity generates harmonics of frequency $2\omega_1$, $4\omega_1$ ... (overtides) and for two primary tides of frequency ω_1 and ω_2, nonlinear interaction generates additional frequencies that are sums $\omega_1 + \omega_2$ (compound tides) and differences $\omega_1 - \omega_2$ (long-period tides). Tides that are higher harmonics of a single astronomical tide are called overtides. M_4 (twice the frequency of M_2) and M_6 (thrice the frequency of M_2) belong to this category. Tides that are generated by the nonlinear interaction of two linear tides are called compound tides (also called mixed or combination tides; see Parker, 1991); for example, MS_4 has a frequency that is the sum of frequencies of M_2 and S_2 and is generated by the interaction of M_2 and S_2. Some overtides and compound tides can reach high amplitudes in seas such as the Yellow Sea and the Persian Gulf and are generally essential to accurate computation of tides in shallow seas and shelves. In rivers and estuaries, it may be essential to consider many shallow water constituents; in some cases, the tide propagates as a bore up the estuary. Tables 6.2.1 and 6.2.2 show the characteristics of the most

TABLE 6.14.1
Shallow Water Tidal Constituents

	Generated by	Angular speeds	°/hour (σ)	Nodal factor (f)	Period hour
Long-period					
*M_{sf}	M_2, S_2	$M_2 - S_2$	1.0159	f(M_2)	
Diurnal					
MP_1	M_2, P_1	$M_2 - P_1$	14.0252	f(M_2)	
SO_1	S_2, O_1	$S_2 - O_1$	16.0570	f(O_1)	
Semidiurnal					
MNS_2	M_2, S_2, N_2	$M_2 + N_2 - S_2$	27.4238	f^2(M_2)	
*$2MS_2$	M_2, S_2	$2M_2 - S_2$	27.9682	f(M_2)	
MSN_2	M_2, S_2, N_2	$M_2 + S_2 - N_2$	30.5444	f^2(M_2)	
$2SM_2$	M_2, S_2	$2S_2 - M_2$ 2S $-$ 2h	31.0159	f(M_2)	11.58
Third-diurnal					
MO_3	M_2, O_1	$M_2 + O_1$	42.9271	f(M_2)*f(O_1)	
MK_3	M_2, K_1	$M_2 + K_1$	44.0252	f(M_2)*f(K_1)	
Fourth-diurnal					
MN_4	M_2, N_2	$M_2 + N_2$	57.4238	f^2(M_2)	
M_4	M_2	$M_2 + M_2 - 4S + 4h$	57.9682	f^2(M_2)	6.21
MS_4	M_2, S_2	$M_2 + S_2 - 2S + 2h$	58.9841	f(M_2)	6.10
MK_4	M_2, K_2	$M_2 + K_1$	59.0662	f(M_2)*f(K_1)	
S_4	S_2	$S_2 + S_2$ 0	60.0000	1.00	6.00
Sixth-diurnal					
M_6	M_2	$M_2 + M_2 + M_2 - 65 + 6h$	86.9523	f^3(M_2)	4.14
$2MS_6$	M_2, S_2	$2M_2 + S_2 - 4s + 4h$	87.9682	f^2(M_2)	4.09
Eighth-diurnal					
M_8	M_2	$4M_2$	115.9364	f^4(M_2)	

* Also contain a significant gravitational component ($2MS_2 \equiv \mu_2$).
From Pugh (1987).

important shallow water tides, including their frequencies and astronomical arguments (Table 6.14.1 from Pugh, 1987, is more detailed).

While shallow water tides at the coast are known from tide gauge records, their distribution on the shelves and shallow seas is not. Precision altimetry holds the promise of providing least the principal ones like M_4 and M_6 in shallow water. However, it is essential to combine this with a numerical model to obtain a global distribution of shallow water tides. The full nonlinear equations need to be solved. An example shown in Figure 6.14.1 is from a regional model of the Yellow Sea from Kantha *et al.* (1996). The distributions of M_4 and MS_4 coamplitude/cophase were obtained by running the full nonlinear regional model for a domain encompassing the Korean seas, with M_2 and S_2 forcing at the boundaries at 1/5 degree resolution for 29 days. The model SSH was saved hourly at each model grid point and the resulting time

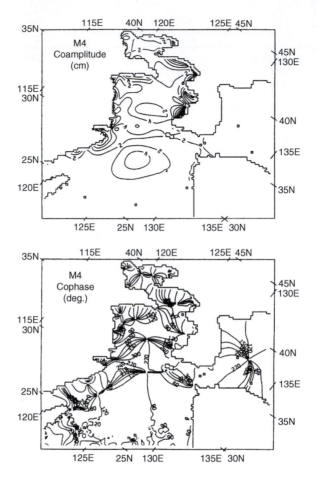

Figure 6.14.1 Coamplitude (top) and cophase (bottom) of M4 shallow water tide in the Yellow Sea from a tidal model. From Kantha *et al.* (1996).

series subjected to harmonic analysis to derive the various components. Note the large values of M_4 near the Korean coast.

Another aspect of shallow water tides is the residual currents induced by tides. These currents arise simply because of the asymmetry between the flooding and ebbing phases of the tide and also because of coastline features that generate vorticity. Consider a uniform channel of depth H with a tidal current of amplitude U and tidal elevation of amplitude of η, such that the ratio (η/H) is not negligible. Then the transport during the flood phase is proportional to U(H+η) and during the ebb phase to U(H-η), and when averaged over a full

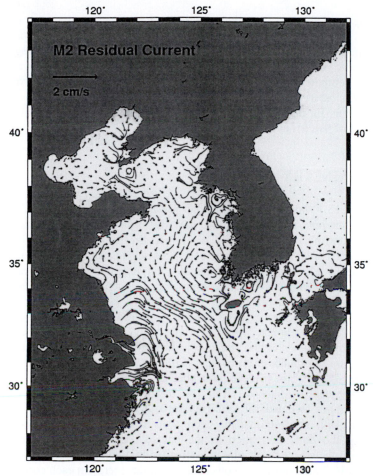

Figure 6.14.2 M_2 tidal residual currents in the Yellow Sea.

tidal cycle, there is therefore a net transport proportional to $2U\eta$ and a residual current of magnitude $2U\eta/H$ in the direction of propagation. Hence, the integrated value of the current over a tidal cycle is not necessarily zero. The strength of the residual current is approximately proportional to the ratio of the tidal amplitude to the mean water depth. In practice, of course, the change in the effective depth makes the tidal characteristics during the flood stage differ from those during the ebb phase, and this expression is approximate at best.

The residual currents generated by primary tides are usually one or two orders of magnitude smaller than that of the primary tidal currents, their typical magnitudes being a few cm s^{-1}. However, they are often similar in magnitude to the background currents, and, because of their persistence, they can transport pollutants and sediments over large distances and hence are quite important to many aspects of coastal engineering. It is rather straightforward to integrate the tidal currents generated from a nonlinear shallow water tidal model over a tidal cycle and compute the residual currents. Normally it is adequate to save hourly values for aposteriori analysis. It is, however, possible to compute them during the model run, for better accuracy. Figure 6.14.2 shows the M_2 residual currents in the Yellow Sea (from Kantha *et al.*, 1996).

6.15 SUMMARY

Figure 6.15.1 shows the flow of tidal power. It can be seen that of the 3.75 TW of power put in by the lunisolar gravitational forces, a negligible amount goes into atmospheric tides (~ 20 GW), either directly or through excitation by ocean tides. About 0.23 TW is dissipated in the solid Earth. Of the remaining 3.5 TW, about 0.6 TW might go into internal tides generated over midocean ridges and topography. Of the remainder, internal tide generation at shelf break, IW solitons, coastal indentations such as the Bay of Fundy, and internal wave generation by tidal currents over roughness on sea bottom may account for another 0.25 TW (Munk, 1997). Of the remaining 2.65 TW, turbulence generated by tidal currents streaming through gaps in island chains and dissipation on the flanks of seamounts (Lueck and Mudge, 1997) may account for another 450 GW, so that the remainder of approximately 2200 GW is most likely dissipated in shallow seas by bottom friction. However, *in situ* observations are essential for confirmation of these energy budget figures. If proven true, this suggests that the deep open ocean accounts for a substantial fraction of the tidal dissipation in the oceans, contrary to the conventional thinking that there is negligible tidal dissipation going on in the deep oceans.

Table 6.10.2 also shows the power input and the energy in each component of the major tides and the baroclinic tides. The total energy in all barotropic tides appears to be about 584 PJ, with M_2 accounting for the lion's share, 390 PJ. The Q factor is around 23 for semidiurnal tides so that the dissipation (or alternatively renewal) timescale for these tides is around 2 days. On the other hand, the Q factor is smaller, about 13 for diurnal tides, with a corresponding dissipation timescale of 2 days also. It is about 5.9 for Mf, and hence the dissipation timescale is 12.8 days. The average energy density in barotropic tides is 1720 J m^{-2}, with M_2 accounting for 1150 J m^{-2}. The energy in internal tides is about 90 PJ, roughly 15% of that in barotropic tides, with an energy density of 300 J m^{-2}, M_2 internal tides accounting for over half of this value. The energy dissipation in internal tides is about 600 GW, about 17% of the total power input into the oceans by the lunisolar gravitational forces. This is certainly more than adequate to maintain (about 200 GW are needed) the observed pelagic

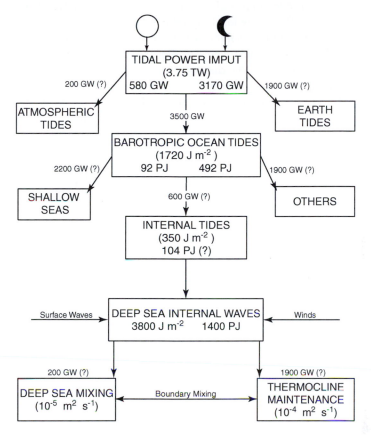

Figure 6.15.1 Tidal power balance.

dissipation rate of 10^{-5} m^2 s^{-1} (Kunze and Sanford, 1996) but not quite sufficient to account entirely for the dissipation rate of 10^{-4} m^2 s^{-1} needed to maintain the permanent thermocline structure in the face of about 25 Sv of dense water formation in subpolar seas (Munk, 1966; Munk and Wunsch, 1998).

The repeatability of tidal phenomenon makes it well suited to a systematic study. Numerical modeling has played a vital role in this task. Most aspects of oceanic tides can be simulated by simple numerical barotropic models, but accurate computation of tidal currents in shallow waters is best done using regional high resolution baroclinic models. Unlike tidal sea level fluctuations, for which a barotropic model is adequate, knowing tidal currents requires knowledge of the vertical current structure. This means that a fully baroclinic

3-D tidal model is needed to model 3-D tidal currents accurately. Also, the tidal currents are greatly affected by local topographic changes; hence, modeling tidal currents accurately requires resolutions generally higher than a model for tidal elevations. Internal tides arising from the density structure in the water column also play an important role in determining the structure of the tidal currents. Data assimilation is valuable and has played a major role in deducing accurate tides over the global oceans. However, accuracy of tides in shallow water still suffers from lack of access to accurate, high resolution digital bathymetric data bases.

Chapter 7

Coastal Dynamics and Barotropic Models

Continental shelves are relatively shallow, and often broad, regions at the margins of the deep ocean basins, with widths ranging from less than 10 to more than 150 kilometers. Over the shelf, the depth increases from zero at the coast to about 200 m at the shelf break, from where there is a rapid increase in depth to abyssal depths of 4 to 5 km over a short distance on the order of 20 to 100 km; this region of steep slope is called the continental slope. The coastal oceans extend all the way from the coastline (including estuaries) to the edge of the abyssal plain of the adjoining basin, encompass both the continental shelf and the slope, and are therefore greatly affected by a variety of processes. The most salient characteristic of the continental shelf is its relatively shallow depth, which causes it to respond vigorously to external forcing. The overlying atmosphere and the adjoining deep ocean are the principal agents of external forcing. The winds (and to a lesser extent, the atmospheric pressure fluctuations) produce vigorous motions in the coastal oceans, whereas tides, strong currents, and eddy motions in the adjacent deep ocean all influence the shelf from across the slope. Fresh water deposited at the coast by rivers and estuaries creates salinity fronts and affects the circulation on the inner shelf. Strong wintertime mixing mixes the water column on the inner shelf generating strong thermal fronts on the shelf. The Louisiana Texas shelf in the Gulf of Mexico provides an excellent example of these and the salinity fronts caused by fresh water from Mississippi/Atchafalaya rivers. Settling of river-transported sediments and resuspension of bottom sediments by vigorous wind and tide-driven mixing is an important process in the coastal oceans. In addition, the shelf acts as an efficient waveguide for subinertial motions generated elsewhere. The resulting coastal circulation is a kaleidoscope of complex motions comprising a broad range of spatial and temporal scales of variability. The time scales involved range from hours to decades and it is a challenge to model these with high fidelity.

The circulation on the continental shelf is affected by Earth's rotation, ambient density stratification, the bottom topography, the offshore ocean, river discharge, and the presence of the coastline. The most important parameter here is the shelf width, the circulation being inherently different for broad and narrow shelves, as evidenced by the circulations along the narrow shelves on the west coast and the broad shelves on the east coast of the United States. Tidal motions tend to get amplified across broad shelves, and the response to atmospheric (both momentum and buoyancy) forcing is also more vigorous. Unlike the open ocean, where strong wind forcing seldom causes strong sea surface height changes, the presence of the coast causes storms to pile up water along the   coast and large setdowns/setups to result during wind-driven coastal upwelling/downwelling. The coastal circulation itself consists of strong persistent but highly variable temperature and/or salinity fronts separating strongly contrasting water masses, well-mixed regions coexisting with strongly vertically stratified waters, narrow intense currents with complex horizontal and vertical structure, propagating coastal trapped waves, and intense mesoscale activity.

Which of the many processes dominate the coastal ocean depends on a variety of factors, including the geometry, geographical location and the adjoining basin. Each coastal ocean is therefore pretty much unique in its characteristics, and the prevailing circulation an aspect of the various forcing functions characteristic of the region. While broad features of the circulation are readily discerned, invariably, concerted efforts by local oceanographers over many years are necessary to sort out the details and decipher the underlying dynamics. Fortunately, coastal oceans are also more easily accessible and hence amenable to *in situ* measurements. Consequently, extensive observations may exist, often as a result of long-term monitoring programs by state and federal agencies that plan and carry out periodic observations often on a well-laid-out observational grid. The CALCOFI program off the California coast and measurements around the Korean peninsula are classic examples. Long-term measurements are also available off the Oregon coast, the Scotian Shelf, and in the Mid-Atlantic Bight (Beardsley and Boicourt, 1981, for example). In general, the coastal oceans off developing countries are not as well studied; this is a major handicap to a well-informed utilization of local resources through fishing and other activities. The recent enactment of the law of the sea makes it hard for anyone but the local authorities to carry out observations in the exclusive economic zones (EEZ) of sovereign countries.

A voluminous literature exists on the circulation in the coastal oceans. It is impossible to do justice to the topic here. Instead, the reader is referred to Allen (1980) and Winant (1980) for dated, but still useful, reviews of wind-driven coastal circulation. Brink (1987) contains a more recent review of physical processes in coastal waters. Results of more recent research can be found in Continental Shelf Research devoted exclusively to coastal oceans, and in the *Journal of Geophysical Research* and the *Journal of Physical Oceanography*. Some journals of the American Society of Civil Engineers focus on the coastal engineering aspects. A recent volume of *The Sea* (Volume 10) edited by Ken Brink and Allan Robinson provides a welcome timely review of coastal ocean processes.

From the point of view of modeling, coastal oceans present a particularly difficult challenge. Complex coastlines and variable shelf widths render the task of horizontal discretization very difficult. Unstructured meshes offer the best hope; however, the fidelity with which they can represent complex processes such as fronts, shelf wave propagations, and coastal upwelling/downwelling has not been fully tested. With structured grids, this geometrical complexity often necessitates the use of high resolution regional models nested in large scale models. The necessity to include the continental slope and the associated processes in many cases makes vertical discretization critical as well. The stepwise discretization employed in z-level models requires a rather large number of vertical levels to simulate processes accurately, while the well-mixed nature of the water column makes isopycnal coordinate models ill suited to the continental shelf. This leaves models with topographically conformal coordinates the best choice, even though because of the large errors in pressure gradient calculations in these models in the presence of steep topography and strong stratification inherent to the continental slope, some caution is necessary. Irrespective of the particular model involved, prescription of conditions on the boundaries open to the ocean basin is a difficult problem and nesting in a basin model is often indispensable.

It is the coastal oceans that most of humanity interacts with. Consequently monitoring, simulating and predicting the state of the coastal oceans has become a high priority, especially in view of their heavy use and abuse, pollution and vanishing fisheries being the two leading issues. A useful coastal ocean prediction system must not only comprise a numerical model capable of faithfully simulating the wide variety of processes prevalent, but also an in-stu monitoring network and a means to rapidly and efficiently assimilate data from such a network to estimate the current state of the coastal ocean and extrapolate into the future. A capability to simulate and predict the chemical, biological and sedimentary state is also essential. There are a variety of efforts to implement such a system in the US, one successful example being the Harvard Ocean Prediction System (HOPS). The core of this system is a PE-based numerical model coupled to a multi-box biological model (Robinson, 1996). The modeling system includes a sophisticated data assimilation scheme (Robinson *et al.*, 1998). A recent overview of HOPS can be found in Robinson (1999). A particularly interesting application of HOPS is to the Massachusetts Bay, where the modeling system was used in real time to assimilate *in-situ* data from various platforms including oceanographic vessels and autonomous underwater vehicles (AUVs) and provide nowcasts and forecasts of the physical/biological state of the Bay. Another notable effort is off the coast of New Jersey, where a long-term environmental observatory (LEO), an observational system comprising a network of telemetering moorings, AUVs and satellite remote sensing integrated with a coastal model, has been pressed into service on the New Jersey shelf to provide real-time estimates of the oceanic state. It is very likely that the above examples will be followed elsewhere in the coastal oceans around the world.

7.1 WIND- AND BUOYANCY-DRIVEN CURRENTS

By far the most important external forcing on the shelf is the alongshore component of the winds. Wind forcing on the shelf involves synoptic timescales and spatial scales of a few hundred to thousand kilometers. There is usually a strong correlation between local alongshore winds and currents on the shelf (see Allen, 1980, and Winant, 1980, for example). However, since the shelf acts as a waveguide, winds elsewhere on the shelf also affect local currents through generation and propagation of continental shelf waves (see Section 7.3), and this correlation is weak at times. Nevertheless, if the winds are known over the entire length of the shelf, it is possible to infer local alongshore currents reasonably well.

A broad continental shelf such as that in the Mid-Atlantic Bight or the Texas-Louisiana shelf can be divided into three parts: inner, middle, and outer (such division is pointless for a narrow shelf such as the one on the west coast of the United States). The inner shelf (< 50 m) is usually well mixed, especially during winter, and driven principally by the alongshore winds (and affected by any fresh water fluxes at the coast), but the outer shelf (100 to 200 m) is affected principally by offshore currents and eddy motions. The midshelf is affected by processes on both the inner and outer shelves. Midshelf fronts are important features of winter-time circulation, because intense cooling and strong storm-induced wind mixing often mix the inner shelf and make it homogeneous, whereas the outer shelf still retains some stratification in the vertical. The differing temperatures lead to a strong thermal front that is often a source of winter-time cyclogenesis in the overlying atmosphere. Midshelf fronts over the Louisiana-Texas shelf in the Gulf of Mexico are often responsible for generation of winter cyclones that impact the eastern United States.

The fresh water flux at the coast causes a density front and strong along-front geostrophic currents. The fresh water often forms a layer of brackish water and inhibits mixing in the vertical. The result is hypoxia in the waters below, a problem of considerable importance in regions such as that in the vicinity of the outflow from Mississippi-Atchafalaya rivers in the Gulf of Mexico during spring flooding.

The importance of mixing on the shelf by winds and buoyancy fluxes cannot be overemphasized. Deep mixing during winter and inhibition of mixing by strong stratification during summer due to solar heating and during spring by river floods affect the density structure of the water column and this in turn has an effect on the circulation on the shelf. Winds also cause upwelling/downwelling, and the associated coastal jets and undercurrents are important to the transport of water masses in both the alongshore and cross-shore directions. In fact, upwelling constitutes an important mechanism for transport of mass and material (pollutants, fresh water) from the shore to the deep sea. Steady upwelling generates a thermal front roughly paralleling the coastline. Figure 7.1.1 shows both thermal IR and ocean color images of upwelling off the California coast. The complexity of the process involving squirts, eddies, meanders, and so on can be seen in both the thermal structure and the chlorophyll concentrations. Upwelling processes are important to the dynamics of shelves such as those off the Oregon and California coasts during summer and off the Arabian and Somali coasts in the Indian Ocean during the summer monsoons.

Finally, processes in the adjacent deep ocean also have an important effect on the shelf circulation. In the Mid-Atlantic Bight, a principal forcing mechanism is the anticyclonic warm-core rings shed by the Gulf Stream and moving south-

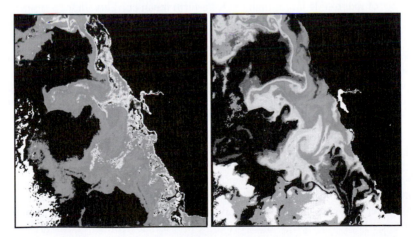

Figure 7.1.1 Ocean color and IR images off the west coast of the United States showing upwelling and the associated front, filaments, eddies, and jets. Courtesy of NASA.

southwest along the continental slope until reabsorbed by the Gulf Stream near Cape Hatteras. These rings transport mass and momentum and have a strong impact on the outer shelf and its circulation. In the South Atlantic Bight, the proximity of the Gulf Stream just off the slope creates frontal cold core eddies that propagate downstream along the shelf edge. On the Texas-Louisiana shelf, it is the eddies shed quasi-periodically by the Loop Current which traverse west, occasionally along the outer slope, that can impact the circulation on the outer shelf. Offshore operations require monitoring of these eddies that pack swirl velocities of as much as 2 m s^{-1}, which can produce large loads on offshore structures and drilling attachments.

The mean gyre scale circulation can also impact the shelf circulation in the form of alongshore currents induced by the sea level slope. On the Mid-Atlantic Bight shelf, there exists a mean current of 5 to 10 cm s^{-1} in the south-southwest direction (Beardsley and Boicourt, 1981) due to the sea level slope induced by the subpolar gyre.

7.2 TIDAL MOTIONS

Tidal sea level fluctuations and tidal currents are an important part of the shelf circulation. We dealt with tides in Chapter 6. Here we will just highlight a few characteristics of tides in coastal seas. Because of the shallow water depths, tides tend to get amplified on the shelf. Shallow water depth also means that nonlinear tides are important on the shelf. Strong tidal currents, which are often an order of magnitude larger than the long-term mean currents, induce significant residual currents, often of the same order as the mean background currents, that are important to the transport of pollutants on the shelf. While tide-induced mixing is very weak and for many practical purposes inconsequential in the deep ocean (except indirectly through internal tides), strong tidal currents are an important source of mixing on many shelves. In the Bering Sea, for example, strong tidal current-driven mixing at the bottom homogenizes the water column on the inner shelf and leads to a thick benthic boundary layer on the outer shelf. The consequence is a single layer density structure on the inner shelf transitioning to a three-layer structure on the outer shelf. Such a density structure has important implications to the biological productivity and management of fisheries in the Bering Sea (Overland *et al.*, 1999). In the Yellow Sea, summertime heating produces a strong two-layer thermal stratification, with upper layers near 28°C and bottom layers near the winter-time values of around 5 to 6 °C. However, spring tides off the Chinese and Korean coasts are strong enough to mix the entire water column in shallower coastal waters so that strong frequent thermal fronts are a fascinating feature of the summertime Yellow Sea circulation. For all these reasons, more often than not, coastal ocean models need to include tidal mixing.

7.3 CONTINENTAL SHELF WAVES

The continental shelf acts as an efficient waveguide for propagation of subinertial motions over long distances along the margins of the ocean basins. These shelf waves have their energy principally in kinetic form and hence have small sea level amplitudes amounting to less than 10 cm. Their period is several days, and they have wavelengths of the order of a thousand kilometers or so. Their motions are confined to the shelf and, like a coastally trapped Kelvin wave, they travel with the coast to the right (left) in the northern (southern) hemisphere. They are rotational wave motions and play an important role in shelf dynamics, since they are a principal means of communicating changes to other locations along the shelf and causing adjustment to changes in wind forcing on the shelf. They were first detected along the Australian coast by Hamon (1962) barely 30 years ago and were explained by Robinson (1964), who coined the term "continental shelf waves" to describe them. They belong to the general category of coastal trapped waves, which also includes nonrotational high frequency edge waves along a beach. Since then a vast literature has built up and these waves have been observed in many parts of the world (for example, Church *et al.,* 1986) becoming an integral part of our understanding of coastal and shelf dynamics. Excellent reviews of the topic exist, for example, Mysak (1980a,b), LeBlond and Mysak (1978, 1979), and Gill (1982), in which can also be found an extensive bibliography on the subject.

Continental shelf waves can be best understood by studying the barotropic shelf waves. Consider therefore the governing equations for unforced barotropic motions in a variable depth ocean (see Figure 7.3.1):

$$\frac{\partial u}{\partial t} - fv + g\frac{\partial \eta}{\partial x} = -u\frac{\partial u}{\partial x} - v\frac{\partial u}{\partial y}$$

$$\frac{\partial v}{\partial t} + fu + g\frac{\partial \eta}{\partial y} = -u\frac{\partial v}{\partial x} - v\frac{\partial v}{\partial y} \qquad (7.3.1)$$

$$\frac{\partial \eta}{\partial t} + \frac{\partial}{\partial x}(Hu) + \frac{\partial}{\partial y}(Hv) = -\frac{\partial}{\partial x}(\eta u) + \frac{\partial}{\partial y}(\eta v)$$

For small amplitude motions of long waves with velocities much smaller than the phase speed of the wave, the nonlinear terms on the RHS of the momentum

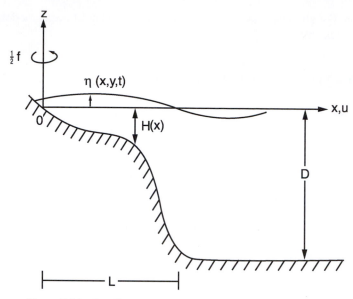

Figure 7.3.1   Coordinate system for barotropic continental shelf waves.

equations can be ignored. Also, the terms on the RHS of the continuity equation are small for $\eta \ll H$. If we linearize the equations by neglecting all terms on the RHS, with some algebraic manipulations this set of equations can be reduced to a single equation in η:

$$
\left(\frac{\partial^2}{\partial t^2}+f^2\right)\left[
\begin{array}{l}
\dfrac{\partial}{\partial t}\left(\dfrac{\partial^2}{\partial x^2}+\dfrac{\partial^2}{\partial y^2}\right)\eta-\dfrac{1}{gH}\left(\dfrac{\partial^2}{\partial t^2}+f^2\right)\dfrac{\partial\eta}{\partial t}+\dfrac{1}{H}\dfrac{\partial H}{\partial x}\left(\dfrac{\partial^2\eta}{\partial x\partial t}+f\dfrac{\partial\eta}{\partial y}\right)\\[3mm]
+\dfrac{1}{H}\dfrac{\partial H}{\partial y}\left(\dfrac{\partial^2\eta}{\partial y\partial t}-f\dfrac{\partial\eta}{\partial x}\right)-\left(\dfrac{\partial f}{\partial y}\dfrac{\partial\eta}{\partial x}-\dfrac{\partial f}{\partial x}\dfrac{\partial\eta}{\partial y}\right)
\end{array}
\right]
$$

$$
+2f^2\left(\frac{\partial f}{\partial y}\frac{\partial\eta}{\partial x}-\frac{\partial f}{\partial x}\frac{\partial\eta}{\partial y}\right)-2f\left(\frac{\partial f}{\partial x}\frac{\partial^2\eta}{\partial x\partial t}+\frac{\partial f}{\partial y}\frac{\partial^2\eta}{\partial y\partial t}\right)=0
$$

$$(7.3.2)$$

This equation of course entertains a vast array of fluid motions including continental shelf waves. Simplifications are preferable before solutions can be sought. The first step is to assume low frequency subinertial motions and to neglect the $\partial^2/\partial t^2$ operator multiplying the square brackets and the last set of terms invoking QG approximation:

$$\frac{\partial}{\partial t}\left(\frac{\partial^2}{\partial x^2}+\frac{\partial^2}{\partial y^2}\right)\eta-\frac{1}{gH}\left(\frac{\partial^2}{\partial t^2}+f^2\right)\frac{\partial\eta}{\partial t}+\frac{1}{H}\frac{\partial H}{\partial x}\left(\frac{\partial^2\eta}{\partial x\partial t}+f\frac{\partial\eta}{\partial y}\right)$$

$$+\frac{1}{H}\frac{\partial H}{\partial y}\left(\frac{\partial^2\eta}{\partial y\partial t}-f\frac{\partial\eta}{\partial x}\right)+\left(\frac{\partial f}{\partial y}\frac{\partial\eta}{\partial x}-\frac{\partial f}{\partial x}\frac{\partial\eta}{\partial y}\right)=0$$

(7.3.3)

This is a generalized form of the equations derived by Mysak (1980a,b), who restricted himself to the f-plane and the case of no depth variations along the y-direction ($\partial H/\partial y = 0$), the direction of wave propagation. Eq. (7.3.3) allows for beta plane solutions as well as an arbitrary orientation of the coastline with respect to the longitude lines (most work on shelf waves considers a coast oriented in the north-south direction). The f-plane dynamics are sufficient for continental shelf wave dynamics (see Kantha, 1984, and Dorr and Grimshaw, 1986, for solutions on a beta plane). We will follow Mysak and put f = constant and $\partial H/\partial y = 0$, and we will seek wavelike solutions of the form $\eta = F(x)\exp i(ky + \sigma t)$. Then Eq. (7.3.3) reduces to

$$\frac{1}{H}\frac{\partial}{\partial x}\left(H\frac{\partial F}{\partial x}\right)+\left(\frac{\sigma^2-f^2}{C^2}-k^2+\frac{fk}{\sigma}\frac{1}{H}\frac{\partial H}{\partial x}\right)F=0 \qquad (7.3.4)$$

This equation along with the boundary conditions Hu = 0 at x = 0 and $\eta \to 0$ as $x \to \infty$ for trapped motions constitutes an eigenvalue problem. The boundary conditions can be written as

$$H\left(fkF+\sigma\frac{\partial F}{\partial x}\right)=0 \quad \text{at } x=0; \ F\to 0 \text{ as } x\to\infty \qquad (7.3.5)$$

The most relevant length scale in the problem is the effective width of the shelf L. The ratio L^2/a^2, called the divergence parameter by Mysak (1980a), where a is the Rossby radius of deformation C/f, is an important nondimensional ratio. Quantities σ/f and kL are two other important nondimensional parameters. For a given off-shore depth profile, it is possible to arrive at a dispersion relationship linking the wavenumber and frequency of the wave motions. However, analytical solutions are possible only for a few very restricted forms of topography, and numerical solutions (Brink and Chapman, 1987) are necessary for an arbitrary profile.

Huthnance (1975) has shown that for any monotonic shelf depth profile that tends to a constant as $x \to \infty$, there exist three types of trapped waves: (1) an infinite set of discrete high frequency waves ($\sigma > f$) that can propagate in either direction, called edge waves; (2) an infinite set of discrete low frequency modes

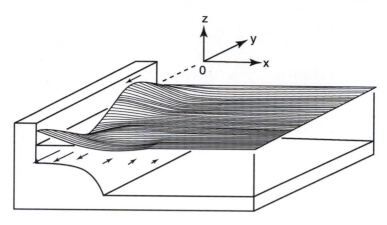

Figure 7.3.2 Continental shelf wave propagating along the coast with the coast to the right (in the northern hemisphere).

($\sigma < f$) that are right-bounded in the northern hemisphere (these are continental shelf waves) (Figure 7.3.2); and (3) a single low frequency Kelvin wave trapped against the continental slope. High frequency superinertial edge waves are not important to shelf dynamics (although they play an important role in processes such as the formation of rip currents, beach cusps, crescentic bars, and surf beat along a beach). At low wave numbers, the edge waves degenerate to leaky Poincare modes, but shelf waves exist at all wave numbers and are important to shelf dynamics. The dispersion relation is continuous for each shelf wave mode. The long waves are nondispersive, but as long as (dH/dx)/H is bounded for all x, the group velocity goes to zero for an intermediate wavenumber. This implies that low wavenumbers have both their phase and energy propagate in the same direction, whereas high wavenumbers have their phase propagate right-bounded, but their energy propagates in the opposite direction.

The right-bounded propagation of shelf waves can be explained by potential vorticity conservation. This elegant explanation was first offered by Longuet-Higgins in 1965 and is presented in Gill (1982). Consider an undisturbed   string of particles parallel to the shoreline located at the depth H_0. The barotropic potential vorticity of these particles is f/H_0. When disturbed in a sinusoidal fashion as shown in Figure 7.3.3, a relative vorticity $\zeta = H\left[(f/H_0) - (f/H)\right]$ is induced to conserve potential vorticity, and hence particles displaced into deeper waters acquire positive (cyclonic) vorticity, while those displaced into shallower waters acquire a negative (anticyclonic) vorticity. These in turn induce velocities that move the particles to positions shown in Figure 7.3.3, the result being that the wave moves with shallow water (or the coast) to the right (in the northern hemisphere).

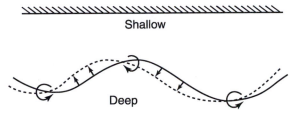

Figure 7.3.3 Conservation of potential vorticity induces vorticities as shown in particles moving offshore and onshore leading to propagation of the wave as shown. From Gill (1982).

It is important to note that a coastal boundary is not essential for the existence of trapped waves. Thus, such waves are possible along midocean ridges. However, the generation of shelf waves by the action of the wind requires the presence of a coastal boundary. The cross-shore Ekman transport produced by alongshore winds requires compensating cross-shore motions below the Ekman layer that force fluid particles to cross the isobaths, the potential vorticity contours in this case. This topographic vortex stretching due to an oscillatory wind stress can give rise to alternating vortices characteristic of shelf waves (Gill, 1982). In mid-latitudes, synoptic disturbances of roughly a week period and with wavelengths of about a 1000 km passing over a shelf are usually responsible for generating shelf waves.

If one is interested primarily in continental shelf waves, it is possible to consider a simpler subset of the governing equations. For these waves, the scale of depth variations on the shelf is much less than the Rossby radius of deformation. In other words, even for barotropic shelf waves, the sea level fluctuations are small (on the order of centimeters); hence, the rigid-lid approximation can be invoked to study their dynamics. Equivalently, these waves have very little potential energy; most of their energy is in kinetic form. This simpler subset can be obtained from Eq. (7.3.3) by ignoring the terms involving gH and obtaining an equation for the sea level perturbation:

$$\frac{\partial}{\partial t}\left(\frac{\partial^2}{\partial x^2}+\frac{\partial^2}{\partial y^2}\right)\eta+\frac{1}{H}\frac{\partial H}{\partial x}\left(\frac{\partial^2\eta}{\partial x\partial t}+f\frac{\partial\eta}{\partial y}\right)$$

$$+\frac{1}{H}\frac{\partial H}{\partial y}\left(\frac{\partial^2\eta}{\partial y\partial t}-f\frac{\partial\eta}{\partial x}\right)+\left(\frac{\partial f}{\partial y}\frac{\partial\eta}{\partial x}-\frac{\partial f}{\partial x}\frac{\partial\eta}{\partial y}\right)=0 \qquad (7.3.6)$$

It is, however, instructive to derive the simpler set from Eq. (7.3.1) itself. Invoking rigid-lid approximation is equivalent to ignoring the $\partial\eta/\partial t$ term in the continuity equation. An equation for vorticity can also be derived from the momentum equations, eliminating the terms containing η in the process:

$$\frac{\partial}{\partial x}(Hu) + \frac{\partial}{\partial y}(Hv) = 0$$

$$\frac{\partial}{\partial t}\left(\frac{\partial v}{\partial x} - \frac{\partial u}{\partial y}\right) + \frac{\partial}{\partial x}(fu) + \frac{\partial}{\partial y}(fv) = 0$$

(7.3.7)

The first equation of this set enables a stream function ψ to be defined: $Hu = -\partial\psi/\partial y; Hv = \partial\psi/\partial x$, so that a single equation governing ψ can be derived:

$$\frac{\partial}{\partial t}\left[\frac{\partial}{\partial x}\left(\frac{1}{H}\frac{\partial\psi}{\partial x}\right) + \frac{\partial}{\partial y}\left(\frac{1}{H}\frac{\partial\psi}{\partial y}\right)\right] + \frac{\partial\psi}{\partial x}\frac{\partial}{\partial y}\left(\frac{f}{H}\right) - \frac{\partial\psi}{\partial y}\frac{\partial}{\partial x}\left(\frac{f}{H}\right) = 0$$

(7.3.8)

Note that Eqs. (7.3.6) and (7.3.8) are equivalent. They are also more general in that since both $\partial f/\partial x$ and $\partial H/\partial y$ are retained, they allow for the more general case of propagation along a coast inclined at an angle to the latitudinal direction on a beta plane to be considered. Note the appearance of the barotropic potential vorticity terms (f/H) in Eq. (7.3.8). This is natural since these waves are rotational. We now invoke the usual approximations f = constant and $\partial H/\partial y=0$, and seek wave solutions of the form $\psi(x,y,t) = \varphi(x)\exp i(ky + \sigma t)$. The eigenvalue problem then reduces to

$$H\frac{\partial}{\partial x}\left(\frac{1}{H}\frac{\partial\varphi}{\partial x}\right) - \left[k^2 + \frac{fk}{\sigma}H\frac{\partial}{\partial x}\left(\frac{1}{H}\right)\right]\varphi = 0 \qquad (0 < x < \infty)$$

(7.3.9)

$$\varphi = 0 \text{ at } x = 0; \ \varphi \to 0 \text{ as } x \to \infty$$

These equations allow only shelf wave solutions; the Kelvin and edge waves have been filtered out by the rigid-lid approximation. In applications of these equations to depth profiles involving discontinuities in H or $\partial H/\partial x$, it is important to assure the continuity of stream function φ and the pressure (or, equivalently, elevation) $(\partial\varphi/\partial x - fk\varphi/\sigma)/H$ across the discontinuity (Mysak, 1980a).

These equations still defy analytical solutions for arbitrary depth profiles. Closed form solutions are possible for two special cases: a step shelf and an exponential shelf. For a step shelf,

$$H(x) = d \qquad 0 < x < L$$

$$H(x) = D \qquad L < x < \infty$$

(7.3.10)

the solution is

$$\varphi = A \ \sin(kx) \qquad\qquad\qquad 0 < x < L$$
$$\varphi = A \ \sin(kL)\exp\left[-k(x-L)\right] \qquad L < x < \infty \tag{7.3.11}$$

and demanding continuity of φ and pressure at $x = L$ gives the dispersion relation

$$\sigma = f(1-r)/(r+\coth \ kL) \tag{7.3.12}$$

where $r = d/D$. Low wavenumbers ($kL \sim 0$) are nondispersive: $\sigma \sim f(1-r)kL$, but high wave numbers ($kL \rightarrow \infty$) are double Kelvin waves trapped along the single-step shelf: $\sigma \sim f(1-r)/(1+r)$. Note also that because of the singular nature of $\partial H/\partial x$, there is only one mode; the higher modes have been filtered out.

This limitation does not exist for the exponential shelf considered originally by Buchwald and Adams (1968), and there exists an infinite discrete set:

$$H(x) = d\exp(2bx) \qquad\qquad 0 < x < L$$
$$H(x) = d\exp(2bL) = D \qquad L < x < \infty \tag{7.3.13}$$

The solution is

$$\varphi = A \ \sin(mx) \ \exp\left[b(x-L)\right] \qquad 0 < x < L$$
$$\varphi = A \ \sin(mL) \ \exp\left[-k(x-L)\right] \qquad L < x < \infty \tag{7.3.14}$$

with the dispersion relation obtained by demanding continuity of pressure:

$$\sigma = \frac{2bfk}{m^2 + k^2 + b^2}; \quad \tan \ (mL) = \frac{-m}{b+k} \tag{7.3.15}$$

There exists, therefore, a countably infinite number of discrete modes, right-bounded. Figure 7.3.4 shows the first five modes for particular values of b, D, and L: L = 80 km, bL = 2.7, and D = 5 km. Note their nondispersive nature at low wavenumbers and zero group velocity at an intermediate wavenumber. The elevation corresponding to each mode n has maximum value at the coast and n zero crossings across the shelf. For waves with wavelengths large compared to the scale of cross-shore depth change, the width of the shelf b^{-1} and $m \sim b$, $\sigma = fk/b$. The phase speed $c \sim f \ b^{-1}$ is therefore a function of the shelf width. Along the west coast where the shelf is narrow, b^{-1} is on the order of 10 to 15

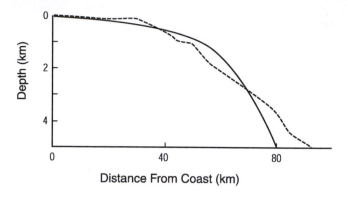

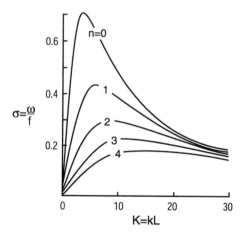

Figure 7.3.4 Dispersion relation of shelf waves for the first five modes (bottom) for an exponential profile (top). From Buchwald and Adams (1968).

km, the phase speed is ~ 1 to 1.5 m s^{-1}, whereas on the east coast with its broad shelves, b^{-1} is of the order of 60 km, so that the phase speed is ~6 m s^{-1}. These are typical of the observed phase speeds of continental shelf waves, 2 to 6 m s^{-1}.

The exponential depth profile considered above captures qualitatively the general behavior of the barotropic shelf waves for a more arbitrary profile. However, for precise quantitative estimates of characteristics such as the phase speed, it is preferable to obtain numerical solutions (Brink and Chapman, 1987).

Mysak (1980a) shows why the atmospheric pressure fluctuations are not very efficient in generating shelf waves (in the absence of resonant interaction). Continental shelf waves are generated on the shelf primarily by fluctuations in the alongshore winds. The generation can be studied by considering the forced form of the momentum equations in Eq. (7.3.1). To a good approximation, the alongshore flow is in geostrophic equilibrium so that the first term in Eq. (7.3.8) can be neglected and Eq. (7.3.8) becomes

$$\frac{\partial}{\partial t \partial x}\left(\frac{1}{H}\frac{\partial \psi}{\partial x}\right) + \frac{\partial \psi}{\partial x}\frac{\partial}{\partial y}\left(\frac{f}{H}\right) - \frac{\partial \psi}{\partial y}\frac{\partial}{\partial x}\left(\frac{f}{H}\right) = -\frac{\tau_y}{H^2}\frac{\partial H}{\partial x} \qquad (7.3.16)$$

The second term is zero for $\partial H / \partial y = 0$ on an f-plane. If one seeks solutions in terms of a sum of shelf wave modes $\psi = \sum A_n(y,t) j_n(x)$, where φ_n are the shelf wave modes, their amplitudes A_n obey a simple first-order forced wave equation,

$$\frac{1}{c_n}\frac{\partial A_n}{\partial t} - \frac{\partial A_n}{\partial y} = \frac{a_n}{f}\tau_y(y,t) \qquad (7.3.17)$$

which can be readily solved by method of characteristics. c_n is the phase speed of each mode, and a_n is a Fourier coefficient. For a periodic moving disturbance, the response is a series of forced disturbances moving with the speed of the disturbance. Because of its forced nature, disturbances can propagate in either direction on the shelf. A similar mechanism applies to coastal trapped waves in a stratified fluid. In practice, it is necessary to add the bottom friction or other dissipation terms to the above equation. Hickey (1979) has looked at generation and propagation of disturbances along the west coast of the US.

Bottom friction is not the only means of dissipation of continental shelf waves. Long waves propagating long distances along a coast can leak their energy into westward propagating Rossby modes. Such shelf waves on a beta plane have been studied by Dorr and Grimshaw (1986). Kantha (1984) studied leaky shelf waves for a generalized exponential depth profile that includes the step shelf and the Buchwald and Adams profiles as subsets:

$$H(x) = d \exp(-2bL) \exp(2bx) \qquad 0 < x < L$$
$$H(x) = D \qquad\qquad\qquad L < x < \infty \qquad (7.3.18)$$

Note that $b = 0$ recovers step shelf and $d = D$ recovers the Buchwald and Adams profile. Starting from Eq. (7.3.8) and seeking wavelike solutions as before (neglecting $\partial H / \partial y$):

$$\frac{\partial^2 \varphi}{\partial x^2} + \left\{ H\frac{\partial}{\partial x}\left(\frac{1}{H}\right) - i\frac{f}{\sigma}\right\}\frac{\partial \varphi}{\partial x} - \left[k_y^2 + \frac{fk_y}{\sigma}H\frac{\partial}{\partial x}\left(\frac{1}{H}\right)\right]\varphi = 0 \quad (0 < x < \infty) \qquad (7.3.19)$$

The equation now involves complex quantities, and we need to allow the wave number to have components in the x and y directions and for the possibility σ is complex. To allow for leaky modes, the open ocean boundary condition needs to be

$$\varphi(x) = \exp(-ik_x x) \quad L < x < \infty \tag{7.3.20}$$

It is straightforward to derive the dispersion relation by matching at $x = L$:

$$\frac{\sigma}{f} = \frac{\left(1 - \dfrac{d}{D}\right) - i\dfrac{\beta}{2fk_y}}{\dfrac{A}{k_y}\coth AL + \dfrac{b}{k_y} + i\dfrac{k_x}{k_y}\dfrac{d}{D}}$$

$$A = \left[k_y^2 - 2b\frac{fk_y}{\sigma} + b^2 - \frac{\beta^2}{4\sigma^2} + i\frac{b\beta}{\sigma}\right]^{1/2} \tag{7.3.21}$$

$$k_x^2 + k_y^2 + \frac{\beta k_x}{\sigma} = 0$$

The last relationship of Eq. (7.3.21) is the Rossby wave dispersion relation for waves with wavenumbers such that their energy is mainly kinetic. If β is put to zero, we get the usual nonleaky shelf wave dispersion relation:

$$\frac{\sigma}{f} = \frac{\left(1 - \dfrac{d}{D}\right)}{\dfrac{A_0}{k_y}\coth A_0 L + \dfrac{b}{k_y} + \dfrac{d}{D}}; \quad A_0 = \left[k_y^2 - 2b\frac{fk_y}{\sigma} + b^2\right]^{1/2} \tag{7.3.22}$$

Putting $b = 0$ recovers the step shelf dispersion relation Eq. (7.3.12), and putting $d = D$ gives $A_0 \coth (A_0 L) + b + k_y = 0$, with $A_0^2 = k_y^2 + b^2 - 2bfk_y/\sigma = -m^2$, the Buchwald and Adams case shown in Eq. (7.3.15).

For nonzero β, closed form solutions are not possible unless we expand in terms of a small parameter $\alpha = \beta L/f$. For small α, σ_r satisfies Eq. (7.3.22), but

$$\sigma_i = \frac{\alpha \sigma_r}{2k_y L} + o(\alpha^2) \tag{7.3.23}$$

For a step shelf ($b = 0$), in the limit $k_y L = 0$,

$$\sigma_r \sim f(1-r)k_y L; \quad \sigma_i \sim \frac{\alpha}{2}(1-r); \quad \frac{1}{fE}\frac{dE}{dt} = -2\frac{\sigma_i}{f} \tag{7.3.24}$$

where E is the energy (kinetic in this case) integrated over a cycle and over x. The rate of decay of energy due to leakage into Rossby waves of the "trapped" wave is proportional to $\beta L/f$.

Stratification and alongshore mean currents modify continental shelf waves. Stratification in particular permits the existence of other subinertial trapped modes such as internal Kelvin waves that can interact effectively with the shelf waves. These aspects can be studied in two idealized cases: (1) a two-layer stratification and (2) continuous but uniform stratification. For the two-layer case, the linear governing equations are given by Eq. (4.1.6), except that the continuity equation for the upper layer is simpler due to the rigid-lid approximation; the term involving the free surface can be ignored, filtering out external surface and Kelvin modes. The important additional parameter is now the internal Rossby radius of deformation $a_i = C_i/f$, and the important nondimensional parameter is the ratio L/a_i, where L is the cross-shore scale of depth changes or equivalently the shelf width. If there is no cross-shore depth variation, but a vertical wall is present at the coast, only a right-bounded nondispersive internal Kelvin wave traveling at a velocity C_i and trapped over a cross-shore scale of the order of a_i can exist. This wave is of course modified by any topographic changes, and stratification in turn affects the shelf waves (Mysak, 1980a).

Two stratification cases are possible (Figure 7.3.5), with deep-sea stratification typical of the east Australian coast and on-shelf stratification typical of the west coast of the United States. In the former case, the stratification tends to enhance considerably the phase speed of the gravest long wavelength shelf wave, whereas in the latter case, the enhancement is quite small (Mysak, 1980a). Thus, the influence of stratification depends on whether it is in the deep or on the shelf. Allen (1977) studied the on-shelf stratification case and the exponential depth profile (considered by Buchwald and Adams, Eq. 7.3.13). Figure 7.3.6a shows the resulting dispersion curves of continental shelf waves. Two aspects are noteworthy. First, the dispersion curves for the shelf modes are substantially the same as those for the nonstratified case. Second, it is easy to see that a close coupling can exist between the shelf modes and the internal Kelvin mode. In fact, the intersection of the shelf mode and the Kelvin mode is somewhat spurious as indicated in Figure 7.3.6b, and instead the modes simply change their nature, the Kelvin mode becoming a shelf mode and vice versa as the intersection point is approached.

Continental shelf wave solutions for uniformly stratified fluid studied in detail by Wang and Mooers (1976) and Huthnance (1978) can be found in Mysak (1980a). The solutions are complex for arbitrary stratification and bottom slope. The governing parameter is the stratification parameter, $S = a_i^2/L^2 = N^2 D^2/(f^2 L^2)$, where a_i is the internal Rossby radius of deformation. Two mechanisms for trapping motions on the shelf are operative: (1) the

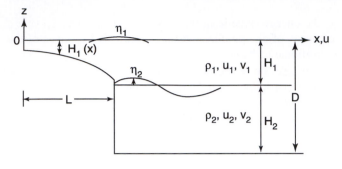

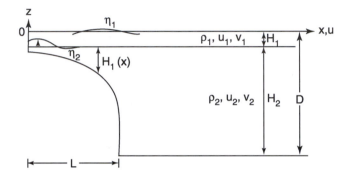

Figure 7.3.5 Two-layer stratification models for baroclinic shelf waves. From Mysak (1980).

presence of the coastal wall, which imposes the constraint of zero normal mass flux, giving rise to motions with a trapping scale of the baroclinic Rossby radius, and (2) the cross-shore bottom slope imposing the constraint of potential vorticity conservation on motions over the sloping bottom, giving rise to topographic Rossby waves with a trapping scale on the order of the shelf width. S is a function of the ratio of these two trapping scales. For weak stratification (S ~ 0), as expected, the shelf waves resemble barotropic shelf waves. For strong stratification, their structure resembles internal Kelvin waves. Short waves become bottom-trapped Rossby waves. In general, for long waves, the flow field tends to be barotropic over the shelf and slightly bottom-trapped over the slope (Mysak, 1980a). No mode coupling was observed for continuous stratification (Huthnance, 1978). Shelf waves in a stratified ocean are usually called coastal trapped waves. There is evidence for equatorial internal Kelvin waves propagating eastward in the equatorial Pacific waveguide to propagate as coastal trapped waves poleward in the coastal waveguide once they encounter the west coast of the Americas.

Coastal trapped waves are also a feature of circulation around islands and

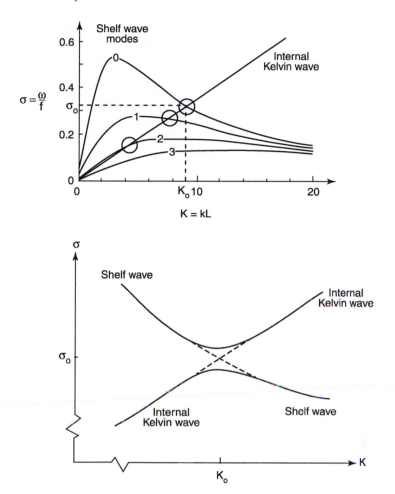

Figure 7.3.6 Superposition of typical dispersion relationship for shelf waves (first four modes) and the internal Kelvin wave (top); details at the transition point between internal Kelvin and shelf wave modes, which shows that the curves corresponding to different wave types do not actually intersect but change families (bottom). From Allen (1975).

submerged topographic features such as seamounts and guyots. In this case, in addition to the dispersion relation relating the wavenumber and the frequency of the trapped wave, there is an additional constraint on the permissible wavenumbers that the number of wavelengths around the island be an integral value. The solutions can be derived as eigenvalues of the inviscid, linear, hydrostatic set of equations in cylindrical coordinates for axisymmetric stratified flow around the feature subject to the usual boundary conditions at the top and the bottom (Brink, 1989; Codiga, 1997a). The result is a discrete set of resonant topographic waves under the influence of stratification and trapped to the bottom. These subinertial waves (n < f) travel around in the anticyclonic direction, with the coast to the right (left) in the northern (southern) hemisphere. The phenomena is of particular importance to flow processes in deep basins such as the Pacific, since the barotropic tidal current appears to be considerably amplified over some seamounts, leading to potentially large dissipation of tidal energy on their flanks. Very high dissipation rates have indeed been observed in microstructure measurements on the flanks of seamounts such as the Cobb and Fieberling seamounts in the north Pacific. Currents at Fieberling Seamount are dominated by strong clockwise mean flow and amplified diurnal tides (Brink, 1995; Eriksen, 1991). Those around Cobb Seamount show similar features. Diurnal tidal currents are amplified by roughly five times the ambient values in the vicinity of the seamount, with clockwise propagation around it and clockwise rotation in time (Codiga and Eriksen, 1997). For an excellent and most recent description and discussion of the physics of subinertial waves trapped around a seamount, see Codiga (1997a,b).

7.4 MODELING SHELF CIRCULATION

With the exception of perhaps the deep western boundary undercurrents, there are very few regions in the deep basins where bottom currents are strong enough to make bottom friction very important to the overall dynamics. This is not true for the coastal oceans. Strong pulses of wind-driven, narrow, swift currents of as much as 50 cm s^{-1} have been observed adjacent to the coast on the LATEX shelf. More often than not, on the inner shelf in waters less than ~25 m deep, the stress applied by the winds at the surface tends to be counterbalanced by the bottom stress. Even on the middle shelf, the bottom stress is important to the shelf dynamics. This has important implications to modeling the coastal oceans. It is important to resolve the bottom boundary layer on the shelf ,and hence a topographically conformal vertical coordinate system is in general preferable, although it is, in principle, possible to use a z-coordinate system, provided very high vertical resolution is provided in the upper 100 to 200 m. This is one reason why despite some of its limitations (see Chapter 9), a sigma coordinate is the coordinate of choice in coastal ocean modeling. It is also important to model accurately the mixing driven by winds, tides, and bottom currents. In addition, the free surface must be explicitly simulated by the model, since it is central to many processes on the shelf, such as barotropic tides, storm surges, and coastal upwelling/downwelling. This rules out rigid lid models for all but studies of coastal processes where free surface dynamics is not essential. An additional complication is that the coastal waveguide must be well resolved, and since the cross-shore spatial scale of the waveguide is the width of the shelf itself, and the coastline shape is seldom simple, this introduces severe constraints on the horizontal coordinate and resolution requirements. This means that it is nearly impossible to resolve the shelf properly in basin scale and global models, especially a narrow shelf, unless one uses a finite element approach. Regional high resolution models are one possibility to circumvent the resolution problem, but they introduce their own complications due to the open lateral boundaries.

Accurate modeling of the circulation on the shelf is much harder than modeling the circulation in the deep basins. The principal reason is the much smaller temporal and spatial scales and the rich variety of processes involved. Since wind and buoyancy flux forcing by the atmosphere are so important to shelf dynamics, the major problem in modeling the shelf is the dearth of accurate, high resolution winds and buoyancy fluxes. For synoptic simulations, the only recourse is to NWP analyses (apart from point measurements by off-shore buoys). At present, these do not have the resolution needed to resolve important atmospheric processes in the coastal oceans (this situation is, however, changing). Nor do they do a good job of simulating the land-sea transition (land-sea breezes) and its feedback on the atmospheric boundary layer. Often a reanalysis is possible using coastal and off-shore buoy measurements (for example, Gordon *et al.*, 1986; Kantha, 1985), but only in a hindcast mode. For forecasting the state of the shelf, the only recourse is to use NWP products. Therefore it is important to remember that the accuracy of any simulation in coastal oceans is a strong function of the accuracy of the atmospheric forcing available.

Finally, a proper treatment of open boundaries and conditions on them is of paramount importance to regional coastal models. Shelf waves propagating into the domain and carrying information on wind-forced events elsewhere on the shelf must be accommodated (Gordon *et al.*, 1986; Kantha, 1985). This is best done by nesting the regional model in a coarser basin scale model, although this is not always feasible, especially in a forecast mode, due to lack of suitable basin scale models. In such circumstances, the best that can be done on cross-shore boundaries is some sort of radiation boundary conditions. The off-shore boundary is usually less of a problem since the slope often acts as the boundary of the waveguide, and except when the deep ocean has a strong effect through mean and eddy motions and such, it is possible to use clamped boundary conditions with essentially tidal forcing prescribed on the off-shore boundary.

7.5 BAROTROPIC MODELS

In Chapter 9 we will describe a comprehensive ocean model that can be used to pattern the coastal ocean and the complex processes involved. Here, we confine ourselves to a simple subset that can nevertheless simulate some coastal processes. If horizontal density gradients are neglected in the governing equations, or alternatively if the ocean is considered to be of uniform density, the velocity distribution in the water column becomes independent of depth (away from regions of frictional influence such as the surface and the bottom). Under these conditions, it is possible to ignore equations involving fluid density (and hence the transport equations for T and S) and to integrate the resulting governing equations for continuity and momentum over the depth of the water column to arrive at a vertically integrated set of equations that govern the sea surface elevation η and the vertically averaged velocity components U_j. In orthogonal curvilinear coordinate system suited to regional models and using tensorial notation, they are

$$\frac{\partial \eta}{\partial t}+\frac{1}{h_1 h_2}\left[\frac{\partial}{\partial \xi_1}(h_2 U_1 D)+\frac{\partial}{\partial \xi_2}(h_1 U_2 D)\right]=0 \tag{7.5.1}$$

$$\frac{\partial}{\partial t}(U_1 D)+\frac{1}{h_1 h_2}\left[\frac{\partial}{\partial \xi_1}(h_2 U_1^2 D)+\frac{\partial}{\partial \xi_2}(h_1 U_1 U_2 D)\right]$$

$$-f U_2 D+\frac{U_1 U_2 D}{h_1 h_2}\frac{\partial h_1}{\partial \xi_2}-\frac{U_2^2 D}{h_1 h_2}\frac{\partial h_2}{\partial \xi_1} \tag{7.5.2}$$

$$=-gD\frac{1}{h_1}\frac{\partial}{\partial \xi_1}\left[\beta(\eta-\eta_a)-\alpha\zeta\right]+(\tau_1-\tau_{b1})+DF_1$$

$$\frac{\partial}{\partial t}(U_2 D)+\frac{1}{h_1 h_2}\left[\frac{\partial}{\partial \xi_1}(h_2 U_1 U_2 D)+\frac{\partial}{\partial \xi_2}(h_1 U_2^2 D)\right]$$

$$+fU_1 D+\frac{U_1 U_2 D}{h_1 h_2}\frac{\partial h_2}{\partial \xi_1}-\frac{U_1^2 D}{h_1 h_2}\frac{\partial h_1}{\partial \xi_2} \tag{7.5.3}$$

$$=-gD\frac{1}{h_2}\frac{\partial}{\partial \xi_2}\left[\beta(\eta-\eta_a)-\alpha\zeta\right]+(\tau_2-\tau_{b2})+DF_2$$

$$U_j=\frac{1}{D}\int_{-H}^{\eta} u_j dz \tag{7.5.4}$$

$$\tau_{bj}=c_d\left|U_j\right|U_j$$

where H is the bottom depth and $D = H + \eta$ is the total depth of the water column. $\overline{F}_j$ denotes the horizontal diffusion terms. η_a represents the atmospheric pressure forcing, where the atmospheric pressure has been written as $P_a = \rho_0\, g\, \eta_a$. Quantity τ_0 is the kinematic wind stress. The bottom friction τ_{bj} (kinematic) in a barotropic model can only be parameterized in terms of the column average velocity U_j, and the form adopted above corresponds to quadratic friction. The bottom friction coefficient c_d is empirically determined; the most commonly used value is around 0.0025. Coefficient β is also put equal to unity in regional models. Note the presence of the gravitational tidal potential terms involving ζ on the right-hand side of the momentum equations that contain astronomical forcing terms due to the gravitational forces of the Moon and the Sun. The astronomical forcing can be prescribed a priori from a knowledge of the ephemerides of the Sun and the Moon (Chapter 6; see also Kantha, 1995b). This forcing is important in marginal seas with restricted access to ocean basins, such as the Mediterranean Sea and the sea of Japan. However, direct astronomical forcing is relatively unimportant in most coastal oceans, where tides tend to be cooscillating, that is, driven by tides in the adjacent basin. We will therefore ignore these terms. Tidal forcing in regional coastal models is prescribed through boundary conditions on the sea surface height at boundaries open to the deep sea.

The atmospheric forcing terms, the wind stress, and pressure can be prescribed as a function of time during the model run. Since astronomical tidal forcing is included also, this set of equations can be used to study the barotropic response of the ocean to both atmospheric and tidal forcing. Thus it can be used to solve for the sea surface height (SSH) and depth-averaged currents due to phenomena such as tides and storm surges. The former was dealt with in Chapter 6. In this chapter, we will consider the response to atmospheric forcing and storm surges.

In the absence of wind and tidal forcing, and frictional influences, the steady solution to Eqs. (7.5.2) and (7.5.3) is $\eta = \eta_a$. This is nothing but a statement of hydrostatic balance in which the sea level adjusts instantaneously to changes in atmospheric pressure at its surface. It implies that the ocean responds to pressure forcing as an inverse barometer with roughly 1 cm (actually 0.993 cm) increase (decrease) for every millibar drop (rise) in atmospheric pressure. This is not always true and the departures from inverse barometer response are quite important to satellite ocean altimetry (Kantha et al., 1994). These departures can be significant at high frequencies and on broad shelves and semienclosed seas. The situation here is akin to considering the tides as equilibrium tides, obtained by neglecting all terms in Eqs. (7.5.2) and (7.5.3), except the sea surface elevation and the tidal potential terms, to obtain $\eta = \alpha\zeta/\beta$. In both cases, to account for the real response of the ocean, it is necessary to solve the full equations with the forcing terms on the RHS due to atmospheric (and tidal) forcing. As in the case of tidal forcing, the oceanic response to pressure forcing may consist of a large resonant response in some regions of the global ocean. The only difference between a barotropic tidal model and a barotropic pressure-forced model is that the tidal forcing is narrow-band and composed of many different discrete frequencies, whereas the pressure forcing is broad-band and has a continuous spectrum. Tidal forcing can be determined a priori centuries in advance, whereas the pressure forcing can only be obtained from numerical weather forecast models, whose forecast skill deteriorates beyond a few days.

It is relatively straightforward to solve Eqs. (7.5.1) to (7.5.4) numerically, using either implicit or explicit, finite element or finite difference schemes. This is the principal reason for the proliferation of regional barotropic coastal ocean and estuarine models. Finite element barotropic models have long been the mainstay of the engineering community for coastal and estuarine research (Pinder and Gray, 1977, and Westerink *et al.*, 1994, for example). In general, the higher the resolution, the better the results, and resolutions in subkilometer range are not unusual. The principal problem lies, however, in resolving the complicated topography and prescribing appropriate lateral open boundary conditions (see Chapter 2). There is a need to fine-tune the bottom friction coefficient, which is essentially an ignorance parameter in this type of model.

Equation set (7.5.1) to (7.5.3) is very much similar to the reduced gravity equations (3.5.1) to (3.5.3), except for the tensorial notation, and the appearance of the surface pressure forcing and bottom friction terms. Therefore, horizontal discretization considerations are also similar. However, because of the high spatial resolutions required and used, the grid of choice in regional coastal ocean models is the C grid.

The finite differencing of Eqs. (7.5.1) to (7.5.4) on the C grid leads to equations very much similar to Eqs. (3.5.12) to (3.5.14) except for additional terms on the RHS due to the presence of bottom friction and atmospheric forcing terms. We will, however, use the control volume approach discussed in Section 3.5 that leads to finite difference equations of the form used in Eqs. (3.5.26) to (3.5.29):

$$
\begin{aligned}
\left(\Delta x_{i,j}\Delta y_{i,j}\right)\frac{d\eta_{i,j}}{dt} = -\frac{1}{4}&\left[
\begin{array}{l}
\left(D_{i+1,j}+D_{i,j}\right)\left(\Delta y_{i,j}+\Delta y_{i+1,j}\right)U_{i+1,j} \\
-\left(D_{i,j}+D_{i-1,j}\right)\left(\Delta y_{i-1,j}+\Delta y_{i,j}\right)U_{i,j}
\end{array}
\right] \\
-\frac{1}{4}&\left[
\begin{array}{l}
\left(D_{i,j+1}+D_{i,j}\right)\left(\Delta x_{i,j}+\Delta x_{i,j+1}\right)V_{i,j+1} \\
-\left(D_{i,j}+D_{i,j-1}\right)\left(\Delta x_{i-1,j}+\Delta x_{i,j}\right)V_{i,j}
\end{array}
\right]
\end{aligned}
\tag{7.5.5}
$$

$$\frac{1}{8}\left(\Delta x_{i-1,j}+\Delta x_{i,j}\right)\left(\Delta y_{i-1,j}+\Delta y_{i,j}\right)\frac{d}{dt}\left[U_{i,j}\left(D_{i-1,j}+D_{i,j}\right)\right]=$$

$$-\frac{1}{4}\left[\left(U_{i+1,j}+U_{i,j}\right)^2\Delta y_{i,j}D_{i,j}-\left(U_{i,j}+U_{i-1,j}\right)^2\Delta y_{i-1,j}D_{i-1,j}\right]$$

$$-\frac{1}{32}\begin{bmatrix}\left(D_{i,j+1}+D_{i,j}+D_{i-1,j+1}+D_{i-1,j}\right)\left(\Delta x_{i,j+1}+\Delta x_{i,j}+\Delta x_{i-1,j+1}+\Delta x_{i-1,j}\right)\\ \left(V_{i,j+1}+V_{i-1,j+1}\right)\left(U_{i,j+1}+U_{i,j}\right)\\ -\left(D_{i,j}+D_{i,j-1}+D_{i-1,j}+D_{i-1,j-1}\right)\left(\Delta x_{i,j}+\Delta x_{i,j-1}+\Delta x_{i-1,j}+\Delta x_{i-1,j-1}\right)\\ \left(V_{i,j}+V_{i-1,j-1}\right)\left(U_{i,j}+U_{i,j-1}\right)\end{bmatrix}$$

$$+\frac{1}{64}\left(\Delta x_{i-1,j}+\Delta x_{i,j}\right)\left(\Delta y_{i-1,j}+\Delta y_{i,j}\right)\left(f_{p\,i-1,j}+f_{p\,i,j}\right)\left(V_{i,j}+V_{i,j+1}+V_{i-1,j}+V_{i-1,j+1}\right)$$

$$\left(D_{i-1,j}+D_{i,j}\right)-\frac{1}{2}g\left(D_{i,j}+D_{i-1,j}\right)\left(\Delta y_{i,j}\eta_{i,j}-\Delta y_{i-1,j}\eta_{i-1,j}\right)$$

$$+\frac{1}{8}\left(\Delta x_{i-1,j}+\Delta x_{i,j}\right)\left(\Delta y_{i-1,j}+\Delta y_{i,j}\right)\left(\tau_{i,j}^x+\tau_{i-1,j}^x\right)$$

$$+0.5g\left(D_{i-1,j}+D_{i,j}\right)\left(\Delta y_{i,j}\eta_{a\,i,j}-\Delta y_{i-1,j}\eta_{a\,i-1,j}\right)$$

$$-\frac{1}{4}c_d\left[U_{i,j}^2+\frac{1}{16}\left(V_{i,j}+V_{i,j+1}+V_{i-1,j}+V_{i-1,j+1}\right)^2\right]^{1/2}U_{i,i}\left(\Delta x_{i-1,j}+\Delta x_{i,j}\right)\left(\Delta y_{i-1,j}+\Delta y_{i,j}\right)$$

$$+\frac{1}{2}\left(D_{i,j}+D_{i-1,j}\right)\left\{A_{i,j}\frac{\Delta y_{i,j}}{\Delta x_{i,j}}\left(U_{i+1,j}-U_{i,j}\right)-A_{i-1,j}\frac{\Delta y_{i-1,j}}{\Delta x_{i-1,j}}\left(U_{i,j}-U_{i-1,j}\right)\right\}$$

$$+\frac{1}{8}\left(D_{i,j}+D_{i-1,j}\right)\begin{Bmatrix}\left(A_{i,j}+A_{i,j+1}+A_{i-1,j+1}+A_{i-1,j}\right)\left(\dfrac{\Delta x_{i,j}+\Delta x_{i,j+1}+\Delta x_{i-1,j+1}+\Delta x_{i-1,j}}{\Delta y_{i,j}+\Delta y_{i,j+1}+\Delta y_{i-1,j+1}+\Delta y_{i-1,j}}\right)\\ \left[U_{i,j+1}-U_{i,j}\right]\\ -\left(A_{i,j-1}+A_{i,j}+A_{i-1,j}+A_{i-1,j-1}\right)\left(\dfrac{\Delta x_{i,j-1}+\Delta x_{i,j}+\Delta x_{i-1,j}+\Delta x_{i-1,j-1}}{\Delta y_{i,j-1}+\Delta y_{i,j}+\Delta y_{i-1,j}+\Delta y_{i-1,j-1}}\right)\\ \left[U_{i,j}-U_{i,j-1}\right]\end{Bmatrix}$$

$$(7.5.6)$$

$$\frac{1}{8}\left(\Delta x_{i,j}+\Delta x_{i,j-1}\right)\left(\Delta y_{i,j}+\Delta y_{i,j-1}\right)\frac{d}{dt}\left[V_{i,j}\left(D_{i,j}+D_{i,j-1}\right)\right]=$$

$$-\frac{1}{4}\left[\left(V_{i,j+1}+V_{i,j}\right)^2\Delta x_{i,j}D_{i,j}-\left(V_{i,j}+V_{i,j-1}\right)^2\Delta x_{i,j-1}D_{i,j-1}\right]$$

$$-\frac{1}{32}\begin{bmatrix}\left(D_{i+1,j}+D_{i,j}+D_{i+1,j-1}+D_{i,j-1}\right)\left(\Delta y_{i+1,j}+\Delta y_{i,j}+\Delta y_{i+1,j-1}+\Delta y_{i,j-1}\right)\\ \left(U_{i+1,j}+U_{i+1,j-1}\right)\left(V_{i,j}+V_{i+1,j}\right)\\ -\left(D_{i,j}+D_{i-1,j}+D_{i,j-1}+D_{i-1,j-1}\right)\left(\Delta y_{i,j}+\Delta y_{i-1,j}+\Delta y_{i,j-1}+\Delta y_{i-1,j-1}\right)\\ \left(U_{i,j}+U_{i,j-1}\right)\left(V_{i-1,j}+V_{i,j}\right)\end{bmatrix}$$

$$-\frac{1}{64}\left(\Delta x_{i,j}+\Delta x_{i,j-1}\right)\left(\Delta y_{i,j}+\Delta y_{i,j-1}\right)\left(f_{p\,i,j}+f_{p\,i,j-1}\right)\left(U_{i,j}+U_{i+1,j}+U_{i,j-1}+U_{i+1,j-1}\right)$$

$$\left(D_{i,j}+D_{i,j-1}\right)-\frac{1}{2}g\left(D_{i,j}+D_{i,j-1}\right)\left(\Delta x_{i,j}\eta_{i,j}-\Delta x_{i,j-1}\eta_{i,j-1}\right)$$

$$+\frac{1}{8}\left(\Delta x_{i,j}+\Delta x_{i,j-1}\right)\left(\Delta y_{i,j}+\Delta y_{i,j-1}\right)\left(\tau^y_{i,j}+\tau^y_{i,j-1}\right)$$

$$+\frac{1}{2}g\left(D_{i,j}+D_{i,j-1}\right)\left(\Delta x_{i,j}\eta_{a\,i,j}-\Delta x_{i,j-1}\eta_{a\,i,j-1}\right)$$

$$-\frac{1}{4}c_d\left[\frac{1}{16}\left(U_{i,j}+U_{i+1,j}+U_{i,j-1}+U_{i+1,j-1}\right)^2+V^2_{i,j}\right]^{1/2}V_{i,j}\left(\Delta x_{i,j}+\Delta x_{i,j-1}\right)\left(\Delta y_{i,j}+\Delta y_{i,j-1}\right)$$

$$+\frac{1}{8}\left(D_{i,j}+D_{i-1,j}\right)\left\{\begin{matrix}\left(A_{i,j}+A_{i+1,j}+A_{i,j-1}+A_{i+1,j-1}\right)\left(\dfrac{\Delta x_{i,j}+\Delta x_{i+1,j}+\Delta x_{i,j-1}+\Delta x_{i+1,j-1}}{\Delta y_{i,j}+\Delta y_{i+1,j}+\Delta y_{i,j-1}+\Delta y_{i+1,j-1}}\right)\\ \left[V_{i+1,j}-V_{i,j}\right]\\ -\left(A_{i-1,j}+A_{i,j}+A_{i-1,j-1}+A_{i,j-1}\right)\left(\dfrac{\Delta x_{i-1,j}+\Delta x_{i,j}+\Delta x_{i-1,j-1}+\Delta x_{i,j-1}}{\Delta y_{i-1,j}+\Delta y_{i,j}+\Delta y_{i-1,j-1}+\Delta y_{i,j-1}}\right)\\ \left[V_{i,j}-V_{i-1,j}\right]\end{matrix}\right\}$$

$$+\frac{1}{2}\left(D_{i,j}+D_{i,j-1}\right)\left\{A_{i,j}\frac{\Delta y_{i,j}}{\Delta x_{i,j}}\left(V_{i+1,j}-V_{i,j}\right)-A_{i,j-1}\frac{\Delta y_{i,j-1}}{\Delta x_{i,j-1}}\left(V_{i,j}-V_{i,j-1}\right)\right\}$$

$$(7.5.7)$$

$$f_{p\ i,j} = f_{i,j} - \frac{1}{4\left(\Delta x_{i,j}\Delta y_{i,j}\right)}\left[\begin{array}{l}\left(U_{i+1,j}+U_{i,j}\right)\left(\Delta x_{i,j+1}-\Delta x_{i,j-1}\right) \\ -\left(V_{i,j}+V_{i,j+1}\right)\left(\Delta y_{i+1,j}-\Delta y_{i-1,j}\right)\end{array}\right] \qquad (7.5.8)$$

Note that D is now the total depth of the water column, and η is the free surface deflection. As before, a conventional second-order leapfrog scheme can be used for time differencing, along with the Asselin-Roberts filter. One salient characteristic of the barotropic model is the higher speed of the gravity wave propagation vis-a-vis that in a reduced gravity model that requires much smaller time steps to be taken in any explicit time stepping scheme. Compared to the baroclinic wave speeds of 1 to 3 m s^{-1}, the external gravity wave speed $(gH)^{1/2}$ can reach values of 50 m s^{-1} near the shelf edge. The CFL condition dictates that the time step Δt be less than the minimum value of 1/2[$(\Delta x^{-2} + \Delta y^{-2})^{-1}(gH)^{-1/2}$] in the domain. When high spatial resolutions are essential, the time step may become too restrictive and semi-implicit schemes may become more attractive.

7.5.1 COASTAL OCEAN RESPONSE TO WIND FORCING

The principal mode of transient response of a coastal ocean to time-varying atmospheric forcing is through generation and propagation of forced and free coastal trapped waves. Production of inertial currents is another. Often during winter, the water column, especially on the inner shelf is well mixed and homogeneous, or nearly so. It may then be possible to ignore stratification and baroclinicity and model the response of the coastal waters to storm passages using a barotropic model.

An idea of the capability of a barotropic model to capture the response of coastal waters to winter storms can be obtained from a case study by Kantha *et al.* (1985) and Gordon *et al.* (1986) who present a simulation of the northeast American shelf to wintertime atmospheric forcing. Figure 7.5.1 shows the model domain and grid, chosen by careful experimentation to adequately resolve the trapped modes on the Mid-Atlantic Bight shelf, where the first mode shelf wave with a phase speed of about 7 m s^{-1} prevails. It also shows the location of current meter moorings from which data are available during the winter of 1980 (January 1 to March 15) for comparison with model results. Six hourly wind stress and atmospheric pressure fields were derived from reanalysis of NMC surface analyses to include pressure data from NDBC buoys. The observed sea level at Halifax was used to prescribe the elevation along the northern boundary, assuming the first mode shelf wave structure in the cross-shelf direction. The off-shore boundary was clamped and radiation conditions were

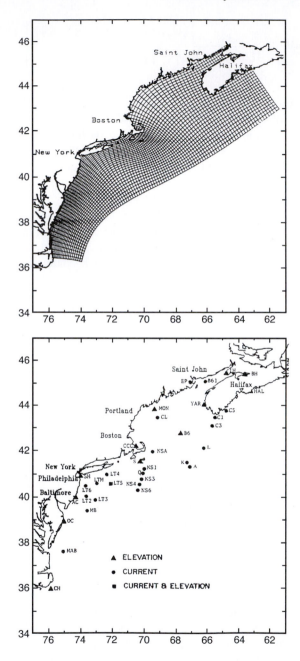

Figure 7.5.1 Barotropic model domain and the orthogonal curvilinear grid used (top) and the locations of the current meter moorings and SSH measurements (bottom).

used at the southern boundary. M_2 tidal forcing from Schwiderski's global model (Schwiderski, 1980, 1983) was prescribed along the open boundaries. Figure 7.5.2 shows comparisons between modeled subtidal currents and observed subtidal currents, along the isobaths. The model appears to capture considerable fraction of the variability in the along-isobath currents, with a correlation coefficient ranging between 0.5 and 0.8. The low frequency sea level response is captured well. While the along-shelf currents generated in response to wind and other forcing on the shelf are generally well simulated by even simple models, the cross-shelf currents and transports are not. These are an order of magnitude smaller and the underlying physical mechanisms are not that well understood. This is true of most simulations of coastal oceans, whether barotropic or baroclinic.

Given the fact that the background currents and density gradients have been ignored, the model simulations are quite skillful. This is simply a reflection of the fact that the barotropic response of a broad shelf to winter storms is a significant fraction of the variance and a simple barotropic model manages to capture some of this variance rather well.

7.5.2 STORM SURGES AND STORM SURGE MODELING

An important application of barotropic models is for prediction of storm surge effects along a coastline due to tropical and extratropical storms. This problem is of great importance in regions like the Gulf of Mexico and the North Sea with broad, shallow coastal oceans that respond vigorously to wind forcing. While summertime tropical storms and hurricanes (called typhoons, cyclones, or willi-willies in other parts of the world) are the principal agents of coastal devastation in the tropics and subtropics, it is the strong wintertime depressions that can cause flooding in the midlatitude regions. The winds are primarily responsible for most of the surge in both cases, the principal difference between the two being the spatial scale of the resulting surge. Tropical storms have strong, intense cores, and therefore they tend to cause intensive but localized surges extending over a few tens of kilometers, usually to the right (in the northern hemisphere) of the point of landfall of the storm center. They also tend to move fast and hence the effect is localized in time as well. In contrast, strong wintertime depressions have low pressure regions associated with strong winds extending over hundreds of kilometers from the low pressure center, and the resulting surge is less intensive but more widespread. They also tend to move slowly and linger longer, and it is the action of the sustained but less intense winds (compared to tropical storms) on a shallow sea that can cause surges on the order of a few meters. In 1953, a strong wintertime depression moving into the North Sea from the Atlantic produced sustained winds of over 25 m s^{-1} and

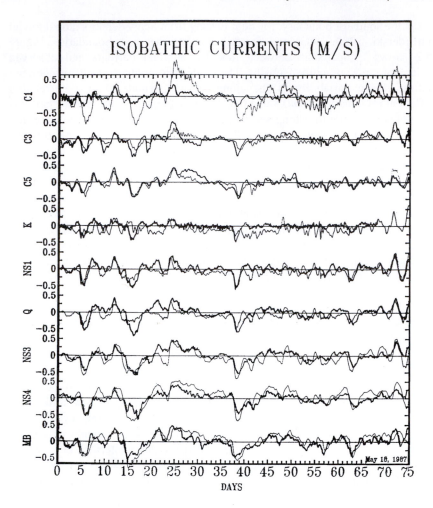

Figure 7.5.2 Comparison of the along-isobath currents from the model and those observed at various locations shown in Figure 7.5.1.

surges of over 2 m causing extensive flooding along the Dutch coast. On the other hand, strong winds in a fast-moving tropical storm (augmented by the pressure drop in the eye of the hurricane) pile up water rather quickly against the coast that often leads to an increase in sea level of several meters and consequent inundation of structures along the coastline. In 1969, hurricane Camille caused a storm surge of nearly 8 m at some points along the Mississippi coast leading to widespread destruction and devastation. Cyclones in the Bay of Bengal have caused frequent flooding and extensive death and destruction, with a loss of 300,000 lives in 1970 alone, due to a storm surge estimated to be nearly 9 m that caused extensive inundation of the heavily populated low-lying areas of the Ganges Delta. Since a likely outcome of the anthropogenic global warming is an increase in the intensity of tropical storms, storm surges are likely to be of even higher concern in the coming century.

Provided the local bathymetry is known accurately and the characteristics of the storm (such as its wind stress distribution and its track) can be deduced reasonably well from NWP forecasts, it is possible to predict the resulting storm surge quite accurately using a barotropic model driven by the wind stress and atmospheric pressure terms on the right hand side. Most often, the storm surge is enhanced by high tides and high surface waves generated by the strong winds. Because of the strong nonlinear interactions between the wind and tide-induced sea surface responses, it is preferable to model the wind and tidal forcing together, although the conventional practice is to model the storm surge separately and simply add the tidal elevations to the resulting surge.

While it is the sea level response to atmospheric forcing that is of concern along the coast, it is the currents induced by the storm (along with the wind- and wave-induced forces) that are of major concern as far as off-shore structures (drilling rigs in the North Sea and in the Gulf of Mexico on the Louisiana-Texas shelf are prime examples) on the continental shelf are concerned. Fail-safe design of these structures requires knowledge of the extreme current and wave environments likely over their designed operational lifetime, whereas operational safety and efficiency are enhanced by any predictive capability. Numerical models run in hindcast and forecast modes respectively can be useful in both cases.

Before modeling the coastal ocean response to storm forcing, it is important to understand how a shallow sea responds to such forcing (see Reid, 1990, for a review). For this, we need to consider its response to pressure and wind effects separately, although the two are often inseparable. It is easy to see that the pressure forcing, given adequate time for the ocean to respond, produces a static inverse barometric response. At high frequencies (typically more than 0.5 cy d^{-1}), the response is muted. At low frequencies, the response is close to the static response. Amplified response is in principle possible if the pressure disturbance travels at a speed comparable to the shallow water gravity wave speed. For a frictionless, nonrotating sea of uniform depth H, this amplification can be shown (see Pugh, 1987) to be $[1 - (V^2/gH)]^{-1}$ and therefore for reasonable values of the depth and storm speed, the amplification is less than 10%. Usually, the change in sea level is less than the static value of -1 centimeter per millibar, usually around -0.7 to -0.85. Since the pressure drop in the center of a tropical storm ranges between 20 and 70 millibars, the sea level rise is less than 70 cm. For an extratropical storm with, say, a 970 mb low pressure center, the sea level rise is about 30 to 40 cm. However, the pressure forcing is barotropic and hence is felt immediately throughout the water column.

The influence of winds on coastal sea level on the other hand depends very much on their strength and duration, the water column depth, the orientation of the coastline, and the width of the shelf. If the sea surface slope were to balance the wind stress exactly, the sea level rise would be $(\tau_w/gH)W$, where W is the width of the shelf. A 20 m s^{-1} wind blowing over a 20 m deep shallow sea of 200 km width could cause a setup of 60 cm. If the wind blows parallel to the shore of a long straight coast, the alongshore currents accelerate until balanced by bottom friction, and the geostrophic balance in the cross-shore direction sets up a sea level gradient on the order of fU/g, where U is the alongshore current. For typical U value of 30 cm s^{-1}, a 100 km shelf width, the setup is on the order of 30 cm or so. These estimates, however, ignore the effects of complex coastline shape and variable bottom depth. Consideration of these effects requires a numerical model.

There are many storm surge models in existence (see, for example, the review by Reid, 1990). These appear to be able to predict the peak surge with an rms error within 20% or so. The major source of error in these models is the inaccuracy in the winds. Because the surge is primarily dependent on the wind effects, it is important to have as accurate winds as possible to force the model, at resolutions sufficient to resolve any small scale features in the wind field. The second major source of error is the inaccuracy in topography. Since the local coastline shape and bottom depth can have a major influence on the resulting surge, it is essential to provide high resolution consistent with the region being modeled and accurate topography consistent with the resolution employed. The third source of error is the bottom friction coefficient; bottom friction is an important component of momentum balance in shallow water and it is therefore necessary to calibrate the coefficient using perhaps the known tidal response and the observed response to earlier storms.

Significant errors are also introduced by neglecting the interaction between the sea level response to atmospheric and tidal forcing. This comes about simply because of the influence of bottom friction, which is considerably affected by any prevailing tidal currents. North Sea is a classic example. This is easily remedied by including tidal forcing in a storm surge model. It is also preferable to estimate the wave field set up by the storm, since this has a bearing on the sea level. According to Pugh (1987), there is usually a setdown in sea level outside of the breaking zone of approximately 0.25 a^2/H, where a is the wave amplitude. Inside the line of breakers, where the water depth is ~2.5 times the wave amplitude a, there is a sea level slope of roughly 0.2 times the slope of the beach. Thus for a 3-m amplitude waves incident on a beach with a slope of 1/30, the line of breakers is in 8 m water, 230 m offshore, leading to a setdown of 0.3 m to the line of breakers and setup of 1.5 m from there to the shoreline (Pugh, 1987). The total wave-induced sea level rise is 1.2 m, on top of the wind- and tide-induced rise, and must be accounted for (Reid, 1990).

In semienclosed seas, bays, and harbors, an additional factor is the seiches produced by storm forcing. Classic examples are the Adriatic Sea and the Gulf of Mexico. Observations at Venice show persistent 22 hr oscillations in sea level often as high as 1 m in amplitude after the passage of depressions across the adriatic. A hurricane entering the Gulf can set off basinwide oscillations with a period of nearly 30 hr, and the sea level can begin to increase long before the arrival of the hurricane itself (Reid, 1990). The Baltic Sea is another semienclosed sea subject to extratropical storms and consequent storm surges and seiches. The surge at Saint Petersburg at the head of the narrow Gulf of Bothnia reached 4 m in 1924 (Pugh, 1987). Typhoon Vera produced a 3.6 m surge at Nagoya off the Japanese coast. Since the shelf is too narrow there for wind action to be able to produce such large surges, it is thought that resonant response of the bay to pressure forcing might have been responsible for the large surge (Pugh, 1987).

Because of the restrictions imposed on the flow by the openings, the sea level rise in one portion of a semi-enclosed basin is often accompanied by a drop in some other portion to conserve mass, and hence while one region is undergoing a strong positive surge, negative surges might occur simultaneously elsewhere. Also the water often driven out of the basin by the storm returns to cause return surges in some regions such as the North Sea. In addition to seiches, inertial oscillations are also an important feature of the oceanic response to storm passages. Strong inertial oscillations are always seen in deep waters away from the coast in the wake of a hurricane.

The magnitude of the storm surge induced by a tropical storm or hurricane is a function of (1) the pressure drop at the center, Δp; (2) the radius to the maximum winds, R; (3) the translational speed of the storm center, V; (4) the orientation of the coastline to the storm track, and (5) the shelf width, W. Jelesnianski (1972), a pioneer in storm surge simulations, was able to construct nomograms involving these parameters that predict surges to within 75% of the observed variance. These nomograms emphasize the overwhelming importance of the shelf width. On the west coast of the United States, where the shelf is nearly nonexistent (tropical storms are also rare), the surge is small and most of the damage is due to high waves. On the east coast and in the northern Gulf of Mexico, where the shelf is broad, hurricanes produce large storm surges. Figure (7.5.3) shows the sea level rise at Galveston, Texas, due to hurricane Carla in 1961.

Numerical models of storm surges are necessarily high resolution regional

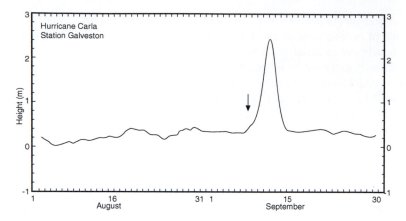

Figure 7.5.3 Sea level rise at Galveston, Texas, from hurricane Carla in 1961. Tides have been removed by 40-hr low pass filter. Arrow indicates the time Carla entered the Gulf of Mexico. From Reid (1990).

models driven by synoptic winds and pressure fields. Finite element barotropic storm surge models have recently become quite popular since they afford local features to be easily resolved (Westerink *et al.,* 1992). However, a nested finite difference approach also enables similar fine resolutions to be realized.

Mehra *et al.* (1996) describe a real-time application of the nested grid approach to combined storm surge/tides modeling to the west coast of the United States. Nesting is one-way. A 5-km resolution model driven by the wind stress and pressure fields from 5 km resolution COAMPS atmospheric model from FNMOC forms the coarse grid model. This model is driven at the boundary by tides from the global model of Kantha (1995b) and six hourly synoptic fields at the surface. The output of the coarse grid model (the sea surface height and the barotropic currents) is saved on the boundaries of a 1-km resolution fine grid model of the San Diego Bay area and used to run the fine grid model for the same duration to provide high resolution currents and sea level.

7.5.3 RESPONSE TO PRESSURE FORCING

Oceanic response to atmospheric pressure forcing can also be studied using a barotropic model (Kantha *et al.,* 1994; Ponte, 1994, Nadiga *et al.,* 1999). The problem is of interest in altimetric data analysis where the response is assumed to be inverse barometric, since any departures from the static response and the

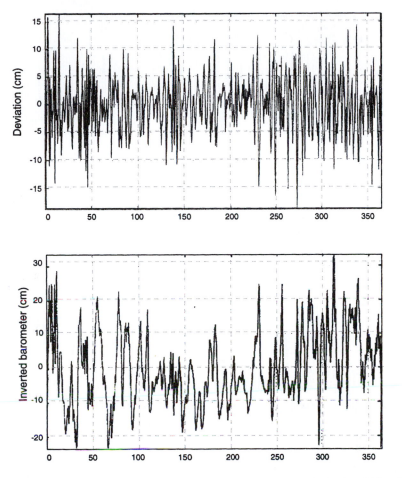

Figure 7.5.4 Sea level response to atmospheric pressure forcing at a point in the middle of the broad shallow Bering Sea Shelf. The bottom panel shows the inverse barometer equivalent to the atmospheric pressure at the point (in cm) and the top panel shows the departure from inverse barometric response.

resulting residual errors can often be mistaken to be genuine mesoscale features. In the equatorial regions, it might be important to include baroclinicity via the use of a baroclinic model. The pressure changes in the tropics are, however, an order of magnitude smaller and on the order of several millibars compared to several tens of millibars in midlatitudes.

Figures 7.5.4 and 7.5.5 show results from a 1/2 degree barotropic model of

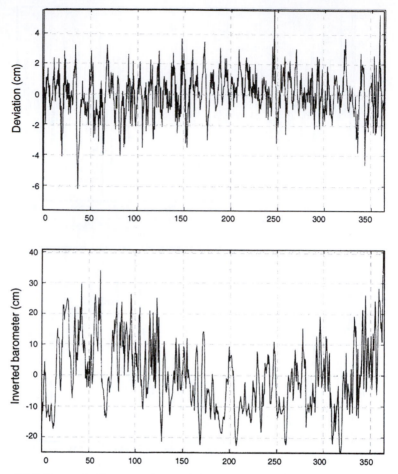

Figure 7.5.5 Same as in Figure 7.5.4 but at a point in the deep North Pacific Ocean. Note smaller departures from inverse barometric response than that in Figure 7.5.4.

the north Pacific, driven by 6 hourly pressure fields from FNMOC. The differing response in the deep oceans and shallow marginal seas is worth noting. The departures from the static response are an order of magnitude smaller in the deep oceans compared to the magnitude of the pressure forcing, whereas they are comparable in magnitude in shallow water and in many semienclosed seas. See Ponte (1994) and Nadiga *et al.* (1999) and references cited therein for a more detailed discussion of departures from inverse barometric response of the ocean to atmospheric pressure forcing.

Chapter 8

Data and Data Processing

So far, we have dealt with simple models that nevertheless capture some of the essential ocean dynamics. These models do not require much input from observations. However, observational data on the oceans (and the lower atmosphere) are an integral part of any numerical ocean modeling effort, when comprehensive models that we are going to describe in the rest of this book are employed, especially for prediction purposes. Preprocessing of the observational data necessary for ingestion into the model or required for verification of its results is a nontrivial task, often more demanding than running the model itself. Once the model is run, its output must be analyzed and transformed into forms suitable for ready comparison with available observations or into forms that lend themselves to easier comprehension. It is not at all unusual for the data preprocessing and model output postprocessing steps to constitute the lion's share of a modeling effort. A practical consequence is that a numerical ocean modeler is also compelled to acquire reasonable familiarity with data sources, data and data processing techniques, and graphics. With this in mind, before we consider more comprehensive ocean models, we will provide a brief overview of oceanic data, their sources and preprocessing, and postprocessing of a model output. However, it would take an entire treatise to cover all these aspects in sufficient detail. We cannot do that. Instead we refer the reader to existing literature on the subject, for example, the recent monograph by Emery and Thomson (1998). Conventional methods such as Fourier transforms and time series analyses can be found in monographs such as this. We will, however, describe in detail two important and modern techniques that the student should be familiar with, wavelet transforms and empirical orthogonal functions (in Appendix B and C, respectively) that are finding increasing use in data analysis in recent years. Here, we will describe data and data sources.

There are two principal uses of ocean models: (1) process studies that shed light on certain aspects of the flow and (2) applications such as hindcasts, nowcasts, and forecasts. In the former, the model is used as a tool (albeit numerical instead of analyical) to explore the dynamical and physical mechanisms underlying specific oceanic processes in an effort to understand the inner workings of the global oceans. The general applicability of numerical models and their ability to deal with complex situations is clearly an advantage in any such investigation, since highly realistic situations can be explored. A unique advantage of numerical models, which observations alone cannot provide, is the ability to ask "what if " questions and investigate hypothetical scenarios (such as, for example, the cessation of the Indonesian throughflow from the tropical Pacific Ocean to the Indian ocean, and the reduced sea level during the glacial) and their impact. They also afford the capability to isolate and study processes, a luxury observationalists seldom have. In addition, the parameter space can be fully explored. In applicational uses, an ocean model simulates the real oceans as closely as feasible so that the model output can be used to provide an estimate of the state of the ocean in the past, present, or sometime in the future. Steadily increasing societal use and abuse of the oceans demands a better understanding of the state of the oceans, and applicational uses of models are therefore steadily increasing.

No matter what the intended use is, all numerical ocean models require observational data for a variety of purposes, including model initialization, model forcing, and model validation, and for providing boundary conditions for limited domain models. For example, one might ask what circulation would be consistent with the observed density field on a particular timescale such as climatological; then the climatological average density field is the needed input. Running models in such a diagnostic mode (where density field is held fixed) and seeking the dynamical state of the ocean corresponding to an observed (albeit almost always incomplete) state requires observational data. When a model is run in a prognostic mode (where the density field is allowed to evolve), it requires data to create a realistic initial state and to provide realistic forcing needed to simulate the desired final state. Only when the initial state is prescribed from observations is the model run called a prediction. When the initial state is idealized, as in process studies, it is called a simulation. In a predictive mode, the model is integrated forward in time subject to a prescribed forcing at its surface by air-sea momentum and buoyancy fluxes, derived directly or indirectly from observations, and the desired final state is then compared to observations for a check on model performance. In any case, observational data are an integral part of a modeling effort and in most cases the success of a modeling effort depends not just on the particular model chosen, but also on the quality and quantity of observational data used in the model.

In most practical applications, observational data are intentionally allowed to influence model results, either through frequent reinitialization (as in NWP initialization, forecast cycle) or through continuous injection of data into the model. This procedure is essential because of the limited predictability (theoretically 2 weeks in the atmosphere and probably 4 to 6 weeks in the oceans) of the highly nonlinear and therefore chaotic geophysical flows. Without the frequent impact from the observed state of the ocean (and the atmosphere), simulations from even the most perfect model would eventually deviate far enough from the real ocean (and the atmosphere) to be of little worth.

Ocean simulation/prediction is an initial value problem and therefore the initial state of the modeled ocean must be prescribed as accurately as possible. This is difficult to do in practice, since unlike the atmosphere (where an extensive observational network exists, at least over land, specifically to provide simultaneous observations over the globe for initialization of NWP models), observational data are insufficient at present to specify the state of the ocean at any given point in time. Although efforts are underway to set up a Global Ocean Observing System (GOOS), the best alternative at present and in the foreseeable future is to prescribe some sort of a climatological average as the initial state and spin up the ocean from that state under prescribed surface forcing. This is usually done using data bases such as Levitus (1982), Levitus *et al.* (1994), Levitus and Boyer (1994), and the U.S. Navy GDEM that contain distributions of climatological average temperature and salinity in the global oceans derived from historical archives of *in situ* observations.

Prescription of surface forcing is itself a major problem. Both momentum and buoyancy fluxes need to be specified at the sea surface, and both are determined by processes in the adjacent atmospheric boundary layer. For determining the long-term average state of the oceans, it is once again possible to appeal to climatological data bases such as that due to Hellerman and Rosenstein (1983) and the Comprehensive Ocean-Atmosphere Data Set (COADS) data base (see Woodruff *et al.,* 1987; the data base has been updated to 1990s) that provide gridded monthly average values of wind stress and other air-sea exchange parameters over the global oceans derived from historical marine surface observations. However, for many applications, including ocean prediction, it is necessary to provide surface forcing on a daily or even a multihourly basis. This is impossible to do from observations alone (although satellite sensed air-sea fluxes might help in the future), and one has little choice but to appeal to data bases such as six-hourly analyses and predictions of NWP centers, even though the accuracy of the surface forcing so derived depends very much on the skill of the particular NWP model and methodology.

Extensive observations during multinational observational campaigns such as

WOCE (*U.S. WOCE Report,* 1998; see also Lindstrom *et al.,* 1998, and the Special Section on WOCE Pacific results in *Journal of Geophysical Research, 103*(C6), 12,897-13,092, 1998) and TOGA (McPhadden *et al.,* 1998) in the recent past have greatly added to the observational data bases for the ocean. These programs are an excellent example of the combined use of *in situ* and remote platforms in a well-integrated campaign to probe and study the oceanic state. They have also helped develop advanced observational techniques such as TAO autonomous arrays, and sophisticated telemetering drifting buoys such as ALACE and PALACE floats. Future programs such as CLIVAR are likely to spawn even more sophisticated technologies such as smart autonomous underwater vehicles (AUVs). Nearly uninterrupted satellite monitoring of the ocean surface is also well underway. The overall outcome will be a much better knowledge of the state of the oceans in the coming decades, which would be a definite help in modeling the oceanic state more accurately.

Salinity measurements in the ocean have been traditionally by *in situ* sampling of the fluid parcel, either through CTDs (or bottles in the past) or thermosalinographs. However, the microwave energy radiated from the sea surface is a function of the SST and to a weaker extent the SSS. A scanning low frequency microwave radiometer (Goodberlet *et al,.* 1997) makes use of this property and measures brightness temperature of the sea surface at 1.4 GHz and the SST, from which SSS can be extracted through Klein and Swift (1977) algorithm. The accuracy depends on the noise level and the temperature and salinity values, but at normal SST and SSS values, the noise level of roughly 1 psu is possible. Increased temporal and spatial sampling/averaging has the potential for reducing this. While the noise level is too high for routine use of this instrument, given the fact that it is at present the only practical means for rapidly and remotely surveying the SSS field (Lagerloef *et al.,* 1995), it is useful for mapping salinity fronts due to river and estuarine outflows in the coastal ocean, where the salinity contrast across the front and hence the signal-to-noise ratio is large. It is therefore a useful alternative to slow *in situ* sampling by thermosalinographs in the coastal ocean. However, until the accuracy of the SSS retrievals attain the 0.1 psu level, and the instrument can be orbited on satellites, the technique will not be of much use in providing the SSS boundary conditions for ocean models in general.

8.1 *IN SITU* OBSERVATIONAL DATA

An important *in situ* data needed for ocean models is the bathymetry. Figure 8.1.1 shows a plot of the bottom topography derived from the Sandwell *et al.* (1996) bathymetric data base. As seen in earlier chapters, bottom depth has an important influence on oceanic processes. The principal data base available at present for this purpose is the Terrainbase data base at NESDIS that contains

Global Topography, TerrainBase

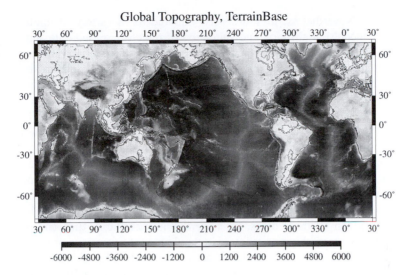

Figure 8.1.1 A bathymetric map of the global oceans generated using the ETOP05 database. The present coastline and the coastline that would result from a 70 m rise and a 140 m drop in sea levels are also shown.

gridded topography over land and under the oceans at 1/12 degree (5') resolution. This data base incorporates DBDB5, the oceanic bathymetric data base derived from ship soundings. It is quite accurate over land and in deep oceans, but it contains large errors over the shelf and in shallow waters. Unfortunately, higher accuracy or higher resolution gridded data bases with global coverage are not yet available. The U.S. Naval Oceanographic Office makes available the DBDBV data base, which supplements DBDB5 with higher resolution (1- and 2-min) over selected areas such as the Mediterranean Sea, and the U.S. coasts. More recently, altimetric data have been used to derive bottom topography over the global oceans (Sandwell *et al.*, 1996). Once again, these are accurate in deep water, but the methodology is not well suited to shallow waters. Limited coverage, high resolution, accurate bathymetric data bases are available from NOS/NOAA, but these are limited to U.S. coastal and EEZ waters. Marine navigation charts are available in most coastal regions of the world, but transferring these to the model grid is tedious at best.

8.1.1 XBT, CTD, CM, ADCP, AND DRIFTER DATA

In situ observations of the ocean are hard and in harsh environments such as the polar oceans, often hazardous to make, and in any case rather expensive. They are therefore inherently sparse with severe temporal and spatial sampling limitations. Nevertheless, they are very valuable for a variety of applications including model initialization and validation of remotely sensed data and model output. The bulk of the *in situ* observations tend to be observations of the temperature of the upper layers, since it is fairly simple to obtain them via expendable bathythermographs (XBTs), to 200 to 800 m depths while underway, without having to stop and make the measurements. A rapid survey of a limited region is also possible with the use of the air-deployed variety (AXBTs), an excellent example being the AXBT survey made in the Eastern Mediterranean for use in operational modeling of the Mediterranean Sea at the Naval Oceanographic Office (Horton *et al.*, 1997). The accuracy is about 0.5°C and suffices for most applications. XBTs are also routinely obtained from naval vessels operating over the global oceans and volunteer commercial shipping of opportunity. TOGA program made extensive use of these voluntary observing ships (VOS). In all these cases, the coverage is sporadic in time and data are obtained only along shipping lanes or in the theater of operations of the vessels. Most of the rest of the oceans have large data voids. Figure 8.1.2 shows the distribution of all XBT observations made in the north Pacific up to 1992 derived from the historical data archives (Levitus *et al.*, 1994; Levitus and Boyer 1994), Figure 8.1.3 shows all observations made in the global oceans that were available in "real time" in individual years, here for example, 1993 to1996. XBT measurements can now be telemetered so that near real-time observations are often available in a limited region.

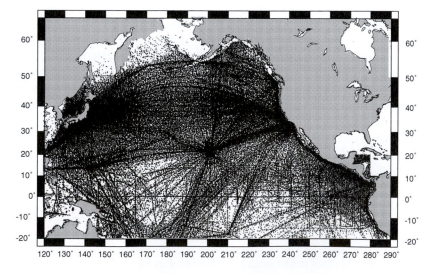

Figure 8.1.2 Distribution of all historical XBT observations to depths greater than 400 m in the North Pacific (to 1992) archived at NODC.

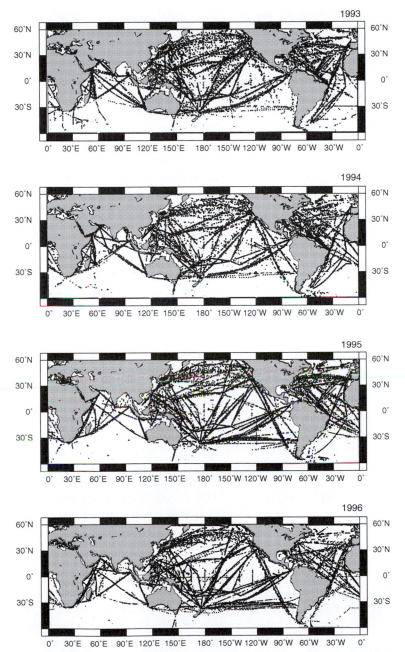

Figure 8.1.3 Distribution of XBT observations that were available in real time in the global oceans archived at NODC for years 1993 to 1996.

Observations of the temperature of the deep oceans used to be made in the past by mechanical bathythermographs (MBTs). This is rarely done at present. Instead conductivity, temperature and depth (CTDs) probes are deployed from ships to measure both the temperature and salinity simultaneously. CTDs rely on measurement of conductivity to infer salinity of the water masses. Normally, salinity needs to be measured to the third or fourth decimal place in most of the oceans to be useful, and the measurement is rather difficult since a wide variety of sources of contamination including biofouling exist. Careful (re)calibration of the sensor and quality checks of data are indispensable. While XCTDs are increasingly being used, conventional tether-deployed CTDs from specially equipped research ships and navy oceanographic vessels are still the mainstay of salinity measurements in the ocean and temperature measurements in the deep oceans. Figure 8.1.4 shows a typical CTD cast in the Sea of Japan taken during the 1993 CREAMS (Circulation Research in East Asian Marginal Seas) cruise. It shows the nearly homogeneous Japan Sea Proper Water below a few hundred meters depth. Since a typical CTD cast to a depth of 3000 to 4000 m takes several hours during which time the vessel must steam in place, these measurements are also rather expensive to make. The expense and difficulty makes salinity observations particularly sparse both in time and space. Even today, there still exist large areas in the southern hemisphere where not even a single salinity profile observation has ever been made. This is unfortunate since salinity is a major factor in the density of water masses and hence ocean dynamics. Fortunately, the changes below a few hundred meters are small in most of the global oceans. Figure 8.1.5 shows the distribution of all CTD observations made in the north Pacific up to 1992 derived from the historical data archives (Levitus *et al.,* 1994). Figures 8.1.6 and 8.1.7 show the sampling grids for one of the very few complete CTD surveys in the Red Sea made in 1993 (Clifford *et al.,* 1997) and one of the only three surveys ever made in the Persian Gulf, this one in 1991 after the Gulf war. The above plots clearly exhibit the principal problem in oceanography and ocean modeling, sparsity of *in situ* observational data. Fortunately the WOCE carried out a systematic one-time survey of the global oceans, measuring not only temperature and salinity but also properties such as dissolved oxygen and carbon content. Figure 8.1.8 shows the sampling grid for the Pacific (*U.S. WOCE Report,* 1998), which took many years to complete.

In situ measurements of SST and SSS are often useful. These are made from ships equipped with thermosalinographs which draw water from intakes a few meters below the surface. Although not strictly surface measurements, they do provide a rapid means of inferring the temperature and, more important, (since accurate remote measurements of SSS is still not possible) the salinity of the upper layers.

Unlike the atmosphere, where ground level measurements and soundings are

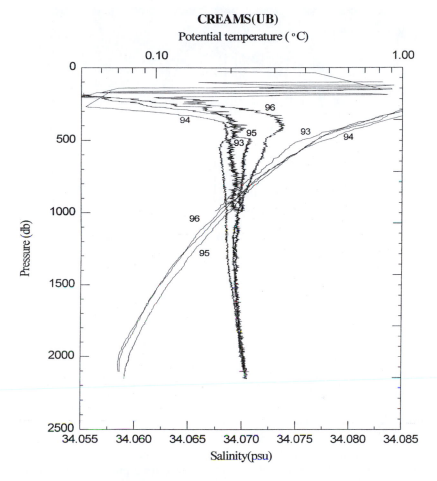

CREAMS(UB)

Figure 8.1.4 CTD casts in the Ulleung Basin of the Sea of Japan, made during the 1990s CREAMS cruise, showing interannual changes. From Kim *et al.* (1997).

made periodically over land and reported routinely for NWP applications, there is no such observational network at present for the oceans. Efforts are underway to set up simple observational networks under Global Ocean Observation System (GOOS) program. For example, there is now a network of sea level stations around the globe (Woodworth *et al.*, 1997) that report fluctuations in global sea level in real time (Figure 8.1.9). Sea level is an important oceanic parameter useful in monitoring processes such as ENSO in the tropical Pacific. There is also a network of coastal tide gage stations that provide usually hourly sea level data and this data base is invaluable in studies of tides and sea level fluctuations in coastal seas.

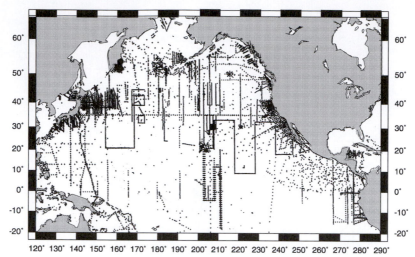

Figure 8.1.5 Distribution of all historical CTD observations to depths greater than 400 m in the North Pacific (to 1992) archived at NODC.

Observations on circulation and currents are even pitifully fewer. Current measurements are even harder to make, are very expensive, and are made only during specific observational campaigns. Traditionally, currents have been measured by current meters strung at various intervals in the vertical on taut moorings anchored to the sea floor with a subsurface float at the top to provide the needed flotation. The data are usually recorded and retrieved when the mooring is recovered, but they can also be telemetered to provide real-time capability. The moorings sway in the presence of strong currents, and pressure sensors are needed on the instruments to determine their location in the vertical at any given time. The instruments are left in place for a predetermined amount of time, days to a year or so. The sampling rate is determined by the capacity of on-board recorders and the duration of deployment, as well as the frequency content of the prevailing processes. The measurable frequencies are delineated at the higher end by the Nyquist frequency $\omega_N = 0.5/\Delta t$, where Δt is the sampling frequency, and at the lower end by $\omega_L = 1/T$, where T is the total duration of the record. The length of the record also determines the frequencies that are separable: to distinguish between two phenomena with a frequency difference of $\Delta\omega$ requires a record of length $1/\Delta\omega$. The bandwidth of measurements is $T/2\Delta t$. Because of the difficulty of simultaneously attaining the high sampling rates needed to record high frequency events and the long record length needed for observing low frequency phenomena, it is often necessary to employ burst sampling techniques, where intensive observations are interspersed with long periods of little or no data.

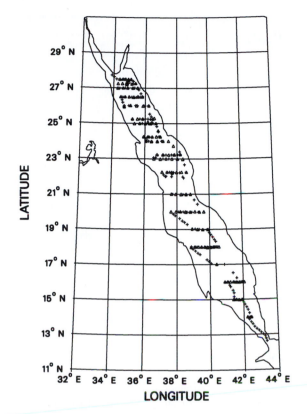

Figure 8.1.6 Sampling grid for the Red Sea (triangles—CTDs, crosses—AXBTs) during the 1994 observational program undertaken by NAVO. From Clifford *et al.* (1997).

The instruments and their recorded data are retrieved by detaching them from the anchor via an acoustic release triggered remotely. The moorings are subject to heavy loads by ocean currents and waves and interference by commercial shipping and fishing activities, and hence losses of instruments and data are quite common. The rotors are also subject to biofouling by marine organisms such as barnacles. Long-term observations therefore require frequent redeployments. It is particularly difficult to make measurements in the upper 30 m or so, not only because of the interference and biofouling but because of also the possibility of contamination by surface waves. In a surface wave field, the vertical motion of the mooring or rotor-pumping of Savonious rotor type instruments leads to erroneous overestimation of currents. Anderaa rotor-type current meters (RCMs), which are of this type, are seldom deployed in the upper 25 m.

Current meters such as RCMs measure the speed and direction of currents. Vector-averaging current meters (VACMs) use burst sampling to measure the

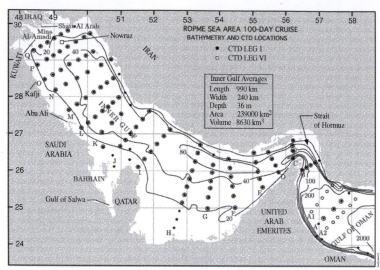

Figure 8.1.7 Sampling grid for the Persian Gulf during the 1992 NOAA observational program immediately after the Gulf War. From Reynolds (1993).

speed and direction and obtain an average over the sampling interval of the two velocity components individually. However, near-surface measurements are most often done with VMCMs, vector measuring current meters, which measure and record the two components of the velocity individually. Current meters may also incorporate sensors to measure the temperature and salinity (and often optical properties) at that depth, Anderaa current meters being typical of this variety. Historically, most of the current meter deployments have been in the coastal oceans either on or close to the shelf, although at some deep ocean stations (for example, Woods Hole site D in the Atlantic), long-term deployments have been made. The recent deployment of TOGA/TAO array in the tropical Pacific as part of a global effort to observe and monitor ENSO (McPhaden, 1995; Piotrowicz, 1997) is a shining example of the kind of measurements needed to properly observe and monitor the oceans (Figure 8.1.10). Some of these arrays also carry current meters. Unfortunately, the effort and expense of deploying and maintaining such an observational network are quite daunting and often beyond the resources of a single nation, even a rich one. Figure 8.1.11 shows a typical current vector stick plot from current meter records deployed off the west coast of the United States (Lentz and Chapman, 1989) during the Coastal Ocean dynamics experiment (CODE). The accuracies of most current meter measurements are typically 1 to 2 cm s^{-1}.

With the advent of acoustic Doppler current profilers (ADCPs) in conjunction with precise positioning possible from differential Global Positioning System

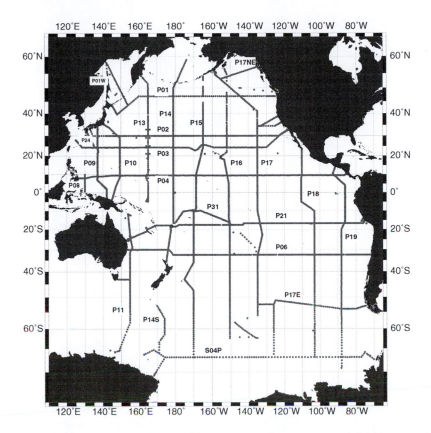

Figure 8.1.8 The sampling grid of the WOCE hydrographic program occupied during the 1990s. Three zonal grids occupied during the 1980s are also included. From Talley (1998).

(GPS), current measurements are becoming more routine in recent years. Shipboard deployments have made it possible to make rapid spatial surveys of oceanic features such as eddies, and mooring-mounted ADCPs provide useful data on currents at isolated places. These instruments rely on the Doppler frequency shift of acoustic pulses by particulate matter in the water column. They measure currents in a small measurement volume, so that currents in typically 2 to 5 m bins can be obtained over the water column. The accuracies are typically 1 to 2 cm s^{-1}, similar to that of conventional current meters. Figure 8.1.12 shows a typical example from measurements made on a warm core ring off the east coast of the United States (Kennelly *et al.,* 1985). ADCPs are also employed on taut moorings to measure currents in the water column. Air-deployed expendable current profilers (AXCPs) have proven useful in probing hurricane-driven oceans. Shore-deployed Coastal Ocean Doppler Radars

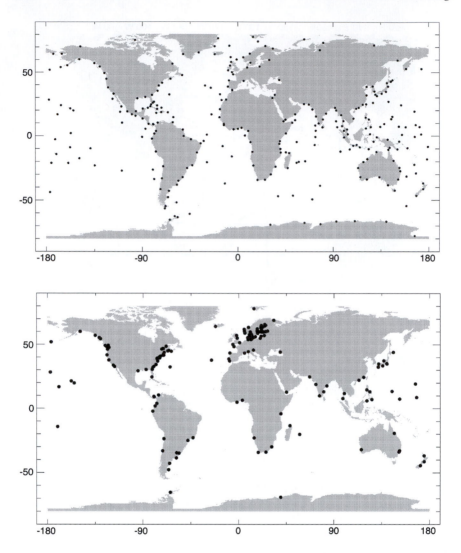

Figure 8.1.9 Global real-time sea level station network under Global Sea Level Observing System (GLOSS) being established under GOOS for monitoring fluctuations in sea level over the globe. The top panel shows 280 core stations, and the bottom level shows the long-term trend set to be equipped with GPS receivers. From Woodworth *et al.* (1997).

(CODARs) measure surface currents in the radial direction over a limited area using microwave energy backscattered by surface waves. The range depends on the frequency used, and is rather limited, 100 to 200 km. Military over-the-horizon radars have a much larger range, typically a 1000 km or so, but are

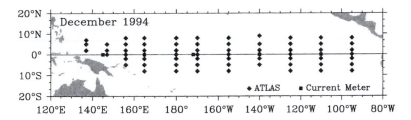

Figure 8.1.10 TOGA/TAO array in the tropical Pacific that enables ENSO and other events to be monitored. From McPhaden (1995).

expensive to operate. Instruments at at least two locations are needed to obtain the complete velocity vector by triangulation.

The third means of inferring circulation is through Lagrangian drifters. Drift bottles and computer punched cards have now been replaced by telemetering buoys drogued to follow currents at a particular depth. Thus real-time measurements have been made possible. Interestingly, some knowledge of the surface circulation in the Gulf Stream has been obtained by Richardson's compilation (Richardson, 1980) of drifting ship wrecks in the area, and an opportunistic spill of a shipment of shoes from a Japanese merchant vessel provided an idea of the prevailing currents in the northeast Pacific. Periodic buoy deployments in the Arctic under the Arctic Buoy program have provided information on the transpolar drift. Drifter deployments were a central part of TOGA observations and are now a routine part of any intensive observations of a region such as the LATEX (Louisiana Texas) program in the Gulf of Mexico. Figure 8.1.13 shows drifters deployed in the East China Sea and the Eulerian currents deduced from these Lagrangian measurements by binning the tracks into lat-lon grids and assuming that the ocean remains in a quasi-steady state (Lie and Choi, 1997). Some of these drifters carry thermistor chains for measurement of

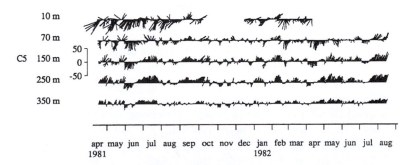

Figure 8.1.11 A typical current vector stick plot showing currents at various depths during the CODE program off the west coast of the United States. From Lentz and Chapman (1989).

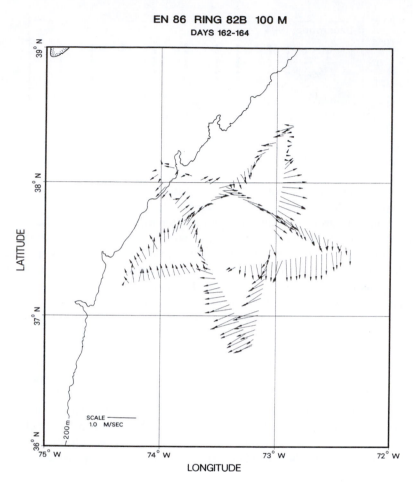

Figure 8.1.12 Typical ADCP current measurements from a ship-deployed current profiler showing currents in a warm-core ring off the east coast of the United States. From Kennelly *et al.* (1985).

ocean temperature profile as well. The drifters can be drogued to follow currents at any particular depth in the upper layers, but windage effects and occasional loss of drogues prevent these floats from following the water masses faithfully. They are therefore quasi-Lagrangian.

For years, SOFAR floats have provided information on the deep currents in the oceans. These have now been replaced by Autonomous Lagrangian Circulation Explorer (ALACE) floats designed to dive to a preprogrammed depth, follow the currents for a predetermined period (~25 days), resurface and telemeter their positions and observations such as temperature. The floats stay on the surface for ~24 hr before returning to their parking depth. From the position

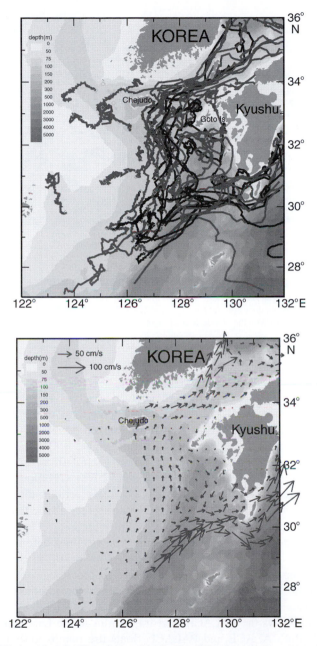

Figure 8.1.13 Trajectories of drifters deployed in the East China Sea and the currents deduced from them by binning the data. Note that this assumes that the currents are quasi-steady over the period. From Lie and Choi (1997).

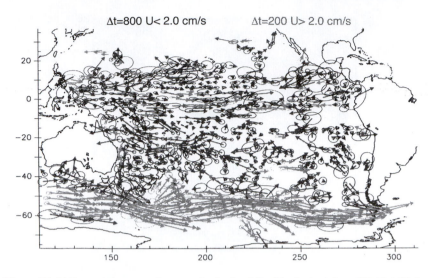

Figure 8.1.14 Space-time averaged currents in the South Pacific obtained from 303 ALACE floats deployed at 900 m depth during 1991–1996 as part of the US WOCE program. The currents are plotted as displacements over 800 days in lighter shade for speeds less than 2 cm s^{-1} and darker for larger values. Ellipses show the standard error. The axes denote longitude and latitude in degrees. From Davis (1998).

of the buoy at the times of surfacing (accurate to 1 to 3 km), it is possible to compute an average speed and direction for the currents at the parking depth. With a lifetime of about 5 years, these floats make it possible to map the long-term circulation in a given piece of the oceans. Figure 8.1.14 shows long-term, space-time averaged circulation in the South Pacific from the U.S. WOCE program (*U.S. WOCE Report,* 1998, Davis, 1998). Profiling Autonomous Lagrangian Circulation Explorer (PALACE) floats periodically dive to their parking depths of 800 to 1600 m, drift with currents for a specified period, then rise to the surface collecting temperature, salinity profiles while ascending, and transmit the collected data to telecommunication satellites before repeating the cycle (roughly once every 10 days). With a lifetime of a few years, these buoyancy-driven devices hold the promise of providing extensive *in-situ* sampling of the water mass structure and the deep currents in the global oceans. There are plans to seed the global oceans with more than 3000 of these floats. Figure 8.1.15 shows the PALACE float trajectories in the Labrador Sea during 1996–1997. Both ALACE and PALACE floats use pumps to actively change their buoyancy in order to ascend or descend, but are otherwise passive drifters. Another *in-situ* instrument that also uses buoyancy adjustments is a glider, but

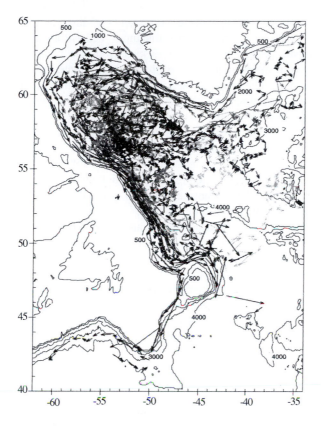

Figure 8.1.15 PALACE float trajectories in the Labrador Sea deployed at 1500 m depth during 1996–1997. The vectors represent averages over 5 to 20 days. Figure provided by R. Davis.

unlike a passive float, it is a winged instrument and therefore capable of gliding limited distances horizontally while ascending/descending. Yet another type of Lagrangian floats use a combination of high drag, nearly neutral buoyancy and compressibility matching that of the seawater to follow water parcels in three dimensions and is very useful in tracking the vertical and horizontal movements of water masses in the upper mixed layer. These are tracked acoustically, and are therefore *in situ* devices, but are helpful in studying the large turbulent eddy motions in the ocean.

Acoustic tomography (Munk *et al.,* 1995) is a technique where a low frequency acoustic signal transmitted from one point and received at another carries information on the intervening water masses. Since the sound speed increases with pressure but decreases with temperature, there exists a minimum in sound speed at a depth of about a kilometer, and this creates a waveguide through which sound can travel large distances with little attenuation. Low frequency acoustic energy travels long distances in the ocean and therefore provides a means of inferring the state of the upper kilometer of the ocean along the propagation path. The propagation time depends essentially on the integrated temperature and therefore provides a convenient means of measuring the variability of temperature along the acoustic path. Acoustic thermometry depends on this principle to infer any long-term increase in the temperature of the upper ocean caused by global warming. Since currents prevailing along the path affect the propagation time, reciprocal acoustic tomography (Munk *et al.,* 1995), in which the acoustic signals are transmitted in both directions using transmitter/receiver combinations at both end points, provides information on ocean currents also, but only along the acoustic path.

Inverted echo sounders (IES) use the acoustic travel time from the sensor to the ocean surface and back to deduce the depth of the main thermocline and its fluctuations. Being quite simple in concept, they provide an inexpensive means of mapping the thermocline depths in a region, provided careful attention is paid to calibration via XBT measurements to reduce the scatter in the measurements (Tracey *et al.,* 1997).

Because of the dearth and expense of observations on currents and circulation, it is traditional to use density measurements and geostrophic approximation to infer the currents in the water column. This method requires questionable assumptions as to the level of no motion but nevertheless is quite useful to obtain an estimate of the currents (more appropriately the vertical shear) in the deep ocean basins.

An exciting new development in oceanic observations is smart autonomous underwater vehicle (AUV) technology. These mini torpedo-size vehicles can be preprogrammed to roam the oceans and collect data along predetermined tracks and depths. The AUV can surface at periodic intervals to telemeter the data collected via communication satellite network to a receiving station and can either resume operation or wait for further instructions. The speed and range are limited, however, by the amount of energy available on board for propulsion and retrieval considerations. However, AUVs are becoming increasingly more versatile and are being increasingly employed for hydrographic and acoustic surveys in the ocean (Levine *et al.,* 1997). While their principal limitation at present is their short endurance, with developing docking and recharging capabilities, there is little doubt that they will be a routine component of future observational networks (Curtin *et al.,* 1993). They are the equivalent of aircraft observations in the atmosphere.

An excellent but dated review of ocean instruments can be found in Baker (1981). A more recent account can be found in Emery and Thomson (1998).

8.1.2 HISTORICAL HYDROGRAPHIC DATA

Oceanographic measurements are routinely archived by government agencies. In the United States, the National Ocean Data Center (NODC, nodc.noaa.gov) is the designated archive in the civilian sector and the U.S. Navy maintains a variety of archives such as MOODS (Master Oceanographic Observational Data Set). Synoptic data for a particular time window with a wide area coverage are very difficult to gather in the oceans. Since density data are indispensable for initialization of ocean models, the only recourse is to appeal to such archived XBT/CTD data. By binning and averaging quality-controlled observations from such archives, it is possible to obtain an estimate of the climatological state of the oceans. Syd Levitus pioneered this approach in 1982 (Levitus, 1982) and his 1° resolution global temperature and salinity fields have been used extensively by ocean modelers since. A recent update includes data up to 1992 (Levitus and Boyer, 1994; Levitus *et al.,* 1994). Distributions of monthly and seasonally averaged temperature and salinity are available at various standard NODC depths. The distributions below 1000 m are annual or overall averages, since there is not enough data to separate the small (if any) seasonal variations. Salinity values in the upper 1000 m are seasonal averages only, while more abundant temperature observations permit monthly averages to be deduced. The vertical levels are closely spaced in the upper 250 m (see Table 8.1.1). Other climatologies are now available, such as the 1/2° resolution U.S. Navy GDEM for the global oceans and the North Atlantic climatology of Lozier *et al.* (1995).

In addition, individual casts (XBTs/CTDs) are also available from NODC and MOODS data bases. It may be necessary, however, to quality control these data and process them for specific use, a tedious though straightforward process. For modeling limited regions such as the Sea of Japan (Bang *et al.,* 1996), it is often preferable to use the Levitus distributions as a first guess and derive suitable averages over a much finer grid making use of such data bases and any other data that are in local but not in the global data archives. While data are never plentiful enough in the oceans to provide completely satisfactory statistical averages, data in some oceanic regions such as the east coast of the United States and the Sea of Japan often suffice to obtain meaningful averages over a much finer grid than that used by Levitus (1982). The 1° resolution of Levitus analysis depended partly on the data availability in data-sparse regions in the southern hemisphere.

8.1.3 HISTORICAL MARINE SURFACE DATA

The wind stress at the surface is the next most important parameter in ocean modeling and wind observations over the ocean are indispensable for deriving this. Such measurements are routinely made and archived on merchant vessels

TABLE 8.1.1
Standard NODC Levels

Level	Depth (m)
1	0
2	10
3	30
4	50
6	75
7	100
8	125
9	150
10	200
11	250
12	300
13	400
14	500
15	600
16	700
17	800
18	900
19	1000
20	1100
21	1200
22	1300
23	1400
24	1500
25	1750
26	2000
27	2500
28	3000
29	3500
30	4000
31	4500
32	5000
33	5500

and navy ships in transit. In the past, these observations were often estimated from periodic (often hourly) visual observations of the sea state (the so-called Beaufort scale), but now they are made continuously from mast-mounted anemometers. However, large errors are usually encountered in mast-mounted instrumental records because of the obstruction of the superstructure, the heat island effect of the ship, and so on. This is the principal reason that research vessels are normally equipped with two anemometers one on either side of the superstructure well above the waterline and as far away from the superstructure as practical. While research vessels can make more accurate and detailed measurements of the wind field, they are not intended for routine monitoring. The process of deducing the wind stress from measured wind itself is fraught with errors. Wind stress is hard to measure accurately and directly with turbulence sensors, except from specially equipped oceanographic research vessels, and is therefore usually deduced from winds measured at anemometric height using bulk formulae. Unless the stability of the lower atmosphere is also measured and accounted for, large errors can result. Nevertheless, wind speed and direction measurements are often the only available information and must be used as best as is feasible.

Marine surface observations such as winds, air temperature, and humidity are also archived at national centers and it is possible to derive some statistical averages from these archives. Once again the coverage is limited to shipping lanes and probably biased toward lower values because of the tendency of ship traffic to avoid regions and periods of high wind and wave state. Hellerman and Rosenstein (1983) were the first to derive monthly averaged wind stresses over global oceans on a $2°$ grid, and these analyses have been heavily used in ocean modeling for climatological type simulations. Figure 8.1.16 shows the distribution of wind stress in the global oceans during the months of January and July. Their analysis does not include parameters needed to calculate air-sea heat and mass fluxes, but the COADS (Comprehensive Ocean and Atmosphere Data Set) data base prepared and archived at NCAR does and is increasingly replacing the Hellerman and Rosenstein data set. This data base provides $2°$ monthly summaries of parameters relevant to air-sea exchange, including wind stress and wind speed, air and water temperatures, and specific humidity. From this data set, it is possible to derive not only monthly averages for the entire data set, but also monthly values (though based on relatively sparse data) for individual months of a year. Thus, model simulations for a particular year are feasible, if high frequency temporal scale variability is not an issue. While this data base extends over more than a century, in terms of accuracy and coverage the past 50 years are perhaps the best. Figure 8.1.17 shows the wind stress distribution for the months of January and July 1994 from COADS.

Long-term monitoring is done by NDBC (NMFS Data Buoy Center) using buoys deployed at selected locations along the U.S. coast. These carry instrumentation to measure the winds and the wave state, as well as parameters to infer the sensible and latent heat fluxes. Some also carry thermistor chains for temperature measurements in the upper layers. Figure 8.1.18 shows a typical observation from 2 NDBC buoys deployed off the east coast of the United States in the Atlantic. Accurate winds from buoys are useful in local analyses including calibration of NWP wind data sets. However, the number of buoys employed at any one time is small.

8 Data and Data Processing

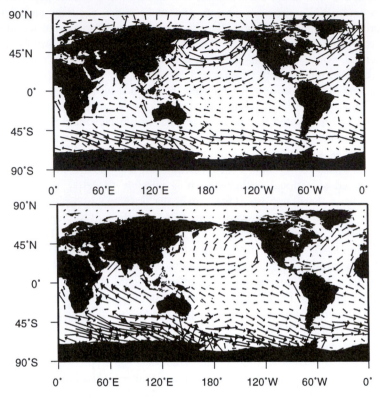

Figure 8.1.16 Long-term mean wind stress distribution over the global ocean during January and July based on data from Hellerman and Rosenstein (1983).

Ocean weather stations and ships, intended primarily for assisting commercial shipping with timely weather and other marine surface status, have also been traditional sources of not only marine surface data but also oceanographic observations, since periodic measurements were made from these platforms while on station. Stations Papa, November, and India are typical examples. Nearly a 25-year time series of oceanic and atmospheric observations are available at Station Papa, including biological parameters. Unfortunately, weather ships are no longer in use. Weather stations along numerous coastal installations provide useful atmospheric data, but they are not usually representative of the state of the atmosphere over the coastal ocean because of the effects of the land-sea interface and the land-sea breeze system.

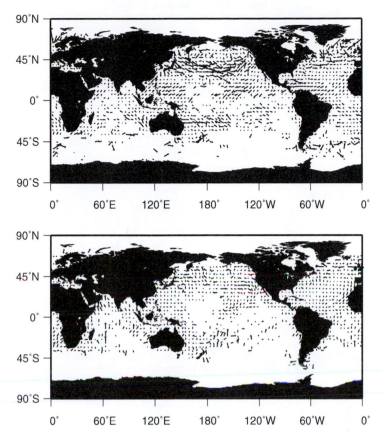

Figure 8.1.17 Wind stress distribution over the global ocean during January and July 1994. Data provided by the NOAA-CIRES Climate Diagnostics Center, Boulder, Colorado, from its Web site at http://www.cdc.noaa.gov/.

Climatological averages, whether for individual months or monthly averages, can also be deduced from NWP center analyses. It is possible to do so because NWP centers now routinely archive their model forecasts and analyses, which are becoming increasingly more reliable. Information on shorter timescales, such as synoptic wind fields at say 6-hourly interval, cannot be generally obtained from any source other than NWP centers; despite problems of accuracy and resolution, they are being increasingly used to drive ocean models in synoptic simulations.

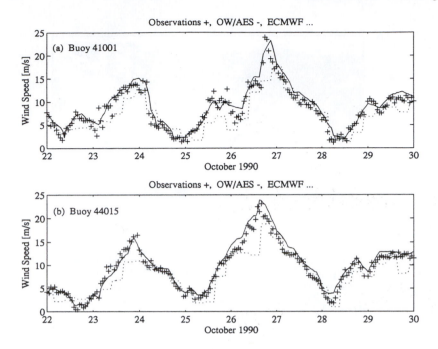

Figure 8.1.18 Wind observations from 2 NDBC buoys off the east coast of the United States compared to ECMWF winds and winds from a special analysis. (From Komen *et al.*, 1994. Reprinted with permission of Cambridge University Press.)

8.2 REMOTELY SENSED DATA

Extensive spatial coverage of the type needed for ocean models is hard to obtain except through satellite-borne instruments. While geosynchronous satellites provide continuous coverage of a particular region, polar-orbiting satellites are the ones that provide coverage of the entire globe. The main limitation is, however, the temporal-spatial sampling. Depending on the orbital pattern, the interval between successive measurements can be anywhere from a day to a month or more. Depending on the footprint or the swath width, the spatial extent of coverage can also be quite variable. It is not our objective here to provide a detailed summary of satellite measurements. Excellent treatises are available (see, for example, Gower, 1981, and Robinson, 1985; see also Emery and Thomson, 1998). Instead we will mention what satellite products are available and are potentially useful for ocean modeling.

8.2.1 SEA SURFACE TEMPERATURE FROM IR SENSORS

The most routinely available near real-time and archival product from remote sensing is the sea surface temperature derived from radiance measurements in the infrared band from a multichannel advanced very high resolution radiometer (AVHRR). AVHRRs are orbited on NOAA satellites in near-polar (~98°), sun-synchronous orbits (meaning the satellite passes over a region at the same local solar time) at an altitude of about 830 km. The cross-track scanning provides a swath width of ~2700 km, which is larger than the track spacing and hence provides for overlap between adjacent passes. Two of these satellites are in orbit at any given time, phased such that they each provide daily mappings but 6 hr apart. Local area coverage (LAC) data broadcast at a pixel size of ~1 km can be captured by a local high resolution picture tube (HRPT) station as the satellite passes overhead, although only data spatially averaged over 4 km × 4 km boxes (global area coverage, GAC) are normally recorded on board and used routinely for global mapping of SST.

The five channels (two IR, one near IR, one reflected IR and one visible) enable more accurate cloud detection and rejection and atmospheric column (principally the water vapor) corrections to be made. A major problem, however, is the effect of atmospheric dust and aerosols. Volcanic eruptions (and large forest fires), for example, can introduce SST biases of as much as −2°C. Such contamination depressed the SST erroneously causing oceanographers to be caught by surprise by the onset of 1982–1983 El Niño (McPhadden *et al.*, 1998). Normally, radiances measured in the three IR bands are combined to derive the SST using a regression formula based on comparison of satellite measured radiances and *in situ* observed temperatures (McClain *et al.*, 1985). These MCSSTs (multichannel SSTs) have a nearly zero bias (0.1 °C) and 0.7 °C variance (Emery *et al.*, 1993; Wick *et al.*, 1996) and are quite useful in data-assimilative models of the ocean. Figure 8.2.1 shows a typical example. The principal problem is the inability (unlike microwave sensors) of the IR sensors to see through the cloud cover. This causes large data voids and, if sufficient care is not taken, cloud contamination of SST data. There are regions of the global oceans that are nearly always covered by clouds, and for these regions even a multiday composite is not of much help. Overall, however, when and where available, MCSSTs provide an alternative means of specifying the heat balance at the ocean surface. Instead of having to prescribe the net heat flux at the ocean surface, which in turn determines the SST in the model, SST can be directly imposed. This often provides a significant improvement in model simulations (Horton *et al.*, 1997).

NOAA geosynchronous GOES satellites also carry multiband IR sensors. The main advantage of these sensors is the more frequent coverage compared to the

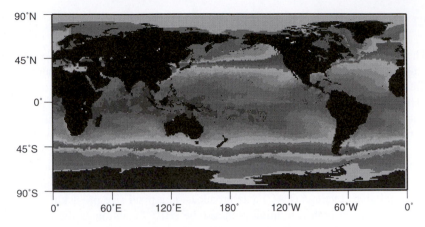

Figure 8.2.1 Composite MCSST over the global oceans during a week in February 1999.

twice daily coverage for the polar orbiting NOAA satellites. Sampling time of half an hour enables diurnal variability in SST to be monitored.

There is, however, a principal problem in converting IR radiance measurements to temperatures. What the IR sensor measures is the radiance related to the temperature of the top few tens of micrometers of the ocean. The sea water is very nearly a black body and over 60% of its emissions in the IR band come from the top 10 μm. MCSST, and other algorithms such as cross product SST (CPSST), and nonlinear SST (NLSST) relate this radiance to the so-called bulk temperature, temperature measured at anywhere between 20 cm to 5 m depths. This is done by regression of satellite-observed radiances against bulk temperature data from drifting buoys. The ocean skin is about 0.3 °C cooler than the temperature just a few millimeters below the surface, owing to the fact that the heat transferred to the atmosphere must occur through a thin molecular layer (see Chapter 4 of Kantha and Clayson, 1999; see also Emery *et al.*, 1993; Wick, 1995; Wick *et al.*, 1996). The temperature in the upper 5 to 10 m of the ocean depends very much on the bulk radiative heating by penetrative solar radiation in the visible band. Strong temperature gradients build up during the day, especially on windless summer days, and this makes the use of MCSSTs problematic, since the regression does not adequately account for strong gradients, which can lead to temperature differences of as much as 2 to 4 °C in the upper few meters. This is the principal reason that nighttime MCSSTs are of better quality for modeling than daytime SSTs. During the night, especially toward sunrise, the upper layers are nearly homogeneous in temperature due to nocturnal cooling and convection in the water column, and the skin temperature is a good indication of the mixed layer temperature below (to zeroth order, one might just add ~0.3 °C to skin temperature to derive the ML temperature or SST as seen by the ocean models, which lack the resolution needed to resolve the cool skin of the ocean). Extraction of the temperature of the cool skin of the ocean itself is not yet practiced in operational circles.

MCSST data are available from various sources including National Environmental Satellite Data and Information Service (NESDIS) and the U.S. Naval Oceanographic Office. NESDIS provides historical as well as real-time 5-day composites of MCSST in the global oceans. Global SST distributions processed operationally from 4 km resolution global area coverage AVHRR satellite data by the Naval Oceanographic Office are available for operational use by civilian and military agencies and NWP centers. Quality-controlled SST retrievals for each day as well as multiday composites are available from 70°S to 80°N over the global oceans (see May *et al.,* 1998, for details on processing and quality control).

The Climate Prediction Center (CPC) at National Centers for Environmental Prediction (NCEP) also provides near-real-time weekly global SST analyses on a 1° grid, obtained by combining the operational satellite AVHRR observations from NESDIS statistically with all available *in situ* data from buoys and ships (within a 20-hr window), using optimal interpolation (Reynolds; 1988, Reynolds and Marsico, 1993; Reynolds and Smith, 1994). In regions of sea-ice cover, as indicated by weekly analyses from U.S. Navy/NOAA Joint Ice Center, the SST is set to −1.8 °C. This data product is available since about 1982 and is extremely useful in real-time nowcasts as well as hindcasts of the global oceans. The principal problem is the coarse resolution.

NASA Jet Propulsion Laboratory (JPL) Physical Oceanography Distributed Active Archive Center (PO.DAAC) makes available a great variety of data sets related to the physical oceanography of the global oceans (podaac.jpl.nasa.gov). Included among them is the 9-km resolution global SST data starting from January 1987 from the NOAA/NASA Pathfinder reanalyses of 4-km resolution global area coverage (GAC) AVHRR data (Smith *et al.,* 1996. Walker and Wilkin, 1998). These data have been very carefully processed, with particular attention to cloud and aerosol contamination, and compared with data from moorings and buoys (see Smith *et al.,* 1996). Walker and Wilkin (1998) have used these data to produce estimates of SST at 10-day intervals by statistical interpolation. These fields are useful to ocean and atmosphere modeling, where gaps due to the presence of clouds in the original data cannot be tolerated.

AVHRRs have been the workhorse of SST measurements since the 1970s and when calibrated with *in situ* buoy data can achieve nominal accuracies of about 0.3 to 0.5°C. However, the requirement that the SST be measured to about 0.1°C for applications to ENSO and global change studies has spurred the development of more accurate sensors such as the along-track scanning radiometer (ASTR), carried aboard ESA ERS-1. Improved atmospheric correction, better internal calibration, and low detector noise enable (see, for example, Donlon and Robinson, 1997) such a radiometer to achieve an rms accuracy of about 0.3 °C, (compared to the ~1°C accuracy of an AVHRR, when not calibrated by buoy measurements). An even more accurate advanced ATSR is planned to be carried on ESA Envisat-1 to be launched in 1999.

8.2.2 SEA SURFACE WINDS FROM MICROWAVE SENSORS

Passive microwave sensors such as SSM/I (special sensor microwave imager) infer marine surface winds from the microwave emissions of a sea surface roughened by the winds. The capillary range of the surface wave field is in close equilibrium with the prevailing winds and therefore the roughness is indicative of the magnitude of the wind stress applied at the ocean surface. It is possible to infer winds to about 2 m s^{-1} accuracy from these measurements. The observations are also unaffected by cloud cover. SSM/I measured winds have been blended with ECMWF analyzed winds to obtain a superior estimate of the wind field prevailing over the global oceans (Atlas *et al.,* 1996). These winds are available from 1989 onward. However, at the time of this writing, such a blended product was not available in real time or near real time for use in real-time nowcasts. Also, SSM/I measurements are not reliable during precipitation events. SSM/I winds in combination with measurements of total liquid water content in the atmospheric column and SST have been used to infer the sensible and latent heat exchanges at the air sea interface (Clayson, 1995; see also Chapter 4 of Kantha and Clayson, 2000). There are two SSM/I sensors currently in orbit on NOAA operational satellites.

Active microwave sensors such as scatterometers use similar physical principles to deduce the surface wind stress magnitude and winds. A scatterometer is currently in orbit on the European ERS-2 satellite (Japanese satellite ADEOS carrying a NASA scatterometer NSCAT failed prematurely in 1997). Microwave energy backscattered from the surface capillary-gravity waves enables scatterometers to estimate the sea surface wind vector, from which surface wind stress can be calculated by the application of bulk aerodynamic formulas and the gridded fields used to drive ocean models. Scatterometers on board ESA ERS-1 and 2 satellite missions and the NASA scatterometer aboard ADEOS have provided proof of this concept. For example, Grima *et al.* (1999) show that the wind stresses estimated from ERS compare favorably with in-situ measurements by wind sensors on the TOGA-TAO array of ATLAS buoys. An example of the NSCAT derived winds is shown in Figure 8.2.2. In combination with altimeters and SSM/I, which also provide estimates of the wind speed (though not direction), the potential for detailed mapping and monitoring of the wind fields over the global oceans exists. The launch in 1999 of

NSCAT wind components and wind speed in M/S 5 day AVE starting 19970613   (WOCE/PO-DAAC VI-NSCAT1)

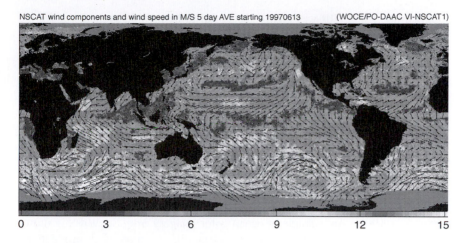

0   3   6   9   12   15

Figure 8.2.2 Typical 5-day average of NSCAT winds. Courtesy of NASA.

NASA QuickSCAT scatterometer expected to provide long-awaited coverage of winds over the global oceans at roughly 3-day intervals, and ESA Envisat, in combination with altimeters in orbit or soon to be launched opens up the possibility of finally realizing this potential.

The principal issue with winds measured from polar-orbiting satellites is the temporal-spatial sampling bias. The spectrum of wind fluctuations contains significant energy at timescales down to a few minutes and it is impossible to sample the winds to a timescale of even a few hours with a single satellite. Similarly, fast moving storm systems and their associated winds cannot be sampled properly in both space and time. The land-sea breeze system along the coast is also difficult to sample. The consequence is that the multiday composite winds over the global ocean derived from satellites are not quite optimal for use in driving a numerical model. The most optimum solution is a combination of satellite-sensed data with NWP wind product.

8.2.3 CHLOROPHYLL AND OPTICAL CLARITY FROM COLOR SENSORS

It is possible to remotely measure the upwelling radiation from the upper ocean and infer its color and hence indirectly its chlorophyll content. Most of the signal seen in certain visible bands is heavily contaminated by the intervening atmosphere, but with appropriate corrections it is possible to infer the slight shift in ocean color from predominantly blue in oligotrophic waters to bluish green in highly productive waters. The Coastal Zone Color Scanner (CZCS) orbited on

GOES in the late 1970s was highly successful in being able to measure, for the very first time, the biological productivity of the upper layers of the global oceans and its spatial and temporal variability. For ocean modeling purposes, ocean color acts as a Lagrangian tracer (albeit imperfect) for the surface circulation and complements quite well the tracer characteristics of the IR sensors. These data are therefore useful for model assessment. However, with the inclusion of a biological (primary productivity) component into an ocean model, the color sensor provides data on phytoplankton concentrations that are useful in assessing biological models and assimilation into biological models. The NASA-sponsored SeaWiFS sensor, successor to the CSCZ on Nimbus-7 from 1978 to 1986, launched on August 1, 1997, is currently in orbit, providing excellent and complete ocean color coverage over the global oceans every 48 hr (the Japanese ADEOS carried the Ocean Color and Temperature Sensor, OCTS, but died prematurely in 1997). An example of ocean color imagery from SeaWiFS is shown in Figure 8.2.3. Not only the spatial variability in biological productivity but also a measure of the prevailing surface circulation can be discerned from color imagery. Images from SeaWiFS are available from seawifs.gsfc.nasa.gov/SEAWIFS.html.

The color sensor is also affected by other factors such as the sediment and yellow stuff (gelbstuff) concentrations in coastal waters. One use of color sensor measurements is to infer the optical clarity of the upper layers. Such data are useful as input to ocean models, since the mixed layer physics is also dependent (not to the same extent as biology) on penetrative radiation in the visible band and the consequent heating of the upper layers. The SeaWiFS sensor is more discriminative and hence more useful for separating the sediment and gelbstuff contributions from that from chlorophyll concentrations.

The spectral characteristics of remotely sensed reflectance (ocean color) are significantly different in coastal waters compared to the open ocean. Higher concentrations of phytoplankton along with suspended sediments and dissolved organic matter in the usually more turbid coastal waters contribute to a much richer coastal spectra, which in turn requires much higher spatial and spectral (especially in the 600-750 nm red region) resolutions than those required to characterize the open ocean waters. Consequently, while multispectral sensors such as SeaWiFS (1.1 km spatial resolution and 8 channels in 402-885 nm band) are adequate for mapping the variability in the open ocean, hyperspectral sensors such as the Coastal Ocean Imaging Spectrometer (COIS, 30 m spatial resolution, 210 channels in 400-2500 nm band)) to be orbited in 2000 under the Navy Earth Map Observer (NEMO) mission would provide the ability to evaluate the contribution of various constituents to the spectral signature of the coastal waters. MODIS to be orbited on the NASA EOS-AM mission also has a higher spectral resolution (36 channels in 405-14385 nm band) than SeaWiFS. For more details on these developments, see Arnone and Gould (1998).

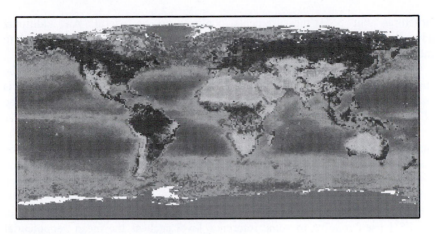

Figure 8.2.3 The global biosphere from SeaWiFS, including the chlorophyll concentrations in the global oceans. Courtesy of NASA.

Both AVHRR and ocean color sensors use the near-visible and visible parts of the electromagnetic spectrum and are therefore "fair-weather" instruments. The images must be "navigated" (accuracy is about 1 km) to properly locate them on the globe, and they must be corrected for off-nadir distortion, and the raw data converted to geophysical parameters such as MCSST and chlorophyll concentrations. This is a complex and tedious process best done at centers that are equipped to handle such tasks. The biggest problem is the ubiquitous cloud cover. The regions with cloud cover must be removed properly by suitable algorithms that can detect their presence. Daytime visible channel on AVHRR helps. Since often, any use of these data in numerical models requires that there be no data-voids or gaps, it is a common practice to composite data from several images to fill in data-voids. A common procedure in deriving void-free real-time global MCSST distributions is to update only those pixels at which the current image has valid data.

8.2.4 SEA SURFACE HEIGHT FROM SATELLITE ALTIMETRY

By far the most valuable measurement from the point of view of ocean modeling is the measurement of the sea surface height (SSH). This is simply because while SST and color measurements are normally observations of the state of the upper mixed layer a few tens of meters deep, SSH is a measure of the state of the entire water column. Due to solar heating of the upper few tens of meters, SST is of little use during spring/summer in inferring subsurface structures, but SSH measurements continue to be useful. An additional advantage is the all-weather capability of active microwave sensors used to measure the SSH, because of their to ability see through cloud cover.

As we saw in Chapter 1, the slope of the sea surface is a measure of the surface currents (minus the Ekman component). Thus measuring the SSH field is equivalent to mapping the geostrophic currents near the surface and, if the horizontal density gradients are known, throughout the water column. A radar altimeter transmits a microwave pulse in the nadir direction and measures the shape and time delay in the return of this pulse. If accurate corrections can be made for propagation delays of the electromagnetic signal in the intervening troposphere and ionosphere, it is then possible to estimate accurately the distance from the satellite to the sensor. If, in addition, the position (ephemeris) of the sensor can be determined independently accurately enough, it is possible to infer the SSH itself relative to some reference level such as the mean geopotential ellipsoid (Figure 8.2.4).

The principal signal in altimetry is from the spatial variations in the Earth's geopotential surface, the geoid, which has undulations principally due to Earth's variable topography (for example, the ridges and trenches in the ocean) on the order of 60 m rms (total) over a wide spectrum of wavenumbers. Superposed on this is the SSH variability with an rms value of less than 1 m over spatial scales ranging from mesoscale eddies to large oceanic gyres. If the oceans were to be rendered motionless, the mean sea surface would conform to the underlying

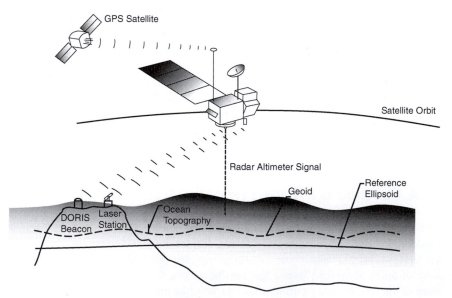

Figure 8.2.4 A sketch of the altimeter showing the various surfaces involved.

geoid. The oceans are in constant motion, however; hence the sea surface undulates, and it is these undulations that are of interest to oceanographers and ocean modelers. If it were possible to infer the invariant geoid, it could be subtracted out of the altimetric signal and one would be able to infer the absolute SSH with respect to a reference level corresponding to motionless oceans. Accurate knowledge of the geoid is important to separating the oceanic SSH signal from the geoid in altimetric measurements (and to applications such as precise orbit determination, tectonics, and geodynamics). However, the geoid is not known accurately at all wavenumbers of interest to oceanography, especially at high wavenumbers (small wavelengths) corresponding to the mesoscale variability in the oceans. It is therefore impossible to separate out the geoid in the altimetric signal. Accurate measurements of Earth's gravity field by a special gravity-mapping mission are needed. Thus what we have in altimetric observations is a large geoid signal plus a small SSH signal due to ocean circulation.

If we could infer the mean sea surface topography (also called SST by geodetists), which is the sum of the geoid and the SSH due to long-term average of the ocean circulation, then deviations of the measured SSH from this mean surface provide a measure of the fluctuations in oceanic circulation about this long-term mean. There are two means to accomplish this. Geodetists have been constructing a mean surface from observations of past satellites (Rapp *et al.*, 1994), and this mean surface is being continually improved. In principle, it is possible to subtract this mean surface from the altimetric signal and arrive at the sea surface deviation and hence the circulation fluctuation from this long-term mean. This method appears to work well for example in the Gulf of Mexico. The major disadvantage of this method is that in many areas the mean surface is not known accurately enough at scales corresponding to mesoscale variability. Also the large inaccuracies in the geoid, especially near oceanic features such as ridges and trenches, contaminate the oceanic signal leading to large errors in the estimated SSH and currents. However, the major advantage is that the altimeter does not have to be in what is known as exact repeat orbit, visiting the same track at precisely the same predetermined lapse of time. Also data from multiple satellites can all be referenced to the same mean surface.

Another means of inferring SSH fluctuations is by subtracting a mean out of the measurements. This is possible if the altimeter is in exact repeat orbit. Since in an exact repeat mission (ERM), the altimeter traverses the same subsatellite track periodically at a predetermined interval (the satellite is maneuvered to repeat any given track to within ± 1 km), it is simple to remove from measurements made during each repeat cycle at any point along track, an average over all cycles. Unfortunately, this removes, along with the geoid, an unknown average oceanographic signal as well. However, the longer the mission, the more feasible the determination of interannual fluctuations becomes.

With the TOPEX/Poseidon precision altimeter in orbit now for 6 years and expected to last beyond 2000, oceanographers will finally be able to obtain a multiyear data set that should be useful for inferring long timescale processes and to drive ocean models to examine interannual variability such as that due to ENSO events. T/P has an approximately 10-day repeat orbit.

Whichever means are used to eliminate the geoid signals, the result is the elimination of some long-term mean oceanographic signal and, for the foreseeable future, altimetry will provide only the SSH anomaly from some mean, not the absolute SSH. However, for a limited number of tracks in a region of interest, it is possible to undertake aircraft underflights and carry out AXBT (or ACTD) surveys along the track as the satellite passes over. Calculating the dynamic height along the track then enables the geoid signal to be recovered and removed to the level of an arbitrary bias. This procedure is often used when it is critical to be able to extract the absolute SSH from altimetric measurements. This strategy is, however, impractical on global or even basin scales. Also errors are inevitable due to contamination of measurements by internal waves and tides.

The feasibility of making SSH measurements from satellites was first demonstrated during the SKYLAB mission in the early seventies (McGoogan *et al.*, 1974). This was followed by NASA's Geodynamics Experimental Ocean Satellite (GEOS-3) in 1975, which used pulse compression to achieve a precision of about 25 cm in SSH measurements. Even though its orbit could be determined to an accuracy of only 1 to 2 m (Parke *et al.*, 1987), GEOS-3 was able to map the geoid for the very first time. A dedicated NASA satellite mission, called SEASAT, was launched in the late 1970s to measure the oceans, and it carried an early generation altimeter on board. However, a catastrophic power failure terminated the mission after only a few months.

The very first successful altimeter, GEOSAT, was launched in 1985 by the U.S. Navy for mapping the mean surface of the Earth (or equivalently the marine geoid) for defense applications (not measuring ocean circulation). However, after the successful 18-month geodetic mission (GM), the satellite was maneuvered into a 17.4-day exact repeat orbit along the same tracks as SEASAT and provided 2-1/2 years of data on ocean circulation (Born *et al.*, 1987). The GM data itself has been recently declassified providing an additional 1-1/2 years of data. The orbit inclination of 72° enabled most of the global oceans to be sampled.

GEOSAT was a single frequency altimeter and hence ionospheric corrections had to be provided by a model. Also its orbit could be determined only to ~20 cm rms (part of the reason was tracking limitations and part of the reason was the low orbit, about 700 km above the Earth's surface, that made orbit determination difficult), and this was a serious handicap in determining the SSH fluctuations. Fortunately most of the orbital error is concentrated in once per revolution frequency and it was possible to eliminate most if not all of the error by removing the appropriate wavenumber. However, the GEOSAT exact repeat period was not optimized for oceanographic applications and hence considerable aliasing of ocean tidal SSH signals into subtidal signals was the result. Because of the satellite stabilization method used, the altimeter tended to lose lock whenever transitioning from a land surface to water causing data outages along the basin margins. Nevertheless GEOSAT was enormously successful in mapping the mesoscale variability in the oceans. The orbit error was recently reduced to ~10 cm as the result of reanalysis (Fu and Cheney, 1995).

The launch of NASA/CNES TOPEX/Poseidon mission in 1992, with a high precision (2 to 3 cm rms) dual frequency altimeter and a GPS receiver on board, has ushered in the era of precision SSH measurements. The dual frequency capability enables accurate ionospheric propagation delay corrections to be made. In addition, the presence on board of a bore-sighted radiometer enables accurate tropospheric corrections to be made (Fu *et al.,* 1994). The high orbit (1343 km semimajor axis; eccentricity 0.0006), the DORIS tracking system based on a network of earth-based laser stations, and the availability of accurate geoid for orbit determination enables the orbit to be calculated very precisely. The GPS capability also enables precision orbit determination. The result is that the distance between the satellite and the sea surface can be determined to a precision of 3 to 4 cm. Overall, the SSH fluctuations can be determined to an accuracy of about 5 cm rms, well within the design specification of 10 cm, and if the time series is long enough they can be determined to an accuracy of 2 to 3 cm rms (Fu and Smith, 1996). The unprecedented accuracy of this altimeter makes residual errors in oceanic tides removed from the altimeter record, departures from inverse barometric response of the ocean to atmospheric pressure fluctuations, and the sea state bias the major sources of error, and not the error in orbit determination as was the case in earlier altimeter missions. Also the 9.9156-day exact repeat (127 revolutions/repeat) was chosen to be optimal for oceanographic measurements, with tidal aliasing of the principal tidal component M_2 a far less serious a problem than in earlier missions (K_1 is still aliased close to semiannual signals and hence there is some mutual contamination). The principal disadvantage is the poorer cross-track resolution, about ~316 km at the equator, compared to ~165 km for GEOSAT. Also the inclination of $66°$ misses a significant portion of the high latitude oceans. However, because of the improved stabilization, the altimetric data remain viable all the way to the coast within the footprint of the radiometer pulse, which depending on the sea state is somewhere between 3 and 7 km (tropospheric correction may be a problem within 30 km of the coast). Also, the precision of the altimeter enables 10 Hz data set with higher along-track resolution (0.7 km versus 7 km for the usual 1 Hz data) to be used to study small scale processes such as internal wave solitons and a better use of the altimetric data to be made on continental shelves.

The Europeans have also orbited altimeters in their ERS-1 and ERS-2 missions at an orbit height of ~780 km, similar to that of GEOSAT. While, they are not as precise and accurate as the TOPEX altimeter, the simultaneous availability of two altimeters has been helpful in the study of oceanographic processes (Choi *et al.,* 1996). ERS-1 and 2, because of their higher inclination, reach farther into polar latitudes (81° compared to 66° for TOPEX) and have been used to map sea ice. ERS-1 has been in 3-day, 35-day, and 135-day repeat orbit configurations, the 35-day configuration being of principal interest to oceanography because of the smaller cross-track spacing (~80 km) that results. By combining the 35-day repeat ERS-1 data and 10-day repeat TOPEX data into a comprehensive circulation model, Choi *et al.* (1996) were able to better simulate the eddies shed by the Loop Current in the Gulf of Mexico during a nowcast/forecast study. With improved orbit determination, ERS-2 has attained accuracies similar to that of TOPEX (Le Traon and Ogor, 1998).

The ground track spacing of a satellite is $2\pi R\cos\theta(n_d T_r)^{-1}$, where n_d is the number of orbits per day $\sim17.06[1+(h/R)]^{-3/2}$, h is the satellite height, T_r is the repeat period in days, R is the earth's radius, and θ is the latitude. Thus there is a trade-off between ground spacing and repeat sampling period. The shorter the repeat period, the wider the track spacing. Thus the 17.4, 9.916, and 35-day repeat periods of GEOSAT, T/P, and ERS-1/2 translate to 165, 315 and 80 km track spacing at the equator.

The U.S. Navy plans to follow up on the GEOSAT mission with the launch of the GEOSAT-follow-on (GFO) series of altimeters. These are designed to follow the same tracks and exact repeat period (17.4 days) as GEOSAT. The first of this series was launched in 1998. NASA and CNES plan to launch a follow-up mission to T/P, Jason1, with the same characteristics (9.916 day repeat period) in 1999. This altimeter is expected to improve the accuracy of SSH measurements even further and to overlap T/P. ERS-2 is to be followed in 1999 by Envisat-1 with the same repeat period of 35 days (Gardini *et al.,* 1995). This opens the possibility of simultaneous operation of 2 or more altimeters, which should enhance applications of altimetry to shallow water, where most of the prevailing variability is primarily due to synoptic wind forcing and hence is far too fast to be meaningfully sampled by a single altimeter.

If there were no intervening matter, the distance from a precision altimeter to the sea surface, the raw range, ζ_{raw}, would be given by $0.5 v_L \Delta t$, where v_L is the speed of light and Δt is the time for the radar pulse to travel to the sea surface and back to the satellite. In practice, this range needs to be corrected for various propagation delays. Free electrons in the ionosphere cause a range delay of ζ_{ion}. Propagation delay in the troposphere is divided into two parts: the wet tropospheric delay ζ_{wet}, due to the water vapor in the atmosphere, and the dry topospheric delay ζ_{dry}, due to the dry mass of the atmosphere. In addition, the fact that the wave troughs reflect the radar pulse better than the wave crests gives rise to an additional correction, called the electromagnetic (EM) or sea state bias ζ_{emb}. The true range is therefore

$$\zeta_{tr} = \zeta_{raw} - \zeta_{ion} - \zeta_{wet} - \zeta_{dry} - \zeta_{emb} \qquad (8.2.1)$$

The accurate determination of these various corrections to the altimetric range (and of the orbital height of the satellite above a reference ellipsoid) is the key to accurate determination of the sea surface height fluctuations. Since the ionospheric delay is a function of frequency, a dual frequency altimeter (such as TOPEX) enables accurate ionospheric corrections to be made. With single frequency altimeters such as Poseidon, GEOSAT, and GFO, this correction has to be obtained from an empirical numerical model of the ionosphere that takes into account the fluctuations in electron density due to day-night, summer-winter, and 11-year solar cycle changes. This correction is typically less than 20 cm. An on-board multichannel microwave radiometer enables the wet tropospheric correction, which is typically less than 40 cm, to be deduced. Otherwise, this information has to be derived from atmospheric models. The dry troposphere correction can be determined if the properties of the atmospheric column beneath the altimeter are known and is typically around 230 cm. For an ideal gas, this correction is proportional to the surface pressure (0.227 cm mb^{-1}), and this information is derived from atmospheric pressure field analyses from NWP centers such as ECMWF. The sea state bias is typically a small fraction of the significant wave height (<5%) and can be inferred if the sea state is known. An added bonus in altimetric measurements is that the return pulse shape contains information on the sea state. It is therefore possible to not only estimate the EM bias correction but also to infer the significant wave height along the satellite subtrack.

Knowing the true range ζ_{tr} and the position of the satellite above a reference ellipsoid (the height of the orbit) ζ_{orb}, the distance between the sea surface and the reference ellipsoid is known:

$$\zeta_{ssh} = \zeta_{orb} - \zeta_{tr} \qquad (8.2.2)$$

Accurate determination of ζ_{orb} is another challenging aspect of altimetry. Precision orbit determination involves computing the ephemeris of the satellite based on an accurate model of the motion of the satellite subject to various forces (such as the atmospheric drag) that minimizes the difference between the computed orbit and observations from tracking systems. An on-board GPS receiver also enables accurate orbit determination.

The major contribution to ζ_{ssh} comes from the static geoid ζ_{geo} (the height of equipotential surface from the reference ellipsoid). Assuming for the moment that ζ_{geo} is known at all points along the subsatellite track, the instantaneous position of the sea surface with respect to the geoid can be determined. This in turn consists of the various tides, the solid Earth (body) tide ζ_{bt}, the ocean tide ζ_{ot} and the associated load tide ζ_{lt}, and the pole tide ζ_{pt}. Of all these tidal corrections, it is the ocean and their associated load tides that are hard to determine; they have to be derived from accurate global ocean tide models (Kantha, 1995b; Le Provost *et al.,* 1994; Tierney *et al.,* 1999; see also Shum *et al.,* 1997). In addition, the changes in atmospheric pressure acting on the surface of the ocean causes the ocean to respond roughly like an inverted barometer rising (falling) roughly 0.996 cm for every millibar drop (rise) in the surface pressure. This correction, called the IB correction ζ_{ib}, is derived once again from atmospheric models. Subtraction of all these from ζ_{ssh} provides the desired sea surface height signal due to subtidal oceanic circulation:

$$\zeta = \zeta_{ssh} - \zeta_{geo} - \zeta_{bt} - \zeta_{ot} - \zeta_{lt} - \zeta_{pt} - \zeta_{ib} \qquad (8.2.3)$$

To this one must add the contribution due to imprecise nature of the altimetric measurements (for example, TOPEX precision is ~2 to 3 cm) and other random errors. Thus, while the determination of the sea surface height from altimetry is quite simple in principle, in practice, the resulting accuracy is very much dependent on the accuracy with which the various quantities mentioned above can be determined. In addition, since the geoid is not known precisely at wavelengths of relevance to mesoscale oceanic variability, ζ_{geo} cannot be determined and therefore one is left with the sum $\zeta_{geo}+\zeta$. The determination of ζ_{geo} requires combination of independent gravity measurements and numerical gravity models. Lemoine *et al.* (1998) describe the most recently developed Earth gravity model, Model 1996, which provides a global map of the gravity anomalies to spherical harmonics of degree and order 360 (30' or 55 km resolution). Based on surface gravity data from around the globe, three decades of satellite tracking data, this is the most accurate geopotential surface deduced to date. Several gravity missions are planned and the next decade should see a more accurate determination of the geoid and hence extraction of the true sea surface height ζ from altimetric measurements. However, it is possible to decompose ζ into a long-term mean ζ_m and a fluctuating component ζ_{ano} (the SSH anomaly from that predefined mean) and to combine ζ_m and ζ_{geo} into ζ_{mss} so that

$$\zeta_{ano} = \zeta_{ssh} - (\zeta_{bt} + \zeta_{ot} + \zeta_{lt} + \zeta_{pt} + \zeta_{ib}) - \zeta_{mss} \qquad (8.2.4)$$

The mean sea surface ζ_{mss} can be derived from previous altimetric measurements over the global oceans. The so-called Rapp mean surface (Rapp *et al.,* 1994) derived in this fashion is available from an ftp site at the Ohio State University (on a $1/16°$ latitude-longitude grid). Another technique is to deduce

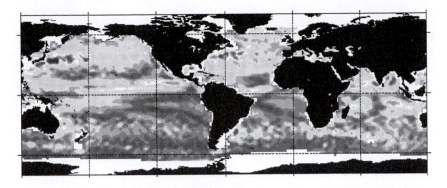

Figure 8.2.5 SSH anomaly from the annual mean from TOPEX altimetry showing the various current systems and mesoscale features.

average observations at each point along the track over many repeat cycles and subtract this long-term mean from the observations at that point. Either way, ζ_{ano} is what is extractable at present from altimetry.

Figure 8.2.5 shows the distribution of ζ_{ano} from a typical TOPEX cycle. The various current systems stand out clearly. The along-track slope of the sea surface height is related by geostrophy to the velocity at the surface (minus the Ekman contribution) and therefore the along-slope gradient of ζ_{ano} gives the surface velocity anomaly. This quantity also highlights regions of strong mesoscale variability in the global oceans.

Altimetric data has been used extensively for comparison with and validation of ocean models (Metzger *et al.*, 1992). For the first time, ocean modelers have observational data with extensive spatial coverage with which to provide a "sanity check" on their model simulations (see Chapters 1 and 11). Also, while it is possible to study aspects such as mesoscale variability and sea level rise from altimetric data alone, a more complete study is possible by assimilating these data into a comprehensive ocean model. Thus other parameters such as currents can be inferred from the resulting oceanic state (Choi *et al.*, 1996, for example). Steady progress is being made in this direction (see the review by Fu and Smith, 1996). Availability of real-time altimetric and MCSST data is the first requirement in establishing nowcast/forecast models for the oceans (Kantha *et al.*, 1999).

8.3 NWP PRODUCTS

Short of using global coupled atmosphere-ocean models, the only way at present to simulate variability on daily-weekly timescales is by driving the ocean model with atmospheric forcing determined from NWP products. Observations alone just do not have the global coverage or the temporal resolution necessary. Satellite-derived sea surface fluxes can in principle be used to provide synoptic type forcing (Clayson *et al.*, 1996), but their efficacy is still largely unproven. The resolution and the skill of NWP models have steadily increased in the past two decades, principally as a result of improved physical parameterizations, increase in quality and quantity of data, improvements in data blending techniques, and significantly improved computer resources. NWP uses forecast-reinitialization methodology to keep the model atmosphere close to the real one, with the cycle repeated every 6 to 12 hr. NWP centers such as NCEP, FNMOC, British Met Office (BMO), ECMWF, and Japan Meteorological Agency (JMA) make routine 2-day forecasts, while centers such as NCEP and ECMWF make extended medium range forecasts with skill up to about 5 days. These centers also archive their model outputs every 6 hr, including analyses from which the predictions are started. Detailed state of the atmosphere is archived, but for ocean modeling purposes, it is adequate to consider only the air-sea exchange parameters. If the NWP model has good boundary layer physics, it is preferable to use the model-indicated fluxes of momentum rather than compute it from model parameters such as winds and stability.

For ocean state forecasts that depend on the surface forcing for adequate skill, it is necessary to drive the ocean model with NWP forecasts. Since the skill of the latter deteriorates steadily, the skill of the ocean forecasts can also be expected to do so. Currently 2 to 3 days are the limit for skillful NWP forecasts, especially with respect to winds and most ocean forecasts are also limited to this duration. Where the surface forcing is not a primary factor as for example in processes such as eddy shedding involving baroclinic instabilities, it may be possible to integrate the model longer. In the hindcast mode of simulations, the period of model simulation is restricted only by the duration of the atmospheric forcing available. NWP centers archive their model output every 6 hr for every forecast cycle. Thus there is some flexibility in the choice of which particular field(s) to use. Since even with current assimilation methods that minimize the shock of data insertion, it takes a few hours for the shock of data insertion to be attenuated, it is preferable to use 12-hr (6-hr) forecasts than either the earlier or later forecasts, if 12-hourly (6-hourly) forcing is adequate. Alternatively, the analyses that blended the observations into the model may be used, although these may not have all the fields needed and the fields have not been brought to dynamical equilibrium by the model.

One problem with NWP products is the aliasing with respect to solar radiation, since hourly sampling is needed to represent diurnal variability. Normally, hourly radiative fluxes are computed using empirical formulae with input such as cloud cover and air temperature from the NWP model interpolated suitably to hourly intervals.

Precipitation is one of the hardest quantities to measure, especially over the oceans, even from a well-equipped research vessel. The skill of the NWP models is highly suspect when it comes to precipitation fields. Yet, precipitation is important to the specification of the surface buoyancy flux in a numerical ocean model. Modern optical rain gauges and microwave radars are beginning to provide better estimates of precipitation over the oceans. NASA has flown a mission focused specifically on precipitation issues, Tropical Rainfall Measurement Mission (TRMM). TRMM is designed to measure precipitation over tropical-subtropical oceans and is expected to revolutionize precipitation prescription in ocean models. There has also been a concerted effort involving operational centers to provide gridded, high quality historical data sets for modeling and other uses by the research community. One such example is the gridded monthly global precipitation data—CPC Merged Analysis of Precipitation (CMAP)—from NOAA/NCEP Global Precipitation Climatology Project (GPCP) on a 2.5° grid for 17 years from 1979 to 1995. This data set combines remotely sensed IR, outgoing long wave (OLR) and MW data and data from NCEP/NCAR reanalyses (Xie and Arkin, 1997). These data show that the annual global precipitation average is 2.69 mm year^{-1} (1.86 over land, 3.01 over water). ITCZ rain band shows up clearly and so does the South Pacific Convergence Zone (SPCZ).

Finally, the Internet and electronic multimedia capabilities have brought a revolution in data exchange and transfer. Instead of having to write to the data center and wait for weeks for the data to arrive on magnetic media by mail, one can now use the Internet or anonymous ftp sites at the data centers to download data on to a workstation. These data are often accompanied by detailed graphics that are well suited to a better comprehension of the coverage and quality. Most federal agencies responsible for collecting and archiving observational data now make their data available in this manner, and so do NWP centers and individual scientists and projects. It is not unusual now to find data collected by satellites and telemetering instrument arrays to be made available almost immediately after collection. Agencies such as NOAA and NASA which are responsible for environmental matters (including the Department of Defense) maintain websites describing their data holdings and the procedures to access them, and any restrictions on their use and distribution. A few of the important ones are listed in Appendix E, but since there are no protocols and standards established yet on indefinite maintenance of the Web sites, and the means to inform the browser when an Internet site is renamed, relocated, or removed, it is best to use a Web browser to locate and access data sources.

8.4 PREPROCESSING OF OBSERVATIONAL DATA AND POSTPROCESSING OF MODEL OUTPUT

Data preprocessing consists usually of several steps: quality control of the data, its display and analysis, and its transformation to a form suitable for ingestion into the model or comparison with model output. The available data fields for ocean model initialization and forcing, even when gridded, seldom conform to the chosen ocean model grid. The data must therefore be prepared in a format that the ocean model requires. It is necessary, for example, to interpolate them onto the model grid. A variety of such interpolation packages are available. For synoptic-type simulations, it is always preferable to animate the surface forcing, especially the surface fluxes, before using them to drive the model to understand what sort of forcing the model is being subjected to. Needless to say that proper attention to units and data formats is essential to avoid costly mistakes.

Data quality is an important issue. While archival and distribution centers pay special attention to this, it may still be necessary to address the quality issue, especially if it becomes necessary to reprocess the data from scratch. Knowledge of data characteristics, error sources, and statistical methods becomes indispensable. One particular problem for numerical modeling is data gaps resulting from sensor malfunction, lack of coverage, or removal of erroneous data. Such data gaps must be filled in by interpolation. Gaps in time series data or vertical profiles are best filled by spline fitting or a similar technique. The model forcing and initialization fields cannot have any data outages either in space or time, and therefore any gaps in coverage must be filled in by surface fitting or an objective analysis technique. Particular attention must be paid to the wind stress applied to the model. Accurate representation of the wind stress curl is particularly important and it is essential to make sure that the interpolation routines do not artificially introduce spurious curl values in the stress fields used to drive the model. See Emery and Thomson (1998) for an in-depth discussion of issues involved in oceanographic data processing and analysis, including time series analysis techniques.

Observational data needed for comparison with model output may also require special handling. First, the modeler must understand what the observations imply in terms of the processes that the model is attempting to simulate. Second, the modeler must examine the model output for similar characteristics. This often involves passing both the data and model output through similar processing steps. Techniques such as empirical orthogonal functions (Appendix C) and Fourier transforms are a part of this process. Finally, a quantitative assessment of the model performance vis-a-vis observations requires statistical techniques such as cross-correlations of temporally and spatially varying property fields. It also requires paying particular attention to an estimation of possible errors in both the data and the model. A detailed discussion of these aspects is beyond the scope of this book.

Before the advent of satellites and remote sensing, processing observational data used to be quite easy, simply because there were not much data to be processed. However, analysis of observational data, even with the aid of modern workstations and high performance computers, has become a nontrivial task. Even this task pales in comparison to processing and analyzing the output of modern three-dimensional models of the atmosphere and the ocean. Unlike observations, which can generate only a limited number of fields with spotty temporal and spatial coverage, a numerical model can generate realizations of 3-D fields of all model prognostic variables every time step for the entire duration of the model simulation. For example, if the model has 10 million grid points and say 10 variables and has a time step of say 30 min, and is run for say 10 yr, it generates 18 trillion pieces of data that would require 72 terabytes of storage space if every time step is saved. This amount of data is rather impractical to save and analyze, and the archival interval of model output has to be carefully chosen so that the data are on the order of a few gigabytes or less. Even then the mere task of processing the model output and arriving at meaningful conclusions is difficult. It is therefore important to preprocess and archive model output with specific goals and hypotheses in mind. This is not always possible, however. A strategy often used in field observations is dense spatiotemporal coverage during preselected intensive observation periods (IOP) with less intensive coverage the rest of the time to merely provide a context for interpreting the IOP data. A similar strategy is also useful in modeling if there exists a particular phenomenon or period worthy of intensive analysis and study.

Before the advent of modern graphics workstations, few preselected model fields were laboriously plotted at some discrete intervals and examined for model performance and conclusions. This is no longer desirable or optimal. Because of the dynamic nature of the model output, better inferences can be drawn by animation of model fields. Animation is also helpful for a quick visual check of the model and forcing fields.

For ocean models, the important fields to plot and study are the sea surface height, the currents at various levels (near-surface, midlevel, and deep suffice for an initial examination), the mass fields, temperature and salinity, the turbulent kinetic energy (and dissipation rate), any chemical and biological variables such as nitrate, and phytoplankton concentrations. It is often advantageous to select a few salient but limited number of the 3-D fields to be visualized on a graphics workstation using a program such as VIS5D for a better understanding of their structure.

Detailed time series should be saved at carefully preselected points. These time series can then be analyzed using standard time series analysis techniques for their spectral content. Modern methods such as wavelet transforms would be a great aid in isolating events in time and spectral space (see Appendix B). Currents and wind vector time series should be plotted as stick plots. Modern statistical methods such as EOFs (see Appendix C) should be employed for a succinct estimate of the variability of temporally varying model fields. All of this requires modern workstations for postprocessing, and graphics workstations for visualization, with adequate speed, disk space, and memory.

8.4.1 GRAPHICS AND VISUALIZATION OF THE MODEL OUTPUT

Conversion to graphics, both static and animated, is the very first step in analyzing the data input into and the output of a numerical model. Many of the strategies detailed below can be used for visualization and analysis of observational data as well. This is particularly important since model checking and skill assessment involve observed variables that need to be visualized and analyzed in a similar format.

Outputs of 1-D and 2-D fields (from either 1-D and 2-D models or a 3-D model) are relatively simpler to analyze and plot since they involve only one and two independent spatial variables, respectively, and time. For example, the evolution in time of, say, the vertical profiles of a scalar such as temperature $T(z,t)$ can be plotted either superimposed on one another or as a waterfall plot or simply contoured (or color-fill mapped) in z-t space. A time series plot at selected points $[T(z_p,t), p = 1,N]$ is often more useful. It is also possible to animate the profiles. Two-dimensional fields of, say, temperature $T(x,y,t)$ on the other hand can be plotted as a 3-D field in x-y-t space or as a series of 2-D (x-y) fields (and animated). Evolution in time is often best seen as a space-time (x-t or y-t) plot often called a Hovmuller diagram. Features oscillating in space/time and features with wavelike propagation characteristics are best visualized by this process. If the model involves nonuniform (including curvilinear and telescoping) grids, either graphics packages designed to handle nonuniform grids must be used or the fields must first be interpolated to a uniform grid overlaid on the original grid.

Time-varying 3-D fields are the hardest to visualize. For layered/isopycnal and sigma-coordinate models, it is necessary to interpolate the fields vertically to preselected horizontal levels, since the influence of the gravitational field tends to make properties on horizontal levels more meaningful, except in cases where it is best to look at properties on predefined surfaces such as isodensity (isopycnal) surfaces. While 3-D animations are feasible, using a package such as VIS5D, and useful, hardcopy presentations necessarily involve 2-D slices. For visualization of the spatial structure of a 3-D field such as, say, $T(x,y,z,t)$, plots of $T(x,y,z_p)$, $T(x,z,y_p)$, or $T(y,z,x_p)$ at selected points at preselected time intervals can provide the visualization needed. Often, it is useful to know the depth of a particular isotherm, say, $z_m(x,y)$ where T=m. This can be done by simple interpolation. Often, it is useful to visualize the property, say, $T(x,y,z)$, on a predefined surface, say, on a particular sigma-t surface ($\sigma_t = m$). This simply requires knowledge of σ_t (x,y,z) from which $z_m(x,y)$ can be determined and $T[x,y, z_m(x,y)]$ evaluated by simple interpolation.

A vector field such as velocity presents even more challenges. Since the vertical velocities are several orders of magnitude smaller than horizontal velocities in the oceans and the atmosphere, the usual practice is to isolate and treat the vertical velocity as a scalar. Horizontal velocities are then two-component vector 3-D fields, and two approaches are possible: (1)Looking at the individual components as scalar fields. This works well if well-defined directions exist, such as the zonal and meridional directions on the globe. (2) Representing vectors as arrows on level surfaces—for example, vector plots of velocity at some depth $V(x,y,z_p)$ or some predefined surface $V[x,y, z_m(x,y)]$. Streak plots, where a particle is released at each grid point and its path plotted over a short preselected interval, are better for defining features such as eddies and retroflections and other complex current patterns. This concept can be carried a step further and tracks of model drifters $x_d(t),y_d(t)$ plotted in x-y space using $V(x,y,z_d,t)$. This is useful for comparison with observed tracks of drifters that are drogued to follow a certain depth that are being used increasingly in the oceans to obtain the spatial structure of prevailing currents. These model drifter tracks can be plotted in x-y-z space if the full 3-D velocity field is used; however, this requires color-coding to indicate the position of the drifter in the vertical. Such tracks are useful for analyzing phenomena such as coastal upwelling and deep convection, which involve significant vertical transport of fluid particles (Moisan *et al.*, 1995).

Temporal evolution of a full 3-D field requires animation of the 3-D field, a time-consuming task for realization and presentation. Simpler subsets are often used for static displays. Space-time plots (Hovmuller diagrams) are often useful if the spatial direction is properly chosen. In equatorial ocean modeling, it is natural to look at propagation along or parallel to the equator. A plot in x-t of a variable such as sea surface height is quite useful for this purpose. It is also possible to plot a time series of a scalar or vector field (stick plots) at selected points in space. These are particularly useful for comparison with observational time series data from instruments such as thermistors and current meters deployed on fixed moorings. Comparison with observed data from polar orbiting satellites requires an entirely different strategy. For example, for comparison with an altimeter, it is necessary to sample the model SSH with a simulated altimeter whose ephemeres are the same as that of the real altimeter. While straightforward to do, this requires integration of such a code into the model code, since the temporal density of archived output is usually inadequate for postsampling by a simulated altimeter.

8.4.2 ANALYSES

Analyses of an ocean model output (and associated observational data) have to be necessarily guided by the modeling goals: model development and checking, broad exploration of resulting physics and dynamics, study of particular phenomena or processes, or prediction of certain oceanic states. It must also be guided by dynamical considerations, available data for comparison/validation, and limitations of the model as well as data. From dynamical considerations, energetics and energy interconversions are quite important. This is the primary reason that the mean kinetic and potential energy computations must be a part of the model output or postprocessing packages. It is a normal practice to distinguish between barotropic and baroclinic energies because of the distinct physics involved. A subset of these (the basin-wide energy levels) is useful for assessing the degree of spin-up of the model from an initial (usually quiescent) state. Since geophysical flows often conserve potential vorticity, tracing potential vorticity of a fluid particle is quite useful. For example, the deep ocean is for the most part adiabatic, and potential vorticity contours highlight advection in the model and can be used for particle tracing. For assessment of turbulent mixing processes, it is necessary to examine the TKE and its dissipation rate, although observational data are seldom available for comparison. However, with more routine deployment of microstructure profilers that infer the dissipation rate in the water column, this situation is changing. Thermodynamic considerations require diagnosis of quantities such as the SST, the mixed layer depth, and the heat content in the upper ocean. Heat and salinity budgets are also often needed. Most of these can be carried out at postprocessing stage, provided the data have been archived appropriately. Some analyses such as term balances needed for deciphering the dynamical balances involved are, however best done during the model run, since they are hard to compute accurately from data archived at intervals much larger than the model stepping time.

Analysis of temporal evolution requires modern time series techniques. The most routine is the analysis of spectral content using FFT techniques for time series analysis. Comparison with observed spectra often reveals any inadequacies in the model. This can be complemented by wavelet transforms, if identifying the spectral contents of isolated events is essential (Appendix B). EOF and SSA techniques (Appendix C) are quite useful for compact quantification of modes of variability of spatially and temporally varying fields, such as SST in the equatorial waveguide, although dynamical interpretations of resulting modes may not always be possible.

Computing correlations between modeled and observed fields is often an excellent way to assess model skill, but this requires similar temporal and spatial coverage.

Chapter 9

Sigma-Coordinate Regional and Coastal Models

A timely review by Haidvogel and Beckmann (1997) discusses the current status of coastal ocean modeling. Table 9.1.1, which was adapted from their paper, provides a succinct summary of the characteristics of various 3D ocean models applicable, in principle at least, to coastal oceans. They cover a wide range of techniques used to solve the same underlying set of 3D, Boussinesq, primitive equations for a stratified fluid. They include hydrostatic and nonhydrostatic versions of the equations: z-level, σ coordinate, s coordinate (hybrid), and isopycnal coordinate systems; rigid lid and free surface formulations; structured (rectangular and curvilinear), and unstructured (triangular and quadrilateral) grids in the horizontal; Arakawa A, B, and C staggering of variables; finite-difference, spectral, finite-element, and spectral element approaches; explicit (leapfrog, third-order Adams-Bashforth), semi-implicit, split-mode, and implicit Crank-Nicolson time-stepping; and centered second-order, first-order upwind, third-order Adams-Bashforth, total variance diminishing (TVD), flux corrected transport (FCT), and piecewise parabolic method (PPM) and other advection schemes (see Chapter 2). The performance of a numerical model depends very much on the particular choices made in its formulation. Unfortunately, few systematic studies of these choices and their effects on model performance have been made (but see Haidvogel and Beckmann, 1997), and it is therefore difficult to make specific recommendations as to the suitability and fidelity of a particular model for a particular flow problem, although some general guidelines can be given. In the following, we

TABLE 9.1.1

Numerical Characteristics of Various Ocean Models (the Last Three, Barotropic)

Model	Eq[1]	Ext[2] md.	Vert[3] discret.	Horiz[4] discret.	Time[5] step	Horiz[6] advn.	Reference	Internet[7] address
DieCAST	H	R	z-FD	R-A/C-FD	LF	C	Dietrichet et al. (1987)	dietrich@cast.msstate.edu
DJM	H	F	σ-S	R-C-FD	S/Si	TVD/PPM	Davies (1987)	amd@pol.ac.uk
GBM	H	R	z-FD	R-C-FD	CN	A/SL	Dippner (1993)	dippner@dkrz.d400.de
GFDL*	H	R	z-FD	R-B-FD	LF	C/U,/FCT	Paconowski (1995)	kd@gfdl.gov
GHREM	H	F	σ-FD	R-C-FD	SpE	TVD	Beckers (1991)	jmb@ocean.oce.ulg.ac.be
HAMSOM	H	F	z-FD	R-C-FD	S/Si	U/A	Backhaus (1985)	backhaus@ifm.uni-hamburg.de
HOPE	H	F	z-FD	R-E-FD	LF	C	Wolff and Maier-Reimer (1997)	
ISPRAMIX	H	R	z-FD	R-C-FD	SpE	U	Eifler and Schrimpf (1992)	walter.eifler@cen.jrc.it
LODYC	H	R	z-FD	C-C-FD	LF	C	Delecluse et al. (1993)	
M3D	H	F	z-FD	C-C-FD	SpE	U	Walker (1996)	walker@ml.csiro.au
MICOM*	H	F	ρ-FD	R-C-FD	SpE, LF	Sm	Bleck et al. (1992)	rbleck@rsmas.miami.edu
MIT*	H/NH	R	z-FD	R-C-FD	LF	C	Marshall et al. (1997)	
NRL	H	F	ρ-FD	R-C-FD	LF	C	Wallcraft (1991)	
POM/CUPOM*	H	F	σ-FD	C-C-FD	SpE, LF	C	Mellor (1996)	glm@splash.princeton.edu
QUODDY*	H	F	σ-FE	U-T-FE	S/Si	C	Ip and Lynch (1994)	daniel.lynch@dartmouth.edu
SCRUM	H	F	s-FE	C-C-FD	SpE/AB3	C/AB3	Song and Haidvogel (1994)	arango@ahab.rutgers.edu
SPEM	H	R	s-S/FD	C-C-FD	LF	C	Hedstrom (1996)	kate@ahab.rutgers.edu
CU	H	F	2D	C-C-FD	LF	C	Kantha (1995)	
FES	H	F	2D	U-T-FE	S/Si	G	Le Provost et al. (1994)	
SEOM*	H	F	2D	U-Q-SE	AB3	G	Iskandarani et al. (1995)	mohamed@ahab.rutgers.edu

* Indicates the model is widely used or has a novel approach.

[1] H—hydrostatic, NH—nonhydrostatic

[2] R—rigid lid, F—free surface dynamics

[3] z—Z coordinate, σ—sigma coordinate, ρ—isopycnal, s—hybrid coordinate, FD—finite difference, S—spectral, FE—finite element, 2D—barotropic

[4] R—regular, C—orthogonal curvilinear, U—unstructured, A, B, C—Arakawa A, B, C grids, T—triangular, Q—quadrilateral, FD—finite difference, FE—finite elements, SE—spectral element

[5] LF—leapfrog, S/Si—synchronous semi-implicit, CN—crank nicolson, SpE—split explicit, AB3—third-order Adams Bashforth

[6] C—centered, TVD—total variance diminishing, PPM—piecewise parabolic method, FCT—flux-corrected transport, U—upwind, G—Galerkin, A—Arakawa enstrophy conserving, SL—semi-Lagrngian, AB3—third-order Adams Bashforth

[7] As known at the time of the writing; the address may or may not stay the same.

attempt to provide a rationale for the formulation of coastal models and discuss, in great detail, one that is widely used. It also happens to be the one we are most familiar with.

9.1 INTRODUCTION

Generally speaking, given the complex coastline and bathymetry, structured grids are the least flexible for coastal ocean modeling, even though orthogonal curvilinear coordinates offer a bit more flexibility and nesting mitigates the problem. Unstructured grids (finite elements) afford essentially unlimited flexibility in tailoring the model grid to a complex coastline geometry, but with some computational and numerical penalties. Rigid lid formulation is, for most purposes, unacceptable in shallow coastal waters, and hence explicit free surface dynamics should be included. This means that split-mode solution techniques (which split the equations into two coupled sets, one involving fast external gravity wave modes, and the other, slow internal gravity waves) have to be used to avoid the computationally costly solution of all the equations at small time steps dictated by the external gravity wave CFL condition (despite the need for temporal filtering to damp a weak instability that may result from split time stepping, see below). To extract maximum efficiency, the more time-consuming internal mode equations are solved less frequently than the external mode equations, with interaction between the two modes at the coupling time step treated carefully. The alternative is synchronous semi-implicit formulation that treats the gravity wave producing terms involving SSH gradients in the external and internal mode equations and the associated terms in continuity equations semi-implicitly (Haidvogel and Beckmann, 1998). This enables both equations to be solved with large time steps, but the penalty is the need for solving a set of elliptic equations at each time step (akin to the rigid lid formulation). The staggered Arakawa C grid is usually the grid of choice for the high grid resolutions normally used in coastal waters, because of its acceptable gravity wave propagation characteristics.

Because of the importance of salinity effects, coastal models must solve for both temperature and salinity and employ the unlinearized equation of state. The presence of nonlinear advective terms in the governing equations requires particular care in handling the advective terms in equations for both momentum and scalars such as temperature and salinity. Ideally, the discretization scheme for advection terms must maintain positive definiteness (no unphysical negative values of scalar concentrations anytime during the solution) and monotonicity (no unphysical oscillations in scalar concentrations). It must also conserve various invariant properties such as energy, enstrophy, and scalar variance. In addition, it must be least diffusive, accurate, and efficient (Haidvogel and Beckmann, 1997). No such advective scheme exists. All are deficient in one way or another (see Rood, 1987, for a thorough review; see also Chapter 2). Low order upwind schemes preserve monotonicity and positive definiteness but at the cost of excessive diffusion. Centered difference schemes, most commonly used, are too dispersive and fail to maintain these two properties without the addition of some explicit diffusivity. Advanced schemes such as TVD and FCT are time-consuming and low order, and hence are less accurate. Higher order spectral schemes are more accurate but may not maintain monotonicity or positive-definiteness. Thus the choice is not straightforward, even though the trend in recent years in ocean modeling has been to use advanced schemes at least for important tracers such as dissolved substances, because their importance to questions about issues such as the ocean circulation and global warming requires that they be simulated as faithfully as possible even if it implies computational penalties. For most coastal ocean models however, centered second-order schemes have sufficed thus far.

Finally, the parameterization of subgrid scales not explicitly simulated by the finite resolution of the model is an important consideration. This includes horizontal and vertical mixing of momentum and scalars. Because of the importance of turbulent mixing in coastal waters, its accurate representation is essential. However, most models avoid having to solve the turbulence equations and instead use simplified parameterization schemes for vertical mixing, such as constant or Richardson number dependent eddy diffusivity coefficients. Convective mixing is often handled using a convective adjustment process that avoids the development of static instability in the water column by iterative adjustment of the density in the water column whenever convective instability develops. This is equivalent to instantaneous vertical mixing and an infinite vertical mixing coefficient, although the use of a large vertical mixing coefficient under convectively unstable conditions appears to give similar results (Marotzke, 1991). However, there is no consensus on whether convective adjustment must be applied to both momentum and density, or just to the density. Models that use turbulence closure (Kantha and Clayson, 1994; Mellor and Yamada, 1982) for parameterizing vertical mixing simulate convective mixing explicitly and therefore avoid this problem altogether.

Horizontal mixing is handled similarly. Most often, Laplacian (harmonic) diffusion terms are used for momentum and scalars, with either constant or mean-deformation-rate-dependent (Smagorinsky, 1963) eddy coefficients. Some models use a higher order (biharmonic) formulation, which is more scale selective and damps out $2\Delta x$ noise efficiently. In either case, horizontal diffusion terms are used primarily for numerical reasons, with the coefficients chosen quite arbitrarily to yield the "best" simulation. Little physics governs this choice. Also most often, this mixing is assumed to act in the horizontal direction (more appropriately along geopotential surfaces). In the interior of the oceans, away from the surface and bottom boundary layers, mixing occurs along isopycnal surfaces, and isopycnal models, or models that implement isopycnal mixing schemes by suitable coordinate rotation when computing mixing terms, are more accurate. In coastal oceans, especially on the shelf, mixing is mostly vertical, and horizontal mixing and diffusion terms are, more often than not, included for numerical reasons, since dispersive advection schemes require the explicit addition of horizontal diffusion.

The layered, the isopycnal, and the z-level ocean models are best suited to applications to deep oceans, especially primary basins, as we shall see shortly (Chapters 10 and 11). When it comes to modeling the shallow coastal oceans and marginal seas such as the Yellow Sea, however, a different type of vertical coordinate is preferable. The distinguishing characteristics of the shallow oceans are the importance of free surface dynamics, vertical mixing, and the bottom boundary layer. Sea level fluctuations due to tides and tidal mixing are important in shallow seas. Sea level changes due to storm surges must often be dealt with. Processes involving wind-induced mixing of the upper layers and current and tide-induced mixing at the bottom must be considered, since often on the inner shelf (for depths less than 20 m), the primary dynamical balance may be between the applied wind stress and the bottom stress. A coastal ocean model must not therefore ignore any of these processes.

The layered models are severely limited in application to the coastal seas, since the topographic variations must be confined to the bottom-most layer. Their close cousin, the isopycnal model, is inherently ill suited to coastal applications, because due to the large vertical mixing that takes place in these waters, isodensity surfaces do not stay intact. These models are best suited to deep basins, where away from the upper mixed layer and the benthic boundary layer (which constitute a small fraction of the water column) the mixing is small and along isopycnal surfaces and hence it makes sense to model the evolution of layers between two well-defined isopycnals, so that spurious diapycnal mixing is avoided and features such as density fronts are well represented. In the coastal oceans, the upper and bottom layers constitute a large fraction of the water column and often overlap on the inner shelf. Under these conditions, neither the layered nor the isopycnal models are satisfactory.

On the other hand, there is no inherent limitation to the application of a z-level model to the coastal ocean (as long as sufficient vertical resolution is provided in the upper 200 m or so), although the bottom boundary layer cannot be as well resolved as in a topographically conformal coordinate system. The primary requirement is that the free surface dynamics be incorporated and vertical mixing well represented. For a long time, the GFDL-developed z-level models were primarily deep basin models because of the rigid lid approximation invoked to filter out external gravity waves and hence free surface dynamics. Newer versions now incorporate free surface dynamics (Killworth *et al.*, 1991) and a better vertical mixing parameterization that makes them applicable to shallow seas (Pacanowski, 1995). Also, the model has been converted from a B grid to a C grid in some versions, to better accommodate high resolutions needed in the coastal oceans (see Chapter 2). However, no version includes tides and tidal mixing yet. Still, the stepwise approximation to the bottom introduces large errors (Tournadre, 1989, for example; see also Chapter 10) and necessitates the use of large number of vertical levels to control these errors; it also appears that excessive horizontal smoothing may be necessary (Haidvogel and Beckmann, 1997).

Haidvogel and Beckmann (1998) have performed a detailed comparison of many coastal models for three test cases:

1. *Gravitational adjustment problem* (Wang, 1984). Here fluids of two different densities are separated initially by a vertical wall. When the wall is removed, the lighter fluid spreads over the other and a two-layer stably stratified system results. The transient adjustment phase, however, involves density currents and sharp density fronts, of which the ability to simulate is a critical test of the model numerics. Depending on the finite difference scheme used, the model can be either excessively diffusive (thus smearing the fronts) or excessively dispersive (producing unphysical oscillations). Advection schemes assuring positive definiteness (for example, the Smolarkiewicz and Grabowski, 1990, scheme used in MICOM) do a much better job in maintaining a sharp, realistic front.

2. *Barotropic flow over a coastal canyon.* Here a time-varying, along-coast wind stress was applied over a coastal shelf/slope with an imbedded canyon. Periodic boundary conditions were used, and the resulting rectified time-mean flow solution was examined. Models using topographically conformal coordinates performed reasonably well here, mainly because of their ability to represent the bottom topography well. Those using z-coordinates represented the flow poorly, unless an excessively large number of vertical levels were employed. The steppy "staircase" nature of the bottom discretization is a major problem in these models, requiring large horizontal diffusion to damp out the resulting noise, which in turn produces weak flow rectification.

3. *Baroclinic flow over a coastal canyon.* This problem combines strong stratification and steep topography and is a hard one for all ocean models, no matter what discretization is used in the vertical. Both z-level and isopycnal models have coordinate surfaces intersecting the bottom that require careful treatment. The sigma coordinates incur large pressure gradient errors. The

conclusions here were less definitive. All three formulations had one problem or another, but the z-level model needed the largest smoothing and produced the weakest flow.

Clearly, the coastal oceans pose a great challenge for ocean modelers. Nevertheless, a large number of modeling studies have been carried out both in the coastal oceans, mostly in the United States and Europe and in some marginal seas (see Haidvogel and Beckmann, 1998, for a partial list). These studies include but are not limited to southeast Australia (Black *et al.*, 1993), the South China Sea (Pohlmann, 1987), the European Shelf (Davies, 1987), the Irish Sea (Davies and Lawrence, 1994), the German Bight (Dippner, 1993; Schrum, 1994), the North Sea (Backhaus and Hainbucher, 1987), the Baltic Sea (Lehmann, 1995), the Mediterranean Sea (Beckers, 1991; Horton *et al.*, 1997), the Red Sea (Clifford *et al.*, 1997), the Black Sea (Stanev *et al.*, 1994), the Bering Sea (Nihoul *et al.*, 1993), the North Atlantic Shelf (Oey and Chen, 1992), the Hudson Bay (Wang *et al.*, 1994), the Gulf of Mexico (Choi *et al.*, 1995; Dietrich and Lin, 1994; Kantha *et al.*, 1999), upwelling off California (Haidvogel *et al.*, 1991; Moisan *et al.*, 1995), the Mid-Atlantic Bight (Keen and Glenn, 1995), submarine canyons (Klinck, 1995), estuarine plumes (Oey and Mellor, 1993), and many others.

There is an inherent advantage to a vertical coordinate system that conforms to the ocean bottom and hence better resolves the bottom frictional charac-teristics so important to inner shelf dynamics, in spite of some limitations that such a coordinate system imposes. Such sigma-coordinate models have their counterparts in the atmosphere also. There are at present two such models, one the popular Princeton Ocean Model (POM), developed by George Mellor's group at the Princeton University (Blumberg and Mellor, 1987; see also Kantha and Piacsek, 1993, 1997; Mellor, 1996), and the other the Semispectral Primitive Equation Model (SPEM) developed by Dale Haidvogel's group at Rutgers University (Haidvogel *et al.*, 1991). Both use sigma vertical coordinates; however, POM uses finite differences in the vertical, whereas SPEM uses the semispectral collocation method based on the use of Chebyshev functions as basis functions (Hedstrom, 1995). Both use orthogonal curvilinear coordinates in the horizontal, second-order leapfrog time differences, with the POM relying on an Asselin filter to remove computational modes, while SPEM relies on a Shapiro smoother. However, the vertical mixing and inclusion of free surface dynamics are more developed and more extensively tested in POM. Published versions of SPEM still use rigid lid approximation and lack turbulence closure. We will therefore concentrate on POM, more specifically, the University of Colorado version of it (CUPOM). For details about SPEM, see Haidvogel *et al.* (1991) and Hedstrom (1995); see Beckmann and Haidvogel (1993, 1997) for its application to circulation around a seamount and, more recently, see Moisan *et al.* (1995) for its application to upwelling off the coast of California. Note that the most recent version of SPEM uses finite-differencing in the vertical on an s-coordinate system and is now called the S-coordinate Primitive Equation Model.

In a sigma-coordinate model, the governing continuity, momentum and scalar conservation equations are cast in a bottom topography following coordinate system by defining a new variable sigma $\sigma = (z - \eta)/(H + \eta)$, so that $\sigma = 0$ defines the free surface and $\sigma = -1$ defines the bottom. σ can also be defined slightly differently as $\sigma = 1 + 2(z - \eta)/(H + \eta)$, so that $\sigma = 1$ defines the free surface (in SPEM, for example). These transformed equations, along with corresponding conservation relations for turbulence quantities, are solved by the model. In this coordinate system, the number of levels is the same everywhere in the ocean, irrespective of the depth of the water column. It is therefore possible to resolve the bottom boundary layer where needed. This set of equations is best suited to modeling the shallow coastal oceans, although there is no inherent limitation to its application to deep basins. The principal problem is in applying it over sharply changing topography such as the continental slope separating the shelf from the deep basin. Here unless the topographic gradients are suitably reduced by a nonlinear smoother, the errors in the calculation of pressure gradients induced by horizontal gradients of density can lead to spurious along-slope currents (Haney, 1991; Mellor *et al.*, 1994). While the problem due to strong topographic changes manifests itself in one form or another in all ocean models, the problem is particularly serious in sigma-coordinate models. It is for this reason that its use requires careful preprocessing of bottom topography to put a bound on topographic gradients yet retain features important to the coastal problem. In practical terms, this means that the continental slope is poorly represented in these models, unless a higher order scheme (and high horizontal resolution) is used to minimize truncation errors involved in calculating the pressure gradient across it.

Many applications of this model can be found in literature (for example, in the *Journal of Physical Oceanography* and the *Journal of Geophysical Research, Oceans*). Users group meetings held in 1996 and 1998, had wide participation by users from around the world, and various applications of the model were presented and discussed. The proceedings of these meetings are available on the Web at www.aos.princeton.edu/htdocs.pom and the model code itself is available from the anonymous ftp site ftp.gfdl.gov on the pub/glm directory. In addition, there exists an excellent users guide, written originally in 1991 but updated for the meeting. The guide, however, details the rectilinear coordinate version of the model and code. Notable applications of POM are the Hudson-Raritan estuary (Oey *et al.*, 1985 a,b), the Mediterranean Sea (Zavatarelli and Mellor, 1995), the Black Sea (Oguz and Malanotte-Rizzoli, 1996), the combined Arctic and North Atlantic oceans at NASA Goddard (Hakkinen, 1997), and the North Atlantic

Ocean at Princeton university (Mellor and Azer, 1995). Over the years, its earlier versions have been applied to the Gulf of Mexico, the South and Mid-Atlantic Bights, the coast of California, and there have been many other coastal applications as well as esturine studies.

This chapter will describe in detail a modified version of POM developed at the University of Colorado, CUPOM (Kantha and Piacsek, 1993, 1997), incorporating an improved mixed layer formulation (Kantha and Clayson, 1994) and involving assimilation of altimetric and other remote sensing as well as *in situ* data. There also exists a version that incorporates astronomical tidal forcing suitable for application to regions with strong tidal mixing. We will also fully describe and document the orthogonal curvilinear version, more consistent with the POM code, partly to illustrate the extremely complex nature of modern ocean models. CUPOM has been applied to the Gulf of Mexico (Choi *et al.*, 1995; Kantha *et al.*, 1999), the Persian Gulf, the Red Sea (Clifford *et al.*, 1997), the Mediterranean Sea (Horton *et al.*, 1997), the Sea of Japan (Bang *et al.*, 1996), the Yellow Sea (Jacobs *et al.*, 1998), the North Pacific Ocean (Engelhardt, 1996), the tropical Pacific (Clayson, 1995), and the North Indian Ocean (Lopez and Kantha, 1999). This version has also been converted to CM5 and applied to the Straits of Sicily and its Cray T3D version to the North Pacific Ocean (Engelhardt, 1996).

9.2 GOVERNING EQUATIONS

We start with the governing equations (see Chapter 1) for an incompressible fluid in a right-handed general orthogonal curvilinear coordinate system, in a frame of reference rotating at an angular velocity $f/2$ ($f = 2\Omega \sin \theta$) about the vertical coordinate z. Invoking hydrostatic and Boussinesq approximations to effect simplifications,

$$\frac{1}{h_1 h_2}\left[\frac{\partial}{\partial x_1}(h_2 u_1) + \frac{\partial}{\partial x_2}(h_1 u_2)\right] + \frac{\partial w}{\partial z} = 0 \tag{9.2.1}$$

$$\frac{\partial u_1}{\partial t} + \frac{1}{h_1 h_2}\left[\frac{\partial}{\partial x_1}(h_2 u_1^2) + \frac{\partial}{\partial x_2}(h_1 u_1 u_2)\right] + \frac{1}{h_1 h_2}\left[u_1 u_2 \frac{\partial h_1}{\partial x_2} - u_2^2 \frac{\partial h_2}{\partial x_1}\right] - f u_2$$

$$+ \frac{\partial}{\partial z}(w u_1) = -\frac{1}{\rho_0 h_1}\frac{\partial p}{\partial x_1} + \frac{1}{h_1 h_2}\left[\frac{\partial}{\partial x_1}(h_2 \tau_{11}) + \frac{\partial}{\partial x_2}(h_1 \tau_{21}) + \tau_{21}\frac{\partial h_1}{\partial x_2} - \tau_{22}\frac{\partial h_2}{\partial x_1}\right]$$

$$+ \frac{\partial}{\partial z}\left(K_M \frac{\partial u_1}{\partial z}\right) \tag{9.2.2}$$

$$\frac{\partial u_2}{\partial t} + \frac{1}{h_1 h_2}\left[\frac{\partial}{\partial x_1}(h_2 u_1 u_2) + \frac{\partial}{\partial x_2}(h_1 u_2^2)\right] + \frac{1}{h_1 h_2}\left[u_1 u_2 \frac{\partial h_2}{\partial x_1} - u_1^2 \frac{\partial h_1}{\partial x_2}\right] + f u_1$$

$$+ \frac{\partial}{\partial z}(w u_2) = -\frac{1}{\rho_0 h_2}\frac{\partial p}{\partial x_2} + \frac{1}{h_1 h_2}\left[\frac{\partial}{\partial x_1}(h_2 \tau_{12}) + \frac{\partial}{\partial x_2}(h_1 \tau_{22}) + \tau_{12}\frac{\partial h_2}{\partial x_1} - \tau_{11}\frac{\partial h_1}{\partial x_2}\right]$$

$$+ \frac{\partial}{\partial z}\left(K_M \frac{\partial u_2}{\partial z}\right) \tag{9.2.3}$$

$$0 = -\frac{1}{\rho_0}\frac{\partial p}{\partial z} - \frac{\rho}{\rho_0}g \tag{9.2.4}$$

$$\frac{\partial \Theta}{\partial t} + \frac{1}{h_1 h_2}\left[\frac{\partial}{\partial x_1}(h_2 u_1 \Theta) + \frac{\partial}{\partial x_2}(h_1 u_2 \Theta)\right] + \frac{\partial}{\partial z}(w\Theta)$$

$$= \frac{\partial Q_S}{\partial z} + \frac{1}{h_1 h_2}\left[\frac{\partial}{\partial x_1}\left(A_H \frac{h_2}{h_1}\frac{\partial \Theta}{\partial x_1}\right) + \frac{\partial}{\partial x_2}\left(A_H \frac{h_1}{h_2}\frac{\partial \Theta}{\partial x_2}\right)\right] + \frac{\partial}{\partial z}\left(K_M \frac{\partial \Theta}{\partial z}\right) \tag{9.2.5}$$

$$\frac{\partial S}{\partial t} + \frac{1}{h_1 h_2}\left[\frac{\partial}{\partial x_1}(h_2 u_1 S) + \frac{\partial}{\partial x_2}(h_1 u_2 S)\right] + \frac{\partial}{\partial z}(wS)$$

$$= \frac{1}{h_1 h_2}\left[\frac{\partial}{\partial x_1}\left(A_H \frac{h_2}{h_1}\frac{\partial S}{\partial x_1}\right) + \frac{\partial}{\partial x_2}\left(A_H \frac{h_1}{h_2}\frac{\partial S}{\partial x_2}\right)\right] + \frac{\partial}{\partial z}\left(K_M \frac{\partial S}{\partial z}\right) \tag{9.2.6}$$

$$\frac{\partial}{\partial t}(q^2) + \frac{1}{h_1 h_2}\left[\frac{\partial}{\partial x_1}(h_2 u_1 q^2) + \frac{\partial}{\partial x_2}(h_1 u_2 q^2)\right] + \frac{\partial}{\partial z}(w q^2)$$

$$= 2K_M\left[\left(\frac{\partial u_1}{\partial z}\right)^2 + \left(\frac{\partial u_2}{\partial z}\right)^2\right] + 2K_H \frac{g}{\rho_0}\frac{\partial \rho}{\partial z} - 2\frac{q^3}{B_1 \ell} + \tag{9.2.7}$$

$$\frac{1}{h_1 h_2}\left[\frac{\partial}{\partial x_1}\left(A_H \frac{h_2}{h_1}\frac{\partial}{\partial x_1}(q^2)\right) + \frac{\partial}{\partial x_2}\left(A_H \frac{h_1}{h_2}\frac{\partial}{\partial x_2}(q^2)\right)\right] + \frac{\partial}{\partial z}\left(K_q \frac{\partial}{\partial z}(q^2)\right)$$

$$\frac{\partial}{\partial t}(q^2 \ell) + \frac{1}{h_1 h_2}\left[\frac{\partial}{\partial x_1}(h_2 u_1 q^2 \ell) + \frac{\partial}{\partial x_2}(h_1 u_2 q^2 \ell)\right] + \frac{\partial}{\partial z}(w q^2 \ell)$$

$$= E_1 \ell\left\{K_M\left[\left(\frac{\partial u_1}{\partial z}\right)^2 + \left(\frac{\partial u_2}{\partial z}\right)^2\right] + E_3 K_H \frac{g}{\rho_0}\frac{\partial \rho}{\partial z}\right\} - \frac{q^3}{B_1}\left[1 + E_2\left(\frac{\ell}{\kappa \ell_w}\right)^2\right] \tag{9.2.8}$$

$$\frac{1}{h_1 h_2}\left[\frac{\partial}{\partial x_1}\left(A_H \frac{h_2}{h_1}\frac{\partial}{\partial x_1}(q^2 \ell)\right) + \frac{\partial}{\partial x_2}\left(A_H \frac{h_1}{h_2}\frac{\partial}{\partial x_2}(q^2 \ell)\right)\right] + \frac{\partial}{\partial z}\left(K_\ell \frac{\partial}{\partial z}(q^2 \ell)\right)$$

$$\rho = \rho(\Theta, S) \tag{9.2.9}$$

Here x_1 and x_2 are the horizontal coordinates, and h_1 and h_2 the corresponding metrics given by

$$ds^2 = h_1^2 dx_1^2 + h_2^2 dx_2^2 + dz^2 \qquad (9.2.10)$$

where ds is the length of a segment in (x_1, x_2, z) space. Quantities u_1, u_2, and w are the velocity components in the x_1, x_2, and z directions; τ_{11}, τ_{12}, and τ_{22} are components of the symmetric Reynolds stress tensor in the horizontal plane:

$$\tau_{11} = 2A_M \left[\frac{1}{h_1} \frac{\partial u_1}{\partial x_1} + \frac{u_2}{h_1 h_2} \frac{\partial h_1}{\partial x_2} \right]$$

$$\tau_{12} = A_M \left[\frac{h_1}{h_2} \frac{\partial}{\partial x_2} \left(\frac{u_1}{h_1} \right) + \frac{h_2}{h_1} \frac{\partial}{\partial x_1} \left(\frac{u_2}{h_2} \right) \right] = \tau_{21} \qquad (9.2.11)$$

$$\tau_{22} = 2A_M \left[\frac{1}{h_2} \frac{\partial u_2}{\partial x_2} + \frac{u_1}{h_1 h_2} \frac{\partial h_2}{\partial x_1} \right]$$

Equation (9.2.1) is the mass conservation equation, Eqs. (9.2.2) and (9.2.3) are the momentum equations in the horizontal direction, and Eq. (9.2.4) is the vertical momentum equation under hydrostatic approximation. Equations (9.2.5) and (9.2.6) are conservation equations for potential temperature and salinity; Eqs. (9.2.7) and (9.2.8) are equations for q^2, twice the TKE and $q^2 l$, the quantity involving the turbulence velocity scale q and length scale l. Equation (9.2.9) is the equation of state relating the density ρ to the potential temperature Θ and salinity S (see Appendix A). Quantities A_M and A_H are coefficients of horizontal viscosity and scalar diffusivity, and K_M, K_H, and K_q are coefficients of vertical diffusivities of momentum, scalars such as temperature and salinity, and turbulence quantities. Quantity ρ_0 is the reference density, g is the gravitational acceleration, and Q_S is the penetrative solar insolation.

The above equations can be transformed to a topographically conformal σ coordinate system in the vertical using the following transformation to go from the (x_1, x_2, z, t) coordinate system to the (x'_1, x'_2, σ, t') system, where $x_1 = x'_1$, $x_2 = x'_2$ and $t = t'$, and

$$\sigma = \frac{z - \eta(x_1, x_2)}{H(x_1, x_2) + \eta(x_1, x_2)} \qquad (9.2.12)$$

Then the derivatives of any dependent variable Φ in the two coordinate systems are related thus

$$\frac{\partial \Phi}{\partial x_{1,2}} = \frac{\partial \Phi}{\partial x'_{1,2}} - \frac{\partial \Phi}{\partial \sigma}\left(\frac{\sigma}{D}\frac{\partial D}{\partial x'_{1,2}} + \frac{1}{D}\frac{\partial \eta}{\partial x'_{1,2}} \right)$$

$$\frac{\partial \Phi}{\partial z} = \frac{1}{D}\frac{\partial \Phi}{\partial \sigma} \qquad\qquad\qquad (9.2.13)$$

$$\frac{\partial \Phi}{\partial t} = \frac{\partial \Phi}{\partial t'} - \frac{\partial \Phi}{\partial \sigma}\left(\frac{\sigma}{D}\frac{\partial D}{\partial t'} + \frac{1}{D}\frac{\partial \eta}{\partial t'} \right)$$

where $D = H + \eta$, the total depth of the water column, H is the bottom depth, and η is the free surface deflection ($\sigma = 0$ at the free surface $z = \eta$ and -1 at the bottom $z = -H$). In the new coordinate system, the governing equations, Eqs. (9.2.1) to (9.2.8), become

$$\frac{\partial \eta}{\partial t} + \frac{1}{h_1 h_2}\left[\frac{\partial}{\partial x_1}(h_2 u_1 D) + \frac{\partial}{\partial x_2}(h_1 u_2 D) \right] + \frac{\partial \omega}{\partial \sigma} = 0 \qquad\qquad (9.2.14)$$

$$\frac{\partial}{\partial t}(u_1 D) + \frac{1}{h_1 h_2}\left[\frac{\partial}{\partial x_1}(h_2 u_1^2 D) + \frac{\partial}{\partial x_2}(h_1 u_1 u_2 D) \right] + \frac{D}{h_1 h_2}\left[u_1 u_2 \frac{\partial h_1}{\partial x_2} - u_2^2 \frac{\partial h_2}{\partial x_1} \right] - f u_2 D$$

$$+ \frac{\partial}{\partial \sigma}(\omega u_1) = -DP_1 + DF_1 + \frac{\partial}{\partial \sigma}\left(\frac{K_M}{D}\frac{\partial u_1}{\partial \sigma} \right) \qquad\qquad (9.2.15)$$

$$\frac{\partial}{\partial t}(u_2 D) + \frac{1}{h_1 h_2}\left[\frac{\partial}{\partial x_1}(h_2 u_1 u_2 D) + \frac{\partial}{\partial x_2}(h_1 u_2^2 D) \right] + \frac{D}{h_1 h_2}\left[u_1 u_2 \frac{\partial h_2}{\partial x_1} - u_1^2 \frac{\partial h_1}{\partial x_2} \right] + f u_1 D$$

$$+ \frac{\partial}{\partial \sigma}(\omega u_2) = -DP_2 + DF_2 + \frac{\partial}{\partial \sigma}\left(\frac{K_M}{D}\frac{\partial u_2}{\partial \sigma} \right) \qquad\qquad (9.2.16)$$

$$\frac{\partial}{\partial t}(\Theta D) + \frac{1}{h_1 h_2}\left[\frac{\partial}{\partial x_1}(h_2 u_1 \Theta D) + \frac{\partial}{\partial x_2}(h_1 u_2 \Theta D) \right] + \frac{\partial}{\partial \sigma}(\omega \Theta)$$

$$= \frac{1}{D}\frac{\partial Q_S}{\partial \sigma} + DF_\Theta + \frac{\partial}{\partial \sigma}\left(\frac{K_M}{D}\frac{\partial \Theta}{\partial \sigma} \right) \qquad\qquad (9.2.17)$$

$$\frac{\partial}{\partial t}(SD) + \frac{1}{h_1 h_2}\left[\frac{\partial}{\partial x_1}(h_2 u_1 SD) + \frac{\partial}{\partial x_2}(h_1 u_2 SD) \right] + \frac{\partial}{\partial \sigma}(\omega S)$$

$$= DF_S + \frac{\partial}{\partial \sigma}\left(\frac{K_M}{D}\frac{\partial S}{\partial \sigma} \right) \qquad\qquad (9.2.18)$$

$$\frac{\partial}{\partial t}\left(q^2 D\right) + \frac{1}{h_1 h_2}\left[\frac{\partial}{\partial x_1}\left(h_2 u_1 q^2 D\right) + \frac{\partial}{\partial x_2}\left(h_1 u_2 q^2 D\right)\right] + \frac{\partial}{\partial \sigma}\left(\omega q^2\right)$$

$$= 2\frac{K_M}{D}\left[\left(\frac{\partial u_1}{\partial \sigma}\right)^2 + \left(\frac{\partial u_2}{\partial \sigma}\right)^2\right] + 2K_H \frac{g}{\rho_0}\frac{\partial \rho}{\partial \sigma} - 2\frac{q^3 D}{B_1 \ell} + DF_q + \frac{\partial}{\partial \sigma}\left(\frac{K_q}{D}\frac{\partial}{\partial \sigma}(q^2)\right)$$

$$(9.2.19)$$

$$\frac{\partial}{\partial t}\left(q^2 \ell D\right) + \frac{1}{h_1 h_2}\left[\frac{\partial}{\partial x_1}\left(h_2 u_1 q^2 \ell D\right) + \frac{\partial}{\partial x_2}\left(h_1 u_2 q^2 \ell D\right)\right] + \frac{\partial}{\partial \sigma}\left(\omega q^2 \ell\right)$$

$$= E_1 \ell\left\{\frac{K_M}{D}\left[\left(\frac{\partial u_1}{\partial \sigma}\right)^2 + \left(\frac{\partial u_2}{\partial \sigma}\right)^2\right] + E_3 K_H \frac{g}{\rho_0}\frac{\partial \rho}{\partial \sigma}\right\} - \frac{q^3 D}{B_1}\left[1 + E_2\left(\frac{\ell}{\kappa L}\right)^2\right]$$

$$+ DF_\ell + \frac{\partial}{\partial \sigma}\left(\frac{K_\ell}{D}\frac{\partial}{\partial \sigma}(q^2 \ell)\right)$$

$$(9.2.20)$$

where ω is the pseudo-vertical velocity in the new coordinate system, and the pressure gradient terms DP_1 and DP_2 have been computed by vertically integrating the hydrostatic equation, Eq. (9.2.4), to derive the pressure:

$$\omega = w - u_1 \sigma \frac{\partial D}{\partial x_1} + \frac{\partial \eta}{\partial x_1} - u_2 \sigma \frac{\partial D}{\partial x_2} + \frac{\partial \eta}{\partial x_2} - \left(\sigma \frac{\partial D}{\partial t} + \frac{\partial \eta}{\partial t}\right) \qquad (9.2.21)$$

$$Dh_1 P_1 = \frac{D}{\rho_0}\frac{\partial p_a}{\partial x_1} + gD\frac{\partial \eta}{\partial x_1} + \frac{gD^2}{\rho_0}\left[\frac{\partial}{\partial x_1}\int_\sigma^0 \rho d\sigma' - \rho\left(\frac{\sigma}{D}\frac{\partial D}{\partial x_1} + \frac{1}{D}\frac{\partial \eta}{\partial x_1}\right)\right]$$

$$- \frac{gD}{\rho_0}\frac{\partial D}{\partial x_1}\int_\sigma^0 \sigma'\frac{\partial \rho}{\partial \sigma'}d\sigma' \qquad (9.2.22)$$

$$Dh_2 P_2 = \frac{D}{\rho_0}\frac{\partial p_a}{\partial x_2} + gD\frac{\partial \eta}{\partial x_2} + \frac{gD^2}{\rho_0}\left[\frac{\partial}{\partial x_2}\int_\sigma^0 \rho d\sigma' - \rho\left(\frac{\sigma}{D}\frac{\partial D}{\partial x_2} + \frac{1}{D}\frac{\partial \eta}{\partial x_2}\right)\right]$$

$$- \frac{gD}{\rho_0}\frac{\partial D}{\partial x_2}\int_\sigma^0 \sigma'\frac{\partial \rho}{\partial \sigma'}d\sigma' \qquad (9.2.23)$$

$$DF_1 = \frac{1}{h_1 h_2}\left\{\begin{array}{l}\frac{\partial}{\partial x_1}\left(h_2 \tau_{11}\right) - h_2\frac{\partial}{\partial \sigma}\left[\left(\frac{\sigma}{D}\frac{\partial D}{\partial x_1} + \frac{1}{D}\frac{\partial \eta}{\partial x_1}\right)\tau_{11}\right] \\[3mm] + \frac{\partial}{\partial x_2}\left(h_1 \tau_{21}\right) - h_1\frac{\partial}{\partial \sigma}\left[\left(\frac{\sigma}{D}\frac{\partial D}{\partial x_2} + \frac{1}{D}\frac{\partial \eta}{\partial x_2}\right)\tau_{21}\right] + \tau_{21}\frac{\partial h_1}{\partial x_2} - \tau_{22}\frac{\partial h_2}{\partial x_1}\end{array}\right\}$$

$$(9.2.24)$$

$$DF_2 = \frac{1}{h_1 h_2} \left\{ \begin{array}{l} \dfrac{\partial}{\partial x_1}(h_2 \tau_{12}) - h_2 \dfrac{\partial}{\partial \sigma}\left[\left(\dfrac{\sigma}{D}\dfrac{\partial D}{\partial x_1} + \dfrac{1}{D}\dfrac{\partial \eta}{\partial x_1} \right)\tau_{12} \right] \\[3ex] + \dfrac{\partial}{\partial x_2}(h_1 \tau_{22}) - h_1 \dfrac{\partial}{\partial \sigma}\left[\left(\dfrac{\sigma}{D}\dfrac{\partial D}{\partial x_2} + \dfrac{1}{D}\dfrac{\partial \eta}{\partial x_2} \right)\tau_{22} \right] + \tau_{12}\dfrac{\partial h_2}{\partial x_1} - \tau_{11}\dfrac{\partial h_1}{\partial x_2} \end{array} \right\}$$

(9.2.25)

$$DF_\Theta = \frac{1}{h_1 h_2} \left\{ \begin{array}{l} \dfrac{\partial}{\partial x_1}(h_2 q_1) - h_2 \dfrac{\partial}{\partial \sigma}\left[\left(\dfrac{\sigma}{D}\dfrac{\partial D}{\partial x_1} + \dfrac{1}{D}\dfrac{\partial \eta}{\partial x_1} \right)q_1 \right] \\[3ex] + \dfrac{\partial}{\partial x_2}(h_1 q_2) - h_1 \dfrac{\partial}{\partial \sigma}\left[\left(\dfrac{\sigma}{D}\dfrac{\partial D}{\partial x_2} + \dfrac{1}{D}\dfrac{\partial \eta}{\partial x_2} \right)q_2 \right] \end{array} \right\}$$

(9.2.26)

where the horizontal Reynolds stresses and heat fluxes are given by

$$\tau_{11} = 2A_M \left\{ \frac{1}{h_1}\left[\frac{\partial}{\partial x_1}(u_1 D) - \frac{\partial}{\partial \sigma}\left(u_1(\sigma\frac{\partial D}{\partial x_1} + \frac{\partial \eta}{\partial x_1}) \right) \right] + \frac{u_2 D}{h_1 h_2}\frac{\partial h_1}{\partial x_2} \right\}$$

$$\tau_{12} = 2A_M \left\{ \begin{array}{l} \dfrac{h_1}{h_2}\left[\dfrac{\partial}{\partial x_2}\left(\dfrac{u_1 D}{h_1} \right) - \dfrac{1}{h_1}\dfrac{\partial}{\partial \sigma}\left(u_1(\sigma\dfrac{\partial D}{\partial x_2} + \dfrac{\partial \eta}{\partial x_2}) \right) \right] \\[3ex] + \dfrac{h_2}{h_1}\left[\dfrac{\partial}{\partial x_1}\left(\dfrac{u_2 D}{h_2} \right) - \dfrac{1}{h_2}\dfrac{\partial}{\partial \sigma}\left(u_2(\sigma\dfrac{\partial D}{\partial x_1} + \dfrac{\partial \eta}{\partial x_1}) \right) \right] \end{array} \right\}$$

(9.2.27)

$$\tau_{22} = 2A_M \left\{ \frac{1}{h_2}\left[\frac{\partial}{\partial x_2}(u_2 D) - \frac{\partial}{\partial \sigma}\left(u_2(\sigma\frac{\partial D}{\partial x_2} + \frac{\partial \eta}{\partial x_2}) \right) \right] + \frac{u_1 D}{h_1 h_2}\frac{\partial h_2}{\partial x_1} \right\}$$

$$q_1 = \frac{A_H}{h_1}\left\{ \frac{\partial}{\partial x_1}(\Theta D) - \frac{\partial}{\partial \sigma}\left[\left(\sigma\frac{\partial D}{\partial x_1} + \frac{\partial \eta}{\partial x_1} \right)\Theta \right] \right\}$$

$$q_2 = \frac{A_H}{h_2}\left\{ \frac{\partial}{\partial x_2}(\Theta D) - \frac{\partial}{\partial \sigma}\left[\left(\sigma\frac{\partial D}{\partial x_2} + \frac{\partial \eta}{\partial x_2} \right)\Theta \right] \right\}$$

(9.2.28)

with similar equations for horizontal diffusion terms F_S, F_q, and F_l in Eqs. (9.2.18) to (9.2.20), except that Θ is replaced by S, q^2, and $q^2 l$, respectively. These most general forms for horizontal diffusion are expensive to compute and most often needlessly complex. Hence they can be simplified considerably:

$$DF_1 = \frac{1}{h_1}\frac{\partial}{\partial x_1}\left[2A_MD\frac{1}{h_1}\frac{\partial u_1}{\partial x_1}\right] + \frac{1}{h_2}\frac{\partial}{\partial x_2}\left[A_MD\left(\frac{1}{h_2}\frac{\partial u_1}{\partial x_2}+\frac{1}{h_1}\frac{\partial u_2}{\partial x_1}\right)\right]$$

$$DF_2 = \frac{1}{h_2}\frac{\partial}{\partial x_2}\left[2A_MD\frac{1}{h_2}\frac{\partial u_2}{\partial x_2}\right] + \frac{1}{h_1}\frac{\partial}{\partial x_1}\left[A_MD\left(\frac{1}{h_1}\frac{\partial u_2}{\partial x_1}+\frac{1}{h_2}\frac{\partial u_1}{\partial x_2}\right)\right] \quad (9.2.29)$$

$$DF_\Theta = \frac{1}{h_1h_2}\left\{\frac{\partial}{\partial x_1}(h_2q_1)+\frac{\partial}{\partial x_2}(h_1q_2)\right\}$$

$$q_1 = \frac{A_H}{h_1}\left[H\frac{\partial\Theta}{\partial x_1}-\left(\frac{\partial H}{\partial x_1}\sigma\frac{\partial\Theta}{\partial\sigma}\right)\right] ; q_2 = \frac{A_H}{h_2}\left[H\frac{\partial\Theta}{\partial x_2}-\left(\frac{\partial H}{\partial x_2}\sigma\frac{\partial\Theta}{\partial\sigma}\right)\right] \quad (9.2.30)$$

The justification is simply that there is a great deal of uncertainty associated with the choice of horizontal mixing coefficients and more often than not, horizontal mixing terms are regarded as a mere numerical necessity to control the subgrid scale energy pileup. Unlike vertical mixing in the oceans, our understanding of horizontal mixing, especially as it relates to subgrid scale processes, is rather primitive in any case and there is generally no need for a complicated, computationally expensive form. However, it is important to avoid artificial, purely numerical diffusion of temperature and salinity along sigma surfaces in the absence of any horizontal gradients. A better strategy still is to implement isopycnal diffusion schemes in regions away from the mixed layers. This prevents spurious vertical mixing resulting from horizontal mixing and sloping isopycnals and removes the threat of significant drift in water mass properties in the deep ocean in long-term integrations.

9.3 VERTICAL MIXING

The vertical mixing processes are relatively better understood and can be calculated from second-moment closure of turbulence (Kantha and Clayson, 1994; see also Mellor and Yamada, 1982). The vertical mixing coefficients K_M, K_H, K_q, and K_ℓ can be regarded as sums of some background values and the turbulent values given by closure theory:

$$K_M = K_{MB} + q\ell S_M ; K_H = K_{HB} + q\ell S_H$$
$$K_q = K_{qB} + q\ell S_q ; K_\ell = K_{\ell B} + q\ell S_\ell \quad (9.3.1)$$

where S_M, S_H, S_q, and S_l are stability functions determined from algebraic relations derived analytically from small perturbation approximations to the full second-moment closure theory (Galperin *et al.*, 1988; Mellor and Yamada, 1974). The following forms are due to Kantha and Clayson (1994):

$$S_M = A_1 \left\{ \frac{\left[\left(1 - \frac{6A_1}{B_1} - 3C_1\right) + 9\left[2A_1 + A_2(1 - C_2)\right]S_H G_H\right]}{(1 - 9A_1 A_2 G_H)} \right\}$$

$$S_H = A_2 \left\{ \frac{\left(1 - \frac{6A_1}{B_1}\right)}{1 - 3A_2 G_H \left[6A_1 + B_2(1 - C_3)\right]} \right\}; \quad S_q = S_\ell = 0.2 \qquad (9.3.2)$$

$$G_H = \frac{\ell^2}{q^2} \frac{g}{\rho_0 D} \frac{\partial \rho}{\partial \sigma}$$

The stability functions are thus a function of the gravitational stratification in the flow. The closure constants A_1, A_2, B_1, B_2, C_1, C_2, and C_3 in Eq. (9.3.2) and E_1, E_2, and E_3 in Eqs. (9.3.19) and (9.3.20) are considered invariant and take the values

$$(A_1, A_2. B_1, B_2, C_1, C_2, C_3, E_1, E_2, E_3)$$

$$= (0.92, 0.74, 16.6, 10.1, 0.08, 0.7, 0.2, 1.8, 1.33, 1.0) \qquad (9.3.3)$$

The term multiplying q^3/B_1 in Eq. (9.2.20) is the wall proximity term inserted empirically to assure log-law behavior near boundaries with L given by $L^{-1} = D^{-1}[\sigma^{-1} + (1 + \sigma)^{-1}]$, and κ (~ 0.4), the well-known von Karman constant. See Kantha and Clayson (1994) and Kantha and Clayson (2000, Chapter 2) for a full description of the second-moment closure.

The background values for mixing coefficients require additional care. Recent observations indicate that the vertical mixing coefficient in the abyssal oceans (Kunze and Sanford, 1996) are remarkably constant at 10^{-5} m^2 s^{-1}, a value 10 times the molecular value for kinematic viscosity. These background values can be used for K_M, K_H, K_q and K_l overall. However, the region immediately below the active mixed layer is a region of strong static stability, but it is also strongly sheared, so that turbulence here is intermittent. It is difficult to model it, but it is unwise to ignore it (Kantha and Clayson, 1994; Large *et al.*, 1994). It is possible to model the mixing immediately below the mixed layer using gradient Richardon number dependent mixing coefficients (see Kantha and Clayson,

1994; Clayson *et al.*, 1997; Large *et al.*, 1994):

$$K_{MB} = K_R + 5.0 \times 10^{-3} \left[1 - \left(Ri_g / 0.7 \right)^2 \right]^3 \qquad (0 < Ri_g < 0.7)$$

$$= K_R \qquad (Ri_g > 0.7) \qquad (9.3.4)$$

$$= K_R + 5.0 \times 10^{-3} \qquad (Ri_g < 0)$$

$$K_{HB} = K_{MB}; \quad K_R = 10^{-4}$$

the values being in $m^2 \, s^{-1}$.

9.4 BOUNDARY CONDITIONS

The flow must also satisfy certain boundary conditions at the top (the free surface), bottom (the ocean bottom), and lateral boundaries. The free surface is a material surface and hence there cannot be any flow across it; this is given by the condition $\omega = 0$ at $\sigma = 0$. The ocean bottom is also impermeable and the condition $\omega = 0$ at $\sigma = -1$ satisfies this requirement. In addition, the fluxes of momentum, net heat, and salt (mass flux is negligible) (or alternatively the velocities, temperature and salinity values) at the free surface and the bottom have to be prescribed. The net heat flux is the sum of the turbulent sensible and latent heat fluxes and the net LW radiative flux at the air-sea interface. Also, either the fluxes of the turbulent quantities must be prescribed, or the quantities themselves must be prescribed. In addition, the SW solar insolation impinging on the ocean surface must be specified, since this provides the volumetric heat source in the upper layers of the ocean. The momentum, heat, and salt fluxes at the free surface are given by

$$K_M \frac{\partial}{D\partial\sigma} (u_1, u_2) = \frac{1}{\rho_0} (\tau_{w1}, \tau_{w2})$$

$$K_H \frac{\partial}{D\partial\sigma} (\Theta, S) = \frac{1}{\rho_0} (q_H, q_S) \qquad (9.4.1)$$

$$q_H = (1 - \bar{\alpha}) L_w - \varepsilon \bar{\sigma} T_S^4 - H_S - H_L$$

$$q_S = S_S (\dot{E} - \dot{P})$$

where τ_w is the wind stress at the sea surface; q_H is the net heat flux, dependent on H_S and H_L, the turbulent sensible and latent heat fluxes at the air-sea interface, the long wave radiative flux L_w and the backradiation by the ocean surface; q_S is the salt flux dependent on the difference between the evaporation rate and the precipitation rate. T_S and S_S are the surface temperature and salinity values, $\bar{\alpha}$ is the albedo and ε the emissivity of the sea surface, and $\bar{\sigma}$ is the Stefan-Boltzmann constant. The parameterization of the air-sea fluxes and the radiative fluxes including the shortwave component S_W has been the object of intense study for decades (see Chapter 4 of Kantha and Clayson, 1999, for example).

The shortwave component that penetrates into the ocean provides a parameterization for Q_S:

$$Q_S = (1-\bar{\alpha})S_w \sum_{n=1}^{N} \beta_n \exp(z/L_n); \quad \sum_{n=1}^{N} \beta_n = 1 \qquad (9.4.2)$$

where S_W is the SW radiative flux, impingent on the sea surface, that is divided into N bands with different extinction scales L_n (β_n is the fraction). Note that unlike LW radiation, which is completely absorbed at the free surface, the SW component penetrates into the ocean, the extent of penetration being a strong function of the optical properties of the water in the visible range of the spectrum (Kantha and Clayson, 1994; see also Kantha and Clayson, 2000, Chapter 2). The near red and infrared components are absorbed within the upper few tens of centimeters and it is the blue-green part of the SW spectrum that penetrates deep into the ocean.

The conditions on the turbulent quantities are

$$q^2 = B_1^{2/3} u_{*0}^2; \quad \ell=0 \qquad (9.4.3)$$

where u_{*0} is the friction velocity at the sea surface $= \left[\left(\tau_{w1}^2 + \tau_{w2}^2 \right)/\rho_0^2 \right]^{1/4}$.
Alternatively, the flux of q^2 related to the flux of TKE can be prescribed. In fact, the latter might be more appropriate, especially since breaking waves under conditions of light winds and shallow mixed layer provide a considerable flux of TKE into the upper layers that cannot be ignored in the upper few meters. The length scale is then related to the wave height. Nevertheless, this wave-generated turbulence is important only in the upper few meters.

Alternative conditions on the heat and salt flux are either to impose the sea surface temperature and salinity or to damp the surface values to the prescribed values with a specific time lag t_l:

$$(\Theta, S) = (T_S, S_S) \text{ or } \frac{\partial}{\partial t}(\Theta, S) = -\frac{1}{t_l}\left[(\Theta - T_S), (S - S_S)\right] \qquad (9.4.4)$$

At the ocean bottom, the heat and salt fluxes are normally zero. The momentum fluxes are determined by the bottom stress, which in turn depends on the surface roughness z_0 and the flow velocity near the bottom:

$$K_M \frac{\partial}{D\partial\sigma}(u_1,u_2) = \frac{1}{\rho_0}(\tau_{b1},\tau_{b2}) = C_D\left(u_{b1}^2 + u_{b2}^2\right)^{1/2}(u_{b1},u_{b2})$$

$$K_H \frac{\partial}{D\partial\sigma}(\Theta,S) = (0,0) \qquad\qquad (9.4.5)$$

$$q^2 = B_1^{2/3}u_{*b}^2; \quad \ell=0$$

where u_{*b} is the friction velocity at the bottom $= \left[\left(\tau_{b1}^2 + \tau_{b2}^2\right)/\rho_0^2\right]^{1/4}$ and C_D is the drag coefficient determined either by matching to the logarithmic law of the wall if resolution is adequate or prescribed at the canonical value of 0.0025:

$$C_D = \text{Max}\left\{0.0025, \left[\frac{1}{\kappa}\ln\left(\frac{1+\sigma_b}{z_0/D}\right)\right]^{-2}\right\} \qquad\qquad (9.4.6)$$

The boundary conditions at closed lateral boundaries are straightforward: zero fluxes. Conditions at open boundaries, however, are difficult to specify, since the state of the medium external to the model domain must be specified but is usually not known. Unlike the atmosphere, where routine global model results are available from NWP centers around the world, no operational global ocean model exists yet, in which the regional model can be nested. For some applications, climatological averages suffice. In most cases, the best that can be done is to prescribe conditions so as to let any disturbances created inside the domain get out through the open boundary. This usually takes the form of Sommerfeld radiation boundary condition on flow quantities, $\frac{\partial\phi}{\partial t} + c\frac{\partial\phi}{\partial n} = 0$, where c is the phase speed of the disturbance of the wave approaching the boundary from the interior and ϕ is any property such as η. It is important to prescribe c correctly so as not to introduce noise at the boundary. Lateral boundary conditions have been the subject of active research for many years (Chapman, 1985; Kantha and Blumberg, 1985; Kantha *et al.*, 1990; Orlanski, 1976; Roed and Cooper, 1987). See Section 2.10 for a more detailed discussion. As far as scalars such as the temperature Θ and salinity S are concerned, if the flow is into the domain, one must specify the values being advected into the domain; if the flow is outward, the values inside the domain need to be advected out:

$$\frac{\partial}{\partial t}(\Theta,S)+u_n\frac{\partial}{\partial n}(\Theta,S)=0 \tag{9.4.7}$$

where n denotes the direction normal to the boundary. As far as the turbulence quantities are concerned, any advection at the lateral boundaries can be assumed to be zero, without much loss of accuracy.

9.5 MODE SPLITTING

Because Eqs. (9.2.14) to (9.2.20) involve free surface dynamics, and the rigid lid approximation is not invoked to filter out fast moving external gravity waves (typical speed 200 m s^{-1}), the solution will involve adjustment due to both these fast external and slow moving internal (typical velocity 2 m s^{-1}) gravity waves. The CFL condition therefore imposes a severe limitation on the time step that can be taken to advance the solutions. To circumvent this, a technique called mode splitting (Madala and Piacsek, 1977) is used. This involves separating the external and internal mode equations and solving the external mode equations at time steps consistent with the fast external gravity waves, while solving the equations for the internal mode at larger time steps consistent with the slow internal gravity wave speeds. Since the barotropic equations are fewer and simpler, this results in considerable savings in computer time, since the ratio of the internal to external time steps permissible ranges from 5 to 10 in estuaries to more than 40 in the deep ocean basins. The external mode equations are obtained by integrating the continuity and momentum equations, Eqs. (9.2.14) to (9.2.16), vertically over the water column:

$$\frac{\partial\eta}{\partial t}+\frac{1}{h_1h_2}\left[\frac{\partial}{\partial x_1}\left(h_2U_1D\right)+\frac{\partial}{\partial x_2}\left(h_1U_2D\right)\right]=0 \tag{9.5.1}$$

$$\frac{\partial}{\partial t}\left(U_1D\right)+\frac{1}{h_1h_2}\left[\frac{\partial}{\partial x_1}\left(h_2U_1^2D\right)+\frac{\partial}{\partial x_2}\left(h_1U_1U_2D\right)\right]+\frac{D}{h_1h_2}\left[U_1U_2\frac{\partial h_1}{\partial x_2}-U_2^2\frac{\partial h_2}{\partial x_1}\right]$$

$$-fU_2D=-\frac{D}{\rho_0}\frac{\partial p_a}{\partial x_1}-gD\frac{\partial\eta}{\partial x_1}+\left(\tau_{w1}-\tau_{b1}\right) \tag{9.5.2}$$

$$+\frac{gD^2}{\rho_0}\left[\frac{\partial}{\partial x_1}\int\limits_{-1}^{0}d\sigma\int\limits_{\sigma}^{0}\rho d\sigma'-\left(\frac{\sigma}{D}\frac{\partial D}{\partial x_1}+\frac{1}{D}\frac{\partial\eta}{\partial x_1}\right)\int\limits_{-1}^{0}\rho d\sigma\right]+\frac{gD}{\rho_0}\frac{\partial D}{\partial x_1}\int\limits_{-1}^{0}d\sigma\int\limits_{\sigma}^{0}\sigma'\frac{\partial\rho}{\partial\sigma'}d\sigma'+D\overline{F}_1$$

$$\frac{\partial}{\partial t}(U_2 D) + \frac{1}{h_1 h_2}\left[\frac{\partial}{\partial x_1}(h_2 U_1 U_2 D) + \frac{\partial}{\partial x_2}(h_1 U_2^2 D)\right] + \frac{D}{h_1 h_2}\left[U_1 U_2 \frac{\partial h_2}{\partial x_1} - U_1^2 \frac{\partial h_1}{\partial x_2}\right]$$

$$+ f U_1 D = -\frac{D}{\rho_0}\frac{\partial p_a}{\partial x_2} - g D \frac{\partial \eta}{\partial x_2} + (\tau_{w2} - \tau_{b2}) \tag{9.5.3}$$

$$+ \frac{g D^2}{\rho_0}\left[\frac{\partial}{\partial x_2}\int_{-1}^{0} d\sigma \int_{\sigma} \rho d\sigma' - \left(\frac{\sigma}{D}\frac{\partial D}{\partial x_2} + \frac{1}{D}\frac{\partial \eta}{\partial x_2}\right)\int_{-1}^{0} \rho d\sigma\right] + \frac{g D}{\rho_0}\frac{\partial D}{\partial x_2}\int_{-1}^{0} d\sigma \int_{\sigma} \sigma' \frac{\partial \rho}{\partial \sigma'} d\sigma' + D\bar{F}_2$$

where

$$(U_1, U_2) = \int_{-1}^{0}(u_1, u_2) d\sigma \tag{9.5.4}$$

and the last terms in each momentum equation contains vertically integrated horizontal diffusion terms as well as the so-called dispersion terms arising out of baroclinicity or equivalently the vertical velocity shear:

$$D\bar{F}_1 = \frac{1}{h_1 h_2}\left\{\frac{\partial}{\partial x_1}(h_2 \bar{\tau}_{11}) + \frac{\partial}{\partial x_2}(h_1 \bar{\tau}_{21}) + \bar{\tau}_{21}\frac{\partial h_1}{\partial x_2} - \bar{\tau}_{22}\frac{\partial h_2}{\partial x_1}\right\}$$

$$-\frac{1}{h_1^2 h_2}\left\{\frac{\partial}{\partial x_1}\left[h_1 h_2 D \int_{-1}^{0}(u_1 - U_1)^2 d\sigma\right] + \frac{\partial}{\partial x_2}\left[h_1^2 D \int_{-1}^{0}(u_1 - U_1)(u_2 - U_2) d\sigma\right]\right\}$$

$$+ \left\{\frac{1}{h_1^2}\frac{\partial h_1}{\partial x_1}\int_{-1}^{0}(u_1 - U_1)^2 d\sigma + \frac{1}{h_1 h_2}\frac{\partial h_2}{\partial x_1}\int_{-1}^{0}(u_2 - U_2)^2 d\sigma\right\} \tag{9.5.5}$$

$$D\bar{F}_2 = \frac{1}{h_1 h_2}\left\{\frac{\partial}{\partial x_1}(h_2 \bar{\tau}_{12}) + \frac{\partial}{\partial x_2}(h_1 \bar{\tau}_{22}) + \bar{\tau}_{12}\frac{\partial h_2}{\partial x_1} - \bar{\tau}_{11}\frac{\partial h_1}{\partial x_2}\right\}$$

$$-\frac{1}{h_1 h_2^2}\left\{\frac{\partial}{\partial x_1}\left[h_2^2 D \int_{-1}^{0}(u_1 - U_1)(u_2 - U_2) d\sigma\right] + \frac{\partial}{\partial x_2}\left[h_1 h_2 D \int_{-1}^{0}(u_2 - U_2)^2 d\sigma\right]\right\}$$

$$+ \left\{\frac{1}{h_1 h_2}\frac{\partial h_1}{\partial x_2}\int_{-1}^{0}(u_1 - U_1)^2 d\sigma + \frac{1}{h_2^2}\frac{\partial h_2}{\partial x_2}\int_{-1}^{0}(u_2 - U_2)^2 d\sigma\right\} \tag{9.5.6}$$

$$\overline{\tau}_{11} = 2A_M \left\{ \frac{h_2}{h_1} \frac{\partial}{\partial x_1} (U_1 D) + \frac{U_2 D}{h_1 h_2} \frac{\partial h_1}{\partial x_2} \right\}$$

$$\overline{\tau}_{12} = 2A_M \left\{ \frac{h_1}{h_2} \frac{\partial}{\partial x_2} \left(\frac{U_1 D}{h_1} \right) + \frac{h_2}{h_1} \frac{\partial}{\partial x_1} \left(\frac{U_2 D}{h_2} \right) \right\} \qquad (9.5.7)$$

$$\overline{\tau}_{22} = 2A_M \left\{ \frac{h_1}{h_2} \frac{\partial}{\partial x_2} (U_2 D) + \frac{U_1 D}{h_1 h_2} \frac{\partial h_2}{\partial x_1} \right\}$$

The dispersion terms involve nonlinear terms in the velocity components and vanish for a homogeneous (barotropic) ocean. However, for a baroclinic ocean, the presence of dispersion terms means that the horizontal diffusion is a function of the vertical coordinate and this is implemented in the model.

The horizontal diffusion parameterization is rather ad hoc. Often, the form and coefficients are chosen solely to control nonlinear instability and aliasing. An efficient scale-selective scheme is the biharmonic diffusion, and this is often used in many global models. Frequent use of Shapiro filters (Shapiro, 1970) during the simulation also provides the needed numerical stability. When Lapacian form is employed as in CUPOM, it is not unusual to use arbitrarily chosen constant values for A_M and A_H equivalent to zeroth order closure. It may, however, be preferable to use values for A_M and A_H that are dependent at least on the deformation rate of the mean flow, the first-moment closure based Smagorinsky formulation (Smagorinsky *et al.*, 1965), where $A_M \sim \Delta^2 (D_{ij} D_{ij})^{1/2}$,

Δ is the grid size, and D_{ij} is the mean deformation rate tensor $\frac{1}{2} \left(\frac{\partial u_i}{\partial x_j} + \frac{\partial u_j}{\partial x_i} \right)$.

This formulation has the advantage that the mixing rate is small if the deformation rate is small, as is required physically and is equivalent to the classical Prandtl mixing length approach (see Chapter 1 of Kantha and Clayson, 1999). It implicitly assumes that the subgrid scales that are being parameterized fall into the Kolmogoroff inertial subrange, which is not always satisfied. It is used extensively in atmospheric models. However, care must be exercised in its use in oceanic models since it tends to smear strong thermal and density fronts because of the strong associated shear, an undesirable feature in ocean models. This can, however, be overcome by choosing a smaller value for the Smagorinsky constant. CUPOM uses a sum of the background and deformation-dependent values and also has provisions for a constant cell Reynolds number approach:

$$A_M = A_{MB} + C_{SM}\left(\Delta x_1 \Delta x_2\right)\left[\left(\frac{\partial u_1}{h_1 \partial x_1}\right)^2 + 0.5\left(\frac{\partial u_1}{h_2 \partial x_2} + \frac{\partial u_2}{h_1 \partial x_1}\right)^2 + \left(\frac{\partial u_2}{h_2 \partial x_2}\right)^2\right]^{1/2}$$

or

$$A_M = A_{MB} + \frac{\left(u_1^2 + u_2^2\right)^{1/2}\left(\Delta x_1 \Delta x_2\right)^{1/2}}{R_c} \tag{9.5.8}$$

$$A_H = A_M / Pr_H$$

where Pr_H is the Prandtl number and C_{SM} is the Smagorinsky constant ($\sim \kappa^2/2$), chosen anywhere from 0.02 to 0.20, the usual value being 0.08. The value of R_C if the cell Reynolds number approach is used, usually is taken to be between 50 and 200. A_{MB} affords some smoothing in regions of little hydrodynamic activity and should be chosen as small as feasible (typical value is 100 m^2 s^{-1} for a 20-km grid). The formulation allows for either constant or deformation-dependent or constant cell Reynolds number values to be used.

9.6 NUMERICS

There is considerable leeway in solving the above governing equations. The solution technique can be explicit or implicit, the former being simpler to implement and somewhat better suited to massively parallel computer architectures. While second-order difference schemes are the norm, higher order schemes are not ruled out, although they involve additional complexities in terms of boundary conditions. The most common practice is to use the lowest order possible and rely on a smaller grid size for better accuracies. In the following, we will lay out one basic scheme: second-order finite difference schemes—implicit in the vertical, explicit in the horizontal—relying on the Thomas algorithm for efficient solution of the resulting tridiagonal matrix equations. Because of the vertical mixing and Laplacian horizontal diffusion terms, the governing equations are second-order PDEs. It is, however, the vertical mixing terms that tend to dominate the solution method, simply because the numerical scheme must allow for flexibility in the choice of the grid size in the vertical. In oceans as well as the atmosphere, it is often essential to resolve processes in mixed layers and the vertical shear due to strong jets such as the EUC. Consequently, the scheme must be implicit in the vertical so as not to impose severe time step limitations that could be brought upon by small grid sizes in the vertical.

The model uses the staggered Arakawa C grid for discretization in the horizontal direction (see Figure 9.6.1). In this grid, quantities such as H and η are at the center of the grid, while the east-west velocity component is displaced half a grid point to the west of the grid center and the north-south component is displaced half a grid to the south of the grid center. C grid has favorable wave propagation properties if the grid size is smaller than the Rossby radius. Since application to coastal oceans implies invariably fine resolution grids, C grid is traditional in coastal modeling. With enormous strides in computing resources, even the global ocean models employed in climate studies are being cast in C grid, to best utilize the fine resolution now affordable.

The internal mode variables Θ, S, ω, q, l, ρ, K_M, and K_H are located at the grid center, while the velocity components u_1 and u_2 are staggered in the horizontal as described above. These variables are staggered in the vertical as well (Figure 9.6.2). While variables associated with turbulence and mixing q, l, K_M, and K_H, as well as the vertical velocity ω are located at the σ levels, the conventional prognostic variables u_1, u_2, Θ, S, and ρ are staggered in the vertical with respect to these and are located between the corresponding σ levels. The shear stress components τ_1 and τ_2 are staggered in the horizontal the same way as u_1 and u_2, but unlike u_1 and u_2 they are located at σ levels (see Figure 9.6.2). The staggered nature of the numerical grid in both horizontal and vertical directions permits higher order accuracies (second-order) vis-a-vis a nonstaggered grid. Similarly, even with problems associated with the spurious computational mode, the leapfrog scheme that uses three time levels is the preferred method for finite differencing in time.

All the governing equations are cast in flux-conservative form before finite differencing is carried out so that mass, momentum, heat, salt, and energy are conserved by the numerical scheme when the equations are discretized. Also a control volume approach is preferable before discretization is attempted. Recognizing that $h_1 \delta x_1 = \Delta x_1$ and $h_2 \delta x_2 = \Delta x_2$, where Δx_1 and Δx_2 are the grid sizes in the x_1 and x_2 directions, we can multiply all the governing equations for both internal and external modes by $h_1 \delta x_1$ and $h_2 \delta x_2$ before finite differencing. The continuity equation, Eq. (9.2.18), then becomes

$$\left(h_1 \delta x_1\right)\left(h_2 \delta x_2\right)\frac{\partial \eta}{\partial t} + \frac{\partial}{\partial x_1}\left(h_2 \delta x_2 \cdot u_1 D\right)\delta x_1$$

$$+ \frac{\partial}{\partial x_2}\left(h_1 \delta x_1 \cdot u_2 D\right)\delta x_2 + \left(h_1 \delta x_1\right)\left(h_2 \delta x_2\right)\frac{\partial \omega}{\partial \sigma} = 0$$

$$(9.6.1)$$

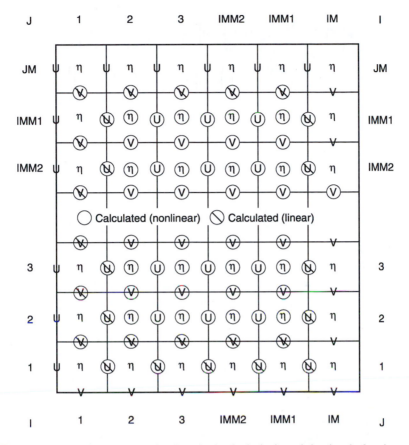

Figure 9.6.1 Arakawa C grid used for discretization in the horizontal showing the locations of the velocities relative to the grid center.

Defining finite difference operators $\delta_t, \delta_\sigma, \Delta_{x_1}, \Delta_{x_2}, \delta_{x_1}, \delta_{x_2}$ for any quantity $\varphi\,(x_1, x_2, \sigma, t)$ as follows:

$$\delta_t \varphi = \frac{1}{2\Delta t}\Big[\varphi\,(x_1, x_2, \sigma, t + \Delta t) - \varphi\,(x_1, x_2, \sigma, t - \Delta t)\Big]$$

$$\delta_\sigma \varphi = \frac{1}{2\Delta t}\left[\varphi\left(x_1, x_2, \sigma + \frac{\Delta\sigma}{2}, t\right) - \varphi\left(x_1, x_2, \sigma - \frac{\Delta\sigma}{2}, t\right)\right]$$

$$\delta_{x_1} \varphi = \frac{1}{\Delta x_1}\left(\Delta_{x_1}\varphi\right); \quad \delta_{x_2}\varphi = \frac{1}{\Delta x_2}\left(\Delta_{x_2}\varphi\right)$$

$$\Delta_{x_1}\varphi = \left[\varphi\left(x_1 + \frac{\Delta x_1}{2}, x_2, \sigma, t\right) - \varphi\left(x_1 - \frac{\Delta x_1}{2}, x_2, \sigma, t\right)\right]$$

$$\Delta_{x_2}\varphi = \left[\varphi\left(x_1, x_2 + \frac{\Delta x_2}{2}, \sigma, t\right) - \varphi\left(x_1, x_2 - \frac{\Delta x_2}{2}, \sigma, t\right)\right]$$

(9.6.2)

one gets an elegant form for the finite difference form of the continuity equation in combined sigma vertical and horizontal orthogonal curvilinear coordinates:

$$\Delta x_1 \Delta x_2 \, \delta_t \eta + \Delta_{x_1}\left(\Delta x_2 u_1 D\right) + \Delta_{x_2}\left(\Delta x_1 u_2 D\right) + \Delta x_1 \Delta x_2 \, \delta_\sigma \omega = 0 \qquad (9.6.3)$$

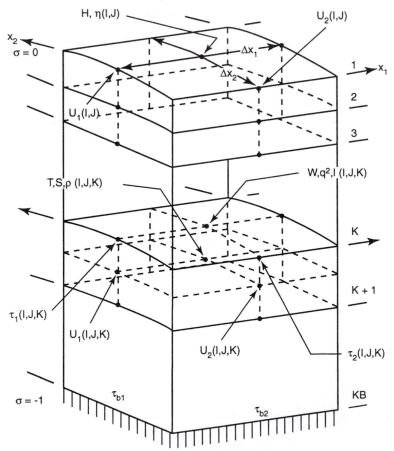

Figure 9.6.2 Sketch of a single computational cell showing the relative locations of various model variables staggered in both the horizontal and the vertical.

Scalar conservation equations can be reduced to their finite difference forms using a similar procedure. For example, the salinity conservation equation becomes

$$\Delta x_1 \Delta x_2 \, \delta_t (SD) + \Delta_{x_1} (\Delta x_2 u_1 SD) + \Delta_{x_2} (\Delta x_1 u_2 SD) + \Delta x_1 \Delta x_2 \, \delta_\sigma (\omega S)$$

$$= \Delta x_1 \Delta x_2 \frac{1}{D} \delta_\sigma \left[K_H \delta_\sigma (S) \right] + \Delta_{x_1} \left[\Delta x_2 A_H D \delta_{x_1} (S) \right] + \Delta_{x_2} \left[\Delta x_1 A_H D \delta_{x_2} (S) \right]$$

$$(9.6.4)$$

The momentum equations are little more complex and for convenience and algebraic clarity, we will omit the density integral and the horizontal diffusion terms in writing the equivalent finite difference form. The momentum equations for u_1 and u_2 become

$$\Delta x_1 \Delta x_2 \, \delta_t (u_1 D) + \Delta_{x_1} \left(\Delta x_2 u_1^2 D \right) + \Delta_{x_2} (\Delta x_1 u_1 u_2 D) + \Delta x_1 \Delta x_2 \, \delta_\sigma (\omega u_1)$$

$$- \Delta x_1 \Delta x_2 \left\{ f + \frac{1}{\Delta x_1 \Delta x_2} \left[u_2 \Delta_{x_1} (\Delta x_2) - u_1 \Delta_{x_2} (\Delta x_1) \right] \right\} u_2 D$$

$$= -\frac{D}{\rho_0} \Delta_{x_1} (\Delta x_2 p_a) - gD\Delta_{x_1} (\Delta x_2 \eta) + \frac{\Delta x_1 \Delta x_2}{D} \delta_\sigma \left[K_M \delta_\sigma (u_1) \right] + \cdots$$

$$\Delta x_1 \Delta x_2 \, \delta_t (u_2 D) + \Delta_{x_1} (\Delta x_2 u_1 u_2 D) + \Delta_{x_2} \left(\Delta x_1 u_2^2 D \right) + \Delta x_1 \Delta x_2 \, \delta_\sigma (\omega u_2)$$

$$+ \Delta x_1 \Delta x_2 \left\{ f + \frac{1}{\Delta x_1 \Delta x_2} \left[u_2 \Delta_{x_1} (\Delta x_2) - u_1 \Delta_{x_2} (\Delta x_1) \right] \right\} u_1 D$$

$$= -\frac{D}{\rho_0} \Delta_{x_2} (\Delta x_1 p_a) - gD\Delta_{x_2} (\Delta x_1 \eta) + \frac{\Delta x_1 \Delta x_2}{D} \delta_\sigma \left[K_M \delta_\sigma (u_2) \right] + \cdots$$

$$(9.6.5)$$

The term in the curly brackets defines a pseudo-Coriolis term in an orthogonal curvilinear coordinate system, which accounts for the acceleration terms arising out of the changes in the grid size from grid point to grid point. The governing equations for the external mode assume similar form and therefore will not be presented.

9.6.1 VERTICAL DIRECTION

The overall solution technique employs explicit schemes in time and horizontal directions, with mode-splitting to ensure efficiency. While it is possible to employ an implicit scheme for at least the external mode solution, the savings realized are not much since the external mode constitutes less than 15% of the overall solution. It is the internal mode solution that is time-consuming, and it is often possible to effect considerable savings by not calculating the turbulence field and prescribing the vertical mixing coefficients suitably by empirical means. The heart of the internal mode calculations is the implicit scheme in the vertical that uses the classical Thomas algorithm to solve the resulting tridiagonal matrices. We will now present details of this numerical solution technique. The transport equations for any scalar quantity such as temperature, salinity, nutrient, or phytoplankton concentration, or TKE, as well as the two momentum equations can be written as

$$\frac{\partial T}{\partial t} + \frac{\partial}{\partial z}(wT) + R = S + \frac{\partial}{\partial z}\left[K(z)\frac{\partial T}{\partial z}\right] \qquad (9.6.6)$$

where T is the prognostic variable, the quantity being modeled and predicted, t is time, and z the vertical coordinate. We will work with the z-coordinate instead of the σ, because the two are equivalent due to the simple relationship between the two: $\frac{\partial}{\partial z} = \frac{1}{D}\frac{\partial}{\partial \sigma}$. The first term is the tendency term or the temporal rate of change. The term R (remainder) denotes all terms in the equation except the vertical mixing, vertical advection, and source term S. This includes horizontal advection terms and horizontal diffusion terms for scalar quantities. For the two momentum equations, these terms include the horizontal pressure gradients, both barotropic (involving sea surface slope) and baroclinic (involving the vertical integrals of horizontal density gradients) terms, and the Coriolis terms. The vertical advection term is normally included as well, but in application to one-dimensional problems, vertical advection has to be prescribed and it is therefore more convenient to separate it out; the second term on the LHS is the vertical advection term, w being the vertical velocity. Similarly, it is more convenient to separate out the source (more appropriately the net source = source − sink) term as well. The source and sink terms are important for quantities such as the nutrients and phytoplankton. In the ocean, the equation for temperature involves penetrative radiative heating term as a prominent source term and is proportional to $\partial I/\partial z$, the vertical divergence of solar insolation, and hence is handled in a manner similar to the vertical advection term. If the source term does not involve vertical derivatives, they are handled in the manner indicated below.

The scheme for time differencing is the second-order three time step leapfrog scheme, because of its higher accuracy, even though the resulting mode splitting problems must be solved by means such as temporal filtering. To make the scheme implicit in the vertical, the variable in the vertical mixing term has to be at time step (n+1) so that

$$\left(\frac{T^{n+1} - T^{n-1}}{2\,\Delta t}\right) + \frac{\partial}{\partial z}\left(w^n T^n\right) + R^n = S^n + \frac{\partial}{\partial z}\left[K^n(z)\frac{\partial T^{n+1}}{\partial z}\right] \quad (9.6.7)$$

Note that while we have used time step n for the remainder terms, any frictional terms, such as bottom friction and horizontal diffusion terms, need to be lagged from stability point of view; in other words, these terms must be evaluated at time step (n−1). Refer to Figure 9.6.3, which indicates the staggered grid in the vertical, in which variables such as the horizontal velocity components and all scalar quantities, such as nutrients, temperature, and so on, are defined at zz points, and quantities involving vertical mixing, such as the turbulence velocity and length scales and mixing coefficients K, and vertical velocity are defined at z points. The very first point at the top is indexed 1, so that as a consequence there are a total of kb z points, but only kb-1 zz points. Although for convenience all arrays are dimensioned kb, care must be taken not to use kb indexed points for quantities defined at zz points, since they are not defined there. With this caveat, Eq. (9.6.7) can be written as

$$\left[\frac{2\Delta t\, K_k^n\left(T_{k-1}^{n+1} - T_k^{n+1}\right)}{zz_{k-1} - zz_k} - \frac{2\Delta t\, K_{k+1}^n\left(T_k^{n+1} - T_{k+1}^{n+1}\right)}{zz_k - zz_{k+1}}\right]\frac{1}{\left(z_k - z_{k+1}\right)} - T_k^{n+1}$$

$$= \left[\frac{2\Delta t\, w_k^n\left(T_{k-1}^n + T_k^n\right)}{z_k - z_{k+1}} - \frac{2\Delta t\, w_{k+1}^n\left(T_k^n + T_{k+1}^n\right)}{z_k - z_{k+1}}\right] \quad (9.6.8)$$

$$+2\Delta t\left(R_k^n - S_k^n\right) - T_k^{n-1}$$

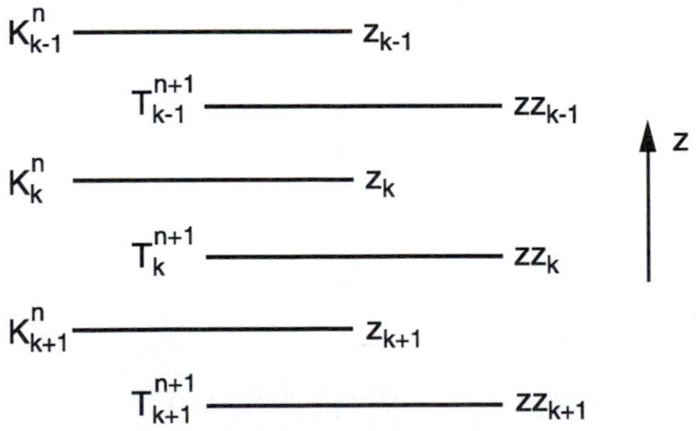

Figure 9.6.3 Staggered vertical grid for temperature (scalar) calculations.

which can be written in a tridiagonal matrix form:

$$
\begin{aligned}
&(-A_k)\,T_{k+1}^{n+1} + (A_k + C_k - 1)\,T_k^{n+1} + (-C_k)\,T_{k-1}^{n+1} \\
&= -D_k + B_k\,(T_{k-1}^n + T_k^n) - E_k\,(T_k^n + T_{k+1}^n)
\end{aligned}
\tag{9.6.9}
$$

where

$$
A_k = \frac{-2\Delta t\ K_{k+1}^n}{(zz_k - zz_{k+1})(z_k - z_{k+1})} = \frac{-2\Delta t\ K_{k+1}^n}{Dzz_k\,Dz_k}
$$

$$
C_k = \frac{-2\Delta t\ K_k^n}{(zz_{k-1} - zz_k)(z_k - z_{k+1})} = \frac{-2\Delta t\ K_k^n}{Dzz_{k-1}\,Dz_k}
$$

$$
D_k = T_k^{n-1} - 2\Delta t\ (R_k^n - S_k^n)
\tag{9.6.10}
$$

$$
B_k = \frac{\Delta t\ w_k^n}{(z_k - z_{k+1})} = \frac{\Delta t\ w_k^n}{Dz_k}
$$

$$
E_k = \frac{\Delta t\ w_{k+1}^n}{(z_k - z_{k+1})} = \frac{\Delta t\ w_{k+1}^n}{Dz_k}
$$

$$
Dz_k = z_k - z_{k+1}, \quad Dzz_k = zz_k - zz_{k+1}
$$

The tridiagonal matrix equation, Eq. (9.6.9), can be solved using the Thomas algorithm, which uses a recursive relationship of the form

$$
T_k^{n+1} = VH_k\,T_{k+1}^{n+1} + VHP_k
\tag{9.6.11}
$$

Changing the index from k to k−1, and substituting for T_{k-1}^{n+1} in Eq. (9.6.9),

$$
\begin{aligned}
T_k^{n+1} &= \frac{A_k}{A_k + C_k(1 - VH_{k-1}) - 1}\,T_{k+1}^{n+1} \\
&+ \frac{C_k\,VHP_{k-1} - D_k + B_k(T_{k-1}^n + T_k^n) - E_k(T_k^n + T_{k+1}^n)}{A_k + C_k(1 - VH_{k-1}) - 1}
\end{aligned}
\tag{9.6.12}
$$

Comparing Eqs. (9.6.11) and (9.6.12), we get recursive relationships for VH_k and VHP_k:

$$
VH_k = \frac{A_k}{A_k + C_k(1 - VH_{k-1}) - 1}
$$

$$
VHP_k = \frac{C_k\,VHP_{k-1} - D_k + B_k(T_{k-1}^n + T_k^n) - E_k(T_k^n + T_{k+1}^n)}{A_k + C_k(1 - VH_{k-1}) - 1}
\tag{9.6.13}
$$

The solution procedure is to compute the coefficients VH_k and VHP_k in the downward sweep from $k = 1$ to $k = kb$ and then use them to compute T_k on the upward sweep. To do this, boundary conditions at the top ($k = 1$ or 2) and bottom ($k = kb-1$ or kb) need to be used. These boundary conditions can be Dirichlet (value of the variable is prescribed) or Neumann (value of the derivative or equivalently the flux of the variable).

Top B.C.

1. If surface value of T is known (Figure 9.6.4), $T_1^{n+1} = T_S$. Using $k=1$ in Eq. (9.6.11),

$$VH_1 = 0, \ VHP_1 = T_S \qquad (9.6.14)$$

2. If the flux of T at the surface is known $\left[K_k^n \dfrac{\partial T}{\partial z} \right]_{z=0} = -\overline{wT}\Big|_S$:

Using $k=1$ in Eq. (9.6.9),

$$\left[2\Delta t(-\overline{wT}\big|_S) - 2\Delta t \, \frac{K_2^n \left(T_1^{n+1} - T_2^{n+1}\right)}{zz_1 - zz_2} \right] \frac{1}{\left(z_1 - z_2\right)} - T_1^{n+1} = \qquad (9.6.15)$$
$$-D_1 + \left[0 - E_1 \left(T_1^n + T_2^n\right) \right]$$

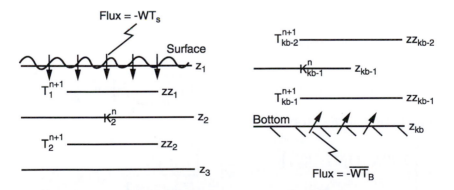

Figure 9.6.4 Boundary conditions at the surface (left) and bottom (right) in solving for the temperature (scalar).

The first term in the second square brackets is zero, because the vertical velocity at the sea surface must vanish. Rearranging Eq. (9.6.15),

$$T_1^{n+1} = \frac{A_1}{A_1 - 1} T_2^{n+1} + \frac{-D_1 - E_1(T_1^n + T_2^n) + \left(\dfrac{2\Delta t \, \overline{wT}\big|_S}{Dz_1} \right)}{A_1 - 1} \tag{9.6.16}$$

so that

$$VH_1 = \frac{A_1}{A_1 - 1}$$

$$VHP_1 = \frac{-D_1 - E_1(T_1^n + T_2^n) + \left(\dfrac{2\Delta t \, \overline{wT}\big|_S}{Dz_1} \right)}{A_1 - 1} \tag{9.6.17}$$

From Eqs. (9.6.13) and (9.6.14) or (9.6.17), it is possible to find all VH_k and VHP_k.

Bottom B.C.

1. If bottom value of T is known (Figure 9.6.4),

$$T_{kb-1}^{n+1} = T_B \tag{9.6.18}$$

and all T_k^{n+1} can be computed.

2. If the flux of T at the bottom is known $\left[K_k^n \dfrac{\partial T}{\partial z} \right]_{z=z_B} = \left[-\overline{wT} \right]_B .$

Using k=kb−1 in Eq. (9.6.9),

$$\left[\frac{2\Delta t K_{kb-1}^n \left(T_{kb-2}^{n+1} - T_{kb-1}^{n+1} \right)}{zz_{kb-1} - zz_{kb-2}} - 2\Delta t(-\overline{wT}\big|_B) \right] \frac{1}{(z_{kb-1} - z_{kb})} - T_{kb-1}^{n+1} =$$
$$-D_{kb-1} + \left[B_{kb-1} \left(T_{kb-2}^{n+1} + T_{kb-1}^{n+1} \right) - 0 \right] \tag{9.6.19}$$

The second term in the second square brackets is zero, because the vertical velocity at the bottom must vanish. Rearranging Eq. (9.6.19),

$$T_{kb-2}^{n+1} = \left[\frac{C_{kb-1}-1}{C_{kb-1}} \right] T_{kb-1}^{n+1} +$$

$$\frac{D_{kb-1} + \dfrac{2\Delta t\, \overline{wT}\big|_B}{Dz_{kb-1}} - B_{kb-1}(T_{kb-2}^n + T_{kb-1}^n)}{C_{kb-1}} \qquad (9.6.20)$$

But putting k=kb−2 in Eq. (9.6.13),

$$T_{kb-2}^{n+1} = VH_{kb-2} T_{kb-1}^{n+1} + VHP_{kb-2} \qquad (9.6.21)$$

From Eqs. (9.6.20) and (9.6.21), eliminating T_{kb-2}^{n+1},

$$T_{kb-1}^{n+1} = \frac{C_{kb-1}VHP_{kb-2} - D_{kb-1} + B_{kb-1}(T_{kb-2}^n + T_{kb-1}^n) - \dfrac{2\Delta t\, \overline{wT}\big|_B}{Dz_{kb-1}}}{C_{kb-1}(1 - VH_{kb-2}) - 1}$$

$$(9.6.22)$$

Equations (9.6.11) and (9.6.18) or (9.6.22) provide all values of T_k on the upward sweep from values of coefficients VH_k and VHP_k determined in the downward sweep.

9.6.2 Horizontal Direction

The explicit scheme in the horizontal direction for both internal and external modes introduces time step constraints from the Courant-Friedrichs-Lewy (CFL) condition. The restrictions due to diffusivity and rotation,

$$\Delta t \leq \Delta x_e (4A_M)^{-1} \text{ and } \Delta t \leq f^{-1} \qquad (9.6.23)$$

are less severe and hence usually not relevant.

The external mode time step has an upper bound defined by

$$\Delta t_e \leq \text{Min}\left[\frac{\Delta x_e}{C_e}\right]; \; \Delta x_e = \left[\frac{1}{(h_1 \delta x_1)^2} + \frac{1}{(h_2 \delta x_2)^2}\right]^{-1/2}; \; C_e = 2(gD)^{1/2} + |V|$$

$$(9.6.24)$$

where the minimum applies over the whole grid. Δx_e is the effective grid size and C_e is the effective external gravity wave speed that accounts for the magnitude of the barotropic component of the flow velocity $|V|$. In the deep oceans, the gravity wave speed is typically 220 m s^{-1} and the flow velocities less than 2 m s^{-1}; hence the contribution of the latter can be neglected and the bound on the external time step depends primarily on the bottom depth and little else. The internal mode calculations are also constrained by the time step:

$$\Delta t_i \leq \text{Min}\left[\frac{\Delta x_e}{c_e}\right]; \; c_e = 2c_i + |v|$$

$$(9.6.25)$$

where c_i is the speed of the gravest internal gravity wave mode, the first baroclinic mode, and $|v|$ is the magnitude of the velocity. As such, the bound on the internal time step and the time split, the ratio of internal to external time steps in the model, depend very much on the stratification in the water column. This is compounded by the fact that the internal gravity wave speeds are of comparable magnitude to oceanic currents, and this also leads to considerable reduction in the time step allowable if the model includes regions of strong western boundary currents. Strong stratification as in estuaries leads to fast internal gravity waves and hence the time split is often limited to as low as 5 to 10. In the deep ocean, the weak prevailing stratification with typical first-mode velocities of 2.5 m s^{-1} or less permits a time split as high as 50.

The second-order accuracy of the leapfrog scheme does not come without some penalties. It is well known that it leads to a spurious computational mode due to the tendency of the solutions at odd and even time steps to diverge from each other (Chapter 2). Remedies include taking occasional Euler forward time steps during the simulation, but the best technique is to use a weak temporal filter to smooth the solution at each time step, the Asselin-Robert filter (Asselin, 1972). Thus a prognostic variable at the current time step is filtered at each time step according to

$$\varphi_f^n = \varphi^n + \nu \left(\varphi^{n+1} - 2\varphi^n + \varphi_f^{n-1}\right)$$

$$(9.6.26)$$

before the values are reset in preparation for the next step. Note that the filter is recursive in the sense that the RHS involves the filtered value from the past time step (subscript f denotes filtered value). Filter coefficient ν is typically 0.05 to 0.1, with the smaller value more prevalent in the ocean community and the larger value more prevalent in the atmosphere. The filter also tends to have a stabilizing influence on the calculations.

9.7 NUMERICAL PROBLEMS

Transformation to sigma-coordinates causes two vexing problems: the artificial purely numerical diffusion along isopycnals, brought about by the form for the horizontal diffusion terms, and spurious currents in the water column along topography, produced by truncation errors in calculating the baroclinic pressure gradients across sharp topographic changes such as the continental slope. The former can produce a spurious vertical transfer of properties by diffusion along sigma surfaces, even when homogeneity prevails in the horizontal direction. To alleviate this problem, only anomalies from a horizontally homogeneous mean background state are retained in the horizontal diffusion terms of temperature and salinity (Mellor and Blumberg, 1985). Thus before these terms are computed, an area-averaged mean value is subtracted, so that horizontal homogeneity would lead to zero diffusion. Nevertheless, spurious diapycnal diffusion due to the inclination of isopycnal surfaces to coordinate surfaces that is present in z-level models also plagues sigma-coordinate models in applications to deep basins. The best remedy is to employ isopycnal diffusion schemes in the interior of the ocean (see Chapter 10), but this is yet to be implemented in a sigma-coordinate model.

The second problem can be illustrated more simply in a 2-D (x–σ) model. Reinstating x' to denote the horizontal coordinate in the sigma-coordinate model and neglecting η terms for simplicity, and using the hydrostatic relationship, the horizontal pressure gradient can be written as

$$\frac{\partial p}{\partial x} = \frac{\partial p}{\partial x'} - \frac{\sigma}{H}\frac{\partial H}{\partial x'}\frac{\partial p}{\partial \sigma} = -\frac{\partial}{\partial x'}(\sigma \rho g H) + \sigma \rho g \frac{\partial H}{\partial x'} \quad (9.7.1)$$

The calculation of the horizontal pressure gradient therefore involves taking the difference between two large terms, leading to severe truncation errors over steep topography, in lower order spatial differencing schemes, a problem well known to atmospheric modelers (Gary, 1973, for example), but highlighted by Haney (1991) for oceanic applications. CUPOM (as well as SPEM and POM) uses staggered grids in both the horizontal and the vertical directions leading to an effectively second-order spatial differencing scheme. The resulting pressure gradient errors cause spurious along-topographic currents even in an ocean basin with horizontally homogeneous density distribution that is expected to be quiescent. In fact, a favorite ploy of modelers is to initialize their sigma-coordinate ocean model with a basin-averaged density profile, so that horizontal density gradients are automatically eliminated, and then perform a long-term integration without any external forcing to watch for any horizontal currents induced by topographic changes in the basin. If the currents induced are "small enough," in magnitude, subsequent calculations with surface forcing and the actual density field are regarded as acceptable. The pressure gradient problem is reduced somewhat, but not eliminated, by subtracting a reference horizontal average value from density before computing the pressure gradient:

$$\frac{\partial p}{\partial x} = -\frac{\partial}{\partial x'}(\sigma\rho'gH) + \sigma\rho'g\frac{\partial H}{\partial x'}; \rho' = [\rho - \bar{\rho}(z)] \qquad (9.7.2)$$

Mellor *et al.* (1994) estimated the errors involved and showed that the pressure gradient error is advectively reduced after a long integration time. Nevertheless, it is essential to limit the topographic gradients in a sigma-coordinate model by the use of a nonlinear filter (a linear Gaussian filter may lead to excessive smoothing of topography and large changes to shelf topography) that caps the ratio of the depths in any two adjacent grids (typical values being 1.2 to 1.5). It is also prudent to test as described above for the magnitude of spurious currents induced. In practice, only the continental slope regions are altered significantly by this filter; the changes in abyssal topographies are small and hence not too consequential. Since it is the calculation of baroclinic pressure gradient that introduces problems, the problem is less severe in shallow water and hence the shelf topography can be kept relatively unchanged without detrimental effects on shelf circulation. Wind forcing tends to dominate the effect of horizontal density gradients on the shelf in many cases.

Higher order schemes tend to reduce the truncation error. McCalpin (1994) used a fourth-order scheme to reduce the pressure gradient errors by an order of magnitude. Chu and Fan (1997) propose an even higher order scheme (sixth-order) and show by application to the seamount problem that it further reduces the error by another order of magnitude. However, for most applications a fourth-order C grid scheme is adequate. Thus instead of the second-order C-grid scheme (for a uniform grid),

$$\left.\frac{\partial p}{\partial x}\right|_i = -\frac{1}{\Delta x'}\Big[(\sigma\rho'gH)_i - (\sigma\rho'gH)_{i-1}\Big] + \frac{\big[(\sigma\rho'g)_i + (\sigma\rho'g)_{i-1}\big]}{2}\left(\frac{H_i - H_{i-1}}{\Delta x'}\right)$$

$$(9.7.3)$$

one can use

$$\left.\frac{\partial p}{\partial x}\right|_i = -\frac{1}{24\Delta x'}\left[(\sigma\rho'gH)_{i+1} - 27(\sigma\rho'gH)_i + 27(\sigma\rho'gH)_{i-1} - (\sigma\rho'gH)_{i-2}\right]$$

$$+\frac{\left[-(\sigma\rho'g)_{i+1} + 9(\sigma\rho'g)_i + 9(\sigma\rho'g)_{i-1} - (\sigma\rho'g)_{i-2}\right]}{16}\left(\frac{H_{i+1} - 27H_i + 27H_{i-1} - H_{i-2}}{24\Delta x'}\right)$$

$$(9.7.4)$$

The expressions are more complicated for a nonuniform orthogonal curvi-linear grid.

Kurihara *et al.* (1998) describe a novel technique for mitigation of pressure gradient errors in the GFDL σ-coordinate hurricane model. Atmospheric modelers have long used the pressure normalized by its surface value as the vertical (σ) coordinate. In this coordinate system, computing the pressure gradient using Eq. (9.7.1) is equivalent to locating the point E at the adjacent grid i+1 that lies on the same coordinate surface as point A at grid point i and then determining the height difference between point E and point B, the intersection with the vertical line at the adjacent grid point i+1 of the isobaric surface passing through point A (Figure 9.7.1). The latter does not use any information on densities between E and B, but only those at A and E. This leads to significant errors. Instead if the slope of the isobaric surface is used to compute the pressure gradient, information on densities between E and B is involved, and no errors result. If pressure at the bottom point G is greater than that at point A, the point B lies above G, and is easily located between σ-coordinate surfaces at points C and D by interpolation. If the pressure at G is less than at A, the point B lies below G, and some assumption as to the state of the fictitious fluid column between G and B is needed. This column is assumed to be in hydrostatic balance at a density that is the average of those at points A and F, the lowest σ-coordinate level at the grid point i+1. Kurihara *et al.* (1998) report no numerical difficulties in pressure gradient computations in steep terrains.

In ocean models, this technique is equivalent to calculating the pressure gradient in z-coordinates. Since the isobaric surface is a level surface, one needs to find the pressure at points A and B by vertical integration of the density and then take the difference. The procedure is very much similar to that used in z-coordinate models.

Arakawa and Suarez (1983) also addressed the problem of vertical finite differencing the primitive equations in sigma-coordinates. They have derived a family of schemes that give exact forms of the hydrostatic equation and the pressure gradient force for particular atmospheres. For example, the pressure gradient force is exact for three-dimensionally isentropic atmospheres. Thus,

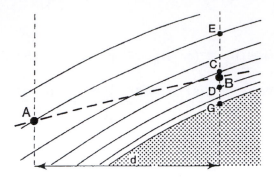

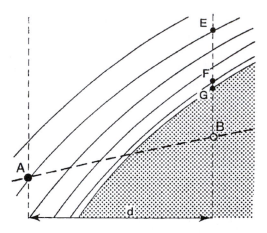

Figure 9.7.1 Position of points in the vertical cross section for pressure gradient calculations (stippled area is land). Dashed line AB indicates the isobar. Point B is (a) in the fluid and (b) below the ground. From Kurihara *et al.* (1998).

while the discretization error cannot be eliminated in general, it can be made to vanish for particular atmospheres.

9.8 APPLICATIONS

We will provide selected examples of application of CUPOM: one basin-scale and the other to semienclosed marginal seas. The North Indian Ocean is under the action of periodically reversing monsoons and exhibits a rich variety of dynamical processes (see, for example, Lopez, 1998). During summer, roughly from June to August, the winds are southwesterly, and during winter, roughly from December to February, they are northeasterly, with the remaining seasons being essentially transition seasons. The reversing winds cause dramatic changes in the circulations of the Arabian Sea and the Bay of Bengal. The seasonally reversing western boundary current, the Somali Current, is a unique feature of the basin circulation. During the southwest monsoon, the Somali Current flows northward along the coast of Somalia and during the northeast monsoons, it reverses and flows southward toward and beyond the equator. The regions off Somalia and Arabia experience intense upwelling during the southwest monsoon. A variety of mesoscale features, such as the Laccadive High that spins off the tip of the Indian subcontinent and drifts westward during NE monsoons, and Socotra Eddy off the island of Socotra during SW monsoons are important features of circulation. The Great Whirl that spins up near the equator off the coast of Africa prior to the full onset of SW monsoons is also noteworthy. In addition, the Indian Monsoon Current below the tip of India and the equatorial current system including the reversing Equatorial Undercurrent are important parts of the circulation in the north Indian ocean. Another unique feature is the high salinities in the Arabian Sea due to excess evaporation, especially during SW monsoons, and the low salinities prevalent in the Bay of Bengal due to the fresh water dumped into the Bay by large rivers draining the Indian subcontinent.

In a basin-scale application of CUPOM to the north Indian Ocean, the model has been run with and without data assimilation, driven by FNMOC windstress and fluxes for 1993 (Lopez, 1998). It has also been run from 1993 to 1998 with altimetric data from TOPEX altimetric data assimilated into it. The horizontal resolution is 1/2 degree, but there are 38 sigma levels in the vertical that enable the upper layer and mixed layer physics to be well resolved. Figure 9.8.1 is a snapshot of the currents at 30 m in the Arabian Sea on August 18, 1993, without (top panel) and with (bottom panel) data assimilation. While the model re-produces various features of the Arabian Sea circulation such as the Somali Current and the Great Whirl that are central to the North Indian Ocean dynamics, these features are more realistic with assistance from altimetric data. Figure 9.8.2 shows the surface circulation during SW and NE monsoons (Lopez, 1998). The model has also been run in a pseudo-operational mode to produce near-real-time nowcasts using real time MCSST and altimetric data streams from the Naval Oceanographic Office. These results will not be presented in any detail here (but see www-ccar.colorado.edu/~kantha/nio/nio.html).

Figure 9.8.3 shows currents from a 10 km resolution model of the Mediterranean Sea (Horton et al., 1997). The model was run using the 40 km resolution Navy Operational Regional Atmospheric Prediction System (NORAPS) winds, orographically modified using a single-layer ABL to provide

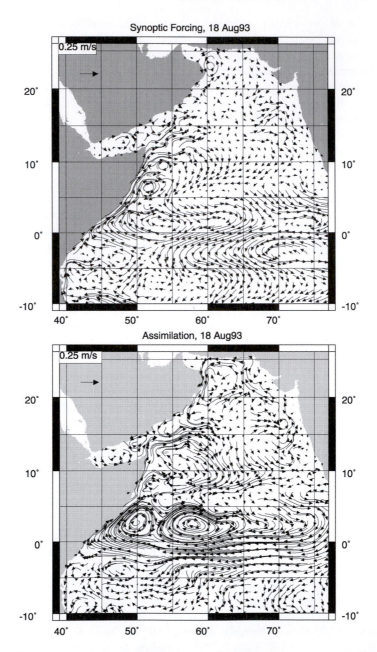

Figure 9.8.1 Snapshots of currents at 30 m depth in the Arabian Sea on August 18, 1993, (top) without and (bottom) with TOPEX altimetric data assimilation. Altimetric data constraint appears to sharpen and intensify the features such as the Great Whirl.

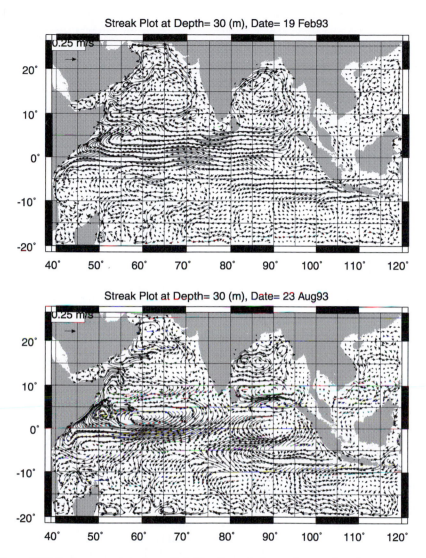

Figure 9.8.2 Surface currents in the north Indian Ocean during northeast and southwest monsoons with MCSST and altimetric assimilation. From Lopez (1998).

orographically steered surface level winds. In addition MCSST was assimilated into the model initialized using a special AXBT survey conducted by NAVO. The model also assimilated data from an XBT/CTD survey during the model run. The data-assimilated model reproduces many of the transient, recurrent, and

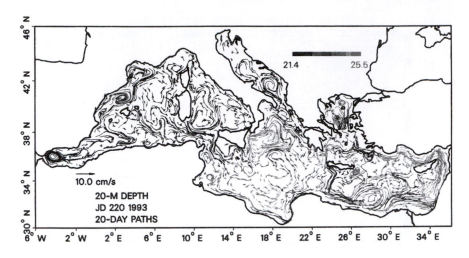

Figure 9.8.3 Currents at 20 m depth from a 10 km resolution model of the Mediterranean Sea showing a wealth of features in both the western and eastern parts. From Horton *et al.* (1997).

permanent eddy and gyre systems such as the Mersa-Metruh, Rhodes gyre, Ierapetra eddy, Peloponnesian gyres, and Shikmona Eddy observed in the Eastern Mediterranean (see Horton *et al.*, 1997). The Alboran gyres and the Algerian Current System in the western Mediterranean are also well simulated. Simulations such as these using an ocean model with comprehensive physics and assimilation of remotely sensed and any available *in situ* data provide an excellent means of not only nowcasting and forecasting the oceanic state, but also of studying using the archived model results, with the dynamical processes realistically modeled in the region.

Finally, Figure 9.8.4 shows the extensive gyre system in the Red Sea as simulated by Clifford *et al.* (1997). Their study underscores the importance of accurate orographically steered winds in simulating the circulation in a marginal or semienclosed sea. This problem is more critical to regional models compared to basin scale models because of the influence of the land-sea interface and the much smaller spatial and temporal scales of variability in wind and other surface forcing.

For examples of application to real-time nowcast/forecasts in a marginal sea (Gulf of Mexico), see Chapter 14, and to hindcast of the atmospheric state using a regional coupled ocean-atmosphere model, see Chapter 13. Both of these applications required accurate simulation of thermodynamic/dynamical processes by the ocean model.

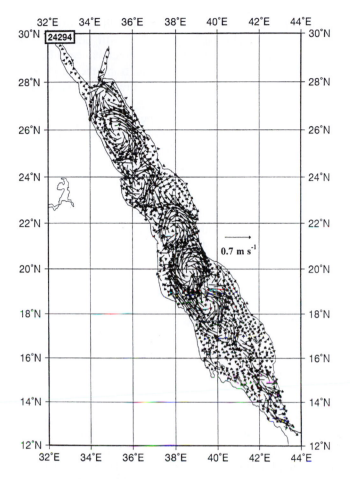

Figure 9.8.4 Near-surface currents in the Red Sea showing a train of principally anticyclonic eddies. From Clifford *et al.* (1997).

9.9 CODE STRUCTURE

The basic computing engine for the POM was set quite early in the 1970s (see Blumberg and Mellor, 1987) and has not changed much over the past two decades. The decisions made then have withstood the test of time well and many routines are therefore common to different derivatives of POM over the years. Major changes are found mostly in routines such as PROFQ, containing the turbulence closure submodel (see, for example Kantha and Clayson, 1994) and

BCOND, which contains the open boundary conditions that are necessarily different for different domains and applications. In addition, one finds in CUPOM additional routines to perform tidal potential calculations and data assimilation. It is here the CUPOM diverges from the original POM. The flowchart of the CUPOM code is shown in Figure 9.9.1. The code structure is similar to that of the original POM (Mellor, 1996) except for some salient differences. The calculation consists of two nested loops, Loop 8000 nested in Loop 9000, the former corresponding to the external mode computations and the latter to the internal mode. The module *main.f* containing the driver program MAIN is the driver module. It initializes fields, reads in all relevant input files and any restart file, and writes out restart and archive files. With the exception of the open boundary condition program BCOND, which is in the module *bcond.f,* and the external mode calculation XMODE, which is in *xmode.f,* the rest of the subroutines do not change much and hence are grouped in the module *rest.f.* The module *wrtpgm.f* contains diagnostic print routines. The common blocks are *Ogcm1.h,* which contain variables that change from domain to domain (such as the size of the 3-D computational grid IM, JM, and KB in x_1, x_2, and z directions), *Ogcm2.h,* which has the various common variables, and *Ogcm3.h,* which defines all the arrays needed. When tides are included, the corresponding variables are found in *Ogcmt.h* and when data assimilation is used, the variables are found in *Ogcm4.h.* The constants are defined in *Ogcmmd.h* and *Ogcmd.h.* The modules are designed such that most of the changes going from one domain to another need to be made for the most part in modules *Ogcm1.h, main.f,* and *bcond.f.*

The internal mode solution is effected by subroutines PROFT, PROFS, PROFU, PROFV, and PROFQ, which solve for Θ, S, u_1, u_2, and q^2 and $q^2 l$ as outlined in 9.6.1. The advection and horizontal diffusion terms needed by these routines are supplied by routines ADVT, ADVU, ADVV, and ADVQ for Θ and S, u_1, u_2, and q^2 and $q^2 l$, respectively. ADVU and ADVV supply the Coriolis and pressure gradient terms as well. This actually constitutes a two-step solution with ADV(X) routines carrying out the first step of explicit time differencing and calculating the advection and other terms, and passing them on to the PROF(X) routines, which then derive the solution at the next time step by an implicit scheme in the vertical. The external mode solution is calculated by XMODE, the advection and horizontal diffusion terms for which are supplied by ADVAVE. For those versions involving the calculation of astronomical tide generating forces, the module *tides.f* contains the routines needed to calculate tides. The routine DENS calculates the density from the equation of state, and BAROPG calculates the pressure gradients arising from horizontal density gradients. Note that the 3-D array RMEAN contains the horizontally averaged values of density interpolated to each 3-D grid point in the sigma-coordinate (hence the need for a 3-D array, when the basic values are horizontally homogeneous) and is

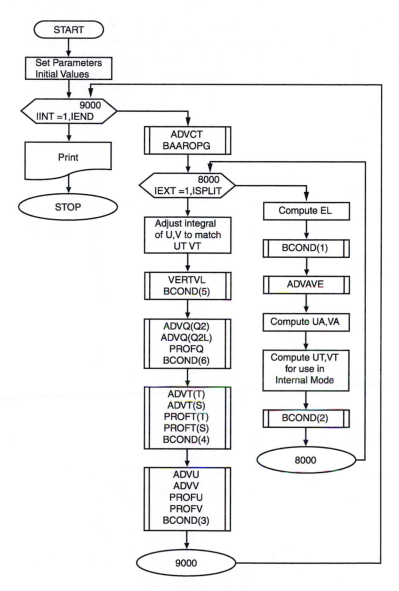

Figure 9.9.1 Flowchart adapted from Mellor (1996).

subtracted out of the density before calculating the pressure gradients and then added back. This procedure reduces the errors in sigma-coordinate model computations of pressure gradients and prevents spurious along-slope currents from destroying the solutions. Note that to extract higher precision, the model salinities are actually (S − 35) psu (and temperatures Θ − 10°C) and the routine DENS adds these offsets to S (and Θ) at the beginning before calculating the density and subtracts them out again toward the end. Thus round-off errors are minimized. Similarly, a constant value of 1.025 is subtracted out of the density (normalized by 1000.0 kg m^{-3}) in all calculations in the model, since only the gradients are relevant and a constant offset is inconsequential. DENS uses the UNESCO equation of state (see Appendix A). Considerable confusion and consternation can be avoided if these aspects are remembered when examining the raw model outputs. Horizontal diffusion calculations for Θ and S in ADVT involve subtracting reference fields TMEAN and SMEAN before these terms are calculated and adding them back afterward (see Mellor and Blumberg, 1985, for justification). Still, there could be a very slow but undesirable drift in the deep ocean fields, the effects of which can be eliminated by restoring the horizontal averages of temperature and salinity in the model domain to predetermined values (from, say, climatology) at predetermined intervals (say, once a year). The uniqueness of this approach vis-a-vis a robust climatology approach where the entire mass field is damped to reference (climatological most often) values is that this permits baroclinicity to evolve yet prevent deleterious water mass drift. This is accomplished in TSDAMP, an additional feature of CUPOM. The routine need not be called if no such damping is deemed desirable or necessary.

While three time levels are involved for each prognostic variable, to save memory requirements, only the past (time step n−1) and present (time step n) time level values are unique to each 3-D variable; the future (time step n+1) time level values are calculated in temporary variables UF and VF for each internal prognostic variable pair Θ and S, u_1 and u_2, and q^2 and q^2l, and immediately after applying Asselin-Robert filter and time level resetting for that pair, used for calculating the future values of the next pair of variables. Therefore, one finds for example only 3-D arrays TB and T defined for temperature, for example, corresponding to n−1 and n time levels. The external mode variables require, however, 2-D arrays only and therefore all three time levels are explicitly defined—for example ELB, EL, and ELF correspond to the three time levels for elevation η (UAB, UA, and UAF for velocity component U_1 and VAB, VA, and VAF for U_2).

The 2-D arrays DX, DY, ALON, and ALAT define the grid spacing and the geophysical locations of each grid point and pretty much define the grid characteristics for any orthogonal curvilinear grid. Although for a rectilinear grid, variables such as DX, DY, ART, ARU, and ARV (the latter three the grid elemental areas) need not be 2-D arrays, for generality, they are retained as 2-D arrays (similarly prognostic variables such as Θ, S, u_1, u_2 defined at only kb−1 levels are still dimensioned kb for convenience). These—along with bottom topography array H and the 3-D arrays related to the water mass properties TB,

SB, TMEAN, SMEAN, RMEAN—pretty much define the problem. It is important to ensure no bugs or unreasonable values in these arrays in order to prevent costly mistakes. It is also essential to ensure that the topography is passed through a nonlinear filter that attenuates strong topographic gradients over the slope before any calculations are ever attempted with the model.

External mode calculations yield sea surface elevation η (EL) and vertically averaged (barotropic) velocity components U_1 and U_2 (UA and VA). They therefore provide the elevation gradients needed for the internal mode calculations. The bottom friction and the vertical integrals of the density gradients (integrated baroclinic forcing) needed to make the external mode calculations come from the internal mode calculations, which yield internal variables Θ, S, u_1, u_2, q^2, and $q^2 l$. This is how the two calculations interact with each other. However, the interaction is tricky because of the time step disparities and care has been taken in constructing the computing engine (Mellor, 1996) to conserve properties and prevent the two calculations from drifting apart in long-term integrations due to inevitable numerical truncation and round off errors, which are not necessarily of the same magnitude in the two sets of calculations.

The internal mode uses time step DTI and at any instant during such calculations, the values at time steps n−1 and n are known (see Figure 9.9.2). The values of the bottom friction and the baroclinic forcing are therefore known from time step n and these are fed (and held constant) to the external mode calculations (which are then stepped forward in time at the smaller time step DTE until time step n+1 is reached). The time-averaged values of vertically-averaged velocities UA and VA, over time intervals n−1 to n (UTB, VTB) and n to n+1 (UTF, VTF) are calculated in the external mode calculations and used to replace the vertical means of internal velocities U and V (values at time level n) by the average of UTB and UTF, and VTB and VTF. This procedure prevents the vertical averages of velocity fields in internal and external mode calculations to drift apart in time from purely numerical causes. Since mass conservation involves the elevations as well, the values of ETB, ET, and ETF (related to ELB, EL, and ELF) from external mode calculations at the three time levels n−1, n, and n+1 are used to ensure conservation of mass and hence conservation of other scalars such as Θ and S. The elevation gradients needed by the internal mode calculations to advance the solution to time level n+1 are provided using the mean of EGB and EGF, which are themselves the averages of EL over time interval n−1 to n and n to n+1, respectively. It is this subtle aspect that makes the external mode CFL irrelevant to internal mode calculations, even though the vertical averaged (barotropic) velocities are retained, *not* subtracted out from the velocity fields in the internal mode calculations (Mellor, 1996).

Each internal time step, MAIN calls XMODE and ADVAVE for NSPLIT external time steps, then ADVT and PROFT to compute Θ, ADVT and PROFS to compute S, ADVU and PROFU to compute U_1, ADVV and PROFV to

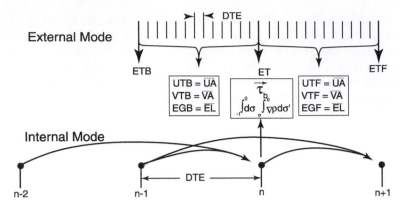

Figure 9.9.2 Split time stepping that computes the barotropic and baroclinic modes separately but couples them frequently.

compute U_2, ADVQ and PROFQ to compute q^2 and q^2l, and DENS and BAROPG to compute the density and integrals of density gradients. ADVAVE computes the advection and horizontal diffusion terms for external mode; XMODE solves the barotropic equations. ADVT computes the horizontal and vertical advection and horizontal diffusion terms needed by PROFT in calculating Θ. ADVT is also called to compute these terms for use by PROFS in calculating S. PROFT and PROFS differ by virtue of the presence of solar insolation terms in the former. Both have provisions for prescribing surface fluxes or surface values or damping to the latter. The bottom boundary conditions are zero flux conditions, but it is also possible to prescribe bottom values. For shallow mixed layers, PROFT can account for the solar insolation absorbed by the bottom and the consequent heat flux to the water column from the bottom. ADVU provides the advection, pressure gradient, Coriolis, and horizontal diffusion terms needed by PROFU in calculating U_1, which is similar to PROFS except for matching to the log law at the bottom. ADVV and PROFV are similar routines but applied to the U_2 component. ADVQ supplies the advection and horizontal diffusion terms needed by PROFQ in calculating q^2 and q^2l. PROFQ embodies turbulence closure and is the most resource-intensive routine of them all in terms of memory and computer time. VERTVL computes the vertical velocity.

Routine BCOND applies conditions on the lateral open boundaries. If the boundary is closed, the masking variables FSM, DUM, and DVM automatically zero out values on land, and because of the C grid, land-ocean fluxes are also automatically set to zero. In this case all BCOND does is masking. If the boundary is open, either the transport normal to the boundary or the sea surface elevation has to be prescribed for the external mode, and the internal mode velocity normal to the boundary, Θ, S, q^2, and q^2l must be prescribed for the internal mode. For time-dependent forcing, the values of surface wind stresses, heat, and salt fluxes are also calculated in BCOND. It is the prescription of the lateral boundary conditions on velocities that is quite demanding. Possible options include radiation boundary conditions, cyclic conditions (for certain simulations), clamped boundary conditions, and so on.

Finally, considerable preprocessing is required to prepare the needed inputs, such as topography H, initial Θ, and S fields, the latter derived from hydrographic archives and interpolated to sigma levels. Surface forcing fields for wind stress, heat, and salt fluxes, as well as SW radiative fluxes, are needed at whatever time intervals permitted by the data, of either a synoptic or a climatological nature. If curvilinear coordinates are used, routines to generate an optimum grid and interpolate the Θ and S fields as well as the surface forcing fields to the model grid points are needed. Postprocessing routines to plot model output in curvilinear coordinates are also essential. Overall, realistic 3-D ocean models such as the one described above are quite effort and resource-intensive.

The code described above runs efficiently on coarse-grained multiple-CPU vector processors of the 1990's, but must be optimized for maximum efficiency on modern massively parallel MIMD machines, such as the CRAY T3E. Massively parallel versions employing high performance Fortran and message passing instructions to coordinate different processors have now been implemented and will be the modeling workhorses in the coming decade.

Chapter 10

Multilevel Basin Scale and Global Models

In this chapter, we will discuss ocean models designed for simulation of basin-scale and global circulation. We will deal principally with models that have had considerable history and user experience around the world. This means inevitably those that have been developed in North America. However, new ocean models have been formulated and are being increasingly used in other parts of the world. One such model is the LODYC finite-difference primitive-equation model developed in France (Delecluse *et al.,* 1993) and applied extensively to investigations of the circulation in the Western Mediterranean Sea (Madec *et al.,* 1991a,b) as well as some global (Madec and Imbard, 1995) and climate (Cassou *et al.,* 1998) studies. The model is similar to the GFDL model discussed in this chapter, in the sense that it is a rigid lid, level model. It uses Arakawa C grid, however, and is formulated in orthogonal curvilinear coordinates. Biharmonic horizontal eddy diffusivity parameterization is used; vertical mixing is based on TKE formulation (Blanke and Delacluse, 1993). The most recent application is by Herbaut *et al.* (1996, 1997), who used the hydrostatic, z-level formulation to study the sensitivity of the Western Mediterranean circulation to atmospheric and other forcing. The LODYC model has also been coupled to a global atmosphere model for climate studies (Cassou *et al.,* 1998).

Another is the Hamburg Ocean Primitive Equation (HOPE) model (Wolff and Maier-Reimer, 1997). It uses hydrostatic and Boussinesq approximations and includes a simple mixed layer. It incorporates a prognostic free surface, uses Arakawa E grid, and is coupled to a comprehensive dynamical (Hibler, 1979) and thermodynamic (Owens and Lemke, 1990) sea-ice model. Kim *et al.* (1998) have applied this model to look at changes in Antarctic climate during the last interglacial period. Coupled to an atmospheric GCM, it has been used in investigations of decadal variability (Latif, 1998; Xu *et al.,* 1998,) and in climate simulations (Schneider, 1998).

The Bryan-Cox-Semtner (GFDL) z-level model (Bryan, 1969; Cox, 1985; Semtner and Chervin, 1992) is the oldest and the most popular global ocean model. Several versions exist, including the Modular Ocean Model (MOM, Pacanowski *et al.,* 1991) from the Geophysical Fluid Dynamics Laboratory (GFDL) in Princeton, New Jersey, the latest version of which (MOM2, Pacanowski, 1995) includes many additional features and options. A version optimized for massively parallel processors called POP (Parallel Ocean Program) has been developed at Los Alamos National Laboratories, Las Alamos, New Mexico (Smith *et al.,* 1992). Other recent developments include inclusion of an explicit (Killworth *et al.,* 1991) and an implicit free surface treatment (Dukowicz and Smith, 1994) in some versions, and the adoption of a C grid. The model has also been reformulated in a general vertical coordinate system that includes the z and σ-coordinate systems as its two limits (Gerdes, 1993). A recent review of the current state of ocean modeling using z-level models, illustrating their current capabilities, can be found in Semtner (1995). A historical perspective can be found in Semtner (1986a, 1995). Discussions of the design philosophy for various features implemented into the original model can be found in Semtner (1986a) and Bryan (1989). McWilliams (1996) has reviewed basin and global scale oceanic general circulation models, concentrating on multilevel and isopycnal models in particular. He discusses their formulations, parameterizations, and the numerical techniques they employ, and attempts to assess how well these models depict the wind and buoyancy-driven circulations in the ocean basins.

10.1 INTRODUCTION

Currents in the oceans are driven by the momentum, heat, and moisture fluxes at the surface from the atmosphere and are greatly affected by the shape and depths of the basins and their interconnections, the Earth's rotation, and the stable stratification that exists in the water column. The oceans are also turbulent. Kirk Bryan was one of the few who realized the necessity for a numerical approach to deal with the complexity of oceanic processes quite early in the history of modern computers. In the late 1960s (Bryan, 1969) and early 1970s, he and his colleagues at GFDL developed a primitive equation-based numerical ocean model, governed by the conservation of mass, momentum, heat, and salt, to deal with these complexities. However, the oceanic circulation consists of swift but narrow currents, such as the Gulf Stream and the Kuroshio, and ubiquitous mesoscale eddies. The width of these currents and the size of these eddies scale with the internal Rossby radius of deformation, which ranges from 5 to 10 km in the subpolar seas to about 40 km in midlatitudes and about 100 to 250 km in the tropics. Most of the kinetic energy in the oceans is in mesoscale variability. Therefore, it is essential to resolve these small features, and it was not until the early 1990s, when computing resources available began to match these requirements, that numerical ocean models began to be more and more realistic (Semtner, 1995). Today, the availability of multigigaflop high performance computers has enabled nearly realistic simulations of the global oceans, and the teraflop computers that will be available in the coming millennium will enable multicentury simulations at realistic resolutions to be carried out. It is very likely that GFDL type models will be at the forefront of such advances. There already exist simulations with GFDL-type models of the Atlantic Ocean (Bryan, Boning, and Holland, 1995), the Southern Ocean (FRAM Group, 1991), and the global ocean (Semtner and Chervin, 1992, Semtner, 1995) that reproduce many of the observed oceanic features. MOM2 (Pacanowski, 1995), the modular version of the original GFDL model, which has over three decades of research and experience behind it, is the most popular and heavily used ocean model at present.

The original GFDL model used a rigid lid for computational efficiency. This method is still useful for long-term climatic simulations. However, this formulation carries no specific advantages when applied to ocean simulations with rapidly changing surface forcing. Also the solution of the barotropic component of the circulation involves the computation of the stream function for the vertically averaged flow (Semtner, 1986b), in which complicated ocean configurations such as multiple islands are hard to implement. The implicit free surface version developed recently at Los Alamos (Dukowicz and Smith, 1994) specifically for massively parallel processing machines, POP, not only circumvents the difficulties of prescribing arbitrary topography but actually speeds up the model by a factor of four (Semtner, 1995). Global simulations have been performed recently with this version at resolution of $1/4°$ to $1/6°$ (Smith *et al.,* 1997). While some problems (such as improper separation latitudes for western boundary currents such as the Gulf Stream) persist, overall, great progress has been achieved with GFDL-type multi-level global models.

10.2 GOVERNING EQUATIONS

For global and basin scale modeling, it is natural to consider the governing equations of motion in a spherical coordinate system fixed to the Earth and hence rotating with it. While the treatment of the Arctic is difficult in this coordinate system, because of the convergence of the meridians at the North Pole, this is not a problem for those models that do not include sea-ice cover. General orthogonal curvilinear coordinates with a specific choice of the coordinate system can overcome this difficulty to some extent (see Chapter 2). The clearest advantage is that unlike the beta plane approximation, the spherical coordinate system allows a more accurate treatment of the effect of the curvature of the Earth's surface, in the sense that the variation of β with latitude and the resulting beta dispersion of Rossby waves can be included. In this section, before describing the z-level model itself, we will show how the simplified equations used in these global and basin scale ocean models result from the more general, complex set of equations governing the motions of a fluid on a rotating Earth. While these aspects were dealt with in Chapter 1 in rather general terms, it is instructive to do so specifically in spherical coordinates applicable to global models. This is best done starting from the inviscid form of the equations, neglecting the viscous terms (or more appropriately the Reynolds or turbulent flux terms that result from statistical averaging of turbulent motions). As shown in Chapter 1, the incompressible, Boussinesq equations for a gravitationally stratified fluid on a rotating sphere are adequate to describe most geophysical flows. These can be written as (Charney and Flierl, 1981)

$$\frac{1}{(R+z)\cos\theta}\frac{\partial u}{\partial\phi}+\frac{1}{(R+z)\cos\theta}\frac{\partial}{\partial\theta}(v\cos\theta)+\frac{1}{(R+z)^2}\frac{\partial}{\partial z}\left[(R+z)^2 w\right]=0$$

$$\frac{du}{dt}-\frac{uv\tan\theta}{(R+z)}-2\Omega v\sin\theta+2\Omega w\cos\theta+\frac{uw}{(R+z)}=-\frac{1}{(R+z)\cos\theta}\frac{\partial p}{\partial\phi}$$

$$+\frac{\partial}{\partial z}\left(K_M\frac{\partial u}{\partial z}\right)+A_M\left[\nabla^2 u+\frac{\left(1-\tan^2\theta\right)u}{(R+z)^2}-\frac{2\sin\theta}{(R+z)^2\cos^2\theta}\frac{\partial v}{\partial\phi}\right]$$

$$\frac{dv}{dt}+\frac{u^2\tan\theta}{(R+z)}+2\Omega u\sin\theta+\frac{vw}{(R+z)}=-\frac{1}{(R+z)}\frac{\partial p}{\partial\theta}+\frac{\partial}{\partial z}\left(K_M\frac{\partial v}{\partial z}\right)$$

$$+A_M\left[\nabla^2 v+\frac{\left(1-\tan^2\theta\right)v}{(R+z)^2}+\frac{2\sin\theta}{(R+z)^2\cos^2\theta}\frac{\partial u}{\partial\phi}\right]$$

$$\frac{dw}{dt}-\frac{u^2}{(R+z)}-\frac{v^2}{(R+z)}+2\Omega u\cos\theta=-\frac{\partial p}{\partial z}-\frac{\rho}{\rho_0}g+\frac{\partial}{\partial z}\left(K_M\frac{\partial w}{\partial z}\right)$$

$$+A_M\left[\nabla^2 w+\frac{\left(1-\tan^2\theta\right)v}{(R+z)^2}+\frac{2\sin\theta}{(R+z)^2\cos^2\theta}\frac{\partial u}{\partial\phi}\right]$$

where

$$\frac{d}{dt}(X) \equiv \frac{\partial X}{\partial t} + \frac{1}{(R+z)\cos\theta}\frac{\partial}{\partial\phi}(uX) + \frac{1}{(R+z)\cos\theta}\frac{\partial}{\partial\theta}(v\cos\theta X) + \frac{\partial}{\partial z}(wX)$$

$$\nabla^2 X \equiv \frac{1}{(R+z)^2\cos^2\theta}\frac{\partial^2 X}{\partial\phi^2} + \frac{1}{(R+z)^2\cos\theta}\frac{\partial}{\partial\theta}\left(\cos\theta\frac{\partial X}{\partial\theta}\right)$$

$$(10.2.1)$$

along with the equation of state and conservation relations for the quantities involved in the equation of state, the potential temperature Θ (temperature attained by a fluid parcel moved isentropically and at constant salinity to a reference pressure), and salinity S:

$$\rho = \rho(\Theta,S,p)$$

$$\frac{d}{dt}(\Theta) = \frac{1}{\rho_0 c_p}\frac{\partial I_s}{\partial z} + \frac{\partial}{\partial z}\left(K_H\frac{\partial}{\partial z}(\Theta)\right) + A_H\nabla^2(\Theta) \qquad (10.2.2)$$

$$\frac{d}{dt}(S) = \frac{\partial}{\partial z}\left(K_H\frac{\partial}{\partial z}(S)\right) + A_H\nabla^2(S)$$

where I_S is the solar insolation. Note that the equations have been written in modified spherical coordinates, where the radial coordinate is replaced by R+z, and R is the mean radius of the Earth quantity. ϕ is the longitude, θ is the latitude, and z is positive vertically up; u, v, and w are velocity components in the zonal, meridional, and vertical directions; p is the kinematic pressure, pressure normalized by average density ρ_0; ρ is the density, and g is the gravitational acceleration; Ω is the Earth's rotation rate (in radians per second). The advection operator is in flux-conservative form. Note the appearance of the terms in the momentum equations involving Ω; these are fictitious acceleration terms and arise from the fact that a rotating frame of reference is not inertial. Therefore unaccelerated straight-line motion must appear curved in a rotating reference frame and hence fictitious forces have to be included in the equations of motion. These are the Coriolis acceleration terms named after the Frenchman Gaspard Gustavede Coriolis who discovered them in 1831. Note also that the frictional terms resulting from Reynolds stresses due to turbulent motions have been separated into horizontal mixing (diffusion) terms involving horizontal diffusion coefficients A_M and A_H, and vertical mixing (diffusion) terms involving vertical mixing (diffusion) coefficients K_M and K_H.

Boussinesq equations are the basis of most ocean circulation models (Bryan, 1969; Kantha and Piacsek, 1997; Semtner, 1986a, b 1974, 1995). These models conserve volume rather than mass. The sea level obtained by some of these models that have a free surface are based on vertical integration of equation for conservation of volume and hence do not account for the sea level changes brought on by expansion and contraction of the water column by heating and cooling, respectively. While the equation of state includes such expansion and contraction, they are not taken into account in the dynamical equations and hence steric sea level changes due to seasonal heating/cooling and global warming are not properly represented. Greatbatch (1994) has shown that this does not represent a serious error, since model-calculated sea level can be adjusted by a spatially uniform, time-varying factor, determined by the net expansion/contraction of the global oceans. Since the correction is spatially uniform, it creates no currents and hence has no dynamical significance. Increase in oceanic mass due to any net fresh water input can be similarly treated. Boussinesq approximation needs to be relaxed if one is interested in problems such as sea level changes brought on by global warming and a full set of flow equations involving variable density needs to be considered. While this is easy to do and Mellor and Ezer (1995) have constructed a non-Boussinesq ocean model, this additional complication and computational expense are unnecessary, since the results from a Boussinesq model are easily corrected for non-Boussinesq effects. Results from models that use a rigid lid and hence calculate only the pressure (equivalently sea level) gradients can also be suitably corrected (see Greatbatch, 1994).

These equations are still more complex than needed for most large-scale circulation models. They can be simplified by making use of the fact that the oceans are very thin compared to their horizontal scales. Let the length scale be L and velocity scale U in the zonal and meridional directions, with corresponding values H and W in the vertical direction. There are two geometric parameters in the problem:

$$\Delta = L / (R \tan \theta); \quad \delta = H / L \qquad (10.2.3)$$

The first refers to the spherical geometry; when this parameter is small it is possible to approximate the spherical surface by its tangent plane, the so-called beta plane approximation. The second one refers to the aspect ratio of the fluid. Since continuity equation demands WL~UH, W/U must be O(δ). For large scale oceanic motions, this parameter is small and therefore it is possible to expand the governing equations in terms of δ and keeping only the zeroth order terms (thin shell approximation) to derive the following hydrostatic form of the governing equations written in a slightly different form (Semtner, 1986b),

$$\Re (1) = 0$$

$$\frac{\partial u}{\partial t} + \Re (u) - \frac{uv \tan \theta}{R} - fv = -\frac{1}{R \cos \theta} \frac{\partial p}{\partial \phi} + \frac{\partial}{\partial z} \left(K_M \frac{\partial u}{\partial z} \right)$$

$$+A_M \left[\bar{\nabla}^2 u + \frac{\left(1-\tan^2 \theta\right)u}{R^2} - \frac{2\sin\theta}{R^2 \cos^2\theta} \frac{\partial v}{\partial\phi} \right]$$

$$\frac{\partial v}{\partial t} + \Re(v) + \frac{u^2 \tan\theta}{R} + fu = -\frac{1}{R}\frac{\partial p}{\partial\theta} + \frac{\partial}{\partial z}\left(K_M \frac{\partial v}{\partial z}\right)$$

$$+A_M \left[\bar{\nabla}^2 v + \frac{\left(1-\tan^2 \theta\right)v}{R^2} + \frac{2\sin\theta}{R^2 \cos^2\theta} \frac{\partial u}{\partial\phi} \right]$$

$$0 = -\frac{\partial p}{\partial z} - \frac{\rho}{\rho_0} g \qquad\qquad (10.2.4)$$

$$\frac{\partial\Theta}{\partial t} + \Re(\Theta) = \frac{1}{\rho_0 c_p}\frac{\partial I_s}{\partial z} + \frac{\partial}{\partial z}\left(K_H \frac{\partial\Theta}{\partial z}\right) + A_H \bar{\nabla}^2\Theta$$

$$\frac{\partial S}{\partial t} + \Re(S) = \frac{\partial}{\partial z}\left(K_H \frac{\partial S}{\partial z}\right) + A_H \bar{\nabla}^2 S$$

$$\rho = \rho(\Theta, S, p)$$

using the flux-conservative form of the advection operator, and the horizontal Laplacian operator,

$$\Re(X) = \frac{1}{R\cos\theta}\left[\frac{\partial}{\partial\phi}(uX) + \frac{\partial}{\partial\theta}(v\cos\theta\, X)\right] + \frac{\partial}{\partial z}(wX)$$

$$\bar{\nabla}^2 X = \frac{1}{R^2 \cos\theta}\left[\frac{1}{\cos\theta}\frac{\partial^2 X}{\partial\phi^2} + \frac{\partial}{\partial\theta}\left(\frac{\partial X}{\partial\theta}\cos\theta\right)\right] \qquad (10.2.5)$$

where $f = 2\Omega\sin\theta$. Note the simplified form the vertical momentum equation takes; hydrostatic balance exists even though the fluid is in motion. This essentially means that the vertical acceleration terms can be ignored compared to the buoyancy terms. Hydrostatic approximation is valid for most large scale motions; exceptions are deep convection in the oceans, where large vertical velocities and accelerations occur when strongly cooled surface water descends in narrow chimney-like plumes (cumulonimbus convection is a similar process in the atmosphere). Also the terms in the zonal and meridional momentum equations involving the vertical velocity are small and have been neglected. Note also the absence of the meridional component of Coriolis acceleration in the equations, required for consistency of energetics when thin shell/hydrostatic approximation is invoked (Section 1.3).

These equations are the starting point for many numerical global OGCMs and basin scale models, including the Bryan-Cox-Semtner type z-level models, and MOM2. Some versions of MOM also include conservation equations for the turbulence velocity and length scales (Rosati and Miyakoda, 1988) for computing the vertical mixing coefficients based on a two-equation model of small scale turbulence (Mellor and Yamada, 1982):

$$\frac{\partial}{\partial t}\left(q^2\right) + \Re\left(q^2\right) = \frac{\partial}{\partial z}\left(K_q \frac{\partial}{\partial z}\left(q^2\right)\right) + A_H \bar{\nabla}^2\left(q^2\right) + S_P + S_B - S_D$$

$$\frac{\partial}{\partial t}\left(q^2\ell\right) + \Re\left(q^2\ell\right) = \frac{\partial}{\partial z}\left(K_q \frac{\partial}{\partial z}\left(q^2\ell\right)\right) + A_H \bar{\nabla}^2\left(q^2\ell\right) + \bar{S}_P + \bar{S}_B - \bar{S}_D$$

$$(10.2.6)$$

where S_P, S_B, and S_D are the source/sink terms in the equation for q^2 are the shear production, buoyancy production (destruction), and dissipation terms, with similar terms in the q^2l equation. The vertical mixing coefficients are given by algebraic relations involving q, l, and gravitational stability factors. The form for these and the source/sink terms can be found in Chapter 9 (see also Rosati and Miyakoda, 1988) and will not be repeated here.

Note the linear form of the horizontal friction terms in the above equations. These terms are necessary in numerical models involving nonlinear advection terms to dissipate the energy that piles up at the grid scale and are designed to represent the subgrid scale mixing and energy dissipation that would act to suppress this catastrophe (were the subgrid scales were to be included in the model). While these are the most popular form, nonlinear mean deformation rate-dependent forms based on the Smagorinsky formulation (Smagorinsky, 1963) are also often used (Rosati and Miyakoda, 1988, Chapter 9). In this case, the horizontal frictional terms in Eqs. (10.4) and (10.8) are replaced by

$$F_\phi = \frac{1}{R\cos\theta}\left[\frac{\partial}{\partial\phi}\left(\tau_{\phi\phi}\right) + \frac{1}{\cos\theta}\frac{\partial}{\partial\theta}\left(\tau_{\phi\theta}\cos^2\theta\right)\right]$$

$$F_\theta = \frac{1}{R\cos\theta}\left[\frac{\partial}{\partial\phi}\left(\tau_{\theta\phi}\right) + \frac{\partial}{\partial\theta}\left(\tau_{\theta\theta}\cos\theta\right) + \tau_{\phi\phi}\sin\theta\right] \qquad (10.2.7)$$

$$F_\chi = \frac{1}{R\cos\theta}\left[\frac{\partial}{\partial\phi}\left(q_\phi\right) + \frac{\partial}{\partial\theta}\left(q_\theta\right)\right]$$

in the zonal (u) and meridional (v) momentum and all scalar (Θ, S, q^2, q^2l) equations. Quantities $\tau_{\phi\phi}$, $\tau_{\phi\theta}$, $\tau_{\theta\phi}$, and $\tau_{\theta\theta}$ are components of the stress tensor, and q_ϕ and q_θ are the components of the flux vector of that particular scalar quantity given by (Rosati and Miyakoda, 1988),

$$\tau_{\phi\phi} = A_{M\phi}D_T \,;\, \tau_{\phi\theta} = A_{M\theta}D_S \,;\, \tau_{\theta\phi} = A_{M\phi}D_S \,;\, \tau_{\theta\theta} = -A_{M\theta}D_T$$

$$q_\phi = \frac{Pr_H}{R\cos\theta}A_{M\phi}\frac{\partial\chi}{\partial\phi}\,;\, q_\theta = \frac{Pr_H}{R\cos\theta}A_{M\theta}\frac{\partial\chi}{\partial\theta}$$

$$D_T = \frac{1}{R\cos\theta}\left[\frac{\partial}{\partial\phi}(u) - \cos^2\theta\frac{\partial}{\partial\theta}\left(\frac{v}{\cos\theta}\right)\right]$$

$$D_S = \frac{1}{R\cos\theta}\left[\frac{\partial}{\partial\phi}(v) + \cos^2\theta\frac{\partial}{\partial\theta}\left(\frac{u}{\cos\theta}\right)\right] \qquad (10.2.8)$$

$$A_{M\phi} = \frac{c_s}{2}\left(\frac{R\Delta\phi}{\cos\theta}\right)^2 |D| \,;\, A_{M\theta} = \frac{c_s}{2}(R\Delta\theta)^2 |D|$$

$$D^2 = \frac{1}{2}\left(D_T^2 + D_S^2\right)$$

where D_T and D_S are the strain rates due to tension and shear and D is the total deformation rate, $\Delta\phi$ and $\Delta\theta$ are the numerical grid sizes in the zonal and meridional directions, and Pr_H is the Prandtl number for horizontal turbulent mixing (most often taken as unity). The constant c_s is the Smagorinsky constant. This form of nonlinear horizontal subgrid scale mixing terms has been extensively used in atmospheric modeling, while the linear form is more prevalent in ocean modeling. A combination of the two is also often used.

In most versions of the Bryan-Cox-Semtner model, the vertical mixing is not computed by second moment closure; instead simpler prescriptions are used, including Richardson number-dependent parameterizations of Pacanowski and Philander (1981),

$$K_M = K_{MB} + K_{MS}\left(1 + 5Ri_g\right)^{-2}$$

$$K_H = K_{HB} + K_M\left(1 + 5Ri_g\right)^{-1} \qquad (10.2.9)$$

$$K_{MB} = 10^{-4}\,m^2 s^{-1}\,;\, K_{HB} = 10^{-5}\,m^2 s^{-1}\,;\, K_{MS} = 5 \times 10^{-3}\,m^2 s^{-1}$$

which is appropriate for the thermocline. Applications to the equatorial oceans have proved useful, even though microstructure observations in the equatorial Pacific by Peters *et al.* (1988) suggest much smaller values:

$$K_M = K_{MB} + K_{MS}\left(1 + 5Ri_g\right)^{-1.5}$$

$$K_H = K_{HB} + K_{MS}\left(1 + 5Ri_g\right)^{-2.5} \qquad (10.2.10)$$

$$K_{MB} = 2 \times 10^{-5}\,m^2 s^{-1}\,;\, K_{HB} = 10^{-6}\,m^2 s^{-1}\,;\, K_{MS} = 5 \times 10^{-4}\,m^2 s^{-1}$$

Since the upper mixed layer is but a small fraction of the water column in the deep basins, these simpler approaches (with an imbedded bulk mixed layer at the top) suffice, as long as the mixed layer model does a good job of handling the air-sea exchanges of momentum and other fluxes. However, such para-meterizations assume that the vertical shear is the principal agent of mixing. They are unable to account for purely convective mixing brought on by surface cooling. Convective mixing is handled instead by the so-called convective adjustment. At each time step, the static stability of the water column is examined and if a layer of greater density is found to overlie one of lower density, the two are allowed to mix and reach the same density. The procedure is applied iteratively until neutral stability is restored to the regions where static instability existed. This is equivalent to assuming infinite vertical mixing coefficients and is a reasonable approach, since convective mixing is highly efficient (Marotzke, 1991). However, care must be taken to mix all properties including momentum.

These equations require boundary conditions at the ocean surface (z = 0) and at the bottom (z = -H) and along lateral boundaries. The ocean surface boundary conditions are standard (see Chapter 9) prescription of momentum flux (wind stress), heat, and salt fluxes (and values of q^2 and l), although damping conditions are also often used for temperature and salinity. The conditions at the ocean bottom are of zero heat and salt fluxes and flow parallel to the bottom, with the momentum flux parameterized by a simple quadratic drag law. At closed lateral boundaries, zero slip and heat and salt flux conditions are imposed. Open boundaries require special treatment (see Chapter 9).

$$K_M\left(\frac{\partial u}{\partial z}, \frac{\partial v}{\partial z}\right) = \left(\tau_{w\phi}, \tau_{w\theta}\right)$$

$$K_H\left(\frac{\partial \Theta}{\partial z}, \frac{\partial S}{\partial z}\right) = \left(q_H, q_S\right) \qquad \text{at } z=0$$

$$q^2 = B_1^{2/3} u_*^2 ; \ell = 0$$

$$w = 0$$

$$K_M\left(\frac{\partial u}{\partial z}, \frac{\partial v}{\partial z}\right) = \frac{1}{2} c_d \left(u_b^2 + v_b^2\right)^{1/2} \left(u_b, v_b\right); \ c_d = 0.0025$$

$$K_H\left(\frac{\partial \Theta}{\partial z}, \frac{\partial S}{\partial z}\right) = 0,0 \qquad \text{at } z=-H$$

$$q^2 = B_1^{2/3} u_{*b}^2 ; \ell = 0$$

$$w = -\frac{1}{R}\left(\frac{u_b}{\cos\theta}\frac{\partial H}{\partial\phi}, v_b\frac{\partial H}{\partial\theta}\right)$$

$$u = v = \frac{\partial\Theta}{\partial n} = \frac{\partial S}{\partial n} = 0 \qquad\qquad \text{at lateral boundaries}$$

$$(10.2.11)$$

where n denotes the direction normal to the boundary.

Most versions use a quadratic drag law for bottom friction. However, Killworth (1997) has recently proposed a bulk benthic boundary layer model that complements the use of bulk mixed layer model at the surface (for example, Large *et al.,* 1994).

The one salient feature in most z-level models is the imposition of a rigid lid by requiring the vertical velocity w to be identically zero at the free surface. This artifact filters out surface gravity waves, not crucial to long-term oceanic circulation dynamics. For models forced by slowly varying climatological-type forcing, this results in considerable saving in computing resources, since large time steps consistent with the internal dynamics can be taken. However, this also leads to the necessity to solve an elliptical Poisson equation that complicates matters in case of the real ocean with its complex shape and numerous islandic features. These solvers are notoriously difficult to vectorize and parallelize. Also the iterative solvers require a large number of iterations to converge when the surface forcing is rapidly changing and the savings realized by imposing a rigid lid diminish. For all these reasons and because free surface dynamics are often central to some studies, these equations are now being solved often with explicit free surface dynamics, requiring other numerical solution techniques, such as that outlined in Chapter 9.

With the imposition of the rigid lid, the free surface height is replaced as a variable in the problem by the pressure against the rigid lid p_S, which is related to the pressure p by

$$p(z) = p_S + \int_z^0 \rho g dz' \qquad\qquad (10.2.12)$$

Terms involving p in the momentum equation can be eliminated by taking the vertical derivative of the u and v equations, interchanging the order of derivatives in the pressure term, and substituting for $\partial p/\partial z$ from the hydrostatic equation:

$$\frac{\partial}{\partial t}\left(\frac{\partial u}{\partial z}\right) - f\frac{\partial v}{\partial z} = -\frac{g}{R\cos\theta}\frac{\partial}{\partial\phi}\left(\frac{\rho}{\rho_0}\right) + \frac{\partial}{\partial z}(\text{O.T.})_\phi$$

$$\frac{\partial}{\partial t}\left(\frac{\partial v}{\partial z}\right) + f\frac{\partial u}{\partial z} = -\frac{g}{R}\frac{\partial}{\partial\theta}\left(\frac{\rho}{\rho_0}\right) + \frac{\partial}{\partial z}(\text{O.T.})_\theta$$

$$(10.2.13)$$

where O.T. stands for other terms in the momentum equations. These equations can predict the vertical shear of the flow and are in fact generalizations of the thermal wind relationship (see Chapter 1). However, the prediction of the velocities requires prediction of their vertical averages or equivalently the vertically integrated transport in the water column. This can be done by deriving an equation for the stream function of the vertically integrated transport. Because of the rigid lid condition imposed, integration of the continuity equation leads to

$$\frac{\partial U}{\partial \phi} + \frac{\partial}{\partial \theta}(V \cos \theta) = 0$$

$$U = \int_{-H}^{0} u dz \; ; V = \int_{-H}^{0} v dz \qquad (10.2.14)$$

from which a stream function ψ can be defined

$$UH = -\frac{1}{a}\frac{\partial \psi}{\partial \theta}; VH = -\frac{1}{a \cos \theta}\frac{\partial \psi}{\partial \phi} \qquad (10.2.15)$$

The momentum equations can also be integrated in the vertical and equations involving the stream function derived:

$$-\frac{1}{HR}\frac{\partial}{\partial \theta}\left(\frac{\partial \psi}{\partial t}\right) - \frac{f}{HR \cos \theta}\left(\frac{\partial \psi}{\partial \phi}\right) = -\frac{1}{R \cos \theta}\frac{\partial p_s}{\partial \phi} + \tau_{w\phi}$$

$$-\frac{g}{HR \cos \theta}\int_{-H}^{0} dz \int_{z}^{0}\frac{\partial}{\partial \phi}\left(\frac{\rho}{\rho_0}\right)dz' + \frac{1}{H}\int_{-H}^{0}(O.T.)_\phi \, dz$$

$$\frac{1}{HR \cos \theta}\frac{\partial}{\partial \phi}\left(\frac{\partial \psi}{\partial t}\right) - \frac{f}{HR}\left(\frac{\partial \psi}{\partial \theta}\right) = -\frac{1}{R}\frac{\partial p_s}{\partial \theta} + \tau_{w\theta} \qquad (10.2.16)$$

$$-\frac{g}{HR}\int_{-H}^{0} dz \int_{z}^{0}\frac{\partial}{\partial \theta}\left(\frac{\rho}{\rho_0}\right)dz' + \frac{1}{H}\int_{-H}^{0}(O.T.)_\theta \, dz$$

However, these involve the unknown p_s. One solution is to eliminate it by cross differentiation of the two equations above (thereby deriving an equation for the vorticity):

$$\left[\frac{\partial}{\partial \phi}\left(\frac{1}{H \cos \theta}\frac{\partial^2 \psi}{\partial \phi \partial t}\right) + \frac{\partial}{\partial \theta}\left(\frac{\cos \theta}{H}\frac{\partial^2 \psi}{\partial \theta \partial t}\right)\right] -$$

$$
\left[\frac{\partial}{\partial \phi} \left(\frac{f}{H} \frac{\partial \psi}{\partial \theta} \right) - \frac{\partial}{\partial \theta} \left(\frac{f}{H} \frac{\partial \psi}{\partial \phi} \right) \right] = R \left[\frac{\partial}{\partial \phi} (\tau_{w\theta}) - \frac{\partial}{\partial \theta} (\tau_{w\phi} \cos \theta) \right]
$$

$$
- \left\{ \frac{\partial}{\partial \phi} \left[\frac{g}{H} \int_{-H}^{0} dz \int_{z}^{0} \frac{\partial}{\partial \theta} \left(\frac{\rho}{\rho_0} \right) dz' \right] - \frac{\partial}{\partial \theta} \left[\frac{g}{H} \int_{-H}^{0} dz \int_{z}^{0} \frac{\partial}{\partial \phi} \left(\frac{\rho}{\rho_0} \right) dz' \right] \right\} +
$$

$$
+ \left\{ \frac{\partial}{\partial \phi} \left[\frac{R}{H} \int_{-H}^{0} (\text{O.T.})_\theta \ dz \right] - \frac{\partial}{\partial \theta} \left[\frac{R \cos \theta}{H} \int_{-H}^{0} (\text{O.T.})_\phi \ dz \right] \right\}
$$

$$
(10.2.17)
$$

The vorticity-stream function formulation is also quite popular in CFD in solutions of full Navier Stokes equations. The approach lends itself to discretization schemes that conserve enstrophy in addition to energy (see Chapter 2). The elliptic equation, Eq. (10.2.17), needs to be solved at each time step subject to conditions imposed on lateral ocean boundaries (which are in general multiply connected, with numerous islands) to obtain the stream function and hence the vertically averaged velocities. Herein lies the principal difficulty with rigid-lid models. While they are efficient, the solution technique is more complicated and not easily adapted to vector and parallel processors. The method is, however, well suited to long-term climatic simulations. When oceanic conditions change more rapidly, as, for example, under synoptic wind forcing, the convergence of the iterative solvers may slow down considerably and the model may become comparable to free surface formulations in terms of computational efficiency.

Islands require careful treatment. The value of the stream function around an island is not known apriori, and must be evaluated at each time step. This is done by requiring the integral of p_S around the island to vanish (Semtner, 1986b). The need to choose a primary continent (such as Eurasia and the Americas) and treat the rest of the land masses as a series of "islands" involves compromises in the land-ocean configuration as well as the precise algorithm used in the elliptic solver. An excellent example is the recent global simulation of Semtner and Chervin (1992), which uses an artificially large interconnection between the Pacific and the Indian Oceans through the Indonesian archipelago. What impact such compromises have on the resulting circulation is hard to assess in the absence of detailed observations in the global oceans.

As mentioned earlier, an alternative approach is not to take the curl of the momentum equations to derive the stream function but to take the divergence to obtain an elliptic equation for the surface pressure p_s on the rigid lid instead (Dukowicz *et al.*, 1993). The main advantage is that the boundary conditions now become Neumann conditions, not Dirichlet, thus avoiding having to solve the island integrals. However, the convergence of the iterative scheme slows down considerably. An implicit free surface approach (Dukowicz and Smith, 1994), in which the rigid lid assumption is replaced by a momentumless free surface layer, leads to considerable acceleration of the convergence rate when solving for the pressure against the rigid lid. Incorporation of an explicit free surface would be an alternative choice but has not yet been implemented in MOM2 and validated (but see Killworth *et al.*, 1991).

Bryan (1969) originally chose the Arakawa B grid for finite differencing the above equations (the model uses flux-conservative form of the governing equations and the numerical scheme therefore conserves various properties such as mass, heat, salt, and energy; see Pacanowski, 1995, and Semtner, 1986b). As long as the computing resources did not permit grid resolutions to fall below the Rossby radius, this was a good choice. With global models now approaching $1/6°$ resolution and beyond (Semtner, 1995), a C grid might be a better choice, since it would lead to less distortion of the wave propagation characteristics on the numerical grid (see Chapter 2). While earlier attempts to convert the model to C grid failed because of the problems associated with the high wave number computational noise in the vertical (Semtner, 1986b), more recent attempts at Los Alamos and elsewhere appear to have succeeded. On the other hand, second-order leap frog has remained the scheme of choice in time differencing, with the occasional use of a backward/forward Euler scheme to prevent solution splitting at alternate time steps.

Ever since the GFDL researchers made the Bryan-Cox-Semtner model modular (modular ocean models MOM and more recently MOM2) and readily available via anonymous ftp (ftp.gfdl.gov, directory pub/GFDL_MOM2), its use has increased exponentially and there are now literally several hundred users around the world. An excellent user's guide and reference manual has been prepared (Pacanowski, 1995) that explains the intricate details of the model, including its numerical implementation and the various options available (some details related to finite-differencing and numerics can also be found in Semtner, 1986b). The model itself is continually and rapidly evolving, and even the originators have difficulty keeping up with all the latest embellishments. We will not therefore undertake the superfluous and futile task of documenting this highly popular model; instead we refer the reader to Pacanowski (1995). We will, however, mention one particular development, the isopycnal mixing scheme, that is of considerable importance in studies of water mass formations and tracer studies in the global oceans and their climatic implications.

10.3 ISOPYCNAL DIFFUSION

From a water mass and tracer transport point of view, the most natural coordinate system is the one aligned with the local, instantaneous isopycnal surfaces. This is simply because the diffusion tensor in this coordinate system is diagonal, and because of the suppression of mixing across isopycnal surfaces by gravitational forces in a stably stratified fluid, the ocean interior is nearly adiabatic, in the sense that there is little mixing across density surfaces, and the cross isopycnal diffusion values are *seven* orders of magnitude smaller than the along isopycnal values (Gent and McWilliams, 1990; Pacanowski, 1995; Redi, 1982). Denoting by prime, the tensor in the isopycnal coordinates,

$$A_{ij}^I = A_I \begin{bmatrix} 1 & 0 & 0 \\ 0 & 1 & 0 \\ 0 & 0 & \varepsilon \end{bmatrix}; \varepsilon = \frac{A_D}{A_I} \sim 10^{-7} \tag{10.3.1}$$

The transformation matrix needed to convert this to traditional (x,y,z) coordinates is

$$T_{ij} = \begin{bmatrix} \dfrac{S_y}{S} & \dfrac{S_x}{S}\left(1+S^2\right)^{-1/2} & -S_x\left(1+S^2\right)^{-1/2} \\ -\dfrac{S_x}{S} & \dfrac{S_y}{S}\left(1+S^2\right)^{-1/2} & -S_y\left(1+S^2\right)^{-1/2} \\ 0 & S\left(1+S^2\right)^{-1/2} & \left(1+S^2\right)^{-1/2} \end{bmatrix} \tag{10.3.2}$$

where $S = (S_x, S_y, 0)$; $S_x = -(\partial\rho/\partial x)(\partial\rho/\partial z)^{-1}$, $S_y = -(\partial\rho/\partial y)(\partial\rho/\partial z)^{-1}$. S is the magnitude of the isopycnal slope. The diffusion tensor in the (x, y, z) coordinate system is

$$A_{ij} = \left(A_{ij}^I\right) T_{ij} \left(A_{ij}^I\right)^T \tag{10.3.3}$$

and therefore

$$A_{ij} = \frac{A_I}{\left(1+S^2\right)} \begin{bmatrix} 1+S_y^2+\varepsilon S_x^2 & -(1-\varepsilon)S_x S_y & (1-\varepsilon)S_x \\ -(1-\varepsilon)S_x S_y & 1+S_x^2+\varepsilon S_y^2 & (1-\varepsilon)S_y \\ (1-\varepsilon)S_x & (1-\varepsilon)S_y & \varepsilon+S^2 \end{bmatrix} \tag{10.3.4}$$

Normally, in the deep oceans away from the mixed layers, the isopycnal slopes are small. For S << 1, the small angle approximation, the tensor simplifies to

(Gent and McWilliams, 1990; Pacanowski, 1995):

$$
A_{ij} = A_I
\begin{bmatrix}
1 & 0 & (1-\varepsilon)S_x \\
0 & 1 & (1-\varepsilon)S_y \\
(1-\varepsilon)S_x & (1-\varepsilon)S_y & \varepsilon + S^2
\end{bmatrix}
\tag{10.3.5}
$$

The small angle approximation does not, however, preserve the diagonal nature of the tensor in isopycnal coordinates as can be seen by transforming A_{ij} in Eq. (10.3.5) back to isopycnal coordinates. But the presence of off-diagonal terms is of minor consequence, and it does preserve the trace (Pacanowski, 1995). It is possible to turn cross isopycnal mixing off totally by choosing $\varepsilon = 0$, thus simulating a completely adiabatic deep ocean and this is one of the advantages of the isopycnal mixing formulation.

The fact that this formulation is different in very important respects from the traditional horizontal-vertical mixing parameterization can be seen from transforming the traditional mixing tensor,

$$
\hat{A}_{ij} = A_H
\begin{bmatrix}
1 & 0 & 0 \\
0 & 1 & 0 \\
0 & 0 & \hat{\varepsilon}
\end{bmatrix}
\;; \hat{\varepsilon} = \frac{A_V}{A_H} \sim 10^{-7}
\tag{10.3.6}
$$

to isopycnal coordinates with the result

$$
\hat{A}_{ij}^I = \frac{A_H}{1+S^2}
\begin{bmatrix}
1+S^2 & 0 & 0 \\
0 & 1+\hat{\varepsilon}S^2 & (1+\hat{\varepsilon})S \\
0 & (1+\hat{\varepsilon})S & \hat{\varepsilon}+S^2
\end{bmatrix}
\tag{10.3.7}
$$

Note the presence of off-diagonal terms in the tensor; the error is of the order of the slope. The error in the first two diagonal terms is of the order of the square of the slope and inconsequential. It is the nonzero $\hat{A}_{33}^I$ term that leads to spurious and detrimental diapycnal mixing in regions of nonzero isopycnal slopes and cannot be tolerated in long-term simulations.

Isopycnal mixing scheme in a z-level model consists of computing the isopycnal slopes and using Eq. (10.3.5) to compute the diffusion coefficients before applying them to compute diffusion terms. The implication that isopycnal slopes are small is generally valid, since slopes of more than 10^{-2} are uncommon, except in and near mixed layers and deep convection regions. This assumption is also implicit in isopycnal models (Chapter 11) which use diffusion tensors that are diagonal. The small angle approximation must be discarded in convection regions. Cox (1987) implemented the small slope isopycnal mixing scheme in the original GFDL model (MOM2 has an option for this), and this indeed led to some improvements such as a more realistic thermocline. However, for reasons of numerical stability, there is a need for a small additional horizontal background diffusion. Its value ($\sim$10–100 m^{2}s^{-1}) is typically 10% of the isopycnal diffusivity, but nevertheless leads to spurious diapycnal diffusivities that are larger than the explicitly modeled values for isopycnal slopes ($\sim$10^{-4}) that are quite common in the oceans. These values are much larger than the 10^{-5} m^{2}s^{-1} canonical values observed in the deep ocean, and they cannot be tolerated in long-term climatic studies. In addition, there exists a bound on the isopycnal slope allowable due to time stepping, $S < [\Delta x\ \Delta z/(4A_H\ \Delta t)]^{1/2}$. However, clipping isopycnal slopes artificially to satisfy this constraint leads to undesirably large diapycnal diffusion. Thus while conceptually simple and straightforward to implement numerically, the actual implementation of isopycnal diffusion has been beset with problems and has led to unsatisfactory results. See Gough (1997) and Griffies (1998) for a discussion of the numerical problems associated with implementation of the isopycnal mixing scheme in z-level models.

The need to very nearly eliminate diapycnal diffusion has led to a careful examination of the numerical implementation for the isopycnal mixing schemes (for example, Griffies, 1998, and Griffies *et al.*, 1998). In addition, in the presence of baroclinic mesoscale eddies, the conversion of the available potential energy to eddy kinetic energy leads to a nearly adiabatic stirring mechanism that reduces isopycnal slopes (Gent and McWilliams, 1990). Numerical models that have resolutions too coarse to resolve mesoscale eddies must parameterize this effect through the addition of a divergence-free eddy-induced advective velocity to the usual tracer velocity (Gent *et al.*, 1995). Model simulations with implementation of this stirring mechanism appear to be more realistic. This eddy-induced advective tracer flux is equivalent to a skew-diffusive tracer flux in the presence of antisymmetric components in the mixing tensor (Griffies, 1998). This mixing is nondissipative and reversible, unlike the conventional dissipative, irreversible mixing due to symmetric components of the tensor, and is more appropriately termed stirring. This antisymmetric stirring tensor is given by (Greatbatch, 1998, see also Griffies, 1998, for an excellent description)

$$\tilde{A}_{ij} = A_s \begin{bmatrix} 0 & 0 & -S_x \\ 0 & 0 & -S_y \\ S_x & S_y & 0 \end{bmatrix} \qquad (10.3.8)$$

The so-called skew-diffusive tracer flux brought on by this tensor is aligned precisely parallel to isopycnal surfaces and leads to downgradient horizontal but upgradient vertical diffusive fluxes. There is no diapycnal component. The off-diagonal terms in Eq. (10.3.8) occur in any coordinate system, z or isopycnal, and choosing to use the z-coordinate to represent eddy-induced stirring, it is possible to combine Eq. (10.3.5) and Eq. (10.3.8) to represent both stirring and mixing effects (Griffies, 1998). This results in considerable savings in computing time. It also avoids numerical difficulties associated with implementing Gent and McWilliams (1990) adiabatic stirring in its advective form (Weaver and Eby, 1997) and the need to employ FCT schemes for advection when computing this advective flux. Often the practice is to set $A_I = A_S$, and in this case, when ε can also be neglected in the off-diagonal terms of Eq. (10.3.5), one gets even a simpler form for combined mixing and stirring, for small angle approximation to the mixing tensor:

$$A_{ij} = A_I \begin{bmatrix} 1 & 0 & 0 \\ 0 & 1 & 0 \\ 0 & 0 & \varepsilon + S^2 \end{bmatrix} \tag{10.3.9}$$

Griffies *et al.* (1998) have implemented a new isopycnal diffusion scheme in the GFDL z-coordinate model that ensures down-the-gradient property for isopycnal diffusion (and hence numerical stability without a background diffusion) and no diapycnal diffusion of locally referenced potential density. This appears to overcome the problems found in Cox's (1987) implementation of the isopycnal mixing scheme.

It must be noted, however, that isopycnal surfaces (isosurfaces of potential density) may not always be appropriate from considerations of transport and mixing. This is simply because neutral surfaces (surfaces along which, when a water mass is transported isentropically without any exchange of heat or salinity with its surroundings, it experiences no buoyancy forces) are not in general parallel to isopycnal surfaces (McDougall, 1987b). This point is of particular importance to deep-ocean mixing considerations and preservation of water mass structure. The use of potential density, the density a water mass would acquire if brought isentropically from its depth z to a reference depth z_r, can lead to errors in computation of the local stability of the water column. This can in turn lead to erroneous mixing in the water column. Traditionally, oceanographers and ocean modelers have used potential density $\rho_{pot}(z,z_r) = \rho\ [\theta(z,z_r), S(z), p(z_r)]$ computed using potential temperature at reference level z_r to compute the buoyancy frequency from $N^2 = -(g/\rho_{pot})d(\rho_{pot})/dz$. This is equivalent to using the thermal and salinity expansion coefficients at the reference depth z_r (which is usually the surface). But this is not necessarily indicative of the local stability of the water column. What is needed in stability and mixing considerations is an assessment of whether a fluid particle when displaced isentropically by an infinitesimal distance in the vertical experiences any restoring buoyancy forces. If the density change due to an infinitesimal isentropic displacement Δz exceeds the existing density change in the water column over the same distance, the water column is locally stable. This condition is equivalent to using thermal and salinity expansion coefficients computed at the local depth.

However, an additional complication is that the concept of local stability involves infinitesimal displacements, and during a finite displacement in a water column that is everywhere locally stable, a water parcel might not experience the same restoring buoyancy forces. This point is of particular importance to ocean modeling since water masses are transported over great distances from their formation regions. Here, however, potential density can be used to assess the stability characteristics as long as z and z_r are the starting and final positions of the water parcel. In other words, one needs to use the final depth of the water parcel as reference depth z_r, not an arbitrary predetermined one such as the surface.

One way to properly assess whether the water column is stable or unstable from a mixing point of view in a numerical model (with discretization in the vertical) is to compute *in situ* density, $\rho_{\text{in-situ}}(z) = \rho\,[T(z), S(z), p(z)]$ at each level using the equation of state. Then to assess if a water column between levels z_1 and z_2 is stable or not, the densities ρ_1 and ρ_2 water parcels from depths z_1 and z_2 would attain if moved isentropically to midlevel $(z_1 + z_2)/2$ can be computed. These values can then be used to compute the local value of N. This is preferable to using $\rho_{\text{pot}}(z_1, z_r)$ and $\rho_{\text{pot}}(z_2, z_r)$, as is the traditional practice. The transport of water masses is still through the use of conservation equations for potential temperature and salinity.

The effect of convection is handled by convective adjustment where successive levels are examined for static stability. If instability is detected, it is removed by homogenizing the layer in between. This is done iteratively so that no instability exists in the water column (Bryan, 1969). Another approach preferred by some is to enhance the vertical diffusivity when instability is detected to simulate convection explicitly.

10.4 ARCHITECTURE AND OTHER MODEL FEATURES

The GFDL model originated in an era when fast access memory was expensive and hence computers had limited core memory (a few MW on CRAY 1S). This meant that for meaningful grid resolutions, the model would not fit in core. Therefore an elaborate scheme was developed to store the model arrays on a fast solid-state disk and read slices of the 3-D arrays into memory as and when needed, perform the needed calculations, and store them back onto the disk. This arrangement is essential to carry out calculations at grid resolutions much higher than a totally in-core model would permit, the major penalty being increased code complexity and the idle time the CPU spends waiting for the values to be brought in and written out to the disk. This necessitates fast I/O routines that are specific to a particular machine architecture, which makes porting the code to a different machine difficult. To minimize the idle time, the values must be brought in well before they are needed. The 3-D model arrays are stored as (IM, KB, JM), where IM denotes the zonal extent, KB the vertical extent and JM, the meridional extent, so that they can be sliced vertically along constant latitude lines into zonal bands (IM, KB).

The model arrays at three time levels (n−1, n, and n+1) reside on separate disks D_{n-1}, D_n and D_{n+1}, with all variables for a given latitude residing together and stacked in the order of latitude. The calculation starts at the southernmost latitude. At any time, values at two time levels (n−1, n) of the variables at three adjacent zonal bands (M−1, M, M+1) needed to calculate the value in the middle zonal band M at time level (n+1) are resident in core. This moving window is shifted north one zonal band at a time as the calculations proceed. To do this, the newly calculated (n+1) level values at zonal band M are written out to the D_{n+1} disk; the (n−1) and n values M−1, M, and M+1 bands resident in memory are shifted south by one, overwriting the M−1 and M bands by the values at M, and M+1, and the n and n+1 values for the next zonal band are read from D_{n-1}, D_n disks into memory at M+1 and the next set of calculations are done. This sequence of steps is continued until the northernmost latitude is reached. The scheme for a multitasking parallel computer with several CPUs is even more complex and will not be repeated here (see Pacanowski, 1995, instead). Needless to say that squeezing the maximum efficiency out of modern computers requires considerable ingenuity and GFDL researchers have painstakingly optimized the MOM2 code for various CRAY architectures. Massively parallel versions exist in the United Kingdom (Webb *et al.,* 1991) and at Los Alamos National Laboratory (Dukowicz *et al.,* 1993), the latter being an in-core version.

Simulations with the GFDL model have shown that it underestimates, by a significant amount, the meridional heat transport in the Atlantic and the Pacific Oceans. This prompted studies into tracer transport in non-eddy-resolving ocean models, and Gent and McWilliams (1990) found that the effective advective velocity of a tracer is considerably higher than the actual value. With their formulation that accounts for the eddy-induced transport (see also Danabasagolu *et al.,* 1994; Gent *et al.,* 1995), the meridional transport of heat appears to attain more realistic values, closer to observed ones. MOM2 has provisions for calculating the eddy-induced advective velocities in tracer transport calculations (Pacanowski, 1995).

A good diagnostics package is a valuable component of a modeling system and essential to deciphering the model biases and sensitivities. As a result of decades of work, the GFDL type models have built up a good diagnostics capability that permits quantities like the eddy kinetic energy and other statistics to be computed and compared with available observations.

10.5 APPLICATIONS

In the past three decades since its development, the z-level Bryan-Cox-Semtner model has greatly improved our understanding of how the oceans work and has relieved oceanographers from the tedium of quasi-geostrophic assumption and flat-bottomed, homogeneous, rectangular ocean basins. The model has been used extensively to study the seasonal, interannual and climatic variations in the global oceans and has been run in both data-assimilative and stand-alone modes. It has also been used for process studies, such as midlatitude ocean dynamics. It is also a central part of the ocean analysis system for the tropical oceans, where a best estimate of the state of these oceans is determined by assimilation of observational data into a tropical ocean version of the model (Leetmaa and Ji, 1989) Numerous applications of its various versions can be found in the literature (for example, in the *Journal of Physical Oceanography* and the *Journal of Geophysical Research, Oceans*). An extensive bibliography can be found in the MOM2 user's guide and reference manual (Pacanowski, 1995).

One of the more recent global applications can be found in Semtner and Chervin (1992). The results from a recent Semtner-Chervin version with a $1/4°$ effective resolution were analyzed by Stammer *et al.* (1996). They show that the model reproduces many of the large scale features of the circulation, including the spatial and temporal structures of variability observed in altimetry. Figure 10.5.1 shows a snapshot of the near-surface circulation from their simulation. This figure illustrates the wealth of details of the structure and evolution of the 3-D circulation that even a relatively coarse resolution but comprehensive-physics numerical ocean model can create. Some of these details are substantially correct, but some are highly inaccurate, a result not necessarily of the deficiencies in the model per se but more due to the compromises that were necessary for tractability and to the inevitable inaccuracies in forcing. The highest resolution global z-level model at present is the $1/6°$ resolution POP model at Los Alamos that is run on a 256-processor CM5 (Fu and Smith, 1996), some results from which were presented in Chapter 1.

Extensive simulations in the 1990s by groups around the world have greatly improved our understanding of the sensitivity of this model to various parameters. A persistent problem in this type of model has been that its circulation is sluggish compared to observations. As anticipated, increased horizontal resolutions have led to more energetic and more realistic mean

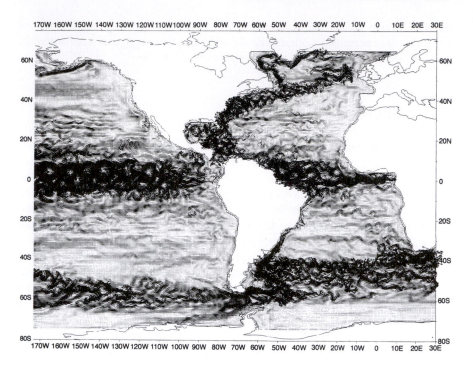

Figure 10.5.1 Surface circulation in the global oceans from the model of Semtner and Chervin (1992). This plot illustrates the wealth of details even a relatively coarse resolution but comprehensive-physics global ocean model can generate, some of which are substantially correct, but some quite inaccurate.

circulation and variability, although the eddy kinetic energy levels are still low, in the 1/4° Semtner-Chervin model, about 1/4 of those indicated by altimetry (Stammer *et al.*, 1996). Comparisons with T/P altimetric data (McClean *et al.*, 1997) have shown that this 1/4° model captures roughly 50% of the altimetric SSH variability, while the higher resolution of the 1/6° POP model increases this to 60%. This is not surprising since observations indicate that the size of the mesoscale eddies scales with the internal Rossby radius, which decreases with latitude. Therefore even the 1/6° resolution achievable at present is woefully inadequate, especially in latitudes higher than about 30°. These models also lack at present a realistic upper mixed layer and are usually driven by synoptic winds, but averaged over a few days. The vertical resolution is also inadequate, about 20 levels at present with about 6 in the upper 200 m, where the variability is high. All these factors tend to produce anemic variability.

Another deficiency has been anemic western boundary currents that do not extend as much as they should into the interior of the basins after their separation from the boundary. Improved horizontal resolution with its concomitant decrease in friction may be needed. Improper separation of the western boundary currents has been a perennial problem in ocean models. The exact cause is still not well known. It is possible that much higher resolutions, both in the horizontal and vertical, are needed to reproduce the JEBAR terms (see Chapter 1) correctly. The decrease in horizontal momentum diffusivity resulting from an increase in horizontal resolution permits the model to be less viscous. Increasing the vertical resolution in the 1/6° POP model from 20 to 37 appears to have been responsible for better simulations of the Gulf Stream separation (Chao *et al.,* 1996). Such high vertical resolutions in concert with a realistic mixed layer model (Kantha and Clayson, 1994; Large *et al.,* 1994) driven by synoptic winds that are not temporally averaged could lead to improved variability as well. More realistic mixed layer models such as Large *et al.* (1994) are increasingly being incorporated into these models (Large and Gent 1999).

McClean *et al.* (1997) made extensive comparisons of Semtner-Chervin 1/4° and POP 1/6° models with T/P altimetric and drifter buoy data, which pretty much illustrate the current status of global ocean modeling. Figures 10.5.2 and 10.5.3 show comparisons of the mesoscale rms SSH and the geostrophic eddy kinetic energy distributions in the global oceans derived from the T/P altimetry and the 1/4° and 1/6° models. While there are glaring deficiencies as detailed above, the agreement that exists is quite remarkable, made possible by the extensive work done with this type of models over the past three decades and the vast increase in available computing resources over the same period.

The GFDL model has also been coupled to atmospheric models for application to coupled ocean-atmosphere simulations of interannual variability (see Chapter 13) and for studies of climate issues such as the impact of anthropogenic CO_2 (Manabe and Stouffer, 1994). More recently, the coupled models have also been used for operational forecasts of ENSO events (Ji *et al.,* 1998).

10.6 HYBRID s-COORDINATE MODELS

In both z and σ-coordinate models, there exists the possibility of spurious diapycnal diffusion resulting from the fact that the diffusion needed for numerical stability of a finite difference scheme is normally implemented along the coordinate surfaces and not along isopycnal surfaces. Since coordinate surfaces seldom coincide with isopycnal surfaces, this causes a component of this diffusion to appear across isopycnal surfaces. This problem can be mitigated in either model by the implementation of isopycnal mixing schemes (Section 10.3) and the use of FCT horizontal advection schemes with very little implicit diffusion (Chapter 2), although the use of isopycnals themselves as coordinate surfaces is a more logical choice from this point of view (see Chapter 11). The principal advantage of a z-coordinate model stems from the fact that the

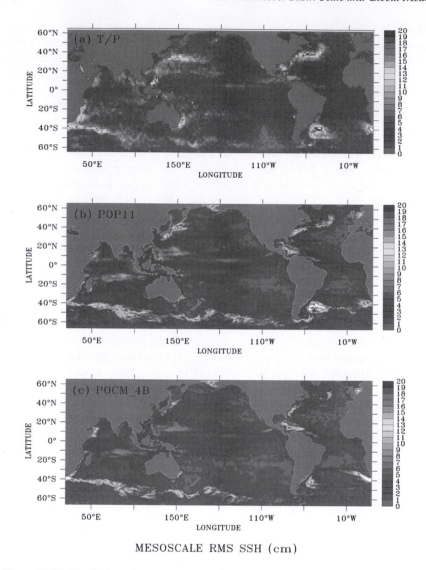

MESOSCALE RMS SSH (cm)

Figure 10.5.2 Distribution of rms SSH from 1/4° Semtner and 1/6° POP model compared to that from T/P altimetry. From McClean *et al.* (1997).

horizontal pressure gradient terms important to the momentum balance can be computed accurately, without the large truncation errors that are possible in regions of strong topographic changes in σ-coordinate models. However, this advantage involves a penalty. In σ-coordinates, the ocean surface and the ocean

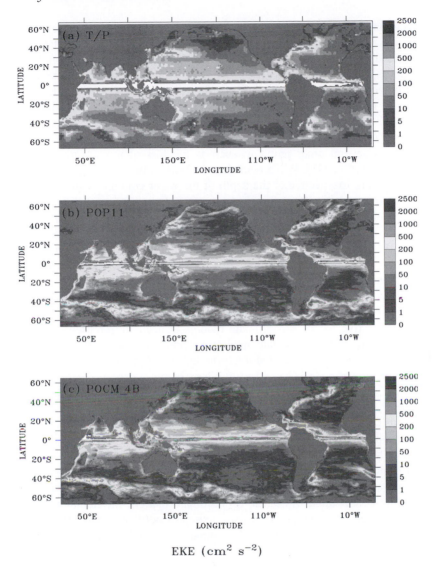

Figure 10.5.3 Distribution of geostrophic eddy KE from 1/4° Semtner and 1/6° POP model compared to that from T/P altimetry. From McClean *et al.* (1997).

bottom are themselves coordinate surfaces, and therefore prescribing the kinematic condition of no mass flow across the bottom is straightforward. The ocean bottom can be represented with accuracy limited only by the grid resolution affordable. In z-coordinates, the ocean bottom cannot be a coordinate surface, unless it is absolutely flat and in fact intersects the model z-levels. This means that the kinematic boundary condition at the piece-wise flat bottom involves truncation errors resulting from discretization of the depth field. This introduces errors in the vertical velocity field, which can be unacceptably large for problems such as the overflow of dense water across a ridge or sill (Gerdes, 1993). In regions of gently sloping bottom, the depth discretization introduces large local depth gradients which in turn cause strong spurious localized vertical motions. Vertical walls form horizontal boundaries in a z-coordinate model, and this causes several other problems such as the loss of momentum by friction (Killworth *et al.*, 1991). While resolving the bottom boundary layer everywhere in the domain, irrespective of the depth of the water column, is simply a matter of providing enough σ-levels near the bottom in a σ-coordinate model, doing so in a z-coordinate model is prohibitive since it requires providing fine vertical resolution over the entire depth range of the model. In regions such as the shelf, where the bottom boundary layer is central to the momentum balance, z-coordinates are inappropriate, unless fine resolution is provided in the top 200 m of the water column, which is not always feasible. This is the primary reason why z-level models have been used primarily for basin-scale and global simulations, where the shallow margins can often be safely ignored.

The attractiveness of σ-coordinate models in accurate representation of the ocean bottom, and z-coordinate models in accurate calculation of horizontal pressure gradients, has motivated the formulation of hybrid s-coordinate models, which permit a σ-coordinate to be employed below a certain depth z_c and a z-coordinate above (Gerdes, 1993; Spall and Robinson, 1990). If it is possible to choose z_c to be below the region of strong prevailing stratification (the permanent thermocline), truncation errors in the pressure gradient calculation due to sloping coordinate surfaces in regions of strong stratification above can be avoided, while simultaneously permitting the bottom to remain a coordinate surface and hence the kinematic condition to be represented accurately below z_c. The pressure gradient errors below z_c are smaller because of the weaker prevailing stratification. As in the σ-coordinate models, the implementation of the hybrid s-coordinate system involves a simple coordinate transformation s = s (x, y, z), which includes z- and σ-coordinates as limiting cases (Bleck, 1978; Gerdes, 1993). The resulting equations look very much similar to the set in Eq. (10.2.4):

$$\Re\,(1) = 0$$

$$\frac{\partial z}{\partial s}\left[\frac{\partial u}{\partial t} - \frac{uv\tan\theta}{R} - fv\right] + \Re\,(u) = -\frac{1}{R\cos\theta}\frac{\partial z}{\partial s}\left(\frac{\partial p}{\partial \phi} + \frac{\rho}{\rho_0}g\frac{\partial z}{\partial \phi}\right) + \frac{\partial}{\partial s}\left[K_M\frac{\partial u}{\partial s}\left(\frac{\partial z}{\partial s}\right)^{-1}\right]$$

$$+\cdots$$

$$\frac{\partial z}{\partial s}\left[\frac{\partial v}{\partial t} + \frac{u^2 \tan \theta}{R} + fu\right] + \Re(v) = -\frac{1}{R}\frac{\partial z}{\partial s}\left(\frac{\partial p}{\partial \theta} + \frac{\rho}{\rho_0}g\frac{\partial z}{\partial \theta}\right) + \frac{\partial}{\partial s}\left[K_M\frac{\partial v}{\partial s}\left(\frac{\partial z}{\partial s}\right)^{-1}\right]$$

$$+ \cdots$$

$$0 = -\frac{\partial p}{\partial s} - \frac{\rho}{\rho_0}g\frac{\partial z}{\partial s}$$

$$\frac{\partial z}{\partial s}\frac{\partial \Theta}{\partial t} + \Re(\Theta) = \frac{1}{\rho_0 c_p}\frac{\partial I_s}{\partial s} + \frac{\partial}{\partial s}\left[K_H\frac{\partial \Theta}{\partial s}\left(\frac{\partial z}{\partial s}\right)^{-1}\right] + \cdots$$

$$\frac{\partial z}{\partial s}\frac{\partial S}{\partial t} + \Re(S) = \frac{\partial}{\partial s}\left[K_H\frac{\partial S}{\partial s}\left(\frac{\partial z}{\partial s}\right)^{-1}\right] + \cdots$$

$$\rho = \rho(\Theta, S, p) \tag{10.6.1}$$

where the advection and the horizontal Laplacian operators are

$$\Re(X) = \frac{1}{R\cos\theta}\left[\frac{\partial}{\partial \phi}\left(u\frac{\partial z}{\partial s}X\right) + \frac{\partial}{\partial \theta}\left(v\frac{\partial z}{\partial s}\cos\theta X\right)\right] + \frac{\partial}{\partial s}(\tilde{w}X)$$

$$\tilde{w} = w + \frac{u}{R\cos\theta}\frac{\partial z}{\partial s}\frac{\partial s}{\partial \phi} + \frac{v}{R}\frac{\partial z}{\partial s}\frac{\partial s}{\partial \theta} \tag{10.6.2}$$

$$\overline{\nabla}^2 X = \frac{1}{R^2\cos\theta}\left[\frac{1}{\cos\theta}\frac{\partial}{\partial \phi}\left(\frac{\partial z}{\partial s}\frac{\partial X}{\partial \phi}\right) + \frac{\partial}{\partial \theta}\left(\frac{\partial z}{\partial s}\frac{\partial X}{\partial \theta}\cos\theta\right)\right]$$

The terms indicating diffusion along coordinate surfaces have been omitted for simplicity.

The boundary conditions become

$$K_M\left(\frac{\partial u}{\partial s}, \frac{\partial v}{\partial s}\right) = \frac{\partial z}{\partial s}(\tau_{w\phi}, \tau_{w\theta})$$

$$K_H\left(\frac{\partial \Theta}{\partial s}, \frac{\partial S}{\partial s}\right) = \frac{\partial z}{\partial s}(q_H, q_s) \qquad\qquad \text{at } z=0$$

$$\tilde{w} = 0$$

$$K_M\left(\frac{\partial u}{\partial s}, \frac{\partial v}{\partial s}\right) = \frac{1}{2}c_d\left(u_b^2 + v_b^2\right)^{1/2}\frac{\partial z}{\partial s}(u_b, v_b); \; c_d = 0.0025$$

$$K_H\left(\frac{\partial\Theta}{\partial s},\frac{\partial S}{\partial s}\right)=(0,0) \qquad\qquad \text{at } z=-H$$

$$\tilde{w}=-\frac{1}{R}\left(\frac{u_b}{\cos\theta}\frac{\partial H}{\partial\phi},v_b\frac{\partial H}{\partial\theta}\right)\frac{\partial z}{\partial s} \qquad\qquad (10.6.3)$$

$$u=v=\frac{\partial\Theta}{\partial n}=\frac{\partial S}{\partial n}=0 \qquad\qquad \text{at lateral boundaries}$$

$\tilde{w}=0$, where the bottom is a coordinate surface. Note that these equations reduce to the forms for z-coordinates when $s=z$ and to σ-coordinates when $s=\sigma$. Defining the gradient operators as in Eq. (3.5.8) and averaging operators as in Eq. (3.5.20) but extended to the s direction as well, the finite-difference equivalents of the advection operator become

$$\Re^v(X)=\frac{1}{R\cos\theta}\left[\delta_\phi\left(\langle u\rangle^\phi\,\delta_s\langle z\rangle^{\phi\theta}\langle X\rangle^\phi\right)+\delta_\theta\left(\langle v\rangle^\theta\,\delta_s\langle z\rangle^{\phi\theta}\cos\theta\langle X\rangle^\theta\right)\right]+\delta_s\left(\tilde{w}\langle X\rangle^s\right)$$

$$(10.6.4)$$

where $\tilde{w}$ is computed from $\Re^v(1)=0$ and the kinematic boundary condition at the surface; the superscript indicates evaluation at velocity grid points (Figure 10.6.1). The momentum equations become

$$\delta_s z\frac{\partial u}{\partial t}=\delta_s\langle z\rangle^{\phi\theta}\left[\frac{uv\tan\theta}{R}+fv\right]-\Re^v(u)-\frac{1}{R\cos\theta}\delta_s\langle z\rangle^{\phi\theta}\left(\delta_\phi\langle\Gamma\rangle^\theta+\frac{\langle\rho\rangle^\phi}{\rho_0}g\delta_\phi\langle z\rangle^{s\theta}\right)$$

$$+\delta_s\left[K_M\delta_s u\left(\delta_s\langle z\rangle^s\right)^{-1}\right]+\frac{1}{R^2\cos\theta}\left[\frac{1}{\cos\theta}\delta_\phi\left(A_M\delta_s\langle z\rangle^\phi\,\delta_\phi u\right)+\delta_\theta\left(A_M\cos\theta\delta_s\langle z\rangle^\theta\,\delta_\theta u\right)\right]$$

$$=G_u$$

$$\delta_s z\frac{\partial v}{\partial t}=-\delta_s\langle z\rangle^{\phi\theta}\left[\frac{u^2\tan\theta}{R}+fu\right]-\Re^v(v)-\frac{1}{R}\delta_s\langle z\rangle^{\phi\theta}\left(\delta_\theta\langle\Gamma\rangle^\phi+\frac{\langle\rho\rangle^\theta}{\rho_0}g\delta_\theta\langle z\rangle^{s\phi}\right)$$

$$+\delta_s\left[K_M\delta_s v\left(\delta_s\langle z\rangle^s\right)^{-1}\right]+\frac{1}{R^2\cos\theta}\left[\frac{1}{\cos\theta}\delta_\phi\left(A_M\delta_s\langle z\rangle^\phi\,\delta_\phi v\right)+\delta_\theta\left(A_M\cos\theta\delta_s\langle z\rangle^\theta\,\delta_\theta v\right)\right]$$

$$=G_v$$

where $\Gamma=\frac{g}{\rho_0}\sum\langle\rho\rangle^s\,\delta_s\langle z\rangle^s$

$$(10.6.5)$$

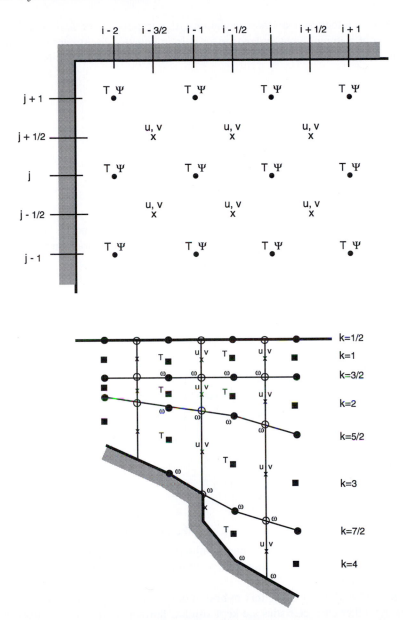

Figure 10.6.1 (Top) Horizontal and (bottom) vertical arrangement of grid points in the B grid s-coordinate model. T indicates tracer grid points, u, v indicates horizontal velocity components. (•) indicates the location of barotropic stream function grid points, and (•) indicates the transformed vertical velocity. Bottom shown is not coincident with a coordinate surface to indicate how intersection of coordinate levels and topography is handled at k = 4.

The vertical integration of these equations, the imposition of a rigid lid and definition of a stream function leads to an elliptic equation for stream function ψ similar in form to Eq. (10.2.17), whose finite-difference equivalent is (Gerdes, 1993)

$$
\left[\delta_\phi \left(\frac{1}{\langle H \rangle^\theta \cos\theta} \delta_\phi \left(\frac{\partial\psi}{\partial t} \right) \right) + \delta_\theta \left(\frac{\cos\theta}{\langle H \rangle^\phi} \delta_\theta \left(\frac{\partial\psi}{\partial t} \right) \right) \right] =
$$

$$
\frac{1}{\Delta\theta} \delta_\phi \left\langle \frac{\Delta\theta}{H} \sum_{k=1}^{K} G_v \right\rangle^\theta - \frac{1}{\Delta\phi} \delta_\theta \left\langle \cos\theta \frac{\Delta\phi}{H} \sum_{k=1}^{K} G_u \right\rangle^\phi
$$

(10.6.6)

where K is the total number of vertical levels. Note that G_u and G_v denote the RHS of momentum equations in Eq. (10.6.5). The tracer equation can be written as

$$
\delta_s z \frac{\partial X}{\partial t} + \Re^x(X) = \delta_s \left[K_H \delta_s X \left(\delta_s \langle z \rangle^s \right)^{-1} \right]
$$

$$
+ \frac{1}{R^2 \cos\theta} \left[\frac{1}{\cos\theta} \delta_\phi \left(A_H \delta_s \langle z \rangle^\phi \delta_\phi X \right) + \delta_\theta \left(A_H \cos\theta \delta_s \langle z \rangle^\theta \delta_\theta X \right) \right]
$$

$$
\Re^x(X) = \frac{1}{R\cos\theta} \left[\delta_\phi \left(\frac{\left\langle u \delta_s \langle z \rangle^{\phi\theta} \Delta\theta \right\rangle^\theta}{\Delta\theta} \langle X \rangle^\phi \right) + \delta_\theta \left(\frac{\left\langle v\cos\theta \delta_s \langle z \rangle^{\phi\theta} \Delta\phi \right\rangle^\phi}{\Delta\phi} \langle X \rangle^\theta \right) \right]
$$

$$
+ \delta_s \left(\tilde{w} \langle X \rangle^s \right)
$$

(10.6.7)

where $\tilde{w}$ is computed from $\Re^x(1) = 0$ and the kinematic boundary condition at the surface; the superscript indicates evaluation at tracer grid points (Figure 10.6.1). The value of A_H is chosen depending on whether implicit or explicit diffusion is used.

Gerdes (1993) formulated this hybrid model along the lines of the GFDL model. All other characteristics are kept similar: horizontal differencing uses the Arakawa B grid, a rigid lid is imposed, convective adjustment is made in unstable regions, the vertical advection is treated explicitly, and the pressure terms are eliminated using the stream function formulation necessitating the solution of an elliptic equation for stream function. The scalars are therefore staggered in the horizontal w.r.t., the horizontal velocities, but not vertically, while the vertical velocities are staggered in the vertical with respect to the horizontal velocities. Thus the finite differencing and other numerical aspects are the same as that of Cox (1984) and Pacanowski (1995). The principal difference is that the FCT algorithm is used for advection of scalars (Zalesak, 1979). This algorithm adds the minimum value of diffusion needed to prevent the development of physically inadmissible negative values for the tracers. There is no explicit diffusion that would be needed for stability of the central difference scheme if the FCT algorithm were not used. Note that by putting $\delta_s \langle z \rangle = 1$, one recovers the original Cox (1984) formulation for the z-coordinate system.

Gerdes (1993) carried out simulations of barotropic and baroclinic topographic Rossby waves with the s-coordinate model and showed that for similar resolutions, the z-coordinate model has considerable difficulty simulating the analytical characteristics of these waves, whereas the s-coordinate does a fairly good job of reproducing not only the dispersion relation but also the spatial structure. This is in spite of the intersection of coordinate surfaces with topography above the depth z_c that tends to degrade the propagation properties of baroclinic Rossby waves. When applied to the problem of the overflow of dense water over a submarine ridge, the flow was advected smoothly along the bottom in the s-coordinate model, whereas the jagged z-level topography caused advective overshoot in the horizontal that led to local static instability and vertical mixing. Such spurious vertical mixing is quite detrimental to the simulation of deep density currents in the ocean, especially since the ambient stratification is rather weak and cannot inhibit such vertical mixing.

However, unless the isopycnal mixing scheme is used, the s-coordinate system also suffers from artificial diapycnal diffusion wherever the isopycnal surfaces intersect the coordinate surfaces at a steep angle. While the FCT algorithm reduces this problem, it does not eliminate it entirely. Nevertheless, for basin scale applications, s-coordinate models hold significant promise, provided z_c can be chosen to minimize the pressure gradient truncation errors. This is not always possible, since the permanent thermocline tends to have a conspicuous slope in the meridional direction, increasing from a few hundred meters in low latitudes to nearly a thousand at mid-high latitudes. Nevertheless, accurate representation of topographic effects such as topographic Rossby waves in the deep (and coastal processes on the shelf) requires careful attention to vertical discretization, just as accurate representation of geostrophic adjustment and gravity waves dictates the choice of horizontal discretization in numerical ocean models. Until very high resolutions become feasible, a choice must be made between z- and σ-coordinates (or equivalently the choice of z_c in s-coordinates) and between Arakawa B and C grids.

Another example of a s-coordinate model, very much similar to the one described earlier, is the S-coordinate Primitive Equation Model (SPEM 5.1)

from Rutgers University (Song and Haidvogel, 1994). It is the current version of
the Semispectral PE Model (SPEM) developed by Haidvogel *et al.* (1991),
which used a spectral decomposition in the vertical in the past. The semispectral
approach has apparently been replaced by the finite-difference approach in the
vertical. This version is a hydrostatic, rigid lid, finite-difference model, with
staggered horizontal and vertical grids, with s-coordinate in the vertical and
orthogonal curvilinear coordinates in the horizontal. Beckmann and Haidvogel
(1997) present high resolution simulations of the flow around the Fieberling
Guyot induced by diurnal barotropic tides, using the latest version of this model.
It uses a stretched grid in the horizontal with maximum change in grid spacing
from one grid cell to another of less than 2.5%. The depth variations from one
grid cell to another is also kept less than 18%, and a fourth-order difference
scheme (McCalpin, 1994) is used to compute the horizontal pressure gradients.
This, combined with the high horizontal resolution, enabled the model to
simulate successfully the observed order of magnitude amplification of tidal
currents around the Guyot.

Proper implementation of isopycnal mixing schemes and the use of terrain-
following hybrid coordinate are making more realistic long term simulations of
ocean circulation possible using these models. However, much remains to be
done. This has to do with our woeful lack of understanding of how the deep
oceans are mixed and the water masses there maintained. The oceanic
thermohaline circulation is a crucial component of climate and climate models
must do a credible job of simulating processes that maintain the water mass
structure in the deep. This in turn may require that some salient deep ocean
mixing processes hitherto ignored in global ocean models, such as the baroclinic
tide-driven mixing in the main thermocline and below in regions of sharp
topographic changes (such as midocean ridges and seamount—see Chapter 6 of
Kantha and Clayson, 1999) be incorporated. Huang (1999) suggests that in
addition to this internal tidal dissipation of perhaps 0.6 TW (some estimates are
as high as twice this value) that elevates the mixing intensities in the deep
oceanic regions, it may be necessary to consider the potential energy generated
by geothermal heating at the ocean bottom, at a rate estimated to be about 0.5
TW. In addition, recent studies have shown that double-diffusive mixing also
needs to be included (Zhang *et al.*, 1998). Thus more accurate simulation of subgrid scale mixing processes appears to hold the key to more realistic simulations
of at least some aspects of global circulation by numerical ocean models.

Finally, the need to accurately simulate important deep convection and
dense/intermediate water formation processes in subpolar seas requires that the
hydrostatic approximation be abandoned and this has recently spurred the
development of nonhydrostatic global ocean models. One example is the M.I.T.
nonhydrostatic model described in Chapter 2 (Marshall *et al.*, 1997a,b). A clever
exploitation of the fact that the oceans are very nearly hydrostatic has permitted

efficient numerical solution to be implemented and global simulations made possible, currently at $1°$ resolution.

10.7 REGIONAL z-LEVEL MODELS

The set of equations in Eq. (10.2.1) are too complicated for many midlatitude regional basin-scale studies. A simplification for small values of Δ, leading to the beta plane equations, is an additional step in many cases. If we consider latitude θ_0, expand in terms of L/R, and retain only terms of order L/R we get, ignoring the frictional terms for simplicity, (Muller, 1995; Veronis, 1981)

$$\left(1+\frac{y}{R}\tan\theta_0\right)\frac{\partial u}{\partial x}+\frac{\partial v}{\partial y}+\frac{\partial w}{\partial z}-\frac{v}{R}\tan\theta_0=0$$

$$\frac{du}{dt}-fv\left(1+\frac{y}{R}\tan\theta_0\right)+\frac{y}{R}\tan\theta_0 u\frac{\partial u}{\partial x}-\frac{uv}{R}\tan\theta_0=-\frac{\partial p}{\partial x}\left(1+\frac{y}{R}\tan\theta_0\right)$$

$$\frac{dv}{dt}+fu\left(1+\frac{y}{R}\tan\theta_0\right)+\frac{y}{R}\tan\theta_0 u\frac{\partial v}{\partial x}+\frac{u^2}{R}\tan\theta_0=-\frac{\partial p}{\partial y}$$

$$0=-\frac{\partial p}{\partial z}-\frac{\rho}{\rho_0}g \qquad\qquad (10.7.1)$$

$$\frac{d\rho}{dt}+\frac{y}{R}\tan\theta_0 u\frac{\partial\rho}{\partial x}=0$$

$$f=2\Omega\sin\theta_0$$

$$\frac{d}{dt}\equiv\frac{\partial}{\partial t}+u\frac{\partial}{\partial x}+v\frac{\partial}{\partial y}+w\frac{\partial}{\partial z}$$

$$\frac{\partial}{\partial x}\equiv\frac{1}{R\cos\theta_0}\frac{\partial}{\partial\phi};\frac{\partial}{\partial y}\equiv\frac{1}{R}\frac{\partial}{\partial\theta}$$

If we now ignore all terms of the order y/R, a simpler subset that is most often used in geophysical flow investigations on a beta plane results:

$$\frac{\partial u}{\partial x}+\frac{\partial v}{\partial y}+\frac{\partial w}{\partial z}=0$$

$$\frac{du}{dt}-fv=-\frac{\partial p}{\partial x} \qquad\qquad (10.7.2)$$

$$\frac{dv}{dt}+fu=-\frac{\partial p}{\partial y}$$

$$0 = -\frac{\partial p}{\partial z} - \frac{\rho}{\rho_0} g$$

$$\frac{d\rho}{dt} = 0$$

(10.7.2)

This set of equations contains the most salient components of midlatitude dynamics such as western intensification, and Rossby wave progation, as we saw in Chapter 4.

Studies of midlatitude gyre scale circulation have benefitted most from eddy resolving models based on this simpler set. It is also applicable to modeling deep marginal seas around the ocean basins.

Chapter 11

Layered and Isopycnal Models

As we have seen, one way to classify ocean models is based on their treatment of the vertical structure. Level models employ the Eulerian approach to deducing properties at horizontal grid points that are also fixed in the vertical space during the simulation. Layered/isopycnal models, on the other hand, employ a semi-Lagrangian approach and monitor the evolution of distinct layers of water that are free to move vertically at any horizontal grid point. The governing primitive equations are integrated vertically between isopycnals (surfaces of equal density). The result is that the ocean is represented by a series of horizontally homogeneous layers of uniform density distributed in the vertical such that any given layer lies below another of lower density but overlies a layer of higher density. In other words, stable stratification prevails. These layers are allowed to deform and the interfaces between them allowed to move vertically, and it is the evolution of these layers at each horizontal grid point that is modeled by layered and isopycnal models.

In truly isopycnal models, outcropping or surfacing of isopycnals, as in frontal regions, is permitted, and the layer thicknesses are allowed to vanish and layers to disappear as a result of thermodynamical processes. Layered models, on the other hand, prevent the outcropping of isopycnals and thinning and vanishing of layers by mixing, most often artificially. While both layered and isopycnal models started as purely dynamical models, both now have thermodynamics incorporated and can therefore be fully dynamical/thermo-dynamic. This means that the surface buoyancy fluxes can be accommodated and diapycnal mixing between adjacent layers permitted. However, considerable ingenuity is involved in parameterizing some of these effects. Also, the use of density in the past as a prognostic variable rather than temperature and salinity in layered/isopycnal models implied a linear equation of state and identical mixing and diffusion characteristics for temperature and salinity. This was a serious limitation near coasts and ice-covered regions, with their broad range of salinity variability, but more recent versions such as the Miami Isopycnic Coordinate Ocean Model (MICOM) (Bleck *et al.*, 1992; New *et al.*, 1995) and the Hamburg isopycnal model (Oberhuber, 1993a,b) now accommodate temperature and salinity as independent thermodynamic variables. It is the difficulty of accommodating complex topography, complex equation of state, and the appearance and disappearance of isopycnal layers that has been responsible for the slow development and acceptance of the isopycnal models compared to their z-level counterparts. The two, however, complement each other quite well for basin scale and global applications.

11.1 LAYERED MODELS

As we saw in Chapters 1, 3, and 4, simple layered dynamical models have always been the favorite tool of theoretical oceanographers, not simply because of tractability, but also because, more often than not, they manage to capture the essence of many dynamical processes. It is not surprising then that with the advent of computers, they also became the favorite tool of some ocean modelers. The single-most advantage of numerical layered dynamical models is that, comparatively speaking, they are less computing resource intensive. This means that they can be run at high horizontal resolutions that enable many dynamical processes to be well represented by virtue of being able to resolve important features of the ocean circulation. In fact, layered dynamical models were the first (QG models being inapplicable to equatorial regions and level and isopycnal models being too resource intensive) to achieve truly global eddy- and narrow western boundary current- resolving capability. This remarkable achievement is not without penalties. Since resource constraints at present dictate that the number of layers be limited (usually three, but a maximum of about six) in order to achieve high horizontal resolutions, layered models sacrifice vertical resolution and often thermodynamics at the altar of eddy resolvability. In contrast, level models invariably include thermodynamics and a comparatively high vertical resolution and hence are restricted to cruder horizontal resolutions, sacrificing eddy resolvability. Irrespective of the computing resources on hand, there has always been a schism in the ocean modeling community between modelers who favor high horizontal resolutions and those who prefer better resolution of the vertical structure.

The most serious drawback of the layered dynamical models is the necessity to "bury" the topography in the bottom-most layer. The ocean bottom in the model is therefore never allowed to be shallower than the depth of the interface between the two bottom-most layers. This is to assure that the horizontal extent of the isopycnal interfaces between the layers in the model does not change during the model simulation. This makes such models inherently limited to deep basins, since the interface between any two layers cannot be allowed to vary in horizontal extent, as might happen if the ocean depths were comparable to the depth of that interface. Another problem is that the layer thicknesses must be maintained at nonzero values, and the density of each layer must be invariant. This restricts their application to regions away from the well-mixed layer at the top. Since thinning of layers and surfacing of interfaces cannot be allowed for numerical reasons, and the density of each layer must be preserved, ingenious devices have to be employed to entrain and detrain water masses across the bounding interfaces of each layer from adjacent layers, and sources and sinks introduced (usually near lateral boundaries). Needless to say, the rest depths of the layers must be carefully chosen to capture the salient dynamics being modeled. That considerable success has indeed been achieved in the application of dynamical layered models to elucidate the dynamics of global and basin circulations (Hogan *et al.*, 1992; Hurlburt and Thompson, 1980; Hurlburt *et al.*, 1992; Kindle and Thompson, 1989; Metzger *et al.*, 1992; Mitchell *et al.*, 1994, Hacker *et al.*, 1998) is a tribute to the remarkable insight and ingenuity of layered modelers.

In layered models, the ocean is divided into several (N) layers in the vertical, and primitive equations for continuity and momentum are integrated over each layer (n=1, N) to obtain expressions for the thickness of and velocity in each layer. In order to do this, hydrostatic approximation is invoked.

The U.S. Navy layered model, developed at the Naval Research Laboratory at Stennis Space Center, Mississippi, is a generalization to N layers of the two-layer model developed originally and applied to Loop Current Eddy shedding processes by Hurlburt and Thompson (1980). Integrating the primitive hydrostatic equations between iso-density surfaces, Wallcraft (1991) shows

$$\frac{\partial}{\partial t}\left(h^n\right)+\frac{\partial}{\partial x_k}\left(h^n U_k^n\right)=w^n-w^{n-1}$$

$$\frac{\partial}{\partial t}\left(h^n U_j^n\right)+\left[\frac{\partial}{\partial x_k}\left(h^n U_k^n\right)+U_k^n\frac{\partial}{\partial x_k}\right]U_j^n+f\varepsilon_{j3k}h^n U_k^n=$$

$$-h^n\sum_{k=1}^{N} G_k^n\frac{\partial}{\partial x_k}\left(h^n-h_0^n\right)+(\tau_j^{n-1}-\tau_j^n)+A_M\frac{\partial^2}{\partial x_k\partial x_k}\left(h^n U_j^n\right) \qquad (11.1.1)$$

$$+\max(0,-w^{n-1})U_j^{n-1}+\max(0,w^k)U_j^{n+1}-[\max(0,-w^n)$$

$$+\max(0,w^{n-1})]U_j^n+\max(0,-c_{de}w^{n-1})(U_j^{n-1}-U_j^n)$$ (11.1.1)

$$+\max(0,-c_{de}w^n)(U_j^{n+1}-U_j^n)\qquad\qquad k=1,2;\ n=1,...,N$$

where h^n is the thickness and U_j^n is the velocity of the nth layer; w^k is the vertical velocity at the kth interface; h^n_o is the layer thickness at rest. The Nth layer contains the model basin topography and its thickness is the total depth of the water column minus the sum of the thicknesses of the remaining layers:

$$G_k^n = \begin{cases} g & k \le n \\ g\left[1-\left(\rho^k-\rho^n\right)/\rho_0\right] & k>n \end{cases}$$ (11.1.2)

$$\tau_j^n = \begin{cases} \tau_w; & n=0 \\ c_{dn}\left|U_j^n-U_j^{n+1}\right|\left(U_j^n-U_j^{n+1}\right); & n=1,N-1 \\ =c_{db}\left|U_j^N\right|U_j^N; & n=N \end{cases}$$

$$w_n = \begin{cases} 0 & n=0 \\ w_n^+ - w_n^- - \Omega_n\hat{w}_n & n=1,\ ...\ N-1 \end{cases}$$

$$w_n^+ = \bar{w}_k\left[\mathrm{Max}(0,h_n^+-h_n)/h_n^+\right]^2$$ (11.1.3)

$$w_n^- = \bar{w}_k\left[\mathrm{Max}(0,h_n-h_n^-)/h_n^+\right]^2$$

$$\hat{w}_n = \frac{\iint\left(w_n^+ - w_n^-\right)}{\iint\Omega_n}$$

The term c_d is the drag coefficient, c_{de} is the drag due to entrainment of fluid from one layer to the adjacent one, τ_w is the wind stress, and ρ^n is the density of the nth layer. Note that the layer densities do not change with time, only their thickness does, at each model grid point. $\bar{w}_n$ is the reference vertical mixing velocity at the nth interface, h_n^+ and h_n^- are nth layer thicknesses where entrainment and detrainment start. Detrainment can be deactivated by making h_n^- large. Ω_n is the nth interface global mixing correction scale factor, a value of 1 for which distributes global mixing uniformly. The conditional statements have to do with entrainment and detrainment at each interface between two adjacent layers, the details of which can be found in Wallcraft (1991). The complex entrainment/detrainment scheme is designed to enable long term integrations without layers "drying up" and surfacing of interfaces and is more a practical than a physical necessity.

The thinning of a layer to vanishing thickness is a major problem in layered models, which leads to numerical difficulties. The traditional solution has been to make each layer thick enough, but this distorts the representation of the oceanic vertical structure. An alternative solution is to entrain fluid into the thinning layer from below to thicken it. Such entrainment has to be balanced by global detrainment from the entraining layer to the detraining layer elsewhere in the model domain or through ports at the lateral boundary, so as to keep the density of both layers constant in space and time.

The equations are finite-differenced using leapfrog scheme on an Arakawa C grid. The scheme is semi-implicit in that the pressure gradient term in the momentum equation and the divergence term in the continuity equation are made implicit. The scheme is unconditionally stable and enables longer time steps to be taken than in explicit methods. The N coupled PDEs are converted into N decoupled Hemholtz equations, which are solved by a fast elliptic solver for rectangular and a capacitance matrix technique for nonrectangular regions (Wallcraft, 1991). At inflow ports, the inflow velocity and angle are prescribed and at outflow ports, Orlanski type radiation boundary conditions are used but the total outflow is adjusted to be equal to the inflow. Wind stress forcing for the model is defined only at alternating points, with the y component defined at u_2 points and x component at u_1 points; the values are interpolated to the rest of the points. This procedure is merely to save space and is permissible since the scale of the winds is usually larger than that of the ocean circulations. Wind stresses can be climatological or synoptic, and values needed at each time step are linearly interpolated from the prescribed wind stress fields. For more details of the model numerics, see Wallcraft (1991).

It is essential to select the number and rest thicknesses of layers carefully in layered models. Since topographic variations are contained in the bottom-most layer only, these models are generally incapable of simulating circulation in coastal and shallow seas. They are, however, excellent at capturing the important lowest order dynamics of the basin circulation and are therefore widely used for process-oriented studies. They are also being increasingly used for a variety of applications. The U.S. Navy layered model has been used extensively for a variety of regions, including the Gulf of Mexico (Hurlburt and Thompson, 1980, 1982), the Mediterranean Sea (Heburn, 1987, 1994), the Atlantic, the Indian (Hacker *et al.*, 1998), and the Pacific Oceans. Fully thermodynamic versions exist for all three ocean basins and the global ocean (Hurlburt *et al.*, 1992; Jacobs *et al.*, 1994; Kindle and Thompson, 1989; Metzger *et al.*, 1992; Mitchell *et al.*, 1994; Thompson and Schmidt, 1988). One example is the six-layer, 1/8° global model, the highest resolution global ocean model at present. Realistic depictions of mesoscale activity—especially in regions of strong ocean currents such as the Gulf Stream in the Atlantic, the Kuroshio in the Pacific, the Brazil/Malvinas Current off Brazil, the Agulhas Current off Africa, and the Circumpolar Current around the continent of Antarctica—are noteworthy. The SSH variability from a layered model like this, driven by synoptic winds from a NWP center such as Fleet Numerical Meteorology and Oceanography Center, compares well with the variability indicated by altimeters such as the U.S. Navy's GEOSAT. Figure 11.1.1 shows evolution of the model SSH variability in the Pacific basin (Hurlburt *et al.*, 1992). The model is quite realistic in depicting the various features of circulation in the basin. Figure 11.1.2 shows the circulation in the top layer in the Indian Ocean from 1/6° NRL six-layer global

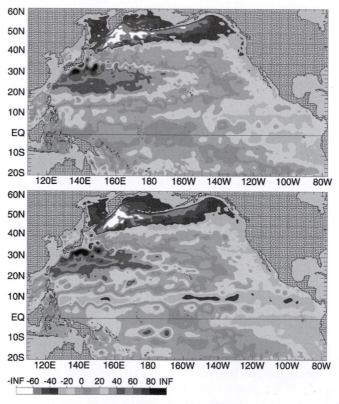

Figure 11.1.1 Sea service height as simulated by 1/8• Pacific version of the NRL Layered Ocean Model. From Hurlburt *et al.* (1992).

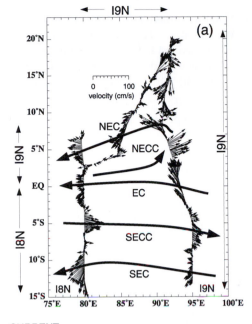

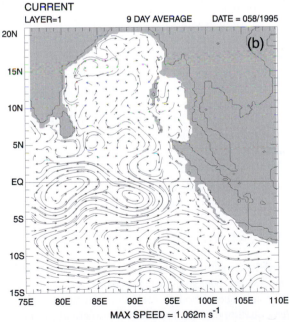

Figure 11.1.2 Comparison of (a) ADCP measurements at 21–75 m depth during 1995 Indian Ocean WOCE cruise with (b) 9-day averaged current in the upper layer (0–115 m thick) of the NRL layered model. From Hacker *et al.* (1998).

model, along with ADCP observations of currents in the upper layer during the 1995 Indian Ocean WOCE cruise. The agreement between the two is quite good.

A simple subset of the layered model is the so-called reduced gravity model (also called N and a half layer model), where the deepest layer is assumed to be infinitely deep and quiescent. A reduced gravity version of the model, with N active layers, has $h^{N+1} \rightarrow \infty, u^{N+1} \rightarrow 0$. The governing equations are similar to that of the N layer model except that h^N_o is constant and

$$G^n_k = \begin{cases} g(\rho^{N+1} - \rho^n)/\rho_o & k \leq n \\ g(\rho^{N+1} - \rho^k)/\rho_o & k > n \end{cases} \tag{11.1.4}$$

$$\tau^n_j = \begin{cases} \tau_w ; & n = 0 \\ c_{dn} \left| U^n_j - U^{n+1}_j \right| \left(U^n_j - U^{n+1}_j \right); & n=1,...N \end{cases}$$

$$w_n = \begin{cases} 0 & n=0 \\ \max\left(0, w^+_n\right) - \max\left(0, w^-_n\right) - h_n \hat{w}_n & n=1, ... N \end{cases} \tag{11.1.5}$$

The reduced gravity model filters out the barotropic mode and is most useful for investigating the internal dynamics.

A particularly popular version is the one-and-a-half layer model (see Chapter 3), where the water column is assumed to consist of two layers: an active top layer of thickness H and a quiescent bottom layer of infinite thickness, with a density interface between the two of intensity $\Delta\rho$. It is remarkable that this very simple model often captures the essential dynamics of the circulation. For example, a reduced gravity model of the Gulf of Mexico demonstrated conclusively that the instability of the Loop Current is responsible for the shedding of the Loop Current Eddies and not the seasonal variability per se (Hurlburt and Thompson, 1980). The governing equations are identical to the barotropic equations, except that the gravity parameter g is replaced by $g' = g\Delta\rho/\rho_o$, the reduced gravity (whose value is two orders of magnitude smaller than g), with H now indicating the rest thickness of the upper layer and η denoting the deflection of the interface.

11.2 ISOPYCNAL MODELS

Isopycnal models are similar to the layered models discussed above, but they do not prevent thinning/vanishing of the isopycnal layers and surfacing of isopycnals. They also accommodate realistic bottom topography and in recent years a complex equation of state. Despite the numerical problems resulting from implementing these capabilities, they are, in principle anyway, ideally suited to simulating the deep basin dynamics. They are also fully dynamical/ thermodynamic. The principal advantage is that these truly isopycnal models are consistent with observations that mixing in the deep oceans is principally along isopycnals, and the cross isopycnal mixing is many orders of magnitude smaller. This enables very long integrations to be carried out without adverse modification of deep water masses. Level models, on the other hand, unless they take advantage of isopycnal mixing schemes (McDougal and Church, 1986; Redi, 1982), introduce artificial diffusion across sloping isopycnals. Even a small slope of the isopycnal implies a large crosspycnal diffusion resulting from large values of horizontal diffusion employed in these models for numerical reasons (the severity of the problem decreases with higher horizontal resolutions). Consequences of such undesirable vertical mixing can be serious in climate-type studies. Isopycnal models started out as simple dynamical wind-driven models, but considerable progress has been made over the past decade in isopycnal modeling, and with the inclusion of adequate upper mixed layer physics and thermodynamics, they are also becoming quite practical. Examples of applications can be found in Oberhuber (1993b), Smith *et al.* (1990), and New *et al.* (1995).

To study wind-driven circulation, it is sufficient to have density be the only variable and keep layer densities constant; essentially a simple dynamical model suffices. For more realistic circulation, including the thermohaline circulation driven by the surface buoyancy fluxes, both temperature and salinity need to be predicted. Inclusion of thermodynamics is therefore a significant complication in these models. Cross-isopycnal mixing is quite intense in the upper mixed layer and it serves no purpose to model this using the isopycnal approach. Instead, isopycnal models represent the interior but superimpose a slab type mixed layer on top, coupled (through entrainment and detrainment) to the isopycnal layers below. This then allows for buoyancy forcing at the ocean surface.

At present, there are only two reasonably mature isopycnal models, since while conceptually simple and elegant, their development is arduous and often involves "engineering" solutions to a variety of technical problems such as the surfacing and vanishing of isopycnals, the presence of two independent thermodynamic variables that determine the layer densities, and the incorporation of thermodynamic forcing and turbulent mixing in the upper layers. It is too early to tell what effect these solutions and the underlying assumptions have on model simulations. Unlike layered models, which avoid the problem of changes in the intersections of coordinate surfaces (isopycnals) with domain boundaries for computational expediency by "burying" the topography in the bottom-most layer, the necessity to handle these changes in isopycnal models requires techniques that are computationally demanding. Handling of mixing processes in the upper ocean, especially the deep convection and intermediate and bottom-dense water formations in subpolar seas, is inherently awkward in an isopycnal model. Nevertheless, isopycnal models form an excellent complement to the traditional z-level models for basin scale and global studies of oceanic circulation.

The first of these models is from the Max-Planck Institute at Hamburg (Oberhuber, 1986, 1993a,b) and the other from the University of Miami, MICOM (Miami Isopycnic Coordinate Ocean Model) (Bleck and Boudra, 1986; Bleck and Smith, 1990; Bleck *et al.,* 1992; New *et al.,* 1995). The Hamburg model uses a B grid and an implicit time-differencing scheme, whereas MICOM uses a C grid and an explicit scheme. Both models have been applied to study the meridional thermohaline circulation in the North Atlantic. But at present, only the Hamburg model incorporates a sea-ice model to account for the influence of ice cover in high latitudes. Both models are fully thermodynamic and solve separate prognostic equations for temperature and salinity. Both have a Kraus-Turner type turbulence kinetic energy-based slab type mixed layer (Kraus and Turner, 1967; see also Chapter 2 of Kantha and Clayson, 1999), driven by surface momentum and buoyancy fluxes, on top of their isopycnal ocean. However, Oberhuber (1993a) has carefully optimized the slab mixed layer model by an inverse-modeling technique and uses a combination of features from different slab models in the Hamburg model. Both handle the interaction of the isopycnal layers at the base of the mixed layer with the mixed layer through layers that are allowed to reach and retain zero thickness. In the Hamburg model, the isopycnal layer immediately below the mixed layer is allowed to lose mass to entrainment during the deepening phase of the mixed layer and ultimately become massless. The procedure is then repeated for the next layer. The shallowing phase is, however, tricky and a complex scheme is employed to accommodate the detrained water into the model isopycnal layers that now reappear as needed (Oberhuber, 1993a). In MICOM also, the isopycnal layers are never allowed to vanish entirely but are retained as massless layers with a tiny residual thickness (Bleck *et al.,* 1992).

Both models use Smagorinsky (1963) formulation for isopycnal mixing. MICOM incorporates diapycnal diffusion. Both retain the barotropic mode, but the numerical scheme to handle it are different. MICOM uses explicit leapfrog time stepping (with an Asselin-Roberts filter to damp out the computational mode), necessitating the use of mode-splitting (see Chapter 9) to avoid computational penalties resulting from the external mode CFL constraint (Bleck *et al.,* 1992). The Hamburg model, on the other hand, uses an unconditionally stable semi-implicit predictor-corrector scheme for time stepping. It is quite difficult to describe the intricate details of the numerical schemes and the "fixes" to the various technical problems inherent to isopycnal models. The reader is referred instead to Bleck and Boudra (1986), Bleck and Smith (1990), and Bleck *et al.* (1992) for the MICOM, and Oberhuber (1986, 1993a) for the Hamburg model, for a more thorough discussion of the various issues.

The governing equations are very much similar to those of the layered models, except that prognostic equations for the two thermodynamic variables Θ and S and an equation of state are also used. By integration of the Boussinesq, hydrostatic primitive equations in the vertical between predefined isopycnal surfaces, one simply gets equations for conservation of integrated mass, momentum, heat, and salt content in each layer, which have, in addition to the usual source/sink terms, those due to entrainment/detrainment from adjacent layers. Because of Boussinesq approximation, density can be normalized out:

$$\frac{\partial}{\partial t}\left(h^n\right)+\frac{\partial}{\partial x_k}\left(h^n U_k^n\right)=(w)^{bot}-(w)^{top}$$

$$\frac{\partial}{\partial t}\left(h^n U_j^n\right)+\left[\frac{\partial}{\partial x_k}\left(h^n U_k^n\right)+U_k^n\frac{\partial}{\partial x_k}\right]U_j^n+f\varepsilon_{j3k}h^n U_k^n=$$

$$-h^n\sum_{k=1}^{N}G_k^n\frac{\partial}{\partial x_k}\left(h^n-h_0^n\right)+(\tau_j^{top}-\tau_j^{bot})+A_M\frac{\partial^2}{\partial x_k \partial x_k}\left(h^n U_j^n\right)$$

$$+\left(wU_j^n\right)^{bot}-\left(wU_j^n\right)^{top}$$

$$\frac{\partial}{\partial t}\left(h^n\Theta^n\right)+\frac{\partial}{\partial x_k}\left(h^n\Theta^n U_k^n\right)=A_H\frac{\partial^2}{\partial x_k\partial x_k}\left(h^n\Theta^n\right)+\left(w\Theta^n\right)^{bot}-\left(w\Theta^n\right)^{top}$$

$$\frac{\partial}{\partial t}\left(h^n S^n\right)+\frac{\partial}{\partial x_k}\left(h^n S^n U_k^n\right)=A_H\frac{\partial^2}{\partial x_k\partial x_k}\left(h^n S^n\right)+\left(wS^n\right)^{bot}-\left(wS^n\right)^{top}$$

$$\rho^n=\rho^n\left(S^n,\Theta^n\right)\qquad\qquad k=1,2;\quad n=1,...,N$$

$$(11.2.1)$$

where the last equation is the equation of state. MICOM uses a simple third-degree polynomial form that can be easily inverted to find T, given ρ and S. Note that only the horizontal derivatives are involved in the above equations. The terms involving w are the entrainment/detrainment terms, which have required considerable ingenuity in formulation so that the densities of upper isopycnal layers are more or less well preserved in the face of mixed layer deepening and shallowing. In the deeper layers, on the other hand, the formulation allows these terms to be zeroed out so that no cross isopycnal transports can take place. The form of the equations in Eq. (11.2.1) is similar to that presented by Oberhuber (1993a) but considerably different from Bleck et al. (1992), who use instead of z, a generalized vertical coordinate s to derive the layered form.

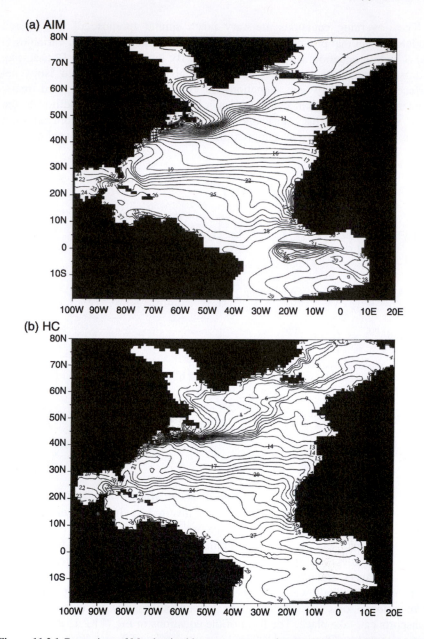

Figure 11.2.1 Comparison of March mixed layer temperatures for year 30 from MICOM isopycnal (top) and MOM z-level (bottom) models; both models suffer from overshoot of the Gulf Stream. From Roberts *et al.* (1996).

The solution of the mass conservation equation requires particular care; MICOM uses a flux corrected transport (FCT) scheme (Zalesak, 1979) to properly handle the problem of vanishing layers, whereas the Hamburg model uses variable lateral boundaries to handle the same problem.

Recent applications of isopycnal models can be found in Oberhuber (1993b), who used the Hamburg model to simulate the seasonal evolution in the North Atlantic, and New *et al.* (1995), who used MICOM to study the formation and evolution of water masses in the North Atlantic. The latter illustrates the promise of isopycnal models, although it may be a while before they are fully tested and validated for such use.

Using 30-year runs of the North Atlantic basin, Roberts *et al.* (1996) and Marsh *et al.* (1996) have made a systematic intercomparison of the Bryan-Cox-Semtner z-level model and MICOM, specifically with respect to the dense overflow over the Greenland-Iceland-Scotland Ridge and the meridional transport of heat in the North Atlantic. Not surprisingly, the isopycnal model better preserves the dense overflow, leading to a better conservation of potential vorticity characteristics, whereas in the z-level model, even with the implementation of an isopycnal mixing scheme (see Chapter 10), the dense overflow mixes vigorously with surrounding waters (apparently due to a limitation on the allowable isopycnal slope in the mixing tensor). While both models developed mean subtropical gyres of similar strength, MICOM transported more heat poleward in the subtropics. These subtle differences are of particular significance to the strength and variability of the meridional circulation and the poleward heat transport, and hence the climate. It is, however, important to realize that some of these conclusions are very much dependent on the grid resolution employed. As grid sizes fall below the Rossby radius of deformation, the two methods might become comparable in terms of preserving the properties of deep oceanic water masses.

Figure 11.2.1 shows a comparison of the MICOM and MOM model SSH for the north Atlantic. Note that both models have difficulty separating the Gulf Stream from the coast at the appropriate point, Cape Hatteras. There are many similarities and dissimilarities in the two results. Overall, this figure illustrates the current state of the art in basin scale ocean modeling.

Chapter 12

Ice-Ocean Coupled Models

There exist very few published "global" ocean models that are truly global, in the sense that they include the polar oceans—namely, the Arctic Ocean and the Southern Ocean around the Antarctic continent—and provide a comprehensive thermodynamic and dynamical treatment of the sea-ice cover; but see Boville and Gent (1998). This has to do with the fact that extension to polar oceans requires coupling to sea ice and a proper representation of the thermodynamics and the dynamics of the sea-ice cover and its interaction with the atmosphere and the underlying ocean. Sea ice is a rather difficult medium to characterize and model, and coupled ice-ocean models on basin scales and larger are still in their infancy. Some ocean models that do extend to and include the subpolar seas, such as the Greenland Sea, avoid treating the sea-ice aspects explicitly and instead choose to put a bound (of about −1.7°C) below which temperatures are not allowed to fall. Coupled atmosphere-ocean global models must and do incorporate sea ice in one fashion or another, but most emphasis is usually placed on accurate treatment of the atmosphere and the ocean, with sea ice being treated in a much simpler fashion. There have also been a few attempts to include the polar oceans in basin scale models (Hakkinen, 1997; Oberhuber, 1993a&b) and to model the Southern Ocean (Hakkinen, 1995). For a comprehensive review of sea-ice models, including the ice-ocean coupled models, see Hakkinen (1990).

Coupled ice-ocean models can be divided into two distinct groups: purely thermodynamically coupled models of usually 1-D variety that couple an ocean mixed layer to a thermodynamic sea-ice model (Hakkinen and Mellor, 1990;

Lemke, 1987; Mellor and Kantha, 1989) and models that couple a fully dynamical/ thermodynamic sea-ice cover to a full ocean, either in 2-D (Heinrichs *et al.,* 1995; Kantha, 1995a; Kantha and Mellor, 1989a) or 3-D (Hakkinen, 1995; Hibler and Bryan, 1987; Oberhuber, 1993a,b; Piacsek *et al.,* 1991; Reidlinger and Preller, 1991). The latter necessarily involve ice rheology and dynamics (Chapter 5). The coupled models can also be regional (Hakkinen *et al.,* 1992; Heinrichs, 1996; Ikeda, 1988) or basin scale (Fleming and Semtner, 1991; Hakkinen, 1995; Hibler and Bryan, 1987; Oberhuber, 1993a,b; Piacsek *et al.,* 1991; Riedlinger and Preller, 1991; Semtner, 1987). Most of the work in this area pertains to the Arctic and the sub-Arctic Seas, but notable exceptions are Parkinson and Washington (1979), who applied a thermodynamic ice model coupled to a fixed depth, nonmoving mixed layer in the Southern Ocean, and Hakkinen (1995), who applied a comprehensive ice-ocean coupled model to the Southern Ocean to study Antarctic Bottom Water formation, among other things. Sea-ice models have been coupled to the Bryan-Cox-Semtner z-level (Hibler and Bryan, 1987), to the Hamburg isopycnal (Oberhuber, 1993a,b), and to sigma-coordinate (Hakkinen *et al.,* 1993; Hakkinen, 1995; Heinrichs, 1996; Kantha and Mellor, 1989a) ocean models.

12.1 SEA-ICE MODELS

As we saw in Chapter 5, dynamics is an important aspect of the sea-ice cover. Since the ice cover mediates the transfer of momentum (and heat and water vapor), the ice motion is what determines the wind-driven part of the circulation in the underlying ocean. To model the ice motion properly, one needs to invoke conservation laws for the concentration, mass, and momentum of sea ice, and incorporate internal stresses through an appropriate ice rheology. The most commonly used rheology is the empirical Hibler plastic-viscous constitutive law connecting the ice internal stresses to the strain rates. This law, while purely empirical, evolves from a long search for a convenient and suitable rheology to characterize the complex ice cover, as we saw in Chapter 5. It accounts for the strong resistance of sea ice to convergence, ridge forming near coasts where the ice motion perpendicular to the coast is strongly inhibited by land, land-fast ice, and grounded ice keels, and free drift in the open ocean away from confining boundaries (especially during summer, when the ice cover is broken up by extensive leads) and during divergent motions of the ice cover under external forcing. It appears to work well for the Arctic and simulates the ice motion and thickness distribution there reasonably well (see Chapter 5).

While it is the ice dynamics that determines, by and large, the circulation in the underlying ocean and the ice thickness near coasts where ridging is important (as well as the distribution of leads and polynyas, where ice can grow in winter and melt laterally during summer), it is the thermodynamic processes that actually control the ice growth and melt. The two are rather intimately interwoven. In the Arctic, accretion at the bottom of the multiyear ice is important to ice growth, while ablation at the top by solar insolation is an important mechanism for ice melt. In addition, ice is created in openings in ice cover such as leads and polynyas during winter and melted laterally there during summer. The surface melting process is rather sensitive to the surface albedo of ice cover, which is affected by the lower albedo of melting ice ponds that cover the multiyear ice extensively during the summer melt season, and the high albedo of any existing snow cover. In addition, there is a net annual export of sea ice (~0.2 Sv) from the Arctic through the Fram Straits into the Atlantic, and this is compensated by ice growth during winter, principally over the broad Siberian shelves. Sea ice growth in whatever form leads to extrusion of brine and hence a salt flux into the underlying ocean that acts to mix it vigorously. Ice growth over the Arctic shelves also leads to the formation of dense water that can spread laterally in the basin below the upper mixed layer and by forming a barrier to exchange of properties with the ocean below, can affect the fate of the perennial ice cover itself.

The interleaving dynamical and thermodynamic processes are similar around the Antarctic, but with an important difference. Unlike the Arctic, where there is perennial ice cover in the central Arctic, perennial ice cover is confined predominantly to the Ross and Weddell Seas, and most of the ice cover is thin (less than a meter thick) first year ice that appears and disappears with the seasons. The mixed layer is deeper than in the Antarctic and provides a significant amount of heat to the ice cover to keep it thin. Frazil ice formation in the bulk of the water column is also important, and as a rule there is more vigorous thermodynamic interaction between the ice and the ocean in the Antarctic than in the Arctic proper.

For many of these reasons, the thermodynamics of sea ice and its interaction with the ocean below are important to modeling the sea-ice cover in the polar oceans. Pioneering contributions were made by Maykut and Untersteiner (1971) to the thermodynamic modeling of sea ice, and most later formulations (Ebert and Curry, 1993; Mellor and Hakkinen, 1995; Mellor and Kantha, 1989; Semtner, 1976) are based on theirs. They considered a comprehensive one-dimensional (horizontally homogeneous) thermodynamic model of the sea ice and its snow cover, accounting for the heat flux from the ocean at the ice bottom, assumed to be known. The radiative and turbulent heat fluxes at the top of the snow/ice are also assumed to be known. They did not, however, account for the leads and thin ice that surround ice floes, and the freezing and lateral melting processes there.

The heat balance at the air-ice/snow interface is quite similar to that at the air-sea interface, except for the components due to conduction through ice and the latent heat required to ablate snow/ice. During winter, when the surface temperatures are low (-20 to $-40\ ^\circ$C), there can be considerable flux of heat from the ocean (temperature typically $-1.7\ ^\circ$C) through ice. Conversely during summer, there can be a heat flux from the atmosphere to the ocean. The fate of the SW solar insolation impinging upon the surface of the ice-snow system depends very much on the albedo of the surface. Snow has typically high albedos, with fresh snow cover having a value of as high as 0.86. On the other hand, the bare ice surface itself has an albedo typically 0.65 or so, but the presence of melt ponds and dirt can decrease these values considerably, with the melt ponds having a typical value of about 0.45. A fraction of the SW radiation impingent upon the top surface penetrates beyond and heats up the interior of ice/snow system but is unavailable for ablation at the surface. The volumetric heating in the ice and snow cover is modeled by exponential decay laws, with the extinction length scale assumed to be constant but different in ice and snow. The properties of ice depend very much on its composition, specifically the presence of brine pockets, and this must be taken into account. The governing equations can be written as

$$(\rho_s c_{ps})\frac{\partial T_s}{\partial t} = \frac{\partial}{\partial z}\left(K_s \frac{\partial T_s}{\partial z}\right) + \frac{\partial I_s}{\partial z}; \quad I_s = Q_{SW}^p \exp(z/l_s)$$

$$(\rho_I c_{pI})\frac{\partial T_I}{\partial t} = \frac{\partial}{\partial z}\left(K_I \frac{\partial T_I}{\partial z}\right) + \frac{\partial I_I}{\partial z}; \quad I_I = Q_{SW}^p \exp(-h_s/l_s)\exp((z+h_s)/l_i)$$

$$\rho_I c_I = \rho_{Ip} c_{Ip} + \gamma_1 S_I(z)/(T_I - 273.15)^2$$

$$K_I = K_{Ip} + \gamma_2 S_I(z)/(T_I - 273.15)$$

$$(12.1.1)$$

where subscripts s, Ip, and I denote snow, pure ice, and ice; T is the temperature (in K), ρ is the density, K is the thermal conductivity, and l is the extinction length scale. γ_1 and γ_2 are constants. Q_{SW}^p is the SW insolation at the bottom of the air-snow/ice interface, $S(z)$ is the salinity of ice and h_s is the snow thickness. ρ and K increase considerably during the melt season in summer, leading to considerable heat storage in brine pockets that is carried over to winter.

Additionally, there has to be a balance of heat flux at the ice-snow and ice-ocean interfaces, accounting for any melting or freezing at the ice-ocean interface:

$$-K_{I/S}\left(\frac{\partial T_{I/S}}{\partial z}\right)_{AI} + (1-\alpha_{I/S})Q_{SW} + Q_{LW} - \sigma\varepsilon_{I/S}T_{AI}^4$$

$$-Q_S - Q_L - Q_{SW}^p = L_{I/S} w_{AI}$$

$$(12.1.2)$$

$$K_I \left(\frac{\partial T_I}{\partial z} \right)_{SI} = K_s \left(\frac{\partial T_s}{\partial z} \right)_{SI}$$

$$-K_I \left(\frac{\partial T_I}{\partial z} \right)_{IO} = Q_O - L_I w_{IO}$$

where subscript I/S refers to ice when the snow cover has melted and snow when it is present. Subscript AI refers to the air-ice/snow interface and $T_{I/S}$ to the surface temperature of snow/ice, SI to the snow-ice interface and IO to the ice-ocean interface. Q_{SW} refers to the short wave and Q_{LW} to long wave radiative fluxes, Q_S and Q_L to sensible and latent heat fluxes from ice/snow to air, and Q_O is the heat flux from the ocean to ice. Q_S and Q_L depend on the characteristics of turbulence in the atmospheric surface layer, and Q_O on those of the oceanic mixed layer. Q_S and Q_L are usually specified using bulk formulae that parameterize the fluxes as functions of the wind speed and the difference in the temperatures and humidities of the air-ice/snow interface, and air at a reference height (Chapter 4 of Kantha and Clayson, 1999). Q_O depends on the turbulent transfer of heat from the ocean to the rough surface underneath ice. α is the albedo, σ is Stefan-Boltzmann constant and ε the emissivity. w_{AI} is the ablation (melting) rate of snow/ice at the air-snow/ice interface, and if it is nonzero, this heat has to be supplied by the heat balance. When $T_{AI} < 273.15$, this term is zero. w_{IO} is the accretion or the melting rate of ice at the ice-ocean interface; if the ocean temperature T_O is less than the freezing value for its salinity S_O, T_f, accretion can occur, and if it is more, melting can occur. Both w_{AI} and w_{IO} are positive for melting. L is the latent heat of fusion.

Maykut and Untersteiner (1971) used monthly average values for the atmospheric and solar forcing, snowfall, and monthly albedo values, and 0.1 m vertical resolution (about 30 ice layers) to simulate the monthly variations in the ice and snow cover in the Arctic (Figure 12.1.1). Their equilibrium thickness of 2.9 m is consistent with the known average thickness of Arctic multiyear ice, but the results do depend sensitively on a variety of parameters, especially the albedo of the snow-ice system and the heat flux at the ice-ocean interface. It is possible to melt the perennial ice cover in a few years by decreasing the average albedo from its value of 0.6 by just 25%. The value for Q_O was taken to be 2 W m^{-2}, in accordance with earlier estimates of the long-term average heat flux in the central Arctic from the warm Atlantic waters a few hundred meters below. However, the equilibrium ice thickness is quite sensitive to the value used.

The resolution of the Maykut and Untersteiner (1971) model is too fine for many large scale ice-only and ice-ocean coupled models. Semtner (1976) explored the possibility of representing the ice thermodynamics with a model of a much cruder resolution. He showed that the results from a three-layer model

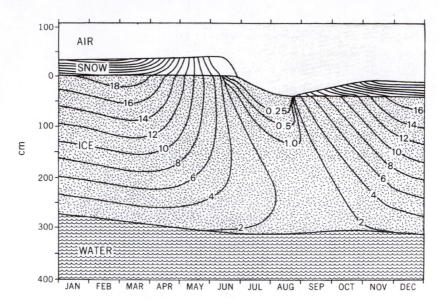

Figure 12.1.1 Seasonal changes in ice thickness and temperature from the model of Maykut and Untersteiner (1971).

with a single snow layer and two ice layers are comparable in accuracy, though slightly worse, to the more detailed Maykut and Unterseiner model. Only the snow temperature T_s and ice temperatures T_{I1} and T_{I2}, and the thicknesses of snow h_s and h_I are solved for. The oceanic part consisted of a 30 m mixed layer with a temperature of $-2°C$. Figure 12.1.2 shows a comparison between ice and snow temperatures from the two models. The maximum error in ice thickness is about 7.5%. This error can, of course, be decreased by using more layers.

Mellor and Kantha (1989; see also Hakkinen, 1995, and Kantha, 1995a) adopted the Semtner model in their ice-ocean coupled model formulation. Figure (12.1.3) shows the configuration of the one ice and one snow layer used by Mellor and Kantha (1989) and Kantha and Mellor (1989b). The snow is assumed to have no heat capacity and the properties of ice are assumed to depend on its salinity S_I taken as constant throughout its thickness for simplicity, even though observations indicate otherwise. A single temperature T_I is also used to characterize the ice temperature; T_0, T_2, and T_3 indicate the temperatures of the ice-ocean (IO), snow-ice (SI), and the snow-air (SA) interfaces. Subscript AI is however used to indicate either the air-ice interface when snow cover is absent and air-snow interface when snow is present.

The heat fluxes at the air-ice and air-ocean interfaces are simply the result of

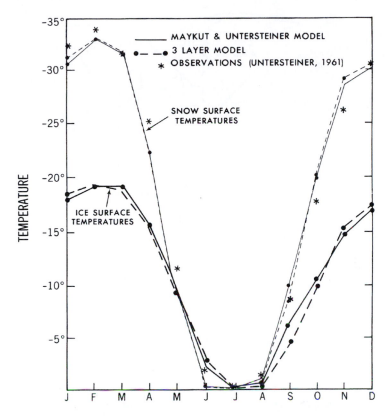

Figure 12.1.2 Comparison of ice and snow temperatures from the simpler three-layer model of Semtner (1976) and the more comprehensive model of Maykut and Untersteiner (1971).

thermodynamic balances at these interfaces:

$$Q_{AI} = Q_{SI} + Q_{LI} - \beta_{AI}(1-\alpha_{AI})Q_{SW} - Q_{LW} + \varepsilon_{AI}\sigma(T_{AI}+273.15)^4$$

$$Q_{AO} = Q_{SO} + Q_{LO} - \beta_{AO}(1-\alpha_{AO})Q_{SW} - Q_{LW} + \varepsilon_{AO}\sigma(T_O+273.15)^4$$

$$(12.1.3)$$

where Q_{SI}, Q_{LI} are sensible and latent heat fluxes from ice/snow to the atmosphere, and Q_{SO}, Q_{LO} are corresponding fluxes from the ocean to the atmosphere; these are parameterized by bulk formulae. Q_{SW} and Q_{LW} are the shortwave and longwave radiative fluxes at the bottom of the atmosphere,

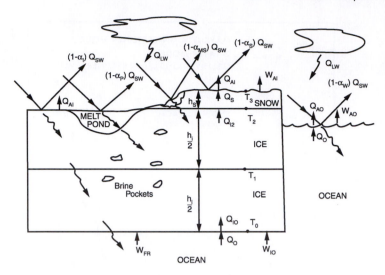

Figure 12.1.3 A sketch of the radiative processes in sea ice and the adjoining lead, illustrating melt ponds, brine packets, and various radiative and conductive fluxes.

dependent on the water vapor content and the cloud cover. α_{AI} is the albedo of the snow/ice surface depending on whether or not snow is present, and β_{AI} is the fraction of the incident shortwave insolation penetrating beyond the interface; α_{AO}, β_{AO} are corresponding values for the ocean. ε_{AI} is the emissivity and T_{AI} the temperature, corresponding to either snow ($T_{AI} = T_3$) or ice ($T_{AI} = T_2$). ε_{AO} is the emissivity and T_O the temperature of the ocean surface. Since the heat capacity of snow is neglected,

$$Q_s = \frac{K_s}{h_s}(T_2 - T_3); Q_{12} = \frac{2K_I}{h_I}(T_I - T_2); Q_s = Q_{12} \qquad (12.1.4)$$

where K_s and K_I are heat conductivities of snow and ice. Equation (12.1.4) can also be written in the form

$$Q_s = \frac{2K_s K_I (T_I - T_3)}{2h_s K_I + h_I K_s} \qquad (12.1.5)$$

When there is no snow ($h_s = 0$), $T_2 = T_3$ and can be eliminated. An auxiliary condition is

$$Q_{AI} = Q_s \quad (T_3 < 273.15) \text{ or}$$

$$T_3 = 273.15 \qquad (12.1.6)$$

When T_3 falls below the freezing point, T_3 is determined by equating Q_{AI} and Q_{I2}. Otherwise, T_3 is held at the freezing point and the difference between Q_{AI} and Q_{I2} is used to melt snow (if present) or ice. The ice melt rate is

$$w_{AI} = \frac{1}{\rho_0 L_F}(Q_{AI} - Q_{I2}) \qquad (12.1.7)$$

The equation for conservation of heat energy in the ice is

$$\rho_I h_I \left(\frac{\partial}{\partial t} + U_{Ij} \frac{\partial}{\partial x_j} \right) \left[(1-r)T_I + \frac{r}{c_{pI}}(L_F + c_{pw}T_I) \right] = Q_{IO} - Q_{I2} \qquad (12.1.8)$$

$$Q_{IO} = \frac{2K_I}{h_I}(T_0 - T_I)$$

where r is the brine fraction, c_{pw} and c_{pI} denote specific heats of water and ice, and L_F is the latent heat of fusion. If the freezing line is approximated by a linear relationship $T_F = mS_F$, then $r = mS_I/T_I$. Q_{IO} is the heat flux just inside the ice-ocean interface. If one chooses to ignore the heat capacity of ice also, the first equation in Eq. (12.1.8) can be replaced simply by $Q_{IO} = Q_{I2}$.

The ocean below is affected by the heat fluxes at both the ice-ocean and air-ocean interfaces. The net oceanic heat flux is given by the weighted combination of the fluxes, Q_{IO} and Q_{AO}, and the melt/freeze rates w_{IO} and w_{AO}:

$$Q_O = \left[A_I Q_{IO} + (1 - A_I)Q_{AO} \right] - w_O L_F \qquad (12.1.9)$$

where

$$w_O = A_I w_{IO} + (1 - A_I)w_{AO} \qquad (12.1.10)$$

where w_O is the total melt/freeze rate of ice due to contributions of both ice cover and open water. The individual contributions w_{IO} and w_{AO} are given by

$$w_{IO} = \frac{1}{\rho_w L_F}(Q_{IO} - Q_O) \; ; \; w_{AO} = \frac{1}{\rho_w L_F}(Q_{AO} - Q_O) \qquad (12.1.11)$$

Q_O, the net oceanic heat flux, and T_O, the temperature of the ice-ocean interface, can be determined by appealing to the laws of heat transfer in a turbulent flow over a rough surface and taking into account the melting/freezing and related processes that can occur at the interface. This requires coupling to an ocean or mixed layer model. If we assume for the moment that T_O and Q_O are known, then assuming $A_I = 1$ (or considering only the ice-covered portion), then Eqs. (12.1.3), (12.1.4), (12.1.5), and (12.1.7) constitute seven equations for seven unknowns T_1, T_2, T_3, Q_{AI}, Q_s, Q_{I2}, and Q_{IO}, from solutions of which w_{AI} and w_{IO} can be determined. Mellor and Kantha (1989) were able to reproduce the monthly changes in ice cover simulated by Maykut and Untersteiner (1972) almost exactly by selecting the snow, wet snow, and ice albedos of 0.82, 0.73, and 0.62. An equilibrium solution with ice ablation at the top of 0.5 m yr^{-1}, balanced by an ice accretion at the bottom of the same amount was attained after 30 years of simulation. Figure (12.1.4) shows the seasonal variations of ice and snow thicknesses and temperatures from the model compared to those of Maykut and Untersteiner (1972).

The presence of a cold halocline in the Arctic between the cold mixed layer and the warm waters below (Aagard *et al.,* 1981) inhibits heat transfer across it to the underside of the perennial ice. Thus unlike the Antarctic, where a large quantity of heat transferred upward into the mixed layer keeps the ice thin, the ice cover and the mixed layer in the Arctic are thermodynamically decoupled from the warm water below, and the upward heat flux is a few W m^{-2} instead of many tens of W m^{-2} on the average in the Antarctic. The majority of the heat for melting ice during summer comes from solar heating through ice and open water, especially the latter. Measurements during AIDJEX (Maykut and McPhee, 1995) show a strong seasonal modulation of this heating with values reaching of 40 to 60 W m^{-2} during August, with the annual average of about 5 W m^{-2}, a value more than twice that of the upward heat flux measured from bottom ablation measurements of multiyear ice. During summer, this large heat source elevates the mixed layer temperature by as much as $0.4°C$ above freezing (Maykut and McPhee, 1995; Perovich and Maykut, 1990). Most ice-ocean models ignore this and assume the water to be at freezing temperature as long as ice is present and tacitly assume that any heat absorbed in open leads/thin ice is used to melt multiyear ice laterally.

Both Maykut and Untersteiner (1971) and Semtner (1986) models ignore the effect of open water adjacent to the ice floes. Even for multiyear ice in the Arctic, ice growth and melt do not entirely produce changes in ice thickness. Lateral melting of ice floes by absorption of solar insolation in leads during summer (Maykut and McPhee, 1995; Maykut and Perovich, 1987) and heat loss and ice production in leads during winter produce changes predominantly in the ice concentration that need to be taken into account. So thermodynamic models of sea ice must solve for both ice concentration A_I as well as the ice thickness h_I. They must also account for the fluxes at the air-ocean interface Q_A. Mellor and Kantha (1989) and Kantha and Mellor (1989b) constructed an ice-ocean coupled model, based on the simpler Semtner thermodynamic ice model, but incorporating the effects of open water/thin ice. The ocean is considered to be covered by open water/thin ice of fractional areal coverage $(1-A_I)$, and equations are written for both the areal ice concentration A_I as well as the average ice

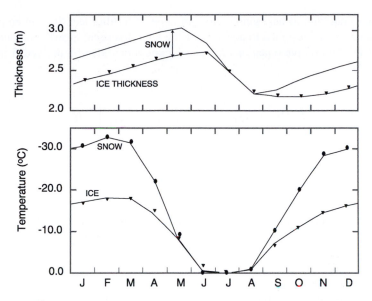

Figure 12.1.4 Comparison of ice and snow thicknesses and temperatures from the Mellor and Kantha (1989) model (lines) with those of Maykut and Untersteiner (1971) (points). From Mellor and Kantha (1989).

thickness $D_I = A_I h_I$, where h_I is the ice thickness:

$$\frac{\partial A_I}{\partial t} + \frac{\partial}{\partial x_j}\left(A_I U_{Ij}\right) = \frac{\rho_0}{\rho_I h_I}\left[\Phi(1 - A_I)w_{AO} + (1 - A_I)w_{FR}\right]$$

$$\frac{\partial D_I}{\partial t} + \frac{\partial}{\partial x_j}\left(D_I U_{Ij}\right) = \frac{\rho_0}{\rho_I}\left[(1 - A_I)w_{AO} + A_I(w_{IO} - w_{AI}) + w_{FR}\right] \qquad (12.1.12)$$

$$0 \leq A_I \leq 1$$

The first equation is purely empirical, first introduced by Nikiferov (1957) without the source/sink terms on the RHS. The second equation is an exact conservation relation for ice mass. The terms w_{AI}, w_{IO}, w_{AO}, and w_{FR} denote melting/freezing rates (in ms^{-1}) at various interfaces: w_{AI} is the melting (positive)/freezing (negative) rate at the ice surface; w_{IO} is the melting (negative)/freezing (positive) rate at the ice bottom related to congelation ice; w_{FR} is the rate of ice growth (positive) due to accretion of frazil ice generated in the water column (due to supercooling) assumed to be deposited instantaneously on the underside of the ice; and w_{AO} is the melting (negative) or freezing (positive) rate due to heat gain or loss in open water/thin ice. Φ is an empirical parameter that is designed to account for the lateral/vertical growth and decay of ice cover. Alternative equations for ice thickness can also be derived from Eq. (12.1.12):

$$\frac{Dh_I}{Dt} = \frac{\rho_0}{\rho_I}\left[\left(\frac{1-A_I}{A_I}\right)(1-\Phi)w_{AO} + (w_{IO} - w_{AI}) + w_{FR}\right] \qquad 0 \le A_I < 1$$

$$\frac{Dh_I}{Dt} + h_I\frac{\partial U_{Ij}}{\partial x_j} = \frac{\rho_0}{\rho_I}\left[(w_{IO} - w_{AI}) + w_{FR}\right] \qquad\qquad A_I = 1$$

(12.1.13)

where D/Dt denotes the substantial derivative. Assume that the source/sink terms on the RHS are zero. Then the second equation in Eq. (12.1.12) is a simple statement of conservation of advected ice mass $\rho_I\, D_I$. The first equation, however, leads to interesting consequences. As long as $A_I < 1$, $Dh_I/Dt = 0$ as can be seen from Eq. (12.1.13). Any convergent/divergent motion of ice does not change the ice thickness but changes the amount of open water. When A_I reaches unity under ice convergence, the first equation is overridden, and Eq. (12.1.13) indicates $Dh_I/Dt = -h_I(\partial U_{Ij}/\partial x_j)$, so that any additional convergence increases the ice thickness, simulating ridging processes.

The source/sink terms on the RHS of the ice concentration equation in Eq. (12.1.12) represent thermodynamic growth/decay of ice cover and hence changes in ice the mass. The role of Φ on the RHS of the first equation can be understood by considering the case when only w_{AO} is nonzero. Because of multiplication by $(1-A_I)$, this term is relevant only when there is open water. If $\Phi = 1$, then Eq. (12.1.13) indicates that any heat loss/gain in open water does not affect the ice thickness; only the ice concentration would change. Thus ice could melt or freeze only laterally. However, summer melting occurs both under and around ice floes, reducing both the ice thickness and its concentration. Analysis by Maykut and Perovich (1987) suggests a value of 0.5 for melting conditions ($w_{AO} < 0$), and Mellor and Kantha (1989) select this value for the ice-ocean coupled model they developed. Under freezing conditions ($w_{AO} > 0$), they select a value of 4.0, based on sensitivity studies in the Arctic to seasonal forcing. Nonzero values of Φ affect both the ice thickness h_I as well as its concentration A_I. The term w_{IO} does not affect the ice concentration, since this represents accretion or melting at the ice undersurface due to its interaction with the ocean.

Hakkinen and Mellor (1992) have modified the source/sink term on the RHS of the concentration equation by replacing the term in the square brackets by $4(1-A_I)w_{AO} + 0.7A_I(w_{AI} - w_{IO})$, where the first term is nonzero only for positive values of w_{AO} and the second term is nonzero only for positive values of

($w_{AI} - w_{IO}$). The first term is to allow leads to freeze over during winter. The second term is zero during freezing, but creates open water when net melting occurs.

Parkinson and Washington (1979) and Hibler (1979) were the first to apply a dynamical-thermodynamic model to simulate the seasonal changes in the polar sea-ice cover. The models were driven by realistic observed (monthly) winds, air temperature, and radiative fluxes. Parkinson and Washington used the Semtner thermodynamic model consisting of a single snow and a single ice layer, overlying a motionless mixed layer of constant depth and coupled to an ice dynamical model without any internal stresses. Their simulations showed the feasibility of reproducing many features of the sea-ice cover in the Arctic and around Antarctica with simple dynamics and thermodynamics. Hibler (1979, 1985) and Hibler and Walsh (1982) used viscous-plastic rheology instead to model the ice cover in the Arctic. Hibler (1979) also formulated equations for ice concentration and thickness, with source and sink terms parameterizing ice growth and decay. For the freezing case, Hibler's formulation is equivalent to that of Mellor and Kantha (1989) if $\Phi = h_I/h_0$, with $h_0 = 0.5$ m; the melting case is equivalent since $\Phi = 0.5$ in his formulation also. He does not, however, consider frazil ice formation. While Hibler (1979) prescribed the observed ice growth rate as function of ice thickness and season from Thorndike *et al.* (1975), Hibler and Walsh (1982) used Semtner's thermodynamic model to estimate ice growth. The salient difference between these simulations and those of Parkinson and Washington (1979) is that, instead of the maximum ice thickness being in the central Arctic, it is found to occur near the Greenland side of the Arctic. Hibler and Walsh (1982) simulated the Arctic ice cover (without snow) for, 1973 to 1975 with daily surface forcing, using both thermodynamics-only and full dynamical-thermodynamic models and showed that the inclusion of a realistic ice rheology is responsible for the more realistic distributions of ice thickness in the Arctic. However, the absence of the warm inflow from the Atlantic through the Fram Strait leads to inaccurate ice extent.

Walsh *et al.* (1982) performed the very first long-term (1951 to 1980) simulation of the Arctic ice cover using the Hibler (1979) model forced by daily winds and air temperature data, with snow cover prescribed from Maykut and Untersteiner (1971). These simulations show the strong interannual variability in the ice cover and the importance of dynamical processes in determining the variability in the thickness of the ice cover in the Arctic.

Most current dynamical-thermodynamic ice models consider only a single ice thickness in addition to open water (or equivalently very thin ice). In reality the ice cover consists of a spectrum of ice thicknesses spanning both multiyear and first-year ice. A single ice thickness is therefore unlikely to represent the observed distribution and all properties relevant to thermodynamics. In fact, Mellor and Kantha (1989) found that as far as heat conduction through ice is concerned, the effective ice thickness is a fraction (0.5 to 0.7) of the average thickness, skewed more toward thinner ice (see Mellor and Hakkinen, 1995). Increasing realization in the 1990s of the complexity of the disposition of solar radiation in ice brought on by ice thickness distribution has prompted explicit radiative modeling of multiple ice thicknesses and categories. Ebert *et al.* (1995) considered 15 different ice thicknesses and two ice categories (multiyear and first year) and accounted for the melt ponds and brine pockets to calculate the amount of SW radiation penetrating through the Arctic ice from March to October. Their results show that the peak in the shortwave radiation transmitted through the ice occurs from July through August and is 10 W m^{-2}, compared to the 6 W m^{-2} for a similar model but with a single thickness distribution (Ebert and Curry, 1993). The incident SW radiation peaks in June at about 300 W m^{-2}. Since the presence of any dry snow cover of significant thickness on sea ice reflects nearly 80% of the incident SW radiation and absorbs nearly the rest, this difference is important only during the summer melt season.

The main problem with this approach is the difficulty of constructing a dynamical model with proper rheological properties for ice with multiple thicknesses, especially in view of the changes in thickness distribution brought on by thermodynamic and dynamical interactions. A more practical strategy might be to use a single ice thickness but to scale it down appropriately when considering its SW radiative properties. An equivalent thickness appears to be about 0.6 times the slab thickness used for dynamical modeling, roughly the same as found by Mellor and Kantha (1989) from conduction point of view. Once again, the SW penetration properties are skewed toward thinner ice.

The complexity of the thermodynamic ice model itself has grown considerably, as attempts are being made to include processes such as melt ponds (Ebert and Curry, 1993; Ebert *et al.*, 1995) and radiative heating in ice more accurately. At the minimum, it is important to distinguish between the albedos of dry fresh snow (~0.8), melted snow (~0.6), melt ponds (~0.4), bare ice (~0.5 to 0.6), and open water (~0.1). Melt ponds alter the surface energy balance for two principal reasons: first, they decrease the albedo of the ice surface; second, they affect the radiation balance in the ice and ocean below. A substantial amount of heating during early summer goes toward creating melt ponds. Melt ponds cover 10 to 25% of the surface of multiyear ice and much higher surface of first-year ice, and their effective albedo depends very much on the albedo of their bottom. The water in melt ponds must freeze first. Podgorny and Grenfell (1996) have constructed a detailed radiative model of melt ponds and their effect on solar energy redistribution in the ice pack. The difference between the albedos of the bottoms of old melt ponds (~0.2) and new ones (~0.2 to 0.6 depending on the wavelength) can affect the thermodynamic evolution of sea ice during summer, especially when the ice surface is covered extensively by melt ponds from early summer melting.

12.2 COUPLED ICE–OCEAN MODELS

Ice-only models either ignore the ocean altogether or assume a motionless ocean. For example, Fichefet and Maqueda (1997) describe a global thermodynamic-dynamic sea-ice model coupled to a one-dimensional slab mixed layer, and show that a prognostic snow layer is important to the Antarctic, whereas the thermal inertia of the snow-ice system is negligible. They also find that internal storage of heat in the arctic ice is determined by internal storage of latent heat in brine pockets. The oceanic heat flux in these models is either unrealistic or has to be tuned to yield correct ice thicknesses. The consequences are rather serious, since the ice thickness is quite sensitive to the heat flux at the bottom of the ice. In simulations of the Arctic ice cover, the errors are greatest in the North Atlantic subpolar seas such as the Greenland and Labrador Seas (and Arctic marginal seas such as the Barents Sea). The ice extends too far during winter. This deficiency is principally due to the lack of thermodynamic interaction of sea ice with the warm Atlantic waters. In the Southern Ocean around Antarctica, the large oceanic heat flux plays a crucial role in keeping the predominantly first-year ice thin (around a meter). In sensible heat polynyas, such as the Weddell Sea polynya and to some extent the North Water polynya in Baffin Bay, where the heat from the upwelled warm waters helps keep the polynya open, the ocean plays a predominant role. It is therefore essential to account for the thermodynamic (and dynamic) interaction of the ice cover with the underlying ocean. This requires that the two media, the ice and the ocean, be coupled both dynamically and thermodynamically. Such coupling demands that proper attention be paid to the turbulent boundary layer flow past the rough bottom surfaces of ice floes.

In modeling the momentum, heat and salt fluxes across an ice-ocean interface, one assumes that the velocity components U_p and V_p and the temperature T_p and salinity S_p are known at a point in the constant flux layer of the turbulent boundary layer. The task then is to relate these quantities to the corresponding quantities U_I, V_I, T_O, and S_O at the undersurface of the ice. If the point is sufficiently close to the surface, but away from the roughness elements, the universal logarithmic laws of the wall (see Kantha and Clayson, 2000) should apply for the profiles so that

$$\frac{1}{\rho_w}\left(\tau_{IOx}, \tau_{IOy}\right) = \frac{\kappa u_*}{\ln\left(-z_p / z_0\right)}\left(U_p - U_I, V_p - V_I\right)$$

$$\rho_w u_*^2 = \left(\tau_{IOx}^2 + \tau_{IOy}^2\right)^{1/2}$$

(12.2.1)

$$\frac{Q_O}{\rho_w c_{pw}} = \frac{\kappa u_*}{Pr_t \ln(-z_p / z_{0T})}(T_p - T_O) = C_T(T_p - T_O) \qquad (12.2.2)$$

$$\frac{Q_O^s}{\rho_w} = \frac{\kappa u_*}{Pr_t \ln(-z_p / z_{0S})}(S_p - S_O) = C_S(S_p - S_O) \qquad (12.2.3)$$

where Q_O and Q_O^s are the heat and salt fluxes from the ocean, and T_O and S_O the temperature and salinity at the ice-ocean interface. u_* is the friction velocity; κ is the von Karman constant. Quantities z_0, z_{0T}, and z_{0S} are the roughness scales for momentum, temperature, and salinity and need to be prescribed. Alternatively, the bulk transfer coefficients c_d, c_h, and c_s for momentum, heat, and salt transfer can be prescribed. Because momentum can be transferred by pressure forces resulting from flow across roughness elements, whereas scalars can only be transferred across molecular sublayers adjacent to the surface, z_{0T} and z_{0S} are different from z_0 (and each other) and are generally much smaller. The ratios of these to z_0 are indicative of the large change in temperature and salinity across these molecular sublayers, and depend very much on the molecular diffusivities, even though the flow is fully turbulent. Analysis of laboratory experiments on heat and mass transfer across rough surfaces by Yaglom and Kader (1974) suggests that

$$\frac{(z_{0T}, z_{0S})}{z_0} \approx \exp\left[-\kappa b \, Pr_t^{-1}\left(\frac{u_* z_0}{\nu}\right)^{1/2}\left(Pr^{2/3}, Sc^{2/3}\right)\right] \qquad (12.2.4)$$

where $Pr \sim 13$ and $Sc \sim 2430$ are the molecular Prandtl and Schmidt numbers corresponding to diffusion of heat and salt; $Pr_t \sim 0.86$, b ranges from approximately 1.5 (McPhee *et al.*, 1987) to 3.0 (Yaglom and Kader, 1974). To a good degree of approximation, z_p/z_{0T} and z_p/z_{0S} terms in Eqs. (12.2.2) and (12.2.3) can be replaced by z_0/z_{0T} and z_0/z_{0S}. This is equivalent to ignoring the change in the logarithmic region and regarding the changes in temperature and salinity (T_p-T_O) and (S_p-S_O) as being solely due to changes across the molecular sublayer. The salt flux is given by

$$Q_o^s = w_O(S_I - S_O) + (1 - A_I)S_O(\dot{P} - \dot{E}) \qquad (12.2.5)$$

where $\dot{P}$ and $\dot{E}$ denote precipitation and evaporation rates in the leads. The momentum roughness scale z_0 itself is a area-weighted average of the values for the open water and ice undersurface. Also T_O and S_O are related to each other, since the interface corresponds to the freezing line, which can be approximated by the linear relationship,

$$T_O = m\, S_O; \quad m \sim -0.054 \qquad (12.2.6)$$

Equations (12.1.9) and (12.2.1) to (12.2.6) can be solved for the five unknowns Q_O, Q_O^S, w_O, T_O, and S_O. w_O can be obtained from

$$w_O = -\frac{1}{2}\left[\hat{b} + \left(\hat{b}^2 - 4\hat{c}\right)^{1/2}\right]$$

$$\hat{b} = -C_S + \frac{\rho_w c_{pw} C_T}{L_F}\left(T_p - m S_I\right) - \frac{1}{L_F}\left[(1-A_I)Q_{AO} + A_I Q_{IO}\right]$$

$$\hat{c} = -\frac{\rho_w c_{pw} C_T C_S}{L_F}\left(T_p - m S_p\right) + \frac{C_S}{L_F}\left[(1-A_I)Q_{AO} + A_I Q_{IO}\right]$$

$$-m\frac{\rho_w c_{pw} C_T}{L_F}\left[(1-A_I)S_O\left(\dot{P} - \dot{E}\right)\right] \qquad (12.2.7)$$

Note that if $A_I = 0$, $w_O = 0$. Q_O and T_O can be obtained from Eqs. (12.1.9) and (12.2.2), and Q_O^S and S_O from Eqs. (12.2.3) and (12.2.5). Quantities w_{AO} and w_{IO} needed in Eq. (12.1.12) are obtained from Eq. (12.1.11).

The ocean underneath is driven by the resulting momentum, heat, and salt fluxes. The boundary conditions at the top of the modeled ocean in a coupled ice-ocean model is

$$K_M\left(\frac{\partial U}{\partial z}, \frac{\partial V}{\partial z}\right)_{z=0} = \frac{1}{\rho_w}\left[A_I\left(\tau_{IOx}, \tau_{IOy}\right) + (1-A_I)\left(\tau_{AOx}, \tau_{AOy}\right)\right]$$

$$K_H\frac{\partial T}{\partial z}\bigg|_{z=0} = -\frac{Q_O}{\rho_w c_{pw}} = C_T\left(T_p - T_O\right) \qquad (12.2.8)$$

$$K_H\frac{\partial S}{\partial z}\bigg|_{z=0} = -\frac{Q_O^S}{\rho_w} = C_S\left(S_p - S_O\right)$$

The ice-ocean coupled model outlined above can be applied to any ocean model, z-level, sigma-coordinate or layered/isopycnal, since it is the boundary conditions [Eq. (12.2.8)] on the surface fluxes that are needed. This ice model coupled to a sigma-coordinate ocean model has been used in studies of the Arctic ice cover by Mellor and Kantha (1989) and Hakkinen and Mellor (1990) using one-dimensional versions, the Bering Sea marginal ice zone by Kantha and Mellor (1989a) and the Arctic leads by Kantha (1995) using two-dimensional versions, and deep convection in the Greenland Sea using a three-dimensional version by Hakkinen *et al.* (1992). Hakkinen and Mellor (1992) used the 3-D coupled model to study the seasonal variability in the Arctic and Hakkinen (1995) used it to study the seasonal variability of ice cover around Antarctica.

Hakkinen and Mellor (1992) have applied the coupled model to the Arctic. The model incorporates a mixed layer based on second-moment closure and ice dynamics. They show that more realistic simulations result when daily surface forcing is used instead of monthly surface forcing. The model simulates well many of the observed features of the Arctic circulation, the Beaufort Gyre, the Transpolar Drift Stream, and the ice pileup in the Greenland sector. Hakkinen (1993) carried long-term simulations with this model for 1955 through 1975 and explored the large ice export from the Arctic as a possible cause for the low salinity anomaly event observed to pass through the Labrador Sea in the 1960s. More recently, Hakkinen (1995) has applied the coupled model to a study of the seasonal variability of ice, circulation, and deep-water formation in the Southern Sea around Antarctica. Hakkinen (1997) models the combined Arctic-north Atlantic oceans on a rotated spherical grid.

Since the sea-ice cover consists of a spectrum of ice thicknesses ranging from thin ice in freshly refreezing leads to multiyear ice several meters thick to heavily ridged ice with deep keels, it is reasonable to suspect that its thermodynamic interaction with the ocean and the atmosphere should account for the ice thickness distribution. But as described earlier, most global climate and regional ice-ocean coupled models consider only two ice thickness categories, thin ice (including open water) and thick ice with a mean thickness (Flato and Hibler, 1992; Hakkinen, 1995; Hibler, 1979, Kantha and Mellor, 1989a; Mellor and Kantha, 1989). Since most of the heat loss and ice growth occurs in thin ice or open water regions during winter, and most of the SW radiation is absorbed there as well during summer, it is the fraction of thin ice and open water that is predominant in thermodynamic interactions of the ice pack. However, it is the dynamics of the ice pack that influences the ice convergence and divergence and therefore the amount of open water and thin ice present at any given time as well as the formation of ridged ice. The sea-ice dynamics and thermodynamics are therefore, unfortunately, inextricably entangled, and dealing with one aspect in great detail while treating the other cursorily is at best likely to lead to a partial answer. Nevertheless, the relative ease of treating the thermodynamic issues vis-a-vis the dynamical ones has led to the proliferation of 1-D ice-only and ice-ocean coupled models (for example, Bjork, 1997; Ebert and Curry, 1993; Hakkinen and Mellor, 1992; Lemke, 1987; Mellor and Kantha, 1989), while models that account for both thermodynamic and dynamical effects are relatively few (Hibler, 1979; Kantha and Mellor, 1989a; Piacsek et al., 1991; Walsh et al., 1995), and the models that couple the atmosphere are even fewer (for example, Kantha and Mellor, 1989b). Given the strong feedback from the ABL over ice, coupled ocean-ice-atmosphere models even of the 1-D variety would be useful.

There have been several attempts to build a very detailed one-dimensional thermodynamic model of sea ice in the Arctic (Bjork, 1992. 1997; Ebert and

Curry, 1993; Ebert *et al.*, 1995; and Schramm *et al.*, 1997) and obtain detailed data from intensive observations (SHEBA, 1994). In these models, the sea-ice model consists of numerous ice categories with different thicknesses and ages and their thermodynamic evolution is very carefully modeled, but the dynamically controlled quantities such as the rate of ice export and ridging are prescribed.

Flato and Hibler (1995) used Thorndike *et al.* (1975) formulation for ice thickness evolution and included 28 ice thickness categories in their model, with highest resolution in thermodynamically important thin ice. Only the area fraction and average properties of each thickness category were modeled. In contrast, Bjork (1992), and more recently Bjork (1997) and Schramm *et al.* (1997), model the evolution of the thicknesses of several categories of ice with different thicknesses, including ridged and thin ice. Distinction between ridged and level ice is retained. In this approach, it is necessary to merge different categories as they grow or decay. Each category is identified by its area fraction, thickness, and temperature, with the area fraction for each controlled by ice export and ridging, while the thickness and temperature are controlled by thermodynamic interactions. The thinner ice categories control the ice growth in winter, and albedo and hence absorption of SW radiation during summer. Bjork (1997) finds that the ice export from the Arctic has more influence on the mean ice thickness than the ridging process, and a 50% reduction in its value to 0.07 Sv would increase the mean ice thickness in the Arctic to 5 m, instead of the prevailing 3 m value, for the same atmospheric forcing. His sensitivity studies underscore the fact that the uncertainty in, for example, the extent of open water often overwhelms any deficiencies in albedo and ridging parameterizations.

Following Bjork's (1992, 1997) approach to dealing with the ice thickness distribution, Ebert and Curry (1993), Schramm *et al.* (1997), and Holland *et al.* (1997) have constructed exceptionally detailed thermodynamic 1-D ice-only and ice-ocean coupled models. The ice in each category is characterized by its surface properties such as albedo, snow cover, and the depth and extent of melt ponds; its interior properties such as its temperature, extent of brine pockets, and its thickness; and whether it is ridged or level, first-year ice with high salinity content, or multiyear ice. Each category is allowed to evolve independently of others (in a comprehensive thermodynamic-dynamical model that would include sea-ice dynamics and rheology, it would be necessary to track each of these ice categories and their mutual dynamical interactions), except that the thin ice is affected by the ridging process, and the categories of similar thickness and ages are occasionally merged. The ice surface temperature and eight interior temperatures are modeled, with the salinity a function of the ice thickness but constant with depth. The extent and depth of melt ponds and the associated albedo are modeled. SW solar insolation is partitioned into four spectral bands and treated separately. Complex surface albedo parameterization is undertaken.

It is difficult to reproduce the complex details here; the reader is instead referred to the above-mentioned references. Not surprisingly, Schramm *et al.* (1997) conclude that it is ice thinner than a meter that needs to be resolved well from the thermodynamic point of view. They also conclude that a minimum of 16 ice categories are needed to accurately reproduce the seasonal fluctuations in the ice cover in the Arctic. Holland *et al.* (1997) find that it is particularly important to account for heat storage by the mixed layer during summer. They also conclude, not surprisingly, that while the exceptionally detailed ice model coupled to a simple slab type mixed layer compares favorably with measurements during AIDJEX, it is difficult to ascertain whether the differences that exist are due to remaining deficiencies in the thermodynamic parameterizations or to the absence of any horizontal advection in the 1-D formulation.

Hibler and Bryan (1987) and Semtner (1987) were the first to use coupled ice-ocean models to study the seasonal changes in ice cover *and* the ocean circulation in the Arctic. Both coupled Hibler's (1979) dynamical-thermodynamic ice model to the GFDL Bryan-Cox-Semtner z-level ocean model. Hibler and Bryan used Newtonian damping to keep the temperature and salinities in the deeper parts close to the prescribed climatological values, and hence their ocean model is nearly diagnostic. The upper ocean was, however, forced by ice melting and freezing processes, driven by data from 1979. Their simulations showed the importance of the North Atlantic warm water in determining the ice extent in the Greenland and Barents Seas. They also observed the transport of ice by the Transpolar Drift Stream and its pileup along the Greenland sector of the Arctic. Semtner (1987), on the other hand, used fully prognostic equations for temperature and salinity but simplified the rheology to retain only the bulk viscosity, which relaxes the time stepping requirements and enables long-term simulations to be more efficient. He also used his simpler three-layer thermodynamic ice model. The monthly mean surface forcing came from Walsh *et al.* (1982). River runoff and flow through Bering Strait were included. While the ice thicknesses were greatly underpredicted, the model did show that the river runoff had only local influence and hence a potential diversions of Siberian rivers would have a rather minimal impact on the Arctic ice as a whole. Fleming and Semtner (1991) used this model to simulate the seasonal and interannual variability during 1971–1980. The results, while more realistic than Semtner's (1987) simulation in some respects, still suffered from overly thin ice cover. None of these models includes a good mixed layer treatment central to ice-ocean interaction processes.

Piacsek *et al.* (1991) coupled the Hibler ice model to the Arctic with values of geostrophic currents, derived from climatological hydrographic fields, prescribed in the underlying ocean. They, however, used a turbulence closure model for the mixed layer and attribute the realistic seasonal variability predicted for 1986 to more realistic mixed layer processes. Riedlinger and Preller (1991) have converted Hibler and Bryan (1987) to operational use by the U.S. Navy as Polar Ice Prediction System (PIPS). The model does a reasonable job of simulating the ice cover, although errors due to available surface forcing and inadequate resolutions often lead to inaccurate simulations in the Greenland and Barents Seas.

Because of the perceived advantages of maintaining the deep oceans adiabatic, isopycnal models are being increasingly used for simulations of the thermohaline circulations in ocean basins. None of these ocean models have been coupled to an ice model, with the exception of one (Oberhuber, 1993b), even though the details of deep-water formation in the Atlantic depend on many things including the ice cover in the subpolar seas. There exist no layered basin or global models coupled to ice, even though there have been some regional models such as that of Hakkinen (1987), who used a two-layer ocean model coupled to sea ice to study processes such as ice edge upwelling. Using the Hibler (1979) model, Holland *et al.* (1993) present a comprehensive study of the thermodynamic/dynamical properties of sea ice and their sensitivity to external forcing. Theirs is an exhaustive study of the Hibler (1979) model and is therefore useful for coupled models incorporating this particular ice model. Oberhuber (1993 a,b) has used the Hamburg isopycnal model coupled to the Hibler dynamic (1979) and Semtner thermodynamic (1976) ice models to simulate the Arctic and the Atlantic (to 30°S) Oceans at a coarse 2° resolution near the equator but higher resolutions toward the pole.

On regional scales, 1-D, 2-D, and 3-D coupled models have been used to study a variety of local processes. 1-D coupled models are essentially ice thermodynamic models coupled to an oceanic mixed layer (Ebert and Curry, 1993; Ebert *et al.,* 1995; Hakkinen and Mellor, 1990; Lemke, 1987; Mellor and Kantha, 1989). Lemke (1987) coupled a Kraus-Turner type bulk mixed layer model to a simple thermodynamic ice model to investigate processes related to Weddell polynya. Mellor and Kantha (1989) and Hakkinen and Mellor (1992) explored the seasonal variability of Arctic sea-ice cover and its sensitivities to external forcing using a mixed layer model based on second-moment closure to Semtner ice model as described earlier. While the incorporation of an active mixed layer leads to definite improvements in the simulation of the ice cover, the ice dynamics partly responsible for the ice thickness and concentration distributions is mostly absent. It requires at least a 2-D model to explore the dynamical aspects. Kantha and Mellor (1989b) applied a 2-D ice-ocean coupled model to study the oceanic temperature and salinity structure at the edge of the Bering Sea MIZ, and its sensitivity to external forcing, and Kantha (1995a) has recently studied the circulation processes underneath a growing winter lead, where convection driven by salt extrusion from growing ice leads to vigorous convection. His simulations compare reasonably well with observations made

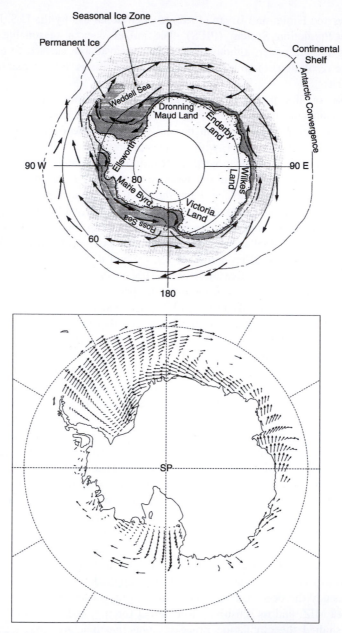

Figure 12.2.1 Sea-ice velocities in September around Antarctica (bottom) from an ice-ocean coupled model of the southern Ocean. Velocities in excess of 15 cm s^{-1} are truncated. Top panel shows a schematic of the current system and the maximum and minimum sea-ice cover. From Hakkinen (1995).

during the 1992 LEADEX studies in the Beaufort Sea. Heinrichs *et al.* (1995) used a 2-D coupled model to show that the heat from the upwelled warm water present at a depth of 100 to 150 m in the Baffin Bay plays an important role in keeping the North Water polynya open.

Nonhydrostatic models have also been applied recently to simulate convective processes under sea ice cover. Smith and Morison (1998) used a nonhydrostatic ocean model without coupling to the ice to simulate convection under refreezing leads. Potts (1998) included coupling to sea ice and frazil ice. Both studies show that in shallow convective layers under leads, nonhydrostaticity is small (though non-negligible) and for all practical purposes, a hydrostatic model suffices. Incorporation of an accurate mixing model is more crucial (Kantha, 1995; Potts, 1998). Nonhydrostaticity cannot, however, be neglected in modeling deep convection in Labrador and Greenland Seas (Marshall *et al.*, 1997a,b; Potts, 1998).

Regional 3-D coupled models have also been used to study processes such as deep convection (Hakkinen *et al.,* 1992). Ikeda (1988) showed the importance of the cross-shore winds in determining the fate of ice in ice-covered coastal oceans. More recently Hakkinen (1995) has applied ice-ocean coupled model to a simulation of the sea ice and circulation in the Southern Ocean (Figure 12.2.1). Heinrichs (1996) has applied it to study the North Water polynya in the Baffin Bay.

To sum up, comprehensive coupled ice-ocean models are essential to the study of the ice cover and its variability in the polar oceans and their potential impact on Earth's climate. It is very likely that in the coming century, we will see more truly global ocean simulations that incorporate both the Arctic Ocean and the Seas around Antarctic. Ultimately, it may be necessary to couple all three media—the atmosphere, the hydrosphere and the cryosphere—in a truly global model of the Earth's climatic system. A step in this direction is the NCAR climate model (Boville and Gent, 1998). However, the excessive resource requirements of such a modeling system, especially at resolutions needed to faithfully represent processes in each of the media and their interactions, await multiteraflop computing capabilities of the coming century. Modeling the Earth's climate system and the anthropogenic impact on it with great fidelity may yet be the grandest challenge of them all, and the stakes in the success or failure of such an enterprise may very well be high for the generations to come.

Chapter 13

Ocean-Atmosphere Coupled Models

Most ocean modeling, whether global or regional, is done with atmospheric forcing assumed to be known a priori and hence prescribable independently. For hindcasts (simulations of past events, for which atmospheric data are available), certain process studies, and climatological type simulations, this does not prove to be a handicap. However, true forecasts, simulations of the state of the ocean at a future time, are severely constrained. Since by definition, no observational data are available, and since NWP model forecast skills deteriorate beyond a few days, the atmospheric surface forcing needed to make dynamical ocean forecasts over timescales of beyond about 3–5 days is currently unavailable. This is a severe restriction on forecasts of the oceanic state, since the timescale for decorrelation from the initial state is several weeks in the oceans (as opposed to several days in the atmosphere). Hence, in principle at least, substantial forecast skill can be retained for a few weeks, if only the surface forcing were known. One solution (the other is a statistical prediction, see Appendix C) is a coupled dynamical ocean-atmosphere model, where the interaction between the two fluids is modeled instead of being guessed or prescribed. The hope is that through accurate simulation of air-sea and air-land interactions, it may be possible to extend the skill of the atmospheric models beyond what is currently feasible without the oceans, before the inevitable sensitivity of nonlinear systems to initial conditions sets in. Such a realization would also make longer ocean forecasts feasible.

In regions where the processes are more deterministic, such as in the tropics, coupled models hold the promise of enabling forecasts of the state of the coupled system to be made months in advance. A classical example is the dynamical forecast of the state of the tropical Pacific Ocean, several months in advance (for ENSO-related applications), which requires a coupled global ocean-atmosphere model or at least a coupled model of the tropical waveguide (Cane, 1992, Ji *et al.*, 1998). Process studies, where the oceanic feedback to the atmosphere is a crucial component—for example, in oceanic studies on timescales longer than a few days in the coastal oceans and a few weeks in the deep oceans—also require coupled models. Studies of Earth's climate on timescales of decades and beyond require the state of the coupled ocean-atmosphere-sea-ice system to be simulated, and this again requires (global) coupled models (Boville and Gent, 1998; Meehl, 1992).

Coupled ocean-atmosphere models can be either global or regional. Regional models are useful if local air-sea and air-land processes dominate the phenomenon being studied. One such example is ENSO, where processes in the tropical waveguide are overwhelmingly important, at least on short enough timescales. Even here, feedback from extra-tropics, through advective and wave propagation processes, requires that global models be employed for better accuracies (Ji *et al.*, 1998). Webster and Palmer (1997) describe the 1997–1998 El Niño, which was correctly forecast by comprehensive coupled air-sea models such as the one at ECMWF (albeit only a few months in advance), whereas the simpler coupled models (Cane, 1992) that enjoyed early success in predicting ENSO events failed to do so. Coastal oceans, with characteristically smaller temporal and spatial scales than the ocean basins, are best studied and forecast using regional coupled models, since processes such as coastal upwelling, which have evolution timescales of mere days, have the potential to significantly alter the state of the ABL immediately above and hence the ability to have a strong feedback on air-sea interactions.

Generally, if local thermodynamic processes related to air-sea (and air-land) interactions are the dominant agents of change, regional coupled models may be useful. Even then, the prescription of conditions on the open boundaries of the atmospheric and oceanic domains requires nesting in global models for the atmosphere and at least basin-scale models for the ocean. If long timescale dynamical processes are involved, global coupled models are indispensable. Studies related to climate, such as investigations of the natural and anthropogenic climate variability over timescales of decades and beyond, require efficient global coupled models that are compatible with available computer resources (for example, Boville and Gent, 1998).

Most (but not all) global atmospheric models employ spectral techniques, while regional models (such as the NCAR MM5 regional model) most often use finite-difference formulations. It is beyond the scope of this book to describe the current state of atmospheric GCMs; instead the reader is referred to the vast literature on the subject in journals such as the *Journal of Atmospheric Sciences,* the *Monthly Weather Review,* the *Journal of Applied Meteorology,* the *Journal of Climate,* and the *Journal of Geophysical Research* (Atmospheres). The concepts are described in, for example, Haltiner and Williams (1980) and Kiehl (1992). AGCMs and their role in climate modeling is described by Ghil and Robertson (1999). The ocean models, whether global or regional, are usually of the finite difference variety. Therefore coupling the global oceans and the atmosphere requires transformation from spectral to physical space of the atmospheric state at each interaction step (this transformation is also needed to run the atmosphere model by itself). Normally (except perhaps in the coastal oceans), the spatial scales of variability in the atmosphere are much larger than those in the ocean and hence the horizontal resolution of the atmospheric component can be coarser than that of the oceanic component. However, the timescales of variability in the atmosphere are much shorter, so that the time stepping requirement for the atmospheric component is a more severe constraint on coupled models. There is generally no need to advance the ocean model at the same time step as the atmospheric model.

Coupling requires continuous or intermittent two-way interactions between the ocean and the atmosphere. It is through the air-sea momentum, heat, and fresh water fluxes that such interaction takes place. Therefore proper representation of turbulent processes in the ABL and OML adjoining the air-sea interface is important. Coupled models are overly sensitive to model biases and errors in parameterization of physical and dynamical processes in each component. For example, an anomalously cold atmosphere or excessive attenuation of the SW radiation by the atmosphere can lead to an unrealistically cold upper ocean. The ocean model also requires SW and LW radiative fluxes at the ocean surface, which are a function principally of the cloud and water vapor content in the atmospheric column and are hence influenced by heat and water vapor exchange across the air-sea interface. Coupled models are well suited to parallel processors since the atmospheric and oceanic components need interact only intermittently and can be solved independently on different processors. At interaction steps, only the information related to fluxes at the air-sea interface need be exchanged, and the bandwidth requirements and computing resource requirements for this intercommunication are rather minimal. However, because of the normally different computing time requirements of the two, synchronization requires considerable care in design so that processor idle time is minimized.

Coupled models are run in either a synchronous mode, where the atmosphere and the ocean interact frequently throughout the simulation, or asynchronously (for example, Manabe *et al.,* 1979), where the deep ocean is spun up to equilibrium first, and only the upper layers interact with the atmosphere during the simulation. The rationale for the latter has to do with the great disparity in timescales of response of the two media to changes in forcing. The lower atmosphere adjusts dynamically to any changes in the conditions at its bottom in a matter of weeks to months, and hence over long timescales, it is in statistical equilibrium with the upper ocean conditions. On the other hand, the ocean responds far more slowly to changes in its surface conditions, some changes taking centuries to percolate through the sluggish deep ocean. It is this inertia (or memory) of the deep ocean that makes asynchronous coupling justifiable, although synchronous coupling is the norm.

Coupled global ocean-atmosphere models got their start in the late 1960s (Manabe and Bryan, 1969), with applications to simulations mostly of potential anthropogenic changes to the atmosphere in mind. These coupled models had initially crude resolutions, highly simplified parameterizations of physical processes such as mixing in the upper ocean, and therefore had great difficulty simulating the climate accurately. Yet over the years they have been improved steadily (Boville and Gent, 1998; Manabe et al., 1979; Washington et al., 1980) and are now able to provide reasonably robust evidence of the global warming likely to result over the coming decades, from the steady increase in atmospheric CO_2 from fossil fuel burning (Gates et al., 1992; Manabe et al., 1990, 1992; Washington and Meehl, 1989). However, more recently the need for accurate forecasts (on a shorter timescale of months to years) of ENSO events has driven the development of atmosphere models coupled to the tropical Pacific or the global oceans specifically to study the internal variability of the coupled dynamical system that leads to anomalous conditions in the tropical waveguide. An overview of climate modeling can be found in Trenberth (1992; see also Gates et al., 1996, and Kattenberg et al., 1996) with many excellent contributed articles on modeling the many components of the climate system, the atmosphere, the hydrosphere, the cryosphere, and the biosphere. A general review of global coupled ocean-atmosphere models can be found in Meehl (1990, 1992). A review of their application to the global warming investigations can be found in Gates et al. (1992, 1996). The subject of climate modeling and the coupled air-sea models is too vast for a thorough coverage here; we instead provide a very brief summary simply for the sake of completeness. We will also provide a brief description of the regional coupled models useful for investigations of local air-sea interaction that nevertheless can have a strong impact on local atmospheric processes, a classical example being the monsoons over the Indian Ocean and the Indian subcontinent. Another, a problem of particular importance to coastal communities, is prediction over a timescale of a few days, of a tropical hurricane intensity and path.

Traditionally, a great deal of attention has been given to modeling the atmospheric component of the hurricane prediction problem, with multiply-nested high resolution atmospheric models. The operational NCEP hurricane model developed at GFDL (Bender et al., 1993; Kurihara et al., 1998) has a triple nest, the outermost grid having a typical resolution of 160 km being fixed, while the two innermost ones having typical resolutions of 80 and 40 km move with the hurricane. The interaction between the nested grids can be either one-way or both and is one of the most complex numerical tasks that can be undertaken. Nevertheless, this system enables thermodynamic processes operative in the hurricane to be modeled more accurately. However, a major limitation is the lack of oceanic feedback, and attempts are being made to correct this with the use of coupled air-sea models (Gines *et al.,* 1997). While the SST can be prescribed at the start and, conceivably, even updated during the simulation, the fuel that powers hurricanes—namely, the latent heat transfer from the ocean to the atmosphere—depends on the total heat content of the upper ocean as well as the SST excess over 26°C. The intense air-sea exchange in the updraft regions in the immediate vicinity of the eye of the hurricane can rapidly deplete the heat stored in the oceanic mixed layer over summer and depress the SST significantly. The entrainment of colder waters from below the seasonal thermocline into the mixed layer by intense wind mixing also depresses SST. This can lead to a decrease in the hurricane intensity as measured by its central pressure. Conversely, the hurricane can traverse over a patch of water with a warm SST and a large heat content, such as a warm core eddy in the Gulf of Mexico or the Gulf Stream off of the Carolinas on the east coast of the United States, which could lead to an increase in the intensity of the hurricane. Hurricanes have a tendency to intensify as they pass over the Gulf Stream, with its very large heat content and warm SST, and wind down as they pass over patches of relatively colder coastal waters. Without accurate knowledge of the initial state of the upper ocean (not just SST) and synchronous coupling to the ocean during the hurricane simulation, it is difficult to accurately forecast hurricane intensities.

Explosive cyclogenesis can occur off eastern sides of continents at midlatitudes due to cold air masses traversing relatively warm water masses during winter. This process is principally driven by the air-sea exchange of heat, as are wintertime storms over the Gulf of Mexico. The thermal front between cold on-shelf water masses and warmer off-shelf ones can have a major impact on the characteristics of wintertime storms over the eastern United States, especially precipitation in the form of snow.

13.1 COUPLING BETWEEN THE OCEAN AND THE ATMOSPHERE

Interaction and hence coupling between the ocean and the atmosphere takes two forms: dynamical and thermodynamic (see Chapter 4 of Kantha and Clayson, 1999, for details of air-sea exchange). Dynamical interaction is most often one way in that it is the atmosphere that drives the ocean. The winds transfer momentum to both ocean surface waves and ocean currents. Accurate modeling of the partitioning of the momentum flux requires therefore a surface wave model to be part of the coupling. This in turn has consequences to thermodynamic interaction, because of the mixing brought about in the upper layers and hence the SST. Thermodynamic interaction, on the other hand, is a two-way process. First of all, the radiation balance at the ocean surface depends very much on the state of the atmospheric column above, especially the cloud cover, which in turn depends on the water vapor transfer from the ocean. The sensible and latent heat exchange between the ocean and the atmosphere depends on the wind stress, the air-sea temperature difference, and the humidity in the surface layer of the atmosphere. The sea state also has an influence in that the roughness felt by the atmosphere, and hence the intensity of air-sea exchange depends on the prevailing wave field. Consequently, accurate modeling of the momentum, heat, and mass transfers across the air-sea (and air-ice and ice-sea) interface is quite complex, but nevertheless crucial to proper coupling of the two media and to prevent introducing systematic errors that lead to biases such as an abnormally warm or cold SST.

It is the SST that the atmosphere "feels" in its interaction with the oceans. This was the rationale for the very first attempts at coupling a global atmospheric model to a "swamp" ocean, with the SST computed from energy balance at the air-sea interface. However, while this avoids assuming the oceans to be an infinite heat source/sink (when SST is prescribed a priori), because it does not allow the oceans to store heat during summer and lose it over the winter, realistic seasonal simulation is impossible. Nevertheless, this type of model enabled problems such as the sensitivity of climate to anthropogenic CO_2 input to be studied (Washington and Meehl, 1989). The next step was the inclusion of an oceanic mixed layer of constant depth (50 to 100 m) to introduce heat storage and thermodynamic memory into the coupled system so that aspects such as seasonal variability could be simulated, albeit crudely. The disadvantage, however, is increased computational time, since unlike the swamp ocean, which adjusts instantly to changes in atmospheric forcing, it takes 10 to 20 years for the coupled system to equilibrate from an arbitrary initial condition (Meehl, 1992). Even then, in the absence of poleward transport of heat from tropics by the ocean current system, the coupled model yields abnormally cold high latitudes and warm tropics, unless artificial heat flux corrections are applied. Nevertheless, such models (with heat flux corrections) improve studies of problems such as the climate sensitivity to increased CO_2 in the atmosphere.

Clearly, proper coupling requires an oceanic GCM that includes effects such as the poleward transport of heat by ocean currents, cold SSTs due to coastal and equatorial upwelling, and variability in upper ocean mixed layer and heat storage. This, however, comes at a considerable increase in computational expense and complexity, since accurate simulation of the above aspects requires an ocean model with resolutions that are hard to realize. Even with a relatively coarse resolution oceans, because of the large timescales (hundreds of years) of response of the deep oceans to changes, "spinup" to equilibrium of a coupled model requires an inordinate amount of time, unless techniques such as artificial (and hence unphysical) acceleration of deep ocean response (Bryan, 1984) or initialization of the ocean from observations are employed. Both have their problems. The former distorts the physics of the deep so that the upper layers and the deep layers still need to equilibrate with each other once the ocean is coupled. The latter suffers from data inadequacies and the fact that the current observed state of the oceans is but a snapshot and hence unsuitable for anything but studies around the present epoch. Another serious problem is a tendency for systematic errors in the state of the coupled media due to even small imbalances and any inadequate or improper representation of physical processes. Nevertheless, a global coupled model is a powerful tool for studying the global climate and its sensitivity to perturbations on a variety of timescales, and as improved understanding of processes and increasing computing resources make climate simulations increasingly more realistic, they are bound to become increasingly valuable as well.

In a coupled model, at the time of coupling, the ocean must provide the atmosphere principally with the SST (and the extent of sea-ice cover), while the atmosphere must provide the ocean with the momentum, heat, mass, and radiative fluxes at the air-sea interface. The sea-ice model imbedded in the ocean model is an essential component of climate studies but is most often highly simplified (for example, only thermodynamics might be included) for practical reasons. It is the current practice to employ synchronous coupling, meaning that both the atmosphere and the ocean models are kept running simultaneously but independently and made to communicate with each other at predetermined intervals (ranging from an hour to a day, depending on the application). In between, each model is run at whatever time step appropriate to it, with the conditions provided to it by the other model held fixed. Depending on the time interval, these conditions may be snapshots at the time of coupling or averages over the previous interval. In the past, the so-called asynchronous coupling was routinely employed in coupled models. This meant that the atmospheric model was run separately for a short period of time (say, a year) with the conditions at the bottom suitably prescribed. The ocean model was then run for a longer period (say, 100 years) because of its longer equilibration timescale, with conditions derived from the atmospheric model at suitable intervals (say, a day and repeated each year). This strategy is seldom used now except perhaps for the spinup phase before coupling. For example, the atmospheric model may be run for a particular period with say observed SST and then the ocean model run with observed atmospheric forcing until it is equilibrated; the two models are then coupled and run to equilibrium before studies of climatic changes to perturbations are performed. Assuring that the two media are in equilibrium with each other is an important and vexing problem in climate sensitivity studies with a coupled model.

A recent trend in coupled models is the use of a flux coupler (Boville and Gent, 1998; Cassou *et al.*, 1998). The component models of the coupled system interact and communicate with one another only through the flux coupler (Figure 13.1.1). The strategy is to pass the instantaneous values of the relevant state variables at each time step from each component model to the coupler, and let the coupler compute the associated fluxes and pass the flux values averaged over the coupling interval. Time synchronization between different models is accomplished through the coupler, with the atmosphere component (including land process and sea-ice models) communicating generally faster with the coupler than the ocean component. The coupler also takes care of the spatial averaging and interpolation needed for the component models that may have different spatial resolutions to interact with one another. The use of the flux coupler permits the flexibility of making changes in one component without affecting others. For example, the ocean model can be replaced by a routine that supplies a predetermined SST; the atmosphere model can be replaced by analyses. The various components of Coupled Ocean-Atmosphere General Circulation Models (COAGCMs) lend themselves readily to running in parallel and relatively independently often on computers at distant sites, principally because their principal interaction is through the air-sea exchange. The SST and the sea-ice extent are written out by the OGCM to the coupler, which provides these fields to the AGCM, and the surface fluxes are written out by the AGCM to the coupler, which interpolates the fields and provides them to the OGCM. This strategy is used in the French coupled climate model (Cassou, 1998) and in NCAR community climate models (Boville and Gent, 1998).

Another important problem is the secular trend, called climate drift, introduced by systematic errors in each component of the model and in the initial fields. More often than not, these errors tend to get amplified by coupling. One such problem is the net heat flux at the air-sea interface which, for a variety of reasons, may not be realistic and hence could lead the coupled model to produce unrealistic SSTs. Obviously, the best strategy is to improve the physics involved so that the coupled model does reproduce realistic SSTs. Careful attention to details and intense efforts over the past decade have made coupled models to be reasonably realistic at present (see also Section 13.2). In the past, however, flux corrections were routinely employed (Meehl, 1992; Neelin *et al.*, 1994) to overcome deficiencies in model resolution and physics. For example, the coarse resolution ocean model, which cannot hope to reproduce the oceanic current system accurately, might be run with SSTs damped to observed values and the mismatch in heat flux needed to produce realistic SSTs saved. The atmospheric model might then be run with observed SSTs and the heat flux values saved. The

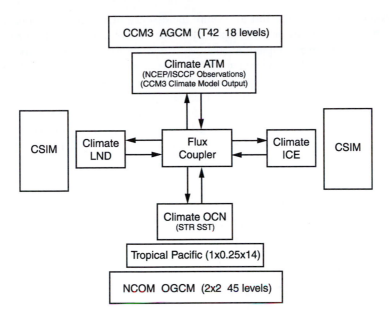

Figure 13.1.1 Flux coupler method of coupling individual components of the coupled NCAR climate model.

coupled model is then run with the corrected heat flux based on the two independent runs. While this ingenious strategy provided acceptable results often, it is highly suspect and the modern practice is to steer away from it. For example, the NCAR climate system model has been successfully integrated for 300 years with remarkably small climate drift (Bryan, 1998), without the use of any artificial flux corrections. This has been achieved by and large by improvements in component models such as better penetrative convection parameterization in the atmosphere and better eddy-mediated heat transport and mixed layer simulation in the ocean model. The careful spinup of the ocean was also required to produce initial conditions that were compatible with the individual component model states (Boville and Gent, 1998). For more details on problems involved in coupled modeling and its applications, see Meehl (1992). For details of successful application of a coupled model to an important problem, ENSO prediction, see Cane (1992) and Ji *et al.* (1998). For successful multicentury simulation of climate, see Boville and Gent (1998).

We have so far discussed coupling comprehensive atmospheric models with simplified to comprehensive ocean models, with climatic studies in mind. However, for certain applications, it is possible to simplify the atmospheric model as well. Some successful coupled models have used highly simplified versions of both in an effort to capture the essence of the phenomenon, without getting bogged down in the complexity of the GCMs. For example, Zebiak and Cane (1987) have coupled a shallow water anomaly (monthly mean state prescribed from observations) model for the tropical Pacific atmosphere to a reduced gravity anomaly model of the equatorial Pacific ocean waveguide for ENSO studies and predictions. For some applications, where only the local influence of surrounding orography and the land-ocean interface need to be taken into account in regional coupled models, the atmospheric model can be simplified to a one- or two-layer model (Horton et al., 1997). It is quite obvious that the power of a comprehensive coupled ocean-atmosphere GCM is not always needed, nor the effort and expense involved always justified.

13.2 COUPLED OCEAN-ATMOSPHERE GENERAL CIRCULATION MODELS

Coupled ocean-atmosphere general circulation models (COAGCM, or CGCM for short) are an essential component of any Earth Climate Modeling System (ECMS) that is designed to study the natural and anthropogenic climate variability. It is for this reason that there has been a great spurt in the development and testing of coupled models in recent years. These efforts have been greatly aided by the rapid increase in available computing resources that has made realistic COAGCMs possible. Neelin et al. (1992; see also Neelin et al., 1994) reviewed the state of the field in the beginning of the 1990s and pointed out the necessity to pay careful attention to modeling the physical processes involved in air-sea interactions so as to avoid the need for artificially correcting the air-sea fluxes to prevent climate drift in the model simulations. They correctly surmised that the coupled feedbacks between the atmosphere and the ocean would amplify any deficiencies in modeling the physical processes in these models. Since then, the progress in diagnosing and alleviating such deficiencies has been quite rapid. The result is that while some deficiencies still exist, it is now possible to model the well-known features of the ocean-atmosphere system such as the annual mean state, the seasonal cycle, and the interannual variability far better without resorting to artificial fixes. The most recent overview by Mechoso et al. (1995) reports on this progress and details the strengths and deficiencies of nearly a dozen COAGCMs. This brief review draws upon this overview as well as a few others such as Neelin et al. (1994).

The realization that an accurate simulation, using COAGCMs, of short-term climatic variability such as ENSO might very well be feasible, and therefore that the potential for successful forecasts of interannual climatic variability exists, has prompted COAGCM efforts to focus principally on the tropical Pacific. Earlier in the 1980s, highly simplified regional coupled ocean-atmosphere models of the tropical Pacific waveguide had demonstrated considerable skill in predicting ENSO events (for example, Zebiak and Cane, 1987). But it was also clear that extratropical influences may need to be taken into account for more accurate predictions. Since ENSO appears to be phase locked and quite sensitive to the seasonal cycle, it appears even more critical to include the seasonal cycle (see the recent review by Latif *et al.,* 1994b) in ENSO prediction models instead of forecasting only the anomalies, as the successful Cane-Zebiak model (Cane, 1992; Cane *et al.,* 1986) does. In other words, realistic prediction of interannual anomalies appears to require realistic prediction of the annual mean state and the seasonal cycle as well (Mechoso *et al.,* 1995). Hence the need for global coupled models, COAGCMs. While the anomaly models appear to excel in predicting the east-west redistribution of heat energy in the tropical Pacific waveguide, simulating the north-south asymmetry about the equator in the seasonal cycle requires a COAGCM. Unqualified success in this endeavor still appears to be elusive (Mechoso *et al.,* 1995).

Neelin *et al.* (1992; see also Neelin *et al.,* 1994) reviewed 17 coupled models (that could be reasonably called GCMs, even though some had limited oceans), out of which 10 did not use artificial flux corrections or damping to climatology so that the mean state could freely evolve. Because of the coarse resolutions possible at the time and inaccuracies in parameterizing physical processes related to air-sea exchange, most of these models exhibited a strong drift of the model mean state or large departures from the expected normal climatological state in the tropical Pacific. The simulated zonal gradients and interannual variability were also poor. In contrast, Mechoso *et al.* (1995) show that most of the 11 COAGCMs they surveyed do a good job in simulating the annual mean east-west SST gradient along the equator without using flux corrections, even though the absolute SST values are considerably underestimated by most (see Figure 13.2.1). Table 13.2.1 from Mechoso *et al.* (1995) provides a detailed description of the characteristics of these models. It is noteworthy that the horizontal (or equivalent horizontal) resolutions of the atmospheric components (AGCM) of these COAGCMs are still rather coarse ranging from $3°$ to $6°$. Roughly half are spectral and the other half finite-difference models. The number of levels in the vertical ranges from 5 to 30. The oceanic components (OGCM) tend to have high meridional resolutions ($1/3°$ to $1/4°$) near the equator, and generally resolutions higher than the atmospheric model elsewhere (roughly $1°$ to $3°$). Only 5 of the OGCMs are near-global, but the rest cover at least the tropical Pacific. None include the Arctic. Most are z-level models, with the number of vertical levels ranging from 7 to 28. A few have an explicit mixed layer, and most use Richardson number-dependent mixing in the ocean, while some use turbulence closure. The details of each of the models can be found in the references cited in the table. Further details on the intercomparison can be found in Mechoso *et al.* (1995).

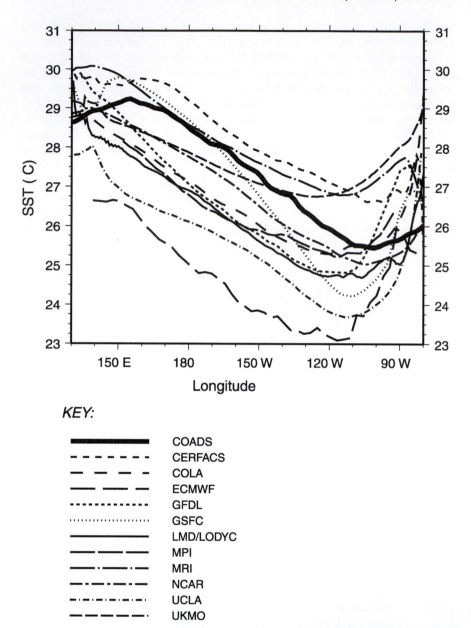

KEY:

▬▬▬▬▬▬▬	COADS
- - - - - - - -	CERFACS
— — — -	COLA
— — — —	ECMWF
··········	GFDL
··················	GSFC
▬▬▬▬▬▬	LMD/LODYC
— — — —	MPI
—·—·—·—	MRI
—·—·—·—	NCAR
—··—··—··—··	UCLA
— — — — —·	UKMO

Figure 13.2.1 Annual mean SST along the equator averaged between 2°S and 2°N from various COAGCMs compared to COADS. Most underestimate the absolute values although the SST gradient appears to be reasonably well predicted. From Mechoso *et al.* (1995).

TABLE 13.2.1
Model Descriptions

Model	CERFACS	COLA	EC	GFDL	GSFC	LMD/LODYC	MPI	MRI	NCAR	UCLA	UKMO
AGCM	Arpege	COLA	EC	GFDL	Aries	LMD	ECHAM	MRI	CCM2	UCLA	UKMO
Resolution	T42 L30	R15 L18	T21 L19	R30 L9	5° × 4° L13	64 × 50 pts L11	T42 L19	5° × 4° L5	T42 L12	5° × 4° L9	3.75° × 2.5° L19
Convection scheme	MF + SC	MCC + SC	MF + SC	MCA	relaxed AS	MCC + MCA	MF + SC	AS	MF	AS	MF + DD
OGCM	OPA T-Pacific 40°S–48°N	GFDL global 70°S–65°N	HOPE global 60°S–60°N	GFDL global 50°S–50°N	Poseidon global 90°S–72°N	OPA T-Pacific 40°S–48°N	HOPE global 60°S–60°N	MRI Ind.–Pac. 40°S–30°N	Gent/Cane T-Pacific 20°S–20°N	GFDL T-Pacific 30°S–50°N	GFDL T-Pacific 30°S–30°N
Resolution	0.75° × 1/3° −1.5° L28	1.5° × 0.5° −1.5° L20	2.8° × 0.5° −2.8° L20	1° × 1/3° −3° L27	1.25° × 2/3° isopyc. L8	0.75° × 1/3° −1.5° L28	2.8° × 0.5° −2.8° L20	2° × 1° L20	1° × 1/4° −1° L7	1° × 1/3° −3° L27	1.5° × 1/3° −1° L16
Ocean mixing	1.5 TKE	Ri	Ri + ML	Ri	Ri + ML	1.5 TKE	Ri + ML	2.0 TKE	Ri	2.5 TKE	Ri + ML
Reference	Terray *et al.* (1994)	Schneider and Kinter (1994)	Stockdale *et al.* (1994)			Barth and Polcher (1994)	Latif *et al.* (1994a)	Nagai *et al.* (1994)	Gent and Tribbia (1993)	Robertson *et al.* (1995)	Ineson and Davey (1994)

Spatial resolution is given as the latitude-longitude grid size or "T" or "R" for triangular or rhomboidal spectral truncation, respectively. The number of model levels/layers is denoted by "L." Key to convection schemes: MF—mass flux, SC—shallow convection, MCC—moisture convergence closure, AS—Arakawa-Schubert, DD—downdrafts. Key to ocean mixing: Ri—Richardson number dependent, ML—explicit mixed layer. See text for further details.

CERFACS: Centre Européende Recherche et de Formation Avancée en Calcul Scientifique
COLA: Center for Ocean-Land-Atmosphere Studies
EC: European Centre for Medium-Range Weather Forecasts
GFDL: Geophysical Fluid Dynamics Laboratory
GSFC: Goddard Space Flight Center

LMD: Laboratoire de Meteorologie Dynamique
MPI: Max-Planck-Institut für Meteorologie
MRI: Meteorological Research Institute, Japan
NCAR: National Center for Atmospheric Research
UCLA: University of California, Los Angeles
UKMO: United Kingdom Meteorological Office

Most models overestimate the SST east of 90°W and west of Peru by as much as 3°C. However, despite the fact that current COAGCMs underestimate the equatorial wind stresses by almost a factor of 2, this tendency is apparently overcompensated by anomalously strong upwelling in OGCMs amplified by the response of the Walker circulation in the equatorial atmosphere, leading to unrealistically large westward extension of the equatorial cold tongue. While the annual mean SST and its gradient along the equator are important to air-sea interactions (and ENSO), and are also good indicators of the skill of the coupled model, the ability to reproduce the expected seasonal cycle of SST in the tropical eastern Pacific (the seasonal cycle is weak in the western Pacific) is crucial to prediction of interannual variability and ENSO. Figure 13.2.2 from Mechoso *et al.* (1995) shows that most of the models still do a poor job as far as the seasonal cycle is concerned. Apparently higher resolution in the AGCM helps better simulate the seasonal cycle. Most did a reasonable job of reproducing the warm SSTs in the western Pacific warm pool, even though the warm pool was too small or too cool in some. Most reproduced the cold SST tongue in the equatorial eastern Pacific, but with varying strengths, widths, and extent. Figure 13.2.3 shows the comparison of the October average SST for the 11 models with observations from COADS data base.

A salient aspect of the atmosphere over the eastern Pacific is the location of the intertropical convergence zone (ITCZ). The ITCZ stays north of the equator there throughout the year. None of the COAGCMs were able to reproduce this behavior. Some simulated an ITCZ that migrated across the equator with seasons or produced a double ITCZ straddling the equator. Because of the poor prediction of the location of the ITCZ, around which most precipitation occurs, the rainfall predictions in the eastern Pacific are also poor in most models. The underlying causes are many and poorly known, but it is suspected that poor simulation of stratus clouds near Peru (see Mechoso *et al.,* 1995), evaporation-wind feedback in the ITCZ, the intensity of coastal upwelling and/or cross-equatorial winds might be responsible.

Therefore, it is clear that while a COAGCM is superior in many ways to the AGCM by itself, coupling exacerbates even small deficiencies in either AGCM or OGCM by amplifying them. In fact, coupling is a rather stringent test of the processes parameterized in both components. Considerable effort needs to be expended in fine-tuning various parameterizations and testing their impact on simulations before these models can be used to simulate or predict interannual (and climatic) fluctuations due to internal variability.

Despite the many persistent problems, Mechoso *et al.* (1995) conclude that considerable progress has been made in modeling the coupled atmosphere-ocean system in the tropical Pacific. Note, however, that this assessment does not extend to other tropical oceans or the extratropics. It is not known how well the models that incorporate the global OGCMs are able to reproduce the very

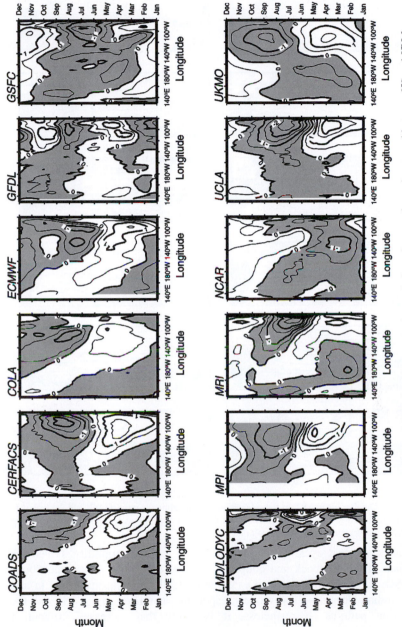

Figure 13.2.2 The seasonal cycle of SST along the equator (anomaly from the annual mean) averaged between 2°S and 2°N from various COAGCMs compared to COADS. Although the trends are there, most do a poor job of simulating the cycle. From Mechoso *et al.* (1995).

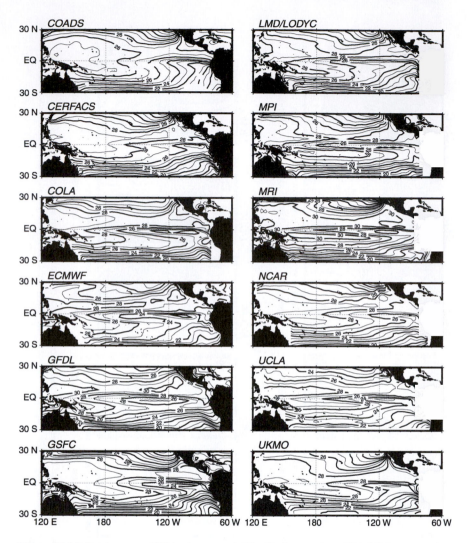

Figure 13.2.3 October mean SST in the equatorial Pacific from various COAGCMs compared to COADS. While broad features are reproduced, the details such as the width and extent of the cold tongue need improvement. From Mechoso *et al.* (1995).

important monsoonal cycle in the Indian Ocean and its variability. Given the importance of the feedback and coupling between the Indian Ocean monsoons and ENSO, it may be important to examine carefully the seasonal cycle in and over the Indian Ocean in these models, especially as related to the SST and the precipitation patterns. Nevertheless, it is likely that increased computing resources and better parameterization of physical processes such as upwelling and stratus clouds will one day enable high resolution truly global COAGCMs to better simulate the seasonal cycle in the tropics, and hence allow the promise of accurate forecasting of the interannual variability of the tropical oceans and atmosphere to be realized.

It is a monumental task to detail the characteristics of all the existing COAGCMs. Instead, we will present some details about one, the University of Los Angeles (UCLA) model, to provide an indication of the level of effort that is involved and refer the reader to the references cited in Table 13.2.1 for others. The development, testing, and application of these models tends to require a group effort with solid institutional support behind it over a number of years and hence beyond the resources of a single individual.

The AGCM in the UCLA model is unique in that unlike most operational AGCMs, which are spectral, it is a grid-point model, developed originally by Akio Arakawa and his coworkers. The resolution is $5°$ longitude by $4°$ latitude, with 15 layers in the vertical at present. The model has evolved over the past, with physical processes better parameterized as deficiencies showed up. It includes the Arakawa and Schubert (1974) cumulus convection parameterization, Katayama (1972) short wave and Harshvardhan *et al.* (1987) long wave, Kim (1996) gravity wave drag and envelope orography, and Suarez *et al.* (1983) PBL processes parameterizations. The PBL is a simple bulk mixed layer type with variable depth. The long wave scheme includes water vapor absorption and the short wave parameterization includes absorption by water vapor and CO_2. Clouds are assumed to form when relative humidity exceeds saturation value or when cumulus convection reaches 500 mb. The prognostic variables are the horizontal velocity, potential temperature, water vapor and ozone mixing ratios, surface pressure and temperature, and PBL depth. Soil moisture, albedo, and sea ice are from monthly climatological values. The numerical differencing scheme is fourth-order accurate and conserves potential enstrophy. For more details, see Mechoso *et al.* (1987) and Robertson *et al.* (1995).

The OGCM in the UCLA model is the GFDL modular ocean model (MOM) (Pacanowski *et al.,* 1991) described in Chapter 10. It covers either the tropical Pacific at 1/3 to $3°$ resolution and 27 layers in the vertical or is near-global at $1°$ resolution and 15 levels. The mixed layer is explicitly simulated using Mellor and Yamada (1982) second-moment closure, and this leads to better thermocline simulations in eastern Pacific than Richardson number-based mixing parameterization. Newtonian damping is applied to temperature and salinity at high latitudes.

The AGCM and OGCM are coupled synchronously, with SST passed from OGCM to AGCM and the momentum, heat, and fresh water fluxes from AGCM to OGCM once per day. The SW and LW radiative fluxes seen by the ocean are, of course, affected by the atmospheric column, and the AGCM provides their surface values to the OGCM.

While the model has probably the very best physical parameterizations known, and simulates well many of the features of the tropical ocean-atmosphere, sensitivity studies and simulations have brought to light many deficiencies that still exist. Many of these are common to other COAGCMs as well. For example, excess surface evaporation in the AGCM leads to an overall cold bias in tropical SST, but a warm bias results off the coasts of California and Peru due to the model's inability to simulate persistent marine stratocumulus there (Ma *et al.*, 1996). Consequently, the low SSTs (by as much as 5°C) observed in these regions due to decreased solar insolation are not reproduced in the model. Excess evaporation in the model appears to be due to low emissivities assigned to high level clouds. In general, representation of cloud-related processes has always been the Achilles' heel of AGCMs and COAGCMs, whose performance is highly sensitive to the distribution and properties of various clouds and cloud types. Recent improvements in cloud parameterizations have apparently made it possible to explore interannual and interdecadal variabilities with this model. As far as the extratropics are concerned, it would be crucial to reproduce accurately the poleward transport of heat by the western boundary currents in the ocean, and its variability. Much higher resolution for the OGCM than is currently feasible will be necessary for accurate depiction of the oceanic state and hence the air-sea exchanges. Nevertheless, along with climate type coupled models, COAGCMs have matured to a stage where meaningful investigations of the natural as well as anthropogenic variability of the Earth's climate system on timescales of months to decades may become possible in the very near future.

An excellent example of quasi-operational application of global coupled ocean-atmosphere general circulation model to seasonal forecasting of the climate system up to 6 months in advance is that of Stockdale *et al.* (1998) at ECMWF. While individual weather events are too chaotic to be predictable more than 7 to 10 days in advance, climate forecasts consisting of an ensemble of integrations of the coupled model with different atmospheric initial conditions to sample the chaotic part of the system hold the promise of predicting the climate several months in advance in a probabilistic sense. This depends on the assumption that while the individual weather events from week to week are uncorrelated, their overall statistics may be biased in a deterministic manner and hence predictable (Palmer, 1993). The ocean is most often the source of such bias and coupled models are therefore essential for long term forecasting of the climate system tendencies (Ji *et al.*, 1996; Livezey *et al.*, 1996; Palmer and Anderson, 1995). Even then, reliable extraction of possible climatic perturbations can be done only in a statistical manner and meaningful statistics requires computationally intensive ensemble of predictions. In the absence of flux corrections, any systematic errors in the component models lead inevitably to a drift in the mean state which must be artificially removed (Stockdale, 1997). One technique is to estimate the drift from a set of forecasts from an earlier period. Stockdale *et al.* (1998) used this method to remove 1-2 °C SST drift in the tropical Pacific to arrive at a seasonal global rainfall forecast consisting of not only the predictable mean climatic shift but also an estimate of the unpredictable random component. For example, the coupled model predicted in the beginning of 1998 that it is 70% certain that southeast China will remain abnormally wet over the coming six months. An operational capability such as this being developed at various NWP centers around the world (Ji *et al.*, 1996; Stockdale, 1997) has been made possible by the availability of more accurate coupled global models and increased computing power.

An example of the application of a coupled model for simulating and studying the global climate system is that of Schneider (1998), who uses the Hamburg coupled model ECHO (Latif *et al.*, 1994) developed at the Max-Planck-Institut fur Meteorologie, Germany. The atmospheric model is the Hamburg version of the ECMWF AGCM at T42 (2.8°) resolution and has 19 levels, and the oceanic model has similar resolution (except near the equator where the resolution is 0.5°) and 20 levels. The coupled model extends to 60° latitude. Sea ice is not included in Schneider's simulations; instead the surface temperature and salinity are damped to climatological values (Levitus, 1982). The coupled model simulation spanning 125 years indicates that while deficiencies such as the waters off South America being too warm exist, overall, given the coarse resolution employed, the model does a credible job of reproducing the principal atmospheric and oceanic features. Schneider (1998) used this model to investigate the impact of a cutoff of the Indonesian Throughflow from the Pacific to the Indian Ocean. He finds that the Throughflow shifts the centroid of the warm pool in the tropical ocean and the center of deep convection in the atmosphere westward and thus exerts a substantial influence on the global climate in general and the tropical climate in particular.

In coupled models, it is the atmospheric component that consumes a majority of the CPU time (~90%). Yet it is the ocean that has longer memory, and for efficient simulations, the atmosphere can be regarded as an adjusted component. One strategy often used in climate simulations is to couple a comprehensive ocean model to a statistical, anomaly atmospheric model. This is known as a hybrid coupled model (HCM). One example is Xu *et al.* (1998), who coupled a statistical atmospheric model based on 100 years of the Hamburg ECHO coupled model (Latif *et al.*, 1994) to the HOPE ocean model (Wolff and Maier-Reimer, 1997) to investigate decadal variability in the North Pacific. Monthly data from the ECHO was used to derive monthly anomalies of SST, net heat flux, freshwater flux, and wind stress. The OGCM was run to equilibrium and the statistical atmosphere model was then coupled to the OGCM by driving the

OGCM with the anomalous fluxes added back to the mean fluxes. Such an HCM is useful in understanding the inherent characteristics of the coupled system in the absence of stochastic forcing by noncoupled variability, such as self-sustained decadal oscillations (Xu et al., 1998).

Another excellent example of the operational use of a coupled model is the use of the NCEP/CPC Pacific basin coupled model (Ji et al., 1998). This model couples the GFDL MOM ocean model to the NCEP global medium range spectral forecast model run at a relatively low resolution of T40. The coupling uses anomalies of surface fluxes from the atmospheric model (since the mean annual cycles of the AGCM are not yet realistic) added to climatological mean annual wind stress and heat fluxes to drive the ocean model. The AGCM is driven by SST from the OGCM. The SSS in the ocean model is relaxed to Levitus et al. (1994) climatology. The details of this coupled model can be found in Ji et al. (1998) and the references therein. This model was more successful in predicting the 1997–1998 ENSO event than most other statistical and dynamical models, but still failed to predict its unexpectedly large intensity and duration accurately (Webster and Palmer, 1997).

Interestingly, none of the ENSO forecast models, both dynamics-based and statistical, including COAGCMs, predicted the intensity of the 1997-1998 event, one of the strongest on record (comparable to the 1982-1983 event), until the rapid warming of the eastern tropical Pacific was well underway (Barnston et al., 1999). Neither did the models correctly predict its spectacular demise caused by a very rapid drop in SST in the eastern Pacific triggered by an abrupt intensification of easterly trade winds, whose cause is unknown (McPhadden, 1999). The most comprehensive coupled ocean-atmosphere dynamical model initialized with in-situ subsurface data from TAO moorings did perform better than simpler dynamical models in predicting the onset. Overall, no dynamical ENSO prediction model simulates well the intraseasonal fluctuations in the tropical atmosphere, the 30–60 day Madden-Julian oscillations that originate over the Indian Ocean and propagate eastward over the Pacific. Neither do statistical ENSO forecast models, trained with seasonal average conditions. McPhadden (1999) points out that the immediate consequence is that ENSO prediction models have difficulty forecasting events and transitions that depend very much on short time scale variability. This is one reason that no model predicted accurately the strength and rapidity of development of the 1997-1998 El Niño, which was triggered by weakening and reversal of the trade winds early in 1997, but heavily influenced by a series of Madden-Julian-oscillation-triggered westerly wind bursts that led to a rapid onset and strengthening of this El Niño. While spectacular progress has been made in ENSO forecasting in the past decade, the lack of understanding and the poor predictability of the variability in intraseasonal oscillations of the tropical atmosphere still limits the skill of these models.

Construction, operation and evaluation of coupled models is an enterprise requiring considerable computing and human resources and hence is beyond the capability of an individual investigator. An idea of the level of effort needed to create a credible model can be found in the Climate Systems Model Special Issue of the *Journal of Climate* (June 1998) that contains a description and evaluation of version one of the NCAR Climate System Model. The title itself indicates that the model is still evolving, but nevertheless, significant accomplishments have been made, such as nearly drift-free multicentury integrations. The atmospheric component is a spectral model with improved hydrologic processes, cloud radiative properties, and convection parameteriza-tions, run at T42 truncation (~2.9°) and 14 levels in the vertical. The ocean component is based on the GFDL z-level model (Gent *et al.*, 1998), with similar overall resolution but 45 levels in the vertical. It uses Gent and McWilliams (1990) eddy flux parameterization, Large *et al.* (1994) mixed layer formulation, and Semtner (1976) thermodynamic and Flato and Hibler (1990) dynamical sea-ice models. Careful sequential integrations of the component atmosphere and ocean models were made to produce compatible conditions for initialization of the coupled model. The overall result is a multicentury climate simulation run with remarkably small climate drift. The reader is referred to the above-mentioned special issue for a good idea of the current state of coupled modeling for climate simulations.

All in all, it appears that coupled ocean-atmosphere models hold out the promise of more accurate short-term deterministic (few weeks) as well as long-term ensemble forecasts (months to years) of the ocean-atmosphere system. The field is rapidly expanding and the next decade should see rapid advances in both capability and skill.

13.3 REGIONAL COUPLED OCEAN-ATMOSPHERE MODELS

A pressing need for predicting the onset and evolution of ENSO events gave the impetus to the development of regional coupled ocean-atmosphere models (RCOAM) in the early 1980s. The success, albeit uneven, of these very simple regional coupled ENSO models (Cane *et al.*, 1986, Zebiak and Cane, 1987) owes a great deal to the validity of the Bjerknes hypothesis (Bjerknes, 1969) that ENSO is mostly due to local air-sea interactions and internal variability in the tropical Pacific waveguide. It is thought that SST anomalies in the waveguide act to either reinforce or weaken the trade winds (and to modulate and shift zonally the Walker circulation driven by the east-west SST gradient), which in turn modify the SST anomalies over certain internal oceanic dynamical timescales, leading to a self-sustained oscillation in the system that is ENSO. In this scenario, extratropics are largely uninfluential. This is approximately correct and makes regional coupled modeling there over timescales of months and beyond meaningful.

A similar situation exists in the coastal oceans, where processes such as coastal upwelling are essentially local but influential to the state of the lower atmosphere across the land-sea interface, and hence regional coupled models are meaningful and useful, but only on short synoptic timescales of a few days. In regions such as the North Indian Ocean, a RCOAM that captures the essence of the monsoonal air-sea and air-land interactions might also provide a means to study monsoon-related processes, without having to resort to an expensive and cumbersome COAGCM.

Starting from simple shallow-water models of the response of the equatorial ocean to wind forcing, Mark Cane and coworkers developed a nonlinear shallow water model of the tropical Pacific Ocean waveguide with an embedded mixed layer coupled to a simple atmosphere over it, also with nonlinear but shallow water dynamics (Cane, 1992; Cane *et al.*, 1986; Zebiak and Cane, 1987). The interactions between the two, namely processes involving SST, are, however, carefully parameterized. Only anomalies from an assumed climatological state are dealt with. Nevertheless, many of the features of ENSO variability exhibited by this simple model are also found in hybrid coupled models that couple an OGCM to a simple atmospheric model (Barnett *et al.*, 1993) and in full-blown COAGCMs (Chao and Philander, 1993; Leetma and Ji, 1989). Even simpler models retaining the essence of delayed-oscillator action exhibit some features of the oscillation in the tropical waveguide (Neelin *et al.*, 1994). The reasons behind the success of these simple models is outlined in detail in Neelin *et al.* (1994). An excellent summary of ENSO can be found in Philander (1990).

The next level in complexity is the full-blown RCOAM, with a comprehensive regional ocean model coupled to a comprehensive regional atmospheric model. One way to look at RCOAM is that it not only facilitates including the oceanic feedback but it also enables high resolution simulations of the local atmosphere to be carried out, which would not be possible with a global AGCM. Models of this nature are useful in the coastal regions and marginal seas, and for studying somewhat localized basin-scale processes such as the monsoons in the Indian Ocean and hurricanes. Efforts are underway at Naval Research Laboratory and Fleet Numerical Meteorology and Oceanography Center at Monterey to develop, test, and put into operational use such a coupled model for the coastal regions, called Coupled Ocean Atmosphere Mesoscale Prediction System (COAMPS). COAMPS is slated to replace the high resolution Navy Operational Regional Atmospheric Prediction System (NORAPS). This model has the sigma coordinate model described in Chapter 9 as the oceanic component and is designed to be nestable and rapidly relocatable (Hodur, 1996).

Mesoscale regional atmospheric models are being increasingly used for both research and real-time weather prediction (Mass and Kuo, 1997). The availability of high-powered workstations and reliable mesoscale models (Cox *et al.*, 1997) capable of running on them, as well as access to real-time analysis and forecast from NWP centers such as NCEP and FNMOC for deriving initial and boundary conditions has made the transition from reliance on central NWP centers to distributed local prediction systems possible. Even ensemble predictions based, for example, on perturbations of the initial conditions are becoming feasible. The regional mesoscale models being increasingly used for this purpose are (1) The PennState/NCAR MM5, (2) Colorado State University Regional Atmospheric Modeling System (RAMS), and (3) Navy Operational Regional Prediction System (NORAPS) being replaced by Coupled Ocean Atmosphere Mesoscale Prediction System (COAMPS). All three are relocatable, three-dimensional, primitive equation-based, regional mesoscale models using staggered grids in the horizontal, and terrain-following coordinates in the vertical, and four-dimensional data assimilation. They have similar subgrid scale mixing, cumulus, and radiation parameterizations. NORAPS (and COAMPS) is specifically designed for oceanic applications, whereas the other two are principally for land applications. Only NORAPS is hydrostatic, whereas MM5, RAMS, and COAMPS have nonhydrostatic options. COAMPS couples an ocean model to the atmosphere, whereas the others do not. See Cox *et al.* (1997) and Mass and Kuo (1997) for a detailed description of the characteristics of these models. Briefly, RAMS uses a leapfrog time-differencing scheme for velocity and pressure, but forward in time for all other prognostic variables, and supports multiple, moving, two-way nesting capability. MM5 is similar (see below) except that a leapfrog scheme is used throughout. Both have a wide range of cumulus and radiation parameterizations available and boundary-layer packages. The performance of both MM5 and RAMS are similar (Cox *et al.*, 1997).

Here we will describe a RCOAM that couples the CU version of the sigma-coordinate POM to the NCAR MM5 regional atmospheric modeland its application to simulating hurricane evolution in the Gulf of Mexico. Marks *et al.* (1997) describe the evolution of hurricane Opal and the troubling questions on the relative roles of internal storm dynamics and the upper ocean posed by its rapid intensification as well as its equally rapid unwinding.

NCAR MM5 is a grid-point atmospheric model developed by Anthes and his coworkers at Penn State and further improved over the years at NCAR by Tom Warner and colleagues. It uses an explicit split-mode time scheme to solve the governing equations and has a built-in Mellor-Yamada Level 2 second-moment turbulence closure to parameterize mixing in the PBL. For applications to the Gulf, the number of levels were 15 in the vertical, and the horizontal resolution was 15 km, roughly the same as that of the ocean model ($1/5°$). The ocean model, CUPOM, has 21 sigma levels in the vertical, with high resolutions in the upper layers to better represent the mixed layer and the Loop Current processes. It has an improved mixed layer formulation due to Kantha and Clayson (1994). The model topography is derived from the 1/12° ETOP05 data base and the model is initialized with historical hydrographic data. For more details on CUPOM, see Chapter 9, and on MM5, see Grell *et al.* (1994).

The need to nest the regional atmospheric component of a RCOAM in a global AGCM requires careful thought as to the domain for the atmospheric model. Either the domain should be such that the lateral boundaries have minimal influence on the results inside or if this is not possible, the boundaries should be such that the internal processes are not dominated by them. In the latter case, it is implicit that the global AGCM in which the RCOAM is imbedded be skillful enough. For example, for the Asian monsoonal problem, it is essential to include the orography around Somalia and the Tibetan Plateau, since in their absence, especially the latter, the monsoonal processes are highly distorted and lateral boundary conditions are unable to overcome this deficiency. The ocean model domain must extend southward beyond the equator. For simulating hurricanes in the Gulf of Mexico, the domain of the atmospheric model should extend beyond the Gulf of Mexico itself, all around. Conditions on the open lateral boundaries of the atmospheric model are derived from the NCEP operational global model analyses. The ocean model is open at the Yucatan to the Caribbean Sea and at the Florida Straits to the Atlantic; monthly average climatological conditions are prescribed at these straits (Choi *et al.*,1995).

The two models are stepped forward synchronously, with the OGCM providing the SST and the AGCM providing the radiative, momentum, heat, and freshwater fluxes at the ocean surface (Bao *et al.*, 1999). Figure 13.3.1 shows how the two models are coupled. Note that a surface wave model mediates the transfer of momentum flux between the atmosphere and the ocean. This is essential to account properly for the transfer of atmospheric momentum to the wave field (see Chapter 5 of Kantha and Clayson, 2000, for a discussion of surface waves and wave models). MM5 is stepped forward at a time step of 15 s, while the ocean model has external/internal time steps of 30 s and 30 minutes. The interaction is half-hourly, fine enough to resolve the diurnal cycle.

Intense, local air-sea exchange of latent heat determines the evolution of a tropical hurricane. This is why a RCOAM is ideally suited to simulations of hurricanes. While operational NWP centers have shown considerable skill in predicting hurricane paths (Ginis *et al.,* 1997), predicting hurricane intensities is still a problem. Heavy reliance is placed instead on periodic observations of central pressure from aircraft penetrating into the eye. A principal problem is the absence of coupling of the hurricane model to the ocean underneath. It is also essential that both the SST and the upper ocean heat content be correctly prescribed at the beginning of the forecast. Hurricane Opal is an excellent example of this need. It developed from a tropical storm that moved into the

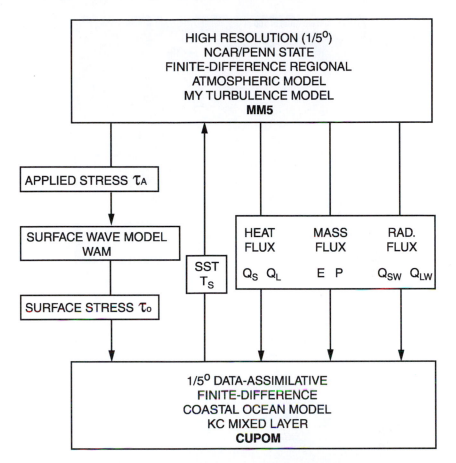

Figure 13.3.1 Coupling between the ocean and atmosphere models for hurricane Opal simulation in the Gulf of Mexico. Note that the atmosphere model domain extends over a much broader domain than that of the ocean.

southwestern part of the Gulf at the end of September 1995. As it evolved and moved along a roughly northward trajectory, it passed over a warm oceanic eddy (eddy Aggie) that had been shed recently by the Loop Current. This is thought to be the reason for its unexpectedly strong and sudden intensification. After its passage over the eddy, Opal unwound rapidly so that at landfall, it did not constitute a major threat to the Gulf coast. Shay *et al.* (1999) describe the possible influence eddy Aggie may have played in Opal's evolution.

Figure 13.3.2 shows the currents and SST from the initial state used to initialize the oceanic component of the coupled model. These were

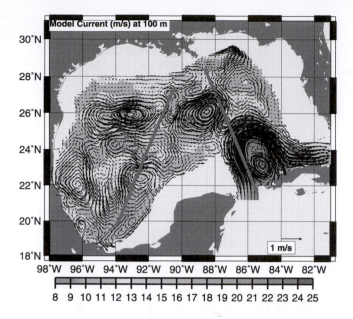

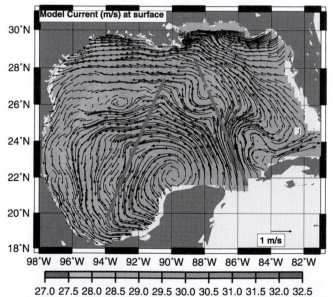

Figure 13.3.2 Currents at 100 m depth (top) showing eddy Aggy and SST and surface current distribution (bottom) in the Gulf at the beginning of the coupled run. These were obtained from a hindcast run of the ocean model alone assimilating altimetric SSH anomalies and MCSST.

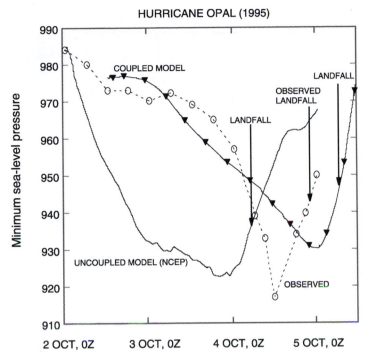

Figure 13.3.3 Center pressure as a function of time without (adapted from Gines *et al.* 1997) and with oceanic feedback compared to observations every 6 hr. Arrows indicate landfall. From Bao *et al.* (1998).

derived   from   a hindcast for 1995 using the ocean model run in a data-assimilative mode driven by FNMOC surface forcing, and assimilating weekly MCSST composites and TOPEX altimetric SSH anomalies (Choi *et al.*, 1995; see Section 14.7). The model simulated eddy Aggy reasonably well, and the eddy was shed from the Loop Current and moved west to roughly the right location at the end of September, when the coupled model simulations began. The SST at the end of September was also quite realistic. The coupled model was then set in motion, with the hurricane prescribed initially as an idealized Rankine vortex (Bao *et al.*, 1999). As observed, hurricane Opal intensified considerably when it passed over Aggy and unwound immediately after. Figure 13.3.3 shows the center pressure as a function of time compared to aircraft observations. It is clear that, in this case at least, prediction of hurricane intensity depended very much on the oceanic feedback and hence coupling was essential to more realistic redictions. Figure 13.3.4 shows the modeled hurricane path compared to observations.

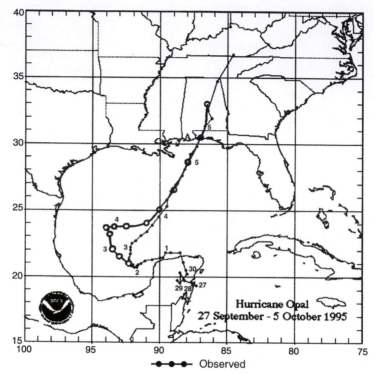

Figure 13.3.4 Observed and predicted paths of Opal every 6 hr. From Bao *et al.* (1998).

Bao *et al.* (1999, see also Andreas and Emanuel, 1999) emphasize the importance of including the effect of sea spray and its evaporation in coupled modeling. At large wind speeds typical of hurricanes, sea spray is an important aspect of the air-sea heat exchange (see Chapter 4 of Clayson and Kantha, 2000). Unfortunately, not much is known about it, although the situation is improving (Andreas, 1992; Fairall *et al.*, 1994; Andreas *et al.*, 1995; Andreas, 1998). Bao *et. al.* (1999) also show the importance of including a wave model in the coupling process, since the stress applied by the atmosphere (τ)is usually partitioned into two components, the wave supported component (τ_w) and turbulence supported component (τ_t), the latter driving the oceans. The ratio τ_w/τ is the wave coupling parameter and is a strong function of the wind wave age. Qunatity τ_w is given by (Makin and Kudryavtsev, 1999; see also Kudryavtsev *et. al.*, 1999)

$$\tau_w = \int_0^\infty \frac{1}{k} T(k) dk = \int_0^\infty \frac{1}{k} dk \int_{-\pi}^{\pi} c^2 B(k,\phi) \beta(k,\phi) \cos\phi \, d\phi \qquad (13.3.1)$$

where $T(k)$ is the omnidirectional spectrum of the momentum flux to waves, $B(k,\phi)$ is the directional wavenumber spectrum, c is the phase velocity, k is the wave number, ϕ is the angle from the wind direction, and $\beta(k,\phi)$ is the growth rate parameter describing the energy flux from wind to the waves (see Chapter 5 of Kantha and Clayson, 2000). Computing τ_w, given τ requires, therefore, a wave model. Following Janssen (1991), Bao *et al.* (1999) take:

$$\frac{u_{*w}}{u_{*t}} = \frac{z_1}{z_0} ; z_0 \sim \frac{u_*^2}{g} \qquad (13.3.2)$$

where z_0 is the roughness scale given by Charnock's relationship, and z_1 is the additional roughness and displacement scale due to waves, so that for neutral stratification, the law of the wall becomes (see Chapter 3 of Kantha and Clayson, 2000) :

$$\frac{U(z)}{u_*} = \frac{1}{\kappa} \ln \frac{z+z_1}{z_0+z_1} \qquad (13.3.3)$$

where κ is the von Karman constant. The atmosphere in MM5 takes $(z_0 + z_1)$ as the roughness scale in computing the total stress τ, with stratification taken into account by Monin-Obukhoff similarity laws.

It is very likely that both COAGCMs and RCOAMs will be useful and heavily used for studying and predicting the state and the internal and externally forced variability of the ocean-atmosphere coupled system in the coming decades. The field has matured to a stage where rapid progress is now possible and inevitable.

Chapter 14

Data Assimilation and Nowcasts/Forecasts

All models of physical and other processes are only approximations to the "truth". Even the best models of complex physical processes might have built-in biases. Inevitable errors in initial conditions, imperfect parameterization of physical processes, and inaccurate forcing make a model ocean diverge rapidly from the real ocean during a forecast. This is simply due to the extreme sensitivity of this system to even minute changes in initial conditions, typical of chaotic nonlinear systems. Even if the physical processes were to be parameterized correctly, the nature of the instability processes is such that the "phase" of events is likely to be incorrect in model simulations, even if the processes are represented well from a statistical point of view. It is therefore essential to employ information derived from actual observations of physical processes, in ocean models to nudge the modeled ocean state close to the real one. The situation is no different from that in modeling the state of the atmosphere for NWP purposes, except that the timescale for the loss of predictability is weeks for the oceans compared to days for the atmosphere. The process of employing observed data from the real ocean (atmosphere) to keep the modeled ocean (atmosphere) realistic is called data assimilation (for example, Anderson and Moore, 1986; Robinson *et al.*, 1998). Data assimilation consists of combining the modeled fields with data observed at various points in the domain to produce the best possible estimate of the real state of the ocean over the entire model domain. Exactly how this is best done has been the subject of considerable research in the atmospheric community over the past few decades (see Bengtsson *et al.*, 1981; Daley, 1991), driven by the need for accurate weather forecasts and, more recently, by an increased need for this information in the oceanic community (Haidvogel and Robinson; 1989, Malanotte-Rizzoli, 1996; Robinson *et. al.*, 1998).

The subject is vast and a voluminous literature exists on the topics of data assimilation and estimating the future state of a dynamical system such as the atmosphere. Most of the work is, however, related to weather prediction. We can only provide a brief review of the topic here and cite appropriate references. A comprehensive review of the status of oceanic data assimilation in the mid-1990s can be found in Malanotte-Rizzoli (1996; see also Ghil and Malanotte-Rizzoli, 1991) and is highly recommended for a broad overview of the field. Articles by Busalacchi (1996) and Leetma and Ji (1996) on data assimilation and forecasts in the tropical oceans, and Aikman *et al.* (1996), Robinson *et al.* (1996), Horton *et al.* (1997), Clifford *et al.* (1997), and Robinson (1999) on operational forecasts in regional oceans are particularly noteworthy. These detail the state of oceanic data assimilation and operational forecasting at present. For a real-time nowcast/forecast capability in the civilian sector, made possible by the timely availability of remotely sensed data such as altimetry, see Kantha *et al.* (1999).

14.1 INTRODUCTION

NWP centers predominantly use the so-called analysis-forecast cycle of assimilation. Here, the current state of the modeled atmosphere as predicted by the previous forecast is combined with observations of the atmosphere such as by radiosondes/surface stations all over the world by an analysis/initialization process to produce initial fields of various model variables suitable for describing the initial state for the next model forecast. The forecast skill depends very much on the accuracy of this initial state so derived, since errors in this initial state tend to get amplified with time during the forecast process. Analysis is the process of combining observational data taken at various points in the model domain with some reference data, which could be the model output from the previous forecast cycle. Initialization, on the other hand, is the process of preparing a grid-point data set from analysis, suitable for a smooth start of the time integration of a numerical model. Analysis and initialization can often be combined into a single process. The main objective of the analysis/initialization process is to bring information obtained from observations to bear upon the initial state of the model fields before embarking on a prediction process. The resulting fields must be a reasonable representation of the true state and be physically consistent with the modeled physics so that the model accepts them "gracefully."

This intermittent data assimilation is the most prevalent method in atmospheric prediction. The main advantage is that the resulting inclusion of a large amount of data tends to suppress random errors. Here the model forecast fields for the current time are used as first guess fields for analysis. Observational data in a time window on the order of 2 to 3 hr around t = 0 are

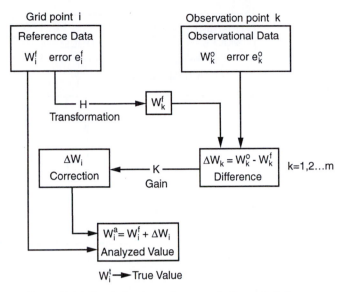

Figure 14.1.1 Sketch showing various steps in data assimilation.

collected, quality-controlled, and processed, and an optimum combination of model "data" and observational data at time $t = 0$ is derived and used to start the next model forecast (Figure 14.1.1). The intervals at which this initialization/forecast cycle is carried out should be long enough to suppress high frequency noise resulting from initialization but short enough to limit error growth due to instabilities and model drift. It is also a function of resources available. This period is 3 to 6 hr at most NWP centers. If a similar method is employed in the oceans, the period would be days, because of the much slower evolution of oceanic fields. However, no such forecast has ever been attempted in the oceans and no definitive answers are available. Model characteristics, data quality and quantity, and dynamical timescales all play a role in this.

An alternative assimilation procedure is the so-called continuous assimilation, where a numerical model is kept running and current by assimilation of observed data as they become available. Thus, the fields are continuously updated and corrected to produce a nowcast (Figure 14.1.2). A forecast can then be initiated at any point by a similar model running forward free without any data assimilation, but initialized from the state of the nowcast model. The principal advantage of this method over the analysis-forecast cycle is that the model derives more benefit from all past data as opposed to a single set of observations at a particular time. In some ways, the method makes the most optimum use of the observational data, although it is more difficult to suppress random errors. Also the shock of data insertion that exists in intermittent

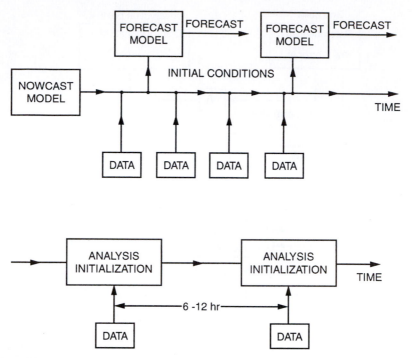

Figure 14.1.2 Comparison of continuous (4-D) assimilation with the analysis-initialization-forecast method.

assimilation due to the inevitable mismatch between the model and analyzed states is less severe provided the insertion interval is smaller. Since such a mismatch can lead to severe "noise" superimposed on the true state of the forecast atmosphere (ocean), often making the forecast unusable, considerable effort has been expended in devising means to minimize such a mismatch, resulting in a procedure called initialization in NWP terminology. Continuous assimilation tends to reduce this shock and is therefore often more preferable. The principal disadvantage is the added expense of running the nowcast model, in addition to the forecast model. Needless to say, the data assimilation scheme must be very efficient for this system to be viable. The reader is referred to Bengtsson *et al.* (1981), Haidvogel and Robinson (1989), Anderson and Willebrand (1989), Ghil and Malanotte-Rizzoli (1991), Brasseur (1995), Malanotte-Rizzoli (1996) and Robinson *et al.* (1998) for a discussion of various assimilation philosophies. For a comprehensive treatment of data assimilation, see also Daly (1991), Bennett (1992) and Wunsch (1996).

The differences between the atmosphere and the oceans that have an impact on data assimilation are worth noting. The spatial scales in the ocean, the scale of the mesoscale eddies, are smaller by two orders of magnitude in the ocean compared to the atmosphere (10-50 km versus 200-1000 km). NWP is well established and an infrastructure exists to collect *in situ* observational data on the state of the atmosphere periodically (00:00 and 12:00 UTC), archive them centrally, and process them efficiently for use in the next forecast cycle. This observational network is quite dense over populated land areas, although it is sparse over the oceans and sparsely inhabited land. Data from commercial aircraft, orbiting satellites, and oceanic drifting buoys are also made immediately available to NWP centers. Such an observing system does not yet exist for the oceans, although plans are underway to establish a similar network (Global Ocean Observing Systems, or GOOS). Only a few drifting buoys and commercial ships provide data at present about the oceanic state at any instant in time. Consequently, observational data are sparser, both in time and in space, by two to three orders of magnitude, especially in the southern hemisphere. However, these disadvantages are offset to some extent by an order of magnitude slower evolution of event scale processes. Still, data assimilation is three to four orders of magnitude harder in the oceans compared to the atmosphere. The advent of satellite altimetry and routine collection of SST data from AVHRR deployed on polar orbiting satellites are now providing more information on the ocean that is increasingly useful for assimilation and nowcast/forecasts.

Initialization does not appear to be a major problem in the oceans. Unlike the atmosphere, which "rings like a bell" and where small departures from dynamical equilibrium of the initial state can cause large problems in the forecast, in most cases, with the exception of perhaps the equatorial waveguide, initial fields have to be quite far from geostrophic equilibrium to cause large unwanted gravity waves. Therefore initialization procedures can be much simpler. Better yet, continuous assimilation can be employed. In the atmosphere, initialization continues to be the subject of intense research and a wholly satisfactory method is yet to be found. Data assimilation is a relatively young field of study in the oceans, barely a decade old, while NWP necessities have made it the subject of intense study in the atmosphere over the past half a century or so. There is still quite a bit of work ahead before a routine nowcast/forecast capability can be established for the oceans, since the method that works best for the atmosphere may not necessarily be optimal for the oceans, and considerable research remains to be done.

The method of combining data into a model can vary from the simplest one, called data insertion in which the model predicted values are just replaced by the observed values, to Kalman filters (Gelb, 1988), which blend the model and observed values optimally, taking into account the model error and observational error statistics, adjoint techniques (Thacker and Long, 1987), and variational methods (Derber and Rosati, 1989). It is also possible to use nudging techniques in which appropriate Newtonian damping terms that damp the variable to the observed value with a predetermined timescale are introduced into the governing equations. In general, the technique can be classified into two groups, optimal and suboptimal, at least in theory. Kalman filter is an example of an optimal technique, and optimal interpolation, of a suboptimal one. The most commonly employed method is optimal interpolation (see Choi and Kantha, 1995, for example), since methods such as Kalman filters and adjoint techniques are computationally intensive and at present still not routinely used for practical applications in NWP and ocean prediction. However, the situation is changing rapidly and a 4-D variational technique has just been introduced operationally at ECMWF. Most sequential assimilation methods replace the model predicted values by a weighted combination of model predicted value and observed value during the assimilation step, the weight determined either a priori by statistical methods such as optimal interpolation or updated at each assimilation step by a method such as Kalman filtering. For examples of oceanic data assimilation, the reader is referred to Derber and Rosati (1989), Glenn and Robinson (1995), Choi and Kantha (1995), Robinson (1996, 1999) and Kantha et al. (1999). An excellent recent summary of data assimilation in ocean models can be found in a collection of articles in Malanotte-Rizzoli (1996). Robinson et al. (1998) is particularly noteworthy. They provide a recent overview, which brings together the various data assimilation methods (used in both the atmosphere and the oceans) under a uniform notation, so that a better appreciation of the interrelationship between various techniques is possible; for too long, this already complex subject has been further complicated needlessly by different notations employed by diverse groups.

A recent example of a data-assimilative ocean model is the tropical Pacific Ocean model based on MOM2 (Seidel and Giese 1999). It assimilates in-stu temperature profiles from expendable probes as well as those from TOGA-TAO array, and TOPEX/Poseidon altimetric data. It is driven by NCEP winds (Kalnay et al., 1996), with SST damped to weekly values and SSS damped to monthly mean climatolgical values from COADS (da Silva et al. 1994). With high resolution in the vertical (15 m layers in the upper 150 m), the model managed to simulate currents in the tropical Pacific reasonably well over a wide range of temporal and spatial scales, including events associated with the 1997-98 ENSO.

14.2 DIRECT INSERTION

In this method, the model predicted values are simply replaced by observed values at grid points where observations are available. The implicit assumption is that the data are perfect, and the model errors are large. This method is seldom used in meteorology because it leads to large dynamical imbalances and corrupts the forecasts in intermittent assimilation and excessive noise in continuous assimilation and nowcasts. It is sometimes used in oceanographic data assimilation, since ocean models are a bit more tolerant to dynamical imbalances that are not too large. A slight variant of this method is replacing the model-generated values at predetermined intervals in a continuous assimilation scheme by a weighted average of the modeled and observed values at the grid points where observations are available. This technique is very much similar to optimal interpolation (OI) schemes, except that model adjustment takes place only at one grid point nearest to the observations and not all points within the decorrelation distance of the observations. Equivalently, the decorrelation scale is assumed to be smaller than the grid size. The model dynamics are relied upon to "spread" the information from observations to nearby grid points. The method is quite simple, very efficient (and hence applicable to models that require large computing resources), and feasible as long as the observations are not far from the state indicated by model dynamics. It has been used successfully by Kantha (1995b) in a high resolution model of the barotropic tides in the global oceans that assimilated tides derived from altimetry and coastal tide gauges. Often in ocean modeling, the data assimilation scheme itself is less influential in producing improved nowcasts than improved model physics (and/or the quality and quantity of data assimilated), and in these situations simple suboptimal schemes such as the modified data insertion turn out to be quite adequate.

14.3 NUDGING

In this method, the model is slowly nudged (or coaxed) toward observations at each time step in the model via a Newtonian damping term in the prognostic equation for the variable:

$$\frac{\partial X}{\partial t} + v \cdot \nabla X + \cdots = -\frac{(X - X_o)}{T_d} \tag{14.3.1}$$

where the model variable X is nudged toward a reference value X_o (in this case the observed value) at a timescale T_d. X_o can itself be a function of time. The smaller the value of T_d, the more rapidly it is nudged toward the reference value, since as $T_d \to 0$, $X \to X_o$. The larger the value of T_d compared to the characteristic timescale of variability in the prognosticated variable, the weaker the influence of assimilated data. Naturally, the value of T_d encodes the confidence in the reference (observed) value vis-a-vis the model value.

Note that this method actually changes the governing "physics" and hence requires some care in use. It is often used in both meteorology and oceanography, even in applications not involving data assimilation, especially where boundary conditions along an open boundary of the model are to be prescribed. For example, in atmospheric simulations, the model velocity can be nudged toward a fixed reference value at the top of the model domain in place of a more sophisticated radiation scheme. In ocean models, a popular use of Newtonian damping is in providing boundary conditions at the surface or a lateral boundary. Instead of prescribing the value for the variable (Dirichlet conditions) or the more difficult task of prescribing the flux of the variable (Neumann conditions), the model is afforded more flexibility by damping the variable toward a reference value. For example, the model SST can be nudged toward, say, its climatological mean, so that it does not deviate too far from climatology yet is not shackled firmly to it. The timescale for this damping is often chosen by trial and error.

A nonstatistical method called the Cressman successive correction (SC) method was used widely in the past at NWP centers prior to the widespread use of optimal interpolation (OI). Here the weights depend only on the absolute magnitude of the distance of the observation point from the grid point:

$$X_a = X_f + \frac{\sum_{k=1}^{M} a'_k \left(X_{o,k} - X_{f,k} \right)}{\sum_{k=1}^{M} a'_k}; \ a'_k \sim \exp\left[-\left(\frac{r_k}{d} \right)^2 \right] \tag{14.3.2}$$

where r_k is the distance of the observation point from the grid point, and d is the decorrelation scale (the influence radius). Subscripts a, f and o refer to analysis, forecast, and observed values. The method is often applied successively by initially choosing a large value of d and then shrinking it gradually for each iteration with the previous analysis value as the new guess value.

14.4 STATISTICAL ASSIMILATION SCHEMES

Statistical assimilation includes simple suboptimal schemes such as optimal interpolation (OI) as well as sophisticated optimal methods such as Kalman filters. It attempts to assign weights to the observations and model values based on their error characteristics, so that a blended value can be obtained that minimizes the analysis error (the difference between the true value and the assimilation result) in a statistical sense. The basics can be illustrated simply by application to a single scalar variable (Ghil and Malanotte-Rizzoli, 1991). Suppose we have two estimates X_1 and X_2 for a quantity X. What is the best possible value for X? Let X_t be its true value and let its unbiased estimate be

$$X_a = a_1 X_1 + a_2 X_2$$

where a_1 and a_2 are the weights given to each value, that is, $E(X_1) = E(X_2) = E(X_t)$, where E is its expectation. Since $E(X_a)$ must equal $E(X_t)$,

$$a_1 + a_2 = 1 \text{ and } X_a = X_1 + a_2(X_2 - X_1) \tag{14.4.1}$$

Let $E(X_1 - X_t)^2 = \sigma_1^2$, $E(X_2 - X_t)^2 = \sigma_2^2$, and $E((X_1 - X_t)(X_2 - X_t)) = 0$. The last condition implies that the two values are uncorrelated. The first two characterize their variability or "error" in a statistical sense. The best linear unbiased estimate for X is obtained by minimizing the estimation error: $\sigma_a^2 = E(X_a - X_t)^2 = a_1^2 \sigma_1^2 + a_2^2 \sigma_2^2$. The minimum occurs when

$$a_1 = \frac{\sigma_a^2}{\sigma_1^2} = \frac{A_1}{A_1 + A_2}; a_2 = \frac{\sigma_a^2}{\sigma_2^2} = \frac{A_2}{A_1 + A_2} \tag{14.4.2}$$

with

$$\frac{1}{\sigma_a^2} = \frac{1}{\sigma_1^2} + \frac{1}{\sigma_2^2} \tag{14.4.3}$$

so that if the weight a_2 is rewritten slightly as K:

$$K = \frac{1}{\sigma_2^2}\left[\frac{1}{\sigma_1^2} + \frac{1}{\sigma_2^2}\right]^{-1} = \frac{A_2}{A_1 + A_2} \tag{14.4.4}$$

$$X_a = X_1 + K(X_2 - X_1) \tag{14.4.5}$$

where A is the inverse of σ^2 and hence can be termed accuracy. Thus if the variance or error of X_1 is very large (or equivalently A_1 is zero), the weight $a_1 = 0$, and $a_2 = 1$, and the best estimation for X is X_2. If both are of equal accuracy, one gets the average of the two values for X.

If we assume X_1 is the model-generated value and X_2 is the observed value, it is easy to see that the best estimate given by Eq. (14.4.5) essentially minimizes the analysis error in a statistical sense, given the model error and the observational error statistics. OI scheme preassigns the weights based on assumed error characteristics, whereas the Kalman filter updates the model error statistics at each assimilation step. The former is less efficient but not very resource intensive, whereas the latter requires prohibitively large resources for practical applications, unless drastic simplifications can be made (Malanotte-Rizzoli *et al.*, 1996). OI is the prevailing scheme in most NWP centers and in operational oceanic data assimilation (for example, Choi *et al.*, 1995; Clifford *et al.*, 1997; Horton *et al.*, 1997; Kantha *et al.*, 1999).

14.4.1 KALMAN FILTER

Kalman filter can be illustrated rather simply by application to a scalar. Let X_f be the model forecast value and let X_o be the observed value. The question is how to combine the two to obtain the best estimate for the analysis value X_a? This can be done by minimizing the mean square estimation error $e_a = \overline{(X_a - X_t)(X_a - X_t)}$. Let $e_f = X_t - X_f$, $e_o = X_o - X_t$ be the forecast error and the observational error respectively. Let $P = \overline{e_f e_f}$; $R = \overline{e_o e_o}$ be the forecast error covariance and the observational error covariance, and let $X_a = X_f + K (X_o - X_f)$, where K is the Kalman gain. Then the analysis error becomes, after some algebra, $e_a = (X_a - X_t) = K(X_o - X_f) - (X_t - X_f) = K (e_o + e_f) - e_f$. The analysis error covariance is

$$P_a = \overline{e_a e_a} = \overline{K(e_o + e_f)(e_o + e_f)K} - \overline{e_f (e_o + e_f)K} - \overline{K(e_o + e_f)e_f} + \overline{e_f e_f}$$

If the observation and the model errors are assumed to be uncorrelated, $\overline{e_f e_o} = 0 = \overline{e_o e_f}$, then the analysis error covariance can be written as

$$P_a = P - P(P+R)^{-1}P + \left[K - P(P+R)^{-1}\right](P+R)\left[K - P(P+R)^{-1}\right] \quad (14.4.6)$$

The minimum value of P_a occurs when the Kalman gain is

$$K = P(P+R)^{-1} \quad (14.4.7)$$

and the analysis error covariance becomes

$$P_a = (1 - K)P \quad (14.4.8)$$

The best statistical estimate in the sense of minimum mean square estimation error becomes

$$X_a = X_f + K (X_o - X_f) \quad (14.4.9)$$

This procedure is of course similar to what we did in deriving Eqs. (14.4.2) to (14.4.5), except that very deliberately, the notation is chosen so that extension to more than one state variable X—that is, when the variances and covariances become matrices—is more decipherable. This can be seen by taking the inverse of P_a:

$$P_a^{-1} = P^{-1} + R^{-1} \tag{14.4.10}$$

This is the same as Eq. (14.4.3), and Eqs. (14.4.4) and (14.4.5) are similar to Eqs. (14.4.7) and (14.4.9). Now, in a time dependent *linear* system,

$$X_{f,n} = L_{n-1} X_{a,n-1} \tag{14.4.11}$$

that is, the forecast value of the variable at time step n depends on the analyzed value at the previous time step n−1. If one assumes that the true value of the variable consists of some additive system noise $e_{t,n-1}$, where $Q_{n-1} = \overline{e_{t,n-1} e_{t,n-1}}$:

$$X_{t,n} = L_{n-1} X_{t,n-1} + e_{t,n-1} \tag{14.4.12}$$

then $e_{f,n} = X_{t,n} - X_{f,n} = e_{t,n-1} - L_{n-1} e_{a,n-1}$. The errors $e_{a,n-1}$ and $e_{t,n-1}$ are correlated, often strongly, but for convenience and simplicity, are assumed to be uncorrelated. Then, the forecast error covariance at time step n becomes

$$P_n = Q_{n-1} + L_{n-1} P_{a,n-1} L_{n-1} \tag{14.4.13}$$

Thus the model error covariance can be updated at time step n using known statistics from the previous step n. The Kalman filter can therefore be looked upon as a sequential recursive linear optimal filter with the following sequential steps:

1. Obtain a forecast for the linear or suitably linearized system:

$$X_{f,n} = L_{n-1} X_{a,n-1} \tag{14.4.14}$$

2. Update the P_n forecast error from statistics from the previous timestep:

$$= Q_{n-1} + L_{n-1} P_{a,n-1} L_{n-1} \tag{14.4.15}$$

3. Compute the Kalman gain:

$$K_n = P_n (P_n + R_n)^{-1} \tag{14.4.16}$$

4. Update the analysis error:

$$P_{a,n} = (1 - K_n) P_n \tag{14.4.17}$$

5. Obtain the new analysis value:

$$X_{a,n} = X_{f,n} + K_n (X_{o,n} - X_{f,n}) \tag{14.4.18}$$

Thus the Kalman filter technique involves updating the forecast error covariance at each assimilation step based on the knowledge of system noise and observational error statistics, as well as the entire past history of the model state. It is the "most optimum least square error fitter."

Because of the notation employed, the above series of steps holds when there is more than one state variable, and the variable becomes a state vector and the variances, the Kalman gain K and the transformation coefficient L, become matrices. Generalizing the analysis to the case when the observations are less numerous, let the forecast and analysis state vectors $\mathbf{X}_f$ and $\mathbf{X}_a$ be of length N and the observation vector $\mathbf{X}_o$ be of length M (M $\ll$ N); then $\mathbf{H}$ is the observation matrix (M $\times$ N), and $\mathbf{K}$ is the Kalman gain matrix (N $\times$ M). The forecast error covariance $\mathbf{P}$ is a (N $\times$ N) matrix, $\mathbf{Q}$ is the system noise covariance matrix (N $\times$ N), and the observational error covariance $\mathbf{R}$ is an (M $\times$ M) matrix, given by

$$\mathbf{P} = \overline{\mathbf{e}_f \mathbf{e}_f^T}; \ \mathbf{e}_f = \mathbf{X}_t - \mathbf{X}_f$$
$$\mathbf{Q} = \overline{\mathbf{e}_t \mathbf{e}_t^T} \qquad (14.4.19)$$
$$\mathbf{R} = \overline{\mathbf{e}_o \mathbf{e}_o^T}; \ \mathbf{e}_o = \mathbf{X}_o - \mathbf{H}\mathbf{X}_t$$

The linear governing equation (with any possible forcing terms not written down explicitly) is

$$\mathbf{X}_{f, n} = \mathbf{L}_{n-1} \, \mathbf{X}_{a, n-1} + \mathbf{e}_{t, n-1} \qquad (14.4.20)$$

The updated forecast error covariance is

$$\mathbf{P}_n = \mathbf{Q}_{n-1} + \mathbf{L}_{n-1} \, \mathbf{P}_{a, n-1} \mathbf{L}^T_{n-1} \qquad (14.4.21)$$

The Kalman gain that minimizes the expected error in analysis, namely $\overline{\left(\mathbf{X}_{a,n} - \mathbf{X}_{t,n}\right)^T \left(\mathbf{X}_{a,n} - \mathbf{X}_{t,n}\right)}$ is

$$\mathbf{K}_n = \mathbf{P}_n \, \mathbf{H}_n^T \, (\mathbf{H}_n \, \mathbf{P}_n \, \mathbf{H}_n^T + \mathbf{R}_n)^{-1} = \mathbf{P}_{a, n} \, \mathbf{H}_n^T \, \mathbf{R}_n^{-1} \qquad (14.4.22)$$

The updated analysis error covariance is

$$\mathbf{P}_{a, n} = (\mathbf{I} - \mathbf{K}_n \, \mathbf{H}_n) \, \mathbf{P}_n \qquad (14.4.23)$$

and the new analyzed value for the state vector is

$$\mathbf{X}_{a, n} = \mathbf{X}_{f, n} + \mathbf{K}_n \, (\mathbf{X}_{o, n} - \mathbf{H}_n \, \mathbf{X}_{f, n}) \qquad (14.4.24)$$

Note that

$$(\mathbf{P}_{a,\,n})^{-1} = (\mathbf{P}_n)^{-1} + \mathbf{H}_n^{\ T}\,\mathbf{R}_n^{\ -1}\,\mathbf{H}_n \qquad (14.4.25)$$

Equations (14.4.22) and (14.4.24) are generalizations of Eqs. (14.4.4) and (14.4.6) for the simple scalar example cited above. Equations (14.4.21) through (14.4.23) are known as the Riccati equations.

Note that the result from a Kalman filter may not be consistent with dynamical constraints, unless imposed as in the variational methods discussed below. Also, while the filter determines the model error covariance $\mathbf{P}_n$ by itself as a part of the solution process, it does require specification of the system noise error covariance $\mathbf{Q}_{n-1}$. This is extremely difficult in practice, although in principle at least it is a more easily determined property of the system independent of data and their assimilation (OI methods essentially require $\mathbf{P}_n$ to be prespecified). Even if $\mathbf{Q}_{n-1}$ were known fully, solving Eq. (14.4.21) at each assimilation step is a formidable computational task, requiring order N^3 multiplications each time. Storage requirements are proportional to N^2. It is therefore impractical for N to be greater than about 10^3. This should be contrasted to modern numerical models, which might have more than half a dozen prognostic variables on a grid with a number of grid points in the range 10^6–10^7.

A Kalman filter provides an unbiased estimate for the current state of the system, that has minimum error variance, and is based on all past observations and dynamics (linear). It can be used for nowcasts of a dynamical system. However, during hindcasts, when "future" data are also available, a Kalman smoother may be preferable, since it makes use of all the data in the assimilation period to make optimal estimates at all points within that period. However, a Kalman smoother is even more resource-intensive than a Kalman filter.

Fukumori *et al.* (1993) and Fu and Fukumori (1996) take advantage of the fact that with regular assimilation of data, when observation locations remain unchanged, $\mathbf{P}_n$ can often approach an asymptotic steady state $\mathbf{P}_\infty$. They use the asymptotic value to assimilate altimetric data from TOPEX/Poseidon into a linear, wind-driven reduced gravity model of the tropical Pacific. $\mathbf{P}_\infty$ can be calculated from the Riccati equations using a technique that approximates a time-evolving system with a time-invariant one (see Anderson and Moore, 1979; Malanotte-Rizzoli *et al.,* 1996) and then used repeatedly to compute the Kalman gain, thus avoiding the need to update the covariance each step. Combining Eqs. (14.4.22) and (14.4.23) one gets

$$\mathbf{P}_n = [\mathbf{P}_\infty^{-1} + \mathbf{H}_n^T \mathbf{R}_n^{-1}\mathbf{H}_n]^{-1} \qquad (14.4.26)$$

which is simply a statement of model state error covariance being the optimum combination of a preselected model forecast error covariance and obserational error covariance (in OI schemes, $\mathbf{P}_\infty$ is empirically estimated; the difference between model simulation and observations can be used). However, computing $\mathbf{P}_\infty$ using Riccati equations is still impractical for large state vectors. Hence, following Fukumori and Malanotte-Rizzoli (1995), Fu and Fukumori (1996) use a grid much coarser than the model grid to reduce the state vector from 12,000 to 831. This is formally equivalent to the reduced state space Kalman filter described below but with an asymptotic error covariance. Malanotte-Rizzoli et al. (1996) have extended the methodology to a fully nonlinear model of an idealized midlatitude zonal jet by linearization of model dynamics around different mean states at successive preselected intervals.

14.4.2 REDUCED STATE SPACE KALMAN FILTERS

While the Kalman filter is arguably the "most optimal least square error fitter" that one can design for data assimilation, it does have some glaring deficiencies. First, it is a linear filter, rigorously formulated only for linear systems. Unlike adjoint methods, which are easy to implement and are quite rigorous for nonlinear systems, the Kalman filter formulation is only approximate, since higher order statistical moments important to nonlinear systems cannot be accounted for. Piecewise (Extended Kalman Filter) linearization around a mean state is needed. While this makes the technique feasible, potential problems such as the possibility of divergence over long periods of time exist (Gelb, 1988). Second, the formidable computational burden needed to update the model error covariance matrix at each assimilation step has so far precluded its application to realistic systems. It is not unusual to have the number of state variables N on the order of 10^7, and updating a $(10^7 \times 10^7)$ matrix every time step is beyond the capabilities of even a modern-day high performance computer. However, the most serious may well be that in practical situations in the atmosphere and the oceans, especially the latter, the error structures needed for optimal filtering are so poorly known that the underlying theoretical bases are violated and hence the filter may not perform well (Dee, 1995). Imperfect knowledge of error covariances makes the use of a full-fledged Kalman filter not only of doubtful utility, especially given the computational expense, but even undesirable in some cases. The task of utilizing poorly known error structures in a computationally feasible filter of a highly complex nonlinear system has spurred the development of reduced state space Kalman filters that appear to hold the promise of making Kalman filters realizable (Cane et al., 1996; Fukumori and Malanotte-Rizzoli, 1995; Malanotte-Rizzoli et al., 1996). Fukumori and Malanotte-Rizzoli (1995) used a grid coarser than the model grid to reduce the state space, guided by EOFs as to its choice. They also used the asymptotic form for the gain matrix instead of updating it (equivalent to OI) and applied it successfully to an idealized eddy-resolving model of a midlatitude jet, whereas Cane *et al.* (1996) used EOFs to simplify the specification of the model system error covaiance $\mathbf{Q}$ and updating the model error covariance $\mathbf{P}$ for a linear tropical sea level model. The reduced state space also prevents local overfitting resulting from misspecified statistics and the solutions are also smoother. Thus the method not only becomes more feasible but also more skillful. We describe the method of Cane *et al.* (1996) for its perceived potential for use in practical data assimilation applications. The whole idea is to reduce the number of state variables from an unmanageable N to more manageable S using multivariate EOFs:

$$\mathbf{X}(x,t) = \mathbf{E}(x)\hat{\mathbf{X}}(t) \rightarrow \hat{\mathbf{X}} = \mathbf{E}^{T}\mathbf{X} \qquad (14.4.27)$$

where $\mathbf{E}$ are the EOFs and $\hat{\mathbf{X}}$ are the principal components governed by the evolution equation:

$$\hat{\mathbf{X}}_{n} = \hat{\mathbf{L}}_{n-1}\hat{\mathbf{X}}_{n-1} + \hat{\mathbf{X}}'_{n-1}; \quad \hat{\mathbf{L}}_{n-1} = \mathbf{E}^{T}\mathbf{L}_{n-1}\mathbf{E} \qquad (14.4.28)$$

Knowing $\mathbf{E}$ and $\mathbf{L}$ enables $\hat{\mathbf{L}}$ to be determined (Cane *et al.,* 1996; Fukumori and Malanotte-Rizzoli, 1995). The updated forecast error covariance of the reduced state variable $\hat{\mathbf{X}}$, $\hat{\mathbf{P}} = \overline{\left(\hat{\mathbf{X}} - \hat{\mathbf{X}}^{t}\right)\left(\hat{\mathbf{X}} - \hat{\mathbf{X}}^{t}\right)^{T}}$ now becomes

$$\hat{\mathbf{P}}_{n} = \hat{\mathbf{Q}}_{n-1} + \hat{\mathbf{L}}_{n-1}\hat{\mathbf{P}}_{a,n-1}\hat{\mathbf{L}}_{n-1}^{T} \qquad (14.4.29)$$

The Kalman gain is

$$\hat{\mathbf{K}}_{n} = \hat{\mathbf{P}}_{n}\hat{\mathbf{H}}_{n}^{T}\left(\hat{\mathbf{H}}_{n}\hat{\mathbf{P}}_{n}\hat{\mathbf{H}}_{n}^{T} + \hat{\mathbf{R}}_{n}\right)^{-1} \qquad (14.4.30)$$

The updated analysis error covariance of the reduced state variable is

$$\hat{\mathbf{P}}_{a,n} = \left(\mathbf{I} - \hat{\mathbf{K}}_{n}\hat{\mathbf{H}}_{n}\right)\hat{\mathbf{P}}_{n} \qquad (14.4.31)$$

where

$$\hat{\mathbf{H}}_{n} = \mathbf{H}_{n}\mathbf{E}_{n} \qquad (14.4.32)$$

The quantities with hats refer to reduced state space S ($\ll$ N). The essence of this approach is to discard higher, unimportant (and often unwelcome) modes of variability through truncation of EOFs. The unreduced space state Kalman gain $\mathbf{K_n}$ can be written as (Cane *et al.*, 1996)

$$\mathbf{K}_n = \mathbf{E}_n \hat{\mathbf{P}}_n \hat{\mathbf{H}}_n^T \left(\hat{\mathbf{H}}_n \hat{\mathbf{P}}_n \hat{\mathbf{H}}_n^T + \hat{\mathbf{R}}_n + \mathbf{R}_a' \right)^{-1} + \mathbf{K}_n'$$

$$\mathbf{R}_n' = \mathbf{H}_n \mathbf{P}_n \mathbf{H}_n^T - \hat{\mathbf{H}}_n \hat{\mathbf{P}}_n \hat{\mathbf{H}}_n^T$$

(14.4.33)

where the $\mathbf{K}_n'$ part of the gain matrix is discarded. $\mathbf{R}_n'$ is the extra noise term arising from discarded modes and this is why the reduced state space gain in Eq. (14.4.29) is different. It can also be regarded as additional sampling error associated with discarded modes (Cane *et al.*, 1996; Fukumori and Malanotte-Rizzoli, 1995).

The implementation of this scheme, however, requires considerable care as to the influence of the discarded modes on system noise. It is difficult to detail the procedures followed; instead we refer the reader to the original references (Cane *et al.*, 1996; Fukumori and Malanotte-Rizzoli, 1995).

Cane *et al.* (1996) applied this method to predicting sea levels in the tropical Pacific by using monthly values of model fields over an approximately 24-year long model run to derive EOFs (Xue *et al.*, 1996), 93 of which retain 99% of the model variance. They used this reduced state space to perform assimilation of data from island gauges from the Pacific and compared the results to that which used the full state space, but a coarser grid (Miller and Cane 1996). They find that the two filters provide very similar results especially at points not assimilated into the model (see Figure 14.4.1), lending support to the notion that poor knowledge of error structures does not let the filter extract useful additional knowledge of the state from data, beyond a limited number of modes.

14.4.3 Optimal (Statistical) Interpolation (OI) Scheme

The OI scheme is a trivial subset of the Kalman filter. Equations (14.4.22) and (14.4.24) are used, but with the error covariance $\mathbf{P}_n$ prescribed and held unchanged in time. Equation (14.4.21), the updating of the error covariance so central to and such a demanding part of Kalman filtering, is not used. Equation (14.4.23) or equivalently Eq. (14.4.25) can then be used, but they are seldom used for diagnostics. Note that if $\mathbf{P}_n$ is made infinity, direct insertion results.

Following the simple example given above, the OI scheme itself can be generalized to account for more than one observational value affecting the analysis. Suppose there are M observations surrounding the grid point value in question: $X_{o, k}$ (k = 1, 2, . . . M) and let the error be $e_{o, k}$. Then the analysis value X_a is

$$X_a = X_f + \sum_{k=1}^{M} a_k \left(X_{o,k} - X_{f,k} \right)$$

(14.4.34)

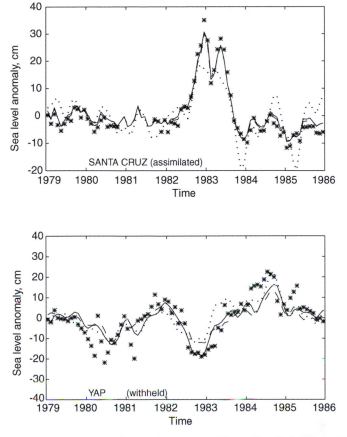

Figure 14.4.1 Comparison of the results of full state space Kalman filter (dashed line) with those of a reduced state filter (solid line). Sea level heights are shown at Santa Cruz where data were assimilated (top) and Yap where they were not (bottom). Stars show observations and dotted line shows unfiltered model output. From Cane *et al.* (1996).

The analysis error e_a can be written as

$$e_a = (X_a - X_t) = e_f + \sum_{k=1}^{M} p_k (e_{o,k} - e_{f,k}); \; e_f = (X_f - X_t) \tag{14.4.35}$$

in terms of the observational errors and the forecast error. The double subscript on X_f implies that the model values have to be interpolated to the locations of the observational data points. Assuming e_f and $e_{o,\,k}$ are not correlated for all k, now we minimize the analysis error by differentiating the analysis error covariance with the weights a_k: $\partial P_a / \partial a_k = 0$:

$$\overline{e_f e_{f,k}} = \sum_{k=1}^{M} a_k \left(\overline{e_{o,k} e_{o,k}} + \overline{e_{f,k} e_{f,k}} \right) \tag{14.4.36}$$

which can be written as

$$p_k = \sum_{l=1}^{M} a_l \left(P_{kl} + R_{kl} \right) \qquad\qquad k=1, 2, \dots M \tag{14.4.37}$$

$$p_k = \overline{e_f e_{f,k}}; \ P_{kl} = \overline{e_{f,k} e_{f,l}}; \ R_{kl} = \overline{e_{o,k} e_{o,l}}$$

a set of M equations for M unknown weights that can be readily solved, if the forecast error structure P_{kl}(and p_k) and the observational error structure R_{kl} are known, and the analysis value can be obtained from Eq. (14.4.34). The trivial scalar example cited above can be obtained by putting k=1. Then $P_{11} = p_1 = \sigma_f^2$; $R_{11} = \sigma_o^2$ and the weight becomes $a_1 = \dfrac{\sigma_f^2}{\sigma_f^2 + \sigma_o^2} = \dfrac{A_o}{A_f + A_o}$ so that

$$X_a = X_f + a_1 \left(X_o - X_f \right) = \left(\frac{A_o}{A_o + A_f} \right) X_o + \left(\frac{A_f}{A_o + A_f} \right) X_o \tag{14.4.38}$$

same as the simple example cited earlier. The weights depend on the relative accuracies of observations and the model forecasts. For $A_o \to \infty$, error-free observations, $X_a = X_o$; for $A_f \to \infty$, error-free model (or no observations), $X_a = X_f$; for $A_o = A_f$, $X_a = 0.5 (X_o + X_f)$. The more usual case of M observational points is a generalization of this concept. Unfortunately it requires the solution of Eq. (14.4.37), a set of M linear equations, or equivalently the inversion of a (M × M) matrix at each assimilation step at each model grid point, where M is the number of observations within the decorrelation distance of the grid point in question and therefore is quite time-consuming to do unless M is artificially bounded. Normally, only the first M most correlated points are assimilated, after sorting the observational points in descending order of correlation, M being typically less than 10 for the oceans and up to 200 in the atmosphere. However, this has been a significant source of noise and error.

Consider two data points, the first one located at a distance 2R and the second R from the grid point, in one case on either side of the grid point, in the second, on the same side. For simplicity, we will consider the observational error to be zero, $R_{kl} = 0$, so that the method yields direct insertion. Let $P_{kl} = \exp [- (r_{kl}/d)^2]$, where r_{kl} is the distance between the two points and let $Y = \exp [- (R/d)^2]$.

Then, for the first case, $P_{11} = P_{22} = 1$, $P_{12} = P_{21} = Y^9$, $p_1 = Y^4$, $p_2 = Y$, so that the OI technique gives

$$X_a = X_f + a_1 \left(X_{o,1} - X_{f,1} \right) + a_2 \left(X_{o,2} - X_{f,2} \right)$$

$$a_1 = \frac{Y^4 (1 + Y^3)}{(1 + Y^9)(1 + Y^3 + Y^6)}; \ a_2 = \frac{(1 + Y^3)(1 + Y^6)}{(1 + Y^9)(1 + Y^3 + Y^6)} \qquad (14.4.39)$$

with more weight given to the second point $a_2 > a_1$. The SC, on the other hand, gives $a_1 = Y^4$, $a_2 = Y$, so that

$$X_a = X_f + a_1 \left(X_{o,1} - X_{f,1} \right) + a_2 \left(X_{o,2} - X_{f,2} \right)$$

$$a_1 = \frac{Y^4}{(Y + Y^4)}; \ a_2 = \frac{Y}{(Y + Y^4)} \qquad (14.4.40)$$

However, for the second case, $P_{12} = P_{21} = Y$, so that OI method gives $a_1 = -Y^2$, $a_2 = Y(1 + Y^2)$, whereas SC gives the same value as before. If for this case, the two points are situated at equal distance R from the grid point, the weights become 1/2 for the SC and $Y/(1+Y^4)$ for the OI. Thus in SC, all observations are given the same weight if they are located the same distance from the grid point and no consideration is given to the observational or model error statistics.

The difference in the case of three observational points, situated at the same distance R from the grid point, but two of them coincident on one side, is even more dramatic. The OI scheme gives

$$X_a = X_f + a_1 \left(X_{o,1} - X_{f,1} \right) + a_2 \left(X_{o,2} - X_{f,2} \right) + a_3 \left(X_{o,3} - X_{f,3} \right)$$

$$a_1 = \frac{Y}{(1 + Y^4)}; \ a_2 = \frac{Y}{2(1 + Y^4)} = a_3 \qquad (14.4.41)$$

with the two coincident points contributing the same together as the third point, whereas SC gives an equal weight of 1/3 to all three points.

The above examples illustrate why OI has replaced SC at NWP centers. However, with the advent of modern multigigaflop computers, efforts are underway to develop and use more optimal data assimilation methods than the OI, such as the Kalman filter, the adjoint technique, and the variational methods.

14.5 VARIATIONAL METHODS

Variational assimilation methods are also optimized data assimilation schemes that attempt to minimize some measure of squared error. They define a functional and use the concepts from the calculus of variations to minimize it subject to certain prescribed constraints. The functional could, for example, be the sum of the squares of the differences between observed and model values of some variable such as the temperature or velocity over the model integration time. The constraint would then be the governing equations of the model, the continuity, momentum, and temperature advection equations. The method has roots in variational calculus and dynamical minimization problems dating back to the time of Euler and Lagrange. It is easily illustrated by a simple one-dimensional problem in variational calculus: Find the function u(x) that makes

the functional $I\{u(x)\} = \int_{x_a}^{x_b} F(x,u,u')dx$ stationary (an extremum) subject to the

constraint $G(x,u,u') = 0$, where the primes denote the x-derivative. Of course, u satisfies boundary conditions at x_a and x_b. Note that integral over the entire interval is involved and therefore the quantities must be well defined over it.

This classical problem can be solved using Lagrange's method of undetermined multipliers (for example, see Daley, 1991), which involves determining the stationary value of a new functional, involving both the functions F and G and an unknown multiplier function $\lambda(x)$:

$I_1\{u(x),G,\lambda(x)\} = \int_{x_a}^{x_b} [F(x,u,u') + \lambda(x)G(x,u,u')]dx$. This requires that the first

variation of I_1, δI_1 be zero and leads to the classic Euler-Lagrange equation

$\dfrac{\partial}{\partial u}[F + \lambda(x)G] - \dfrac{d}{dx}\dfrac{\partial}{\partial u'}[F + \lambda(x)G] = 0$ that along with the constraint is adequate

to determine the function u(x). The second variation $\delta^2 I_1$ determines if the stationary value is also an extremum. If we regard G as the governing equations of the numerical model and F as a quadratic quantity to be minimized over the integration interval, the application of the variational methods to the data assimilation problem becomes clear.

Consider the cost function

$$J = \int_V \left[\alpha_1^2 (p - p_a)^2 + \alpha_2^2 (U - U_a)^2 \right] dv \qquad (14.5.1)$$

where subscript a denotes analyzed (observed) values and α_I are preassigned weights. The variables p and U are functions of space and some independent variable S. It is required to find S that minimizes the errors in variables p and U over the domain subject to the constraint f (U,p) = 0, usually the governing equation such as the geostrophic relationship. The problem can be tackled by the Lagrange multiplier method if the constraint is required to be satisfied exactly, the so-called strong constraint approach (Sasaki, 1970). The problem then is to find the stationary value of the augmented functional:

$$J = \int_V \left[\alpha_1^2 (p - p_a)^2 + \alpha_2^2 (U - U_a)^2 + \lambda f \right] dv \qquad (14.5.2)$$

By solving the resulting Euler-Lagrange equations, one can determine the variable S needed. Often it is desirable to satisfy the constraint only approximately. This could be the case when, for example, the model or the forcing fields in the model are imperfect. Then there is no advantage to insisting that the model-governed constraints be satisfied exactly. In this weak constraint approach (Sasaki, 1970), one finds the stationary value of a modified functional instead:

$$J = \int_V \left[\alpha_1^2 (p - p_a)^2 + \alpha_2^2 (U - U_a)^2 + \beta(f)^2 \right] dv \qquad (14.5.3)$$

where β is a prespecified weight, which if chosen small makes the constraint weak in that it is only satisfied approximately. The first two terms constitute the cost function, whereas the last one is the penalty function, since the constraint violations are penalized in the process of minimizing the cost function. Examples of strong and weak constraint variational methods applied to practical problems in atmospheric analysis can be found in Daley (1991).

Four-dimensional variational analysis extends this technique to include both spatial and temporal dimensions. The idea is to try and minimize the model forecast error over the entire forecast cycle by determining, say, the initial conditions (the classical initialization problem in weather forecast) that would do so. This requires variational approaches (Derber, 1987, for example).

Derber and Rosati (1989) used a variational formulation of OI in their global model to assimilate XBT data. Their scheme attempts to minimize the functional:

$$J = T^T P^{-1} T + \left[D(T) - T_o \right]^T R^{-1} \left[D(T) - T_o \right] \qquad (14.5.4)$$

where T is an N component vector of corrections to temperature, **P** is the (N × N) first-guess error covariance matrix, modeled as a Gaussian (correlation of error between any two points being ~ a exp $(-r^2/b^2)$, where a and b are prescribed constants and r is the distance between the points), **R** is the (M × M) observational error covariance matrix, assumed traditionally to be diagonal (no correlation between individual observations), and D is an interpolation operator. T_o is an M component vector containing the difference between the observations and the interpolated guess value. The functional J is minimized using a preconditioned conjugate gradient technique.

14.5.1 ADJOINT MODELS

Adjoint models and methods (Lewis and Derber, 1985; Long and Thacker, 1989a,b; see also Bennett, 1992) are a formal way of minimizing a quadratic cost function subject to the constraints of the nonlinear governing equations in a time-dependent model forecast. The procedure begins with a cost function defined as a measure of the misfit between model forecasts and observations. The goal is to minimize this cost function under the constraint that the governing equations are satisfied by adjusting variables such as model parameters or initial conditions. This is done by integrating the governing equations forward in time (forward model) with an initial guess for the control variables and estimating the cost function in the process. Then the adjoint equations (adjoint model) are integrated backward in time, forced by the model-data misfit. This permits the gradient of the cost function with respect to the control variables to be computed from this backward integration, from which an improved estimate of the control variables can be made, and the procedure repeated until the cost function is minimized. The adjoint method is therefore an efficient iterative technique for solving the minimization problem. The adjoint equations can be readily derived from the governing equations by the use of the Lagrange multiplier method (Long and Thacker, 1989a,b). In practice, however, it is important for numerical consistency to construct the adjoint of the discretized form of the forward model, rather than discretizing the continuous form of the adjoint model. Fortunately, automated software (Giering and Kaminski, 1996) is available to do this.

Suppose we integrate the model for a time period and define a quantity, which is an integral over that time period. We wish to find either the initial state of the model or a model parameter that satisfies a constraint on this quantity. For example, we may want to minimize mean squared error relative to some preanalysis. One way to do this is to make a number of model runs varying the parameter of interest (or initial condition), determine the quantity for each integration, and arrive at the desired value that gives the optimum value of the quantity. This brute force approach is acceptable if the number of parameters is few but is obviously prohibitively expensive for large number of parameters. The same objective can be achieved in two integrations, one forward integration of the model and one backward integration of the corresponding adjoint model to determine the gradient with respect to the parameters, which is then followed by an iteration process to convergence. Let the governing equation (not necessarily linear) be

$$\frac{\partial X}{\partial t} = f(X, \alpha) \tag{14.5.5}$$

where X is the state vector and α is the vector of model parameters. Define a functional R, which can be obtained from $X(t)$ determined from integration from $X = X_0$ at $t = t_0$ to $X = X_1$ at $t = t_1$. R is a function of X_0 and α. Now construct an adjoint model, governed by the operator adjoint to f,

$$\frac{\partial X}{\partial t} = f_a(X, \alpha) \qquad (14.5.6)$$

and integrate it backward from t_1 to t_0, starting from $X = X_1$. From the outputs of the forward and the adjoint model, an equation can be derived that relates the variation of R to that of X_0 and α. And it is possible to derive the value of say X_0 to give the desired value of R. These methods are complex and, because of the large state vectors involved in typical oceanographic (and atmospheric) problems, quite expensive to implement. Nevertheless, they are quite appealing in concept and have been vigorously pursued in the past few years. The following is a description based on Miller and Cane (1996) of the adjoint formulation that attempts to determine the initial conditions for optimum model forecast.

Consider a model state **u** governed by the evolution equation:

$$\frac{\partial \mathbf{u}}{\partial t} = L(\mathbf{u}) + F(\mathbf{x}, t) + \mathbf{q} \qquad (14.5.7)$$

where L is a partial differential operator (linear or nonlinear), F is the forcing function, and **q** is the system noise. Let a set of measurements **d** be related to the state vector and the measurement noise by $\mathbf{d} = H\mathbf{u} + \mathbf{e}$, where

$$H\mathbf{u} = \int_0^T \int_S G(\mathbf{x}_1, \mathbf{x}_2, t_1, t_2) \mathbf{u}(\mathbf{x}_2, t_2) d\mathbf{x}_2 dt_2 \qquad (14.5.8)$$

where S denotes the spatial domain. Define a cost (objective) function:

$$J(\mathbf{u}) = \frac{1}{2} \int_0^T \int_0^T \int_S \int_S \mathbf{q}^T(\mathbf{x}_1, t_1) W(\mathbf{x}_1, \mathbf{x}_2, t_1, t_2) \mathbf{q}(\mathbf{x}_2, t_2) d\mathbf{x}_1 d\mathbf{x}_2 dt_1 dt_2$$

$$+ \frac{1}{2} \int_S \int_S [\mathbf{u}(\mathbf{x}_1, 0) - \mathbf{u}_0(\mathbf{x}_1)]^T V(\mathbf{x}_1, \mathbf{x}_2) [\mathbf{u}(\mathbf{x}_2, 0) - \mathbf{u}_0(\mathbf{x}_2)] d\mathbf{x}_1 d\mathbf{x}_2 \qquad (14.5.9)$$

$$+ \frac{1}{2} \mathbf{e}^T \mathbf{w} \mathbf{e}$$

where $\mathbf{u}_0(\mathbf{x})$ is the desired initial condition. The first term on the RHS is the model error, the second is the first guess, and the last one corresponds to the observations. W, w, and V are model error covariance, observational error covariance, and first guess error covariance, corresponding to model forecast, data, and initial conditions. If these are chosen to be inverses of corresponding error covariance functions, then minimizing J results in a least square estimate, equivalent to fixed interval Kalman smoother. Applying the method of calculus of variations and requiring the gradient of the cost function J w.r.t. both $\mathbf{u}$ $(\mathbf{x},t)$ and $\mathbf{u}$ $(\mathbf{x},0)$ vanish yields

$$\frac{\partial \mathbf{v}}{\partial t} + L^*(\mathbf{v}) = -\mathbf{e}^T \mathbf{w} G \qquad (14.5.10)$$

$$\mathbf{v}(\mathbf{x},0) = \left[\mathbf{u}(\mathbf{x},0) - \mathbf{u}_0(\mathbf{x})\right]^T V \qquad (14.5.11)$$

where L^* is the adjoint operator of L. For a nonlinear model, L^* is the adjoint of the operator L, linearized about the current state vector of the model. Equation (14.5.10) along with the final condition $\mathbf{v}(\mathbf{x},T) = 0$, and definition of the adjoint variable $\mathbf{v}$,

$$\frac{\partial \mathbf{u}}{\partial t} = L(\mathbf{u}) + F(\mathbf{x},t) + W^{-1}\mathbf{v}^T \qquad (14.5.12)$$

constitute the Euler-Lagrange equations for this variational problem. Integration of Eq. (14.5.10) backward in time from its final condition provides the initial value of the adjoint variable $\mathbf{v}(\mathbf{x},0)$, which then gives the gradient of the cost function w.r.t. the initial state vector $\mathbf{u}$ $(\mathbf{x},0)$:

$$\frac{\partial J}{\partial\left[\mathbf{u}(\mathbf{x},0)\right]} = \left[\mathbf{u}(\mathbf{x},0) - \mathbf{u}_0(\mathbf{x})\right]^T V - \mathbf{v}(\mathbf{x},0) \qquad (14.5.13)$$

Finding a stationary value of this by a conjugate gradient technique (see Chapter 2) completes the problem solution.

Putting $\mathbf{q} = 0$ in the above formulation is equivalent to requiring a strong constraint approach (Sasaki, 1970), since the model (and its forcing) is then assumed to be perfect and can be obtained by putting $W^{-1} = 0$. The only free parameter is then the initial state vector. However, the model parameters such as drag or diffusion coefficients or wave speeds can be added to the cost function. In situations where the system retains little memory of the initial conditions after a while, such as the tropical ocean, a parameter such as the wave speed may be the appropriate independent control variable. Smedstad and O'Brien (1991) used the Kelvin wave in their tropical Pacific Ocean assimilation model.

The major appeal of this strong constraint approach is the resulting considerable computational simplification, since otherwise the forward integrations using Eq. (14.5.12) and backward integration in time using the adjoint equation, Eq. (14.5.10), are fully coupled (Miller and Cane, 1996). Substituting for $\mathbf{q}$ from Eq. (14.5.7) in Eq. (14.5.9) and putting $\mathbf{v} = \mathbf{q}^T W$ shows that the adjoint variable is just a Lagrange multiplier.

In addition to providing weighting based on the confidence, the second integral in the cost function prevents the high wavenumber "noise" components of the initial field $\mathbf{u}$ $(\mathbf{x},0)$ from degrading the solutions and making the search for optimal solution poorly conditioned (Miller and Cane, 1996). A similar effect can be achieved by adding a smoothing term $\int_S (\nabla \mathbf{u})^2 d\mathbf{x}$ to the cost function, even though this does not perform the function of weighting. Either way, it is very important to impose a smoothness constraint on the solutions (Busalacchi, 1996). Examples and oceanographic applications of the adjoint technique can be found in Long and Thacker (1989a,b) and Sheinbaum and Anderson (1990a,b).

To summarize, the practice of assimilating observed data into numerical models for the purpose of constraining the model results to remain close to the observed "real" state is extremely complex, continuously evolving, and is by no means a fully solved problem, even for the atmosphere where decades of experience have been accumulated from routine day-to-day applications. The above synopsis does not do justice to the subject, which is vast and varied, and the reader is encouraged to explore it further; the references cited should help in this task.

14.6 PREDICTABILITY OF NONLINEAR SYSTEMS: LOW ORDER PARADIGMS

Once Newton formulated the laws of motion (and gravitation) in 1687, it became possible to determine the positions of heavenly bodies in the solar system from first principles and to apply the laws successfully to explain earthly phenomena such as surface gravity waves and oceanic tides. The success led some dynamicists such as Marques de Laplace to assume that in a deterministic system governed by well-known laws, given the initial state of the system, it should be possible to determine the future state of the system in perpetuity and hence provide an unlimited forecasting capability. Little was known then about the extraordinary sensitivity of nonlinear systems to even infinitesimal errors in their initial conditions that would make any forecast rapidly diverge from reality due to explosive growth of any errors in prescribing the initial state. There is evidence, however, that other dynamicists, such as Laplace's contemporary, Henri Poincare, were aware of this aspect of nonlinear dynamics. Poincare clearly alluded to this sensitivity that makes forecasting beyond a certain finite interval of time impractical, but he did not pursue the topic in any greater detail. There the subject stood for more than half a century. It was only in 1963 that an American meteorologist at the Massachusetts Institute of Technology, Ed Lorenz, rediscovered this extraordinary sensitivity to initial conditions in the context of weather forecasting and launched the science of (low order) nonlinear systems and the theory of chaos. While error growth can also be a feature of a linear system, it is particularly problematic in nonlinear sytems.

Many of the characteristics of a chaotic dynamical system can be explored using low order systems governed by the corresponding number of coupled nonlinear ODEs desribing their evolution in time. While they provide valuable experience in applying ODE solvers, they also provide insight into the behavior of dynamical systems that exhibit extreme sensitivity to initial conditions. Predictability issues permeate the field of numerical prediction of real-life flow systems such as the atmosphere and the oceans, which are, however, far too complex, and insight gained therefore from simple low order time-dependent systems might prove valuable in the study of atmospheric and oceanic predictability as well. It is for this reason, we will explore the realm of low order chaotic systems.

In his attempt to find a tractable low order paradigm for the complex highly nonlinear fluid motion related to two-dimensional Rayleigh-Benard convection, Lorenz (1963) used Saltzman's (1962) technique of series expansions to convert the governing nonlinear partial differential equations to a system of nonlinear ordinary differential equations, and by truncating the series, derived a low order analog of the convective fluid motion, widely known as the Lorenz system:

$$\frac{dX}{dt} = -\alpha X + \alpha Y$$

$$\frac{dY}{dt} = rX - Y - XZ \qquad\qquad (14.6.1)$$

$$\frac{dZ}{dt} = XY - bZ$$

where α represents the Prandtl number, b is the aspect ratio, and r represents the Rayleigh number. In a fluid subjected to convection (heated from below or cooled from above in the presence of gravitational body forces), it is well known that as the Rayleigh number is increased, the initially laminar motions become unstable and eventually lead to fully developed turbulent convection. Exactly how this transition to chaotic turbulent motions takes place has been studied for over a century. However, the nonlinear nature and the complexity of the problem has made progress very difficult. Thus any simplification that can lead to a better understanding of the route to turbulence (loosely speaking, chaos) has been welcome.

Lorenz's equations constitute the lowest order system that can be representative of the much higher order, much more complex phenomenon of convection. This system of three first-order coupled nonlinear differential equations is the most well-known low order nonlinear dynamical system and has been investigated extensively by nonlinear dynamicists interested in the subject of chaotic systems and predictability theory. The system possesses two states of

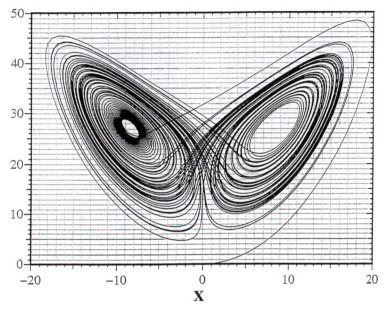

Figure 14.6.1 Lorenz "butterfly." The trajectories of the system in X-Z projection of the phase space showing the two strange attractors and abrupt transitions between the two.

equilibria. For certain values of the Lorenz parameters, α, b, and r, the system is highly predictable and executes oscillations around one or the other of these two states. However, for other values of α, b, and r, the system exhibits chaotic, unpredictable transitions between these two states. For $\alpha = 10$, b = 8/3, and r = 28, the parameter values explored by Lorenz and extensively since then, in the three-dimensional phase space (X,Y,Z) of the dynamical system, the trajectory of the state vector describes the famous Lorenz attractor. This aesthetically pleasing "Lorenz Butterfly" (Figure 14.6.1) consists of two wing-shaped segments centered around the two dynamical states called the attractors. The trajectory of the system started from an arbitrary initial state executes looping motions around the two attractors, with frequent highly unpredictable (chaotic) transitions between the two.

The most fascinating aspect of this system is that no matter how infinitesimally close two initial states are in the phase space, their trajectories diverge eventually. If one regards one of the two initial states as reality and the other as a forecast, and the difference between the two initial states as an "error" in prescribing the initial state, the explosive growth in error in time makes the forecast depart radically from reality and therefore all predictability is lost after a finite amount of time. The trajectories of a cluster of points, no matter how closely spaced in phase space at $t = 0$, will evolve such that at a future time, the

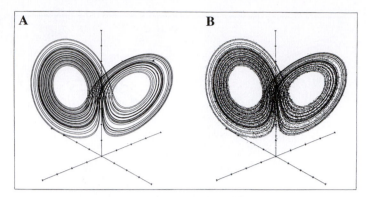

Figure 14.6.2   In (A), all trajectories will eventually converge on this structure. Trajectories originally close in location can, however, end up anywhere on the structure. This is shown in (B): 10,000 measurements, initially located in the dot in (A), produce varying states within the structure. This illustrates the basic nature of the predictability problem for a nonlinear system. From Tsonis and Elsner (1989).

points end up scattered all over the phase space around the two attractors (Figure 14.6.2). In fact, if the number of points is large enough, their final state will define the two butterfly wings quite well. If the system were not chaotic, the points would stay clustered for all times, defining, at most, a small volume in the phase space of the system.

Two timescales are associated with the Lorenz system: one is the typical residence time of the state vector within one of the butterfly wings, and the second describes the evolution of the system around the unstable fixed point at the center of each butterfly wing (Palmer, 1993). These two timescales can be seen clearly in plots of any of the state variable with time. Figure 14.6.3 shows the evolution of the state variable Y with time. The two timescales are evident from this plot, and so is the fact that predictability of a nonlinear system is limited. The limitation on predictability depends critically on where in the phase space the initial conditions lie. A cluster of points near the most unstable point, the stagnant state (X=Y=Z=0), can disperse quite rapidly over the phase space, whereas another elsewhere on the attractor might stay together for a longer but finite period of time (see Palmer, 1993).

The discovery of chaos in nonlinear systems has profound implications to the long-term predictability of these systems. It means that because of the sensitive dependence on initial conditions, no matter how accurately one can hope to measure and specify the initial state of the system, predictability would be lost after a finite time. This places an upper bound on the time interval over which complex systems like weather and climate can be predicted in advance. The positions of planetary bodies in the solar system can be predicted highly accurately centuries in advance, but even this highly predictable system is chaotic over longer timescales (Ruelle, 1994) in the sense that the positions of

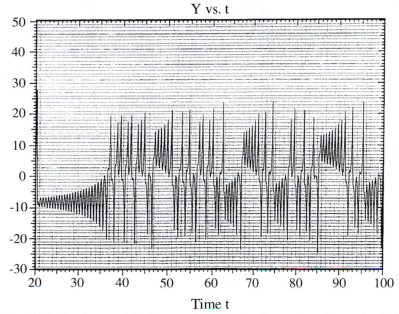

Figure 14.6.3 Plot of state variable Y versus time t, showing the two timescales associated with the Lorenz system, the timescale of orbit around each attractor, and the timescale for transitions between the two attractors.

the planets can be predicted accurately for only a period of 20 Myr (Laskar, 1990). See Murray (1998) for a discussion of chaos in the solar system. Secular changes in the Earth's ellipticity brought about by geological processes, such as postglacial rebound due to surface mass load variations associated with ice-age cycles, and mass redistribution due to convective flow in the viscous mantle can affect the Earth's precession and obliquity (Sussman and Wisdom, 1992; Forte and Mitrovica, 1997).

There is evidence that even the behavior of more complex higher order dynamical systems such as the atmosphere bears some resemblance to the simple Lorenz system, although the chaotic behavior is much more complex. For example, the truncated equations describing atmospheric motions often exhibit two equilibrium states (Palmer, 1993) corresponding to zonal and blocked flows (Weeks *et al.*, 1997), with the two characteristic times (determined by baroclinic instabilities) being the residence time in each of these states and the time for transitions between them. Palmer (1994) offers a variant of the Lorenz system as a low order paradigm to describe the active and break phases of the Indian monsoons and the transition between them. Palmer (1993) describes an extension of the Lorenz system coupled to an oscillator as a possible paradigm to describe the tropical influence on midlatitude weather.

Huang and Dewar (1996) describe a low order paradigm for the haline circulation in the oceans. Meridional thermohaline circulation in the global oceans (Stommel, 1961) plays an important role in the Earth's climate on timescales of millenia and beyond and many studies have focused on its possible chaotic nature. Periodic reversals in the Earth's magnetic field exhibit chaotic behavior and have been studied using a low order nonlinear model of the Earth's dynamo (see Roberts, 1987). Zeldovich and Ruzmaikin (1983) have derived a model for the solar dynamo that is identical to the Lorenz system, and Weiss (1988), for example, has proposed a fifth order system that appears to be capable of simulating the chaotic behavior of sunspot cycle, including the Maunder minimum observed in the 17th century. Low order chaotic systems have also been demonstrated in other fields such as biology (for example, insect populations, heartbeats) and chemistry (Belousov-Zhabotinsky chemical reactions). Nonlinearity and hence chaotic behavior is ubiquitous in nature and we present a concise review here in the context of geophysical flows and ocean modeling, simply because the student must be aware of its relevance to the behavior and prediction of the oceans and the atmosphere. It is also a fascinating subject worthy of study by itself.

Lorenz's rediscovery of chaotic behavior of nonlinear dynamical systems went unnoticed till the mid-1970s, despite his insistence that it was a dynamical consequence and not an arcane numerical artifact associated with computational errors. There is now literally many hundreds of papers and books describing chaotic behavior in a wide variety of nonlinear systems. For a good summary of chaos, the reader is referred to Ruelle (1989) and Baker and Gollub (1990). Sparrow (1982) describes the Lorenz system in great detail. Holden and Muhamad (1988) describe the properties of many low order chaotic systems. There is also considerable interest in deriving some order out of chaos and hence in predicting a chaotic time series (Casdagli, 1989; Farmer and Siderowich, 1987; Henderson and Wells, 1988; Mundt et al., 1991; Ottino et al., 1992). The phenomenon of turbulence and how an ordered laminar flow transitions to a random chaotic turbulent flow has always preoccupied fluid dynamicists and the discovery of chaos has reshaped many of the underlying concepts such as bifurcation and other routes to chaos and transition to turbulence (Swinney and Gollub, 1981).

The behavior of dynamical systems can be described well in the state space of their dependent variables. For example, while the motion of a pendulum can be described as a time series of the angle (with respect to the vertical) versus time, it can also be described in the two-dimensional state space of the angle versus the angular velocity. In this space, the origin describes the state of rest. A damped pendulum describes a trajectory in this space that spirals inward toward this point and eventually ends up there. This point is called an attractor. If the dissipation of the pendulum's motion is compensated exactly (equivalently in the absence of friction), its trajectory is simply a circle around the origin. This is called a limit cycle. It indicates periodic motion (Figure 14.6.4). The attractor

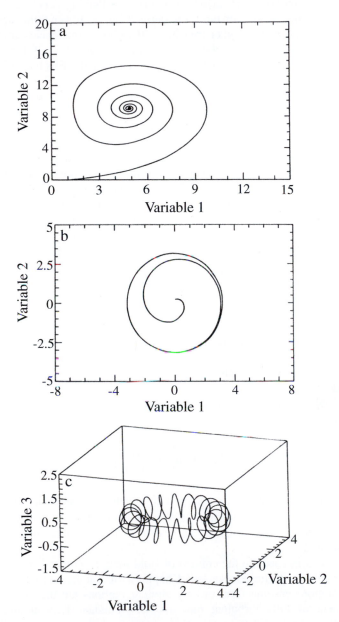

Figure 14.6.4 Different normal attractors of a dynamical system: (top) fixed point, (middle) stable orbit (circle), and (bottom) a torus for motions involving two frequencies. From Henderson and Wells (1988).

for quasi-periodic motions, consisting of two or more frequencies, is a torus. For dual frequency motions, the trajectory describes a looping curve on the surface of the torus, with the size of the torus defined by the lower frequency and the radius of the tube by the higher frequency. If the two frequencies are multiples of each other, the curve is closed and endlessly repeated. If the ratio is a nonrational number, the curve is not closed and eventually fills the surface of the torus. Nonchaotic systems have regular attractors such as fixed points, limit cycles, and tori, and their long-term predictability is assured, since two points in state space close to each other at any given time will remain close to each other for all time. They are not sensitive to initial conditions. Their attractors are characterized by an integer dimension (limit cycle has dimension of one, for example). Chaotic systems such as the Lorenz system, however, possess strange attractors. They are exquisitely sensitive to initial conditions and therefore two neighboring trajectories will not stay close to each other but will diverge rapidly, making long-term predictability unattainable. Strange attractors have fractal dimension larger than 2. The fractal dimension is defined as d = (log n/ log m), when an object is increased by a factor of m it is found to contain n original objects within it (for example, for a geometrical shape of a square, if m = 2, n = 4 and therefore d = 2, an integer, but for a fractal shape, the dimension is not an integer). For example, the Lorenz attractor has a dimension of 2.06. Its motion is deterministic (not random), but aperiodic and unpredictable except for a brief time span.

The sensitivity of a nonlinear system to initial conditions depends on at least one of its Lyapunov exponents being positive. To define the Lyapunov exponent, consider an infinitesimal hypersphere in the N-dimensional phase space of the system (3 for the Lorenz system) at time zero. This hypersphere will be deformed into a hyper-ellipsoid at a later point in time as the system evolves. The nth Lyapunov exponent is related to the length of the nth principal axis of the ellipsoid:

$$\lambda_n = \lim_{t \to \infty} \left[\frac{1}{t} \ln \frac{p_n(t)}{p_n(0)} \right] \tag{14.6.2}$$

with the exponents ordered from largest to smallest. A necessary condition for chaos is that at least λ_1 be positive. It is possible to extract Lyapunov exponents from the observed time series of any system, if the record is long enough and noise is limited.

The interest in chaos in the context of fluid motions comes from the point of view of transition to turbulence (see Chapter 1 of Kantha and Clayson, 2000). Chaotic motions ensuing from orderly laminar motions are the first step in the development of fully turbulent motions. Lev Landau (Landau and Lifshitz, 1959) once suggested that as a particular parameter is increased (for example Rayleigh number in convective flows), the initially steady flow would become unstable to any perturbations and a periodic motion would appear in the flow.

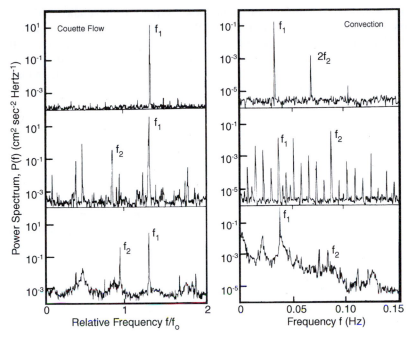

Figure 14.6.5 Transition to turbulence in (left) Couette flow and (right) convection. Note the rapid degeneration into turbulence after the appearance of a few discrete frequencies and their harmonics. From Gollub and Swinney (1978).

As the parameter is increased further, this motion in turn would become unstable, and more and more frequencies would appear until there would be infinitely many. In spectral space, a single peak would appear first, followed by the appearance of another and yet another until the peaks would be so closely spaced together as to yield almost a continuous broad band-looking spectrum characteristic of turbulent flows. A complex and confused motion would result. Careful observations in flows like the Taylor Couette flow between two rotating cylinders (with the inner one rotating faster) and the Rayleigh-Benard convection (Gollub and Swinney, 1978) have shown that while initially the flow follows this scenario, turbulence sets in quite rapidly (Figure 14.6.5). Starting from a steady flow, the spectrum develops a single peak, then one or two additional unrelated frequencies appear, but chaotic motion ensues shortly thereafter resulting in a broad band spectrum with a few weak peaks.

Chaos theory has made contributions in explaining this transition. Under the Ruelle-Takens quasi-periodic route to chaos of a low order system, chaos should set in once three distinct unrelated frequencies have appeared. Other routes have also been proposed. Under the Figenbaum route to chaos, period-doubling marks the route, where successive frequencies 1/2, 1/4, 1/8 . . . of the original

frequency appear. When infinitely many new frequencies have appeared, chaos
ensues. Yet another proposed route is the Pomeau and Manneville intermittency
route. Here, long periods of regular motion are punctuated by bursts of chaotic
motion, with the bursts becoming more and more frequent until complete
chaotic motion ensues. The Lorenz system exhibits such intermittent chaos for
certain values of the parameters, and intermittent turbulence is common in fluid
flows. However, turbulence and transition to turbulence even in the simple
Rayleigh-Benard convection are much more complicated than what chaos
theories suggest. For example, the appearance of as many as four to five distinct
frequencies or as few as none can precede turbulence. Spatial and temporal
intermittency and hysterisis complicate the picture further. This no doubt has
something to do with the fact that real fluid flows are higher order systems and
the turbulence has both spatial and temporal variability. Chaos deals essentially
with temporal variability. Nevertheless, a study of chaos provides some insights
into the nature of turbulent flows, and this is attractive because of the basically
intractable nature of the subject.

While the Lorenz system can be derived rigorously from equations for a two-
dimensional thermal convection in a box of aspect ratio b, whether this highly
truncated system truly represents the low order aspects of convection is often
doubted. On the other hand, loop models have been useful in investigating
thermal, haline, and thermohaline convection (Welander, 1967, for example),
and it is possible to derive a low order truncated set of equations to describe
some of the flow behavior. Here one considers a circular loop of radius R,
formed from a tube of radius r_0 and filled with saline (or fresh) water. If the loop
is heated at the bottom and cooled at the top, thermal convection ensues. If, in
addition, there is evaporation and freshwater fluxes around the loop containing
saline water, thermohaline convection occurs. If the loop is filled with saline
water and driven only by variations in the evaporation and precipitation rates,
haline convection ensues. There are, of course, two equilibrium states of motion,
one for each direction of the flow in the loop. Under certain conditions, the flow
executes random, chaotic reversals in flow direction. It is instructive to study the
characteristics of the low order system governing haline convection as a
prototypical chaotic system. Here we will follow Huang and Dewar (1996)
closely.

If the loop is forced by the difference in evaporation and precipitation
through the skin according to $\dot{E} - \dot{P} = -E\cos\theta$, with θ measured from the loop
bottom, the momentum and the salinity conservation are

$$\frac{\partial \Omega}{\partial t} = -\varepsilon\Omega - \frac{\beta g}{R}\langle S\sin\theta\rangle$$

$$\frac{\partial S}{\partial t} + \frac{\partial}{\partial \theta}\left(\Omega + \frac{2E}{r_0}\sin\theta\right) = \frac{A_s}{R^2}\frac{\partial^2 S}{\partial \theta^2} - E\bar{S} \qquad (14.6.3)$$

where Ω is the angular rotation rate, ε is the friction coefficient, β is the expansion coefficient, and A_s is the diffusion coefficient. $\overline{S}$ is the reference salinity and the angle brackets denote the average over the entire loop. These equations can be normalized:

$$\frac{\partial \Omega}{\partial t} = -\alpha \Omega - \langle S \sin \theta \rangle$$

$$\frac{\partial S}{\partial t} + \Omega \frac{\partial S}{\partial \theta} = k \frac{\partial^2 S}{\partial \theta^2} - \lambda \cos \theta \qquad (14.6.4)$$

$$\tau = \left(\frac{R}{\beta g \overline{S}} \right)^{1/2}, \alpha = \varepsilon \tau, k = \frac{A_s \tau}{R^2}, \lambda = \frac{E \tau}{2 r_0}$$

The various nondimensional timescales in the problem are τ, the basic timescale, α, the spindown timescale, k, the diffusive timescale and λ, the freshwater renewal timescale. The equations can be solved by expanding salinity in a Fourier series:

$$S(\theta) = 1 + 2 \sum_{1}^{\infty} \left[a_n \sin(n\theta) + b_n \cos(n\theta) \right] \qquad (14.6.5)$$

One thus gets a series of equations for coefficients a_n and b_n instead of salinity. The different salinity modes are decoupled, because the salinity equation is linear in S. However, because of the appearance of Ω in the second term of the salinity equation, the salinity equation is coupled to the angular velocity equation and is nonlinear. The higher modes decay more rapidly and hence for looking at the long-term evolution, one needs to look at only the first mode:

$$\frac{d\Omega}{dt} = -\alpha \Omega - a_1$$

$$\frac{da_1}{dt} = \Omega b_1 - k a_1 \qquad (14.6.6)$$

$$\frac{db_1}{dt} = -\Omega a_1 - k b_1 - \frac{\lambda}{2}$$

Transformations

$$r = \frac{\lambda}{2\alpha k}, X = \Omega, Y = -\frac{a_1}{\alpha}, Z = \frac{b_1}{\alpha} + \frac{\lambda}{2k\alpha} \qquad (14.6.7)$$

reduce the system to the classical water wheel equation, where the buckets are leaky but are also being filled at a certain rate (Lorenz, 1979; Sparrow, 1982):

$$\frac{dX}{dt} = -\alpha X + \alpha Y$$

$$\frac{dY}{dt} = rX - kY - XZ \qquad (14.6.8)$$

$$\frac{dZ}{dt} = XY - kZ$$

The water wheel has two stable equilibrium modes because it can rotate in either direction. However, under certain parameter conditions, it undergoes chaotic transitions between these two states. If $k = 1$, the water wheel equations reduce to the Lorenz system but with $b = 1$.

The water wheel equations have three stationary solutions

$$(0,0,0), \left(\hat{\Omega}, \hat{\Omega}, \hat{\Omega}^2 k^{-1}\right), \left(-\hat{\Omega}, -\hat{\Omega}, \hat{\Omega}^2 k^{-1}\right)$$
$$\hat{\Omega} = \left[k(r-k)\right]^{1/2} \qquad (14.6.9)$$

corresponding to one stagnant and two steady advection states. The stability of the system can be explored by putting the equations in Eq. (14.6.8) in matrix form:

$$\frac{\partial \mathbf{X}}{\partial t} = \mathbf{A} \cdot \mathbf{X}$$

$$\mathbf{X} = (X, Y, Z)^t; \mathbf{A} = \begin{bmatrix} -\alpha & \alpha & 0 \\ r & -k & -X_p \\ Y_p & X_p & -k \end{bmatrix} \qquad (14.6.10)$$

and examining the eigenvalues of the matrix $\mathbf{A}$ (X_p, Y_p denote the stationary points). The stagnant state is unstable when $r = k$ leading to steady advection through pitchfork bifurcation. The governing equation for the eigen values is a cubic for the advective states,

$$\sigma^3 + (\alpha + 2k)\sigma^2 + \left(\alpha k + k^2 + \bar{\Omega}^2\right)\sigma + 2\alpha\bar{\Omega}^2 = 0 \qquad (14.6.11)$$

The advective states have therefore 1 real and 2 complex conjugate roots. The real part can go from a negative value through zero to a positive value, and this zero-crossing value is the Hopf bifurcation point. At the bifurcation point, the real part is zero so that the three roots are $(c, +id, -id)$ and hence it is possible to eliminate c and d and get a relation that describes the critical condition for Hopf bifurcation. In dimensional quantities this means

$$\lambda_c = 2\alpha^2 k (\alpha + 4k)(\alpha - 2k)^{-1} \qquad (14.6.12)$$

If the freshwater flux increases beyond this critical value, limit cycle or chaotic solutions result. The investigation of the resulting bifurcation structure, however, requires numerical techniques (Doedel, 1986). Numerical solutions can be obtained for any arbitrary values of the parameters, and Figure 14.6.6 shows the solution for $\alpha = 0.5$, $\lambda = 0.02$ and $k = 0.01$. More details can be found on the behavior of this system and more complex 4-mode and 64-mode systems related to haline convection in Huang and Dewar (1996).

A few more examples of low order chaotic systems are in order. A paradigm for chaotic monsoon system is the nudged Lorenz system (Palmer 1996):

$$\frac{dX}{dt} = -\alpha X + \alpha Y + f$$

$$\frac{dY}{dt} = rX - XZ + f \qquad (14.6.13)$$

$$\frac{dZ}{dt} = XY - bZ$$

Unlike the Lorenz equations, where the system spends roughly the same amount of time near both the attractors, in the modified equations, the system spends more time near one or other attractor depending on the value of the parameter f. If one thinks of f as an external influence on the system, it is possible that external influences can nudge the system to one or other of the attractors, depending on its value. ENSO events might thus have some influence on the monsoons in the North Indian Ocean. Figure 14.6.7 shows the situation. A similar paradigm is the Lorenz system coupled to a linear oscillator. The former is representative of the more chaotic midlatitude weather, whereas the latter is representative of the more regular tropical pattern (Palmer, 1993):

$$\frac{dX}{dt} = -aX + aY + V$$

$$\frac{dY}{dt} = rX - Y - XZ + W$$

$$\frac{dZ}{dt} = XY - bZ \qquad (14.6.14)$$

$$\frac{dW}{dt} = -cV - k(W-W^*) - Y$$

$$\frac{dV}{dt} = c(W-W^*) - kV - X$$

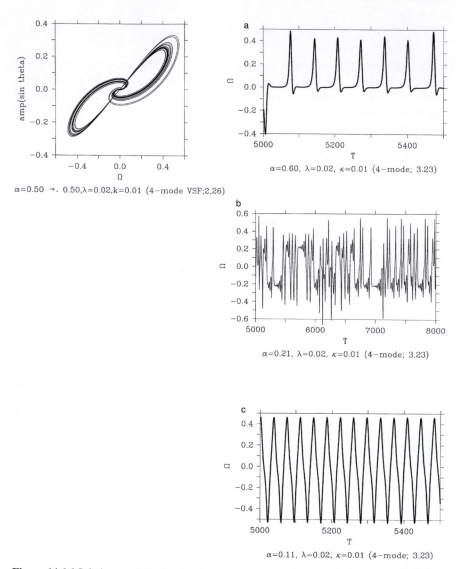

Figure 14.6.6 Solutions to the haline circulation showing periodic and chaotic motions at different parametric values (right) and the phase diagram showing chaotic motions (left). From Huang and Dewar (1996).

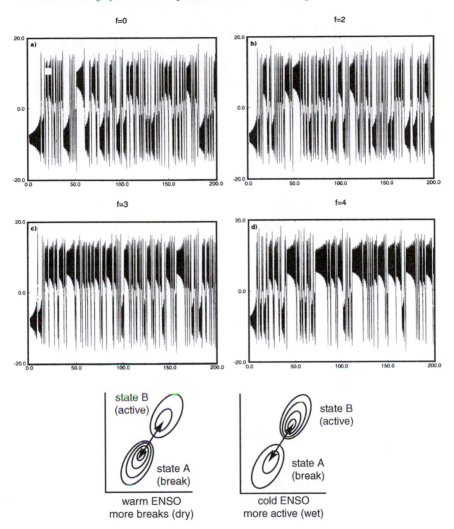

Figure 14.6.7 Behavior of a nudged chaotic system that spends more time preferentially around one regime of an attractor depending on the degree of nudging. Hypothesized to represent the active and break phases of the southwest monsoons over the Indian Ocean. Bottom panels from P. Webster, based on Palmer (1994); time series from Palmer (1996).

Here c (~1.5) is the frequency of the dominant mode of variability in the tropics, k (~0.1) is a damping coefficient. The parameter W^* is the SST anomaly. For certain values of the parameters, this set of equations forces the system to spend more time on one of the attractors than the other. Figure 14.6.8 shows the probability distribution function for the Lorenz and the coupled models for W = 4, representing the El Niño state. Clearly the distribution is asymmetric. This has

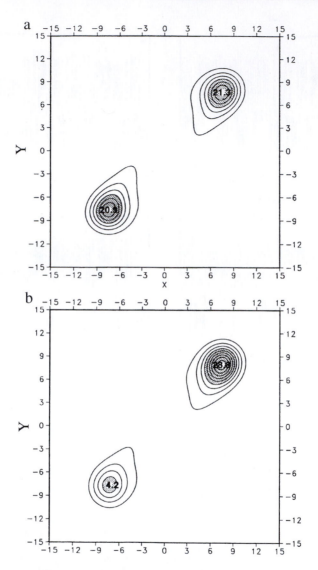

Figure 14.6.8 Probability distribution function of the coupled Lorenz-oscillator system (a) without and (b) with coupling. In the former case, the probability of finding the state at either attractor is the same, whereas in the latter state, the system spends more time around one of the attractors. From Palmer (1993).

implications to predictability; a system that spends more time on the average around one of the two attractors is a bit more predictable (on the average) than the one that spends roughly the same amount of time around either.

14.7 NOWCASTS/FORECASTS IN THE GULF OF MEXICO

An important application of ocean models is in prediction of the past (hindcast), current (nowcast) and future (forecast) state of the ocean. Given the fact that more than 50% of the burgeoning human population lives within 100 miles of a coastline and hence heavily uses/abuses the coastal oceans, such predictions, especially in the coastal and marginal seas, are particularly useful for societal needs such as sea level predictions, mapping of currents, and pollution tracking. We will provide one such example from a marginal semienclosed sea in the North Atlantic, the Gulf of Mexico. The offshore oil fields of this "miniocean basin" account for roughly half the U.S. domestic oil production and the Louisiana-Texas (LATEX) continental shelf is dotted with thousands of oil platforms. Exploration and production is expanding steadily into deeper waters of the Gulf, waters as deep as 1000 to 2000 m.

A major source of oceanic variability in the Gulf is the Loop Current. About 28 Sv of subtropical waters enter the Gulf through the Yucatan Straits between Mexico and Cuba and leave it through the Florida Straits between Florida and Cuba to eventually become the Gulf Stream. The extent of penetration of this so-called Loop Current into the Gulf is highly variable. Occasionally the Loop Current becomes unstable and sheds off a huge anticyclonic eddy, anywhere from 100 to 350 km in diameter, that pinches off the Loop Current and moves into the western Gulf (Hurlburt and Thompson, 1980). The Loop Current Eddy (LCE) is the principal mechanism for renewal of water masses in the western Gulf. The path of LCEs is also highly variable, and occasionally a LCE traverses the Gulf in close proximity to the LATEX continental shelf. Because of the strong currents (often as much as 4 kts, 2 m s^{-1}, in magnitude) associated with LCEs, this is the second major source of operational concern (the first being hurricanes in late summer through fall) to production and exploration activities in the Gulf. A capability to accurately forecast the shedding and movement of a LCE, and the locations of the Loop Current front and LCEs, along with the associated currents in the water column, is valuable to the offshore oil industry.

A forecast of the path an LCE takes is possible with the use of a numerical model of the Gulf. However, accurate information on the initial location of the LCE once it is shed and the corresponding Gulf-wide oceanic state is rather crucial to the forecast skill. An accurate nowcast is therefore essential, and this requires a data-assimilative numerical model. Since *in situ* data, even in the Gulf, are sparse and often nonexistent, remotely sensed data need to be relied upon for this purpose. Since sea surface temperature from IR sensors is not always useful in locating a LCE (especially in summer), and since altimetry can almost always detect such an eddy if it happens to straddle its track, altimetric SSH anomalies assimilated into an ocean model provide a reliable nowcast of not only the eddy location but also the initial state of the Gulf. Forecasts can then be made from this nowcast and the path of the LCE predicted.

The methodology employed (Choi *et al.*, 1995, Kantha *et al.*, 1999) for producing such a nowcast is continuous assimilation. The model is run from a time in the past to the present assimilating altimetric data track by track, and MCSST data from satellite-borne IR sensors. Altimetric SSH anomalies are not injected directly (since they would otherwise spin up a barotropic mode), but converted to anomalies in the temperature of the water column (following Carnes *et al.*, 1990), which are then assimilated into the model, in this case, using simple optimal interpolation. The conversion is done by using either historical CTD data or long-term simulations with the model run free but forced synoptically to derive a statistical relationship between dynamic height anomalies and the subsurface temperature anomalies, using EOFs (see Appendix C for a description of EOFs). Altimetric SSH anomalies are treated as dynamic height anomalies (valid only in deep waters and not on the continental shelf), and regressed to temperature anomalies for assimilation into the model. A nowcast produced thus is then used as the basis of a forecast.

Traditionally, salinity fields in ocean models are initialized with climatological values, sparse as they are, but absence of information on salinity profiles prevents salinity from being assimilated into the model during nowcasts. This can lead to significant errors in the estimate of the state of the upper ocean, especially in regions like the equatorial Pacific, where large errors in modeled geopotential surfaces and currents can result (Cooper, 1988; Acro-Shertzer *et al.*, 1997; Reynolds *et al.*, 1998). It is therefore preferable to assimilate salinity profiles also when estimating the state of the upper ocean. Since they are seldom measured, inferring them from temperature profiles using the prevailing T-S relationships is the only possibility at present. However, the large scatter in the near-surface layers in the T-S relationship is a source of major error. But it appears that knowledge of surface salinity can help reduce errors in estimation of the salinity profiles in the upper layers (Hansen and Thacker 1999). This is especially attractive because of the possibility of estimating the surface salinity remotely (Lagerloef *et al.*, 1995; Miller *et al.*, 1998).

Real-time nowcast/forecasts in the Gulf of Mexico have been made possible by the timely availability of high quality real-time satellite data streams produced and disseminated by the Naval Oceanographic Office. Kantha *et al.* (1999) use real-time altimetric SSH anomalies derived from NASA/CNES TOPEX and ESA ERS-2 altimeters, and composite MCSST data derived from NOAA AVHRR in a continuous data assimilation mode to produce a nowcast and a 4-week forecast (see www-ccar.colorado.edu/~jkchoi/gomforecast.html). Figure 14.7.1 shows a real-time nowcast and a 3-week forecast of eddy El Dorado that was shed from the Loop Current in the fall of 1997. Kantha *et al.* (1999) suggest that the forecasts retain considerable skill to about 1-2 weeks, beyond which the forecasts begin to deviate increasingly from reality. Winds are not very important to Loop Current variablilty, and hence forecasts of the Loop

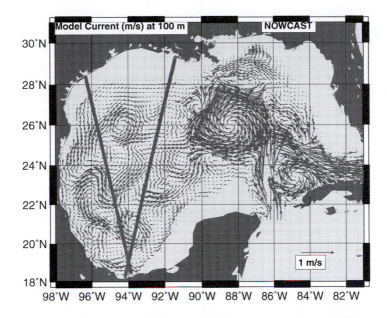

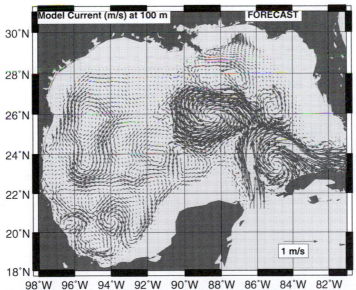

Figure 14.7.1 Currents at 100 m depth from real-time nowcast on October 2, 1997, and 3-week forecast on October 23, 1997, of eddy El Dorado in the Gulf of Mexico. Only TOPEX and ERS-1 altimetric SSH anomaly data (tracks shown) were assimilated into the model. Real-time MCSST data were not available for assimilation. From Kantha *et al.* (1999).

Current and its eddies are not limited by the skill of the forecast winds, as would be expected otherwise. Since altimetric data are available within a few hours of its collection by the sensor, this suggests that forecasts with some skill can be made roughly 1-2 weeks in advance. If this is proven correct, such a nowcast/forecast capability would be useful to drilling/exploration activities in the Gulf. It is in applications such as this that an ocean model, acting in concert with routine ocean monitoring via satellite-borne (and any *in situ*) sensors, can prove to be useful to societal needs.

Sparsity of *in situ* observational data, especially on currents, makes it difficult to assess the skill of such oceanic nowcast/forecasts. The best recourse is to wait for the satellite to collect data for the time period of interest and compare it with the forecast for that period. Another is to compare the current nowcast for the period with a previous forecast. Three factors are essential for accurate forecasts: an accurate nowcast to initialize the model running free, accurate forcing for the forecast run, and a model with accurate physics and free from blatant biases and errors.

In regional forecasts, it is often important to reproduce certain local features, such as eddies and fronts faithfully. The Loop Current front and eddies are one such example. Since *in situ* observational resources are almost always limited, if the observational strategy can be adapted to sample the feature of interest more thoroughly and the resulting data used in the model, it is possible to better forecast the feature of interest. This can be done by adaptive sampling (Kantha *et al.*, 1999), an iterative procedure where observational assets are dispatched to the location of the feature of interest as forecast by a model (albeit not very accurately, for whatever reasons). These data are then assimilated into the nowcast/forecast model to produce a better nowcast/forecast of the feature, which can then be used, if need be, to refine the sampling strategy. This iterative process can be continued as long as needed. Figures 14.7.2 and 14.7.3 show an example of adaptive sampling for eddy Deviant in the Gulf of Mexico in the summer of 1997. This small eddy (<150 km in diameter) was not adequately sampled by the altimeters and therefore was not accurately nowcast or forecast by the model. While the location of the eddy was simulated reasonably well by the model, the currents predicted were anemic (Figure 14.7.2). *In situ* XBT sampling of the eddy and the northwest edge of the Loop Current enabled the model to correct its nowcast/forecast so that the currents in the eddy were more realistic at about 1 m s^{-1} (Figure 14.7.3).

Ocean nowcasting and forecasting is a rapidly evolving field, with origins dating back to the 1980s (for example, see INO, 1986; JOI, 1990). Interest in skillful nowcasts and forecasts for use by offshore industry and coastal and navy communities has remained high (for example, see the 1992 Special Issues of *Oceanography* on Ocean Prediction, and *Marine Technology Society Journal* on Oceanic and Atmospheric Nowcasting and Forecasting; see also Malonette-Rizzoli and Tziperman, 1996). Extensive feasibility studies have been conducted in recent years in one of the most dynamic and therefore hard-to-predict regions

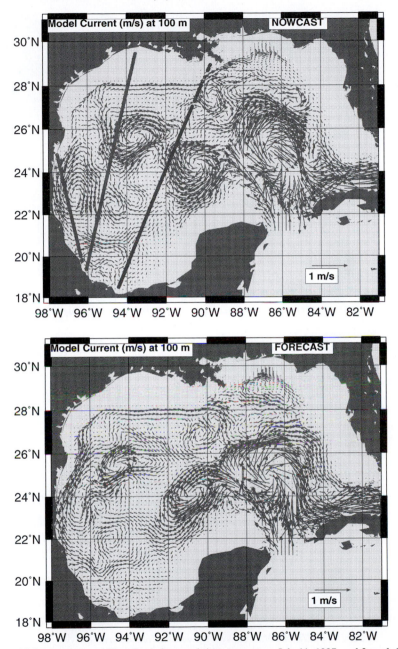

Figure 14.7.2 Currents at 100 m depth from real-time nowcast on July 11, 1997, and 2-week forecast on July 25, 1997, for eddy Deviant in the Gulf of Mexico. Only TOPEX and ERS-1 altimetric SSH anomaly data (tracks shown) were assimilated into the model. From Kantha *et al.* (1999).

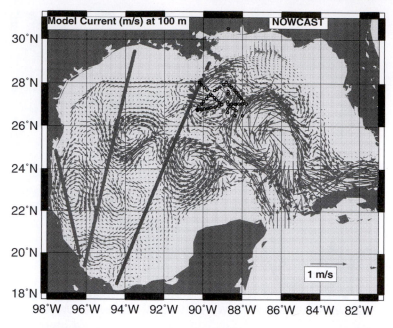

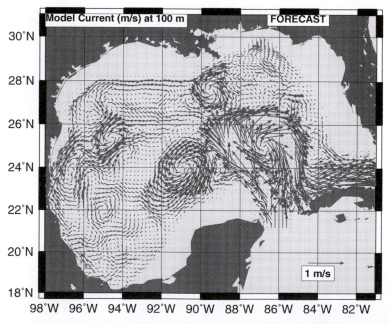

Figure 14.7.3 Currents at 100 m depth from real-time nowcast on July 11, 1997, and 2-week forecast on July 25, 1997, for eddy Deviant in the Gulf of Mexico with *in situ* XBT data (shown as circles) assimilated in the nowcast. TOPEX and ERS-1 altimetric SSH anomaly data (tracks shown) were also assimilated into the model. From Kantha *et al.* (1999).

of the ocean, the Gulf Stream system off Cape Hatteras in the western North Atlantic, under the auspices of Data Assimilation and Model Evaluation Experiment (DAMEE; see Willems *et al.,* 1994). Different kinds of models and assimilation methodologies have been tested (for example, Ezer and Mellor, 1994; Fox *et al.,* 1992) against high quality observational data sets (Lai *et al.,* 1994) to assess model forecast skills against persistence (persistence assumes the forecast state is no different from the initial state). Once the Gulf Stream departs the coast at cape Hatteras, it is subject to growing meanders and pinch-off of warm core and cold core rings, with timescales for growth of instabilities ranging from 1 to 3 weeks. It is therefore not surprising that even with very careful initialization and comprehensive physics, successful forecasts of the Gulf Stream position are limited to about 2 weeks. The DAMEE does, however, demonstrate the importance of assimilating observational data to produce as realistic an initial state as possible so that the forecast can be successful in beating persistence to a significant degree. This is consistent with similar experiences in atmospheric forecasting, where a large emphasis is placed on obtaining an optimum initial state during the analysis-initialization cycle. The initial state of the ocean must be dynamically well-balanced (Gangopadhyay and Robinson, 1997b; Hurlburt *et al.,* 1990) so that its future evolution can be accurately tracked by a model, without interference from growing perturbations generated by initial imbalances. This is particularly important in regions subject to vigorous flow instabilities.

Since observational data are seldom adequate to describe the three-dimensional synoptic state of the mesoscale features at the start of a model forecast, an alternative strategy is essential. A useful technique pioneered by the Harvard University ocean modelers is the use of feature models (Robinson, 1996; Robinson *et al.,* 1988) that assume that characteristics such as the velocity and temperature distributions of mesoscale features such as the rings can be described by parametric models, whose few parameters can be readily inferred from sparse observations. The procedure is described by Gangopadhyay and Robinson (1997a,b). Based on historical observations, multiscale feature models of the Gulf Stream jet, warm and cold core rings, and the western boundary undercurrent are constructed. While it is seldom possible to obtain observational data that define these features in 3-D space, it is often possible to infer, from remotely sensed and other observational data, parameters such as the sizes and locations of these features in the model domain. Using these locations and feature models (this process is also called bogusing), the resulting initial field is adjusted to balance dynamically using simple quasi-geostrophic dynamics, and this balanced state is then used to initialize the forecast model and start the forecast. This strategy appears to be quite successful in retaining forecast skill vis-a-vis persistence for up to about 2 weeks. See Gangopadhyay and Robinson (1997a,b) for a detailed description and testing of this strategy with Harvard ocean models.

An accurate nowcast of the oceanic state is also a necessary ingredient in accurate atmospheric forecasts using a regional coupled ocean-atmosphere model. It is especially important to depict the state of the upper ocean, both the SST and the heat content, correctly so that the model can better simulate events, such as hurricane evolution during summer and wintertime storms, that involve intense air-sea interaction processes. See Chapter 13 for a coupled model simulation of hurricane Opal that illustrates this aspect (see also Bao *et al.*, 1999).

Finally, there is a tendency in the west to focus almost exclusively on the North Pacific and the North Atlantic oceans in both observational and modeling campaigns. The oceans in the southern hemisphere and the Indian Ocean have received far less attention so far than they truly deserve. The summer monsoons in the Indian Ocean and the adjoining seas in the Pacific (Webster *et al.*, 1998, 1999b) play a crucial role in the welfare of more than half the global population (for example, Webster *et al.*, 1999a), and yet our knowledge of the coupled ocean-atmosphere system there is skimpy at best. Hopefully, the coming decades will see as much attention paid to the Indian Ocean and its summer monsoons as that paid to ENSO in the 1990s.

We stop here in our rather overly ambitious and certainly very difficult task of providing an overview of *Numerical Models of Oceans and Oceanic Processes*. We hope we have succeeded in, if nothing else, conveying the complexity of the tasks faced by ocean modelers. The oceans are not only fascinating from an intellectual point of view, but are also vital to our survival in the coming millenium. We therefore encourage the reader to become actively involved in the study of the oceans and contribute to a better understanding of our *blue* planet.

Note added in proof: Genetic algorithms are also proving useful in time series prediction. See, for example, G. G. Szpiro, 1997. Forecasting chaotic time series with genetic algorithms, *Phys. Rev. E.,* **55,** 2557–2568.

Appendix A

Equations of State

A.1 EQUATION OF STATE FOR THE OCEAN

The equation of state that is now commonly used is presented by Millero and Poisson (1981) and given in UNESCO Technical Paper in Marine Science Number 36 (UNESCO, 1981). The density of sea water (in kg m^{-3}) as a function of temperature T (°C), salinity S (psu), and pressure p (bars) over the typical oceanic range of −2° to 40 °C temperature, 0 to 42 psu salinity, and 0 to 1000 bars pressure is given to a precision of 3.5×10^{-3} kg m^{-3} by

$$\rho(T, S, p) = \rho(T, S, 0)\left[1 - p / K(T, S, p)\right]^{-1} \qquad (A.1)$$

where K(T,S,p) is the secant bulk modulus and the density of sea water at one standard atmosphere pressure (p = 0) is given by (Gill, 1982; Pond and Pickard, 1989)

$$
\begin{aligned}
\rho(T, S, 0) = \quad & \\
& +\ 999.842\ 594 & & +\ 6.793\ 952 \quad \times 10^{-2}\ T \\
& -\ 9.095\ 290 \ \times 10^{-3}\ T^2 & & +\ 1.001\ 685 \quad \times 10^{-4}\ T^3 \\
& -\ 1.120083 \ \ \times 10^{-6}\ T^4 & & +\ 6.536\ 332 \quad \times 10^{-9}\ T^5 \\
& +\ 8.244\ 93 \ \ \times 10^{-1}\ S & & -\ 4.089\ 9 \qquad \times 10^{-3}\ TS \\
& +\ 7.643\ 80 \ \ \times 10^{-5}\ T^2 S & & -\ 8.246\ 7 \qquad \times 10^{-7}\ T^3 S \\
& +\ 5.387\ 50 \ \ \times 10^{-9}\ T^4 S & & -\ 5.724\ 66 \quad \times 10^{-3}\ S^{1.5} \\
& +\ 1.022\ 70 \ \ \times 10^{-4}\ TS^{1.5} & & -\ 1.654\ 6 \qquad \times 10^{-6}\ T^2 S^{1.5} \\
& +\ 4.831\ 40 \ \ \times 10^{-4}\ S^2 & &
\end{aligned}
$$

$$(A.2)$$

K (T,S,p) =

+ 19 652.21

+ 148.420 6		T	- 2.327 105	T^2
+	1.360 477	$\times 10^{-2}\ T^3$	- 5.155 288	$\times 10^{-5}\ T^4$
+	3.239 908	p	+ 1.437 13	$\times 10^{-3}\ Tp$
+	1.160 92	$\times 10^{-4}\ T^2p$	- 5.779 05	$\times 10^{-7}\ T^3p$
+	8.509 35	$\times 10^{-5}\ p^2$	- 6.122 93	$\times 10^{-6}\ Tp^2$
+	5.278 7	$\times 10^{-8}\ T^2p^2$		

+ 54.674 6		S	- 0.603 459	TS
+	1.099 87	$\times 10^{-2}\ T^2S$	- 6.167 0	$\times 10^{-5}\ T^3S$
+	7.944	$\times 10^{-2}\ S^{1.5}$	+ 1.648 3	$\times 10^{-2}\ TS^{1.5}$
-	5.300 9	$\times 10^{-4}\ T^2S^{1.5}$	+ 2.283 8	$\times 10^{-3}\ pS$
-	1.098 1	$\times 10^{-5}\ TpS$	- 1.607 8	$\times 10^{-6}\ T^2pS$
+	1.910 75	$\times 10^{-4}\ pS^{1.5}$	- 9.934 8	$\times 10^{-7}\ p^2S$
+	2.081 6	$\times 10^{-8}\ Tp^2S$	+ 9.169 7	$\times 10^{-10}T^2p^2S$ (A.3)

Test values for various quantities above (Pond and Pickard, 1989):

At T = 5 °C, S = 0, p = 0 bars: ρ = 999.96675 kg m^{-3}
At T = 5 °C, S = 0, p = 1000 bars: ρ = 1044.12802 kg m^{-3}
At T = 25 °C, S = 0, p = 0 bars: ρ = 997.04796 kg m^{-3}
At T = 25 °C, S = 0, p = 1000 bars: ρ = 1037.90204 kg m^{-3}
At T = 5 °C, S = 35, p = 0 bars: ρ = 1027.67547 kg m^{-3}
At T = 5 °C, S = 35, p = 1000 bars: ρ = 1069.48914 kg m^{-3}
At T = 25 °C, S = 35, p = 0 bars: ρ = 1023.34306 kg m^{-3}
At T = 25 °C, S = 35, p = 1000 bars: ρ = 1062.53817 kg m^{-3}

The specific volume is given by

$$\alpha\left(T,S,p\right)=\alpha\left(T,S,0\right)\left[1-p/K\left(T,S,p\right)\right]^{-1} \tag{A.4}$$

The adiabatic lapse rate can be calculated using

$$\Gamma = g\alpha T / c_p \tag{A.5}$$

The static stability can be calculated by

$$N^2 = g\alpha\left(\Gamma+\frac{dT}{dz}\right)-g\beta\frac{dS}{dz}=g^2\alpha^2T/c_p+g\alpha\frac{dT}{dz}-g\beta\frac{dS}{dz} \tag{A.6}$$

where N is the Brunt-Vaisala (buoyancy) frequency and g the gravitational constant and

$$\alpha = \frac{1}{\rho}\left[\frac{\partial \rho}{\partial T}\right]_{p,S} \quad ;\beta = \frac{1}{\rho}\left[\frac{\partial \rho}{\partial S}\right]_{p,T} . \tag{A.7}$$

Potential temperature $\theta(T,S,p)$ to 0.001 °C precision in the range 2 to 30°C, 30 to 40 psu, 0 to 1000 bars is given by

$$\theta(T,S,p) = T$$

$$
\begin{array}{llll}
- & 3.650\ 4 & \times 10^{-4}\ p & -8.319\ 8 & \times 10^{-5}\ Tp \\
+ & 5.406\ 5 & \times 10^{-7}\ T^2p & -4.027\ 4 & \times 10^{-9}\ T^3p \\
- & 8.930\ 9 & \times 10^{-7}\ p^2 & +3.162\ 8 & \times 10^{-8}\ Tp^2 \\
- & 2.198\ 7 & \times 10^{-10}T^2p^2 & & \\
+ & 1.605\ 6 & \times 10^{-10}p^3 & -5.048\ 4 & \times 10^{-12}Tp^3 \\
- & 1.743\ 9 & \times 10^{-5}\ p(S-35) & & \\
+ & 2.977\ 8 & \times 10^{-7}\ Tp(S-35) & & \\
+ & 4.105\ 7 & \times 10^{-9}\ p^2\ (S-35) & &
\end{array}
\tag{A.8}
$$

A test value for $T = 10°C$, $S = 25$ psu, $p = 1000$ bars: θ is 8.467 851 6.

The dependence of freezing point (°C) on salinity (0 to 40 psu) and pressure (0 to 50 bars) to 0.004 °C precision is

$$T_f(S,p) = -0.0575\ S + 1.710523 \times 10^{-3}\ S^{1.5} - 2.154996 \times 10^{-4}\ S^2$$
$$- 7.53 \times 10^{-3}\ p \tag{A.9}$$

This equation will be of particular interest in studying the formation of super-cooled waters and ice formation on the bottom of ice shelves.

A.2 EQUATION OF STATE FOR THE ATMOSPHERE

Air is a mixture of gases, the primary constituents being in nearly constant proportions: N_2 (78.1% by volume), O_2 (21%), and Ar (0.9%). Dry air can be treated as an ideal gas:

$$p_{da} = \rho_{da} R_a T \tag{A.10}$$

where $R_a = 287.04$ J kg^{-1} K^{-1}. T is the air temperature, ρ_{da} is the density and p_{da} is the pressure of dry air. Water vapor can also be treated as an ideal gas:

$$p_v = \rho_v R_v T \tag{A.11}$$

where $R_v = 461.50$ J kg^{-1} K^{-1}. T is the temperature, ρ_v is the density of water vapor (also called absolute humidity, the mass of vapor per unit volume of moist air), and p_v is the partial pressure of the water vapor. The total pressure is the sum of the two partial pressures, assuming air is a mixture of ideal gases:

$$p = p_{da} + p_v \tag{A.12}$$

If q is the specific humidity, mass of water vapor per unit mass of moist air,

$$\rho_v = q\rho_a \,; \ \rho_a = \rho_{da} + \rho_v \tag{A.13}$$

$$\frac{p_v}{p} = \frac{q}{\varepsilon + (1-\varepsilon)q} = \frac{r}{r+\varepsilon} \tag{A.14}$$

where $\varepsilon = \dfrac{R_a}{R_v} = 0.62197$, $r = \dfrac{q}{1-q}$ is the mixing ratio. The saturation value q_s, the maximum possible value for q, depends strongly on temperature: 0.0038 (r = 0.0038) at 0°C and 0.04 (r = 0.042) at 37°C.

The equation of state for the moist air can therefore be written as

$$\rho_a = \frac{p}{RT_v} \tag{A.15}$$

where T_v is the virtual temperature given by

$$T_V = T(1-q-q/\varepsilon) = T(1+0.6078q) \tag{A.16}$$

Since air is diatomic mostly ($c_{pda} = 3.5\ R_{da}$) and water vapor is triatomic ($c_{pv} = 4\ R_v$),

$$c_p = c_{pda}(1-q+8q/7\varepsilon) = 1004.6(1+0.8375q) \tag{A.17}$$

$$c_V = c_{vda}(1-q+6q/5\varepsilon) = 714.3(1+0.9294q) \tag{A.18}$$

The Clausius-Clapeyron equation provides the rate of change of saturation vapor pressure p_{vs} over a water surface with temperature:

$$\frac{dp_{vs}}{dT} = \frac{L_v}{T\left(\dfrac{1}{\rho_v} - \dfrac{1}{\rho_w}\right)} \sim \frac{L_v \rho_v}{T} = \frac{L_v p_{vs}}{R_v T^2} \tag{A.19}$$

where ρ_V is the density of water vapor and ρ_W is the density of liquid water. Expressions for L_v, the latent heat of vaporization (in J kg^{-1}), and saturation vapor pressure p_{vs} (mb) of pure water vapor over a water surface (to 0.2% precision between $-40\,°C$ and $+40\,°C$) can be written as

$$L_v\,(T) = 2.5008 \times 10^6 - 2.3 \times 10^3\,T \tag{A.20}$$

$$p_{vs}\,(T) = 10^{(0.7859+0.03477T)/(1+0.00412T)} \tag{A.21}$$

In air, the saturation vapor pressure is slightly higher by a factor of 1 to 1.006:

$$p_{VS} = p'_{VS}\left[1 + 10^{-6}\,p(4.5 + 0.0006T^2)\right], \tag{A.22}$$

where p is the total pressure in mb. Over sea water, this value must be multiplied by a factor of 0.98. Knowing T and p, one can therefore compute p_{vs} and hence the saturation specific humidity q_s, the saturation mixing ratio r_s, and the relative humidity RH:

$$RH = \frac{r}{r_s} = \frac{q(1-q_s)}{q_s(1-q)}, \tag{A.23}$$

For evaporation over ice, the expressions are different, and to 0.3% precision in $-40°C$ and $0°C$ range, the saturation vapor pressure is given by

$$p'_{vsi}\,(T) = p'_{vs}\,(T)10^{0.00422T} \tag{A.24}$$

and

$$L_{vi}\,(T) = 2.839 \times 10^6 - 3.6\,(T+35)^2 \tag{A.25}$$

The standard psychrometer equation that relates wet-bulb temperature T_{wb} to air temperature is

$$T_{wb} = T_a - \frac{L_E k_v}{c_{pa} k_H}[q_s\,(T_{wb}) - q_a] \tag{A.26}$$

where k_v and k_H are water vapor and heat diffusivities and q_s the saturation specific humidity.

To within 0.3%, the dry lapse rate for the atmosphere is $\Gamma = g/c_p$, about 10 K km^{-1}. The moist adiabatic lapse rate (in K km^{-1}) is given by

$$\Gamma_m = 6.4 - 0.12T + 2.5 \times 10^{-5}\,T^3 + [-2.4 + 10^{-3}(T-5)^2][1 - 10^{-3}p] \tag{A.27}$$

Appendix B
Wavelet Transforms

Interpreting observational data and, increasingly, the output of modern ocean models often requires an analysis of the characteristics of the temporal variations of forcing and response. The traditional and time-honored method is the time series analysis via Fourier transforms, which provide the characteristics of a temporally varying signal in frequency space. The widespread use of computers and the discovery of the Fast Fourier Transform (FFT) algorithm have made Fourier transforms the centerpiece of any time series analysis since the 1960s.

However, Fourier transforms provide only a limited amount of information on the characteristics of a temporally varying signal. They provide only its frequency content over the entire record length. All phase information related to individual events comprising the time series is lost in the transformation from temporal to frequency space. Thus while it is possible to identify the magnitude and frequency (or equivalently the period) of various events embedded in the time series, it is impossible to localize each of the events in the time domain; in other words, it is impossible to identify when a particular event is occurring in the time series and what its characteristics are. Wavelet transforms (WT) provide a unique capability for localizing and quantifying the variability of a signal in both frequency and time domains simultaneously and are one of the major breakthroughs in time series analysis since the advent of FFTs.

Lau and Weng (1995) provide an excellent illustration of the utility of wavelet transforms in time series analysis by application to a record of the oxygen isotope ($\delta^{18}O$) in the sediments at the ocean bottom, which contains information related to variations in the global ice volume (Raymo *et al.*, 1990). Even a cursory examination of this time series (Figure B.1) of $\delta^{18}O$ shows that a significant change occurred in the characteristics of ice volume fluctuations around 0.7 Myr BP (before present). The ice volume started varying at a 100 Kyr period instead of the 40 Kyr period before then. These two cycles, 40 Kyr and 100 Kyr have approximately the same periods as two of the major components of the variations in the Earth's orbit around the Sun. They are related to the variations in the obliquity and eccentricity of the Earth's orbit,

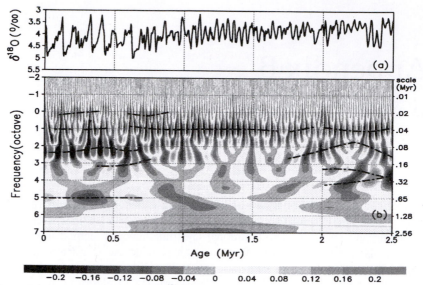

Figure B.1 Time series of deep sea $\delta^{18}O$ at a site in the North Atlantic and the corresponding Morlet wavelet transform; note the transition from 40 kyr cycle to 100 kyr cycle about 0.7 Myr BP. From Lau and Weng (1995).

while the precession of the equinoxes occurs at roughly 20 Kyr cycles (Berger, 1978). Similar cycles often show up in climate records such as the oxygen isotope recorded in deep-sea sediments. While the precise dynamical causes underlying such variability in Earth's climate are still being sorted out, it is the belief that these climatic variations are a response of the Earth to variations in the solar constant brought on by the orbital variations, as first proposed by Milankovic (Berger, 1978).

In addition to the 40 Kyr and 100 Kyr cycles, there are other cycles embedded in the signal that are not that readily apparent. Extraction of such signals usually meant subjecting the time series to FFT analysis. The resulting spectrum will show dominant peaks at 40 and 100 Kyrs and minor ones at various other periods. However, all time (phase) information is lost and individual events cannot be identified. While it is clear that 40 Kyr and 100 Kyr cycles are embedded in the climate record, the change that occurred at 0.7 Myr BP that was readily discernible is lost. The need for a technique that preserves the phase information (pertaining to when a particular event of certain frequency characteristics occurred in the time series) is quite obvious.

It can be seen from the wavelet transform in Figure B.1 that it is not only possible to identify the 40 Kyr and 100 Kyr cycles in the sediment record, but it is also possible to say that a major change occurred in the Earth's climate at around 0.7 Myr BP. The abrupt steplike transition between the two frequency regimes is clearly seen in WT. The frequency and amplitude modulation involving splitting and merging of the harmonics and subharmonics of the basic 40 Kyr period, the 20, 80, 160, and 640 Kyr cycles, can be readily identified in the WT. The 40 Kyr cycle is overwhelmingly predominant over the subharmonics between 1 Myr and 1.7 Myr BP, while the 100 Kyr cycle dominates the record from 0.7 Myr BP to the present, even though the 40 Kyr cycle has not disappeared. The 100 Kyr cycle is present between 1.7 Myr and 2.5 Myr BP, but it is modulated by even a lower frequency variation during that period. The 40 Kyr cycle is present in the entire record except for some noticeable gaps in the 0.7 Myr BP to the present. The 20 Kyr cycles are present throughout the record but are significantly enhanced during the past 1 Myr when the 80 to 100 Kyr cycles are also strong. The occurrences of the 20 Kyr, 160 Kyr, and 640 Kyr cycles at various times in the past and their relative strengths compared to the dominant 40 and 100 Kyr cycles are readily apparent in the WT. The longest resolvable cycle, 2.5 Myr, represents the trend in the data and it can be seen that the most rapid change in the trend occurs around approximately 1.2 Myr BP, when the 2.5 Myr cycle changes sign. This coincides with the abrupt bifurcation and revival of strong 80 and 160 Kyr cycles and the appearance of a 640 Kyr cycle in the record. It is clear that a WT is a far more complete characterization of the time series than a traditional FFT and in fact provides a complete picture of scale (equivalently frequency)-time evolution of various events comprising the signal. Unlike an FFT, it is well suited for analysis of nonstationary time series as well.

By summing up all the WT coefficients scale by scale, it is possible to compute the approximate Fourier spectrum of a signal and thus derive information equivalent to that provided by Fourier transforms. Figure B.2 from Lau and Weng (1995) shows the mean wavelet spectrum for three time periods, the entire 2.5 Myr record, for the past 0.7 Myr, and between 1.0 and 2.5 Myr. The dominance of the 40 Kyr cycle over the 100 Kyr cycle over the entire period can be seen from the relative magnitudes of the two peaks in the plots that preserve the variance in the signal. Other cycles are not apparent. In the past 0.7 Myr, the 100 Kyr cycle is overwhelmingly dominant, although the 20, 40, and 640 Kyr peaks can be seen. Between 1.0 and 2.5 Myr BP, the 40 Kyr cycle is the most dominant.

By plotting the various WT coefficients as a function of time, the variability in each of the cycles over time can be seen (Figure B.3). Figure B.3 provides another example, time series of northern hemisphere temperatures. The WT clearly shows the variability on a variety of time scales. It is also possible to reconstitute the signal including (or excluding) certain frequency components. This technique not only provides a means of decomposing and retaining only the most significant components in the time series record, but it also enables very efficient data compression to be effected. Consequently WTs can be used not only for time series analyses but also for data compression.

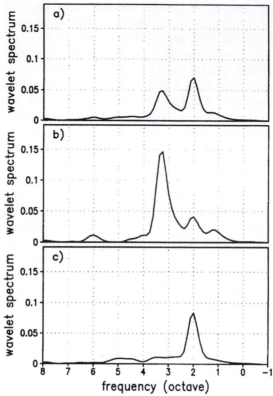

Figure B.2 Wavelet spectra of the time series in Figure B.1, averaged over the past 2.5 Myr (top), 0.7 Myr (middle), and 1-2.5 Myr (bottom). From Lau and Weng (1995).

B.1 INTRODUCTION

Wavelet transform is a powerful time series analysis (and data compression) tool that is capable of localizing the signal variability simultaneously in both time and scale (frequency) domains. It is ideally suited for analyzing records containing episodic events and multiple scales, an excellent example being the proxy climate record discussed above. It complements the traditional Fourier transforms and empirical orthogonal functions in time series analysis. Since their introduction by Morlet (1983) in the 1980s for the analysis of seismic signals, wavelet transforms have enjoyed an explosive growth and have been used in a wide variety of fields, including physics, geophysics, optics, quantum mechanics, seismology, and in various engineering fields, especially electrical engineering. Rigorous treatments of wavelets and their mathematical foundations can be

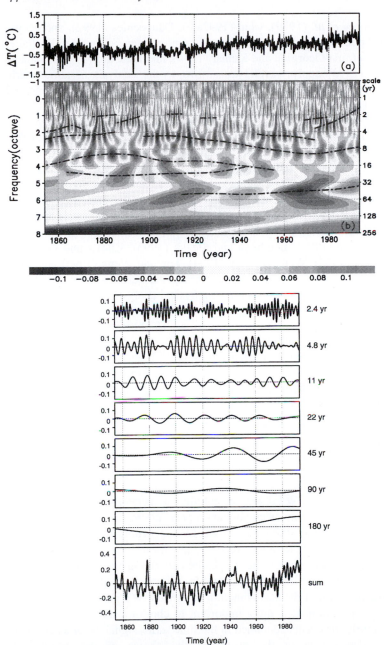

Figure B.3 Time series of northern hemisphere temperature, with its Morlet wavelet transforms, (top). Reconstitution of the signal obtained by summing up fluctuations from 2.4 years and up (bottom). From Lau and Weng (1995).

found in Combes *et al.* (1989), Chui (1992a,b), Daubechies (1992), and Meyer (1993a, b). Applications to a wide variety of problems can be found in Ruskai *et al.* (1992), Foufoula-Georgiou and Kumar (1994), and Hubbard (1996). Gamage and Blumen (1993) apply wavelet transforms to the analysis of atmospheric fronts; Meyers *et al.* (1993) to Yanai waves in the equatorial oceans; Weng and Lau (1994) to tropical convection; Liandrat and Moret-Bailly (1990), Farge (1992), Bendjoya and Slezak (1993), Collineau and Brunet (1993), Gao and Li (1993), Hudgins *et al.* (1993), Howell and Mahrt (1994), and Farge *et al.* (1996) to analysis of turbulence; Liu (1994) to wind waves; Mak (1995) to interannual variability in sea surface temperature; Lau and Weng (1995) to climate and weather records; and Torrence and Compo (1997) to ENSO analysis in the topical Pacific SST data. Kantha *et al.* (1996) use wavelet transforms to elucidate the deviations in the inverse barometric response of ocean basins to atmospheric forcing, a topic of considerable interest to altimetric applications.

Of these, Farge (1992), Bendjoya and Slezak (1993), Kumar and Foufoula-Georgiou (1994, 1997), Weng and Lau (1994), Lau and Weng (1995), and Mak (1995) are particularly lucid and useful to grasp the bare essentials needed to appreciate and use wavelets as a tool in geophysical analysis without getting lost in the underlying mathematical complexities, for which it is suggested that the reader consult mathematical treatises such as Daubechies (1992) or Chui (1992a,b). Kumar and Foufoula-Georgiou (1997) should be consulted for an excellent review of geophysical applications of WT and an up-to-date list of references. The following summary of wavelet transforms is principally derived from the above seven references. However, an attempt has been made to use uniform notation, as this is the first and foremost obstacle to a novice.

B.1.1 THEORY

Wavelet transforms (WT) are essentially a generalization of Fourier transforms (FT) and are somewhat similar in principle to the windowed Fourier transforms (WFT). In fact, the heavily used FFT algorithm (Bergland, 1969) is easily adapted to WT and provides a very efficient means of computing WT. The principal difference between FT and WT is the underlying basis function used to decompose the time series into its individual frequencies or scales.

The particular form of the basis function to use in a WT depends critically on the intended application. It is quite important to choose a mother wavelet that resembles closely the events contained in the time series signal as far as possible, since some dependence of the result on the choice of the wavelets is unavoidable. In practice, Meyer (and Haar) orthogonal discrete wavelets and Morlet (and DOG—derivative of a Gaussian) continuous wavelets are the two that are widely used in geophysics. It is important to remember that continuous wavelets such as Morlet are best suited for scale analysis and event identification, while orthogonal wavelets such as Meyer wavelets are best at signal decomposition, reconstitution, and data compression applications. However, in practice, the distinction may not be that important since it is often possible to subsample the continuous wavelets at quasi-orthogonal discrete intervals and interpolate the orthogonal WT coefficients at octave intervals to suitable voice intervals. Thus the desirable time localization property of continuous WT and the accurate reconstruction property of discrete orthogonal WT can be combined to some extent. In any case, it is preferable to be able to go back and forth between the two. We will use only Meyer and Morlet wavelets in the examples, since they are the most commonly used wavelet basis functions.

B.1.1.1 Fourier Transforms

Fourier transform (FT) uses sines and cosines as basis functions for signal decomposition, functions that are invariant with and extend to infinity in time and hence are global in nature. For a real valued signal f(t), its FT is given by a complex-valued function:

$$S(\omega) = \int_{-\infty}^{\infty} f(t)e^{-i\omega t}dt \qquad (B.1)$$

The transform preserves energy,

$$E = \frac{1}{2}\int_{-\infty}^{\infty} f(t)f^*(t)dt = \frac{1}{2}\int_{-\infty}^{\infty} S(\omega)S^*(\omega)d\omega \qquad (B.2)$$

and is invertible; the asterisk denotes the complex conjugate. Signal f(t) is related to S(ω) by

$$f(t) = \frac{1}{2\pi}\int_{-\infty}^{\infty} S(\omega)e^{i\omega t}d\omega \qquad (B.3)$$

In this form, it is easy to see that f(t) is just a weighted sum of elementary functions $e^{i\omega t}$, with S(ω) denoting the weight, which is a function of only frequency ω and is independent of time t. The square of its modulus $|S(\omega)|^2$ is the Fourier energy density (power) spectrum E(ω). Thus it provides information on only the frequency content of a signal and no information on where these frequencies are located in the time domain. Because of the form of the basis function (a delta function in frequency space and extending to ±∞ in time domain), the resulting transform contains no time information. All phase information of individual events in the time series has been lost, and hence FT

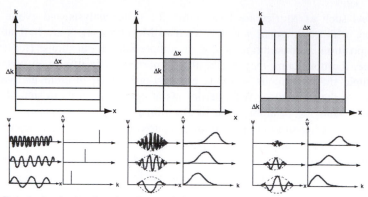

Figure B.4 A sketch showing the characteristics of Fourier, windowed Fourier, and wavelet transforms. Courtesy of Marie Farge and Alexandre Azzalini, from Farge (1992) with permission from the Annual Review of Fluid Mechanics, Volume 24. Copyright 1992 by Annual Reviews http://www.AnnualReviews.org.

provides no capability for localization in time. The FT of a stationary sine wave is simply a line spectrum (Figure B.4) and that of a delta function contains an infinite number of components in frequency space. A reversal of time leads to no changes in the FT, even though the time of occurrence of events could be vastly different. If a periodic signal changes frequency suddenly, the FT consists of two line spectra and has no information on when the change occurred. Retaining such information is often critical to the analysis of geophysical records, such as the climatic record mentioned above, and the analysis of numerical ocean model outputs. Signals containing episodic events occur in many other fields, music, speech, electronics, and optics being among them.

Figure B.5 shows two time series entirely different in character: one has two superimposed signals of different frequencies for the entire duration of the record, while the other has the same signal change its frequency. Yet the Fourier transforms are identical (compare with the corresponding WT) as shown by the Fourier spectrum for each case, except for the spurious frequencies in the spectrum of the latter that arise from the inability of FT to handle discontinuities gracefully. Note that the area under the spectrum is the energy content of the signal. Figure B.6 shows a signal that consists of three constant frequency segments with two abrupt changes in frequency, with the middle segment lasting for only a brief lifetime. The corresponding spectrum (plotted using a linear scale for the ordinate, called variance preserving plot) illustrates the nature of the FT quite well. It consists of three peaks, one of which is much shorter than it should be; this corresponds to the briefer middle segment, which the FT has difficulty resolving. The discontinuous changes in the signal give rise to spurious oscillations all over the spectrum. No information on the duration or the time of occurrence of each segment exists in the spectrum.

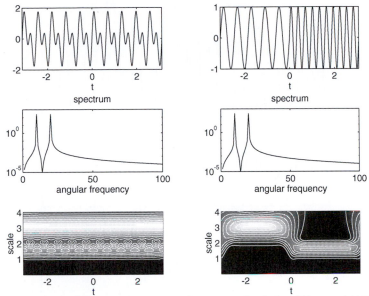

Figure B.5 Conventional Fourier and wavelet analysis of two signals that differ, yet yield the same Fourier transform. From Kumar and Foufoula-Georgiou (1994).

It is worth pointing out that a FT assumes that the signal is stationary. It is therefore essential to remove the underlying trend before subjecting a time series to FT. Also the finiteness of the time series and the infinite nature of the basis functions used to obtain FT cause leakage problems. The two ends of the record must be properly altered for an accurate FT. The most favored treatment is to smooth or taper the ends of the record using a cosine taper—in other words, multiplying it by a cosine squared term to bring the function smoothly to zero. However, some distortion in the result is quite unavoidable, no matter what technique is used to condition the ends.

Since in practice, the FT is discrete and digital, it is also necessary to filter a continuous analog signal to remove all frequencies higher than the Nyquist frequency (twice the sampling frequency) before sampling it to produce a sequence of points to be subjected to the discrete Fourier transform. Otherwise the frequencies higher than the Nyquist will be aliased or folded into lower frequencies leading to serious errors. Also the quality of the resulting spectra depends critically on how well a particular frequency is sampled in time. For those frequencies that are sufficiently long lived in the original time series, the corresponding peaks are sharp and well defined. For those that are not, the peak is poorly defined and the energy is spread over a broad range of neighboring spectral space. The Fast Fourier Transform algorithm developed by Cooley and Tuckey is the most favored technique for obtaining FT of a digitized signal

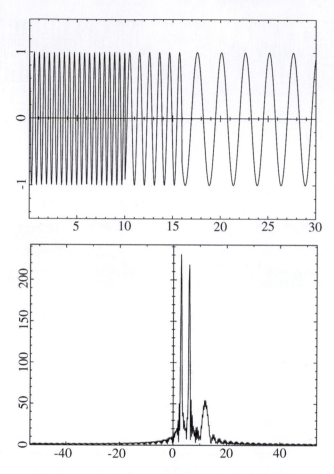

Figure B.6 Fourier transform of a signal consisting of three frequencies and the corresponding FT. (From Bendjoya, Ph. and Slezak, E., *Celestial Mechanics and Dynamic Astronomy*, figures 1 and 2, pp. 233–234, © 1993 with permission from Kluwer Academic Publishers.)

and can in general be adapted to a record with the number of points not a multiple of 2. However, it is traditional and more efficient to use a record of length 2^N, and this is accomplished simply by padding the record with zeros. Note that the FT does not contain any information at frequencies higher than that corresponding to the sampling rate ($2\pi/2\Delta t$), and lower than that corresponding to the record length ($2\pi/T_1$). Naturally, the lower frequencies are less well sampled than higher ones and hence statistically the confidence level decreases with the decrease in frequency. For details of FT and FFT, see Bergland (1969; see also Jenkins and Watts, 1968).

B.1.1.2 Windowed Fourier Transforms

Windowed Fourier Transform (WFT), also called short-time FT (Gabor, 1946) is an attempt to overcome the problem of loss of temporal information in an FT and uses a sliding window g(t) of length T_w centered around time t (Figure B.4b). The window width is constant in both time and frequency domains:

$$S_s(\omega,t) = \int_{-\infty}^{\infty} f(t')g^*(t'-t)e^{-i\omega t'}dt' \qquad (B.4)$$

where g^* denotes the complex conjugate of g. WFT measures the amplitude of a sine wave of frequency ω locally around a point t in time. Since the support (the extent over which the basis function has nonzero values) is the same for all ω and t, the number of cycles contained in the window is a function of frequency (Figure B.7). However, the constancy of the window in time makes it include a large number of high frequency cycles and only few or part of low frequency cycles comprising a signal, resulting in nonuniform representation of different frequency components. Constancy in frequency space means the high frequencies are overemphasized. Since all $(T_w/2\Delta t)$ frequencies are analyzed regardless of the frequency content of the signal, the result is wasted computational effort. Aliasing of frequencies, both higher and lower than that corresponding to the window width, results in spurious spectral peaks (Kaiser, 1994; Torrence and Compo, 1997).

As long as $\int_{-\infty}^{\infty} |g(t)|^2 dt = 1$, WFT preserves energy contained in the signal and the function f(t) can be reconstituted from the windowed transform by using

$$f(t) = \int_{-\infty}^{\infty} dt' \int_{-\infty}^{\infty} S_s(\omega,t')g(t-t')e^{-i\omega t'}d\omega \qquad (B.5)$$

When $g(t) = \pi^{-1/4} \exp(-t^2/2)$, the WFT is called the Gabor transform. Its use of a Gaussian window (Figure B.7) provides the best possible localization in both frequency and time. However, the number of oscillations within the window is a function of the location of the window in frequency space and, unlike WT, is not a constant.

B.1.1.3 Wavelet Transforms (WT)

The uniqueness of WT is that the time-frequency window is appropriately dilated (stretched) as it is moved (translated) around in the time-frequency phase space, so as to provide appropriate support at all frequencies within the limitation of the uncertainty principle that prevents arbitrarily high precision to be achieved simultaneously in both frequency and time for any time-frequency localization scheme. This is made possible through the use of a generalized and localized two-parameter family of basis functions (for example, shaped like the Mexican hat) called wavelets, instead of global sines and cosines. The window narrows when moved to the high frequency region of the wavelet domain and broadens in the low frequency portions of the time-frequency space (Figure B.4c). This way, both high and low frequencies are analyzed with optimal resolution and the least number of basis functions. This zoom-in capability permits localization in time of short-lived, high frequency signals, including abrupt changes, without losing the ability to resolve the low frequency components.

However, the width and height of the window in the time-frequency wavelet domain (WD) cannot be arbitrarily chosen and is in fact constrained by the uncertainty principle similar to the Heisenberg uncertainty principle in quantum mechanics. More precise localization in time in the high frequency part of the WD can only be achieved at the expense of reduced frequency resolution, and higher frequency resolution in the low frequency portion can be achieved only by sacrificing the time localization. Simultaneous, arbitrarily fine characterization of a signal in both time and frequency domains is impossible to achieve. WT enables sharp changes and other small scale features to be well localized in the time domain with an accompanying uncertainty in their frequency content; similarly large scale features are well resolved in the frequency domain, but there is an uncertainty in their localization in the time domain. Note that FT achieves precise localization of all events in frequency space by sacrificing completely any localization in time. Figure B.4 summarizes the characteristics of basis functions and the resulting time-frequency windows for FT, WFT, and WT. Note that unlike WFT, the windows (called Heisenberg cells) shorten in the time and lengthen in the frequency domain as frequency analyzed is increased.

B.1.2 CONTINUOUS WAVELET TRANSFORMS (CWT)

Consider first continuous wavelet transforms (CWT). Mathematically, the CWT $S_w(\lambda, \tau)$ of a real signal f(t) is the convolution integral:

$$S_w(\lambda, \tau) = \lambda^{-1/2} \int_{-\infty}^{\infty} f(t) \psi^* \left(\frac{t - \tau}{\lambda} \right) dt \quad (b>0) \tag{B.6}$$

where ψ^* is the complex conjugate of a mother wavelet ψ. Note that the CWT of a real function of a single variable f(t) is a complex-valued function of two variables τ and λ. A CWT decomposes a time series signal f(t) in terms of elementary basis functions $\psi_{\lambda, \tau}(t)$, derived from an analyzing (mother) wavelet

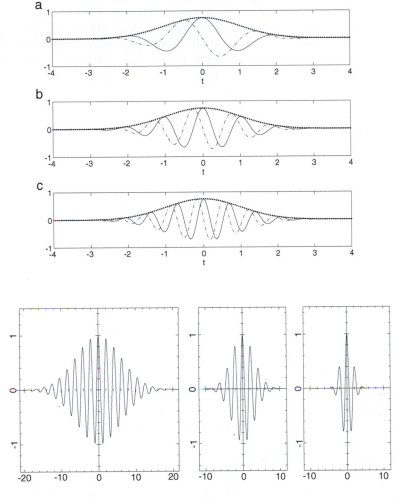

Figure B.7 Windowed FT, using a Gaussian envelope, for different frequencies. (Top panel from Kumar and Foufoula-Georgios, 1994; bottom panel from Bendjoya, Ph., and Slezak, E. *Celestial Mechanics and Dynamic Astronomy,* figure 3, p. 235, © 1993, with kind permission from Kluwer Academic Publishers.)

$\psi(t)$, by dilation and translation:

$$\psi_{\lambda,\tau}(t) = \lambda^{-1/2}\psi\left(\frac{t-\tau}{\lambda}\right) \tag{B.7}$$

where τ denotes the position (the translation parameter), and λ (> 0) denotes the scale (the dilation parameter) of the mother wavelet. $\psi_{\lambda,\tau}(t)$ are called daughter wavelets. If $\psi(t)$ is complex valued, it allows for analysis of both the amplitude and phase of the original signal. Complex wavelets (such as Morlet) are well suited to analyzing oscillatory signals. If it is a real function (for example, DOG), only the amplitude information can be retrieved. Real wavelets are best suited to analyzing discontinuous and peaky signals. Thus CWT provides a picture of the signal content of the original time series in the time (τ)- scale (λ) domain, permitting localization in both time and space. The normalization factor $\lambda^{-1/2}$ is quite important in that it keeps the energy content of all daughter wavelets for all scales the same as that of the mother wavelet.

While it is possible to compute WT using the time-domain definition in Eq. (B.6) and evaluating the integral on the RHS for various values of scale (equivalently frequency) λ and τ, it is more efficient to use Parseval's theorem to define it in the frequency-domain and FFTs to do so:

$$S_w(\lambda, \tau) = \sqrt{\lambda} \int_{-\infty}^{\infty} S(\omega) S_\psi^*(\lambda\omega) d\omega \qquad (B.8)$$

where $S_\psi(\omega)$ is the Fourier transform of the mother wavelet $\psi(t)$. To do this, the Fourier transform of the mother wavelet must be known analytically. Then the various steps in calculating WT of a function is to choose the mother wavelet (with known FT or determine it), find FT of the time series using FFT after appropriate signal conditioning, choose the scales for analysis, multiply the FT of the time series by the complex conjugate of the FT of the daughter wavelet at that scale, and use inverse FT to obtain the WT (Torrence and Compo, 1997).

While any number of suitable wavelet bases can be constructed and used, there are a few restrictions on the wavelet basis. It must be compactly supported, meaning it must vanish beyond a finite time interval or decay sufficiently rapidly (at least quadratically) for $t \rightarrow -\infty$ and $+\infty$. This is essential for localization in the time domain (remember, sine and cosine functions used in FT do not satisfy this condition). It must also satisfy the admissibility condition that its integral with respect to time must be zero:

$$\int_{-\infty}^{\infty} \psi(t) dt = 0 \qquad (B.9)$$

This requirement for a zero mean implies that the wavelet must have positive and negative segments so that it integrates to zero; it must be "wavelike." The daughter wavelets also have a zero mean. Higher order moments may also be required to be zero in some applications, but this is not normally necessary. A multiplicative constant is usually chosen so that the function has unit energy $\int_{-\infty}^{\infty} |\psi(t)|^2 dt = 1$. From the FT of the mother wavelet $\psi(t)$,

$$S_\psi(\omega) = \int\limits_{-\infty}^{\infty} \psi(t)e^{-i\omega t}dt \qquad (B.10)$$

it is easy to see that the admissibility condition for a real-valued wavelet implies that $S_\psi(0)$ must be zero. The wavelet can therefore be viewed as a notch filter that removes the mean. Figure B.8 shows this for a mother wavelet in the form of a Mexican hat function (Sadowsky, 1994):

$$\psi(t) \sim (1-t^2)e^{-t^2/2}; S_\psi(\omega) \sim \omega^2 e^{-\omega^2/2} \qquad (B.11)$$

It is clear that this wavelet is a low pass in the time domain and a band pass in the frequency domain. The filtering property of the CWT can be seen from Eq. (B.8) (Sadowsky, 1994). A CWT is therefore equivalent to filtering of the signal in the frequency domain by the dilated filter $\sqrt{\lambda}S_\psi(\lambda\omega)$ with a phase lag $\tau\omega$. Figure B.8 shows the effect of dilation for the Mexican hat wavelet. For $\lambda \gg 1$, the low frequency part of the phase space, the filter has a small bandwidth and hence fine resolution; for $\lambda \ll 1$, the high frequency part, the bandwidth is large and the resolution is coarse. Unlike the Gabor windowed transform, where the frequency resolution or the bandwidth $\Delta\omega$ is constant, in CWT it is variable such that the ratio of the frequency resolution to frequency $\Delta\omega/\omega$ is constant. This provides the zoom-in/zoom-out capability of the CWT that is often referred to as a microscope for examination in phase space. The admissibility condition ensures that the gain of the entire family of wavelet filters is finite. Any basis function can be chosen as long as the admissibility condition is satisfied.

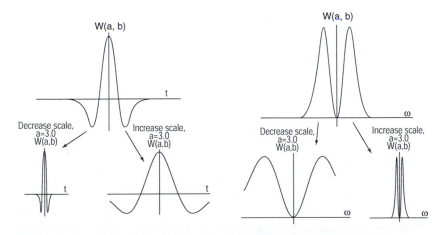

Figure B.8 Mexican hat mother and daughter wavelets. From Sadowsky (1994).

The original signal can be rebuilt from the wavelet coefficients:

$$f(t) = \frac{1}{C_\psi} \int\limits_{0}^{\infty} \int\limits_{-\infty}^{\infty} \lambda^{-2} S_w(\lambda, \tau) \psi_{\lambda, \tau} d\lambda d\tau = \frac{1}{C_\psi} \int\limits_{0}^{\infty} \lambda^{-2} d\lambda \int\limits_{-\infty}^{\infty} S_w(\lambda, \tau) \frac{1}{\sqrt{\lambda}} \psi\left(\frac{t - \tau}{\lambda}\right) d\tau \qquad (B.12)$$

where

$$C_\psi = 2\pi \int\limits_{-\infty}^{\infty} \frac{1}{\omega} S_\psi(\omega) S_\psi^*(\omega) d\omega = 2\pi \int\limits_{0}^{\infty} \frac{|S_\psi(\omega)|^2}{\omega} d\omega \qquad (B.13)$$

is finite-valued, which follows from the admissibility condition for any integrable functional form for ψ. The second equality of Eq. (B.13) implies that the function is required to be regular or progressive, meaning that its FT vanish for negative frequencies. This is not restrictive for real-valued wavelets, since any square integrable real-valued function can be considered a real part of a complex-valued square integrable function, whose FT is zero for negative frequencies. This just implies that the CWT is energy-preserving for any square-integrable function:

$$E = \frac{1}{2} \int\limits_{-\infty}^{\infty} f(t) f^*(t) dt = \int\limits_{0}^{\infty} \left[\frac{1}{2C_\psi} \int\limits_{-\infty}^{\infty} \lambda^{-2} S_w(\lambda, \tau) S_w^*(\lambda, \tau) d\tau\right] d\lambda \qquad (B.14)$$

This property has important implications in the applications of CWT to turbulence research, since the quantity within square brackets denotes the energy at each scale λ and the total energy over all scales is preserved in a CWT. If, for example, a Morlet wavelet, which bears a remarkable similarity to a turbulent eddy (Liandrat and Moret-Bailly, 1990; Tennekes and Lumley, 1972), is used to analyze a turbulent velocity signal, such a wavelet decomposition provides a much more powerful means of understanding the turbulence dynamics than the traditional Fourier spectrum approach.

The principal problem, however, is the redundancy of the CWT coefficients, which are not independent of one another as in an orthogonal discrete wavelet transform, but are related in the following fashion:

$$S_w(\lambda_0, \tau_0) = \int\limits_{0}^{\infty} \frac{d\lambda}{\lambda^2} \int\limits_{-\infty}^{\infty} S_w(\lambda, \tau) \bar{S}_w\left(\frac{\tau_0 - \tau}{\lambda}, \frac{\lambda_0}{\lambda}\right) d\tau \qquad (B.15)$$

where

$$\bar{S}_w(\lambda, \tau) = \lambda^{-1/2} \frac{1}{C_\psi} \int\limits_{-\infty}^{\infty} \psi(t) \psi^*\left(\frac{t - \tau}{\lambda}\right) dt \qquad (B.16)$$

Thus any square-integrable signal $f(t)$ can then be decomposed into a linear combination of wavelets. Unlike a FT, the analysis is applicable to nonstationary

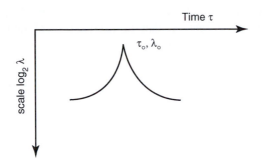

Figure B.9 Cone of influence of a point in a Morlet WT. From Lau and Weng (1995).

processes as well. Needless to say, the choice of the basis function, the mother wavelet, is quite important. For continuous wavelets, there are no restrictions on the values of the dilation and translation parameters. Many continuous wavelets such as the Morlet wavelet and the Mexican hat function are symmetric, although asymmetric ones can be used as long as the admissibility condition is satisfied. The Morlet wavelet is widely used in geophysics and is a plane wave modulated by a Gaussian envelope, whereas the Mexican hat is the second derivative of a Gaussian function. Continuous wavelets are particularly effective in identifying and localizing events in a time series and have been used extensively in geophysical (Gamage and Blumen, 1993; Lau and Weng, 1995; Liu, 1994; Meyers *et al.*, 1993; Weng and Lau, 1994) and engineering applications. The major disadvantage is that the continuous wavelets are not an orthogonal set and hence they cannot be used to accurately quantify the relative contribution of different frequency components to the total variability of the signal. One needs orthogonal wavelet bases for this purpose.

Unlike a FT, which maps a 1-D time series, a function of time, into a 1-D spectrum, a function of frequency, a WT yields a 2-D timescale map. For continuous wavelets, the map (see Figure B.1) is usually presented as a contour plot of the wavelet coefficients with a linear time (τ) abscissa, with time increasing to the right and a logarithmic scale (λ) ordinate, with scale (frequency) increasing downward (upward). The WT is usually complex, but the modulus (amplitude) and phase (or real and imaginary parts) can be plotted separately, although most often only the real part of the complex function is presented. The mother wavelet is chosen to vanish outside a time interval (t_1 - t_2) and this means that a point in (τ_1, λ_1) space influences regions only within the cone of influence: $|\tau - \tau_1| = (\lambda - \lambda_1)\Lambda$, where $\Lambda = \sqrt{2}$ for a Morlet wavelet (Figure B.9).

In practical computations of a CWT (just like FT), it is necessary to deal with the discretized version of Eq. (B.8). This is done by discretizing both the scale and location parameters usually in octaves (integer powers of 2): $\lambda = 2^m$ and $\tau = \lambda n$, where m and n are integers, so that the discrete form of the wavelet becomes

$$\psi_{m,n} = 2^{-m/2} \psi\left(2^{-m}t - n\right) \tag{B.17}$$

and the discrete form of the CWT becomes

$$S_w(m,n) = 2^{-m/2} \int_{-\infty}^{\infty} f(t)\psi^*(2^{-m}t - n)dt \tag{B.18}$$

and

$$S_w(m,n) = 2^{m/2} \int_{-\infty}^{\infty} S(2^m \omega)S_\psi^*(\omega)d\omega \tag{B.19}$$

The only difference between this discrete form of CWT and an orthogonal discrete WT is in the constraint on the wavelet in the latter to ensure orthogonality of every daughter wavelet to all others:

$$\int \psi_{m,n}(t)\psi_{m',n'}(t)dt = \delta_{mm'}\delta_{nn'} \tag{B.20}$$

where δ_{ij} is the Kronecker delta (equal to 1 if i = j; 0 otherwise). This renders the wavelets orthogonal to their dilates and translates, and enables the best possible decomposition to be accomplished. Also, in an orthogonal WT, the scales are analyzed only at octaves (powers of 2). While this permits a broad range of scales in the frequency domain to be efficiently analyzed (linear octave scale is logarithmic in frequency space), continuous wavelets allow analysis at fractional values of each octave (called voice) and are even better at localizing and characterizing events in the time domain, albeit at the expense of accurate quantification of the variance content. The subdivision of each octave into voices is usually such that the ratio of two consecutive voices is constant so that the scale at octave j and voice i is simply

$$\lambda_{i,j} = \lambda_0 2^{j+i/n} \tag{B.21}$$

where λ_0 can be chosen arbitrarily.

While brute-force numerical integration of Eq. (B.6) can provide wavelet transforms, the computational time required is proportional to NM^2, where N is the number of scales λ being analyzed and M is the logarithm of the number of points in time. Efficiency requires computation in spectral space as a convolution or filtering operation using Eq. (B.8). This allows the use of algorithms similar to FFT (computational time is only $NM\log_2 M$). The reader is referred to Sadowsky (1994), who describes an algorithm developed by Holschneider *et al.*

(1989) that explicitly makes use of the fact that the various filters needed to derive a CWT using the discretized version of Eq. (B.11) can be realized by dilations of a single filter. Thus, the number of sampled points is kept the same for all wavelets and the operation count is the same as an FFT and algorithm similar to the Mallat algorithm for orthogonal wavelets, if discretization is by octaves (Farge, 1992).

It is, however, important to precondition the signal appropriately, just as in the use of FFT routines, to prevent aliasing and other errors. However, the techniques used for FFT, such as detrending or cosine tapering, are inappropriate for wavelet analysis. Instead, it is preferable to buffer the data on both ends by a sufficient number of zero values and discard these tail regions in the resulting transforms. This technique appears to give acceptable results (Meyers *et al.*, 1993), even though more sophisticated buffer regions that match derivative values at the ends may yield better results.

B.1.2.1 Morlet Wavelet

The Morlet wavelet is the most commonly used continuous wavelet and is a complex-valued function approximating a simple wave modulated by a Gaussian envelope:

$$\psi(t) = \pi^{-1/4} e^{i\omega_0 t} e^{-t^2/2} \qquad \omega_0 \geq 5 \qquad (B.22)$$

The constant is chosen to make $|\psi|^2 = 1$. The restriction on the range is essential to assure $\psi \approx 0$ and satisfy the admissibility condition. The more precise form is

$$\psi(t) = \pi^{-1/4} e^{i\omega_0 t - e^{-\omega_0^2/2}} e^{-t^2/2} \qquad (B.23)$$

but the second term in the first exponential is negligible in the above range. Morlet proposed the range $5 < \omega_0 < 6$. The FT of a Morlet wavelet is

$$S_\psi(\omega) = \pi^{-1/4} e^{-(\omega - \omega_0)^2/2} \qquad \omega > 0 \qquad (B.24)$$

It is zero for negative ω. The complex-valued nature of the Morlet wavelet enables the detection of both amplitude and phase of the events in a time series and is well suited for scale analysis of time series describing wavelike processes. Its Gaussian envelope makes it optimal for localization in time and frequency space. Its shape for various scales are shown in Figure B.10. The real and imaginary parts of the Morlet mother wavelet and its FT are also shown in unit scale for $\omega_0 = 5$. Note the absence of the spectrum in the negative portion. The daughter Morlet wavelet has a spread of λ in the time domain, while its FT is centered at ω_0/λ, with a spread of $1/\lambda$.

Strictly speaking, a CWT provides a timescale, not a time-frequency localization. However, for some wavelets such as Morlet wavelets, it is possible to relate the scale to a frequency. Meyers *et al.* (1993) derive such a relationship for Morlet wavelets,

$$\lambda_0 = \frac{1}{2\omega_0}\left[\omega_0 + \left(2 + \omega_0^2\right)^{1/2}\right] \tag{B.25}$$

so that any signal that is a superposition of waves with different frequencies ω_{1-N} will have a WT with maximum moduli at scale values τ_{1-N} given by the above equation. The reader is referred to Meyers *et al.* (1993) for application of Morlet wavelets to an oceanographic problem and to Weng and Lau (1994) and Lau and Weng (1995) for application to geophysical time series analysis.

B.1.2.2 Mexican Hat (Maar Wavelet)

This belongs to the class of real-valued wavelets. One such class corresponds to the mth derivative of the Gaussian (DOG) function. DOG and its Fourier transform are

$$\psi(t) = \frac{(-1)^{m+1}}{\sqrt{\Gamma(m + (1/2))}}\pi^{-1/4}\frac{d^m}{dt^m}\left(e^{-t^2/2}\right)$$

$$S_\psi(\omega) = \frac{(i\omega)^m}{\sqrt{\Gamma(m + (1/2))}}\pi^{-1/4}\left(e^{-\omega^2/2}\right) \tag{B.26}$$

The larger the value of m, the farther apart the two peaks in frequency of the transform are spaced. Maar wavelet is the second derivative of the Gaussian function, DOG (m=2),

$$\psi(t) = \frac{2}{\sqrt{3}}\pi^{-1/4}(1 - t^2)e^{-t^2/2} \tag{B.27}$$

with applications to edge detection. The wavelet coefficients derived using this wavelet can be interpreted to yield the second derivative or the curvature of the signal at scale λ. Thus zero values correspond to inflection points, or zero second derivatives, whereas high values indicate strong curvature. Discontinuities are then characterized by two local increases in the modulus mediated by a steep zero crossing (Liandrat and Moret-Bailly, 1990).

A variant of the Mexican hat called the French hat wavelet is piece-wise continuous and similar except for its boxy shape in both the positive and negative lobes.

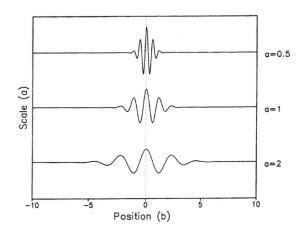

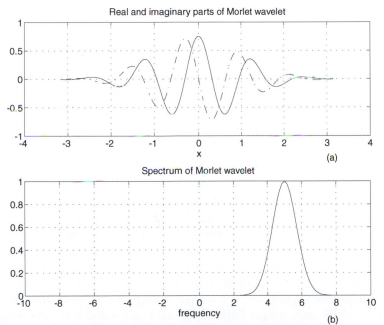

Figure B.10 Morlet wavelet of different scales (top, from Lau and Weng, 1995), its real and imaginary parts, and its Fourier transform (bottom, from Kumar and Foufoula-Georgiou, 1994).

B.1.2.3 Paul Wavelet

This is another complex-valued wavelet with use similar to Morlet wavelet:

$$\psi(t) = \frac{2^m i^m m!}{\sqrt{\pi(2m)!}} \pi^{-1/4} (1-it)^{-(m+1)}$$

$$S_\psi(\omega) = \frac{2^m}{\sqrt{m(2m-1)!}} \pi^{-1/4} \omega^m e^{-\omega} \qquad (\omega > 0) \qquad\qquad (B.28)$$

$$= 0 \qquad\qquad\qquad (\omega \le 0)$$

   In addition to these traditionally used wavelets, it is also possible to construct one's own, subject to admissibility constraints. For example, a real-valued antisymmetric quadratic spline or a symmetric cubic spline wavelet could be used. For a time series with relatively flat segments and abrupt changes between the segments, Hagelberg and Gamage (1994) have shown that the amplitude of WT is large in flat regions for antisymmetric wavelets thus emphasizing these regions of the time series. Symmetric ones emphasize the transition regions.

   While the choice the wavelet basis requires some care as to the degree of match between the basis and the time series (smooth series are best analyzed by smooth basis), roughly similar basis functions give roughly similar WT, qualitatively speaking. However, the power spectrum may look different, since real valued ones transform positive and negative oscillations in the original series into separate peaks in wavelet power, whereas complex-valued ones tend to blend them.

B.1.3 DISCRETE WAVELET TRANSFORMS (DWT)

   In general, there are two types of wavelets, continuous and discrete. Discrete wavelets allow only discrete values for the translation and dilatation parameters and can be made orthogonal. The orthogonality property of a discrete WT makes it very useful in quantifying the variance contained in any particular frequency range for a particular time interval. It is also excellent for decomposing and reconstituting a signal with minimum basis functions and hence quite useful for data compression applications. If a wavelet is such that the time series behavior at infinity is inconsequential (strictly speaking, the wavelet is zero outside a given finite interval), the wavelet is said to be compactly supported; Daubechies' wavelet is compactly supported and the Haar wavelet is a subset of it. However, most wavelets (all continuous ones) have infinitely supported basis functions, although the time series behavior at infinity exerts only a very small influence on the transform. Haar basis functions (Haar, 1910) dating back well before the rigorous development of WT in the 1980s have been used to characterize fronts (Gamage and Hagelberg, 1993) and boundary layer turbulence (Collinau and Brunet, 1993; Howell and Mahrt, 1994). They have also been used extensively in electrical engineering for data compression applications. However, the Haar basis has a boxcar shape that involves discontinuities and hence leads to poor localization in frequency space. Fortunately, better orthogonal wavelets such as Meyer wavelets (Meyer, 1993a) are available and have been used for analyzing seismic signals and interannual variability in oceanic SST (Mak, 1995). The Meyer wavelet is continuous, easy to use, and involves a high degree of symmetry. There also exist other orthogonal wavelets such as Daubechies' orthogonal wavelets (Daubechies, 1992) and semiorthogonal ones such as spline wavelets (Chui, 1992 a,b).

Orthogonal wavelets are defined by Eqs. (B.17) and (B.20). Any real-valued function f(t) can be expanded therefore in terms of its wavelet series expansion:

$$f(t) = \sum_{m=1}^{M} \sum_{n=1}^{N} \alpha_{m,n} \psi_{m,n}(t) \tag{B.29}$$

where $N = 2^M$ denotes the length of the record in terms of number of points in the time series, and M denotes the number of octaves in the decomposition. The wavelet coefficients are given by

$$\alpha_{m,n} = \int \psi_{m,n}(t) f(t) dt = \frac{1}{2\pi} \int 2^{-m/2} e^{i\omega n} S_\psi^*(2^{-m}\omega) S(\omega) d\omega \tag{B.30}$$

and the integrated wavelet spectrum, the counterpart of Fourier spectrum, is given by

$$\phi_m = \sum_n \alpha_{m,n}^2 \tag{B.31}$$

The summation should really be carried over only the N independent sets, not the MN coefficients obtained from Eq. (B.29), or the coefficient $\alpha_{m,n}$ should be multiplied by $\sqrt{2^{m-M}}$ to obtain the true spectral values. The spectrum consisting of these weighted coefficients is referred to as the weighted spectrum.

B.1.3.1 Haar Wavelets

The Haar wavelet is the simplest orthogonal discrete wavelet and is shaped like a boxcar:

$$\psi(t) = \begin{cases} 1 & 0 \le t \le 1/2 \\ -1 & 1/2 \le t \le 1 \\ 0 & \text{elsewhere} \end{cases} \tag{B.32}$$

Weng and Lau (1994) describe an excellent application of Haar wavelets to the analysis of period doubling in a chaotic dynamical system. The principal problem encountered in its use for geophysical signal analysis is that in geophysics signals contain continuous phases and scales, not dyadic ones (multiples of 2). Thus the scale decomposition provided by Haar wavelets can be distorted. See Weng and Lau (1994) for an illustration of this problem in application to satellite IR radiance data and comparison to Morlet WT. While the complex-valued Morlet wavelet provides both amplitude and phase information, the real-valued Haar wavelet provides only information on amplitude with possible aliasing problems. Fortunately, better orthogonal wavelets are available for the analysis of geophysical signals.

B.1.3.2 Meyer Wavelets

The Meyer wavelet is an excellent example of an orthogonal discrete wavelet and is widely used in geophysical analysis. It is a continuous function, defined in terms of its FT (Mak, 1995):

$$S_w(\omega) = e^{i\omega/2} \sin\left[\frac{\pi}{2} A\left(\frac{3|\omega|}{2\pi} - 1\right)\right] \qquad \frac{2\pi}{3} \le |\omega| \le \frac{4\pi}{3}$$

$$S_w(\omega) = e^{i\omega/2} \cos\left[\frac{\pi}{2} A\left(\frac{3|\omega|}{4\pi} - 1\right)\right] \qquad \frac{4\pi}{3} \le |\omega| \le \frac{8\pi}{3} \tag{B.33}$$

$$S_w(\omega) = 0 \qquad\qquad\qquad\qquad\qquad \text{otherwise}$$

where

$$A(x) = 0 \qquad\qquad\qquad\qquad\qquad\qquad x \le 0$$

$$A(x) = (35 - 84x + 70x^2 - 20x^3)x^4 \qquad 0 \le x \le 1$$

$$A(x) = 1 \qquad\qquad\qquad\qquad\qquad\qquad x \ge 1$$

with the admissibility condition

$$A(x) + A(1-x) = 1 \tag{B.35}$$

A(x) can be defined differently in the domain 0<x<1, as long as the admissibility condition is satisfied. Figure B.11 shows the structure of this Meyer wavelet for different values of j (k=0) both in temporal and frequency space. From its FT, it is easy to see the zoom-in/zoom-out capability of Meyer wavelets. Mak (1995)

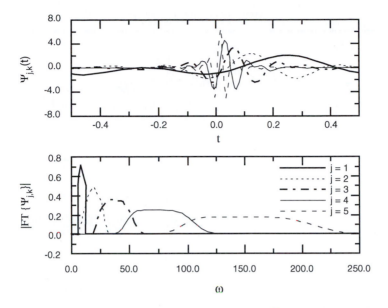

Figure B.11 Meyer wavelet for different values of j. From Mak (1995).

illustrates its application to SST data for analysis of El Niño and La Niña events. Of the 44-year monthly values of SST from 1949 to 1992, he used 2^9 (512) points for deriving the Meyer DWT. Figure B.12 shows the weighted Meyer spectrum for SST in the tropical Central Pacific. Note the prominent features at m = 4 and 5, corresponding to periods of 5.3 and 2.6 years. They are consistent with the occurrence of El Niño/La Niña events at various times in the past 40 years.

While the orthogonality of the DWT used has desirable attributes, this comes at the expense of spectral resolution, since there are only nine octaves. Figure B.13 shows the integrated Meyer wavelet spectrum, which highlights this feature. Most of the variance (56%) is contained at m = 4 and 5, the El Niño timescales, whereas the longer timescales and the annual/semiannual ones contain far less. Subjecting the series to a FT yields the Fourier spectrum shown in Figure B.14, plotted in variance-preserving linear ordinate against a logarithm scale in frequency as the abscissa. It is clear that the Fourier spectrum has a far better spectral resolution, with prominent peaks at 2 and 4.5 years. Therefore the DWT and the FT complement each other in analyses of geophysical signals. When the event is too short, however, the FT may have trouble indicating its presence, whereas a DWT would not (see Meyers and O'Brien, 1994, for such an application to detection of 29-day oscillations in the Indian Ocean).

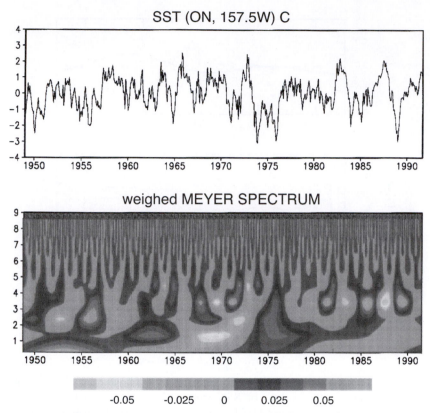

Figure B.12 Weighted Meyer cross-spectrum for SST in the Central Pacific. From Mak (1995).

B.2 EXAMPLES

The following four examples from Lau and Weng (1995) illustrate the utility of Morlet wavelets for the analysis of geophysical signals. Here octave 0 is defined as corresponding to 8 time units, and therefore octave 1 corresponds to 16 and 2 to 32 time units. Only the real part of the Morlet WT is shown, with dimensionless time units t (normalized by T, where T is the period corresponding to the fundamental frequency ω) along the abscissa and octaves along the ordinate.

1. Amplitude-Modulated Simple Wave:

$$f(t) = A \cos(\omega t + a \sin \sigma t) \qquad (B.36)$$

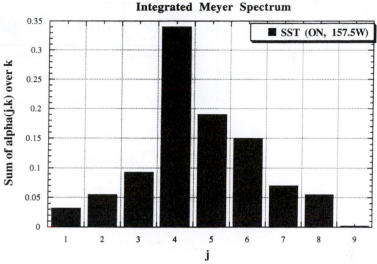

Figure B.13 Integrated Meyer spectrum of the SST in Figure B.12. From Mak (1995).

Figure B.15 shows the time series, its FT, and the real part of its WT, for a wave with a period of 32 time units (octave 2), modulated in amplitude by a wave of longer period of 256 units. The FT shows a peak at the fundamental frequency ω and its side bands $\omega \pm \alpha$, since the signal represents a superposition of three waves with frequencies ω, $\omega + \alpha$, and $\omega - \alpha$. The corresponding WT shows

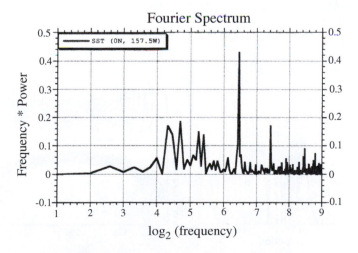

Figure B.14 Fourier transform of the SST in Figure B.12. From Mak (1995).

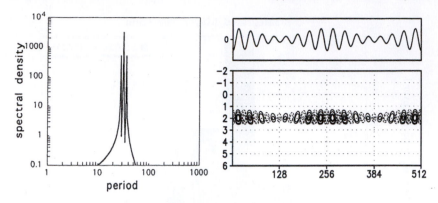

Figure B.15 FT and Morlet WT of an amplitude-modulated signal. From Lau and Weng (1995).

it concentrated along octave 2, with positive and negative phases corresponding to those of the original signal. The modulation corresponds to the amplitude modulation in the original signal and the slight tilt in phase shows the existence of side bands to the fundamental frequency.

2. Frequency-Modulated Simple Wave:

$$f(t) = A\cos(\omega t + a\sin\sigma t) \qquad (B.37)$$

Figure B.16 shows a basic signal with a period of 32 units (octave 2) undergoing frequency modulation with frequency gradually increasing such that the period falls to 8 units (octave 0) at t = 256 and decreasing back to the original value. Since there are no distinct frequencies involved, the cor-

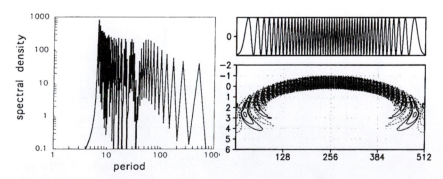

Figure B.16 FT and Morlet WT of a frequency-modulated signal. From Lau and Weng (1995).

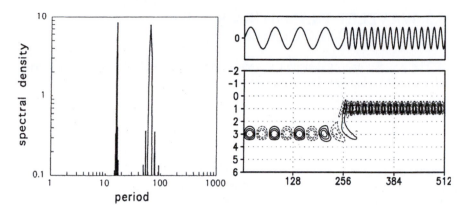

Figure B.17 FT and Morlet WT of an abrupt frequency change in the signal. From Lau and Weng (1995).

responding FT looks like a noisy signal and has no distinct peaks. The WT, however, reflects the original signal and appears arched with values starting near octave 2, spanning octave 0, before falling back to octave 2. This is an excellent illustration of the vastly superior information content of a WT compared to a regular FT.

3. Abrupt Change in the Frequency of a Simple Wave:

Here the signal period changes from 64 units (octave 3) to 8 (octave 1) at t = 256. Such changes are often characteristic of natural systems, whose natural frequency often changes abruptly. The climatic series described in the introduction is a good example. The corresponding FT simply shows two distinct peaks but has no information on when or even if a change occurred. It also contains spurious peaks arising from the use of infinite basis functions for a finite time series. The WT (Figure B.17) on the other hand shows the abrupt change rather well by a shift in scale from octave 3 to octave 1 at t = 256.

4. Abrupt Localized Disturbance:

$$f(t) = A\left[(t-256)/a\right]\exp\left[-(t-256)^2/2a^2\right] \qquad (B.38)$$

Here, a transient disturbance occurs at t = 256. It is a brief disturbance well localized in time. Its FT is quite uninformative, since it consists of a broad spectrum of frequencies needed to represent the abrupt change. The WT (Figure B.18), on the other hand, localizes the change in the time domain, with a distinct cone of influence with converging phase lines. Sudden transient disturbances are another frequent characteristic of natural systems, for example, the climatic perturbation produced by a volcanic eruption or the impact of a comet.

B.3 WAVELET TRANSFORMS AND STOCHASTIC PROCESSES

The wavelet energy density (power) spectrum or simply the wavelet spectrum (also called scalogram) can be written as

$$E_w(m,n) = S_w(m,n)S_w^*(m,n) = \left|S_w(m,n)\right|^2 \tag{B.39}$$

It is the localized energy spectrum of a time series, a generalization of the Fourier spectrum. As Kumar and Foufouls-Georgiou (1994) point out, the utility and popularity of the Fourier transform in the analysis of stochastic (and stationary) processes in a wide variety of fields are mostly due to the fact that the energy density (power) spectrum of a stochastic time series is simply the Fourier transform of the autocovariance function

$$
\begin{array}{ccc}
f(t) & \Leftrightarrow & S(\omega) \\
\downarrow & \downarrow & \\
R(\tau) = \langle f(t)f(t-\tau)\rangle & \Leftrightarrow & E(\omega) = \left|S(\omega)\right|^2
\end{array}
\tag{B.40}
$$

A similar and even more versatile relationship exists for wavelet transforms:

$$
\begin{array}{ccc}
f(t) & \Leftrightarrow & S_w(\lambda,t) \\
\downarrow & \downarrow & \\
R(t,\tau) = \left\langle f(t-\tfrac{\tau}{2})f(t+\tfrac{\tau}{2})\right\rangle & \Rightarrow & E(\lambda,t) = \left|S(\lambda,t)\right|^2
\end{array}
\tag{B.41}
$$

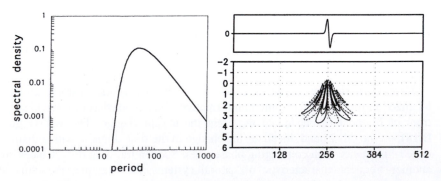

Figure B.18 FT and Morlet WT of a localized disturbance. From Lau and Weng (1995).

The greater utility of WT arises from the fact that this relationship is no longer restricted to stationary stochastic processes. The wavelet energy density spectrum is also known as Wigner-Ville spectrum.

B.4 TWO-DIMENSIONAL WAVELET TRANSFORMS

We have so far dealt with only 1-D time series and 1-D WT. Extension to higher dimensions is straightforward (Kumar and Foufoula-Georgiou, 1994, 1997) but is messy and involves wavelets that are not only dilated and translated but also rotated in higher dimensional space. The principal problem in higher dimensional WT is the difficulty of displaying the results, since the parameter space is three dimensional or higher. We will illustrate the concepts using extension of CWT to two dimensions. This is effected by considering the independent variables as vectors instead of scalars, $t = (t_1, t_2)$, $T = (\tau_1, \tau_2)$:

$$S_w(\lambda, T) = \lambda^{-1} \int_{-\infty}^{\infty} \int_{-\infty}^{\infty} f(T)\psi^*\left(\frac{t-T}{\lambda}\right)dt \qquad (B.42)$$

$$S_w(\lambda, T) = \int_{-\infty}^{\infty} \int_{-\infty}^{\infty} S(w)\left[\lambda\psi^*(\lambda w)\right]e^{iTw}dw \qquad (B.43)$$

$$f(t) = \frac{1}{C_\psi}\int_0^{\infty}\frac{d\lambda}{\lambda^3} \int_{-\infty}^{\infty} \int_{-\infty}^{\infty} S_w(\lambda, T)\psi\left(\frac{t-w}{\lambda}\right)dT \qquad (B.44)$$

The condition of admissibility remains unchanged: compact support or sufficiently fast decay in the higher dimensional space and zero mean over the higher dimensional space. The two-dimensional version of continuous Morlet wavelet is

$$\psi^\theta(t) = \pi^{-1/2}e^{iw_0 \cdot t}e^{-|t|^2/2} \qquad |w_0| \geq 5 \qquad (B.45)$$

Its Fourier transform is

$$S_\psi^\theta(w) = \pi^{-1/2}e^{-|w-w_0|^2/2} \qquad w > 0 \qquad (B.46)$$
$$S_\psi^\theta(w) = 0 \qquad w \leq 0$$

where $w = (\omega_1, \omega_2)$ is a point in the 2-D frequency space and $w_0 = (\omega_1, \omega_2)_0$ is a constant, by choosing which the directional selectivity of the transform can be changed. The superscript θ indicates the wavelet direction:

$$\theta = \tan^{-1}\left(\omega_2 / \omega_1\right)_0 \tag{B.47}$$

Figure B.19 shows the spectrum of the 2-D Morlet wavelet; note that it is not progressive, meaning it is not entirely confined to the positive quadrant. By fixing the scale a and traversing along θ, directional information can be deduced from the WT, and by fixing θ and traversing along a, scale information can be extracted. Thus, more complete information can be extracted from 2-D variables than 2-D Fourier transforms alone. A useful property of the Morlet wavelet is its angular selectivity when extended to higher dimensions, which improves as scale decreases, but at the expense of spatial selectivity.

Directional selectivity can be suppressed if desired using a modified Morlet wavelet, called Halo wavelet because of its shape in the Fourier space:

$$S_\psi^\theta(w) = \pi^{-1/2} e^{-[|w| - w_0]^2 / 2} \tag{B.48}$$

Two-dimensional WTs are ideally suited to analyses of time series of vector quantities such as the horizontal component of currents in the ocean and directional properties of wind waves. They are being used in the study of turbulence and its coherent structures (Farge et al., 1996). See Kumar and

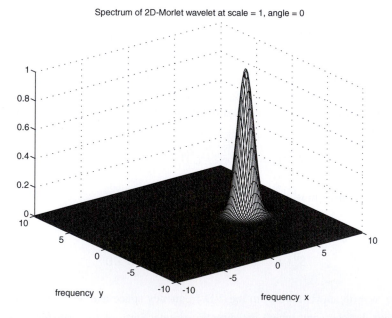

Figure B.19 Two-dimensional Morlet WT. From Kumar and Foufoula-Georgios (1994).

Foufoula-Georgiou (1994, 1997) for a fuller description of 2-D WT and their applications to a variety of problems. The reader is also referred to Ostrovskii (1995) for application of 2-D wavelet transforms of SST in the Sea of Japan to investigate mixing processes in summer and winter.

B.5 CROSS WAVELET TRANSFORMS (CRWT)

Many problems in geophysics involve deciphering and understanding the response of a geophysical system to external forcing, and therefore it is necessary to examine how the forcing and the response are related to each other. Here a WT is useful in characterizing the forcing and response and relating them to each other in the time-frequency space. For example, the response of the ocean basins to fluctuating atmospheric pressure forcing at various frequencies is of particular interest to geophysicists. Cross wavelet transforms are even more useful in such analyses and are very much similar in approach to the cross Fourier transforms in time series analysis (Liu, 1994). In cross Fourier transforms, the cross-correlation between two time series $f_1(t)$ and $f_2(t)$ is transformed into Fourier space. The cross-correlation is normally at zero lag. The Fourier cross-spectra, the co- and quad-spectrum, contain information on how the two time series are related to each other. Once again all information

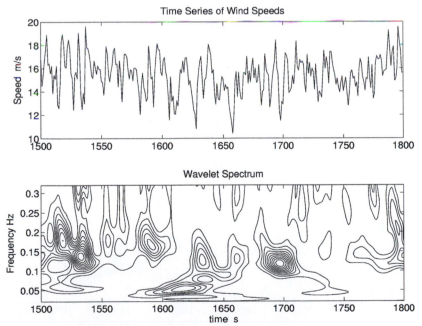

Figure B.20 WT of wind speed time series. From Liu (1994).

content pertaining to time localization of events comprising the two time series is lost. The cross wavelet spectrum provides the means to localize the events in both time and frequency domains.

The complex-valued cross wavelet spectrum of two time series $f_1(t)$ and $f_2(t)$, with corresponding wavelet transforms S_{w1} and S_{w2}, is

$$E_{w_1 w_2}(m,n) = S_{w_1}(m,n) S_{w_2}^*(m,n) \tag{B.49}$$

Its real part is the co-wavelet spectrum and the imaginary part is the quad-wavelet spectrum. The wavelet coherency is also complex valued,

$$\Gamma(m,n) = \frac{E_{w_1 w_2}(m,n)}{\sqrt{E_{w_1}(m,n) E_{w_2}(m,n)}} \tag{B.50}$$

and the square of its modulus is the real-valued wavelet coherence between the two time series. This has information on how the two time series are related to each other, similar to Fourier coherence, but localizes information in temporal space as well. Figures B.20 to B.22 from Liu (1994), who applied continuous cross wavelet transform s to wind wave observations, illustrate this quite well.

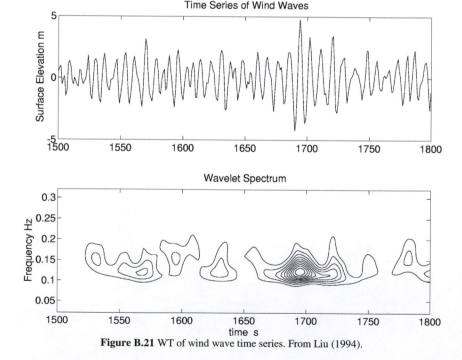

Figure B.21 WT of wind wave time series. From Liu (1994).

Figure B.20 shows the wind speed time series and its wavelet transform, and Figure B.21 shows the corresponding quantities for the resulting wind-driven surface waves. It is clear, both from an examination of the time series and the transforms, that the wind and wave bursts are strongly correlated. However, Figure B.22 quantifies the coherence between the two and localizes it in time-frequency space.

Mak (1995) presents an application of the discrete cross wavelet transform to the analysis of SST in the ocean. Following the same methodology as for DWT, one can define

$$f_1(t) = \sum_{m=1}^{M} \sum_{n=1}^{N} \alpha_{1m,n} \psi_{m,n}(t) \tag{B.51}$$

$$f_2(t) = \sum_{m=1}^{M} \sum_{n=1}^{N} \alpha_{2m,n} \psi_{m,n}(t) \tag{B.52}$$

and makecross-spectrum use of the orthonormality property to define the integrated wavelet

$$\phi_m = \sum_{n} \alpha_{1m,n} \alpha_{2m,n} \tag{B.53}$$

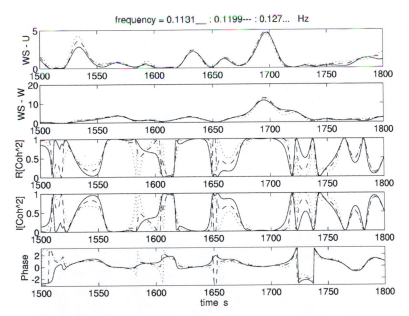

Figure B.22 CWT of wind speed–wind wave time series. From Liu (1994).

which is related to the covariance of the two signals. Note that the scale factor is needed to weight the spectrum because of the redundancy, just as in WT. Note also that since the cross spectrum is related to the covariance of the two stochastic time series, it is possible to derive this at any time lag between the two series. Thus, strictly speaking, the lag is an additional variable, even though most cross-spectra (including Fourier ones) are derived for zero lag. In order to do this, it is necessary to redefine one of the two time series by shifting it in time by the desired lag and deriving a cross spectrum using the modified series.

B.6 ERROR ANALYSIS

Proper quantitative interpretation of WT requires computation of various statistical measures. This is especially important given the wide flexibility in the choice of wavelet basis and the qualitative nature of most comparisons. We present the following from Torrence and Compo (1997), who have provided theoretical wavelet spectra of red and white noise processes so that these can be used to estimate the statistical measures of confidence in the transform results. The discrete FT containing N points, of a red noise process modeled as a first-order autoregressive (Markov) process $x_{m+1} = a\ x_m + z_m$, where a is the autocorrelation and z_m is Gaussian white noise, is

$$P_n = \frac{1-a^2}{1-2a\cos(2\pi n/N)+a^2} \qquad (n=0,1,\,...\,N/2) \qquad (B.54)$$

If the analyzed process consisted of random red noise, a slice of the WT of the process at any given time should resemble Eq. (B.54). Thus any power above the level of the red noise would indicate statistical significance. See Torrence and Compo (1997) for more details.

Appendix C

Empirical Orthogonal Functions and Empirical Normal Modes

Like Fourier transforms and wavelet transforms, Empirical Orthogonal Functions (EOFs) are a means of decomposing a signal into its constituents, in this case into a set of mutually orthogonal components. They are also a form of principal component analysis (PCA) (Preisendorfer, 1988; see also Mitchum, 1993) of data. The method depends on finding the eigenfunctions of the correlation function of the variable. When applied properly to a stochastic variable, it is also called proper orthogonal decomposition (POD; Berkooz *et al.,* 1993; Lumley, 1971). While Fourier transforms use universal global and wavelet transforms, universal local basis functions for decomposition, EOFs use specific basis functions that are unique to that particular signal or data set and therefore do not always apply to a different signal or data set. In fact, often, small changes in the signal or composition of a data set can cause significant changes in the EOFs. EOFs constitute the most optimum means of decomposing (and reconstituting) a signal, made possible by the fact that they are orthogonal to one another. They can also be looked upon as the most efficient way of characterizing a time series or data set. The EOFs are an ordered set in that the first one contains the most variance, the second one the second-most, and so on, so that by truncation at a particular level, it is possible to retain only the most essential components of the signal (and reconstitute it appropriately). Thus they can be used for data compression purposes as well. They are called empirical, since they are derived from the data set itself. They are the most efficient approximation to a data set possible, in the absence of any a priori knowledge of the actual principal components of the data set. In fact, the prime advantage of EOFs is they do not require any a priori knowledge on the pattern of variability. Examples of application of EOFs to oceanic data are Hendricks *et al.* (1996) and Nerem *et al.* (1997).

While the EOF analysis is applicable to the time-dependent spatial fields of a single variable, coupled pattern analyses (CPA) are a means of decomposing the time dependent spatial fields of two variables so that the two resulting spatial patterns have the maximum possible temporal correlation for any particular mode. They are useful for studying the temporal variability of two separate but related geophysical fields (Bretherton *et al.*, 1992; Hsu, 1994; Leuliette, 1998; Peng and Fyfe, 1996; Wallace *et al.*, 1992).

Empirical normal modes (ENMs) are solutions to a homogeneous linear first-order stochastic equation and form the basis of linear statistical prediction methods (Davis, 1976; Penland and Sardeshmukh, 1995). They are also called principal oscillation patterns (POPs).

C.1 EMPIRICAL ORTHOGONAL FUNCTIONS

Obukhov (1947) and Lorenz (1956) were the first to make use of EOF analysis in meteorology. The method has become increasingly popular in recent years and has found a central place in statistical forecasting (Penland and Matrosova, 1994; Penland and Sardeshmukh, 1995) and analysis of observational data (Hendricks *et al.*, 1996; Zou, 1993; Zou and Latif, 1994). It has become an increasingly important analysis tool for both observations and output of numerical models.

Consider a scalar variable $f(x,t)$, defined at a discrete set of M points in physical space x and N points in time. There is no restriction as to the dimension in physical space as long as the values are defined in some order in the spatial domain. Thus, for example, the function could be the spatial distribution of SST or SSH in the ocean (Penland and Sardeshmukh, 1995; Zou, 1993; Zou and Latif, 1994), and hence two-dimensional in physical space. It could even be a function of three-dimensional space as it is when analyzing and synthesizing the characteristics of human facial features. The sequence

$$f(x_i, t_n), i = 1, M; \ n = 1, N \qquad (C.1)$$

can be written also as

$$F = f(x_i, t_n) = \begin{bmatrix} f_{x_1,t_1} & f_{x_2,t_1} & \cdots & f_{x_{M-1},t_1} & f_{x_M,t_1} \\ f_{x_1,t_2} & f_{x_2,t_2} & \cdots & f_{x_{M-1},t_2} & f_{x_M,t_2} \\ \vdots & & & & \vdots \\ f_{x_1,t_{N-1}} & f_{x_2,t_{N-1}} & \cdots & f_{x_{M-1},t_{N-1}} & f_{x_M,t_{N-1}} \\ f_{x_1,t_N} & f_{x_2,t_N} & \cdots & f_{x_{M-1},t_N} & f_{x_M,t_N} \end{bmatrix}$$

This matrix has dimensions $N \times M$. Let $\mathbf{R}$ be its covariance matrix $\mathbf{F^T F}$:

$$\mathbf{R}(x_i, x_j) = \sum_{n=1}^{N} f(x_i, t_n) f(x_j, t_n) \tag{C.2}$$

This temporal covariance matrix is real, symmetric, positive-definite, and of dimension $M \times M$. It is possible to map the physical space represented by x into an orthogonal space represented by the eigenvectors of the covariance matrix $\mathbf{R}$. Let $u_k(x_j)$ be the eigenvectors and λ_k be the corresponding eigenvalues, which form an ordered set of positive values, ordered such that they decrease in value: $\lambda_1 \ge \lambda_2 \ldots \ge \lambda_M$:

$$\sum_{i=1}^{M} R(x_i, x_j) u_k(x_i) = \lambda_k u_k(x_j) \qquad (k = 1, M) \tag{C.3}$$

The eigenvectors $u_k(x_j)$ satisfy the orthogonality condition,

$$\sum_{i=1}^{M} u_k(x_i) u_l(x_i) = \delta_{kl} \sum_{i=1}^{M} u_k^2(x_i) \tag{C.4}$$

and therefore can be regarded as uncorrelated modes of variability in the data set (Davis, 1976). These eigenvectors are called empirical orthogonal functions (EOFs) or empirical orthogonal modes. These form the empirical basis functions for a decomposition of the original data set, which can be written in terms of EOFs as

$$f(x_i, t_n) = \sum_{k=1}^{M} \alpha_k(t_n) u_k(x_i) \tag{C.5}$$

Each mode is analogous to a standing wave whose spatial pattern is given by $u_k(x_i)$ and whose temporal changes are denoted by $\alpha_k(t_n)$. The modal amplitudes are related to the original data set:

$$\alpha_k(t_n) = \sum_{i=1}^{M} f(x_i, t_n) u_k(x_i) \tag{C.6}$$

The most remarkable property of these eigenmode amplitudes is that they are uncorrelated over the data set, that is, uncorrelated in time:

$$\sum_{n=1}^{N} \alpha_k(t_n) \alpha_l(t_n) = \delta_{kl} \lambda_k \tag{C.7}$$

where λ_k is the portion of the total variance of the original signal associated with mode k:

$$\sigma^2 = \sum_{k=1}^{M}\lambda_k = \sum_{i=1}^{M}\sum_{n=1}^{N} f(x_i,t_n)f(x_i,t_n) = \sum_{k=1}^{M}\sum_{n=1}^{N} \alpha_k(t_n)\alpha_k(t_n) \qquad (C.8)$$

They also indicate the time average variance in each of the various modes; their sum is the total variance. Since the amplitude of the mode decreases with mode number, the significance of each mode's contribution to total variance decreases as the mode number increases. Therefore, for real data, only a few leading modes of the possible total of M modes are relevant to the dynamical processes represented in the data set. The higher modes are related to the "noise" in the data set and can often be safely ignored and the original series reconstituted with a truncated set of eigenvectors.

What makes EOFs useful in analyses of large data sets is the property that they are the most efficient way of characterizing the data using a smaller number of parameters than that in the original data set. By this we mean that for any J < M, no series approximation

$$\hat{f}(x_i,t_n) = \sum_{i=1}^{J} \alpha_k(t_n)v_k(x_i) \qquad (C.9)$$

can produce a lower mean square error,

$$\sum_{i=1}^{M}\sum_{n=1}^{N}\left[f(x_i,t_n)-\hat{f}(x_i,t_n)\right]\left[f(x_i,t_n)-\hat{f}(x_i,t_n)\right] \qquad (C.10)$$

than using $v_k(x_i) = u_k(x_i)$. See Davis (1976) for a proof of this statement.

Another advantage of EOF analysis is that it is possible to truncate the series summation at or before the noise level in the data set (if known a priori) since it is possible to derive the mean square error associated with estimating the amplitude of each EOF (Davis, 1976):

$$\varepsilon_k^2 = \lambda_k\left[1-\frac{\lambda_k}{\lambda_k+\varepsilon_n^2}\right] \qquad (C.11)$$

where ε_n^2 is the variance of the uncorrelated noise in the data set. This estimate is quite useful for determining how many EOFs should be retained to represent the data set. When the variance associated with the EOF mode is comparable to the noise variance, then its estimate becomes unreliable, and that mode and all higher modes can be deleted without affecting the signal content of the original data set.

In practice, the truncation can be effected more easily, albeit, approximately by plotting the $\ln(\lambda_k)$ against mode number k. Because of the tendency of the variance in principal components to become exponential for less important components, the curve becomes linear at higher modes and truncation can be effected at the knee of the curve.

Overland and Preisendorfer (1982) point out a Monte Carlo technique for truncating the eigenvalues at the noise level. The principle is simple. One compares each of the normalized eigenvalue: $\bar{\lambda}_k = \lambda_k \left(\sum_{k=1}^{M} \lambda_k \right)^{-1}$ (derived from N data sets) with that derived from a spatially and temporally uncorrelated random process. Monte Carlo technique is used to create a data set randomly drawn from a population of uncorrelated Gaussian variables. A random number generator is used to generate independent sequences of length N for M independent Gaussian variables with zero mean and unit variance, and M eigenvalues of the correlation matrix are computed. This Monte Carlo experiment is repeated 100 times (l =1, 100) and the average value of each eigenvalue is computed over this data set,

$$\bar{\hat{\lambda}}_k^1 = \hat{\lambda}_k^1 \left(\sum_{k=1}^{M} \hat{\lambda}_k^1 \right)^{-1} , \text{ and ordered so that } \bar{\hat{\lambda}}_k^1 \le \bar{\hat{\lambda}}_k^2 \le \cdots \le \bar{\hat{\lambda}}_k^{100} . \text{ Compare } \bar{\lambda}_k \text{ with}$$

$\bar{\hat{\lambda}}_k^{95}$ and terminate the sequence of eigenvalues λ_k at $k = \tilde{k}$ when $\bar{\lambda}_k < \bar{\hat{\lambda}}_k^{95}$. The values of $\bar{\hat{\lambda}}_k^{95}$ are tabulated for different values of M and N (from Overland and Preisendorfer, 1982).

The name principal components to describe EOFs arises from analogy to moment of inertia and stress tensors in a solid object. Both these tensors are, of course, defined in an arbitrary Cartesian coordinate system (for example, the moment of inertia tensor I_{ij}). However, there exists a principal axis about which the moment of inertia of the solid is maximum. Determining this direction requires diagonalizing the matrix I_{ij}. Then, the moment of inertia is maximized about this principal axis, minimized along another.

If one regards each realization in time of the variable $f(x_i,t_n)$ as denoting a single point in M-dimensional physical space, there are a total of N such points in this space and the question is which direction in this space maximizes the variance or the energy (Kundu *et al.*, 1975). The real and symmetric covariance matrix **R**, Eq. (C.2), takes the place of the inertia tensor, and the new set of orthogonal axes are defined by the eigenvectors of this matrix. The transformed matrix is the α_{ij} matrix, Eq. (C.6), which, of course, does not have any off-diagonal terms, Eq. (C.7), and the eigenvalues are the contributions to the total variance. One of the diagonal terms of the diagonalized α_{ij} matrix is maximized and one other is minimized, with the rest reaching stationary values. Since the modes are ordered from maximum to minimum, the first eigenvector extracts the maximum variance out of the data set, the second one out of the "remainder," and so on until the last one extracts the least. The first one is the least square straight-line fit to data in M-dimensional space, the second one to whatever is left and so on. Since each successive EOF mode is efficient at extracting the variance in the remaining data, it follows that expansion in terms of EOFs leads to the fastest convergence in a series expansion of the original variable. This is the principal attractiveness of EOF analysis, since it not only permits examination of the most important parts of variability, but also enables more effective statistical prediction methods to be devised, based on the best features of variability of available data. Of course, truncation using EOF analysis and reconstitution using only the most essential lowest order modes also leads to efficient data compression.

Interpretation of univariate EOF patterns is straightforward. Positive and negative lobes are present in the spatial EOF patterns. In regions with the same sign, temporal variations are in phase; in those with opposite signs they are out of phase. Often two successive EOFs show similar patterns, but they are out of phase in one compared to the other. These indicate oscillating features. Naturally, areas of large amplitude are also regions of large temporal variability.

Although Fourier transforms use universal (trigonometric) basis functions and EOFs those very specific to the data set, there is considerable similarity between the two. In particular, when there is stationarity in the spatial variable x, EOFs are trigonometric functions and the eigenvalues are nothing but Fourier spectral estimates multiplied by the bandwidth (Davis, 1976).

Extension to multivariate EOFs is straightforward, but the interpretation of the resulting EOFs should take into account the fact that there is no orthogonality at times other than at zero lag, because the covariance matrix for the EOF analysis is defined at zero lag. It is, however, possible to derive EOFs at nonzero lag.

Finally, it is worth pointing out that since EOF analysis deals with variability, the data must have the mean removed. However, if the data also contain a dominant cycle such as the seasonal or diurnal cycle, and its characteristics are known a priori, it might be better to least-square fit that cycle and remove it before subjecting the data to EOF analysis so as to highlight the other modes of variability. Otherwise, important variability such as interannual changes might look like noise. However, in practice, because of the imperfectness associated with defining such a cycle, it may be better to let the EOF analysis least-square fit the signal out. Also there are questions regarding whether the matrix should be a covariance matrix or a correlation matrix. The use of the latter implies normalization. In a multivariate analysis, it is prudent to normalize, to prevent biasing the results toward the variable with largest variance. In a univariate situation, it depends on the variable being considered . For a nonconserved

Mode 1 Spatial Field

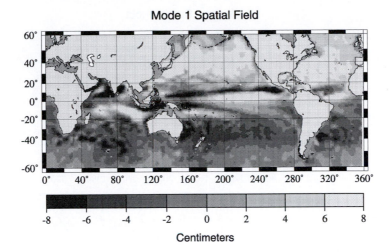

Centimeters

Mode 1 Amplitude Time Series (Normalized)

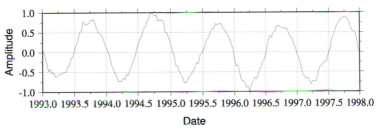

Date

Figure C.1 The first EOF mode from six years of TOPEX/Poseidon altimetric SSH anomalies. Courtesy of C. Fox.

quantity such as SST, normalization might be fine, but for a globally conserved quantity such as SSH, normalization may not ensure certain conservation properties.

In practice, it is much simpler to derive EOF modes and their amplitudes by singular value decomposition (SVD) of the original matrix $\mathbf{F}$ itself (Eq. C.1), rather than dealing with its covariance matrix $\mathbf{R}$ (Eq. C.2). SVD decomposes the matrix $\mathbf{F}$ such that $\mathbf{F}=\mathbf{U}\cdot\mathbf{S}\cdot\mathbf{V}$, where matrix $\mathbf{U}$ (dimension NxM) contains the eigenvectors and $\mathbf{S}\cdot\mathbf{V}^{\mathbf{T}}$ contains their amplitudes, where $\mathbf{S}$ and $\mathbf{V}$ are square matrices with dimension (MxM) . $\mathbf{S}$ is a diagonal matrix, with $S_{n,n}$ corresponding to the nth EOF mode, so that the diagonal elements of $\mathbf{S}\cdot\mathbf{S}^{\mathbf{T}}$, when normalized by the trace of $\mathbf{S}\cdot\mathbf{S}^{\mathbf{T}}$, denote the fractional variance in each mode. Standard, optimized routines are available to do this [see Press *et al.*, 1992 for details and the code; MATLAB instruction to do SVD is simply [U,S,V] = SVD (F)].

Mode 2 Spatial Field

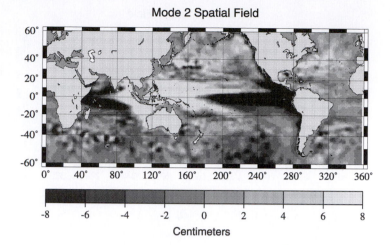

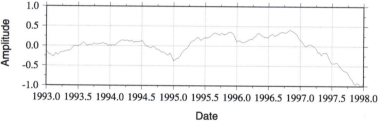

Figure C.2 The second EOF mode. Courtesy of C. Fox.

Figures C.1 to C.4 show the first four EOF modes from the analysis of six years of TOPEX/Poseidon SSH anomaly data (see Hendricks *et al.*, 1996 for an earlier analysis with two years of data). These plots are remarkable for the variability they capture. Modes 1, 3, and 4 show the seasonal cycle that principally consists of the North-South hemisphere oscillations. Mode 2 shows the prominent interannual variability; the ENSO-related oscillation in the equatorial Pacific is quite apparent.

Figure C.5 shows the amplitudes of the first seven EOF modes from nine year-long run (from 1987 to 1995) of a 1/5° resolution numerical model of the Gulf of Mexico (Choi *et al.*, 1995) driven by synoptic winds from FNMOC and running free with no assimilation of data. They were obtained using the 657 snapshots of the SSH fields from the model, archived at 5-day intervals. These modes were obtained using the SVD decomposition of the time series of the SSH fields. Of the possible 657 modes, the first and second EOF modes contain

Mode 3 Spatial Field

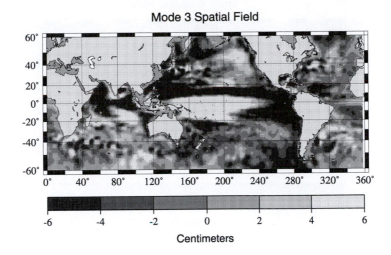

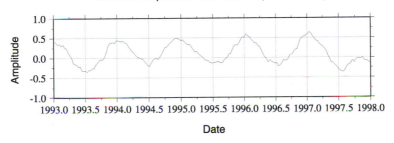

Figure C.3 The third EOF mode. Courtesy of C. Fox.

about 60% of the variance. They can be identified with the quasi-periodic shedding of Loop Current eddies, an important dynamical process in the Gulf, indicating that the Loop Current eddy shedding process accounts for a majority of the variability in the Gulf of Mexico model. The peaks in the second mode time series are associated with eddy shedding events. The eigenvectors or the spatial patterns corresponding to modes 1 and 2 are shown in Figure C.6.

Of the first two EOF modes derived from 40 years of 3-month running mean SST data in the tropical Pacific (Penland and Sardeshmukh, 1995), the first mode shows the characteristic ENSO-like east-west oscillation. The appearance of warm water off the coast of Peru during an ENSO event is depicted quite well. The first EOF accounts for 31% of the variance and the second for more than 7%. Figure C.7 shows the first two EOFs. Compare these with Figure C.16 that shows the first three EOF modes derived from 3 years of 4-month low pass filtered SSH anomalies in the tropical Pacific derived from the Geosat altimeter

Mode 4 Spatial Field

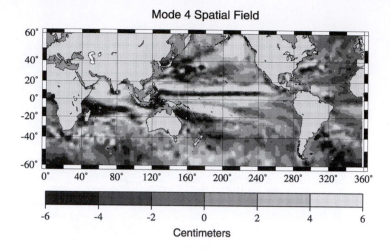

Mode 4 Amplitude Time Series (Normalized)

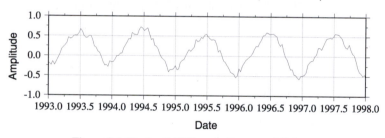

Figure C.4 The fourth EOF mode. Courtesy of C. Fox.

(Zou, 1993). Notice the remarkable resemblance between the first two EOF modes in Figures C.7 and C.16, in spite of the brevity of the Geosat data record compared to SST data. The first EOF mode accounts for 36% of the variance, the second mode accounts for 21%, and the third mode accounts for 13%. The characteristic east-west oscillation pattern associated with ENSO is captured in the short Geosat record as well.

C.1.1 COMPLEX EOFs

Extension to vector fields, such as two-dimensional currents, is straightforward (Kundu and Allen, 1976). Instead of dealing with a real-valued function, we need to consider a complex-valued function:

$$f(x_i, t_n) = f_1(x_i, t_n) + i f_2(x_i, t_n) \tag{C.12}$$

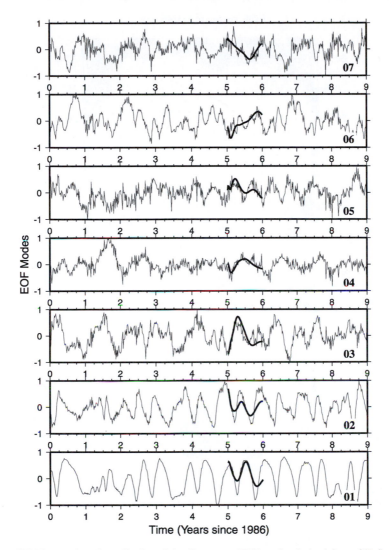

Figure C.5 Time series of amplitudes of the first seven EOF modes derived from 657 five-day snapshots of the SSH fields from a nine year-long run (1987–1995) of the 1/5 degree Gulf of Mexico model. The first seven modes capture nearly 90% of the variance. Time series predicted using statistical methods (see Section C.3) is shown by a thick line.

The covariance matrix R is now complex, but Hermitian,

$$R(x_i, x_j) = \frac{1}{N} \sum_{n=1}^{N} f^*(x_i, t_n) f(x_j, t_n) \qquad (C.13)$$

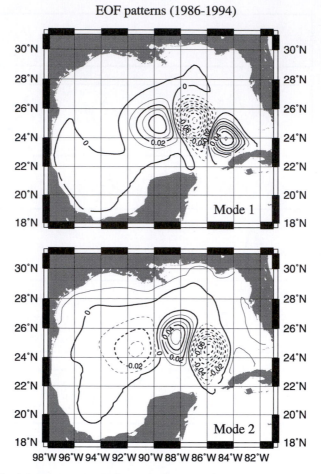

Figure C.6 Spatial patterns corresponding to the first two EOF modes. The peaks of the amplitude time series of the second mode correspond to the Loop Current eddy shedding events in the Gulf.

$$R(x_i, x_j) = R^*(x_j, x_i)$$

where asterisks denote the complex conjugate. The eigenvectors $u_k(x_j)$ are now complex-valued, but the corresponding eigenvalues λ_k are still real and form an ordered set of positive values:

$$\sum_{i=1}^{M} R(x_i, x_j) u_k(x_i) = \lambda_k u_k(x_j) \qquad (k=1, M) \qquad (C.14)$$

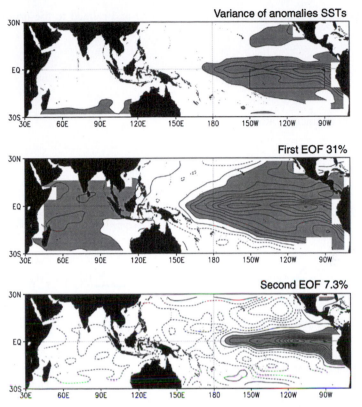

Figure C.7 The first two EOFs of SST in the tropical Pacific. From Penland and Sardeshmukh (1995).

The complex eigenvectors $u_k(x_j)$ satisfy the orthogonality condition:

$$\sum_{i=1}^{M} u_k^*(x_i)u_1(x_i) = \delta_{kl} \sum_{i=1}^{M} \left| u_k(x_i) \right|^2 \tag{C.15}$$

These eigenvectors are called complex empirical orthogonal functions (CEOF) or complex empirical orthogonal modes. These form the empirical complex basis functions for decomposition of the original data set, which can be written in terms of EOFs as

$$f(x_i, t_n) = \sum_{k=1}^{M} \alpha_k(t_n)u_k(x_i) \tag{C.16}$$

The modal amplitudes are now complex-valued and related to the original data set by

$$\alpha_k(t_n) = \sum_{i=1}^{M} f(x_i, t_n) u_k^*(x_i) \tag{C.17}$$

They form an uncorrelated set

$$\sum_{n=1}^{N} \alpha_k^*(t_n) \alpha_l(t_n) = \delta_{kl} \lambda_k \tag{C.18}$$

where λ_k is the portion of the total variance of the original signal associated with mode k:

$$\sigma^2 = \sum_{k=1}^{M} \lambda_k = \sum_{i=1}^{M} \sum_{n=1}^{N} f^*(x_i, t_n) f(x_i, t_n) = \sum_{k=1}^{M} \sum_{n=1}^{N} \alpha_k^*(t_n) \alpha_k(t_n) \tag{C.19}$$

We will not provide any examples of CEOFs; the reader is referred instead to Kundu and Allen (1976) for a typical application.

When applied to the correlation function of a random field, EOFs are known as proper orthogonal decompositions (POD) (Lumley, 1971). This technique is useful in analysis of turbulence (for example, Wilson, 1996). The state vector can consist of any combination of random variables, such as three components of velocity and temperature. We will illustrate it here for a single random variable, say the streamwise component of velocity $v(x_i, t)$. The correlation function is

$$R(x_i, \hat{x}_i) = \overline{v(x_i) v(\hat{x}_i)} \tag{C.20}$$

and the EOFs are defined by

$$\iiint R(x_i, \hat{x}_i)\, u_k(\hat{x}_i) d\hat{x}_i = \lambda_k u_k(x_i)$$
$$\iiint u_l(x_i)\, u_m(x_i) dx_i = \delta_{lm} \tag{C.21}$$

with the latter indicating orthogonality of the modes. For statistically horizontally homogeneous situations, the analysis is usually done in wavenumber space rather than in physical space. Then eigenfunctions become harmonic, since there is no information on spatial location and only on spatial scales. Statistical orthogonality or uncorrelated nature of the eigenmode amplitudes in POD analysis can be used to great advantage in identifying flow structures that are mutually uncorrelated and the reader is referred to Wilson (1996) for details and an example of application to LES simulation of a convective ABL.

EOF analysis can also be applied to time series data to detect oscillations and propagating signals. Instead of a sequence comprising of sampled data sets in time, one uses sequences lagged in time. When used this way, the technique is called singular spectrum analysis (SSA). SSA is well suited to detecting weak, anharmonic oscillations in short noisy time series (Unal and Ghil, 1995; Vautard *et al.*, 1992). When the technique is applied to temporally varying spatial patterns, it is called extended EOF (EEOF) or multichannel SSA (MSSA) and is well suited to detecting propagating and standing oscillatory patterns in a noisy space-time data sequence.

C.1.2 SINGULAR SPECTRUM ANALYSIS

SSA applied to a time series can resolve a weak anharmonic oscillation embedded in short noisy data, even when the variance of the white noise is twice that of the signal. This it does through near-equality of the variances associated with two eigen elements of the lagged covariance matrix of the time series. Unlike classical spectral analysis, which uses predetermined sines and cosines, SSA uses data-adaptive orthogonal basis functions and is therefore more effective. It can detect a nonlinear, anharmonic oscillation by only two EOFs, while a Fourier analysis would require a large number of sine and cosine pairs to represent the same (Unal and Ghil, 1995). It is therefore very useful as a data-adaptive filter to extract low frequency signals from a time series that has higher frequencies and noise superimposed. Unal and Ghil (1995) give an excellent example using sea level data, which has directly forced semiannual and annual cycles superimposed on sampling noise, to extract interannual and interdecadal variability. SSA is very similar to regular EOF (PCA), except that it is applied to a single time series in delay coordinates, rather than to a sequence of distinct time series, indexed by spatial coordinates. The ensuing treatment follows Unal and Ghil (1995).

Consider a time series $\hat{f}_n = \hat{f}(t_n); n = 1, N$ with equal spacing Δt, with zero mean and unit variance. This time series is then embedded in an M-dimensional space by taking N–M+1 consecutive lagged sequences of length M:

$$f(x_i, t_n) = \begin{bmatrix} \hat{f}_1 & \hat{f}_2 & \cdots & \hat{f}_{M-1} & \hat{f}_M \\ \hat{f}_2 & \hat{f}_3 & \cdots & \hat{f}_M & \hat{f}_{M+1} \\ \vdots & & & & \vdots \\ \hat{f}_{N-M} & \hat{f}_{N-M+1} & \cdots & \hat{f}_{N-2} & \hat{f}_{N-1} \\ \hat{f}_{N-M+1} & \hat{f}_{N-M+2} & \cdots & \hat{f}_{N-1} & \hat{f}_N \end{bmatrix} \qquad (C.22)$$

The matrix is (N–M+1) × M in dimension. Once this sequence is formed, the rest follows the usual EOF decomposition methodology (see Eq. C.1). Singular value decomposition of this matrix provides the needed M eigenfunctions and N eigenvectors.

Naturally, the sampling interval Δt and the maximum lag (or equivalently window length) ΔMt form the lower and upper bounds on the timescale of information extracted from the signal and hence require considerable care in selection. Δt should be such that the phenomenon is sampled well and care must be taken to prevent aliasing by prefiltering to remove all components above the Nyquist frequency. M should be chosen based on the timescale of variability, not to exceed N/3 to prevent excessive sampling error in computing covariances at large lags (Unal and Ghil, 1995). The time series $\hat{f}_n = \hat{f}(t_n)$ can be expanded using orthonormal basis u_k:

$$\hat{f}_{n+p-1} = \sum_{k=1}^{M} \alpha_{k,n} u_{k,p}; \ n=1, N-M+1, \ p=1, M \qquad (C.23)$$

where u and α have the same meaning as before; they are the eigenvector (of length M) and its modal amplitude (of length N–M+1), with λ being the variance or eigenvalue associated with each eigenmode or EOF (the modes ordered in decreasing order of variance). Thus the EOFs constitute a moving-average filter of length M, which is data-adaptive! The technique lends itself to selective reconstitution of the original time series by elimination of unwanted components or EOFs. Then if a single element F_{ij} can be written as

$$F_{n,p} = \sum_{k} \alpha_{k,n} u_{k,p}; \ n=1, N-M+1, \ p=1, M \qquad (C.24)$$

where summation over k is only over the EOF modes retained, then the elements of the reconstituted time series are given by

$$\tilde{f}_n = \frac{1}{n} \sum_{p=1}^{n} F(n-p+1, p) \qquad n=1, M-1$$

$$= \frac{1}{M} \sum_{p=1}^{M} F(n-p+1, p) \qquad n=M, N-M+1 \qquad (C.25)$$

$$= \frac{1}{(N-n+1)} \sum_{p=n-N+M}^{M} F(n-p+1, p) \qquad n=N-M+2, N$$

The Monte Carlo test described earlier can be used to test the significance of the The EOFs. Visual inspection of eigenvalue spectrum can also help eliminate

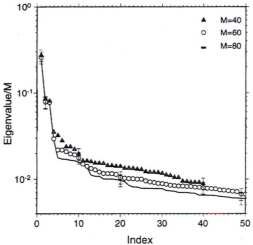

Figure C.8 The first 50 normalized eigenvalues for three different window lengths. From Unal and Ghil (1995).

insignificant modes beyond the cutoff point where the spectral slope becomes nearly horizontal.

Unal and Ghil (1995) analyze the monthly sea level data at Brest spanning more than 180 years from 1807 to 1987 (N=2160) using SSA. Figure C.8 shows the first 50 normalized eigenvalues for window widths (in months) of M = 40, 60, and 80. Only the first 19 modes, which capture 66.66% of the variance, are considered significant. Figure C.9 shows the first nine eigenmodes, and Figure C.10 shows the corresponding modal amplitudes for M=60. EOF mode 1 has a nonzero mean and is the data-adaptive running mean. Modes 2 and 3 form an oscillating quadrature pair and correspond to the annual cycle (plotted only for 20 years to show the quadrature relationship). EOF modes 8 and 9 also form an oscillating pair corresponding to the semiannual cycle. Modes 5 and 6 correspond to a less regular 3-year interannual oscillations (also plotted for 20 years only). EOFs 4 and 7 contain low frequency variability. Figure C.11 shows the power spectra of the modal amplitude time series for modes 2, 5, and 9, which show they peak at 3 years, 1 year, and 1/2 year.

C.1.3 EXTENDED EOFs

MSSA or EEOFs, introduced by Weare and Nasstrom (1982) allow systematic identification of standing and propagating oscillatory patterns in terms of their spatial and temporal behavior over a wide range of timescales (Unal and Ghil, 1995), whereas regular EOFs allow identification of standing oscillatory patterns only. The method is superior to complex PCA, if there are

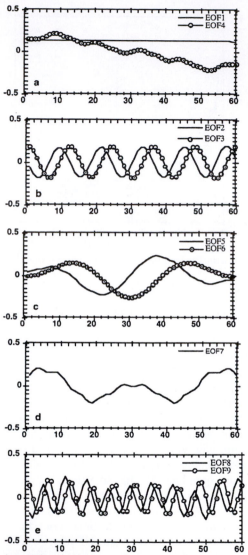

Figure C.9 First nine eigenmodes of 180-year sea level series at Brest. From Unal and Ghil (1995).

several waves propagating with irregular forms. EEOFs are better at identifying a larger number of oscillatory patterns than EOFs (POPs) and according to Unal and Ghil (1995), they have been shown to be superior in analyzing interannual signals in the tropical Pacific. As in SSA, if two consecutive eigenvalues are nearly equal and the corresponding eigenvectors are in (space- and time-) quadrature, they form an oscillation, with the spatial pattern described by

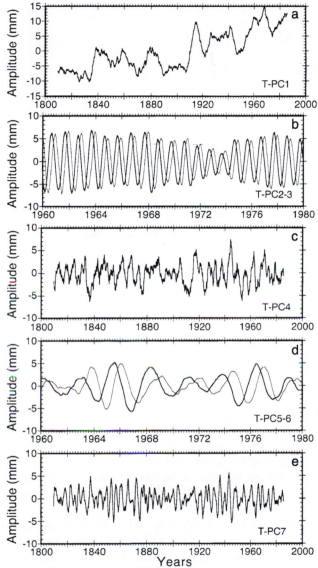

Figure C.10 Modal amplitudes for the nine eigenmodes shown in Figure C.9. From Unal and Ghil (1995).

alternation of the two associated EOFs (Unal and Ghil, 1995). Similar to SSA, the reconstituted time series has elements given by

$$F = F_{n,p} = \sum_{k}^{M \cdot L} \alpha_{k,n} u_{k,p}; \quad n=1, N-M+1, \ p=1, M \cdot L \tag{C.26}$$

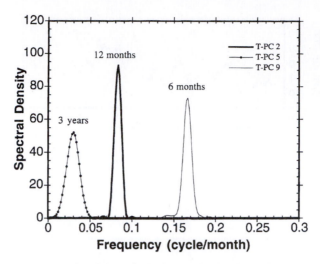

Figure C.11 Power spectra of modal amplitudes 2, 5, and 9 showing peaks at 3 years, 1 year, and 1/2 year. From Unal and Ghil (1995).

where L is the number of data points in space. All lagged sequences for each spatial point are written first, before going to the next spatial point. The M eigenfunctions of the spatio-temporal covariance matrix $\mathbf{F^TF}$, u_k are now spatial patterns (and functions of lag), but the modal amplitudes are functions of time only as in SSA or EOF analysis.

Unal and Ghil (1995) have analyzed sea level data in the western Pacific using EEOFs, and Figures C.12 and C.13 show the sea level patterns reconstructed from low frequency modes of interest to ENSO and the corresponding temporal evolution. ENSO events during 1969, 1972–1973, 1982–1983, and 1986 are manifest in the time series as drops in sea level in the tropical Pacific. Fox (1997) has used EEOFs to extract Rossby wave signals in the TOPEX/Poseidon altimetric data.

A note of caution might be worthwhile. While the EOF analysis is a power tool for analyzing and characterizing the variability in a time series record, as the above examples show, it is a statistical tool and has no dynamical underpinnings. Therefore the interpretation of the meaning of the modes is often difficult. The hope is that the data is representative enough of the dynamical processes embedded in other extraneous "noise" so that the EOFs capture what is essential and relevant to the dynamics. Second, there is some sensitivity to sampling errors (North *et al.,* 1982) that need to be kept in mind and, of course, there is always that need to relate what the EOFs extract to the salient dynamical processes embedded in the time series.

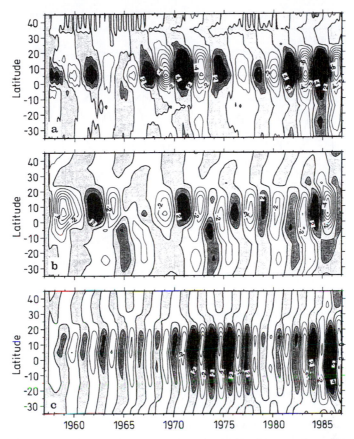

Figure C.12 Reconstructed sea level in the western Pacific using EOFs. From Unal and Ghil (1995).

Finally, it is often necessary to "rotate" the EOFs to obtain meaningful patterns of variability (Enfield and Meyer, 1997; Horel, 1981; Kawamura, 1994; Richman, 1986). The rotated EOFs (REOFs) are also temporally orthogonal (uncorrelated) to one another; however, they do not retain complete orthogonality of the spatial patterns. The most commonly used rotation algorithm is called the varimax-orthogonal rotation (Kawamura, 1994; Richman, 1986), which tries to minimize the number of principal modes that are highly correlated with the original field (Leuliette, 1998). To minimize computational resources needed, the rotation is often accomplished by linear transformations of only a truncated subset of the most dominant EOFs, not the complete set. While the sum of the total variance in the modes so rotated is preserved, there is a redistribution of the variance among the different modes. Kawamura (1994) presents REOF analysis of global monthly SST anomalies (seasonal cycle

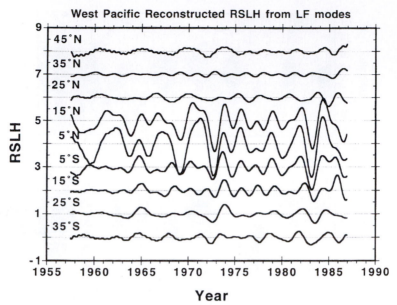

Figure C.13 Corresponding sea level variability showing the various ENSO events. From Unal and Ghil (1995).

removed) from 1955 to 1988. He was able to identify the first REOF as the ENSO mode but without contamination by interdecadal variability, while the remaining modes were dominated by interdecadal variability, but without any contamination by the ENSO signal. This is a definite advantage when compared to unrotated EOFs, where the first mode is the ENSO mode, but it also contains some interdecadal variability. REOFs are also able to identify localized modes confined to certain subdomains: the third and fourth REOFs indicated variability that is confined primarily to the Atlantic. It is more difficult to deduce such localized modes of variability for the whole domain with unrotated EOFs. See Richman (1986) for more details on REOFs.

C.1.4 COUPLED PATTERN ANALYSIS

CPA provides another means of matching patterns of variability of two separate time-dependent fields such as the SST and the SSH (Leuliette, 1998, for example), provided the processes underlying the two are simply and directly related. In essence, they are a means of deducing matched pairs of spatial patterns, one for each field, that are highly correlated temporally. This property makes them preferable to deriving the EOF modes of each of the two fields separately and comparing them mode by mode (for example, Nerem *et al.,* 1997), since the purely statistical nature of the analysis and the empirical nature of the basis functions for the two fields means that the EOF modes of the two fields may not be necessarily related to each other. CPA has been used in analyzing climate data (Bretherten *et al.,* 1992), and matching the patterns of variability of SST and atmospheric pressure (Peng and Fyfe, 1996; Wallace *et al.,* 1992).

CPA analysis is similar to EOF analysis (Leuliette, 1998) in that the two fields are decomposed into orthonormal modes:

$$f(x_i, t_n) = \sum_{k=1}^{M} \alpha_k(t_n) u_k(x_i)$$

$$g(x_i, t_n) = \sum_{k=1}^{M} \beta_k(t_n) v_k(x_i)$$

(C.27)

However, while the eigenvectors of each field form an orthonormal set

$$\sum_{i=1}^{M} u_k(x_i) u_l(x_i) = \delta_{kl} \sum_{i=1}^{M} u_k^2(x_i)$$

$$\sum_{i=1}^{M} v_k(x_i) v_l(x_i) = \delta_{kl} \sum_{i=1}^{M} v_k^2(x_i)$$

(C.28)

the eigenvalues do not (unlike in EOF analysis):

$$\sum_{n=1}^{N} \alpha_k(t_n) \alpha_l(t_n) \neq \delta_{kl} \lambda_k$$

$$\sum_{n=1}^{N} \beta_k(t_n) \beta_l(t_n) \neq \delta_{kl} \pi_k$$

(C.29)

Instead, the eigenvalues (or the CPA expansion coefficients) of one field are orthogonal to those of the other:

$$\sum_{n=1}^{N} \alpha_k(t_n) \beta_l(t_n) \neq \delta_{kl} \sigma_k$$

(C.30)

If the cross-covariance matrix is defined,

$$\hat{R}(x_i, x_j) = \frac{1}{N} \sum_{n=1}^{N} f(x_i, t_n) g(x_j, t_n)$$

(C.31)

$u_k(x_i)$ is the eigenvector of $\hat{\mathbf{R}}\hat{\mathbf{R}}^T$ and $v_k(x_i)$ is the eigenvector of $\hat{\mathbf{R}}^T\hat{\mathbf{R}}$. Leuliette (1998) has been successful in applying CPA to global SSH fields from TOPEX/Poseidon altimeter and SST fields from NCEP analyses (see the above-cited references for more details on the technique).

C.2 EMPIRICAL NORMAL MODES

Empirical normal modes (ENMs) form the basis of linear statistical prediction methods (Davis, 1976; Penland and Sardeshmukh, 1995; Zou, 1993), which belong to the class of inverse models in which the properties of the dynamical system are extracted from observations instead of being derived from dynamical equations.

The simplest possible model for a stochastic system is a first-order multivariate linear Markov process governed by the following linear autoregressive stochastic equation where the forcing is by Gaussian white noise:

$$\frac{dy}{dt} = \mathbf{B}y + \xi \qquad (C.32)$$

where y(t) is the state vector that characterizes the system and ξ is the Gaussian white noise forcing. Matrix $\mathbf{B}$ characterizes the system and can be determined from observations. It is real but nonsymmetric in general. If the relationship in Eq. (C.32) is valid, then given the state vector y(t) at any time t, the most probable vector $\hat{y}(t+\tau_0)$ at time $t+\tau_0$ is given by (Penland and Sardeshmukh, 1995)

$$\hat{y}(t+\tau_0) = e^{\mathbf{B}t_0}y(t) \qquad (C.33)$$

provided the matrix $\mathbf{B}$ is known. Since the system is Gaussian, $\mathbf{B}$ can be estimated from the observed state vector y(t) by an error minimization process:

$$\mathbf{B} = \frac{1}{\tau_0}\ln\left(\frac{C(\tau_0)}{C(0)}\right) \qquad (C.34)$$

where $C(\tau_0)$ and $\mathbf{C}(0)$ are covariance matrices of the state vector at lag τ_0 and zero lag:

$$C(\tau_0) = \left\langle y(t+\tau_0)y^T(t)\right\rangle \qquad (C.35)$$

$$C(0) = \left\langle y(t)y^T(t)\right\rangle \qquad (C.36)$$

where T denotes the transpose. The angular brackets indicate ensemble averages, but through ergodicity, time averages can be used for a stationary time series. The eigenvalues of matrix **B** are complex-valued but have negative real parts, meaning that the modes decay. Also, the EOFs are eigenfunctions of **C**(0). Now the state vector can be predicted at any arbitrary lead time τ by

$$\hat{y}(t+\tau) = e^{B\tau}y(t) = G(\tau)y(t) \tag{C.37}$$

The predicted behavior of the system depends critically on the characteristics of the homogeneous version of Eq. (C.32), that is, without the white noise forcing term:

$$\frac{dy}{dt} = \mathbf{B}y \tag{C.38}$$

Empirical normal modes (ENMs) are solutions of this equation. They are also called POPs, but they are not always oscillatory and their ordering in terms of their damping rate does not necessarily reflect their true importance. As Penland and Sardeshmukh (1995) show, it is not just the behavior of one ENM, which is always a decaying mode, but the constructive interference between several ENMs that can lead to growth in the system.

The ENMs can also be viewed as the eigenmodes of the matrix $A = C(\tau_0)/C(0) = e^{Bt_0}$ [EOFs are eigenmodes of the covariance matrix **C**(0)]:

$$\sum_{i=1}^{M} A(x_i, x_j)v'_k(x_i) = \Lambda'_k v'_k(x_i) \quad (k=1,M) \tag{C.39}$$

The eigenvalues $\Lambda_k(t)$ can be real (positive or negative) but are in general complex. In this form it is easier to understand the relationships between EOFs and ENMs. Since EOFs are based on autocovariance, they codify the current state of the state vector, whereas ENMs based on normalized covariance in time codify the behavior in temporal space and can form the basis of a linear statistical prediction method.

If u_k is the kth complex-valued eigenvector of the real matrix **B**:

$$\sum_{i=1}^{M} B(x_i, x_j)v_k(x_i) = \Lambda_k v_k(x_i) \quad (k=1,M) \tag{C.40}$$

the state vector y can be expressed as

$$y(x_i, t_n) = \sum_{i=1}^{M} \beta_k(t_n)v_k(x_i) \tag{C.41}$$

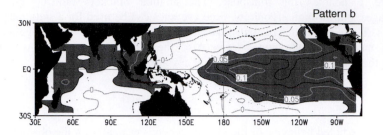

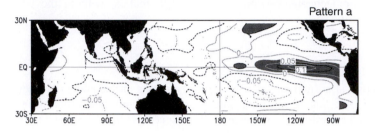

Figure C.14 Mode 6/7 ENMs derived from monthly SST data. From Penland and Sardeshmukh (1995).

where β_k is the corresponding complex-valued amplitude of the eigenmode. ENMs occur in complex-conjugate pairs (and single real-valued ones) so that for each such pair,

$$y_k(t) \approx e^{\Lambda_k^r t}\left[v_k^r(x)\cos(\Lambda_k^i t) - v_k^i(x)\sin(\Lambda_k^i t) \right] \qquad (C.42)$$

Note that Λ_k^r is negative so that this describes a single damped oscillatory mode:

$$y_k(t) \approx \left[a(t)v_k^r(x) + b(t)v_k^i(x) \right] \qquad (C.43)$$

describing oscillation between the real and imaginary parts of the ENM complex conjugate pair. The effect is an oscillation expressed by

$$v_k^r \to -v_k^i \to -v_k^r \to v_k^i \to v_k^r \qquad (C.44)$$

where each state represented above is visited for a quarter of the period of the oscillation. The terms a(t) and b(t) are, of course, damped sines and cosines with

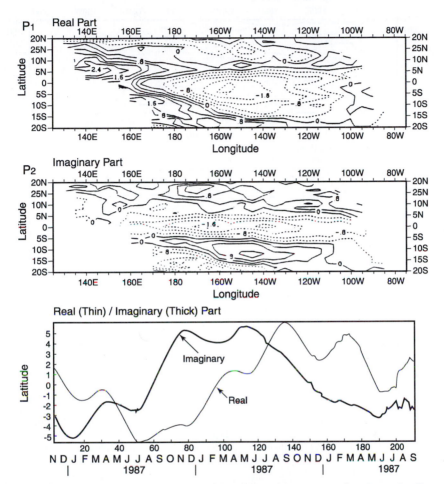

Figure C.15 First POP pair derived from GEOSAT data and the corresponding time series. From Zou (1993).

a phase difference of 90^0 between them. Also, each ENM is orthogonal to every other so that ENMs constitute the best possible decomposition of the time behavior embedded in the data and hence form an excellent basis for time series prediction assuming the linear autoregressive Markov model.

Figure C.14 shows the ENM 6/7 derived from the SST time series by Penland and Sardeshmukh (1995), which shows a remarkable resemblance to the first two EOF patterns derived from the same data (Figure C.7). The oscillation period is 46 months and the decay period is 8.5 months. Note that arranged in the order of decay rates, these ENMs are not the slowest-decaying ones, yet these and the ENM pair 8/9 with a period of 25 months and decay time of 7.9

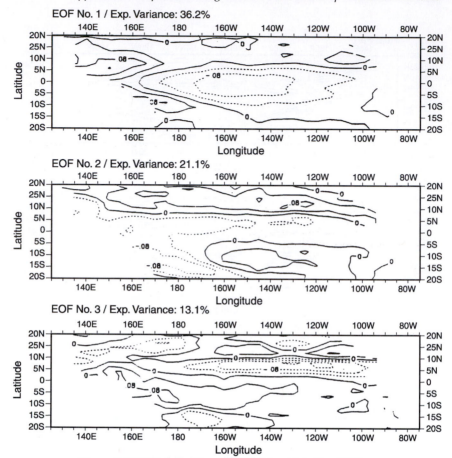

Figure C.16 EOFs derived from GEOSAT data. From Zou (1993).

months, and ENM pair 14/15 with a period of 72 months and a decay rate of 3.7 months are the most relevant to ENSO processes. Therefore unlike EOFs, the order of ENMs do not necessarily reflect their importance in the timewise behavior of the dynamical system.

Figure C.15 shows the dominant ENM pair derived from Geosat SSH anomaly data by Zou (1993). This pair has a period of 3 years and a decay time of roughly the same. It has 25% of the variance. The resemblance of the real-imaginary parts to the first two EOFs (Figure C.16) is striking. Figure C.15 also shows the amplitudes of the resulting oscillation. The exact nature of the physical processes behind the oscillation cannot, of course, be inferred from the ENM analysis; it is hypothesized that the generation, propagation, and reflection of Rossby and Kelvin waves in the equatorial waveguide are responsible for this

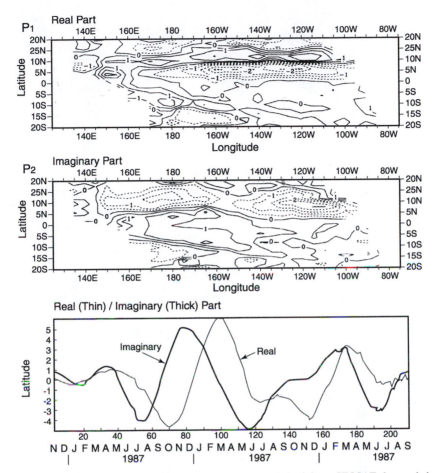

Figure C.17 POP pair corresponding to the annual cycle derived from GEOSAT data and the corresponding time series. From Zou (1993).

kind of oscillation. Figure C.17 shows the corresponding quantities for the next ENM pair. This pair, with a period of 395 days, is associated with the seasonal cycle in the SSH variability.

The utility of statistical prediction is illustrated by Figures C.18 and C.19. Here the 5-day SSH time series from 9 years of Gulf of Mexico model run mentioned in Section C.1 is used as the basis of a linear autoregressive statistical prediction model. These data were used to obtain the EOFs (Figures C.5 and C.6) and ENMs, and these formed the basis of prediction starting from the 8th year. Figure C.5 also shows the first 7 predicted EOF modes. Figures C.18 and C.19 show the comparison between the predicted SSH field (statistical

SSH (meter) on January 29, 1991

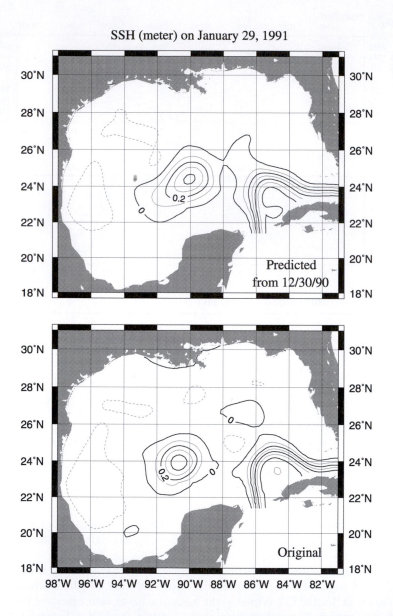

Figure C.18 Predicted SSH field a month into the forecast reconstituted from the modal amplitudes, and their corresponding spatial patterns, of the first 40 EOF modes (top) compared to the original SSH field on that day (bottom).

SSH (meter) on June 28, 1991

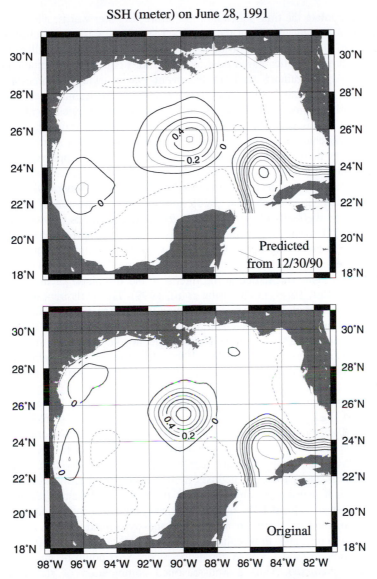

Figure C.19 Predicted SSH field 6 months into the forecast reconstituted from the modal amplitudes, and their corresponding spatial patterns, of the first 40 EOF modes (top) compared to the original SSH field on that day (bottom).

prediction) and the original SSH distribution 30 and 180 days from the start of the prediction. The remarkable similarity between the two is noteworthy.

The example cited above illustrates the best possible scenario, and therefore, certain caveats are in order. Normally, methods such as this are useful when used to extrapolate the time series into the future, in other words, when used for forecasting future events. Then the predicted time series is not a subset of the "training set" time series used to derive the EOFs and ENMs. Needless to say, the skill of these statistics-based prediction methods depends very much on how representative the training set is of the event one is trying to forecast. If a signature of the event being forecast is present in the training set, then the method exhibits some skill. If not, it fails. This is the Achiles' heel of these methods and this is why, the longer the time series used in training, the better, since then the chances of adequate and meaningful statistical sampling of the type of events being forecast is more likely. In real world applications, more often than not, the available time series is too short. Nevertheless, statistical prediction methods such as this (and those based on neural networks) are a very useful complement to dynamical prediction methods. Barnston *et al.* (1999) have evaluated the skills of statistical and dynamical methods in predicting the 1997-98 ENSO event. While the ENM-based statistical prediction had been quite skillful in predicting earlier ENSO events, it did not fare well in predicting the 1997-98 event. In fact, almost all the prediction methods failed in one way or other, when it came to predicting the intensity and the phase of the onset and demise of the 1997-98 event (Barnston *et al.*, 1999). This is thought to be due to the absence of skill in dynamical models when it comes to reproducing faithfully the intra-seasonal variability (such as Madden-Julian oscillations in the tropical waveguide), and the absence of such high frequency signals in the training sets used in statistical models (McPhaden, 1999). Comprehensive dynamical ocean models did perform somewhat better than the simple ones (Webster and Palmer, 1998).

While statistical prediction methods have so far been used in only a stand-alone mode, their use in conjunction with dynamical models may be more promising. Dynamics-based forecast methods are known to have only a short-range (a few weeks in the oceans) predictive capability, due to uncontrollable error growth brought on by the nonlinearity of the dynamical system. When and where statistics-based forecast methods demonstrate useful medium/long range forecast skill, it may be possible to utilize their predicted fields (reconstituted from the most significant EOF modes) to slowly nudge a dynamical model to the state indicated by them, so as to keep the dynamical system around the "correct" attractor, and thus enhance the forecast range and utility of the dynamical forecast model. For example, the model SSH could be nudged gently to the SSH predicted. Thus a well-reasoned combination of statistical and dynamical prediction methods may be better than either by itself, although this idea remains to be explored further.

Appendix D

Units and Constants

D.1 USEFUL QUANTITIES

D.1.1 SI (INTERNATIONAL SYSTEM OF UNITS) UNITS AND CONVENTIONS

The basic SI units for our applications are

Length	meter (m)
Mass	kilogram (kg)
Time	second (s)
Temperature	kelvin (K)

From these, units for all other quantities can be derived. The derived units are

$1 \text{ rad} = 1 \text{ m m}^{-1}$	angle
$1 \text{ Hz} = 1 \text{ s}^{-1}$	frequency
$1 \text{ N} = 1 \text{ kg m s}^{-2}$	force
$1 \text{ Pa} = 1 \text{ N m}^{-2}$	pressure (also stress)
$1 \text{ J} = 1 \text{ N m} = 1 \text{ kg m}^2 \text{ s}^{-2}$	energy (also work, heat)
$1 \text{ W} = 1 \text{ J s}^{-1} = 1 \text{ kg m}^2 \text{ s}^{-3}$	power (also heat flux)

Prefixes

10^{-1}	deci	(d)	10^{1}	deka	(da)
10^{-2}	centi	(c)	10^{2}	hecto	(h)
10^{-3}	milli	(m)	10^{3}	kilo	(k)
10^{-6}	micro	(u)	10^{6}	mega	(M)
10^{-9}	nano	(n)	10^{9}	giga	(G)
10^{-12}	pico	(p)	10^{12}	tera	(T)
10^{-15}	femto	(f)	10^{15}	peta	(P)
10^{-18}	atto	(a)	10^{18}	exa	(E)
10^{-21}	zepto	(z)	10^{21}	zetta	(Z)
10^{-24}	yocto	(y)	10^{24}	yotta	(Y)

Symbols for physical quantities are italicized, but units are not (for example, temperature T in degrees Celcius °C). Unit symbols are lowercase, unless

851

derived from proper names (for example, 1 watt is 1 W, 1 kilogram is 1 kg), not pluralized (for example, 3 seconds is 3 s), and not followed by a period. When prefixes are used, no space is allowed in between (for example, 3 nano-seconds is 3 ns , not 3 n s); compound prefixes are to be avoided (for example, 1 terawatt is 1 TW, not 1 MMW), and when written out fully should begin with a lowercase letter (for example, millijoules). Multiplica-tion of units may be indicated by the use of a space in between, and division by a negative exponent (for example, 1 newton meter is 1 N m and 1 newton per meter squared per second is 1 N m^{-2} s^{-1}). When written out in full, no mathematical operations should be comingled (for example, 1 watt per meters squared, not 1 watt/meter^2), and a product should be denoted by a space in between (for example, 1 newton meter). Large numbers are to be divided by spaces between each 3 digit groups (for example, 237 865 540) and decimal marker is to be always preceded if unpreceded by a zero (for example, 0.8 kg, not .8 kg). Units and the numbers are to be divided by a space (for example, 10 kg, not 10kg). Prefixes are to be used whenever possible to bring the numerical value between 0.1 and 1000 (for example 180 kW, not 180 000 W). The word *degree* or its symbol is not used in conjunction with the unit kelvin (for example, 215 kelvin or 215 K, not 215 degrees kelvin or 215 °K), but only with Celsius (degrees Celsius or °C).

Following the above conventions (Nelson, 1982) avoids needless proofreading corrections by copy editors. Also in the era of electronic self-publishing on the Internet, it is important to adhere strictly to standards that have been painstakingly arrived at after decades of experimentation.

D.1.2 USEFUL CONVERSION FACTORS

1 pound (mass) = 0.453 593 kg
1 inch = 2.54 cm
1 foot = 0.3048 m
1 fathom = 6 ft = 1.8288 m
1 mile = 1.609 344 km
1 nautical mile = 1.85318 km
1 acre = 4047 m^2
1 gal = 3.786 liter
1 bbl = 31.5 gal = 119.26 liter
1 mile hr^{-1} = 0.447041 m s^{-1} = 1.609 344 km hr^{-1}
1 ft s^{-1} = 0.3048 m s^{-1}
1 knot = 0.51477 m s^{-1} = 1.853 172 km hr^{-1}
1 pound (force) = 4.4482 N
1 bar = 10^5 Pa = 10^5 N m^{-2} = 10^6 dyn cm^{-2}

1 psi = 6894.724 Pa
1 horsepower = 746 W
°F = 32.0 + 1.8 × °C
1 solar day = 86400 s
1 sidereal day = 86164 s
1 degree = 0.01745 (= $2\pi/360$) rad
1 Hz = 2π rad s^{-1}
1 cm s^{-1} = 0.864 km day^{-1}
1 N = 10^5 dynes
1 Pa = 1 N m^{-2} = 10 dyn cm^{-2} = 10^{-2} mb
1 mb = 10^2 Pa; 1 m of water ~ 10.1 kPa
1 standard atmosphere = 101.325 kPa (14.7 psi, 76 cm Hg)
1 J = 1 N m = 1 W s = 10^7 ergs = 6.24 × 10^{18} ev = 0.2389 cal
1 W = 1 J s^{-1}
1 petawatt = 10^{15} W
1 ly day^{-1} = 0.484 W m^{-2}
1 W m^{-2} = 4.15 μE m^{-2} s^{-1}
1 metric ton (t) = 1000 kg
1 hectare (ha) = 10^4 m^2
0 °C = 273.15 K

D.1.3 USEFUL UNIVERSAL CONSTANTS:

π = 3.141 592 653 589 793 238 462 643
e = 2.718 281 828 459 045 235 360 287
γ = 0.577 215 664 901 532 860 606 512 (Euler constant)

Universal gas constant R = 8.314 51 J mol^{-1} K^{-1}
Gas constant for dry air = 287.05 J kg^{-1} K^{-1}
Gas constant for water vapor = 461.53 J kg^{-1} K^{-1}
Molecular weight of dry air = 28.97
Molecular weight of water vapor = 18.016
Avagadro constant = 6.02252 × 10^{23} mol^{-1}
Speed of light = 2.9979246 × 10^8 m s^{-1}
Newton's gravitational constant G = 6.672 590 × 10^{-11} N m^2 kg^{-2}
Solar constant = 1376 W m^{-2} (in 200 to 4000 nm range; 40% of it in visible
    400 – 670 nm range, 1.78 × 10^{17} W)
Stefan-Boltzmann constant σ = 5.670 51 × 10^{-8} W m^{-2} K^{-4}
Boltzmann constant k = 1.38 × 10^{-23} J K^{-4}
Planck's constant = 6.6262 × 10^{-34} J s

D.1.4 USEFUL GEODETIC CONSTANTS:

Earth

Radius a = 6371 km (6378.139 km – equatorial, 6356.754 km – polar, hydrostatic flattening – 1/299.638)

Radius of the core (equatorial) = 3486 km (flattening = 1/392.7)

Mass of Earth = 5.977×10^{24} kg (mantle = 4.05×10^{24} kg, core = 1.90×10^{24} kg, crust = 2.5×10^{22} kg)

Average density $\rho_E = 5517$ kg m^{-3}

Average density of the upper mantle = 3330 kg m^{-3}

Average density of the crust = 2850 kg m^{-3}

Polar moment of inertia of the entire Earth C ~ 8.0376×10^{37} kg m^2

Equatorial moment of inertia of the entire Earth A ~ 8.0115×10^{37} kg m^2 (C–A = 2.610×10^{35} kg m^2)

Polar moment of inertia of the core C_c ~ 0.9140×10^{37} kg m^2

Equatorial moment of inertia of the core A_c ~ 0.9117×10^{37} kg m^2 (C_c–A_c = 2.328×10^{34} kg m^2)

Polar moment of inertia of the mantle C_m ~ 7.1236×10^{37} kg m^2

Equatorial moment of inertia of the mantle A_m ~ 7.1000×10^{37} kg m^2 (C_m–A_m = 2.377×10^{34} kg m^2)

Orbital dates: spring (vernal) equinox = March 20, autumnal equinox = September 22, summer solstice = June 20, winter solstice = December 21, perihelion = January 3, aphelion = July 4

Obliquity of Ecliptic = 23° 26'

Rotation rate = 7.292115×10^{-5} rad s^{-1}

Mean orbital velocity = 29.77 km s^{-1}

Chandler wobble period = 433.3 days

Average surface temperature = 288 K

Average albedo = 0.33

Sidereal day = 23 hr 56 m 4.09 s

Sidereal year = 365.25 day

Tropical year = 365.2422 day

Current angular momentum of the Earth-Moon System = 3.5×0^{34} Kg m^2 s^{-1}

Current energy in the Earth-Moon System = 2.5412×10^{29} J

Current energy dissipation rate in the Earth-Moon System = 3.17×10^{12} W

Moon

Mean radius = 1738 km = 0.273 a

Mass = 7.35×10^{22} kg = 0.01235 M_E

Mean density = 3342 kg m^{-3} = 0.606 ρ_E

Mean moment of inertia = 8.665×10^{34} kg m^2

Mean gravity = 1.623 m s^{-2}

Mean orbital velocity = 1.01 km s^{-1}

Inclination to the ecliptic = $0.09°$

Mean rotation rate = 2.662×10^{-6} rad s^{-1}

Mean distance between Earth and Moon = $0.384\ 405 \times 10^6$ km = 60.3 a
  (apogee = 363 263 km, perigee = 405 547 km, eccentricity = 0.0549)

Surface temperature = 120–390 K

Albedo = 0.073

Roache limit for the Moon: R ~ 2.456 a $(\rho_E / \rho_M)^{1/3}$ ~ 2.9 a

Sidereal month = 27.3217 day

Synodic month = 29.5306 day

Anomalistic month = 27.5545 day

Sun

Mean radius = 6.96×10^5 km = 109.245 a

Mass = 1.991×10^{30} kg = 3.33311×10^5 M$_E$

Density = 1410 kg m^{-3} = 0.256 ρ_E

Rotation period of the Sun (15°) = 25.5 day

Orbital velocity = 220 km s^{-1}

Surface temperature = 5785 K

Mean distance between Earth and Sun = 149.5×10^6 km = 23465.7 a
  (aphelion = 147.1×10^6 km, perihelion = 152.1×10^6 km, eccentricity =0.016718)

D.1.5 USEFUL PHYSICAL QUANTITIES

Mean mass of the atmosphere = 5.1352×10^{18} kg,
  (dry air = 5.122×10^{18} kg, water vapor = 1.32×10^{16} kg)

Mass of the ocean =1.35×10^{21} kg

Mass of ice = 2.2×10^{19} kg

Mass of water in lakes and rivers = 5.0×10^{17} kg
  (in sediments = 2×10^{20} kg)

Surface area of Earth = 5.10×10^{14} m^2

Ocean surface area = 3.611×10^{14} m^2 (71%, includes sea ice)

Deep ocean (>1000 m) surface area = 3.05×10^{14} m^2

Land surface area = 1.489×10^{14} m^2

Surface area of sea ice = 0.261×10^{14} m^2 (7% of ocean area) average,
  0.11×10^{14} km^2 in the northern (8–15 $\times 10^6$ km^2, Arctic, sub-Arctic Seas), 0.14×10^{14} km^2 in the Southern hemisphere (4–21 $\times 10^6$ km^2)

Mean ocean depth = 3795 m (~ 4.2 km excluding the shelf)

Mean sea ice thickness = 3 m (Arctic), 0.7 m (Antarctic)

Standard sea level pressure = 101.325 kPa

Mean sea level pressure = 98.44 kPa (dry air = 98.19, water vapor = 0.25 kPa)

Standard sea level temperature = 288.15 K

Typical density of dry air at 20 °C and 100 kPa (1 bar) = 1.210 kg m^{-3}

Typical density of sea water = 1026 kg m^{-3}

Typical density of fresh water = 1000 kg m^{-3}

Typical density of ice = 917 kg m^{-3}

Typical density of snow = 330 kg m^{-3}

Average density of sea water = 1035 kg m^{-3}

Latent heat of vaporization of water at 0 °C = 2.501 ×10^6 J kg^{-1}

Latent heat of vaporization of water at 100 °C = 2.250 × 10^6 J kg^{-1}

Latent heat of sublimation of ice at 0 °C = 2.835 × 10^6 J kg^{-1}

Latent heat of fusion for water at 0 °C = 3.347 × 10^5 J kg^{-1}

Specific heat at constant pressure for dry air = 1005.6 J kg^{-1} K^{-1}

Specific heat of water vapor at 0 °C = 1859 J kg^{-1} K^{-1}

Specific heat at constant pressure for water at 0 °C = 4217.4 J kg^{-1} K^{-1}

Specific heat of ice and snow at 0 °C = 2099 J kg^{-1} K^{-1}

Ratio of specific heats at constant pressure and volume for dry air = 1.4

Kinematic viscosity of sea water at 10 °C, 35 psu = 1.8 × 10^{-6} m^2 s^{-1}

Molecular heat diffusivity for sea water = 1.39 × 10^{-7} m^2 s^{-1}

Molecular salt diffusivity for sea water = 9.0 × 10^{-10} m^2 s^{-1}

Kinematic viscosity of air = 1.53 × 10^{-5} m^2 s^{-1} [1.35 × 10^{-5}+ 10^{-7} (T_a–273.15)]

Heat diffusivity for air = 2.16 × 10^{-5} m^2 s^{-1} [1.9 × 10^{-5}+ 1.26 × 10^{-7} (T_a–273.15)]

Water vapor diffusivity for air = 2.57 × 10^{-5} m^2 s^{-1} [2.26 × 10^{-5}+ 1.51 × 10^{-7} (T_a–273.15)]

Molecular salt diffusivity for sea water = 9.0 × 10^{-10} m^2 s^{-1}

Prandtl number for air = 0.71

Schmidt number for water vapor in air =0.595

Prandtl number for water = 8 to 13

Schmidt number for salt diffusion in water ~ 2000

Typical emissivity of water, ice and snow surfaces = 0.97

Typical albedo of water surface = 0.1

Typical albedo of ice surface = 0.7

Typical albedo of fresh snow = 0.85

Typical thermal conductivity of air = 0.024 W m^{-1} K^{-1} [equal to 0.0238+7.12 × 10^{-5}(T_a–273.15) at temperature T_a in K]

Typical thermal conductivity of water = 0.6 W m^{-1} K^{-1}

Typical thermal conductivity of ice = 2.034 W m^{-1} K^{-1}

Typical thermal conductivity of snow = 0.3097 W m^{-1}K^{-1}

Typical sound speed at sea level in air = 320 m s^{-1}

Typical sound speed at sea level in water = 1500 m s^{-1}

Typical sound speed in ice = 4000 m s^{-1}

Typical volumetric expansion coefficient of sea water for heat ~ 2×10^{-4} K^{-1}

Typical volumetric expansion coefficient of sea water for salinity ~ 8 × 10^{-4} psu^{-1}

Typical surface tension for water = 7.4 × 10^{-5} N m^{-1}

Typical geothermal heat flux at sea bottom = 0.05 W m^{-2} (at midocean spreading centers, this value can be an order of magnitude or more higher)

D.1.6 USEFUL DYNAMICAL QUANTITIES

von Karman constant k = 0.40-0.41

Charnock constant = 0.011

Kolmogoroff constant 1-D (3-D) = 0.50 (1.62)

Batchelor constant 1-D (3-D) = 0.42 (1.36)

Phillips constant = 0.015

Toba constant = 0.11

Acceleration due to gravity g = 9.806 65 m s^{-2} (midlatitude), 9.780 32 (equator) and 9.832 04 (poles) at sea level, 9.797 6 (surface average), but at arbitrary latitude θ and altitude z,

$$g = \left(9.78032 + 5.172 \times 10^{-2} \sin^2 \theta - 6.0 \times 10^{-5} \sin^2 2\theta\right)\left(1 + z/a\right)^{-2}$$

Earth's rotation rate Ω = 7.292 116 × 10^{-5} s^{-1}

1 inertial period (IP) = T /(2 sin θ) ; θ is the latitude; T is the rotation period, 1/2 sidereal day at the poles, 1 sidereal day at 30° latitude

Value of f = 2Ω sin θ: Equatorial–0; midlatitude (30°)–7.29 × 10^{-5} s^{-1}; poles (90°)–1.458 × 10^{-4} s^{-1}

Beta (β) = (2Ω/a) cos θ; a is Earth's radius; θ is the latitude is 2.289 159 × 10^{-11} cos θ m^{-1} s^{-1}: 2.2891 × 10^{-11} m^{-1} s^{-1} at 0°, 2.0 × 10^{-11} m^{-1} s^{-1} at 30°, 1.618 68 × 10^{-11} at 45°, 0 at 90° latitude

Rossby radius of deformation a = C/f for the first baroclinic mode: 330 km at 3°, 100 km at 10°, 40 km at 30°, 20 km at 45°, 5 km at the poles

Equatorial Rossby radius of deformation a = (C/β)$^{1/2}$ for the first baroclinic mode: 330 km, second baroclinic mode: 256 km

Equatorial Rossby radius = 1200 km (atmosphere)

Extra-tropical long Rossby wave speed C = βa^2: 22 cm s^{-1} at 10°, 1.5 cm s^{-1} at 30°

Equatorial Kelvin wave speed = 2.5 m s^{-1} (first baroclinic mode), 1.5 m s^{-1} (second), 30 m s^{-1} (atmosphere)

Equatorial Rossby wave speed: Mode 1 of the first baroclinic mode ~ 0.8 m s^{-1}, Mode 2 ~0.5 m s^{-1}

Baroclinic midlatitude gravity/Kelvin wave speed: First mode ~ 2.5 m s^{-1}, second mode ~ 1.5 m s^{-1}

[*] The values quoted above for some physical properties as typical values are suitable as rough estimates. For more precise values, which are often functions of many other parameters such as temperature, see appropriate tables.

D.2 IMPORTANT SCALES AND QUANTITIES

D.2.1 LENGTH SCALES

Deep convection length scale, $l_{dc} = (B_0/f^3)^{1/2}$, where B_0 is the buoyancy flux, characterizes the size of the convective plumes in rotation-dominated deep convection process. It also denotes the depth beyond which rotational effects become important in the convective layer and the plumes assume a sinewy rope-like shapes.

Ekman scale, $\ell_E \sim \left(\dfrac{u_*}{f}\right)$, characterizes the size of the friction-influenced layer adjacent to a boundary due to an applied shear stress. It sets the upper bound on the depth of the OML and ABL under neutrally stratified and stably stratified conditions. It is often defined as $(2K_M/f)^{1/2}$.

Intermediate geostrophic radius, $a_i = a (R/a)^{1/3}$: where R is the planetary radius and a is the Rossby radius of deformation: characterizes the scale at which the curvature of the planet is felt by fluid motions.

Rhines length scale, $L_z = (U/2\beta)^{1/2}$, where U is the velocity of the jet, is the scale at which nonlinearity effects in a zonal jet become comparable to planetary beta effects. It is the meridional scale of features associated with zonal jets.

Rossby radius of deformation (midlatitude), $a \sim (C/f)$ which characterizes the horizontal scale of response in a rotating fluid, perhaps the most fundamental length scale in midlatitude oceanic and atmospheric dynamics. The most important Rossby radius is that corresponding to the first baroclinic mode. A radius of deformation based on the barotropic shallow water gravity wave speed (~ 200 m s^{-1}) can be defined, $a_e = (gD)^{1/2}/ f$, and this is called the external Rossby radius of deformation or Rossby-Obukhoff radius. For uniformly stratified fluids, with depth D and buoyancy frequency N, there exist a sequence of internal Rossby radii: $a_n = C_n/ f$, where $C_n = ND/(n\pi)$, n = 1, 2, . . .

Rossby radius of deformation (equatorial), $a \sim (2C/\beta)^{1/2}$ which characterizes the horizontal scale of response in a rotating fluid, perhaps the most fundamental length scale in equatorial oceanic and atmospheric dynamics. The most important are those corresponding to the first baroclinic mode ($C \sim 2.4$ m s^{-1}.

Rossby radius of deformation (deep convection), $a \sim (B_0 D^2/f^3)^{1/4}$, where B_0 is the buoyancy flux that characterizes the horizontal scale of the convective patches resulting from deep convection.

Western boundary layer scale: $\delta_S \sim c/\beta$; $\delta_M \sim (A_M/\beta)^{1/3}$; $\delta_I \sim (U/\beta)^{1/2}$ are the Stommel, Munk, and inertial scales, about 100 km.

D.2.2 TIMESCALES

Buoyancy timescale, $T_b = \dfrac{2\pi}{N}$ is the timescale of oscillation of a vertically displaced parcel in a stably stratified fluid.

Rotational timescale, $T_I = 2\pi/f$, the inertial period, characteristic period of inertial oscillations in midlatitudes. The equivalent period in the equatorial waveguide $\sim (\beta C)^{-1/2}$

Baroclinic instability timescale, $T_{bc} = \dfrac{N}{f|\partial U/\partial z|} = \dfrac{Ri^{1/2}}{f}$ is the Eady-Charney timescale for growth of baroclinic instabilities.

Rossby wave dispersion timescale: $T_\beta = 2\pi(\beta a)^{-1}$ is the dispersion timescale for a linear Rossby wave packet with a characteristic radius equal to the Rossby radius.

D.2.3 VELOCITY SCALES

Deep convection velocity scale, $u_{dc} = (B_o/f)^{1/2}$, where B_0 is the buoyancy flux, characterizes the vertical velocity associated with convective plumes in rotation-dominated deep convection process.

(Reduced) Gravity wave speed, $C = (gH_e)^{1/2} = (g'H)^{1/2}$, where g' is the reduced gravity and H_e is the equivalent depth, is the most important velocity scale in the oceans and the atmosphere. Usually the first baroclinic mode is the most dominant.

Rossby drift speed, $V_c = \beta a^2$, is the westward propagation speed of long nondispersive planetary (Rossby) waves. The drift speed of mesoscale features scales with this velocity scale.

Sverdrup velocity scale, $U_I = \dfrac{\pi}{\beta L H} \dfrac{\tau_0}{\rho_0}$ is the scale of the wind stress curl-induced meridional velocity in the basin.

Sverdrup transport scale, $M_I = \dfrac{\pi}{\beta L} \dfrac{\tau_0}{\rho_0}$ is the scale of the wind stress curl-induced meridional transport in the basin.

D.2.4 Nondimensional Quantities

Burger number $Bu = (a/L)^2$ is the square of the ratio of the Rossby radius of deformation to the horizontal scale of the flow.

Rayleigh number, $Ra = \Delta b D^3/(\nu k)$, where Δb is the buoyancy change across the layer and D the layer depth, is the ratio of the destabilizing buoyancy forces to stabilizing viscous forces in free convection.

(Deep convection) Rayleigh number, $Ra = B_0 D^4/(\nu k^2)$, where B_0 is the buoyancy flux and D the convective layer depth, characterizes the Rayleigh number associated with the deep convection process.

Deep convection Rossby number, $R_{dc} = l_{cd}/D = [B_0/(f^3 D^2)]^{1/2}$ where B_0 is the buoyancy flux and D the convective layer depth, denotes the importance of rotation. For small values of R_{dc}, the rotational effects are important.

Ekman number, $E = \left(\dfrac{K_M, A_M}{f L^2} \right)$ is the parameter denoting the importance of frictional layers adjacent to the boundary.

Froude number, $F = \left(\dfrac{U}{NH} \right)$ is the ratio of the timescale associated with stable stratification to the flow timescale.

Nonhydrostaticity number, $N_H = U/(NL)$, is the ratio of the characteristic advection timescale to the buoyancy timescale and determines whether the flow is hydrostatic or nonhydrostatic.

(Flow) Reynolds number, $R_N = \left(\dfrac{UL}{\nu}\right)$ is the ratio of inertial forces to viscous forces in a flow. Low R_N flows are laminar. Transition occurs at a certain value of R_N in a particular sheared flow.

(Western boundary) Reynolds number, $Re = \left(\dfrac{\delta_I}{\delta}\right)^2$ is the ratio of inertial forces to viscous forces in a flow in the western current; much larger than unity.

(Bulk) Richardson number, $Ri_b = \left(\dfrac{D\Delta b}{u_*^2}\right)$ denotes the stability of a buoyancy interface and hence its resistance to turbulent entrainment across it.

(Gradient) Richardson number,

$$Ri_g = \left(\frac{g}{\rho}\frac{\partial\overline{\rho}}{\partial z}\right)\left(\frac{\partial U_i}{\partial z}\frac{\partial U_i}{\partial z}\right)^{-1} = \left(\frac{N}{f_s}\right)^2 = \left(\frac{t_s}{t_b}\right)^2$$ is indicative of static stability of

a stratified sheared flow. The classical Miles-Howard theorem states that the flow is stable if $Ri_g > 0.25$. It can be thought of as the square of the ratio of the timescale associated with mean shear and the buoyancy period.

Rossby number, $Ro = \left(\dfrac{U}{fL}\right)$ characterizes the relative importance of rotation on flow dynamics. For small Rossby numbers typical of large scale geophysical flows, rotational effects dominate. It is also a measure of the nonlinearity as indicated by the ratio of relative to planetary vorticities, and of the vorticity of the fluid relative to the planetary vorticity.

(Temporal) Rossby number, $Ro_t = \dfrac{\omega}{f} = \dfrac{T_I}{T}$ is the ratio of the inertial period to the flow timescale, useful for periodic flow processes.

Taylor number, $Ta = \dfrac{f^2 D^4}{\nu^2}$, where D is the layer depth, denotes the effect of rotation in fluid instability. The onset of instability in a rotating fluid depends on this number.

D.3 USEFUL WEBSITES

The following oceanography-related websites may be useful. Note that the government and organizational sites are probably stable, whereas the same cannot be said about others. An extensive list of oceanography-related web pages and links can be found on:

Oceanography Resources: **www.whoi.edu/resources/oceanography.html** and **www.esdim.noaa.gov/ocean_page.html**

Directory of Oceanographic Institutions: **scilib.ucsd.edu/sio/inst**

Naturally, with the continuing explosive growth of the internet, many more resources will be made available in the future to oceanographers and ocean modelers. Therefore, it may be useful to employ standard web browsers to locate a particular resource of interest.

U.S. GOVERNMENT AGENCIES

www.nasa.gov: NASA home page with links to its various centers and laboratories
podaac-www.jpl.nasa.gov: NASA Jet Propulsion Laboratory Physical Oceanography Distributed Active Archive Center; a NASA data distribution center
daac.gsfc.nasa.gov: NASA Goddard data archive and distribution center
stardust.jpl.nasa.gov: Planetary data site
imagine.gsfc.nasa.gov: NASA educational site
www.noaa.gov: NOAA home page with links to National Environmental Satellite, Data and Information Service (NESDIS), National Weather Service (NWS), Office of Ocean and Atmospheric Research (OAR), NOAA Marine Fisheries Service (NMFS), and National Ocean Service (NOS)
www.ncdc.noaa.gov: NOAA Climatic Data Center archives all climate-related data
www.nodc.noaa.gov: NOAA Oceanographic Data Center archives ocean-related data, including hydrography
www.ngdc.noaa.gov: NOAA Geophysical Data Center archives atmosphere-related data
www.noaa.gov/nws/nws_intro.html: National Weather Service website
www.ncep.noaa.gov: NCEP website
www.cdc.noaa.gov: NCEP Climate Diagnostics Center website

www.cpc.noaa.gov: NCEP Climate p\Prediction Center website

www.pmel.noaa.gov: NOAA Pacific Marine Environmental Laboratory home page

www.aoml.noaa.gov: NOAA Atlantic Oceanographic and Meteorological Laboratory home page

www.navo.navy.mil: Naval Oceanographic Office products and support website

www.onr.navy.mil: Office of Naval Research website

www.nrl.navy.mil: Naval Research Laboratory website

www.nrlssc.navy.mil: Naval Research Laboratory, Stennis Space Center Branch website

www.mms.gov: Minerals Management Service, Department of the Interior, website

www.epa.gov: Environmental Protection Agency website

www.doe.gov: Department of Energy website

www.nsf.gov: National Science Foundation website

OTHERS

www.ecmwf.int: Homepage for ECMWF and its products

www.wmo.ch: World Meteorological Organization home page

shark1.esrin.esa.it: European Space Agency home page with links to AVHRR, ATSR, and Ocean Color data

www.ncar.ucar.edu: NCAR home page

www.whoi.edu: Woods Hole Oceanographic Institution home page

www.sio.ucsd.edu: Scripps Institute of Oceanography home page

www.ldeo.colombia.edu: Lamont Doherty Earth Observatory of Columbia University

earth.agu.org: American Geophysical Union home page. You can link to its various publications from here. Of most interest are the *Journal of Geophysical Research, Geophysical Research Letters, Reviews of Geophysics,* and EOS. Information on upcoming meetings can also be found here.

www.ametsoc.org/AMS: American Meteorological Society home page. You can link to its various publications from here. Of most interest are the *Journal of Physical Oceanography, Journal of Atmospheric Sciences, Journal of Oceanic and Atmospheric Technology,* and *Journal of Climate.* Information on meetings can also be found here.

xxx.lanl.gov/archive/ao-sci: Index to recent publications in atmospheric and oceanic sciences

www.carl.org: Colorado Association of Research Libraries website; useful for locating books, publications, and journal articles.

www.amazon.com: On-line book search and purchase service

DATA

Altimetry data products (real time) from University of Colorado Colorado
   Center for Astrodynamics Research (CCAR) home page: **www-
   ccar.colorado.edu**
AVHRR data from NASA/NOAA Pathfinder: **podaac-www.jpl.nasa.gov/sst**
Bathymetry data: **www.ngdc.noaa.gov/mgg/bathymetry/relief.html**
Carbon Dioxide Information Analysis Center: **cdiac.esd.ornl.gov/cdiac**
Computational Science Education Project (DoE): **csep1.phy.ornl.gov/csep.html**
Ferret data visualization and analysis tool (PMEL): **ferret.wrc.noaa.gov**
Global Change master directory: **gcmd.gsfc.nasa.gov**
NCAR data archives: **www.scd.ucar.edu/dss/catalogs/index.html**
NCAR Graphics documentation: **ngwww.ucar.edu**
Ocean Color Imagery (from NASA SeaWIFS):
   seawifs.gsfc.nasa.gov/SEAWIFS/ IMAGES/IMAGES.html
Precipitation data (monthly global) from NOAA/NCEP global precipitation
   climatology project: **www.ncdc.noaa.gov/opg**
Recent highlights on global ocean models: **vislab-www.nps.navy.mil/~braccio**
Sea Level (global) network data: **www.soest.hawaii.edu/UHSLC**
TOGA-COARE data: **www.ncdc.noaa.gov/coare/index.html**
TOGA-TAO home page: **www.pmel.noaa.gov/toga-tao/home.html**
Weather forecasts from a variety of weather centers from Purdue University
   weather processor: **thunder.atms.purdue.edu**
WOCE (U.S.) project home page: **www-ocean.tamu.edu/WOCE/uswoce.html**

References

Aagaard, K. 1989. A synthesis of the Arctic Ocean circulation. *Rapp. P.-V. Reun. Cons. Int. Explor. Mer.* **188,** 11–22.

Aagaard, K., L. K. Coachman, and E. C. Carmack. 1981. On the halocline of the Arctic Ocean. *Deep-Sea Res. A* **28,** 529–545.

Aagaard, K., and E. C. Carmack. 1989. The role of sea ice and other fresh water in the Arctic circulation. *J. Geophys. Res.* **94,** 14485–14498.

Aberson, S. D., and J. L. Franklin. 1999. Impact on hurricane track and intensity forecasts of GPS dropwindsonde observations from the first season flights of the NOAA Gulfstream-IV jet aircraft, *Bull. Amer. Meteorol. Soc.,* **80,** 421–427.

Accad, Y., and C. L. Pekeris. 1978. Solution of the tidal equations for the M_2 and S_2 tides in the world oceans from a knowledge of the tidal potential alone. *Philos. Trans. R. Soc. London Ser. A* **290,** 235–266.

Acro-Shertzer, C. E., D. V. Hansen, and M. S. Swenson. 1997. Evaluation and diagnosis of surface currents in the National Centers for Environmental Prediction's ocean analyses., *J. Geophys. Res.,* **102,** 21,037–21,048.

Adkins, J. F., H. Cheng, E. A. Boyle, E. R. M. Druffel, and R. L. Edwards. 1998. Deep-sea coral evidence for rapid change in ventilation of the deep North Atlantic 15,400 years ago. *Science* **280,** 725–728.

Agnew, D. C., and W. E. Farrell. 1978. Self-consistent equilibrium ocean tides. *Geophys. J. R. Astron. Soc.* **55,** 171–181.

Aikman, F., III, G. L. Mellor, T. Ezer, D. Sheinin, P. Chen, L. Breaker, and D. B. Rao. 1996. Towards an operational nowcast/forecast system for the U.S. east Coast. In P. Malanotte-Rizzoli (Ed.), *Modern Approaches to Data Assimilation in Ocean Modeling,* pp. 347–376. Elsevier, Amsterdam/New York.

Alcock, G. A., and D. E. Cartwright. 1978. Some experiments with 'orthotides.' *Geophys. J. R. Astron. Soc.* **54,** 681–696.

Allen, J. S. 1977. Coastal trapped waves in a stratified ocean. *J. Phys. Oceanogr.* **5,** 300–325.

Allen, J. S. 1980. Models of wind-driven currents on the continental shelf. *Annu. Rev. Fluid Mech.* **12,** 389–433.

Allen, J. S., J. A. Barth, and P. A. Newberger. 1990. On intermediate models for barotropic continental shelf and slope flow studies. Part I. Formulation and comparison of exact solutions. *J. Phys. Oceanogr.* **20,** 1017–1042.

Amone, R. A., and R. W. Gould. 1998. Coastal monitoring using ocean color. *Sea Technology*, 18–27, September 1998.

Andersen, O. B. 1994. Ocean tides in the northern North Atlantic and adjacent seas from ERS 1 altimetry. *J. Geophys. Res.* **99**, 22557–22573.

Andersen, O. B. 1999. Shallow water tides in the northwest European shelf region from TOPEX/POSEIDON altimetry., *J. Geophys. Res.*, **104**, 7729–7742.

Andersen, O. B., P. L. Woodworth, and R. A. Flather. 1995. Intercomparison of recent ocean tidal models. *J. Geophys. Res.* **100**, 25261–25282.

Anderson, B. D. O., and J. N. Moore. 1979. *Optimal Filtering*. Prentice-Hall, New York. 357 pp.

Anderson, D., and J. Willebrand, (Eds.). 1989. *Oceanic Circulation Models: Combining Data and Dynamics.*, Kluwer, Dordrecht, Netherlands.

Anderson, D. L. T., and A. M. Moore. 1986. Data assimilation. In J. J. O'Brien (Ed.), *Advanced Physical Oceanographic Numerical Modeling*, pp. 437–464, Reidel, Dordrecht.

Anderson, J. D., Jr. 1995. *Computational Fluid Dynamics: The Basics with Applications.* McGraw Hill, New York. 547 pp.

Andreas, E. L., 1992, Sea spray and the turbulent air-sea heat fluxes, *J. Geophys. Res.* **97**, 11,429–11,441.

Andreas, E. L., 1998., A new sea spray generation function for wind speeds up to 32 m/s, *J. Phys. Oceanogr.* **28**, 2175–2184.

Andreas, E. L., J. B. Edson, E. C. Monahan, M. P. Rouault, and S. D. Smith. 1995. The spray contribution to net evaporation from the sea: A review of recent progress, *Boundary-Layer Meteorol.*, **72**, 3–52.

Andreas, E. L., and K. A. Emanuel. 1999. Effects of sea spray on tropical cyclone intensity, *23rd Conference on Hurricane and Tropical Meteorology*, 79th AMS annual meeting, Dallas, TX, 10-15 January 10–15, 1999.

Anthes, R. A. 1974. Data assimilation and initialization of hurricane prediction models. *J. Atmos. Sci.* **31**, 702–719.

Apel, J. R. 1987. *Principles of Ocean Physics*. Academic Press, San Diego. 634 pp.

Apel, J. R., J. R. Holbrook, A. K. Liu, and J. J. Tsai. 1985. The Sulu Sea Internal Soliton Experiment. *J. Phys. Oceanogr.* **15**, 1625–1651.

Arakawa, A. 1966. Computational design for long-term numerical integration of the equations of fluid motion: Two-dimensional incompressible flow. Part I. *J. Comput. Phys.* **1**, 119–143.

Arakawa, A. 1970. Numerical simulation of large-scale atmospheric motions. In G. Birkhoff and S. Varga (Eds.), *Numerical Solution of Field Problems in Continuum Physics*, Vol. 2, pp. 24–40. SIAM-AMS Proc., Am. Math. Soc., Providence.

Arakawa, A., and W. H. Schubert. 1974. Interaction of a cumulous ensemble with the large scale environment, Part I. *J. Atmos. Sci.* **31**, 674–701.

Arakawa, A., and V. R. Lamb. 1977. Computational design of the basic dynamical processes of the UCLA General Circulation Model. In *Methods in Computational Physics*, Vol. 17, pp. 174–265. Academic Press, San Diego.

Arakawa, A., and V. R. Lamb. 1981. A potential enstrophy and energy conserving scheme for the shallow water equations. *Mon. Weather Rev.* **109**, 18–36.

Arakawa, A., and M. J. Suarez. 1983. Vertical differencing of the primitive equations in sigma coordinates. *Mon. Weather Rev.* **111**, 34–45.

Arakawa, A., and S. Moorthi. 1988. Baroclinic instability in vertically discrete systems. *J. Atmos. Sci.* **45**, 1688–1707.

Arakawa, A., and Y.-J. G. Hsu. 1990. Energy conserving and potential-enstrophy dissipating schemes for the shallow water equations. *Mon. Weather Rev.* **118**, 1960–1969.

Asselin, R. 1972. Frequency filters for time integrations. *Mon. Weather Rev.* **100**, 487–490.

Atlas, R., R. N. Hoffman, S. C. Bloom, J. C. Jusem, and J. Ardizzone. 1996. A multi-year global surface wind velocity data using SSM/I wind observations. *Bull. Am. Meteorol. Soc.* **77**, 869–882.

Backhaus, J. O. 1985. A three-dimensional model for simulation of shelf sea dynamics. *Dtsch. Hydrograph. Z.* **38**, 165–187.

Backhaus, J. O., and D. Hainbucher. 1987a. A finite-difference general circulation numerical model for shelf seas and its application to low frequency variability on the North European Shelf. In J. C. J. Nihoul and B. M. Jamart (Eds.), *Three-Dimensional Models of Marine and Estuarine Dynamics*, pp. 221–244. Elsevier, New York.

Backhaus, J. O., and D. Hainbucher. 1987b. On the application of a three dimensional numerical model to the waters between Vancouver Island and the mainland coast of British Columbia and Washington State. In N. Heaps (Ed.), *Three-Dimensional Coastal Ocean Models,* pp. 149–176. American Geophysical Union, Washington, DC.

Baer, F., 1972. An alternate scale representation of atmospheric energy spectra. *J. Atmos. Sci.* **29**, 649-664.

Baines, P. G. 1982. On internal tide generation models. *Deep-Sea Res.* **29**, 307–338.

Baker, D. J. 1981. Ocean instruments and experiment design. In B. A. Warren and C. Wunsch (Eds.), *Evolution of Physical Oceanography,* pp. 396–433. MIT Press, Cambridge, MA.

Baker, G. L., and J. P. Gollub. 1990. *Chaotic Dynamics, an Introduction.* Cambridge Univ. Press, Cambridge.

Bang, I.-K., J.-K. Choi, L. Kantha, C. Horton, M. Clifford, M.-S. Suk, K.-I. Chang, S.-Y. Nam, and H.-J. Lie. 1996. A hindcast experiment in the East Sea (Sea of Japan). *La Mer,* **34**, 108–130.

Bao, J.-W., J. M. Wilczak, J.-K. Choi, and L. H. Kantha. 1999. Numerical simulations of air-sea interaction under high wind conditions using a coupled model: A study of hurricane development., *Mon. Wea. Rev.*

Barnes, R. T. H., R. Hide, A. A. White, and C. A. Wilson. 1983. Atmospheric angular momentum fluctuations, length-of-day changes and polar motion. *Proc. Roy. Soc. London, Ser. A* **387**, 31–73.

Barnett, T. P., M. Latif, N. Graham, M. Flugel, S. Pazan, and W. White. 1993. ENSO and ENSO-related predictability. Part I. Prediction of equatorial Pacific sea surface temperature with hybrid coupled ocean-atmosphere model. *J. Clim.* **6**, 1545–1566.

Barnston, A. G., M. H. Glantz, and Y. He, 1999. Predictive skill of statistical and dynamical climate models in SST forecasts during the 1997–98 El Niño episode and the 1998 La Niña onset. *Bull. Amer. Meteorol. Soc.* **80**, 217–243.

Barry, R. G., M. C. Serreze, J. A. Maslanik, and R. H. Preller. 1993. The Arctic sea ice-climate system: Observations and modeling. *Rev. Geophys.* **31**, 397–422.

Barth, J. A., J. S. Allen, and P. A. Newberger. 1990. On intermediate models for barotropic continental shelf and slope flow studies. Part II. Comparison of numerical model solutions in doubly periodic domains. *J. Phys. Oceanogr.* **20**, 1044–1076.

Barth, N., and J. Polcher. 1994. *Two Ocean/Atmosphere Simulations Using OPA6 and LMD5bis.* Laboratoire de Meteorologie Dynamique Tech. Report.

Batchelor, G. K. *Introduction to Fluid Mechanics.* Cambridge Univ. Press, Cambridge. 615 pp.

Bates, J. R., S. Moorthi, and R. W. Higgins. 1993. A global multilevel atmospheric model using a vector semi-Lagrangian finite difference scheme. Part I. Adiabatic formulaion. *Mon. Weather Rev.* **121**, 244–263.

Beardsley, R. C., and W. C. Boicourt. 1981. On estuarine and continental-shelf circulation in the Middle Atlantic Bight. In B. Warren and C. Wunsch (Eds.), *Evolution of Physical Oceanography,* 198–233. MIT Press, Cambridge, MA.

Beckers, J. M. 1991. Application of a 3D model to the western Mediterranean. *J. Mar. Syst.* **1**, 315–322.

Beckers, J. M., and J. C. J. Nihoul. 1992. Model of the Algerian Current's instability. *J. Mar. Syst.* **3**, 441–451.

Beckmann, A., and D. B. Haidvogel. 1993. Numerical simulation of flow around a tall isolated seamount. Part I. Problem formulation and model accuracy. *J. Phys. Oceanogr.* **23**, 1736–1753.

Beckmann, A., and D. B. Haidvogel. 1997. A numerical simulation of flow at Fieberling Guyot. *J. Geophys. Res.* **102**, 5595–5614.

Behrenfeld, M. J. and Z. S. Kolber. 1999. Widespread iron limitation of phytoplankton in the South Pacific Ocean. *Science* **283,** 840–843.

Behringer, D. W., M. Ji, and A. Leetmaa. 1998. An improved coupled model for ENSO prediction and implications for ocean initialization. Part I. The ocean data assimilation system. *Mon. Weather Rev.* **126,** 1013–1021.

Bender, L. C. 1996. Modification of the physics and numerics in a third generation ocean wave model. *J. Atmos. Oceanic Technol.* **13,** 726–750.

Bender, M. A., I. Ginis, and Y. Kurihara. 1993. Numerical simulations of tropical cyclone-ocean interaction with a high resolution coupled model. *J. Geophys. Res.* **98,** 23-245–23,263.

Bendjoya, Ph., and E. Slezak. 1993. Wavelet analysis and applications to some dynamical systems. *Celestial Mech. Dynamical Astron.* **56,** 231–262.

Bengtsson, L., M. Ghil, and E. Kallen (Eds.). 1981. *Dynamic Meteorology: Data Assimilation Methods.* Springer-Verlag, New York. 330 pp.

Benilov, E. S. 1996. Beta-induced translation of strong isolated eddies. *J. Phys. Oceanogr.* **26,** 2223–2229.

Bennett, A. F. 1992. *Inverse Methods in Physical Oceanography.*, Cambridge Univ. Press, Cambridge.

Bereger, A. 1978. Long-term variations of caloric insolation resulting from the earth's orbital elements. *Quat. Res.* **9,** 139–167.

Berger, A., *et al.* (Eds.). 1984. *Milankovitch and Climate.* Reidel, Dordrecht, Netherlands.

Bergland, G. D. 1969. A guided tour of the fast Fourier transform. *IEEE Spectrum* **6,** 41–51.

Berkooz, G., P. Holmes, and J. L. Lumley. 1993. The proper orthogonal decomposition in the analysis of turbulent flows. *Annu. Rev. Fluid Mech.* **25,** 539–575.

Bevington, P. R. 1969. *Data Reduction and Error Analysis for the Physical Sciences.* McGraw Hill, New York. 336 pp.

Bindschadler, R. 1998. Monitoring ice sheet behavior from space. *Rev. Geophys.* **36,** 79–104.

Bjork, G. 1992. On the response of the equilibrium thickness distribution of sea ice to ice export, mechanical deformation, and thermal forcing with application to the Arctic Ocean. *J. Geophys. Res.* **97,** 11287–11298.

Bjork, G. 1997. The relation between ice deformation, oceanic heat flux, and the ice thickness distribution in the Arctic Ocean. *J. Geophys. Res.* **102,** 18681–18698.

Black, K., D. Hatton, and M. Rosenberg. 1993. Locally and externally driven dynamics of a large semi-enclosed bay in southeastern Australia. *J. Coastal Res.* **9,** 509–538.

Blandford, R. R. 1971. Boundary conditions in homogeneous ocean models. *Deep-Sea Res.* **18,** 739–751.

Blanke, B., and P. Delacluse. 1993. Variability of the tropical Atlantic Ocean simulated by a general circulation model with two different mixed layer physics. *J. Phys. Oceanogr.* **23,** 1363–1388.

Bleck, R. 1978. Finite difference equations in generalized vertical coordinates. P1. Total energy conservation. *Beitr. Phys. Atmos.* **51,** 360–372.

Bleck, R., and D. B. Boudra. 1986. Wind-driven spin-up in eddy-resolving ocean models formulated in isopycnic and isobaric coordinates. *J. Geophys. Res.* **91,** 7611–7621.

Bleck, R., and L. T. Smith. 1990. A wind-driven isopycnic coordinate model of the north and equatorial Atlantic Ocean. 1. Model development and supporting experiments. *J. Geophys. Res.* **95,** 3273–3285.

Bleck, R., C. Rooth, D. Hu, and L. T. Smith. 1992. Salinity-driven thermohaline transients in a wind- and thermohaline-forced isopycnic coordinate model of the North Atlantic. *J. Phys. Oceanogr.* **22,** 1486–1505.

Blumberg, A. F., and L. H. Kantha. 1985. Open boundary condition for circulation models. *J. Hydraul. Engineer.* **111,** 237–255.

Blumberg, A. F., and G. L. Mellor. 1987. A description of a three-dimensional coastal ocean circulation model. In N. Heaps (Ed.), *Three-Dimensional Coastal Ocean Models,* pp. 1–16. American Geophysical Union, Washington, DC.

Boer, G. J., K. Arpe, M. Blackburn, M. De'Que, W. L. Gates, T. L. Hart, H. Le Treut, E. Roeckner, D. A. Sheirin, I. Simmonds, R. N. B. Smith, T. Tokioka, R. T. Wetherald, and D. Williamson. 1992. Some results from intercomparison of the climates simulated by 14 atmospheric general circulation models. *J. Geophys. Res.* **97,** 12771–12786.

Boris, J. P., and D. L. Book. 1973. Flux corrected transport. I. SHASTA, a fluid transport algorithm that works. *J. Comput. Phys.* **11,** 38–69.

Boris, J. P., and D. L. Book. 1975. Solution of the continuity equations by the method of flux-corrected transport. *Methods Comput. Phys.* **16,** 85–129.

Born, G. H., J. L. Mitchell, and G. A. Heyler. 1987. Geosat ERM—mission design. *J. Astron. Sci.* **35,** 119–134.

Bourke, R. H. and R. P. Garrett, 1987. Sea ice thickness distribution in the Arctic ocean. *Cold Regions Science and Technology* **13,** 259-280.

Boville, B. A., and P. R. Gent. 1998. The NCAR climate system model, version one. *J. Clim.* **11,** 1115–1130.

Brandt, A. 1977. *Math. Comput.* **31,** 333–390.

Brasseur, P. (Ed.). 1995, Data Assimilation in Marine Science., *J. Mar. Syst.* **6,** 175 pp., 1995.

Bretherton, C. S., C. Smith, and J. M. Wallace. 1992. An intercomparison of methods for finding coupled patterns in climate data. *J. Clim.* **5,** 541–560.

Briggs, W. L. 1987. *A Multigrid Tutorial.* SIAM, Philadelphia.

Brink, K. H. 1987. Coastal ocean physical processes. *Rev. Geophys.* **25,** 204–216.

Brink, K. H. 1989. The effect of stratification on seamount-trapped waves. *Deep-Sea Res. A* **36,** 825–844.

Brink, K. H. 1990. On the generation of seamount-trapped waves. *Deep-Sea Res. A* **37,** 1569–1582.

Brink, K. H. 1995. Tidal and lower frequency currents above Fieberling Guyot. *J. Geophys. Res.* **100,** 10817–10832.

Brink, K. H., and D. C. Chapman. 1987. *Programs for Computing Properties of Coastal-Trapped Waves and Wind-Driven Motions over the Continental Shelf and Slope.* WHOI Report 85-17.

Brink, K., R. Arnone, P. Coble, C. Flagg, B. Jones, J. Kindle, C. Lee, D. Phinney, M. Wood, C. Yentsch, and D. Young. 1998. Monsoons boost biological productivity in Arabian Sea. *EOS Trans.* **79,** 165.

Broecker, W. S. 1987. The biggest chill. *Nat. Hist. Mag.* **97,** 74–82.

Broecker, W. S. 1991. The great ocean conveyor. *Oceanography* **4,** 79–89.

Broecker, W. S. 1992. The great ocean conveyor. In B. G. Levi, D. Hafemeister, and R. Scribner (Eds.),*Global Warming: Physics and Facts,* pp. 129–161. Am. Inst. of Phys., New York.

Broecker, W. S. 1997. Thermohaline circulation, the Achilles heel of our climate system: Will man-made CO_2 upset the current balance? *Science* **278,** 1582–1588.

Brown, R. A., 1990. Meteorology, in *Polar Oceanography*, ed. W. O. Smith, Jr., Academic Press, pp. 1-46.

Bruce, J. G., J. C. Kindle, L. Kantha, J. L. Kerling, and J. F. Baily. 1998. A note on recent observations in the Arabian Sea Laccadive Eddy region. *J. Geophys. Res.* **103,** 7593–7600.

Bryan, F. O. 1987. Parameter sensitivity of primitive equation ocean general circulation models. *J. Phys. Oceanogr.* **17,** 970–985.

Bryan, F. O. 1998. Climate drift in a multicentury integration of the NCAR climate system model. *J. Clim.* **11,** 1455–1471.

Bryan, F. O., C. W. Boning, and W. R. Holland. 1995. On the midlatitude circulation in a high-resolution model of the North Atlantic. *J. Phys. Oceanogr.* **25,** 289–301.

Bryan, K. 1969. A numerical model for the study of the circulation of the world oceans. *J. Comput. Phys.* **4,** 347–359.

Bryan, K. 1984. Accelerating the convergence to equilibrium of ocean climate models. *J. Phys. Oceanogr.* **14,** 666–673.

Bryan, K. 1989. The design of numerical models of the ocean circulation. In D. L. T. Anderson and J. Willebrand (Eds.),*Ocean Circulation Models: Combining Data and Dynamics*, pp. 465–500. Kluwer Academic, Dordrecht/Norwell, MA.

Bryan, K. 1991. Poleward heat transport in the ocean. *Tellus* **43**, 104–115.

Bryan, K., and L. J. Lewis. 1979. A water mass model of the world ocean. *J. Geophys. Res.* **84**, 2503–2517.

Bryan, K., and J. L. Sarmiento. 1985. Modeling ocean circulation. *Adv. Geophys.* **28A**, 433–459.

Bryden, H. L., and M. M. Hall. 1980. Heat transport by currents across 25° N latitude in the Atlantic Ocean. *Science* **207**, 884–886.

Bryden, H. L., D. H. Roemmich, and J. A. Church. 1991. Ocean heat transport across 24° N in the Pacific. *Deep-Sea Res.* **38**, 297–324.

Buchwald, V. T., and J. K. Adams. 1968. The propagation of continental shelf waves. *Proc. R. Soc. Sec. A* **305**, 235–250.

Bursa, M. 1990. *Tidal Friction and Earth's Rotation.* Final report of IAG Special Study Group 5-99, 19th IUGG, IAG General Assembly.

Busalacchi, A. J. 1996. Data assimilation in support of tropical ocean circulation studies. In P. Malanotte-Rizzoli (Ed.), *Modern Approaches to Data Assimilation in Ocean Modeling,* pp. 235–270. Elsevier, Amsterdam/New York.

Campbell, W. J. 1965. The wind driven circulation of the ice and water in a polar ocean. *J. Geophys. Res.* **70**, 3279–3301.

Cane, M. A. 1992. Tropical Pacific ENSO models: ENSO as a mode of the coupled system. In K. E. Trenberth (Ed.),*Climate System Modeling*, pp. 583–616. Cambridge Univ. Press, New York.

Cane, M. A., S. E. Zebiak, and S. C. Dolan. 1986. Experimental forecasts of El Niño, *Nature* **321**, 827–832.

Cane, M. A., A. Kaplan, R. N. Miller, B. Tang, E. C. Hackert, and A. Busalacchi. 1996. Mapping tropical Pacific sea level: Data assimilation via a reduced state space Kalman filter. *J. Geophys. Res.* **101**, 22599–22617.

Cane, M. A., V. M. Kamenkovich, and A. Krupitsky. 1998. On the utility and disutility of JEBAR. *J. Phys. Oceanogr.* **28**, 519–526.

Canuto, C., M. Y. Hussaini, A. Quarteroni, and T. A. Zang. 1988. *Spectral Methods in Fluid Dynamics.* Springer-Verlag, Berlin/New York. 567 pp.

Carnes, M. R., J. L. Mitchell, and P. W. deWitt. 1990. Synthetic temperature profiles derived from satellite altimetry: Comparison with air-dropped expendable bathythermograph profiles. *J. Geophys. Res.* **95**, 17,979–17,992.

Cartwright, D. E. 1977. Oceanic tides. *Rep. Prog. Phys.* **40**, 665–708.

Cartwright, D. E. 1991. Detection of tides from artificial satellites (review). In B. B. Parker (Ed.), *Tidal Hydrodynamics,* pp. 547–568. Wiley, New York.

Cartwright, D. E. 1993. Theory of ocean tides with application to altimetry. In *Satellite Altimetry in Geodesy and Oceanography,* pp. 99–141. Springer-Verlag, Berlin/New York.

Cartwright, D. E. 1997. Some thoughts on the spring-neap cycle of tidal dissipation. *Prog. Oceanogr.* **40**, 125–133.

Cartwright, D. E. 1998. *In Quest of Tides.* Cambridge Univ. Press, Cambridge.

Cartwright, D. E., and R. J. Taylor. 1971. New computations of the tide-generating potential. *Geophys. J. R. Astron. Soc.* **23**, 45–74.

Cartwright, D. E., and A. C. Edden. 1973. Corrected tables of tidal harmonics. *Geophys. J. R. Astron. Soc.* **33**, 253–264.

Cartwright, D. E., and R. D. Ray. 1989. New estimates of oceanic tidal energy dissipation from satellite altimetry. *Geophys. Res. Lett.* **16**, 73–76.

Cartwright, D. E., and R. D. Ray. 1990. Oceanic tides from Geosat altimetry. *J. Geophys. Res.* **95**, 3069–3090.

Cartwright, D. E., and R. D. Ray. 1991. Energetics of global ocean tides from Geosat altimetry. *J. Geophys. Res.* **96**, 16897–16912.

Cartwright, D. E., R. D. Ray, and B. V. Sanchez. 1991. *Oceanic tide maps and spherical harmonic coefficients from Geosat altimetry.* NASA Tech. Memo. 104544.

Casdagli, M. 1989. Nonlinear prediction of chaotic time series. *Physica* **35D**, 335–338.

Cassou, C., P. Noyret, E. Sevault, O. Thual, L. Terray, D. Beaucourt, and M. Imbard. 1998. Distributed ocean-atmosphere modeling and sensitivity to the coupling flux precision: CATHODe Project. *Mon. Weather Rev.* **126**, 1035–1053.

Chang, S. C. 1995. The method of space-time conservation element and solution element—A new approach for solving Navier-Stokes and Euler equations *J. Comput. Phys.* **191**, 295–324.

Chao, B. F. 1994. The geoid and Earth rotation. In P. Vanicek and N. Christou (Eds.),*Geophysical Interpretations of Geoid*. CRC Press, Boca Raton, FL.

Chao, B. F., R. D. Ray, and G. D. Egbert. 1996a. Diurnal/semidiurnal oceanic tidal angular momentum: Topex/Poseidon models in comparison with Earth's rotation rate. *Geophys. Res. Lett.* **22**, 1993–1996.

Chao, B. F., R. D. Ray, J. M. Gipson, G. D. Egbert, and C. Ma. 1996b. Diurnal/semidiurnal polar motion excited by variations in oceanic tidal angular momentum. *J. Geophys. Res.* **101**, 20151–20163.

Chao, B. F., and R. D. Ray. 1997. Oceanic tidal angular momentum and Earth's rotation variations. *Prog. Oceanogr.* **40**, 399–421.

Chao, B. F., R. D. Ray, and J. M. Gipson. 1997. Non-tidal high-frequency signals in Earth's rotational variations during VLBI intensive measurement campaigns. *EOS Trans. AGU* **78**, 150.

Chao, Y., and S. G. H. Philander. 1993. On the structure of the Southern Oscillation. *J. Clim.* **6**, 450–469.

Chao, Y., A. Gangopadhyay, F. O. Bryan, and W. R. Holland. 1996. Modelling the Gulf Stream system: How far from reality? *Geophys. Res. Lett.* **23**, 3155–3158.

Chapman, D. C. 1985. Numerical treatment of cross-shelf open boundaries in a barotropic coastal ocean model. *J. Phys. Oceanogr.* **11**, 355–375.

Chapman, S., and R. S. Lindzen. 1970. *Atmospheric Tides*. Reidel, Dordrecht.

Charney, J. G. 1947. The dynamics of long waves in a baroclinic westerly current. *J. Meteorol.* **4**, 135–163.

Charney, J. G. 1955. The Gulf Stream as an inertial boundary layer. *Proc. Natl. Acad. Sci.* **41**, 731–740.

Charney, J. G., and G. R. Flierl. 1981. Oceanic analogues of large-scale atmospheric motions. In B. Warren and C. Wunsch (Eds.), *Evolution of Physical Oceanography*, pp. 504–549. MIT Press, Cambridge, MA.

Charnock, H., and S. G. H. Philander. 1989. The dynamics of the coupled atmosphere and ocean. In *Proc. Roy. Soc. Meeting, Dec. 13–14, 1988*, pp. 160.

Chatfield, C. 1989. *The Analysis of Time Series*. Chapman and Hall, London/New York.

Chelton, D. B. and M. G. Schlax. 1996. Global observations of oceanic Rossby waves. *Science* **272**, 234–238.

Chelton, D. B., R. A. DeSzoeke, M. G. Schlax, K. El Naggar, and N. Siwertz. 1998. Geographical variability of the first baroclinic Rossby radius of deformation *J. Phys. Oceanogr.* **28**, 433–460.

Chiswell, S. M. 1994. Vertical structure of the baroclinic tides in the central North Pacific subtropical gyre. *J. Phys. Oceanogr.* **24**, 2032–2039.

Choi, J. K., L. H. Kantha, and R. R. Leben. 1995a. A nowcast/forecast experiment using TOPEX/POSEIDON and ERS-1 altimetric data assimilation into a three-dimensional circulation model of the Gulf of Mexico. *IAPSO Abstract, XXI General Assembly, Honolulu, Hawaii, August 1995*.

Choi, J.-K., L. H. Kantha, and R. R. Leben. 1995b. A nowcast/forecast experiment using TOPEX/Poseidon and ERS-1 altimetric data assimilation into a three-dimensional circulation model of the Gulf of Mexico. IUGG, XXI General Assembly, Boulder, Colorado, July 1995.

Choi, J.-K., L. H. Kantha, and C. Penland. 1996. Prediction of eddy shedding events in the Gulf of Mexico using the empirical normal mode analysis of SSH data. Unpublished manuscript.

Chow, J. H. S., and W. R. Holland. 1986. *Description of a quasi-geostrophic multi-layer box ocean model*. NCAR report, NCAR, Boulder, CO.

Chu, P. C., and C. Fan. 1997. Sixth-order difference scheme for sigma coordinate ocean models. *J. Phys. Oceanogr.* **27**, 2064–2071.

Chu, P. C., Y. Chen, and S. Lu. 1998. On Haney-type surface thermal boundary conditions for ocean circulation models. *J. Phys. Oceanogr.* **28,** 890–901.

Chui, F. K. (Ed.). 1992a. *An Introduction to Wavelets: Wavelet Analysis and Its Applications.* Academic Press, San Diego. 266 pp.

Chui, F. K. (Ed.). 1992b. *Wavelets: A Tutorial in Theory and Applications.* Academic Press, San Diego. 266 pp.

Church, J. A., H. J. Freeland, and R. L. Smith. 1986a. Coastal trapped waves on the east Australian continental shelf: Part 1, Propagation modes. *J. Phys. Oceanogr.* **16,** 1929–1943.

Church, J. A., N. J. White, A. J. Clarke, H. J. Freeland, and R. L. Smith. 1986a. Coastal trapped waves on the east Australian continental shelf: Part 2, Model verification. *J. Phys. Oceanogr.* **16,** 1945–1957.

Church, *et al.* 1986. Coastal trapped waves on the east Australian continental shelf: Parts I & II. *J. Phys. Oceanogr.* **16,** 1929–1957.

Ciyuan, L., and K. K. C. Yau. 1990. Application of early Chinese records of lunar occultations and close approaches. In P. Brosche and J. Sundermann (Eds.), *Earth's Rotation from Eons to Days*, pp. 33–39. Springer-Verlag, Berlin/New York.

Clark, T. A., C. Ma, J. W. Ryan, B. F. Chao, J. M. Gipson, D. S. MacMillan, N. R. Vandenberg, T. M. Eubanks, and A. E. Niell. 1998. Earth rotation measurement yields valuable information about the dynamics of the Earth system. *EOS Trans. AGU* **79,** 205.

Clayson, C. A. 1995. Modeling the mixed layer in the tropical Pacific Ocean and air-sea interactions: Application to a westerly wind burst. Ph.D. dissertation, University of Colorado, Boulder.

Clayson, C. A., and J. A. Curry. 1996. Determination of surface turbulent fluxes for the Tropical Ocean-Global Atmosphere Coupled Ocean-Atmosphere response experiment: Comparison of satellite retrievals and *in-situ* measurements. *J. Geophys. Res.* **101,** 28,515–28,528.

Clayson, C. A., C. W. Fairall, and J. A. Curry. 1996. Evaluation of turbulent fluxes at the ocean surface using surface renewal theory. *J. Geophys. Res.* **101,** 28,503–28,513.

Clayson, C. A., A. Chen, L. H. Kantha, and P. J. Webster. 1997. Numerical simulations of the equatorial Pacific during the TOGA/COARE IOP, Abstract, 22nd Conference on Hurricanes and Tropical Meteorology, May 19–23. Fort Collins, CO, 600–601.

Clifford, M., C. Horton, J. Schmitz, and L. Kantha. 1997. An oceanographic nowcast/forecast system for the Red Sea. *J. Geophys. Res.* **102,** 25101–25122.

Coachman, L. K., and K. Aagaard. 1975. Physical oceanography of Arctic and sub-Arctic seas. In Y. Herman (Ed.), *Marine Geology and Oceanography of the Arctic Seas,* pp. 1–72. Springer-Verlag, New York.

Cocke, S. 1998. Case study of Erin using the FSU nested regional spectral model. *Mon. Weather Rev.* **126,** 1337–1346.

Codiga, D. L. 1997a. Physics and observational signatures of free, forced, and frictional stratified seamount-trapped waves. *J. Geophys. Res.* **102,** 23009–23024.

Codiga, D. L. 1997b. Trapped wave modification and critical surface formation by mean flow at a seamount with applications at Fieberling Guyot. *J. Geophys. Res.* **102,** 23025–23040.

Codiga, D. L., and C. C. Eriksen. 1997. Observations of low-frequency circulation and amplified subinertial tidal currents at Cobb Seamount. *J. Geophys. Res.* **102,** 22993–23008.

Cole, J. D. 1968. *Perturbation Methods in Applied Mathematics.* Blaisdell, Waltham, MA.

Collineau, S., and Y. Brunet. 1993. Detection of turbulent coherent motions in a forest canopy. Part I. Wavelet analysis. *Boundary Layer Meteor.* **65,** 357–379.

Colony, R., and D. A. Rothrock. 1980. A perspective of the time-dependent response of the AIDJEX model. In R. S. Pritchard (Ed.), *Sea Ice Processes and Models*, pp. 124–133. Univ. Washington Press, Seattle.

Combes, J. M., A. Grossman, and Ph. Tchamitchian. 1989. Wavelets: Time-frequency methods and phase space. *Proc. Int. Conf.,* Marseille, France. 331 pp.

Coon, M. D. 1980. A review of AIDJEX modeling. In R. S. Pritchard (Ed.), *Sea Ice Processes and Models*, pp. 12–27. Univ. Washington Press, Seattle.

Coon, M. D., S. A. Maykut, R. S. Pritchard, D. A. Rothock, and A. S. 1974. Thorndike, modeling the pack ice as an elastic-plastic material. A*IDJEX Bull.* **24**, 1–105.

Coon, M. D., G. S. Knoke, D. C. Echert, and R. S. Pritchard. 1998. The architecture of an anisotropic elastic-plastic sea ice mechanics constitutive law. *J. Geophys. Res.* **103**, 21915–21925.

Cooper, N. S. 1988. The effect of salinity on tropical ocean models. *J. Phys. Oceanogr.* **18**, 697–707.

Courant, R., K. O. Friedrichs, and H. Lewy. 1928. Uber die partiellen differenzgleichungen der mathematischen physik. *Math. Annalen* **100**, 32–74. (English translation, 1967. *IBM J.* 215-34.)

Cox, M. D. 1984. *A Primitive Equation, 3-Dimensional Model of the Ocean.* GFDL Ocean Group Tech. Rep. 1, Geophys. Fluid Dyn. Lab., Princeton Univ., Princeton, NJ.

Cox, M. D. 1985. An eddy-resolving numerical model of the ventilated thermocline. *J. Phys. Oceanogr.* **15**, 1312–1324.

Cox, M. D. 1987. Isopycnal diffusion in a z-coordinate model. *Ocean Modell.* **74**, 1–5.

Cox, R., B. L. Bauer, and T. Smith. 1997. A mesoscale model intercomparison. *Bull. Am. Meteorol. Soc.* **79**, 265–284.

CREAMS, 1994. Proceedings of the Third Circulation Research of the East Asian Marginal Seas (CREAMS'94) Workshop, Seoul National University, Seoul, Korea, Nov 7-8, 1994.

CREAMS, 1996. Proceedings of the fourth Circulation Research of the East Asian Marginal Seas (CREAMS'96) Workshop, Vladivostok, February 12-13, 1996.

Csanady, G. T. 1985. "Pycnobathic" currents over the upper continental slope. *J. Phys. Oceanogr.* **15**, 306–315.

Curtin, T., J. G. Bellingham, J. Catipovic, and D. Webb. 1993. Autonomous oceanographic sampling networks. *Oceanography* **6**, 86–94.

Cushman-Roisin, B. 1994. *Geophysical Fluid Dynamics.* Prentice Hall, Upper Saddle River, NJ.

Da Silva, A. M., C. C. Young, and S. Levitus. 1994a. *Atlas of Surface Marine Data 1994,* Vol. 1: *Algorithm and Procedures.* NOAA Atlas NESDIS 6, NOAA, Wash. D.C.

Da Silva, A. M., C. C. Young, and S. Levitus. 1994b. *Atlas of Surface Marine Data 1994,* Vol. 3: *Anomalies of Heat and Momentum Fluxes.* NOAA Atlas NESDIS 8, NOAA.

Dahlen, F. A. 1976. The passive influence of the oceans upon the rotation of the Earth. *Geophys. J. R. Astron. Soc.* **46**, 363–406.

Daley, R. 1991. *Atmospheric Data Analysis.* Cambridge Univ. Press, Cambridge. 457 pp.

Danabasoglu, G., J. C. McWilliams, and P. R. Gent. 1994. The role of mesoscale tracer transports in the global ocean circulation. *Science* **264**, 1123–1126.

Danabasoglu, G., J. C. McWilliams, and W. C. Large. 1996. Approach to equilibrium in accelerated global oceanic models. *J. Clim.* **9**, 1092–1110.

Darwin, G. H. 1886. On the dynamical theory of the tides of long period. *Proc. R. Soc. London* **41**, 337–342.

Daubechies, I. 1992. *Ten Lectures on Wavelets.* CBMS-NSF Regional Conf. Ser. in Applied Mathematics. Society for Industrial and Applied Mathematics, Capital City Press. 357 pp.

Davies, A. M. 1987. A three-dimensional numerical model of semidiurnal tides on the European continental shelf. In J. C. J. Nihoul and B. M. Jamart (Eds.), *Three-Dimensional Models of Marine and Estuarine Dynamics,* pp. 573–590. Elsevier, New York.

Davies, A. M., and J. Lawrence. 1994. The response of the Irish Sea to boundary and wind forcing: Results from a three-dimensional hydrodynamic model. *J. Geophys. Res.* **99**, 22665–22687.

Davies, H. C. 1983. Limitations of some common lateral boundary schemes used in regional NWP models. *Mon. Weather Rev.* **111**, 1002–1012.

Davis, R. E. 1976. Predictability of sea surface temperature and sea level pressure anomalies over the North Pacific Ocean. *J. Phys. Oceanogr.* **6**, 249–266.

Davis, R. E. 1998. Preliminary results from directly measuring middepth circulation in the tropical and South Pacific. *J. Geophys. Res.* **103**, 24619–24640.

Deardorff, J. W. 1972. Parameterization of the planetary boundary layer for use in general circulation models. *Mon. Weather Rev.* **100**, 93–106.

Dee, D. P. 1995. On-line estimation of error covariance parameters for atmospheric data assimilation. *Mon. Weather Rev.* **123**, 1128–1145.

Defant, A. 1961. Tides in the Mediterranean and adjacent seas. Observations and discussion. *Phys. Oceanogr.* **1**, 364–456.

Dehant, V., C. R. Wilson, D. A. Salstein, B. F. Chao, R. S. Gross, C. Le-Provost, and R. M. Ponte. 1997. Study of Earth's rotation and geophysical fluids progresses. *EOS Trans. AGU* **78**, 357, 360.

Delecluse, P., G. Madec, M. Imbard, and C. Levy. 1993. *OPA Version 7 Ocean General Circulation Model Reference Manual.* Laboratoire d'Oceanographie Dynamique et de Climatologie (LODYC) internal report 93/05. 90 pp.

Deleersnijder, E., and J. M. Beckers. 1992. On the use of the σ-coordinate system in regions of large bathymetric variations. *J. Mar. Syst.* **3**, 381–390.

Denton, G. H., and T. J. Hughes. 1981. *The Last Great Ice Sheets.* Wiley, New York. 484 pp.

Derber, J. 1987. Variational four-dimensional analysis using quasi-geostrophic constraints. *Mon. Weather Rev.* **115**, 998–1008.

Derber, J. 1989. A variational continuous assimilation technique. *Mon. Weather Rev.* **117**, 2437–2446.

Derber, J., and A. Rosati. 1989. A global oceanic data assimilation system. *J. Phys. Oceanogr.* **19**, 1333–1347.

Desai, S. D. 1996. Ocean tides from TOPEX/POSEIDON altimetry with some geophysical applications. Ph.D. dissertation submitted to the Department of Aerospace Engineering Sciences, University of Colorado, Boulder, CO.

Desai, S. D., and J. M. Wahr. 1994. Another ocean tide model derived from TOPEX/POSEIDON satellite altimetry (abstract). *Eos Trans. AGU* **75**(44), 57.

Desai, S. D., and J. M. Wahr. 1995. Empirical ocean tide models estimated from TOPEX/POSEIDON altimetry. *J. Geophys. Res.* **100**, 25205–25228.

Dickey, J. O. 1995. Earth rotation variations from hours to centuries. In I. Appenzeller (Ed.), *Highlights in Astronomy*, Vol., 10, pp. 17–44.

Dickey, J. O., *et al.* 1994. Lunar laser ranging: A continuing legacy of the Apollo program. *Science* **265**, 482–490.

Dickman, S. R. 1993. Dynamic ocean-tide effects on Earth's rotation. *Geophys. J. Int.* **112**, 448–470.

Dickman, S. R., and Y. S. Nam. 1995. Revised predictions of long-period ocean tidal effects on Earth's rotation rate. *J. Geophys. Res.* **100**, 8233–8243.

Dickman, S. R. 1998. Determination of oceanic dynamic barometer corrections to atmospheric excitation of Earth rotation. *J. Geophys. Res.* **103**, 15,127–15,143.

Dietrich, D. E., M. G. Marietta, and P. J. Roach. 1987. An ocean modeling system with turbulent boundary layers and topography. *Int. J. Numer. Methods Fluids* **7**, 833–855.

Dietrich, D. E., and C. A. Lin. 1994. Numerical studies of eddy shedding in the Gulf of Mexico. *J. Geophys. Res.* **99**, 7599–7615.

Dietrich, G., K. Kalle, W. Krauss, and G. Siedler. 1980. Tidal phenomenon. In *General Oceanography*, pp. 407–459. Wiley, New York.

Dippner, J. W. 1993. A front-resolving model of the German Bight. *Cont. Shelf Res.* **13**, 49–66.

Doedel, E. J. 1986. AUTO: A program for the automatic bifurcation analysis of autonomous systems. *Proc. 10th Manitoba Conf. Numerical Math. Comput.* **30**, 265–274.

Dongarra, J. L. 1999. Performance of various computers using standard linear equations software. University of Tennessee Computer Science Department report CS-89-85, 1999 (available also as www.netlib.org/benchmark/performance.ps).

Donlon, C. J., and I. S. Robinson. 1997. Observations of the oceanic thermal skin in the Atlantic Ocean. *J. Geophys. Res.* **102**, 18585–18606.

Doodson, A. T. 1921. The harmonic development of the tide-generating potential. *Proc. R. Soc. London A* **100**, 305–329.

Doodson, A. T. 1927. The analysis of tidal observations. *Philos. Trans. R. Soc. London Ser. A* **227**, 223–279.

Dorr and R. Grimshaw. 1986. Barotropic continental shelf waves on a beta plane. *J. Phys. Oceanogr.* **16**, 1345–1358.

Dukowicz, J. K., R. D. Smith, and R. C. Malone. 1993. A reformulation and implementation of the Bryan-Cox-Semtner ocean model on the Connection Machine. *J. Atmos. Ocean. Technol.* **10**, 195.

Dukowicz, J. K., and Smith, R. D. 1994. Implicit free-surface model for the Bryan-Cox-Semtner ocean model. *J. Geophys. Res.* **99**, 7991.

Durran, D. R. 1991. The third-order Adams-Bashforth method: An attractive alternative to leapfrog time differencing. *Mon. Weather Rev.* **119**, 702–720.

Dushaw, B. D., B. D. Cornuelle, P. F. Worcester, B. M. Howe, and D. S. Luther. 1995. Barotropic and baroclinic tides in the central north Pacific Ocean determined from long-range reciprocal acoustic transmissions. *J. Phys. Oceanogr.* **25**, 631–647.

Eady, E. T. 1949. Long waves and cyclone waves. *Tellus* **1**, 33–52.

Eanes, R. J. 1994. Diurnal and semidiurnal tides from TOPEX/POSEIDON altimetry (abstract). *Eos Trans. AGU* **75**(16, Spring Meeting Suppl.), 108.

Ebert, E. E., and J. A. Curry. 1993. An intermediate one-dimensional thermodynamic sea ice model for investigation of ice-atmosphere interactions. *J. Geophys. Res.* **98**, 10085–10109.

Ebert, E. E., J. L. Schramm, and J. A. Curry. 1995. Disposition of solar radiation in sea ice and the upper ocean. *J. Geophys. Res.* **100**, 15965–15975.

Eby, M., and G. Holloway. 1994. Grid transformation for incorporating the Arctic in a global ocean model *Clim. Dynamics* **10**, 241.

Egbert, G. D. 1997. Tidal data inversion: interpolation and inference. *Prog. Oceanogr.* **40**, 53–80.

Egbert, G. D., A. F. Bennett, and M. G. G. Foreman. 1994. TOPEX/POSEIDON tides estimated using a global inverse model. *J. Geophys. Res.* **99**, 24821–24852.

Egbert, G. D., and A. F. Bennett. 1996. Data assimilation methods for ocean tides. In P. Malanotte-Rizzoli (Ed.), *Modern Approaches to Data Assimilation in Ocean Modelling*, pp. 147–179. Elsevier, New York.

Eifler, W., and W. Schrimpf. 1992. *ISPRAMIX, a Hydrodynamic Program for Computing Regional Sea Circulation Patterns and Transfer Processes*. Report EUR 14856EN, European Commission.

Ekman, V. W. 1905. On the influence of the earth's rotation on ocean currents. *Arkiv f. Matem. Astr. a. Fysik (Stockholm) Bd.* 2, no. 11, 53 pp.

Elfouhaily, T., D. Thompson, D. Vandemark, and B. Chapron. 1999. Weakly nonlinear theory and sea state bias estimations. *J. Geophys. Res.* **104**, 7641–7648.

Emery, W. J., L. Kantha, G. A. Wick, and P. Schluessel. 1993. The relationship between skin and bulk sea surface temperatures. In I. S. F. Jones, Y. Sigimori, and R. W. Stewart (Eds.), *Satellite Remote Sensing of Ocean Environment*, 25–40. Seibutsu Kenkyusha, Tokyo.

Emery, W. J., W. G. Lee, and L. Magaard. 1984. Geographic and seasonal distributions of Brunt-Vaisala frequency and Rossby radii in the north Pacific and north Atlantic. *J. Phys. Oceanogr.* **14**, 294–317.

Emery, W. J., and R. E. Thomson. 1998. *Data Analysis Methods in Physical Oceanography*. Pergamon, Elmsford, NY. 634 pp.

Enfield, D. B., and D. A. Meyer. 1997. Tropical Atlantic sea surface temperature variability and its relation to El Niño-southern oscillation. *J. Geophys. Res.* **102**, 929–945.

Engelhardt, D. B. 1996. Estimation of north Pacific ocean dynamics and heat transfer from TOPEX/Poseidon satellite altimetry and a primitive equation ocean model. Ph.D. dissertation, Department of Aerospace Engineering Sciences, Univ. of Colorado, Boulder. 132 pp.

Engquist, B., and A. Majda. 1977. Absorbing boundary conditions for the numerical simulation of waves. *Math. Comput.* **31**, 629–651.

Eriksen, C. C. 1982. Observations of internal wave reflection off sloping bottoms. *J. Geophys. Res.* **87**, 525–538.

Eriksen, C. C. 1991. Observations of amplified flows atop a large seamount. *J. Geophys. Res.* **96,** 15227–15236.

Eubanks, T. M. 1993. Variations in the orientation of the Earth. In D. E. Smith and D. L. Turcott (Eds.), *Contributions of Space Geodesy to Geodynamics: Earth Dynamics,* pp. 1–54. AGU, Washington, DC.

Ezer, T., and G. L. Mellor. 1994. Continuous assimilation of GEOSAT altimeter data into a three-dimensional primitive equation Gulf Stream model. *J. Phys. Oceanogr.* **24,** 832–847.

Fairall, C. W., J. D. Kepert, and G. J. Holland. 1994. The effect of sea spray on surface energy transports over the ocean. *The Global Atmosphere and Ocean System* **2,** 121–142.

Falkowski, P. G., R. T. Barber, and V. Smetacek. 1998. Biogeochemical controls and feedbacks onocean primary production. *Science* **281,** 200–206.

Faller, A. J. 1981. The origin and development of laboratory models and analogues of the ocean circulation. In B. Warren and C. Wunsch (Eds.), *Evolution of Physical Oceanography*, pp. 462–480. MIT Press, Cambridge, MA.

Farge, M. 1992. Wavelet transforms and their applications to turbulence. *Annu. Rev. Fluid Mech.* **24,** 395–457.

Farge, M., N. Kevlahan, V. Perrier, and E. Goirand. 1996. Wavelets and turbulence. *Proc. IEEE* **84,** 639–669.

Farmer, J. D., and J. J. Sidorowich. 1987. Predicting chaotic time series. *Phys. Rev. Lett.* **59,** 845–848.

Farrel, W. E. 1972. Deformation of the Earth by surface loads. *Rev. Geophys.* **10,** 761–797.

Farrow, D. E., and D. P. Stevens. 1995. A new tracer advection scheme for Bryan and Cox type ocean general circulation models. *J. Phys. Oceanogr.* **25,** 1731–1741.

Fichefet, T., and M. A. M. Maqueda. 1997. Sensitivity of a global sea ice model to the treatment of ice thermodynamics and dynamics. *J. Geophys. Res.* **102,** 12609–12646.

Field, C. B., M. J. Behrenfeld, J. T. Randerson, and P. Falkowski. 1998. Primary production of the biosphere: Integrating terrestrial and oceanic components. *Science* **281,** 237–240.

Findlater, J. 1971. Mean monthly air flow at low levels over the western Indian Ocean. *Geophysical. Memoirs,* Vol. 115, Her Majesty's Stn. Off., London. 53 pp.

Fischer, P. F. 1989. Spectral element solution of the Navier-Stokes equations on high performance distributed-memory parallel processors. Ph.D. thesis, MIT, Cambridge, MA. 133 pp.

Flato, G. M., and W. D. Hibler III. 1990. On a simple sea-ice dynamics model of climate studies. *Ann. Glaciol.* **14,** 72–77.

Flato, G. M., and W. D. Hibler III. 1992. Modeling pack ice as a cavitating fluid. *J. Phys. Oceanogr.* **22,** 626–651.

Flato, G. M., and W. D. Hibler III. 1995. Ridging and strength in modeling the thickness distribution of Arctic sea ice. *J. Geophys. Res.* **100,** 18611–18626.

Fleming, G. H., and A. J. Semtner. 1991. A numerical study of interannual ocean forcing on Arctic ice. *J. Geophys. Res.* **96,** 4589–4603.

Fletcher, C. A. J. 1988a. *Computational Techniques for Fluid Dynamics*, Vol. 1: *Fundamental and General Techniques*, Springer-Verlag, Berlin/New York.

Fletcher, C. A. J. 1988b. *Computational Techniques for Fluid Dynamics*, Vol. 2: *Specific Techniques for Different Flow Categories*. Springer-Verlag, Berlin/New York.

Forbes, J. M., and M. E. Hagan. 1988. Diurnal propagating tide in the presence of mean winds and dissipation: A numerical investigation. *Planet. Space Sci.* **36,** 579–590.

Foreman, M. G. G. 1977. *Manual for Tidal Heights Analysis and Prediction.* Pac. Mar. Sci. Rep. 77-10, Inst. for Ocean Sci., Sidney, B.C., Canada.

Foreman, M. G. G., R. F. Henry, R. A. Walters, and V. A. Ballantyne. 1993. A finite element model for tides and resonance along the north coast of British Columbia. *J. Geophys. Res.* **98,** 2509–2532.

Forte, A. M., and J. X. Mitrovica. 1997. A resonance in the Earth's obliquity and precession over the past 20 Myr driven by mantle convection. *Nature* **390,** 676–680.

Foster, T. D. 1972. An analysis of the cabbeling instability in sea water. *J. Phys. Oceanogr.* **2,** 294–301.

Foufoula-Georgiou, E., and P. Kumar (Eds.). 1994. *Wavelets in Geophysics.* Academic Press, San Diego. 372 pp.

Fox, A. D., and S. J. Maskell. 1995. Two-way interactive nesting of primitive equation ocean models with topography. *J. Phys. Oceanogr.* **25,** 2977–2996.

Fox, C. 1997. Estimation of mid-latitude Rossby waves using a simple ocean model and kalman filtering with TOPEX/Poseidon altimeter data. Ph.D. dissertation, Department of Aerospace Engineering Sciences, University of Colorado, Boulder, CO. 129 pp.

Fox, D. N., M. R. Carnes, and J. L. Mitchell. 1992. Characterizing major frontal systems: A nowcast/forecast system for the northwest Atlantic. *Oceanography* **5,** 49–54.

Francis, O., and P. Mazzega. 1990. Global charts of ocean tide loading effects. *J. Geophys. Res.* **95,** 11411–11424.

Fu, L. L., E. J. Christensen, C. A. Yamarone, Jr., M. Lefebvre, Y. Menard, M. Dorrer, and P. Escudier. 1994. TOPEX/POSEIDON mission overview. *J. Geophys. Res.* **99,** 24369–24382.

Fu, L.-L., and R. E. Cheney. 1995. Application of satellite altimetry to ocean circulation studies: 1987–1994. *Rev. Geophys.* **33**(Suppl.), 213–223.

Fu, L.-L., and I. Fukumori. 1996. A case study of the effects of errors in satellite altimetry on data assimilation. In P. Malanotte-Rizzoli (Ed.), *Modern Approaches to Data Assimilation in Ocean Modeling,* pp. 77–96. Elsevier, Amsterdam/New York.

Fu, L.-L., and R. D. Smith. 1996. Global ocean circulation from satellite altimetry and high-resolution computer simulation. *Bull. Am. Meteorol. Soc.* **77,** 2625–2636.

Fukumori, I., J. Benveniste, C. Wunsch, and D. B. Haidvogel. 1993. Assimilation of sea surface topography into an ocean circulation model using a steady-state smoother. *J. Phys. Oceanogr.* **23,** 1831–1855.

Fukumori, I., and P. Malanotte-Rizzoli. 1995. An approximate Kalman filter for ocean data assimilation: An example with an idealized Gulf Stream model. *J. Geophys. Res.* **100,** 6777–6793.

Gabor, D. 1946. Theory of communications. *J. IEEE (London)* **93,** 429–457.

Galperin, B., L. H. Kantha, S. Hassid, and A. Rosati. 1988. A quasi-equilibrium turbulent energy model for geophysical flows. *J. Atmos. Sci.* **45,** 55–62.

Gamage, N., and W. Blumen. 1993. Comparative analysis of low level cold fronts: Wavelet, Fourier, and empirical orthogonal function decompositions. *Mon. Weather Rev.* **121,** 2867–2878.

Gamage, N. K. K., and C. Hagelberg. 1993. Detection and analysis of microfronts and associated coherent events using localized transforms. *J. Atmos. Sci.* **50,** 750–756.

Gambis, D. 1992. Wavelet transform analysis of the length of the day and El Niño/Southern Oscillation variations at interseasonal and interannual time scales. *Ann. Geophys.* **10,** 429–437.

Gangopadhyay, A., and A. R. Robinson. 1997a. Circulation and dynamics of the western North Atlantic. Part I. Multiscale feature models. *J. Atmos. Oceanic Technol.* **14,** 1314–1332.

Gangopadhyay, A., and A. R. Robinson. 1997b. Circulation and dynamics of the western North Atlantic. Part III. Forecasting the meanders and rings. *J. Atmos. Oceanic Technol.* **14,** 1352–1365.

Gao, W., and B. L. Li. 1993. Wavelet analysis of coherent structures at the atmosphere-forest interface. *J. Appl. Meteorol.* **32,** 1717–1725.

Gardini, B., G. Graf, and G. Ratier. 1995. The instruments on ENVISAT. *Acta Astronaut.* **37,** 301–312.

Gargett, A., P. Cummins, and G. Halloway. 1989. Effects of variable vertical diffusivity in the GFDL model. In P. Muller and D. Henderson (Eds.), *Parameterization of Small Scale Processes,* pp. 11–20. Proc. Aha Hulikoa Workshop, Hawaii Institute of Geophysics Special Publication.

Gargett, A., and G. Halloway. 1992. Sensitivity of the GFDL ocean model to different diffusivities for heat and salt. *J. Phys. Oceanogr.* **22,** 1158–1177.

Garwood, R., Jr. 1991. Enhancements to deep turbulent entrainment. In P. C. Chu and J. C. Gascard (Eds.), *Deep Convection and Deep Water Formation in the Oceans,* pp. 197–273. Elsevier, New York.

Garwood, R., Jr., S. Isakari, and P. Gallacher. 1994. Thermobaric convection. In O. M. Johannesen, R. D. Muench, and J. E. Overland (Eds.), *The Polar Oceans and Their Role in Shaping the Global Environment,* pp. 199–209. AGU, Washington, DC.

Gary, J. M. 1973. Estimate of truncation error in transformed coordinate primitive equation atmospheric models. *J. Atmos. Sci.* **30,** 223–233.

Gates, W. L. 1992. AMIP: The atmospheric model intercomparison project. *Bull. Am. Meteorol. Soc.* **73,** 1962–1970.

Gates, W. L., J. F. B. Mitchell, G. Boer, U. Cubasch, and V. P. Meleshko. 1992. Climate modeling, climate prediction and model validation. In J. T. Houghton, B. A. Callander, and S. K. Varney (Ed.), *Climate Change 1992: The Supplementary Report to the Intergovernmental Panel on Climate Change Scientific Assessment,* pp. 97–133. Cambridge Univ. Press, Cambridge.

Gates, W. L., *et al.* 1996. Climate models—Evaluation. In J. T. Houghton, L. G. Meira Filho, B. A. Callander, N. Harris, A. Kattenberg, and K. Maskell (Eds.), *Climate Change 1995: The Science of Climate Change,* pp. 229–284. Cambridge Univ. Press, Cambridge.

Gear, W. C. 1971. *Numerical Initial Value Problems in Ordinary Differenetial Equations.* Prentice-Hall, New York. 253 pp.

Geiger, C. A., W. D. Hibler III, and S. F. Ackley. 1998. Large scale sea ice drift and deformation: Comparison between models and observations in the Western Weddell Sea during 1992. *J. Geophys. Res.* **103,** 21893–21914.

Giering, R., and T. Kamainski. 1996. *Recipes for Adjoint Code Construction.* Tech. Report 212, Max-Planck Institut fur Meteorologie, Hamburg.

Gelb. A. (Ed.). 1988. *Applied Optimal Estimation.* MIT Press, Cambridge, MA. 374 pp.

Gent, P. R., and J. C. McWilliams. 1982. Intermediate model solutions to the Lorenz equations: Strange attractors and other phenomena. *J. Atmos. Sci.* **39,** 3–13.

Gent, P. R., and J. C. McWilliams. 1990. Isopycnal mixing in ocean circulation models. *J. Phys. Oceanogr.* **20,** 150–155.

Gent, P. R., and J. J. Tribbia. 1993. Simulation and predictability in a coupled TOGA model. *J. Clim.* **6,** 1843–1858.

Gent, P. R., J. Willebrand, T. McDougall, and J. C. McWilliams. 1995. Parameterizing eddy-induced tracer transports in ocean circulation models. *J. Phys. Oceanogr.* **25,** 463–474.

Gent, P. R., F. O. Bryan, G. Danabasoglu, S. C. Doney, W. R. Holland, W. G. Large, and J. C. McWilliams. 1998. The NCAR climate system model global ocean component. *J. Clim.* **11,** 1287–1306.

Gerdes, R. 1993. A primitive equation ocean circulation model using a general vertical coordinate transformation. 1. Description and testing of the model. *J. Geophys. Res.* **98,** 14683–14701.

Gerdes, R., C. Koberle, and J. Willebrand. 1991. The influence of numerical advection schemes on the results of ocean general circulation models. *Clim. Dyn.* **5,** 211–226.

Gerstenkorn, H. 1955. Uber Gezeitenreibung beim zweikorperproblem. *Z. Astrophys. B* **36,** 245–274.

Ghil, M., and P. Malanotte-Rizzoli. 1991. Data assimilation in meteorology and oceanography. *Advances in Geophysics,* Vol. 33, pp. 141–266. Academic Press, San Diego, 141-266.

Ghil, M., and A. W. Robertson. 1999. Solving problems with GCMs: General circulation models and their role in the climate modeling hierarchy. In D. Randall (Ed.), *General Circulation Model Development: Past, Present and Future.* Academic Press, San Diego.

Gill, A. E. 1973. Circulation and bottom water production in the Weddel Sea. *Deep-Sea Res.* **20,** 111–140.

Gill, A. E. 1982. *Atmosphere-Ocean Dynamics.* Academic Press, New York. 666 pp.

Ginis, I., M. A. Bender, and Y. Kurihara. 1997. Development of a coupled hurricane-ocean forecast system in the north Atlantic. AMS abstract, 22nd Conference on Hurricanes and Tropical

Meteorology, May 19–23, Fort Collins, Colorado, 443–444.

Ginis, I., R. A. Richardson, and L. M. Rothstein. 1998. Design of a multiply nested primitive equation ocean model. *Mon. Weather Rev.* **126,** 1054–1079.

Gipson, J. M. 1996. VLBI determination of neglected terms in high frequency Earth orientation parameter variation. *J. Geophys. Res.* **101,** 28051–28064.

Glass, L., and M. C. Mackey. 1988. *From Clocks to Chaos.* Princeton Univ. Press, Princeton, NJ. 248 pp.

Glenn, S. M., and A. R. Robinson. 1995. Verification of an operational Gulf Stream forecasting model. In D. R. Lynch and A. M. Davies (Eds.), *Quantitative Skill Assessment for Coastal Ocean Models,* pp. 469–499. American Geophysical Union, Washington, DC.

Godfrey, J. S. 1989. A Sverdrup model of the depth-integrated flow for the world ocean allowing for island circulations. *Geophys. Astrophys. Fluid Dyn.* **45,** 89–112.

Godfrey, J. S. 1996. The effect of the Indonesian throughflow on Indian Ocean circulation and heat exchange with the atmosphere: A review. *J. Geophys. Res.* **101,** 12217–12237.

Goodberlet, M., C. Swift, K. Kiley, J. Miller, and J. Zaitzeff. 1997. Microwave remote sensing of coastal zone salinity. *J. Coastal Res.* **13,** 363–372.

Gordon, R. B., R. A. Flather, J. Wolf, L. H. Kantha, and H. J. Herring. 1986. Modeling the current response of continental shelf waters to winter storms: Comparisons with data. *Offshore Technology Conference OTC 5417,* pp. 503–512.

Gough, W. A. 1997. Numerical diffusion in an isopycnal ocean general circulation model. *Ocean Modell.* **115,** 1–4.

Gower, J. F. R. (Ed.). 1981. *Oceanography from Space.* Plenum, New York.

Graef, F. 1998. On the westward translation of isolated eddies. *J. Phys. Oceanogr.* **28,** 740–745.

Grassberger, P., and I. Procaccia. 1983. Measuring the strangeness of strange attractors. *Physica* **9D,** 189–193.

Gray, W. G. 1982. Some inadequacies of finite element models as simulators of two-dimensional circulation. *Adv. Water Resources* **5,** 171–177.

Greatbatch, R. J. 1994. A note on the representation of steric sea level in models that conserve volume rather than mass. *J. Geophys. Res.* **99,** 12767–12771.

Greatbatch, R. J. 1998. Exploring the relationship between eddy-induced transport velocity, vertical momentum transfer, and the isopycnal flux of potential vorticity. *J. Phys. Oceanogr.* **28,** 422–432.

Greatbatch, R. J., A. F. Fanning, A. D. Goulding, and S. Levitus. 1991. A diagnosis of interpentadal circulation changes in the North Atlantic. *J. Geophys. Res.* **96,** 22009–22023.

Green, J. S. A. 1970. Transfer properties of the large-scale eddies and the general circulation of the atmosphere. *Quart. J. R. Meteorol. Soc.* **96,** 157–185.

Greenberg, D. A., F. E. Werner, and D. R. Lynch. 1998. A diagnostic finite-element ocean circulation model in spherical-polar coordinates. *J. Atmos. Oceanic Technol.* **15,** 942–958.

Gregg, M. C., and T. B. Sanford. 1988. The dependence of turbulent dissipation on stratification in a diffusively stable thermocline. *J. Geophys. Res.* **93,** 12381–12392.

Grell, G. A., J. Dudhia, and D. R. Stauffer. 1994. *A Description of the Fifth-Generation Penn State/NCAR Mesoscale Model (MM5).* NCAR Technical Note 398, National Center for Atmospheric Research, Boulder, CO. 138 pp.

Grejner-Brzezinska, D. A., and C. C. Goad. 1996. Subdaily earth rotation determined from GPS. *Geophys. Res. Lett.* **23,** 2701–2704.

Griffies, S. M. 1998. The Gent-McWilliams skew flux. *J. Phys. Oceanogr.* **28,** 831–841.

Griffies, S. M., A. Gnanadesikan, R. C. Pacanowski, V. Larichev, J. K. Dukowicz, and R. D. Smith. 1998. Isoneutral diffusion in a z-coordinate ocean model. *J. Phys. Oceanogr.* **28,** 805–830.

Grima, N., A. Bentamy, K. Katsaros, Y. Quilfen, P. Delecluse, and C. Levy. 1999. Sensitivity of an oceanic general circulation model forced by satellite wind stress fields. *J. Geophys. Res.* **104,** 7967–7989.

Gross, R. S. 1993. The effect of ocean tides on Earth's rotation as predicted by the results of an ocean tide model. *Geophys. Res. Lett.* **20,** 293–296.

Gross, R. S., B. F. Chao, and S. D. Desai. 1997. Effect of long-period ocean tides on the Earth's polar motion. *Prog. Oceanogr.* **40,** 385–398.

Groves, G. W., and R. W. Reynolds. 1975. An orthogonalized convolution method of tide prediction. *J. Geophys. Res.* **80,** 4131–4138.

Gunther, H., S. Hasselmann, and P. A. E. M. Janssen. 1992. *The WAM Model Cycle 4*, DKRZ Tech. Rep. No. 4. Hamburg, Germany, October 1992.

Gustafson, T., and B. Kullenberg. 1933. Tragheitsstromungen in der Ostsee Medd. Goteborgs Oceanogr. Inst., no. 5, Goteborg.

Gustafsson, N. 1990. Sensitivity of limited area model data assimilation to lateral boundary condition fields. *Tellus* **42A,** 109–115.

Gutfraind, R., and S. B. Savage. 1997. Marginal ice zone rheology: Comparison of results from continuum-plastic models and discrete-particle simulations. *J. Geophys. Res.* **102,** 12647–12661.

Haar, A. 1910. Zur theorie der orthogonalen functionensysteme. *Math. Ann.* **69,** 331–371.

Hackbusch, W. 1985. *Multigrid Methods and Applications*. Springer-Verlag, Berlin/New York.

Hacker, P., E. Firing, and J. C. Kindle, 1998. Bay of Bengal currents during the northeast monsoon. *Geophys. Res. Letters*, **25,** 2769-2773.

Hagelberg, C., and N. K. K. Gamage. 1994. Applications of structure preserving wavelet decomposition to intermittent turbulence. In E. Foufoula-Georgiou and P. Kumar (Eds.), *Wavelets in Geophysics*, pp. 45–80. Academic Press, San Diego.

Haidvogel, D. B., and Robinson, A. R. 1989. Data assimilation. *Dyn. Atmos. Oceans* **13**(Special issue), 171–515.

Haidvogel, D. B., A. Beckmann, and K. S. Hedstrom. 1991a. Dynamical simulations of filament formation and evolution in the coastal transition zone. *J. Geophys. Res.* **96,** 15017–15040.

Haidvogel, D. B., J. Wilkin, and R. Young. 1991b. A semi-spectral primitive equation ocean circulation model using vertical sigma and orthogonal curvilinear horizontal coordinates. *J. Comp. Phys.* **94,** 151–185.

Haidvogel, D. B., and F. O. Bryan. 1993. Ocean circulation modeling. In K. E. Trenberth (Ed.), *Climate System Modeling,* pp. 371, 412. Cambridge University Press, Cambridge.

Haidvogel, D. B., and A. Beckmann. 1997. Numerical modeling of the coastal ocean. In K. H. Brink and A. R. Robinson (Eds.), *The Sea*, Vol. 10, pp. 457–482. Wiley, New York.

Haidvogel, D. B., E. Curchitser, M. Iskandarani, R. Hughes, and M. Taylor, 1997. Global Modelling Using the Spectral Element Method, in *Numerical Methods in Atmospheric and Oceanic Modelling*: The André J. Robert Memorial Volume (companion volume to Atmosphere-Ocean), C.A. Lin, R. Laprise and H. Ritchie editors, pp 505-531; co-published by Canadian Meteorological and Oceanographic Society and National Research Council of Canada.

Hakkinen, S. 1987a. A constitutive law for sea ice and some applications. *Math. Modell.* **9,** 9469–9478.

Hakkinen, S. 1987b. A coupled dynamic-thermodynamic model of an ice-ocean system in the marginal ice zone. *J. Geophys. Res.* **92,** 9469–9478.

Hakkinen, S. 1990. Models and their applications to polar oceanography. In W. O. Smith, Jr. (Ed.), *Polar Oceanography,* Part A: *Physical Science*, pp. 335–384. Academic Press, San Diego.

Hakkinen, S. 1993. An Arctic source for the Great Salinity Anomaly: A simulation of the Arctic ice-ocean system for 1955–1975. *J. Geophys. Res.* **98,** 16397–16410.

Hakkinen, S. 1995. Seasonal simulation of the Southern Ocean coupled ice-ocean system. *J. Geophys. Res.* **100,** 22733–22748.

Hakkinen, S. 1997. Personal communication.

Hakkinen, S., and G. L. Mellor. 1990. One hundred years of Arctic ice cover variations as simulated by a one dimensional coupled ice-ocean model. *J. Geophys. Res.* **95,** 15959–15969.

Hakkinen, S., and G. L. Mellor. 1992. Modeling the seasonal variability of a coupled Arctic ice-ocean system. *J. Geophys. Res.* **97,** 5389–5408.

Hakkinen, S., G. L. Mellor, and L. H. Kantha. 1992. Modeling deep convection in the Greenland Sea. *J. Geophys. Res.* **97**, 5389–5408.

Haltiner, G. J., and R. T. Williams. 1980. *Numerical Prediction and Dynamic Meteorology*, 2nd ed. Wiley, New York. 477 pp.

Hamilton, P. 1990. Deep currents in the Gulf of Mexico. *J. Phys. Oceanogr.* **20**, 1087–1104.

Hamon, B. V. 1962. The spectrums of mean sea level at Sydney, Coff's Harbour, and Lord Howe Island. *J. Geophys. Res.* **67**, 5147–5155.

Haney, R. L. 1971. Surface thermal boundary condition for ocean circulation models. *J. Phys. Oceanogr.* **1**, 156–167.

Haney, R. L. 1991. On the pressure gradient force over steep topography in sigma-coordinate ocean models. *J. Phys. Oceanogr.* **21**, 610–619.

Hansen, D. V., and W. C. Thacker. 1999. Evaluation of salinity profiles in the upper ocean, *J. Geophys. Res.*, **104**, 7921–-7933.

Hansen, K. S. 1982. Secular effects of oceanic tidal dissipation on the lunar orbit and the Earth's rotation. *Rev. Geophys. Space Phys.* **20**, 457–480.

Harder, M., P. Lemke, and M. Hilmer. 1998. Simulation of sea ice transport through Fram Strait: Natural variability and sensitivity to forcing. *J. Geophys. Res.* **103**, 5595–5606.

Harrison, E. J., and R. L. Elsberry. 1972. A method for incorporating nested grids in the solutions of systems of geophysical equations. *J. Atmos. Sci.* **29**, 1235–1245.

Harshvardhan, R. D., D. A. Randall, and T. G. Corsetti. 1987. A fast radiation parameterization for general circulation models. *J. Geophys. Res.* **92**, 1009–1016.

Harten, A. 1983. High resolution schemes for hyperbolic conservation laws. *J. Comput. Phys.* **49**, 357–393.

Hasselmann, K. 1976. PIPs and POPs—A general formalism for the reduction of dynamical systems in terms of principal interaction patterns and principal oscillation patterns. *J. Geophys. Res.* **93**, 11015–11020.

Haurwitz, B., 1956. The geographical distribution of the solar semidiurnal pressure oscillation. *Meteorol. Pap.*, **2**.

Haurwitz, B., and A. D. Cowley. 1973. The diurnal and semidiurnal barometric oscillations, global distribution and annual variations. *Pure Appl. Geophys.* **102**, 193–222.

Heaps, N. (Ed.). 1987. *Three-Dimensional Coastal Ocean Models*. American Geophysical Union, Washington, DC. 208 pp.

Heburn, G. W. 1987. The dynamics of the western Mediterranean Sea: A wind-forced case study. *Ann. Geophys. Ser. B* **5**, 61–74.

Heburn, G. W. 1994. The dynamics of the seasonal variability of the western Mediterranean circulation. In P. E. LaViolette (Ed.), *Seasonal and Interannual Variability of the Western Mediterranean Sea*, pp. 249–285.

Hecht, M. W., W. R. Holland, and P. J. Rasch. 1995. Upwind-weighted advection schemes for ocean tracer transport: An evaluation in a passive tracer context. *J. Geophys. Res.* **100**, 29763–29778.

Hecht, M. W., F. O. Bryan, and W. R. Holland. 1998. A consideration of tracer advection schemes in a primitive equation ocean model. *J. Geophys. Res.* **103**, 3301–3321.

Hedley, M., and M. K. Yau. 1988. Radiation boundary conditions in numerical modeling. *Mon. Weather Rev.* **116**, 1721–1736.

Hedstrom, K. 1995. *User's Manual for a Semi-spectral Primitive Equation Ocean Circulation Model Version 3.9.* Institute of Marine and Coastal Sciences, Rutgers University, New Brunswick, NJ. 131 pp.

Heinrichs, J. F. 1996. Coupled ice/ocean modeling of Baffin bay and the formation of the North Water Polynyas. Ph.D. dissertation, Department of Geography, University of Colorado, Boulder, CO.

Heinrichs, J. F., K. Steffen, and L. Kantha. 1995. Warm water upwelling and polynya formation in the North Water area of Baffin Bay—A study using a two-dimensional coupled ice/ocean model.

Presented at XXI IUGG meeting, Boulder, CO, July 2–14, 1995.

Held, I., and V. D. Larichev. 1998. A scaling theory for horizontally homogeneous, baroclinically unstable flow on a beta-plane. *J. Atmos. Sci.* **53,** 946–952.

Hellerman, S., and M. Rosenstein. 1983. Normal monthly wind stress over the world ocean with error estimates. *J. Phys. Oceanogr.* **13,** 1093–1104.

Hendershott, M. C. 1977. Numerical models of ocean tides. In E. D. Goldberg, I. N. McCave, J. J. O'Brien, and J. H. Steele (Eds.), *The Sea,* Vol. 6, *Marine Modelling,* pp. 47–95. Wiley, New York.

Hendershott, M. 1981. Long waves and ocean tides. In B. Warren and C. Wunsch (Eds.), *Evolution of Physical Oceanography,* pp. 292–341. MIT Press, Cambridge, MA.

Henderson, H. W., and R. Wells. 1988. Obtaining attractor dimensions from meteorological time series. *Adv. Geophys.* **30,** 205–237.

Hendricks, J. R., R. R. Leben, G. H. Born, and C. J. Koblinsky. 1996. Empirical orthogonal function analysis of global TOPEX/POSEIDON altimeter data and implications for detection of global sea level rise. *J. Geophys. Res.* **101,** 14131–14145.

Hendry, R. M. 1977. Observations of the semidiurnal internal tide in the western North Atlantic Ocean. *Philos. Trans. R. Soc. London Ser. A* **286,** 1–24.

Herbaut, C., L. Mortier, and M. Crepon. 1996. A sensitivity study of the general circulation of the western Mediterranean Sea. *J. Phys. Oceanogr.* **26,** 65–84.

Herbaut, C., F. Martel, and M. Crepon. 1997. A sensitivity study of the general circulation of the western Mediterranean Sea. Part II: The response to atmospheric forcing. *J. Phys. Oceanogr.* **27,** 2126–2145.

Herring, T. 1991. The rotation of the Earth. *Rev. Geophys.* **29**(Suppl.), 172–175.

Herring, T. A., and D. Dong. 1994. Measurement of diurnal and semidiurnal rotational variations and tidal parameters of Earth. *J. Geophys. Res.* **99,** 18051–18071.

Hibler, W. D., III. 1979. A dynamic thermodynamic sea ice model. *J. Phys. Oceanogr.* **9,** 815–846.

Hibler, W. D., III. 1980a. Modeling a variable thickness sea ice cover. *Mon. Weather Rev.* **108,** 1943–1973.

Hibler, W. D., III. 1980b. Modeling pack ice as a viscous-plastic continuum: Some preliminary results. In R. S. Pritchard (Ed.), *Sea Ice Processes and Models*, pp. 163–176. Univ. Washington Press, Seattle.

Hibler, W. D., III. 1980c. Sea ice growth, drift, and decay. In *Dynamics of Snow and Ice Masses,* pp. 141–209. Academic Press, San Diego, CA.

Hibler, W. D., III. 1985. Modeling sea-ice dynamics. *Adv. Geophys.* **28,** 549–580.

Hibler, W. D., III. 1986. Ice dynamics. In N. Untersteiner (Ed.), *The Geophysics of Sea Ice*, pp. 577–640. Plenum, New York.

Hibler, W. D., III, and J. E. Walsh. 1982. On modeling seasonal and interannual fluctuations of Arctic sea ice. *J. Phys. Oceanogr.* **12,** 1514–1523.

Hibler, W. D., III, and K. Bryan. 1987. A diagnostic ice-ocean model. *J. Phys. Oceanogr.* **17,** 987–1015.

Hibler, W. D., III, and C. F. Ip. 1995. The effect of sea ice rheology on Arctic buoy drift. In J. P. Dempsey and Y. D. S. Rajapakse (Eds.), *Ice Mechanics,* Vol. 204, pp. pp.255–264. Am. Soc. Mech. Eng., New York.

Hibler, W. D., III, and E. M. Schulson. 1997. On modeling sea ice fracture and flow in numerical investigations of climate. *Ann. Glaciol.* **25,** 26–32.

Hickey, B., M. 1979. The California current system—Hypotheses and facts. *Prog. Oceanogr.* **8,** 191–279.

Hickey, B. M., and N. E. Pola. 1983. The seasonal alongshore pressure gradient on the west coast of the United States. *J. Geophys. Res.* **88,** 7623–7633.

Hide, R., and J. O. Dickey. 1991. Earth's variable rotation. *Science* **253,** 629–637.

Hirsch, C. 1988. *Numerical Computation of Internal and External Flows.* Wiley, New York. 515 pp.

Hirst, A. C., and T. J. McDougall. 1998. Meridional overturning and dianeutral transport in a z-coordinate ocean model including eddy-induced advection. *J. Phys. Oceanogr.* **28,** 1205–1223.

Hodur, R. M. 1996. The Naval Research Laboratory's Coupled Ocean/Atmosphere Mesoscale Prediction System (COAMPS). *Mon. Wea. Rev.* **125,** 1414–1430.

Hoffmann, K. A., and S. T. Chiang. 1993. *Computational Fluid Dynamics for Engineers,* Vols. 1 and 2. Engineering Education System, Kansas.

Hogan, P. J., H. E. Hurlburt, G. A. Jacobs, A. J. Wallcraft, W. J. Teague, and J. L. Mitchell. 1992. Simulation of GEOSAT, TOPEX/Poseidon, and ERS-1 altimeter data from 1/8° Pacific Ocean model: Effects of space-time resolution on mesoscale sea surface height variability. *Mar. Technol. Soc. J.* **26,** 98–107.

Holden, A. V., and M. A. Muhamad. 1986. A graphical zoo of strange and peculiar attractors, 15–35.

Holland, D. M., L. A. Mysak, D. K. Manak, and J. M. Oberhuber. 1993. Sensitivity study of a dynamic thermodynamic sea ice model. *J. Geophys. Res.* **98,** 2561–2586.

Holland, M. M., J. A. Curry, and J. L. Schramm. 1997. Modeling the thermodynamics of a sea ice thickness distribution. 2. Sea ice/ocean interactions. *J. Geophys. Res.* **102,** 23093–23108.

Holland, W. R. 1973. Baroclinic and topographic influences on the transport in western boundary currents. *Geophys. Fluid Dyn.* **4,** 187–210.

Holland, W. R. 1978. The role of mesoscale eddies in the general circulation of the ocean; numerical experiments using a wind-driven quasigeostrophic model. *J. Phys. Oceanogr.* **8,** 363–392.

Holland, W. R. 1985. Quasi-geostrophic modeling of eddy-resolved ocean circulation. In J. J. O'Brien (Ed.), *Advanced Physical Oceanographic Numerical Modeling*, pp. 203–231. Reidel, Dordrecht.

Holland, W. R. 1989. Experiences with various parameterizations of sub-grid scale dissipation and diffusion in numerical models of ocean circulation. In P. Muller and D. Henderson (Eds.), *Parameterization of Small Scale Processes*, pp. 1–9. Proc. Aha Hulikoa Workshop, Hawaii Institute of Geophysics Special Publication.

Holland, W. R., and A. D. Hirschman. 1972. A numerical calculation of the circulation of the North Atlantic Ocean. *J. Phys. Oceanogr.* **2,** 336–352.

Holland, W. R., and L. B. Lin. 1975a. Generation of mesoscale eddies and their contribution to the oceanic general circulation. Part 1. Preliminary numerical experiment. *J. Phys. Oceanogr.* **5,** 642–657.

Holland, W. R., and L. B. Lin. 1975b. Generation of mesoscale eddies and their contribution to the oceanic general circulation. Part 2. Parameter study. *J. Phys. Oceanogr.* **5,** 642–657.

Holland, W. R., J. C. Chow, and F. O. Bryan. 1998. Application of a third order upwind scheme in the NCAR ocean model *J. Clim.* **11,** 1487–1493.

Holschneider, M., R. Kronland_Martinet, J. Morlet, and Ph. Tchamitchian. 1989. A real-time algorithm for signal analysis with the help of the wavelet transform. In J. Combes, A. Grossmann, and Ph. Tchamitchian (Eds.), *Wavelets: Time-Frequency Methods and Phase Space*, pp. 286–297. Springer-Verlag, New York.

Holt, M. 1988. *Numerical Methods in Fluid Dnamics.* Springer-Verlag, Berlin/New York.

Holton, J. R. 1992. *An Introduction to Dynamic Meteorology,* 2nd ed. Academic Press, San Diego. 511 pp.

Horel, J. D. 1981. A rotated principal component analysis of the interannual variability of the northern hemisphere 500-mb height field. *Mon. Weather Rev.* **109,** 2080–2092.

Horton, C., M. Clifford, D. Cole, J. Schmitz, and L. Kantha. 1991. Water circulation modeling system for the Persian Gulf. In *Proceedings of 1991 Marine Technology Society Meeting, New Orleans.*

Horton, C., M. Clifford, D. Cole, J. Schmitz, and L. Kantha. 1992. Operational modeling: Semi-enclosed basin modeling at the Naval Oceanographic Office. *Oceanography Magazine of the Oceanography Society.*

Horton, C., M. Clifford, J. Schmitz, and L. Kantha. 1997. A real-time oceanographic nowcast/forecast system for the Mediterranean Sea. *J. Geophys. Res.* **102,** 25123–25156.

Howard, L. N. Note on the paper of John W. Miles. *J. Fluid Mech.* **13,** 158–160.

Howell, J. F., and L. Mahrt. 1994. An adaptive decomposition application to turbulence. In E. Foufoula-Georgiou and P. Kumar (Eds.), *Wavelets in Geophysics*, pp. 107–128. Academic Press, San Diego.

Hsu, H. 1994. Relationship between tropical heating and global circulation: Interannual variability. *J. Geophys. Res.* **99,** 10473–10489.

Huang, R. X. 1998. Mixing and available potential energy in a Boussinesq ocean. *J. Phys. Oceanogr.* **28,** 669–678.

Huang, R. X. 1999. Mixing and energetics of the oceanic thermohaline circulation. *J. Phys. Oceanogr.* **29,** 727–746.

Huang, R. X., and W. K. Dewar. 1996. Haline circulation: Bifurcation and chaos. *J. Phys. Oceanogr.* **26,** 2093–2106.

Hubbard, B. B. 1996. *The World According to Wavelets*. Peters, Wellseley, MA. 264 pp.

Hudgins, L. H., C. A. Friehe, and M. E. Mayer. 1993. Wavelet transform and atmospheric turbulence. *Phys. Rev. Lett.* **71,** 3279–3282.

Hunke, E. C., and J. K. Dukowicz. 1997. An elastic-viscous-plastic model for sea ice dynamics. *J. Phys. Oceanogr.* **27,** 1849–1867.

Hurlburt, H. E., and J. D. Thompson. 1980. A numerical study of the Loop Current intrusions and eddy shedding. *J. Phys. Oceanogr.* **10,** 1611–1631.

Hurlburt, H. E., and J. D. Thompson. 1982. The dynamics of the Loop Current and shed eddies in a numerical model of the Gulf of Mexico. In J. C. J. Nihoul (Ed.), *Hydrodynamics of Semi-enclosed Seas*, pp. 243–298. Elsevier, New York.

Hurlburt, H. E., D. N. Fox, and E. J. Metzger. 1990. Statistical inference of weakly correlated subthermocline fields from satellite altimeter data. *J. Geophys. Res.* **95,** 11375–11409.

Hurlburt, H. E., A. J. Wallcraft, Z. Sirkes, and E. J. Metzger. 1992. Modeling of the global and Pacific oceans: On the path to eddy-resolving ocean prediction. *Oceanography* **5,** 9–18.

Huthnance, J. M. 1977. On trapped waves over a continental shelf. *J. Fluid Mech.* **69,** 689–704.

Huthnance, J. M. 1978. On coastal trapped waves: Analysis and numerical calculation by inverse iteration. *J. Phys. Oceanogr.* **8,** 74–92.

Huthnance, J. M. 1984. Slope currents and "JEBAR." *J. Phys. Oceanogr.* **14,** 795–810.

Ida, S., R. M. Canup, and G. R. Stewart. 1997. Lunar accretion from an impact-generated disk. *Nature* **389,** 353–357.

Ierley, G. R. 1987. On the onset of recirculation in barotropic general circulation models. *J. Phys. Oceanogr.* **17,** 2366–2374.

Ikeda, M. 1988. A three-dimensional coupled ice-ocean model of coastal circulation. *J. Geophys. Res.* **93,** 10731–10748.

Ineson, S., and M. K. Davey. 1994. Some results from a coupled TOGA model. In *Proc. IOC WestPac III Meeting*.

Ingraham, W. J., Jr., C. C. Ebbesmeyer, and R. A. Hinrichsen. 1998. Imminent climate and circulation shift in the northeast Pacific Ocean could have major impact on marine resources. *EOS Trans. AGU* **79,** 197.

INO. 1986. *Ocean Prediction Workshop, a Status and Prospectus Report on the Scientific Basis and the Navy's Needs*. Institute of Naval Oceanography Report, Stennis Space Center, MS.

International Hydrographic Organization. 1979. Tidal Constituent Bank, station catalogue, Ocean and Aquatic Sci., Dept. of Fish. and Oceans, Ottawa, Ontario.

Ip, C. F., W. D. Hibler III, and G. M. Flato. 1991. On the effect of rheology on seasonal sea ice simulations. *Ann. Glaciol.* **15,** 17–25.

Ip, J. T. C., and D. Lynch. 1994. *Three-Dimensional Shallow Water Hydrodynamics on Finite Elements: Nonlinear Time-Stepping Prognostic Model*. Report NML-94-1, Dartmouth College, Hanover, NH.

Iskandarani, M., D. B. Haidvogel, and J. P. Boyd. 1995. A staggered spectral element model with applications to the oceanic shallow water equations. *Int. J. Numer. Methods Fluids* **20,** 393–414.

Israeli, M., and S. A. Orzag. 1981. Approximation of radiation boundary conditions. *J. Comput. Phys.* **41,** 115–135.

Jacobs, G. A., W. J. Teague, J. L. Mitchell, and H. E. Hurlburt. 1996. An examination of the north Pacific Ocean in the spectral domain using GEOSAT altimeter data and a 1/8° 6-layer Pacific Ocean model. *J. Geophys. Res.* **101,** 1025–1044.

Jacobs, G. A., W. J. Teague, S. K. Riedlinger, and R. H. Preller. 1998. Sea surface height variations in the Yellow and East China Seas 2. SSH variability in the weekly and semiweekly bands. *J. Geophys. Res.* **103,** 18,479–18,496.

Janssen, P. A. E. M. 1991. The quasi-linear theory of wind wave generation applied to wave forecasting. *J. Phys. Oceanogr.* **21,** 1631–1642.

Jeffreys, H. 1920. Tidal friction in shallow seas. *Philos. Trans. R. Soc. London Ser. A* **221,** 239.

Jelesnianski, C. P. 1972. *SPLASH (Special Programs to List Amplitudes of Surges from Hurricanes); Part II: General Track and Variant Storm Conditions.* NOAA Tech. Mem. NWS TDL-52, National Weather Service, Silver Spring, MD.

Jenkins, G. M., and D. G. Watts. 1968. *Spectral Analysis and Its Applications.* Holden-Day, Oakland, CA. 523 pp.

Ji, M., A. Leetmaa, and V. E. Kousky. 1996. Coupled model predictions of ENSO during the 1980s and the 1990s at the National Centers for Environmental Prediction. *J. Clim.* **9,** 3105–3120.

Ji, M., D. W. Behringer, and A. Leetmaa. 1998. An improved coupled model for ENSO prediction and implications for ocean initialization. Part II. The coupled model. *Mon. Weather Rev.* **126,** 1022–1034.

Johannessen, O. M., R. D. Muench, and J. E. Overland (Eds.). 1995. *The Polar Oceans and Their Role in Shaping the Global Environment.* The Nansen Centennial Volume, American Geophysical Union, Washington, DC. 525 pp.

JOI. 1990. *Coastal Ocean Prediction Systems Program: Understanding and Managing Our Coastal Ocean.* Joint Oceanographic Institutions Report, Washington, DC.

Jones, H., and J. Marshall. 1993. Convection with rotation in a neutral ocean: A study of open-ocean deep convection. *J. Phys. Oceanogr.* **23,** 1009–1039.

Kagan, B. A. 1997. Earth-Moon tidal evolution: Model results and observational evidence. *Prog. Oceanogr.* **40,** 109–124.

Kagan, B. A., and J. Sundermann. 1996. Dissipation of tidal energy, paleotides, and the evolution of the Earth-Moon system. *Adv. Geophys.* **38,** 179–266.

Kaiser, G. 1994. *A Friendly Guide to Wavelets.* Birkhauser, Cambridge, MA. 300 pp.

Kalman, R. 1960. A new approach to linear fitting and prediction problems. *Trans. ASME Ser. D, J. Basic Eng.* **82,** 35–45.

Kalnay, E., *et al.* 1996. The NCEP/NCAR 40-year reanalysis project. *Bull. Amer. Meteorol. Soc.* **77,** 437–471.

Kalnay-Rivas, E., A. Bayliss, and J. Storch. 1977. The 4th order GISS model of the global atmosphere. *Contrib. Atmos. Phys.* **50,** 306–311.

Kampf, J., and J. O. Backhaus. 1998. Shallow, brine-driven free convection in polar oceans: Nonhydrostatic numerical process studies. *J. Geophys. Res.* **103,** 5577–5594.

Kantha, L. H. 1984. On leaky coastal trapped waves. *Ocean Modell.* **60,** 9–12.

Kantha, L. H. 1985. Comments on "On tidal motion in a stratified inlet, with particular reference to boundary conditions." *J. Phys. Oceanogr.* **15,** 1608–1609.

Kantha, L. H. 1995a. A numerical model of Arctic leads. *J. Geophys. Res.* **100,** 4653–4672.

Kantha, L. H. 1995b. Barotropic tides in the global oceans from a nonlinear tidal model assimilating altimetric tides. 1. Model description and results. *J. Geophys. Res.* **100,** 25283–25308.

Kantha, L. H. 1998. Tides—A modern perspective. *Mar. Geodesy* **21,** 275–297.

Kantha, L. H., H. J. Herring, and G. L. Mellor. 1985. Investigation and simulation of storm current events in the Mid-Atlantic Bight, Dynalysis of Princeton Report December 1985. Princeton, NJ. 89 pp.

Kantha, L. H., G. L. Mellor, and A. F. Blumberg. 1982. A diagnostic calculation of the general circulation in the South Atlantic Bight. *J. Phys. Oceanogr.* **12,** 805–819.

Kantha, L. H., and G. L. Mellor. 1989a. A numerical model of the atmospheric boundary layer over a marginal ice zone. *J. Geophys. Res.* **94,** 4959–4970.

Kantha, L. H., and G. L. Mellor. 1989b. Application of a two-dimensional coupled ocean-ice model to the Bering Sea marginal ice zone. *J. Geophys. Res.* **94,** 10921–10935.

Kantha, L. H., A. F. Blumberg, and G. L. Mellor. 1990. Computing phase speeds at an open boundary. *J. Hydraul. Engineer.* **116,** 592–597.

Kantha, L. H., and S. Piacsek. 1993. Ocean models. In *Computational Science Education Project*, pp. 273–361. Dept. of Energy Electronic Book (http://csep1.phy.ornl.gov/ csep.html).

Kantha, L. H., and C. A. Clayson. 1994. An improved mixed layer model for geophysical applications. *J. Geophys. Res.* **99,** 25235–25266.

Kantha, L., K. Whitmer, and G. Born. 1994. The inverted barometer effect in altimetry: A study in the North Pacific. *TOPEX/Poseidon Res. News* **2,** 18–23.

Kantha, L. H., C. Tierney, J. W. Lopez, S. D. Desai, M. E. Parke, and L. Drexler. 1995. Barotropic tides in the global oceans from a nonlinear tidal model assimilating altimetric tides. 2. Altimetric and geophysical implications. *J. Geophys. Res.* **100,** 25309–25317.

Kantha, L. H., I.-K. Bang, J.-K. Choi, and M.-S. Suk. 1996. Shallow water tides in the Yellow Sea. *J. Korean Soc. Oceanogr.* **31,** 123–133.

Kantha, L. H., and J.-K. Choi. 1997. A real-time nowcast/forecast system for the Gulf of Mexico. www-ccar.colorado.edu/~jkchoi/gomforecast.html.

Kantha, L. H., and S. Piacsek. 1997. Computational ocean modeling. In A. B. Tucker, Jr. (Ed.), *The Computer Science and Engineering Handbook*, pp. 934–958. CRC Press, Boca Raton, FL.

Kantha, L. H., and C. C. Tierney. 1997. Global baroclinic tides. *Prog. Oceanogr.* **40,** 163–188.

Kantha, L. H., J. S. Stewart, and S. D. Desai. 1998. Long period lunar fortnightly and monthly ocean tides. *J. Geophys. Res.* **103,** 12639–12648.

Kantha, L. H., and C. A. Clayson. 2000. *Small Scale Processes in Geophysical Flows*. Academic Press, San Diego, CA.

Kantha, L. H., J.-K. Choi, R. R. Leben, C. Cooper, K. Schaudt, M. Vogel, and J. Feeney. 1999. Hindcasts and real-time nowcast/forecasts of currents in the Gulf of Mexico, Offshore Technology Conference, Houston, TX, May 3–6, 1999.

Katayama, A. 1972. A simplified scheme for computing radiative transfer in the troposphere. In *Numerical Simulation of Weather and Climate*. Dept. Meteorol. Tech Rep. 6, University of Los Angeles, Los Angeles.

Kattenberg, A., *et al.* 1996. Climate models—Projections of future climate. In J. T. Houghton, L. G. Meira Filho, B. A. Callander, N. Harris, A. Kattenberg, and K. Maskell (Eds.), *Climate Change 1995: The Science of Climate Change*, pp. 229–284. Cambridge Univ. Press, Cambridge.

Kawamura, R. 1994. A rotated EOF analysis of global sea surface temperature variability with interannual and interdecadal scales. *J. Phys. Oceanogr.* **24,** 707–715.

Keen, T. R., and S. M. Glenn. 1995. A coupled hydrodynamic-bottom boundary layer model of storm and tidal flow in the Middle Atlantic Bight of North America. *J. Phys. Oceanogr.* **25,** 391–406.

Kennelly, M. A., Evans, R. H., and T. N. Joyce, 1985. Small-scale cyclones at the periphery of warm-core ring, *J. Geophys. Res.*, **90,** 8845-8857.

Kessler, W. S. 1990. Observations of long Rossby waves in the northern tropical Pacific. *J. Geophys. Res.* **95,** 5183–5217.

Khedouri, E., C. Szczechowski, and R. E. Cheney. 1983. Potential oceanographic applications of satellite altimetry for inferring subsurface thermal structure, Oceans 83. *Proc. Mar. Technol. Soc.* **1,** 274–280.

Kiehl, J. T. 1992. Atmospheric general circulation modeling. In K. E. Trenberth (Ed.), *Climate System Modeling*, pp. 319–370. Cambridge Univ. Press, New York.

Killworth, P. D. 1977. Mixing on the Weddel Sea continental slope. *Deep-Sea Res.* **24,** 427–448.

Killworth, P. D. 1979. On chimney formation in the ocean. *J. Phys. Oceanogr.* **9,** 531–554.

Killworth, P. D. 1983. Deep convection in the world ocean. *Rev. Geophys.* **21,** 1–26.

Killworth, P. D., D. Stainforth, D. J. Webb, and S. M. Paterson. 1991. The development of a free surface Bryan-Cox-Semtner ocean model. *J. Phys. Oceanogr.* **21,** 1333–1348.

Killworth, P. D., and N. R. Edwards. 1997. A turbulent bottom boundry layer code for use in numerical ocean models. *Ocean Modell.* **114,** 6–9.

Kim, K., Y.-G. Kim, Y.-K. Cho, S. C. Hwang, M. Danchenkov, A. Scherbinin, and S. Yarosh. 1994. CTD Observation of CREAMS. *Third Workshop of Circulation Research of the East Asian Marginal Seas (CREAMS),* pp. 23–30.

Kim, K., *et al.* 1997. Preliminary report on physical observations during CREAMS'96 summer expedition. In *Proc. CREAMS'97 International Symposium, Fukuoka, Japan, 28–30 Jan. 1997,* pp. 47–50.

Kim, S.-J., T. J. Crowley, and A. Stossel. 1998. Local orbital forcing of Antarctic climate change during the last interglacial. *Science* **280,** 728–730.

Kim, Y.-J. 1996. Representation of subgrid-scale orographic effects in a general circulation model. Part I. Impact on the dynamics of simulated January climate. *J. Clim.* **9,** 2698–2717.

Kindle, J. C., and J. D. Thompson. 1989. The 26- and 50-day oscillations in the western Indian Ocean: Model results, *J. Geophys. Res.* **94,** 4721–4736.

Klemp, J. B., and D. R. Durran. 1983. An upper boundary condition permitting internal gravity wave radiation in numerical mesoscale models. *Mon. Weather Rev.* **111,** 430–445.

Klinck, J. M. 1995. Circulation near submarine canyons: A modeling study. *J. Geophys. Res.* **101,** 1211–1223.

Kline, L., and C. Swift. 1977. An improved model for the dieelectric constant of sea water at microwave frequencies. *IEEE J. Oceanic Eng.* **OE-2,** 104–111.

Komen, G. J., L. Cavaleri, M. Donelan, K. Hasselmann, S. Hasselmann and P. A. E. M. Janssen, 1994. *Dynamics and Modelling of Ocean Waves.* Cambridge Univ. Press, 532 pp.

Kowalik, Z., and T. S. Murty. 1993. *Numerical Modeling of Ocean Dynamics.* World Scientific, Singapore. 481 pp.

Kowalik, Z., and A. Yu. Proshutinsky. 1993. Diurnal tides in the Arctic Ocean. *J. Geophys. Res.* **98,** 16449–16468.

Kowalik, Z., and A. Yu. Proshutinsky. 1995. The Arctic Ocean tides. In O. M. Johannessen, R. D. Muench, and J. E. Overland (Eds.), *The Polar Oceans and Their Role in Shaping the Global Environment,* pp. 137–158. American Geophysical Union, Washington, DC.

Kowalik, Z., and I. Polyakov. 1998. Tides in the Sea of Okhotsk. *J. Phys. Oceanogr.* **28,** 1389–1409.

Kraus, E. B., and J. S. Turner. 1967. A one-dimensional model of the seasonal thermocline II: The general theory and its consequences. *Tellus* **19,** 98–106.

Kudryavtsev, V. N., V. K. Makin, and B. Chapron. 1999. Coupled sea surface-atmosphere model, 2, Spectrum of short wind waves. *J. Geophys. Res.* **104,** 7625–7640.

Kumar, P., and E. Foufoula-Georgiou. 1994. Wavelet analysis in geophysics: An introduction. In E. Foufoula-Georgiou and P. Kumar (Eds.), *Wavelets in Geophysics,* pp. 1–43. Academic Press, San Diego.

Kumar, P., and E. Foufoula-Georgiou. 1997. Wavelet analysis for geophysical applications. *Rev. Geophys.* **35,** 385–412.

Kundu, P. K., J. S. Allen, and R. L. Smith. 1975. Modal decomposition of the velocity field near the Oregon coast. *J. Phys. Oceanogr.* **5,** 683–704.

Kundu, P. K., and J. S. Allen. 1976. Some three-dimensional characteristics of low-frequency current fluctuations near the Oregon coast. *J. Phys. Oceanogr.* **6,** 181–199.

Kunze, E., and T. B. Sanford. 1996. Abyssal mixing: Where it is not. *J. Phys. Oceanogr.* **26,** 2286–2296.

Kurihara, Y. 1965. On the use of implicit and iterative methods for the time integration of the wave equation. *Mon. Weather Rev.* **93,** 33–46.

Kurihara, Y., G. J. Tripoli, and M. A. Bender. 1979. Design of a movable nested-mesh primitive equation model. *Mon. Weather Rev.* **107,** 239–249.

Kurihara, Y., C. L. Kerr, and M. A. Bender. 1989. An improved numerical scheme to treat the open lateral boundary of a regional model. *Mon. Weather Rev.* **117,** 2714–2722.

Kurihara, Y., R. E. Tuleya, and M. A. Bender. 1998. The GFDL hurricane prediction system and its performance in the 1995 hurricane season. *Mon. Weather Rev.* **126,** 1306–1322.

Lab Sea Group. 1997. *The Labrador Sea Deep Convection Experiment.* Center for Global Change Science Report, MIT, Cambridge, MA.

Lagerloef, G. S. E., C. T. Swift, and D. M. Le Vine. 1995. Sea surface salinity: The next remote sensing challenge. *Oceanography* **8,** 44–50.

Lai, C. A., W. Qian, and S. M. Glenn. 1994. Data assimilation and model evaluation experiment data sets., *Bull. Am. Meteorol. Soc.*, **75,** 793–810.

Lambeck, K. 1980. *The Earth's Variable Rotation.* Cambridge University Press, Cambridge, U.K. pp. 449.

Lambeck, K. 1988. *Geophysical Geodesy.* Oxford Univ. Press, London. 718 pp.

Landau, L. D., and E. M. Lifshitz. 1959. *Fluid Mechanics.* Pergamon, Elmsford, NY.

Laplace, P. S. 1776. Recherches sur plusierurs points du systeme du monde. *Memoires de l'Academie Royale des Sciences de Paris*, **88,** 75–182.

Large, W. G., J. C. McWilliams, and S. Doney. 1994. Oceanic vertical mixing: A review and a model with a nonlocal boundary layer parameterization. *Rev. Geophys.* **32,** 363–403.

Laskar, J. 1990. The chaotic motion of the solar system: A numerical estimate of the size of the chaotic zones. *Icarus* **88,** 266–291.

Latif, M. 1998. Dynamics of interdecadal variability in coupled ocean-atmosphere models. *J. Clim.* **11,** 602–624.

Latif, M., T. Stockdale, J. Wolff, G. Burgers, E. Maier-Reimer, M. M. Junge, K. Arpe, and L. Bengtsson. 1994a. Climatology and variability in the ECHO coupled GCM. *Tellus* **46A,** 351–366.

Latif, M., T. P. Barnett, M. A. Cane, M. Flugel, N. E. Graham, H. von Storch, J.-S. Xu, and S. E. Zebiak. 1994b. A review of ENSO prediction studies. *Clim. Dyn.* **9,** 167–179.

Lau, K.-M., and H. Weng. 1995. Climate signal detection using wavelet transform: How to make a time series sing. *Bull. Am. Meteorol. Soc.* **76,** 2391–2402.

Leben, R. R., G. H. Born, and J. D. Thompson. 1990. Mean sea surface and variability of the Gulf of Mexico using Geosat altimetry data. *J. Geophys. Res.* **95,** 3025.

LeBlond, P. H., and L. A. Mysak. 1978. *Waves in the Ocean.* Elsevier, Amsterdam/New York.

LeBlond, P. H., and L. A. Mysak. 1979. Ocean waves: A survey of some recent results. *SIAM Rev.* **21,** 289–328.

Lee, D. C., A. N. Halliday, G. A. Snyder, and L. A. Taylor. 1997. Age and origin of the Moon. *Science* **278,** 1098–1103.

Leetma, A., and M. Ji. 1989. Operational hindcasting of the tropical Pacific. *Dyn. Atmos. Oceans* **13,** 465–490.

Leetmaa, A., and M. Ji. 1996. Ocean data assimilation as a component of a climate forecast system. In P. Malanotte-Rizzoli (Ed.), Modern Approaches to Data Assimilation in Ocean Modeling, pp. 271–295.ed. by P. Malanotte-Elsevier, Amsterdam/New York.271–295, 1996.

Legg, S., J. McWilliams, and J. Gao. 1998. Localization of deep ocean convection by a mesoscale eddy. *J. Phys. Oceanogr.* **28,** 944–970.

Lehmann, A. 1995. A three-dimensional baroclinic eddy resolving model of the Baltic Sea. *Tellus* **47A,** 1013–1031.

Lemke, P. 1987. A coupled one-dimensional sea ice-mixed layer model. *J. Geophys. Res.* **92,** 13164–13172.

Lemke. P., W. D. Hibler III, G. M. Flato, M. Harder, and M. Kreyscher. 1998. On the improvement of sea-ice models for climate simulations: The sea-ice model intercomparison project. *Ann. Glaciol.*, **25.**

Lemoine, F. G., N. K. Pavlis, S. C. Kenyon, R. H. Rapp, E. C. Pavlis, and B. F. Chao. 1998. New high-resolution model developed for Earth's gravitational field. *EOS Trans. AGU* **79,** 113.

Lentz, S. J., and D. C. Chapman. 1989. Seasonal differences in the current and temperature variability over the northern california shelf during the Coastal Ocean Dynamics Experiment. *J. Geophys. Res.* **94,** 12571–12592.

Leonard, B. P. 1979. A stable and accurate convective modeling procedure based upon quadratic upstream interpolation. *Comput. Methods Appl. Mech.* **19,** 59–98.

Le Provost, C., and P. Vincent. 1986. Some tests of precision for finite element model of ocean tides. *J. Computational Phys.* **65,** 273–291.

Le Provost, C., F. Lyard, and J.-M. Molines. 1991. Improving ocean tide predictions by using additional semidiurnal constituents from spline interpolation in the frequency domain. *Geophys. Res. Lett.* **18,** 845–848.

Le Provost, C., M. L. Genco, F. Lyard, P. Vincent, and P. Canceil. 1994. Spectroscopy of the world tides from a finite element hydrodynamic model. *J. Geophys. Res.* **99,** 24777–24797.

Le Provost, C., F. Lyard, J. M. Molines, M. L. Genco, and F. Rabilloud. 1998. A hydrodynamic ocean tide model improved by assimilating a satellite altimeter-derived data set. *J. Geophys. Res.* **103,** 5513–5529.

Le Traon, P.-Y., and F. Ogor. 1998. ERS-1/2 orbit improvement using TOPEX/Poseidon: The 2 cm challenge. *J. Geophys. Res.* **103,** 8045–8057.

Leuliette, E. W. 1998. Two TOPEX/POSEIDON studies: Steric and tidal ocean modes. Doctoral dissertation, Dept. Physics, University of Colorado, Boulder. 153 pp.

Levine, E. R., D. N. Connors, R. S. Shell, and R. C. Hanson. 1997. Autonomous underwater vehicle-based hydrographic sampling. *J. Atmos. Oceanic Tech.* **14,** 1444–1454.

Levitus, S. 1982. *Climatological Atlas of the World Ocean.* NOAA Professional Paper 13, Geophys. Fluid Dyn. Lab., Princeton, NJ. 173 pp.

Levitus, S., and T. P. Boyer. 1994. *World Ocean Atlas 1994 Volume 2: Temperature.* U.S. Department of Commerce, Washington, DC. 117 pp.

Levitus, S., R. Burgett, and T. P. Boyer. 1994. *World Ocean Atlas 1994 Volume 3: Salinity.* U.S. Department of Commerce, Washington, DC. 99 pp.

Lewis, J., and J. Derber. 1985. The use of adjoint equations to solve a variational adjustment problem with advective constraints. *Tellus* **37,** 309–327.

Liandrat, J., and F. Moret-Bailly. 1990. The wavelet transform: Some applications to fluid dynamics and turbulence. *Eur. J. Mech. B Fluids* **9,** 1–19.

Lie, H.-J., and C.-H. Cho. 1997. Surface current field in the eastern East China Sea. In *Proc. CREAMS'97 International Symposium, Fukuoka, Japan, 28–30 Jan. 1997,* pp. 133–136.

Lindstrom, E., F. Bingham, J. Kindle, P. Saunders, and S. Wiffels. 1998. Preface(to special section on world ocean circulation experiment (WOCE): Pacific results). *J. Geophys. Res.* **103,** 12897.

Lindzen, R. S. 1979. Atmospheric tides. *Annu. Rev. Earth Planet. Sci.,* 199–225.

Lindzen, R. S. 1990. *Dynamics in Atmospheric Physics.* Cambridge Univ. Press, Cambridge. 310 pp.

Lissauer, J. L. 1997. It's not easy to make the Moon. *Nature* **389,** 327–328.

Lithgow-Bertelloni, C., and M. A. Richards. 1998. The dynamics of Cenozoic and Mesozoic plate motions. *Rev. Geophys.* **36,** 27–78.

Liu, P. C. 1994. Wavelet spectrum analysis and ocean wind waves. In E. Foufoula and P. Kumar (Eds.), *Wavelets in Geophysics,* Vol. 4, pp. 151–166, Academic Press, San Diego.

Liu, S. K., and J. J. Leenderste. 1982. A three-dimensional shelf model of the Bering and Chuckchi seas. In *Proceedings, 18th Conference on Coastal Engineering,* pp. 598–616. American Society of Civil Engineers, New York.

Livezey, R. E., M. Masutani, and M. Ji. 1996. SST-forced seasonal simulation and prediction skill for versions of the NCEP/NRF model. *Bull. Am. Meteorol. Soc.* **77,** 507–517.

Llubes, M., and P. Mazzega. 1997. Testing recent global ocean tide models with loading gravimetric data. *Prog. Oceanogr.* **40,** 369–383.

Long, R. B., and W. C. Thacker. 1989a. Data assimilation into a numerical equatorial ocean model. I. The model and the assimilation algorithm. *Dyn. Atmos. Oceans* **13**, 379–412.

Long, R. B., and W. C. Thacker. 1989b. Data assimilation into a numerical equatorial ocean model. II. Assimilation experiments. *Dyn. Atmos. Oceans* **13**, 413–440.

Lopez, J. W. 1998. A study of physical processes of the northern Indian Ocean using a comprehensive primitive equation numerical model. Ph.D. dissertation, Department of Aerospace Engineering Sciences, University of Colorado, Boulder, CO. 144 pp.

Lopez, J. W., and L. H. Kantha. 1998. Results from a numerical model of the northern Indian Ocean: Circulation in the South Arabian Sea. *J. Mar. Syst.* (in press).

Lorenc, A. 1986. Analysis methods for numerical weather prediction. *Q. J. R. Meteorol. Soc.* **112**, 1177–1194.

Lorenz, E. N. 1955. Available potential energy and maintenance of general circulation. *Tellus* **7**, 157–167.

Lorenz, E. N. 1956. *Empirical Orthogonal Functions and Statistical Weather Prediction.* Report no. 1, Statistical Forecasting Project, Dept. Meteorology, MIT, Cambridge, MA. 49 pp.

Lorenz, E. N. 1963. Deterministic nonperiodic flow. *J. Atmos. Sci.* **20**, 130–141.

Lorenz, E. N. 1979. On the prevalence of aperiodicity in simple systems. In M. Gremland and J. E. Marsden (Eds.), *Global Analysis, Lecture Notes in Mathematics*, pp. 53–75. Springer-Verlag, Berlin/New York.

Lozier, M. S., W. B. Owens, and R. G. Curry. 1995. The climatology of the North Atlantic. *Prog. Oceanogr.* **36**, 1–44.

Loyning, T. B., and J. E. Weber. 1997. Thermobaric effect on buoyancy-driven convection in cold seawater. *J. Geophys. Res.* **102**, 27875–27885.

Lueck, R., and R. Reid. 1984. On the production and dissipation of mechanical energy in the ocean. *J. Geophys. Res.* **89**, 3439–3445.

Lueck, R. G., and T. D. Mudge. 1997. Topographically induced mixing around a shallow seamount. *Science* **276**, 1831–1833.

Lumley, J. L. 1971. *Stochastic Tools in Turbulence.* Academic Press, New York. 194 pp.

Lyard, F. H. 1997. The tides in the Arctic Ocean from a finite element model. *J. Geophys. Res.* **102**, 15611–15638.

Lynch, D. R. 1983. Progress in hydrodynamic modeling. *Rev. Geophys.* **21**, 741–754.

Lynch, D. R., and W. G. Gray. 1979. A wave equation model for finite element tidal computations. *Comput. Fluids* **7**, 207–228.

Lynch, D. R., and A. M. Davies (Eds.). 1995. *Quantitative Skill Assessment for Coastal Ocean Models.* American Geophysical Union, Washington, DC.

Lynch, D. R., J. T. C. Ip, C. E. Naimie, and F. E. Werner. 1996. Comprehensive coastal circulation model with application to the Gulf of Maine. *Cont. Shelf Res.* **16**, 875–906.

Lynn, R. J., and J. L. Reid. 1968. Characteristics and circulation of deep and abyssal waters. *Deep-Sea Res.* **15**, 577–598.

Ma, C.-C., C. R. Mechoso, A. W. Robertson, and A. Arakawa. 1996. Peruvian stratus clouds and the tropical Pacific circulation: A coupled ocean-atmosphere GCM study. *J. Clim.* **9**, 1635–1645.

Ma, H. 1993. A spectral element basin model for the shallow water equations. *J. Comput. Phys.* **109**, 133–149.

Ma, X. C., C. K. Shum, R. J. Eanes, and B. D. Tapley. 1994. Determination of ocean tides from the first year of TOPEX/POSEIDON altimeter measurements. *J. Geophys. Res.* **99**, 24809–24820.

McCalpin, J. D. 1994. A comparison of second-order and fourth-order pressure gradient algorithms in a σ-coordinate ocean model. *Int. J. Numer. Methods Fluid* **18**, 361–383.

McClain, E. P., W. G. Pichel, and C. C. Walton. 1985. Comparative performance of AVHRR-based multichannel sea surface temperatures. *J. Geophys. Res.* **90**, 11587–11601.

McClean, J. L., A. J. Semtner, and V. Zlotnicki. 1997. Comparisons of mesoscale variability in the Semtner-Chervin 1/4° model, the Los Alamos Parallel Ocean Program 1/6° model, and TOPEX/POSEIDON data. *J. Geophys. Res.* **102**, 25203–25226.

McCormick, S. F.(Ed.). 1987. *Multigrid Methods*. SIAM, Philadelphia.

McCreary, J. 1981a. A linear stratified ocean model of the equatorial undercurrent. *Philos. Trans. Roy. Soc. London* **298**, 603–605.

McCreary, J. 1981b. A linear stratified ocean model of the coastal undercurrent. *Philos. Trans. Roy. Soc. London* **302**, 385–413.

McDougall, T. J. 1984. The relative roles of diapycnal and isopycnal mixing on subsurface water mass conversion. *J. Phys. Oceanogr.* **14**, 1577–1589.

McDougall, T. J. 1987a. Neutral surfaces. *J. Phys. Oceanogr.* **17**, 1950–1964.

McDougall, T. J. 1987b. Thermobaricity, cabelling, and water-mass conversion. *J. Geophys. Res.* **92**, 5448–5464.

McDougall, T. J., and J. A. Church. 1986. Pitfalls with the numerical representation of isopycnal and diapycnal mixing. *J. Phys. Oceanogr.* **16**, 196–199.

McGoogan, J. T., L. S. Miller, G. S. Brown, and G. S. Hayne. 1974. The S-193 radar altimeter experiment. *Proc. IEEE* **62**, 793–803.

McPhaden, M. J. 1995. The tropical atmosphere ocean array is completed. *Bull. Am. Meteorol. Soc.* **76**, 739–741.

McPhadden, M. J. 1999. Genesis and evolution of the 1997–98 El Niño. *Science* **283**, 950–954.

McPhaden, M. J., A. J. Busalacchi, R. Cheney, J.-R. Donguy, K. S. Gage, D. Halpern, M. Ji, P. Julian, G. Meyers, G. T. Mitchum, P. P. Niiler, J. Picaut, R. W. Reynolds, N. Smith, and K. Takeuchi. 1998. The Tropical Ocean-Global Atmosphere observing system: A decade of progress. *J. Geophys. Res.* **103**, 14169–14240.

McPhee, M. G. 1980. An analysis of pack ice drift in summer. In R. S. Pritchard (Ed.), *Sea Ice Processes and Models*, pp. 62–75. Univ. Washington Press, Seattle.

McPhee, M. G. 1992. Turbulent heat flux in the upper ocean under sea ice. *J. Geophys. Res.* **97**, 5365–5379.

McPhee, M. G., G. A. Maykut, and J. H. Morrison. 1987. Dynamics and thermodynamics of the ice/upper ocean systems in the marginal ice zone of the Greenland Sea. *J. Geophys. Res.* **92**, 7017–7031.

McWilliams, J. C. 1996. Modeling the oceanic general circulation. *Annu. Rev. Fluid Mech.* **28**, 215–248.

McWilliams, J. C., and P. R. Gent. 1980. Intermediate models of planetary circulations in the atmophere and ocean. *J. Atmos. Sci.* **37**, 1657–1678.

Madala, R. V., and S. A. Piacsek. 1977. A semi-implicit numerical model for baroclinic oceans. *J. Comput. Phys.* **23**, 167–178.

Madec, G., M. Chartier, and M. Crepon. 1991a. Effect of thermohaline forcing variability on deep water formation in the western Mediterranean Sea. *Dyn. Atmos. Oceans* **15**, 301–332.

Madec, G., M. Chartier, P. Delacluse, and M. Crepon. 1991b. Numerical study of deep water formation in the northwestern Mediterranean Sea: A high resolution three dimensional numerical study. *J. Phys. Oceanogr.* **21**, 1349–1371.

Madec, G., and M. Imbard. 1995. A global ocean mesh to overcome the North Pole singularity. *Clim. Dyn.* **12**, 381–388.

Mak, M. 1995. Orthogonal wavelet analysis: Interannual variability in the sea surface temperature. *Bull. Am. Meteorol. Soc.* **76**, 2179–2186.

Makin, V. K., and V. N. Kudryavtsev 1999. Coupled sea surface-atmosphere model, 1, Wind over waves coupling. *J. Geophys. Res.* **104**, 7613–7624.

Malanotte-Rizzoli, P. (Ed.). 1996. *Modern Approaches to Data Assimilation in Ocean Modeling*. Elsevier, Amsterdam/New York. 455 pp.

Malanotte-Rizzoli, P., I. Fukumori, and R. E. Young. 1996. A methodology for the construction of a hierarchy of Kalman filters for nonlinear primitive equation models. In P. Malanotte-Rizzoli

(Ed.), *Modern Approaches to Data Assimilation in Ocean Modeling,* pp. 297–317. Elsevier, Amsterdam/New York.

Malanotte-Rizzoli, P., and E. Tziperman. 1996. The oceanographic data assimilation problem: Overview, motivation and purposes. In P. Malanotte-Rizzoli (Ed.), *Modern Approaches to Data Assimilation in Ocean Modeling,* pp. 3–17. Elsevier, Amsterdam/New York.

Manabe, S., and K. Bryan. 1969. Climate calculations with a combined ocean-atmosphere model. *J. Atmos. Sci.* **26,** 786–789.

Manabe, S., K. Bryan, and M. J. Spelman. 1979. A global ocean-atmosphere climate model with seasonal variation for future studies of climate sensitivity. *Dyn. Atmos. Oceans* **3,** 393–426.

Manabe, S., M. J. Spelman, and R. J. Stouffer. 1990. Transient response of a coupled ocean-atmosphere model to a doubling of atmospheric carbon dioxide. *J. Phys. Oceanogr.* **20,** 722–749.

Manabe, S., K. Bryan, and M. J. Spelman. 1992. Transient response of a global ocean-atmosphere model to gradual changes of atmospheric carbon dioxide. Part II. Seasonal response. *J. Clim.* **5,** 105–126.

Manabe, S., and R. J. Stouffer. 1994. Multiple-century response of a coupled ocean-atmosphere model to an increase of atmospheric carbon dioxide. *J. Clim.* **7,** 5–23.

Mantyla, A. W., and J. L. Reid. 1983. Abyssal characteristics of the world ocean waters. *Deep-Sea Res.* **30,** 805–833.

Marcus, S. L., Yi Chao, J. O. Dickey, and P. Gegout. 1998. Detection and modeling of nontidal oceanic effects on Earth's rotation rate. *Science* **281,** 1656–1659.

Marks, F. D., L. K. Shay, and PDT-5. 1997. Landfalling tropical cyclones: Forecast problems and associated research opportunities. *Bull. Am. Meteorol. Soc.* **79,** 305–323.

Marotzke, J. 1991. Influence of convective adjustment on the stability of the thermohaline circulation. *J. Phys. Oceanogr.* **21,** 903–907.

Marotzke, J. 1994. Ocean models in climate problems. In P. Malanotte-Rizzoli and A. R. Robinson (Eds.), *Ocean Processes in Climate Dynamics: Global and Mediterranean Examples,* pp. 79–109. Kluwer, Dordrecht/Norwell, MA.

Marotzke, J. 1996. Analysis of thermohaline feedbacks. In D. L. T. Andersen and J. Willebrand (Eds.), *Climate Variability: Dynamics and Predictability.* NATO ASI Series.

Marsh, J. G., *et al.* 1990. The GRM-T2 gravitational model. *J. Geophys. Res.* **95,** 22043–22071.

Marsh, R., M. J. Roberts, R. A. Wood, and A. L. New. 1996. An intercomparison of a Bryan-Cox type ocean model and an isopycnic ocean model. Part II. The subtropical gyre and meridional heat transport. *J. Phys. Oceanogr.* **26,** 1528–1551.

Marshall, J., A. Adcroft, C. Hill, L. Perelman, and C. Heisey. 1997a. A finite-volume, incompressible Navier-Stokes model for studies of the ocean on parallel computers. *J. Geophys. Res.* **102,** 5753–5766.

Marshall, J., C. Hill, L. Perelman, and A. Adcroft. 1997b. Hydrostatic, quasi-hydrostatic, and nonhydrostatic ocean modeling. *J. Geophys. Res.* **102,** 5733–5752.

Marshall, J., and F. Schott. 1998. *Open-Ocean Convection: Observations, Theory and Models.* Center for Global Change Science Report no. 52, MIT, Cambridge, MA.

Mass, C. F., and Y.-H. Kuo. 1997. Regional real-time numerical weather prediction: Current status and future potential. *Bull. Am. Meteorol. Soc.* **79,** 253–264.

May, D. G., M. M. Parmeter, D. S. Olszewski, and B. D. McKenzie. 1998. Operational processing of satellite sea surface temperature retrievals at the Naval Oceanographic Office. *Bull. Am. Meteorol. Soc.* **79,** 397–408.

Maykut, G. A., and N. Untersteiner. 1972. Some results from a time-dependent thermodynamic model of sea ice. *J. Geophys. Res.* **76,** 1550–1575.

Maykut, G. A., and D. K. Perovich. 1987. The role of short wave radiation in the summer decay of a sea ice cover. *J. Geophys. Res.* **92,** 7032–7044.

Maykut, G. A., and M. G. McPhee. 1995. Solar heating of the Arctic mixed layer. *J. Geophys. Res.* **100,** 24691–24703.

Mazzega, P., and M. Berge. 1994. Ocean tides in the Asian semienclosed seas from TOPEX/POSEIDON. *J. Geophys. Res.* **99**, 24867–24881.

Mechoso, C. R., A. Kitoh, S. Moorthi, and A. Arakawa. 1987. Numerical simulations of the atmospheric response to a sea surface temperature anomaly over the equatorial eastern Pacific Ocean. *Mon. Weather Rev.* **115**, 2936–2956.

Mechoso, C. R., A. W. Robertson, N. Barth, M. K. Davey, P. Delecluse, P. R. Gent, S. Ineson, B. Kirtman, M. Latif, H. Le Treut, T. Nagai, J. D. Neelin, S. G. H. Philander, J. Polcher, P. S. Schopf, T. Stockdale, M. J. Suarez, L. Terray, O. Thual, and J. J. Tribbia. 1995. The seasonal cycle over the tropical Pacific in coupled ocean-atmosphere general circulation models. *Mon. Weather Rev.* **123**, 2825–2838.

Meehl, G. A. 1990. Development of global coupled ocean-atmosphere general circulation models. *Clim. Dyn.* **5**, 19–33.

Meehl, G. A. 1992. Global coupled models: Atmosphere, ocean, sea ice. In K. E. Trenberth (Ed.), *Climate System Modeling*, pp. 555–582. Cambridge Univ. Press, New York.

Mehra, A., V. Anantharaj, S. Payne, and L. Kantha. 1996. *Demonstration of a Real Time Capability to Produce Tidal Heights and Currents for Naval Operational Use: A Case Study for the West Coasts of Africa and the United States.* Mississippi State University Center for Air-Sea Technology Technical Note 96-2. 45 pp.

Melchior, P. 1981. *The Tides of the Planet Earth.* Pergamon, Tarrytown, NY. 665 pp.

Mellor, G. L. 1996a. *Introduction to Physical Oceanography.* Am. Inst. Phys. Press, New York. 260 pp.

Mellor, G. L. 1996b. *Users Guide for a Three-Dimensional, Primitive Equation, Numerical Ocean Model,* Rev. ed. Princeton University Report, Princeton, NJ. 39 pp.

Mellor, G. L., and T. Yamada. 1974. A hierarchy of turbulence closure models for planetary boundary layers. *J. Atmos. Sci.* **31**, 1791–1805.

Mellor, G. L., and P. A. Durbin. 1975. The structure and dynamics of the ocean surface mixed layer. *J. Phys. Oceanogr.* **5**, 718–728.

Mellor, G. L., C. R. Mechoso, and E. Keto. 1982. A diagnostic calculation of the general circulation of the Atlantic Ocean. *Deep-Sea Res.* **29**, 1171–1192.

Mellor, G. L., and T. Yamada. 1982. Development of a turbulence closure model for geophysical fluid problems. *Rev. Geophys. Space Phys.* **20**, 851–875.

Mellor, G. L., and A. F. Blumberg. 1985. Modeling vertical and horizontal diffusivities with the sigma-coordinate system. *Mon. Weather Rev.* **113**, 1380–1383.

Mellor, G. L., and L. H. Kantha. 1989. An ice-ocean coupled model. *J. Geophys. Res.* **94**, 10937–10954.

Mellor, G. L., T. Ezer, and L.-Y. Oey. 1994. The pressure gradient conundrum of sigma coordinate ocean models. *J. Atmos. Oceanic Technol.* **11**, 1126–1134.

Mellor, G. L., and T. Azer. 1995. Sea level variations induced by heating and cooling: An evaluation of the Boussinesq approximation in ocean models. *J. Geophys. Res.* **100**, 20565–20578.

Mellor, G. L., and S. Hakkinen. 1995. A review of coupled ice-ocean models. In *Polar Oceans and Their Role in Shaping the Global Environment.* Geophysical Monograph 85, American Geophysical Union, Washington, DC.

Mertens, C., and F. Schott. 1998. Interannual variability of deep-water formation in the northwest Mediterranean. *J. Phys. Oceanogr.* **28**, 1410–1424.

Mertz, G., and D. G. Wright. 1992. Interpretations of the JEBAR term. *J. Phys. Oceanogr.* **22**, 301–305.

Mesinger, F., and A. Arakawa. 1976. *Numerical Methods Used in Atmospheric Models,* Vol. 1. Global Atmospheric Research Program Publication 17, WMO, Geneva. 64 pp.

Metzger, E. J., Hurlburt, H. E., Kindle, J. C., Serkes, Z., and Pringle, J. M. 1992. Hindcasting of wind-driven anomalies using a reduced-gravity global ocean model. *Mar. Technol. Soc. J.* **26**, 23–32.

Meyer, Y. 1993a. *Wavelets, Algorithms and Applications.* SIAM, Philadelphia. 133 pp.

Meyer, Y. 1993b. *Wavelets and Operators*. Cambridge Univ. Press, Cambridge. 238 pp.

Meyers, S. D., B. G. Kelly, and J. J. O'Brien. 1993. An introduction to wavelet analysis in oceanography and meteorology: With applications to the dispersion of Yanai waves. *Mon. Weather Rev.* **121,** 2858–2866.

Meyers, S. D., and J. J. O'Brien. 1994. Spatial and temporal 26-day SST variations in the equatorial Indian Ocean using wavelet analysis. *Geophys. Res. Lett.* **21,** 777–780.

Miller, A. J., D. S. Luther, and M. Hendershott. 1993. The fortnightly and monthly tides: Resonant Rossby waves or nearly equilibrium gravity waves? *J. Phys. Oceanogr.* **23,** 879–899.

Miller, M. J., and A. J. Thorpe. 1981. Radiation conditions for the lateral boundaries of limited-area numerical models. *Q. J. R. Meteorol. Soc.* **107,** 615–628.

Miller, R. N., and M. A. Cane. 1996. Tropical data assimilation: Theoretical concepts. In P. Malanotte-Rizzoli (Ed.), *Modern Approaches to Data Assimilation in Ocean Modeling,* pp. 207–233. Elsevier, Amsterdam/New York.

Millero, F. J., and A. Poisson. 1981. International one-atmosphere equation of state of seawater. *Deep-Sea Res.* **27A,** 255–264.

Mitchell, J. L., W. J. Teague, G. A. Jacobs, and H. E. Hurlburt. 1996. Kuroshia Extension dynamics from satellite altimetry and a model simulation. *J. Geophys. Res.* **101,** 1045–1058.

Mitchum, G. T. 1993. Principal component analysis: Basic methods and extensions. In P. Muller and D. Henderson (Eds.), *Statistical Methods in Physical Oceanography, Proc. 'Aha Huliko'a Hawaiian Winter Workshop, Hawaii, Jan 12–15, 1993,* pp. 185–199. School of Ocean and Earth Science Technology, Honolulu Special Publication.

Miura, R. M. 1976. The Korteweg-deVries equation: A survey of results. *SIAM Rev.* **18,** 412–459.

Miyakoda, K., and A. Rosati. 1977. One-way nested grid models: The interface conditions and the numerical accuracy. *Mon. Weather Rev.* **105,** 1092–1107.

MODE Group. 1978. The Mid-Ocean Dynamics Experiment. *Deep-Sea Res.* **25,** 859–910.

Moeng, C.-H., and P. P. Sullivan. 1994. A comparison of shear- and buoyancy-driven planetary boundary layer flows. *J. Atmos. Sci.* **51,** 999–1022.

Mofjeld, H. O. 1986. Observed tides on the northeastern Bering Sea shelf. *J. Geophys. Res.* **91,** 2593–2606.

Moisan, J. R., E. E. Hofmann, and D. B. Haidvogel. 1995. Modeling nutrient and plankton processes in the California coastal transition zone. 2. A three-dimensional physical-bio-optical model. *J. Geophys. Res.* **101,** 22677–22692.

Molines, J. M., C. Le Provost, F. Lyard, R. D. Ray, C. K. Shum, and R. J. Eanes. 1994. Tidal corrections in the TOPEX/POSEIDON geophysical records. *J. Geophys. Res.* **99,** 24749–24760.

Moody, J. A., *et al.* 1984. Atlas of tidal elevation and current observations on the northeast American Continental shelf and slope. *U.S. Geol. Surv. Bull.* **1611.**

Moon, F. C. 1987. *Chaotic Vibrations.* Wiley, New York.

Moore, D. W., and S. G. H. Philander. 1977. Modeling of the tropical oceanic circulation. In E. D. Goldberg, I. N. McCave, J. J. O'Brien, and J. H. Steele (Eds.), *The Sea,* Vol. 6, pp. 319–362. Wiley, New York.

Morlet, J. 1983. Sampling theory and wave propagation. In F. H. Chen (Ed.), *Issues in Acoustic Signal/Image Processing and Recognition,* Vol. 1, pp. 233–261. NATO ASI Series, Springer-Verlag, Berlin/New York.

Morozov, E. G. 1995. Semidiurnal internal wave global field. *Deep-Sea Res.* **42,** 135–148.

Morrison, L. V. 1978. Tidal decelerations of the Earth's rotation deduced from astronomical observations in the period AD 1600 to the present. In P. Brosche and J. Sundermann (Eds.), *Tidal Friction and the Earth's Rotation,* pp. 22–27. Springer-Verlag, Berlin/New York.

Muench, R. D., K. Jezek, and L. H. Kantha. 1991. Introduction: Third marginal ice zone research collection. *J. Geophys. Res.* **96,** 4529–4530.

Muller, P. 1995. Ertel's potential vorticity theorem in physical oceanography. *Rev. Geophys.* **33,** 67–98.

Muller, P., and G. Halloway. 1989. Parameterization of small scale processes, EOS. American

Geophysical Union, Washington, DC.

Muller, P., and D. Henderson (Eds.). 1989. *Parameterization of Small Scale Processes, Proc. Aha Hulikoa Workshop.* Hawaii Institute of Geophysics Special Publication. Univ. Hawaii, Manoa. 354 pp.

Mullineaux, C. W. 1999. The plankton and the planet. *Science* **283**, 801–802.

Mundt, M. D., W. B. Maguire II, and R. R. P. Chase. 1991. Chaos in the sunspot cycle: Analysis and prediction. *J. Geophys. Res.* **96**, 1705–1716.

Munk, W. H. 1950. On the wind-driven ocean circulation. *J. Meteorol.* **7**, 79–93.

Munk, W. H. 1966. Abyssal recipes. *Deep-Sea Res.* **13**, 707–730.

Munk, W. H. 1968. Once again—Tidal friction, *Q. J. R. Astron. Soc.* **9**, 352–375.

Munk, W. H. 1997. Once again—Once again—Tidal friction. *Prog. Oceanogr.* **40**, 7–36.

Munk, W. H., and G. J. F. MacDonald. 1960. *The Rotation of the Earth, a Geophysical Discussion.* Cambridge Univ. Press, New York. 323 pp.

Munk, W. H., and D. E. Cartwright. 1966. Tidal spectroscopy and prediction. *Proc. R. Soc. London A* **259**, 533–581.

Munk, W. H., and C. Wunsch. 1998. The Moon and mixing: Abyssal recipes II. Energetics of tidal and wind mixing. *Deep-Sea Res.* **145**, 1977–2010.

Munk, W., B. Zetler, J. Clark, S. Gill, D. Porter, J. Spiesberger, and R. Spindel. 1981. Tidal effects on long-range sound transmission. *J. Geophys. Res.* **86**, 6399–6410.

Munk, W., P. Worcester, and C. Wunsch. 1995. *Ocean Acoustic Tomography.* Cambridge University Press, Cambridge. 433 pp.

Murray, C. D. 1998. Chaotic motion in the solar system. In P. R. Weissman, L.-A. McFadden, and T. V. Johnson (Eds.), *Encyclopedia of the Solar System*, pp. 825–844. Academic Press, San Diego.

Murray, R. J. 1996. Explicit generation of orthogonal grids for ocean models. *J. Comput. Phys.* **126**, 251–273.

Murty, T. S., and M. I. El-Sabh. 1986. The age of tides. *Oceanogr. Mar. Biol. Annu. Rev.* **23**, 11–103.

Myers, P. G., A. F. Fanning, and A. J. Weaver. 1996. JEBAR, bottom pressure torque, and the Gulf Stream separation. *J. Phys. Oceanogr.* **26**, 671–705.

Mysak, L. A. 1980a. Recent advances in shelf wave dynamics. *Rev. Geophys. Space Phys.* **18**, 211–241.

Mysak, L. A. 1980b. Topographically trapped waves. *Annu. Rev. Fluid Mech.* **12**, 45–76.

Nagai, T., Y. Kitamura, M. Endoh, and T. Tokoika. 1995. Coupled atmosphere-ocean model simulations of El Niño-Southern Oscillation with and without an active Indian Ocean. *J. Clim.* **8**, 3–14.

Nayfeh, A. 1973. *Perturbation Methods.* Wiley, New York. 425 pp.

Neelin, J. D., *et al.* 1992. Tropical air-sea interaction in general circulation models. *Clim. Dyn.* **7**, 73–104.

Neelin, J. D., F.-F. Jin, and M. Latif. 1994. Dynamics of coupled ocean-atmosphere models: The tropical problem. *Annu. Rev. Fluid Mech.* **26**, 617–659.

Nelson, R. A. 1982. *SI: The International System of Units,* 2nd ed. American Association of Physics Teachers, College Park, MD.

Nerem, R. S., K. E. Rachlin, and B. D. Beckley. 1997. Characterization of global mean sea level variations observed by TOPEX/POSEIDON using empirical orthogonal functions. *Surv. Geophys.* **18**, 293–302.

Neta, B., and R. T. Williams. 1989. Rossby wave frequencies and group velocities for finite element and finite difference approximations to the vorticity-divergence and primitive forms of the shallow water equations. *Mon. Weather Rev.* **117**, 1439–1457.

New, A. L., and R. Bleck. 1995. An isopycnal model study of the North Atlantic. Part II. Interdecadal variability of the subtropical gyre. *J. Phys. Oceanogr.* **25**, 2700–2714.

New, A. L., R. Bleck, Y. Jia, R. Marsh, M. Huddleston, and S. Barnard. 1995. An isopycnal model study of the North Atlantic. Part I. Model experiment. *J. Phys. Oceanogr.* **25**, 2667–2699.

Newman, M., and P. D. Sardeshmukh. 1995. A caveat concerning singular value decomposition. *J. Clim.* **8**, 352–360.

Nezlin, M. V., and E. N. Snezhkin. 1993. *Rossby Vortices, Spiral Structures, Solitons.* Springer-Verlag, Berlin. 240 pp.

Nicolis, C., and G. Nicolis. (Eds.). 1987. *Irreversible Phenomena and Dynamical Systems Analysis in Geosciences.* Reidel, Dordrecht.

Nihoul, J. C. J., P. Adam, P. Brasseur, E. Deleersnijder, S. Djenidi, and J. Haus. 1993. Three-dimensional general circulation model of the northern Bering Sea's summer ecohydrodynamics. *Cont. Shelf Res.* **13**, 509–542.

Nikiferov, E. G. 1957. Variations in ice cover compaction due to its dynamics. In *Problemy Arktiki,* Vol. 2. Morskoi Transport Press, St. Petersburg.

NOAA. 1986. *ETOP05 Digital Relief of the Surface of the Earth.* Data Announcement 86-MGG-07, National Geophysical Data Center, Washington, DC.

Nof, D. 1981. On the beta-induced movement of isolated baroclinic eddies. *J. Phys. Oceanogr.* **11**, 1662–1672.

Nof, D. 1983. On the migration of isolated eddies with application to Gulf Stream rings. *J. Mar. Res.* **41**, 399.

North, G. R., T. L. Bell, R. F. Cahalan, and F. J. Moeng. 1982. Sampling errors in the estimation of empirical orthogonal functions. *Mon. Weather Rev.* **110**, 699–706.

Oberhuber, J. M. 1986. About some numerical methods used in an ocean general circulation model with isopycnic coordinates. In J. J. O'Brien (Ed.), *Advanced Physical Oceanographic Numerical Modelling,* pp. 511–522. NATO ASI Series. Reidel, Dordrecht, Holland.

Oberhuber, J. M. 1993a. Simulation of the Atlantic circulation with a coupled sea ice-mixed layer-isopycnal general circulation model. Part I. Model description. *J. Phys. Oceanogr.* **23**, 808–829.

Oberhuber, J. M. 1993b. Simulation of the Atlantic circulation with a coupled sea ice-mixed layer-isopycnal general circulation model. Part II. Model experiment. *J. Phys. Oceanogr.* **23**, 830–845.

O'Brien, J. J. 1985. *Advanced Physical Oceanographic Numerical Modeling.* Reidel, New York.

Obukhov, A. M. 1947. Statistically homogeneous fields on a sphere. *Usp. Mat. Nauk* **2**, 196–198.

Oey, L.-Y., G. L. Mellor, and R. I. Hires. 1985a. A three-dimensional simulation of the Hudson-Raritan estuary. Part I. Description of the model and model simulations. *J. Phys. Oceanogr.* **15**, 1676–1692.

Oey, L.-Y., G. L. Mellor, and R. I. Hires. 1985b. A three-dimensional simulation of the Hudson-Raritan estuary. Part II. Comparison with observation. *J. Phys. Oceanogr.* **15**, 1693–1709.

Oey, L. Y., and P. Chen. 1992. A nested-grid ocean model: With application to the simulation of meanders and eddies in the Norwegian Coastal Current. *J. Geophys. Res.* **97**, 20063–20086.

Oey, L. -Y., and G. L. Mellor. 1993. Subtidal variability of estuarine outflow, plume and coastal current: A model study. *J. Phys. Oceanogr.* 23, 164–171.

Oguz, T., and P. Malanotte-Rizzoli. 1996. Seasonal variability of wind and thermohaline-driven circulation in the Black Sea: Modeling studies. *J. Geophys. Res.* **101**, 16,551–16,570.

Olson, D. B. 1991. Rings in the ocean. *Annu. Rev. Earth Planet. Sci.* **19**, 283–311.

Ookochi, Y. 1972. A computational scheme for the nesting fine mesh in the primitive equation model. *J. Meteorol. Soc. Japan.* **50**, 37–47.

Oort, A. H., S. C. Ascher, S. Levitus, and J. P. Peixoto. 1989. New estimates of the available potential energy in the world ocean. *J. Geophys. Res.* **94**, *3187–3200.*

Oort, A. H., L. A. Anderson, and J. P. Peixoto. 1994. Estimates of the energy cycle of the oceans. *J. Geophys. Res.* **99**, 7665–7688.

Orlanski, I. 1976. A simple boundary condition for unbounded hyperbolic flows. *J. Comput. Phys.* **21**, 251–269.

Ostrovskii, A. G. 1995. Signatures of stirring and mixing in the Japan sea surface temperature patterns in autumn 1993 and spring 1994. *Geophys. Lett.* **22**, 2357–2360.

Ottino, J. M., F. J. Muzzio, M. Tjahjadi, J. G. Franjione, S. C. Jana, and H. A. Kusch. 1992. Chaos, symmetry and self-similarity: Exploiting order and disorder in mixing processes. *Science* **257**, 754–760.

Overland, J. E. 1985. Atmospheric boundary layer structure and drag coefficients over sea ice. *J. Geophys. Res.* **90**, 9029–9049.

Overland, J. E., and R. W. Preisendorfer. 1982. A significance test for principal components applied to a cyclone climatology. *Mon. Weather Rev.* **110**, 1–4.

Overland, J. E., and C. H. Pease. 1988. Modeling ice dynamics of coastal seas. *J. Geophys. Res.* **93**, 15619–15637.

Overland, J. E., S. L. McNutt, S. Salo, J. Groves, and S. Li. 1998. Arctic ice as a granular plastic. *J. Geophys. Res.* **103**, 21845–21868.

Overland, J. E., S. Solo, L. H. Kantha, and C. A. Clayson. 1999. Thermal stratification and mixing on the Bering Shelf. In *The Bering Sea: Physical, Chemical and Biological Dynamics, PICES Bering Sea Volume* (in press).

Owens, W. B., and P. Lemke. 1990. Sensitivity studies with a sea ice-mixed layer-pycnocline model in the Weddell Sea. *J. Geophys. Res.* **95**, 9527.

Pacanowski, R. C. 1995. *MOM2 Documentation, User's Guide and Refernce Manual.* Ocean Technical Report 3, Geophysical Fluid Dynamics Laboratory, Princeton, NJ. 233 pp.

Pacanowski, R. C., and G. Philander. 1981. Parameterization of vertical mixing in numerical models of the tropical ocean. *J. Phys. Oceanogr.* **11**, 1442–1451.

Pacanowski, R. C., K. Dixon, and A. Rosati. 1991. *The GFDL Modular Ocean Model User Guide.* Ocean Technical Report 2, Geophysical Fluid dynamics Laboratory, Princeton, NJ.

Pagiatakis, S. D. 1990. The response of a realistic Earth to ocean tide loading. *Geophys. J. Int.* **103**, 541–1560.

Palmer, T. N. 1993. Extended range atmospheric prediction and the Lorenz model. *Bull. Am. Meteorol. Soc.* **74**, 49–65.

Palmer, T. N. 1994. Chaos and the predictability in forecasting the monsoons. *Proc. Indian Nat. Sci. Acad.* **60A**, 57–66.

Palmer, T. N., 1996. Predictability of the atmosphere and oceans: from days to decades. In *Decadal Climate Variability: Dynamics and Predictability,* D. L. T. Anderson and J. Willebrand (Eds.), Springer-Verlag, Berlin.

Palmer, T. N., and D. L. T. Anderson. 1995. The prospects for seasonal forecasting: A review. *Q. J. R. Meteorol. Soc.* **120**, 755–793.

Park, Y.-G. 1996. Rotating convection driven by differential bottom heating and its application. Ph.D. dissertation, MIT and Woods Hole Oceanographic Institution Joint program. 137 pp.

Parke, M. E., and M. C. Hendershott. 1980. M_2, S_2 and K_1 models of the global ocean tide on an elastic Earth. *Mar. Geodesy* **3**, 379–408.

Parke, M. E., R. H. Stewart, and D. L. Farless. 1987. On the choice of orbits for an altimetric satellite to study ocean circulation and tides. *J. Geophys. Res.* **92**, 11693–11707.

Parker, B. B. (Ed.). 1991. *Tidal Hydrodynamics.* Wiley, New York. 883 pp.

Parkinson, C. L., and W. M. Washington. 1979. A large scale numerical model of sea ice. *J. Geophys. Res.* **84**, 311–336.

Pedlosky, J. 1987. *Geophysical Fluid Dynamics,* 2nd ed. Springer Verlag, New York. 710 pp.

Pedlosky, J. 1996. *Ocean Circulation Theory.* Springer, New York. 453 pp.

Peggion, G. 1994. Numerical inaccuracies across the interface of a nested grid. *Numer. Methods Partial Differential Equations* **10**, 455–473.

Pekeris, C. L., and Y. Accad. 1969. Solution of Laplace's equations for the M_2 tide in the world oceans. *Philos. Trans. R. Soc. London A* **265**, 413–436.

Peng, S., and J. Fyfe. 1996. The coupled patterns between sea level pressure and sea surface temperature in the midlatitude North Atlantic. *J. Clim.* **9**, 1824–1839.

Penland, C., and L. Matrosova. 1994. A balance condition for stochastic numerical models with application to the El Niño-Southern Oscillation. *J. Clim.* **7**, 1352–1372.

Penland, C., and P. D. Sardeshmukh. 1995. The optimal growth of tropical sea surface temperature anomalies. *J. Clim.* **8,** 1999–2024.

Perkey, D. J., and R. A. Maddox. 1985. A numerical investigation of a mesoscale convective system. *Mon. Weather Rev.* **113,** 553–566.

Perkins, A. L., L. F. Smedstad, D. W. Blake, G. W. Heburn, and A. J. Wallcraft. 1997. A new nested boundary condition for a primitive equation ocean model. *J. Geophys. Res.* **102,** 3483–3500.

Perovich, D. K., and G. A. Maykut. 1990. Solar heating of a stratified ocean in the presence of a static ice cover. *J. Geophys. Res.* **95,** 18233–18245.

Peyret, R., and T. D. Taylor. 1983. *Computational Methods for Fluid Flow.* Springer-Verlag, Berlin/New York. 358 pp.

Pfirman, S. L., R. Colony, D. Nurnberg, H. Eicken, and I. Rigor. 1997. Reconstructing the origin and trajectory of drifting Arctic sea ice. *J. Geophys. Res.* **102,** 12575–12586.

Philander, S. G. H. 1990. *El Niño, La Niña, and the Southern Oscillation.* Academic Press, San Diego. 293 pp.

Phillips, N. A. 1957. A coordinate system having some special advantages for numerical forecasting. *J. Meteorol.* **14,** 184–185.

Phillips, N. A. 1966. The equation of motion for a shallow rotating atmosphere and the "traditional approximation." *J. Atmos. Sci.* **23,** 626–628.

Phillips, N. A. 1968. Reply to G. Veronis's comments on Phillips (1966). *J. Atmos. Sci.* **25,** 1156–1158.

Phillips, N. A. 1973. Principles of large-scale numerical weather prediction. In P. Morel (Ed.), *Dynamic Meteorology,* pp. 1–96. Reidel, Norwell, MA.

Phillips, N. A. 1979. *The Nested Grid Model.* Tech. Report NWS 22, NMC, NOAA, MD.

Phillips, N., and J. Shukla. 1973. On the strategy of combining coarse and fine meshes in numerical weather prediction. *J. Appl. Meteorol.* **12,** 763–770.

Piacsek, S., R. Allard, and A. Warn-Varnas. 1991. Studies of the Arctic ice cover and upper ocean with a coupled ice-ocean model. *J. Geophys. Res.* **96,** 4631–4650.

Pickard, G. L., and W. J. Emery. 1982. *Descriptive Physical Oceanography.* Pergamon, Oxford. 249 pp.

Pietrzak, J. D. 1998. The use of TVD flux limiters for forward-in-time upstream-biased advection schemes in ocean modeling. *Mon. Weather Rev.* **126,** 812–830.

Pinder, G. F., and W. G. Gray. 1977. *Finite Element Simulation in Surface and Subsurface Hydrology.* Academic Press, San Diego.

Pinkel, R., M. Merrifield, M. McPhaden, J. Picaut, S. Rutledge, D. Siegel, and L. Washburn. 1996. Solitary waves in the western equatorial Pacific Ocean.

Piotrowicz, S. R. 1997. TOGA observing system and GOOS. *Sea Technol.,* 39–44.

Platzman, G. W. 1984. Planetary energy balance for tidal dissipation. *Rev. Geophys. Space Phys.* **22,** 73–84.

Platzman, G. W. 1991. An observational study of energy balance in the atmospheric lunar tide. *Pure Appl. Geophys.* **137,** 1–33.

Platzman, G. W., G. A. Curtis, K. S. Hansen, and R. D. Slater. 1981. Normal modes of the world ocean. Part II. Description of modes in the period range 8–80 hours. *J. Phys. Oceanogr.* **11,** 579–603.

Podgorny, I. A., and T. C. Grenfell. 1996. Partitioning of solar energy in melt ponds from measurements of pond albedo and depth. *J. Geophys. Res.* **101,** 22737–22748.

Pohlmann, T. 1987. A three-dimensional circulation model of the South China Sea. In J. C. J. Nihoul and B. M. Jamart (Eds.), *Three-Dimensional Models of Marine and Estuarine Dynamics,* pp. 245–268. Elsevier, New York.

Pohlmann, T. 1995. Predicting of the thermocline in a circulation model of the North Sea. I. Model description, calibration and verification. *Cont. Shelf Res.* **16,** 131–146.

Polito, P. S., and P. Cornillon. 1997. Long baroclinic Rossby waves detected by TOPEX/POSEIDON. *J. Geophys. Res.* **102,** 3215–3235.

Pond, S., and G. L. Pickard. 1989. *Introductory Dynamical Oceanography*, 2nd ed. Pergamon, New York. 329 pp.

Ponte, R. M. 1994. Understanding the relation between wind driven sea level variability. *J. Geophys. Res.* **99,** 8033–8040.

Ponte, R. M. 1997. Nonequilibrium response of the global ocean to the 5-day Rossby-Haurwitz wave in atmospheric surface pressure. *J. Phys. Oceanogr.* **27,** 2158–2168.

Potemra, J. T., M. E. Luther, and J. J. O'Brien. 1991. The seasonal circulation of the upper ocean in the Bay of Bengal. *J. Geophys. Res.* **99,** 25127–25141.

Potts, M. A. 1998. A study of convective processes in polar and subpolar seas using mon-hydrostatic models. Doctoral dissertation, Department of Aerospace Engineering Sciences, University of Colorado, Boulder.

Preisendorfer, R. W. 1988. *Principal Component Analysis in Meteorology and Oceanography, Developments in Atmospheric Science,* Vol. 17. Elsevier, Amsterdam/New York. 425 pp.

Press, W. H., S. A. Teukolsky, W. T. Vetterling, and B. P. Flannery. 1992. *Numerical Recipes in FORTRAN, the Art of Scientific Computing*, 2nd ed. Cambridge Univ. Press, Cambridge. 963 pp.

Price, J. F., R. A. Weller, and R. R. Schudlich. 1987. Wind-driven ocean currents and Ekman transport. *Science* **238,** 1534–1538.

Pritchard, R. S. (Ed.). 1980. *Sea Ice Processes and Models*. Univ. Washington Press, Seattle. 474 pp.

Proudman, J. 1959. The condition that long-period tide shall follow equilibrium law. *Geophys. J.* 244–249.

Pugh, D. T. 1987. *Tides, Surges and Mean Sea-Level*. Wiley, New York.

Randall, D. A. 1994. Geostrophic adjustment and the finite-difference shallow-water equations. *Mon. Weather Rev.* **122,** 1371–1377.

Randall, D., J. Curry, D. Battisti, G. Flato, R. Gumbine, S. Hakkinen, D. Martinson, R. Preller, J. Walsh, and J. Weatherly. 1998. Status of and outlook for large-scale modeling of atmosphere-ice-ocean interactions in the Arctic. *Bull. Am. Meteorol. Soc.* **79,** 197–219.

Rapp, R. H., Y. Yi, and Y. M. Wang. 1994. Mean sea surface and geoid gradient comparisons with TOPEX altimeter data. *J. Geophys. Res.* **99,** 24,657–24,668.

Rattray, M., Jr., J. G. Dworski, and P. E. Kovala. 1969. Generation of long internal waves at the continental slope. *Deep-Sea Res.* **16**(Suppl.), 179–195.

Ray, R. D. 1998. Ocean self-attraction and loading in numerical tidal models. *Mar. Geodesy* **21,** 181–192.

Ray, R. D., and B. V. Sanchez. 1989. Radial deformation of the Earth by oceanic tidal loading. NASA Technical Report 100743, Goddard Space Flight Center, Greenbelt, MD.

Ray, R. D. 1993. Global ocean tide models on the eve of TOPEX/POSEIDON. *IEEE Trans. Geosci. Remote Sens.* **31,** 355–364.

Ray, R. D. 1998. Ocean self-attraction and loading in numerical tidal models. *Mar. Geodesy* **21,** 181–192.

Ray, R. D., and D. E. Cartwright. 1991. Satellite altimeter observations of the Mf and Mm ocean tides, with simultaneous orbit corrections. In *Proc. XX General Assembly IUGG, Vienna, Austria.*

Ray, R. D., B. Sanchez, and D. E. Cartwright. 1994a. Some extensions to the response method of tidal analysis applied to TOPEX/POSEIDON (abstract). *EOS Trans. AGU* **75**(16), 108.

Ray, R. D., D. J. Steinberg, B. F. Chao, and D. E. Cartwright. 1994b. Diurnal and semidiurnal variations in the Earth's rotation rate induced by oceanic tides. *Science* **264,** 830–832.

Ray, R. D., R. J. Eanes, and B. F. Chao. 1996. Detection of tidal dissipation in the solid Earth by satellite tracking and altimetry. *Nature* **381,** 595–597.

Ray, R. D., and G. T. Mitchum. 1996. Surface manifestation of internal tides generated near Hawaii. *Geophys. Res. Lett.* **23,** 2101–2104.

Ray, R. D., B. F. Chao, Z. Kowalik, and A. Y. Proshutinsky. 1997. Angular momentum of Arctic

Ocean tides. *J. Geodesy* **71**, 344–350.

Ray, R. D., and G. T. Mitchum. 1997. Surface manifestation of internal tides in the deep ocean: Observations from altimetry and island gauges. *Prog. Oceanogr.* **40**, 135–162.

Raymo, M. E., W. F. Ruddiman, N. J. Shackleton, and D. W. Oppo. 1990. Evolution of Atlantic-Pacific C gradients over the last 2.5 m.y. *Earth Planet. Sci. Lett.* **97**, 353–368.

Raymond, W. H., and H. L. Kuo. 1984. A radiation boundary condition for multi-dimensional flows. *Q. J. R. Meteorol. Soc.* **110**, 535–551.

Redi, M. H. 1982. Oceanic isopycnal mixing coordinate rotation. *J. Phys. Oceanogr.* **12**, 1154–1158.

Reid, J. L. 1981. On the mid-depth circulation of the World Ocean. In B. A. Warren and C. Wunsch (Eds.), *Evolution of Physical Oceanography, Scientific Surveys in Honor of Henry Stommel,* pp. 70–111. The MIT Press, Cambridge, MA.

Reid, R. O. 1990. Tides and storm surges. In J. B. Herbich (Ed.), *Handbook of Coastal and Ocean Engineering,* Vol. 1, *Wave Phenomena and Coastal Structures,* Chap. 9, pp. 533–589. Gulf, Houston, TX.

Reynolds, R. M. 1993. Overview of physical oceanographic measurements taken during the Mt. Mitchell cruise to the ROPME sea area. In *Proceedings of the First Scientific Workshop on the Results of the R/V Mt. Mitchell Cruise.*

Reynolds, R. W., M. Ji, and A. Leetmaa. 1998. Use of salinity to improve ocean modeling. *Phys. Chem. Earth* **23**, 543–553.

Reynolds, R. W. 1988. A real-time global sea surface temperature analysis. *J. Clim.* **1**, 75–86.

Reynolds, R. W., and D. C. Marsico. 1993. An improved real-time global sea surface temperature analysis. *J. Clim.* **6**, 114–119.

Reynolds, R. W., and T. M. Smith. 1994. Improved global sea surface temperature analyses using optimal interpolation. *J. Clim.* **7**, 929–948.

Richardson, L. F. [1922] 1965. *Weather Prediction by Numerical Process.* Dover, New York.

Richardson, P. L. 1984. Drifting derelict trajectories in the North Atlantic. *EOS Trans. AGU.* **65**, 730–731.

Richman, M. B. 1986. Rotation of principal components. *J. Climatol.* **6**, 293–335.

Richtmyer, R. D., and K. W. Morton. 1967. *Difference Methods for Initial Value Problems,* 2nd ed. Interscience, New York.

Riedlinger, S. H., and R. H. Preller. 1991. The development of a coupled ice-ocean model for forecasting ice conditions in the Arctic. *J. Geophys. Res.* **96**, 16955–16977.

Riser, S. C. 1998. An examination of the North Atlantic circulation using PALACE floats. 1998 U.S. WOCE Report 22–25.

Robert, A. J. 1966. The integration of a low-order spectral form of the primitive meteorological equations. *J. Meteorol. Soc. Japan.* **44**, 237–244.

Roberts, M. J., R. Marsh, A. L. New, and R. A. Wood. 1996. An intercomparison of a Bryan-Cox type ocean model and an isopycnic ocean model. Part I. The subtropical gyre and high latitude processes. *J. Phys. Oceanogr.* **26**, 1495–1527.

Roberts, P. H. 1987. Dynamo theory. In C. Nicolis and G. Nicolis (Eds.), *Irreversible Phenomena and Dynamical Systems Analysis in Geosciences,* pp. 73–133. Reidel, Dordrecht/Norwell, MA.

Robertson, A. W., C.-C. Ma, C. R. Mechoso, and M. Ghil. 1995. Simulation of the tropical Pacific climate with a coupled ocean-atmosphere general circulation model. Part I. The seasonal cycle. *J. Clim.* **8**, 1178–1198.

Robertson, D. S., J. R. Ray, and W. E. Carter. 1994. Tidal variations in UT1 observed with very long baseline interferometry. *J. Geophys. Res.* **99**, 621–636.

Robinson, A. R. 1964. Continental shelf waves and the response of the sea level to weather systems. *J. Geophys. Res.* **69**, 367–368.

Robinson, A. R. (Ed.). 1983. *Eddies in Marine Sciences.* Springer, New York.

Robinson, A. R., P. F. J. Lermusiaux, and N. Q. Sloan III. 1998. Data assimilation. In K. H. Brink and A. R. Robinson (Eds.), *The Sea,* Vol. 10, pp. 541–594. John Wiley and Sons, New York.

Robinson, A. R. 1996. Physical processes, field estimation and an approach to interdisciplinary ocean modeling. *Earth Science Reviews* **40**, 3–54.

Robinson, A. R. 1999. Forecasting and simulating coastal ocean processes and variabilities with the Harvard Ocean Prediction System, *Coastal Ocean Prediction*. Amer. Geophys. Union, 77–99.

Robinson, A. R., P. F. J. Lermusiaux, and N. Q. Sloan III. 1998. Data assimilation. In K. H. Brink, A. R. Robinson (Eds.), *The Sea*, Vol. 10, John Wiley, New York, pp. 541–594.

Robinson, A. R., M. A. Spall, and N. Pinardi. 1988. Gulf Stream simulations and the dynamics of ring and meander processes. *J. Phys. Oceanogr.* **18**, 1811–1853.

Robinson, A. R., H. G. Arango, A. Warn-varnas, W. Leslie, A. J. Miller, P. J. Haley, and C. J. Lozano. 1996. Real-time regional forecasting. In P. Malanotte-Rizzoli (Ed.), *Modern Approaches to Data Assimilation in Ocean Modeling*, pp. 377–412. Elsevier, Amsterdam/New York.

Robinson, I. S. 1985. *Satellite Oceanography*. Ellis Horwood, Chichester.

Roe, P. L. 1986. Characteristic-based schemes for the Euler equations. *Annu. Rev. Fluid Mech.* **18**, 337–365.

Roe, P. L. 1989. A survey of upwind differencing techniques. In *Lecture Notes in Physics*, Vol. 323, *Proceedings, Eleventh International Conference on Numerical Methods in Fluid Dynamics, 1988*.

Rosati, A., and K. Miyakoda. 1988. General circulation model for upper ocean simulation. *J. Phys. Oceanogr.* **18**, 1601–1626.

Sadourny, R., A. Arakawa, and Y. Mintz. 1968. Integration of the nondivergent barotropic equation with an icosahedral hexagonal grid for the sphere. *Mon. Wea. Rev.* **96**, 351–356.

Sadowsky, J. 1994. The continuous wavelet transform: A tool for signal investigation and understanding. *Johns Hopkins APL Technical Digest* **15**, 306–318.

Salstein, D. A., D. M. Kann, A. J. Miller, and R. D. Rosen. 1993. The sub-bureau for atmospheric angular momentum of the international earth rotation service: A meteorological data center with geodetic applications. *Bull. Am. Meteorol. Soc.* **74**, 67–80.

Saltzman, B. 1962. Finite amplitude free convection as an initial value problem. *J. Atmos. Sci.* **19**, 329.

Sanchez, B. V., R. D. Ray, and D. E. Cartwright. 1992. A Proudman-function expansion of the M_2 tide in the Mediterranean Sea from satellite altimetry and coastal gages. *Oceanol. Acta* **15**, 325–337.

Sanchez, B. V., and N. K. Pavlis. 1996. Estimation of main tidal constituents from TOPEX altimetry using a Proudman function expansion. *J. Geophys. Res.* **100**, 25229–25248.

Sanderson, B. G. 1998. Order and resolution for computational ocean dynamics. *J. Phys. Oceanogr.* **28**, 1271–1286.

Sandstrom, H., and N. S. Oakey. 1995. Dissipation in internal tides and solitary waves. *J. Phys. Oceanogr.* **25**, 604–614.

Sarkisyan, A. S. and V. F. Ivanov. 1971. The combined effect of baroclinicity and bottom topography as an important factor in the dynamics of ocean currents. *Izv. Acad. Nauka, USSR, Atmos. Ocean. Phys.* **1**, 173–188.

Sarmiento, J. L. 1992. Biogeochemical ocean models. In K. E. Trenberth (Ed.), *Climate System Modeling*, pp. 519–551. Cambridge Univ. Press, Cambridge.

Sasaki, Y. 1970. Some basic formalisms in numerical variational analysis. *Mon. Weather Rev.* **98**, 875–883.

Sasao, T., and J. M. Wahr. 1981. An excitation mechanism for the free "core nutation." *Geophys. J. R. Astron. Soc.* **64**, 729–746.

Schastok, J., M. Soffel, and H. Ruder. 1994. A contribution to the study of fortnightly and monthly zonal tides in UT1. *Astron. Astrophys.* **283**, 650–654.

Schlax, M. G., and D. B. Chelton. 1994. Aliased tidal errors in TOPEX/POSEIDON sea surface height data. *J. Geophys. Res.* **99**, 24761–24776.

Schlichting, H. 1978. *Boundary Layer Theory*, 7th ed. McGraw Hill, New York. 747 pp.

Schmitz, W. J., Jr. 1995. On the interbasin-scale thermohaline circulation. *Rev. Geophys.* **33,** 151–174.

Schmitz, W. J., Jr. 1996a. *On the World Ocean Circulation: Volume I. Some Global Features/North Atlantic Circulation.* Woods Hole Oceanographic Institution TR WHOI-96-08. 240 pp.

Schmitz, W. J., Jr. 1996b. *On the World Ocean Circulation: Volume II. The Pacific and Indian Oceans/A Global Update.* Woods Hole Oceanographic Institution TR WHOI-96-03. 141 pp.

Schneider, E. K., and J. L. Kinter III. 1994. An examination of the internally generated variability in long climate simulations. *Clim. Dyn.* **10,** 181–204.

Schneider, N. 1998. The Indonesian throughflow and the global climate system. *J. Clim.* **11,** 676–689.

Schott, F. 1977. On the energetics of baroclinic tides in the North Atlantic. *Ann. Geophys.* **33,** 41–62.

Schrama, E. J. O., and R. D. Ray. 1994. A preliminary tidal analysis of TOPEX/POSEIDON altimetry. *J. Geophys. Res.* **99,** 24799–24808.

Schramm, J. L., M. M. Holland, J. A. Curry, and E. E. Ebert. 1997. Modeling the thermodynamics of a sea ice thickness distribution. 1. Sensitivity to ice thickness resolution. *J. Geophys. Res.* **102,** 23079–23092.

Schrum, C. 1994. Numerische simulation thermodynamischer prozesse in der Deutschen Bucht. *Ber. Zentrum Meeres-Klimaforsch. B* **15.**

Schudlich, R. R., and J. F. Price. 1998. Observations of seasonal variation in the Ekman layer. *J. Phys. Oceanogr.* **28,** 1187–1204.

Schureman, P. 1941. *Manual of Harmonic Analysis and Prediction of Tides.* U.S. Dept. of Commerce Spec. Publ. 98, U.S. Govt. Printing Office, Washington, DC. 313 pp.

Schwiderski, E. W. 1980. On charting global ocean tides. *Rev. Geophys.* **18,** 243–268.

Schwiderski, E. W. 1982. *Global Ocean Tides, 10, the Fortnightly Lunar Tide (Mf), Atlas of Tidal Charts and Maps.* Rep. TR 82-151, Naval Surface Weapons Center, Dahlgren, VA.

Schwiderski, E. W. 1983. Atlas of ocean tidal charts and maps. I. The semidiurnal principal lunar tide M$_2$. *Mar. Geodesy* **6,** 219–265.

Seidel, H. F., and B. S. Giese. 1999. Equatorial currents in the Pacific Ocean 1992–1997. *J. Geophys. Res.* **104,** 7849–7863.

Seiler, U. 1991. Periodic changes of the angular momentum budget due to tides of the world ocean. *J. Geophys. Res.* **96,** 10287–10300.

Semtner, A. J., Jr. 1976. A model for the thermodynamic growth of sea ice in numerical investigation of climate. *J. Phys. Oceanogr.* **6,** 379–389.

Semtner, A. J. 1986a. History and methodology of modelling the circulation of the world ocean. In J. J. O'Brien (Ed.), *Advanced Physical Oceanographic Numerical Modeling*, pp. 23–32. Reidel, Dordrecht, Holland.

Semtner, A. J. 1986b. Finite-difference formulation of a world ocean model. In J. J. O'Brien (Ed.), *Advanced Physical Oceanographic Numerical Modeling*, pp. 187–202. Reidel, Dordrecht, Holland.

Semtner, A. J. 1987. A numerical study of sea ice and ocean circulation in the Arctic. *J. Phys. Oceanogr.* **17,** 1077–1099.

Semtner, A. J. 1995. Modeling ocean circulation. *Science* **269,** 1379–1385.

Semtner, A. J., and W. R. Holland. 1978. Intercomparison of quasigeostrophic simulations of the western North Atlantic circulation with primitive equation results. *J. Phys. Oceanogr.* **8,** 735–754.

Semtner, A. J., Jr., and R. M. Chervin. 1992. Ocean general circulation from a global eddy-resolving model. *J. Geophys. Res.* **97,** 5493–5550.

Servain, J., A. J. Busalacchi, M. J. McPhaden, A. D. Moura, G. Reverdin, M. Vianna, and S. E. Zebiak. 1998. A pilot research moored array in the tropical Atlantic (PIRATA). *Bull. Am. Meteorol. Soc.* **79,** 2019–2031.

Shapiro, M. A., and J. J. O'Brien. 1970. Boundary conditions for fine-mesh limited area forecasts. *J. Appl. Meteorol.* **9**, 345–349.

Shapiro, R. 1970. Smoothing, filtering and boundary effects. *Rev. Geophys. Space Phys.* **8**, 359–387.

Shay, L. K., G. J. Goni, and P. G. Black. 1999. Effects of Warm Oceanic Feature on Hurricane Opal, *Mon. Wea. Rev.* (in press).

SHEBA Science Working Group. 1994. New program to research issues of global climate in the Arctic. *EOS Trans. AGU* **75**, 249–252.

Sheinbaum, J., and D. L. T. Anderson. 1990a. Variational assimilation of XBT data. Part I. *J. Phys. Oceanogr.* **20**, 672–688.

Sheinbaum, J., and D. L. T. Anderson. 1990a. Variational assimilation of XBT data. Part II. Sensitivity studies and use of smoothing constraints. *J. Phys. Oceanogr.* **20**, 689–704.

Shum, C. K., P. L. Woodworth, O. B. Andersen, E. Egbert, O. Francis, C. King, S. Klosko, C. Le Provost, X. Li, J. Molines, M. Parke, R. Ray, M. Sclax, D. Stammer, C. Tierney, P. Vincent, and C. Wunsch. 1997. Accuracy assessment of recent ocean tide models. *J. Geophys. Res.* **102**, 25173–25194.

Simmons, A. J. and L. Bengtsson, Atmospheric general circulation models: their design and use for climate studies. In J. T. Houghton (ed.), *The Global Climate*, Cambridge University Press, London, 37-62.

Skamarock, W., J. Oliger, and R. L. Street. 1989. Adaptive grid refinement for numerical weather prediction. *J. Comput. Phys.* **80**, 27–60.

Smagorinsky, J. S. 1963. General circulation experiments with the primitive equations. I. The basic experiment. *Mon. Weather Rev.* **91**, 99–164.

Smagorinsky, J., S. Manabe, and J. L. Holloway, Jr. 1965. Numerical results from a nine-level general circulation model of the atmosphere. *Mon. Weather Rev.* **93**, 727–768.

Smedstad, O. M., and J. J. O'Brien. 1991. Variational data assimilation and parameter estimation in an equatorial Pacific Ocean model. *Prog. Oceanogr.* **26**, 179–241.

Smith, D., and J. Morison. 1998. Numerical Study of Haline Convection Beneath Leads in Sea Ice. *J. Geophys. Res.* **103**, 10069–10083.

Smith, E., J. Vazquez, A. Tran, and R. Sumagaysay. 1996. Satellite-derived sea surface temperature data available from the NOAA/NASA Pathfinder program. *EOS Trans. AGU Electron.* Suppl., April 2.

Smith, J. A. 1998. Evolution of Langmuir circulation during a storm. *J. Geophys. Res.* **103**, 12649–12668.

Smith, L. T., D. B. Boudra, and R. Bleck. 1990. A wind-driven isopycnic coordinate model of the north and equatorial Atlantic Ocean. 2. The Atlantic basin experiments. *J. Geophys. Res.* **95**, 13105–13128.

Smith, R. D., Dukowicz, J. K., and Malone, R. C. 1992. Parallel ocean general circulation modeling. *Physica D* **60**, 38–61.

Smith, R. D., S. Kortas, and B. Meltz. 1997a. *Curvilinear Coordinates for Global Ocean Models*. Los Alamos National Laboratory Tech. Note LA-UR-95-1146. 38 pp.

Smith, W. O. (Ed.). 1990. *Polar Oceanography, Part A: Physical Science; Part B: Biological Science*. Academic Press, San Diego. 406 and 405 pp.

Smolarkiewicz, P. K. 1984. A fully multidimensional positive definite advection transport algorithm with small implicit diffusion. *J. Comput. Phys.* **54**, 325–362.

Smolarkiewicz, P. K., and W. W. Grabowski. 1990. The multidimensional positive definite advection transport algorithm: Non-oscillatory option. *J. Comput. Phys.* **86**, 355–375.

Sommerfeld, A. 1979. *Partial Differential Equations: Lectures on Theoretical Physics*, Vol. 6. Academic Press, San Diego, CA.

Sonett, C. P., 1996. Late Proterozoic and Paleozoic tides, retreat of the Moon, and rotation of the Earth. *Science* **273**, 100–103.

Song, X., and C. A. Friehe. 1997. Surface air-sea fluxes and upper ocean heat budget at 156° E, 4° S during the Tropical Ocean-Global Atmosphere Coupled Ocean Atmosphere Response Experiment. *J. Geophys. Res.* **102,** 23109–23130.

Song, Y., and D. Haidvogel. 1994. A semi-implicit ocean circulation model using a generalized topography-following coordinate system. *J. Comput. Phys.* **115,** 228–244.

Sovers, O. J., C. S. Jacobs, and R. S. Gross. 1993. Measuring rapid ocean tidal Earth orientation variations with very long baseline interferometry. *J. Geophys. Res.* **98,** 19959–19971.

Spall, M. A., and A. R. Robinson. 1990. Regional primitive equation studies of the Gulf Stream meander and ring formation region. *J. Phys. Oceanogr.* **20,** 985–1016.

Spall, M. A., and W. R. Holland. 1991. A nested primitive equation model for oceanic application. *J. Phys. Oceanogr.* **21,** 205–220.

Sparrow, C. 1982. *The Lorenz Equations: Bifurcations, Chaos and Stange Attractors,* pp. 110–133. Springer-Verlag, Berlin/New York.

Spencer Jones, H. 1939. The rotation of the Earth and the secular acceleration of the sun, moon and planets. *Mon. Not. R. Astronom. Soc.* **39,** 541–558.

Stammer, D. 1998. On eddy characteristics, eddy transports, and mean flow proerties. *J. Phys. Oceanogr.* **28,** 727–739.

Stammer, D., R. Tokmakian, A. Semtner, and C. Wunsch. 1996. How well does a 1/4° global circulation model simulate large-scale oceanic observations? *J. Geophys. Res.* **101,** 25799–25832.

Stanev, E. V., V. M. Roussenov, N. H. Rachev, and J. V. Staneva. 1994. Sea response to atmospheric variability: Model study for the Black Sea. *J. Mar. Syst.* **6,** 241–267.

Steele, M., J. Zhang, D. Rothrock, and H. Stern. 1997. The force balance of sea ice in a numerical model of the Arctic Ocean. *J. Geophys. Res.* **102,** 21061–21079.

Stern, H. L., D. A. Rothrock, and R. K. Kwok. 1995. Open water production in Arctic sea ice: Satellite measurements and model parameterizations. *J. Geophys. Res.* **100,** 20601–20612.

Stockdale, T. N. 1997. Coupled ocean-atmosphere forecasts in the presence of climate drift. *Mon. Weather Rev.* **125,** 809–818.

Stockdale, T., M. Latif, G. Burgers, and J.-O. Wolff. 1994. Some sensitivities of coupled ocean-atmosphere GCM. *Tellus,* **46A,** 367–380.

Stockdale, T. N., D. L. T. Anderson, J. O. S. Alves, and M. A. Balmaseda. 1998. Global seasonal rainfall forecasts using a coupled ocean-atmosphere model. *Nature* **392,** 370–373.

Stommel, H. M. 1948. The westward intensification of wind-driven ocean currents. *Trans. Am. Geophys. Union* **29,** 202–206.

Stommel, H. M. 1961. Thermohaline convection with two stable regimes of flow. *Tellus* **13,** 224–230.

Stone, P. 1972. A simplified radiative-dynamical model for the static stability of the rotating atmosphere. *J. Atmos. Sci.* **29,** 405–418.

Strang, G. 1994. Wavelets. *Am. Scientist* **82,** 250–255.

Suarez, M. J., A. Arakawa, and D. A. Randall. 1983. The parameterization of the planetary boundary layer in the UCLA general circulation model: Formulation and results. *Mon. Weather Rev.* **111,** 2224–2243.

Sundermann, J. 1977. The semidiurnal principal lunar tide M$_2$ in the Bering Sea. *Dtsch. Hydrogr. Z.* **30,** 91–101.

Sussman, S. J., and J. Wisdom. 1992. Chaotic evolution of the solar system. *Science* **257,** 56–62.

Sverdrup, H. U. 1947. Wind-driven currents in a baroclinic ocean; with application to the equatorial currents of the eastern Pacific. *Proc. Natl. Acad. Sci.* **33,** 318–326.

Swinney, H. L., and J. P. Gollub. 1978. The transition to turbulence. *Physcis Today* **31,** 41–43.

Swinney, H. L., and J. P. Gollub (Eds.). 1981. *Hydrodynamic Instabilities and the Transition to Turbulence.* Springer-Verlag, Berlin.

Talagrand, O. 1981. On the mathematics of data assimilation. *Tellus* **33,** 321–329.

Talley, L. D. 1998. WOCE hydrographic programme atlas for the Pacific Ocean. 1998 U.S. WOCE Report. 15–17.

Tamura, Y. 1987. A harmonic development of the tide-generating potential. *Bull. d'Informations Marees Terrestres* **99**, 6813–6855.

Taylor, G. I. 1919. Tidal friction in the Irish Sea. *Philos. Trans. R. Soc. Lond. A* **220**, 1–93.

Tennekes, H., and J. L. Lumley. 1972. *A First Course in Turbulence*. MIT Press, Cambridge, MA.

Terray, L., O. Thual, S. Belamari, M. Deque, P. Dandin, C. Levy, and P. Delecluse. 1994. *Climatology and Interannual Variability Simulated by the ARPEGE-OPA Model*. CERFACS Tech. Rep. TR/CMGC/94-05. 33 pp.

Thacker, W. C., and R. B. Long. 1987. Fitting dynamics to data. *J. Geophys. Res.* **93**, 1227–1240.

Thompson, J. D., and W. J. Schmidt, Jr. 1988. A limited area model of the Gulf Stream: Design, initial experiments, and model-data intercomparison. *J. Phys. Oceanogr.* **19**, 792–814.

Thompson, J. F., Z. U. A. Warsi, and C. W. Mastin. 1985. *Numerical Grid Generation, Foundations and Applications*. North-Holland, Amsterdam.

Thorndike, A. S., D. A. Rothrock, G. A. Maykut, and R. Colony. 1975. The thickness distribution of sea ice. *J. Geophys. Res.* **80**, 4501–4513.

Thorndike, A. S., and R. Colony. 1980. *Arctic Ocean Buoy Program Data Report*. Polar Science Center, Univ. Washington, Seattle.

Thorpe, S. A. 1975. The excitation, dissipation and interaction of internal waves in the deep ocean. *J. Geophys. Res.* **80**, 328–338.

Thual, O., and J. C. McWilliams. 1992. The catastrophe structure of thermohaline convection in a two-dimensional model and a comparison with low-order box models. *Geophys. Astrophys. Fluid Dyn.* **64**, 67–95.

Thuburn, J. 1997. A PV-based shallow water model on a hexagonal-icosahedral grid. *Mon. Weather Rev.* **125**, 2328–2347.

Tierney, C. C., L. H. Kantha, and G. H. Born. 1999. Shallow and deep water global ocean tides from altimetry and numerical modeling. *J. Geophys. Res.* (in press).

Toggweiler, J. R. 1994. The ocean's overturning circulation. *Phys. Today* **47**, 45–50.

Toggweiler, J. R., and B. Samuels. 1998. On the ocean's large scale circulation near the limit of no vertical mixing. *J. Phys. Oceanogr.* **28**, 1832–1852.

Tomczak, M., and J. S. Godfrey. 1994. *Regional Oceanography: An Introduction*. Pergamon, Elmsford, NY. 422 pp.

Toole, J. M., R. W. Schmitt, K. L. Polzin, and E. Kunze. 1997. Near-boundary mixing above the flanks of a mid-latitude seamount. *J. Geophys. Res.* **102**, 947–959.

Torrence, C., and G. P. Compo. 1998. A practical guide to wavelet analysis. *Bull. Am. Meteorol. Soc.* **79**, 61–78.

Tournadre, J. 1989. *Traitement de la topographie dans le modele aux equations primitives du GFDL*. Document de Travail U. A. 710, IFREMER, Brest, France.

Tracey, K. L., S. D. Howden, and D. R. Watts. 1997. IES calibration and mapping procedures. *J. Atmos. Oceanic Tech.* **14**, 1483–1493.

Trenberth, K. E. (Ed.). 1992. *Climate System Modeling*. Cambridge Univ. Press, New York. 817 pp.

Trenberth, K. E., and A. Solomon. 1994. The global heat balance: Heat transport in the atmosphere and the oceans. *Clim. Dyn.* **10**, 107–134.

Tsonis, A. A., and J. B. Elsner. 1989. Chaos, strange attractors and weather. *Bull. Am. Meteorol. Soc.* **70**, 14–23.

Turcotte, D. L. 1988. Fractals in fluid mechanics. *Annu. Rev. Fluid Mech.* **20**, 5–16.

Unal, Y. S., and M. Ghil. 1995. Interannual and interdecadal oscillation patterns in sea level. *Clim. Dyn.* **11**, 255–278.

UNESCO. 1981. *Tenth Report of the Joint Panel on Oceanographic Tables and Standards*. UNESCO Technical Papers in Marine Science no. 36. 24 pp.

Untersteiner, N. (Ed.). 1986. *The Geophysics of Sea Ice*. Plenum, New York. 1196 pp.

Untersteiner, N., and A. S. Thorndike. 1982. Arctic data buoy program. *Polar Rec.* **21**, 127–135.

U.S. WOCE Office. 1998. *U.S. WOCE Report 1998*. U.S. World Ocean Circulation Experiment Report no. 10, U.S. WOCE Office, College Station, TX. 56 pp.

Van der Steen, A. J. 1999. Overview of recent supercomputers (available on www.netlib.org/benchmark).

Van Dyke, M. 1964. *Perturbation Methods in Fluid Mechanics.* Academic Press, New York.

Vautard, R., P. Yiou, and M. Ghil. 1992. Singular spectrum analysis: A toolkit for short noisy chaotic time series. *Physica D* **58,** 95–126.

Veronis, G. 1968. Comments on Phillips's (1966) proposed simplification of the equation of motion for a shallow rotating atmosphere. *J. Atmos. Sci.* **25,** 1154–1155.

Veronis, G. 1981. Dynamics of large-scale ocean circulation. In B. Warren and C. Wunsch (Eds.), *Evolution of Physical Oceanography,* pp. 140–183. MIT Press, Cambridge, MA.

Vinje, T. E., and O. Finnekasa. 1986. *The Ice Transport through Fram Strait.* Rep. NR 186, Nor. Polarinst., Oslo. 39 pp.

Wagner, C. A., C. K. Tai, and J. M. Kuhn. 1994. Improved M_2 ocean tide from TOPEX/POSEIDON and Geosat altimetry. *J. Geophys. Res.* **99,** 24853–24866.

Wahr, J. M. 1981. Body tides on an elliptical, rotating, elastic, oceanless Earth. *Geophys. J. R. Astron. Soc.* **64,** 677–703.

Wahr, J. M. 1982. The effects of the atmosphere and oceans on the Earth's wobble. I. Theory. *Geophys. J. R. Astron. Soc.* **70,** 349–372.

Wahr, J. M. 1983. The effects of the atmosphere and oceans on the Earth's wobble. II. Results. *Geophys. J. R. Astron. Soc.* **74,** 451–487.

Wahr, J. M. 1986. Geophysical aspects of polar motion, variations in the length of day, and the luni-solar nutations. In *Space Geodesy and Geodynamics,* pp. 281–313. Academic Press, San Diego.

Wahr, J. M. 1988. The Earth's rotation. *Annu. Rev. Earth Planet. Sci.* **16,** 231–249.

Wahr, J. M., and Z. Bergen. 1986. The effects of mantle anelasticity on nutations, Earth tides and tidal variations in rotation rate. *Geophys. J. R. Astron. Soc.* **87,** 633–668.

Walker, A. E., and J. L. Wilkin. 1998. Optimal averaging of NOAA/NASA Pathfinder satellite sea surface temperature data. *J. Geophys. Res.* **103,** 12869–12883.

Walker, S. J. 1996. *A 3-Dimensional Non-linear Variable Density Hydrodynamic Model with Curvilinear Coordinates.* Technical Report OMR-60/00, CSIRO Division of Oceanography, Hobart, Australia.

Wallace, J. M., C. Smith, and C. S. Bretherton. 1992. Singular value decomposition of wintertime sea surface temperature and 500-mb height anomalies. *J. Clim.* **5,** 561–576.

Wallcraft, A. J. 1991. *The Navy Layered Ocean Model Users Guide.* Naval Oceanographic and Atmospheric Research Laboratory Report 35. NRL Stennis Space Center, MS. 21 pp.

Walsh, J. E., W. D. Hibler III, and B. Ross. 1982. Numerical simulation of northern hemisphere sea ice variability, 1951–1980. *J. Geophys. Res.* **90,** 4847–4865.

Walsh, J. J., D. A. Dieterle, F. E. Muller-Karger, R. Bohrer, W. P. Bissett, R. J. Varela, R. Aparicio, R. Diaz, R. Thunell, G. T. Taylor, M. I. Scranton, K. A. Fanning, and E. T. Peltzer. 1999. Simulation of carbon-nitrogen cycling during spring upwelling in the Cariaco Basin. *J. Geophys. Res.* **104,** 7807–7825.

Wang, B., and Y. Wang. 1996. Temporal structure of the southern oscillation as revealed by waveform and wavelet analysis. *J. Clim.* **9,** 1586–1598.

Wang, D.-P. 1984. Mutual intrusion of a gravity current and density front formation. *J. Phys. Oceanogr.* **14,** 1191–1199.

Wang, D.-P., and C. N. K. Mooers. 1976. Coastal trapped waves in a continuously stratified ocean. *J. Phys. Oceanogr.* **6,** 853–863.

Wang, J. L., L. A. Mysak, and R. G. Ingram. 1994. A three-dimensional numerical simulation of Hudson Bay summer ocean circulation: Topographic gyres, separations and coastal jets. *J. Phys. Oceanogr.* **24,** 2496–2514.

Wang, X.-Y. 1995. Computational fluid dynamics based on the method of space-time conservation element and solution element. Ph.D. dissertation, Department of Aerospace Engineering Sciences, University of Colorado, Boulder. 180 pp.

Warner, T. T., R. A. Peterson, and R. E. Treadon. 1997. A tutorial on lateral boundary conditions as a basic and potentially serious limitation to regional numerical weather prediction. *Bull. Am. Meteorol. Soc.* **78,** 2599–2617.

Warren, B. A., and C. Wunsch (Eds.). 1981. *Evolution of Physical Oceanography.* MIT Press, Cambridge, MA. 623 pp.

Warren, S. G., C. J. Hahn, J. London, R. M. Chervin, and R. L. Jenne. 1988. *Globl Distribution of Total Cloud Cover and Cloud Type Amounts over the Ocean.* NCAR TN-317+STR, Boulder, CO. 42 pp.

Washington, W. M., and C. L. Parkinson. 1986. *An Introduction to Three-Dimensional Climate Modeling.* University Science Books, Sausalito, CA.

Washington, W. M., and G. A. Meehl. 1989. Climate sensitivity due to increased CO_2: Experiments with a coupled atmosphere and ocean general circulation model. *Clim. Dyn.* **4,** 1–38.

Washington, W. M., A. J. Semtner, Jr., G. A. Meehl, D. J. Knight, and T. A. Mayer. 1980. A general circulation experiment with a coupled atmosphere, ocean, and sea ice model. *J. Phys. Oceanogr.* **10,** 1887–1908.

Weare, B. F., A. R. Navato, and R. E. Newell. 1976. Empirical orthogonal analysis of Pacific sea surface temperatures. *J. Phys. Oceanogr.* **6,** 671–678.

Weaver, A. J., and T. M. C. Hughes. 1992. Stability and variability of the thermohaline circulation and its link to climate. *Trends Phys. Oceanogr.* **1,** 15–70.

Weaver, A. J., and M. Eby. 1997. On the numerical implementation of advection schemes for use in conjunction with various mixing parameterizations in the GFDL ocean model. *J. Phys. Oceanogr.* **27,** 369–377.

Webb, D. J. 1982. Tides and the evolution of the Earth-Moon system. *Geophys. J. R. Astron. Soc.* **70,** 261–271.

Webb, D. J. 1996. An ocean code for array processor computers. *Computers Geosci.* **22,** 569–578.

Webb, D. J., *et al.* (FRAM Group). 1991. Using an eddy resolving model to study the Southern Ocean. *EOS Trans. AGU* **72,** 169, 174–175.

Webster, P. J., and T. N. Palmer. 1997. The past and future of El Niño. *Nature* **390,** 562–564.

Webster, P. J., T. Palmer, M. Yanai, V. Magana, J. Shukla, and A. Yasunari. 1998. Monsoons: Processes, predictability and the prospects for prediction. *J. Geophys. Res.* **103,** 14451–14510.

Webster, P. J., L. H. Kantha, J. W. Lopez, and H.-R. Chang. 1999a. The sea surface height and Bangladesh floods. *Bull. Amer. Meteorol. Soc.* (in review).

Webster, P. J., J. Loschnigg, A. Moore, and R. Leben, 1999b. The great 1997–1998 warming of the Indian Ocean: Evidence for coupled ocean-atmosphere instabilities. *Nature.* (in press).

Weeks, E. R., Y. Tian, J. S. Urbach, K. Ide, H. L. Swinney, and M. Ghil. 1997. Transitions between blocked and zonal flows in a rotating annulus with topography. *Science* **278,** 1598–1601.

Weeks, W. F., and G. Timco. 1998. Preface. *J. Geophys. Res.* **103,** 21737–21738.

Wehner, M. F., and C. Covey. 1995. Description and validation of the LLNL/UCLA parallel atmospheric GCM. Lawrence Livermore National Laboratory Report UCRL-ID-123223. 21 pp.

Weiss, N. O. 1988. Is the solar cycle an example of deterministic chaos? In F. R. Stephenson and A. W. Wolfendale (Eds.), *Secular Solar and Geomagnetic Variations in the Last 10,000 Years,* pp. 69–78. Kluwer, Norwell, MA.

Welander, P. 1967. On the oscillatory instability of a differentially heated fluid loop. *J. Fluid Mech.* **29,** 17–30.

Welander, P. 1986. Thermocline effects in the ocean circulation and related simple model. In J. Willebrand and D. L. T. Anderson (Eds.), *Large Scale Transport Processes in Oceans and Atmosphere,* pp. 163–200. Reidel, Dordrecht/Norwell, MA.

Wendt, J. F. (Ed.). 1992. *Computational Fluid Dynamics, an Introduction.* Springer-Verlag, Berlin/New York. 291 pp.

Weng, H., and K.-M. Lau. 1994. Wavelets, period doubling, and time frequency localization with application to organization of convection over the tropical western Pacific. *J. Atmos. Sci.* **51,** 2523–2541.

Wenzel, H.-G., and W. Zurn. 1990. Errors of the Cartwright-Taylor-Edden 1973 tidal potential displayed by gravimetric Earth tide observations at BFO Schlitach. *Marees Terrestres Bull. d'Informations* **107,** 7559–7574.

Wessel, P., and W. H. F. Smith. 1991. Free software helps map and display data. *EOS Trans. AGU* **72,** 441.

Westerink, J. J., and W. G. Gray. 1991. Progress in surface water modeling. *Rev. Geophys. Suppl.,* 210–217.

Westerink, J. J., R. A. Luettich, A. M. Bapista, N. M. Scheffner, and P. Farrar. 1992. Tide and storm surge predictions using finite element model. *J. Hydraul. Engineer.* **118,** 1373–1390.

Westerink, J. J., R. A. Luettich, and J. C. Muccino. 1994. Modeling tides in the western North Atlantic using unstructured graded grids. *Tellus* **46A,** 178–199.

Wettlaufer, J. S. 1991. Heat flux at the ice-ocean interface. *J. Geophys. Res.* **96,** 7215–7236.

White, A. A., and R. A. Bromley. 1995. Dynamically consistent, quasi-hydrostatic equations for global models with a complete representation of the Coriolis force. *Q. J. R. Meteorol. Soc.* **121,** 399–418.

Whitehead, J. A. 1995. Thermohaline ocean processes and models. *Annu. Rev. Fluid Mech.* **27,** 89–113.

Whiteman, C. D., and X. Bian. 1996. Solar semidiurnal tides in the troposphere: Detection by radar profilers. *Bull. Am. Meteorol. Soc.* **77,** 529–542.

Wick, G. 1995. Evaluation of the variability and predictability of the bulk-skin sea surface temperature difference with applications to satellite-measured sea surface temperature. Ph.D. dissertation, University of Colorado, Boulder. 140 pp.

Wick, G. A., W. J. Emery, L. H. Kantha, and P. Schluessel. 1996. The behavior of the bulk-skin sea surface temperature difference under varying wind speed and heat flux. *J. Phys. Oceanogr.* **26,** 1969–1988.

Willems, R. C., *et al.* 1994. Experiment evaluates ocean models and data assimilation in the Gulf Stream. *EOS Trans. AGU* **75,** 385–394.

Willett, C. S. 1996. A study of anticyclonic eddies in the eastern tropical Pacific Ocean with integrated satellite, in-situ, and modelled data. Ph.D. dissertation, Aerospace Engineering Sciences, University of Colorado, Boulder, CO.

Wilson, C. R. 1995. Earth rotation and global change, U.S. National Report to IUGG, 1991–1994. *Rev. Geophys. Space Phys.* **33**(Suppl.), 225–229.

Wilson, C. R. 1998. Oceanic effects on Earth's rotation rate. *Science* **281,** 1623–1624.

Wilson, D. K. 1996. Empirical orthogonal function analysis of the weakly convective atmospheric boundary layer. Part I. Eddy structures. *J. Atmos. Sci.* **53,** 801–823.

Winant, C. D. 1980. Coastal circulation and wind-induced currents. *Annu. Rev. Fluid Mech.* **12,** 271–301.

Wolf, A. 1986. Quantifying chaos with Lyapunov exponents. 273–290.

Wolff, J.-O., and E. Maier-Reimer. 1997. *HOPE: The Hamburg Ocean Primitive Equation Model.* Deutsches Klimarechenzentrum TR 13, Max-Planck-Institut fur Meteorologie, Hamburg, Germany. 103 pp.

Woodruff, S. D., R. J. Slutz, R. L. Jenne, and P. M. Streurer. 1987. A comprehensive ocean-atmosphere data set. *Bull. Am. Meteorol. Soc.* **68,** 1239.

Woodworth, P. L., A. Tolkatchev, and D. T. Pugh. 1997. The global sea level observing system (GLOSS) [Abstract]. In *International Symposium on Monitoring the Oceans in the 2000s: An Integrated Approach, October 15–17, 1997, Biarritz, France.*

World Climate Research Program. 1995. *CLIVAR, a Study of Climate Variability and Predictability.* Tech. Doc. WMO/TD-690, World Meteorol. Organ., Geneva, 157 pp.

Wunsch, C. 1967. The long-period tides. *Rev. Geophys. Space Phys.* **5,** 447–475.

Wunsch, C. 1975. Internal tides in the ocean. *Rev. Geophys. Space Phys.* **13,** 167–182.

Wunsch, C. 1981. Low-frequency variability of the sea. In B. Warren and C. Wunsch (Eds.), *Evolution of Physical Oceanography,* pp. 342–375. MIT Press, Cambridge, MA.

Wunsch, C. 1996. *The Ocean Circulation Inverse Problem.* Cambridge University Press,

Cambridge.

Wunsch, C. 1998. The work done by wind on the oceanic general circulation. *J. Phys. Oceanogr.* **28,** 2332–2340.

Wunsch, C., and D. Stammer. 1997. Atmospheric loading and the oceanic "inverted barometer" effect. *Rev. Geophys.* **35,** 79–107.

Wunsch, C., and D. Stammer. 1998. Satellite altimetry, the marine geoid, and the oceanic circulation. *Annu. Rev. Earth Planet. Sci.* **26,** 219–253.

Wunsch, C., D. B. Haidvogel, M. Iskandarani, and R. Hughes. 1997. Dynamics of the long period tides. *Prog. Oceanogr.* **40,** 81–108.

Xie, P., and P. A. Arkin. 1997. Global precipitation: A 17-year monthly analysis based on gauge observations, satellite estimates, and numerical model outputs. *Bull. Am. Meteorol. Soc.* **78,** 2539–2558.

Xu, W., T. P. Barnett, and M. Latif. 1998. Decadal variability in the North Pacific as simulated by a hybrid coupled model. *J. Clim.* **11,** 297–312.

Xue, Y., M. A. Cane, and S. E. Zebiak. 1997. Predictability of a coupled model of ENSO using singular vector analysis. Part II: Optimal growth and forecast skill. *J. Geophys. Res.* **125,** 2057–2073.

Yaglom, A. M., and B. A. Kader. 1974. Heat and mass transfer between a rough wall and turbulent flow at high Reynolds and Peclet numbers. *J. Fluid Mech.* **62,** 601–623.

Yorke, J. A., and E. D. Yorke. 1981. Chaotic behavior and fluid dynamics. In *Hydrodynamic Instabilities and the Transition to Turbulence*, pp. 77–95. Springer-Verlag, Berlin/New York.

Zahel, W. 1977. A global hydrodynamical numerical 1° model of the ocean tides, *Ann. Geophys.* **33,** 31–40.

Zahel, W. 1991. Modeling ocean tides with and without assimilating data. *J. Geophys. Res.* **96,** 20379–20391.

Zalesak, S. T. 1979. Fully multidimensional flux-corrected transport algorithms for fluids. *J. Comput. Phys.* **31,** 335–362.

Zavatarelli, M., and G. L. Mellor. 1995. A numerical study of the Mediterranean Sea circulation. *J. Phys. Oceanogr.* **25,** 1384–1414.

Zebiak, S. E., and M. A. Cane. 1987. A model El Niño Southern Oscillation. *Mon. Weather Rev.* **115,** 2262–2278.

Zeldovich, Ya. B., A. A. Ruzmaikin, and D. D. Sokoloff. 1983. *Magnetic Fields in Astrophysics.* Gordon and Breach, 365 pp.

Zhang, J., and W. D. Hibler III. 1997. On an efficient numerical method for modeling sea ice dynamics. *J. Geophys. Res.* **102,** 8691–8702.

Zhang, J., R. W. Schmitt, and R. X. Huang. 1998. Sensitivity of the GFDL Modular Ocean Model to parameterization of double-diffusive processes. *J. Phys. Oceanogr.* **28,** 589–605.

Zheng, D.-L., H.-R. Chang, N. L. Seaman, T. T. Warner, and J. M. Frisch. 1986. A two-way interactive nesting procedure with variable terrain resolution. *Mon. Weather Rev.* **114,** 1330–1339.

Zou, J. 1993. Principal oscillation pattern analysis of sea level variations in the tropical Pacific measured by the Geosat altimeter. I. S. F. Jones, Y. Sugimori, and R. Stewart (Eds.), *Satellite Remote Sensing of the Oceanic Environment,* pp. 382–392. Seibutsu Kenkyusha, Tokyo.

Zou, J., and M. Latif. 1994. Modes of ocean variability in the tropical Pacific as derived from Geosat altimetry. *J. Geophys. Res.* **99,** 9963–9975.

Biographies

There are many names associated with numerical ocean modeling: Kirk Bryan, James O'Brien, Allan Robinson, Rainer Bleck, George Mellor, Dale Haidvogel, Bert Semtner, Bill Holland, Mike Cox, and several others. Here, we mention a few that stand out.

Dr. Kirk Bryan regarded by many as the founding father of numerical ocean modeling. At a time when computers were still very primitive, he had the vision to formulate a comprehensive baroclinic model necessary for studying the complex physical processes in our global oceans. Only recently, with the advent of high-performance multi-gigaflop (and very soon multi-teraflop) computers, is his vision of accurately modeling oceanic processes numerically coming to pass. The modular ocean model (MOM) that is widely used in the oceanic community originates from the model that he and colleagues Mike Cox and Bert Semtner developed. Many ocean models employed around the world have been derived from it. Kirk has spent most of his career at the NOAA Geophysical Fluid Dynamics Laboratory at Princeton University and is widely known in the ocean modeling community for his many contributions to the field. He has been instrumental in introducing many to ocean modeling and training them in this difficult art. Dr. Bryan has received numerous awards and honors for this work. Now retired from GFDL, he is still active in the field.

Professor Allan R. Robinson is the Gordon McKay Professor of Geophysical Fluid Dynamics at the Harvard University. He has been active in the field for more than four decades. Principally a theoretician, he helped advance the dynamics of rotating and stratified fluids, the theory and modeling of ocean circulation, and the assimilation of data into ocean models. In the arena of ocean modeling, he is well known for his contributions to comprehensive, data-assimilative, predictive, coupled physical-biological ecosystem models of the ocean. His Harvard Ocean Prediction System (HOPS), an ecosystem model, is used for real-time nowcast/forecasts of the oceanic physical-biological state in regions such as the Massachusetts Bay and the Mediterranean Sea. He is the recipient of numerous awards and honors. In 1998, a special symposium was convened at the AGU-ASLO meeting in San Diego to honor his many contributions to the theory and modeling of ocean circulation, data assimilation, and interdisciplinary applications. He is currently the editor-in-chief of *The Sea* and the journal *Dynamics of Atmospheres and Oceans*.

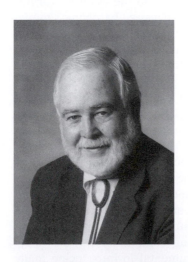

James J. O'Brien, Secretary of the Navy Professor in Oceanography at the Florida State University (FSU), has made many contributions to the understanding of oceanic processes and has helped train an entire generation of numerical ocean modelers. Dr. O'Brien is the director of the FSU Center for Ocean-Atmospheric Prediction studies, recipient of the Sverdrup Gold Medal, and a Fellow of the American Meteorology Society, the American Geophysical Union, and the Royal Meteorological Society. Jim is particularly noted for contributions of his ideas and students to the U.S. Navy's modeling efforts. A passionate advocate of layered models, he, along with his former students and colleagues, has been instrumental in the development and application of eddy-resolving dynamical models to the global oceans.

George L. Mellor of Princeton University, began his research career in the field of aerodynamics, studying turbulence and boundary layers under neutral stratification. His association with GFDL began around 1970 with the establishment of the Geophysical Fluid Dynamics Program (now the Atmospheric and Oceanic Sciences Program), which he founded. Dr. Mellor headed the group in Princeton that developed and fine-tuned the widely used sigma coordinate model, Princeton Ocean Model (POM), for application to coastal ocean problems. Now retired from Princeton University, Dr. Mellor is still active in numerical modeling of estuarine and oceanic processes. He is a Fellow of the American Meteorological Society and the American Geophysical Union.

Index

International Geophysics Series

EDITED BY

RENATA DMOWSKA

Division of Applied Science
Harvard University
Cambridge, Massachusetts

JAMES R. HOLTON

Department of Atmospheric Sciences
University of Washington
Seattle, Washington

H. THOMAS ROSSBY

Graduate School of Oceanography
University of Rhode Island
Narragansett, Rhode Island

* Out of print.

937